Lecture Notes in Computer Science 10455

FoLLI Publications on Logic, Language and Information

Subline of Lectures Notes in Computer Science

More information about this series at http://www.springer.com/series/7407

Alexandru Baltag · Jeremy Seligman
Tomoyuki Yamada (Eds.)

Logic, Rationality, and Interaction

6th International Workshop, LORI 2017
Sapporo, Japan, September 11–14, 2017
Proceedings

Springer

Editors
Alexandru Baltag
Institute for Logic, Language
and Computation
University of Amsterdam
Amsterdam, Noord-Holland
The Netherlands

Jeremy Seligman
Department of Philosophy
The University of Auckland
Auckland, Auckland
New Zealand

Tomoyuki Yamada
Graduate School of Letters
Hokkaido University
Sapporo
Japan

ISSN 0302-9743 ISSN 1611-3349 (electronic)
Lecture Notes in Computer Science
ISBN 978-3-662-55664-1 ISBN 978-3-662-55665-8 (eBook)
DOI 10.1007/978-3-662-55665-8

Library of Congress Control Number: 2017949521

LNCS Sublibrary: SL1 – Theoretical Computer Science and General Issues

Printed on acid-free paper

This Springer imprint is published by Springer Nature
The registered company is Springer-Verlag GmbH Germany
The registered company address is: Heidelberger Platz 3, 14197 Berlin, Germany

Preface

This volume collects the papers presented at LORI-6, the 6th International Workshop on Logic, Rationality, and Interaction, held in Sapporo, Japan, during September 11–14 and hosted by the Philosophy Department of Hokkaido University.

The workshop received 73 submissions and the final program consisted of 44 full papers and 13 short papers, although four of these were withdrawn from the proceedings. Each paper was selected on the basis of two or more reviews. The number of submissions to the LORI series is growing, with 42 at LORI-4 and 66 at LORI-5. Moreover, many are of high quality. We took the decision this year to have a number of parallel sessions at the workshop, thus allowing us to accept more papers.

The topics of papers contributed to the workshop covered the spectrum of topics within the scope of the LORI series. There were also invited talks by J. Michael Dunn (Indiana University, USA), Nina Gierasimczuk (Technical University of Denmark, Denmark), Alan Hájek (Australian National University, Australia), Natasha Alechina (University of Nottingham, UK), Marta Bílková (Charles University in Prague, Czech Republic), and Hiroakira Ono (JAIST, Japan).

The LORI series was started in China, with the first event (LORI-1) in August 2007, hosted by Beijing Normal University. This was judged to be a great success in both providing a focus for relevant research in East Asia and as a means of attracting scholars from outside the region to interact and exchange ideas. From then on LORI workshops have been held every two years. The next three were all in mainland China: LORI-2 at Southwest University, Chongqing; LORI-3 at Sun Yet-sen University, Guangzhou; and LORI-4 at Zhejiang University, Hangzhou. The most recent event was LORI-5, hosted jointly by the National Taiwan University and Yang-Ming University in Taipei, Taiwan. LORI-6 is the first of the series to be held in Japan.

More details about the LORI conference series can be found at www.golori.org.

As Organizing and Program Committee (PC) chairs we would like to thank the PC members and all the additional reviewers for working so hard and efficiently within extremely tight time constraints. The program is greatly indebted to their contribution. We owe further thanks to the LORI Standing Committee, Fenrong Liu and Johan van Benthem, together with a number of former LORI PC chairs, who offered timely and insightful advice and support. We are grateful for the generous support of The Graduate School of Letters and the Department of Philosophy at Hokkaido University, the continued support of the Tsinghua University - University of Amsterdam Joint Research Center for Logic, and the Association of Symbolic Logic. We would also like to acknowledge the use of EasyChair, for both organizing the reviewing process and creating these proceedings. The final thanks should go to all those colleagues and students on the ground at Hokkaido University, for their effort in making the workshop happen.

July 2017

Alexandru Baltag
Jeremy Seligman
Tomoyuki Yamada

Organization

Organizing Committee

Masahiro Matsuo	Department of Natural History Sciences, Hokkaido University, Japan
Shigehiro Kato	Department of Linguistic Sciences, Hokkaido University, Japan
Katsuhiko Sano	Department of Philosophy, Hokkaido University, Japan
Tomoyuki Yamada	Department of Philosophy, Hokkaido University, Japan

Program Committee

Natasha Alechina	University of Nottingham, UK
Alexandru Baltag	University of Amsterdam, The Netherlands
Thomas Bolander	Technical University of Denmark, Denmark
Zoé Christoff	University of Bayreuth, Germany
Branden Fitelson	University of California-Berkeley, USA
Nina Gierasimczuk	Technical University of Denmark, Denmark
Davide Grossi	University of Liverpool, UK
Jiahong Guo	Beijing Normal University, China
Meiyun Guo	Southwest University, China
Andreas Herzig	INRIT-CNRS, France
Wesley Holliday	UC Berkeley, USA
Thomas Icard	Stanford University, USA
Makoto Kanazawa	National Institute of Informatics, Japan
Willemien Kets	Northwestern University, USA
Kohei Kishida	University of Oxford, UK
Dominik Klein	University of Bamberg, Germany
Hidenori Kurokawa	Kobe University, Japan
Emiliano Lorini	INRIT-CNRS, France
Xudong Luo	Guangxi Normal University, China
Minghui Ma	Sun Yat-Sen University, China
Edwin Mares	Victoria University of Wellington, New Zealand
Sara Negri	University of Helsinki, Finland
Hiroakira Ono	JAIST, Japan
Eric Pacuit	University of Maryland, College Park, USA
Gabriella Pigozzi	Université Paris-Dauphine, France
Soroush Rafiee Rad	University of Amsterdam, The Netherlands
R. Ramanujam	Institute of Mathematical Sciences, Chennai, India
Joshua Sack	California State University Long Beach, USA
Katsuhiko Sano	Hokkaido University, Japan
Jeremy Seligman	University of Auckland, New Zealand

Kaile Su	Griffith University, Australia
Paolo Turrini	Imperial College London, UK
Fernando R. Velázquez-Quesada	University of Amsterdam, The Netherlands
Ren-June Wang	National Chung-Cheng University, Taiwan
Xuefeng Wen	Sun Yat-Sen University; National University of Defence Technology, China
Yì N. Wáng	Zhejiang University, China
Junhua Yu	Tsinghua University, China
Thomas Ågotnes	University of Bergen, Norway

Additional Reviewers

Anantha Padmanabha, M.S.
Charrier, Tristan
Chen, Gong
Chen, Qingliang
Ciardelli, Ivano
Cinà, Giovanni
Eva, Benjamin
Fernández-Duque, David
Fjellstad, Andreas
Fujita, Ken-etsu
Galimullin, Rustam
Genin, Konstantin
Girard, Patrick
Hansen, Jens Ulrik
He, Shunan
Jing, Xiaoxin
Liang, Zhen
Liberman, Andrés Occhipinti
Lin, Yuanlei
Liu, Chanjuan
Ma, Wenjun
Paoli, Francesco
Pedersen, Truls
Picollo, Lavinia
Rennela, Mathys
Suresh, S.P.
Verdée, Peter
von Plato, Jan
Xiong, Zuojun
Zamansky, Anna

Contents

Long Papers

Short Papers

Long Papers

A Logical Framework for Graded Predicates

Petr Cintula[1(✉)], Carles Noguera[2], and Nicholas J. J. Smith[3]

[1] Institute of Computer Science, Czech Academy of Sciences, Pod Vodárenskou věží 271/2, 182 07 Prague, Czech Republic
cintula@cs.cas.cz

[2] Institute of Information Theory and Automation, Czech Academy of Sciences, Pod Vodárenskou věží 4, 182 08 Prague, Czech Republic
noguera@utia.cas.cz

[3] Department of Philosophy, University of Sydney, Main Quadrangle A14, Sydney, NSW 2006, Australia
njjsmith@sydney.edu.au

Abstract. In this position paper we present a logical framework for modelling reasoning with graded predicates. We distinguish several types of graded predicates and discuss their ubiquity in rational interaction and the logical challenges they pose. We present mathematical fuzzy logic as a set of logical tools that can be used to model reasoning with graded predicates, and discuss a philosophical account of vagueness that makes use of these tools. This approach is then generalized to other kinds of graded predicates. Finally, we propose a general research program towards a logic-based account of reasoning with graded predicates.

Keywords: Graded predicates · Vagueness · Mathematical fuzzy logic

1 Introduction

A contemporary view of reasoning goes beyond the analysis of argumentation in discourse and takes rationality as a broad phenomenon that encompasses competence not only at organizing discourse, but also at making decisions, and taking actions towards goals, in light of our beliefs and knowledge. A logic-based account of reasoning should recognize that all these aspects of rationality involve heavy use of graded properties. Indeed, predicates that are a matter of more-or-less (such as *red*, *old*, *tall*, or *rich*) are ubiquitous in most domains of discourse and everyday reasoning scenarios. They include vague predicates (such as the examples just mentioned), but also predicates with sharply-defined boundaries. Take, for example, the predicate *acute angle* whose extension is exactly the set of angles strictly smaller than a right angle, and yet its instances admit mutual comparison: if α and β are angles of, respectively, 30° and 89°, it is true that both are acute angles, and it also makes sense to assert that α is strictly more acute than β. Non-graded all-or-nothing properties are actually rare in everyday communication and usually belong to quite restricted domains of discourse (typically, mathematics and other sciences as well as legal discourse).

A. Baltag et al. (Eds.): LORI 2017, LNCS 10455, pp. 3–16, 2017.
DOI: 10.1007/978-3-662-55665-8_1

Arguably, graded properties are epistemologically necessary, as they gather together many similar notions that would otherwise collapse the conceptual system (and the language) with too many properties (and predicates). It is necessary for reasons of economy to reason with one predicate *red*, instead of having infinitely many predicates, for each possible level in the colour spectrum (or as many as the human eye can distinguish). Reasoning with such graded properties is successfully and correctly carried out in many contexts (notwithstanding the fact that natural language has enough devices to provide higher levels of precision whenever necessary).

We view Logic as the science of correct reasoning and, as such, we expect it to provide us with the formal means to deal with all forms of valid consequence that can potentially be carried out by rational beings. During most of its history as a formal science, Logic has tried to explain correct reasoning by means of the classical paradigm based on the bivalence principle. Despite its many merits and achievements, this approach does violence to many properties, forcing sharp all-or-nothing definitions (splitting graded properties into many binary ones) in contexts that normally do not require them. Indeed, natural language allows satisfactory communication and correct reasoning using graded predicates. The classical logical analysis seems, therefore, too artificial—too detached from actual reasoning.

On the one hand, there have been several attempts in philosophical logic and analytic philosophy at understanding vague predicates and their potential for generating logical paradoxes—although most of these attempts do not treat vague predicates as graded. On the other hand, Mathematical Fuzzy Logic (MFL) was proposed [17] as a study of many-valued logical systems able to handle graded properties (and related notions of partial truth, vagueness, fuzziness, imprecision, etc.). It has attracted a considerable number of researchers who have mostly disregarded its original motivations and focused on developing a deep and extensive corpus of mathematical results (see e.g. [6]), covering all technical aspects of such logical systems. Philosophers of vagueness have often attacked MFL as an inadequate framework for dealing with vagueness, based on allegations that usually disregard most of the mathematical development of MFL and focus on a few characteristics of the logical systems that were proposed at the beginning of the field. However, there has been a recent philosophical account of vague predicates [33] that treats them as graded, employs the modern logical machinery of MFL, and offers good answers to the traditional arguments against degree-based approaches.

In this paper we will defend a logical approach to reasoning with graded predicates that goes beyond that offered in [33] by considering other graded predicates besides vague ones. The structure of the paper is as follows. After this Introduction, Sect. 2 discusses the different kinds of graded predicates that we want to model. Section 3 presents a brief up-to-date account of mathematical fuzzy logic and Sect. 4 shows how it can be used to provide a satisfactory explanation of vague predicates and a solution to the paradoxes they generate. Section 5 discusses possible ways of modelling other kinds of graded predicates by means of the tools of MFL and Sect. 6 proposes a general program, extending the first steps outlined in this paper, to develop a full logic-based account of reasoning with graded predicates.

2 Graded Predicates

The essential feature of graded predicates is that they may apply with different intensities to different objects. If F is a graded unary predicate and a and b are objects in the domain of discourse relevant for F, it may happen that a is strictly more F than b, i.e. the degree of F-ness of a is greater than that of b; it could also be the other way around; or a and b could be equal; or they could be incomparable. All these possible comparisons do not entail the existence of any numerical scale, but only a purely ordinal notion of degree. Also, there are graded predicates of any higher arity, characterized in the same way, that apply to tuples instead of single individuals.

We may distinguish the following different kinds of graded predicates:[1]

1. **Classical predicates:** As an extreme case of our classification we must include the classical predicates. They obey the bivalence and excluded middle principles and hence yield a perfect division of the domain into the elements that satisfy the predicate and those that do not. They are a limit case of graded predicates that admit only two degrees. Classical predicates correspond to sharply-defined all-or-nothing properties and are ideal for analysing reasoning in domains that typically employ such notions, for example mathematics or legal discourse. However, they have often been abused to model other kinds of graded properties in an unnatural way.
2. **Vague predicates:** Vague predicates exhibit three surface characteristics: (a) their extension has blurry boundaries, (b) they have borderline cases (objects such that we can neither confidently assert nor confidently deny that they fall under the predicate), and (c) they generate *sorites* paradoxes, as follows. A sorites series for a predicate F is a series of objects $x_0, x_1, \ldots, x_n$ such that:

   - F definitely applies to x_0
   - F definitely does not apply to x_n
   - for each $i < n$, the objects x_i and x_{i+1} are extremely similar in all respects relevant to the application of F.

Such a series generates the following argument: (1) x_0 is F; (2) for each $i < n$, if x_i is F, then so is x_{i+1}; therefore x_n is F. When F is vague, this argument becomes a logical paradox, because it has the form of a valid argument whose first premise is clearly true and whose second premise also seems true, and yet its conclusion is clearly false.

Typical examples of vague predicates are those mentioned in the Introduction: *red*, *old*, *tall* and *rich*. Vague predicates can be subdivided into **linear**

[1] This classification is a modification of that presented by Paoli in [26,27].

(or **unidimensional)** and **nonlinear** vague predicates. The application of a linear vague predicate to an object depends only on the extent to which the object possesses some underlying attribute, which varies along a single dimension. For example, (once we fix a context) whether someone is 'tall' depends only on her height (and heights vary along a single dimension) and whether someone is 'old' depends only on his age (and ages vary along a single dimension). By contrast, the application of a nonlinear vague predicate to an object does not depend only on the position of that object along a single dimension. Some (perhaps all) nonlinear vague predicates are **multidimensional**: for example, whether an object is 'red' depends on its position in a three-dimensional colour space—that is, it depends on its position along three different dimensions (e.g. hue, saturation and brightness). Arguably, there is also a second kind of nonlinear vague predicate: one whose application conditions *cannot* be factored into a series of linear dimensions. For example, some might argue that 'beautiful' is such a predicate. We do not take a position either way on whether the nonlinear vague predicates are completely exhausted by the multidimensional vague predicates (i.e. on whether the class of nonlinear nonmultidimensional vague predicates is empty). Note that nonlinear vague predicates may have *incomparable* instances: for example, it may be possible to come up with two individuals such that there is no way to determine who is more clever, because they are clever in different ways.

Vagueness has been clearly distinguished from other phenomena (such as uncertainty, context sensitivity, ambiguity, and generality) and has been addressed by several competing theories (see e.g. [14,21,22,28,29,33,34,36]). In Sect. 4 we will summarize a degree-based treatment of vague predicates.

3. **Graded precise predicates:** These are predicates that have sharply defined limits but that, unlike classical predicates, admit more than two degrees of application. An example, already mentioned in the Introduction, is the unary predicate *acute angle*: it is sharply defined (as applying to angles strictly smaller than a right angle) but it applies with different intensities to different acute angles (an angle of 30° is more acute than an angle of 89°, although both are acute) and also to different non-acute angles (an angle of 170° is less acute than an angle of 91°, although both are non-acute). Other sciences also employ graded precise predicates: for example *acid* and *base* in chemistry, defined as having a pH smaller (resp. greater) than 7. Finally, to give an example from legal language, consider the predicate *guilty*. The judicial system does not want borderline cases, and will do everything it takes to prevent them and always declare an accused person either guilty or not. However, there are different degrees of guilt, which translate into more or less severe sentences.

Their well-defined limits save graded precise predicates from the difficulties of vagueness (in particular, the generation of sorites paradoxes), but they are still quite different from classical predicates and require a different logical treatment.

3 Mathematical Fuzzy Logic

Petr Hájek founded MFL [17] as an attempt to provide solid logical foundations for fuzzy set theory and its engineering applications. Among other motivations, fuzzy set theory had been explicitly proposed as a mathematical apparatus for dealing with vagueness and imprecision, but it lacked a focus on syntactical formalization of discourse and a notion of logical consequence, thus keeping it far from being a logical study of reasoning under vagueness. Hájek and his collaborators developed MFL as a genuine subdiscipline of Mathematical Logic, specializing in the study of certain many-valued logics.

The first examples of fuzzy logics were two many-valued propositional systems that had been studied already for quite some time before the inception of fuzzy sets: Łukasiewicz [24] and Gödel–Dummett logics [11]. Both were considered fuzzy logics because—similar to the definition of membership functions in fuzzy sets—they were semantically defined as infinitely-valued logics taking truth-values in the real unit interval $[0, 1]$. But they had more characteristics in common: a language with conjunction $\wedge$ and disjunction $\vee$ respectively interpreted as the operations minimum and maximum, constants for (total) falsity and truth $\overline{0}$ and $\overline{1}$ respectively interpreted as the values 0 and 1, an implication $\rightarrow$, and, in the case of Łukasiewicz, another conjunction connective (fusion) & satisfying the following residuation law with respect to the implication, for each $a, b, c \in [0, 1]$ (in the case of Gödel–Dummett logic it is satisfied by $\wedge$):

$$a \mathbin{\&} b \leq c \text{ if, and only if, } a \leq b \rightarrow c.$$

Both these operations, used to interpret conjunctions, are particular instances of binary functions called *triangular norms* (or *t-norms* for short): binary commutative, associative, monotone functions on $[0, 1]$; and moreover they are both continuous, which guarantees the existence of a binary function satisfying the residuation law. Therefore, Hájek and other MFL researchers started proposing alternative $[0, 1]$-valued logics by keeping the interpretation of $\wedge$, $\vee$, $\overline{0}$ and $\overline{1}$ as in the previous systems, but taking other continuous t-norms for & and their corresponding residuum for $\rightarrow$ [5,17]. It was later observed that the necessary and sufficient condition for a t-norm to have a residuum was not continuity, but just left-continuity. This motivated the introduction, by Esteva and Godo, of MTL [12], a weaker logic that was later proved to be complete w.r.t. the semantics given by all left-continuous t-norms and their residua [19]. Therefore, MTL was proposed as a basic fuzzy logic upon which other fuzzy logics could be obtained as axiomatic extensions.

Besides their intended t-norm-based semantics over $[0, 1]$ (also called *standard* semantics), all these fuzzy logics were also given an algebraic semantics based on classes of MTL-algebras, that is, structures of the form $\boldsymbol{A} = \langle A, \wedge, \vee, \&, \rightarrow, \overline{0}, \overline{1}\rangle$ such that

- $\langle A, \wedge, \vee, \overline{0}, \overline{1}\rangle$ is a bounded lattice
- $\langle A, \&, \overline{1}\rangle$ is a commutative monoid

- for each $a, b, c \in A$ we have

$$a \mathbin{\&} b \leq c \quad \text{iff} \quad b \leq a \to c \qquad \text{(residuation)}$$
$$(a \to b) \vee (b \to a) = \overline{1} \qquad \text{(prelinearity)}$$

We say that an MTL-algebra is:

- *Linearly ordered* (or an MTL-chain) if its lattice order is total.
- *Standard* if its lattice reduct is the real unit interval $[0, 1]$ with its usual order.

Note that in a standard MTL-algebra & is interpreted by a left-continuous t-norm and $\to$ by its residuum—and vice versa: each left-continuous t-norm fully determines its corresponding standard MTL-algebra.

MTL is an algebraizable logic in the sense of [3] and the variety of MTL-algebras is its equivalent algebraic semantics. Thus each finitary extension of MTL (like all the other logics mentioned so far) also has an equivalent algebraic semantics which is a corresponding subquasivariety of MTL-algebras. Conversely, given any subquasivariety $\mathbb{K}$ of MTL-algebras, the corresponding finitary extension L of MTL is obtained by setting that for each set of formulas Γ and each formula φ, $\Gamma \vdash_{\mathrm{L}} \varphi$ iff for each algebra $\boldsymbol{A} = \langle A, \wedge, \vee, \&, \to, \overline{0}, \overline{1}\rangle \in \mathbb{K}$ and each $\boldsymbol{A}$-evaluation e we have: if $e(\psi) = \overline{1}$ for each $\psi \in \Gamma$, then $e(\varphi) = \overline{1}$.

It soon became clear that fuzzy logics were closely related to substructural logics; indeed it was proven that MTL is the axiomatic extension of the logic $\mathrm{FL_{ew}}$ (the full Lambek logic with exchange and weakening, see e.g. [15]) obtained by adding the prelinearity axiom $(\varphi \to \psi) \vee (\psi \to \varphi)$ [13]. Several papers have considered weaker fuzzy logics as extensions of other substructural logics:

(a) By dropping *commutativity* of conjunction Petr Hájek obtained a system, psMTL^r [18], which is an axiomatic extension of $\mathrm{FL_w}$ and was proven to be complete with respect to the semantics on non-commutative residuated t-norms [20].
(b) By removing *integrality* (i.e. not requiring the neutral element of conjunction to be maximum of the order) Metcalfe and Montagna proposed the logic UL which is an axiomatic extension of $\mathrm{FL_e}$ with bounds and was, in turn, proven to be complete with respect to left-continuous uninorms (that is, a generalization of t-norms that allows the neutral element to be any element $u \in [0, 1]$) [25].
(c) By removing *associativity* (i.e. not requiring conjunction to be interpreted by an associative operation) as well as commutativity and integrality, one obtains a very weak fuzzy logic SL^ℓ which extends the non-associative Lambek logic [16,23] and is still complete with respect to models over $[0, 1]$. The axiomatization and completeness theorems for this logic and for systems obtained with other combinations of the properties (associativity, integrality, and commutativity) are presented in [7,8].

All these fuzzy logics, weaker than MTL, are still algebraizable in the sense of [3] and their algebraic counterparts are classes of lattice-ordered residuated unital groupoids (not necessarily associative, commutative, or integral) in which the semantical consequence relation has to be defined in a more general way than before. More precisely, if $\mathbb{K}$ is a class of such algebras, Γ is a set of formulas and φ is a formula, $\Gamma \vdash_{\mathrm{L}} \varphi$ iff for each algebra $\boldsymbol{A} \in \mathbb{K}$ and each $\boldsymbol{A}$-evaluation e we have: if $e(\psi) \geq \overline{1}$ for each $\psi \in \Gamma$, then $e(\varphi) \geq \overline{1}$. Therefore, in these fuzzy logics the interpretation of the constant $\overline{1}$ is not the only relevant truth-value when it comes to defining consequence, but all the elements greater than $\overline{1}$; that is, the truth-preserving definition of consequence (usual in algebraic logic) uses in any algebra $\boldsymbol{A}$ the following set of *designated truth-degrees*: $D = \{a \in A \mid a \geq \overline{1}\}$.

A general property shared by all the mentioned algebraic semantics for fuzzy logics, from Łukasiewicz and Gödel–Dummett logics to these weaker systems, is that each algebra can be represented as a subdirect product of chains, i.e. can be embedded into a product of linearly ordered algebras in such a way that each projection is surjective. Therefore, all these logics are complete with respect to the semantics given just by linearly ordered algebras (and, for many prominent logics this gets even better, as we have mentioned, because they are complete w.r.t. standard chains). Based on this fact it has been argued that the only essential feature of fuzzy logics is that they are the logics of chains [2,8].

The field of research determined by this wide family of logics has attracted many researchers who have extensively carried out for MFL a typical agenda of mathematical logic: proof theory, model theory, modalities, first and higher order formalisms, axiomatic fuzzy set and fuzzy class theories, recursion and complexity, functional representation, different kinds of semantics, connections with other areas of Mathematics, applications to Philosophy, etc.; see e.g. the handbook series [6] and references therein.

4 A Degree-Based Account of Vagueness

The fundamental questions that a theory of vagueness should answer are: (1) What is the meaning (semantic value) of a vague predicate? and (2) How should we reason in the presence of vagueness? As part of answering these questions, a theory of vagueness should solve the sorites paradox—and this in turn involves two tasks: (i) Locate the error in the sorites argument: the premise that isn't true or the step of reasoning that is incorrect. This part of the solution should fall out of the answers to (1) and (2) above. (ii) Explain why the sorites argument for a vague predicate is a paradox rather than a simple fallacy: that is, provide an explanation of why competent speakers find the argument compelling but not convincing—why they initially go along with the reasoning but are still not inclined to accept the conclusion.

The simplest fuzzy answer to (1) is that the meaning of a vague predicate is a fuzzy set. However, this simple answer is inadequate. For it is generally accepted that language is a human artefact: the sounds we make mean what they do because of the kinds of situations in which we (and earlier speakers) have made (and would make) those sounds. This generates a constraint on any theory of vagueness: if the theory says that vague predicates have meanings of such-and-such a kind (e.g. fuzzy sets), then we must be able to satisfy ourselves that our past and present usage (and usage dispositions) could indeed determine such meanings for actual vague predicates. However it seems that usage and usage dispositions do not suffice to pick out a single fuzzy set—a particular function from objects to $[0, 1]$—as the extension of 'is tall' (and similarly for other vague predicates). For this reason, Smith [33] proposed *fuzzy plurivaluationism.* Instead of each vague discourse being associated with a unique intended fuzzy model, the plurivaluationist idea is that each vague discourse is associated with multiple acceptable fuzzy models. The acceptable models are all those that our usage and usage dispositions do *not* rule out as being incorrect interpretations of our language (e.g. an interpretation that does not map persons generally agreed to be paradigmatic instances of 'tall' to 1 is incorrect). On this view, a fuzzy set is the right *kind* of thing to be the meaning of a vague predicate—but there is not, in general, just one fuzzy set that is the uniquely correct meaning of a vague predicate in ordinary discourse. Rather, there are many fuzzy sets—one in each acceptable model—each of which is an equally correct meaning.

The answer that we propose to question (2) is that we should reason in the presence of vagueness in accordance with some system of MFL—although not necessarily the same system in every context: different reasoning scenarios may require different logics (see Sect. 6 below for more details).

Suppose we have a sorites series $x_0, \ldots, x_n$ for the predicate F and the associated sorites argument:

$$Fx_0, \quad Fx_0 \to Fx_1,\ Fx_1 \to Fx_2,\ \ldots,\ Fx_{n-1} \to Fx_n \qquad \therefore Fx_n.$$

If we employ Łukasiewicz logic with a definition of consequence as preservation of degree 1, then we get the following solution to the sorites. (i) The problem with the argument is that, although it is valid, it is unsound: it is not the case that every premise is true to degree 1. (ii) The argument is nevertheless compelling because all the premises are either true to degree 1 or *very nearly* true to degree 1, and in ordinary reasoning contexts we tend to apply a useful approximation heuristic that involves rounding very small differences up or down—hence we go along with the premisses, even though they are not all, strictly speaking, fully true.

Two key arguments in favour of this theory of vagueness are as follows. First, no other theory can solve the sorites in an equally satisfactory way: all other extant theories are forced to attribute ad hoc, implausible mistakes to ordinary reasoners to explain why they go along with the sorites reasoning [30]. Second, no other theory fits with our best understanding of what vagueness fundamentally consists in. In Sect. 2 we introduced vague predicates via three characteristics: blurry boundaries, borderline cases and sorites susceptibility. This can be compared to explaining what water is by saying it's a clear potable liquid that falls as rain and boils at 100 °C: this helps someone who doesn't know what water is to identify samples of it, but it still leaves open the question of the underlying nature or essence of water—of what water fundamentally *is*, that explains why it has these characteristics. The same goes for vagueness: it would be desirable to understand its fundamental nature and explain *why* it has the three surface characteristics. Smith [32,33] has argued that a predicate F is vague iff it satisfies the following Closeness principle:

> If x and y are very similar in respects relevant to the application of F, then Fx and Fy are very similar in respect of truth.

This yields explanations of why vague predicates have their surface characteristics: i.e. assuming only that a predicate P satisfies Closeness, we can derive that P must have blurry boundaries and borderline cases and generate sorites paradoxes. Furthermore, only theories of vagueness that admit degrees of truth can allow that there exist predicates that satisfy Closeness. This, then, is a strong reason for accepting fuzzy theories of vagueness, which do admit degrees of truth.

Two key arguments against fuzzy theories of vagueness are as follows. First, there is the artificial precision objection, that it is implausible to associate each vague predicate in natural language with a *particular* function that assigns a unique real number to each object. Fuzzy plurivaluationism avoids this objection, however, as it associates each vague predicate with many such functions (one per admissible model). Second, there is the truth-functionality objection, that fuzzy theories are incompatible with ordinary usage of compound propositions in the presence of borderline cases. However, this objection is based on an outdated understanding of fuzzy logics as having only very limited resources—for example, minimum and maximum as the only possible interpretations of conjunction and disjunction [31].

5 Modelling Graded Predicates in MFL

In the previous section we have seen that the algebraic semantics of many prominent fuzzy logics (such as Łukasiewicz, Gödel–Dummett, and other t-norm-based logics), though it has only one designated element on each algebra, is already powerful enough to provide a model of vagueness. This unique designated element (the maximum value of the lattice order, the number 1 on $[0,1]$-valued models) represents full truth and plays two important roles: it is the value used for the truth-preserving definition of logical consequence, and it is the neutral element of the operation that interprets the conjunction &.

We now sketch a proposal for modelling reasoning with graded predicates that exploits a greater part of the power of MFL. Indeed, we consider models for the weaker fuzzy logics mentioned in Sect. 3 where the neutral element of conjunction is an element $u \leq 1$, and the set of designated values that define the truth-preserving notion of consequence is $D = \{a \mid a \geq u\}$, not necessarily a singleton.

Vague Predicates: The usage of non-integral algebras of truth-values gives an interesting complement to the theory of vagueness explained in Sect. 4, that allows one to distinguish between different clear instances of a vague predicate. Take, for example, the predicate *tall* (and assume that we have already fixed a particular context of application). From a fuzzy plurivaluationist perspective, such a predicate admits many models which should all agree on clear instances and clear non-instances, and may assign degrees to borderline cases in different ways. Take individuals a, b, c, d, and e with respective heights of 2.1, 1.87, 1.78, 1.63, and 1.58 m. All models will agree that a and b are tall and that d and e are not tall, while c is a borderline case that will receive different degrees of tallness in different models. Algebraic models of MTL (and its extensions) have only one designated value for truth (1) and one for falsity (0), so the mentioned clear cases will only take these values. If T is a unary predicate symbol for *tall*, then the formulas Ta and Tb will be evaluated to 1, while Td and Te will be evaluated to 0, in symbols: $\|Ta\| = \|Tb\| = 1$ and $\|Td\| = \|Te\| = 0$. However, if we take instead an algebraic model of UL, defined for example by a left-continuous uninorm, then the set of designated elements is the interval $[u, 1]$, where u is the neutral element (the interpretation of $\overline{1}$). This provides a finer model for the vague predicate that allows one to make distinctions among clear cases, and clear non-cases, that is, both a and b are definitely tall, hence $\|Ta\|, \|Tb\| \in [u, 1]$, but a is much taller than b, which can be captured in the model by requiring $\|Ta\| > \|Tb\|$; hence the identification of all clear cases in one truth-value enforced by MTL and its extensions is no longer necessary. Similarly with the cases that are definitely not tall. Where f is the interpretation of $\overline{0}$, the set of degrees $[0, f]$ gives a whole range to interpret clear non-cases, in particular: $\|Td\|, \|Te\| \in [0, f]$ with $\|Te\| < \|Td\|$. This suggests, as already pointed out by Paoli [26], the following revision of the Closeness principle:

> x and y are very similar in F-relevant respects if, *and only if*, Fx and Fy are very similar in respect of truth.

Paoli uses this revised principle to argue that vague predicates can be better interpreted in models that have more than one truth-degree for clear cases, and more than one for clear non-cases. However, his proposal is restricted to the algebraic models of Casari's comparative logic [4]. We believe that such a restriction is not flexible enough, because for example it excludes non-commutative or non-associative interpretations of residuated conjunction, which are necessary in some reasoning scenarios.

On the other hand, algebraic models of fuzzy logics can always be decomposed into linearly ordered components (technically, by means of subdirect representation). This property allows us to account for the fact that many (maybe all) vague predicates depend on underlying parameters that vary on linear scales. Furthermore, if there are any vague predicates that cannot be explained from a set of parameters that vary on a linear scale of degrees—that is, if there are nonlinear nonmultidimensional vague predicates—they can still be modelled in a degree-based approach if we enhance somewhat the logical framework and allow for systems that do not enjoy completeness (and subdirect decomposition) w.r.t. chains. Algebraic logic offers methods to build algebraic models for any non-classical logic. In particular, if the logic has a reasonable implication connective, it induces an order relation in its algebraic models [9,10] and, hence, such algebras can be seen as (not necessarily linearly ordered) scales of degrees adequate for modelling such predicates.

Graded Precise Predicates: These predicates also admit models on algebras of fuzzy logics, but with an important restriction of the evaluation functions to account for the fact that there are no borderline cases. Take, for instance, the predicate *acute angle*, represented by the unary predicate symbol A. Consider again a model defined by a left-continuous uninorm with neutral element $u < 1$, and let f be the interpretation of $\overline{0}$. Then, an admissible evaluation would be given by:

$$\|Ax\| = \begin{cases} (u-1)x/90 + 1, & \text{if } 0 \leq x < 90 \\ -fx/270 + 4f/3, & \text{if } 90 \leq x < 360 \end{cases}$$

that is, a piecewise linear map that maps all acute angles to the interval of designated elements $[u, 1]$ (in particular it maps 0 to 1, because 0° is the most acute angle), and maps all non-acute angles to the interval $[0, f]$ (in particular, it maps 90° to f, because it is the least non-acute angle). Observe that no angle is mapped to the interval (f, u) of intermediate truth-values. This will be a common characteristic of all models for graded precise predicates, because, unlike vague predicates, they have no borderline cases. Again, MFL offers a wealth of logical systems to model a multitude of graded precise predicates depending on the needs of each context.

This semantical treatment of graded precise predicates is inspired by Paoli's proposal [26,27] where the interpretations were given on algebraic models of Casari's comparative logic.

6 A General Program

We propose a research program of correct reasoning with graded properties, done from the point of view of Logic and based on the following three layers of analysis:

1. *Natural language and natural reasoning scenarios:* Interdisciplinary research relating Logic to Cognitive Science, Psychology and Linguistics in order to understand how correct reasoning is actually carried out in natural language with graded properties.
2. *Formal interpreted languages and artificial reasoning scenarios:* The application of tools of mathematical logic (level 3, below) to natural reasoning scenarios (level 1, above) requires the introduction of a middle level, in which the logical formalisms come with specific interpretations of graded properties in specific contexts. This has the potential for applications to Computer Science that require handling graded predicates.
3. *Formal abstract languages and mathematical logic:* The systematic mathematical study of non-classical logical systems with a graded semantics upon which the study of the previous levels can be based.

The ideas sketched in the previous section illustrate how the mathematical machinery developed in the study of non-classical logics can be used in modelling graded predicates and reasoning with them. The study of such logical systems in layer three is done in a completely abstract way, as free mathematical research unconstrained by the possible interpretations of the formal language.

However, by requiring specific behaviour of (part of) the formal language, motivated by particular reasoning scenarios, we move to the second layer: for example, studying logics with additional modalities aimed at modelling agents' partial knowledge or belief, or their degrees of preference—or other specific fragments of first-order or higher-order logics with good representational power and complexity properties. The potential for applications is suggested by the fact that some areas of Computer Science use several kinds of *weighted* notions, i.e. graded properties, for example in valued constraint satisfaction problems, weighted graphs, weighted automata, and so on. The application of MFL and other algebraic logical tools to these areas is still very much under-explored.

Finally, the first layer can be seen as a proposal in the spirit of Stenning and van Lambalgen's endeavour to bring Logic back to the study of reasoning [35], after years of evolving in separate directions. They have convincingly argued that Logic is very much domain-dependent: valid forms of inference depend on the domain of discourse. Accordingly, they claim that each instance of reasoning requires two stages:

1. *reasoning to an interpretation:* in which one has to decide what are the appropriate formal tools for the particular reasoning scenario (language, models, notion of consequence)
2. *reasoning from an interpretation:* in which, having established the previous parameters, one can reason according to the chosen form of inference.

This approach is compatible with the kind of plurivaluationism defended in [33] and with the idea, advocated in [1] and mentioned in Sect. 4 above, that we need not pick one particular logic from the MFL family and then use it in every reasoning context that involves graded notions: there are differences among graded notions (e.g. vague vs graded precise—and within the vague predicates, linear vs nonlinear; etc.) and different contexts may well require different logics. In general, our broader aim is to apply the full suite of MFL tools, the plurivaluationism of [33], and the methodology proposed in [35] to reasoning scenarios involving graded properties.

Acknowledgments. Petr Cintula and Carles Noguera are supported by the project GA17-04630S of the Czech Science Foundation (GAČR); both authors have also received funding from the European Union's Horizon 2020 research and innovation programme under the Marie Sklodowska-Curie grant agreement No 689176 (SYSMICS project). Petr Cintula also acknowledges the support of RVO 67985807. Nicholas Smith is supported by the SOPHI Research Support Scheme at the University of Sydney.

References

1. Běhounek, L.: In which sense is fuzzy logic a logic for vagueness? In: Lukasiewicz, T., Peñaloza, R., Turhan, A.Y. (eds.) Proceedings of the First Workshop on Logics for Reasoning about Preferences, Uncertainty, and Vagueness (PRUV 2014), Vienna, pp. 26–39 (2014)
2. Běhounek, L., Cintula, P.: Fuzzy logics as the logics of chains. Fuzzy Sets Syst. **157**(5), 604–610 (2006)
3. Blok, W.J., Pigozzi, D.L.: Algebraizable Logics, Memoirs of the American Mathematical Society, vol. 396. American Mathematical Society, Providence (1989). http://orion.math.iastate.edu/dpigozzi/
4. Casari, E.: Comparative logics and Abelian ℓ-groups. In: Logic Colloquium 1988, Padova. Studies in Logic and the Foundations of Mathematics, vol. 127, pp. 161–190. North-Holland, Amsterdam (1989)
5. Cignoli, R., Esteva, F., Godo, L., Torrens, A.: Basic fuzzy logic is the logic of continuous t-norms and their residua. Soft. Comput. **4**(2), 106–112 (2000)
6. Cintula, P., Fermüller, C.G., Hájek, P., Noguera, C. (eds.): Handbook of Mathematical Fuzzy Logic (in three volumes). Studies in Logic, Mathematical Logic and Foundations, vols. 37, 38, and 58. College Publications (2011, 2015)
7. Cintula, P., Horčík, R., Noguera, C.: Non-associative substructural logics and their semilinear extensions: axiomatization and completeness properties. Rev. Symbolic Logic **6**(3), 394–423 (2013)
8. Cintula, P., Horčík, R., Noguera, C.: The quest for the basic fuzzy logic. In: Montagna, F. (ed.) Petr Hájek on Mathematical Fuzzy Logic. OCL, vol. 6, pp. 245–290. Springer, Cham (2015). doi:10.1007/978-3-319-06233-4_12
9. Cintula, P., Noguera, C.: Implicational (semilinear) logics I: a new hierarchy. Arch. Math. Logic **49**(4), 417–446 (2010)
10. Cintula, P., Noguera, C.: A Henkin-style proof of completeness for first-order algebraizable logics. J. Symb. Log. **80**(1), 341–358 (2015)
11. Dummett, M.: A propositional calculus with denumerable matrix. J. Symb. Log. **24**(2), 97–106 (1959)
12. Esteva, F., Godo, L.: Monoidal t-norm based logic: towards a logic for left-continuous t-norms. Fuzzy Sets Syst. **124**(3), 271–288 (2001)
13. Esteva, F., Godo, L., García-Cerdaña, À.: On the hierarchy of t-norm based residuated fuzzy logics. In: Fitting, M.C., Orlowska, E. (eds.) Beyond Two: Theory and Applications of Multiple-Valued Logic, Studies in Fuzziness and Soft Computing, vol. 114, pp. 251–272. Springer, Heidelberg (2003)
14. Fine, K.: Vagueness, truth and logic. Synthese **30**, 265–300 (1975). Reprinted with corrections in [22], pp. 119–150
15. Galatos, N., Jipsen, P., Kowalski, T., Ono, H.: Residuated Lattices: An Algebraic Glimpse at Substructural Logics. Studies in Logic and the Foundations of Mathematics, vol. 151. Elsevier, Amsterdam (2007)
16. Galatos, N., Ono, H.: Cut elimination and strong separation for substructural logics: an algebraic approach. Ann. Pure Appl. Logic **161**(9), 1097–1133 (2010)
17. Hájek, P.: Metamathematics of Fuzzy Logic, Trends in Logic, vol. 4. Kluwer, Dordrecht (1998)
18. Hájek, P.: Fuzzy logics with noncommutative conjunctions. J. Log. Comput. **13**(4), 469–479 (2003)
19. Jenei, S., Montagna, F.: A proof of standard completeness for Esteva and Godo's logic MTL. Stud. Logica. **70**(2), 183–192 (2002)

20. Jenei, S., Montagna, F.: A proof of standard completeness for non-commutative monoidal t-norm logic. Neural Netw. World **13**(5), 481–489 (2003)
21. Keefe, R.: Theories of Vagueness. Cambridge University Press, Cambridge (2000)
22. Keefe, R., Smith, P. (eds.): Vagueness: A Reader. MIT Press, Cambridge (1997)
23. Lambek, J.: On the calculus of syntactic types. In: Jakobson, R. (ed.) Structure of Language and Its Mathematical Aspects, pp. 166–178. American Mathematical Society, Providence (1961)
24. Łukasiewicz, J., Tarski, A.: Untersuchungen über den Aussagenkalkül. Comptes Rendus des Séances de la Société des Sciences et des Lettres de Varsovie, cl. III **23**(iii), 30–50 (1930)
25. Metcalfe, G., Montagna, F.: Substructural fuzzy logics. J. Symb. Log. **72**(3), 834–864 (2007)
26. Paoli, F.: Truth degrees, closeness, and the sorites. In: Bueno, O., Abasnezhad, A. (eds.) On the Sorites Paradox. Springer (to appear)
27. Paoli, F.: Comparative logic as an approach to comparison in natural language. J. Semant. **16**, 67–96 (1999)
28. Raffman, D.: Vagueness without paradox. Philos. Rev. **103**, 41–74 (1994)
29. Shapiro, S.: Vagueness in Context. Oxford University Press, Oxford (2006)
30. Smith, N.J.J.: Consonance and dissonance in solutions to the sorites. In: Bueno, O., Abasnezhad, A. (eds.) On the Sorites Paradox. Springer (to appear)
31. Smith, N.J.J.: Undead argument: the truth-functionality objection to fuzzy theories of vagueness. To appear in Synthese. doi:10.1007/s11229-014-0651-7
32. Smith, N.J.J.: Vagueness as closeness. Australas. J. Philos. **83**, 157–183 (2005)
33. Smith, N.J.J.: Vagueness and Degrees of Truth. Oxford University Press, Oxford (2008). Paperback 2013
34. Sorensen, R.: Vagueness and Contradiction. Oxford University Press, Oxford (2001)
35. Stenning, K., van Lambalgen, M.: Human Reasoning and Cognitive Science. MIT Press, Cambridge (2008)
36. Williamson, T.: Vagueness. Routledge, London (1994)

Evidence Logics with Relational Evidence

Alexandru Baltag[1] and Andrés Occhipinti[2(✉)]

[1] University of Amsterdam, Amsterdam, Netherlands
A.Baltag@uva.nl
[2] DTU Compute, Technical University of Denmark, Copenhagen, Denmark
aocc@dtu.dk

Abstract. We introduce a family of logics for reasoning about *relational evidence*: evidence that involves an ordering of states in terms of their relative plausibility. We provide sound and complete axiomatizations for the logics. We also present several evidential actions and prove soundness and completeness for the associated dynamic logics.

Keywords: Evidence logic · Dynamic epistemic logic · Belief revision

Dynamic evidence logics [2,14–17] are logics for reasoning about the evidence and evidence-based beliefs of agents in a dynamic environment. Evidence logics are concerned with scenarios in which an agent collects several pieces of evidence about a situation of interest, from a number of sources, and uses this evidence to form and revise her beliefs about this situation. The agent is typically uncertain about the actual state of affairs, and as a result takes several alternative descriptions of this state as possible (these descriptions are typically called *possible worlds* or *possible states*). The existing evidence logics, i.e., *neighborhood evidence logics* (NEL) [2,14–17], have the following features:

1. *All evidence is 'binary'.* Each piece of evidence is modeled as a set of possible states. This set indicates which states are good candidates for the actual state, and which ones are not, according to the source. Hence the name binary; every state is either a good candidate ('in'), or a bad candidate ('out').
2. *All evidence is equally reliable.* The agent treats all evidence pieces on a par. There is no explicit modeling of the relative *reliability* of pieces of evidence.
3. *One procedure to combine evidence.* The logics developed so far study the evidence and beliefs held by an agent relying on one specific procedure for combining evidence.

This work presents a family of dynamic evidence logics which we call *relational evidence logics* (REL). Relational evidence logics aim to contribute to the existing work on evidence logic as follows.

1. *Relax the assumption that all evidence is binary.* This is accomplished by modeling pieces of evidence by *evidence relations*. Evidence relations are preorders over the set of possible states. The ordering is meant to represent the

A. Baltag et al. (Eds.): LORI 2017, LNCS 10455, pp. 17–32, 2017.
DOI: 10.1007/978-3-662-55665-8_2

relative plausibility of states based on the evidence. While a special type of evidence relation – *dichotomous order* – can be used to model binary evidence, less 'black-and-white' forms of evidence can also be encoded in REL models.

2. *Model levels of evidence reliability.* In general, not all evidence is equally reliable. Expert advice and gossip provide very different grounds for belief, and a rational agent should weight the evidence that it is exposed to accordingly. To model evidence reliability, we equipped our models with *priority orders*, i.e., orderings of the family of evidence relations according to their relative reliability. Priority orders were introduced in [1], and have already been used in other DEL logics (see, e.g. [9,12]). Here, we use them to model the relative reliability of pieces of evidence.
3. *Explore alternative evidence aggregation rules.* Our evidence models come equipped with an aggregator, which merges the available evidence relations into a single relation representing the combined plausibility of the possible states. The beliefs of the agent are then defined on the basis of this combined plausibility order. By focusing on different classes of evidence models, given by their underlying aggregator, we can then compare the logics of belief arising from different approaches to combining evidence.

1 Relational Evidence Models

Relational evidence. We call *relational evidence* any type of evidence that induces an ordering of states in terms of their relative plausibility. A suitable representation for relational evidence, which we adopt, is given by the class of *preorders*. We call preorders representing relational evidence, *evidence relations*, or *evidence orders*. As is well-known, preorders can represent several meaningful types of orderings, including those that feature incomparable or tied alternatives.

Definition 1 (Preorder). *A* preorder *is a binary relation that is reflexive and transitive. We denote the set of all preorders on a set* X *by* $Pre(X)$. *For a preorder* R *on* X *and an element* $x \in X$, *we define the following associated notions:* $R[x] := \{y \in X \mid Rxy\}$; $R^{<} := \{(x,y) \in X^2 \mid Rxy \text{ and } Ryx\}$; $R^{\sim} := \{(x,y) \in X^2 \mid Rxy \text{ and } \neg Ryx\}$.

Evidence reliability. In general, not all sources are equally trustworthy, so an agent combining evidence may be justified in giving priority to some evidence items over others. As suggested in [17], a next reasonable step in evidence logics is modeling levels of reliability of evidence. One general format for this is given by the *priority graphs* of [1], which have already been used extensively in dynamic epistemic logic (see, e.g., [9,12]). In this work, we will use the related, yet simpler format of a 'priority order', as used in [5,6], to represent hierarchy among pieces of evidence. Our definition of a priority order is as follows:

Definition 2 (Priority order). *Let* $\mathscr{R}$ *be a family of evidence orders over* W. *A priority order for* $\mathscr{R}$ *is a preorder* $\preceq$ *on* $\mathscr{R}$. *For* $R, R' \in \mathscr{R}$, $R \preceq R'$ *reads as: "the evidence order* R' *has at least the same priority as evidence order* R*".*

Intuitively, priority orders tell us which pieces of evidence are more reliable according to the agent. They give the agent a natural way to break stalemates when faced with inconsistent evidence.

Evidence aggregators. We are interested in modeling a situation in which an agent integrates evidence obtained from multiple sources to obtain and update a combined plausibility ordering, and forms beliefs based on this ordering. When we consider relational evidence with varying levels of priority, a natural way model the process of evidence combination is to define a function that takes as input a family of evidence orders $\mathscr{R}$ together with a priority order $\preceq$ defined on them, and combines them into a plausibility order. The agent's beliefs can then be defined in terms of this output.

Definition 3 (Evidence aggregator). *Let W be a set of alternatives. Let $\mathcal{W}$ be the set of preorders on W. An* evidence aggregator *for W is a function Ag mapping any preordered family $P = \langle \mathscr{R}, \preceq \rangle$ to a preorder $Ag(P)$ on W, where $\emptyset \notin \mathscr{R} \subseteq \mathcal{W}$ and $\preceq$ is a preorder on $\mathscr{R}$. $\mathscr{R}$ is seen here as a family of evidence orders over W, $\preceq$ as a priority order for $\mathscr{R}$, and $Ag(P)$ as an evidence-based plausibility order on W.*

At first glance, our definition of an aggregator may seem to impose mild constraints that are met by most natural aggregation functions. However, as it is well-known, the output of some common rules, like the majority rule, may not be transitive (thus not a preorder), and hence they don't count as aggregators. A specific aggregator that *does* satisfy the constraints is the *lexicographic rule.* This aggregator was extensively studied in [1], where it was shown to satisfy several nice aggregative properties. The definition of the aggregator is the following:

Definition 4. *The (anti-)lexicographic rule is the aggregator* lex *given by*

$$(w,v) \in \mathsf{lex}(\langle \mathscr{R}, \preceq \rangle) \text{ iff } \forall R' \in \mathscr{R}\ (R'wv \ \vee\ \exists R \in \mathscr{R}(R' \prec R \wedge R^{<}wv))$$

Intuitively, the lexicographic rule works as follows. Given a particular hierarchy $\preceq$ over a family of evidence $\mathscr{R}$, aggregation is done by giving priority to the evidence orders further up the hierarchy in a compensating way: the agent follows what all evidence orders agree on, if it can, or follows more influential pieces of evidence, in case of disagreement. Other well-known aggregators that satisfy the constraints, but don't make use of the priority structure, are the intersection rule (defined below), or the Borda rule.

Definition 5. *The intersection rule is the aggregator $Ag_{\cap}$ given by $(w,v) \in Ag_{\cap}(\langle \mathscr{R}, \preceq \rangle)$ iff $(w,v) \in \bigcap \mathscr{R}$.*

The models. Having defined relational evidence and evidence aggregators, we are now ready to introduce relational evidence models.

Definition 6 (Relational evidence model). *Let* P *be a set of propositional variables. A* relational evidence model *(*REL *model, for short) is a tuple*

$M = \langle W, \langle \mathscr{R}, \preceq \rangle, V, Ag \rangle$ where W is a non-empty set of states; *$\langle \mathscr{R}, \preceq \rangle$ is an ordered family of evidence, where: $\mathscr{R}$ is a set of evidence orders on W with $W^2 \in \mathscr{R}$ and $\preceq$ is a priority order for $\mathscr{R}$; $V : \mathsf{P} \to 2^W$ is a valuation function; Ag is an evidence aggregator for W.*

$W^2 \in \mathscr{R}$ is called the *trivial evidence order*. It represents the evidence stating that "the actual state is in W", which is taken to be always available to the agent as a starting point. $M = \langle W, \langle \mathscr{R}, \preceq \rangle, V, Ag \rangle$ is called an *f-model* iff $Ag = f$.

Syntax and semantics. We now introduce a *static* language for reasoning about relational evidence, which we call $\mathscr{L}$. In [2], this language is interpreted over NEL models (there, the language is called $\mathscr{L}_{\forall \Box \Box_0}$).

Definition 7 ($\mathscr{L}$). *Let P be a countably infinite set of propositional variables. The language $\mathscr{L}$ is defined by:*

$$\varphi ::= p \mid \neg \varphi \mid \varphi \wedge \varphi \mid \Box_0 \varphi \mid \Box \varphi \mid \forall \varphi \quad (p \in \mathsf{P})$$

The intended interpretation of the modalities is as follows. $\Box_0 \varphi$ reads as: 'the agent has basic, factive evidence for φ'; $\Box \varphi$ reads as: 'the agent has combined, factive evidence for φ'. The language $\mathscr{L}$ is interpreted over REL models as follows.

Definition 8 (Satisfaction). *Let $M = \langle W, \langle \mathscr{R}, \preceq \rangle, V, Ag \rangle$ be an REL model and $w \in W$. The satisfaction relation $\models$ between pairs (M, w) and formulas $\varphi \in \mathscr{L}$ is defined as follows (the propositional clauses are as usual):*

$M, w \models \Box_0 \varphi$ *iff there is $R \in \mathscr{R}$ such that, for all $v \in W$, Rwv implies $M, v \models \varphi$*
$M, w \models \Box \varphi$ *iff for all $v \in W$, $Ag(\langle \mathscr{R}, \preceq \rangle)wv$ implies $M, v \models \varphi$*
$M, w \models \forall \varphi$ *iff for all $v \in W$, $M, v \models \varphi$*

Definition 9 (Truth map). *Let $M = \langle W, \langle \mathscr{R}, \preceq \rangle, V, Ag \rangle$ be a REL model. We define a truth map $[\![\cdot]\!]_M : \mathscr{L} \to 2^W$ given by: $[\![\varphi]\!]_M = \{w \in W \mid M, w \models \varphi\}$.*

Next, we introduce some definable notions of evidence and belief over REL models, illustrated below with an example. Fix a model $M = \langle W, \langle \mathscr{R}, \preceq \rangle, V, Ag \rangle$.

Basic (factive) Evidence. We say that a piece of evidence $R \in \mathscr{R}$ *supports* φ at $w \in W$ iff $R[w] \subseteq [\![\varphi]\!]_M$. That is, every world that is at least as plausible as w under R satisfies φ. Using this notion of support, we say that the agent has basic, factive evidence for φ at $w \in W$ if there is a piece of evidence $R \in \mathscr{R}$ that supports φ at w. That is: 'the agent has basic evidence for φ at $w \in W$' iff $\exists R \in \mathscr{R}(R[w] \subseteq [\![\varphi]\!]_M)$ iff $M, w \models \Box_0 \varphi$. We also have a non-factive version of this notion, which says that the agent has basic evidence for φ if there is a piece of evidence R that supports φ at *some* state, i.e.: 'the agent has basic evidence for φ (at any state)' iff $\exists w(\exists R \in \mathscr{R}(R[w] \subseteq [\![\varphi]\!]_M))$ iff $M, w \models \exists \Box_0 \varphi$. We can also have a *conditional* version of basic evidence: 'the agent has basic, factive evidence for ψ at w, conditional on φ being true'. Putting $\Box_0^\varphi \psi := \Box_0(\varphi \to \psi)$, we have: 'the agent has basic, factive evidence for ψ at w, conditional on φ being true' iff $\exists R \in \mathscr{R}(\forall v(Rwv \Rightarrow (v \in [\![\varphi]\!]_M \Rightarrow v \in [\![\psi]\!]_M)))$ iff $M, w \models \Box_0^\varphi \psi$. The notion of conditional evidence reduces to that of plain evidence when $\varphi = \top$.

Aggregated (factive) Evidence. We propose a notion of aggregated evidence based on the output of the aggregator: the agent has aggregated, factive evidence for φ at $w \in W$ iff $Ag(\langle \mathscr{R}, \preceq \rangle)[w] \subseteq [\![\varphi]\!]_M$ iff $M, w \models \Box\varphi$. The non-factive version of the previous notion is as follows: the agent has aggregated evidence for φ (at any state) iff $\exists w(Ag(\langle \mathscr{R}, \preceq \rangle)[w] \subseteq [\![\varphi]\!]_M)$ iff $M, w \models \exists\Box\varphi$. As we did with basic evidence, we can define a conditional notion of aggregated evidence in φ by putting $\Box^{\varphi}\psi := \Box(\varphi \to \psi)$. The unconditional version is given by $\varphi = \top$.

Evidence-Based Belief. The notion of belief we will work with is based on the agent's plausibility order, which in REL models corresponds to the output of the aggregator. As we don't require the plausibility order to be converse-well founded, it may have no maximal elements, which means that Grove's definition of belief may yield inconsistent beliefs. For this reason, we adopt a usual generalization of Grove's definition, which defines beliefs in terms of truth in all 'plausible enough' worlds (see, e.g., [3,16]). Putting $B\varphi := \forall\Diamond\Box\varphi$, we have: the agent believes φ (at any state) iff $\forall w(\exists v((w,v) \in Ag(\langle \mathscr{R}, \preceq \rangle)$ and $Ag(\langle \mathscr{R}, \preceq \rangle)[v] \subseteq [\![\varphi]\!]_M))$ iff $M, w \models \forall\Diamond\Box\varphi$. That is, the agent believes φ iff for every state $w \in W$, we can always find a more plausible state $v \in [\![\varphi]\!]_M$, all whose successors are also in $[\![\varphi]\!]_M$. When the plausibility relation is indeed converse well-founded, this notion of belief coincides with Grove's one, while ensuring consistency of belief otherwise. We can also define a notion of conditional belief. Putting $B^{\varphi}\psi := \forall(\varphi \to \Diamond(\varphi \to (\Box\varphi \to \psi)))$, we have: 'the agent believes ψ conditional on φ iff $\forall w(w \in [\![\varphi]\!]_M \Rightarrow \exists v(Ag(\langle \mathscr{R}, \preceq \rangle)$ wv and $v \in [\![\varphi]\!]_M$ and $Ag(\langle \mathscr{R}, \preceq \rangle)[v] \cap [\![\varphi]\!]_M \subseteq [\![\psi]\!]_M))$ iff $M, w \models B^{\varphi}\psi$. As before, this conditional notion reduces to that of absolute belief when $\varphi = \top$.

Example 1 (The diagnosis). Consider an agent seeking medical advice on an ongoing health issue. To keep thing simple, assume that there are four possible diseases: asthma (a), allergy (al), cold (c), and flu (f). This can be described by a set W consisting of four possible worlds, $\{w_a, w_{al}, w_c, w_f\}$ and a set of atomic formulas $\{a, al, c, f\}$ (each true at the corresponding world). The agent consults three sources, a medical intern (IN), a family doctor (FD) and an allergist (AL). The doctors inspect the patient, observing fairly non-specific symptoms: cough, no fever, and some inconclusive swelling at an allergen test spot. Given the non-specificity of the symptoms, the doctors can't single out a condition that best explains all they observed. Instead, comparing the diseases in terms of how well they explain the observed symptoms, and drawing on their experience, each doctor arrives at a ranking of the possible diseases. Let us denote by R_{IN}, R_{FD} and R_{AL} the evidence orders representing the judgment of the intern, family doctor and allergist, respectively, which we assume to be as depicted below. If the agent has no information about how reliable each doctor is, she may just trust them all equally. We can model this by a priority order $\preceq$ over the evidence orders $R_{IN} \sim R_{FD} \sim R_{AL}$ that puts all evidence as equally likely. On the other hand, if the agent knows that the intern is the least experienced of the doctors, she may consider his evidence as strictly less reliable than the one provided by the other doctors. Similarly, if the allergist has a strong reputation, the agent may wish to give the allergist's judgment strict priority over the rest. We can model this by a different priority order $\preceq'$ given by $R_{IN} \prec' R_{FD} \prec' R_{AL}$ (note that this is meant to be reflexive and transitive). If, e.g., the agent uses the lexicographic rule, we arrive at the following scenarios, with different aggregated evidence depending on the priority order used:

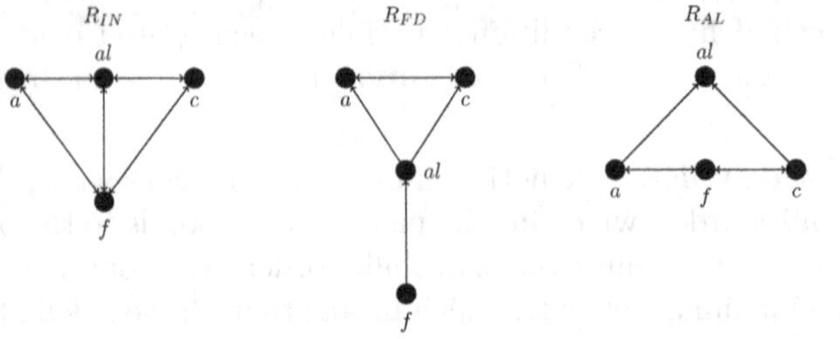

Lexicographic aggregation based on $\preceq$

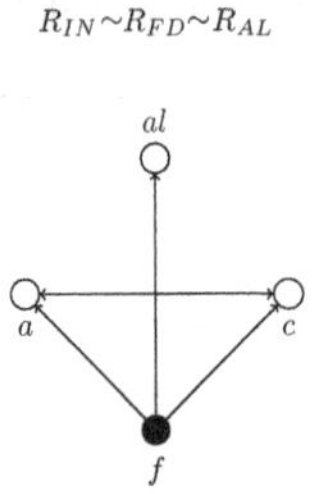

Lexicographic aggregation based on $\preceq'$

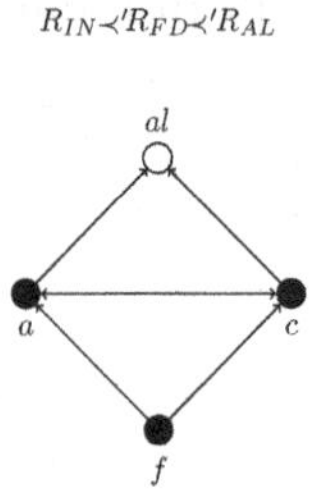

The best candidates for the actual disease, in each case, are depicted in white. Note that, e.g., the agent has basic evidence for $a \vee al \vee c$, but she doesn't have evidence for f. Moreover, in the scenario based on $\preceq'$, the agent believes that the allergy is the actual disease, but she doesn't in the scenario based on $\preceq$.

A PDL language for relational evidence. Later in this work, we will discuss *evidential actions* by which the agent, upon receiving a new piece of relational evidence, revises its existing body of evidence. To encode syntactically the evidence pieces featured in evidential actions, we will enrich our basic language $\mathscr{L}$ with formulas that stand for specific evidence relations. A natural way to introduce relation-defining expressions, in a modal setting such as ours, is to employ suitable program expressions from Propositional Dynamic Logic (PDL). We will follow this approach, augmenting $\mathscr{L}$ with PDL-style *evidence programs* that define pieces of relational evidence. As evidence orders are preorders, we will employ a set of program expressions whose terms are guaranteed to always define preorders. An natural fragment of PDL meeting this condition is the one provided by programs of the form π^*, which always define the reflexive transitive closure of some relation.

Definition 10 (Evidence programs). *The set Π has all program symbols π defined as follows:*

$$\pi ::= A \mid ?\varphi \mid \pi \cup \pi \mid \pi;\pi \mid \pi^*$$

where $\varphi \in \mathscr{L}$*. Here* A *denotes the universal program, while the rest of the programs have their usual* PDL *meanings (see, e.g., [10]). We call* $\Pi_* := \{\pi^* \mid \pi \in \Pi\}$ *the set of* evidence programs.

To interpret evidence programs in REL models, we extend the truth map:

Definition 11 (Truth map). *Let* $M = \langle W, \langle \mathscr{R}, \prec \rangle, V, Ag \rangle$ *be an REL model. We define an extended truth map* $[\![\cdot]\!]_M : \mathscr{L} \cup \Pi \to 2^W \cup 2^{W^2}$ *given by:* $[\![\varphi]\!]_M = \{w \in W \mid M, w \models \varphi\}$; $[\![A]\!]_M = W^2$; $[\![?\varphi]\!]_M = \{(w, w) \in W^2 \mid w \in [\![\varphi]\!]^M\}$; $[\![\pi \cup \pi']\!]_M = [\![\pi]\!]_M \cup [\![\pi']\!]_M$; $[\![\pi; \pi']\!]_M = [\![\pi]\!]_M \circ [\![\pi']\!]_M$; $[\![\pi^*]\!]_M = [\![\pi]\!]_M^*$.

Some Examples of Definable Evidence Programs. Here are some natural types of relational evidence that can be constructed with programs from Π_*.

Dichotomous evidence. For a formula φ, let $\pi_\varphi := (A; ?\varphi) \cup (?\neg\varphi; A; ?\neg\varphi)$. π_φ puts the φ worlds strictly above the $\neg\varphi$ worlds, and makes every world equally plausible within each of these two regions. It is easy to see that π_φ always defines a preorder, and therefore $(\pi_\varphi)^*$ is an evidence program equivalent to π_φ (Fig. 1).

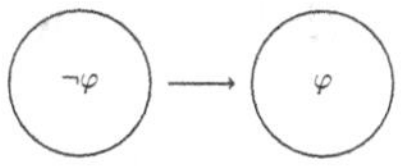

Fig. 1. The dichotomous order defined by π_φ

Totally ordered evidence. Several programs can be used to define total preorders. For example, for formulas $\varphi_1, \ldots, \varphi_n$, we can define the program

$$\begin{aligned}\pi_{\varphi_1,\ldots,\varphi_n} := & (A; ?\varphi_1) \cup (?\neg\varphi_1; A; ?\neg\varphi_1; ?\varphi_2) \\ & \cup (?\neg\varphi_1; \neg\varphi_2; A; ?\neg\varphi_1; ?\neg\varphi_2; ?\varphi_3) \\ & \cup \ldots \\ & \cup (?\neg\varphi_1; \ldots; ?\neg\varphi_n; A; ?\neg\varphi_1; \ldots; ?\neg\varphi_{n-1}; ?\varphi_n) \cup (?\top)\end{aligned}$$

This type of program, described in [18], puts the φ_1 worlds above everything else, the $\neg\varphi_1 \wedge \varphi_2$ worlds above the $\neg\varphi_1 \wedge \neg\varphi_2$ worlds, and so on, and the $\neg\varphi_1 \wedge \neg\varphi_2 \wedge \cdots \wedge \neg\varphi_{n-1} \wedge \varphi_n$ above the $\neg\varphi_1 \wedge \neg\varphi_2 \wedge \cdots \wedge \neg\varphi_n$ worlds. $\pi^t(\varphi_1, \ldots, \varphi_n)$ always defines a preorder, so the evidence program $(\pi^t(\varphi_1, \ldots, \varphi_n))^*$ is equivalent to it (Fig. 2).

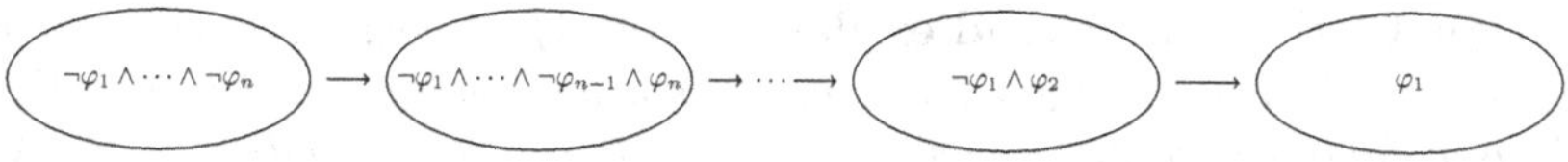

Fig. 2. The total preorder defined by $\pi_{\varphi_1,\ldots,\varphi_n}$

Incompletely ordered evidence. Several programs can be used to define evidence orders featuring incomparabilities. To illustrate this, let us consider the program $\pi_{\varphi\wedge\psi} := (A;?\varphi\wedge\psi)\cup(?\neg\varphi\wedge\neg\psi;A;?\varphi\vee\psi)\cup(?\neg\varphi\wedge\psi;A;?\neg\varphi\wedge\psi)\cup(?\varphi\wedge\neg\psi;A;?\varphi\wedge\neg\psi)\cup(?\top)$. As depicted in Fig. 3, this program puts the $\varphi\wedge\psi$ worlds above everything else, the $\neg\varphi\wedge\psi$ and $\varphi\wedge\neg\psi$ as incomparable 'second-best' worlds, and the $\neg\varphi\wedge\neg\psi$ below everything else. As with the other programs $\pi_{\varphi\wedge\psi}$ always defines a preorder, so $(\pi_{\varphi\wedge\psi})^*$ is an equivalent evidence program.

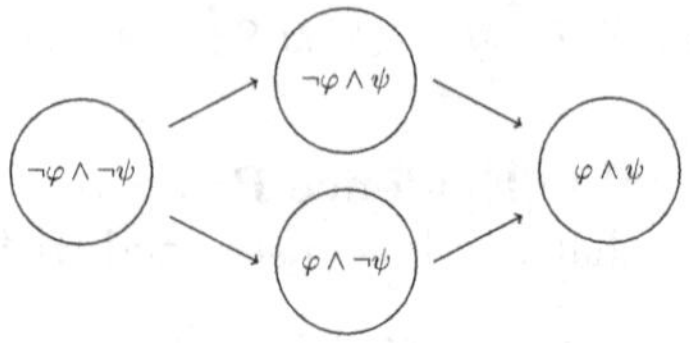

Fig. 3. The incomplete preorder defined by $\pi_{\varphi\wedge\psi}$

2 The Logics of $Ag_\cap$-Models and lex-Models

We initiate here our logical study of the statics of belief and evidence in the REL setting. We first zoom into two specific classes of REL models, the classes of $Ag_\cap$-models and lex-models, and study the static logics for belief and evidence based on these models. In particular, we introduce systems $\mathsf{L}_\cap$ and $\mathsf{L}_{\mathsf{lex}}$ that axiomatize the class of $Ag_\cap$-models and the class of lex-models, respectively. (To simplify notation, we write $\cap$-models instead of $Ag_\cap$-models hereafter). In later sections, we will 'zoom out' and study the class of all REL models. Our choice to study $\cap$ and lex models in some detail is motivated as follows. The class of $\cap$-models is interesting because it links our relational evidence setting back to the NEL setting that inspired it. Indeed, as we show right below, given any NEL model with finitely many pieces of evidence, we can always find a $\cap$-model that is modally equivalent to it (with respect to language $\mathscr{L}$). This $\cap$-model represents binary evidence in a relational way, thereby encoding the same information presented in the NEL model. lex-models, on the other hand, provide a good study case for the REL setting, as they exemplify its main novel features: non-binary evidence and reliability-sensitive aggregation. We recall here the definition of a NEL model to compare them to $\cap$-models. The definition given for these models follows the one in [2]. For a more general notion, see [15], where the models we consider are called *uniform* models.

Definition 12 (Neighborhood evidence model). *A* neighborhood evidence model *is a tuple $M=\langle W,E_0,V\rangle$ where: W is a non-empty set of states; $E_0\subseteq\mathscr{P}(W)$ is a family of basic evidence sets, such that $\emptyset\notin E_0$ and $W\in E_0$; $V:\mathsf{P}\to\mathscr{P}(W)$ is a valuation function. A model is called* feasible *if E_0 is finite. A* body of evidence *is a family $F\subseteq E_0$ such that every non-empty finite subfamily*

$F' \subseteq F$ is consistent, i.e., $\bigcap F' \neq \emptyset$. A piece of combined evidence *is any non-empty intersection of finitely many pieces of basic evidence. We denote by $E := \{\bigcap F \mid F \subseteq E_0, |F| \in \mathbb{N}\}$ the family of all combined evidence.*

Definition 13 (Satisfaction). *Let $M = \langle W, E_0, V \rangle$ be an NEL model and $w \in W$. The satisfaction relation $\models$ between pairs (M, w) and formulas $\varphi \in \mathscr{L}$ is:*

$M, w \models \Box_0\varphi$ *iff there is $e \in E_0$ such that $w \in e \subseteq [\![\varphi]\!]_M$*
$M, w \models \Box\varphi$ *iff there is $e \in E$ such that $w \in e \subseteq [\![\varphi]\!]_M$*
$M, w \models \forall\varphi$ *iff $W = [\![\varphi]\!]_M$*

We now present a way to 'transform' a NEL model into a matching REL model. To do that, we first encode binary evidence, the type of evidence considered in NEL models, as relational evidence.

Definition 14. *Let W be a set. For each $e \subseteq W$, we denote by R_e the relation given by: $(w, v) \in R_e$ iff $w \in e \Rightarrow v \in e$.*

That is, R_e is a preorder with at most two indifference classes (i.e., a dichotomous weak order) of 'good' and 'bad' candidates for the actual state, which puts all the 'bad' candidates strictly below the 'good' ones (Fig. 4).

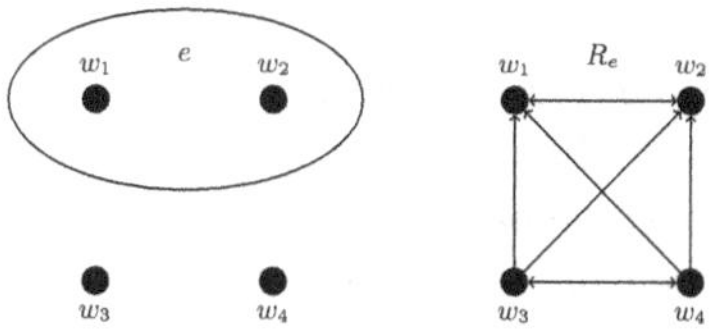

Fig. 4. A piece of binary evidence, represented as an evidence set e (left) and as a dichotomous evidence order R_e (right).

Having fixed this connection between evidence sets and evidence orders, we can now consider a natural way to transform every NEL into a $\cap$-model in which each evidence order is dichotomous. To fix this link, we define a mapping between NEL and REL models.

Definition 15. *Let Rel be a map from NEL to REL models given by:*

$$\langle W, E_0, V \rangle \mapsto \langle Rel(W), \langle Rel(E_0), \preceq \rangle, Rel(V), Ag_\cap \rangle$$

where $Rel(W) := W$, $Rel(V) := V$ $Rel(E_0) := \{R_e \mid e \in E_0\}$ and $\preceq = Rel(E_0)^2$.

We can then observe that feasible NEL models and their images under *Rel* are modally equivalent, in the sense of having point-wise equivalent modal theories.

Proposition 1. *Let $M = \langle W, E_0, V \rangle$ be a feasible NEL model. For any $\varphi \in \mathscr{L}$ and any $w \in W$, we have: $M, w \models \varphi$* iff *$Rel(M), w \models \varphi$.*

That is, feasible NEL models can be seen as 'special cases' of REL models in which all evidence is dichotomous and equally reliable. As the following proposition shows, the modal equivalence result does not extend to non-feasible **NEL** models. This is because, in models with infinitely many pieces of evidence, the notion of combined evidence presented in [2] differs from the one proposed here for **REL** models. To clarify this, consider a **NEL** model $M = \langle W, E_0, V\rangle$. Recall that the agent has combined evidence for a proposition φ at w if there is a *finite* body of evidence whose combination contains w and supports φ, i.e., if there is some finite $F \subseteq E_0$ such that $w \in \bigcap F \subseteq [\![\varphi]\!]_M$. Suppose M is a non-feasible model in which we have $w \in \bigcap E_0 \subseteq [\![\varphi]\!]_M$, while no finite family $F \subseteq E_0$ is such that $w \in \bigcap F \subseteq [\![\varphi]\!]_M$. That is, the combination of *all* the evidence supports φ at w, but no combination of a finite subfamily of E_0 does. In a **NEL** model like this, the agent does *not* have combined evidence for φ at w. That is, $M, w \not\models \Box\varphi$. However, our proposed notion of aggregated evidence for **REL** models is based on combining *all* the available evidence, and as a result in $Rel(M)$ the agent does have aggregated evidence for φ (i.e., $Rel(M), w \models \Box\varphi$). A concrete example of such a model is $M = \langle W, E_0, V\rangle$ with $W = \mathbb{N}$, $E_0 = \{\mathbb{N} \setminus \{2n+1\} \mid n \in \mathbb{N}\}$ and $V(p) = \{2n \mid n \in \mathbb{N}\}$. It is easy to verify that $M, 0 \not\models p$, while $Rel(M), 0 \models p$. *The proofs for all the results presented in this paper can be found in an extended version of it that will constitute the basis for a journal version. This extended version can be found in* [4].

Proposition 2. *Non-feasible* ***NEL*** *models need not be modally equivalent to their images under Rel. In particular, the left-to-right direction of Proposition 1 holds for non-feasible evidence models, but the right-to-left direction doesn't: there are non-feasible neighborhood models M s.t. $Rel(M), w \models \Box\psi$ but $M, w \not\models \Box\psi$.*

Having motivated our interest in $\cap$-models via their connection to neighborhood evidence logics, we now focus again on the static logics of $\cap$- and lex-models. Table 1 lists the axioms and rules in $\mathsf{L}_\cap$ and $\mathsf{L}_{\mathsf{lex}}$.

As stated in Theorem 1, these two systems completely axiomatize the logics of $\cap$ and lex models, respectively.

Theorem 1. *$\mathsf{L}_\cap$ and $\mathsf{L}_{\mathsf{lex}}$ are sound and strongly complete with respect to $\cap$-models and* lex*-models, respectively.*

Evidence dynamics for $\cap$-models. Having established the soundness and completeness of the static logics, we now turn to evidence dynamics, starting with $\cap$-models. In line with the work on NEL, we consider update, evidence addition and evidence upgrade actions for $\cap$-models. As the intersection rule is insensitive to the priority order, when we consider $\cap$-models, it is convenient to treat the models as if they came with a family of evidence orders $\mathscr{R}$ only, instead of an ordered family $\langle \mathscr{R}, \preceq\rangle$. Accordingly, hereafter we will write $\cap$-models as follows: $M = \langle W, \mathscr{R}, V, Ag_\cap\rangle$. Let us fix a $\cap$-model $M = \langle W, \mathscr{R}, V, Ag_\cap\rangle$, some proposition $P \subseteq W$ and some evidence order $R \in Pre(W)$.

Table 1. The systems $\mathsf{L}_\cap$ and L_{lex}

Axioms and inference rules	System(s)
All tautologies of propositional logic	both
S5 axioms for $\forall$, S4 axioms for $\Box$, axiom 4 for $\Box_0$	both
$\forall\varphi \to \Box_0\varphi$	both
$(\Box_0\varphi \wedge \forall\psi) \to \Box_0(\varphi \wedge \forall\psi)$	$\mathsf{L}_\cap$
$(\Box_0\varphi \wedge \forall\psi) \leftrightarrow \Box_0(\varphi \wedge \forall\psi)$	L_{lex}
$\Box_0\varphi \to \Box\varphi$	$\mathsf{L}_\cap$
Axioms T and N for $\Box_0$	L_{lex}
$\forall\varphi \to \Box\varphi$	L_{lex}
Modus ponens	both
Necessitation Rule for $\bullet \in \{\forall, \Box\}$: from φ infer $\bullet\varphi$	both
Monotonicity Rule for $\Box_0$: from $\varphi \to \psi$ infer $\Box_0\varphi \to \Box_0\psi$	both

Update. We first consider updates that involve learning a new fact P with absolute certainty. Upon learning P, the agent rules out all possible states that are incompatible with it. For REL models, this means keeping only the worlds in $[\![P]\!]_M$ and restricting each evidence order accordingly.

Definition 16 (Update). *The model* $M^{!P} = \langle W^{!P}, \mathscr{R}^{!P}, V^{!P}, Ag_\cap^{!P}\rangle$ *has* $W^{!P} := P$, $\mathscr{R}^{!P} := \{R \cap P^2 \mid R \in \mathscr{R}\}$, $Ag_\cap^{!P} := Ag_\cap$ *restricted to* P, *and for all* $p \in \mathsf{P}$, $V^{!P}(p) := V(p) \cap P$.

Evidence addition. Unlike update, which is standardly defined in terms of an incoming proposition $P \subseteq W$, our proposed notion of evidence addition for $\cap$-models involves accepting a new piece of *relational evidence* R from a trusted source. That is, relational evidence addition consists of adding a new piece of relational evidence $R \subseteq Pre(W)$ to the family $\mathscr{R}$.

Definition 17 (Evidence addition). *The model* $M^{+R} = \langle W^{+R}, \mathscr{R}^{+R}, V^{+R}, Ag_\cap^{+R}\rangle$ *has* $W^{+R} := W$, $\mathscr{R}^{+R} := \mathscr{R} \cup \{R\}$, $V^{+R} := V$ *and* $Ag_\cap^{+R} := Ag_\cap$.

Evidence upgrade. Finally, we consider an action of upgrade with a piece of relational evidence R. This upgrade action is based on the notion of *binary lexicographic merge* from Andréka et al. [1].

Definition 18 (Evidence upgrade). *The model* $M^{\Uparrow R} = \langle W^{\Uparrow R}, \mathscr{R}^{\Uparrow R}, V^{\Uparrow R}, Ag_\cap^{\Uparrow R}\rangle$ *has* $W^{\Uparrow R} := W$, $\mathscr{R}^{\Uparrow R} := \{R^< \cup (R \cap R') \mid R' \in \mathscr{R}\}$, $V^{\Uparrow R} := V$ *and* $Ag_\cap^{\Uparrow R} := Ag_\cap$.

Intuitively, this operation modifies each existing piece of evidence R' with R following the rule: "keep whatever R and R' agree on, and where they conflict, give priority to R". To encode syntactically the evidential actions described above, we present extensions of $\mathscr{L}$, obtained by adding to $\mathscr{L}$ dynamic modalities for update, evidence addition and evidence upgrade. The modalities for update will be standard, i.e., modalities of the form $[!\varphi]\psi$. The new formulas of the form $[!\varphi]\psi$ are used to express the statement: "ψ is true after φ is publicly announced".

Definition 19 **($\mathscr{L}^!$).** *The language $\mathscr{L}^!$ is defined recursively by:*

$$\varphi ::= p \mid \neg\varphi \mid \varphi \wedge \varphi \mid \Box_0\varphi \mid \Box\varphi \mid \forall\varphi \mid [!\varphi]\varphi \quad (p \in \mathsf{P})$$

Satisfaction for formulas $[!\varphi]\psi \in \mathscr{L}^!$ is given by: $M, w \models [!\varphi]\psi$ iff $M, w \models \varphi$ implies $M^{![\![\varphi]\!]_M}, w \models \psi$. For the remaining actions, we extend $\mathscr{L}$ with dynamic modalities of the form $[+\pi]\psi$ for addition and $[\Uparrow \pi]\psi$ for upgrade, where the symbol π occurring inside the modality is an evidence program.

Definition 20 **($\mathscr{L}^\bullet$).** *Let $\bullet \in \{+, \Uparrow\}$. The language $\mathscr{L}^\bullet$ is defined by:*

$$\varphi ::= p \mid \neg\varphi \mid \varphi \wedge \varphi \mid \Box_0\varphi \mid \Box\varphi \mid \forall\varphi \mid [\bullet\pi^*]\varphi \quad (p \in \mathsf{P})$$
$$\pi ::= A \mid ?\varphi \mid \pi \cup \pi \mid \pi ; \pi \mid \pi^*$$

The new formulas of the form $[+\pi]\varphi$ are used to express the statement: "φ is true after the evidence order defined by π is added as a piece of evidence", while the $[\Uparrow \pi]\varphi$ are used to express: "φ is true after the existing evidence is upgraded with the relation defined by π". We extend the satisfaction relation $\models$ to cover formulas of the form $[\bullet\pi]\varphi$ as follows: for a formula $[\bullet\pi]\varphi \in \mathscr{L}^\bullet$, we have $M, w \models [\bullet\pi]\varphi$ iff $M^{\bullet[\![\pi]\!]_M}, w \models \varphi$.

The next natural step is to introduce proof systems for the languages $\mathscr{L}^!$, $\mathscr{L}^+$ and $\mathscr{L}^\Uparrow$ with respect to $\cap$-models. A standard approach to obtain soundness and completeness proofs is via a reductive analysis, appealing to *reduction axioms*. We refer to [11] for an extensive explanation of this technique. Taking this route, we obtained complete proof systems for the dynamic logics. The reduction axioms and the completeness proofs can be found in the Extended version of this paper (see Definitions 12, 13 and 21, Lemma 1 and Theorem 2 therein).

Theorem 2. *There exist proof systems for $\mathscr{L}^!$, $\mathscr{L}^+$ and $\mathscr{L}^\Uparrow$ that are sound and complete with respect to $\cap$-models.*

Evidence dynamics for lex-models. We now have a first look at the dynamics of evidence over lex models. In the REL setting, evidential actions can be seen as complex actions involving two possible transformations on the initial model: (i) modifying the stock of evidence, $\mathscr{R}$, perhaps by adding a new evidence relation R to it, or modifying the existing evidence with R; and (ii) updating the priority order, $\preceq$, e.g. to 'place' a new evidence item where it fits, according to its reliability. We may also have actions involving evidence, not about the world, but about evidence itself or its sources (sometimes called 'higher-order evidence' [7]), which trigger a reevaluation of the priority order without changing the stock of evidence (for instance, upon learning that a specific source is less reliable than we initially thought, we may want to lower the priority of the evidence provided by this source). To illustrate the type of actions that can be explored in this setting, here we study an action of *prioritized addition* over lex models. For the sake of generality, we describe this action over REL models.

Prioritized addition. Let $M = \langle W, \langle \mathscr{R}, \preceq \rangle, V, Ag\rangle$ be a REL model and $R \in Pre(W)$ a piece of relational evidence. The prioritized addition of R adds R to the set of available evidence $\mathscr{R}$, giving the highest priority to the new evidence.

Definition 21 (Prioritized addition). *The model* $M^{\oplus R} = \langle W^{\oplus R}, \langle \mathscr{R}^{\oplus R}, \preceq^{\oplus R}\rangle, V^{\oplus R}, Ag^{\oplus R}\rangle$ *has* $W^{\oplus R} := W$, $\mathscr{R}^{\oplus R} := \mathscr{R} \cup \{R\}$, $V^{\oplus R} := V$, $Ag^{\oplus R} := Ag$ *and* $\preceq^{\oplus R} := \preceq \cup \{(R', R) \mid R' \in \mathscr{R}\}$.

To encode this action, we add formulas $[\oplus\pi]\varphi$, used to express the statement that φ is true after the prioritized addition of the evidence order defined by π.

Definition 22 ($\mathscr{L}^{\oplus}$). *The language* $\mathscr{L}^{\oplus}$ *is given by:*

$$\begin{aligned}\varphi &::= p \mid \neg\varphi \mid \varphi \wedge \varphi \mid \Box_0\varphi \mid \Box\varphi \mid \forall\varphi \mid [\oplus\pi^*]\varphi \quad (p \in \mathsf{P})\\ \pi &::= A \mid ?\varphi \mid \pi \cup \pi \mid \pi;\pi \mid \pi^*\end{aligned}$$

Satisfaction for formulas $[\oplus\pi]\varphi \in \mathscr{L}^{\oplus}$ is given by: $M, w \models [\oplus\pi]\varphi$ iff $M^{\oplus[\![\pi]\!]_M}, w \models \varphi$. As we did with the dynamic extensions presented for actions in $\cap$-models, we wish to obtain a matching proof system for our dynamic language $\mathscr{L}^{\oplus}$. We do this via reduction axioms; the axioms and the completeness proof can be found in the Extended version of this paper (see Definition 29 and Theorem 3 there).

Theorem 3. *There exists a proof system for* $\mathscr{L}^{\oplus}$ *that is sound and complete with respect to* lex *models.*

3 The Logic of REL Models

In this section, we briefly study the logic of evidence and belief based on some abstract aggregator. That is, instead of fixing an aggregator, we are now interested in reasoning about the beliefs that an agent would form, based on her evidence, *irrespective* of the aggregator used. With respect to dynamics, we will focus on the action of *prioritized addition* introduced for lex-models, considering an *iterated* version of prioritized addition, defined with a (possibly empty) sequence of evidence orders $\boldsymbol{R} = \langle R_1, \ldots, R_n\rangle$ as input.

Definition 23 (Iterated prioritized addition). *Let* $M = \langle W, \langle \mathscr{R}, \preceq\rangle, V, Ag\rangle$ *be a REL model and* $\boldsymbol{R} = \langle R_1, \ldots, R_n\rangle$ *be a sequence of evidence orders. The model* $M^{\oplus \boldsymbol{R}} = \langle W^{\oplus \boldsymbol{R}}, \langle \mathscr{R}^{\oplus \boldsymbol{R}}, \preceq^{\oplus \boldsymbol{R}}\rangle, V^{\oplus \boldsymbol{R}}, Ag^{\oplus \boldsymbol{R}}\rangle$ *has* $W^{\oplus \boldsymbol{R}} := W$, $\mathscr{R}^{\oplus \boldsymbol{R}} := \mathscr{R} \cup \{R_i \mid i \in \{1, \ldots n\}\}$, $V^{\oplus \boldsymbol{R}} := V$, $Ag^{\oplus \boldsymbol{R}} := Ag$ *and*

$$\begin{aligned}\preceq^{\oplus \boldsymbol{R}} := \preceq &\cup \{(R, R_1) \mid R \in \mathscr{R}\} \cup \{(R, R_2) \mid R \in \mathscr{R} \cup \{R_1\}\}\\ &\cup \ldots\\ &\cup \{(R, R_n) \mid R \in \mathscr{R} \cup \{R_j \mid j \in \{1, \ldots, n-1\}\}\}\end{aligned}$$

That is, first R_1 is added as the highest priority evidence, then R_2 is added as the highest priority evidence, on top of every other evidence (including R_1), and so on, up to R_n. When $\boldsymbol{R}$ has one element, we get the basic notion of addition.

Syntax and semantics. To pre-encode part of the dynamics of iterated prioritized addition, we will modify our basic language $\mathscr{L}$ with *conditional modalities* of the form $\Box^{\boldsymbol{\pi}}$, where $\boldsymbol{\pi}$ is a finite, possibly empty sequence of evidence programs $\pi_1, \ldots, \pi_n$. The intended interpretation of $\Box^{\boldsymbol{\pi}}\varphi$ is "the agent would have aggregated evidence for φ, after the iterated prioritized addition of $\boldsymbol{\pi}$".

Definition 24 ($\mathscr{L}_c$). *The language $\mathscr{L}_c$ is defined as follows:*

$$\varphi ::= p \mid \neg\varphi \mid \varphi \wedge \varphi \mid \Box_0\varphi \mid \Box^{\boldsymbol{\pi}}\varphi \mid \forall\varphi \quad (p \in \mathsf{P})$$
$$\pi ::= A \mid ?\varphi \mid \pi \cup \pi \mid \pi ; \pi \mid \pi^*$$

*where $\boldsymbol{\pi}$ is a (possibly empty) finite sequence of evidence programs (i.e. *-programs).*

Notation 1. We abuse the notation for the truth map $[\![\cdot]\!]_M$ and write $[\![\boldsymbol{\pi}]\!]_M$ to denote $\langle [\![\pi_1]\!]_M, \ldots, [\![\pi_n]\!]_M \rangle$, where $\boldsymbol{\pi} = \langle \pi_1, \ldots, \pi_n \rangle$.

As we allow $\boldsymbol{\pi}$ to be empty, $\Box^{\boldsymbol{\pi}}$ reduces to the $\Box\varphi$ from $\mathscr{L}$ when $\boldsymbol{\pi}$ is the empty sequence, giving us a fully *static* sub-language. Satisfaction for formulas $\Box^{\boldsymbol{\pi}}\varphi \in \mathscr{L}_c$ is given by: $M, w \models \Box^{\boldsymbol{\pi}}\varphi$ iff $Ag(\langle \mathscr{R}^{\oplus[\![\boldsymbol{\pi}]\!]_M}, \preceq^{\oplus[\![\boldsymbol{\pi}]\!]_M} \rangle)[w] \subseteq [\![\varphi]\!]_M$. Next, we introduce a complete proof system for the language with conditional modalities (proof of completeness in the Extended Version).

Definition 25 (L_c). *The system L_c includes the same axioms and inference rules as $\mathsf{L}_{\mathsf{lex}}$, with axioms and inference rules for $\Box$ in $\mathsf{L}_{\mathsf{lex}}$ applying to $\Box^{\boldsymbol{\pi}}$ in L_c.*

Theorem 4. *L_c is sound and strongly complete with respect to* REL *models.*

Evidence dynamics for REL models. Having established the soundness and completeness of the static logic, we now turn to evidence dynamics, focusing on prioritized evidence addition. To encode prioritized addition, we add formulas of the form $[\oplus\boldsymbol{\pi}]\varphi$, used to express the statement that φ is true after the prioritized addition of the sequence of evidence orders defined by $\boldsymbol{\pi}$.

Definition 26 ($\mathscr{L}_c^{\oplus}$). *The language $\mathscr{L}_c^{\oplus}$ is given by:*

$$\varphi ::= p \mid \neg\varphi \mid \varphi \wedge \varphi \mid \Box_0\varphi \mid \Box^{\boldsymbol{\pi}}\varphi \mid \forall\varphi \mid [\oplus\boldsymbol{\pi}]\varphi \quad (p \in \mathsf{P})$$
$$\pi ::= A \mid ?\varphi \mid \pi \cup \pi \mid \pi ; \pi \mid \pi^*$$

*where $\boldsymbol{\pi}$ is a (possibly empty) finite sequence of evidence programs (i.e. *-programs).*

The satisfaction for these formulas is given by $M, w \models [\oplus\boldsymbol{\pi}]\varphi$ iff $M^{\oplus[\![\boldsymbol{\pi}]\!]_M}, w \models \varphi$. A complete system for $\mathscr{L}_c^{\oplus}$ can be found in the extended version of this paper (see Definition 36).

Theorem 5. *There is a proof system for $\mathscr{L}_c^{\oplus}$ that is complete w.r.t* REL *models.*

4 Conclusions and Future Work

We have presented evidence logics that use a novel, non-binary representation for evidence and consider reliability-sensitive forms of evidence aggregation. Here are a few avenues for future research. *Additional aggregators*: we studied two of them. An interesting extension to this work involves developing logics based on other well-known rules. *Additional actions*: in a setting with ordered evidence, evidence actions are complex transformations, both of the stock of evidence and the priority order. For the lexicographic case, we studied a form of prioritized addition. More general actions, e.g., transforming the priority order (re-evaluation of reliability) without affecting the stock of evidence, can be explored. *Probabilistic evidence*: we moved from the binary evidence case to the relational evidence case. *Probabilistic opinion pooling* [8] and *pure inductive logic* [13] study the aggregation of probability functions, but a dynamic-logic analysis is missing.

Acknowledgement. We thank the two anonymous referees for their useful comments. We also thank Johan van Benthem, Benedikt Löwe, Ulle Endriss and Aybüke Özgün for their valuable feedback on a previous version of this paper.

References

1. Andréka, H., Ryan, M., Schobbens, P.: Operators and laws for combining preference relations. J. Logic Comput. **12**(1), 13–53 (2002)
2. Baltag, A., Bezhanishvili, N., Özgün, A., Smets, S.: Justified belief and the topology of evidence. In: Väänänen, J., Hirvonen, Å., de Queiroz, R. (eds.) WoLLIC 2016. LNCS, vol. 9803, pp. 83–103. Springer, Heidelberg (2016). doi:10.1007/978-3-662-52921-8_6
3. Baltag, A., Fiutek, V., Smets, S.: DDL as an "Internalization" of dynamic belief revision. In: Trypuz, R. (ed.) Krister Segerberg on Logic of Actions, vol. 1, pp. 253–283. Springer, Dordrecht (2014). doi:10.1007/978-94-007-7046-1_12
4. Baltag, A., Occhipinti Liberman, A.: Evidence logics with relational evidence. Preprint arxiv:1706.05905 (2017)
5. Baltag, A., Smets, S.: Protocols for belief merge: reaching agreement via communication. Logic J. IGPL **21**(3), 468–487 (2013)
6. Baltag, A., Smets, S., et al.: Talking your way into agreement: belief merge by persuasive communication. In: MALLOW (2009)
7. Christensen, D.: Higher-order evidence. Philos. Phenomenolog. Res. **81**(1), 185–215 (2010)
8. Dietrich, F., List, C.: Probabilistic opinion pooling generalized. Part one: general agendas. Soc. Choice Welfare **48**(4), 747–786 (2017)
9. Girard, P.: Modal logic for lexicographic preference aggregation. In: van Benthem, J., Gupta, A., Pacuit, E. (eds.) Games, Norms and Reasons. Synthese Library (Studies in Epistemology, Logic, Methodology, and Philosophy of Science), vol. 353, pp. 97–117. Springer, Dordrecht (2011). doi:10.1007/978-94-007-0714-6_6
10. Harel, D., Kozen, D., Tiuryn, J.: Dynamic Logic (Foundations of Computing). The MIT Press, Cambridge (2000)

11. Kooi, B., van Benthem, J.: Reduction axioms for epistemic actions. AiML-2004: Advances in Modal Logic, UMCS-04-9-1 in Technical Report Series, pp. 197–211 (2004)
12. Liu, F.: Reasoning about Preference Dynamics. Springer Nature, New York (2011)
13. Paris, J., Vencovská, A.: Pure Inductive Logic. Cambridge University Press, Cambridge (2015)
14. van Benthem, J.: Belief update as social choice. In: Girard, P., Roy, O., Marion, M. (eds.) Dynamic Formal Epistemology. Synthese Library (Studies in Epistemology, Logic, Methodology, and Philosophy of Science), vol. 351, pp. 151–160. Springer, Dordrecht (2011). doi:10.1007/978-94-007-0074-1_8
15. van Benthem, J., Fernández-Duque, D., Pacuit, E.: Evidence and plausibility in neighborhood structures. Ann. Pure Appl. Logic **165**(1), 106–133 (2014)
16. van Benthem, J., Pacuit, E.: Dynamic logics of evidence-based beliefs. Stud. Logica **99**(1–3), 61–92 (2011)
17. Benthem, J., Pacuit, E.: Logical dynamics of evidence. In: Ditmarsch, H., Lang, J., Ju, S. (eds.) LORI 2011. LNCS (LNAI), vol. 6953, pp. 1–27. Springer, Heidelberg (2011). doi:10.1007/978-3-642-24130-7_1
18. van Eijck, J.: Yet more modal logics of preference change and belief revision. In: New Perspectives on Games and Interaction. Amsterdam University Press (2009)

Rational Coordination with no Communication or Conventions

Valentin Goranko[1,2(✉)], Antti Kuusisto[3], and Raine Rönnholm[4]

[1] Stockholm University, Stockholm, Sweden
valentin.goranko@philosophy.su.se
[2] Visiting professor, University of Johannesburg, Johannesburg, South Africa
[3] University of Bremen, Bremen, Germany
[4] University of Tampere, Tampere, Finland

Abstract. We study pure coordination games where in every outcome, all players have identical payoffs, 'win' or 'lose'. We identify and discuss a range of 'purely rational principles' guiding the reasoning of rational players in such games and analyse which classes of coordination games can be solved by such players with no preplay communication or conventions. We observe that it is highly nontrivial to delineate a boundary between purely rational principles and other decision methods, such as conventions, for solving such coordination games.

1 Introduction

Pure coordination games ([11]), aka *games of common payoffs* ([12]), are strategic form games in which all players receive the same payoffs and thus all players have fully aligned preferences to coordinate in order to reach the best possible outcome for everyone. Here we study one-step *pure win-lose coordination games* (WLC games) in which all payoffs are either 1 (i.e., *win*) or 0 (i.e., *lose*).

Clearly, if players can communicate when playing a pure coordination game with at least one winning outcome, then they can simply agree on a winning strategy profile, so the game is trivialised. What makes such games non-trivial is the limited, or no possibility of communication before the game is presented to the players. In this paper we assume *no preplay communication*[1] *at all*, meaning that the players must make their choices by reasoning individually, without any contact with the other players before (or during) playing the game.

There are many natural real-life situations where such coordination scenarios occur. For example, (A) two cars driving towards each other on a narrow street such that they can avoid a collision by swerving either to the right or to the left. Or, (B) a group of n people who get separated in a city and they must each decide on a place where to get together ('regroup'), supposing they do not have any way of contacting each other.

[1] Note that, unlike the common use of 'preplay communication' in game theory to mean communication before the given game is played, here we mean communication before the players are even presented with the game.

A. Baltag et al. (Eds.): LORI 2017, LNCS 10455, pp. 33–48, 2017.
DOI: 10.1007/978-3-662-55665-8_3

Even if no preplay communication is possible, players may still share some *conventions* ([8,11,18]) which they believe everyone to follow. In (A), a collision could be avoided by using the convention (or rule) that cars should always swerve to the right (or, to the left). In (B), everyone could go to a famous meeting spot in the city, e.g., the main railway station. Conventions need not be explicit agreements, but they can also naturally emerge as so-called *focal points*, for example. The theory of focal points, originating from Schelling [17], has been further developed in the context of coordination games, e.g. in [13,20].

In this paper we assume that the players share no conventions, either. Thus, in our setting players play independently of each other. They can be assumed to come from completely different cultures, or even from different galaxies, for that matter. However, we assume that *it is common belief* among the players that:

(1) *every player knows the structure of the game*;
(2) *all players have the same goal*, viz. selecting together a winning profile;

Initially in this paper we will only assume *individual rationality*, i.e. that every player acts with the aim to win the game. Later we will assume in addition *common belief in rationality*, i.e. that every player is individually rational and that it is commonly believed amongst all players that every player is rational.

Our main objective is to analyse what kinds of reasoning can be accepted as 'purely rational' and what kinds of WLC games can be solved by such reasoning. Thus, we try to identify '*purely rational principles*' that *every* rational player ought to follow in *every* WLC game. We also study the hierarchy of such principles based on classes of WLC games that can be won by following different reasoning principles. It is easy to see that coordination by pure rationality is not possible in the example situations (A) and (B) above. However, we will see that there are many natural pure coordination scenarios in which it seems clear that rational players can coordinate successfully.

One of the principal findings of our study is that it is highly nontrivial to demarcate the "purely rational" principles from the rest[2]. Indeed, this seems to be an open-ended question and its answer depends on different background assumptions. Still, we identify a hierarchy of principles that can be regarded as rational and we also provide justifications for them. However, these justifications have varying levels of common acceptability and a more in-depth discussion would be needed to settle some of the issues arising there. Due to space constraints, a more detailed discussion on these issues is deferred to a follow-up work.

Coordination and rationality are natural and interesting topics that have been studied in various contexts in, e.g., [3,6–8,19]. We note the close conceptual relationship of the present study with the notion of *rationalisability* of strategies [2,5,15], which is particularly important in epistemic game theory. We also mention two recent relevant works related to logic to which the observations and results in the present paper could be directly applied: in [10], two-player

[2] Schelling shares this view on pure coordination games (see [17], p. 283, footnote 16).

coordination games were related to a variant of *Coalition Logic*[3], and in [1], coordination was analysed with respect to the game-theoretic semantics of *Independence Friendly Logic*.

An extended version of this paper, containing more examples and technical details, is available as a companion technical report [9]. In addition to the theoretical work presented here, we have also run some empirical experiments on people's behaviour in certain WLC games. One of our tests can be accessed from the link given in the technical report [9].

2 Pure Win-Lose Coordination Games

2.1 The Setting

A *pure win-lose coordination game* G is a strategic form game with n players $(1, \dots, n)$ whose available *choices* (*moves, actions*) are given by sets $\{C_i\}_{i \leq n}$. The set of winning *choice profiles* is presented by an n-ary *winning relation* W_G. For technical convenience and simplification of some definitions, we present these games as *relational structures* (see, e.g., [4]). A formal definition follows.

Definition 1. *An n-player* **win-lose coordination game** *(*WLC **game**) *is a relational structure* $G = (A, C_1, \dots, C_n, W_G)$ *where A is a finite domain of* **choices**, *each* $C_i \neq \emptyset$ *is a unary predicate, representing the choices of player i, s.t.* $C_1 \cup \dots \cup C_n = A$, *and* W_G *is an n-ary relation s.t.* $W_G \subseteq C_1 \times \dots \times C_n$. *Here we also assume that the players have pairwise disjoint choice sets, i.e.,* $C_i \cap C_j = \emptyset$ *for every* $i, j \leq n$ *s.t.* $i \neq j$. *A tuple* $\sigma \in C_1 \times \dots \times C_n$ *is called a* **choice profile** *for G and the choice profiles in* W_G *are called* **winning**.

We use the following terminology for any WLC game $G = (A, C_1, \dots, C_n, W_G)$.

- Let $A_i \subseteq C_i$ for every $i \leq n$. The **restriction** of G to $(A_1, \dots, A_n)$ is the game $G \upharpoonright (A_1, \dots, A_n) := (A_1 \cup \dots \cup A_n, A_1, \dots, A_n, W_G \upharpoonright A_1 \times \dots \times A_n)$.
- For every choice $c \in C_i$ of a player i, the **winning extension of** c **in** G is the set $W_G^i(c)$ of all tuples $\tau \in C_1 \times \dots \times C_{i-1} \times C_{i+1} \times \dots \times C_n$ such that the choice profile obtained from τ by adding c to the i-th position is winning.
- A choice $c \in C_i$ of a player i is **(surely) winning**, respectively **(surely) losing**, if it is guaranteed to produce a winning (respectively losing) choice profile regardless of what choices the other player(s) make. Note that c is a winning choice if $W_G^i(c) = C_1 \times \dots \times C_{i-1} \times C_{i+1} \times \dots \times C_n$. Similarly, c is a losing choice if $W_G^i(c) = \emptyset$.
- A choice $c \in C_i$ is **at least as good as** (respectively, **better than**) a choice $c' \in C_i$ if $W_G^i(c') \subseteq W_G^i(c)$ (respectively, $W_G^i(c') \subsetneq W_G^i(c)$). A choice $c \in C_i$ is **optimal** for a player i if it is at least as good as any other choice of i.

[3] In fact, the initial motivation for the present work came from concerns with the semantics of Alternating time temporal logic ATL, extending Coalition Logic.

Note that a choice $c \in C_i$ is better than a choice $c' \in C_i$ precisely when c *weakly dominates* c' in the usual game-theoretic sense (see e.g. [12,16]), and a choice $c \in C_i$ is an optimal choice of player i when it is a weakly dominant choice. Note also that c *strictly dominates* c' (*ibid.*) if and only if c is surely winning and c' is surely losing. Thus, strict domination is a too strong concept in WLC games. Also the concept of *Nash equilibrium* is not very useful here.

Example 1. We present here a 3-player coordination story which will be used as a running example hereafter. The three robbers Casper, Jesper and Jonathan[4] are planning to quickly steal a cake from the bakery of Cardamom Town while the baker is out. They have two possible plans to enter the bakery: either (a) to break in through the front door or (b) to sneak in through a dark open basement. For (a) they need a *crowbar* and for (b) a *lantern*. The baker keeps the cake on top of a high cupboard, and the robbers can only reach it by using a *ladder*.

When approaching the bakery, Casper is carrying a crowbar, Jesper is carrying a ladder and Jonathan is carrying a lantern. However, the robbers cannot agree whether they should follow plan (a) or plan (b). While the robbers are quarreling, suddenly Constable Bastian appears and the robbers all flee to different directions. After this the robbers have to individually decide whether to go to the front door (by plan (a)) or to the basement entrance (by plan (b)). They must do the right decision fast before the baker returns.

The scenario we described here can naturally be modeled as a WLC game. We relate Casper, Jesper and Jonathan with players 1, 2 and 3, respectively. Each player i has two choices a_i and b_i which correspond to either going to the front door or to the basement entrance, respectively. The robbers succeed in obtaining the cake if both Casper and Jesper go to the front door (whence it does not matter what Jonathan does). Or, alternatively, they succeed if both Jonathan and Jesper go to the basement (whence the choice of Casper is irrelevant). Hence this coordination scenario corresponds to the following WLC game $G^* = (\{a_1, b_1, a_2, b_2, a_3, b_3\}, C_1, C_2, C_3, W_{G^*})$, where for each player i, $C_i = \{a_i, b_i\}$ and $W_{G^*} = \{(a_1, a_2, a_3), (a_1, a_2, b_3), (a_1, b_2, b_3), (b_1, b_2, b_3)\}$. (For a graphical presentation of this game, see Example 2 below.)

2.2 Presenting WLC Games as Hypergraphs

The n-ary winning relation W_G of an n-player WLC game G defines a *hypergraph* on the set of all choices. We give visual presentations of hypergraphs corresponding to WLC games as follows: The choices of each player are displayed as columns of nodes starting from the choices of player 1 on the left and ending with the column with choices of player n. The winning relation consists of lines that go through some choice of each player[5]. This kind of graphical presentation of a WLC game G will be called a *game graph (drawing) of G*.

[4] This example is based on the children's book *When the Robbers Came to Cardamom Town* by Thorbjørn Egner, featuring the characters Casper, Jesper and Jonathan.

[5] In pictures these lines can be drawn in different styles or colours, to tell them apart.

Example 2. The WLC game G^* in Example 1 has the following game graph:

G^* :

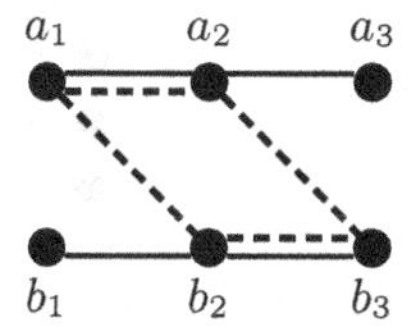

We now introduce a uniform notation for certain simple classes of WLC games. Let $k_1, \ldots, k_n \in \mathbb{N}$.

- $G(k_1 \times \cdots \times k_n)$ is the n-player WLC game where the player i has k_i choices and the winning relation is the *universal relation* $C_1 \times \cdots \times C_n$.
- $G(\overline{k_1 \times \cdots \times k_n})$ is the n-player WLC game where the player i has k_i choices and the winning relation is the *empty relation*. Some examples:

$G(2 \times 3)$

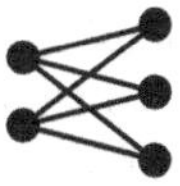

$G(\overline{1 \times 3 \times 1})$

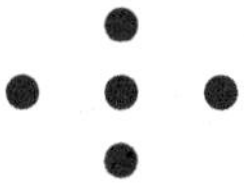

$G(1 \times 1 \times 2)$

- Let $k \in \mathbb{N}$. We write $G(Z_k)$ for the *2-player* WLC game in which both players have k choices and the winning relation forms a single path that goes through all the choices (see below for an example). Similarly, $G(O_k)$, where $k \geq 2$, denotes the 2-player WLC game where the winning relation forms a $2k$-cycle that goes through all the choices. These are exemplified by the following:

$G(Z_2)$

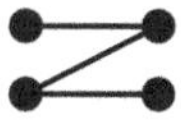

$G(Z_3)$

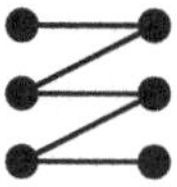

$G(O_2) = G(2 \times 2)$

$G(O_3)$

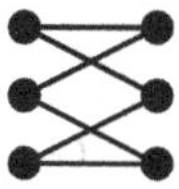

- Suppose that $G(A)$ and $G(B)$ have been defined, both having the same number of players. Then $G(A + B)$ is the *disjoint union* of $G(A)$ and $G(B)$, i.e., the game obtained by assigning to each player a disjoint union of her choices in $G(A)$ and $G(B)$, and where the winning relation for $G(A+B)$ is the union of the winning relations in $G(A)$ and $G(B)$. Some examples:

$G(1 \times 2 + 1 \times 0)$

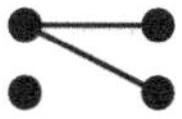

$G(2 \times 1 + \overline{1 \times 2})$

$G(1 \times 1 + 2 \times 2)$

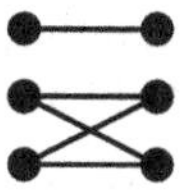

$G(Z_2 + \overline{1 \times 1})$

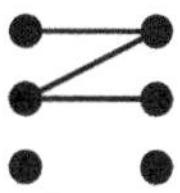

- Let $m \in \mathbb{N}$. Then $G(mA) := G(A + \cdots + A)$ (m times). Examples:

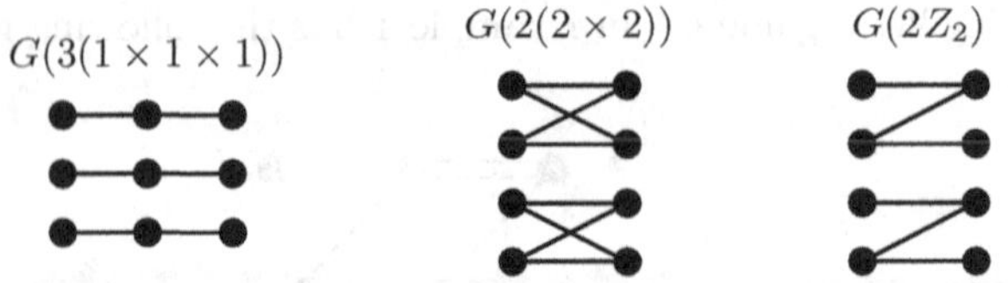

- Recall our "regrouping scenario" (B) from the introduction. If there are n people in the group and there are m possible meeting spots in the city, then the game is of the form $G(m(1^n))$, where $1^n := 1 \times \cdots \times 1$ (n times).

2.3 Symmetries of WLC Games and Structural Protocols

A **protocol** is a mapping Σ that assigns to every pair (G, i), where G is a WLC game and i a player in G, a nonempty set $\Sigma(G, i) \subseteq C_i$ of choices. Thus a protocol gives global nondeterministic strategy for playing any WLC game in the role of any player. Intuitively, a protocol represents a global mode of acting in any situation that involves playing WLC games. Hence, protocols can be informally regarded as global "reasoning styles" or "behaviour modes". Thus, a protocol can also be identified with an agent who acts according to that protocol in all situations that involve playing different WLC games in different player roles.

Assuming a setting based on pure rationality with no special conventions or preplay communication, a protocol will only take into account the *structural properties* of the game and its winning relation. Thus the names of the choices and the names (or ordering) of the players should be of no relevance. In this section we make this issue precise. (For more details and examples, see [9].)

Definition 2. *An isomorphism*[6] *between games G and G' is called a* ***choice-renaming****. An automorphism of G is called a* ***choice-renaming of G****.*

Let $G = (A, C_1, \ldots, C_n, W_G)$ be a WLC *game. For a player i, we say that the choices $c, c' \in C_i$ are* ***i-equivalent****, denoted by $c \simeq_i c'$, if there is a choice-renaming of G that maps c to c'. For each $i \leq n$, the relation $\simeq_i$ is an equivalence relation on the set C_i. We denote the equivalence class of $c \in C_i$ by $[\![c]\!]_i$.*

Definition 3. *Consider n-player* WLC *games $G = (A, C_1, \ldots, C_n, W_G)$ and $G' = (A, C'_1, \ldots, C'_n, W'_G)$. A permutation $\beta : \{1, ..., n\} \to \{1, ..., n\}$ is called a* ***player-renaming*** *between G and G' if the following conditions hold:*

(1) $C_{\beta(i)} = C'_i$ for each $i \leq n$.
(2) $W'_G = \{ (c_{\beta(1)}, \ldots, c_{\beta(n)}) \mid (c_1, \ldots, c_n) \in W_G \}$.

If there is a player-renaming between two WLC games, the games are essentially the same, the only difference being the ordering of the players.

Definition 4. *Consider* WLC *games G and G'. A pair (β, π) is a* ***full renaming*** *between G and G' if there is a* WLC *game G'' such that β is a player-renaming between G and G'' and π is a choice-renaming between G'' and G'.*

[6] Isomorphism is defined as usual for relational structures (see, e.g., [4]).

If $G = G'$, we say that (β, π) is a ***full renaming of*** *G. We say that choices $c \in C_i$ and $c' \in C_j$ in the same game are* ***structurally equivalent****, denoted by $c \sim c'$, if there is a full renaming (β, π) of G such that $\beta(i) = j$ and $\pi(c) = c'$. It is quite easy to see that $\sim$ is an equivalence relation on the set A of all choices. We denote the equivalence class of a choice c by $[c]$.*

Example 3. Consider a WLC game of the form $G(1 \times 2 + 2 \times 1)$.

$G(1 \times 2 + 2 \times 1)$: 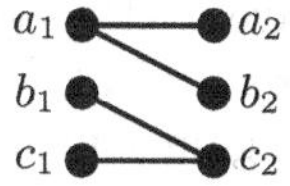

It is easy to see that $\simeq_1$ has the equivalence classes $\{a_1\}$ and $\{b_1, c_1\}$, and similarly, $\simeq_2$ has the equivalence classes $\{c_2\}$ and $\{a_2, b_2\}$. Furthermore, $\sim$ has the equivalence classes $\{a_1, c_2\}$ and $\{b_1, c_1, a_2, b_2\}$. Likewise, in the game G^* from Example 1 the relation $\sim$ has the equivalence classes $\{a_1, b_3\}$, $\{b_1, a_3\}$, $\{a_2, b_2\}$.

We say that a protocol Σ is **structural** if it is "indifferent" with respect to full renamings, which means that, given any WLC games G, G' for which there exists a full renaming (β, π) between G and G', for any i and any choice $c \in C_i$, it must hold that $c \in \Sigma(G, i)$ iff $\pi(c) \in \Sigma(G', \beta(i))$. Intuitively, this reflects the idea that when following a structural protocol, one acts independently of the names of choices and names (or ordering) of player roles.

It is worth noting that if we considered a framework where WLC games were presented so that the names of the choices and players could be used to define an ordering (of the players and their choices), things would trivialize because it would be easy to win all games by the prenegotiated agreement to always choose the lexicographically least tuple from the winning relation.

3 Purely Rational Principles in WLC Games

By a **principle** we mean any nonempty class of protocols. Intuitively, these are the protocols "complying" with that principle. If protocols are regarded as "reasoning styles" (or "behaviour modes"), then principles are *properties* of such reasoning styles (or behaviour modes). Principles that contain only structural protocols are called **structural principles.**

A player i **follows a principle P** in a WLC game G if she plays according to some protocol in P. We are mainly interested in structural principles which describe "purely rational" reasoning that involves neither preplay communication nor conventions and which are rational to follow in *every* WLC game. Such principles will be called **purely rational principles**. Intuitively, purely rational principles should always be followed by all rational players. Consider:

$$\mathrm{P}_1 := \{\Sigma \mid \Sigma(G, i) \text{ does not contain any surely losing choices when } W_G \neq \emptyset\},$$

$$\mathrm{P}_2 := \{\Sigma \mid \Sigma(G, i) \text{ contains all choices } c \in C_i \text{ such that } |W_G^i(c)|$$
$$\text{is a prime number. If there are no such choices, } \Sigma(G, i) = C_i.\}.$$

If player i follows P_1, then she always uses some protocol which does not select surely losing choices, if possible. This seems a principle that any rational agent would follow. If player i follows P_2, then she always plays choices whose degree (in the game graph) is a prime number, if possible. Note that both principles are structural, but P_1 can be seen as a purely rational principle, while P_2 seems arbitrary; it could possibly be some seemingly odd convention, for example.

We say that a **principle P solves** a WLC game G (or G **is P-solvable**), if G is won whenever every player follows some protocol that belongs to P. Formally, this means that $\Sigma_1(G,1)\times\cdots\times\Sigma_n(G,n)\subseteq W_G$ for all protocols $\Sigma_1,\ldots,\Sigma_n\in$ P. The class of all P-solvable games is denoted by $s(\mathrm{P})$.

In this paper we try to identify (a hierarchy of) principles that can be considered to be purely rational and analyse the classes of games that they solve.

3.1 Basic Individual Rationality

Hereafter we describe principles by the properties of protocols that they determine. We begin by considering the case where players are individually rational, but there is no common knowledge about this being the case. It is safe to assume that any individually rational player would follow at least the following principle.

Fundamental individual rationality (FIR):
Never play a strictly dominated choice.[7]

As noted before, strict domination is a very weak concept with WLC games. Following FIR simply means that a player should never prefer a losing choice to a winning one. Therefore FIR is a very weak principle that can solve only some quite trivial types of games such as $G(1\times 2+1\times 0)$. In general, FIR-solvable games have a simple description: at least one of the players has (at least one) winning choice, and all non-winning choices of that player are losing. FIR has two natural strengthenings which can be considered purely rational:

1. **Non-losing principle (NL):** Never play a losing choice, if possible.
2. **Sure winning principle (SW):** Always play a winning choice, if possible.

Since losing choices cannot be winning choices, these principles can naturally be put together (by taking the intersection of these principles):

Basic individual rationality (BIR): NL $\cap$ SW.

When following BIR, a player plays a winning choice if she has one, and else she plays a non-losing choice. We make the following observations. (For a more detailed justification of these claims, see the technical report [9].)

1. NL and SW do not imply each other and neither of them follows from FIR. This can be seen by the following examples.
   - The game $G(1\times 1+\overline{1\times 1})$ is NL-solvable but not SW-solvable.

[7] Recall, that a choice a is strictly dominated by a choice b if the choice b guarantees a strictly higher payoff than the choice a in every play of the game (see e.g. [12,16]).

   - The game $G(Z_2)$ is SW-solvable but not NL-solvable.
2. FIR-solvable games are solvable by *both* SW and NL.
3. Every BIR-solvable game is *either* NL or SW-solvable.

Therefore we can see that the sets of games solvable by FIR, NL, SW, BIR form the following lattice:

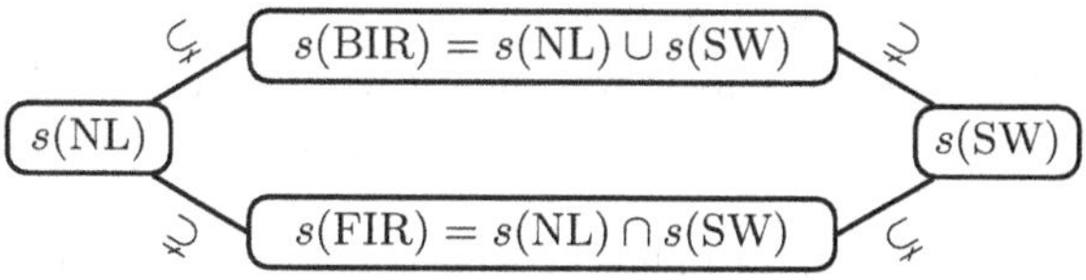

SW-solvable and NL-solvable games have simple descriptions: In SW-solvable games, at least one player has a surely winning choice. In NL-solvable games, the winning relation forms a nonempty *Cartesian product* between all non-losing choices. BIR-solvable games have (at least) one of these two properties.

3.2 Common Beliefs in Rationality and Iterated Reasoning

In contrast to individual rationality, collective rationality allows players to make assumptions on each other's rationality. Let P be a (purely rational) principle. When *all players believe that everyone follows* P, they can reason as follows:

($\star$) Suppose that by following P each player i must play a choice from $A_i \subseteq C_i$ (that is, A_i is the smallest set such that $\Sigma(G,i) \subseteq A_i$ for every $\Sigma \in \text{P}$). By this assumption, the players may collectively assume that the game that is played is actually $G' := G \upharpoonright (A_1, \ldots, A_n)$, and therefore all P-compliant protocols should only prescribe choices in G'.

If players have *common belief* in P being followed, then the reasoning ($\star$) above can be repeated for the game G' and this iteration can be continued until a fixed point is reached. By $\mathsf{cir}(\text{P})$ we denote the principle of **collective iterated reasoning of P** which prescribes that P is followed in the reduced game obtained by the iterated reasoning of ($\star$). Since every iteration of ($\star$) only reduces the players' sets of acceptable choices (yet, keeps them nonempty), it is easy to see that $s(\text{P}) \subseteq s(\mathsf{cir}(\text{P}))$ for any principle P (see [9] for more details.)

When considering principles of *collective* rationality, we will apply collective iterated reasoning. It may be debated whether such reasoning counts as purely rational, so a question arises: if P is a purely rational principle, is $\mathsf{cir}(\text{P})$ always purely rational as well? For the lack of space we will not discuss this issue here. We note, however, the extensive literature relating common beliefs and knowledge with individual and collective rationality, see e.g. [5,11,14,21].

3.3 Basic Collective Rationality

Here we extend individually rational principles of Sect. 3.1 by adding common belief in the principles (as described in Sect. 3.2) to the picture. We first analyse what happens with principles NL and SW. It is easy to see that the collective iterated reasoning of NL reaches a fixed point in a single step by simply removing the losing choices of every player. Hence $s(\mathrm{NL}) = s(\mathsf{cir}(\mathrm{NL}))$. Collective iterated reasoning of SW also reaches a fixed point in a single step by eliminating all non-winning choices of every player who has a winning choice. But if even one player has a winning choice, then the game is already SW-solvable. Therefore $s(\mathrm{SW}) = s(\mathsf{cir}(\mathrm{SW}))$.

However, assuming common belief in BIR, some games which are not BIR-solvable may become solvable. See the following example.

Example 4. The game $G(Z_2 + \overline{1 \times 1})$ cannot be solved with NL or SW. However, if the players can assume that neither of them selects a losing choice (by NL) and eliminate those choices from the game, then they (both) have a winning choice in the reduced game and can win in it by SW.

Thus, we define the following principle:

Basic collective rationality (BCR): cir(BIR).

The above example shows that $s(\mathrm{BIR}) \subsetneq s(\mathrm{BCR})$, i.e. BCR is *stronger* than BIR. The games solvable by BCR have the following characterisation: *after removing all surely losing choices of every player, at least one of the players has a surely winning choice.* It is worth noting that common belief in SW is not needed for solving games with BCR because a *single* iteration of cir(NL) suffices. Thus, players could solve BCR-solvable games simply by following BIR and believing that everyone follows NL. We also point out that the principle BCR is equivalent to the principle applied in [10] for Strategic Coordination Logic.

3.4 Principles Using Optimal Choices

If a rational player has optimal choices (that are at least as good as all other choices), it is natural to assume that she selects such a choice.

Individual optimal choices (IOC): Play an optimal choice, if possible.

Example 5. Recall the WLC game G^* from Example 1. For Casper (who is carrying the crowbar) it is a better choice to go to the front door than to the basement. Likewise, for Jonathan (who is carrying the lantern) it is a better choice to go to the basement than to the front door. Therefore the choice a_1 is (the only) optimal choice for player 1 and b_3 is (the only) optimal choice for the player 3. The player 2 (Jesper) does not have any optimal choices, but if both 1 and 3 play their optimal choices, then the game is won regardless of the choice of 2. Therefore, the game G^* is solvable with IOC. But since no player has winning or losing choices in this game, it is easy to see that it is not BCR-solvable.

By the description of BIR-solvable games, it is easy to see that they are IOC-solvable. We will show that IOC is *incomparable* with BCR (in terms of their sets of solvable games). As explained above, the game G^* is IOC-solvable but not BCR-solvable. Furthermore, the following BCR-solvable game G_Σ is not IOC-solvable since player 1 does not have any optimal choices and so might end up playing a losing choice.

G_Σ:

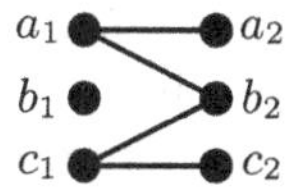

In order to avoid pathological cases like this we can add NL to IOC.

Improved basic individual rationality (BIR$^+$): IOC $\cap$ NL

This principle is stronger than BCR (see [9]) even though it is based only on *individual* reasoning. We now consider the collective version of IOC:

Collective optimal choices (COC): cir(IOC)

We can show that COC is stronger than BIR$^+$ and therefore also stronger than BCR (see [9]). Finally, observe that in a 2-player WLC game G where $W_G \neq \emptyset$ *the only optimal choices are those that are winning against all non-losing choices of the other player.* Therefore, in the special case of 2-player WLC games, it is easy to see that the hierarchy collapses as $s(\text{BCR}) = s(\text{BIR}^+) = s(\text{COC})$.

3.5 Elimination of Weakly Dominated Choices

Usually in game-theory, rationality is associated with the elimination of strictly or weakly dominated strategies. As noted in Sect. 3.1, *strict* domination is a too strong concept for WLC games. Weak domination, on the other hand, gives the following principle when applied individually.

Individually rational choices (IRC): Do not play a choice a when there is a better choice b available, i.e., if $W_G^i(a) \subsetneq W_G^i(b)$, then i does not play a.

Note that by this definition, when a player follows IRC, she also follows NL and IOC, and therefore $s(\text{BIR}^+) \subseteq s(\text{IRC})$. The inclusion is, in fact, proper since the following WLC game $G_\#$ is solvable with IRC but not with BIR$^+$.

$G_\#$:

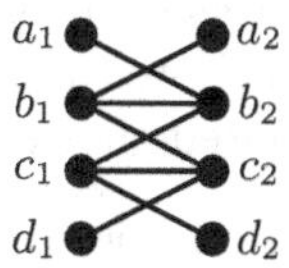

We show in [9] that IRC is incomparable with COC. However, based on the observations above, in the 2-player case $s(\text{COC}) = s(\text{BIR}^+) \subsetneq s(\text{IRC})$.

We next assume common belief in IRC. As commonly known (see e.g. [14]), *iterated elimination of weakly dominated strategies* eventually stabilises in some reduced game but different elimination orders may produce different results. However, when applying cir(IRC), the process will stabilise to a unique reduced game since all weakly dominated choices are always removed *simultaneously*. By following the next principle, players will play a choice within this reduced game.

Collective rational choices (CRC): cir(IRC)

For example, $G(Z_3)$ is not solvable with IRC, but can be solved with CRC by doing two collective iterations of IRC. Thus $s(\text{IRC}) \subsetneq s(\text{CRC})$. This observation can be generalized as follows: to solve a game of the form $G(Z_n)$, the players need $n-1$ iterations of IRC. Therefore different numbers of iterations of IRC form a proper hierarchy of CRC-solvable 2-player WLC games within $s(\text{CRC})$.

3.6 Symmetry-Based Principles

By only following the concept of rationality from game-theory, one could argue that CRC reaches the border of purely rational principles. However, we now define more principles which are incomparable with CRC but can still be regarded as purely rational. These principles are based on *symmetries* in WLC games and the assumption that players follow only structural protocols is central here.

We begin with auxiliary definitions. We say that a choice profile $(c_1, \ldots, c_n)$ **exhibits a bad choice symmetry** if $[\![c_1]\!]_1 \times \cdots \times [\![c_n]\!]_n \not\subseteq W_G$ (recall Definition 2), and that a choice c **generates a bad choice symmetry** if σ_c exhibits bad choice symmetry for *every* choice profile σ_c that contains c.

Elimination of bad choice symmetries (ECS):
Never play choices that generate a bad choice symmetry, if possible.

Why should this principle be considered rational? Suppose that a player i plays a choice c_i which generates a bad choice symmetry. It is now possible to win only if some tuple $(c_1, \ldots, c_{i-1}, c_i, c_{i+1}, \ldots, c_n) \in W_G$ is eventually chosen. However, the players have *exactly the same reason* (based on structural principles) to play so that any other tuple in $[\![c_1]\!]_1 \times \cdots \times [\![c_n]\!]_n$ is selected, and such other tuple may possibly be a losing one since $[\![c_1]\!]_1 \times \cdots \times [\![c_n]\!]_n \not\subseteq W_G$.

Example 6. Here is a typical example of using ECS. Suppose that the game graph of G has two (or more) connected components that are isomorphic to each other. Since no player can see a difference between those components, all players should avoid playing choices from them. With this reasoning, games like $G(1 \times 1 + 2(1 \times 2))$ can be solved. Note that this game is not CRC-solvable since no player has any weakly dominated choices.

While ECS only considers symmetries between similar choices, the next principle takes symmetries *between players* into account. Consider a choice profile $\mathbf{c} = (c_1, \ldots, c_n)$ and let $S_i^p(\mathbf{c}) := \{c_i\} \cup (C_i \cap \bigcup_{j \neq i}[c_j])$ for each i (recall Definition 4). We say that $(c_1, ..., c_n)$ **exhibits a bad player symmetry** if $S_1^p(\mathbf{c}) \times \cdots \times S_n^p(\mathbf{c}) \not\subseteq W_G$ and a choice c **generates a bad player symmetry** if σ_c exhibits a bad player symmetry for every choice profile σ_c that contains c.

Elimination of bad player symmetries (EPS):
Never play choices that generate bad player symmetries, if possible.

Here the players assume that all players reason similarly, or alternatively, each player wants to play so that she would at least coordinate with herself in the case she was to use her protocol to make a choice in each player role of a WLC game. Suppose that the players have some reasons to select a choice profile $(c_1, \ldots, c_n)$. Now, if there are players $i \neq j$ and a choice $c'_j \in C_j$ such that $c'_j \sim c_i$, then the player j should have the same reason to play c'_j as i has for playing c_i. Hence, if the players have their reasons to play $(c_1, \ldots, c_n)$, they should have the same reasons to play any choice profile in $S^p_1(\boldsymbol{c}) \times \cdots \times S^p_n(\boldsymbol{c})$. Winning is not guaranteed if $S^p_1(\boldsymbol{c}) \times \cdots \times S^p_n(\boldsymbol{c}) \nsubseteq W_G$.

Example 7. Consider EPS in the case of a two-player game WLC game G. If for a given choice $c \in C_1$, there is a structurally equivalent choice $c' \in C_2$ such that $(c, c') \notin W_G$, then by following EPS, player 1 does not play the choice c (and likewise player 2 does not play the choice c'). With this kind of reasoning, some CRC-unsolvable games like $G(1 \times 1 + 1 \times 2 + 2 \times 1)$ become solvable.

Note also that the game G^* (recall Example 1) is EPS-solvable since both choices b_1 and a_3 generate a bad player symmetry.

Finally, we introduce a principle that takes both types of symmetries into account. For a choice profile $\boldsymbol{c} = (c_1, ..., c_n)$ let $S_i(\boldsymbol{c}) := C_i \cap \bigcup_j [c_j]$ for each i. We say that $(c_1, ..., c_n)$ **exhibits a bad symmetry** if $S_1(\boldsymbol{c}) \times \cdots \times S_n(\boldsymbol{c}) \nsubseteq W_G$, and a choice c **generates a bad symmetry** if σ_c exhibits a bad symmetry for every choice profile σ_c that contains c.

Elimination of bad symmetries (ES):
Never play choices that generate bad symmetries, if possible.

It is easy to show that ECS and EPS are not comparable and that they are both weaker than ES. Furthermore, all symmetry based principles can clearly solve NL-solvable games, but they are incomparable with SW and all the stronger principles. For proofs of these claims and further examples, see [9].

In a follow-up work we will address questions about compatibility of the symmetry principles ECS and EPS with each other and with the other principles considered so far, in particular with CRC which is the strongest of them.

3.7 Hierarchy of the Principles Presented so Far

The partially ordered diagram below presents the hierarchy of solvable games with the principles we have presented in this paper. The principles that only use individual reasoning have normal frames and the ones that use collective reasoning have double frames.

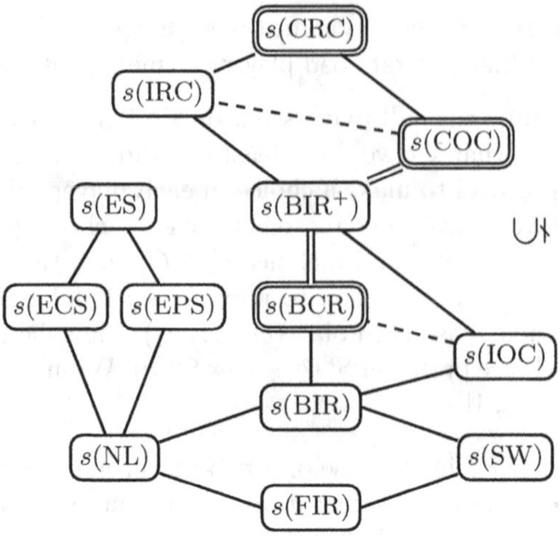

- Normal lines represent proper inclusions in both the general *and* 2-player case.
- *Double* lines represent proper inclusions in the general case. In the 2-player case there is an identity.
- *Dashed* lines represent proper inclusions in the 2-player case. In the general case the two sets are not comparable.

3.8 Beyond the Limits of Pure Rationality

How far can we go up the hierarchy of rational principles? This seems a genuinely difficult question to answer. We now mention—without providing precise formal definitions—two structural principles for which it would seem somewhat controversial to claim them rational in our sense, but they are definitely meaningful and natural nevertheless.

The first one is the **principle of probabilistically optimal reasoning (PR)**. Informally put, this principle prescribes to always play a choice that have as large winning extension as possible. These choices have the highest probability of winning, supposing that all the other players play randomly (but *not* if the others follow PR, too: consider e.g. $G(1 \times 2 + 2 \times 1)$).

With PR one can solve games like $G(1\times 1+2\times 2)$ that are unsolvable with all other principles presented here. However, in $G(1\times 1+2\times 2)$ one could also reason (perhaps less convincingly) that both players should pick their choices from the subgame $G(1\times 1)$ since that is the '*simplest*' (and, also the only '*unique*') winning choice profile. We call this kind of reasoning the **Occam razor principle (OR)**. In fact, it generalises the idea of *focal point* [13,17,20].

Note that $G(1 \times 1 + 2 \times 2)$ can be won if both players follow PR or if both follow OR, but not if one follows PR while the other follows OR. Moreover, in this game it is impossible for a player to follow *both* PR and OR. Hence, *at least one* of these principles is not purely rational. Actually, it can be argued that *none of them* is purely rational. It is also interesting to note that following PR can violate the symmetry principles, as demonstrated by the game $G(2(2\times 2)+1\times 1)$.

3.9 Characterising Structurally Unsolvable Games

So far we have characterised several principles with different levels of justification for being purely rational. It seems difficult to pinpoint a single strongest principle of pure rationality, but even if such a principle existed, certain games would nevertheless be unsolvable (assuming that purely rational principles must be *structural*). The simplest nontrivial example of such a game is $G(2(1 \times 1))$.

We now characterise the class of WLC games that are **structurally unsolvable**, i.e., *unsolvable by any structural principle.* We say that G is **structurally indeterminate** if all choice profiles in W_G exhibit a bad symmetry (recall the definition of the principle ES). For an example the game $G(1 \times 2 + 2 \times 1)$ is structurally indeterminate, whereas the game $G(1 \times 1 + 2 \times 2)$ is not.

Claim I. *No structural principle can solve a structurally indeterminate game.*

For a proof for this claim, see [9]. This characterisation is optimal in the sense that all games that are not structurally indeterminate, can be solved by some structural principle. This follows from the following even stronger claim.

Claim II. *There exists a protocol Σ such that the principle $\{\Sigma\}$ can solve all* WLC *games that are not structurally indeterminate.*

For a proof, see [9]. There are many games that are not structurally unsolvable, but in order to solve them, the players need to follow structural principles that seem arbitrary and certainly cannot be considered purely rational. We call such principles *structural conventions.* However, it is difficult to separate some rational principles from structural conventions. This and other related conceptual issues will be discussed in an extended version of this paper.

4 Concluding Remarks

In this paper we have focused on scenarios where players look for choices that guarantee winning if a suitable rational principle is followed. But it is very natural to ask how players should act in a game which seems not solvable by any purely rational principle. If players cannot guarantee a win, it is natural to assume that they should at least try to maximize somehow their collective chances of winning, say, by considering protocols involving some probability distribution between their choices. Another natural extension of our framework is to consider non-structural principles based on limited preplay communication and use of various types of conventions. Also, studying pure dis-coordination games and combinations of coordination/dis-coordination are major lines for further work.

Acknowledgements. The work of Valentin Goranko was partly supported by a research grant 2015-04388 of the Swedish Research Council. The work of Antti Kuusisto was supported by the ERC grant 647289 "CODA." We thank the reviewers of this paper, as well as those of the Strategic Reasoning 2017 abstract for valuable remarks.

References

1. Barbero, F.: Cooperation in games and epistemic readings of independence-friendly sentences. J. Logic Lang. Inform. 1–40 (2017). https://link.springer.com/article/10.1007/s10849-017-9255-1
2. Bernheim, B.D.: Rationalizable strategic behavior. Econometrica **52**, 1007–1028 (1984)
3. Bicchieri, C.: Rationality and Coordination. Cambridge University Press, London (1994)
4. Ebbinghaus, H., Flum, J., Thomas, W.: Mathematical logic, 2nd edn. Undergraduate Texts in Mathematics, Springer, New York (1994). doi:10.1007/978-1-4757-2355-7
5. Fudenberg, D., Tirole, J.: Game theory. MIT Press, Cambridge (1991)
6. Gauthier, D.: Coordination. Dialogue **14**(02), 195–221 (1975)
7. Genesereth, M.R., Ginsberg, M.L., Rosenschein, J.S.: Cooperation without communication. In: Proceedings of AAAI 1986, vol. 1, pp. 51–57 (1986)
8. Gilbert, M.: Rationality, coordination, and convention. Synthese **84**(1), 1–21 (1990)
9. Goranko, V., Kuusisto, A., Rönnholm, R.: Rational coordination with no communication or conventions. Technical report arXiv:1706.07412 (2017)
10. Hawke, P.: The logic of joint ability in two-player tacit games. Rev. Symbolic Logic, 1–28 (2017). https://www.cambridge.org/core/journals/review-of-symbolic-logic/article/logic-of-joint-ability-in-twoplayer-tacit-games/1E5DFEC4C52DDCA2F3D0B4614AA74DCE
11. Lewis, D.: Convention, A Philosophical Study. Harvard University Press, Cambridge (1969)
12. Leyton-Brown, K., Shoham, Y.: Essentials of Game Theory: A Concise Multidisciplinary Introduction. Morgan & Claypool Publishers, San Rafael (2008)
13. Mehta, J., Starmer, C., Sugden, R.: Focal points in pure coordination games: an experimental investigation. Theor. Decis. **36**(2), 163–185 (1994)
14. Osborne, M., Rubinstein, A.: A Course in Game Theory. MIT Press, Cambridge (1994)
15. Pearce, D.G.: Rationalizable strategic behavior and the problem of perfection. Econometrica. **52**, 1029–1050 (1984)
16. Peters, H.: Game Theory: A Multi-Leveled Approach. Springer, Heidelberg (2008). doi:10.1007/978-3-540-69291-1
17. Schelling, T.C.: The Strategy Of Conflict. Harvard University Press, Cambridge (1960)
18. Sugden, R.: Spontaneous order. J. Econ. Perspect. **3**(4), 85–97 (1989)
19. Sugden, R.: Rational Choice: A Survey of Contributions from Economics and Philosophy. Econ. J. **101**(407), 751–785 (1991)
20. Sugden, R.: A theory of focal points. Econ. J. **105**, 533–550 (1995)
21. Syverson, P.: Logic, Convention, and Common Knowledge: A Conventionalist Account of Logic. CSLI Lecture Notes 142, CSLI Publications, California (2002)

Towards a Logic of Tweeting

Zuojun Xiong[1,2](✉), Thomas Ågotnes[2,3], Jeremy Seligman[4], and Rui Zhu[4]

[1] Institute of Logic and Intelligence, Southwest University, Chongqing, China
xiongzuojun@gmail.com
[2] Department of Information Science and Media Studies, University of Bergen, Bergen, Norway
thomas.agotnes@infomedia.uib.no
[3] Center for the Study of Language and Cognition, Zhejiang University, Hangzhou, China
[4] Department of Philosophy, University of Auckland, Auckland, New Zealand
j.seligman@auckland.ac.nz, zrui956@aucklanduni.ac.nz

Abstract. In this paper we study the logical principles of a common type of network communication events that haven't been studied from a logical perspective before, namely *network announcements*, or *tweeting*, i.e., simultaneously sending a message to all your friends in a social network. In particular, we develop and study a minimal modal logic for reasoning about propositional network announcements. The logical formalisation helps elucidate core logical principles of network announcements, as well as a number of assumptions that must be made in such reasoning. The main results are sound and complete axiomatisations.

1 Introduction

Formalising reasoning about different types of interaction between multiple agents is an active research topic in logic, artificial intelligence, knowledge representation and reasoning, multi-agent systems, formal specification and verification, and other fields. Most modern approaches are based on modal logic. Despite their ubiquitousness, relatively little attention has been given to formalising reasoning about interaction in *social networks* – with some notable exceptions [2–5, 8–14].

In this paper we deal with an issue that has not yet been studied in the sparse but growing literature on logics for social networks. Existing works broadly speaking fall in two main categories; those using formal logic to characterise "global" network phenomena such as cascades (e.g., [2]), and those using formal logic to capture the often subtle details of "local" social network events such as message passing (e.g., [11]). Works in the latter category, in which the current paper falls, have mostly been motivated by capturing events typical in *Facebook*-like applications, such as privately sending a message to a friend (one-to-one messaging). In this paper we formalise reasoning about (what we call) *network announcements* in social networks, the primary communication event on, e.g., *Twitter*: the sending by one agent of a message which received simultaneously

A. Baltag et al. (Eds.): LORI 2017, LNCS 10455, pp. 49–64, 2017.
DOI: 10.1007/978-3-662-55665-8_4

by a number of other agents (the sender's *followers*), determined by the network structure. We will also refer to the act of making network announcements as *tweeting*.

We introduce a minimal modal logic for reasoning about network announcements, having expressions of the form $\langle a : \theta \rangle \varphi$ with the intuitive meaning that if agent a tweets θ, φ will become true. This tweeting operator is not completely original; a very similar operator $[F!\varphi]$ with the meaning "after I announce φ to my friends" was defined semantically in [10] (and also mentioned in [11]), but not systematically studied. In particular, the *logic of network announcements*, their logical principles, axiomatic basis, and so on, has not been studied. Certain aspects of what we call network announcements have been studied in computer science under the term (one-to-many) *multi-cast messaging* [6,7], but not using formal logic.

In this paper we restrict the beliefs of the agents, and thus also the messages they can tweet, to be about basic propositional facts only, as opposed to higher-order beliefs, beliefs about who follows whom, beliefs about who said what, etc. We also make a number of additional idealising assumptions:

Sincerity	Agents only tweet what they believe.
Credulity	Agents believe the messages they receive.
Conservatism	Agents never stop believing what they believed before.
Network stability	Who follows whom after a tweet is the same as before.
Rationality	Agents only believe what follows logically from their previous beliefs and the messages they receive.
Doxastic Omniscience	Agents believe all the logical consequences of what they believe.

These assumptions limit the applicability of the logic, but also allow us to focus on the core concepts of network announcement epistemology. These need to be understood before other more complex issues are addressed. We will see that already several interesting phenomena emerge. In Sect. 6 we discuss the prospect for extensions.

One remaining, natural assumption is the consistency of each agent's beliefs. Instead of building this into our models from the beginning, we develop the logic without any assumption of consistency and then characterise classes of models in which various consistency assumptions hold. In this paper, we will consider two:

Weak Coherence	Each agent has consistent beliefs.
Global Coherence	Agents have mutually consistent beliefs.

The paper is structured as follows. In the next section we introduce the syntax and semantics of the logic, and illustrate what it can express and discuss some of its properties. In Sect. 3 we study logical properties in the form of valid formulas, and the relationship between them, enabling us to form a Hilbert-style axiomatic system that is shown to be complete in Sect. 4. A red thread, and indeed the crux of the completeness proof, is how formulas expressing some agents' beliefs or ignorance after some tweet implicitly contains information about the network structure. In Sect. 5 we look at completeness results for some variants of the logic, and we conclude in Sect. 6.

2 Propositional Network Announcements

Let Agnt and Prop be non-empty sets of agent names and atomic propositional letters, respectively.

Definition 1 (Language). The language of *propositional network announcement logic* is defined by the following grammar, where $p \in \mathsf{Prop}$ and $a \in \mathsf{Agnt}$:

$$\theta ::= p \mid \neg\theta \mid \theta \wedge \theta \qquad \varphi ::= B_a\theta \mid \neg\varphi \mid \varphi \wedge \varphi \mid \langle a : \theta\rangle\varphi.$$

Expressions of type θ are called *message formulas* or just *messages*; those of type φ are called *formulas*. The usual derived propositional connectives are used, as well as $[a : \theta]$ for $\neg\langle a : \theta\rangle\neg$. The intended meaning of $B_a\theta$ is that agent a believes θ, while $\langle a : \theta\rangle\varphi$ means that a can tweet θ, after which φ is the case. Formulas of the form $B_a\theta$ are called *belief formulas*.

A model for our language has two parts: an assignment of belief states to each agent and a "following" relation between agents. Recall that a propositional logic *valuation* is a function from Prop to truth values. We denote the set of all valuations Val. We model an agent's belief state by a subset of Val, with no further restrictions. Each message θ determines the set $[\![\theta]\!]$ of those valuations that make it true, according to the usual semantics of propositional logic.

Definition 2 (Models). A *propositional network announcement model* over Agnt and Prop is a pair (F, ω), where the *following relation* F is a binary relation on Agnt and the *belief state function* $\omega\colon \mathsf{Agnt} \to \mathrm{pow}(\mathsf{Val})$ assigns each agent a (possibly empty) set of valuations. We write Fa for the set $\{b \mid bFa\}$ of followers of a.

Note that the subset ordering of belief states is inverse to the strength of the state, so we define $\omega_1 \leq \omega_2$ iff $\omega_2(a) \subseteq \omega_1(a)$ for every a. Any belief of a's in state ω_1 is also a belief in state ω_2.

Definition 3 (Updates). When (only) agent a's belief state is updated with θ, the result is the belief state function $[a \uparrow \theta]\omega$. More generally, the result of simultaneously updating all the agents in a set C of agents with θ is $[C \uparrow \theta]\omega$, where

$$[C \uparrow \theta]\omega(b) = \begin{cases} \omega(b) \cap [\![\theta]\!] & \text{if } b \in C \\ \omega(b) & \text{otherwise} \end{cases}$$

Note that updating is monotonic: $\omega \leq [C \uparrow \theta]\omega$. The language is interpreted in these models as follows.

Definition 4 (Satisfaction). A formula φ of the language of propositional network announcement logic is *satisfied* by a model (F, ω), written $F, \omega \models \varphi$, as follows:

$$\begin{array}{lll} F,\omega \models B_a\theta & \text{iff} & \omega(a) \subseteq [\![\theta]\!] \\ F,\omega \models \neg\varphi & \text{iff} & F,\omega \not\models \varphi \\ F,\omega \models \varphi \wedge \psi & \text{iff} & F,\omega \models \varphi \text{ and } F,\omega \models \psi \\ F,\omega \models \langle a:\theta\rangle\varphi & \text{iff} & F,\omega \models B_a\theta \text{ and } F,[Fa\!\uparrow\!\theta]\omega \models \varphi \end{array}$$

As usual we say that a formula is *valid* iff it is satisfied by every model. We will also be interested in the class of models in which agents' beliefs are mutually consistent.

Definition 5. A model (F, ω) is *weakly coherent* iff $\omega(a) \neq \emptyset$ for every $a \in \mathsf{Agnt}$. It is *globally coherent* iff $\bigcap_{a\in\mathsf{Agnt}} \omega(a) \neq \emptyset$.

Clearly global coherence implies weak coherence but not vice versa.

2.1 A Simple Example

Figure 1 shows a model of three agents with the following beliefs (this is *all* they believe, modulo logical consequence): Claire believes that the party will be at Anna's place (q); Bill believes that Anna's mother is in town (p); and Anna believes that if her mother is in town the party will not be at her (Anna's) place ($p \to \neg q$). Note that the beliefs are mutually inconsistent (the model is not globally coherent).

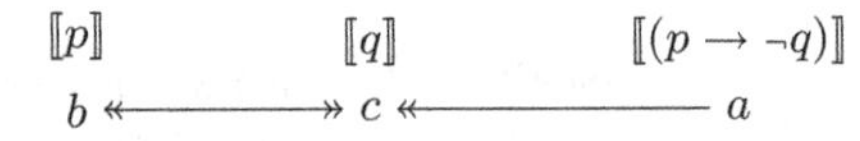

Fig. 1. A simple model. An arrow from a to b means that a follows b.

At this point, each of the three friends have a consistent belief state (the model is *weakly* coherent). Claire can tweet q, after which Anna and Bill will also believe the party is at Anna's place. That's described by the formula $\langle c : q\rangle(B_a q \wedge B_b q)$. Moreover, Bill can tweet p, but the formula $\langle b : p\rangle\neg B_a p$ tells us that still Anna would not believe p. That's because only Claire is following Bill, and so only she will get the message, and before she does she cannot tweet it to Anna: $\neg\langle c : p\rangle B_a p$.

After Claire receives Bill's tweet she can retweet it, and since Anna is following her, she will then believe her mother is in town: $\langle b : p\rangle\langle c : p\rangle B_a p$. Anna also believes $(p \to \neg q)$, so when she receives this message she will also believe $\neg q$, that the party is not going to be at her place. Problems arise for Anna when she receives both of Claire's tweets: one indicating that the party will be at her place, and the other that it will not. Since she is completely credulous, this will leave her in an inconsistent state: $\langle c : q\rangle\langle b : p\rangle\langle c : p\rangle B_a\bot$. Neither Bill nor Claire suffer the same fate. In fact, both will still believe that the party is at Anna's place: $\langle c : q\rangle\langle b : p\rangle\langle c : p\rangle(B_b q \wedge B_c q)$.

The possibility of inconsistent belief states can be regarded as a limitation of our model due to the sometimes unrealistic assumption of credulity: our agents have no way of revising their beliefs. But it can also be regarded as a feature. Even if the agents' beliefs are globally inconsistent, the network structure will allow that inconsistency to emerge in some places but not in others, and this can be described by formulas of our language.

2.2 Conditional Tweeting and Relational Semantics

The dual tweeting operator $[a : \theta]$, can be seen to be a conditional:

Proposition 1. *$F, \omega \models [a : \theta]\varphi$ iff if $F, \omega \models B_a\theta$ then $F, [Fa \uparrow \theta]\omega \models \varphi$.*

Our notation for $\langle a : \theta\rangle$ and $[a : \theta]$ is no accident. Just like the diamond and box of ordinary modal logic, they can be given a relational ("Kripke") semantics. (We omit the straightforward proof of the following.)

Proposition 2 (Relational Semantics). *Define the relation F^a_θ between belief state functions by: $F^a_\theta(\omega_1, \omega_2)$ iff $\omega_1(a) \subseteq [\![\theta]\!]$ and $\omega_2 = [Fa \uparrow \theta]\omega_1$. Then:*

$$F, \omega \models \langle a : \theta\rangle\varphi \quad \textit{iff} \quad F, \nu \models \varphi \textit{ for some } \nu \textit{ such that } F^a_\theta(\omega, \nu)$$
$$F, \omega \models [a : \theta]\varphi \quad \textit{iff} \quad F, \nu \models \varphi \textit{ for every } \nu \textit{ such that } F^a_\theta(\omega, \nu)$$

Define a function V from belief formulas to sets of belief state functions, by $V(B_a\theta) = \{\omega \mid \omega(a) \subseteq [\![\theta]\!]\}$. Let W be the set of belief state functions and let $M(F)$ be the (multi-)modal model (W, F, V) and take our language to be a language of propositional modal logic, with each belief formula considered as a propositional variable and each announcement operator as a modal operator. Then $F, \omega \models \varphi$ iff $M(F), \omega \models \varphi$.

2.3 Potential Belief, and Tracking Ghosts

An agent's *potential beliefs* are those she may acquire as a result of communications from other agents. To describe these clearly we need some notation. Given agents $c_0, \ldots, c_n$ and messages $\theta_0, \ldots, \theta_n$, let $\langle c_0 : \theta_0, \ldots, c_n : \theta_n\rangle$ be an abbreviation for the sequence of tweets $\langle c_0 : \theta_0\rangle \ldots \langle c_n : \theta_n\rangle$. Let $\vec{c}$ be a variable over expressions of the form $c_0 : \theta_0, \ldots, c_n : \theta_n$, so we can also write the (possibly empty) sequence of tweets as $\langle\vec{c}\rangle$. As for the basic language, we define $[\vec{c}]$ as $\neg\langle\vec{c}\rangle\neg$. The *reversal* of $\vec{c}$, denoted $\overleftarrow{c}$, is the reverse sequence of tweets $c_n : \theta_n, \ldots, c_0 : \theta_0$.

Definition 6. A formula of the form $\langle\vec{c}\rangle B_a\theta$ is a *potential belief formula*. Agent a has a *potential belief* that θ iff $F, \omega \models \langle\vec{c}\rangle B_a\theta$ for some $\vec{c}$. (F, ω) is *potentially equivalent* to (F', ω') iff every agent has the same potential beliefs in (F, ω) as in (F', ω').

In the example, Bill's and Claire's beliefs are always consistent, whereas Anna's are not. She potentially believes a contradiction. The formula $\langle c : q\rangle\langle b : p\rangle\langle c : p\rangle B_a\bot$ that expresses a way in which Anna's beliefs can become inconsistent is an example of a potential belief formula. The concept of potential inconsistency gives a finer-grained picture of the initial distribution of beliefs than merely saying they are mutually inconsistent.

Lemma 1. *If* (F, ω) *is potentially equivalent to* (F', ω') *and* $\omega(a) \subseteq [\![\theta]\!]$*, then* $(F, [F_a \uparrow \theta]\omega)$ *is potentially equivalent to* $(F', [F'a \uparrow \theta]\omega')$*.*

Proof. Assume that (F, ω) is potentially equivalent to (F', ω') and $\omega(a) \subseteq [\![\theta]\!]$. By Definition 6, $\omega'(a) \subseteq [\![\theta]\!]$. Then for any $\vec{c}$, χ, and agent b, the following are equivalent:

$F, [Fa \uparrow \theta]\omega \models \langle\vec{c}\rangle B_b\chi$
$F, \omega \models \langle a : \theta\rangle\langle\vec{c}\rangle B_b\chi$ by Definition 4 and $\omega(a) \subseteq [\![\theta]\!]$
$F', \omega' \models \langle a : \theta\rangle\langle\vec{c}\rangle B_b\chi$ by Definition 6 and assumption
$F', [F'a \uparrow \theta]\omega' \models \langle\vec{c}\rangle B_b\chi$ by Definition 4 and $\omega'(a) \subseteq [\![\theta]\!]$.

Theorem 1. *Propositional network announcement models satisfying the same potential belief formulas are indistinguishable: they satisfy all the same formulas.*

Proof. We prove that for any pair of potentially equivalent models (F, ω) and (F', ω'), $F, \omega \models \varphi$ iff $F', \omega' \models \varphi$ for any φ by induction on φ. (i). The atomic case, $F, \omega \models B_a\theta$ iff $F', \omega' \models B_a\theta$, follows directly from Definition 6. (ii). The boolean cases are straightforward. (iii). For $\varphi = \langle a : \theta\rangle\psi$, the induction hypothesis is that for any pair of potentially equivalent models (F, ω) and (F', ω'), $F, \omega \models \chi$ iff $F', \omega' \models \chi$ for any subformula χ of φ – in particular for $\chi = \psi$. The following are equivalent:

$F, \omega \models \langle a : \theta\rangle\psi$
$\omega(a) \subseteq [\![\theta]\!]$ and $F, [Fa \uparrow \theta]\omega \models \psi$ by Definition 4
$\omega'(a) \subseteq [\![\theta]\!]$ and $F', [F'a \uparrow \theta]\omega' \models \psi$ by (i), i.h. and Lemma 1
$F', \omega' \models \langle a : \theta\rangle\psi$ by Definition 4.

A closely related idea is that of "tracking".

Definition 7. Agent b *tracks* agent a in a model (F, ω) iff $F, \omega \models \langle\vec{c}\rangle B_a\theta \rightarrow \langle\vec{c}\rangle B_b\theta$ for every $\vec{c}$ and θ.

If b tracks a then any potential belief of a is also a potential belief of b, but more than that, their potential beliefs are synchronised, in the sense that whenever a acquires a belief, b either acquires it at the same time or already has it.

The interesting point about tracking is that it obscures the following relation. It is possible for an agent to track without following, perhaps by coincidence or just because she already believes every one of some other agent's potential beliefs. If agent b tracks agent a, it is impossible to detect, using the logical language, whether b follows a. We say that b is a *ghost follower* of a if b tracks a without following a. Ghost followers are indistinguishable from real followers.

3 The Logic

We consider various valid principles and properties of propositional network announcement logic, working towards an axiomatisation (shown to be complete in the next section): the system pNAL, shown in Fig. 3. We use $\vdash$ to represent derivability in pNAL. We now introduce the axioms step by step, and along the way present some additional properties that are derivable by the axioms we have introduced "so far".

3.1 The Modal Base

We have observed that the announcement operators have a relational semantics (Proposition 2). It follows that their logic must be an extension of the modal logic K. The belief operator B also has most of the properties of a normal modal operator, except that substitution of propositional variables is restricted to message formulas. By the *modal base* of our logic, we mean the system in Fig. 2. The following is straightforward.

Taut	$\vdash \varphi$ φ subst. inst. of prop. tautology	MP	if $\vdash \varphi \to \psi$ and $\vdash \varphi$ then $\vdash \psi$
K_B	$\vdash B_a(\theta \to \chi) \to (B_a\theta \to B_a\chi)$	Nec_B	if $\vdash_0 \theta$ then $\vdash B_a\theta$
$\mathsf{K}_:$	$\vdash [a:\theta](\varphi \to \psi) \to ([a:\theta]\varphi \to [a:\theta]\psi)$	$\mathsf{Nec}_:$	if $\vdash \varphi$ then $\vdash [a:\theta]\varphi$

Fig. 2. The modal base of pNAL. $\vdash_0$ represents derivability in classical propositional logic.

Proposition 3. *The modal base is sound: every derivable formula is valid.*

Because of this modal base, we can do standard normal modal logic reasoning in our logic (with the syntactic restriction that θ must be propositional in $B_a\theta$), which we will make frequent use of in the following. The base is also enough to show that equivalent formulas can be swapped. The proof (omitted here) is a standard induction on formulas.

Proposition 4 (Replacement of Logical Equivalents).

RLE   *if* $\vdash_0 \theta \leftrightarrow \chi$ *then* $\vdash \Theta(\theta) \leftrightarrow \Theta(\chi)$ *if* $\vdash \varphi \leftrightarrow \psi$ *then* $\vdash \Phi(\varphi) \leftrightarrow \Phi(\psi)$

where $\Theta(\chi)$ *is the formula obtained from* $\Theta(\theta)$ *by replacing some instances of* θ *by* χ *or vice versa, and similarly for formulas* $\Phi(\varphi)$ *and* $\Phi(\psi)$.

3.2 Duality and Sincerity

The following dualities are derivable using propositional logic (and replacement of equivalents) alone:

Proposition 5. Dual $\vdash [a:\theta]\neg\varphi \leftrightarrow \neg\langle a:\theta\rangle\varphi \qquad \vdash \neg[a:\theta]\varphi \leftrightarrow \langle a:\theta\rangle\neg\varphi$

But the operators are also linked by our assumption of sincerity, which says that $B_a\theta$ is presupposed by $\langle a:\theta\rangle$ and serves as the antecedent of $[a:\theta]$:

$$\mathsf{Sinc} \quad [a:\theta]\varphi \leftrightarrow (B_a\theta \rightarrow \langle a:\theta\rangle\varphi) \qquad \langle a:\theta\rangle\varphi \leftrightarrow (B_a\theta \wedge [a:\theta]\varphi)$$

These are inter-derivable using the modal base; the first is included as an axiom in pNAL. Validity can be easily checked directly from the semantical definitions.

Proposition 6. Sinc *is valid.*

Sinc implies that when the precondition is satisfied, the two operators are equivalent:

Proposition 7. Swap $\vdash B_a\theta \rightarrow ([a:\theta]\varphi \leftrightarrow \langle a:\theta\rangle\varphi)$

Using only the precondition principle Sinc and the modal base, we can show the existence of normal forms.

Theorem 2 (Normal form). *Every formula is provably equivalent to one in conjunctive normal form or disjunctive normal form, with atoms generated by*

$$\varphi ::= B_a\theta \mid [a:\theta]\varphi \mid \langle a:\theta\rangle\varphi$$

which is to say: sequences of diamonds and boxes ending in a belief formula. Moreover, the modal depth of the normal form is no greater than the depth of the original formula.

Proof. Given an arbitrary formula, first rewrite (by expanding or introducing abbreviations) so that the only operators are $\neg$, $\wedge$, B_a and $[a:\theta]$. Then, given RLE, we only need note that:

$$\mathsf{Red}_\neg \quad \vdash [a:\theta]\neg\varphi \leftrightarrow \neg(B_a\theta \wedge [a:\theta]\varphi)$$
$$\mathsf{Red}_\wedge \quad \vdash [a:\theta](\varphi \wedge \psi) \leftrightarrow ([a:\theta]\varphi \wedge [a:\theta]\psi)$$

($\mathsf{Red}_\neg$ is tautologically equivalent to an instance of Sinc and $\mathsf{Red}_\wedge$ is just the modally derivable distribution of box over conjunction.)

To see that there is no increase of modal depth (nesting of tweets), it is enough to note that the formulas on either side of the equivalences $\mathsf{Red}_\neg$ and $\mathsf{Red}_\wedge$ are of the same depth.

3.3 Rational Conservative Updating

The direct effect of a tweet is captured by two axioms:

$$\mathsf{Cnsv} \quad B_b\chi \rightarrow [a:\theta]B_b\chi \qquad \mathsf{Rat} \quad \langle a:\theta\rangle B_b\chi \rightarrow B_b(\theta \rightarrow \chi)$$

Conservatism (Cnsv) is our assumption that old beliefs are retained when receiving new information. Rationality (Rat) is our assumption that agents only believe what follows logically from their old beliefs and the content of the tweets they receive. Note that the soundness of these axioms rely on the fact that there are no higher-order beliefs.

Proposition 8. *Both* Cnsv *and* Rat *are valid.*

Proof. For Cnsv recall the previously mentioned monotonicity of updating: $\omega \leq [C \uparrow \theta]\omega$. Since tweeting is a special case of updating, it is also monotonic. For Rat consider two cases. If b is a follower of a, then b's updated state $[F_a \uparrow \theta]\omega(b) = \omega(b) \cap \theta$. But $\omega(b) \cap \theta \subseteq [\![\chi]\!]$ iff $\omega(b) \subseteq [\![\theta \to \chi]\!]$. If b is not a follower of a then $[F_a \uparrow \theta]\omega(b) = \omega(b)$ and $\omega(b) \subseteq [\![\chi]\!]$ implies $\omega(b) \subseteq [\![\theta \to \chi]\!]$.

The implication in Rat can be turned into an equivalence under the assumption that b believes what a tweets. The following can be proved using Rat and Cnsv.

Proposition 9. Up $\vdash \langle a : \theta \rangle B_b\theta \to (\langle a : \theta \rangle B_b\chi \leftrightarrow B_b(\theta \to \chi))$

Proof. One half of the equivalence is a weakening of Rat. For the other, note that $[a : \theta]B_b(\theta \to \chi) \to (\langle a : \theta \rangle B_b\theta \to \langle a : \theta \rangle B_b\chi)$ is derivable by purely modal reasoning. And $B_b(\theta \to \chi) \to [a : \theta]B_b(\theta \to \chi)$ is an instance of Cnsv.

3.4 Following

Given the problem of ghosts described in Sect. 2.3, there can be no formula that exactly defines the relation of following. Nonetheless, some sufficient and necessary conditions are expressible.

Proposition 10. *Given a and b, for any $\vec{c}$ and any θ,*

(Sufficient) If $F, \omega \models \langle \vec{c} \rangle (\neg B_b\chi \wedge \langle a : \theta \rangle B_b\chi)$ then bFa.
(Necessary) If bFa then $F, \omega \models [\vec{c}][a : \theta]B_b\theta$

Proof. For the sufficient condition, suppose $F, \omega \models \langle \vec{c} \rangle (\neg B_b\chi \wedge \langle a : \theta \rangle B_b\chi)$. Let ω' be the result of updating ω according to $\langle \vec{c} \rangle$. Then $F, \omega' \not\models B_b\chi$ but $F, [Fa \uparrow \theta]\omega' \models B_b\chi$. So $\omega'(b) \neq [Fa \uparrow \theta]\omega'(b)$ and so $b \in Fa$.

For the necessary condition, suppose bFa. Then either one of the preconditions in evaluating $[\vec{c}]$ fails, and in which case the formula is satisfied, or they all succeed. In that case, let ω' be as before. Since $b \in Fa$, $[Fa \uparrow \theta]\omega'(b) = \omega'(b) \cap [\![\theta]\!]$ and so $F, [Fa \uparrow \theta]\omega' \models B_b\theta$. Thus $F, \omega \models [\vec{c}][a : \theta]B_b\theta$.

Our approach to the logic, then, is to include an axiom saying that the sufficient condition implies the necessary condition:

$$\text{Foll} \quad \langle \vec{c} \rangle (\neg B_b\chi \wedge \langle a : \chi' \rangle B_b\chi) \to [\vec{e}][a : \theta]B_b\theta$$

3.5 Network Stability

The following axiom captures the assumption that the only thing that changes anything is tweeting events, and that an "empty" tweet (of a tautology) changes nothing.

$$\text{Null} \quad \text{if} \vdash_0 \theta \text{ then} \vdash \varphi \leftrightarrow \langle a : \theta \rangle \varphi$$

Null is in fact the last axiom we need in order to get a complete axiomatic system, as we shall see in the next section. We end this section with mentioning two additional properties related to network stability. They are both already derivable (follows from the completeness result in the next section).

First, tweeting does not affect the network structure; in fact, the following relation is kept fixed. That's our assumption of network stability. Consider the following property:

$$\text{Stab} \quad \langle b : \chi \rangle B_c\delta \rightarrow [a : \theta]\langle b : \chi \rangle B_c\delta$$

Stab is a necessary but insufficient condition of network stability; and that's the best we can do. There is no sufficient condition available. Fortunately, we don't need one to get a complete axiomatisation, as we shall see in the next section.

Network stability is needed for many principles involving the iteration of tweets; in particular, for moving a conditional tweet to the beginning of a sequence of tweets:

$$\text{Perm} \quad \langle \vec{a} \rangle [b : \chi]\varphi \rightarrow [b : \chi]\langle \vec{a} \rangle \varphi$$

Proposition 11. Null, Stab *and* Perm *are valid.*

Proof. The case for Null is trivial. We show the case for Stab, Perm is (also) straightforward. Suppose $F, \omega \models \langle b : \chi \rangle B_c\delta$ then (1) $F, \omega \models B_b\chi$ and (2) $F, [Fb \uparrow \chi]\omega \models B_c\delta$. Now suppose $F, \omega \models B_a\theta$. Then (3) $F, [Fa \uparrow \theta]\omega \models B_b\chi$ since (1) and $[Fa \uparrow \theta]\omega(b) \subseteq \omega(b)$. We also have (4) $F, [Fb \uparrow \chi]([Fa \uparrow \theta]\omega) \models B_c\delta$ since $[Fb \uparrow \chi]([Fa \uparrow \theta]\omega) = [Fa \uparrow \theta]([Fb \uparrow \chi]\omega) \subseteq [Fb \uparrow \chi]\omega$ and (2). Thus by (3) and (4), $F, \omega \models [a : \theta]\langle b : \chi \rangle B_c\delta$.

Taut	if $\vdash_0 \varphi$ then $\vdash \varphi$	MP	if $\vdash \varphi \rightarrow \psi$ and $\vdash \varphi$ then $\vdash \psi$
K_B	$\vdash B_a(\theta \rightarrow \chi) \rightarrow (B_a\theta \rightarrow B_a\chi)$	$\mathsf{K}_:$	$\vdash [a : \theta](\varphi \rightarrow \psi) \rightarrow ([a : \theta]\varphi \rightarrow [a : \theta]\psi)$
Nec_B	if $\vdash_0 \theta$ then $\vdash B_a\theta$	$\mathsf{Nec}_:$	if $\vdash \varphi$ then $\vdash [a : \theta]\varphi$
Sinc	$\vdash [a : \theta]\varphi \leftrightarrow (B_a\theta \rightarrow \langle a : \theta \rangle \varphi)$	Cnsv	$\vdash B_b\chi \rightarrow [a : \theta]B_b\chi$
Rat	$\vdash \langle a : \theta \rangle B_b\chi \rightarrow B_b(\theta \rightarrow \chi)$	Foll	$\vdash \langle \vec{c} \rangle(\neg B_b\chi \wedge \langle a : \chi' \rangle B_b\chi) \rightarrow [\vec{e}][a : \theta]B_b\theta$
Null	if $\vdash_0 \theta$ then $\vdash \varphi \leftrightarrow \langle a : \theta \rangle \varphi$		

Fig. 3. Axioms and rules of pNAL.

4 Completeness

We show that the system pNAL, displayed in Fig. 3 is (strongly) complete. We assume the usual concepts of consistency, maximal consistency, and logical closure. The Lindenbaum result that any consistent set of formulas can be extended to a maximal consistent set holds for standard reasons. We have shown the derivability of various additional principles (Dual, Swap, and Up) that will be used below (Propositions 5, 7 and 9). All that remains is to use a maximal consistent set of formulas Γ to construct a following relation F_Γ and a belief state function ω_Γ for which we can prove a Truth Lemma (that F_Γ, ω_Γ satisfies every formula in Γ). So we define:

$$bF_\Gamma a \text{ iff } [\vec{c}][a:\theta]B_b\theta \in \Gamma \text{ for all } \vec{c} \text{ and } \theta \qquad \omega_\Gamma(a) = \bigcap\{[\![\theta]\!] \mid B_a\theta \in \Gamma\}.$$

Note that the definition of F_Γ uses the set of all formulas providing *necessary* conditions for following, as identified in Proposition 10. In a given model, any ghost follower of a will also satisfy all these conditions. Our approach is to build a model in which all trackers of a are taken to be followers.

Regarding belief formulas, the definition of ω_Γ does it's job.

Lemma 2. *$\omega_\Gamma(a) \subseteq [\![\theta]\!]$ iff $B_a\theta \in \Gamma$ for any a and θ,*

Proof. Right-to-left is immediate since $\omega_\Gamma(a) = \bigcap\{[\![\chi]\!] \mid B_a\chi \in \Gamma\}$. For the other direction, by completeness of propositional logic, $\vdash_0 (\chi_1 \wedge \ldots \wedge \chi_n) \to \theta$ for some $B_a\chi_1, \ldots B_a\chi_n \in \Gamma$. Then $\vdash B_a((\chi_1 \wedge \ldots \wedge \chi_n) \to \theta)$ by Nec_B and so $\vdash (B_a\chi_1 \wedge \ldots \wedge B_a\chi_n) \to B_a\theta$ by more modal reasoning using K_B and Nec_B. Thus $B_a\theta \in \Gamma$.

Lemma 2 is the obvious base case of an attempt to prove the Truth Lemma by induction on the structure of formulas. But such a direct approach doesn't work because the clause for $\langle a:\theta\rangle$ requires us to update the model. There are several options here. A first thought is to try to construct the set of formulas satisfied by the updated model, i.e., to find a maximal consistent set Γ' such that $F_{\Gamma'} = F_\Gamma$ and $\omega_{\Gamma'} = [F_\Gamma a \uparrow \theta]\omega_\Gamma$. But the search for something satisfying the first of these conditions is plagued by ghosts: each time the model is updated, new ghost followers may appear. Instead, we'll construct a Γ' to meet only the second condition, using a syntactic update operation:

$$\langle a:\theta\rangle\Gamma = \{\varphi \mid \langle a:\theta\rangle\varphi \in \Gamma\}.$$

Our proof of the Truth Lemma (Lemma 5) will involve a strengthening of it that quantifies over sets obtained by repeated applications of syntactic update. This requires new notation and a some technical lemmas.

Definition 8. Define the relation $\trianglelefteq$ between sets of formulas as follows: $\Gamma \trianglelefteq \Gamma'$ iff $B_a\theta \in \Gamma$ and $\Gamma' = \langle a:\theta\rangle\Gamma$ for some a and θ. Let $\leq$ be the transitive closure of $\trianglelefteq$.

Lemma 3. *If Γ is a maximal consistent set and $\Gamma \leq \Gamma'$ then*

1. *Γ' is also a maximal consistent set and*
2. *there is a $\vec{c}$ such that: (a) $\Gamma' = \langle\vec{c}\rangle\Gamma$, and (b) $[\overleftarrow{c}]\varphi \in \Gamma$ iff $\varphi \in \Gamma'$ for all φ, where $\overleftarrow{c}$ is the reversal of $\vec{c}$.*

Proof. By induction on the length of the shortest chain $\Gamma \trianglelefteq \ldots \trianglelefteq \Gamma'$. In the base case, $\Gamma' = \Gamma$ and we can take $\vec{c}$ to be the empty sequence. So now suppose the length of the shortest chain is strictly positive. Then there is a Γ'' such that $\Gamma \trianglelefteq \Gamma'' \leq \Gamma'$. Note that the chain $\Gamma'' \trianglelefteq \ldots \trianglelefteq \Gamma'$ is shorter. By definition of $\trianglelefteq$, there are a and θ such that $B_a\theta \in \Gamma$ and $\Gamma'' = \langle a : \theta \rangle \Gamma$.

1. We first show that Γ'' is a maximal consistent set: $\neg\varphi \in \Gamma''$ iff $\neg\varphi \in \langle a : \theta\rangle\Gamma$ (by $\Gamma'' = \langle a : \theta\rangle\Gamma$) iff $\langle a : \theta\rangle\neg\varphi \in \Gamma$ (by def. of $\langle a : \theta\rangle\Gamma$) iff $\neg[a : \theta]\varphi \in \Gamma$ (by Dual) iff $\neg\langle a : \theta\rangle\varphi \in \Gamma$ (by Swap and $B_a\theta \in \Gamma$) iff $\langle a : \theta\rangle\varphi \notin \Gamma$ iff $\varphi \notin \langle a : \theta\rangle\Gamma$ (by def. of $\langle a : \theta\rangle\Gamma$) iff $\varphi \notin \Gamma''$ ($\Gamma'' = \langle a : \theta\rangle\Gamma$). Now, since there is a shorter chain from Γ'' to Γ' and Γ'' is a maximal consistent set we can apply the induction hypothesis, to get that Γ' is a maximal consistent set.
2. Also from the induction hypothesis: there is $\vec{c}$ such that (1) $\Gamma' = \langle\vec{c}\rangle\Gamma''$ and (2) $[\overleftarrow{c}]\varphi \in \Gamma''$ iff $\varphi \in \Gamma'$ for all φ. So let $\vec{e} = \vec{c}, a : \theta$. Then from (1): $\Gamma' = \langle\vec{c}\rangle\Gamma'' = \langle\vec{c}\rangle\langle a : \theta\rangle\Gamma = \langle\vec{c}, a : \theta\rangle\Gamma = \langle\vec{e}\rangle\Gamma$. We get the following equivalences: $[\overleftarrow{e}]\varphi \in \Gamma$ iff $[a : \theta, \overleftarrow{c}]\varphi \in \Gamma$ (def. of $\overleftarrow{e}$) iff $[a : \theta][\overleftarrow{c}]\varphi \in \Gamma$ (definition of $[a : \theta, \overleftarrow{c}]$) iff $\langle a : \theta\rangle[\overleftarrow{c}]\varphi \in \Gamma$ (Swap, $B_a\theta \in \Gamma$) iff $[\overleftarrow{c}]\varphi \in \langle a : \theta\rangle\Gamma$ (def. of $\langle a : \theta\rangle\Gamma$) iff $[\overleftarrow{c}]\varphi \in \Gamma''$ ($\Gamma'' = \langle a : \theta\rangle\Gamma$) iff $\varphi \in \Gamma'$ (2).

Lemma 3 enables us to show that belief state functions behave properly under updates.

Lemma 4. *For any m.c.s. Γ, if $\Gamma \leq \Gamma'$ and $B_a\theta \in \Gamma'$ then $[F_\Gamma a \uparrow \theta]\omega_{\Gamma'} = \omega_{\langle a:\theta\rangle\Gamma'}$.*

Proof. Suppose $\Gamma \leq \Gamma'$ and $B_a\theta \in \Gamma'$. We first need a fact about propositional logic:

Claim: $\bigcap\{[\![\chi]\!] \mid \theta \to \chi \in \Delta\} = \bigcap\{[\![\chi]\!] \mid \chi \in \Delta\} \cap [\![\theta]\!]$
for any formula θ and any logically closed set Δ of formulas (of propositional logic). It can be proved easily from the deduction theorem.

By Lemma 3, $\Gamma' = \langle\vec{c}\rangle\Gamma$ for some $\vec{c}$ and this is a maximal consistent set. Let b be in Agnt. We will show that $\omega_{\langle a:\theta\rangle\Gamma'}(b) = [F_\Gamma a \uparrow \theta]\omega_{\Gamma'}(b)$. We have two cases depending on whether or not b is tracking a:

$b \in F_\Gamma a$ Then $[\overleftarrow{c}][a : \theta]B_b\theta$ is in Γ. By Lemma 3.2, $[a : \theta]B_b\theta$ is in $\langle\vec{c}\rangle\Gamma$. But $B_a\theta$ is in Γ' and so by Swap so is $\langle a : \theta\rangle B_b\theta$. From this, Up tells us that $\langle a : \theta\rangle B_b\chi \leftrightarrow B_b(\theta \to \chi)$ is in Γ' for any χ, but also by definition of $\langle a : \theta\rangle\Gamma'$, we know that $\langle a : \theta\rangle B_b\chi \in \Gamma'$ iff $B_b\chi \in \langle a : \theta\rangle\Gamma'$. Putting these together: $B_b\chi \in \langle a : \theta\rangle\Gamma'$ iff $B_b(\theta \to \chi) \in \Gamma'$.

$$\text{So } \bigcap\{[\![\chi]\!] \mid B_b\chi \in \langle a : \theta\rangle\Gamma'\} = \bigcap\{[\![\chi]\!] \mid B_b(\theta \to \chi) \in \Gamma'\}$$
$$= \bigcap\{[\![\chi]\!] \mid B_b\chi \in \Gamma'\} \cap [\![\theta]\!] \quad \text{by Claim, with } \Delta = \{\chi \mid B_b\chi \in \Gamma'\}$$

$$\text{Thus} \quad \omega_{\langle a:\theta\rangle\Gamma'}(b) = [F_\Gamma a \uparrow \theta]\omega_{\Gamma'}(b)$$

$b \notin F_\Gamma a$ Then there is some $\vec{e}$ and some θ' for which $[\overleftarrow{e}][a : \theta']B_b\theta$ is not in Γ. Foll then tells us that $\langle\overleftarrow{c}\rangle(\neg B_b\chi \wedge \langle a : \theta\rangle B_b\chi)$ is not in Γ for any χ. But Γ is a maximal consistent set so it does contain $[\overleftarrow{c}](\langle a : \theta\rangle B_b\chi \to B_b\chi)$. So by Lemma 2, $\langle a : \theta\rangle B_b\chi \to B_b\chi$ is in $\langle\vec{c}\rangle\Gamma = \Gamma'$. We also have that $B_b\chi \to \langle a : \theta\rangle B_b\chi$ is in Γ'. (This is by Cnsv and Swap since $B_a\theta \in \Gamma'$.) Thus:

$$
\begin{aligned}
B_b\chi \in \Gamma' &\text{ iff } \langle a:\theta\rangle B_b\chi \in \Gamma' \\
&\text{ iff } B_b\chi \in \langle a:\theta\rangle\Gamma' \quad \text{Defn. } \langle a:\theta\rangle\Gamma'
\end{aligned}
$$

Hence $\bigcap\{[\![\chi]\!] \mid B_b\chi \in \langle a:\theta\rangle\Gamma'\} = \bigcap\{[\![\chi]\!] \mid B_b\chi \in \Gamma'\}$

Thus $\qquad \omega_{\langle a:\theta\rangle\Gamma'}(b) = [F_\Gamma a\!\uparrow\!\theta]\omega_{\Gamma'}(b)$

Lemma 5 (Truth Lemma). *$F_\Gamma, \omega_\Gamma \models \varphi$ iff $\varphi \in \Gamma$, for any formula φ and m.c.s. Γ.*

Proof. Let Γ be a maximal consistent set. We prove that for any Γ', if $\Gamma \leq \Gamma'$ then $F_\Gamma, \omega_{\Gamma'} \models \psi$ iff $\psi \in \Gamma'$, by induction on ψ. The base case $\psi = B_a\theta$ follows from Lemma 2, and the cases for negation and conjunction are straightforward.

Consider the case that $\psi = \langle a : \theta\rangle\varphi$. Note that Γ' is a maximal consistent set by Lemma 3.1. The following are equivalent:

$F_\Gamma, \omega_{\Gamma'} \models \langle a:\theta\rangle\varphi$	
$F_\Gamma, \omega_{\Gamma'} \models B_a\theta$ and $F_\Gamma, [F_\Gamma a\!\uparrow\!\theta]\omega_{\Gamma'} \models \varphi$	by semantics (Definition 4)
$B_a\theta \in \Gamma'$ and $F_\Gamma, [F_\Gamma a\!\uparrow\!\theta]\omega_{\Gamma'} \models \varphi$	by the $B_a\theta$ case, above
$B_a\theta \in \Gamma'$ and $F_\Gamma, \omega_{\langle a:\theta\rangle\Gamma'} \models \varphi$	by Lemma 4, since $\Gamma \leq \Gamma'$
$B_a\theta \in \Gamma'$ and $\varphi \in \langle a:\theta\rangle\Gamma'$	by I.H., since $\Gamma \leq \langle a:\theta\rangle\Gamma'$
$B_a\theta \in \Gamma'$ and $\langle a:\theta\rangle\varphi \in \Gamma'$	by definition of $\langle a:\theta\rangle\Gamma'$
$\langle a:\theta\rangle\varphi \in \Gamma'$	by Sinc and closure of Γ'

This completes the induction. Finally, let $\top$ be any tautology. Then $B_a\top$ by Nec$_B$ and $\Gamma = \langle a : \top\rangle\Gamma$ by Null. Thus $\Gamma \trianglelefteq \Gamma$ and so $\Gamma \leq \Gamma$, and the result follows.

When Γ is a set of formulas and φ a formula, we write $\Gamma \models \varphi$ to mean that any model satisfying Γ also satisfies φ, and $\Gamma \vdash \varphi$ to mean that there is a theorem $(\varphi_1 \wedge \ldots \wedge \varphi_n) \to \varphi$ of pNAL for some finite sequence of formulas $\varphi_1, \ldots, \varphi_n$ in Γ.

Theorem 3 (Soundness and Completeness). *For any Γ and φ, $\Gamma \models \varphi$ iff $\Gamma \vdash \varphi$.*

Proof. Soundness follows from validity of the axioms and rules (Propositions 3, 6, 8, 10 and 11). (Strong) completeness follows from the Truth Lemma.

5 Variants

We have a brief look at some natural variants of the logic.

5.1 Irreflexivity

Do agents follow themselves? We have not assumed that they *don't*; we allow models were agents do follow themselves. Indeed, the canonical model in the previous section is reflexive – all agents always follow themselves. However, it is easy to see that self-following cannot be expressed in our logical language. I.e., we have the following property.

Proposition 12. *For any formula* φ *and any model* (F, ω), $F, \omega \models \varphi$ *iff* $F^{-}, \omega \models \varphi$, *where* F^{-} *is the largest irreflexive submodel of* F *(i.e.,* $F^{-} = F \setminus \{(a, a) : a \in \mathsf{Agnt}\}$*).*

We thus immediately get the following corollary of the results in the previous section (by taking F to be the (reflexive) canonical following relation).

Corollary 1. pNAL *is sound and strongly complete with respect to the class of models with an irreflexive following relation.*

5.2 Coherence

Our logic can easily be extended to axiomatise the class of weakly or globally coherent models. Consider the following schemata (n ranges over positive natural numbers):

WCoh if $\vdash_0 \neg\theta$ then $\vdash \neg B_a \theta$

GCoh if $\vdash_0 \neg(\theta_1 \wedge \ldots \wedge \theta_n)$ then $\vdash \neg(B_{a_1}\theta_1 \wedge \cdots \wedge B_{a_n}\theta_n)$

Let the wCpNAL be the axiom system pNAL extended with WCoh and let gCpNAL be the axiom system pNAL extended with GCoh. The following can be easily checked.

Lemma 6. *A model is weakly (globally) coherent iff it satisfies all inst. of* WCoh *(*GCoh*).*

From this, and the fact that the rules of the logic preserve validity on the classes of weakly and globally coherent models, respectively, we immediately get the following.

Theorem 4. wCpNAL *and* gCpNAL *are sound and strongly complete with respect to the classes of weakly and globally coherent models, respectively.*

6 Discussion

In this paper we laid the groundwork for formal reasoning about network announcements in social networks ("tweeting"). We defined a minimal modal logic based on (not necessarily consistent) propositional beliefs and a "tweeting" modality $\langle a : \theta \rangle$, and studied the logic in detail. We believe that this detailed study lays a solid foundation for richer frameworks to be studied in the future. For example, the technique we used for encoding network structure using logical formulas in the completeness proof is general. We made several assumptions clear in the beginning of the paper, some of which showed up again as axioms of the logic. It could be interesting to investigate a weakening of some of these assumptions starting from a syntactic angle, by weakening the axioms. Regarding coherence, there is a third, natural, form that we haven't considered in this paper: no agent can enter an inconsistent belief state as a result of network announcements ("local coherence").

There are two main and orthogonal directions for future work. The first is extending the *semantics* to model agents with higher-order beliefs and possibly even beliefs about the network structure. Higher-order beliefs would introduce a number of subtleties and complications and would require a number of assumptions. For example, with higher order beliefs assumptions would have to be made about different agents' beliefs about the possibility of tweeting events taking place, belief states are would no longer be monotonic under tweeting, the belief state of the tweeter would not be static under tweeting, and so on. With incomplete information about the network structure, new beliefs about that structure could actually be formed as a result of receiving tweets in certain situations. The second direction is extending the *syntax*. One natural and interesting possibility for enriching the language is to add modalities of the form $\langle a \rangle$ (where a is an agent) *quantifying* over tweets, known from group announcement logic [1], where a formula of the form $\langle a \rangle \varphi$ would intuitively mean that a can make φ true by tweeting some message. Such operators can potentially be used to capture many interesting phenomena related to the information flow in social networks.

Acknowledgments. The first author is supported by Projects of the National Social Science Foundation of China under research no. 15AZX020.

References

1. Ågotnes, T., Balbiani, P., van Ditmarsch, H., Seban, P.: Group announcement logic. J. Appl. Logic **8**(1), 62–81 (2010)
2. Baltag, A., Christoff, Z., Ulrik Hansen, J., Smets, S.: Logical Models of Informational Cascades. Studies in Logic. College Publications, London (2013)
3. Christoff, Z.: Dynamic Logics of Networks. Ph.D. thesis, University of Amsterdam (2016)
4. Christoff, Z., Ulrik Hansen, J.: A logic for diffusion in social networks. J. Appl. Logic **13**(1), 48–77 (2015)
5. Ulrik Hansen, J.: Reasoning about opinion dynamics in social networks. J. Logic Comput. (2015). https://doi.org/10.1093/logcom/exv083
6. Harte, L.: Introduction to Data Multicasting. Althos Publishing, Raleigh (2008)
7. Hosszú, G.: Mediacommunication based on application-layer multi-cast. In: Encyclopedia of Virtual Communities and Technologies, pp. 302–307. IGI Global (2006)
8. Pacuit, E., Parikh, R.: The logic of communication graphs. In: Leite, J., Omicini, A., Torroni, P., Yolum, I. (eds.) DALT 2004. LNCS, vol. 3476, pp. 256–269. Springer, Heidelberg (2005). doi:10.1007/11493402_15
9. Ruan, J., Thielscher, M.: A logic for knowledge flow in social networks. In: Wang, D., Reynolds, M. (eds.) AI 2011. LNCS, vol. 7106, pp. 511–520. Springer, Heidelberg (2011). doi:10.1007/978-3-642-25832-9_52
10. Seligman, J., Liu, F., Girard, P.: Logic in the community. In: Banerjee, M., Seth, A. (eds.) ICLA 2011. LNCS, vol. 6521, pp. 178–188. Springer, Heidelberg (2011). doi:10.1007/978-3-642-18026-2_15
11. Seligman, J., Liu, F., Girard, P.: Facebook and the epistemic logic of friendship. In: Schipper, B.C. (ed.) Proceedings of the 14th Conference on Theoretical Aspects of Rationality and Knowledge, Chennai, India, pp. 229–238 (2013)

12. van Ditmarsch, H., van Eijck, J., Pardo, P., Ramezanian, R., Schwarzentruber, F.: Gossip in dynamic networks. Liber Amicorum Alberti. A Tribute to Albert Visser, pp. 91–98. College Publications, London (2016)
13. van Eijck, J., Sietsma, F.: Message-generated kripke semantics. In: The 10th International Conference on Autonomous Agents and Multiagent Systems, vol. 3, pp. 1183–1184 (2011)
14. Wang, Y., Sietsma, F., Eijck, J.: Logic of information flow on communication channels. In: Omicini, A., Sardina, S., Vasconcelos, W. (eds.) DALT 2010. LNCS, vol. 6619, pp. 130–147. Springer, Heidelberg (2011). doi:10.1007/978-3-642-20715-0_8

Multi-Path vs. Single-Path Replies to Skepticism

Wen-fang Wang(✉)

Institute of Philosophy of Mind and Cognition,
National Yang Ming University, Taipei City, Taiwan
wfwang@ym.edu.tw

Abstract. In order to reply to the contemporary skeptic's argument for the conclusion that we don't have any empirical knowledge about the external world, several authors have suggested different fallibilist theories of knowledge that reject the epistemic closure principle. Holliday [8], however, shows that almost all of them suffer from either the problem of containment or the problem of vacuous knowledge. Furthermore, Holliday [9] suggests that the fallibilist should allow a proposition to have multiple sets of relevant alternatives, each of which is sufficient while none is necessary, if all its members are eliminated, for knowing that proposition. Not completely satisfied with Holliday's multi-path reply to the skeptic, the author suggests a new single-path relevant alternative theory of knowledge and argues that it can avoid both the problem of containment and the problem of vacuous knowledge while rejecting skepticism.

1 The Skeptic's Argument

The contemporary skeptic has no intention to deny that we have reflective or conceptual knowledge. After all, s/he has tried to convince us by *logical* reasonings that we do not have knowledge of a certain kind. What the contemporary skeptic does want to deny is that we have any empirical knowledge about the external world. If you claim that you do know, say, that you have two hands, here then is a simple argument from the contemporary skeptic trying to convince you that you are absolutely wrong about this:[1]

($\mathbf{P_1}$) You don't know that you are not a brain in a vat.
($\mathbf{P_2}$) You know that that you have two hands implies that you are not a brain in a vat.
($\mathbf{P_3}$) If you know that that you have two hands implies that you are not a brain in a vat and you know that you have two hands, then you know that you are not a brain in a vat.
(**C**) Therefore, you don't know that you have two hands.

In symbols (take ψ to be 'you are not a brain in a vat' and 'ϕ' to be 'you have two hands'), the argument ($\mathbf{P_1}$)–(**C**) has the following valid form:

[1] This is, in essence but not in form, Unger's argument in Chap. 1 of [12].

A. Baltag et al. (Eds.): LORI 2017, LNCS 10455, pp. 65–78, 2017.
DOI: 10.1007/978-3-662-55665-8_5

(P$_1$) $\neg K\psi$
(P$_2$) $K(\phi \rightarrow \psi)$
(P$_3$) $(K(\phi \rightarrow \psi) \wedge K\phi) \rightarrow K\psi$
(C) $\neg K\phi$

For the contemporary skeptic, a similar argument of the same form is always available for any specific claim ϕ about the external world about which you want to claim that you have knowledge. This argument pattern is, then, a general weapon that the contemporary skeptic often uses to frustrate his/her optimistic enemies. Note that the premise (**P$_3$**) is an instance of a version of the so-called 'epistemic closure principle' (ECP). Since the argument is valid, it is sound if all its premises are true. The important question is, of course, whether (**P$_1$**)–(**P$_3$**) are true; and, if not, which premise is to be blamed.

2 Single-path Fallibilism and Its Problems

There are different replies to this question; actually, there are too many to be discussed in a short paper. Holliday, however, has done an admirably extensive research on quite a substantial part of these replies and points out several serious problems about them. To see what the main problems of these replies are, let us pick just one of them out as an example: Dreske's 'relevant alternative theory', as it is now called, in his [2–5]. For Dretske, what is wrong in the argument is (**P$_3$**) and ECP that backs it up. Dretske argues that epistemic operators in general, and the knowledge operator 'S knows that' in particular, are not fully penetrating operators; rather, they are semi-penetrating operators. Here, an operator 'O' is fully penetrating iff, for whatever *types* of sentences ϕ and ψ, $O\phi$ implies $O\psi$ if one knows that ϕ implies ψ. On the other hand, an operator 'O' is semi-penetrating iff, for some but not all *types* of sentences ϕ and ψ, $O\phi$ implies $O\psi$ if one knows that ϕ implies ψ. Dretske [2] gives several examples endeavoring to convince us that epistemic operators are not fully penetrating but only semi-penetrating and hence ECP is not universally valid.

Why is the knowledge operator only semi-penetrating according to Dretske? The answer has something to do with a consequence of his theory of knowledge in [3]. Call a $\neg\phi$-possibility w, i.e., a possibility in which ϕ is false, 'an alternative to ϕ'. Then, one consequence of Dretske's theory of knowledge in [3] is that, in order to know that ϕ, a subject S's total evidence does not have to be so strong as to exclude[2] every alternative to ϕ; all s/he needs in order to know that ϕ is only to have his/her evidence strong enough to exclude every 'relevant' alternative to ϕ. This consequence of Dretske's theory of knowledge is actually shared with several other theories of knowledge, such as those of Heller [6,7], Lewis [10], Sosa [11], and DeRose [1]. Indeed, if we define 'fallibilism' to be the thesis that one can know an empirical proposition ϕ even if s/he has not ruled out every way that ϕ could be false, it is easy to see that every theory mentioned here is a sort of

[2] Or rule out, or eliminate. In what follows, 'eliminate', 'rule out', and 'exclude' will be used as synonyms.

fallibilism. Given this fallibilist conception of knowledge, it is then possible that a $(\neg\phi \wedge \neg\psi)$-possibility w is a 'relevant' but unexcluded alternative to ψ though not a 'relevant' alternative to ϕ even if one knows that ϕ implies ψ. Hence it is also possible that the total evidence that a subject S has, though being strong enough to exclude every relevant alternative to ϕ (so that S knows that ϕ), is not good enough to exclude *every* relevant alternative to ψ (so that S does not know that ψ) even if S knows that ϕ implies ψ. This is why the knowledge operator is not fully penetrating according to Dretske. However, the operator is still semi-penetrating, for, to Dretske, it is trivial, at least for an astute logician, to say that whenever a subject S knows that ϕ, s/he therefore knows that $(\phi \vee \psi)$ as well (giving that s/he also understand ψ) and that whenever S knows that $(\phi \wedge \psi)$, s/he therefore knows that ϕ (and knows that ψ too).

Borrowing from Heller [6] the idea that we treat the relevance relation mentioned here in a way similar to that logicians treat the similarity relation in conditional logic, Holliday [8] suggests that we formalize Dretske's theory of knowledge by what he calls 'RA models' and 'D-semantics'. An RA model (short for 'a relevant alternatives model') $\mathfrak{M}$ is a tuple $<\mathrm{W}_{\mathfrak{M}}, \Rightarrow_{\mathfrak{M}}, \leq_{\mathfrak{M}}, \mathrm{V}_{\mathfrak{M}}>$ that satisfies conditions 1–4 (where At is the set of all atomic sentences in a formal propositional language with a modal operator 'K'):

1. $\mathrm{W}_{\mathfrak{M}}$ is a non-empty set;
2. $\Rightarrow_{\mathfrak{M}}$ is a reflexive binary relation on $\mathrm{W}_{\mathfrak{M}}$;
3. $\leq_{\mathfrak{M}}$ assigns to each $w \in \mathrm{W}$ a binary relation $\leq^w_{\mathfrak{M}}$ on some field $\mathrm{W}^w_{\mathfrak{M}} \subseteq \mathrm{W}_{\mathfrak{M}}$ such that:
   3.1 $\leq^w_{\mathfrak{M}}$ is reflexive and transitive in $\mathrm{W}^w_{\mathfrak{M}}$ (preorder), and
   3.2 $w \in \mathrm{W}^w_{\mathfrak{M}}$, and for all $v \in \mathrm{W}^w_{\mathfrak{M}}$, $w \leq^w_{\mathfrak{M}} v$ (weak centering);
4. $\mathrm{V}_{\mathfrak{M}}$: At $\to \mathcal{P}(\mathrm{W}_{\mathfrak{M}})$.

Here, '$w \Rightarrow_{\mathfrak{M}} v$' is intended to mean that v is not eliminated by the subject S's total evidence at w, while the preorder relation $\leq^w_{\mathfrak{M}}$ orders worlds in $\mathrm{W}^w_{\mathfrak{M}}$ according to how relevant they are to w. Given a model $\mathfrak{M}$ and a world w of it, the truth condition for sentences of the form 'Kϕ' is defined as follows (This is what Holliday [8] calls 'D-semantics'):

$$\mathfrak{M}, w \models \mathrm{K}\,\phi \text{ iff } \forall v \in \mathrm{Min}_{\leq^w_{\mathfrak{M}}}[\neg\phi]_{\mathfrak{M}}\text{: not } w \Rightarrow_{\mathfrak{M}} v;$$

where $[\neg\phi]_{\mathfrak{M}} = \{v \in \mathrm{W}_{\mathfrak{M}} \mid \mathfrak{M}, v \models \neg\phi\}$ and $\mathrm{Min}_{\leq^w_{\mathfrak{M}}}[\neg\phi]_{\mathfrak{M}} = \{v \in [\neg\phi]_{\mathfrak{M}} \cap \mathrm{W}^w_{\mathfrak{M}} \mid \neg\exists u(u \in [\neg\phi]_{\mathfrak{M}} \wedge u \leq^w_{\mathfrak{M}} v \wedge \neg v \leq^w_{\mathfrak{M}} u)\}$. Thus, the clause for 'Kϕ' in effect says shat 'Kϕ' is true at a world w in a model $\mathfrak{M}$ iff all of the most relevant $\neg\phi$-worlds in $\mathrm{W}^w_{\mathfrak{M}}$ are eliminated by S's total evidence at w. Validity ('D-validity') is then defined as truth-preserving in all worlds of all models.

With the help of these definitions, Holliday [8] proves several interesting and surprising results about D-semantics. In particular, he proves that '$(\mathrm{K}\phi \wedge \mathrm{K}(\phi \to \psi)) \to \mathrm{K}\phi$' is not D-valid. This is supposed to be a good news for Dretske, for RA models and D-semantics are supposed to reflect the essential features of Dretske's relevant alternative theory of knowledge. This result shows that Dretskes knowledge operator 'S knows that' is indeed not fully penetrating.

There is, however, another related result: it can also be shown that many simple epistemic closure principles, including 'K($\phi \wedge \psi$) $\rightarrow$ Kϕ' and 'K$\phi \rightarrow$ K($\phi \vee \psi$)', are not D-valid either. Holliday calls this 'the problem of containment' of D-semantics, i.e., the problem that invalidity tends to spread out all over the realm of epistemic closure principles[3] even to cover some intuitively very plausible closure principles, and he thinks that this should be a bad news for Dretske: it shows that Dretske's knowledge operator is, contrary to what Dretske thinks, not semi-penetrating either. These results therefore point to a dilemma for a relevant theory like Dretske's: either skepticism or the problem of containment.

Dretske's relevant alternative theory of knowledge is by no means the only victim of this problem. Holliday [8] goes on to show that each of the fallibilist theories – Lewis's, Dretske's, Heller's, Sosa's, and DeRose's – that he investigates in [8] faces a similar trilemma: either skepticism, or the problem of containment, or the problem of vacuous knowledge, where a theory of knowledge has the problem of vacuous knowledge if it allows a subject S to have knowledge about a proposition ϕ without requiring S's total evidence to be able to eliminate *any* alternative to ϕ. It is then a natural guess that these negative results may apply quite generally to fallibilist theories of a certain sort. Indeed, all these negative results about the cited theories and the possible generalization are nicely put together by a theorem that Holliday proves in [9]. Call a fallibilist theory of knowledge 'a standard single-path theory' (or 'a standard theory' or 'a single-path theory' for short) if it associates a *unique* set of 'relevant alternatives' with each proposition-world pair $<\phi, w>$ such that the necessary and sufficient condition for S to know ϕ at w is for S to evidentially eliminate the associated set of relevant alternatives. In terms of the notion of a function, an essential element of a single-path fallibilist theory of knowledge is a two-place function r that assigns a unique set of possibilities, $\mathrm{r}_{\mathfrak{M}}(<\phi, w>) \subseteq \mathrm{W}_{\mathfrak{M}}$, to each proposition-world pair in a model $\mathfrak{M}$.[4] Holliday [9, Sect. 2] then proves the following proposition:

[3] Following Holliday, we call a sentence of the form 'an epistemic closure principle':

$$\phi_0 \wedge \mathrm{K}\phi_1 \wedge \ldots \wedge \mathrm{K}\phi_n \rightarrow \mathrm{K}\psi_1 \vee \ldots \vee \mathrm{K}\psi_m,$$

where 'ϕ_0' is a propositional conjunction, i.e., a conjunction none of whose conjuncts contains an occurrence of 'K'. When n is equal to 1, we call such an epistemic closure principle 'a single-premise epistemic closure principle'. When n is greater than 1, we call such an epistemic closure principle 'a multiple-premise epistemic closure principle'.

[4] For example, in Holliday's formalization of Dretske's relevant alternative theory, the function r is such that, for every model $\mathfrak{M}$, world w, and proposition ϕ, $\mathrm{r}_{\mathfrak{M}}(w, \phi) = \mathrm{Min}_{\leq^w_{\mathfrak{M}}}[\neg\phi]_{\mathfrak{M}} = \{v \in [\neg\phi]_{\mathfrak{M}} \cap \mathrm{W}^w_{\mathfrak{M}} \mid \neg\exists u(u \in [\neg\phi]_{\mathfrak{M}} \wedge u \leq^w_{\mathfrak{M}} v \wedge \neg v \leq^w_{\mathfrak{M}} u)\}$.

Proposition 1. For any model $\mathfrak{M}$, world w, and area Σ,[5] the following principles are jointly inconsistent in the 'standard' fallibilist picture:[6]

contrast/enough$_\Sigma$ – $\forall\phi \in \Sigma$: $r_{\mathfrak{M}}(\phi, w) \subseteq (W_{\mathfrak{M}} - [\phi]_{\mathfrak{M}})$;
e-fallibilism$_\Sigma$ – $\exists\phi \in \Sigma\ \exists\psi \in \Sigma$: $r_{\mathfrak{M}}(\phi, w) \subseteq [\psi]_{\mathfrak{M}}$ and it is not the case that $(W^w_{\mathfrak{M}} - [\phi]_{\mathfrak{M}}) \subseteq [\psi]_{\mathfrak{M}}$;
noVK$_\Sigma$ – $\forall\phi \in \Sigma$: $(W^w_{\mathfrak{M}} \cap [\phi]_{\mathfrak{M}}) \neq W^w_{\mathfrak{M}}$ implies $r_{\mathfrak{M}}(\phi, w) \neq \emptyset$;
TF-cover$_\Sigma$ – $\forall\phi \in \Sigma\ \forall\psi \in \Sigma$: if ψ is a TF-consequence of ϕ, then $r_{\mathfrak{M}}(\psi, w) \subseteq r_{\mathfrak{M}}(\phi, w)$.

Here, **contrast/enough**$_\Sigma$ says that, for every sentence ϕ (in Σ), every alternative to ϕ at w is a $\neg\phi$-world of the model $\mathfrak{M}$, while **e-fallibilism**$_\Sigma$ in effect says that, for at least some sentence ϕ, not every $\neg\phi$-world in $W^w_{\mathfrak{M}}$ is a relevant alternative to ϕ at w. **NoVK**$_\Sigma$ demands that, for every sentence ϕ and world w, the set of relevant alternatives for $<\phi, w>$ must not be empty if ϕ is not true in every world in $W^w_{\mathfrak{M}}$, while **TF-cover**$_\Sigma$ ensures us that all single-premise epistemic closure principles of the form '$(\phi_0 \wedge K\phi) \rightarrow K\psi$' will be valid if ψ is a truth-functional consequence of ϕ. Proposition 1 then asserts that no standard fallibilist theory can satisfy all of these four principles. In short, what Proposition 1 says is that a single-path e-fallibilist theory of knowledge (i.e., a standard fallibilist theory of knowledge that satisfies **e-fallibilism**$_\Sigma$) that identifies every relevant alternative to ϕ with some $\neg\phi$-possibility cannot avoid both the problem of vacuous knowledge and the problem of containment. This seems to be a terribly bad news for the fallibilist. What can the fallibilist do with it?

3 Holliday's Multi-path Approach

For Holliday, what the fallibilist should learn from the bad news is that s/he should abandon *both* **contrast/enough**$_\Sigma$ *and* the 'single-path assumption' if s/he is to maintain both **noVK**$_\Sigma$ and **TF-cover**$_\Sigma$ and therefore avoid the problem of vacuous knowledge and that of containment. That is, s/he should abandon *both* the idea that a relevant alternative to a proposition ϕ (at w) is always some $\neg\phi$-possibility *and* the assumption that for every proposition-world pair $<\phi, w>$ there is a *unique* associated set $\mathfrak{S}$ of relevant alternatives such that the necessary and sufficient condition for a person to know ϕ at w is for him/her to evidentially eliminate relevant alternatives in $\mathfrak{S}$.

Holliday [9] further thinks that there are independent reasons for abandoning **contrast/enough**$_\Sigma$ and the single-path assumption, but what are these? He

[5] An area Σ is a set of sentences such that if $\phi \in \Sigma$ and ψ is truth-functional consequence of ϕ, then $\psi \in \Sigma$. Note that where I talk about a model $\mathfrak{M}$ and a world w, Holliday [9] talks about a context C and a scenario w; the terminological difference here is unimportant.

[6] Where I use the phrase 'standard fallibilist picture' Holliday [9] uses the phrase 'standard relevant picture'. Again, I think the terminological difference here is unimportant.

asks us to consider a disjunction 'p ∨ q' as an example. There seem, Holliday says, to be at least three 'paths' to know it: one could start by eliminating the set of all relevant ¬p-alternatives and thereby knowing 'p ∨ q' via knowing 'p', or by eliminating the set of all relevant ¬q-alternatives and thereby knowing it via knowing 'q', or by eliminating the set of all relevant ¬(p ∨ q)-alternatives and thereby knowing it directly. More importantly, these three sets (the set of relevant ¬p-alternatives, that of relevant ¬q-alternatives, and that of relevant ¬(p ∨ q)-alternatives) may all be different, and this is the reason why the fallibilist should give up the single-path assumption. Furthermore, all or some of the ¬p-alternatives (or ¬q-alternatives) may also be q-scenarios (or p-alternatives) and therefore be (p ∨ q)-alternatives as well. If this is the case, then one can know 'p ∨ q' by eliminating some or all (p ∨ q)-alternatives, and this is the reason why the fallibilist should give up $\textbf{contrast/enough}_{\Sigma}$.

Assuming that Holliday is right about this (I will come back to this in Sects. 4 and 5), the fallibilist has no choice but to accept what Holliday calls 'the multi-path picture of knowledge' according to which neither single-path assumption nor $\textbf{contrast/enough}_{\Sigma}$ principle should be retained, if s/he is to maintain both $\textbf{noVK}_{\Sigma}$ and $\textbf{TF-cover}_{\Sigma}$. Indeed, Holliday [9] goes on to propose one such theory. A few preliminaries is, however, needed in order to understand his proposal. Let $\mathfrak{L}$ be an ordinary propositional language with a non-truth-functional connective 'K'. Call a sentence of $\mathfrak{L}$ 'a TF-atomic' if it is either a propositional symbol or a sentence whose main connective is 'K'. Call any TF-atomic or its negation 'a TF-basic' and call a disjunction of any number of TF-basics 'a clause'. A sentence ϕ is in canonical conjunctive normal form (CCNF) iff ϕ is a conjunction of any number of nontrivial clauses (a nontrivial clause is a clause that does not include both 'p' and '¬p' as its disjuncts) such that, for each TF-atomic $p \in at(\phi)$ ($at(\phi)$ is the set of TF-atomics that occur in ϕ), each clause of ϕ contains either 'p' or '¬p'. Given these definitions, each sentence ϕ that is not a tautology is truth-functionally equivalent to a sentence ϕ' in CCNF with $at(\phi) = at(\phi')$ (call the latter 'CCNF(ϕ)') that is unique up to reordering of the conjuncts and disjuncts. Now, if ϕ is a sentence in CCNF, we let $\mathfrak{c}(\phi)$ to be the set of all subclauses C of conjuncts in ϕ such that every nontrivial superclause C′ of C, with $at(C') = at(\phi)$, is a conjunct of ϕ.[7] It is provable that $\mathfrak{c}(\phi)$ is the set of all nontrivial clauses C with $at(C) \subseteq at(\phi)$ that are TF-consequences (truth-functional-consequences) of ϕ.

We can now explain how multiple sets of relevant alternatives to a pair $<\phi, w>$ are to be determined according to Holliday.[8] Given an RA model $\mathfrak{M}$ and the alternatives function $r_{\mathfrak{M}}$ defined in it,[9] we define inductively a multi-path relevant alternative function $\mathbf{r}^{r}_{\mathfrak{M}}$ for sentences in CCNF:

7 If a clause C′ can be obtained by adding zero or more disjuncts to C, then C′ is a superclause of C, and C is a subclause of C′: e.g., '(p ∨ ¬q ∨ r)' is a superclause of '(p ∨ ¬q)' and a subclause of '(p ∨ ¬q ∨¬s ∨ r)'.

8 What follows is not the only way that multiple sets of relevant alternatives can be assigned to a proposition at a world, but it seems to be a quite natural way to do so.

9 See footnote 5 for how it is to be defined.

(1) For any clause C standing alone, $\mathbf{r}^r_{\mathfrak{M}}(\mathrm{C}, w) = \{\mathrm{r}_{\mathfrak{M}}(\mathrm{C}', w) \mid \mathrm{C}'$ is a subclause of C$\}$;
(2) For any CCNF which is a conjunction '$\mathrm{C}_1 \wedge \ldots \wedge \mathrm{C}_n$' of clauses with $\mathfrak{c}(\mathrm{C}_1 \wedge \mathrm{C}_n) = \{\psi_1, \ldots, \psi_m\}$; we define $\mathbf{r}^r_{\mathfrak{M}}(\mathrm{C}_1 \wedge \ldots \wedge \mathrm{C}_n, w) = \{\mathrm{A} \subseteq \mathrm{W}_{\mathfrak{M}} \mid \exists \mathrm{A}_1 \in \mathbf{r}^r_{\mathfrak{M}}(\psi_1, w) \ldots \exists \mathrm{A}_m \in \mathbf{r}^r_{\mathfrak{M}}(\psi_m, w)\colon \mathrm{A} = \cup_{1 \leq i \leq m} \mathrm{A}_i\}$;

We then define $\mathbf{r}^r_{\mathfrak{M}}$ for sentences in general as follows:

- $\mathbf{r}^r_{\mathfrak{M}}(\phi, w) = \mathbf{r}^r_{\mathfrak{M}}(\mathrm{CCNF}(\phi), w)$, for every sentence ϕ of $\mathfrak{L}$ that is not a truth-functional tautology.[10]

The idea behind the definition of $\mathbf{r}^r_{\mathfrak{M}}$ is simple: (1) says that any path to know a subclause of a clause is a path to know the clause, which is a generalization of the idea that any path to know a disjunct of a disjunction is a path to know the disjunction; and (2) says that knowing a conjunction of clauses requires doing enough epistemic work to know each of the clauses that are TF-consequences of the conjunction. With the multi-path relevant alternative function $\mathbf{r}^r_{\mathfrak{M}}$ in hand, Holliday [9] proposes the following truth-condition for sentences of the form '$\mathrm{K}\phi$':

- $\mathfrak{M}, w \models K\phi$ iff $\exists \mathrm{A} \in \mathbf{r}^{\mathbf{r}}_{\mathfrak{M}}(\phi, \mathbf{w})\colon \mathrm{A} \cap \{v \mid w \Rightarrow_{\mathfrak{M}} v\} = \emptyset$.

Call any member of $\mathbf{r}^{\mathbf{r}}_{\mathfrak{M}}(\phi, \mathbf{w})$ 'a relevant alternative set for ϕ at w', the above condition then says that 'S knows that ϕ' is true at w in $\mathfrak{M}$ iff S eliminates all alternatives in at least one of the relevant alternative sets for ϕ at w.

Holliday [9] then proves beautifully that the following conditions are jointly consistent[11] for the multi-path relevant alternative function $\mathbf{r}^r_{\mathfrak{M}}$:

e-fallibilism^{multi} – $\exists\phi\exists\psi\exists \mathrm{A}(\mathrm{A} \in \mathbf{r}^{\mathbf{r}}_{\mathfrak{M}}(\phi, w) \wedge \mathrm{A} \subseteq [\psi]_{\mathfrak{M}})$ and it is not the case that $(\mathrm{W}^w_{\mathfrak{M}} - [\phi]_{\mathfrak{M}}) \subseteq [\psi]_{\mathfrak{M}}$;
noVK^{multi} – $\forall\phi$: $(\mathrm{W}^w_{\mathfrak{M}} \cap [\phi]_{\mathfrak{M}}) \neq \mathrm{W}^w_{\mathfrak{M}}$ implies that $\emptyset \notin \mathbf{r}^{\mathbf{r}}_{\mathfrak{M}}(\phi, w)$;
TF-cover^{multi} – $\forall\phi\forall\psi$: if ψ is a TF-consequence of ϕ, then $\forall \mathrm{A}(\mathrm{A} \in \mathbf{r}^{\mathbf{r}}_{\mathfrak{M}}(\phi, w) \rightarrow \exists \mathrm{B}(\mathrm{B} \in \mathbf{r}^{\mathbf{r}}_{\mathfrak{M}}(\psi, w) \wedge \mathrm{B} \subseteq \mathrm{A}))$.

Here, the principles **e-fallibilism**^{multi}, **noVK**^{multi}, and **TF-cover**^{multi} are multipath versions generalized from **e-fallibilism**$_\Sigma$, **NoVK**$_\Sigma$, and **TF-cover**$_\Sigma$ that we saw earlier in Proposition 1 of Sect. 2. In effect, **e-fallibilism**^{multi} says that, at least for some sentence ϕ and some relevant alternative set A for ϕ at w, not every $\neg\phi$-world in $\mathrm{W}^w_{\mathfrak{M}}$ is included in A, and this amounts to saying that eliminating all alternatives in $\mathrm{W}^w_{\mathfrak{M}}$ to the proposition ϕ is not a necessarily condition for knowing that ϕ at w. **NoVK**^{multi} demands that, for every sentence ϕ, no relevant alternative set for ϕ at w should be empty if ϕ is not true at every world in $\mathrm{W}^w_{\mathfrak{M}}$, while **TF-cover**^{multi} ensures us that all single-premise epistemic closure principles of the form '$(\phi_0 \wedge \mathrm{K}\phi) \rightarrow \mathrm{K}\psi$' will be valid if ψ is a truth-functional consequence of ϕ. Moreover, the above semantics allows the fallibilist to prove

[10] If ϕ is a truth-functional tautology, we define $\mathbf{r}^r_{\mathfrak{M}}(\phi, w) = \mathbf{r}^r_{\mathfrak{M}}((\mathrm{p} \vee \neg\mathrm{p}), w)$.
[11] More correctly, the following conditions *and several others* are jointly consistent. But these extra ones are not important for my purpose in this paper.

that the epistemic closure principle ECP that lies behind the skeptic's argument is actually invalid and therefore is able to avoid the skeptic's conclusion. In short, the above result assures the fallibilist that s/he can reject skepticism and at the same time avoid both the problem of containment and the problem of vacuous knowledge – what a release for the fallibilist! What else can the fallibilist expect?

4 Single-path Approach Revisited

Holliday's proposal is indeed a very nice one, yet, for several reasons that cannot be explained here due to the limit of space, I have some qualms about it. One reason for my uneasiness about his proposal is the worry that Holliday may actually require the reliabilist to abandon too much to escape the impossibility result, i.e., Proposition 1 of Sect. 2. Recall that the impossibility result says that it is impossible for the reliabilist to consistently embrace the following four principles at once: **contrast/enough**$_\Sigma$, **e-fallibilism**$_\Sigma$, **noVK**$_\Sigma$, and **TF-cover**$_\Sigma$. I believe that Holliday's proof of the result is not questionable. However, the reliabilist may wonder why, in the face of the result, s/he could not simply drop one of the four principles out of the picture and retain the central idea that, for each sentence of the language, there is a unique set of relevant alternatives for that sentence. Since Holliday has not proved that this position is impossible, it is at least worthwhile to try to work it out. And this is exactly what I am going to do in this section.

I will begin with a few interesting ideas proposed by Heller. Heller [7, p. 201] proposes the following necessary condition (the so-called 'ERA' condition, short for 'expanded relevant alternatives') for S to know that ϕ:

> (ERA) S knows that ϕ only if S doesn't believe ϕ in any of the closest $\neg\phi$-world or any more distant $\neg\phi$-worlds that are still close enough.

Note that Heller actually uses the phrase 'S doesn't believe ϕ in a relevant $\neg\phi$-world w' to paraphrase Dretske's phrase 'S can rule out the $\neg\phi$-world w', so, for the sake of brevity and easy comparison, we can also write ERA as:

> (ERA*): S knows that ϕ only if S can rule out both the closest $\neg\phi$-worlds and all $\neg\phi$-worlds that are close enough.

To make Heller's idea more precise, we may take an epistemic model for Heller, or an H-model, $\mathfrak{M}$ to be a tuple $<W_{\mathfrak{M}}, \$_{\mathfrak{M}}, \Rightarrow_{\mathfrak{M}}, V_{\mathfrak{M}}>$ that satisfies:[12]

1. $W_{\mathfrak{M}}$ is a non-empty set.
2. $\$_{\mathfrak{M}}$ is a function from $W_{\mathfrak{M}}$ to $\mathcal{P}(\mathcal{P}(W_{\mathfrak{M}}))$ that is weakly centered, nested, closed under unions and nonempty intersection, and satisfies Limit Assumption.

[12] Holliday [8] formalizes Heller's ideas in [6,7] differently. Due to limit of space, however, I will not explain how he formalizes it, nor compare mine formalization with his.

3. $\Rightarrow_{\mathfrak{M}}$ is a reflexive binary relation on $W_{\mathfrak{M}}$ and contains every such pair $<w, v>$ that $v \in (W_{\mathfrak{M}} - \cup\$^w_{\mathfrak{M}})$. (We will write '$\cup\$_{\mathfrak{M}}(w)$' as '$\cup\$^w_{\mathfrak{M}}$'.)
4. $V_{\mathfrak{M}}$: At $\rightarrow \mathcal{P}(W_{\mathfrak{M}})$.

Here, for each world w, the 'sphere function' \$ orders a part of worlds, $\cup\$^w_{\mathfrak{M}}$, by putting them into 'spheres' that are nested, closed under unions and nonempty intersection relations with a smallest sphere inside. The smaller a sphere is, the more relevant worlds in it are to w. Again, '$w \Rightarrow_{\mathfrak{M}} v$ means that v is not eliminated or not ruled out by the subject S's total evidence at w. The stipulation that $\Rightarrow_{\mathfrak{M}}$ contains every pair $< w, v >$ such that $v \in (W^w_{\mathfrak{M}} - \cup\$^w_{\mathfrak{M}})$ embodies the idea that those worlds in $W_{\mathfrak{M}} - \cup\$^w_{\mathfrak{M}}$ are all too remote to be elimin*able* by the total evidence of the subject at w. (I will have more to say about the justification of this stipulation at the end of this paper.)

Given an H-model $\mathfrak{M} = <W_{\mathfrak{M}}, \$_{\mathfrak{M}}, \Rightarrow_{\mathfrak{M}}, V_{\mathfrak{M}}>$, a world $w \in W_{\mathfrak{M}}$, a formula ϕ in the epistemic language $\mathfrak{L}$, we define the truth-condition for $\mathfrak{M}, w \models K\phi$ as follows (call this 'H*-semantics'[13]):[14]

$$\mathfrak{M}, w \models K\phi \text{ iff } r_{\mathfrak{M}}(\phi, w) \cap \{v \mid w \Rightarrow_{\mathfrak{M}} v\} = \emptyset.$$

The condition says that the subject knows that ϕ iff s/he rules out all relevant alternatives to ϕ. H*-validity is then defined in the usual way. What is crucial here is how should we define the single-path relevant function $r_{\mathfrak{M}}$. In view of (ERA*), it is tempting to define $r_{\mathfrak{M}}(\phi, w)$ as the union of the set of the closest $\neg\phi$-worlds to w and the set of those $\neg\phi$-worlds that are close enough to w *for whatever sentence* ϕ and world w. However, if we defined $r_{\mathfrak{M}}$ in this uniformed way, the resultant semantics will not be able to avoid the problem of containment. So, how are we going to define $r_{\mathfrak{M}}$?

This is how: we define $r_{\mathfrak{M}}(\phi, w)$ in such a way that different types of sentences get different definitions. To be a bit more precise: we preserve Heller's suggestion (ERA*) for TF-basics *only* (recall that a sentence of $\mathfrak{L}$ is TF-atomic if it is either a propositional symbol or a sentence whose main connective is 'K' and that any TF-atomic or its negation is a TF-basic), accommodate a part of Holliday's insights into our treatment of disjunctions, adjust a bit Holliday's treatment of conjunctions, and then follow Holliday's suggestion to identify $r_{\mathfrak{M}}(\phi, w)$ with $r_{\mathfrak{M}}(\text{CCNF}(\phi), w)$ for every sentence ϕ. The following definitions then shows the details of these ideas. First, we define $\text{Min}_{\leq^w_{\mathfrak{M}}}[\phi]_{\mathfrak{M}}$ to be the empty set $\emptyset$ if $[\phi]_{\mathfrak{M}} = \emptyset$. However, if $[\phi]_{\mathfrak{M}} \neq \emptyset$, we define $\text{Min}_{\leq^w_{\mathfrak{M}}}[\phi]_{\mathfrak{M}}$ to be the intersection of $[\phi]_{\mathfrak{M}}$ and the smallest sphere S of $\$_{\mathfrak{M}}(w)$, if there is one, such that $[\phi]_{\mathfrak{M}} \cap S \neq \emptyset$, and we define it to be $[\phi]_{\mathfrak{M}}$ if otherwise. In words, $\text{Min}_{\leq^w_{\mathfrak{M}}}[\phi]_{\mathfrak{M}}$ (the set of the closest ϕ-worlds) is empty if ϕ is impossible, and it is the set of closest ϕ-worlds to w in

[13] I call the semantics 'H*-semantics' in order not to confuse it with Holliday's H-semantics in [8]. Readers should be able to see the similarity between H*-semantics as proposed here and D-semantics that we discussed in Sect. 2.

[14] Holiday [8] gives a different semantics (what he calls 'H-semantics') for Heller [6,7]. Again, due to the limit of space, however, I will not explain how he formalizes it, nor compare my formalization with his.

$\cup\$^w_{\mathfrak{M}}$ if there is some ϕ-world in $\cup\$^w_{\mathfrak{M}}$; otherwise it is just the set of all ϕ-worlds. Given an H-model $\mathfrak{M}$ and a world w, we then define 'the' relevant set for any sentence in CCNF inductively as follows:

(r_1) $r_{\mathfrak{M}}(p, w) = \text{Min}_{\leq^w_{\mathfrak{M}}}[\neg p]_{\mathfrak{M}} \cup (\cup\$^w_{\mathfrak{M}} \cap [\neg p]_{\mathfrak{M}})$ if p is a TF-basic;
(r_2) if $[p_1 \vee \ldots \vee p_n]_{\mathfrak{M}} = W_{\mathfrak{M}}$, then $r_{\mathfrak{M}}(p_1 \vee \ldots \vee p_n, w) = \emptyset$;[15]
(r_3) if $[p_1 \vee \ldots \vee p_n]_{\mathfrak{M}} \neq W_{\mathfrak{M}}$, then:[16]
  (a) $r_{\mathfrak{M}}(p_1 \vee \ldots \vee p_n, w) = \cup\$^w_{\mathfrak{M}} \cap ([\neg p_{i_1}]_{\mathfrak{M}} \cap \ldots \cap [\neg p_{i_m}]_{\mathfrak{M}})$, where $1 \leq i_j \leq n$ for each j between 1 and m, if 'p_{i_1}' ... 'p_{i_m}' are those and only those disjuncts in $(p_1 \vee \ldots \vee p_n)$ such that $(\cup\$_{\mathfrak{M}}(w) \cap [\neg p_{i_j}]_{\mathfrak{M}}) \neq \emptyset$ for each i_j; and
  (b) $r_{\mathfrak{M}}(p_1 \vee \ldots \vee p_n, w) = \cap\{r_{\mathfrak{M}}(p_i, w) \mid 1 \leq i \leq n\}$, if there is no such disjunct;
(r_4) $r_{\mathfrak{M}}(C_1 \wedge \ldots \wedge C_n, w) = \cup\{r_{\mathfrak{M}}(A, w) \mid A \in \mathfrak{c}(C_1 \wedge \ldots \wedge C_n)\}$.

(r_1) asserts that (ERA*) is a correct principle for all TF-basics: the subject S knows a TF-basic ϕ at w iff[17] S has ruled out both the closest $\neg$p-worlds (i.e., worlds in $\text{Min}_{\leq^w_{\mathfrak{M}}}[\neg p]_{\mathfrak{M}}$) and all $\neg$p-worlds that are close enough to w (i.e., worlds in $\cup\$^w_{\mathfrak{M}} \cap [\neg p]_{\mathfrak{M}}$). ($r_2$) says that, if a disjunction is a necessary truth in $\mathfrak{M}$ (call such a clause 'an r_2-clause in $\mathfrak{M}$'), then no $\neg\phi$-world in $\mathfrak{M}$ is relevant for knowing it at w (since there is no $\neg\phi$-world in $\mathfrak{M}$ at all). (r_{3a}) says that, if a clause '$(p_1 \vee \ldots \vee p_n)$' is not a necessary truth in $\mathfrak{M}$ and if 'p_{i_1}' ... 'p_{i_m}' are all those and only those disjuncts in it that are not true all over $\cup\$_{\mathfrak{M}}(w)$ (call such a clause 'an r_{3a}-clause in $\mathfrak{M}$'), then the set of relevant alternatives to '$(p_1 \vee \ldots \vee p_n)$' at w is the set of worlds in $\cup\$_{\mathfrak{M}}(w)$ that falsify all of these 'p_{i_1}' ... 'p_{i_m}'. However, if every disjunct in '$(p_1 \vee \ldots \vee p_n)$' is true all over $\cup\$_{\mathfrak{M}}(w)$ (call such a clause 'an r_{3b}-clause in $\mathfrak{M}$'), then (r_{3b}) says that the set of relevant alternatives to '$(p_1 \vee \ldots \vee p_n)$' is the intersection of all relevant alternative sets of its disjuncts, i.e., the set of worlds in $W_{\mathfrak{M}}$ that falsify all of its disjuncts. (r_4) says that knowing a conjunction of clauses requires doing enough epistemic work to know each of the clauses that are TF-consequences of the conjunction. Finally, for any H-model $\mathfrak{M}$ and any world w, we define $r_{\mathfrak{M}}(\phi, w)$ to be $r_{\mathfrak{M}}(p \vee \neg p, w)$ if ϕ is a tautology and define $r_{\mathfrak{M}}(\phi, w)$ to be $r_{\mathfrak{M}}(\text{CCNF}(\phi), w)$ if ϕ is not a tautology (recall that CCNF(ϕ) is a sentence ϕ' of CCNF that is truth-functionally equivalent to ϕ with $\text{at}(\phi) = \text{at}(\phi')$).

What will the fallibilist get from the above definition for $r_{\mathfrak{M}}(\phi, w)$? We can prove that the epistemic closure principle (ECP) is invalid in H*-semantics while the following two conditions holds for every ϕ and ψ at every world w in every H*-model $\mathfrak{M}$:

[15] When n = 1, this case reduces to case (r_1).
[16] When n = 1, this case also reduces to case (r_1).
[17] Heller takes (ERA*) to be merely a necessary condition, yet, for the sake of simplicity, I take it to be both a necessary and a sufficient condition. As far as I can tell, nothing important hinges on this difference for the purpose of this paper.

noVK$^{H^*}$ – $\forall\phi$: $(\cup\$^{w}_{\mathfrak{M}}\cap[\phi]_{\mathfrak{M}}) \neq \cup\$^{w}_{\mathfrak{M}}$ implies $\mathrm{r}_{\mathfrak{M}}(\phi, w) \neq \emptyset$;
TF-cover$^{H^*}$ – $\forall\phi\forall\psi$: if ψ is a TF-consequence of ϕ and $\mathrm{r}_{\mathfrak{M}}(\phi, w) \subset \cup\$^{w}_{\mathfrak{M}}$, then $\mathrm{r}_{\mathfrak{M}}(\psi, w) \subseteq \mathrm{r}_{\mathfrak{M}}(\phi, w)$.

noVK$^{H^*}$ demands that, for every sentence ϕ of $\mathfrak{L}$, the set of relevant alternatives for $< \phi, w >$ must not be empty if ϕ is not true at every world that is close enough to w, while **TF-cover**$^{H^*}$ ensures us, whose proof I omit, that all single-premise epistemic closure principles of the form '$(\phi_0 \wedge \mathrm{K}\phi) \rightarrow \phi$', where ψ is a TF-consequence of ϕ, will be valid. Since $\mathrm{r}_{\mathfrak{M}}(\phi, w)$, as defined in this section, can satisfy both **noVK**$^{H^*}$ and **TF-cover**$^{H^*}$ while invalidates ECP, the above results therefore show that H*-semantics with its single-path relevant function $\mathrm{r}_{\mathfrak{M}}(\phi, w)$ enables the fallibilist to reject skepticism and to avoid the problem of containment and the problem of vacuous knowledge at the same time. Due to the limit of space, however, I will leave the proof of these claims to a fuller version.

5 Comparison and Discussion

We are now in a position to briefly compare my proposal with Holliday's of Sect. 3. There are some similarities as well as dissimilarities between the two proposals. Some similarities between them are easily seen: both proposals rely upon identifying $\mathrm{r}_{\mathfrak{M}}(\phi, w)$ with $\mathrm{r}_{\mathfrak{M}}(\mathrm{CCNF}(\phi), w)$, both avoid the skeptic's conclusion by invalidating the multi-premise epistemic closure principle ECP, both avoid the problem of vacuous knowledge[18], and both avoid the problem of containment. Besides, there is one more thing common in both proposals: both agree that, in some cases, one can know a sentence ϕ by just ruling out relevant alternatives at which ϕ is true, and both thereby reject **contrast/enough**$_\Sigma$. This is possible for H*-semantics mainly because of (r_{3a}), as illustrated by the following example: if 'p' is not but 'q' is a 'heavy-weight proposition', i.e., a proposition that is true all over $\cup\$^{w}_{\mathfrak{M}}$, then the set of relevant alternatives of 'p $\vee$ q', according to (r_{3a}), will be those $\neg$p-worlds within $\cup\$^{w}_{\mathfrak{M}}$, and these might well all be q-worlds and therefore all be (p $\vee$ q)-worlds as well. In case that they are, one thereby knows 'p $\vee$ q' by just ruling out (p $\vee$ q)-worlds within $\cup\$^{w}_{\mathfrak{M}}$.

To be sure, the more significant thing to do is to compare the dissimilarities between the two proposals. Again, some dissimilarities between them are easily seen. For example, my proposal is, while Holliday's is not, a single-path relevant alternative theory in the sense that it proposes that, for each sentence of the language, there is a unique set of relevant alternatives associated with that sentence in such a way that one needs to eliminate all of them in order to know that sentence is true. I take this to be a major advantage of my proposal over Holliday's, in view of the potentially chaotic proliferation of relevant alternative sets in his proposal. For another example, while both proposals suggest that the

[18] Compare Holliday's **noVK**^{multi} with my **noVK**$^{H^*}$. If we identify $\mathrm{W}^{w}_{\mathfrak{M}}$ in **noVK**^{multi} with $\cup\$^{w}_{\mathfrak{M}}$ in **noVK**$^{H^*}$, it can be seen vividly that the two results are essentially the same.

content of $r_{\mathfrak{M}}(\phi, w)$ depends on the content of $r_{\mathfrak{M}}(\text{CCNF}(\phi), w)$, my proposal does not, while Holliday's does, demand that *the number of members* in $r_{\mathfrak{M}}(\phi, w)$ also depends on $r_{\mathfrak{M}}(\text{CCNF}(\phi), w)$. I take this feature of Holliday's proposal to be an *ad hoc* one, and my proposal certainly avoids it.

There is, however, a difference between the two proposals that some people might think that it indicates an important advantage of Hollidays proposal over mine (though I will argue shortly that this is not the case). Consider the principle **e-fallibilism**^{multi} that Holliday shows it to be compatible with **noVK**^{multi} and **TF-cover**^{multi}:

e-fallibilism^{multi} – $\exists\phi\exists\psi\exists A(A\in \mathbf{r}^{\mathbf{r}}_{\mathfrak{M}}(\phi, w) \wedge A\subseteq [\psi]_{\mathfrak{M}})$ and it is not the case that $(W^{w}_{\mathfrak{M}} - [\phi]_{\mathfrak{M}}) \subseteq [\psi]_{\mathfrak{M}}$

Since my proposal does not allow a variant of **e-fallibilism**^{multi} to be compatible with the other two principles that my proposal validates, viz., **noVK**^{H*} and **TF-cover**^{H*}, some people may take this difference as indicating an important merit of Holliday's proposal and a serious defect of mine. This conclusion, however, is an over-statement. While my proposal, indeed, does not allow **noVK**^{H*} and **TF-cover**^{H*} to be compatible with a variant of **e-fallibilism**^{multi}, it nonetheless allows them to be compatible with the following principle:

fallibilism$^{H^*}$ – $\exists\phi\exists\psi\exists w$: $r_{\mathfrak{M}}(\phi, w) \subseteq [\psi]_{\mathfrak{M}}$ and it is not the case that $(W_{\mathfrak{M}} - [\phi]_{\mathfrak{M}}) \subseteq [\psi]_{\mathfrak{M}}$.

fallibilism$^{H^*}$ allows the possibility that an epistemic subject S knows some proposition ϕ without ruling out every $\neg\phi$-possibility in a model. Actually, any model that invalidates ECP can be turned into a model that makes **fallibilism**$^{H^*}$ true. Even though **fallibilism**$^{H^*}$ is weaker than **e-fallibilism**^{multi}, I don't see any reason why we should not count the former as a kind of fallibilism as well. I therefore see that the difference between my proposal and Holliday's over the principle **e-fallibilism**^{multi} indicates no special favor for Holliday's proposal. At most, it shows only that Holliday's proposal allows a stronger form of reliabilism that my proposal does not.

Finally, it may be said that the difference between Holliday's suggested semantics in Sect. 3 and my H*-semantic in Sect. 4 favors his proposal in that mine is *ad hoc* in a way that his semantics is not. '*Ad hoc* in what way?', I ask? And I suggest that the correct answer to this question is 'In no way'.

(r_1), as I explained in Sect. 4, is just Heller's [7] ERA (or ERA*) restricted to TF-basics. I don't think that ERA can be plausibly generalized to apply to all sentences. Consider, for example, a disjunction 'p ∨ q', where 'p' is not, while 'q' is, a heavy-weight proposition. (Recall that a heavy-weight proposition is such a proposition that any possibility that falsifies it is too remote to be reachable by the total evidence of the subject S at w. In other words, a heavy-weight proposition is a proposition that is true all over $\cup\$^{w}_{\mathfrak{M}}$.) If we apply Heller's ERA to such a disjunction, the set of relevant alternatives for it will be the set of the closet ($\neg$p $\wedge$ $\neg$q)-worlds that fall outside the range of $\cup\$^w_{\mathfrak{M}}$. As a result, we should conclude that 'p $\vee$ q' is actually not known or even not knowable to the subject S. Yet, of course the sentence 'p $\vee$ q' is knowable or even known to S: if S knows that p, then S knows that p $\vee$ q as well. This example shows that, even if ERA is the correct principle for TF-basics, still some other principle should rule disjunctions; this is why disjunctions and TF-basics are treated differently in my proposed semantics. The important question, then, is how we should define the relevant set of a disjunction, and this is described in (r_2) and (r_3). (r_2) is a quite natural semantics for any necessary truth, no matter it is a disjunction or not. If ϕ is necessarily true (in a model $\mathfrak{M}$), then no 'alternative' to ϕ needs to be eliminated in order to know ϕ for the simple reason that there is no alternative to ϕ at all (in the model $\mathfrak{M}$). As to (r_{3a}), the rationale behind it has already been hinted by the example we illustrate twice in this section. If 'p' is not, while 'q' is, a heavy-weight proposition, then intuitively the disjunction 'p $\vee$ q' can still be known if and only if one can know that p. Generalizing this example, you then reach (r_{3a}): a disjunction of multiple TF-basics can be known if and only if one knows its non-heavy-weight part, i.e., its non-heavy-weight subclause. However, if every disjunct of a disjunction is a heavy-weight proposition, then there seems to be no way to know the disjunction as a whole. (r_{3b}) therefore requests that the subject should rule out the intersection of the relevant alternative sets of all these heavy-weight propositions. Since this intersection, if not empty, will surely fall out of $\cup\$^w_{\mathfrak{M}}$, this request is in effect a request of an impossible mission for the subject to accomplish. Needless to say, my (r_4) is simply Holliday's rule for CCNF but restricted to the case that each conjunct has only a unique set of relevant alternatives. If Holliday's rule for CCNF is not *ad hoc*, neither is mine. This shows that H*-semantics is not an *ad hoc* semantics.

After this brief comparison between my and Holliday's proposals and a dense defense of my semantics, I now leave the evaluation about which one of us should win the debate to readers.

References

1. DeRose, K.: Solving the skeptical problem. Philos. Rev. **104**(1), 1–52 (1995)
2. Dretske, F.: Epistemic operators. J. Philos. **67**(24), 1007–1023 (1970)
3. Dretske, F.: Conclusive reasons. Australas. J. Philos. **49**(1), 1–22 (1971)
4. Dretske, F.: The pragmatic dimension of knowledge. Philos. Stud. **40**(3), 363–378 (1981)
5. Dretske, F.: The case against closure. In: Steup, M., Sosa, E. (eds.) Contemporary Debates in Epistemology, pp. 13–26. Blackwell Publishing Ltd., Oxford (2005)
6. Heller, M.: Relevant alternatives. Philos. Stud. **55**(1), 23–40 (1987)
7. Heller, M.: Relevant alternatives and closure. Australas. J. Philos. **77**(2), 196–208 (1999)
8. Holliday, W.H.: Epistemic closure and epistemic logic I: relevant alternatives and subjunctivism. J. Philos. Logic **44**(1), 1–62 (2014)
9. Holliday, W.H.: Fallibilism and Multiple Paths to Knowledge. Oxford Studies in Epistemology, vol. 15. Oxford University Press, Oxford (2015)

10. Lewis, D.: Elusive knowledge. Australas. J. Philos. **74**(4), 549–567 (1996)
11. Sosa, E.: How to defeat opposition to Moore. Philos. Perspect. **13**, 141–153 (1999)
12. Unger, P.: Ignorance: A Case for Scepticism. Oxford University Press, Oxford (1975)

An Extended First-Order Belnap-Dunn Logic with Classical Negation

Norihiro Kamide[1(✉)] and Hitoshi Omori[2]

[1] Teikyo University, Toyosatodai 1-1, Utsunomiya, Tochigi 320-8551, Japan
drnkamide08@kpd.biglobe.ne.jp
[2] Kyoto University, Yoshida Honmachi, Sakyo-ku, Kyoto 606-8501, Japan
hitoshiomori@gmail.com

Abstract. In this paper, we investigate an extended first-order Belnap-Dunn logic with classical negation. We introduce a Gentzen-type sequent calculus FBD+ for this logic and prove theorems for syntactically and semantically embedding FBD+ into a Gentzen-type sequent calculus for first-order classical logic. Moreover, we show the cut-elimination theorem for FBD+ and prove the completeness theorems with respect to both valuation and many-valued semantics for FBD+.

1 Introduction

The main aim of this paper is to devise a well-behaving Gentzen-type sequent calculus for an extended first-order *Belnap-Dunn logic* (cf. [2,3,6]) with classical negation. The target logic is a first-order extension of De and Omori's axiomatic propositional expansion BD+ [5] of Belnap-Dunn logic. The authors in [5] showed that BD+ is essentially equivalent to *Béziau's four-valued modal logic* PM4N [4] and *Zaitsev's paraconsistent logic* FDEP [22].[1] As yet, no first-order extension of BD+ or Gentzen-type sequent calculi for such a first-order extension or BD+ has been considered in the literature, although a somewhat similar first-order extension of Belnap-Dunn logic with an additional unary connective $\triangle$, known as the Baaz' delta operator in the literature of fuzzy logic, has been studied on the basis of a Gentzen-type natural deduction system [16].[2]

The logic BD+, introduced as a Hilbert-style axiomatic system in [5], is obtained from a Hilbert-style axiomatic system for propositional classical logic with the standard language $\{\wedge, \vee, \rightarrow, \neg\}$ by adding the following axiom schemes with a paraconsistent negation connective $\sim$: $\sim\sim\alpha \leftrightarrow \alpha$, $\sim\neg\alpha \leftrightarrow \neg\sim\alpha$, $\sim(\alpha \wedge \beta) \leftrightarrow \sim\alpha \vee \sim\beta$, $\sim(\alpha \vee \beta) \leftrightarrow \sim\alpha \wedge \sim\beta$, and $\sim(\alpha \rightarrow \beta) \leftrightarrow \neg\sim\alpha \wedge \sim\beta$. Note here that the characteristic axiom schemes of BD+ are the following ones: $\sim(\alpha \rightarrow \beta) \leftrightarrow \neg\sim\alpha \wedge \sim\beta$ and $\sim\neg\alpha \leftrightarrow \neg\sim\alpha$. These axiom schemes are discussed to be very natural and plausible from the point of view of many-valued semantics in [5].

[1] Another system which is equivalent to BD+ is PŁ4 of Méndez and Robles (cf. [11]).
[2] Belnap-Dunn logic with $\triangle$ is equivalent to the expansion of Belnap-Dunn logic by what is sometimes called *exclusion negation* (cf. [5, p. 829]).

A. Baltag et al. (Eds.): LORI 2017, LNCS 10455, pp. 79–93, 2017.
DOI: 10.1007/978-3-662-55665-8_6

In this paper, we show that the above axiom schemes have some advantages in order to develop a simple cut-free Gentzen-type sequent calculus, as well as some nice properties such as completeness and strong equivalence substitution.[3] The substitution property is a particularly novel property, since some typical paraconsistent logics such as *Nelson's logic* N4 [1,12,19] lacks this property.[4]

We can summarize the contributions of this paper as follows: We introduce a natural first-order extension of BD+ with the strong equivalence substitution property, develop a simple Gentzen-type sequent calculus FBD+ for this extension, and prove several theorems for syntactically and semantically embedding FBD+ into a Gentzen-type sequent calculus FLK for first-order classical logic. These embedding results show that the existing standard algorithms for the automated theorem proving based on first-order classical logic are also applicable to FBD+. We also prove the cut-elimination theorem for FBD+ as well as the completeness theorems with respect to both valuation and many-valued semantics for FBD+. These fundamental results give us a proof-theoretic justification for both the proposed extended first-order Belnap-Dunn logic and the original propositional logic BD+.

The rest of this paper is structured as follows. In Sect. 2, we introduce FBD+ and show the substitution property with respect to the strong equivalence for it. This will be followed by Sect. 3 in which we first establish a syntactical embedding from FBD+ into FLK and second prove the cut-elimination theorem for FBD+ based on the syntactical embedding. Then, in Sect. 4, we first establish a semantical embedding from FBD+ into FLK and second prove the completeness theorem with respect to a valuation semantics using both the syntactical and semantical embedding theorems. In Sect. 5, we prove the completeness theorem with respect to a many-valued semantics for FBD+, and Sect. 6 concludes the paper with a summary and future directions.

2 Sequent Calculus and Strong Equivalence

To begin with, we introduce the first-order language (without individual constants and function symbols) $\mathcal{L}_{\text{FBD+}}$ of an extended first-order *Belnap-Dunn logic* with classical negation. This is denoted simply as $\mathcal{L}$ when no confusion can arise. Thus, *formulas* of $\mathcal{L}$ are constructed from countably many predicate symbols, countably many individual variables, and the logical connectives $\wedge$ (conjunction), $\vee$ (disjunction), $\rightarrow$ (implication), $\sim$ (paraconsistent negation), $\neg$ (classical negation), $\forall$ (universal quantifier) and $\exists$ (existential quantifier). We use small letters $p, q, ...$ to denote predicate symbols, small letters $x, y, ...$ to denote

[3] By using the strong equivalence substitution property, we can show the Herbrand theorem for FBD+, although we omit the details due to space limitations.

[4] Another interesting property of BD+ is the maximality with respect to the set of theorems (but *not* with respect to the rules of inference) which is proved in [5, Sect. 3.3]. Maximality does not hold for Nelson logics (even for "classical" extensions) since there are extensions, obtained by adding some axioms, that are not classical logic.

individual variables, Greek small letters $\alpha, \beta, \ldots$ to denote formulas, and Greek capital letters $\Gamma, \Delta, \ldots$ to represent finite (possibly empty) sets of formulas. An expression $\alpha[y/x]$ means the formula which is obtained from the formula α by replacing all free occurrences of the individual variable x in α with the individual variable y, but avoiding a clash of variables by a suitable renaming of bound variables. A 0-ary predicate is regarded as a propositional variable. If Φ is the set of all atomic formulas of $\mathcal{L}$, we then say that $\mathcal{L}$ is based on Φ. The symbol $\equiv$ is used to denote the equality of symbols. A *sequent* is an expression of the form $\Gamma \Rightarrow \Delta$. An expression $\alpha \Leftrightarrow \beta$ is used to represent the abbreviation of the sequents $\alpha \Rightarrow \beta$ and $\beta \Rightarrow \alpha$. An expression $L \vdash S$ means that a sequent S is provable in a sequent calculus L. If L of $L \vdash S$ is clear from the context, we omit L in it. A rule R of inference is said to be *admissible* in a sequent calculus L if the following condition is satisfied: For any instance $\dfrac{S_1 \cdots S_n}{S}$ of R, if $L \vdash S_i$ for all i, then $L \vdash S$.

We now introduce a Gentzen-type sequent calculus FBD+ for the first-order extension of De and Omori's extended Belnap-Dunn logic BD+ with classical negation as follows.[5]

Definition 1 (FBD+). *The initial sequents of* FBD+ *are of the following form, for any atomic formula* p,

$$p \Rightarrow p \qquad {\sim}p \Rightarrow {\sim}p.$$

The structural inference rules of FBD+ *are of the form:*

$$\frac{\Gamma \Rightarrow \Delta, \alpha \quad \alpha, \Sigma \Rightarrow \Pi}{\Gamma, \Sigma \Rightarrow \Delta, \Pi}\ (\text{cut}) \qquad \frac{\Gamma \Rightarrow \Delta}{\alpha, \Gamma \Rightarrow \Delta}\ (\text{we-left}) \qquad \frac{\Gamma \Rightarrow \Delta}{\Gamma \Rightarrow \Delta, \alpha}\ (\text{we-right}).$$

The pure logical inference rules of FBD+ *are of the form:*

$$\frac{\alpha, \beta, \Gamma \Rightarrow \Delta}{\alpha \wedge \beta, \Gamma \Rightarrow \Delta}\ (\wedge\text{left}) \qquad \frac{\Gamma \Rightarrow \Delta, \alpha \quad \Gamma \Rightarrow \Delta, \beta}{\Gamma \Rightarrow \Delta, \alpha \wedge \beta}\ (\wedge\text{right})$$

$$\frac{\alpha, \Gamma \Rightarrow \Delta \quad \beta, \Gamma \Rightarrow \Delta}{\alpha \vee \beta, \Gamma \Rightarrow \Delta}\ (\vee\text{left}) \qquad \frac{\Gamma \Rightarrow \Delta, \alpha, \beta}{\Gamma \Rightarrow \Delta, \alpha \vee \beta}\ (\vee\text{right})$$

$$\frac{\Gamma \Rightarrow \Delta, \alpha \quad \beta, \Sigma \Rightarrow \Pi}{\alpha \to \beta, \Gamma, \Sigma \Rightarrow \Delta, \Pi}\ (\to\text{left}) \qquad \frac{\alpha, \Gamma \Rightarrow \Delta, \beta}{\Gamma \Rightarrow \Delta, \alpha \to \beta}\ (\to\text{right})$$

$$\frac{\Gamma \Rightarrow \Delta, \alpha}{\neg\alpha, \Gamma \Rightarrow \Delta}\ (\neg\text{left}) \qquad \frac{\alpha, \Gamma \Rightarrow \Delta}{\Gamma \Rightarrow \Delta, \neg\alpha}\ (\neg\text{right})$$

$$\frac{\alpha[y/x], \Gamma \Rightarrow \Delta}{\forall x\alpha, \Gamma \Rightarrow \Delta}\ (\forall\text{left}) \qquad \frac{\Gamma \Rightarrow \Delta, \alpha[z/x]}{\Gamma \Rightarrow \Delta, \forall x\alpha}\ (\forall\text{right})$$

[5] In FBD+, we can replace the multiplicative (context splitting) type inference rules (cut) and ($\to$left) with their additive (non context splitting) type modifications. But, we adopt the multiplicative type inference rules since these are compatible in the system LK (for the classical logic) presented in [17].

$$\frac{\alpha[z/x], \Gamma \Rightarrow \Delta}{\exists x\alpha, \Gamma \Rightarrow \Delta}\ (\exists\text{left}) \qquad \frac{\Gamma \Rightarrow \Delta, \alpha[y/x]}{\Gamma \Rightarrow \Delta, \exists x\alpha}\ (\exists\text{right})$$

where y is an arbitrary individual variable, and z is an individual variable which has the eigenvariable condition, i.e., z does not occur as a free individual variable in the lower sequent of the rule.

The $\sim$*-combined logical inference rules of* FBD+ *are of the form:*

$$\frac{\alpha, \Gamma \Rightarrow \Delta}{\sim\sim\alpha, \Gamma \Rightarrow \Delta}\ (\sim\sim\text{left}) \qquad \frac{\Gamma \Rightarrow \Delta, \alpha}{\Gamma \Rightarrow \Delta, \sim\sim\alpha}\ (\sim\sim\text{right})$$

$$\frac{\sim\alpha, \Gamma \Rightarrow \Delta \quad \sim\beta, \Gamma \Rightarrow \Delta}{\sim(\alpha \wedge \beta), \Gamma \Rightarrow \Delta}\ (\sim\wedge\text{left}) \qquad \frac{\Gamma \Rightarrow \Delta, \sim\alpha, \sim\beta}{\Gamma \Rightarrow \Delta, \sim(\alpha \wedge \beta)}\ (\sim\wedge\text{right})$$

$$\frac{\sim\alpha, \sim\beta, \Gamma \Rightarrow \Delta}{\sim(\alpha \vee \beta), \Gamma \Rightarrow \Delta}\ (\sim\vee\text{left}) \qquad \frac{\Gamma \Rightarrow \Delta, \sim\alpha \quad \Gamma \Rightarrow \Delta, \sim\beta}{\Gamma \Rightarrow \Delta, \sim(\alpha \vee \beta)}\ (\sim\vee\text{right})$$

$$\frac{\sim\beta, \Gamma \Rightarrow \Delta, \sim\alpha}{\sim(\alpha \rightarrow \beta), \Gamma \Rightarrow \Delta}\ (\sim\rightarrow\text{left}) \qquad \frac{\sim\alpha, \Gamma \Rightarrow \Delta \quad \Gamma \Rightarrow \Delta, \sim\beta}{\Gamma \Rightarrow \Delta, \sim(\alpha \rightarrow \beta)}\ (\sim\rightarrow\text{right})$$

$$\frac{\Gamma \Rightarrow \Delta, \sim\alpha}{\sim\neg\alpha, \Gamma \Rightarrow \Delta}\ (\sim\neg\text{left}) \qquad \frac{\sim\alpha, \Gamma \Rightarrow \Delta}{\Gamma \Rightarrow \Delta, \sim\neg\alpha}\ (\sim\neg\text{right})$$

$$\frac{\sim\alpha[z/x], \Gamma \Rightarrow \Delta}{\sim\forall x\alpha, \Gamma \Rightarrow \Delta}\ (\sim\forall\text{left}) \qquad \frac{\Gamma \Rightarrow \Delta, \sim\alpha[y/x]}{\Gamma \Rightarrow \Delta, \sim\forall x\alpha}\ (\sim\forall\text{right})$$

$$\frac{\sim\alpha[y/x], \Gamma \Rightarrow \Delta}{\sim\exists x\alpha, \Gamma \Rightarrow \Delta}\ (\sim\exists\text{left}) \qquad \frac{\Gamma \Rightarrow \Delta, \sim\alpha[z/x]}{\Gamma \Rightarrow \Delta, \sim\exists x\alpha}\ (\sim\exists\text{right})$$

where y is an arbitrary individual variable, and z is an individual variable which has the eigenvariable condition, i.e., z does not occur as a free individual variable in the lower sequent of the rule.

In order to show some syntactical embedding theorems, we introduce a Gentzen-type sequent calculus FLK for first-order classical logic. The first-order language $\mathcal{L}_{\text{FLK}}$ is obtained from $\mathcal{L}_{\text{FBD+}}$ by deleting $\sim$. This language will also be denoted simply by $\mathcal{L}$ when no confusion can arise.

Definition 2 (FLK). FLK *is the* $\sim$*-free part of* FBD+*, i.e., it is obtained from* FBD+ *by deleting the initial sequents of the form* $\sim p \Rightarrow \sim p$ *and the* $\sim$*-combined logical inference rules.*

Remark 3. *Here are some remarks which are simple but useful.*

1. *Let L be FBD+ or FLK. Sequents of the form* $\alpha \Rightarrow \alpha$ *for any formula* α *are provable in cut-free L. This fact can be shown by induction on* α.
2. *The following sequents are provable in cut-free* FBD+*:* $\sim\sim\alpha \Leftrightarrow \alpha$, $\sim(\alpha\wedge\beta) \Leftrightarrow \sim\alpha \vee \sim\beta$, $\sim(\alpha \vee \beta) \Leftrightarrow \sim\alpha \wedge \sim\beta$, $\sim(\alpha\rightarrow\beta) \Leftrightarrow \neg\sim\alpha \wedge \sim\beta$, $\sim\neg\alpha \Leftrightarrow \neg\sim\alpha$, $\sim\forall x\alpha \Leftrightarrow \exists x\sim\alpha$ *and* $\sim\exists x\alpha \Leftrightarrow \forall x\sim\alpha$.

3. *The inference rules ($\sim\!\rightarrow$left) and ($\sim\!\rightarrow$right) just correspond to the Hilbert-style axiom scheme* $\sim(\alpha\rightarrow\beta) \leftrightarrow \neg\!\sim\!\alpha \wedge \sim\!\beta$. *The inference rules ($\sim\!\neg$left) and ($\sim\!\neg$right) just correspond to the Hilbert-style axiom scheme* $\sim\!\neg\alpha \leftrightarrow \sim\!\neg\alpha$. *These axiom schemes were introduced by De and Omori [5] in order to axiomatize the extended Belnap-Dunn logic BD+ with classical negation.*
4. *As well-known, the cut-elimination theorem holds for FLK (see e.g., [7,17]).*

In order to prove Theorem 8 (weak syntactical embedding from FBD+ into FLK), we need the following proposition.

Proposition 4. *The following rules are admissible in cut-free* FBD+*:*

$$\frac{\sim\!\neg\alpha, \Gamma \Rightarrow \Delta}{\Gamma \Rightarrow \Delta, \sim\!\alpha}\ (\sim\!\neg\text{left}^{-1}) \qquad \frac{\Gamma \Rightarrow \Delta, \sim\!\neg\alpha}{\sim\!\alpha, \Gamma \Rightarrow \Delta}\ (\sim\!\neg\text{right}^{-1}).$$

An expression $\alpha \leftrightarrow_s \beta$ for any formulas α and β, called a *strong equivalence* between α and β, is defined by FBD+ $\vdash \alpha \Leftrightarrow \beta$ and FBD+ $\vdash \sim\!\alpha \Leftrightarrow \sim\!\beta$.

Proposition 5 (Strong equivalence). *We have:* $\sim(\alpha\wedge\beta) \leftrightarrow_s \sim\!\alpha\vee\sim\!\beta$, $\sim(\alpha\vee\beta) \leftrightarrow_s \sim\!\alpha\wedge\sim\!\beta$, $\sim(\alpha\rightarrow\beta) \leftrightarrow_s \neg\!\sim\!\alpha\wedge\sim\!\beta$, $\sim\sim\!\alpha \leftrightarrow_s \alpha$, $\sim\!\neg\alpha \leftrightarrow_s \neg\!\sim\!\alpha$, $\sim(\forall x\alpha) \leftrightarrow_s \exists x\!\sim\!\alpha$ *and* $\sim(\exists x\alpha) \leftrightarrow_s \forall x\!\sim\!\alpha$.

It is known that $\sim(\alpha\rightarrow\beta) \leftrightarrow_s \alpha\wedge\sim\!\beta$ does not hold for the standard Gentzen-type sequent calculus for Nelson's paraconsistent logic N4. Indeed, $\sim(\alpha\rightarrow\beta) \leftrightarrow \alpha \wedge \sim\!\beta$ is a characteristic axiom scheme for N4, but $\sim\sim(\alpha\rightarrow\beta) \leftrightarrow \sim(\alpha \wedge \sim\!\beta)$ is not a theorem for N4. For more discussions on strong equivalence in some variants of N4, see e.g., [19,20].

Proposition 6 (Substitution for strong equivalence). *Let α be a subformula of a formula γ, and $\gamma^\star$ be the formula obtained from γ by replacing a occurrence of α with that of β. Then, we have: If $\alpha \leftrightarrow_s \beta$, then $\gamma \leftrightarrow_s \gamma^\star$.*

3 Syntactical Embedding and Cut-Elimination

We introduce an FLK-translation function for formulas of FBD+, and by using this translation, we show several theorems for embedding FBD+ into FLK. A similar translation has been used by Gurevich [8], Rautenberg [15] and Vorob'ev [18] to embed Nelson's constructive logic [1,12] into intuitionistic logic. More recently, some similar translations have been applied, for example, in [9,10] to embed some paraconsistent logics into classical logic.

Definition 7. *We fix a set Φ of atomic formulas, and define the set $\Phi' := \{p' \mid p \in \Phi\}$ of atomic formulas. Let languages $\mathcal{L}_{\text{FBD+}}$ and $\mathcal{L}_{\text{FLK}}$, defined as above, be based on sets Φ and $\Phi \cup \Phi'$, respectively. A mapping f from $\mathcal{L}_{\text{FBD+}}$ to $\mathcal{L}_{\text{FLK}}$ is defined inductively by:*

1. $f(p) := p$, $f({\sim}p) := p' \in \Phi'$ *for* $p \in \Phi$,
2. $f(\alpha \wedge \beta) := f(\alpha) \wedge f(\beta)$,
3. $f(\alpha \vee \beta) := f(\alpha) \vee f(\beta)$,
4. $f(\alpha \rightarrow \beta) := f(\alpha) \rightarrow f(\beta)$,
5. $f(\neg\alpha) := \neg f(\alpha)$,
6. $f(\forall x\alpha) := \forall x f(\alpha)$,
7. $f(\exists x\alpha) := \exists x f(\alpha)$,
8. $f({\sim}(\alpha \wedge \beta)) := f({\sim}\alpha) \vee f({\sim}\beta)$,
9. $f({\sim}(\alpha \vee \beta)) := f({\sim}\alpha) \wedge f({\sim}\beta)$,
10. $f({\sim}(\alpha \rightarrow \beta)) := \neg f({\sim}\alpha) \wedge f({\sim}\beta)$,
11. $f({\sim}{\sim}\alpha) := f(\alpha)$,
12. $f({\sim}\neg\alpha) := \neg f({\sim}\alpha)$,
13. $f({\sim}\forall x\alpha) := \exists x f({\sim}\alpha)$,
14. $f({\sim}\exists x\alpha) := \forall x f({\sim}\alpha)$.

An expression $f(\Gamma)$ denotes the result of replacing every occurrence of a formula α in Γ by an occurrence of $f(\alpha)$.

Theorem 8 (Weak syntactical embedding from FBD+ into FLK). *Let* Γ, Δ *be sets of formulas in* $\mathcal{L}_{\mathrm{FBD+}}$, *and* f *be the mapping defined in Definition 7.*

1. *If* FBD+ $\vdash \Gamma \Rightarrow \Delta$, *then* FLK $\vdash f(\Gamma) \Rightarrow f(\Delta)$.
2. *If* FLK $-$ (cut) $\vdash f(\Gamma) \Rightarrow f(\Delta)$, *then* FBD+ $-$ (cut) $\vdash \Gamma \Rightarrow \Delta$.

Proof. • (1): By induction on the proofs P of $\Gamma \Rightarrow \Delta$ in FBD+. We distinguish the cases according to the last inference of P, and show some cases.

1. Case ${\sim}p \Rightarrow {\sim}p$: The last inference of P is of the form: ${\sim}p \Rightarrow {\sim}p$ for any $p \in \Phi$. In this case, we obtain FLK $\vdash f({\sim}p) \Rightarrow f({\sim}p)$, i.e., FLK $\vdash p' \Rightarrow p'$ ($p' \in \Phi'$), by the definition of f.
2. Case (${\sim}\neg$left): The last inference of P is of the form:

$$\frac{\Gamma \Rightarrow \Delta, {\sim}\alpha}{{\sim}\neg\alpha, \Gamma \Rightarrow \Delta}\ ({\sim}\neg\text{left}).$$

By induction hypothesis, we have FLK $\vdash f(\Gamma) \Rightarrow f(\Delta), f({\sim}\alpha)$. Then, we obtain the required fact:

$$\cfrac{\begin{matrix}\vdots \\ f(\Gamma) \Rightarrow f(\Delta), f({\sim}\alpha)\end{matrix}}{\neg f({\sim}\alpha), f(\Gamma) \Rightarrow f(\Delta)}\ (\neg\text{left})$$

where $\neg f({\sim}\alpha)$ coincides with $f({\sim}\neg\alpha)$ by the definition of f.
3. Case (${\sim}{\rightarrow}$right): The last inference of P is of the form:

$$\frac{{\sim}\alpha, \Gamma \Rightarrow \Delta \quad \Gamma \Rightarrow \Delta, {\sim}\beta}{\Gamma \Rightarrow \Delta, {\sim}(\alpha \rightarrow \beta)}\ ({\sim}{\rightarrow}\text{right}).$$

By induction hypothesis, we have FLK $\vdash f({\sim}\alpha), f(\Gamma) \Rightarrow f(\Delta)$ and FLK $\vdash f(\Gamma) \Rightarrow f(\Delta), f({\sim}\beta)$. Then, we obtain the required fact:

$$\cfrac{\cfrac{\begin{matrix}\vdots \\ f({\sim}\alpha), f(\Gamma) \Rightarrow f(\Delta)\end{matrix}}{f(\Gamma) \Rightarrow f(\Delta), \neg f({\sim}\alpha)}\ (\neg\text{right}) \qquad \begin{matrix}\vdots \\ f(\Gamma) \Rightarrow f(\Delta), f({\sim}\beta)\end{matrix}}{f(\Gamma) \Rightarrow f(\Delta), \neg f({\sim}\alpha) \wedge f({\sim}\beta)}\ (\wedge\text{right})$$

where $\neg f({\sim}\alpha) \wedge f({\sim}\beta)$ coincides with $f({\sim}(\alpha \rightarrow \beta))$ by the definition of f.

4. Case (∼∀left): The last inference of P is of the form:

$$\frac{\sim\alpha[z/x], \Gamma \Rightarrow \Delta}{\sim\forall x\alpha, \Gamma \Rightarrow \Delta}\ (\sim\forall\text{left}).$$

By induction hypothesis, we have FLK $\vdash f(\sim\alpha[z/x]), f(\Gamma) \Rightarrow f(\Delta)$. Then, we obtain the required fact:

$$\frac{\begin{matrix}\vdots\\ f(\sim\alpha[z/x]), f(\Gamma) \Rightarrow f(\Delta)\end{matrix}}{\exists x f(\sim\alpha), f(\Gamma) \Rightarrow f(\Delta)}\ (\exists\text{left})$$

where $\exists x f(\sim\alpha)$ coincides with $f(\sim\forall x\alpha)$ by the definition of f.

• (2): By induction on the proofs Q of $f(\Gamma) \Rightarrow f(\Delta)$ in FLK − (cut). We distinguish the cases according to the last inference of Q, and show only the following case.

Case (∧right): The last inference of Q is (∧right).

1. Subcase (1): The last inference of Q is of the form:

$$\frac{f(\Gamma) \Rightarrow f(\Delta), f(\alpha) \quad f(\Gamma) \Rightarrow f(\Delta), f(\beta)}{f(\Gamma) \Rightarrow f(\Delta), f(\alpha \wedge \beta)}\ (\wedge\text{right})$$

where $f(\alpha \wedge \beta)$ coincides with $f(\alpha) \wedge f(\beta)$ by the definition of f. This case can straightforwardly be shown.
2. Subcase (2): The last inference of Q is of the form:

$$\frac{f(\Gamma) \Rightarrow f(\Delta), f(\sim\alpha) \quad f(\Gamma) \Rightarrow f(\Delta), f(\sim\beta)}{f(\Gamma) \Rightarrow f(\Delta), f(\sim(\alpha \vee \beta))}\ (\wedge\text{right})$$

where $f(\sim(\alpha \vee \beta))$ coincides with $f(\sim\alpha) \wedge f(\sim\beta)$ by the definition of f. This case can straightforwardly be shown.
3. Subcase (3): The last inference of Q is of the form:

$$\frac{f(\Gamma) \Rightarrow f(\Delta), f(\sim\neg\alpha) \quad f(\Gamma) \Rightarrow f(\Delta), f(\sim\beta)}{f(\Gamma) \Rightarrow f(\Delta), f(\sim(\alpha\rightarrow\beta))}\ (\wedge\text{right})$$

where $f(\sim(\alpha\rightarrow\beta))$ and $f(\sim\neg\alpha)$ respectively coincide with $\neg f(\sim\alpha) \wedge f(\sim\beta)$ and $\neg f(\sim\alpha)$ by the definition of f. By induction hypothesis, we have FBD+ − (cut) $\vdash \Gamma \Rightarrow \Delta, \sim\neg\alpha$ and FBD+ − (cut) $\vdash \Gamma \Rightarrow \Delta, \sim\beta$. We thus obtain the required fact:

$$\frac{\dfrac{\begin{matrix}\vdots\\ \Gamma \Rightarrow \Delta, \sim\neg\alpha\end{matrix}}{\sim\alpha, \Gamma \Rightarrow \Delta}\ (\sim\neg\text{right}^{-1}) \quad \begin{matrix}\vdots\\ \Gamma \Rightarrow \Delta, \sim\beta\end{matrix}}{\Gamma \Rightarrow \Delta, \sim(\alpha\rightarrow\beta)}\ (\sim\rightarrow\text{right})$$

where $(\sim\neg\text{right}^{-1})$ is admissible in cut-free FBD+ by Proposition 4. ∎

Theorem 9 (Cut-elimination for FBD+). *The rule* (cut) *is admissible in cut-free* FBD+.

Proof. Suppose FBD+ $\vdash \Gamma \Rightarrow \Delta$. Then, we have FLK $\vdash f(\Gamma) \Rightarrow f(\Delta)$ by Theorem 8(1), and hence FLK $-$ (cut) $\vdash f(\Gamma) \Rightarrow f(\Delta)$ by the cut-elimination theorem for FLK. By Theorem 8(2), we obtain FBD+ $-$ (cut) $\vdash \Gamma \Rightarrow \Delta$. ∎

Theorem 10 (Syntactical embedding from FBD+ into FLK). *Let Γ, Δ be sets of formulas in $\mathcal{L}_{\text{FBD+}}$, and f be the mapping defined in Definition 7.*

1. FBD+ $\vdash \Gamma \Rightarrow \Delta$ *iff* FLK $\vdash f(\Gamma) \Rightarrow f(\Delta)$.
2. FBD+ $-$ (cut) $\vdash \Gamma \Rightarrow \Delta$ *iff* FLK $-$ (cut) $\vdash f(\Gamma) \Rightarrow f(\Delta)$.

Proof. • (1): ($\Longrightarrow$): By Theorem 8(1). ($\Longleftarrow$): Suppose FLK $\vdash f(\Gamma) \Rightarrow f(\Delta)$. Then we have FLK $-$ (cut) $\vdash f(\Gamma) \Rightarrow f(\Delta)$ by the cut-elimination theorem for FLK. We thus obtain FBD+ $-$ (cut) $\vdash \Gamma \Rightarrow \Delta$ by Theorem 8(2). Therefore we have FBD+ $\vdash \Gamma \Rightarrow \Delta$.

• (2): ($\Longrightarrow$): Suppose FBD+ $-$ (cut) $\vdash \Gamma \Rightarrow \Delta$. Then we have FBD+ $\vdash \Gamma \Rightarrow \Delta$. We then obtain FLK $\vdash f(\Gamma) \Rightarrow f(\Delta)$ by Theorem 8(1). Therefore we obtain FLK $-$ (cut) $\vdash f(\Gamma) \Rightarrow f(\Delta)$ by the cut-elimination theorem for FLK. ($\Longleftarrow$): By Theorem 8(2). ∎

4 Semantical Embedding and Completeness for Valuation Semantics

Definition 11. *A structure $\mathcal{A} := \langle U, I^* \rangle$ is called a* paraconsistent model *if the following conditions hold:*

1. *U is a non-empty set,*
2. *p^{I^*} and $(\sim p)^{I^*}$ are mappings such that $p^{I^*}, (\sim p)^{I^*} \subseteq U^n$ (i.e., p^{I^*} and $(\sim p)^{I^*}$ are n-ary relations on U) for an n-ary predicate symbol p.*

We introduce the notation $\underline{u}$ as the name of $u \in U$, and we denote as $\mathcal{L}[\mathcal{A}]$ the language obtained from $\mathcal{L}$ by adding the names of all the elements of U. A formula α is called a closed formula *if α has no free individual variable. A formula of the form $\forall x_1 \cdots \forall x_m \alpha$ is called the* universal closure *of α if the free variables of α are $x_1, ..., x_m$. We write $cl(\alpha)$ for the universal closure of α.*

Definition 12 (Valuation semantics for FBD+). *Let $\mathcal{A} := \langle U, I^* \rangle$ be a paraconsistent model. The paraconsistent satisfaction relation $\mathcal{A} \models^* \alpha$ for any closed formula α of $\mathcal{L}[\mathcal{A}]$ are defined inductively by:*

1. *$[\mathcal{A} \models^* p(\underline{x}_1, ..., \underline{x}_n)$ iff $(x_1, ..., x_n) \in p^{I^*}]$ for any n-ary atomic formula $p(\underline{x}_1, ..., \underline{x}_n)$,*
2. *$[\mathcal{A} \models^* {\sim}p(\underline{x}_1, ..., \underline{x}_n)$ iff $(x_1, ..., x_n) \in (\sim p)^{I^*}]$ for any n-ary negated atomic formula ${\sim}p(\underline{x}_1, ..., \underline{x}_n)$,*

3. $\mathcal{A} \models^* \alpha \wedge \beta$ *iff* $\mathcal{A} \models^* \alpha$ *and* $\mathcal{A} \models^* \beta$,
4. $\mathcal{A} \models^* \alpha \vee \beta$ *iff* $\mathcal{A} \models^* \alpha$ *or* $\mathcal{A} \models^* \beta$,
5. $\mathcal{A} \models^* \alpha \rightarrow \beta$ *iff* $\mathcal{A} \not\models^* \alpha$ *or* $\mathcal{A} \models^* \beta$,
6. $\mathcal{A} \models^* \neg\alpha$ *iff* $\mathcal{A} \not\models^* \alpha$,
7. $\mathcal{A} \models^* \forall x\alpha$ *iff* $\mathcal{A} \models^* \alpha[\underline{u}/x]$ *for all* $u \in U$,
8. $\mathcal{A} \models^* \exists x\alpha$ *iff* $\mathcal{A} \models^* \alpha[\underline{u}/x]$ *for some* $u \in U$,
9. $\mathcal{A} \models^* {\sim}(\alpha \wedge \beta)$ *iff* $\mathcal{A} \models^* {\sim}\alpha$ *or* $\mathcal{A} \models^* {\sim}\beta$,
10. $\mathcal{A} \models^* {\sim}(\alpha \vee \beta)$ *iff* $\mathcal{A} \models^* {\sim}\alpha$ *and* $\mathcal{A} \models^* {\sim}\beta$,
11. $\mathcal{A} \models^* {\sim}(\alpha \rightarrow \beta)$ *iff* $\mathcal{A} \not\models^* {\sim}\alpha$ *and* $\mathcal{A} \models^* {\sim}\beta$,
12. $\mathcal{A} \models^* {\sim}{\sim}\alpha$ *iff* $\mathcal{A} \models^* \alpha$,
13. $\mathcal{A} \models^* {\sim}\neg\alpha$ *iff* $\mathcal{A} \not\models^* {\sim}\alpha$,
14. $\mathcal{A} \models^* {\sim}\forall x\alpha$ *iff* $\mathcal{A} \models^* {\sim}\alpha[\underline{u}/x]$ *for some* $u \in U$,
15. $\mathcal{A} \models^* {\sim}\exists x\alpha$ *iff* $\mathcal{A} \models^* {\sim}\alpha[\underline{u}/x]$ *for all* $u \in U$.

The paraconsistent satisfaction relation $\mathcal{A} \models^* \alpha$ *for any formula* α *of* $\mathcal{L}$ *is defined by* ($\mathcal{A} \models^* \alpha$ *iff* $\mathcal{A} \models^* cl(\alpha)$). *A formula* α *of* $\mathcal{L}$ *is called FBD+-*valid *iff* $\mathcal{A} \models^* \alpha$ *holds for any paraconsistent model* $\mathcal{A}$.

Definition 13 (Valuation semantics for FLK). *A structure* $\mathcal{A} := \langle U, I \rangle$, *called a* model, *is defined in a similar way as in Definition 11. Then, the satisfaction relation* $\mathcal{A} \models \alpha$ *for any closed formula* α *of* $\mathcal{L}[\mathcal{A}]$ *is defined inductively by:*

1. $[\mathcal{A} \models p(\underline{x}_1, ..., \underline{x}_n)$ *iff* $(x_1, ..., x_n) \in p^I]$ *for any n-ary atomic formula* $p(\underline{x}_1, ..., \underline{x}_n)$,
2. $\mathcal{A} \models \alpha \wedge \beta$ *iff* $\mathcal{A} \models \alpha$ *and* $\mathcal{A} \models \beta$,
3. $\mathcal{A} \models \alpha \vee \beta$ *iff* $\mathcal{A} \models \alpha$ *or* $\mathcal{A} \models \beta$,
4. $\mathcal{A} \models \alpha \rightarrow \beta$ *iff* $\mathcal{A} \not\models \alpha$ *or* $\mathcal{A} \models \beta$,
5. $\mathcal{A} \models \neg\alpha$ *iff* $\mathcal{A} \not\models \alpha$,
6. $\mathcal{A} \models \forall x\alpha$ *iff* $\mathcal{A} \models \alpha[\underline{u}/x]$ *for all* $u \in U$,
7. $\mathcal{A} \models \exists x\alpha$ *iff* $\mathcal{A} \models \alpha[\underline{u}/x]$ *for some* $u \in U$.

The satisfaction relation $\mathcal{A} \models \alpha$ *for any formula* α *of* $\mathcal{L}$ *is defined in a similar way as in Definition 12. The notion of FLK-validity for a formula is also defined in a similar way as in Definition 12.*

The following completeness theorem for FLK is well-known: For any formula α, FLK $\vdash \Rightarrow \alpha$ iff α is FLK-valid.

Lemma 14. *Let* f *be the mapping defined in Definition 7 and* $\mathcal{A} := \langle U, I^* \rangle$ *be a paraconsistent model. For any paraconsistent satisfaction relation* $\models^*$ *on* $\mathcal{A}$, *we can construct a satisfaction relation* $\models$ *on a model* $\mathcal{A}' = \langle U, I \rangle$ *such that for any formula* α, $\mathcal{A} \models^* \alpha$ *iff* $\mathcal{A}' \models f(\alpha)$.

Proof. Let Φ be a set of atomic formulas, Φ' be the set $\{p' \mid p \in \Phi\}$ of atomic formulas, and that $\models^*$ is a paraconsistent satisfaction relation on $\mathcal{A}$. Suppose that $\models$ is a satisfaction relation on $\mathcal{A}'$ such that, for any atomic formula $p \in \Phi$,

1. $\mathcal{A} \models^* p$ iff $\mathcal{A}' \models p$,
2. $\mathcal{A} \models^* {\sim}p$ iff $\mathcal{A}' \models p'$.

Then, the lemma is proved by induction on α. For the base case:

1. Case $\alpha \equiv p$ where p is a propositional variable: $\mathcal{A} \models^* p$ iff $\mathcal{A}' \models p$ (by the assumption) iff $\mathcal{A}' \models f(p)$ (by the definition of f).
2. Case $\alpha \equiv {\sim}p$ where p is a propositional variable: $\mathcal{A} \models^* {\sim}p$ iff $\mathcal{A}' \models p'$ (by the assumption) iff $\mathcal{A}' \models f({\sim}p)$ (by the definition of f).

For induction step, we show some of the cases.

1. Case $\alpha \equiv \forall x\beta$: $\mathcal{A} \models^* \forall x\beta$ iff $\mathcal{A} \models^* \beta[\underline{u}/x]$ for all $u \in U$ iff $\mathcal{A}' \models f(\beta[\underline{u}/x])$ for all $u \in U$ iff (by induction hypothesis) iff $\mathcal{A}' \models \forall x f(\beta)$ iff $\mathcal{A}' \models f(\forall x\beta)$.
2. Case $\alpha \equiv {\sim}{\sim}\beta$: $\mathcal{A} \models^* {\sim}{\sim}\beta$ iff $\mathcal{A} \models^* \beta$ iff $\mathcal{A}' \models f(\beta)$ (by induction hypothesis) iff $\mathcal{A}' \models f({\sim}{\sim}\beta)$.
3. Case $\alpha \equiv {\sim}\neg\beta$: $\mathcal{A} \models^* {\sim}\neg\beta$ iff $\mathcal{A} \not\models^* {\sim}\beta$ iff $\mathcal{A}' \not\models f({\sim}\beta)$ (by induction hypothesis) iff $\mathcal{A}' \models \neg f({\sim}\beta)$ iff $\mathcal{A}' \models f({\sim}\neg\beta)$.
4. Case $\alpha \equiv {\sim}(\beta \rightarrow \gamma)$: $\mathcal{A} \models^* {\sim}(\beta \rightarrow \gamma)$ iff $\mathcal{A} \not\models^* {\sim}\beta$ and $\mathcal{A} \models^* {\sim}\gamma$ iff $\mathcal{A}' \not\models f({\sim}\beta)$ and $\mathcal{A}' \models f({\sim}\gamma)$ (by induction hypothesis) iff $\mathcal{A}' \models \neg f({\sim}\beta)$ and $\mathcal{A}' \models f({\sim}\gamma)$ iff $\mathcal{A}' \models \neg f({\sim}\beta) \wedge f({\sim}\gamma)$ iff $\mathcal{A}' \models f({\sim}(\beta \rightarrow \gamma))$.
5. Case $\alpha \equiv {\sim}\forall x\beta$: $\mathcal{A} \models^* {\sim}\forall x\beta$ iff $\mathcal{A} \models^* {\sim}\beta[\underline{u}/x]$ for some $u \in U$ iff $\mathcal{A}' \models f({\sim}\beta[\underline{u}/x])$ for some $u \in U$ (by induction hypothesis) iff $\mathcal{A}' \models \exists x f({\sim}\beta)$ iff $\mathcal{A}' \models f({\sim}\forall x\beta)$.

Note here that the last steps are by the definition of f. ∎

Lemma 15. *Let f be the mapping defined in Definition 7 and let $\mathcal{A} := \langle U, I\rangle$ be a model. For any satisfaction relation $\models$ on $\mathcal{A}$, we can construct a paraconsistent satisfaction relation $\models^*$ on a paraconsistent model $\mathcal{A}' = \langle U, I^*\rangle$ such that for any formula α, $\mathcal{A} \models f(\alpha)$ iff $\mathcal{A}' \models^* \alpha$.*

Proof. Similar to the proof of Lemma 14. ∎

Theorem 16 (Semantical embedding from FBD+ into FLK). *Let f be the mapping defined in Definition 7. For all formula α, α is FBD+-valid iff $f(\alpha)$ is FLK-valid.*

Proof. By Lemmas 14 and 15. ∎

Theorem 17 (Completeness for FBD+). *For all formula α,* FBD+ $\vdash \Rightarrow \alpha$ *iff α is FBD+-valid.*

Proof. We have: FBD+ $\vdash \Rightarrow \alpha$ iff FLK $\vdash \Rightarrow f(\alpha)$ (by Theorem 10) iff $f(\alpha)$ is FLK-valid (by the completeness theorem for FLK) iff α is FBD+-valid (by Theorem 16). ∎

5 Completeness for Many-Valued Semantics

We assume the Hilbert-style axiomatic system for FBD+, and also use the same name for the system as FBD+. The Hilbert-style system FBD+ is obtained from BD+ [5] by adding the standard quantifier axiom schemes and inference rules for first-order classical logic and the axiom schemes of the form: ${\sim}\forall x\alpha \leftrightarrow \exists x{\sim}\alpha$ and ${\sim}\exists x\alpha \leftrightarrow \forall x{\sim}\alpha$.

Definition 18. *An* interpretation $\mathcal{I}$ *is a pair* $\langle U, v\rangle$ *where* U *is a non-empty set and we assign* $v(c) \in U$ *for every constant* c*, assign both the* extension $v^+(p) \subseteq U^n$ *and the* anti-extension $v^-(p) \subseteq U^n$ *for every* n*-ary predicate symbol* p*. Given any interpretation* $\langle U, v\rangle$*, we can define FBD+-valuation* $\overline{v}$ *for all the sentences of* $\mathcal{L}$ *expanded by* $\{\underline{u} : u \in U\}$ *inductively as follows: as for the atomic sentences,*

$$1 \in \overline{v}(p(t_1, ..., t_n)) \text{ iff } \langle v(t_1), \ldots, v(t_n)\rangle \in v^+(p),$$
$$0 \in \overline{v}(p(t_1, ..., t_n)) \text{ iff } \langle v(t_1), \ldots, v(t_n)\rangle \in v^-(p).$$

The rest of the clauses are as follows:

$1 \in \overline{v}({\sim}\alpha)$	*iff* $0 \in \overline{v}(\alpha)$,	$0 \in \overline{v}({\sim}\alpha)$	*iff* $1 \in \overline{v}(\alpha)$,
$1 \in \overline{v}(\neg\alpha)$	*iff* $1 \notin \overline{v}(\alpha)$,	$0 \in \overline{v}(\neg\alpha)$	*iff* $0 \notin \overline{v}(\alpha)$,
$1 \in \overline{v}(\alpha \wedge \beta)$	*iff* $1 \in \overline{v}(\alpha)$ *and* $1 \in \overline{v}(\beta)$,	$0 \in \overline{v}(\alpha \wedge \beta)$	*iff* $0 \in \overline{v}(\alpha)$ *or* $0 \in \overline{v}(\beta)$,
$1 \in \overline{v}(\alpha \vee \beta)$	*iff* $1 \in \overline{v}(\alpha)$ *or* $1 \in \overline{v}(\beta)$,	$0 \in \overline{v}(\alpha \vee \beta)$	*iff* $0 \in \overline{v}(\alpha)$ *and* $0 \in \overline{v}(\beta)$,
$1 \in \overline{v}(\alpha{\rightarrow}\beta)$	*iff* $1 \notin \overline{v}(\alpha)$ *or* $1 \in \overline{v}(\beta)$,	$0 \in \overline{v}(\alpha{\rightarrow}\beta)$	*iff* $0 \notin \overline{v}(\alpha)$ *and* $0 \in \overline{v}(\beta)$,
$1 \in \overline{v}(\forall x\alpha)$	*iff* $1 \in \overline{v}(\alpha[\underline{u}/x])$, *for all* $u \in U$,	$0 \in \overline{v}(\forall x\alpha)$	*iff* $0 \in \overline{v}(\alpha[\underline{u}/x])$, *for some* $u \in U$,
$1 \in \overline{v}(\exists x\alpha)$	*iff* $1 \in \overline{v}(\alpha[\underline{u}/x])$, *for some* $u \in U$,	$0 \in \overline{v}(\exists x\alpha)$	*iff* $0 \in \overline{v}(\alpha[\underline{u}/x])$, *for all* $u \in U$.

Finally, let $\Gamma \cup \{\alpha\}$ *be any set of sentences. Then,* α *is a* FBD+-semantic consequence *from* Γ $(\Gamma \models_{FBD+} \alpha)$ *iff for every interpretation* $\mathcal{I} = \langle U, v\rangle$ *and for every FBD+-valuation* $\overline{v}$ *(recall that* $\overline{v}$ *is determined by the given interpretation* $\langle U, v\rangle$*),* $1 \in \overline{v}(\alpha)$ *if* $1 \in \overline{v}(\gamma)$ *for all* $\gamma \in \Gamma$.

Remark 19. *The truth tables for the propositional connectives are as follows:*

α	${\sim}\alpha$	$\neg\alpha$
t	f	f
b	b	n
n	n	b
f	t	t

$\alpha \wedge \beta$	t	b	n	f
t	t	b	n	f
b	b	b	f	f
n	n	f	n	f
f	f	f	f	f

$\alpha \vee \beta$	t	b	n	f
t	t	t	t	t
b	t	b	t	b
n	t	t	n	n
f	t	b	n	f

$\alpha{\rightarrow}\beta$	t	b	n	f
t	t	b	n	f
b	t	t	n	n
n	t	b	t	b
f	t	t	t	t

For a discussion on the relation between many-valued semantics and Dunn semantics for expansions of Belnap-Dunn logic, see [14].

Proposition 20 (Soundness for FBD+). *If* $\Gamma \vdash_{\text{FBD+}} \alpha$ *then* $\Gamma \models_{\text{FBD+}} \alpha$.

Proof. By induction on the derivation $\Gamma \vdash_{\text{FBD+}} \alpha$, as usual. ∎

For the completeness, we need the following standard notions.

Definition 21. *Let* Σ *be a set of formulas. Then,*

1. Σ *is a* theory *iff* $\Sigma \vdash_{\text{FBD+}} \alpha$ *implies that* $\alpha \in \Sigma$ *for all* α*;*
2. Σ *is* prime *iff* $\alpha \vee \beta \in \Sigma$ *implies that* $\alpha \in \Sigma$ *or* $\beta \in \Sigma$ *for all* α *and* β*;*
3. Σ *is* non-trivial *iff for some* α, $\alpha \notin \Sigma$*;*
4. Σ *is* saturated *iff the following holds:*
   (a) $\forall x\alpha \in \Gamma$ *iff* $\alpha[c/x] \in \Gamma$ *for all constant* c*, and*
   (b) $\exists x\alpha \in \Gamma$ *iff* $\alpha[c/x] \in \Gamma$ *for some constant* c*.*

The rest of the proof is quite standard.

Lemma 22. *If Γ is a prime theory, then $\alpha \rightarrow \beta \in \Gamma$ iff ($\alpha \notin \Gamma$ or $\beta \in \Gamma$).*

Lemma 23. *If Γ is a non-trivial prime theory, then $\neg\alpha \in \Gamma$ iff $\alpha \notin \Gamma$.*

Lemma 24. *Let $\Gamma \cup \{\alpha\}$ be any set of sentences. If $\Gamma \nvdash_{\mathrm{FBD+}} \alpha$, then by adding countably new constant symbols, we can extend $\langle \Gamma, \{\alpha\} \rangle$ to $\langle \Gamma^+, \Pi^+ \rangle$ such that $\Gamma \subseteq \Gamma^+, \alpha \in \Pi^+, \Gamma^+ \nvdash_{\mathrm{FBD+}} \Pi^+$, either $\beta \in \Gamma^+$ or $\beta \in \Pi^+$ holds for all β, and Γ^+ is a prime and saturated theory.*[6]

Proof. Let us expand our language with a countable set $E := \{e_n : n \in \omega\}$ of fresh constant symbols. Moreover, let $(\alpha_n)_{n \geq 1}$ be an enumeration of all formulas in the expanded syntax. We inductively define the sequence $(\langle \Gamma_n, \Pi_n \rangle)_{n \in \omega}$ such that $\Gamma_n \nvdash_{\mathrm{FBD+}} \Pi_n$ as follows:

1. $\Gamma_0 := \Gamma$ and $\Pi_0 := \{\alpha\}$.
2. Suppose that we have constructed $\langle \Gamma_{n-1}, \Pi_{n-1} \rangle$ such that $\Gamma_{n-1} \nvdash_{\mathrm{FBD+}} \Pi_{n-1}$. We have the following two cases:
   (a) if $\Gamma_{n-1} \cup \{\alpha_n\} \nvdash_{\mathrm{FBD+}} \Pi_{n-1}$, then we split the case depending on the form of α_n:
      i. If $\alpha_n = \exists x \beta$, we define $\Gamma_n := \Gamma_{n-1} \cup \{\alpha_n, \beta[e/x]\}$ and $\Pi_n := \Pi_{n-1}$, where e is the first constant in the enumeration of E such that it is fresh in Γ_{n-1}, Π_{n-1} and α_n.
      ii. Otherwise, $\Gamma_n := \Gamma_{n-1} \cup \{\alpha_n\}$ and $\Pi_n := \Pi_{n-1}$.
   (b) If $\Gamma_{n-1} \cup \{\alpha_n\} \vdash_{\mathrm{FBD+}} \Pi_{n-1}$, then we again split the case depending on the form of α_n:
      i. If $\alpha_n = \forall x \beta$, we define $\Gamma_n := \Gamma_{n-1}$ and $\Pi_n := \Pi_{n-1} \cup \{\alpha_n, \beta[e/x]\}$, where e is the first constant in the enumeration of E such that it is fresh in Γ_{n-1}, Π_{n-1} and α_n.
      ii. Otherwise, we put $\Gamma_n := \Gamma_{n-1}$ and $\Pi_n := \Pi_{n-1} \cup \{\alpha_n\}$.

In both cases, it is easy to see that $\Gamma_n \nvdash_{\mathrm{FBD+}} \Pi_n$.

We define the limit of the sequence $(\langle \Gamma_n, \Pi_n \rangle)_{n \in \omega}$ as $\Gamma^+ := \bigcup_{n \in \omega} \Gamma_n$ and $\Pi^+ := \bigcup_{n \in \omega} \Pi_n$. It is clear that $\Gamma^+ \nvdash_{\mathrm{FBD+}} \Pi^+$. By construction, $\alpha \in \Gamma^+$ or $\alpha \in \Pi^+$ for every α. Moreover, Γ^+ is a prime and saturated theory. Here we only show the saturation requirement for $\forall$ of Γ^+: $\forall x \alpha \in \Gamma^+$ iff $\alpha[c/x] \in \Gamma^+$ for any constant c. The left-to-right direction is easy once we establish that Γ^+ is a theory. For the other direction, we show the contrapositive. Assume that $\forall x \alpha \notin \Gamma^+$. Since $\forall x \alpha \in \Gamma^+$ or $\forall x \alpha \in \Pi^+$, we have $\forall x \alpha \in \Pi^+$, which implies $\alpha[e/x] \in \Pi^+$ for some e by construction. Then, we obtain $\alpha[e/x] \notin \Gamma^+$. Indeed, suppose for reductio that $\alpha[e/x] \in \Gamma^+$. Then, it follows from $\alpha[e/x] \in \Pi^+$ that $\Gamma^+ \vdash_{\mathrm{FBD+}} \Pi^+$, which is a contradiction. ∎

Theorem 25 (Completeness for FBD+). *$\Gamma \models_{\mathrm{FBD+}} \alpha$ iff $\Gamma \vdash_{\mathrm{FBD+}} \alpha$.*

Proof. Since we have already observed the soundness, we prove the completeness part. To this end, we prove the contrapositive. Assume that $\Gamma \nvdash_{\mathrm{FBD+}} \alpha$.

[6] Note that $\Gamma \nvdash_{\mathrm{FBD+}} \Pi$ is defined as $\Gamma \nvdash_{\mathrm{FBD+}} \alpha_1 \vee \cdots \vee \alpha_n$ for some $\alpha_1, \ldots, \alpha_n \in \Pi$.

Then by Lemma 24, there is a prime and saturated theory Γ^+ such that $\Gamma \subseteq \Gamma^+$ and $\Gamma^+ \nvdash_{\text{FBD+}} \alpha$. Now, define an interpretation $\mathcal{I}_{\Gamma^+} = \langle U, v \rangle$ as follows: $U=\{c : c$ is a constant symbol$\}$ and, for every n-ary predicate symbol p:

$$v^+(p) := \{\langle t_1, \dots, t_n \rangle : p(t_1, \dots, t_n) \in \Gamma^+\},$$
$$v^-(p) := \{\langle t_1, \dots, t_n \rangle : {\sim}p(t_1, \dots, t_n) \in \Gamma^+\},$$

and, for every constant symbol c, $v(c) = c$. Then, the following holds for every sentence α.

$$1 \in \overline{v}(\alpha) \text{ iff } \alpha \in \Gamma^+,$$
$$0 \in \overline{v}(\alpha) \text{ iff } {\sim}\alpha \in \Gamma^+.$$

This can be proved by induction on α. We will here only check two cases in which α is of the form $\beta \to \gamma$ and $\forall x\beta$. For the positive case for the conditional,

$$\begin{aligned} 1 \in \overline{v}(\beta{\to}\gamma) &\text{ iff } 1 \notin \overline{v}(\beta) \text{ or } 1 \in \overline{v}(\gamma) \\ &\text{ iff } \beta \notin \Gamma^+ \text{ or } \gamma \in \Gamma^+ && \text{(by induction hypothesis)} \\ &\text{ iff } \beta{\to}\gamma \in \Gamma^+. && \text{(by Lemma 22)} \end{aligned}$$

For the negative case for the conditional,

$$\begin{aligned} 0 \in \overline{v}(\beta{\to}\gamma) &\text{ iff } 0 \notin \overline{v}(\beta) \text{ and } 0 \in \overline{v}(\gamma) \\ &\text{ iff } {\sim}\beta \notin \Gamma^+ \text{ and } {\sim}\gamma \in \Gamma^+ && \text{(by induction hypothesis)} \\ &\text{ iff } \neg{\sim}\beta \in \Gamma^+ \text{ and } {\sim}\gamma \in \Gamma^+ && \text{(by Lemma 23)} \\ &\text{ iff } {\sim}(\beta{\to}\gamma) \in \Gamma^+. && \text{(by } {\sim}(\beta{\to}\gamma) \leftrightarrow \neg{\sim}\beta \wedge {\sim}\gamma) \end{aligned}$$

For the positive case for the universal quantifier,

$$\begin{aligned} 1 \in \overline{v}(\forall x\beta) &\text{ iff } 1 \in \overline{v}(\beta[\underline{u}/x]), \text{for all } u \in U \\ &\text{ iff } \beta[\underline{u}/x] \in \Gamma^+, \text{for all } u \in U && \text{(by induction hypothesis)} \\ &\text{ iff } \forall x\beta \in \Gamma^+. && \text{(by } \Gamma^+ \text{ being saturated)} \end{aligned}$$

For the negative case for the universal quantifier,

$$\begin{aligned} 0 \in \overline{v}(\forall x\beta) &\text{ iff } 0 \in \overline{v}(\beta[\underline{u}/x]), \text{for some } u \in U \\ &\text{ iff } {\sim}\beta[\underline{u}/x] \in \Gamma^+, \text{for some } u \in U && \text{(by induction hypothesis)} \\ &\text{ iff } \exists x{\sim}\beta \in \Gamma^+ && \text{(by } \Gamma^+ \text{ being saturated)} \\ &\text{ iff } {\sim}\forall x\beta \in \Gamma^+. && \text{(by } {\sim}\forall x\beta \leftrightarrow \exists x{\sim}\beta) \end{aligned}$$

Therefore we obtain the desired result since we have that $1 \in \overline{v}(\gamma)$ for all $\gamma \in \Gamma^+$ and that $1 \notin \overline{v}(\alpha)$ (since $\Gamma^+ \nvdash_{\text{FBD+}} \alpha$, i.e. $\alpha \notin \Gamma^+$), that is, $\Gamma \not\models_{\text{FBD+}} \alpha$. ∎

By making use of the above completeness results, we establish the equivalence of Gentzen and Hilbert systems.

Proposition 26. *For any finite set $\Gamma \cup \{\alpha\}$ of formulas in $\mathcal{L}_{\text{FBD+}}$, if $\Gamma \vdash_{\text{FBD+}} \alpha$ then* FBD+ $\vdash \Gamma \Rightarrow \alpha$.

Proof. By induction on the length of derivations in the Hilbert style system. ∎

Proposition 27. *For any finite set $\Gamma \cup \{\alpha\}$ of formulas in $\mathcal{L}_{\mathrm{FBD+}}$, if* FBD+ $\vdash$ $\Gamma \Rightarrow \alpha$ *then* $\Gamma \vdash_{\mathrm{FBD+}} \alpha$.

Proof. Let $\tau(\Gamma \Rightarrow \Delta) = \bigwedge \Gamma \to \bigvee \Delta$, where $\bigwedge \varnothing = (p \to p)$ and $\bigvee \varnothing = (p \wedge \neg p)$ for some fixed atomic formula p. We first note that FBD+ $\vdash \Gamma \Rightarrow \alpha$ iff FBD+ $\vdash \varnothing \Rightarrow \tau(\Gamma \Rightarrow \alpha)$. In view of the completeness of the Hilbert style system with respect to the semantics from Definition 18, it is enough to show that for every sequent rule $\frac{S_1 \ldots S_n}{S}$ of FBD+, we have $\{\tau(S_1), \ldots, \tau(S_n)\} \models_{\mathrm{FBD+}} \tau(S)$. For axiomatic sequents, this is obvious, and since (cut) is eliminable, (cut) need not be considered. The remaining cases are relatively straightforward, and leave the details to the reader. This completes the proof. ∎

6 Conclusion

Here is a brief summary of our results presented in this paper. We developed a Gentzen-type sequent calculus FBD+ for the proposed first-order extension of BD+ and proved several theorems for embedding FBD+ into FLK (a Gentzen-type sequent calculus for first-order classical logic) both syntactically and semantically. These embedding results show that the existing standard algorithms for the automated theorem proving based on first-order classical logic are also applicable to FBD+. We also proved the cut-elimination theorem for FBD+ as well as the completeness theorems with respect to both valuation and many-valued semantics for FBD+. These fundamental results give us a proof-theoretic justification for both the proposed extended first-order Belnap-Dunn logic and the original propositional logic BD+.

For future directions, we only mention three of them among others. First, we may apply the same technique to the connexive variant of BD+, called dBD in [13].[7] In particular, this variant has the following falsity condition for the conditional:

$$0 \in \overline{v}(\alpha \to \beta) \text{ iff } (1 \notin \overline{v}(\alpha) \text{ or } 0 \in \overline{v}(\beta))$$

The details are kept for another occasion. Second, we may also consider a constructive variant of BD+, and apply the same technique to establish some interesting results. Finally, we may consider a modal expansion of BD+ as well. We keep these topics, together with others that are not mentioned here, for a subsequent paper.

Acknowledgments. We would like to thank the anonymous referees for their valuable comments. Norihiro Kamide was partially supported by JSPS KAKENHI Grant Number JP26330263. Hitoshi Omori is a Postdoctoral Research Fellow of Japan Society for the Promotion of Science (JSPS), and was partially supported by JSPS KAKENHI Grant Number JP16K16684.

[7] For connexive logic in general, see [21].

References

1. Almukdad, A., Nelson, D.: Constructible falsity and inexact predicates. J. Symbol. Logic **49**(1), 231–233 (1984)
2. Belnap, N.D.: A useful four-valued logic. In: Epstein, G., Dunn, J.M. (eds.) Modern Uses of Multiple-Valued Logic, pp. 5–37. Reidel, Dordrecht (1977)
3. Belnap, N.D.: How a computer should think. In: Ryle, G. (ed.) Contemporary Aspects of Philosophy, pp. 30–56. Oriel Press, Stocksfield (1977)
4. Béziau, J.Y.: A new four-valued approach to modal logic. Logique et Analyse **54**(213), 109–121 (2011)
5. De, M., Omori, H.: Classical negation and expansions of Belnap-Dunn logic. Stud. Logica **103**(4), 825–851 (2015)
6. Dunn, J.M.: Intuitive semantics for first-degree entailment and 'coupled trees'. Philos. Stud. **29**(3), 149–168 (1976)
7. Gentzen, G.: Collected papers of Gerhard Gentzen. In: Szabo, M.E. (ed.) Studies in Logic and the Foundations of Mathematics. North-Holland (English translation) (1969)
8. Gurevich, Y.: Intuitionistic logic with strong negation. Stud. Logica **36**, 49–59 (1977)
9. Kamide, N.: Paraconsistent double negation that can simulate classical negation. In: Proceedings of the 46th IEEE International Symposium on Multiple-Valued Logic (ISMVL 2016), pp. 131–136 (2016)
10. Kamide, N., Shramko, Y.: Embedding from multilattice logic into classical logic and vice versa. J. Logic Comput. **27**(5), 1549–1575 (2017)
11. Méndez, J.M., Robles, G.: A strong and rich 4-valued modal logic without Łukasiewicz-type paradoxes. Log. Univers. **9**(4), 501–522 (2015)
12. Nelson, D.: Constructible falsity. J. Symbol. Logic **14**, 16–26 (1949)
13. Omori, H.: From paraconsistent logic to dialetheic logic. In: Andreas, H., Verdée, P. (eds.) Logical Studies of Paraconsistent Reasoning in Science and Mathematics. TL, vol. 45, pp. 111–134. Springer, Cham (2016). doi:10.1007/978-3-319-40220-8_8
14. Omori, H., Sano, K.: Generalizing functional completeness in Belnap-Dunn logic. Stud. Logica **103**(5), 883–917 (2015)
15. Rautenberg, W.: Klassische und nicht-klassische Aussagenlogik. Vieweg, Braunschweig (1979)
16. Sano, K., Omori, H.: An expansion of first-order Belnap-Dunn logic. Logic J. IGPL **22**(3), 458–481 (2014)
17. Takeuti, G.: Proof Theory, 2nd edn. Dover Publications, Inc., Mineola (2013)
18. Vorob'ev, N.N.: A constructive propositional calculus with strong negation. Dokl. Akad. Nauk SSSR **85**, 465–468 (1952). (in Russian)
19. Wansing, H.: The Logic of Information Structures. LNCS, vol. 681, 163 pages. Springer, Heidelberg (1993)
20. Wansing, H.: Informational interpretation of substructural propositional logics. J. Logic Lang. Inform. **2**(4), 285–308 (1993)
21. Wansing, H.: Connexive logic, Stanford Encyclopedia of Philosophy (2014). http://plato.stanford.edu/entries/logic-connexive/
22. Zaitsev, D.: Generalized relevant logic and models of reasoning. Moscow State Lomonosov University doctoral dissertation (2012)

A Characterization Theorem for Trackable Updates

Giovanni Cinà(✉)

ILLC, University of Amsterdam, Amsterdam, The Netherlands
giovanni.cina88@gmail.com

Abstract. The information available to some agents can be represented with several mathematical models, depending on one's purpose. These models differ not only in their level of precision, but also in how they evolve when the agents receive new data. The notion of tracking was introduced to describe the matching of information dynamics, or 'updates', on different structures.

We expand on the topic of tracking, focusing on the example of plausibility and evidence models, two central structures in the literature on formal epistemology. Our main result is a characterization of the trackable updates of a certain class, that is, we give the exact condition for an update on evidence models to be trackable by a an update on plausibility models. For the positive cases we offer a procedure to compute the other update, while for the negative cases we give a recipe to construct a counterexample to tracking. To our knowledge, this is the first result of this kind in the literature.

1 Introduction

Given a model representing the epistemic or doxastic state of some agents, the signature theme of dynamic epistemic logic (DEL henceforth) is the study of how such model is transformed by new incoming information (see among others [1,3,4,6]). The possible transformations of a model are generally called 'updates', to highlight the idea that the agents are revising their beliefs or knowledge in light of new data. The work on updates originally revolved around Kripke structures, but in recent years several scholars exported the idea to other kinds of models, most importantly evidence models [7] and probabilistic models [12].

Beside re-proposing the host of DEL problems and techniques for new structures, this move opened up the possibility for a new direction of enquiry. These different kinds of models do not live in separated compartments: on the contrary, they are connected by various constructions, some of which have been studied extensively in other fields such as Duality Theory. Piecing together the existence of updates on different structures with the possibility to transform one kind of structure into the other, scholars were presented with diagrams of the following shape.

A. Baltag et al. (Eds.): LORI 2017, LNCS 10455, pp. 94–107, 2017.
DOI: 10.1007/978-3-662-55665-8_7

$$\begin{array}{ccc}
\text{structure}_1 & \xrightarrow{\text{update}_1} & \text{structure}_1 \\
{\scriptstyle\text{construction}}\uparrow & & \uparrow{\scriptstyle\text{construction}} \\
\text{structure}_2 & \xrightarrow[\text{update}_2]{} & \text{structure}_2
\end{array}$$

The commutation of such diagrams, namely the validity of the equation $\text{update}_2 \circ \text{construction} = \text{construction} \circ \text{update}_1$, captures the intuition that update_1 *matches*, or *tracks*, update_2. This correspondence between updates is especially significant if the structures at the top level of the diagram, structure_1 in this case, are poorer than the ones at the bottom level: this means that we can *reduce* update_2, which in general may be rather complicated, to an update on a poorer structure. On the other hand the lack of tracking is also informative, as it highlights the structural differences between different levels. Tracking therefore plays a central role in the comparability of diverse approaches to modeling agency.

The topic of tracking was first addressed by [12] in the context of probabilistic models, while [5] set up and discussed the problem in the setting of evidence and plausibility models. The former paper fixed the two updates (top and bottom of the diagram) and studied the constructions that enabled the commutation, while the latter article fixed the construction (left and right of the diagram) and searched for the pairs of updates which gave a successful commutation. This paper addresses some open problems raised by [5]; therefore we frame the issue of tracking in the second fashion.

From this second point of view, we argue that the optimal answer would be a characterization giving:

- the shape of all and only the trackable updates;
- for any update U of that shape, an algorithm that manipulates the definition of U and produces the definition of its tracking companion;
- for the updates that are not of that shape, a procedure to construct a counterexample to tracking.

This may be a highly non-trivial result, for example if the updates at the richer level are cast in a second order language and the ones at the poorer level are given in a first order signature.

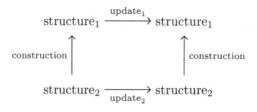

This paper explore this issue in the setting of evidence and plausibility models. Our main result is a characterization result of this kind for a restricted class of updates on evidence models: we give the exact condition for an update on evidence models belonging to the class to be trackable by an update on plausibility models. For the positive cases we offer a procedure to compute the other update, while for the negative cases we give a recipe to construct a counterexample to tracking. To our knowledge, this is the first result of this kind in the literature. As for the implications of this result, our focus on the definability aspects of tracking underscores the possibility to study this problem with logical techniques and suggests the possibility to apply the same methodology to other examples.

2 Plausibility and Evidence Models

We begin by introducing the two main actors featuring in our play, evidence and plausibility models. Plausibility models are widely used in formal epistemology [2,4]; their introduction can be traced back at least to [11]. They consist of a carrier, to be understood as a collection of possible worlds, and a preorder representing how an agent ranks the possible scenarios in terms of plausibility.[1]

Definition 1 (Plausibility model). *A* plausibility model *is a tuple* $\mathcal{M} = \langle W, \leq, V\rangle$ *with* W *a non-empty set of worlds, a reflexive and transitive relation* $\leq \subseteq W \times W$ *and* $V : W \to \wp(At)$ *a valuation function. We denote with* **PM** *the class of plausibility models.*

Introduced in [7], evidence models are structures capturing the evidence available to an agent. Such evidence is represented via a family of sets of possible worlds: intuitively each set in the family constitutes a piece of evidence that the agent can use to draw conclusions.[2]

Definition 2 (Evidence model). *An* evidence model *is a tuple* $\mathcal{M} = \langle W, E, V\rangle$ *with* W *a non-empty set of worlds, a function* $E : W \to \wp(\wp(W))$ *and* $V : W \to \wp(At)$ *a valuation function. We furthermore assume* $W \in E(w)$ *and* $\emptyset \notin E(w)$ *for all* $w \in W$. *A* uniform evidence model *is an evidence model where* E *is a constant function, we denote with* $\mathcal{E}$ *this fixed collection of evidence sets.*

The requirements on evidence ensure that the agent has trivial evidence, namely the whole set W, and does not have inconsistent evidence, i.e. the empty set. Notice that the structure on evidence models is world-dependent, therefore to match with plausibility models from now on we only consider uniform evidence models.[3] We denote with **EM** the class of uniform evidence models. We now describe how to construct a plausibility model starting from an evidence model.

---

[1] In the literature plausibility models are sometimes assumed to have a complete and/or well-founded relation, we drop these assumptions here, adopting the definition of [4].

[2] Evidence models contain more information than plausibility models; such information is captured by operators such as the evidence modality. See [4,5] for a discussion on the relationship between these models. The sphere systems of [9] also constitute an example of neighborhood models with a close tie to relational structures.

[3] Alternatively one could consider plausibility models with a different relation for each world, this generalization does not add much depth to our results, so we employ the simpler models.

Definition 3 (Construction *ORD*, **[7]).** *Given a uniform evidence model* $\mathcal{M} = \langle W, \mathcal{E}, V\rangle$ *construct the plausibility model* $ORD(M) = \langle W, \leq_{\mathcal{E}}, V\rangle$ *where* $\leq_{\mathcal{E}}$ *is defined as follows.*

$$w \leq_{\mathcal{E}} v \textit{ iff } \forall X \in \mathcal{E}, v \in X \textit{ implies } w \in X$$

A reader with some knowledge in Topology or Duality Theory will recognize this construction as (the converse of) the *specialization preorder* obtainable from a neighborhood structure. Conversely, we can reconstruct an evidence model from a plausibility model, as follows.

Definition 4 (Construction *EV*, **[7]).** *Call a set* $X \subseteq W$ downward closed *with respect to* $\leq$ *if* $w \in X$ *and* $v \leq w$ *entail* $v \in X$. *A set with such property is called a* down-set.

Given a plausibility model $\mathcal{M} = \langle W, \leq, V\rangle$ *construct the evidence model* $EV(M) = \langle W, \mathcal{E}_{\leq}, V\rangle$ *where* $\mathcal{E}_{\leq}$ *is the set of non-empty downward closed subsets of* W.

Lemma 1. *Performing EV and ORD in this order on a plausibility model M one obtains the same model M, that is,* $ORD \circ EV = Id_{\mathbf{PM}}$.

The converse is not the case, in general $EV \circ ORD \neq Id_{\mathbf{EM}}$. Starting from an evidence model M and performing $EV \circ ORD$ one obtains a copy of M where all the intersections of the evidence sets have been added. This observation together with the Lemma, which is just folklore, captures the idea that the level of plausibility models is 'poorer', or in other words contains less information compared to the level of evidence models.

3 Tracking

As we mentioned, the issue of tracking in the case of evidence and plausibility models was addressed in [5], where the author investigated when an update on evidence models is mirrored by another update at the level of plausibility models. As we mentioned in the introduction, an update is a model-changing operation that is meant to represent the change of an agent's internal state after new information is taken into account. The new available data is typically encoded in a formula of a static language, thus an update is parametric on a formula. For plausibility models some well-studied examples are the updates 'public announcement of ϕ', where all the worlds satisfying $\neg\phi$ are removed, and 'radical upgrade with ϕ', where all the worlds satisfying $\neg\phi$ are ranked as less plausible than worlds satisfying ϕ, all the rest being equal. In [5] an update is regarded as a purely semantical operation, meaning that a set is used as the parameter for the update instead of the extension of a formula.

Definition 5 (Tracking, [5]). *We indicate with* $U(X)$ *the update that uses as a parameter the set* X. *A function* $U(X) : \mathbf{PM} \to \mathbf{PM}$ tracks *a function* $U'(X) : \mathbf{EM} \to \mathbf{EM}$ *if* $U(X)(ORD(M)) = ORD(U'(X)(M)$ *for all* X, *or equivalently if the following square of functions commutes:*

$$\begin{array}{ccc}
\mathbf{PM} & \xrightarrow{U(X)} & \mathbf{PM} \\
{\scriptstyle ORD}\uparrow & & \uparrow{\scriptstyle ORD} \\
\mathbf{EM} & \xrightarrow[U'(X)]{} & \mathbf{EM}
\end{array}$$

We sometimes omit the parameter X when it is clear from the context.

The concept of tracking can be explored in a richer categorical setting, where we equip **PM** and **EM** with suitable notions of morphism; indeed the constructions ORD and EV can be extended to adjoint functors that preserve some categorical constructions. We leave aside these concerns for now, since our result can be cast without the categorical machinery, but direct the reader to [8] where these issues are studied systematically. The definition of tracking highlights the fact that we are interested in tracking updates on the richer structures (evidence models) with updates on the poorer structures (plausibility models). The other direction, from poor to rich structures, is less interesting since every update on plausibility models has a canonical counterpart on evidence models, as the next proposition shows.

Proposition 1 (See [5]). *For every update $U(X) : \mathbf{PM} \to \mathbf{PM}$ there is an update $U'(X) : \mathbf{EM} \to \mathbf{EM}$ that is tracked by $U(X)$.*

The next proposition provides an equivalent condition for tracking.

Proposition 2. *The existence of an update U tracking an update U' is equivalent to the following: for every pair of evidence models M_1, M_2, if $ORD(M_1) = ORD(M_2)$ then $ORD(U'(M_1)) = ORD(U'(M_2))$.*

The left-to-right direction of this proposition suggests how to prove that an update on evidence models *cannot* be tracked: it is sufficient to find two models for which the condition of Proposition 2 fails. This strategy is adopted in [5] to prove that some updates cannot be tracked.

The other direction of the proposition may at first glance seem to trivialize the problem of tracking: given an update U' on evidence models, we can just verify that the condition of Proposition 2 is fulfilled and then we immediately have an update $U := ORD \circ U' \circ EV$ that tracks U'. Such definition, however, is only partially satisfactory: even though it fits the bill from a semantic perspective, the interest of tracking lies in the possibility to rewrite an update on a complex structure in the language of a poorer structure. We expand on this matter in Sect. 3.2. The definition $U := ORD \circ U' \circ EV$ circumvents this problem altogether and is therefore not very informative. For this reason it is still noteworthy to obtain positive tracking results.

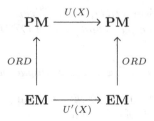

3.1 Examples of Tracking

The aforementioned paper [5] contains a few examples of tracking results; we refer the reader to said article for details on definitions and proofs.

Proposition 3 (See [5]). *The following statements hold:*

- *Public announcement at the plausibility level tracks public announcement on evidence models.*
- *Suggestion tracks evidence addition.*
- *Radical upgrade tracks the upgrade called "up".*

We want however to show a concrete instance of a tracking proof, for the reader to understand what these arguments look like. This result is actually a direct consequence of the more general Theorem 2, proved later, but it serves for explanatory purposes. Our example is the tracking of the update 'evidence weakening'.

Definition 6. *Given a set X,* evidence weakening *is a construction of type* $\cup(X) : \mathbf{EM} \to \mathbf{EM}$. *For an evidence model $M = \langle W, E, V\rangle$ the update returns a model $\cup(X)(M)$ is defined as:*

- $W^{\cup X} = W$,
- $V^{\cup X} = V$,
- $\mathcal{E}^{\cup X} = \{Y \cup X | Y \in \mathcal{E}\}$,

We define the update on plausibility models that tracks evidence weakening, we call it 'collapse of ϕ'.

Definition 7. *Given a formula ϕ in the language,* collapse of ϕ *is a construction of type $coll(\phi) : \mathbf{PM} \to \mathbf{PM}$. For a plausibility model $M = \langle W, \leq, V\rangle$ the action on objects $coll(\phi)(M)$ is defined as:*

- $coll(W) = W$
- $coll(V)(p) = V(p)$
- *$coll(\leq)$ is defined via a case distinction:*
  1. *If $w, v \in [\![\phi]\!]$ then $(w, v) \in coll(\leq)$;*
     all ϕ-worlds are equi-plausible.
  2. *If $w, v \in [\![\neg\phi]\!]$ then $(w, v) \in coll(\leq)$ iff $w \leq v$;*
     the relation is unaltered on $\neg\phi$-worlds.
  3. *If $w \in [\![\neg\phi]\!]$ and $v \in [\![\phi]\!]$ then $(w, v) \in coll(\leq)$ iff $\forall k \in W\, w \leq k$;*
     a $\neg\phi$-world is at least as plausible as a ϕ-world iff the former was the bottom element of $\leq$.
  4. *if $v \in [\![\neg\phi]\!]$ and $w \in [\![\phi]\!]$ then $(w, v) \in coll(\leq)$;*
     all ϕ-worlds are at least as plausible as $\neg\phi$-worlds.

Theorem 1 (Tracking of evidence weakening). *The evidence weakening update on evidence models is tracked by the collapse update, making the following diagram commute on objects:*

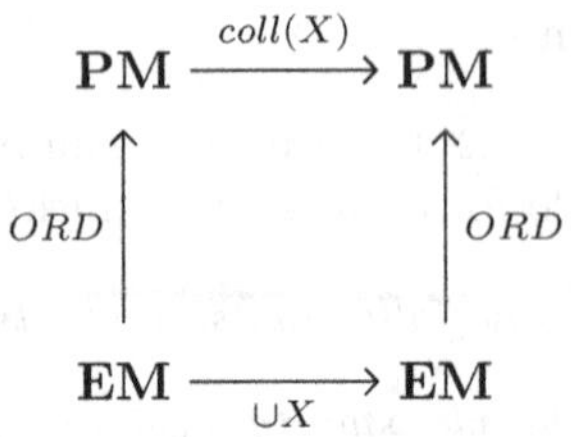

Proof. Consider an evidence model $\mathcal{M} = \langle W, \mathcal{E}, V\rangle$. The functor ORD and the two updates leave the set of worlds W and the valuation V unaltered, thus there is nothing to check there. Applying first ORD and then $coll(X)$ we obtain the relation $coll(\leq_{\mathcal{E}})$, while applying the update $\cup X$ and then ORD we get the relation $\leq_{\cup X(\mathcal{E})}$: we need to show that the two coincide, that is

$$(w, v) \in coll(\leq_{\mathcal{E}}) \quad \text{iff} \quad w \leq_{\cup X(\mathcal{E})} v$$

We do so by a case distinction, using $\overline{X}$ to denote the complement of X.:

- Suppose $w, v \in X$. Then by definition $(w, v) \in coll(\leq_{\mathcal{E}})$ is always the case. But also $w \leq_{\cup X(\mathcal{E})} v$ must always be the case: since $w, v \in X$ the condition $\forall Y \in \mathcal{E}$ if $v \in Y \cup X$ then $w \in Y \cup X$ is always fulfilled.
- Assume $w, v \in \overline{X}$. Then $(w, v) \in coll(\leq_{\mathcal{E}})$ iff, by definition, $w \leq_{\mathcal{E}} v$, which means that for all $Y \in \mathcal{E}$ if $v \in Y$ then $w \in Y$. Since we assumed $w, v \in \overline{X}$, the last condition is equivalent to the following: for all $Y \in \mathcal{E}$ if $v \in Y \cup X$ then $w \in Y \cup X$. But this is just the definition of $w \leq_{\cup X(\mathcal{E})} v$.
- Suppose now that $w \in \overline{X}$ and $v \in X$. By the definition of collapse, $(w, v) \in coll(\leq_{\mathcal{E}})$ is the case iff w is below every element in W with respect to relation $\leq_{\mathcal{E}}$. This latter condition is the case iff w is contained in all the evidence sets in $\mathcal{E}$: if it does then clearly it is below every other element by the definition of $\leq_{\mathcal{E}}$; for the other direction consider that every evidence set Y is not empty (by definition of evidence model $\emptyset \notin \mathcal{E}$) so there is $k \in Y$ but because $w \leq_{\mathcal{E}} k$ we get $w \in Y$.
  If for all $Y \in \mathcal{E}$ we have $w \in Y$ then for all $Y \in \mathcal{E}$ we have that if $v \in Y \cup \overline{X}$ then $w \in Y \cup \overline{X}$, because the consequent always holds. Hence $w \leq_{\cup X(\mathcal{E})} v$. Conversely, under the assumption $v \in X$ and $w \in \overline{X}$, the condition $\forall Y \in \mathcal{E}$ if $v \in Y \cup X$ then $w \in Y \cup X$ entails that $w \in Y$ for all $Y \in \mathcal{E}$.
- For the last case assume that $v \in \overline{X}$ and $w \in X$. Then $(w, v) \in coll(\leq_{\mathcal{E}})$ is always the case by definition. Note that the same holds for $w \leq_{\cup X(\mathcal{E})} v$: since $w \in X$, we have $w \in Y \cup X$ for every $Y \in \mathcal{E}$.

Of course one of the main difficulties, when dealing with problems such as tracking, is to *find* the right candidate for a tracking companion. In general one would like to have a constructive procedure to compute the second update from the first, instead of having to rely on guesswork. Our result, presented later, also addresses this concern.

There are also plenty of examples of updates on evidence models that are *not* trackable. One case mentioned in [5] is evidence deleting, an update that removes all the evidence sets implying the negation of the new data.

3.2 Tracking as a Definability Problem

As we mentioned previously, the interesting part of tracking is the reduction of updates cast in a complex language to updates cast in a poor language, typically a fragment. In other words, tracking is ultimately a *definability* issue. We begin by making explicit what we mean by saying that an update is defined in a certain language. We focus exclusively on updates that preserve the carrier of the structure.

Definition 8 (Definability). *An n-ary relation R in a model M is definable in a language $\mathcal{L}$ iff there is a formula $\phi(\overline{x}) \in \mathcal{L}$ with n open variables such that:*

$$R = \{(a_1, \ldots, a_n) | M \vDash \phi(\overline{x})[(a_1, \ldots, a_n)]\}$$

The signature of plausibility models is FOL with a binary relational symbol $\leq$ which is meant to be interpreted on the plausibility relation. For evidence models we need a stronger language in order to quantify over evidence sets.

Definition 9 (Evidence language). *Consider the grammar*

$$\phi ::= \mathcal{E}(n) \,|\, x \in n \,|\, \neg\phi \,|\, \phi \wedge \phi \,|\, \forall x\, \phi \,|\, \forall n\, \phi$$

where variables $n, n', \ldots$ are for subsets and variables $x, y, \ldots$ are for elements. To the signature we add a unary predicate $\mathcal{E}$ on subsets, denoting whether a subset is a piece of evidence, and a binary relation $\in$ denoting elementhood. We adopt the standard conventions for free and bound variables, as well as the classical abbreviations for defined propositional connectives.

We will use 'plausibility language' or 'evidence language' to refer to such languages. The semantics of these languages are just the standard first and second-order semantics; the former language is meant to be interpreted over the class of plausibility models, while the second over the class of evidence models.

In order to define an update we define its intended output via a formula containing the suitable parameters. On plausibility models for example, given an update U, a plausibility model M and a set P, we define U by defining the plausibility relation on the updated model $U(P)(M)$ with a formula $\beta(x, y, P, \leq)$ such that:

- it depends on P, a unary predicate interpreted on the set P;[4]
- it depends on $\leq$, a binary relational symbol interpreted on the relation $\leq$ in the model;
- it has two open variables in order to define a binary relation;
- it is in the signature of plausibility models.

In the case of evidence models we define an update $U'(P)$ with a formula $\alpha(n, P, \mathcal{E})$, where n is an open variable of sort 'subset', P is again a unary

[4] We ambiguously use the same symbol for the corresponding semantic and the syntactic objects; the context will disambiguate.

predicate for worlds and $\mathcal{E}$ is the aforementioned unary predicate for subsets; a formula $\alpha(n, P, \mathcal{E})$ will denote the evidence sets of the updated model. We can now state precisely what the problem of tracking amounts to in the case of evidence and plausibility models.

Question 1 (Tracking). Given an update U' on evidence models defined by a formula $\alpha(n, P, \mathcal{E})$ in the evidence language, can we find an update U on plausibility models that tracks U' and is defined by a formula $\beta(x, y, P, \leq)$ in the plausibility language?

The optimal answer to this problem is a characterization result giving:

- the syntactic shape of all and only the trackable updates;
- for the updates of that shape, an algorithm that manipulates syntactically the corresponding formulas α and produces the definitions β of their tracking companion;
- for the updates that are not of that shape, a procedure to construct a counter-example to tracking.

Our main theorem is a result of this kind, for a specific class of updates.

Notice that the evidence language is strong enough to express the action of the function ORD: given an evidence collection $\mathcal{E}$ the relation $x \leq_{\mathcal{E}} y$ is defined by the following formula with two open variables.

$$x \leq_{\mathcal{E}} y := \forall n(\mathcal{E}(n) \rightarrow (y \in n \rightarrow x \in n))$$

Call $FOL(P, \leq_{\mathcal{E}})$ the language of FOL enriched with two additional symbols for P and $\leq_{\mathcal{E}}$. Note that this is a fragment of the evidence language (enriched with the unary predicate P), namely a fragment where the quantification over subsets occurs only within $\leq_{\mathcal{E}}$. The next proposition points to the fact that any update in the language of evidence models that is definable with a formula of $FOL(P, \leq_{\mathcal{E}})$ is in fact trackable.

Proposition 4 ($FOL(P, \leq_{\mathcal{E}})$-definability entails tracking). *Given an update U' on evidence models and a set P, assume $U'(P)$ preserves the domain of the models. Suppose that for any given model M the relation $ORD(U'(P)(\mathcal{E}))$, namely the plausibility relation in the model $ORD(U'(P)(M))$, is definable by a formula $\beta(x, y, P, \leq_{\mathcal{E}}) \in FOL(P, \leq_{\mathcal{E}})$. Then $U'(P)$ is tracked by an update $U(P)$ defined by $\beta(x, y, P, \leq)$.*

If an update $U'(P)$ on evidence models is defined by $\alpha(n, P, \mathcal{E})$ in the evidence language (that is, such formula denotes the subsets that are pieces of evidence after the update) then the relation $ORD(U'(P)(\mathcal{E}))$ is also defined by

$$\forall n(\alpha(n, P, \mathcal{E}) \rightarrow (y \in n \rightarrow x \in n)) \tag{1}$$

Therefore Proposition 4 guarantees that if we can reduce 1 to the fragment $FOL(P, \leq_{\mathcal{E}})$ then we know that U' is trackable.

4 Characterizing Trackable Updates

Now for the main course: we isolate a class of updates for which we can prove a characterization result. We begin with a preliminary definition and some notation.

Definition 10 (Simple formulas). *Given a predicate P on elements, a formula $\psi(n, x, P)$ in the evidence language is* simple *if it is built from the atomic formulas $x \in n$ and Px using only negations, conjunctions and disjunctions.*

Simple formulas are essentially just boolean combinations of the two atomic formulas $x \in n$ and Px.

Notation. We use the following abbreviations in the evidence language:

- $n \subseteq n' := \forall y\, [y \in n \to y \in n']$
- $n = n' := n \subseteq n' \wedge n' \subseteq n$
- $n \subset n' := n \subseteq n' \wedge \neg(n' \subseteq n)$
- $n \subseteq P := \forall y\, [y \in n \to Py]$
- $n = P := n \subseteq P \wedge P \subseteq n$
- $n \subset P := n \subseteq P \wedge \neg(P \subseteq n)$
- $n \subseteq \phi(n', P) := \forall y\, [y \in n \to \phi(n', y, P)]$
- $n = \phi(n', P) := (n \subseteq \phi(n', P)) \wedge (\phi(n', P) \subseteq n)$
- $n \subset \phi(n', P) := (n \subseteq \phi(n', P)) \wedge \neg(\phi(n', P) \subseteq n)$

Note how we remove the variable y from $\phi(n', P)$ to stress that this variable has been quantified over. We use the same notation with other formulas such as $\theta(n', x, P)$ in the same fashion.

A first observation is that all updates that are defined with a formula of the following shape

$$\alpha(n, P) := \exists n'(\mathcal{E}(n') \wedge n = \phi(n', P))$$

turn out to be trackable. In these cases all evidence sets are modified uniformly by ϕ. An example of such updates is evidence weakening, in which case $\phi(n', y, P) := x \in n' \vee Px$. We thus seek to enlarge this class of updates to a more diverse one, including some non-trackable updates. As witnessed by some examples treated in [5], counterexamples to tracking seem to occur when we break this uniformity, that is, we modify some evidence sets while we leave some other unchanged. This suggests the introduction of a 'precondition' θ, which may be triggered or not by an evidence set; to keep things under control we require θ to also be a simple formulas. This leads us to the definition of simple updates.

Definition 11 (Simple updates). *An update on evidence models is* simple *if it is definable with a formula of the following shape:*

$$\begin{aligned}\alpha(n, P) := \exists n'(&\mathcal{E}(n') \wedge \exists x\, \theta(n', x, P) \wedge n = \phi(n', P))\\ \vee\ (&\mathcal{E}(n') \wedge \neg\exists x\, \theta(n', x, P) \wedge n = n')\end{aligned}$$

where both $\theta(n, x, P)$ and $\phi(n', y, P)$ are simple formulas.

Simple updates can be intuitively understood as follows: the new collection of evidences keeps all the old evidence sets n' for which the condition θ 'fails', namely when $\neg\exists x\, \theta(n', x, P)$ is the case, while it replaces with $\phi(n', P)$ all the old evidence sets n' for which the precondition $\exists x\, \theta(n', x, P)$ holds. If $\theta(n, x, P)$ is equivalent to $\top$ then a simple update performs a uniform operation on all evidence sets, thus we recover all updates of the form $\alpha(n, P) := \exists n'(\mathcal{E}(n') \wedge n = \phi(n', P))$. If θ has more structure then it can be exploited to select the evidence sets that we intend to manipulate.

The class of simple updates contains both examples and counterexamples to tracking, therefore it is amenable for a characterization result as described in the previous section. Despite being defined in terms of simple formulas, simple updates already exhibit a complex behaviour due to the interaction between the 'precondition' θ and the 'effect' of the update ϕ. Now for some further terminology. The elements that belong to a subset n but do not belong to its updated version, namely $\phi(n, x, P)$, are called *separated.* The elements that do not belong to a subset n but belong to $\phi(n, x, P)$ are instead called *adopted.* We encode these notions in the following formulas:

- $Sep(n, x, P) := x \in n \wedge \neg\phi(n, x, P)$
- $Ado(n, x, P) := x \notin n \wedge \phi(n, x, P)$

With this terminology in place we can state our characterization result for tracking of simple updates.

Theorem 2. *A simple update U' is trackable if and only if one of the following conditions hold.*

1. *All separated points and all adopted points are witnesses:* $\forall n\, [Sep(n, P) \subseteq \theta(n, P) \wedge Ado(n, P) \subseteq \theta(n, P)]$ *is a tautology.*
2. *The formula* $\forall n\, \mathcal{E}(n) \rightarrow \forall x(\gamma(n, x, P) \rightarrow \theta(n, x, P))$ *is a tautology on evidence models, where* $\gamma(n, x, P)$ *is one of the following formulas:*
   - $x \in n$
   - $x \notin n$
   - Px
   - $\neg Px$
3. $\exists x\, \theta(n, x, P)$ *is equivalent to* $\bot$.

If one of the aforementioned conditions holds then we have a procedure to construct the tracking companion of U'; if they all fail we have a procedure to construct a counterexample to tracking.

We provide an example in the Appendix A to show an application of our result; we refer the reader to [8] for the full proof.

5 Conclusions

In this article we dived into the upcoming topic of tracking, namely the matching of information dynamics on different structures. We focused on the case study of plausibility and evidence models. After setting up the stage and discussing some examples, we described the sense in which tracking is primarily a definability issue. We then presented our main result, an if and only if characterization of the trackable updates in the class of simple updates. For the trackable updates the proof of the theorem provides a procedure to construct the corresponding update on plausibility models, while for the non-trackable updates we offer a strategy to build a counterexample to tracking.

We have seen the sense in which tracking, at least in the case study we analyzed, connects to the reduction of second-order formulas to first-order ones. Updates whose definition can be reduced in such a way are, loosely speaking, treating a second order structure as if it were first-order. Indeed trackable updates give the "same" treatment to evidence models that generate the same preorder, where "same" means from the point of view of a first-order signature (see Proposition 2). This perspective becomes particularly interesting if we consider that evidence models are examples of neighborhood models. Even though our results are tailored to work on the class of evidence models specifically, the techniques employed in this chapter could be tested in the general case, namely the tracking of operations on neighborhood models by operations on preorders.

The groundwork put forward in this article provides a basis to further study the issues connected to dynamic updates and tracking, as well as a methodology that can be adapted to other classes of models different from the ones covered here. A prominent example would be probabilistic models: one may want to apply our techniques to study their interaction with the poorer plausibility models. The interplay between probabilities and plausibility relations is a classic theme in epistemology and it has seen some renewed interest in recent years, witness for example [10]. On the other hand, neighborhood models have their own history of connections with probabilistic models, (see e.g. [13]), and may serve as a bridge between the coarse-grained plausibility representation and the probabilistic realm.

A Appendix: Application of Theorem 2

In this appendix we show how our main result can be applied, deriving the update 'collapse of X' (Definition 7) from the update 'evidence weakening of X' (Definition 6). This will showcase how our characterization result solves the problem of finding the tracking companion for a given update.

We start by encoding evidence weakening in a formula α of the evidence language, using P as the unary predicate that is meant to be interpreted on X:

$$\alpha(n, P, \mathcal{E}) := \exists n'(\mathcal{E}(n') \wedge \forall x\, x \in n \leftrightarrow (x \in n' \vee Px))$$

Intuitively this formula is saying 'a set n is a piece of evidence in the updated model iff there exists an evidence set n' in the original model such that $n = n' \cup X$'. Note that this is a simple update, where $\phi(x, n', P) = (x \in n' \vee Px)$ and $\theta(x, n', P) = \top$.

Evidence weakening satisfies the premises of our Theorem; this ensures that we can apply the procedure to find its tracking companion. As we know from Proposition 4, it is sufficient to show that, for $\alpha(n, P, \mathcal{E})$ as above, the formula

$$\forall n(\alpha(n, P, \mathcal{E}) \to (y \in n \to x \in n))$$

is equivalent to a formula $\beta(x, y, P, \leq_{\mathcal{E}}) \in FOL(P, \leq_{\mathcal{E}})$. Thus we proceed manipulate the formula syntactically exploiting various first-order and propositional laws. Starting from

$$\forall n(\alpha(n, P, \mathcal{E}) \to (y \in n \to x \in n))$$

we plug in α and obtain

$$\forall n(\exists n'(\mathcal{E}(n') \wedge \forall z\, z \in n \leftrightarrow (z \in n' \vee Pz)) \to (y \in n \to x \in n))$$

From this formula we extract the existential quantifier:

$$\forall n, \forall n'(\mathcal{E}(n') \wedge \forall z\, z \in n \leftrightarrow (z \in n' \vee Pz)) \to (y \in n \to x \in n)$$

Now we eliminate the quantifier over n exploiting the bi-conditional, substituting the instances of $x \in n$ and $y \in n$:

$$\forall n'(\mathcal{E}(n') \to ((y \in n' \vee Py) \to (x \in n' \vee Px)))$$

We proceed to split the first disjunction into two implications.

$$\forall n'(\mathcal{E}(n') \to ([(y \in n') \to (x \in n' \vee Px)] \wedge [(Py) \to (x \in n' \vee Px)]))$$

We can then distribute the outermost implication and the quantifier over the conjunction, obtaining

$$\begin{array}{l}\forall n'(\mathcal{E}(n') \to [(y \in n') \to (x \in n' \vee Px)]) \wedge \\ \forall n'(\mathcal{E}(n') \to [(Py) \to (x \in n' \vee Px)])\end{array}$$

The next step is to rewrite the innermost disjunctions as implications.

$$\begin{array}{l}\forall n'(\mathcal{E}(n') \to [(y \in n') \to (\neg Px \to x \in n')]) \wedge \\ \forall n'(\mathcal{E}(n') \to [(Py) \to (\neg Px \to x \in n')])\end{array}$$

Since n' does not feature in the literals Px or Py or their negations, we can pull out these formulas from the quantifications.

$$\begin{array}{l}[\neg Px \to \forall n'(\mathcal{E}(n') \to ((y \in n') \to (x \in n')))] \wedge \\ [(Py \wedge \neg Px) \to \forall n'(\mathcal{E}(n') \to x \in n')]\end{array}$$

Finally, using the definition of $\leq_{\mathcal{E}}$, we can rewrite the last formula as

$$\beta(x, y, P, \leq_{\mathcal{E}}) := [\neg Px \rightarrow x \leq_{\mathcal{E}} y] \wedge [(Py \wedge \neg Px) \rightarrow \forall z\, x \leq_{\mathcal{E}} z]$$

The equivalence of $\forall z\, x \leq_{\mathcal{E}} z$ and $\forall n'(\mathcal{E}(n') \rightarrow x \in n')$ over evidence models is an easy check.

Proposition 4 argued that the tracking companion of evidence weakening is defined by $\beta(x, y, P, \leq)$ (which is just $\beta(x, y, P, \leq_{\mathcal{E}})$ where we have substituted the relation $\leq_{\mathcal{E}}$ with $\leq$). In this case this means that two elements a and b are in the new, updated relation if and only if they satisfy $\beta(x, y, P, \leq)$. Is is easy to see that $\beta(x, y, P, \leq)$ defines exactly the collapse update of Definition 7: $(a, b) \in coll(\leq)$ if and only if $\beta(x, y, P, \leq)$ is true when instantiated to (a, b) if and only if one of the following is the case

- $a \in X$ (recall that P is interpreted on the set X);
- $a \notin X$, $b \notin X$ and $a \leq b$;
- $a \notin X$, $b \in X$ and $\forall z\, a \leq z$.

The reader is invited to contrast this with Definition 7. The manipulations showed in this section are part of a general strategy to reduce the updates of the right shape into formulas of the fragment $FOL(P, \leq_{\mathcal{E}})$, giving us a procedure to compute the tracking companion.

References

1. Baltag, A., Moss, L.S.: Logics for epistemic programs. Synthese **139**, 165–224 (2004)
2. Baltag, A., Smets, S.: Conditional doxastic models: a qualitative approach to dynamic belief revision. Electron. Notes Theor. Comput. Sci. **165**, 5–21 (2006)
3. Baltag, A., Smets, S.: A qualitative theory of dynamic interactive belief revision. In: Woolridge, M., Bonanno, G., van der Hoek, W. (eds.) Texts in Logic and Games, vol. 3, pp. 9–58. Amsterdam University Press, Amsterdam (2008)
4. van Benthem, J.: Dynamic logic for belief revision. J. Appl. Non-Class. Logics **17**(2), 129–155 (2007)
5. van Benthem, J.: Tracking information. In: Bimbo, K. (ed.) J. Michael Dunn on Information Based Logics, pp. 363–389. Springer, Cham (2016). doi:10.1007/978-3-319-29300-4
6. van Benthem, J., van Eijck, J., Kooi, B.: Logics of communication and change. Inf. Comput. **204**(11), 1620–1662 (2006)
7. van Benthem, J., Pacuit, E.: Dynamic logics of evidence-based beliefs. Studia Logica **99**(1–3), 61–92 (2011)
8. Cinà, G.: Categories for the working modal logician. Ph.D. thesis, Institute for Logic, Language and Computation, University of Amsterdam (2017)
9. Grove, A.: Two modellings for theory change. J. Philos. Logic **17**(2), 157–170 (1988)
10. Leitgeb, H.: The stability theory of belief. Philos. Rev. **123**(2), 131–171 (2014)
11. Lewis, D.: Counterfactuals. Harvard University Press, Cambridge (1973)
12. Lin, H., Kelly, K.T.: Propositional reasoning that tracks probabilistic reasoning. J. Philos. Logic **41**, 957–981 (2012)
13. van Eijck, J., Renne, B.: Belief as willingness to bet. arXiv preprint arXiv:1412.5090 (2014)

Convergence, Continuity and Recurrence in Dynamic Epistemic Logic

Dominik Klein[1,2(✉)] and Rasmus K. Rendsvig[3,4(✉)]

[1] Department of Philosophy, Bayreuth University, Bayreuth, Germany
dominik.klein@uni-bayreuth.de
[2] Department of Political Science, University of Bamberg, Bamberg, Germany
[3] Theoretical Philosophy, Lund University, Lund, Sweden
rendsvig@gmail.com
[4] Center for Information and Bubble Studies, University of Copenhagen, Copenhagen, Denmark

Abstract. The paper analyzes dynamic epistemic logic from a topological perspective. The main contribution consists of a framework in which dynamic epistemic logic satisfies the requirements for being a topological dynamical system thus interfacing discrete dynamic logics with continuous mappings of dynamical systems. The setting is based on a notion of logical convergence, demonstratively equivalent with convergence in Stone topology. Presented is a flexible, parametrized family of metrics inducing the latter, used as an analytical aid. We show maps induced by action model transformations continuous with respect to the Stone topology and present results on the recurrent behavior of said maps.

Keywords: Dynamic epistemic logic · Limit behavior · Convergence · Recurrence · Dynamical systems · Metric spaces · General topology · Modal logic

1 Introduction

Dynamic epistemic logic is a framework for modeling information dynamics. In it, systematic change of Kripke models are punctiliously investigated through model transformers mapping Kripke models to Kripke models. The iterated application of such a map may constitute a model of information dynamics, or be may be analyzed purely for its mathematical properties [6,8,10,11,13,16,18,40–43].

Dynamical systems theory is a mathematical field studying the long-term behavior of spaces under the action of a continuous function. In case of discrete time, this amounts to investigating the space under the iterations of a continuous map. The field is rich in concepts, methodologies and results developed with the aim of understanding general dynamics.

The two fields find common ground in the iterated application of maps. With dynamic epistemic logic analyzing very specific map types, the hope is that general results from dynamical systems theory may shed light on properties

A. Baltag et al. (Eds.): LORI 2017, LNCS 10455, pp. 108–122, 2017.
DOI: 10.1007/978-3-662-55665-8_8

of the former. There is, however, a chasm between the two: Dynamical systems theory revolves around spaces imbued with metrical or topological structure with respect to which maps are continuous. No such structure is found in dynamic epistemic logic. This chasm has not gone unappreciated: In his 2011 *Logical Dynamics of Information and Interaction* [10], van Benthem writes

> **From discrete dynamic logics to continuous dynamical systems** "We conclude with what we see as a major challenge. Van Benthem [7,8] pointed out how update evolution suggests a long-term perspective that is like the evolutionary dynamics found in dynamical systems. [...] Interfacing current dynamic and temporal logics with the continuous realm is a major issue, also for logic in general." [10, Sect. 4.8. Emph. is org. heading]

This paper takes on the challenge and attempts to bridge this chasm.

We proceed as follows. Section 2 presents what we consider natural spaces when working with modal logic, namely sets of pointed Kripke models *modulo* logical equivalence. These are referred to as *modal spaces*. A natural notion of "logical convergence" on modal spaces is provided. Section 3 seeks a topology on modal spaces for which topological convergence coincides with logical convergence. We consider a metric topology based on n-bisimulation and prove it insufficient, but show an adapted Stone topology satisfactory. Saddened by the loss of a useful aid, the metric inducing the n-bisimulation topology, a family of metrics is introduced that all induce the Stone topology, yet allow a variety of subtle modelling choices. Sets of pointed Kripke models are thus equipped with a structure of compact metric spaces. Section 4 considers maps on modal spaces based on multi-pointed action models using product update. Restrictions are imposed to ensure totality, and the resulting *clean maps* are shown continuous with respect to the Stone topology. With that, we present our main contribution: A modal space under the action of a clean map satisfies the standard requirements for being a topological dynamical system. Section 5 applies the now-suited terminology from dynamical systems theory, and present some initial results pertaining to the recurrent behavior of clean maps on modal spaces. Section 6 concludes the paper by pointing out a variety of future research venues. Throughout, we situate our work in the literature.

Remark 1. To make explicit what may be apparent, note that the primary concern is the *semantics* of dynamic epistemic logic, i.e., its models and model transformation. Syntactical considerations are briefly touched upon in Sect. 6.

Remark 2. The paper is not self-contained. For notions from modal logic that remain undefined here, refer to e.g. [14,27]. For topological notions, refer to e.g. [37]. For more on dynamic and epistemic logic than the bare minimum of standard notions and notations rehearsed, see e.g. [2–5,10,20,22,30,38,39]. Finally, a background document containing generalizations and omitted proofs is our [31].

2 Modal Spaces and Logical Convergence

Let there be given a countable set Φ of **atoms** and a finite set I of **agents**. Where $p \in \Phi$ and $i \in I$, define the **language** $\mathcal{L}$ by

$$\varphi := \top \mid p \mid \neg\varphi \mid \varphi \wedge \varphi \mid \Box_i\varphi.$$

Modal logics may be formulated in $\mathcal{L}$. By a **logic** Λ we refer only to extensions of the **minimal normal modal logic** K over the language $\mathcal{L}$. With Λ given by context, let $\boldsymbol{\varphi}$ be the set of formulas Λ-provably equivalent to φ. Denote the resulting partition $\{\boldsymbol{\varphi} : \varphi \in \mathcal{L}\}$ of $\mathcal{L}$ by $\boldsymbol{\mathcal{L}}_\Lambda$.[1] Call $\boldsymbol{\mathcal{L}}_\Lambda$'s elements **$\Lambda$-propositions**.

We use relational semantics to evaluate formulas. A **Kripke model** for $\mathcal{L}$ is a tuple $M = (\llbracket M \rrbracket, R, \llbracket\cdot\rrbracket)$ where $\llbracket M \rrbracket$ is a countable, non-empty set of **states**, $R : I \longrightarrow \mathcal{P}(\llbracket M \rrbracket \times \llbracket M \rrbracket)$ assigns to each $i \in I$ an **accessibility relation** R_i, and $\llbracket\cdot\rrbracket : \Phi \longrightarrow \mathcal{P}(\llbracket M \rrbracket)$ is a **valuation**, assigning to each atom a set of states. With $s \in \llbracket M \rrbracket$, call $Ms = (\llbracket M \rrbracket, R, \llbracket\cdot\rrbracket, s)$ a **pointed Kripke model**. The used semantics are standard, including the modal clause:

$$Ms \vDash \Box_i\varphi \text{ iff for all } t : sR_it \text{ implies } Mt \vDash \varphi.$$

Throughout, we work with pointed Kripke models. Working with modal logics, we find it natural to identify pointed Kripke models that are considered equivalent by the logic used. The domains of interest are thus the following type of quotient spaces:

Definition 1. *The $\boldsymbol{\mathcal{L}}_\Lambda$ **modal space** of a set of pointed Kripke models X is the set $\boldsymbol{X} = \{\boldsymbol{x} : x \in X\}$ for $\boldsymbol{x} = \{y \in X : y \vDash \varphi$ iff $x \vDash \varphi$ for all $\boldsymbol{\varphi} \in \boldsymbol{\mathcal{L}}_\Lambda\}$.*

Working with an $\boldsymbol{\mathcal{L}}_\Lambda$ modal space portrays that we only are interested in differences between pointed Kripke models insofar as these are modally expressible and are considered differences by Λ.

In a modal space, how may we conceptualize that a sequence $\boldsymbol{x}_1, \boldsymbol{x}_2, ...$ converges to some point $\boldsymbol{x}$? Focusing on the concept from which we derive the notion of identity in modal spaces, namely Λ-propositions, we find it natural to think of $\boldsymbol{x}_1, \boldsymbol{x}_2, ...$ as converging to $\boldsymbol{x}$ just in case $\boldsymbol{x}_n$ moves towards satisfying all the same Λ-propositions as $\boldsymbol{x}$ as n goes to infinity. We thus offer the following definition:

Definition 2. *A sequence of points $\boldsymbol{x}_1, \boldsymbol{x}_2, ...$ in an $\boldsymbol{\mathcal{L}}_\Lambda$ modal space $\boldsymbol{X}$ is said to **logically converge** to the point $\boldsymbol{x}$ in $\boldsymbol{X}$ iff for every $\boldsymbol{\varphi} \in \boldsymbol{\mathcal{L}}_\Lambda$ for which $x \vDash \varphi$, there is an $N \in \mathbb{N}$ such that $x_n \vDash \varphi$ for all $n \geq N$.*

To avoid re-proving useful results concerning this notion of convergence, we next turn to seeking a topology for which logical convergence coincides with topological convergence. Recall that for a topology $\mathcal{T}$ on a set X, a sequence of points $x_1, x_2, ...$ is said to **converge** to x in the **topological space** $(X, \mathcal{T})$ iff for every open set $U \in \mathcal{T}$ containing x, there is an $N \in \mathbb{N}$ such that $x_n \in U$ for all $n \geq N$.

[1] $\boldsymbol{\mathcal{L}}_\Lambda$ is isomorphic to the domain of the **Lindenbaum algebra** of Λ.

3 Topologies on Modal Spaces

One way of obtaining a **topology** on a space is to define a **metric** for said space. Several metrics have been suggested for sets of pointed Kripke models [1,17]. These metrics are only defined for finite pointed Kripke models, but incorporating ideas from the metrics of [36] on **shift spaces** and [26] on sets of **first-order logical theories** allows us to simultaneously generalize and simplify the n-**Bisimulation-based Distance** of [17] to the degree of applicability:

Let $\boldsymbol{X}$ be a modal space for which modal equivalence and **bisimilarity** coincide[2] and let $\leftrightarroweq_n$ relate $x, y \in X$ iff x and y are n-**bisimilar**. Then proving

$$d_B(\boldsymbol{x}, \boldsymbol{y}) = \begin{cases} 0 & \text{if } x \leftrightarroweq_n y \text{ for all } n \\ \frac{1}{2^n} & \text{if } n \text{ is the least intenger such that } x \not\leftrightarroweq_n y \end{cases}$$

a metric on $\boldsymbol{X}$ is trivial. We refer to d_B as the n-**bisimulation metric**, and to the induced **metric topology** as the n-**bisimulation topology**, denoted $\mathcal{T}_B$. A basis of the topology $\mathcal{T}_B$ is given by the set of elements $B_{\boldsymbol{x}n} = \{\boldsymbol{y} \in \boldsymbol{X} : y \leftrightarroweq_n x\}$.

Considering the intimate link between modal logic and bisimulation, we consider both n-bisimulation metric and topology highly natural.[3] Alas, logical convergence does not:

Proposition 1. *Logical convergence in arbitrary modal space* $\boldsymbol{X}$ *does not imply convergence in the topological space* $(\boldsymbol{X}, \mathcal{T}_B)$.

Proof. Let $\boldsymbol{X}$ be an $\boldsymbol{\mathcal{L}}_\Lambda$ modal space with $\mathcal{L}$ based on the atoms $\Phi = \{p_k : k \in \mathbb{N}\}$. Let $x \in X$ satisfy $\Box\bot$ and p_k for all $k \in \mathbb{N}$. Let $\boldsymbol{x_1}, \boldsymbol{x_2}, ...$ be a sequence in $\boldsymbol{X}$ such that for all $k \in \mathbb{N}$, x_k satisfies $\Box\bot$, p_m for all $m \leq k$, and $\neg p_l$ for all $l > k$. Then for all $\varphi \in \boldsymbol{\mathcal{L}}_\Lambda$ for which $x \vDash \varphi$, there is an N such that $x_n \vDash \varphi$ for all $n \geq N$, hence the sequence $\boldsymbol{x_1}, \boldsymbol{x_2}, ...$ converges to $\boldsymbol{x}$. There does not, however, exist any N' such that $\boldsymbol{x}_{n'} \in B_{\boldsymbol{x}0}$ for all $n' \geq N'$. Hence $\boldsymbol{x_1}, \boldsymbol{x_2}, ...$ does not converge to $\boldsymbol{x}$ in $\mathcal{T}_B$. □

Proposition 1 implies that the n-bisimulation topology may not straightforwardly be used to establish negative results concerning logical convergence. That it may be used for positive cases is a corollary to Propositions 2 and 6 below. On the upside, logical convergence coincides with convergence in the n-bisimulation topology – i.e. Proposition 1 fails – when $\mathcal{L}$ has finite atoms. This is a corollary to Proposition 5.

An alternative to a metric-based approach to topologies is to construct the set of all open sets directly. Comparing the definition of logical convergence with that of convergence in topological spaces is highly suggestive: Replacing every occurrence of the formula φ with an open set U while replacing satisfaction $\vDash$ with inclusion $\in$ transforms the former definition into the latter. Hence the

[2] That all models in X are **image-finite** is a sufficient condition, cf. the Hennessy-Milner Theorem. See e.g. [14] or [27].

[3] Space does not allow for a discussion of the remaining metrics of [1,17], but see [31].

collection of sets $U_\varphi = \{\boldsymbol{x} \in \boldsymbol{X} : x \vDash \varphi\}$, $\varphi \in \mathcal{L}_\Lambda$, seems a reasonable candidate for a topology. Alas, this collection is not closed under arbitrary unions, as all formulas are finite. Hence it is not a topology. It does however constitute the basis for a topology, in fact the somewhat influential **Stone topology**, $\mathcal{T}_S$.

The Stone topology is traditionally defined on the collection of complete theories for some propositional, first-order or modal logic, but is straightforwardly applicable to modal spaces. Moreover, it satisfies our *desideratum*:

Proposition 2. *For any $\mathcal{L}_\Lambda$ modal space $\boldsymbol{X}$, a sequence $\boldsymbol{x}_1, \boldsymbol{x}_2, \ldots$ logically converges to the point $\boldsymbol{x}$ if, and only if, it converges to $\boldsymbol{x}$ in $(\boldsymbol{X}, \mathcal{T}_S)$.*

Proof. Assume $\boldsymbol{x}_1, \boldsymbol{x}_2, \ldots$ logically converges to $\boldsymbol{x}$ in $\boldsymbol{X}$ and that U containing $\boldsymbol{x}$ is open in $\mathcal{T}_S$. Then there is a basis element $U_\varphi \subseteq U$ with $\boldsymbol{x} \in U_\varphi$. So $x \vDash \varphi$. By assumption, there exists an N such that $x_n \vDash \varphi$ for all $n \geq N$. Hence $\boldsymbol{x}_n \in U_\varphi \subseteq U$ for all $n \geq N$.

Assume $\boldsymbol{x}_1, \boldsymbol{x}_2, \ldots$ converges to $\boldsymbol{x}$ in $(\boldsymbol{X}, \mathcal{T}_S)$ and let $x \vDash \varphi$. Then $\boldsymbol{x} \in U_\varphi$, which is open. As the sequence converges, there exists an N such that $\boldsymbol{x}_n \in U_\varphi$ for all $n \geq N$. Hence $x_n \vDash \varphi$ for all $n \geq N$. □

Apart from its attractive characteristic concerning convergence, working on the basis of a logic, the Stone topology imposes a natural structure. As is evident from its basis, every subset of $\boldsymbol{X}$ characterizable by a single Λ-proposition $\varphi \in \mathcal{L}_\Lambda$ is clopen. If the logic Λ is compact and $\boldsymbol{X}$ saturated (see footnote 7), also the converse is true: every clopen set is of the form U_φ for some φ. We refer to [31] for proofs and a precise characterization result. In this case, a subset is open, but not closed, iff it is characterizable only by an infinitary disjunction of Λ-propositions, and a subset if closed, but not open, iff it is characterizable only by an infinitary conjunction of Λ-propositions. The Stone topology thus transparently reflects the properties of logic, language and topology. Moreover, it enjoys practical topological properties:

Proposition 3. *For any $\mathcal{L}_\Lambda$ modal space $\boldsymbol{X}$, $(\boldsymbol{X}, \mathcal{T}_S)$ is **Hausdorff** and **totally disconnected**. If Λ is (logically) **compact**[4] and $\boldsymbol{X}$ is **saturated**[5], then $(\boldsymbol{X}, \mathcal{T}_S)$ is also (topologically) **compact**.*

Proof. These properties are well-known for the Stone topology applied to complete theories. For the topology applied to modal spaces, we defer to [31]. □

One may interject that, as having a metric may facilitate obtaining results, it may cause a loss of tools to move away from the n-bisimulation topology. The Stone topology, however, is **metrizable**. A family of metrics inducing it, generalizing the Hamming distance to infinite strings by using weighted sums, was introduced in [31]. We here present a sub-family, suited for modal spaces:

[4] A logic Λ is logically compact if any arbitrary set A of formulas is Λ-consistent iff every finite subset of A is Λ-consistent.

[5] An $\mathcal{L}_\Lambda$ modal space $\boldsymbol{X}$ is saturated iff for each Λ-consistent set of formulas A, there is an $\boldsymbol{x} \in \boldsymbol{X}$ such that $x \vDash A$. Saturation relates to the notion of **strong completeness**, cf. e.g. [14, Proposition 4.12]. See [31] for its use in a more general context.

Definition 3. *Let $D \subseteq \mathcal{L}_\Lambda$ contain for every $\psi \in \mathcal{L}_\Lambda$ some $\{\varphi_i\}_{i \in I}$ that Λ-entails either ψ or $\neg\psi$, and let $\varphi_1, \varphi_2, \ldots$ be an enumeration of D.*

Let $\boldsymbol{X}$ be an $\mathcal{L}_\Lambda$ modal space. For all $\boldsymbol{x}, \boldsymbol{y} \in \boldsymbol{X}$, for all $k \in \mathbb{N}$, let

$$d_k(\boldsymbol{x}, \boldsymbol{y}) = \begin{cases} 0 & \text{if } x \vDash \varphi \text{ iff } y \vDash \varphi \text{ for } \varphi \in \varphi_k \\ 1 & \text{else} \end{cases}$$

*Let $w : D \longrightarrow \mathbb{R}_{>0}$ assign strictly positive **weight** to each φ_k in D such that $(w(\varphi_n))$ forms a convergent series. Define the function $d_w : \boldsymbol{X}^2 \longrightarrow \mathbb{R}$ by*

$$d_w(\boldsymbol{x}, \boldsymbol{y}) = \sum_{k=0}^{\infty} w(\varphi_k) d_k(\boldsymbol{x}, \boldsymbol{y})$$

for all $\boldsymbol{x}, \boldsymbol{y} \in \boldsymbol{X}$. The set of these functions is denoted $D_{\boldsymbol{X}}$. Let $\mathcal{D}_{D,\boldsymbol{X}} = \bigcup_{D \subseteq \mathcal{L}_\Lambda} D_{\boldsymbol{X}}$.

We refer to [31] for the proof establishing the following proposition:

Proposition 4. *Let $\boldsymbol{X}$ be an $\mathcal{L}_\Lambda$ modal space and d_w belong to $\mathcal{D}_{\boldsymbol{X}}$. Then d_w is a metric on $\boldsymbol{X}$ and the metric topology $\mathcal{T}_w$ induced by d_w on $\boldsymbol{X}$ is the Stone topology of Λ.*

For a metric space $(\boldsymbol{X}, d)$, we will also write $\boldsymbol{X}_d$.

With variable parameters D and w, $\mathcal{D}_{\boldsymbol{X}}$ allows one to vary the choice of metric with the problem under consideration. E.g., if the n-bisimulation metric seems apt, one could choose that, with one restriction:

Proposition 5. *If $\boldsymbol{X}$ is an $\mathcal{L}_\Lambda$ modal space with $\mathcal{L}$ based on a finite atom set, then $\mathcal{D}_{\boldsymbol{X}}$ contains a topological equivalent to the n-bisimulation metric.*

Proof (sketch). As $\mathcal{L}$ is based on a finite set of atoms, for each $\boldsymbol{x} \in \boldsymbol{X}, n \in \mathbb{N}_0$, there exists a characteristic formula $\varphi_{x,n}$ such that $y \vDash \varphi_{x,n}$ iff $y \leftrightarroweq_n x$, cf. [27]. Let $D_n = \{\varphi_{\boldsymbol{x},n} : \boldsymbol{x} \in \boldsymbol{X}\}$ and $D = \bigcup_{n \in \mathbb{N}_0} D_n$. Then each D_n is finite and D satisfies Definition 3. Finally, let $w(\varphi) = \frac{1}{|D_n|} \cdot \frac{1}{2^{n+1}}$ for $\varphi \in D_n$. Then $d_w \in \mathcal{D}_{\boldsymbol{X}}$ and is equivalent to the n-bisimulation metric d_b. □

As a corollary to Proposition 5, it follows that, for finite atom languages, the n-bisimulation topology is the Stone topology. This is not true in general, as witnessed by Proposition 1 and the following:

Proposition 6. *If $\boldsymbol{X}$ is an $\mathcal{L}_\Lambda$ modal space with $\mathcal{L}$ based on a countably infinite atom set, then the n-bisimulation metric topology on $\boldsymbol{X}$ is strictly finer than the Stone topology on $\boldsymbol{X}$.*

Proof (sketch). We refer to [31] for details, but for $\mathcal{T}_B \not\subseteq \mathcal{T}_S$, note that the set $B_{\boldsymbol{x}0}$ used in the proof of Proposition 1, is open in $\mathcal{T}_B$, but not in $\mathcal{T}_S$. □

With this comparison, we end our exposition of topologies on modal spaces.

4 Clean Maps on Modal Spaces

We focus on a class of maps induced by action models applied using product update. Action models are a popular and widely applicable class of model transformers, generalizing important constructions such as public announcements. An especially general version of action models is *multi-pointed* action models with *postconditions*. Postconditions allow action states in an action model to change the valuation of atoms [12,19], thereby also allowing the representation of information dynamics concerning situations that are not factually static. Permitting multiple points allows the actual action states executed to depend on the pointed Kripke model to be transformed, thus generalizing single-pointed action models.[6]

A **multi-pointed action model** is a tuple $\Sigma\Gamma = (\llbracket\Sigma\rrbracket, \mathsf{R}, pre, post, \Gamma)$ where $\llbracket\Sigma\rrbracket$ is a countable, non-empty set of **actions**. The map $\mathsf{R} : I \to \mathcal{P}(\llbracket\Sigma\rrbracket \times \llbracket\Sigma\rrbracket)$ assigns an **accessibility relation** R_i on $\llbracket\Sigma\rrbracket$ to each agent $i \in I$. The map $pre : \llbracket\Sigma\rrbracket \to \mathcal{L}$ assigns to each action a **precondition**, and the map $post : \llbracket\Sigma\rrbracket \to \mathcal{L}$ assigns to each action a **postcondition**,[7] which must be $\top$ or a conjunctive clause[8] over Φ. Finally, $\emptyset \neq \Gamma \subseteq \llbracket\Sigma\rrbracket$ is the set of designated actions.

To obtain well-behaved total maps on a modal spaces, we must invoke a set of mild, but non-standard, requirements: Let X be a set of pointed Kripke models. Call $\Sigma\Gamma$ **precondition finite** if the set $\{\boldsymbol{pre(\sigma)} \in \boldsymbol{\mathcal{L}}_\Lambda : \sigma \in \llbracket\Sigma\rrbracket\}$ is finite. This is needed for our proof of continuity. Call $\Sigma\Gamma$ **exhaustive over** X if for all $x \in X$, there is a $\sigma \in \Gamma$ such that $x \vDash pre(\sigma)$. This conditions ensures that the action model $\Sigma\Gamma$ is universally applicable on X. Finally, call $\Sigma\Gamma$ **deterministic over** X if $X \vDash pre(\sigma) \wedge pre(\sigma') \to \bot$ for each $\sigma \neq \sigma' \in \Gamma$. Together with exhaustivity, this condition ensures that the product of $\Sigma\Gamma$ and any $Ms \in X$ is a (single-)pointed Kripke model, i.e., that the actual state after the updates is well-defined and unique.

Let $\Sigma\Gamma$ be exhaustive and deterministic over X and let $Ms \in X$. Then the **product update** of Ms with $\Sigma\Gamma$, denoted $Ms \otimes \Sigma\Gamma$, is the pointed Kripke model $(\llbracket M\Sigma\rrbracket, R', \llbracket\cdot\rrbracket', s')$ with

$$\begin{aligned}
\llbracket M\Sigma\rrbracket &= \{(s,\sigma) \in \llbracket M\rrbracket \times \llbracket\Sigma\rrbracket : (M,s) \vDash pre(\sigma)\} \\
R' &= \{((s,\sigma),(t,\tau)) : (s,t) \in R_i \text{ and } (\sigma,\tau) \in \mathsf{R}_i\}\,, \text{ for all } i \in N \\
\llbracket p\rrbracket' &= \{(s,\sigma) : s \in \llbracket p\rrbracket, post(\sigma) \nvDash \neg p\} \cup \{(s,\sigma) : post(\sigma) \vDash p\}\,, \text{ for all } p \in \Phi \\
s' &= (s,\sigma) : \sigma \in \Gamma \text{ and } Ms \vDash pre(\sigma)
\end{aligned}$$

Call $\Sigma\Gamma$ **closing over** X if for all $x \in X$, $x \otimes \Sigma\Gamma \in X$. With $\Sigma\Gamma$ exhaustive and deterministic, $\Sigma\Gamma$ and $\otimes$ induce a well-defined total map on X.

The class of maps of interest in the present is then the following:

[6] Multi-pointed action models are also referred to as *epistemic programs* in [2], and allow encodings akin to *knowledge-based programs* [22] of interpreted systems, cf. [42].

[7] The precondition of σ specify the conditions under which σ is executable, while its postcondition may dictate the posterior values of a finite, possibly empty, set of atoms.

[8] I.e. a conjuction of literals, where a literal is an atom or a negated atom.

Definition 4. *Let $\boldsymbol{X}$ be an $\mathcal{L}_\Lambda$ modal space. A map $\boldsymbol{f} : \boldsymbol{X} \to \boldsymbol{X}$ is called* ***clean*** *if there exists a precondition finite, multi-pointed action model $\Sigma\Gamma$ closing, deterministic and exhaustive over X such that $\boldsymbol{f}(\boldsymbol{x}) = \boldsymbol{y}$ iff $x \otimes \Sigma\Gamma \in \boldsymbol{y}$ for all $\boldsymbol{x} \in \boldsymbol{X}$.*

Clean maps are total by the assumptions of being closing and exhaustive. They are well-defined as $\boldsymbol{f}(\boldsymbol{x})$ is independent of the choice of representative for $\boldsymbol{x}$: If $x' \in \boldsymbol{x}$, then $x' \otimes \Sigma\Gamma$ and $x \otimes \Sigma\Gamma$ are modally equivalent and hence define the same point in $\boldsymbol{X}$. The latter follows as multi-pointed action models applied using product update preserve bisimulation [2], which implies modal equivalence. Clean maps moreover play nicely with the Stone topology:

Proposition 7. *Let $\boldsymbol{f}$ be a clean map on an $\mathcal{L}_\Lambda$ modal space $\boldsymbol{X}$. Then $\boldsymbol{f}$ is continuous with respect to the Stone topology of Λ.*

Proof (sketch). We defer to [31] for details, but offer a sketch: The map $\boldsymbol{f}$ is shown uniformly continuous using the ε-δ formulation of continuity. The proof relies on a lemma stating that for every $d_w \in \mathcal{D}_{\boldsymbol{X}}$ and every $\epsilon > 0$, there are formulas $\chi_1, \ldots, \chi_l \in \mathcal{L}$ such that every $x \in X$ satisfies some χ_i and whenever $y \vDash \chi_i$ and $z \vDash \chi_i$ for some $i \leq l$, then $d_w(y, z) < \epsilon$. The main part of the proof establishes the claim that there is a function $\boldsymbol{\delta} : \mathcal{L} \to (0, \infty)$ such that for any $\varphi \in \mathcal{L}$, if $f(x) \vDash \varphi$ and $d_{\boldsymbol{a}}(x, y) < \boldsymbol{\delta}(\varphi)$, then $f(y) \vDash \varphi$. Setting $\delta = \min\{\boldsymbol{\delta}(\chi_i) \colon i \leq l\}$ then yields a δ with the desired property. □

With Proposition 7, we are positioned to state our main theorem:

Theorem 1. *Let $\boldsymbol{f}$ be a clean map on a saturated $\mathcal{L}_\Lambda$ modal space $\boldsymbol{X}$ with Λ compact and let $d \in \mathcal{D}_{\boldsymbol{X}}$. Then $(\boldsymbol{X}_d, \boldsymbol{f})$ is a* ***topological dynamical system.***

Proof. Propositions 2, 3, 4 and 7 jointly imply that $\boldsymbol{X}_d$ is a compact metric space on which $\boldsymbol{f}$ is continuous, thus satisfying the requirements of e.g. [21,29,44]. □

With Theorem 1, we have, in what we consider a natural manner, situated dynamic epistemic logic in the mathematical discipline of dynamical systems. A core topic in this discipline is to understand the long-term, qualitative behavior of maps on spaces. Central to this endeavor is the concept of *recurrence*, i.e., understanding when a system returns to previous states as time goes to infinity.

5 Recurrence in the Limit Behavior of Clean Maps

We represent results concerning the limit behavior of clean maps on modal spaces. In establishing the required terminology, we follow [29]: Let f be a continuous map on a metric space X_d and $x \in X_d$. A point $y \in X$ is a **limit point**[9] for x under f if there is a strictly increasing sequence $n_1, n_2, \ldots$ such that the subsequence $f^{n_1}(x), f^{n_2}(x), \ldots$ of $(f^n(x))_{n \in \mathbb{N}_0}$ converges to y. The **limit set** of x under f is the set of all limit points for x, denoted $\omega_f(x)$. Notably, $\omega_f(x)$ is closed under f: For $y \in \omega_f(x)$ also $f(y) \in \omega_f(x)$. We immediately obtain that any modal system satisfying Theorem 1 has a nonempty limit set:

[9] Or ω-limit point. The ω is everywhere omitted as time here only moves forward.

Proposition 8. *Let $(\boldsymbol{X}_d, \boldsymbol{f})$ be as in Theorem 1. For any point $\boldsymbol{x} \in \boldsymbol{X}$, the limit set of $\boldsymbol{x}$ under $\boldsymbol{f}$ is non-empty.*

Proof. Since $\boldsymbol{X}$ is is compact, every sequence in $\boldsymbol{X}$ has a convergent subsequence, cf. e.g. [37, Theorem 28.2].

Proposition 8 does not inform us of the *structure* of said limit set. In the study of dynamical systems, such structure is often sought through classifying the possible repetitive behavior of a system, i.e., through the system's *recurrence* properties. For such studies, a point x is called (positively) **recurrent** if $x \in \omega_f(x)$, i.e., if it is a limit point of itself.

The simplest structural form of recurrence is *periodicity*: For a point $x \in X$, call the set $\mathcal{O}_f(x) = \{f^n(x) : n \in \mathbb{N}_0\}$ its **orbit**. The orbit $\mathcal{O}_f(x)$ is **periodic** if $f^{n+k}(x) = f^n(x)$ for some $n \geq 0, k > 0$; the least such k is the **period** of $\mathcal{O}_f(x)$. Periodicity is thus equivalent to $\mathcal{O}_f(x)$ being finite. Related is the notion of a **limit cycle**: a periodic orbit $\mathcal{O}_f(x)$ is a limit cycle if it is the limit set of some y not in the period, i.e., if $\mathcal{O}_f(x) = \omega_f(y)$ for some $y \notin \mathcal{O}_f(x)$.

It was conjectured by van Benthem that certain clean maps—those based on finite action models and without postconditions—would, whenever applied to a finite x, have a periodic orbit $\mathcal{O}_f(x)$. I.e., after finite iterations, the map would oscillate between a finite number of states. This was the content of van Benthem's "*Finite Evolution Conjecture*" [8]. The conjecture was refuted using a counterexample by Sadzik in his 2006 paper, [43].[10] The example provided by Sadzik (his Example 33) uses an action model with only Boolean preconditions. Interestingly, the orbit of the corresponding clean map terminates in a limit cycle. This is a corollary to Proposition 9 below.

Before we can state the proposition, we need to introduce some terminology. Call a multi-pointed action model $\Sigma\Gamma$ **finite** if $[\![\Sigma]\!]$ is finite, **Boolean** if $pre(\sigma)$ is a Boolean formula for all $\sigma \in [\![\Sigma]\!]$, and **static** if $post(\sigma) = \top$ for all $\sigma \in [\![\Sigma]\!]$. We apply the same terms to a clean map $\boldsymbol{f}$ based on $\Sigma\Gamma$. In this terminology, Sadzik showed that for any finite, Boolean, and static clean map $\boldsymbol{f} : \boldsymbol{X} \to \boldsymbol{X}$, if the orbit $\mathcal{O}_{\boldsymbol{f}}(\boldsymbol{x})$ is periodic, then it has period 1.[11] This insightful result immediates the following:

Proposition 9. *Let $(\boldsymbol{X}_d, \boldsymbol{f})$ be as in Theorem 1 with $\boldsymbol{f}$ finite, Boolean, and static. For all $\boldsymbol{x} \in \boldsymbol{X}$, the orbit $\mathcal{O}_{\boldsymbol{f}}(\boldsymbol{x})$ is periodic with period* 1.

Proof. By Proposition 8, the limit set $\omega_{\boldsymbol{f}}(\boldsymbol{x})$ of $\boldsymbol{x}$ under $\boldsymbol{f}$ is non-empty. Sadzik's result shows that it contains a single point. Hence $(\boldsymbol{f}^n(\boldsymbol{x}))_{n \in \mathbb{N}_0}$ converges to this point. As the limit set $\omega_{\boldsymbol{f}}(\boldsymbol{x})$ is closed under $\boldsymbol{f}$, its unique point is a fix-point. □

Proposition 9 may be seen as a partial vindication of van Benthem's conjecture: Forgoing the requirement of reaching the limit set in finite time and the possibility of modal preconditions, the conjecture holds, even if the initial state has an infinite set of worlds $[\![x]\!]$. This simple recurrent behavior is, however, not

[10] We paraphrase van Benthem and Sadzik using the terminology introduced.

[11] See [16] for an elegant and generalizing exposition.

the general case. More complex clean maps may exhibit **nontrivial recurrence**, i.e., produce non-periodic orbits with recurrent points:

Proposition 10. *There exist finite, static, but non-Boolean, clean maps that exhibit nontrivial recurrence.*

Proposition 11. *There exist finite, Boolean, but non-static, clean maps that exhibit nontrivial recurrence.*

We show these propositions below, building a clean map which, from a selected initial state, has uncountably many limit points, despite the orbit being only countable. Moreover, said orbit also contains infinitely many recurrent points. In fact, every element of the orbit is recurrent. We rely on Lemma 1 in the proof. A proof of Lemma 1.1 may be found in [32], a proof of Lemma 1.2 in [15].

Lemma 1. *Any Turing machine can be emulated using a set X of $S5$ pointed Kripke models for finite atoms and a finite multi-pointed action model $\Sigma\Gamma$ deterministic over X. Moreover, $\Sigma\Gamma$ may be chosen 1. static, but non-Boolean, or 2. Boolean, but non-static.*

Proof (of Propositions 10 and 11). For both propositions, we use a Turing machine *ad infinitum* iterating the successor function on the natural numbers. Numbers are represented in mirrored base-2, i.e., with the *leftmost* digit the *lowest*. Such a machine may be build with alphabet $\{\triangleright, 0, 1, \sqcup\}$, where the symbol $\triangleright$ is used to mark the starting cell and $\sqcup$ is the *blank* symbol. We omit the exact description of the machine here. Of importance is the content of the tape: Omitting blank ($\sqcup$) cells, natural numbers are represented as illustrated in Fig. 1.

0 :	▷ 0	2 :	▷ 0 1	4 :	▷ 0 0 1	6 :	▷ 0 1 1	8 :	▷ 0 0 0 1
1 :	▷ 1	3 :	▷ 1 1	5 :	▷ 1 0 1	7 :	▷ 1 1 1	9 :	▷ 1 0 0 1

Fig. 1. Mirrored base-2 Turing tape representation of $0, .., 9 \in \mathbb{N}_0$, blank cells omitted. Notice that the mirrored notation causes perpetual change close to the start cell, $\triangleright$.

Initiated with its read-write head on the cell with the start symbol $\triangleright$ of a tape with content n, the machine will go through a number of configurations before returning the read-write head to the start cell with the tape now having content $n+1$. Auto-iterating, the machine will thus, over time, produce a tape that will have contained every natural number in order.

This Turing machine may be emulated by a finite $\Sigma\Gamma$ on a set X cf. Lemma 1. Omitting the details[12], the idea is that the Turing tape, or a finite fragment,

[12] The details differ depending on whether $\Sigma\Gamma$ must be static, but non-Boolean for Proposition 10, or Boolean, but non-static for Proposition 11. See resp. [15,32].

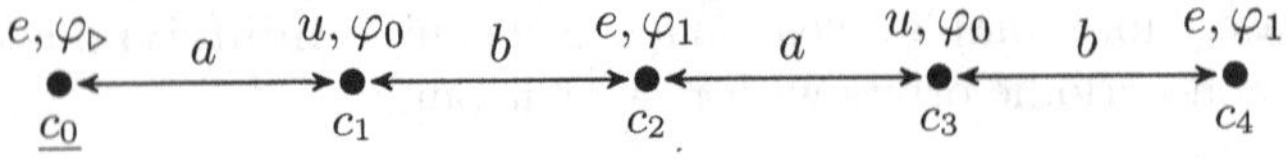

Fig. 2. A pointed Kripke model emulating the configuration of the Turing machine with cell content representing the number 10. The designated state is the underlined c_0. Each state is labeled with a formula $\varphi_\triangleright$, φ_0 or φ_1 expressing its content. Relations a and b allow expressing distance of cells: That c_0 satisfies $\Diamond_a(u \wedge \Diamond_b(e \wedge \varphi_1))$ exactly expresses that cell c_2 contains a 1. Omitted are reflexive loops for relations, and the additional structure marking cell content and read-write head position.

thereof may be encoded as a pointed Kripke model: Each cell of the tape corresponds to a state, with the cell's content encoded by additional structure,[13] which is modally expressible. By structuring the cell states with two equivalence relations and atoms u and e true at cells with odd (even) index respectively, (cf. Fig. 2), also the position of a cell is expressible. The designated state corresponds to the start cell, marked $\triangleright$.

Let $(c_n)_{n\in\mathbb{N}_0}$ be the sequence of configurations of the machine when initiated on a tape with content 0. Each c_n may be represented by a pointed Kripke model, obtaining a sequence $(x_n)_{n\in\mathbb{N}_0}$. By Lemma 1, there thus exists a $\Sigma\Gamma$ such that for all n, $x_n \otimes \Sigma\Gamma = x_{n+1}$. Hence, moving to the full modal space $\boldsymbol{X}$ for the language used, a clean map $\boldsymbol{f}\colon \boldsymbol{X} \to \boldsymbol{X}$ based on $\Sigma\Gamma$ will satisfy $\boldsymbol{f}(\boldsymbol{x}_n) = \boldsymbol{x}_{n+1}$ for all n. The Turing machine's run is thus emulated by $(\boldsymbol{f}^k(\boldsymbol{x}_0))_{k\in\mathbb{N}_0}$.

Let $(c'_n)_{n\in\mathbb{N}_0}$ be the subsequence of $(c_n)_{n\in\mathbb{N}_0}$ where the machine has finished the successor operation and returned its read write head to its starting position $\triangleright$, ready to embark on the next successor step. The tape of the first 9 of these c'_n are depicted in Fig. 1. Let $(\boldsymbol{x}'_n)_{n\in\mathbb{N}_0}$ be the corresponding subsequence of $(\boldsymbol{f}^k(\boldsymbol{x}_0))_{k\in\mathbb{N}_0}$. We show that $(\boldsymbol{x}'_n)_{n\in\mathbb{N}_0}$ has uncountably many limit points:

For each subset Z of $\mathbb{N}$, let c^Z be a tape with content 1 on cell i iff $i \in Z$ and 0 else. On the Kripke model side, let the corresponding $x^Z \in X$ be a model structurally identical to those of $(x'_n)_{n\in\mathbb{N}_0}$, but satisfying φ_1 on all "cell states" distance $i \in Z$ from the designated "$\triangleright$" state, and φ_0 on all other.[14] The set $\{x^Z : Z \subseteq \mathbb{N}\}$ is uncountable, and each $\boldsymbol{x}^Z$ is a limit point of $\overline{\boldsymbol{x}}$: For each $Z \subseteq \mathbb{N}$ and $n \in \mathbb{N}$, there are infinitely many k for which $x_k \vDash \varphi$ iff $x^Z \vDash \varphi$ for all φ of modal depth at most n. Hence, for every n, the set $\{\boldsymbol{x}_k : d_b(\boldsymbol{x}_k, \boldsymbol{x}^Z) < 2^{-n}\}$ is infinite, with d_b the equivalent of the n-bisimulation metric, cf. Proposition 5. Hence, for each of the uncountably many $Z \subseteq \mathbb{N}$, $\boldsymbol{x}^Z$ is a limit point of the sequence $\overline{\boldsymbol{x}}$.

Finally, every $\boldsymbol{x}'_k \in (\boldsymbol{x}'_n)_{n\in\mathbb{N}_0}$ is recurrent: That $\boldsymbol{x}'_k \in \omega_{\boldsymbol{f}}(\boldsymbol{x}'_k)$ follows from $\boldsymbol{x}'_k$ being a limit point of $(\boldsymbol{x}'_n)_{n\in\mathbb{N}_0}$, which it is as $\boldsymbol{x}'_k = \boldsymbol{x}^Z$ for some $Z \subseteq \mathbb{N}$.[15] As the set of recurrent points is thus infinite, it cannot be periodic. □

[13] For Proposition 11, tape cell content may be encoded using atomic propositions, changeable through postconditions, cf. [15]; for Proposition 10, cell content is written by adding and removing additional states, cf. [32].

[14] The exact form is straightforward from the constructions used in [15,32].

[15] A similar argument shows that all x^Z with $Z \subseteq \mathbb{N}$ co-infinite are recurrent points. Hence $\omega_{\boldsymbol{f}}(\boldsymbol{x}'_k)$ for any $\boldsymbol{x}'_k \in (\boldsymbol{x}'_n)_{n\in\mathbb{N}_0}$ contains uncountably many recurrent points.

As a final result on the orbits of clean maps, we answer an open question: After having exemplified a period 2 system, Sadzik [43] notes that it is unknown whether finite, static, but non-Boolean, clean maps exhibiting longer periods exist. They do:

Proposition 12. *For any $n \in \mathbb{N}$, there exists finite, static, but non-Boolean clean maps with periodic orbits of period n. This is also true for finite Boolean, but non-static, clean maps.*

Proof. For the given n, find a Turing machine that, from some configuration, loops with period n. From here, Lemma 1 does the job. □

Finally, we note that brute force determination of a clean map's orbit properties is not in general a feasible option:

Proposition 13. *The problems of determining whether a Boolean and non-static, or a static and non-Boolean, clean map, a) has a periodic orbit or not, and b) contains a limit cycle or not, are both undecidable.*

Proof. The constructions from the proofs of Lemma 1 allows encoding the halting problem into either question. □

6 Discussion and Future Venues

We consider Theorem 1 our main contribution. With it, an interface between the discrete semantics of dynamic epistemic logic with dynamical systems have been provided; thus the former has been situated in the mathematical field of the latter. This paves the way for the application of results from dynamical systems theory and related fields to the information dynamics of dynamic epistemic logic.

The term *nontrivial recurrence* is adopted from Hasselblatt and Katok, [29]. They remark that "[nontrivial recurrence] is the first indication of complicated asymptotic behavior." Propositions 10 and 11 indicate that the dynamics of action models and product update may not be an easy landscape to map. Hasselblatt and Katok continue: "In certain low-dimensional situations [...] it is possible to give a comprehensive description of the nontrivial recurrence that can appear." [29, p. 24]. That the Stone topology is zero-dimensional fuels the hope that general topology and dynamical systems theory yet has perspectives to offer on dynamic epistemic logic. One possible direction is seeking a finer parametrization of clean maps combined with results specific to zero-dimensional spaces, as found, e.g., in the field of symbolic dynamics [36]. But also other venues are possible: The introduction of [29] is an inspiration.

The approach presented furthermore applies to model transformations beyond multi-pointed action models and product update. Given the equivalence shown in [33] between single-pointed action model product update and *general arrow updates*, we see no reason to suspect that "clean maps" based on the latter should not be continuous on modal spaces. A further conjecture is that the *action-priority update* of [5] on plausibility models[16] yields "clean maps"

[16] Hence also the multi-agent belief revision policies *lexicographic upgrade* and *elite change*, also known as *radical* and *conservative upgrade*, introduced in [9], cf. [5].

continuous w.r.t. the suited Stone topology, and that this may be shown using a variant of our proof of the continuity of clean maps. A more difficult case is the *PDL-transformations* of *General Dynamic Dynamic Logic* [25] given the signature change the operation involves.

There is a possible clinch between the suggested approach and epistemic logic with common knowledge. The state space of a dynamical system is compact. The Stone topology for languages including a common knowledge operator is non-compact. Hence, it cannot constitute the space of a dynamical system—but its *one-point compactification* may. We are currently working on this clinch, the consequences of compactification, and relations to the problem of attaining common knowledge, cf. [28].

Questions also arise concerning the *dynamic logic* of dynamic epistemic logic. Propositions 10 and 11 indicate that there is more to the semantic dynamics of dynamic epistemic logic than is representable by finite compositional dynamic modalities—even when including a Kleene star. An open question still stands on how to reason about limit behavior. One interesting venue stems from van Benthem [10]. He notes[17] that the reduction axioms of dynamic epistemic logic could possibly be viewed on par with differential equations of quantitative dynamical systems. As modal spaces are zero-dimensional, they are imbeddable in $\mathbb{R}$ cf. [37, Theorem 50.5], turning clean maps into functions from $\mathbb{R}$ to $\mathbb{R}$, possibly representable as discrete-time difference equations.

An alternative approach is possible given by consulting Theorem 1. With Theorem 1, a connection arises between dynamic epistemic logic and *dynamic topological logic* (see e.g. [23,24,34,35]): Each system $(\boldsymbol{X}_d, \boldsymbol{f})$ may be considered a dynamic topological model with atom set $\boldsymbol{\mathcal{L}}_\Lambda$ and the 'next' operator's semantics given by an application of $\boldsymbol{f}$, equivalent to a $\langle \boldsymbol{f} \rangle$ dynamic modality of DEL. The topological 'interior' operator has yet no DEL parallel. A 'henceforth' operator allows for a limited characterization of recurrence [35]. We are wondering about and wandering around the connections between a limit set operator with semantics $x \vDash [\omega_{\boldsymbol{f}}]\varphi$ iff $y \vDash \varphi$ for all $\boldsymbol{y} \in \omega_{\boldsymbol{f}}(\boldsymbol{x})$, dynamic topological logic and the study of oscillations suggested by van Benthem [11].

With the focal point on pointed Kripke models and action model transformations, we have only considered a special case of logical dynamics. It is our firm belief that much of the methodology here suggested is generalizable: With structures described logically using a countable language, the notion of logical convergence will coincide with topological convergence in the Stone topology on the quotient space *modulo* logical equivalence, and the metrics introduced will, *mutatis mutandis*, be applicable to said space [31]. The continuity of maps and compactness of course depends on what the specifics of the chosen model transformations and the compactness of the logic amount to.

Acknowledgements. The contribution of R.K. Rendsvig was funded by the Swedish Research Council through the Knowledge in a Digital World project and by The Center for Information and Bubble Studies, sponsored by The Carlsberg Foundation.

[17] In the omitted part of the quotation from the introduction.

The contribution of D. Klein was partially supported by the Deutsche Forschungsgemeinschaft (DFG) and Grantová agentura České republiky (GAČR) as part of the joint project From Shared Evidence to Group Attitudes [RO 4548/6-1].

References

1. Aucher, G.: Generalizing AGM to a multi-agent setting. Logic J. IGPL **18**(4), 530–558 (2010)
2. Baltag, A., Moss, L.S.: Logics for epistemic programs. Synthese **139**(2), 165–224 (2004)
3. Baltag, A., Moss, L.S., Solecki, S.: The logic of public announcements, common knowledge, and private suspicions. In: TARK 1998. Morgan Kaufmann (1998)
4. Baltag, A., Renne, B.: Dynamic epistemic logic. In: The Stanford Encyclopedia of Philosophy (2016). Fall 2016th Edition
5. Baltag, A., Smets, S.: A qualitative theory of dynamic interactive belief revision. In: Proceedings of LOFT 7. Amsterdam University Press (2008)
6. Baltag, A., Smets, S.: Group belief dynamics under iterated revision: fixed points and cycles of joint upgrades. In: TARK 2009. ACM (2009)
7. van Benthem, J.: Games in dynamic-epistemic logic. Bull. Econ. Res. **53**(1), 219–249 (2001)
8. van Benthem, J.: "One is a Lonely Number": logic and communication. In: Logic Colloquium 2002. Lecture Notes in Logic, vol. 27. Association for Symbolic Logic (2006)
9. van Benthem, J.: Dynamic logic for belief revision. J. Appl. Non-Class. Logics **17**(2), 129–155 (2007)
10. van Benthem, J.: Logical Dynamics of Information and Interaction. Cambridge University Press, Cambridge (2011)
11. van Benthem, J.: Oscillations, logic, and dynamical systems. In: Ghosh, S., Szymanik, J. (eds.) The Facts Matter. College Publications, London (2016)
12. van Benthem, J., van Eijck, J., Kooi, B.: Logics of communication and change. Inf. Comput. **204**(11), 1620–1662 (2006)
13. van Benthem, J., Gerbrandy, J., Hoshi, T., Pacuit, E.: Merging frameworks for interaction. J. Philos. Logic **38**(5), 491–526 (2009)
14. Blackburn, P., de Rijke, M., Venema, Y.: Modal Logic. Cambridge University Press, Cambridge (2001)
15. Bolander, T., Birkegaard, M.: Epistemic planning for single- and multi-agent systems. J. Appl. Non-Class. Logics **21**(1), 9–34 (2011)
16. Bolander, T., Jensen, M., Schwarzentruber, F.: Complexity results in epistemic planning. In: Proceedings of IJCAI 2015. AAAI Press (2015)
17. Caridroit, T., Konieczny, S., de Lima, T., Marquis, P.: On distances between KD45n Kripke models and their use for belief revision. In: ECAI 2016. IOS Press (2016)
18. Dégremont, C.: The temporal mind: observations on the logic of belief change in interactive systems. Ph.D. thesis, University of Amsterdam (2010)
19. van Ditmarsch, H., Kooi, B.: Semantic results for ontic and epistemic change. In: Logic and the Foundations of Game and Decision Theory (LOFT 7). Texts in Logic and Games, vol. 3. Amsterdam University Press (2008)
20. van Ditmarsch, H., van der Hoek, W., Kooi, B.: Dynamic Epistemic Logic. Springer, Dordrecht (2008). doi:10.1007/978-1-4020-5839-4

21. Eisner, T., Farkas, B., Haase, M., Nagel, R.: Operator Theoretic Aspects of Ergodic Theory. Springer, Heidelberg (2015). doi:10.1007/978-3-319-16898-2
22. Fagin, R., Halpern, J.Y., Moses, Y., Vardi, M.Y.: Reasoning About Knowledge. The MIT Press, Cambridge (1995)
23. Fernández-Duque, D.: A sound and complete axiomatization for dynamic topological logic. J. Symb. Logic **77**(3), 1–26 (2012)
24. Fernández-Duque, D.: Dynamic topological logic of metric spaces. J. Symb. Logic **77**(1), 308–328 (2012)
25. Girard, P., Seligman, J., Liu, F.: General dynamic dynamic logic. In: Bolander, T., Brauner, T., Ghilardi, S., Moss, L. (eds.) Advances in modal logics, vol. 9. College Publications, London (2012)
26. Goranko, V.: Logical topologies and semantic completeness. In: Logic Colloquium 1999. Lecture Notes in Logic, vol. 17. AK Peters (2004)
27. Goranko, V., Otto, M.: Model theory of modal logic. In: Blackburn, P., van Benthem, J., Wolter, F. (eds.) Handbook of Modal Logic. Elsevier, Amsterdam (2007)
28. Halpern, Y.J., Moses, Y.: Knowledge and common knowledge in a distributed environment. J. ACM **37**(3), 549–587 (1990)
29. Hasselblatt, B., Katok, A.: Principal structures. In: Hasselblatt, B., Katok, A. (eds.) Handbook of Dynamical Systems, vol. 1A. Elsevier, Amsterdam (2002)
30. Hintikka, J.: Knowledge and Belief: An Introduction to the Logic of the Two Notions. College Publications, London (1962). 2nd, 2005 Edition
31. Klein, D., Rendsvig, R.K.: Metrics for Formal Structures, with an Application to Kripke Models and their Dynamics. arXiv:1704.00977 (2017)
32. Klein, D., Rendsvig, R.K.: Turing Completeness of Finite, Epistemic Programs. arXiv:1706.06845 (2017)
33. Kooi, B., Renne, B.: Generalized arrow update logic. In: TARK 2011. ACM, New York (2011)
34. Kremer, P., Mints, G.: Dynamical topological logic. Bull. Symb. Logic **3**, 371–372 (1997)
35. Kremer, P., Mints, G.: Dynamic Topological Logic. In: Aiello, M., Pratt-Hartmann, I., Van Benthem, J. (eds.) Handbook of Spatial Logics. Springer, Dordrecht (2007). doi:10.1007/978-1-4020-5587-4_10
36. Lind, D., Marcus, B.: An Introduction to Symbolic Dynamics and Coding. Cambridge University Press, Cambridge (1995)
37. Munkres, J.R.: Topology, 2nd edn. Prentice-Hall, Englewood Cliffs (2000)
38. Plaza, J.A.: Logics of public communications. In: Proceedings of the 4th International Symposium on Methodologies for Intelligent Systems (1989)
39. Rendsvig, R.K.: Towards a theory of semantic competence. Master's thesis, Department of Philosophy and Science Studies and Department of Mathematics, Roskilde University (2011)
40. Rendsvig, R.K.: Diffusion, influence and best-response dynamics in networks: an action model approach. In: Proceedings of ESSLLI 2014 Student Session arXiv:1708.01477 (2014)
41. Rendsvig, R.K.: Pluralistic ignorance in the bystander effect: Informational dynamics of unresponsive witnesses in situations calling for intervention. Synthese **191**(11), 2471–2498 (2014)
42. Rendsvig, R.K.: Model transformers for dynamical systems of dynamic epistemic logic. In: Hoek, W., Holliday, W.H., Wang, W. (eds.) LORI 2015. LNCS, vol. 9394, pp. 316–327. Springer, Heidelberg (2015). doi:10.1007/978-3-662-48561-3_26
43. Sadzik, T.: Exploring the iterated update universe. ILLC PP-2006-26 (2006)
44. de Vries, J.: Topological Dynamical Systems. de Gruyter, Berlin (2014)

Dynamic Logic of Power and Immunity

Huimin Dong(✉) and Olivier Roy(✉)

Universität Bayreuth, Bayreuth, Germany
{huimin.dong,olivier.roy}@uni-bayreuth.de

Abstract. We present a dynamic logic for modelling legal competences, and in particular for the Hohfeldian categories of power and immunity. We argue that this logic improves on existing models by explicitly capturing the norm-changing character of legal competences, while at the same time providing a sophisticated reduction of the latter to static normative positions. The logic is shown to be completely axiomatizable; an analysis of its resulting dynamic normative positions is provided; and it is finally applied to a concrete case in German contract law to illustrate how the logic can distinguish legal ability and legal permissibility.

The Hohfeldian [7] typology of rights distinguishes what one might call *static* and *dynamic* rights. Static basic rights encompass claims and privileges, as well as their respective correlatives of duties and no-claim. On the dynamic side one finds power and immunity together with the correlatives of liability and no-power. See Table 1 for the classical presentation.

What we call here static and dynamic rights have been labelled in various ways in the formal literature. Kanger called static rights the "type of the states of affairs" and dynamic ones the "type of influence" [9]. Makinson instead uses the "deontic family" and the "legally capacitative family" for static and dynamic rights, respectively [12]. Bentham, von Wright and Hart on the other hand use "legal validity" and "norm-creating action" [11], while Lindahl [11] followed this action viewpoint, and call it "the range of action."

Although logical approaches to legal competences are scarcer than for static normative positions, existing theories can be divided into two broad families. The first formalizes power and immunity as (legal) permissibility, or absence thereof, to see to it that a certain normative position obtains [10,11]. Lindahl [11], for instance, captures j's power to make it the case that i ought to see to it that φ using a combination of action and embedded deontic modalities:

$$P\ Do_j O(Do_i \varphi)$$

We call such an approach reductive because it takes power and immunity as definable in the language of obligations, permissions, and action, where claims and privileges are also defined. Non-reductive approaches, on the other hand, view power and immunity as position-changing actions that are not reducible to static normative positions [8,12]. A typical example of this is Jones and Sergot [8], who capture legal power through "counts as" conditionals.

A. Baltag et al. (Eds.): LORI 2017, LNCS 10455, pp. 123–136, 2017.
DOI: 10.1007/978-3-662-55665-8_9

Table 1. Legal Rights [14]

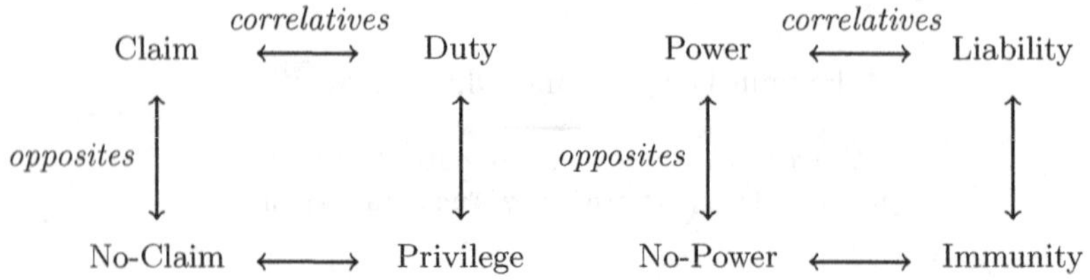

Each of these families have assets and drawbacks. Reductive approaches come with a rich logical theory of the relationship between static normative conditions and legal competences, with the latter inheriting its logic from the former. Defining power as above, however, obfuscates the dynamic character of legal competence by reducing it to permissibility, a simple static legal relation. This dynamic aspect was arguably crucial for Hohfeld who defined power as the ability to "*change* legal relations" [7, p. 44–45]. The formalization above furthermore conflates legal ability (*rechtliches Können*) with legal permissibility (*rechtliches Dürfen*), although these two concepts are distinct [8,12]. Non-reductive approaches, on the other hand, do better justice to the dynamic character of legal competences by taking norm-changing actions as first class citizens in the logic. This allows, by the same token, to distinguish legal ability and legal permissibility. The cost of this is a relatively weak logic of legal competences, which is at least at the outset completely independent from the logic of the static normative position.

The dynamic logic that we present in this paper provides a plausible middle ground between these two types of approaches. It is reductive, and as such comes with a rich set of principles of interaction between static and dynamic rights. It does so, however, while retaining both the dynamic character of legal competences and the distinction between legal ability and legal permissibility.

The reader familiar with dynamic *epistemic* logic [15] will recognize both the modelling methodology and many of the canonical results (axiomatization, bisimulation invariance) that we present here. What we propose is a deontic reinterpretation of this framework. We show that this yields interesting insights for the theory of legal competences, and can be applied to the concrete question of the distinction between legal ability and legal permissibility.

The rest of the paper is structured as follows. Our main contribution being the dynamic part of the model, for the static part we follow a fairly standard approach for conditional obligations [19]. We present it briefly in Sect. 1, and then move to dynamic modalities and legal competence in Sect. 2. We show how the two put together capture the four Hohfeldian basic types of right, present a complete axiomatization, and study its model theory. We then turn to the combinatorics of dynamic normative positions in Sect. 3. Finally, we apply it to a concrete case in the German civil code in order to show that legal ability and legal permissibility can be naturally distinguished in this logic.

1 Static Rights

Our starting point is a conditional version of the Kangerian model of claims and privileges [9,11,12]. The latter is the standard in current theories of normative positions [14], and hence comes with well-studied models of claims and privilege. The conditional version we propose follows the one developed in [19], but goes back at least to [6]. Little, however, rests on this modelling choice in the sense that the dynamic methodology that we present later is fairly modular. In other contexts it has been successfully used to extend very different static logical systems [16,17]. The same could be done here.

On the surface the language we use differs from classical Kangerian approaches in that it contains a single Kripke modality on an underlying preference relation, along with the usual "seeing to it that" modality. We use this language instead of the classical deontic one for technical reasons. It facilitates the axiomatization of the dynamic modalities. It is well known, however, this language can define the conditional obligations and permissions [3,20]. We come back to this at the end of the section.

Definition 1. *Let Prop be a countable set of propositions and $\mathcal{I}$ a set of agents. The language $\mathcal{L}$ is defined as follows:*

$$\varphi := p \in Prop \mid \neg\varphi \mid \varphi \wedge \varphi \mid [\leq]\varphi \mid A\varphi \mid Do_i\varphi$$

where $i \in \mathcal{I}$.

We write $\langle\leq\rangle\varphi$ for $\neg[\leq]\neg\varphi$, and $E\varphi$ for $\neg A\neg\varphi$. A formula $A\varphi$ is read as "it is necessary that φ." $Do_i\varphi$ indicates a *non-deontic* or *ontic* [21] action of agent i, and should be read in the usual sense of "i sees to it that φ".

The semantics of this language is provided in standard preferential models augmented with a Kripke relation for the Do_i operators. We do not assume that the preference ordering is connected. This will give rise to slight differences from, e.g. [18,19].

Definition 2. *Let Prop and $\mathcal{I}$ be as above. A* **preference-action model** *$\mathcal{M}$ is a tuple $\langle W, \leq, \{\sim_i\}_{i\in\mathcal{I}}, V\rangle$ where:*

- *W is a non-empty set of states.*
- *$\leq$ is a converse well-founded, reflexive and transitive relation on W.*
- *for each $i \in \mathcal{I}$, $\sim_i$ is an equivalence relation.*
- *$V : Prop \to \mathcal{P}(W)$ is a valuation function.*

Preference-action frames are models minus the valuation. The assumption that the relations $\sim_i$ are equivalence relations is present only to simplify the treatment of static rights and thus put the emphasis on our dynamic extension. As above, this assumption could be lifted. We then can interpret the sentences in language $\mathcal{L}$ according to a preference-action model as follows.

Definition 3. *The truth conditions for sentence $\varphi \in \mathcal{L}$ are defined in the following:*

- $\mathcal{M}, w \models p$ *iff* $w \in V(p)$.
- $\mathcal{M}, w \models \neg\varphi$ *iff* $\mathcal{M}, w \not\models \varphi$.
- $\mathcal{M}, w \models \varphi \wedge \psi$ *iff* $\mathcal{M}, w \models \varphi$ *and* $\mathcal{M}, w \models \psi$.
- $\mathcal{M}, w \models [\leq]\varphi$ *iff* $\mathcal{M}, w' \models \varphi$ *for all* $w' \geq w$.[1]
- $\mathcal{M}, w \models A\varphi$ *iff* $\mathcal{M}, w' \models \varphi$ *for all* $w' \in W$.
- $\mathcal{M}, w \models Do_i\varphi$ *iff* $\mathcal{M}, w' \models \varphi$ *for all* $w' \sim_i w$.

Validity on models and frames, and classes thereof, is defined as usual. We define $||\varphi||$, the truth set of φ, as $\{w : \mathcal{M}, w \models \varphi\}$.

As mentioned, conditional obligation, understood in terms of "truth in all the most preferred worlds," is definable in this language. The argument is standard. Let $O(\psi/\varphi)$ be defined as follows:

- $\mathcal{M}, w \models O(\psi/\varphi)$ iff $\mathcal{M}, w' \models \psi$ for all $w' \in max_{\geq}(||\varphi||)$.

with $max_{\geq}(X) = \{w \in X : \neg\exists w' \in X \ s.t. \ w' > w\}$. Then conditional obligation is definable as:

$$O(\psi/\varphi) \leftrightarrow A(\varphi \rightarrow \langle\leq\rangle(\varphi \wedge [\leq](\varphi \rightarrow \psi))),$$

Unconditional obligations $O\varphi$ are defined as $O(\varphi/\top)$, and permission as "weak permissions", i.e. $P(\varphi/\psi)$ iff $\neg O(\neg\varphi/\psi)$. With this in hand we have the machinery required to define claims and privileges, which we do again using the standard Kangerian approach:

- Given ψ, agent i has a claim against j regarding φ: $O(Do_j\varphi/\psi)$
- Given ψ, agent i has a privilege against j regarding φ: $\neg O(Do_i\neg\varphi/\psi)$ iff $P(\neg Do_i\neg\varphi/\psi)$

We close this section with a short example, to which we will return later.

Example 1. Ivy has parked her car but she forgot to put the mandatory parking permit in her windshield. Parking (p), with or without a parking permit, is a non-deontic action. It can of course have deontic consequences, but only if a city clerk with the *power* to *issue* parking tickets passes by. Absent this deontic action of issuing a ticket, the city has no claim against Ivy regarding the payment of a fine ($\neg O(Do_{Ivy}f/p)$). Possessing a permit being mandatory for parking, Ivy is forbidden to park. In other words, the city has a claim against her not to park her car where she did ($O(Do_{Ivy}\neg p)$). This is illustrated in Fig. 1.

[1] $\geq$ is the inverse relation of $\leq$. More precisely, $w' \geq w$ iff $w \leq w'$. There is a similar case for the preference order of actions later.

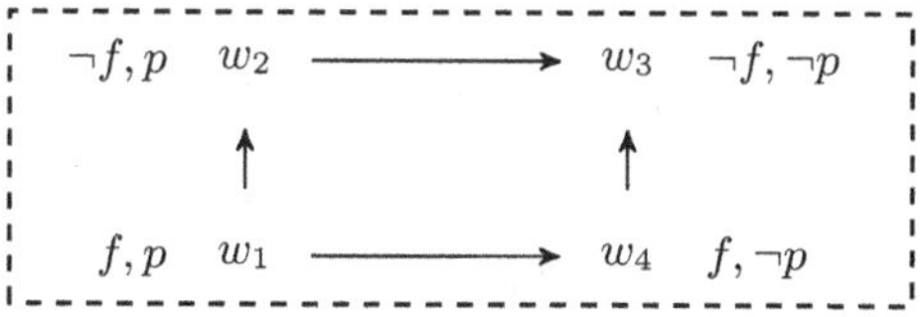

Fig. 1. A static model of Ivy's example. The arrows → represent the preference order $\leq$ between states. Reflexive loops are omitted.

2 Dynamic Rights

2.1 Core Model

Our modeling of legal competence follows the so-called "event models" methodology developed in [1] for epistemic modalities. See [15] for details. Transposed into our deontic context, the proposal is to model explicitly the structure of deontic action or legal competences using what we call *deontic action models.* These are agent-indexed to capture the fact that different agents will have different legal competences.

Definition 4. *A* **deontic action model for agent i** $\mathscr{A}_i$ *is a tuple* $\langle A, \geq^{\mathscr{A}_i}, Pre\rangle$ *where:*

- *A is a non-empty, finite set of acts.*
- $\geq^{\mathscr{A}_i}$ *is a converse well-founded reflexive and transitive relation on A.*
- $Pre : A \to \mathcal{L}$ *is a precondition function.*

We write $a \cong^{\mathscr{A}_i} a'$ *whenever* $a \geq^{\mathscr{A}_i} a'$ *and* $a' \geq^{\mathscr{A}_i} a$.

Each act $a \in A$ should be seen as a deontic action or a legal ability. It encodes an action that agent i can take in order to bring about changes in obligations and permissions or, in more Hohfeldian terminology, changes in underlying legal relations. These acts also come in different levels of ideality, which is encoded by the preference order $\geq^{\mathscr{A}_i}$. Finally, the preconditions function Pre specifies for each act a the conditions in the underlying static models that need to obtain for a to be executable in the first place.

Example 2. John is city clerk. He can confirm a violation of the parking regulations (v), or not (n). He confirms a violation if a fine applies (f), otherwise ($\neg f$) not. Given that Ivy's car has no permit in the windshield, the preferred situation is one where the fine indeed applies. This is illustrated in Fig. 2.

The effect of executing a deontic action in a particular situation is computed by the so-called lexicographic update.

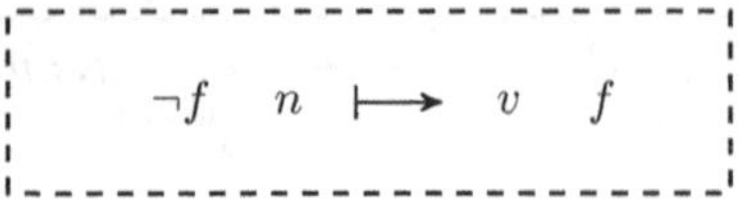

Fig. 2. The deontic action model $\mathscr{A}_{John}$ for John the city clerk. The arrow $\mapsto$ represents the preference order $\leq^{\mathbf{A}_j}$, with reflexive loops omitted. The precondition of n and v are written down to the left and the right, respectively.

Definition 5. *Let $\mathcal{M}$ be a preference-action model and $\mathscr{A}_i$ be a deontic action model. The preference-action model $\mathcal{M} \otimes \mathscr{A}_i = \langle W', \geq', \{\sim'_i\}_{i \in \mathcal{I}}, V' \rangle$ is defined as follows:*

- $W' = \{(w, a) \mid \mathcal{M}, w \models Pre(a),$ *where* $a \in A\}$.
- $(w, a) \geq' (w', a')$ *iff either* $a >^{\mathscr{A}_i} a'$ *or* $a \cong^{\mathscr{A}_i} a'$ *and* $w \geq w'$.
- $(w, a) \sim' (w', a')$ *iff* $w \sim_i w'$.
- $(w, a) \in V'(p) \Leftrightarrow w \in V(p)$.

The lexicographic update takes pairs of preference-action models and event models and returns an updated model $\mathcal{M} \otimes \mathscr{A}_i$. The adjective "lexicographic" comes from the update rule for the pre-orders $\geq$, which gives priority to the deontic action. The domain of that new model is the set of pairs (w, a) such that $\mathcal{M}, w$ satisfies the pre-condition of a, written $\mathcal{M}, w \models Pre(a)$.

Lexicographic updates capture what we call *pure* deontic actions. These are actions that *only* change legal relations. This is encoded in the condition defining the valuation V' in the updated model: $(w, a) \in V'(p) \Leftrightarrow w \in V(p)$. One can take pure deontic action to be acts that are explicitly defined by the legislator, for instance entering into a contract or getting married. Of course non-deontic action might change the legal relation too. By breaking your neighbour's window you create a claim for her against you to cover the repair costs. Such mixed deontic and non-deontic action are the object of [8]. A full comparison between their and our models of deontic actions and legal competences is left for future work.

Example 3. John notices that Ivy's car doesn't have a permit. He issues a parking ticket, which results in the city having a claim against Ivy regarding the payment of a fine. This is represented by updating the model in Fig. 1 with the one in

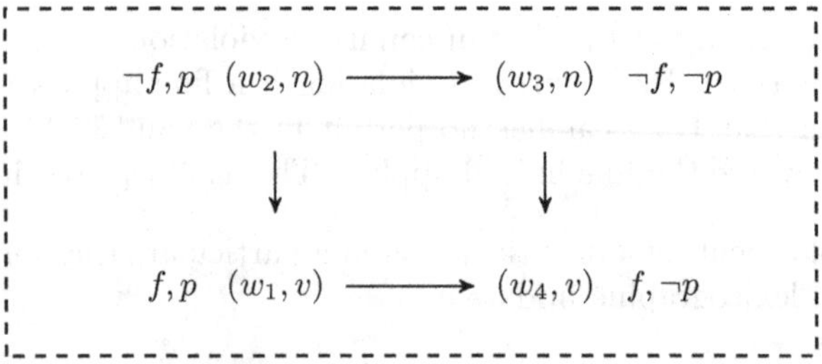

Fig. 3. The model $\mathcal{M} \otimes \mathscr{A}_{John}$ resulting from John's execution of a deontic action to issue a parking ticket.

Fig. 2. The result is in Fig. 3. After the ticket has been issued, Ivy still ought not to park there, but she now ought to pay a fine.

Of course, executing different deontic actions will have different effects on the same initial legal relations. This notion of "different deontic action" can be made precise using the standard notions of bisimulation [2] and action emulation [22], but we leave this out for reasons of space, and refer the reader to [5] for details.

To express the effect of deontic action the language $\mathcal{L}$ is extended with a dynamic, unary operator $[\mathscr{A}_i, a]$, with the following semantics:

- $\mathcal{M}, w \models [\mathscr{A}_i, a]\varphi$ iff if $\mathcal{M}, w \models Pre(a)$ then $\mathcal{M} \otimes \mathscr{A}_i, (w, a) \models \varphi$.

A formula $[\mathscr{A}_i, a]\varphi$ thus reads "if i's deontic action a is executable, then doing so results in φ". Dynamic modalities allow us to introduce our key notions, powers and immunity. Let $T(i, j, \psi/\varphi)$ denote an arbitrary (conditional) normative position definable in the static language $\mathcal{L}$. Then:

- i has a *power* against j regarding $T(i, j, \psi/\varphi)$:

$$\bigvee_{a \in \mathscr{A}_i} [\mathscr{A}_i, a]T(i, j, \psi/\varphi)$$

- i has an *immunity* against j regarding $T(i, j, \psi/\varphi)$:

$$\neg \bigvee_{a \in \mathscr{A}_j} [\mathscr{A}_j, a]T(i, j, \psi/\varphi)$$

In words, i has a power against j regarding the normative position $T(i, j, \psi/\varphi)$ whenever there is a deontic action that i can be executed which results in $T(i, j, \psi/\varphi)$. Similarly, i has an immunity against j regarding $T(i, j, \psi/\varphi)$ if j doesn't have a power against i regarding that position. A quick check of the example above reveals that, as expected, John has a power against Ivy regarding her paying a fine.

This formalization of dynamic rights has two assets in comparison with classical, reductive approaches. First, it explicitly captures, both semantically and syntactically, the dynamic character of power and immunity. Second, as we will see below, this clear static-dynamic distinction allows for a natural distinction between legal ability and legal permissibility. This analysis of power and immunity does so, however, while staying reductive. This is what we show now.

2.2 Axiomatization and Reduction to Static Positions

Axiomatizing the set of validities for the frames, models and updates just defined proceeds in two modules: one for the static modalities of $\mathcal{L}$ and one for the dynamic extension. For the static part the axiomatization proceeds in a standard manner. We use K for $[\leq]$, together with the Löb axiom for well-foundedness and transitivity. A and Do_i are S5 modalities. Interaction between $[\leq]$, Do_i and A can be captured by standard inclusion axioms.

Table 2. Reduction axioms for lexicographic update

$$[\mathscr{A}_i, a]p \leftrightarrow (Pre(a) \rightarrow p)$$
$$[\mathscr{A}_i, a]\neg\varphi \leftrightarrow (Pre(a) \rightarrow \neg[\mathscr{A}_i, a]\varphi)$$
$$[\mathscr{A}_i, a](\varphi \wedge \psi) \leftrightarrow [\mathscr{A}_i, a]\varphi \wedge [\mathscr{A}_i, a]\psi$$
$$[\mathscr{A}_i, a]A\varphi \leftrightarrow (Pre(a) \rightarrow A[\mathscr{A}_i, a]\varphi)$$
$$[\mathscr{A}_i, a][\leq]\varphi \leftrightarrow (Pre(a) \rightarrow \bigwedge_{c>^{\mathscr{A}_i}a} A[\mathscr{A}_i, c]\varphi \wedge \bigwedge_{c\cong^{\mathscr{A}_i}a}[\leq][\mathscr{A}_i, c]\varphi)$$
$$[\mathscr{A}_i, a]Do_j\varphi \leftrightarrow (Pre(a) \rightarrow Do_j[\mathscr{A}_i, a]\varphi)$$

Axiomatizing the dynamic part uses the well-known "reduction axioms" methodology [1,19]. Formulas containing dynamic modalities are shown to be semantically equivalent to formulas of $\mathcal{L}$, that is without dynamic modalities. The formulas in Table 2 indeed show how to "push" dynamic modalities inside the various connectives and modal operators of the static language, until they range over atomic propositions where they can be eliminated. These formulas are sound with respect to the lexicographic update over preference-action models. Taking them as axioms thus makes formulas containing dynamic modalities provably equivalent to formulas of $\mathcal{L}$. Completeness for the extended language then follows from completeness of the static part with respect to the class of preference-action models.

Through the soundness of the reduction axioms, together with the fact that conditional obligations are definable in $\mathcal{L}$, we obtain that power and immunity are also reducible to static normative positions. So the approach presented here is reductive. This reduction, however, is more complex than the simple reduction proposed for instance in [11]. Here is the valid reduction validity for conditional obligation:

$$[\mathscr{A}_i, a]O(\psi/\varphi) \leftrightarrow [Pre(a) \rightarrow$$
$$A \bigwedge_{d\in A} (((\langle\mathscr{A}_i, d\rangle\varphi \wedge A \bigwedge_{c>^{\mathscr{A}_i}d} [\mathscr{A}_i, c]\neg\varphi) \rightarrow O([\mathscr{A}_i, d]\psi / \bigvee_{c\cong^{\mathscr{A}_i}d} \langle\mathscr{A}_i, c\rangle\varphi))]$$

with $\langle\mathscr{A}_i, d\rangle\varphi$ being the dual of $[\mathscr{A}_i, a]$. The complexity of this formula results from it essentially encoding syntactically the lexicographic update rule in combination with the specific semantic definition of obligations as truth in all the *most* ideal worlds. For unconditional obligations, however, it simplifies to the following:

$$[\mathscr{A}_i, a]O(\psi) \leftrightarrow [Pre(a) \rightarrow$$
$$A \bigwedge_{d\in A} ((pre(d) \wedge A \bigwedge_{c>^{\mathscr{A}_i}d} \neg pre(c)) \rightarrow O([\mathscr{A}_i, d]\psi / \bigvee_{c\cong^{\mathscr{A}_i}d} pre(c)))]$$

The effect of changes in legal relations are thus reducible to statements describing legal relations holding *before* the deontic action takes place. In particular, the latter formula states that executing an action a would result in an obligation to ψ exactly when, if a is executable in the first place, for any maximally ideal and executable action d, it ought to be the case before executing d that ψ would hold after d.

The reduction allows to distinguish the legal permissibility of a deontic action a and its legal ability. The latter boils down to a being executable in a particular situation, which in turn reduces to the preconditions of a obtaining. This, however, is *not* equivalent to the execution of a being permitted. Defining permissibility of deontic action requires some additional machinery. We return to this question in Sect. 4.

3 Dynamic Normative Positions

We now turn to the study of the dynamic normative positions that are generated by our theory of legal competences. By analogy with the study of static normative positions [14], we answer now the question of how many distinct, atomic legal competences there are. It is well known that for static normative positions this number increases substantially with the number of agents. This is also the case here, but there is an additional complication. As hinted at by the valid reduction law for conditional obligation, the number of dynamic normative positions will also grow with the size of the deontic action model. The theory of normative positions for conditional static rights is less well studied than for unconditional ones [14]. In brief, there are 256 static normative positions $T(i, j, \psi/\varphi))$ in $[\![\pm O(\pm\binom{Do_i}{Do_j} \pm \psi/\varphi)]\!]$ using Makinson's [12] notation, in which all the elements are consistent and logically independent, and their disjunction is a tautology in the given logic. We omit the argument for reasons of space, but refer to [5] for details.

Let us define the set of *atomic legal competences* for two agents i, j and a given static normative position $T(i, j, \psi/\varphi))$:

$$[\![\pm \bigvee_{i \in \mathcal{I}} \bigvee_{a \in \mathscr{A}_i} [\mathscr{A}_i, a] \pm T(i, j, \psi/\varphi)]\!]$$

The question we ask now is thus how large is that set for a given action model of size at most n? In this paper we restrict ourselves to the two-agents cases. Our base result establishes the number of combinations of claims and no-claims rights that can result from a given deontic action. There are 16 total maximally consistent elements in $[\mathscr{A}_i, a][\![\pm O(\pm\binom{Do_i}{Do_j} \pm \psi/\varphi)]\!]$. We refer again to [5] for details.

With this in hand we can get our calculation going. There are 256 maximally consistent elements in $\bigvee_{a \in \mathscr{A}_i} [\mathscr{A}_i, a][\![\pm O(\pm\binom{Do_i}{Do_j} \pm \psi/ \pm \varphi)]\!]$ when $|\mathscr{A}_i| = 1$. When $|\mathscr{A}_i| = 2$, the total number of the maximally consistent elements of that set is 2^{256}. This number happens to grow linearly. With $|\mathscr{A}_i| = n$, we get n^{256} possibilities. On the other hand, when $|\mathscr{A}_i| = 2$, the total number of the maximally consistent elements in $\neg \bigvee_{a \in \mathscr{A}_i} [\mathscr{A}_i, a][\![\pm O(\pm\binom{Do_i}{Do_j} \pm \psi/ \pm \varphi)]\!]$ is 2^{256} This number is also n^{256} for $|\mathscr{A}_i| = n$. Furthermore, when $|\mathscr{A}_i| = 1$, for each element in the first set above there are 255 elements in the second. The total number of elements in

$$[\![\pm \bigvee_{a\in\mathscr{A}_i} [\mathscr{A}_i, a][\![\pm O(\pm\binom{Do_i}{Do_j}) \pm \psi / \pm \varphi)]\!]]\!]$$

is thus $256 \times 255 = 65280$, when $|\mathscr{A}_i| = 1$, and the total number is $n^{256} \times (n^{255})$, when $|\mathscr{A}_i| = n$.

4 Legal Ability and Legal Permissibility

Although legal ability and legal permissibility often go together, they are conceptually distinct notions [8,12]. The German Civil Code (*Bürgerliches Gesetzbuch*) offers a concrete case where they can come apart.

Article 179 of this code regulates the contractual delegation of the right to auction one's land to a third party. The article allows for the following case. Suppose that A contracts B to auction her land. B sells to C, but C is not the highest bidder. The sale being a deontic action, it is still considered valid. It transfers property rights from A to C. B selling to C is, however, not legally permissible, because C is not the highest bidder. B can consequently be asked to compensate A for the difference between the selling price and the highest bid.

In this example B has the legal ability to sell the land to C. She is legally capable of executing a deontic action which transfers the set of static and dynamic property rights with regard to the land from A to C. This action, however, is impermissible. It cannot be executed without B incurring a sanction. We now show that our logical model of legal competences can capture this case in a simple manner.

4.1 Legal Ability

Let c be the fact that C is in possession of the land's property titles, and S be that B compensate A for the price difference. Before the selling, A owns her land, and A contracts B to auction her land. B, so entrusted by A, has a privilege to transfer the land's property titles, which we represent here simply as transferring them to c or to someone else $\neg c$. Hence, even on the condition S, the states with $Do_B\neg c$ are the most ideal, the states with $Do_B c$ are the least ideal, and the others are as ideal as each other. This situation is illustrated in Fig. 4. There, as before, the arrows $\rightarrow$ representing the preference ordering $\leq$ between distinct states, and the dashed arc labelled B represents the relations $\sim_B$. The reflexive loops for both relations are everywhere omitted. In this model the following atomic type holds for B: $O(\neg Do_B c)$, $\neg O(Do_B\neg c)$, and $\neg O(\neg Do_B\neg c)$. It also contains $\neg O(c)$. Here do_B is the dual of Do_B, i.e. $do_B\varphi := \neg Do_B\neg\varphi$.

B's legal ability can be modelled in the deontic action model of Fig. 5. As before, the arrow $\mapsto$ represents the preference order $\leq^{\mathsf{S}_B}$ with the reflexive loops omitted. In this particular model, the precondition of action s is $Do_B c$, and that of action n is $\neg Do_B c$.

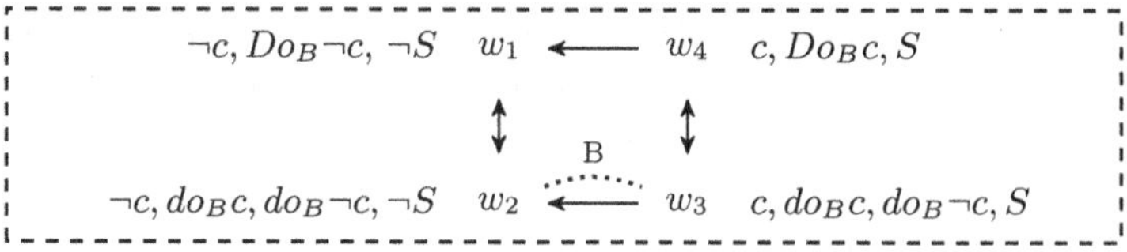

Fig. 4. The preference-action model **C** for the normative positions before the sale

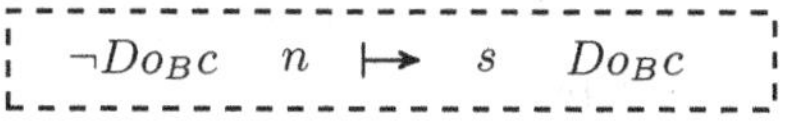

Fig. 5. The deontic action model $\mathbf{S}_B$ for B's sale to C

The result of the lexicographic update $\mathbf{C} \otimes \mathbf{S}_B$ is presented in Fig. 6. In this model the normative position of B has changed. Now we have $O(Do_B c)$, and $O(c)$ as well. The action B selling the land to C of course changes A's normative position as well, but we omit those here.

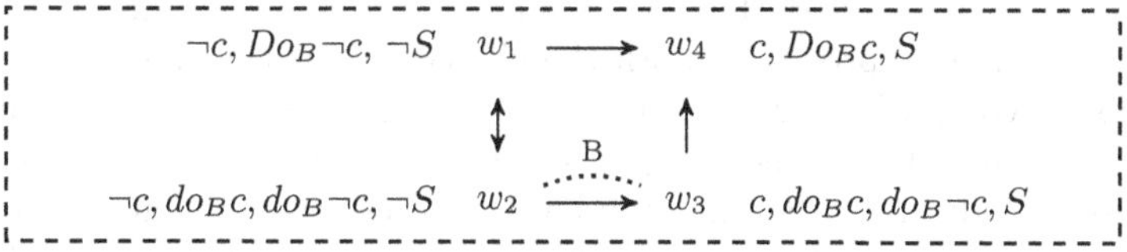

Fig. 6. The preference-action model **C** for the normative positions after the sale

This simple example shows how to model B's legal ability to sell A's land to C. Since C is not the highest bidder, however, B's action is not legally permissible. This can also be easily expressed here.

4.2 Legal Permissibility

As mentioned, the language $\mathcal{L}$, even extended with the dynamic modalities, cannot directly express the notion of legal permissibility. This language is designed to describe the effects of deontic action or, by having deontic modalities scoping over dynamic expressions, the normative status of those effects. But this is still different from saying that a certain deontic action is obligatory, permitted or forbidden. In our example it is arguably the case that it ought to be that C owns the land after B has sold it to her, even though this sale is not legally permitted.

To express legal permissibility we instead use a type of Anderson-Kanger reduction that is also present in dynamic deontic logic [13]. Let S, as above, represent the constant that a sanction will occur, viz. in our particular case that B must compensate A. Then we define "action a is legally permitted" as follows.

$$P(a) := [\mathscr{A}_i, a]\neg S$$

This is the standard definition of strong permission in dynamic deontic logic introduced by Dignum et al. [4]. In our case $Do_{BC} \rightarrow S$. B incurs a sanction upon selling the land to C. Since $\mathbf{C}, w_4 \models \langle \mathbf{S}_B, s\rangle S$, we get that $\mathbf{C}, w_4 \models \neg P(s)$ and, together with our analysis of legal ability $\mathbf{C}, w_4 \models \langle \mathbf{S}_B, s\rangle \top \wedge \neg P(s)$. Selling to C is legally possible but not permissible.

5 Conclusion

This paper can be seen as a test drive for a new way of representing the structure of legal competences, and of deontic action more generally: (deontic) action models and the related update mechanism. This methodology is well-established in epistemic and doxastic logic, and many of the results presented here are direct imports from that literature. Our main contribution is to bring it to bear on the theory of dynamic rights.

Indeed, we have argued that the model of Hohfeldian power and immunity developed here improves on both the classical reductive and non-reductive approaches. In comparison with for instance Lindahl's [11] approach, our model explicitly captures the norm-changing or dynamic character of legal competences. It does so both at the semantic level, through the explicit update mechanism, and at the semantic level, by using an explicit dynamic modality to express the effects of deontic actions. The approach we propose here is however still reductive in the sense that formulas with dynamic modalities are semantically and provably reducible to formulas without. As a result it comes with a rich set of interaction principles between static and dynamic rights. From that point of view it improves on Jones and Sergot's non-reductive approach to dynamic rights through "count as" conditionals [8]. Finally, we have shown that this system can capture the distinction between legal ability and legal permissibility in a more auspicious way then in reductive approaches, without paying the price of full-blown non-reductionism.

We take this to be a promising starting point for the methodology we propose, but of course it also raises a number of questions that could not be addressed in this paper. Probably the most important next step will be to enrich the model to cover not only pure but also the combination of deontic and non-deontic actions, for instance breaking someone's window or crossing someone's property. This would allow for a closer comparison between ours and the Jones and Sergot approach just mentioned. Equally important in our view is to study the theory of legal competences that would result from extending a static base that is different from Standard Deontic Logic. In the epistemic context a wide variety of static logics of knowledge and belief have been "dynamified" using the action model methodology. The proposal here already slightly deviates from the classical Kangerian approach in that it includes conditional obligations. Studying the normative positions stemming from that addition would have taken us too far from the core proposal of this paper, and we again refer the reader to the forthcoming PhD thesis [5] for details. But more radical departures from SDL have been proposed to capture actual legal reasoning, and the question remains whether they would yield a plausible theory of power and immunity once augmented with a dynamic module as we have done here.

Acknowledgments. The research reported in this paper has been supported by the PIOTR project (RO 4548/4-1] of the German Research Foundation (DFG). The first author is supported by the MOE Project of Key Research Institute of Humanities and Social Sciences in Universities [No. 12JJD720005], by the China Scholarship Council grant [CSC No. 201306380078], and by the Chair of Philosophy I, Universität Bayreuth.

References

1. Baltag, A., Smets, S.: A qualitative theory of dynamic interactive belief revision. Log. Found. Game Decis. Theory **3**, 9–58 (2008)
2. Blackburn, P., De Rijke, M., Venema, Y.: Modal Logic, vol. 53. Cambridge University Press, Chennai (2002)
3. Boutilier, C.: Toward a logic for qualitative decision theory. KR **94**, 75–86 (1994)
4. Dignum, F., Meyer, J.-J.C., Wieringa, R.: Free choice and contextually permitted actions. Stud. Log. **57**(1), 193–220 (1996)
5. Dong, H.: Permission in Non-Monotonic Normative Reasoning. PhD thesis, Universität Bayreuth (2017)
6. Hansson, B.: An analysis of some deontic logics. Nous **33**, 373–398 (1969)
7. Hohfeld, W.N.: Some fundamental legal conceptions as applied in judicial reasoning. Yale Law J. **23**(1), 16–59 (1913)
8. Jones, A.J., Sergot, M.: A formal characterisation of institutionalised power. Log. J. IGPL **4**(3), 427–443 (1996)
9. Kanger, S.: Law and logic. Theoria **38**(3), 105–132 (1972)
10. Kanger, S., Kanger, H.: Rights and parliamentarism. Theoria **32**(2), 85–115 (1966)
11. Lindahl, L.: Position and Change: A Study in Law and Logic, vol. 112. Springer Science and Business Media, Dordrecht (1977)
12. Makinson, D.: On the formal representation of rights relations. J. philos. Log. **15**(4), 403–425 (1986)
13. Meyer, J.-J.C.: A different approach to Deontic Logic: Deontic Logic Viewed as a Variant of Dynamic Logic. Notre Dame J. Formal Log. **29**, 109–136 (1988)
14. Sergot, M.: Normative positions. In: Gabbay, D., Horty, J., Parent, X., van der Meyden, R., van der Torre, L. (eds.) Handbook of Deontic Logic and Normative Systems, vol. 1. College Publication, London (2013)
15. Van Benthem, J.: Dynamic logic for belief revision. J. Appl. Non-classical log. **17**(2), 129–155 (2007)
16. van Benthem, J.: Modal Logic for Open Minds. Center for the Study of Language and Information (CSLI), California (2010)
17. van Benthem, J.: Logical Dynamics of Information and Interaction. Cambridge University Press, Chennai (2011)
18. Benthem, J., Grossi, D., Liu, F.: Deontics = Betterness + Priority. In: Governatori, G., Sartor, G. (eds.) DEON 2010. LNCS, vol. 6181, pp. 50–65. Springer, Heidelberg (2010). doi:10.1007/978-3-642-14183-6_6

19. van Benthem, J., Grossi, D., Liu, F.: Priority structures in deontic logic. Theoria **80**(2), 116–152 (2014)
20. van Benthem, J., van Otterloo, S., Roy, O.: Preference logic, conditionals, and solution concepts in games. In: Festschrift for Krister Segerberg (2005)
21. van Ditmarsch, H., Kooi, B.: Semantic results for ontic and epistemic change. Log. Found. Game Decis. Theory **3**, 87–117 (2008)
22. van Eijck, J., Ruan, J., Sadzik, T.: Action emulation. Synthese **185**, 131–151 (2012)

A Propositional Dynamic Logic for Instantial Neighborhood Models

Johan van Benthem[1], Nick Bezhanishvili[1], and Sebastian Enqvist[1,2](✉)

[1] ILLC, University of Amsterdam, Amsterdam, The Netherlands
{J.vanBenthem,N.Bezhanishvili}@uva.nl
[2] Department of Philosophy, Stockholm University, Stockholm, Sweden
thesebastianenqvist@gmail.com

Abstract. We propose a new perspective on logics of computation by combining instantial neighborhood logic INL with bisimulation safe operations adapted from PDL and dynamic game logic. INL is a recently proposed modal logic, based on a richer extension of neighborhood semantics which permits both universal and existential quantification over individual neighborhoods. We show that a number of game constructors from game logic can be adapted to this setting to ensure invariance for instantial neighborhood bisimulations, which give the appropriate bisimulation concept for INL. We also prove that our extended logic IPDL is a conservative extension of dual-free game logic, and its semantics generalizes the monotone neighborhood semantics of game logic. Finally, we provide a sound and complete system of axioms for IPDL, and establish its finite model property and decidability.

1 Introduction

In this paper, we introduce a new modal logic of computation, in the style of propositional dynamic logic, based on *instantial neighborhood logic* INL [3]. The logic INL is based on a recent variant of monotone neighborhood semantics for modal logics, called instantial neighborhood semantics. In the standard neighborhood semantics, the box operator has the interpretation: $\Box p$ is true at a point if *there exists* a neighborhood in which *all* the elements satisfy the proposition p. So the box operator has a built-in fixed existential-universal quantifier pattern. In instantial neighborhood logic, we allow both universal and existential quantification over individual neighborhoods, so the basic modality has the form $\Box(p_1, ..., p_n; q)$. This formula is true at a point if *there exists* a neighborhood N in which *all* the elements satisfy the proposition q, and furthermore each of the propositions $p_1, ..., p_n$ are satisfied by *some* elements of N. INL is more expressive than monotone neighborhood logic, and comes with a natural associated notion of bisimulation together with a Hennessy-Milner theorem for finite models. It has a complete system of axioms, has the finite model property, is decidable and PSpace-complete.

Formally, our proposal is to consider an extension of the base language INL by bisimulation safe "program constructors", as in the standard propositional

A. Baltag et al. (Eds.): LORI 2017, LNCS 10455, pp. 137–150, 2017.
DOI: 10.1007/978-3-662-55665-8_10

dynamic logic of sequential programs (PDL). The usual repertoire here consists of choice, test, sequential composition and a Kleene star for program iteration. Similar additions have already been studied extensively for the standard (monotone) neighborhood semantics, where the constructors are interpreted as methods of constructing complex *games* (this idea dates back to [13]). In the neighborhood setting, some additional operations are available, including the *dual* construction. This is a very powerful construction, and it is well known that dynamic game logic is not contained in any fixed level of the μ-calculus alternation hierarchy [4].

We think of our extended logic, which we call instantial PDL (IPDL for short), as a dynamic logic for a richer notion of computation than sequential programs. We consider a computational process as an agent acting in an uncertain environment that affects the outcome of each action. This is similar to the thinking behind the alternating-time temporal logic ATL of Alur et al. [1]. Dynamic game logic can be interpreted in a similar way, thinking of processes as "games against the environment". Instantial neighborhood semantics introduces a more fine-grained perspective to this setting, with a more expressive language and a finer bisimulation concept than standard neighborhood bisimilarity, namely the instantial neighborhood bisimulations of [3].

We generalize operations from game logic in the setting of instantial neighborhood logic, with the implicit desiderata that the extended language should be bisimulation invariant, and that the operations should be reasonably simple. Note that bisimulation invariance now has a new meaning, since we are working with instantial neighborhood bisimulations. This means that setting up the program constructors correctly is a non-trivial task, and the constructors known from game logic need to be revised in order to ensure bisimulation invariance. The case for sequential composition of programs is particularly subtle, and a naive generalization of the composition operation from game logic could easily break bisimulation safety. In particular, the standard definition from game logic is not bisimulation safe in our sense. One of our key contributions in this paper is to provide a bisimulation safe sequential composition operation. We also find natural analogues of test, choice and Kleene star. As opposed to the case of dynamic game logic, we cannot see any obvious candidate for a dual constructor. However, a dual to the choice operator can be defined, generalizing "demonic choice" in game logic and bearing a similarity to the parallel game composition operation considered in [15]. We show that our logic is in fact a conservative extension of dual-free game logic, and the instantial neighborhood semantics can be seen as a generalization of the semantics for dual-free game logic over monotone neighborhood structures, in a sense that will be made precise in Sect. 4.2.

We provide sound and complete axioms for our instantial propositional dynamic logic IPDL, prove decidability via finite model property, and establish bisimulation invariance. The latter amounts to bisimulation safety for our program constructors. The completeness proof for the language IPDL, including all the program constructors that we consider, is based on the standard completeness proof for PDL (see [5] for an exposition), but involves some non-trivial new features. In particular, the axiom system requires two distinct induction rules, corresponding to a nested least fixpoint induction, and the model construction makes heavy use of a normal form for INL-formulas established in [3].

2 Instantial Neighborhood Logic

2.1 Syntax and Semantics

We start by reviewing the basic language for instantial neighborhood semantics. The only difference with our first paper on instantial neighborhood logic is that we are interpreting the language over *labelled* neighborhood structures, where the labels play the same role as "atomic programs" in PDL or "atomic games" in game logic.

The syntax of INL is given by the following grammar:

$$\varphi := p \in \mathsf{Prop} \mid \varphi \wedge \varphi \mid \neg\varphi \mid [a](\Psi;\varphi)$$

where a ranges over a fixed set $\mathcal{A}$ of *atomic labels*, and Ψ ranges over finite sets of formulas of INL. We have deviated a bit from the syntax of [3] here in allowing Ψ to be a finite *set* rather than a tuple of formulas. We shall sometimes write $[a](\psi_1, ..., \psi_n; \varphi)$ rather than $[a](\{\psi_1, ..., \psi_n\}; \varphi)$, in particular we write $[a](\psi;\varphi)$ rather than $[a](\{\psi\};\varphi)$, and $[a]\varphi$ rather than $[a](\emptyset;\varphi)$.

Formulas in INL will be interpreted over neighborhood structures.

Definition 1. *A* neighborhood *frame is a structure* (W, R) *where* W *is a set and* R *associates with each* $a \in \mathcal{A}$ *a binary relation* $R_a \subseteq W \times \mathcal{P}W$*. A* neighborhood model (W, R, V) *is a neighborhood frame together with a valuation* $V : \mathsf{Prop} \to \mathcal{P}W$.

We define the interpretations of all formulas in a neighborhood model $\mathfrak{M} = (W, R, V)$ as follows:

- $[\![p]\!] = V(p)$.
- $[\![\varphi \wedge \psi]\!] = [\![\varphi]\!] \cap [\![\psi]\!]$.
- $[\![\neg\varphi]\!] = W \setminus [\![\varphi]\!]$.
- $u \in [\![[a](\psi_1, ..., \psi_k;\varphi)]\!]$ iff there is some $Z \subseteq W$ such that:

$(u, Z) \in R_a$ and $Z \subseteq [\![\varphi]\!]$, $Z \cap [\![\psi_i]\!] \neq \emptyset$ for $i \in \{1, ..., k\}$

We write $\mathfrak{M}, v \Vdash \varphi$ for $v \in [\![\varphi]\!]$, and we write $\Vdash \varphi$ and say that φ is *valid* if, for every game model $\mathfrak{M}$ and $v \in W$, we have $\mathfrak{M}, v \Vdash \varphi$. We allow the notation $[\![-]\!]_{\mathfrak{M}}$ to make explicit reference to the model in the background.

Neighborhood models come with a natural notion of bisimulation, introduced in a more general setting in [3]. For this definition, the so called *Egli-Milner lifting* of a binary relation will play an important role:

Definition 1. *The* Egli-Milner lifting *of a binary relation* $R \subseteq X \times Y$*, denoted* $\overline{R}$*, is a relation from* $\mathcal{P}X$ *to* $\mathcal{P}Y$ *defined by:* $Z\overline{R}Z'$ *iff:*

1. *For all* $z \in Z$ *there is some* $z' \in Z'$ *such that* zRz'.
2. *For all* $z' \in Z'$ *there is some* $z \in Z$ *such that* zRz'.

We write $R;S$ for the composition of relations R and S. It is well known that the Egli-Milner lifting preserves relation composition:

$$\overline{R;S} = \overline{R};\overline{S}$$

Definition 2. *Let* $\mathfrak{M} = (W, R, V)$ *and* $\mathfrak{M}' = (W', R', V')$ *be any neighborhood models. The relation* $B \subseteq W \times W'$ *is said to be an* instantial neighborhood bisimulation *if for all* uBu' *and all atomic labels* a *we have:*

Atomic *For all* p, $u \in V(p)$ *iff* $u' \in V'(p)$.
Forth *For all* Z *such that* uR_aZ, *there is some* Z' *such that* $u'R'_aZ'$ *and* $Z\overline{B}Z'$.
Back *For all* Z' *such that* $u'R'_aZ'$ *there is some* Z *such that* uR_aZ *and* $Z\overline{B}Z'$.

We say that pointed models $\mathfrak{M}, w$ *and* $\mathfrak{N}, v$ *are* bisimilar, *written* $\mathfrak{M}, w \underline{\longleftrightarrow} \mathfrak{N}, v$, *if there is an instantial neighborhood bisimulation* B *between* $\mathfrak{M}$ *and* $\mathfrak{N}$ *such that* wBv.

It is easy to check that all formulas of INL are invariant for instantial neighborhood bisimilarity:

Proposition 1. *If* $\mathfrak{M}, w \underline{\longleftrightarrow} \mathfrak{N}, v$ *then* $\mathfrak{M}, w \Vdash \varphi$ *iff* $\mathfrak{N}, v \Vdash \varphi$, *for each formula* φ *of INL.*

2.2 Axiomatization

We now turn to the task of axiomatizing the valid formulas of INL. Our system of axioms is a gentle modification of the axiom system for instantial neighborhood logic presented in [3].

INL Axioms

Mon: $[a](\psi_1, ..., \psi_n; \varphi) \to [a](\psi_1 \vee \alpha_1, ..., \psi_n \vee \alpha_n; \varphi \vee \beta)$
Weak: $[a](\Psi; \varphi) \to [a](\Psi'; \varphi)$ for $\Psi' \subseteq \Psi$
Un: $[a](\psi_1, ..., \psi_n; \varphi) \to [a](\psi_1 \wedge \varphi, ..., \psi_n \wedge \varphi; \varphi)$
Lem: $[a](\Psi; \varphi) \to [a](\Psi \cup \{\gamma\}; \varphi) \vee [a](\Psi; \varphi \wedge \neg\gamma)$
Bot: $\neg[a](\bot; \varphi)$

Rules

MP:

$$\frac{\varphi \to \psi \qquad \varphi}{\psi}$$

RE:

$$\frac{\varphi \leftrightarrow \psi \qquad \theta}{\theta[\varphi/\psi]}$$

where $\theta[\varphi/\psi]$ is the result of substituting some occurrences of the formula ψ by φ in θ.

We denote this system of axioms by Ax1 and write Ax1 $\vdash \varphi$ to say that the formula φ is provable in this axiom system. We also write $\varphi \vdash_{\mathsf{Ax1}} \psi$ for Ax1 $\vdash \varphi \to \psi$, and say that φ *provably entails* ψ.

Theorem 1. *The system* Ax1 *is sound and complete for validity on neighborhood models.*

The proof of this result is essentially the same as in [3], and will not be repeated here.

Since the proof in [3] constructs a finite model for each consistent formula, we also get:

Theorem 2. *The logic INL is decidable and has the finite model property.*

3 Test, Choice, Parallel Composition and Sequential Composition

We now extend the language INL with four basic PDL-style operations: test, choice, parallel composition and sequential composition. The resulting language will be called *dynamic instantial neighborhood logic*, or (DINL). The syntax of DINL is defined by the following dual grammar.

$$\varphi := p \in \mathsf{Prop} \mid \varphi \wedge \varphi \mid \neg\varphi \mid [\pi](\Psi; \varphi)$$

$$\pi := a \in \mathcal{A} \mid \varphi? \mid \pi \cup \pi \mid \pi \cap \pi \mid \pi \circ \pi$$

We define the interpretation $[\![o]\!]$ of each operation $o \in \{\cup, \cap, \circ\}$ in a neighborhood model $\mathfrak{M}$ as a binary map from pairs of neighborhood relations to neighborhood relations, as follows:

- $R_1[\![\cup]\!]R_2 = R_1 \cup R_2$
- $R_1[\![\cap]\!]R_2 = \{(w, Z_1 \cup Z_2) \mid (w, Z_1) \in R_1 \;\&\; (w, Z_2) \in R_2\}$
- $(w, Z) \in R_1[\![\circ]\!]R_2$ iff there is some set Y and some family of sets F such that $(w, Y) \in R_1$, $(Y, F) \in \overline{R}_2$ and $Z = \bigcup F$.

The interpretation $[\![?]\!]$ of the test operator will be a map $[\![?]\!]$ assigning a neighborhood relation to each subset Z of W, defined by:

$$[\![?]\!]Z := \{(u, \{u\}) \mid u \in Z\}$$

Note that $[\![?]\!]$ is monotone in the sense that $Z \subseteq Z'$ implies $[\![?]\!]Z \subseteq [\![?]\!]Z'$. Each operator $o \in \{\cup, \cap, \circ\}$ is also monotone, in the sense that $R_1[\![o]\!]R_2 \subseteq R'_1[\![o]\!]R'_2$

whenever $R_1 \subseteq R_1'$ and $R_2 \subseteq R_2'$. For the sequential composition operator, this uses the well known fact that the Egli-Milner lifting is monotone, i.e. $\overline{R} \subseteq \overline{R'}$ whenever $R \subseteq R'$.

We can now define the semantic interpretations of all formulas, and the neighborhood relations corresponding to all complex labels π, by the following mutual recursion:

- $[\![p]\!] = V(p)$.
- $[\![\varphi \wedge \psi]\!] = [\![\varphi]\!] \cap [\![\psi]\!]$.
- $[\![\neg\varphi]\!] = W \setminus [\![\varphi]\!]$.
- $u \in [\![[\pi](\psi_1, ..., \psi_k; \varphi)]\!]$ iff there is some $Z \subseteq W$ such that:

$(u, Z) \in R_\pi$ and $Z \subseteq [\![\varphi]\!]$, $Z \cap [\![\psi_i]\!] \neq \emptyset$ for $i \in \{1, ..., k\}$.

- $R_{\pi_1 o \pi_2} = R_{\pi_1} [\![o]\!] R_{\pi_2}$ for $o \in \{\cup, \cap, \circ\}$.
- $R_{\varphi?} = [\![?]\!][\![\varphi]\!]$

The definitions of the dynamic operations are tailored towards obtaining the following result:

Proposition 2. *All formulas of DINL are invariant for instantial neighborhood bisimulations.*

3.1 Axiomatization

Our axiom system for DINL will take the sound and complete axioms for INL as its foundation, and extend it with reduction axioms for the test, choice, parallel composition and sequential composition operators. The axioms and rules are listed below; note that the INL axioms and the axioms for frame constraints are now stated for arbitrary complex labels π rather than just atoms a.

INL Axioms: (Mon), (Weak), (Un), (Lem) and (Bot)

Reduction Axioms:

Test: $[\gamma?](\Psi; \varphi) \leftrightarrow \gamma \wedge \bigwedge \Psi \wedge \varphi$
Ch: $[\pi_1 \cup \pi_2](\Psi; \varphi) \leftrightarrow [\pi_1](\Psi; \varphi) \vee [\pi_2](\Psi; \varphi)$
Pa: $[\pi_1 \cap \pi_2](\Psi; \varphi) \leftrightarrow \bigvee\{[\pi_1](\Theta_1; \varphi) \wedge [\pi_2](\Theta_2; \varphi) \mid \Psi = \Theta_1 \cup \Theta_2\}$
Cmp: $[\pi_1 \circ \pi_2](\psi_1, ..., \psi_n; \varphi) \leftrightarrow [\pi_1]([\pi_2](\psi_1; \varphi), ..., [\pi_2](\psi_n; \varphi); [\pi_2]\varphi)$

Rules: (MP) and (RE)

We denote this system of axioms by Ax2 and write Ax2 $\vdash \varphi$ to say that the formula φ is provable in this axiom system. We also write $\varphi \vdash_{\mathsf{Ax2}} \psi$ for Ax2 $\vdash \varphi \rightarrow \psi$. We shall sometimes drop the reference to Ax2 to keep notation cleaner.

Proposition 3 (Soundness). *If* $\mathsf{Ax2} \vdash \varphi$*, then* φ *is valid on all neighborhood models.*

By applying soundness of the reduction axioms, we can use a standard argument to obtain for every consistent formula φ of DINL a provably (and hence semantically) equivalent formula φ^t in INL, which is then satisfiable by Theorem 1. For example, the formula $[\gamma?](\psi_1, ..., \psi_n; \varphi)^t$ is defined to be $\gamma^t \wedge \psi_1^t \wedge ... \wedge_n^t \wedge \varphi$.

We get:

Theorem 3 (Completeness). *A formula* φ *of DINL is valid on all neighborhood models iff* $\mathsf{Ax2} \vdash \varphi$.

Furthermore, the finite model property and decidability clearly carry over from INL:

Theorem 4. *The logic DINL is decidable and has the finite model property.*

4 Iteration

4.1 The Language IPDL

We now introduce the final operation that we consider here, a Kleene star for finite iteration. This operation will be set up to generalize the game iteration operation from game logic. The corresponding language will be denoted by IPDL, read "instantial PDL", and is given by the following dual grammar:

$$\varphi := p \in \mathsf{Prop} \mid \varphi \wedge \varphi \mid \neg\varphi \mid [\pi](\Psi; \varphi)$$

$$\pi := a \in \mathcal{A} \mid \varphi? \mid \pi \cup \pi \mid \pi \cap \pi \mid \pi \circ \pi \mid \pi^*$$

For the semantic interpretation of the Kleene star, it will be useful to first define the relation skip by:

$$\mathsf{skip} := \{(w, \{w\}) \mid w \in W\}$$

We now define a relation $R^{[\xi]}$ for each ordinal ξ by induction as follows.

- $R^{[0]} = \emptyset$
- $R^{[\xi+1]} = \mathsf{skip}[\![\cup]\!](R[\![\circ]\!]R^{[\xi]})$
- $R^{\kappa} = \bigcup_{\xi<\kappa} R^{[\xi]}$ if κ is a limit ordinal.

We define $[\![*]\!]R$ to be equal to $R^{[\xi]}$, where ξ is the smallest ordinal satisfying $R^{[\xi]} = R^{[\xi+1]}$. It is easy to see that this is a standard least fixpoint construction, in particular we have:

Proposition 4. *Let* W *be a finite set and* $R \subseteq W \times \mathcal{P}(W)$*. Then:*

$$[\![*]\!]R = \bigcup_{n\in\omega} R^{[n]}$$

Semantics of IPDL-formulas in a neighborhood model $\mathfrak{M} = (W, R, V)$ are now defined as follows:

- $[\![p]\!] = V(p)$.
- $[\![\varphi \wedge \psi]\!] = [\![\varphi]\!] \cap [\![\psi]\!]$.
- $[\![\neg\varphi]\!] = W \setminus [\![\varphi]\!]$.
- $u \in [\![[\pi](\psi_1, ..., \psi_k; \varphi)]\!]$ iff there is some $Z \subseteq W$ such that:

$(u, Z) \in R_\pi$ and $Z \subseteq [\![\varphi]\!]$, $Z \cap [\![\psi_i]\!] \neq \emptyset$ for $i \in \{1, ..., k\}$.

- $R_{\pi_1 o \pi_2} = R_{\pi_1} [\![o]\!] R_{\pi_2}$ for $o \in \{\cup, \cap, \circ\}$.
- $R_{\varphi?} = [\![?]\!] [\![\varphi]\!]$.
- $R_{\pi^*} = [\![*]\!] R_\pi$.

Proposition 5. *All formulas of IPDL are invariant for instantial neighborhood bisimulations.*

The proof of this is a bisimulation safety argument, and the step for the Kleene star involves using the bisimulation safety of union and sequential composition to prove the appropriate back-and-forth conditions for each approximant $R_\pi^{[\xi]}$ of the least fixpoint $R_{\pi^*} = [\![*]\!] R_\pi$. We omit the details.

4.2 Comparison with Dual-Free Game Logic

We now show that IPDL can, in a precise sense, be viewed as a language extension of dual-free game logic. We shall denote this language simply by GL, for "game logic", although the full dynamic game logic also includes a dual constructor. Formally, formulas of GL and game terms are defined by the following dual grammar:

$$\varphi := p \in \mathsf{Prop} \mid \varphi \wedge \varphi \mid \neg\varphi \mid [\pi]\varphi$$

$$\pi := a \in \mathcal{A} \mid \varphi? \mid \pi \circ \pi \mid \pi \cup \pi \mid \pi \cap \pi \mid \pi^*$$

where Prop is a fixed set of propositional variables and $\mathcal{A}$ is a set of atomic games, both assumed to be countably infinite. Note that GL is a syntactic fragment of IPDL. Here, $\cup$ is interpreted as "angelic choice" (choice for Player I), $\cap$ is interpreted as "demonic choice" (choice for Player II), $\circ$ is sequential game composition and $*$ is finite game iteration (controlled by Player I).

Semantics of game logic formulas are again given by neighborhood frames, with the extra constraint that neighborhoods associated with a world are upwards closed under subsethood:

Definition 3. *A neighborhood frame (W, R) is said to be a* monotonic power frame *if the following condition holds for each $a \in \mathcal{A}$:*
(Monotonicity) For all $u \in W$, if $(u, Z) \in R_a$ and $Z \subseteq Z'$ then $(u, Z') \in R_a$.
A monotonic power model is a neighborhood model whose underlying frame is a monotonic power frame.

In order to provide the semantic interpretations of formulas in a model, we need to provide semantic interpretations of the game constructors. We shall use double vertical lines $\|-\|$ to refer to semantic interpretations of formulas in GL and game constructors in monotonic neighborhood models, in order to distinguish it from the semantics given for IPDL, where we use square brackets $[\![-]\!]$. We follow the definitions in [2]. Formally, we define operations on the lattice $\mathcal{N}W = \mathcal{P}(W \times \mathcal{P}(W))$ of *neighborhood relations* over W as follows:

- $R\|\cup\|R' = R \cup R'$
- $R\|\cap\|R' = R \cap R'$
- $(u, Z') \in R\|\circ\|R'$ iff there is some $Z \subseteq W$ with $(u, Z) \in R$ and $(v, Z') \in R'$ for all $v \in Z$.
- $\|?\|(Z) = \{(w, Z') \in W \times \mathcal{P}(W) \mid w \in Z \cap Z'\}$

Finally, we define $\|*\|R$ to be the least fixpoint in the lattice $\mathcal{N}W$ of the monotone map F defined by:

$$FS = \mathsf{skip}^{\uparrow}\|\cup\|(R\|\circ\|S)$$

where $\mathsf{skip}^{\uparrow} = \{(w, Z) \in W \times \mathcal{P}(W) \mid w \in Z\}$. We can now set up the semantics of GL. Fixing a monotonic power model $\mathfrak{M}$, we define the interpretation of every formula φ and the neighborhood relations R_π corresponding to each game term π in the obvious way, so that in particular we have $R_{\pi_1 \cup \pi_2} = R_{\pi_1}\|\cup\|R_{\pi_2}$, $R_{\pi_1 \cap \pi_2} = R_{\pi_1}\|\cap\|R_{\pi_2}$ etc., and $u \in \|[\pi]\varphi\|$ iff $(u, \|\varphi\|) \in R_a$. For a monotonic power model $\mathfrak{M} = (W, R, V)$ and $u \in W$ we shall also write $\mathfrak{M}, u \vDash \varphi$ for $u \in \|\varphi\|$. Since semantic interpretations are always defined relative to a model, if necessary we shall use the notation $\|-\|_{\mathfrak{M}}$ rather than $\|-\|$ to make it clear which model $\mathfrak{M}$ is being referred to. We write $\vDash \varphi$ if $\mathfrak{M}, u \vDash \varphi$ for every pointed monotone power model $(\mathfrak{M}, u)$. We get the following result, showing in what sense IPDL indeed generalizes the semantics of GL:

Proposition 6. *For any GL-formula φ, and any monotonic power model $\mathfrak{M}$, we have $\|\varphi\|_{\mathfrak{M}} = [\![\varphi]\!]_{\mathfrak{M}}$.*

From this proposition, we get the following result:

Theorem 5. *IPDL is a conservative extension of GL. That is, for every GL-formula φ, we have*

$$\vDash \varphi \textit{ iff } \Vdash \varphi$$

In other words: the formulas of IPDL that are valid on arbitrary neighborhood frames form a conservative extension of the GL-formulas that are valid over monotonic power frames.

4.3 Axiomatization

Our axiomatization for IPDL is given below.

INL Axioms: (Mon), (Weak), (Un), (Lem) and (Bot).

Reduction Axioms from DINL: (Test), (Ch), (Pa) and (Cmp).

Basic Rules: (MP) and (RE).

Kleene Star: Finally we add axioms and rules for iteration. The Kleene star is a least fixpoint construction, and a standard approach to axiomatizing least fixpoints is to use one *fixpoint axiom* and one *induction rule* (see [10]). The fixpoint axiom **Fix** is stated as follows:

$$[\pi^*](\Psi;\varphi) \leftrightarrow (\bigwedge \Psi \wedge \varphi) \vee [\pi \circ \pi^*](\Psi;\varphi)$$

We will actually need *two* induction rules:

Ind1:

$$\frac{\varphi \to \gamma \qquad [\pi]\gamma \to \gamma}{[\pi^*]\varphi \to \gamma}$$

Ind2:

$$\frac{(\psi \wedge \varphi) \to \gamma \qquad [\pi](\gamma;[\pi^*]\varphi) \to \gamma}{[\pi^*](\psi;\varphi) \to \gamma}$$

Remark 1. The reason that we require two distinct induction rules can be seen as follows: the reduction axioms for IPDL should be interpreted as encoding a recursive translation of the language IPDL into the modal μ-calculus (interpeted on instantial neighborhood models). When we pass by formulas involving the Kleene-star in this translation, the translation will not surprisingly involve least fixpoint operators, and the induction rules then correspond to the Kozen-Park induction rules for least fixpoint operators. This step of the translation is trickier than the step for the Kleene star in a translation of PDL into the μ-calculus (see [6]), and requires use of nested least fixpoint variables.

Note also that the second induction axiom only involves a single instantial formula ψ. This is because we can "pre-process" an arbitrary formula $[\pi^*](\psi_1,...,\psi_n;\varphi)$ by applying the axiom **Fix**, and then applying the composition axiom (Cmp) to the formula $[\pi \circ \pi^*](\psi_1,...,\psi_n;\varphi)$ to obtain the formula:

$$[\pi]([\pi^*](\psi_1;\varphi),....,[\pi^*](\psi_n;\varphi);[\pi^*]\varphi)$$

Here, each occurrence of the operator $[\pi^*]$ is followed by at most one instantial formula.

We denote this axiom system as Ax3 and write $\varphi \vdash_{\mathsf{Ax3}} \psi$ to say that $\mathsf{Ax3} \vdash \varphi \to \psi$. We will also sometimes drop the explicit reference to the system Ax3, simply writing $\vdash \varphi$ or $\varphi \vdash \psi$.

Theorem 6. *The axiom system* Ax3 *is sound and complete for validity over neighborhood models.*

The soundness part of this theorem is a fairly straightforward check. For the completeness proof, we shall rely heavily on the following lemma, which was proved (in a slightly different formulation) in [3]: fix a finite and subformula closed set of formulas Σ. An *atom* over Σ is a maximal consistent subset of Σ, and we denote the set of atoms over Σ by $\mathsf{At}(\Sigma)$. Given any atom $w \in \mathsf{At}(\Sigma)$, let $\widehat{w}$ be its conjunction, and let $\widehat{Z} = \{\widehat{w} \mid w \in Z\}$ for a set of atoms Z.

Lemma 1. *Let $[\pi](\Psi; \varphi)$ be any formula such that each formula in $\Psi \cup \{\varphi\}$ is a boolean combination of formulas in Σ. Then $[\pi](\Psi; \varphi)$ is provably equivalent to a disjunction of formulas of the form $[\pi](\widehat{Z}; \bigvee \widehat{Z})$ for $Z \subseteq \mathsf{At}(\Sigma)$ being some set of atoms with $w \vdash \varphi$ for each $w \in Z$ and for all $\psi \in \Psi$ there is some $v \in Z$ with $v \vdash \psi$.*

We shall also need an adapted concept of Fischer-Ladner closure:

Definition 4. *A set Σ of formulas is said to be* Fischer-Ladner closed *if the following clauses hold:*

- *If $\varphi \in \Sigma$, and the main connective of φ is not $\neg$, then the formula $\neg\varphi$ is in Σ.*
- *Any subformula of a formula in Σ is in Σ.*
- *If $[\gamma?](\Psi; \varphi)$ is in Σ then so is $\gamma \wedge \bigwedge \Psi \wedge \varphi$.*
- *If $[\pi_1 \circ \pi_2](\psi_1, ..., \psi_n; \varphi) \in \Sigma$, then $[\pi_1]([\pi_2](\psi_1; \varphi), ..., [\pi_1](\psi_n; \varphi); [\pi_2]\varphi)$ is in Σ too.*
- *If $[\pi_1 \cup \pi_2](\Psi; \varphi) \in \Sigma$ then $[\pi_1](\Psi; \varphi) \vee [\pi_2](\Psi; \varphi) \in \Sigma$ too.*
- *If $[\pi_1 \cap \pi_2](\Psi; \varphi) \in \Sigma$ then the formula:*

$$\bigvee \{[\pi_1](\Theta_1; \varphi) \wedge [\pi_2](\Theta_2; \varphi) \mid \Psi = \Theta_1 \cup \Theta_2\}$$

  is in Σ too.
- *If $[\pi^*](\Psi; \varphi) \in \Sigma$ then $(\bigwedge \Psi \wedge \varphi) \vee [\pi \circ \pi^*](\Psi; \varphi)$ is in Σ too.*

Lemma 2. *Every formula φ is a member of some finite Fischer-Ladner closed set of formulas.*

Proof. Standard, see for example [5].

Fix a finite and Fischer-Ladner closed set of formulas Σ. An *atom* over Σ is a maximal consistent subset of Σ, and we denote the set of atoms over Σ by $\mathsf{At}(\Sigma)$. Given any atom $w \in \mathsf{At}(\Sigma)$, let $\widehat{w}$ be its conjunction, and let $\widehat{Z} = \{\widehat{w} \mid w \in Z\}$ for a set of atoms Z.

Lemma 3. *Let $[\pi](\Psi; \varphi)$ be any formula such that each formula in $\Psi \cup \{\varphi\}$ is a boolean combination of formulas in Σ. Then $[\pi](\Psi; \varphi)$ is provably equivalent to a disjunction of formulas of the form $[\pi](\widehat{Z}; \bigvee \widehat{Z})$ for $Z \subseteq \mathsf{At}(\Sigma)$ being some set of atoms with $w \vdash \varphi$ for each $w \in Z$ and for all $\psi \in \Psi$ there is some $v \in Z$ with $v \vdash \psi$.*

Definition 5. *Given any label* π, *we define the relation* $S^{\Sigma}_{\pi} \subseteq \mathsf{At}(\Sigma) \times \mathcal{P}(\mathsf{At}(\Sigma))$ *by setting* $(w, Z) \in S^{\Sigma}_{\pi}$ *iff* $\widehat{w} \wedge [\pi](\widehat{Z}; \bigvee \widehat{Z})$ *is consistent with respect to the system* Ax3.

The canonical neighborhood model *over* Σ *denoted* $\mathfrak{C}^{\Sigma}$ *is defined as the triple* $(W^{\Sigma}, R^{\Sigma}, V^{\Sigma})$ *where* W^{Σ} *is the set of atoms over* Σ, $R^{\Sigma}_a = S^{\Sigma}_a$ *for each atomic label* a, *and* $V^{\Sigma}(p) = \{w \in W^{\Sigma} \mid p \in w\}$.

The key lemma in the completeness proof, which is proved using the induction rules for the Kleene star, is the following:

Lemma 4. *For each label* π, *we have* $S^{\Sigma}_{\pi^*} \subseteq [\![*]\!](S^{\Sigma}_{\pi})$.

Lemma 4 is needed to prove Lemma 5 below, by induction on the complexity of program terms. Say that a label π is *safe* if, for every formula γ such that the term $\gamma?$ appears in π, we have $\gamma \in \Sigma$ and furthermore, $\gamma \in w$ iff $\mathfrak{C}^{\Sigma}, w \Vdash \gamma$ for each $w \in \mathsf{At}(\Sigma)$.

Lemma 5. *For every safe label* π, *we have* $S^{\Sigma}_{\pi} \subseteq R^{\Sigma}_{\pi}$.

Using Lemma 5 we can prove a truth lemma for the canonical model:

Lemma 6. *For every atom* w *and any* $\psi \in \Sigma$, *we have* $(\mathfrak{C}^{\Sigma}, w) \Vdash \psi$ *if and only if* $\psi \in w$.

Finally, we can now prove Theorem 6: suppose the formula φ is not provable, so that $\neg\varphi$ is consistent. By Lemma 2, $\neg\varphi$ belongs to some finite Fischer-Ladner closed set Σ and since $\neg\varphi$ is consistent it belongs to some atom w. Hence $\varphi \notin w$ and by Lemma 6 we have $\mathfrak{C}^{\Sigma}, w \nVdash \varphi$. So φ is not valid.

As a corollary to the completeness proof, which produces a finite model for a consistent formula, we get:

Theorem 7. *IPDL has the finite model property and is decidable.*

5 Concluding Remarks

We have explored a propositional dynamic logic defined over instantial neighborhood logic. A language extension that is clearly related to the framework of this paper is the addition to the base language of least and greatest fixpoint operators, which for standard modal logic results in the *modal* μ*-calculus*. It is well known that PDL can be viewed as a fragment of the modal μ-calculus. In fact, our logic IPDL can also be translated into the analogous extension of INL with fixpoints. The translation is not straightforward though, and in fact the best translation we have found so far even causes an exponential blowup in formula size. We have omitted this material here due to lack of space. The fixpoint extension of INL is a very well behaved language: as shown in [3], INL is a coalgebraic modal logic corresponding to a weak pullback preserving functor - the double covariant powerset functor - that additionally preserves finite sets.

(This should be contrasted with the monotone neighborhood functor, which is the appropriate functor for monotone modal logic and is known *not* to preserve weak pullbacks - see [12]. The monotone neighborhood functor is not suitable for INL since INL-formulas are not invariant for the behavioural equivalence associated with this functor.) This means that the μ-calculus extension of INL will inherit a number properties that hold in much wider generality: the language has the finite model property and is decidable [16], a sound and complete system of axioms is available [8] and the uniform interpolation property holds [11]. Note however that it does *not* mean that we obtain our completeness result (and hence decidability and finite model property) for free, since completeness for fragments of modal μ-calculi does not generally follow easily from completeness of the full languages. Witnessing examples are Reynold's highly non-trivial completeness proof for CTL* [14] (which is a fragment of the μ-calculus [7]), or Parikh's game logic, which still lacks a complete system of axioms.

There is a growing body of work on PDL-like coalgebraic logics, with generic results on axiomatizability, see for example [9]. This setting is clearly related to the present work, however our system IPDL is not covered by this framework as it stands: while the covariant powerset functor is a monad, the *double* covariant powerset functor is not, which would be a requirement for existing work on coalgebraic PDL-logics to readily apply[1]. Perhaps the framework can be modified to capture IPDL as an instance – we offer this as a challenge and an interesting direction for future research.

References

1. Alur, R., Henzinger, T.A., Kupferman, O.: Alternating-time temporal logic. J. ACM (JACM) **49**(5), 672–713 (2002)
2. van Benthem, J.: Logic in Games. MIT Press, Cambridge (2014)
3. van Benthem, J., Bezhanishvili, N., Enqvist, S., Yu, J.: Instantial neighborhood logic. Rev. Symbol. Logic **10**(1), 116–144 (2017)
4. Berwanger, D.: Game logic is strong enough for parity games. Stud. Logica **75**(2), 205–219 (2003)
5. Blackburn, P., de Rijke, M., Venema, Y.: Modal Logic. Cambridge Tracts in Theoretical Computer Science. Cambridge University Press, Cambridge (2001)
6. Carreiro, F., Venema, Y.: PDL inside the μ-calculus: a syntactic and an automata-theoretic characterization. Adv. Modal Logic **10**, 74–93 (2014)
7. Dam, M.: CTL* and ECTL* as fragments of the modal μ-calculus. Theor. Comput. Sci. **126**(1), 77–96 (1994)
8. Enqvist, S., Seifan, F., Venema, Y.: Completeness for coalgebraic fixpoint logic. In: Proceedings of the 25th EACSL Annual Conference on Computer Science Logic (CSL 2016), LIPIcs, vol. 62, pp. 7:1–7:19 (2016)
9. Hansen, H.H., Kupke, C.: Weak completeness of coalgebraic dynamic logics. arXiv preprint arXiv:1509.03017 (2015)
10. Kozen, D.: Results on the propositional μ-calculus. Theor. Comput. Sci. **27**, 333–354 (1983)

[1] We are thankful to Helle Hansen for pointing this out to us.

11. Marti, J., Seifan, F., Venema, Y.: Uniform interpolation for coalgebraic fixpoint logic. In: Proceedings of the Sixth Conference on Algebra and Coalgebra in Computer Science (CALCO 2015), pp. 238–252 (2015)
12. Marti, J., Venema, Y.: Lax extensions of coalgebra functors. In: Pattinson, D., Schröder, L. (eds.) CMCS 2012. LNCS, vol. 7399, pp. 150–169. Springer, Heidelberg (2012). doi:10.1007/978-3-642-32784-1_9
13. Parikh, R.: The logic of games and its applications. Ann. Discrete Math. **24**, 111–139 (1985)
14. Reynolds, M.: An axiomatization of full computation tree logic. J. Symbol. Logic **66**, 1011–1057 (2001)
15. van Benthem, J., Ghosh, S., Liu, F.: Modelling simultaneous games in dynamic logic. Synthese **165**(2), 247–268 (2008)
16. Venema, Y.: Automata and fixed point logic: a coalgebraic perspective. Inform. Comput. **204**, 637–678 (2006)

Contradictory Information as a Basis for Rational Belief

Adam Přenosil[1,2(✉)]

[1] Department of Logic, Faculty of Arts, Charles University, Prague, Czech Republic
adam.prenosil@gmail.com

[2] Institute of Computer Science, Czech Academy of Sciences, Prague, Czech Republic

Abstract. As agents faced with fallible information, we frequently find ourselves in situations where we are forced to base our beliefs on evidence which is in some way or another contradictory. We nevertheless want these beliefs to be rational. This paper presents a simple probabilistic model of what it means for a belief based on a contradictory body of evidence to be rational. In this approach, we model contradictions in the evidence available to us as resulting from random noise, and we model our task as rational agents as reconstructing the most likely states of affairs given the evidence available to us. Our main result consists in providing several equivalent descriptions of the non-reflexive and non-monotonic consequence relation which formalizes the notion that it is reasonable to accept that a proposition is true given good evidence supporting some set of propositions.

Keywords: Paraconsistent logic · Belnap–Dunn logic · Non-monotonic logic · Non-reflexive logic · Belief revision

1 Introduction

It is an indisputable fact that we as rational agents are frequently faced with information which, although generally trustworthy, is fallible. Our trust in such information is limited at least by logic: when faced with two logically conflicting pieces of information, we conclude that at least one of them must be wrong, rather than concluding that a contradiction is true. That much is clear. It is less clear what exactly it means for a belief to be rationally supported by such trustworthy but fallible (briefly: *good*) evidence. The present paper is intended as a contribution towards answering this question in a certain idealized case. In particular, we shall provide a formal answer to this question in the form a consequence relation $\Gamma \vDash \varphi$ which is to be interpreted as: it is reasonable to accept the belief that φ is true given good evidence supporting Γ.

Such a consequence relation needs to tolerate inconsistency in the sense that logical contradictions ought not imply everything (as they have the unfortunate habit of doing in classical logic). It is, after all, not reasonable to conclude in the face of contradictory information that every proposition is true. Since such

A. Baltag et al. (Eds.): LORI 2017, LNCS 10455, pp. 151–165, 2017.
DOI: 10.1007/978-3-662-55665-8_11

logics have long been studied, it would at first glance seem reasonable to look for the consequence relation among the well-known examples of these so-called *paraconsistent* logics. However, this impression is mistaken, since we want the consequence relation $\Gamma \vDash \varphi$ to be non-reflexive and non-monotonic.

For example, given good (but contradictory) evidence supporting $p \wedge \neg p$, it is not reasonable to conclude that $p \wedge \neg p$ is true, since it is never reasonable to conclude that a contradiction is true.[1] Likewise, although it is reasonable to conclude that p is true given good evidence supporting p, it is not reasonable to conclude that p is true given both good evidence supporting p and good evidence supporting $\neg p$. In other words, we want to reject both reflexivity (since $p \wedge \neg p \nvDash p \wedge \neg p$) and monotonicity (since $p \vDash p$ but $p, \neg p \nvDash p$).

To illustrate what kind of consequence relation we seek and what its potential use is, let us compare it to one of the simplest and best motivated paraconsistent logics, namely the four-valued Belnap–Dunn logic. This logic, first introduced by Dunn [8], was proposed by Belnap [6,7] as a logic of how a computer should deal with inconsistent information. Its idea is very simple: in addition to the two truth values of classical logic, we allow for two additional truth values: being both true and false and being neither true nor false. These are interpreted epistemically rather than ontologically: a proposition is true if we have information supporting its truth, and it is false if we have information supporting its falsity. (Belnap aptly calls these values "told true" and "told false".) This, of course, allows for the possibility that we have information supporting both the truth and the falsity of a proposition, as well as for the possibility that we have neither.

A computer then simply collects the information it is presented with. Although such a computer can record contradictions in a non-trivial manner and infer e.g. that information supporting p and $q \vee r$ also supports $(p \wedge q) \vee r$, it leaves the handling of the contradictions to the user. The user is presented with information that e.g. the computer has both information supporting p and information supporting $\neg p$, but is given no guidance on how to respond to such information. Our ambition, on the other hand, is to potentially automatize a reasonable response to contradictory information. That is, rather than being content with having a computer *record* contradictory information, we want it to go further and to *sidestep* the contradictions in this information and present the user with consistent data, or even to circumvent the user altogether and take decisions on its own based on the contradictory data it is given.

Moreover, we wish to achieve this goal in a purely logical way, without using any non-logical information about the relative plausibility of different states of affairs, or the relative strength of different pieces of evidence, or indeed without considering such objects as distinct pieces of evidence. While a theory which involves such notions is no doubt worth exploring, we claim that part of the interest of the results presented here is that they do not rely on any kind of

[1] At least if we disregard issues relating to logical non-omniscience and dialetheism. Our logic is intended to model reasoning with ordinary empirical propositions where such issues are not prominent.

extralogical structure, i.e. that a purely logical theory specifying which beliefs it is reasonable to adopt in the face of contradictory evidence is possible.

We will not offer here any substantial explanation of what constitutes good evidence, apart from whatever explanation will be implicit in our probabilistic model layed out below. By good evidence we will simply mean evidence which is convincing in the absence of contrary evidence but may become unconvincing in the presence of contrary evidence. That is, while it is reasonable to conclude that φ is true on the basis of some good evidence, we may be forced to abandon this belief if we come into possession of contrary evidence. This not mean that our evidence was not good (in our sense), only that good evidence is still fallible.

The structure of the paper is as follows. In Sect. 2 we lay out a simplified model of how contradictions arise in the evidence available to us. Based on this model, we then define a consequence relation which determines whether it is reasonable to accept a given belief based on some body of evidence, taking inspiration from the probabilistic semantics of classical logic due to Adams [1]. We then describe this consequence relation syntactically. Finally, in Sect. 3 we discuss the potential use of this consequence relation in belief revision theory. In particular, we propose that what we ought to revise when faced with new evidence are what we call evidence sets rather than belief sets.

The consequence relation studied in the present paper is a close relative of the so-called minimally inconsistent Logic of Paradox introduced by Priest [10], the distance-based paraconsistent logics of Arieli [3], and it is also related to the adaptive logics of Batens [5] and his school. These connections will be spelled out in more detail in the course of the paper.

2 The Noise and the Signal

We propose to analyze the notion of a rational belief based on an inconsistent body of information in probabilistic terms, using an analogy from information theory. A fundamental task of information theory is to reconstruct the original form of a message which was in some way distorted by passing through a noisy channel. In our analogy, the original message is the actual state of affairs (or some relevant aspect of it), the noise is some mechanism which introduces false evidence into the body of evidence available to us, and the distorted message is precisely this body of evidence. Our goal is then to reconstruct the most probable states of affairs from this distorted evidence.

More precisely, we shall imagine that even though p ($\neg p$) is false, there is some small probability ε_p ($\varepsilon_{\neg p}$) that we will actually come into possession of good evidence that p ($\neg p$) is true. This probability, of course, is difficult to quantify precisely, but that need not concern us: we will be concerned with the limit case where this probability approaches zero. On the other hand, the probability that we will come into possession of good evidence that p ($\neg p$) is true given that p

($\neg p$) is true, denoted τ_p ($\tau_{\neg p}$), will be a quantity whose precise value is again irrelevant but which remains far enough from zero.[2]

The above will be our rudimentary working model of how inconsistencies arise in our data. It may be no doubt elaborated further in various ways, but the fundamental point of the approach should be clear enough: we are given some probabilistic mechanism which corrupts our evidence by introducing some amount of falsehood into it, and our task is to reconstruct the most probable states of affairs which give rise to this corrupted body of evidence.

Let us now formalize these notions more precisely. Consider a finite set of atomic propositions $At = \{p, q, r, \ldots\}$. The set Fm of all *formulas* generated by At in the signature $\{\wedge, \vee, \neg\}$ is defined in the obvious way. By a *state of affairs* we shall mean a classical valuation of formulas into the two truth values of classical logic. By a *state of evidence* we shall mean a valuation of formulas into the four truth values of the Belnap–Dunn logic. Such valuations follows precisely the classical truth and falsity conditions:

$$\begin{array}{ll} \neg\varphi \text{ is true} \Leftrightarrow \varphi \text{ is false} & \neg\varphi \text{ is false} \Leftrightarrow \varphi \text{ is true} \\ \varphi \wedge \psi \text{ is true} \Leftrightarrow \varphi \text{ is true } \& \ \psi \text{ is true} & \varphi \vee \psi \text{ is true} \Leftrightarrow \varphi \text{ is true or } \psi \text{ is true} \\ \varphi \wedge \psi \text{ is false} \Leftrightarrow \varphi \text{ is false or } \psi \text{ is false} & \varphi \vee \psi \text{ is false} \Leftrightarrow \varphi \text{ is false } \& \ \psi \text{ is false} \end{array}$$

except we allow for the possibility of being both true and false or being neither true nor false. The consequence relation $\Gamma \vdash_{\mathcal{B}} \varphi$ of the Belnap–Dunn logic is defined as follows: φ is true in each state of evidence in which each $\gamma \in \Gamma$ is true. Being "true" or "false" in a state of evidence corresponds to having good evidence for or against a given proposition.

Note that we could restrict to *complete* states evidence here, i.e. to the three-valued Logic of Paradox rather than the four-valued Belnap–Dunn logic. In such states of evidence we have evidence for or against each proposition, possibly both. This would not substantially change the model layed out below. However, we prefer to stick to the four-valued setting, since it is a better approximation of the epistemic situations that we are normally faced with.

States of affairs and states of evidence combine to form *full states*, which are pairs (w, e) consisting of a state of affairs w and a state of evidence e. In the following, "φ true" shall denote the set of full states (w, e) such that φ is true in w, "φ told" shall denote the set of full states (w, e) such that φ is true in e, and "Γ told" shall denote the intersection of the sets "γ told" for $\gamma \in \Gamma$.

Let P be a probability measure on the finite set of all states of affairs. We only impose one condition on P, namely that it is *non-excluding* in the sense that $P(w) \in (0, 1)$ for each state of affairs w. Given a choice of constants ε_p,

[2] We have no particular story to tell about how one comes into possession of good evidence, or what precisely constitutes good evidence. It is entirely up to the user of the logic to supply such a story. We simply imagine that the user of the logic collects some information about the world and at the end of this process he ends up with a certain probability with information supporting or contradicting p (given that p is true, or given that p is false).

$\varepsilon_{\neg p}$, τ_p, $\tau_{\neg p} \in (0,1)$ for each $p \in At$, briefly denoted $\boldsymbol{\varepsilon}$ and $\boldsymbol{\tau}$, we extend P to a probability measure $P_{\boldsymbol{\varepsilon\tau}}$ on the finite set of all full states as follows (here l ranges over literals):

$$\frac{P_{\boldsymbol{\varepsilon\tau}}(w,e)}{P(w)} = \prod_{\substack{l \text{ true in } w \\ l \text{ told in } e}} \tau_l \cdot \prod_{\substack{l \text{ true in } w \\ l \text{ not told in } e}} (1-\tau_l) \cdot \prod_{\substack{\neg l \text{ true in } w \\ l \text{ told in } e}} \varepsilon_l \cdot \prod_{\substack{\neg l \text{ true in } w \\ l \text{ not told in } e}} (1-\varepsilon_l)$$

That is, we are assuming that in each state of affairs w the events of being told l and being told l' are independent for distinct literals l and l' and moreover they are determined by the parameters $\boldsymbol{\tau}$ and $\boldsymbol{\varepsilon}$ as outlined in the equation above. This equation is intended to model the connection between the actual state of affairs and the evidence available to us. In particular, we are assuming here that evidence may be only distorted at the atomic level.

The above model is schematically represented by Fig. 1, where on the left we have the four possible states of affairs over two propositional variables and on the right have have some of the sixteen possible states of evidence. The numbers next to the arrows represent the probabilities of the transitions between the states of affairs and the states of evidence, e.g. the probability that we will end up in the state of evidence $p\overline{p}q$ given that $\overline{p}q$ is the case is 0.1. If the state of evidence does not conflict with the state of affairs, the transition will (at least in the limit case) have a relatively high probability, even if not all aspects of the state of affairs are captured by the state of evidence. This is the case with the transitions $pq \rightarrow pq$ and $pq \rightarrow p$. On the other hand, the more conflict there is between the two, the less likely the transition is. For example, the transition $\overline{p}q \rightarrow p\overline{p}q$ is relatively unlikely due to the conflict between p and $\overline{p}$, and the transition $\overline{pq}$ is even less likely due to the additional conflict betwen q and $\overline{q}$.

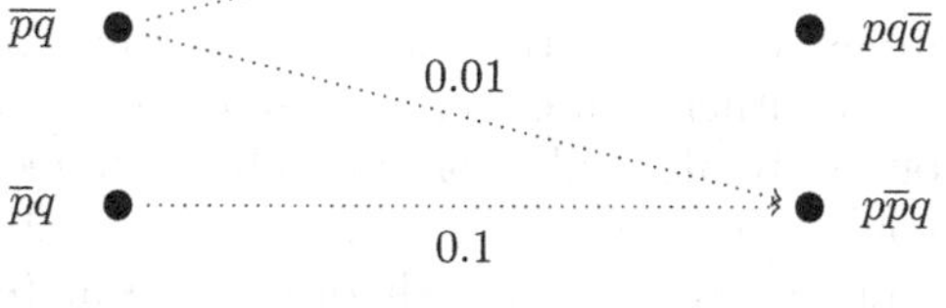

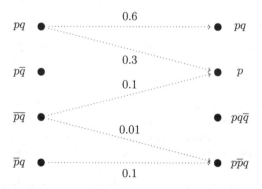

Fig. 1. Transitions between states of affairs and states of evidence

Recall that the conditional probability $P_{\boldsymbol{\varepsilon\tau}}(\varphi \text{ true} \,|\, \Gamma \text{ told})$ is given by the formula $P_{\boldsymbol{\varepsilon\tau}}(\varphi \text{ true } \& \ \Gamma \text{ told})/P_{\boldsymbol{\varepsilon\tau}}(\Gamma \text{ told})$. Once we have fixed the values of the parameters $\boldsymbol{\varepsilon}$ and $\boldsymbol{\tau}$, we can compute this conditional probability for each

φ and Γ. The consequence relation that we seek now admits a simple definition in terms of this conditional probability. At first glance, the definition relies on the parameters $\boldsymbol{\tau}$ and the probability measure P, but we shall see later that in fact it does not depend on these. The main result of the present paper consists in providing a syntactic description of this consequence relation, i.e. in proving a completeness theorem for it.

Definition 1. *We shall say that φ is a fallible consequence of Γ, symbolically $\Gamma \vDash \varphi$, if $\lim_{\boldsymbol{\varepsilon}\to 0} P_{\boldsymbol{\varepsilon\tau}}(\varphi \text{ true} \mid \Gamma \text{ told}) = 1$.*

It is worth recalling here the probabilistic characterization of classical logic due to Adams [1] for the sake of comparison. Adams proved that $\Gamma \vdash \varphi$ holds in classical logic if and only if for each $\varepsilon > 0$ there is some $\delta > 0$ such that $P(\varphi) \geq 1 - \varepsilon$ for each probability measure P such that $P(\gamma) \geq 1 - \delta$ for all $\gamma \in \Gamma$. In other words, Adams proved that classical logic is applicable not only in the idealized case where the premises are known with certainty, but also in the more realistic case where the premises are only known to a high degree of confidence. This is fine as long as it is possible to have a high degree of confidence in our premises. However, if this is not possible due to their inconsistency, we have to take into account the fact that some of our premises are false and reason instead in a way which strives to minimize the distance between the evidence available to us and the actual state of affairs.

Some basic properties of fallible consequence may immediately be inferred. Namely, the premises are governed by the Belnap–Dunn logic $\mathcal{B}$, whereas the conclusion is governed by classical logic $\mathcal{CL}$, in the following sense. Recall that $\Gamma \dashv\vdash_{\mathcal{B}} \Delta$ denotes that Γ and Δ are equivalent in $\mathcal{B}$, i.e. true in the same states of evidence. $\Gamma \vdash_{\mathcal{CL}} \varphi$ denotes that φ is a classical consequence of Γ.

Proposition 1. *If $\Gamma \dashv\vdash_{\mathcal{B}} \Delta$ and $\varphi \vdash_{\mathcal{CL}} \psi$, then $\Gamma \vDash \varphi$ implies $\Delta \vDash \psi$.*

Proof. If $\Gamma \dashv\vdash_{\mathcal{B}} \Delta$, then "$\Gamma$ told" and "Δ told" are the same event, hence $\Gamma \vDash \varphi$ implies $\Delta \vDash \varphi$. If $\varphi \vdash_{\mathcal{CL}} \psi$, then the event "$\varphi$ true" is a subset of the event "ψ true", hence $\lim_{\boldsymbol{\varepsilon}\to 0} P_{\boldsymbol{\varepsilon\tau}}(\psi \text{ true} \mid \Delta \text{ told}) \geq \lim_{\boldsymbol{\varepsilon}\to 0} P_{\boldsymbol{\varepsilon\tau}}(\varphi \text{ true} \mid \Delta \text{ told}) = 1$.

Note that since we are considering languages with only finitely many atomic propositions and there are up to equivalence in $\mathcal{B}$ only finitely many formulas over finitely many atoms, we may without loss of generality restrict to Γ finite.

Recall that a *literal* is either an atom or a negated atom, a *conjunctive clause* is a finite conjunction of literals, and a formula in *disjunctive normal form* is a finite disjunction of conjunctive clauses. We shall use e.g. $p\overline{p}q\overline{r}$ as an abbreviation for the conjunctive clause $p \wedge \neg p \wedge q \wedge \neg r$. Each formula has an (essentially unique) equivalent formula in disjunctive normal form in $\mathcal{B}$.

Let us now introduce some useful notions relating to conjunctive clauses. By the *consistent part* of a conjunctive clause φ, denoted cp φ, we mean the conjunction of all atoms $p \in \varphi$ such that $\neg p \notin \varphi$ and all negated atoms $\neg p \in \varphi$ such that $p \notin \varphi$. By the *inconsistent part* of a conjunctive clause φ, denoted ip φ, we mean the conjunction of all atoms $p \in \varphi$ such that $\neg p \in \varphi$ and all negated atoms $\neg p \in \varphi$ such that $p \in \varphi$. Clearly $\varphi = \mathrm{cp}\ \varphi \wedge \mathrm{ip}\ \varphi$ and moreover $\mathrm{cp}\ \varphi$ and $\mathrm{ip}\ \varphi$ are disjoint as sets of literals.

Conjuctive clauses may be ordered according to how consistent they are. We shall say that a conjunctive clause φ is *less consistent* than a conjunctive clause ψ if $\mathrm{ip}\ \varphi \vdash_{\mathcal{B}} \mathrm{ip}\ \psi$. Given a set of conjunctive clauses C, by the *minimally inconsistent* elements of C we shall mean those $\varphi \in C$ such that no $\psi \in C$ is strictly more consistent than φ. For example, among the clauses $p\overline{p}q\overline{r}$, $pq\overline{q}r\overline{r}s$, $p\overline{p}rs\overline{s}$, the first two are minimally inconsistent while the last one is not, being strictly less consistent than the first clause.

Conjunctive clauses will be identified with sets of literals in the obvious way. In particular, given two conjunctive clauses φ and ψ, we shall use $\varphi \setminus \psi$ to denote the conjunction of all literals which occur in φ but not in ψ. Since each state of affairs α corresponds to a unique conjunctive clause, $\varphi \setminus \alpha$ will denote the conjunction of all literals in φ which do not hold at α. The conjunctive clause $\varphi \setminus \alpha$ is thus a measure of how far α is from what φ claims to be the case. Likewise, $\varphi \cap \psi$ (in particular, $\varphi \cap \alpha$) will denote the conjunction of all literals which occur in both φ and ψ (in particular, all literals in φ which hold at α).

With this notation in hand, our definition of $P_{\varepsilon\tau}(\varphi \text{ told} \mid \alpha \text{ true})$, where φ is a conjunctive clause and α is a state of affairs, may be written as

$$P_{\varepsilon\tau}(\varphi \text{ told} \mid \alpha \text{ true}) = \prod_{l \in \varphi \setminus \alpha} \varepsilon_l \cdot \prod_{l \in \varphi \cap \alpha} \tau_l \cdot \prod_{l \in \alpha \setminus \varphi} (1 - \tau_l) \cdot \prod_{l \notin \alpha \,\&\, l \notin \varphi} (1 - \varepsilon_l). \quad (*)$$

Lemma 1. $\lim_{\varepsilon \to 0} P_{\varepsilon\tau}(\varphi$ *told* $\mid \alpha$ *true*$) = 0$ *if and only if* $\alpha \wedge \varphi$ *is inconsistent.*

Proof. If φ is a conjunctive clause, then the claim holds because the conditional probability $P_{\varepsilon\tau}(\varphi \text{ told} \mid \alpha \text{ true})$ contains a factor of the form ε_l if and only if some atom occurs negated (unnegated) in φ but unnegated (negated) in α, i.e. if and only if $\alpha \wedge \varphi$ is inconsistent.

Now let $\bigvee_{i \in I} \varphi_i$ be the disjunctive normal form of φ. If $\alpha \wedge \varphi$ is inconsistent, then $\alpha \wedge \varphi_i$ is inconsistent for each $i \in I$, therefore $\lim_{\varepsilon \to 0} P_{\varepsilon\tau}(\varphi \text{ told} \mid \alpha \text{ true}) = \lim_{\varepsilon \to 0} P_{\varepsilon\tau}(\bigvee_{i \in I} \varphi_i \text{ told} \mid \alpha \text{ true}) \leq \lim_{\varepsilon \to 0} \sum_{i \in I} P_{\varepsilon\tau}(\varphi_i \text{ told} \mid \alpha \text{ true}) \leq \sum_{i \in I} \lim_{\varepsilon \to 0} P_{\varepsilon\tau}(\varphi_i \text{ told} \mid \alpha \text{ true}) \leq 0$.

Conversely, if $\alpha \wedge \varphi$ is consistent, then $\alpha \wedge \varphi_i$ is consistent for some $i \in I$. But $\lim_{\varepsilon \to 0} P_{\varepsilon\tau}(\varphi \text{ told} \mid \alpha \text{ true}) = 0$ only if $\lim_{\varepsilon \to 0} P_{\varepsilon\tau}(\varphi_i \text{ told} \mid \alpha \text{ true}) = 0$, which is not the case, since the lemma holds for φ_i.

Lemma 2. *Let* φ *and* ψ *be conjunctive clauses and let* α *be a state of affairs. Then* $\lim_{\varepsilon \to 0} \frac{P_{\varepsilon\tau}(\varphi \text{ told} \mid \alpha \text{ true})}{P_{\varepsilon\tau}(\psi \text{ told} \mid \alpha \text{ true})} = 0$ *if and only if* $\varphi \setminus \alpha \supsetneq \psi \setminus \alpha$.

Proof. This claim follows immediately from $(*)$, since $\tau_l \in (0, 1)$.

Theorem 1 (Completeness theorem). *Let* $\{\gamma_i \mid i \in I\}$ *be a set of conjunctive clauses and* $\{\gamma_j \mid j \in J\}$ *be the subset of all minimally inconsistent clauses among these. Then* $\bigvee_{i \in I} \gamma_i \vDash \varphi$ *if and only if* $\bigvee_{j \in J} \mathrm{cp}\ \gamma_j \vdash_{\mathcal{CL}} \varphi$.

Proof. Let $\gamma = \bigvee_{i \in I} \gamma_i$. Then by Bayes' theorem

$$P_{\varepsilon\tau}(\varphi \text{ true} \,|\, \gamma \text{ told}) = \frac{P_{\varepsilon\tau}(\gamma \text{ told} \,|\, \varphi \text{ true}) P_{\varepsilon\tau}(\varphi \text{ true})}{P_{\varepsilon\tau}(\gamma \text{ told})} = \frac{\sum_{\alpha \models \varphi} P_{\varepsilon\tau}(\gamma \text{ told} \,|\, \alpha \text{ true}) P_{\varepsilon\tau}(\alpha \text{ true})}{\sum_{\alpha} P_{\varepsilon\tau}(\gamma \text{ told} \,|\, \alpha \text{ true}) P_{\varepsilon\tau}(\alpha \text{ true})}$$

where α ranges over all states of affairs in the denominator, and over all states of affairs where φ holds in the numerator. Recall that $P_{\varepsilon\tau}(\gamma \text{ told} \,|\, \alpha \text{ true})$ is determined by the constants ε and τ via $(*)$. Clearly $P_{\varepsilon\tau}(\alpha \text{ true}) \in (0,1)$. In fact, $P_{\varepsilon\tau}(\alpha \text{ true}) = P(\alpha)$, even though we will not need this observation.

Suppose first that γ is classically consistent. By Lemma 1 summation in both the numerator and the denominator may be restricted to summation over α such that $\alpha \wedge \gamma$ is consistent, i.e. α such that γ holds at α. Moreover, each such summand contributes a strictly positive number in the limit. Therefore $\lim_{\varepsilon \to 0} P_{\varepsilon\tau}(\varphi \text{ true} \,|\, \gamma \text{ told}) = 1$ if and only if no such summands occur in the denominator but not the numerator, or equivalently if and only if γ does not hold at α whenever φ does not hold at α, or equivalently if and only if $\gamma \vdash_{\mathcal{CL}} \varphi$. But if γ is consistent, then γ and $\bigvee_{j \in J} \mathrm{cp}\ \gamma_j$ are classically equivalent.

Suppose on the other hand that γ is not classically consistent. We shall first group together those γ_i which have the same inconsistent part. That is, there is a finite set $K \subseteq J$ and formulas δ_k such that each such formula δ_k is a disjunction of all formulas γ_i which have the same inconsistent part, and moreover each γ_j occurs as a disjunct in some δ_k. There is also a subset $L \subseteq K$ of those δ_k which are disjunctions of formulas γ_j for $j \in J$.

Let us define $\mathrm{cp}\ \delta_k$ as the disjunction of the consistent parts of those γ_i which occur as disjuncts in δ_k, and $\mathrm{ip}\ \delta_k$ as the inconsistent part of each γ_i which occurs as a disjunct in δ_k. Moreover, we may define $\mathrm{ip}\,(\delta_k \wedge \delta_{k'})$ as $\mathrm{ip}\ \delta_k \wedge \mathrm{ip}\ \delta_{k'}$. Then the conjunctions $\delta_k \wedge \delta_{k'}$ for distinct $k, k' \in K$ are strictly more inconsistent than each $\delta_{k''}$ for $k'' \in K$, in the sense that $\mathrm{ip}\,(\delta_k \wedge \delta_{k'}) \vdash_{\mathcal{B}} \mathrm{ip}\ \delta_{k''}$ for each $\delta_{k''}$, but $\mathrm{ip}\ \delta_{k''} \nvdash_{\mathcal{B}} \mathrm{ip}\,(\delta_k \wedge \delta_{k'})$ for each $\delta_{k''}$. Observe that $\bigvee_{i \in I} \gamma_i$ is equivalent in $\mathcal{B}$ to $\bigvee_{k \in K} \delta_k$, and $\bigvee_{j \in J} \gamma_j$ is equivalent in $\mathcal{B}$ to $\bigvee_{l \in L} \delta_l$.

Clearly $P_{\varepsilon\tau}(\bigvee_{i \in I} \gamma_i \text{ told} \,|\, \alpha \text{ true}) = P_{\varepsilon\tau}(\bigvee_{k \in K} \gamma_k \text{ told} \,|\, \alpha \text{ true})$. Lemma 2 now implies that $P_{\varepsilon\tau}(\bigvee_{k \in K} \delta_k \text{ told} \,|\, \alpha \text{ true}) \approx P_{\varepsilon\tau}(\bigvee_{l \in L} \delta_l \text{ told} \,|\, \alpha \text{ true})$, disregarding some summands which tend to zero faster than the right-hand side. Moreover, $P_{\varepsilon\tau}(\bigvee_{l \in L} \delta_l \text{ told} \,|\, \alpha \text{ true}) \approx \sum_{l \in L} P_{\varepsilon\tau}(\delta_l \text{ told} \,|\, \alpha \text{ true})$ in the same sense, using the observations from the previous paragraph.

Appealing to Lemma 2 again yields that $\sum_{\alpha} \sum_{l \in L} P_{\varepsilon\tau}(\delta_l \text{ told} \,|\, \alpha \text{ true}) \approx \sum_{l \in L} \sum_{\alpha \models \mathrm{cp}\ \delta_l} P_{\varepsilon\tau}(\delta_l \text{ told} \,|\, \alpha \text{ true})$ and also $\sum_{\alpha \models \varphi} \sum_{l \in L} P_{\varepsilon\tau}(\delta_l \text{ told} \,|\, \alpha \text{ true}) \approx \sum_{l \in L} \sum_{\alpha \models \varphi \wedge \mathrm{cp}\ \delta_l} P_{\varepsilon\tau}(\delta_l \text{ told} \,|\, \alpha \text{ true})$. We therefore obtain that

$$P_{\varepsilon\tau}(\varphi \text{ true} \,|\, \gamma \text{ told}) \approx \frac{\sum_{l \in L} \sum_{\alpha \models \varphi \wedge \mathrm{cp}\ \delta_l} P_{\varepsilon\tau}(\delta_l \text{ told} \,|\, \alpha \text{ true}) P_{\varepsilon\tau}(\alpha \text{ true})}{\sum_{l \in L} \sum_{\alpha \models \mathrm{cp}\ \delta_l} P_{\varepsilon\tau}(\delta_l \text{ told} \,|\, \alpha \text{ true}) P_{\varepsilon\tau}(\alpha \text{ true})} \qquad (\dagger)$$

in the same sense as above. The right-to-left direction of the theorem is now immediate: setting $\varphi = \bigvee_{l \in L} \mathrm{cp}\ \delta_l$ in $(\dagger)$, which is equivalent to $\bigvee_{j \in J} \gamma_j$ in $\mathcal{B}$, yields a fraction where the numerator and the denominator are equal (since $\varphi \wedge \text{cp } \delta_l$ is equivalent to cp δ_l if $\varphi = \bigvee_{l \in L} \text{cp } \delta_l$).

To prove the left-to-right implication, suppose that $\bigvee_{j \in J} \text{cp } \gamma_j \nvdash_{\mathcal{CL}} \varphi$, i.e. $\bigvee_{l \in L} \text{cp } \delta_l \nvdash_{\mathcal{CL}} \varphi$. Then there is some state of affairs β where cp δ_l holds for some $l \in L$ but φ does not. Then this state of affairs β is summed over in the denominator of (†) but not in the numerator. Thus $\gamma \vDash \varphi$ can hold only if the term $P_{\varepsilon\mathcal{T}}(\delta_l \text{ told} \mid \beta \text{ true}) P_{\varepsilon\mathcal{T}}(\beta \text{ true})$ tends to zero faster than the remaining terms in the denominator of (†). But this is not the case by the left-to-right direction of Lemma 2, since δ_l is minimally inconsistent among the disjuncts δ_k for $k \in K$ and cp δ_l holds at β.

Corollary 1. *If Γ is classically consistent, then $\Gamma \vDash \varphi$ if and only if $\Gamma \vdash_{\mathcal{CL}} \varphi$.*

Recall that each formula is equivalent in $\mathcal{B}$ to a formula in disjunctive normal form. Theorem 1 therefore suffices to fully describe the fallible consequence relation. Observe also that Theorem 1 justifies calling the formula $\bigvee_{j \in J} \text{cp } \gamma_j$ the *classical approximation* to the formula $\gamma = \bigvee_{i \in I} \gamma_i$. Theorem 1 may then be rephrased as saying that it is reasonable (in a precise technical sense) to accept the belief that φ is true on the basis of good evidence supporting γ if and only if φ is classically entailed by this classical approximation to γ.

To illustrate how fallible consequence works, let us consider a simple example due to Dunn [9]. A fire has broken out in the building that you are in and you are trying to escape to safety. You find yourself at a crossroads with three ways out: to your left, to your right, and straight ahead. Two fireman are there to provide you with assistance. Unfortunately, their guidance is contradictory. The first one says that the only safe way out is left. The second one says that the only safe way out is right. How should you respond to such a situation?

Formalizing the first fireman's advice as $l \wedge \neg r \wedge \neg s$ and the second fireman's advice as $r \wedge \neg l \wedge \neg s$, the best classical approximation to this body of evidence by Theorem 1 is $\neg s$, i.e. the theory stating that the way straight ahead is not safe. This agrees with Dunn's answer and presumably with our intuition.

Theorem 1 may in fact be rephrased in a more semantic way. Let Val_4 be the set of all states of evidence, let $Val_3 \subseteq Val_4$ be the set of all *complete* states of evidence, i.e. of all states of evidence where each atom, or equivalently each formula, is either true or false (possibly both), and let $Val_2 \subseteq Val_4$ be the set of all *classical* states of evidence, i.e. of all states of evidence where each atom, or equivalently each formula, is either true and not false or it is false and not true.

The states of evidence may be ordered according to their information content and according to how inconsistent they are. The information order is a partial order on Val_4 such that $u \sqsubseteq v$ if and only if each atom which is true in u is true in v and each atom which is false in u is false in v (equivalently, each formula which is true in u is true in v). The inconsistency order is a preorder on Val_4 such that u is *less inconsistent* than v if and only if each atom which is both true and false in u is both true and false in v. This inconsistency order was introduced and used for a similar purpose already by Arieli and Avron [4].

In the following, we shall use the notation $||\varphi|| = \{w \in Val_4 \mid \varphi \text{ is true at } w\}$ and $||\Gamma|| = \bigcap_{\gamma \in \Gamma} ||\gamma||$. If $S \subseteq Val_4$ is a set of complete states of evidence, then

$\text{Min}_{\text{inc}}\, S$ will denote the set of all minimal states of evidence in S with respect to the ordering by *inconsistency*.

Theorem 2 (Completeness theorem). *$\Gamma \vDash \varphi$ if and only if $w \in ||\varphi||$ for each classical state of evidence w such that $w \sqsubseteq v$ for some $v \in \text{Min}_{\text{inc}}(||\Gamma|| \cap Val_3)$.*

Proof. Let $\gamma = \bigvee_{i \in I} \gamma_i$ be a formula in disjunctive normal form equivalent to Γ, and let $\{\gamma_j \mid j \in J\}$ be the set of all minimally inconsistent disjuncts of γ. In particular, $w \in ||\Gamma||$ if and only if there is some $i \in I$ such that $w \in ||\gamma_i||$ and each atom which occurs negated (unnegated) in cp γ_i is false but not true (true but not false) in w. If $w \in \text{Min}_{\text{inc}}(||\Gamma|| \cap Val_3)$, then in fact we may take $i \in J$. Otherwise there is some $j \in J$ such that γ_j is strictly more consistent than γ_i, in which case each complete state of evidence which only assigns the value both true and false to the atoms which occur both negated and unnegated in γ_j is strictly below w in the inconsistency order.

If w is a classical state of evidence and v is a complete state of evidence, then $w \sqsubseteq v$ if and only if each atom (or equivalently, each formula) which is true but not false in v is true in w and each atom (or equivalently, each formula) which is false but not true in v is false in w. That is, there is some $v \in \text{Min}_{\text{inc}}(||\Gamma|| \cap Val_3)$ such that $w \sqsubseteq v$ if and only if there is some $j \in J$ such that cp γ_j is true in w.

Of course, Theorem 2 is simply a reformulation of Theorem 1 using a slightly different language. Nevertheless, we believe that it may help to have Fig. 2 in mind when thinking about fallible consequence.

Figure 2 schematically represents how the classical approximation to Γ is computed based on Theorem 2. First, the set $||\Gamma||$ of all models which satisfy Γ is computed and intersected with Val_3. Then, the minimaly inconsistent elements of this set are determined. In Fig. 2, this is represented by the black area. Finally, the downset of this set is determined and intersected with the set of all classical valuation Val_2, which yields the hatched area of Fig. 2. This set is then the classical approximation to the original set of states $||\Gamma||$.

The close relation between the fallible consequence relation introduced above, the "minimally inconsistent Logic of Paradox" $Mi\mathcal{LP}$ introduced by Priest [10], and the "consequence relation for preserving consistency" $\vDash^4_{\mathcal{I}_1}$ introduced by Arieli and Avron [4] deserves noticing. Indeed, Priest's definition of $Mi\mathcal{LP}$ and Arieli and Avron's definition of $\vDash^4_{\mathcal{I}_1}$ (Definition 2) is very similar to Theorem 2, except that no projection onto the classical states is performed. In the case of $Mi\mathcal{LP}$ this is, of course, due to the fact that Priest believes that some contradictions are true, therefore he has no reason to perform such a projection. We recall the definitions of these logics below, rephrased in our terminology.

Definition 2. *$\Gamma \vdash_{Mi\mathcal{LP}} \varphi$ if and only if $\text{Min}_{\text{inc}}(||\Gamma|| \cap Val_3) \subseteq ||\varphi||$. $\Gamma \vDash^4_{\mathcal{I}_1} \varphi$ if and only if $\text{Min}_{\text{inc}}\, ||\Gamma|| \subseteq ||\varphi||$.*

In fact, there is a very simple explicit relationship between our logic and a variant of the logic $Mi\mathcal{LP}$. Let δ denote the so-called conflation operation on

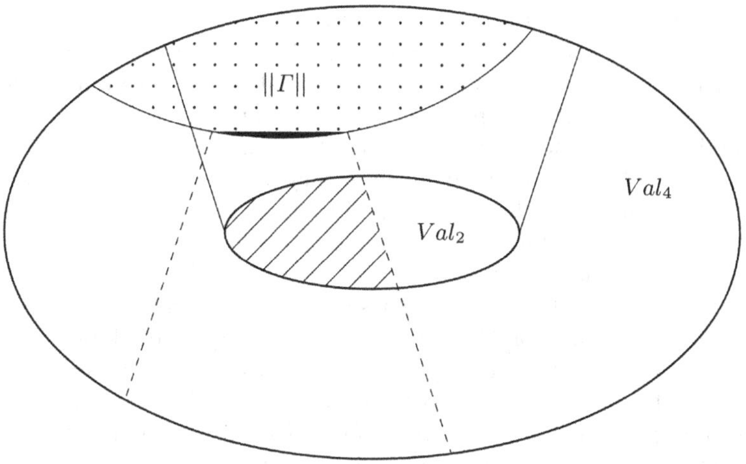

Fig. 2. Computing classical approximations

the four truth values, defined by $\delta\mathsf{t} = \mathsf{t}$, $\delta\mathsf{f} = \mathsf{f}$, $\delta\mathsf{n} = \mathsf{b}$, and $\delta\mathsf{b} = \mathsf{n}$. That is, $\delta\varphi$ is true (false) if and only if φ is not false (not true). The logic $Mi\mathcal{LP}$ does not include δ in its signature, and moreover its standard three-valued semantics with the truth values $\{\mathsf{f}, \mathsf{b}, \mathsf{t}\}$ cannot be expanded by δ, as $\delta\mathsf{b} = \mathsf{n} \notin \{\mathsf{f}, \mathsf{b}, \mathsf{t}\}$. However, if we use the *four-valued semantics* for $Mi\mathcal{LP}$ from Definition 2, we can easily expand the logic $Mi\mathcal{LP}$ by the operator δ, obtaning the logic $Mi\mathcal{LP}\delta$. The relationship between fallible consequence and $Mi\mathcal{LP}\delta$ can now be expressed as follows.[3]

Proposition 2. *$\Gamma \vDash \varphi$ if and only if $\Gamma \vdash_{Mi\mathcal{LP}\delta} \delta\varphi$.*

Proof. Suppose that $\Gamma \vdash_{Mi\mathcal{LP}\delta} \delta\varphi$. Then $\neg\varphi$ does not hold in any world in $\mathrm{Min_{inc}}(||\Gamma|| \cap Val_3)$. Therefore it does not hold in any classical w such that $w \sqsubseteq v$ for some $v \in \mathrm{Min_{inc}}(||\Gamma|| \cap Val_3)$. But then φ does hold in w whenever $w \sqsubseteq v \in \mathrm{Min_{inc}}(||\Gamma|| \cap Val_3)$ for w classical, hence $\Gamma \vDash \varphi$. Conversely, suppose that $\Gamma \nvdash_{Mi\mathcal{LP}\delta} \delta\varphi$. Since each φ is equivalent to a conjunction of disjunctive clauses, we may assume without loss of generality that $\varphi = \bigvee_{k\in K} l_k$, where l_k are literals. Then there is some $w \in \mathrm{Min_{inc}}(||\Gamma|| \cap Val_3)$ such that $\neg\varphi$ holds in w. Let $\gamma = \bigvee_{i\in I} \gamma_i$ be a formal in disjunctive normal form equivalent in $\mathcal{B}$ to Γ. Then there is some disjunct γ_j, minimally inconsistent among the conjunctive clauses γ_i, such that $\gamma_j \vdash_{\mathcal{B}} \neg\varphi$. Thus $\gamma_j \vdash_{\mathcal{B}} \neg l_k$ for each $k \in K$. It now follows that there is some classical $w \sqsubseteq v$ where $\gamma_j \vdash_{\mathcal{B}} \neg l_k$ for each $k \in K$. But then $\neg\varphi$ is true in w, hence φ is not and $\Gamma \nvDash \varphi$.

Finally, let us note that a variant of the fallible consequence relation may be obtained by tinkering with the definition of the probability measure $P_{\boldsymbol{\varepsilon}\mathcal{T}}$. Instead of having a tuple of parameters $\boldsymbol{\varepsilon}$, we may suppose that a single parameter ε

[3] The following proposition was suggested to the author by one of the referees.

suffices, i.e. that $\varepsilon_l = \varepsilon_{l'}$ for all literals l and l'. This yields a *stronger* logic, since we have more control over the way in which ε approaches zero. (Notice that the definition of fallible consequence essentially quantifies universally over all possible ways in which the parameters ε may approach zero.)

To obtain an analogue of Theorem 1 for this consequence relation, it suffices to modify the notion of minimal inconsistency by saying that a conjunctive clause φ is less consistent than a conjunctive clause ψ if $|\,\mathrm{ip}\;\varphi| \geq |\,\mathrm{ip}\;\psi|$ rather than $\mathrm{ip}\;\varphi \supseteq \mathrm{ip}\;\psi$. Accordingly, the inclusion $\varphi \setminus \alpha \supsetneq \psi \setminus \alpha$ in Lemma 2 would be replaced by the inequality $|\varphi \setminus \alpha| > |\psi \setminus \alpha|$. With these modifications, the statement and proof of Theorem 1 carries over to this modified relation.

To obtain an analogue of Theorem 2 for this consequence relation, it suffices to modify the inconsistency order by saying that u is less inconsistent than v if less atoms (in the sense of cardinality) are both true and false in u then in v. The statement and proof of Theorem 2 then carries over to the modified relation.

The distance-based paraconsistent logic of Arieli [3] based on the Hamming distance is similar in spirit to this variant of the fallible consequence relation. However, there are some differences between the framework used here, where consequence is a relation between a set of formulas and a formula, and the framework of Arieli, whose consequence relation is a relation between a *multiset* of formulas and a formula. This corresponds to distinguishing between how well supported different formulas are, which we disregard in our framework. Moreover, while Arieli allows for inconsistent multisets of premises, he assumes that each individual formula in such a multiset is classically consistent, or equivalently in his framework, that inconsistent states are infinitely far away from consistent ones. Due to these differences, Arieli's framework is reflexive, while ours is not.

The idea of handling inconsistent sets of premises by restricting to minimally inconsistent models is also known from the adaptive logic program of Batens [5] and others. In particular, given a Tarskian consequence relation and a set of "abnormal" formulas, one can uniformly obtain an adaptive logic which also tries to minimize the amount of inconsistency or abnormality forced on us by the premises. One advantage of the adaptive logic approach is that a dynamic proof theory is available for such logics, a topic which is left untouched in the present paper.

3 Beliefs Bases vs. Evidence Sets

In this final section, we shall briefly discuss the possible applications of the fallible consequence relation in the field of belief revision. In particular, this consequence relation allows us to replace revisions of belief bases by revisions of *evidence sets*, by which we mean theories of the Belnap–Dunn logic.

In the standard AGM model of belief revision [2], a belief set B is revised in the face of new information supporting φ to form a new belief set $B * \varphi$ such that $\varphi \in B * \varphi$. There are several problematic aspects to this, one of which is that such an approach forces us to accept the primacy of new information, i.e. new information is *a priori* taken to be more reliable than old information.

Such problems are often circumvented by revising *belief bases* rather than belief sets, where a belief base is simply a set of formulas (not necessarily closed under classical consequence) which we have independent reasons to believe.

In that approach, belief bases perform much the same task as evidence sets. That is, they allow us to distinguish between different inconsistent bodies of information. Both belief bases and evidence sets form a potential basis for our beliefs in the sense that we can compute a consistent belief set from either of them. In the case of belief bases, this process of extracting consistent information from an inconsistent belief base is called consolidation. In the case of evidence sets, a consistent belief set B may be obtained from an inconsistent evidence set E with the help of the fallible consequence relation as $B = \{\varphi \mid E \vDash \varphi\}$. Of course, studying the fallible consequence relation $\vDash$ is equivalent to studying the assignment $B \mapsto E$, just like describing a Tarskian consequence relation is the same as studying the behaviour of the associated consequence operator.

Even though belief bases and evidence sets perform the same task within the theory of belief revision, there is, however, a major difference between the two notions. Belief bases are a very fine-grained, syntactic tool for representing inconsistent bodies of information, whereas evidence sets are more coarse-grained and more neutral between syntax and semantics. Accordingly, each notion has its own advantages and shortcomings which make it appropriate in certain contexts but not in others. From one point of view, evidence sets are insufficient to capture the distinctions made by belief bases. From another, belief bases make distinctions for which there may be no basis in our evidence.

Generating an evidence set E from a belief base B is easy: simply take the deductive closure of B in the Belnap–Dunn logic. Every evidence set may be generated in this way. However, observe that some information is lost in this transition to evidence sets. For example, the distinct belief bases $\{p, q\}$ and $\{p \wedge q\}$, which behave differently under belief revision by $\neg p$, both generate the same evidence set. Evidence sets therefore do not fully represent the information present in belief bases. Whether this is a desirable feature of evidence sets or not depends entirely on whether we view the fine-grained, syntactic character of belief bases as something we wish to avoid or embrace.

The approach based on belief bases is more appropriate whenever we can break down our evidence into minimal independent chunks. But it is often not clear whether this is in fact achievable. If our belief in $p \wedge q$ derives from being told $p \wedge q$ by a trusted colleague, then we do indeed have a single piece of evidence supporting $p \wedge q$, namely our colleague's claim. But if we have no knowledge of how our colleague arrived at this belief (perhaps he simply combined two independent pieces of evidence for p and q), it is far from clear whether the appropriate reaction to learning that p is false is to rescind our belief in q as well, as we would if we formalized our belief base as $\{p \wedge q\}$, or whether we ought to retain our belief in q, as we would if we formalized our belief base as $\{p, q\}$. In order to be applicable to in actual reasoning, the approach based on belief bases needs to provide some guidance for distinguishing these two cases.

The approach based on evidence sets, by contrast, breaks down the available evidence as thoroughly as possible. For example, in the above situation we have $\neg p, p \wedge q \vDash q$. That is, even though it is given no information on how exactly we obtained the evidence that $p \wedge q$ is true, the consequence relation $\vDash$ behaves as if our evidence for $p \wedge q$ came from two independent pieces of evidence for p and for q and then discards the evidence for p while keeping the evidence for q.

4 Conclusion

We have obtained a consequence relation with a probabilistic flavour which determines whether it is reasonable to accept a certain belief based on a potentially inconsistent body of information. Moreover, this consequence relation is purely logical in the sense that no extralogical choice of selection functions or distance functions or plausibility measures was needed to define it. How such extralogical information, if available, can be used to refine this consequence relation is yet to be seen. However, we find it worth noting that a reasonable consequence relation is available even in the absence of such extralogical information.

On the other hand, a major limitation of the results presented above is the assumption that each proposition arises from applying certain connectives to some independent atomic propositions and that the random noise in our evidence acts only on these atomic propositions. A more general approach would start from a De Morgan algebra of propositions which does not single out any of them as atomic. It is not difficult to devise a relational semantics for a consequence relation similar to $\vDash$ which does not suffer from this limitation: in Theorem 2 we may simply restrict to states of evidence which satisfy some given extralogical constraints. Instead, the difficulty lies in providing a reasonable probabilistic justification/interpretation for such a relation along the lines of Theorem 1.

In conclusion, recall that the justification of the fallible consequence relation consisted in a certain simpified model of how contradictions arise in our evidence: *given that* contradictions in our evidence arise as a result of random noise with certain characteristics, φ is almost sure to be true given evidence supporting Γ. Of course, one may consider different models of how contradictions arise in our evidence, leading to different consequence relations. Therefore, we do not claim that this is the unique correct consequence relation for this purpose. We would, however, like to suggest that in the absence of some such model, it appears difficult to specify what it means for a given belief or a given decision to be rational in the face of some contradictory body of evidence.

Acknowledgments. This research was supported by the project 16-07954J SEGA "From shared attitudes to group agency" of the Czech Science Foundation and DFG. The author would also like to thank the two anonymous referees for their helpful comments, in particular for suggesting that Proposition 2 might hold.

References

1. Adams, E.W.: The Logic of Conditionals: An Application of Probability to Deductive Logic. Springer Science and Business Media, Netherlands (1975)
2. Alchourrón, C.E., Gärdenfors, P., Makinson, D.: On the logic of theory change: Partial meet contraction and revision functions. J. Symb. Logic **50**, 510–530 (1985)
3. Arieli, O.: Distance-based paraconsistent logics. Int. J. Approx. Reason. **48**(3), 766–783 (2008)
4. Arieli, O., Avron, A.: The value of four values. Artif. Intell. **102**(1), 97–141 (1998)
5. Batens, D.: A universal logic approach to adaptive logics. Logic. Universalis **1**(1), 221–242 (2007)
6. Belnap, N.D.: How a computer should think. In: Contemporary Aspects of Philosophy, Oriel Press Ltd., pp. 30–56 (1977)
7. Belnap, N.D.: A useful four-valued logic. In: Dunn, J.M., Epstein, G. (eds.) Modern Uses of Multiple-Valued Logic of Episteme, vol. 2, pp. 5–37. Springer, Netherlands (1977)
8. Michael, J.: Dunn. Intuitive semantics for first-degree entailments and 'coupled trees'. Philos. Stud. **29**(3), 149–168 (1976)
9. Dunn, J.M.: Contradictory information can be better than nothing: The example of the two firemen. Logica 2012, (2012)
10. Priest, G.: Minimally inconsistent LP. Studia Logic. **50**(2), 321–331 (1991)

Stability in Binary Opinion Diffusion

Zoé Christoff[1(✉)] and Davide Grossi[2]

[1] Department of Philosophy, University of Bayreuth, Bayreuth, Germany
zoe.christoff@gmail.com
[2] Department of Computer Science, University of Liverpool, Liverpool, England

Abstract. The paper studies the stabilization of the process of diffusion of binary opinions on networks. It first shows how such dynamics can be modeled and studied via techniques from binary aggregation, which directly relate to neighborhood frames. It then characterizes stabilization in terms of such neighborhood structures, and shows how the monotone μ-calculus can express relevant properties of them. Finally, it illustrates the scope of these results by applying them to specific diffusion models.

1 Introduction

The paper establishes necessary and sufficient conditions for stabilizing behavior in the dynamics of binary opinions on networks, and shows how they can be expressed in a known modal fixpoint logic.

Context. This paper brings together tools and methods from three different fields: network theory, judgment aggregation, and logic. The most well-known models for opinion diffusion, such as the stochastic linear averaging models [11] and the threshold models [19] were developed some decades ago and their stabilization conditions have been extensively studied (e.g., [16], cf. [22] for an overview). In the meantime, judgment aggregation theory [13,20] has developed the formal tools to analyse collective opinion formation. In particular, binary judgment aggregation [17] focuses on opinions consisting of the acceptance or rejection of a given set of issues. The recent 'propositional opinion diffusion' setting from [18]—but see also the very related framework developed in [8]—, as well as the analysis of liquid democracy from [10], blended the two above traditions studying forms of iterated aggregation on networks. Our paper is a contribution to this very recent line of research. Furthermore, in the past few years, logicians have also designed different systems to reason about diffusion in networks [9], ranging from the most specific 'Facebook logic' setting [23], to logics for threshold diffusion [1–3], to the more abstract discussion of the general laws of oscillations [4].

Contribution. We look at opinion diffusion as a synchronous process of iterated binary aggregation on networks: all agents take up the opinion resulting from aggregating the individual opinions of their network neighbors (their influencers). The two key parameters for binary opinion diffusion are the underlying network and the 'rule' each agent uses to aggregate its neighbors' opinions. It is known that in binary aggregation, rules (under mild assumptions) can be represented

A. Baltag et al. (Eds.): LORI 2017, LNCS 10455, pp. 166–180, 2017.
DOI: 10.1007/978-3-662-55665-8_12

through sets of winning and veto coalitions. The same applies also when such rules are 'restricted' to sets of neighbors on a network. So the above diffusion process can alternatively be represented as follows: *each agent accepts an issue p* whenever one of its winning coalitions accepts p, and rejects it whenever one of its veto coalitions rejects$p$. Sets of winning and veto coalitions offer therefore a useful level of abstraction to study diffusion processes of binary opinions, leaving aside the specifics of each underlying network and aggregation rule. As such sets of winning and veto coalitions are nothing but neighborhoods, in the modal logic sense [7], we show how neighborhood logics enhanced with fixpoint operators [14] can be used to logically characterize key properties of those processes, such as their stability and their stabilization.

Outline. Section 2 introduces some preliminaries on binary aggregation and presents the basic model of opinion diffusion as iterated aggregation on networks. Section 3 contains the paper's main results: it establishes the correspondence between diffusion as binary aggregation on networks and neighborhood structures, exploits this correspondence to obtain general stabilization results, and gives a logical characterization of stability and stabilization in the μ-calculus on monotone neighborhood structures. Section 4 illustrates these general results via examples of two specific diffusion models. Section 5 concludes.

2 Preliminaries

In this section we first recall standard notions from binary aggregation (mainly from [13,17,20]) and the theory of winning coalitions (mainly from [12]), and then introduce the model of binary opinion diffusion used in the paper.

2.1 Binary Aggregation

A binary aggregation structure (*BA structure*) is a tuple $\mathcal{A} = \langle N, \mathbf{P} \rangle$ where: $N = \{1, \ldots, n\}$ is a non-empty finite set individuals with $|N| = n \in \mathbb{N}$; and $\mathbf{P} = \{p_1, \ldots, p_m\}$ is a non-empty finite set of issues with $|\mathbf{P}| = m \in \mathbb{N}$, each represented by a propositional atom. An *opinion* function O is an assignment of acceptance/rejection values (or, truth values) to the issues in $\mathbf{P}$. Thus, $O(p) = \mathbf{0}$ (respectively, $O(p) = \mathbf{1}$) indicates that opinion O rejects (respectively, accepts) the issue p. Syntactically, the two opinions correspond to the truth of the literals p or $\neg p$. For $p \in \mathbf{P}$ we write $\pm p$ to denote one element from $\{p, \neg p\}$, and $\pm \mathbf{P}$ to denote $\bigcup_{p \in \mathbf{P}} \{p, \neg p\}$, which we will refer to as the *agenda* of $\mathcal{A}$. The set of opinions is denoted $\mathcal{O}$. An *opinion profile* $\mathbf{O} = (O_1, \ldots, O_n)$ is a tuple recording the opinion of each individual in N.

We need to introduce some further terminology and notation. We sometimes treat opinion profiles as $n \times m$ binary matrices recording the opinion of n agents (rows) over m issues (columns). So given a profile $\mathbf{O}$ and $i \in N$ the ith projection of $\mathbf{O}$ is denoted O_i (i.e., the opinion of agent i on each issue), and the pth projection of $\mathbf{O}$ is denoted O^p (i.e., the vector consisting of the opinion of each agent on issue p). The sub-profile consisting of the agents in $C \subset N$ is denoted

$\mathbf{O}_C$. So $\mathbf{O}_C = \langle O_i \rangle_{i \in C}$. We also denote by $\mathbf{O}(p) = \{i \in N \mid O_i(p) = \mathbf{1}\}$ the set of agents accepting issue p in profile $\mathbf{O}$, and by $\mathbf{O}(\neg p) = \{i \in N \mid O_i(p) = \mathbf{0}\}$ the set of agents rejecting issue p in profile $\mathbf{O}$. The latter notation is naturally extended to sub-profiles.

Given a BA structure $\mathcal{A}$, a (resolute) *aggregator* (for $\mathcal{A}$) is a function $F : \mathcal{O}^N \to \mathcal{O}$, mapping every profile of individual opinions to one collective opinion. $F(\mathbf{O})(p)$ denotes the outcome of the aggregation of profile $\mathbf{O}$ on issue p. In this paper we will work only with *independent* aggregators, that is, aggregators such that the collective opinion on each issue is a function only of the individual opinions on that issue.[1] An independent aggregator $F : \mathcal{O}^N \to \mathcal{O}$ can be represented as a tuple $\langle F^p \rangle_{p \in \mathbf{P}}$ of functions $F^p : \{\mathbf{0}, \mathbf{1}\}^n \to \{\mathbf{0}, \mathbf{1}\}$ mapping binary vectors (the opinions on a given issue p) to acceptance or rejection (of that issue). If for all $p, q \in \mathbf{P}$ $F^p = F^q$ then the resulting independent aggregator F is said to be *neutral*, in the sense that all issues are aggregated in the same way.

Example 1 (Aggregators). The benchmark aggregator is the so-called *issue-wise strict majority rule*, which we will refer to simply as majority (maj). The rule accepts an issue p if and only if a majority of voters accept it, formally:

$$\mathsf{maj}(\mathbf{O})(p) = \mathbf{1} \Longleftrightarrow |\mathbf{O}(p)| \geq \frac{|N| + 1}{2}. \tag{1}$$

So-called *quota rules* are aggregators that generalize majority. They accept an issue if the number of voters accepting it exceeds a given quota, possibly different for each issue. Formally, a quota rule is defined via a function $q : \mathbf{P} \to (2^N \to N)$, associating a natural number (between 1 and $|N|$) to each issue and subset of agents such that $q(p)(C) \leq |C|$. Intuitively, a quota tells you how many agents, in a given set C, should accept p in order for p to be collectively accepted. The quota aggregator (F_q) is defined as follows: $F_q(\mathbf{O}_C)(p) = \mathbf{1} \Longleftrightarrow |\mathbf{O}_C(p)| \geq q(p)$. A quota rule F_q is called uniform in case q is a constant function. Issue-wise majority is a uniform quota rule, with quota $q = \lceil \frac{N+1}{2} \rceil$.[2] Finally, the *dictatorship* of i (d_i) is the aggregator so defined: $d_i(\mathbf{O})(p) = O_i(p)$, for any opinion profile $\mathbf{O}$.

2.2 Winning and Veto Coalitions

Given an aggregator F, a set of voters $C \subseteq N$ is a *winning coalition* for issue $p \in \mathbf{P}$ if for every profile $\mathbf{O}$ we have that *if* $O_i(p) = \mathbf{1}$ for all $i \in C$ and $O_i(p) = \mathbf{0}$ for all $i \notin C$, *then* $F(\mathbf{O})(p) = \mathbf{1}$. Furthermore, we call a set of voters $C \subseteq N$ a *veto coalition* for issue $p \in \mathbf{P}$ if for every profile $\mathbf{O}$ we have that *if* $O_i(p) = \mathbf{0}$ for

[1] Formally, an aggregator F is independent iff, for all $p \in \mathbf{P}$: for any profiles $\mathbf{O}, \mathbf{O}'$ such that for all $i \in N, O_i(p) = O'_i(p)$, $F(\mathbf{O})(p) = F(\mathbf{O}')(p)$. Independence is a natural assumption in settings like ours, where issues are assumed not to be logically interrelated.

[2] Recall that the ceiling function $\lceil x \rceil$ denotes the smallest integer larger than x.

all $i \in C$ and $O_i(p) = \mathbf{1}$ for all $i \notin C$, *then* $F(\mathbf{O})(p) = \mathbf{0}$. Let $\mathcal{W}_p$ denote the set of winning coalitions and $\mathcal{V}_p$ the set of veto ones, for issue p. As aggregators are of type $\mathcal{O}^N \to \mathcal{O}$, and by the independence assumption, one can show that winning and veto coalitions (for any issue p) are dual notions, i.e., for any $C \subseteq N$:

$$C \in \mathcal{W}_p \iff N \backslash C \notin \mathcal{V}_p. \tag{2}$$

It is well known that further properties imposed on an independent aggregator F induce extra structure on its set of winning (and veto) coalitions (cf. [13, Lemma 17.1]). In particular, an independent aggregator F is *monotonic* iff for each $p \in \mathbf{P}$ and for any $C \in \mathcal{W}_p$ (respectively, $C \in \mathcal{V}_p$), if $C \subseteq C'$ then $C' \in \mathcal{W}_p$ (respectively, $C' \in \mathcal{V}_p$), i.e., winning and veto coalitions are closed under supersets. Moreover, an independent aggregator is *responsive* iff for each $p \in \mathbf{P}$ $\emptyset \notin \mathcal{W}_p$ and $\emptyset \notin \mathcal{V}_p$, that is, the aggregator is not a constant function. In this paper we focus on aggregators that are independent, monotonic and responsive. All aggregators in Example 1 have these properties.

Given a set of winning coalitions $\mathcal{W}_p$ for a given issue, a *dummy agent* is an agent $i \in N$ such that: if $C \in \mathcal{W}_p$, then $C \cup \{i\} \in \mathcal{W}_p$ (trivial if $\mathcal{W}_p$ is closed under supersets), and if $i \in C \in \mathcal{W}_p$ then $C \backslash \{i\} \in \mathcal{W}_p$. Intuitively, dummy agents are agents whose opinions are irrelevant for $\mathcal{W}_p$ (and, dually, $\mathcal{V}_p$).

Example 2. The winning and veto coalitions for the majority aggregator are (for issue p): $\mathcal{W}_p = \left\{C \subseteq N \mid |C| \geq \frac{|N|+1}{2}\right\}$ and $\mathcal{V}_p = \left\{C \subseteq N \mid |C| \geq \frac{|N|}{2}\right\}$. The majority aggregator induces a set of winning coalitions with no dummy agents. A dictatorship, instead, induces a set of winning coalitions where all agents except the dictator are dummy.

2.3 Opinion Diffusion as Binary Aggregation on Networks

Let $\mathcal{G} = \left\langle N, \{R_p\}_{p \in \mathbf{P}} \right\rangle$ be a multi-relational structure where each $\langle N, R_p \rangle$ is a directed graph over N, with R_p serial. We call these graphs (multi-issues) *networks*. Intuitively, $\langle N, R_p \rangle$ represents the network of influence among agents in N about issue p. So we write iR_pj to denote that i's opinion on p is influenced by j's opinion on p. The set of influencers (for issue p) of agent i is denoted $R_p(i) = \{j \in N \mid iR_pj\}$. Note that i may belong to such set.

Recall that an independent aggregator $F : \mathcal{O}^N \to \mathcal{O}$ can be represented as a tuple $\langle F^p \rangle_{p \in \mathbf{P}}$ of functions $F^p : \{\mathbf{0}, \mathbf{1}\}^{|N|} \to \{\mathbf{0}, \mathbf{1}\}$, one for each issue. So fixing a network $\mathcal{G}$, we can associate to each agent i and issue p a function F^p mapping the opinions of the influencers of i on p to an opinion on p. We call this an *allocation*, and denote it with $\mathbf{F}$.[3] We denote with $\mathbf{F}(i)$ the independent aggregator $\langle \mathbf{F}(i,p) \rangle_{p \in \mathbf{P}}$ obtained by collating the functions $\mathbf{F}(i,p) : \{\mathbf{0}, \mathbf{1}\}^{|R_p(i)|} \to \{\mathbf{0}, \mathbf{1}\}$

[3] More precisely, $\mathbf{F} : N \to \left(\mathbf{P} \to \bigcup_{X \subseteq N} \{\mathbf{0}, \mathbf{1}\}^{\{\mathbf{0},\mathbf{1}\}^{|X|}}\right)$.

modeling how i aggregates its neighbors' opinions over p. Each $\mathbf{F}(i)$ is furthermore assumed to be monotonic and responsive in $R_p(i)$ (cf. Sect. 2.2 above).[4]

The key idea is that the opinion of each agent i is the result of the aggregation, via an independent, monotonic and responsive aggregator, of the opinions of its influencers.

Definition 1. *Let a BA structure* $\mathcal{A} = \langle N, \mathbf{P} \rangle$, *a network* $\mathcal{G} = \left\langle N, \{R_p\}_{p \in \mathbf{P}} \right\rangle$ *and an allocation* $\mathbf{F}$ *be given. Now let* $\mathbf{O}$ *be an opinion profile (for* $\mathcal{A}$*). The binary opinion diffusion process induced by* $\mathcal{G}$ *and* $\mathbf{F}$ *on* $\mathbf{O}$ *is the stream of opinion profiles* $\mathbf{O}^0, \mathbf{O}^1, \ldots, \mathbf{O}^n, \ldots$ *defined inductively as follows:* $\boxed{base}$ $\mathbf{O}^0 = \mathbf{O}$; $\boxed{step}$ $O_i^{n+1}(p) = \mathbf{F}(i)(\mathbf{O}^n)(p)$, $\forall i \in N$, $\forall p \in \mathbf{P}$.

Section 4 will illustrate the above definition extensively, so we refrain from giving an example at this point. Finally, we say that the stream of opinion profiles $\mathbf{O}^0, \mathbf{O}^1, \ldots, \mathbf{O}^n, \ldots$ *stabilizes* (or converges) if there exists $n \in \mathbb{N}$ such that $\mathbf{O}^n = \mathbf{O}^{n+1}$. We will also say that a stream of opinion profiles stabilizes *for issue* p if there exists $n \in \mathbb{N}$ such that $\mathbf{O}^n(p) = \mathbf{O}^{n+1}(p)$. We say that agent $i \in N$ stabilizes for issue p if there exists $n \in \mathbb{N}$ such that $O_i^n(p) = O_i^m(p)$, for any $m > n$. If such an n is reached, we say that agent i is *stable* for issue p.

3 Networks, Aggregators and Neighborhoods

In this section we first show how the opinion diffusion model of Definition 1 can be recast in terms of neighborhood structures from modal logic [7], we then use such structures to establish novel general stabilization theorems, and finally show how a known variant of the μ-calculus [14] can be used to express properties of such structures, which are relevant for their stabilizing behavior.

3.1 Neighborhood Structures for Aggregation on Networks

Lemma 1. *Let* $\mathcal{A} = \langle N, \mathbf{P} \rangle$ *be given, and fix an issue* $p \in \mathbf{P}$*:*

1. *For each serial directed graph* $\langle N, R_p \rangle$ *and each allocation* $\mathbf{F}$, *there exists* $\{\mathcal{W}_i\}_{i \in N}$ *such that* $\mathbf{F}(i)(\mathbf{O})(p) = \mathbf{1}$ *iff* $\mathbf{O}(p) \in \mathcal{W}_i$*;*
2. *For each collection* $\{\mathcal{W}_i\}_{i \in N}$ *of winning coalitions, there exists a graph* $\langle N, R_p \rangle$ *and an allocation* $\mathbf{F}$ *such that* $\mathbf{F}(i)(\mathbf{O})(p) = \mathbf{1}$ *iff* $\mathbf{O}(p) \in \mathcal{W}_i$.

Proof. The proof is by construction. $\boxed{1}$ By the definition of $\mathbf{F}$, $\mathbf{F}(i)$ is an independent aggregator for the aggregation structure $\langle R(i), \mathbf{P} \rangle$. It follows that $\mathbf{F}(i)$ can be represented by a set of winning coalitions $\mathcal{W}_i', \mathcal{V}_i' \subseteq \wp(R_p(i))$. We therefore have that, with respect to the agents in $R(i)$, $\mathbf{F}(i)$ can be represented through

[4] It is worth noticing that $\mathbf{F}(i)$ is not an aggregator in the strict sense, as the set of individuals whose opinions are aggregated varies from issue to issue. However, it can be represented by an aggregator on N where $N \backslash R_p(i)$ are dummy agents, as shown later in Lemma 1. We will therefore slightly abuse terminology and still refer to such functions as aggregators.

those sets, that is, for any profile $\mathbf{O}$, $\mathbf{F}(i)(\mathbf{O}_{R_p(i)})(p) = \mathbf{1}$ iff $\mathbf{O}_{R_p(i)}(p) \in \mathcal{W}'_i$. These sets need then to be lifted from the set of influencers $R(i)$ to the full set of agents N, as follows: $\mathcal{W}_i = \{X \subseteq N \mid \exists Y \in \mathcal{W}'_i \ s.t. \ Y \subseteq X\}$. To establish the claim then it suffices to observe that the construction guarantees the following property $X \cap R_p(i) \in \mathcal{W}'_i$ iff $X \in \mathcal{W}_i$. [2] Trivial, as it suffices to take $R_p = N^2$ and each $\mathbf{F}(i)$ to be the aggregator induced by $\mathcal{W}_i$. □

What Lemma 1 tells us is that we can think of a network $\mathcal{G} = \left\langle N, \{R_p\}_{p \in \mathbf{P}} \right\rangle$ where each node i's opinion on the issues is determined by i's neighbors (the set $R_p(i)$ for each issue) through an independent aggregator $\mathbf{F}(i)$ essentially as a neighborhood frame of winning (and, dually, veto) coalitions $\mathcal{C} = \left\langle N, \{\mathcal{W}_p\}_{p \in \mathbf{P}} \right\rangle$ where $\mathcal{W}_p : N \to 2^{2^N}$ assigns, for each issue, the winning coalitions $\mathcal{W}_p(i)$ of each agent on that issue.[5] The bottom line is that binary opinion dynamics on networks can equivalently be studied through neighborhood (multi-)frames.

One final piece of notation: for $C \subseteq N$ and issue p, let $\mathcal{C}_p(C)$ be the set defined as follows: [base] $C^0 = C$; [step] $C^{n+1}_{\mathcal{W}} = \{i \in N \mid C^n \in \mathcal{W}_p(i)\}$ and $C^{n+1}_{\mathcal{V}} = \{i \in N \mid C^n \in \mathcal{V}_p(i)\}$; $\mathcal{C}_p(C) = \bigcup C^n_{\mathcal{W}} \cup \bigcup C^n_{\mathcal{V}}$. Intuitively, $\mathcal{C}_p(C)$ denotes the set of individuals whose opinion on p can be influenced, directly or indirectly, by C once the individuals in C all accept or reject issue p.

Example 3. Consider a given $\mathcal{A} = \langle N, \mathbf{P} \rangle$, and assume that the opinion of all agents in a given network structure $\mathcal{G} = \left\langle N, \{R_p\}_{p \in \mathbf{P}} \right\rangle$ is determined by aggregating the opinions of their network neighbors via the same aggregator F. If F is strict majority maj, then the corresponding neighborhood structure $\mathcal{C} = \left\langle N, \{\mathcal{W}_p\}_{p \in \mathbf{P}} \right\rangle$ is defined as follows (for each $p \in \mathbf{P}$, each $i \in N$): $\mathcal{W}_i = \left\{ C \subseteq N \mid \exists C' \subseteq C \ s.t. \ C' \subseteq R_p(i) \ and \ |C'| \geq \frac{|R_p(i)|+1}{2} \right\}$. In general, for any quota rule F_q, the corresponding neighborhood structure is given by: $\mathcal{W}_i = \{C \subseteq N \mid \exists C' \subseteq C \ s.t. \ C' \subseteq R_p(i) \ and \ |C'| \geq q(p)(R_p(i))\}$. And if F is a dictatorship of a neighbor $j \in R_p(i)$, then all and only the sets containing the dictator matter: $\mathcal{W}_i = \{C \subseteq N \mid j \in C\}$.

3.2 Stabilization in Neighborhood Structures

In this section we assume that network $\mathcal{G} = \left\langle N, \{R_p\}_{p \in \mathbf{P}} \right\rangle$ and allocation $\mathbf{F}$ are given, and we exploit their representation as a system of neighborhoods $\mathcal{C} = \left\langle N, \{\mathcal{W}_p\}_{p \in \mathbf{P}} \right\rangle$.

[5] Note that the construction in the proof of Lemma 1 is such that each agent $j \notin R_p(i)$ participates to i's set of winning and veto coalitions only as a 'dummy' agent who can be added or removed to a winning (or veto) coalition without changing the status of that coalition.

Dependence, Consensus and Gurus. A coalition $C \subseteq N$ is said to have a *win dependence* (w.r.t. a given issue p) on a coalition $D \subseteq N$ whenever there exists a sequence $D = C_0, \ldots, C_m = C$ such that $C_i \in \mathcal{W}_p(i)$ for all $i \in C_{k+1}$ for all $0 \leq k < m$. Similarly, it is said to have a *veto dependence* (w.r.t. a given issue p) on a coalition $D \subseteq N$ whenever there exists a sequence $D = C_0, \ldots, C_m = C$ such that $C_i \in \mathcal{V}_p(k)$ for all $j \in C_{k+1}$ for all $0 \leq k < m$. Intuitively, a win (resp., veto) dependence of C on D means that whenever the agents in D all agree on accepting (resp. rejecting) p, this acceptance (resp. rejection) will 'reach' the agents in C in m steps. We refer to m as the *dependency length* of C on D.

In the above definition, when $C = D$ we say that C is a win (resp. veto) *self-dependent* coalition, that is, $C \in \mathcal{W}_p(i)$ for all $i \in C$ (resp. $C \in \mathcal{V}_p(i)$ for all $i \in C$). In the first case we refer to C as a *winning consensus* (C can force the acceptance of p among its own members), and in the second case we refer to it as a *veto consensus* (C can force the rejection of p among its own members). One same coalition can be both a winning and a veto consensus, in which case we will refer to it simply as a *consensus*. A singleton consensus, that is a set $\{i\}$ such that $\{i\} \in \mathcal{W}_p(i) \cap \mathcal{V}_p(i)$, is called a *guru* (for issue p). That is, an individual that has itself as smallest winning and veto coalition.

We will also need the following terminology. We say that a given profile *win-locks* a coalition C whenever for some D on which C has a win dependence $D \subseteq \mathbf{O}(p)$. It *veto-locks* a coalition C whenever for some D on which C has a veto dependence $D \subseteq \mathbf{O}(\neg p)$. If $D = C$, we say that the profile *directly win-locks* (resp. *directly veto-locks*) C. Intuitively, a profile *locks* a coalition whenever it triggers the acceptance or rejection, at some future moment, of the issue by all agents in the coalition.

Interlocking. A coalition $C \subseteq N$ is said to be *interlocked* (on issue p) if there exists a sequence $C_0, C_1, \ldots, C_m$ ($m \geq 1$) of subcoalitions of C, where $C_m = C_0$ and $C_\ell \neq C_{\ell+1}$ for $0 \leq \ell < m$, and such that: for all ℓ, $0 \leq \ell < m$: $C_\ell \in \mathcal{W}_p(i)$ for all $i \in C_{\ell+1}$ and $C \backslash C_\ell \in \mathcal{V}_p(i)$ for all $i \in C \backslash C_{\ell+1}$. That is, a coalition is interlocked whenever it can be partitioned in two cells one of which is win dependent on itself and the other of which is veto dependent on itself with a same dependency length. Intuitively, a coalition is interlocked whenever it can be split in two sub-coalitions whose winning and veto coalitions exhibit cyclical dependencies. The sequence $C_0, C_1, \ldots, C_m$ is called an *interlocking* of C. Integer m is referred to as the *interlocking length* of C. Each subcoalition C_ℓ in an interlocking is called a *lock* of C. Observe that if C is a winning (resp. veto) consensus, then C is trivially interlocked with an interlocking of length 1 consisting of $C_0 = C = C_1$ for a winning consensus (resp. $C_0 = \emptyset = C_1$, for a veto consensus).

Example 4 (Interlockings of length at most 1). Let $\{a, b\} = N$ and $p \in \mathbf{P}$. Let $\mathcal{W}_p(a) = \{\{a\}, \{b\}, \{a, b\}\}$ and $\mathcal{V}_p(a) = \{\{a, b\}\}$. Let $\mathcal{W}_p(b) = \{\{a, b\}\}$ and $\mathcal{V}_p(b) = \{\{a\}, \{b\}.\{a, b\}\}$. Coalition $\{a, b\}$ has the following interlockings, all of length 1:

- $C_0 = \{a, b\} = C_1$, since $\{a, b\}$ is a winning coalition both for a and for b and $\emptyset$ is a veto coalition for itself (trivially).
- $C_0 = \emptyset = C_1$, since $\emptyset$ is a winning coalition (trivially) for itself and $\{a, b\}$ is a veto coalition both for a and for b;[6]
- $C_0 = \{a\} = C_1$, since $\{a\}$ is a winning coalition for a and $\{b\}$ is a veto coalition for b.

Notice that this example can be instantiated by a complete network where b has a rejection bias (uses the standard unanimity aggregator: accepts an issue only when all influencers accept it) while a has an acceptance bias (rejects an issue only when all its influencers reject it).

Example 5 (Interlockings of length at most 2). Let $\{a, b\} \subseteq N$ and $p \in \mathbf{P}$. Let $\mathcal{W}_p(a) = \mathcal{V}_p(a) = \{\{b\}, \{a, b\}\}$. Let $\mathcal{W}_p(b) = \mathcal{V}_p(b) = \{\{a\}, \{a, b\}\}$. Coalition $\{a, b\}$ has the following interlockings

- *of length 1*: $C_0 = \{a, b\} = C_1$, since $\{a, b\}$ is a winning coalition both for a and b and $\emptyset$ is a veto coalition for itself. $C_0 = \emptyset = C_1$, since $\emptyset$ is a winning coalition for itself and $\{a, b\}$ is a veto coalition for a and for b;
- *of length 2*: $C_0 = C_2 = \{a\}, C_1 = \{b\}$, since $\{a\}$ is a winning coalition for b, and $\{b\}$ is a veto coalition for a, and since $\{b\}$ is a winning coalition for a and $\{a\}$ is a veto coalition for b. $C_0 = C_2 = \{b\}, C_1 = \{a\}$, for the same reasons.

This example can be instantiated either by a symmetric and irreflexive network between two agents both using any responsive monotonic and independent rule; or by a complete graph where the agents are each other's dictator. Note that any opinion profile where a and b disagree triggers an oscillating behavior.

Stabilization Results. Coalition interlocking is a useful abstraction to study stabilizing and oscillating behavior in neighborhood structures and thereby, indirectly, on networks.

Lemma 2. *Let* $\mathcal{A} = \langle N, \mathbf{P} \rangle$, $\mathcal{G} = \left\langle N, \{R_p\}_{p \in \mathbf{P}} \right\rangle$ *and* $\mathbf{F}$ *be given.* C *is a consensus coalition on issue* p *if and only if all agents in* C *are stable on the same value of* p *in any opinion profile* $\mathbf{O}$ *that directly win- or veto-locks* C *w.r.t.* p.

Proof. Left-to-right Recall that a consensus is defined as both a winning and a veto consensus. W.l.o.g. we assume $\mathbf{O}$ directly win-locks C, that is, $C \subseteq \mathbf{O}(p)$. So, $\forall i, j \in C$ $O_i(p) = O_j(p) = \mathbf{1}$. Since C is assumed to be a winning consensus, by the definition of winning coalition and Definition 1, it follows that $\forall i, j \in C, \mathbf{F}(i)(\mathbf{O})(p) = \mathbf{F}(j)(\mathbf{O})(p) = \mathbf{1}$, and therefore all agents in C are stable on p.

Right-to-left We proceed by contraposition, assuming that C is not a winning consensus (the case for veto consensus is symmetric). There exists $j \in C$ s.t. $C \notin \mathcal{W}_p(j)$. By (2), it follows that $N \backslash C \in \mathcal{V}_p(j)$. Let now $\mathbf{O}$ be such that $\mathbf{O}(p) = C$ (hence $\mathbf{O}(\neg p) = N \backslash C$). Clearly $\mathbf{O}$ directly locks C, but $\mathbf{F}(j)(\mathbf{O})(p) = \mathbf{0}$. Not all agents in C are therefore stable on the same value of p in $\mathbf{O}$. □

[6] Note that $\{a, b\}$ and $\emptyset$ are consensuses.

Intuitively, the lemma states that the members of a coalition that depends on itself for the collective acceptance or rejection of an issue hold stable opinions (on this issue) in every profile where they all agree (with respect to this issue).

Corollary 1. *Let* $\mathcal{A} = \langle N, \mathbf{P}\rangle$, $\mathcal{G} = \left\langle N, \{R_p\}_{p\in\mathbf{P}}\right\rangle$ *and* $\mathbf{F}$ *be given. If* C *is a consensus coalition on issue* p*, then all agents in* $\mathcal{C}_p(C)$ *stabilize to the same value of* p *from any opinion profile* $\mathbf{O}$ *that win- or veto-locks* C *w.r.t.* p*.*

Proof. By Lemma 2 we know that if C is a consensus coalition, and profile $\mathbf{O}$ locks C w.r.t. p, then all agents in C are stable and agree on p. We can then prove by induction on n that each individual in $\mathcal{C}_p(C) = \bigcup C^n$ eventually stabilizes. Base All individuals in $C^0 = C$ are stable on the same value of p by the above argument. Step Assume (IH) that all individuals in C^n are stable on the same value of p. We show that all individuals in $C^{n+1} = \{i \in N \mid C^n \in \mathcal{W}_p(i) \textit{ and } C^n \in \mathcal{V}_p(i)\}$ are. Consider such an individual. By construction $C^n \in \mathcal{W}_p(i)$ and $C^n \in \mathcal{V}_p(i)$. By IH and Definition 1, i's opinion on p will therefore stabilize in (at most) n steps. □

Lemma 3. *Let* $\mathcal{A} = \langle N, \mathbf{P}\rangle$, $\mathcal{G} = \left\langle N, \{R_p\}_{p\in\mathbf{P}}\right\rangle$, $\mathbf{F}$ *and issue* p *be given. All agents in* N *stabilize on* p *from a given profile* $\mathbf{O}$ *if and only if there exists no* $C \subseteq N$ *such that* C *is interlocked on* p *with an interlocking* $C_0, C_1, \ldots, C_m$ $(m > 1)$ *and for some* C_ℓ $(0 \leq \ell \leq m)$ $\mathbf{O}$ *win-locks* C_ℓ *and veto-locks* $C \backslash C_\ell$*.*

Proof. Left-to-right We prove the claim by contraposition. Assume that there exists an interlocking $C_0, C_1, \ldots, C_m$ of C of length $m \geq 2$. By the assumption that $\mathbf{O}$ win-locks C_ℓ and veto-locks $C \backslash C_\ell$ (with $0 \leq \ell < m$), we have that the dynamics reaches a profile $\mathbf{O}_\star$ such that $C_\ell \subseteq \mathbf{O}_\star(p)$ and $C\backslash C_\ell \subseteq \mathbf{O}_\star(\neg p)$. Now we consider the dynamics of opinions starting from $\mathbf{O}_\star$, showing that it cycles. We focus on the opinions of agents in C. Now, at each step $\mathbf{O}^k_\star$, with $0 < k \leq m$, we have by the definition of interlocking and of winning (and veto) coalitions that $C_{(\ell+k) \bmod m} \subseteq \mathbf{O}^k_\star(p)$ and $C\backslash C_{(\ell+k) \bmod m} \subseteq \mathbf{O}^k_\star(\neg p)$. Since, by the definition of interlocking, each $C_{\ell+1} \neq C_\ell$ (for $0 \leq \ell \leq m-1$), it follows that at each step at least one individual in C changes its opinion. Therefore, not all agents in C (and therefore in N) stabilize.

Right-to-left Again, we proceed by contraposition and assume that not all agents in N stabilize from $\mathbf{O}$. There exists therefore a cycle of profiles $\mathbf{O}_\star \neq \mathbf{O}^1_\star \neq \ldots \neq \mathbf{O}^m_\star = \mathbf{O}_\star$. We show that N is interlocked. Observe that, for any $0 \leq k < m$, $\mathbf{O}^k_\star(p) \in \mathcal{W}_i(p)$ for all $i \in \mathbf{O}^{k+1}_\star(p)$. This holds because each agent establishes its opinion on p at $k+1$ by aggregating the opinions of its neighbors on p at k via independent aggregators (Definition 1). Similarly, for any $0 \leq k < m$, $\mathbf{O}^k_\star(\neg p) \in \mathcal{V}_i(p)$ for all $i \in \mathbf{O}^{k+1}_\star(\neg p)$. It follows that $\mathbf{O}_\star(p) \neq \mathbf{O}^1_\star(p) \neq \ldots \neq \mathbf{O}^m_\star(p)$ constitutes an interlocking. As the dynamics has been assumed to end up in cycle $\mathbf{O}_\star \neq \mathbf{O}^1_\star \neq \ldots \neq \mathbf{O}^m_\star = \mathbf{O}_\star$ there exists $0 \leq k < m$ s.t. $\mathbf{O}$ win-locks $\mathbf{O}^k_\star(p)$ and veto-locks $\mathbf{O}^k_\star(\neg p)$. □

Lemma 3 is a central result. Its underpinning intuition is that stabilization occurs whenever it is not possible to drive the opinion diffusion dynamics into a state where, because of the way winning and veto coalitions are intertwined—that is our formal notion of interlocking—a cyclical behavior is bound to occur.

3.3 A Fixpoint Logic for Stability

Monotone μ-calculus. We use the syntax and semantics for the monotone (multi) modal *mu*-calculus from [14].[7] The language $\mathcal{L}$ is given by:

$$\varphi ::= p \mid \bot \mid \neg p \mid \varphi \wedge \varphi \mid \varphi \vee \varphi \mid \langle p \rangle \varphi \mid [p] \varphi \mid U\varphi \mid E\varphi \mid \mu p.\varphi \mid \nu p.\varphi$$

where $p \in \mathbf{P}$ and in formulas $\mu p.\varphi$ and $\nu p.\varphi$, p does not appear in the scope of a negation. This language can be directly interpreted on models $\mathcal{M} = \langle \mathcal{C}, \mathbf{O} \rangle$ consisting of a winning coalitions structure $\mathcal{C} = \left\langle N, \{\mathcal{W}_p\}_{p \in \mathbf{P}} \right\rangle$ plus an opinion profile $\mathbf{O}$. Recall that sets of winning coalitions are here assumed to be closed under supersets and do not contain the empty set (cf. Sect. 2).[8] Let $i \in N$, $p \in \mathbf{P}$ and $\varphi \in \mathcal{L}$. The satisfaction of φ in a model $\mathcal{M} = \langle \mathcal{C}, \mathbf{O} \rangle$ is inductively defined as follows (Boolean clauses omitted):[9]

$$\begin{array}{ll} \|[p]\varphi\|_{\mathcal{M}} = \{i \mid \|\varphi\|_{\mathcal{M}} \in \mathcal{W}_p(i)\} & \|\langle p \rangle \varphi\|_{\mathcal{M}} = \{i \mid (N \backslash \|\varphi\|_{\mathcal{M}}) \notin \mathcal{W}_p(i)\} \\ \|U\varphi\|_{\mathcal{M}} = \{i \mid \|\varphi\|_{\mathcal{M}} = N\} & \|E\varphi\|_{\mathcal{M}} = \{i \mid \|\varphi\|_{\mathcal{M}} \neq \emptyset\} \\ \|\mu p.\varphi\|_{\mathcal{M}} = \bigcap \{Z \mid \|\varphi\|_{\mathcal{M}_{[p \mapsto Z]}} \subseteq Z\} & \|\nu p.\varphi\|_{\mathcal{M}} = \bigcup \{Z \mid Z \subseteq \|\varphi\|_{\mathcal{M}_{[p \mapsto Z]}}\} \end{array}$$

where $\mathcal{M}_{[p \mapsto Z]}$ denotes model $\mathcal{M} = \langle \mathcal{C}, \mathbf{O} \rangle$ where $\mathbf{O}(p)$ is set to be Z.

Some Validities. To fix intuitions about the above logic it is worth mentioning some simple validities of the class of neighborhood models corresponding to winning coalitions structures. As winning coalitions are closed under supersets (because of the monotonicity of the underlying aggregators), we have that $\mathcal{M}, i \models U(\varphi \to \psi) \to ([p]\varphi \to [p]\psi)$ and $\mathcal{M}, i \models U(\varphi \to \psi) \to (\langle p \rangle \varphi \to \langle p \rangle \psi)$, for any model $\mathcal{M} = \langle \mathcal{C}, \mathbf{O} \rangle$, issue p, and agent i. As $\emptyset$ is neither a winning nor a veto coalition (because of the responsiveness of the underlying aggregators), we have that $\mathcal{M} \models [p]\top \wedge \langle p \rangle \top$ for any model $\mathcal{M}$, issue p and agent i.

Expressing Stabilization Conditions. We give a glimpse of how the above logic can be used to express properties of neighborhood structures induced by binary aggregation on networks, which are relevant to the stabilizing behavior

[7] The monotone μ-calculus was already used in [3] to model threshold-based diffusion.

[8] These properties force the resulting class of structures to validate specific formulae expressed in the above language. We refer the reader to [21] for an overview of the logics induced by monotonic neighborhood structures and subclasses thereof.

[9] We alternatively write $\mathcal{M}, i \models \varphi$ whenever $i \in \|\varphi\|$.

of the resulting opinion diffusion processes. As an example, we provide a formalization of the properties involved in Corollary 1. First of all, observe that the property "coalition C is a consensus coalition (w.r.t. p)" can be expressed as follows, using a dedicated atom C:

$$\mathsf{Con}(C) := U(C \rightarrow [p]\, C) \wedge U(C \rightarrow \langle p \rangle\, C)$$

The notion of "win-locking" and "veto-locking" of a coalition C (w.r.t. p) by a given profile can be expressed as:

$$\mathsf{wLock}(C) := U(C \rightarrow \mu q.p \vee [p]\, q) \qquad \mathsf{vLock}(C) := U(C \rightarrow \mu q.\neg p \vee \langle p \rangle\, q)$$

Given the above, the condition for stabilization identified in Corollary 1 can be expressed as follows:

$$\bigvee_{C \subseteq N} \mathsf{Con}(C) \wedge (((\mu q.C \vee [p]\, q) \wedge \mathsf{wLock}(C)) \vee ((\mu q.C \vee \langle p \rangle\, q) \wedge \mathsf{vLock}(C))) \quad (3)$$

In words, there exists a consensus coalition C—$\mathsf{Con}(C)$—and, either C is win-locked by the current profile—$\mathsf{wLock}(C)$—and is reachable from the agent (at the evaluation point) through a chain of winning coalitions—$\mu q.C \vee [p]\, q$—, or it is veto-locked by the current profile and is reachable from the agent through a chain of veto coalitions—$\mu q.\neg p \vee \langle p \rangle\, q$. By Corollary 1, on the class of neighborhood models $\mathcal{M} = \langle \mathcal{C}, \mathbf{O} \rangle$ induced by a network $\mathcal{G}$ and an allocation $\mathbf{F}$ of independent, responsive and monotonic aggregators, Formula (3) expresses a sufficient condition for the stabilization of an individual's opinion.

4 Instantiations

We illustrate the above general framework, and the reach of the stabilization results we established, via two examples of binary opinion dynamics on networks.

4.1 Boolean DeGroot Processes

In [11], DeGroot models step by step opinion change under social influence, using stochastic matrices both for representing opinions and influences. The opinion of each agent at the next time step is obtained through linear averaging (see [22] for a comprehensive exposition of the model). In this section, we focus on the "Boolean extreme" of these matrices, when both opinions and influence are taken to be binary. In such case, each agent has exactly one influencer, or "guru", whose opinions it repeatedly copies. We call these processes, which we first introduced in [10] as a model of liquid democracy, *Boolean DeGroot processes* (BDPs).

Binary Influence. Opinions are defined over a given BA structure with issues $\mathbf{P}$ and are therefore binary. Similarly, we take influence to be of an "all-or-nothing" type too, each agent is therefore taken to be influenced by exactly one agent, possibly itself. A binary influence matrix (for issue p) induces the *influence graph*

$\langle N, R_p \rangle$ where iR_pj still denotes that "i is influenced by j on issue p" but here amounts to "j is i's guru on issue p". Each R_p is serial ($\forall i \in N, \exists j \in N : iRj$) and functional ($\forall i, j, k \in N$ if iRj and iRk then $j = k$). So each agent i has exactly one successor (influencer or 'guru'), possibly itself, which we denote $R(i)$ slightly abusing notation. Combining the influence graphs of the issues in $\mathbf{P}$ one therefore obtains then a multi-issue influence network $\mathcal{G} = \left\langle N, \{R_p\}_{p \in \mathbf{P}} \right\rangle$.

BDP Dynamics. Given the above properties, applying any responsive and monotonic aggregator to the opinion of one unique neighbor results in making that neighbor effectively a dictator of the opinions of the agents it influences. So fixing any multi-issue influence network $\mathcal{G} = \left\langle N, \{R_p\}_{p \in \mathbf{P}} \right\rangle$, and any allocation $\mathbf{F}$ which assigns n independent responsive and monotonic aggregator to each agent, Definition 1 induces the following type of opinion dynamics for BDPs:[10] $\boxed{\text{base}}$ $\mathbf{O}_0 := \mathbf{O}$; $\boxed{\text{step}}$ $O_i^{n+1}(p) := O^n_{R_p(i)}(p)$ for all $i \in N$, $p \in \mathbf{P}$. That is, at each step, each agent simply copies the opinion of its guru.[11]

Stabilization in BDPs. We can derive the following result using Lemma 3:

Theorem 1. *Let $\mathcal{G} = \left\langle N, \{R_p\}_{p \in \mathbf{P}} \right\rangle$ be an influence network and $\mathbf{F}$ allocate to each agent an independent, monotonic and responsive aggregator. Then the following statements are equivalent: (1) Each agent stabilizes on issue p from $\mathbf{O}$. (2) There is no coalition $C \subseteq N$ such that: C is a cycle in G_p and there are two agents $i, j \in C$ such that $O_i(p) \neq O_j(p)$.*

Proof. We first show that under the BDP assumptions over $\mathcal{G}$ and $\mathbf{F}$, (2) holds if and only if there exists no $C \subseteq N$ such that C is interlocked with interlocking of length greater than 1 and $\mathbf{O}$ win-locks a coalition involved in the interlocking, and veto-locks its complement. The result then follows directly from Lemma 3. $\boxed{\text{Left-to-right}}$ Assume there is no coalition $C \subseteq N$ such that: C is a cycle in R_p and there are two agents $i, j \in C$ such that $O_i(p) \neq O_j(p)$. So there are no cycles of length ≥ 2 containing two agents that disagree on p in $\mathbf{O}$. Observe that each cycle $i_0, \ldots, i_m$ in an influence graph $\langle N, R_p \rangle$ (under the assumptions over $\mathbf{F}$) trivially defines an interlocked coalition $\{i_0, \ldots, i_m\}$ with interlocking $\{i_0\}, \ldots, \{i_m\}$. As no disagreement occurs in $\mathbf{O}$ among members

[10] Note that this dynamics is the extreme case of linear averaging applied on binary opinions and binary influence.

[11] BDPs are also limit cases of *propositional opinion diffusion* processes recently proposed by [18], i.e., cases where (1) the aggregation rule is the unanimity rule (an agent *changes* its opinion if and only if all her influencers disagree with it), and (2) each agent has exactly one influencer. Note that, in general, the 'unanimity rule' from the setting of propositional opinion diffusion differs from what we call the unanimity rule, which prescribes not only to 'change' your opinion if all your influencers have the opposite opinion, but to adopt their opinion no matter what opinion you currently hold. In the limit case of BDPs, those two notions trivially coincide.

of an interlocked coalitions, it follows that all interlocked coalitions are locked by $\mathbf{O}$. Right-to-left The argument is analogous. □

Observe that, if R_p contains only cycles of length 1 then, trivially, no two agents in a cycle can disagree (second statement in Theorem 1).

4.2 Unanimity Processes

Up Dynamics. We consider now diffusion processes based on the unanimity rule. Let us call this type of diffusion "Unanimity Processes" (UPs) and define their dynamics in the obvious way: Fix an opinion profile $\mathbf{O}$ and a serial (non-necessarily functional) influence profile $\mathcal{G}$. Consider the stream $\mathbf{O}^0, \mathbf{O}^1, \ldots, \mathbf{O}^n, \ldots$ of opinion profiles recursively defined as follows: base $\mathbf{O}^0 := \mathbf{O}$; and step for all $i \in N$ and all $p \in \mathbf{P}$, $O_i^{n+1}(p) = p$ iff for all $i \in R_p(i), O_i(p) = \mathbf{1}$. That is, an agent accepts p next if p is accepted by all its neighbors, and otherwise rejects it. Formally, we are assuming an allocation $\mathbf{F}$ of aggregators assigning to each agent i a quota rule based on a quota q_p for issue p which is equal to $|R_p(i)|$.

Stabilization of UPs. Again, we can leverage Lemma 3 to establish stabilization results:

Theorem 2. *Let $\mathcal{G} = \left\langle N, \{R_p\}_{p \in \mathbf{P}} \right\rangle$ be symmetric and serial;*[12] *let $\mathbf{F}$ allocate to each agent the unanimity aggregator (as in the above definition of UPs) and let $\mathbf{O}$ be an opinion profile. The following are equivalent: (1) Each agent stabilizes on issue p from $\mathbf{O}$; (2) For all connected component C of G_p, one of the two conditions hold: (i) there exist $i, j \in C$, such that: $O_i(p) = O_j(p) = \mathbf{0}$ and there is an R_p-path of odd length from i to j; (ii) for all $i \in C$, $O_i(p) = \mathbf{1}$.*

Proof. 1) ⇒ 2) By contraposition: assume that for some $p \in \mathbf{P}$ and some connected component C of G_p, neither i nor ii hold: Then for any $i, j \in C$, such that $O_i(p) = O_j(p) = \mathbf{0}$, the distance n from i to j is ≥ 2. W.l.o.g, let i be any $i \in C$ such that $O_i(p) = \mathbf{0}$. By definition of UP and by symmetry, there exists an interlocking of length 2 where $C_1 = \{j \in C \mid jR^n i \textit{ and } n \textit{ is odd}\}$ and $C_2 = \{j \in C \mid jR^n i \textit{ and } n \textit{ is even}\}$, and $C_2 \subseteq \mathbf{O}(p)$. Hence, $\mathbf{O}$ win-locks C_2 and veto-locks C_1. By Lemma 3, $\mathbf{O}$ does not converge. 2) ⇒ 1) Assume that i holds for some $p \in \mathbf{P}$, and some connected component C of G_p: there exist $i, j \in C$, such that $O_i(p) = O_j(p) = \mathbf{0}$ with a path of odd length n from i to j in R_p. By definition of UPs and by symmetry, this implies that there are two agents k, l at distance $d = \frac{n-1}{2}$ such that $O_k^d(p) = O_l^d(p) = \mathbf{0}$. Therefore, $O_k^d(p) = O_l^d(p) = \mathbf{0}$ and O_k^d and O_l^d are stable. It follows by the definition of UP that any agents within any distance m from k or l will be stable (after at most $d + m$ steps). Assume that ii holds for some $p \in \mathbf{P}$, and some connected component C of G_p: for all $i \in C$, $O_i(p) = \mathbf{1}$. Then, for all $i \in C$, $O_i(p)$ is stable. □

[12] These correspond to the typical case of 'friendship' networks (cf. [23]).

We illustrated our setup with Boolean DeGroot processes and unanimity processes but there are diverse additional instantiations to consider, such as diffusion under a unique numerical threshold (the typical 'threshold models'), and under majority, for instance. It should be clear that these cases can all be captured by our setting. We leave the additional results out for space reasons.

5 Conclusions

We have characterized stabilization conditions for opinion diffusion modeled as iterated binary aggregation on networks, and showed how existing modal fixpoint logics are well-suited to express properties of this setting. These investigations open up several lines of research. First, although our opinion diffusion model is directly related to the propositional opinion diffusion model of [18], the two are not identical, and a precise comparison is worth closer inspection and may lead to a yet more general theory of binary opinion diffusion. Second, we have handled opinion dynamics on independent issues, sidestepping the hard problems involved in the aggregation of logically interdependent issues. Opinion dynamics on interdependent issues is only since very recently receiving attention (e.g., [6,15]) and is a promising, challenging line for future research within our framework. Third, the paper has given only a glimpse of the sort of logical languages which, in light of our results, appear to be relevant for the study of opinion dynamics on networks. Expanding the logical side of our paper is a natural direction for future work. In particular, the study of fragments and variants of the monotone μ-calculus from [14] is a promising line of research (see the recent [5]). Moreover, we have focused here on stabilization, which is nothing but the limit case of oscillating behavior, where oscillations have size one. We believe that a richer logic to capture long-term behavior—beyond stabilization alone—in binary opinion diffusion on networks should blend some monotone μ-calculus with the oscillatory fixpoint operators introduced in [4]. The study of such logical systems would without doubt be an interesting line for further research.

Acknowledgments. Zoé Christoff and Davide Grossi acknowledge support for this research by EPSRC (grant EP/M015815/1, "Foundations of Opinion Formation in Autonomous Systems"). Zoé Christoff also acknowledges support from the Deutsche Forschungsgemeinschaft (DFG) and Grantová agentura České republiky (GAČR) joint project RO 4548/6–1.

References

1. Azimipour, S., Naumov, P.: Lighthouse principle for diffusion in social networks. J. Appl. Logic. (2017, to appear)
2. Baltag, A., Christoff, Z., Rendsvig, R.K., Smets, S.: Dynamic epistemic logic of diffusion and prediction in social networks. In: Twelfth Conference on Logic and the Foundations of Game and Decision Theory (LOFT 2016) (2016)

3. Baltag, A., Sonja, S.: Logic goes viral - modalities for social networks. Presented at the workshop Trends in Logic - Presenting the Tsinghua-UvA Joint Research Center, Tsinghua University, 2 July 2014
4. van Benthem, J.: Oscillations, logic, and dynamical systems. In: Ghosh, S., Szymanik, J. (eds.) The Facts Matter. Essays on Logic and Cognition in Honour of Rineke Verbrugge, pp. 9–22. College Publications (2015)
5. van Benthem, J., Bezhanishvili, N., Enqvist, S., Junhua, Y.: Instantial neighbourhood logic. Rev. Symbol. Logic **10**(1), 116144 (2017)
6. Botan, S.: Propositional opinion diffusion with constraints. ILLC Master of Logic Thesis (2016)
7. Chellas, B.F.: Modal Logic. An Introduction. Cambridge University Press, Cambridge (1980)
8. Cholvy, L.: Influence-based opinion diffusion (extended abstract). In: Proceedings of AAMAS 2016, pp. 1355–1356. IFAAMAS (2016)
9. Christoff, Z.: Dynamic logics of networks: information flow and the spread of opinion. Ph.D thesis, Institute for logic, Language and Computation, University of Amsterdam, Amsterdam, The Netherlands. ILLC Dissertation Series DS-2016-02 (2016)
10. Christoff, Z., Grossi, D.: Binary aggregation with delegable proxy: an analysis of liquid democracy. In: Proceedings of TARK 2017, EPTCS, vol. 251 (2017)
11. Degroot, M.H.: Reaching a consensus. J. Am. Stat. Assoc. **69**(345), 118–121 (1974)
12. Dokow, E., Holzman, R.: Aggregation of binary evaluations. J. Econ. Theory **145**(2), 495–511 (2010)
13. Endriss, U.: Judgment aggregation. In: Brandt, F., Conitzer, V., Endriss, U., Lang, J., Procaccia, A.D. (eds.) Handbook of Computational Social Choice, chap. 17. Cambridge University Press (2016)
14. Enqvist, S., Seifan, F., Venema, Y.: Expressiveness of the modal mu-calculus on monotone neighborhood structures. Technical report, arXiv: 1502.07889 (2015)
15. Friedkin, N.E., Proskurnikov, A.V., Tempo, R., Parsegov, S.E.: Network science on belief system dynamics under logic constraints. Science **354**(6310), 321–326 (2016)
16. Goles, E.: Periodic behavior of generalized threshold functions. Discrete Math. **30**, 187–189 (1980)
17. Grandi, U., Endriss, U.: Lifting integrity constraints in binary aggregation. Artif. Intell. **199**, 45–66 (2013)
18. Grandi, U., Lorini, E., Perrussel, L.: Propositional opinion diffusion. In: Proceedings of the 2015 International Conference on Autonomous Agents and Multiagent Systems, AAMAS 2015, pp. 989–997. International Foundation for Autonomous Agents and Multiagent Systems, Richland (2015)
19. Granovetter, M.: Threshold models of collective behavior. Am. J. Sociol. **83**(6), 1420–1443 (1978)
20. Grossi, D., Pigozzi, G.: Judgment aggregation: a primer. Synth. Lect. Artif. Intell. Mach. Learn. **8**(2), 1–151 (2014)
21. Hansen, H.H.: Monotonic modal logics. Technical Report PP-2003-24, ILLC (2003)
22. Jackson, M.O.: Social and Economic Networks. Princeton University Press, Princeton (2008)
23. Liu, F., Seligman, J., Girard, P.: Logical dynamics of belief change in the community. Synthese **191**(11), 2403–2431 (2014)

Quotient Dynamics: The Logic of Abstraction

Alexandru Baltag[1], Nick Bezhanishvili[1], Julia Ilin[1], and Aybüke Özgün[1,2](✉)

[1] University of Amsterdam, Amsterdam, The Netherlands
[2] LORIA, CNRS - Université de Lorraine, Nancy, France
ozgunaybuke@gmail.com

Abstract. We propose a Logic of Abstraction, meant to formalize the act of "abstracting away" the irrelevant features of a model. We give complete axiomatizations for a number of variants of this formalism, and explore their expressivity. As a special case, we consider the "logics of filtration".

1 Introduction

In this work, we aim to formalize the process of *abstraction*, in the specific sense of "abstracting away", i.e. disregarding all 'irrelevant' distinctions. Since reality is potentially infinitely complex, abstraction is essential for scientific modeling. In principle, a model should represent *all* the facts, but in practice the model is always tailored to the relevant issues under discussion. In particular, this phenomenon is all-pervasive in the formal epistemology literature: when modeling epistemic scenarios, the modeler focuses on a set of relevant issues, and identifies situations that agree on all these issues, thus reducing the size and complexity of the model to manageable proportions. A well-known example is the Muddy Children puzzle [10]. A standard relational model for the n-children puzzle has 2^n states, but with this we disregard all irrelevant facts (e.g. the color of each kid's clothes, etc.), focusing only on whether each of the n children is dirty or not. However, the same situation may be analyzed at various levels of abstraction, depending on the particular application. Rather than modeling again every new application from scratch, a good modeler develops the art of simplifying older models in order to reuse them in new situations, by again "abstracting away" some of the issues.

We develop technical tools to formalize this concept of abstraction as a dynamic process. We do this in a modal framework based on the standard Kripke models, by introducing *dynamic abstraction modalities*, similar to the update operators of Dynamic Epistemic Logic [1,9]. The "relevant issues" may be given syntactically, as a set of formulas, inducing an equivalence relation on worlds that satisfy the same relevant formulas; or we may give them semantically, by starting directly with an equivalence relation on possible worlds (the so-called *issue relation* of [3,15], telling us which worlds agree on all the relevant issues). For most of this paper, we focus on the first (syntactic) option, but we also consider the second option in Sect. 5. Roughly speaking, we represent the

A. Baltag et al. (Eds.): LORI 2017, LNCS 10455, pp. 181–194, 2017.
DOI: 10.1007/978-3-662-55665-8_13

process of abstraction as a *model transformation*, that maps any given model to a *quotient model*. While the *states* of the quotient model can be defined in a natural and canonical way (as *equivalence classes* with respect to the relevant equivalence relation, there are many different ways to define the valuation function, and more interestingly the accessibility relation(s), of the quotient model. This problem is already known from Modal Logic, where it occurs when an appropriate notion of *filtration* is needed for a given logic. Defining the quotient relations corresponds to *lifting* the relation(s) of the initial model (maybe after first performing a *relational transformation*) to some relation(s) between the induced equivalence classes. Depending on the context, different such liftings can be used. In this work, we focus on what is called the $(\exists,\exists)$-lifting, which corresponds to the so-called *minimal filtration* in Modal Logic.[1]

In Sect. 2, we start with the (single-agent) basic modal language, then generalize it to all **PDL**-definable relations [7,11,14]. Here, **PDL**-programs play a meta-syntactic role: they are used to specify a *relational transformer*. We define a logic for each such transformer, by applying it to the original relation of the model, then applying the $(\exists,\exists)$-lifting to obtain the quotient relation. We investigate the expressivity of these logics, and prove completeness using reduction axioms in the style of Public Announcement Logic (PAL) [2,12,16,17]. In Sect. 3, we apply this to the special case of modal filtrations. As an added benefit, we show that these logics internalize the so-called Filtration Theorem [7]: while usually stated as meta-logical result, it becomes a plain logical theorem in our proof systems. In Sect. 4, we move to a "multi-agent" (multi-relational) framework, but also increase the expressivity (by including all **PDL** programs into the syntax), thus obtaining "the Logic of Abstraction": a general logical formalism that can treat and compare various types of quotient-taking operations in a unified formalism. We give a complete axiomatization via reduction axioms. In contrast to PAL (where adding common knowledge operators increases expressivity), the addition of Kleene star (iteration) on programs is innocuous: this logic is co-expressive with a version of **PDL** (with a "universal program" 1). Finally, in Sect. 5 we discuss two further generalizations and variations of our setting: by considering other relation liftings than the $(\exists,\exists)$-lifting; and by taking the above-mentioned "semantic option", of starting with an issue relation on worlds, and investigating the corresponding logic of quotients.

Due to page restrictions, we omitted the long proofs in Sect. 4 from this submission. The proofs can be found in the extended version of this paper at https://sites.google.com/site/ozgunaybuke/publications.

2 Quotient-Taking as a Model Transformer

In this section, we explain the main ideas behind the formalism developed in this paper and fix some notations. In particular, we provide a detailed description of our *quotient models* (defined for a specific modal language through a finite

[1] However, we'll show that, in combination with applying relational transformers described by regular PDL programs, this lifting can capture other filtrations.

set of formulas), and introduce the so-called *abstraction modalities*. Our quotient models are similar to *filtrations* from modal logic (see [7, Sect. 2.3] for an overview of filtrations), but our notion is more general[2]. We then introduce our formal dynamic language including the abstraction modalities, and provide sound and complete axiomatizations of a specific family of dynamic abstraction logics.

We start the section by introducing the static language we work with throughout the section. By $\mathcal{L}_E$ we denote the language of *basic modal logic* enriched with the *universal modality* defined by the grammar

$$\varphi ::= p \mid \neg\varphi \mid \varphi \wedge \varphi \mid E\varphi \mid \Diamond\varphi,$$

where p is a propositional variable, and E stands for the (dual) of the universal modality. We employ the usual definitions for $\vee$, $\rightarrow$, $\leftrightarrow$, $\top$, $\bot$, and $\Box$. The fragment of $\mathcal{L}_E$ without the modality E is denoted by $\mathcal{L}$. Formulas of $\mathcal{L}_E$ are interpreted on Kripke models $\mathfrak{M} = (W, R, V)$ in a standard way (see, e.g., [7, Chap. 1]). In particular, $\mathfrak{M}, w \models E\varphi$ iff there is $v \in W$ with $\mathfrak{M}, v \models \varphi$.

In the following let a Kripke model $\mathfrak{M} = (W, R, V)$ be fixed. Our aim is to define a *quotient model* $\mathfrak{M}_\Sigma = (W_\Sigma, R_\Sigma, V_\Sigma)$ of $\mathfrak{M}$ wrt a finite[3] set of formulas $\Sigma \subseteq \mathcal{L}_E$.

The set $\Sigma \subseteq \mathcal{L}_E$ induces an *equivalence relation* $\sim_\Sigma$ on W: for $w, v \in W$

$$w \sim_\Sigma v \quad \text{iff} \quad \text{for all } \varphi \in \Sigma\, (\mathfrak{M}, w \models \varphi \text{ iff } \mathfrak{M}, v \models \varphi). \tag{1}$$

In other words, two worlds are *Σ-equivalent* iff they satisfy the same formulas from Σ. We denote by $|w|_\Sigma$ the equivalence class of w with respect to $\sim_\Sigma$, i.e., $|w|_\Sigma := \{v \in W \mid w \sim_\Sigma v\}$. The domain of our quotient model will be the set of equivalence classes with respect to $\sim_\Sigma$, i.e. $W_\Sigma = \{|w|_\Sigma \mid w \in W\}$.

Concerning the valuation V_Σ, for any propositional letter p, we set

$$V_\Sigma(p) := \{|w|_\Sigma \mid \text{ there is } w' \in |w|_\Sigma \text{ with } w' \in V(p)\}.$$

While this generalizes the definition of the valuation used in filtrations, (see, e.g., [7, Chap. 2.3], it also constitutes the minimal valuation that preserves the truth value of *true* propositional letters in each world, in the sense that if $w \models p$ then $|w|_\Sigma \models p$[4].

Finally, we get to the most important defintion, namely, the definition the relation R_Σ. The relation R_Σ is determined by two factors: the first factor is a prescription on how to transfer a relation on W to a relation on W_Σ. We refer to such a prescription as a *lifting* of the relation R from W to W_Σ (similar to

[2] In Sect. 3, we will show precisely how filtrations fit into our framework.

[3] The finiteness of Σ is in fact irrelevant for the definition of quotient models, however, this will be required in order to be able to provide reduction axioms for our new dynamic modalities introduced later in this section. This is why we keep the setting simple and work only with finite Σs.

[4] Note that two Σ-equivalent worlds may disagree on the propositional variables that are not in the set Σ.

relation liftings studied theoretical computer science). As an example consider the definition

$$|w|_\Sigma R_\Sigma |v|_\Sigma \text{ iff there exists } w' \in |w|_\Sigma, \text{ and there exists } v' \in |v|_\Sigma \text{ such that } w'Rv'. \quad (2)$$

We call this the $(\exists, \exists)$-lifting of R for obvious reasons[5]. In a similar manner, we can also define $(\exists, \forall)$-, $(\forall, \exists)$- and $(\forall, \forall)$-liftings of R. However, in this paper, we work with the $(\exists, \exists)$-lifting, and briefly mention the other options in Sect. 5.

The second factor to characterize R_Σ consists in deciding which relation to lift from W to W_Σ. For example, in (2), the relation R is lifted (as maybe the most obvious choice). In our framework though, we will allow more flexibility by considering liftings of the so-called **PDL**$_{-*}$-*definable relations* (à la van Benthem and Liu [4]). More formally, the programs in the language of *star-free Propositional Dynamic Logic* (**PDL**$_{-*}$) are defined by the grammar

$$\pi ::= r \mid ?\varphi \mid 1 \mid \pi ; \pi \mid \pi \cup \pi,$$

where r is the (only) basic program[6] and φ is a formula in the language $\mathcal{L}_E$. The program 1 stands for the *universal program*. As usual, a program π determines a relation R_π on the model $\mathfrak{M}$ recursively defined as: $R_r := R$, $R_1 := W \times W$, and $R_{?\varphi} := \{(x, x) \mid \mathfrak{M}, x \models \varphi\}$, for some $\varphi \in \mathcal{L}_E$, and for any two programs π and π', we have $R_{\pi;\pi'} := R_\pi ; R_{\pi'}$, and $R_{\pi \cup \pi'} := R_\pi \cup R_{\pi'}$, where $R_\pi ; R_{\pi'}$ and $R_\pi \cup R_{\pi'}$ are the composition and the union of the relations R_π and $R_{\pi'}$, respectively. A binary relation Q on W is called **PDL**$_{-*}$-*definable* iff $Q = R_\pi$ for some program π of **PDL**$_{-*}$.

In this section, any **PDL**$_{-*}$-definable relation can be used to determine the relations on our quotient models. In detail, in our framework each program π leads to a model transformation function that takes a Kripke model $\mathfrak{M}$ and a finite $\Sigma \subseteq \mathcal{L}_E$, and returns the quotient model $\mathfrak{M}_\Sigma$ *whose relation* R_Σ *is determined by the* $(\exists, \exists)$-*lifting of the relation* R_π. As a consequence, each program π will lead to a π-dependent dynamic logic.

Definition 1 (Quotient model wrt π). *Let $\mathfrak{M} = (W, R, V)$ be a Kripke model. For every finite $\Sigma \subseteq \mathcal{L}_E$, the quotient model of $\mathfrak{M}$ with respect to Σ is $\mathfrak{M}_\Sigma = (W_\Sigma, R_\Sigma, V_\Sigma)$, where $W_\Sigma := \{|w|_\Sigma \mid w \in W\}$, $V_\Sigma(p) := \{|w|_\Sigma \mid$ there is $w' \in |w|_\Sigma$ with $w' \in V(p)\}$, and*

$$|w|_\Sigma R_\Sigma |v|_\Sigma \textit{ iff there is } w' \in |w|_\Sigma \textit{ and there is } v' \in |v|_\Sigma \textit{ such that } w' R_\pi v'.$$

Therefore, each π describes a particular type of model transformation whose arguments vary over finite subsets Σ of the language $\mathcal{L}_E$. As usual in dynamic

---

[5] This definition is known to modal logicians under the name of *smallest filtration* (see, e.g., [7, Chap. 2.3]).

[6] In this section—since the formalism is based on Kripke models with a single relation—we have only one basic program r in our syntax. In Sect. 4, we work with multi-relational Kripke models allowing for more than one basic programs, as standard in **PDL**.

epistemic logics [9], we introduce dynamic modalities, denoted by $[\Sigma]$, capturing this type of model change and call them the *abstraction modalities.* Before we formally define the dynamic language and the semantics of the abstraction modalities, we point out some observations concerning their expressive power. Unlike e.g. the public announcement operator (see, e.g., [16,17]), the abstraction modality adds expressivity to the basic modal language $\mathcal{L}$:

Fact 1. *The abstraction modality adds expressivity to the basic modal language* $\mathcal{L}$.

Indeed, let $\pi = r$ be the basic program, i.e. the relation R_Σ on the quotient model $\mathfrak{M}_\Sigma$ is defined as in (2). Using the abstraction modality we can e.g. express the existential statements $\Psi :=$ "$\exists x, y \in W$ with xRy", or $\Psi' :=$ "$\exists x \in W$ with $\mathfrak{M}, x \models p$.", namely by $[\{\top\}]\Diamond\top$, and $[\{\top\}]p$, respectively. It is well-known that neither Ψ nor Ψ' are expressible in the basic modal language $\mathcal{L}$. Note, however, that the statements *are* expressible in $\mathcal{L}_E$, that is, when the universal modality is added to $\mathcal{L}$. On the other hand, the universal modality can express statements that are not expressible via the abstraction modality.

Fact 2. *The universal modality and the abstraction modality are not equally expressive.*

For example, the statement $\chi :=$ "$\exists x \in W$ with $\mathfrak{M}, x \models \neg p$" for some propositional letter p is not expressible with the abstraction modality. To illustrate, consider the two models $\mathfrak{M}$ and $\mathfrak{M}'$.

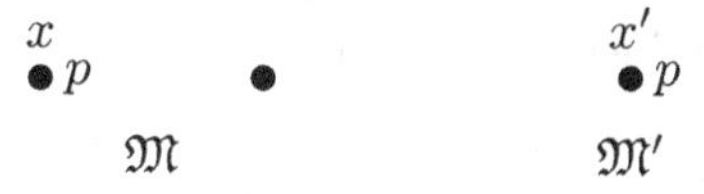

Then $\mathfrak{M}, x$ satisfies χ but $\mathfrak{M}', x'$ does not satisfy χ. Since x and x' are bisimilar for $\mathcal{L}$, they satisfy the same formulas in the language $\mathcal{L}$. Now for every finite $\Sigma \subseteq \mathcal{L}$, either ($\mathfrak{M}_\Sigma = \mathfrak{M}$ and $\mathfrak{M}'_\Sigma = \mathfrak{M}'$) or $\mathfrak{M}_\Sigma = \mathfrak{M}'_\Sigma = \mathfrak{M}'$. Therefore, x and x' agree on all formulas in the language $\mathcal{L}$ extended by the abstraction modality. Thus, χ is not expressible via $[\Sigma]$. We point out that these examples of course depend on the program π we choose for the quotient model.

The above expressivity results imply that the basic modal language with the abstraction modality is not reducible to basic modal language. This motivates why we work with the language $\mathcal{L}_E$ (but not with the simpler basic modal language $\mathcal{L}$) as our static language. In fact, we will show that $\mathcal{L}_E$ together with the abstraction modality is co-expressive with $\mathcal{L}_E$.

Formally, our dynamic language $\mathcal{L}_{E,[\Sigma]}$ is defined by the grammar

$$\varphi ::= p \mid \neg\varphi \mid \varphi \wedge \varphi \mid E\varphi \mid \Diamond\varphi \mid [\Sigma]\varphi$$

where Σ is a finite subset of $\mathcal{L}_E$. For a fixed program π, we evaluate formulas of $\mathcal{L}_{E,[\Sigma]}$ as follows:

Definition 2 (Semantics for $[\Sigma]\varphi$ wrt π). *Given a Kripke model $\mathfrak{M} = (X, V, R)$ and a state $w \in W$, the truth of $\mathcal{L}_{E,[\Sigma]}$-formulas is defined for Boolean cases, and the modalities $\Diamond$ and E as usual. The semantics for the abstraction modality $[\Sigma]\varphi$ is given by*

$$\mathfrak{M}, w \models [\Sigma]\varphi \textit{ iff } \mathfrak{M}_\Sigma, |w|_\Sigma \models \varphi,$$

where $\mathfrak{M}_\Sigma$ is the quotient model built wrt the program π.

In the rest of this section we will define a family of logics $\mathbf{K}_{E,\Sigma}(\pi)$—one for each program π of $\mathbf{PDL}_{-*}$— and show their soundness and completeness wrt to our semantics. While the soundness proof is standard, the completeness is established via reducing the dynamic logic to its underlying static base through a set of so-called *reduction axioms*. The reduction axioms (given in Table 1) describe a recursive rewriting algorithm that converts the formulas in $\mathcal{L}_{E,[\Sigma]}$ to semantically and provably equivalent formulas in $\mathcal{L}_E$. The key property that allows us to obtain reduction axioms in this particular setting is that—by finiteness of Σ and the presence of the universal modality—the equivalence relation $\sim_\Sigma$ becomes *definable* in our language in the sense of Lemma 1.

We fix the following notation: for every finite $\Sigma \subseteq \mathcal{L}_E$, and for every formula $\chi \in \mathcal{L}_{E,[\Sigma]}$ let

$$\langle\sim_\Sigma\rangle\chi := \bigvee_{\Psi\subseteq\Sigma} \left(\hat{\Psi} \wedge E\left(\hat{\Psi} \wedge \chi\right)\right), \tag{3}$$

where $\hat{\Psi} = \bigwedge\Psi \wedge \bigwedge\neg(\Sigma \setminus \Psi)$. The modality $\langle\sim_\Sigma\rangle$ is the diamond modality of the equivalence relation induced by Σ, thus $\sim_\Sigma$ is definable in $\mathcal{L}_{E,[\Sigma]}$:

Lemma 1. *Let $\mathfrak{M} = (W, R, V)$ be a model and let Σ be a finite set of formulas of $\mathcal{L}_{E,[\Sigma]}$. Then $\mathfrak{M}, x \models \langle\sim_\Sigma\rangle\chi$ iff there is $x' \sim_\Sigma x$ with $\mathfrak{M}, x' \models \chi$.*

Proof. Let $\mathfrak{M} = (W, R, V)$ be a Kripke model, Σ a finite subset of $\mathcal{L}_E$ and $\chi \in \mathcal{L}_{E,[\Sigma]}$.

($\Rightarrow$) Suppose $\mathfrak{M}, x \models \bigvee_{\Psi\subseteq\Sigma}\left(\hat{\Psi} \wedge E\left(\hat{\Psi} \wedge \chi\right)\right)$. This means that $\mathfrak{M}, x \models \hat{\Psi} \wedge E\left(\hat{\Psi} \wedge \chi\right)$ for some $\Psi \subseteq \Sigma$. I.e., we have $\mathfrak{M}, x \models \hat{\Psi}$ and $\mathfrak{M}, x \models E\left(\hat{\Psi} \wedge \chi\right)$. The latter implies that there is $x' \in W$ such that $\mathfrak{M}, x' \models \hat{\Psi} \wedge \chi$. Since $\mathfrak{M}, x' \models \hat{\Psi}$, we obtain $x \sim_\Sigma x'$, therefore the result follows.

($\Leftarrow$) Suppose there is $x' \in W$ such that $x \sim_\Sigma x'$ and $\mathfrak{M}, x' \models \chi$. As $x \sim_\Sigma x'$, the states x and x' make exactly the same formulas in Σ true. Therefore, we obtain that $\mathfrak{M}, x \models \hat{\Psi}$ and $\mathfrak{M}, x' \models \hat{\Psi}$ for some $\Psi \subseteq \Sigma$. The latter together with the assumption $\mathfrak{M}, x' \models \chi$ implies that $\mathfrak{M}, x \models E(\hat{\Psi} \wedge \chi)$. We therefore obtain $\mathfrak{M}, x \models \hat{\Psi} \wedge E(\hat{\Psi} \wedge \chi)$. Thus, $\mathfrak{M}, x \models \bigvee_{\Psi\subseteq\Sigma}\left(\hat{\Psi} \wedge E\left(\hat{\Psi} \wedge \chi\right)\right)$.

Table 1 contains reduction axioms and rules of the logic $\mathbf{K}_{E,\Sigma}(\pi)$. Note that the axiom (Ax-$\Diamond_\pi$) contains the symbol $\langle\pi\rangle$ which is *not* part of the language

$\mathcal{L}_{E,[\Sigma]}$. Recall that the programs used to build π do not contain the star-operator. Since the language of star-free-**PDL** (with the universal program) is as expressive as the language $\mathcal{L}_E$, we can legitimately use the axiom (Ax-$\Diamond_\pi$) as an abbreviation for a formula in the language $\mathcal{L}_{E,[\Sigma]}$ (cf. [4]). To be precise, we employ the following abbreviations: $\langle r\rangle\psi := \Diamond\psi$, $\langle 1\rangle\psi := E\psi$, $\langle ?\varphi\rangle\psi := \psi \wedge \varphi$, $\langle \pi;\pi'\rangle\psi := \langle\pi\rangle\langle\pi'\rangle\psi$, and $\langle\pi \cup \pi'\rangle\psi := \langle\pi\rangle\psi \vee \langle\pi'\rangle\psi$ for formulas $\psi \in \mathcal{L}_{E,[\Sigma]}$, $\varphi \in \mathcal{L}_E$ and programs π, π' of $\mathbf{PDL}_{-*}$.

Table 1. The logic $\mathbf{K}_{E,\Sigma}(\pi)$

(**K**)	Axioms and rules of the basic modal logic **K**
(**E**)	**S5**-axioms and rules for E, $\Diamond\varphi \rightarrow E\varphi$
(Ax-p)	$[\Sigma]\,p \leftrightarrow \langle\sim_\Sigma\rangle p$
(Ax-$\neg$)	$[\Sigma]\neg\varphi \leftrightarrow \neg[\Sigma]\varphi$
(Ax-$\wedge$)	$[\Sigma](\varphi \wedge \psi) \leftrightarrow [\Sigma]\varphi \wedge [\Sigma]\psi$
(Ax-E)	$[\Sigma]E\varphi \leftrightarrow E[\Sigma]\varphi$
(Ax-$\Diamond_\pi$)	$[\Sigma]\Diamond\varphi \leftrightarrow \bigvee_{\Psi\subseteq\Sigma}\left(\hat{\Psi} \wedge E\left(\hat{\Psi} \wedge \langle\pi\rangle[\Sigma]\varphi\right)\right)$
(Nec$_{[\Sigma]}$)	From φ infer $[\Sigma]\varphi$

Completeness of $\mathbf{K}_{E,\Sigma}(\pi)$ is shown by defining a translation $t_\pi : \mathcal{L}_{E,[\Sigma]} \rightarrow \mathcal{L}_E$ that transforms each formula in the language $\mathcal{L}_{E,[\Sigma]}$ to a $\mathbf{K}_{E,\Sigma}(\pi)$-provably equivalent formula in the language $\mathcal{L}_E$. We will skip the details of this translation since we will later discuss a similar translation in Sect. 4. We then obtain:

Theorem 1 (Expressivity). *Let π be a $\mathbf{PDL}_{-*}$-program. For every $\varphi \in \mathcal{L}_{E,[\Sigma]}$, $\vdash_{\mathbf{K}_{E,\Sigma}(\pi)} \varphi \leftrightarrow t_\pi(\varphi)$.*

We can now derive completeness results by standard arguments from the completeness of the basic modal logic with the universal modality $\mathbf{K}_E$ (see [13] for the completeness of $\mathbf{K}_E$) and the soundness of $\mathbf{K}_{E,\Sigma}(\pi)$.

Theorem 2 (Completeness). *Let π be a $\mathbf{PDL}_{-*}$-program. The logic $\mathbf{K}_{E,\Sigma}(\pi)$ is sound and complete wrt to the class of all Kripke models, where the quotient models are taken wrt the program π.*

3 Special Case: Logics of Filtrations

Thinking of quotient models, *filtrations* may be the first thing coming to the mind of a modal logician. Filtrations are used in order to prove the finite model property of some modal logics (see e.g. [7, Sect. 2.3] and [8, Sect. 5.3]). Roughly speaking, they turn a (refutation) model into a finite one by forming a quotient. In order to preserve some relational properties of Kripke models (such as transitivity, reflexivity etc.) there are several ways to define quotient models, leading to several notions of filtrations. In this section, we show how some well-known filtrations can be captured by the quotient models given in Definition 1. More precisely, for a filtration f—where f stands for the *smallest*, the *largest*, the *transitive* or the *smallest-transitive* filtration—we will define a program π_f of $\mathbf{PDL}_{-*}$ such that the quotient model wrt the program π_f corresponds exactly to an f-filtration. In this sense, we can say that the logic of the filtration f is the logic $\mathbf{K}_{E,\Sigma}(\pi_f)$ axiomatized in Table 1. We will also comment on the possibility of adding additional axioms to these logics. Roughly, an axiom χ of the basic modal language $\mathcal{L}$ can be safely added to the logic $\mathbf{K}_{E,\Sigma}(\pi_f)$, whenever the basic modal logic axiomatized by χ *admits f-filtrations* (see, e.g., [8, Chap. 5.3]).

We will refer to the smallest, the largest, the transitive (a.k.a. Lemmon filtrations) and the smallest-transitive filtration, by s, l, t, and st, respectively. For the definitions of the first three filtrations, see e.g. [7, Sect. 2.3]. The smallest transitive filtration is obtained by first taking the smallest filtration and then replacing the resulting relation Q with its transitive closure Q^+ (see e.g. [8, Chap. 5.3])[7].

We will now define programs π_f in the language of $\mathbf{PDL}_{-*}$ whose corresponding quotient models coincide with that of the f-filtration for $f \in \{s, l, t, st\}$. Let Σ be a finite set of formulas in the language $\mathcal{L}_E$. For $\Psi \subseteq \Sigma$, we set $\Psi_\Diamond = \bigwedge_{\Diamond\varphi\in\Sigma,\varphi\in\Psi} \Diamond\varphi$, $\Psi_{\Diamond,\vee} = \bigwedge_{\Diamond\varphi\in\Sigma,\varphi\in\Psi}(\Diamond\varphi \vee \varphi)$, and $\neg\Psi = \{\neg\varphi \mid \varphi \in \Psi\}$. We then define the following programs: let $\pi_\Sigma = \bigcup_{\Psi\subseteq\Sigma}(?\hat{\Psi}; 1; ?\hat{\Psi})$, and for $k \in \mathbb{N}$, let $\pi_1 = r$ and $\pi_{k+1} = r; \pi_\Sigma; \pi_k$, then define

$$\pi_s := r, \quad \pi_l := \bigcup_{\Psi\subseteq\Sigma} (?\Psi_\Diamond; 1; ?\hat{\Psi}), \quad \pi_t := \bigcup_{\Psi\subseteq\Sigma} (?\Psi_{\Diamond,\vee}; 1; ?\hat{\Psi}), \text{ and } \quad \pi_{st} := \bigcup_{1\leq k\leq 2^{|\Sigma|}} \pi_k.$$

It is easy to see that the quotient model wrt the program π_f corresponds exactly to an f-filtration for $f \in \{s, l, t, st\}$. To prove this for the smallest-transitive filtration, observe that by finiteness of Σ, the size of W_Σ is bounded by $2^{|\Sigma|}$. Thus, the transitive closure of a relation on W_Σ is reached by at most $2^{|\Sigma|}$ many iterations.

Proposition 1. *Let $f \in \{s, l, t, st\}$. For every finite and subformula closed*[8] *set $\Sigma \subseteq \mathcal{L}_E$, the model $\mathfrak{M}_\Sigma^{\pi_f}$ is an f-filtration of $\mathfrak{M}$ through Σ.*

The quotient models resulting from the transitive and the smallest-transitive filtrations are always transitive. To a modal logician, these filtrations in fact become interesting only when applied to transitive Kripke models, since otherwise the Filtration Theorem does not hold (see, e.g., [8, Theorem 5.23]). The transitivity of the quotient models implies that the (4)-Axiom ($\Diamond\Diamond\varphi \to \Diamond\varphi$) is valid on these models. Therefore, if the (4)-Axiom is added to the logics $\mathbf{K}_{E,\Sigma}(\pi_t)$

[7] The filtrations in the aforementioned sources are defined for a language without the universal modality. However, as observed in [13, Sect. 5.2], the universal modality does not cause any problems in the theory of filtrations.

[8] Since filtrations are usually only defined for subformula closed sets—the reason being that the Filtration Theorem can only be proved in this case—we add this as an additional condition.

and $\mathbf{K}_{E,\Sigma}(\pi_{st})$, the necessitation rule ($\mathrm{Nec}_{[\Sigma]}$) for $[\Sigma]$ remains sound. We can therefore extend the logics $\mathbf{K}_{E,\Sigma}(\pi_t)$ and $\mathbf{K}_{E,\Sigma}(\pi_{st})$ by the (4)-axiom and obtain sound systems. In more general terms, whenever a quotient model wrt a filtration type f preserves the validity of a certain axiom, this axiom can be safely added to the dynamic logic $\mathbf{K}_{E,\Sigma}(\pi_f)$ without affecting soundness and completeness. In fact, the same is true under slightly weaker assumptions. Let χ be an axiom that characterizes the class $\mathcal{K}$ of Kripke models. Sometimes the validity of χ is not preserved in all quotient models (wrt filtration type f) of the class $\mathcal{K}$, but it is preserved in quotient models of a smaller class of models $\mathcal{K}' \subseteq \mathcal{K}$ (e.g., the filtration st preserves the validity of the (.2)-axiom only on *rooted*[9] transitive models, but not on arbitrary transitive models (see [8, Theorem 5.33])). If the smaller class $\mathcal{K}'$ is "big enough", meaning that the logic axiomatized by χ is complete wrt $\mathcal{K}'$, then its dynamic extension is also complete wrt the class $\mathcal{K}'$, where quotient models are taken wrt π_f. To modal logicians such considerations run under the name of *admitting filtration*, see [8, Sect. 5.2].

For a normal modal logic $\mathbf{L}$ (e.g., $\mathbf{T}, \mathbf{KB}$ or $\mathbf{K4}$ etc., see [8] for our notational convention for the normal modal logics.), by $\mathbf{L}_{E,\Sigma}(\pi_f)$, we denote the logic that is obtained from the axioms and rules of $\mathbf{L}$ and from Table 1 for $f \in \{s, l, t, st\}$. Using results explored in [8, Chap. 5.2], Proposition 1 and Theorem 2, we obtain the following:

Corollary 1. *1. For $f \in \{s, l, t, st\}$, the logics $\mathbf{D}_{E,\Sigma}(\pi_f)$ and $\mathbf{T}_{E,\Sigma}(\pi_f)$ are sound and complete wrt the class of serial Kripke models and reflexive Kripke models, respectively (where the quotient models are taken wrt π_f).*

2. $\mathbf{KB}_{E,\Sigma}(\pi_s)$ is sound and complete wrt symmetric Kripke models.

3. For $f \in \{t, st\}$, the logics $\mathbf{K4}_{E,\Sigma}(\pi_f)$, $\mathbf{D4}_{E,\Sigma}(\pi_f)$, and $\mathbf{S4}_{E,\Sigma}(\pi_f)$ are sound and complete wrt transitive, transitive serial, and reflexive transitive models, respectively (where the quotient models are taken wrt π_f).

4. $\mathbf{K4.2}_{E,\Sigma}(\pi_{st})$ and $\mathbf{K4.3}_{E,\Sigma}(\pi_{st})$ are sound and complete wrt the class of rooted transitive directed models, and rooted transitive connected models, respectively. Moreover, $\mathbf{S4.2}_{E,\Sigma}(\pi_{st})$ and $\mathbf{S4.3}_{E,\Sigma}(\pi_{st})$ are sound and complete wrt the class of Kripke models based on rooted directed quasi-orders, and rooted linear quasi-orders, respectively.

5. For $f \in \{s, l, t, st\}$, the logic $\mathbf{S5}_{E,\Sigma}(\pi_f)$ is sound and complete wrt the class of Kripke models based on clusters. In fact, $\mathbf{S5}_{E,\Sigma}(\pi_{st})$ is sound and complete wrt the class of Kripke models based on equivalence relations.

Remark 1. We note that the above corollary can be proved for a larger class of stable and transitive stable logics of [5,6]. These are logics that are sound and complete with respect to classes of rooted frames closed under graph homomorphisms. In other words, these are the logics admitting all filtrations and all transitive filtrations, respectively. In this respect, stable logics play a similar role to the abstraction modality that subframe logics—logics whose frames are closed under subframes [8, Chap. 11.3]—play for the public announcement operator.

[9] Recall that a transitive Kripke model $\mathfrak{M}$ is called *rooted* if there is $s \in W$ such that sRw for all $w \in \mathfrak{M}$.

Finally, we comment on the meaning of the Filtration Theorem in our context (see, e.g., [7, Theorem 2.39] for the filtration theorem). Due to the completeness result stated in Theorem 2, the Filtration Theorem can be proven syntactically in our logics $\mathbf{K}_{E,\Sigma}(\pi_f)$, i.e., it can be *internalized* as a theorem of these systems:

Corollary 2 (Internalized Filtration Theorem). *For every finite subformula closed set $\Sigma \subseteq \mathcal{L}_E$ and all $\varphi \in \Sigma$, we have the following:*

1. $\vdash_{\mathbf{K}_{E,\Sigma}(\pi_f)} [\Sigma]\varphi \leftrightarrow \varphi$, *for $f \in \{s, l\}$;*
2. $\vdash_{\mathbf{K4}_{E,\Sigma}(\pi_f)} [\Sigma]\varphi \leftrightarrow \varphi$, *for $f \in \{t, st\}$.*

4 The Logic of Abstraction

This section generalizes the setting presented in Sects. 2 and 3 in many ways. To start with, we move to a multi-relational setting, also called multi-agent setting, allowing for many basic programs in a given **PDL**-language. Secondly—and more importantly—we generalise the abstraction modalities in such a way that the **PDL**-programs become a component of these modalities. More precisely, an abstraction modality contains a sequence of programs $\overrightarrow{\pi}$ that are indexed by the set of agents as a parameter. The program π_r corresponding to agent r determines the relation of the same agent in the quotient model. Another generalization over the previous setting is that we allow programs in the (full) **PDL**-language, i.e. the language including the star-operator. In this section, we introduce semantics for this extended language on multi-relational Kripke models and provide a sound and complete axiomatization for the logic of abstraction $\mathbf{PDL}_{\mathbf{\Sigma}}$. Since the star operator properly adds expressivity to the static (multi-) modal language, our resulting dynamic logic $\mathbf{PDL}_{\mathbf{\Sigma}}$ will not be reducible to basic modal logic. Instead, we will employ the language of **PDL** as our base language.

We would like to stress the two different uses of the language of propositional dynamic logic: while, in the previous sections, the language of $\mathbf{PDL}_{-*}$ was only used as a "meta-language" for abbreviations of formulas in $\mathcal{L}_E$, the language of **PDL** here becomes an essential part of our logical language. Our dynamic language $\mathsf{PDL}_{[\overrightarrow{\pi}/\Sigma]}$ is defined by extending the language of propositional dynamic logic **PDL** with the abstraction modalities $[\overrightarrow{\pi}/\Sigma]\varphi$. More precisely, $\mathsf{PDL}_{[\overrightarrow{\pi}/\Sigma]}$ is defined by the grammar:

$$\pi ::= r \mid ?\psi \mid 1 \mid \pi;\pi \mid \pi \cup \pi \mid \pi^*, \text{ and } \quad \varphi ::= p \mid \neg\varphi \mid \varphi \wedge \varphi \mid \langle\pi\rangle\varphi \mid [\overrightarrow{\pi}/\Sigma]\varphi,$$

where r is an element of the set of basic programs Π_0, $\psi \in \mathsf{PDL}$, $\overrightarrow{\pi} = (\pi_r)_{r\in\Pi_0}$ is a sequence of PDL-programs, and Σ is a *finite*[10] subset of PDL (the language $\mathsf{PDL}_{[\overrightarrow{\pi}/\Sigma]}$ without $[\overrightarrow{\pi}/\Sigma]\varphi$).

[10] Similar to the case in Sect. 2, the sets Σ being finite is essential in order to obtain reduction axioms for the corresponding dynamic logic.

Given a (multi-relational) Kripke model $\mathfrak{M} = (W, (R_r)_{r\in\Pi_0}, V)$, we interpret programs as relations on $\mathfrak{M}$ as usual and denote the relation corresponding to the program π by R_π. Recall that the relation R_π is defined recursively on the structure of π. In particular, $R_1 := W \times W$, and $R_{?\psi} := \{(x,x) \mid \mathfrak{M}, x \models \psi\}$, for some $\psi \in \mathsf{PDL}$. Just as in (1) (see Sect. 2), a finite subset $\Sigma \subseteq \mathsf{PDL}$ induces an equivalence relation $\sim_\Sigma$ on W by relating two worlds that satisfy the same formulas of Σ. We denote by $|w|_\Sigma$ the equivalence class of $w \in W$ with respect to $\sim_\Sigma$.

Next we define the *(multi-relational) quotient models.* Recall that in Definition 1 we defined quotient models wrt a fixed program π. In the current setting, the sequence of programs $\overrightarrow{\pi}$ becomes a parameter of the quotient models, thus receives a similar status as the set Σ. This is reflected in the shape of the abstraction modalities $[\overrightarrow{\pi}/\Sigma]\varphi$.

Definition 3 (Quotient model). *Let $\mathfrak{M} = (W, (R_r)_{r\in\Pi_0}, V)$ be a Kripke model. For every finite $\Sigma \subseteq \mathsf{PDL}$ and every sequence $\overrightarrow{\pi} = (\pi_r)_{r\in\Pi_0}$ of programs, the* quotient model $\mathfrak{M}^{\overrightarrow{\pi}}_\Sigma$, *is $\mathfrak{M}^{\overrightarrow{\pi}}_\Sigma = (W_\Sigma, (R^{\pi_r}_\Sigma)_{r\in\Pi_0}, V_\Sigma)$, where $W_\Sigma := \{|w|_\Sigma \mid w \in W\}$, $V_\Sigma(p) := \{|w|_\Sigma \mid$ there is $w' \sim_\Sigma w$ with $w' \in V(p)\}$, and for each $r \in \Pi_0$*

$$|w|_\Sigma R^{\pi_r}_\Sigma |v|_\Sigma \text{ iff there is } w' \sim_\Sigma w \text{ and there is } v' \sim_\Sigma v \text{ with } w' R_{\pi_r} v'.$$

In other words, using the terminology of Sect. 2, the quotient model $\mathfrak{M}^{\overrightarrow{\pi}}_\Sigma$ arises from $\mathfrak{M}$ by interpreting a basic program $r \in \Pi_0$ via the $(\exists, \exists)$-lifting of the relation R_{π_r} from W to W_Σ.

Definition 4 (Semantics for $\mathsf{PDL}_{[\overrightarrow{\pi}/\Sigma]}$). *Given a Kripke model $\mathfrak{M} = (W, (R_r)_{r\in\Pi_0}, V)$ and a state w in W, the truth of $\mathsf{PDL}_{[\overrightarrow{\pi}/\Sigma]}$-formulas at a world w in $\mathfrak{M}$ is defined recursively as for PDL with the additional clause:*

$$\mathfrak{M}, w \models [\overrightarrow{\pi}/\Sigma]\varphi \text{ iff } \mathfrak{M}^{\overrightarrow{\pi}}_\Sigma, |w|_\Sigma \models \varphi$$

where $\mathfrak{M}^{\overrightarrow{\pi}}_\Sigma$ is as given in Definition 3.

Next we introduce reduction axioms that allow us to convert a formula of $\mathsf{PDL}_{[\overrightarrow{\pi}/\Sigma]}$ to a provably equivalent formula in PDL. In the current setting, there are two key properties that allow us to obtain reduction axioms. Firstly, the equivalence relation $\sim_\Sigma$ is *definable* in the language $\mathsf{PDL}_{[\pi/\Sigma]}$ similar to the case in Sect. 2. Secondly, Σ being finite ensures that the model $\mathfrak{M}^{\pi}_\Sigma$ is not only finite but its size is bounded in terms of the size of Σ. In fact, the size of $\mathfrak{M}^{\pi}_\Sigma$ is at most $2^{|\Sigma|}$. For this reason we can obtain reduction axioms for the star-operator. As in (3), for every formula $\chi \in \mathsf{PDL}_{[\overrightarrow{\pi}/\Sigma]}$ and finite $\Sigma \subseteq \mathsf{PDL}$ we fix the following notation:

$$\langle\sim_\Sigma\rangle\chi := \bigvee_{\Psi\subseteq\Sigma} \left(\hat{\Psi} \wedge \langle 1\rangle \left(\hat{\Psi} \wedge \chi\right)\right).$$

The modality $\langle\sim_\Sigma\rangle$ is the diamond modality of the relation $\sim_\Sigma$, as can be shown analogously to Lemma 1.

For an axiomatization of **PDL**, see [7, Sect. 4.8] or [14]. The universal program 1 requires the **S5** axioms and rules, and $\langle\pi\rangle p \to \langle 1\rangle p$ for every program π. The logic $\mathbf{PDL_\Sigma}$ is defined by the axioms and rules given in Table 2.

Table 2. The logic $\mathbf{PDL_\Sigma}$

(**PDL**)	Axiom-schemes and rules of **PDL**
(Ax-p)	$[\overrightarrow{\pi}/\Sigma]\,p \leftrightarrow \langle\sim_\Sigma\rangle p$
(Ax-$\neg$)	$[\overrightarrow{\pi}/\Sigma]\neg\varphi \leftrightarrow \neg[\overrightarrow{\pi}/\Sigma]\varphi$
(Ax-$\wedge$)	$[\overrightarrow{\pi}/\Sigma](\varphi\wedge\psi) \leftrightarrow [\overrightarrow{\pi}/\Sigma]\varphi \wedge [\overrightarrow{\pi}/\Sigma]\psi$
(Ax-$\langle 1\rangle$)	$[\overrightarrow{\pi}/\Sigma]\langle 1\rangle\varphi \leftrightarrow \langle 1\rangle[\overrightarrow{\pi}/\Sigma]\varphi$
(Ax-$\langle r\rangle$)	$[\overrightarrow{\pi}/\Sigma]\langle r\rangle\varphi \leftrightarrow \langle\sim_\Sigma\rangle\langle\pi_r\rangle[\overrightarrow{\pi}/\Sigma]\varphi$ for all $r \in \Pi_0$
(Ax-$*$)	$[\overrightarrow{\pi}/\Sigma]\langle\alpha^*\rangle\varphi \leftrightarrow [\overrightarrow{\pi}/\Sigma]\bigvee_{n\leq 2^{\vert\Sigma\vert}}\langle\alpha\rangle^n\varphi$
(Nec$_{[\overrightarrow{\pi}/\Sigma]}$)	From φ infer $[\overrightarrow{\pi}/\Sigma]\varphi$

The reduction axioms enables us to show that every formula in $\mathsf{PDL}_{[\overrightarrow{\pi}/\Sigma]}$ is provably equivalent (in the system $\mathbf{PDL_\Sigma}$) to a formula in the language PDL.

Theorem 3 (Expressivity). *For every $\varphi \in \mathsf{PDL}_{[\overrightarrow{\pi}/\Sigma]}$ there is a $\psi \in \mathsf{PDL}$ such that $\vdash_{\mathbf{PDL_\Sigma}} \varphi \leftrightarrow \psi$.*

Using Theorem 3, the completeness of $\mathbf{PDL_\Sigma}$ is a consequence of the completeness theorem for **PDL** and the soundness of the system $\mathbf{PDL_\Sigma}$.

Theorem 4 (Completeness). $\mathbf{PDL_\Sigma}$ *is sound and complete.*

5 Further Generalizations and Variations

In this final section, we outline some further results and alternatives.

Other Liftings: We used $(\exists,\exists)$-lifting to build the quotient models in Definition 3, but we can use other liftings as discussed in Sect. 2. However, we conjecture that reduction axioms for the $(\forall,\forall)$- and the $(\exists,\forall)$-lifts are *not available* in our setting. Though such reduction axioms might become available if we extend the base language by nominals as in hybrid logics. On the other hand, the setting using the $(\forall,\exists)$-lift of the relation R_π admits reduction axioms, obtained by replacing Ax-$\langle r\rangle$ from Table 2 by:

$$(\text{Ax-}\langle r\rangle)\quad [\overrightarrow{\pi}/\Sigma]\langle r\rangle\varphi \leftrightarrow \bigvee_{\Psi\subseteq\Sigma}\bigvee_{\Phi\subseteq\Sigma}\left(\hat{\Psi}\wedge\langle\pi_r\rangle\left(\hat{\Phi}\wedge[\overrightarrow{\pi}/\Sigma]\varphi\right)\wedge[1]\left(\hat{\Psi}\to\langle\pi_r\rangle\hat{\Phi}\right)\right)$$

The 'Semantic Option': While in this paper we focused on the 'syntactic option' (issues given by a set of formulas), we are also investigating the semantic option: each model comes with its own equivalence "issue" relation Q. In this setup, models are of the shape $\mathfrak{M} = (W, (R_r)_{r\in\Pi_0}, Q, V)$, where $(W, (R_r)_{r\in\Pi_0}, V)$

is a Kripke model and Q is an equivalence relation on W. We then define a language $\mathsf{PDL}_{Q,\vec{\pi}/Q}$ as:

$$\pi := r \mid Q \mid ?\psi \mid 1 \mid \pi;\pi \mid \pi \cup \pi \mid \pi^*, \text{ and } \quad \varphi := p \mid \neg\varphi \mid \varphi \wedge \varphi \mid \langle\pi\rangle\varphi \mid [\vec{\pi}/Q]\varphi,$$

where r is an element of the set of the basic programs Π_0 and $\psi \in \mathsf{PDL}$ (the language $\mathsf{PDL}_{Q,\vec{\pi}/Q}$ without $[\vec{\pi}/Q]\varphi$). Note that we add a symbol Q to the basic programs whose intended interpretation is the equivalence relation Q. Its modality $[Q]$ is the so-called *issue modality* from [3]. For a model $\mathfrak{M} = (W, (R_r)_{r\in\Pi_0}, Q, V)$ and a sequence of programs $\vec{\pi}$, we define a model $\mathfrak{M}_Q^{\vec{\pi}} := (W_Q, (R_Q^{\pi_r})_{r\in\Pi_0}, \mathsf{Id}, V_Q)$, where $W_Q := \{|w| \mid \text{there is } w'Qw \text{ with } w' \in V(p)\}$, $V_Q(p) := \{|w| \mid w \in V(p)\}$, Id denotes the identity relation, and

$$|w|R_Q^{\pi_r}|v| \text{ iff there is } w'Qw \text{ and there is } v'Qv \text{ such that } w'R_{\pi_r}v',$$

where $|w|$ is the equivalence class of w wrt Q. The crucial step in the semantics is:

$$\mathfrak{M}, x \models [\vec{\pi}/Q]\varphi \text{ iff } \mathfrak{M}_Q^{\vec{\pi}}, |x| \models \varphi.$$

To get a convenient representation of the reduction axioms, we define functions $f_{Q,\vec{\pi}}$ on programs by $f_{\vec{\pi},Q}(Q) = ?\top$, $f_{\vec{\pi},Q}(r) = Q;\vec{\pi}$, $f_{\vec{\pi},Q}(\alpha_1 \circ \alpha_2) = f_{\vec{\pi},Q}(\alpha_1) \circ f_{\vec{\pi};Q}(\alpha_2)$ for $\circ \in \{\cup, ;\}$ and $f_{\vec{\pi},Q}(\pi^*) = (f_{\vec{\pi},Q}(\pi))^*$. Here is the full list of reduction axioms (Table 3):

Table 3. The logic $\mathbf{PDL}_Q$

(**PDL**)	Axiom-schemes and rules of **PDL**
(**Q**)	**S5**-axioms and rules for Q
(Ax-p)	$[\vec{\pi}/Q]p \leftrightarrow \langle Q\rangle p$
(Ax-$\neg$)	$[\vec{\pi}/Q]\neg\varphi \leftrightarrow \neg[\vec{\pi}/Q]\varphi$
(Ax-$\wedge$)	$[\vec{\pi}/Q](\varphi \wedge \psi) \leftrightarrow [\vec{\pi}/Q]\varphi \wedge [\vec{\pi}/Q]\psi$
(Ax-$\langle\alpha\rangle$)	$[\vec{\pi}/Q]\langle\alpha\rangle\varphi \leftrightarrow \langle f_{Q,\vec{\pi}}(\alpha)\rangle[\vec{\pi}/Q]\varphi$
(Ax-$\langle Q\rangle$)	$[\vec{\pi}/Q]\langle Q\rangle\varphi \leftrightarrow [\vec{\pi}/Q]\varphi$
(DR-Nec)	From φ infer $[\vec{\pi}/Q]\varphi$

Note that in our earlier versions, the analogue of the modality $\langle Q\rangle$ was *definable* in the language $\mathsf{PDL}_{[\vec{\pi}/\Sigma]}$ (cf. Sect. 2, Lemma 1), thus was not needed in the syntax.

Acknowledgments. A. Özgün acknowledges financial support from European Research Council grant EPS 313360.

References

1. Baltag, A., Renne, B.: Dynamic epistemic logic. In: Zalta, E.N. (ed.), The Stanford Encyclopedia of Philosophy, Metaphysics Research Lab, Stanford University (2016)
2. van Benthem, J.: Logical Dynamics of Information and Interaction. Cambridge University Press, New York (2014)
3. van Benthem, J., Minică, Ş.: Toward a dynamic logic of questions. J. Philos. Logic **41**, 633–669 (2012)
4. van Benthem, J., Liu, F.: Dynamic logic of preference upgrade. J. Appl. Non Class. Logics **17**, 157–182 (2007)
5. Bezhanishvili, G., Bezhanishvili, N., Iemhoff, R.: Stable canonical rules. J. Symb. Logic **81**, 284–315 (2016)
6. Bezhanishvili, G., Bezhanishvili, N., Ilin, J.: Stable modal logics. https://www.illc.uva.nl/Research/Publications/Reports/PP-2016-11.text.pdf
7. Blackburn, P., de Rijke, M., Venema, Y.: Modal Logic. Cambridge University Press, New York (2001)
8. Chagrov, A.V., Zakharyaschev, M.: Modal Logic. Oxford Logic Guides, vol. 35. Oxford University Press, Oxford (1997)
9. van Ditmarsch, H., van der Hoek, W., Kooi, B.: Dynamic Epistemic Logic, 1st edn. Springer Publishing Company, Incorporated, Netherlands (2007)
10. Fagin, R., Halpern, J.Y., Moses, Y., Vardi, M.Y.: Reasoning About Knowledge. MIT Press, Cambridge (1995)
11. Fischer, M.J., Ladner, R.E.: Propositional dynamic logic of regular programs. J. Comput. Syst. Sci. **18**, 194–211 (1979)
12. Gerbrandy, J., Groeneveld, W.: Reasoning about information change. J. Logic Lang. Inf. **6**, 147–169 (1997)
13. Goranko, V., Passy, S.: Using the universal modality: gains and questions. J. Logic Comput. **2**, 5–30 (1992)
14. Harel, D., Kozen, D., Tiuryn, J.: Dynamic Logic. MIT Press, Cambridge (2000)
15. Minică, Ş.: Dynamic Logic of Questions. Ph.D. thesis, ILLC, University of Amsterdam (2011)
16. Plaza, J.: Logics of public communications. In: Proceedings of the 4th International Symposium on Methodologies for Intelligent Systems, pp. 201–216 (1989)
17. Plaza, J.: Logics of public communications. Synthese **158**, 165–179 (2007)

The Dynamics of Group Polarization

Carlo Proietti(✉)

Lund University, Lund, Sweden
carlo.proietti@fil.lu.se

Abstract. Exchange of arguments in a discussion often makes individuals more radical about their initial opinion. This phenomenon is known as Group-induced Attitude Polarization. A byproduct of it are bipolarization effects, where the distance between the attitudes of two groups of individuals increases after the discussion. This paper is a first attempt to analyse the building blocks of information exchange and information update that induce polarization. I use Argumentation Frameworks as a tool for encoding the information of agents in a debate relative to a given issue a. I then adapt a specific measure of the degree of acceptability of an opinion (Matt and Toni 2008). Changes in the degree of acceptability of a, prior and posterior to information exchange, serve here as an indicator of polarization. I finally show that the way agents transmit and update information has a decisive impact on polarization and bipolarization.

1 Introduction

Almost sixty years ago MIT student J.A. Stoner observed and studied a strange group phenomenon that he classified as "risky shift". This term categorizes the tendency of a group to make decisions that are riskier than the average of the individual decisions of members before the group met [27]. Subsequent research in social psychology showed that a similar pattern applies more generally to change of attitude and opinion after debate. This phenomenon is nowadays famous as *Group-induced attitude polarization.* Understanding the dynamics that lead to polarization is particularly relevant in the era of social networks, because of the dramatic global effects they may cause. Indeed, virtual forums and political debate seem to accrue so-called *bipolarization effects*, i.e. the tendency of different subgroups to radicalize their opinions towards opposite directions [28].

A long tradition in social psychology has regarded polarization and bipolarization as a byproduct of social influence in groups [26], where the main explanatory mechanism is *social comparison* [11].[1] An alternative explanation is

[1] According to social comparison explanations, such as [26], polarization may arise in a group because individuals are motivated to perceive and present themselves in a favorable light in their social environment. To this end, they take a position which is similar to everyone else but a bit more extreme. This kind of explanation assumes a lot. Indeed, models that explain bipolarization effects by social comparison mechanisms usually postulate both positive influence by ingroup members and negative influence by outgroup members [12,16]. However, a number of criticisms have been addressed towards the accuracy of empirical research showing the presence of negative influence in social interaction [19].

A. Baltag et al. (Eds.): LORI 2017, LNCS 10455, pp. 195–208, 2017.
DOI: 10.1007/978-3-662-55665-8_14

provided by *persuasive arguments theory*, which was developed and tested in a number of lab experiments in the 1970s [30].[2]

Both social comparison and persuasive arguments theory provide interesting clues for explaining polarization phenomena. However, much more is hidden behind the mechanisms of *information* transmission and update among agents. The present work is a first attempt towards the formal description of such mechanisms. The aim is to unravel all the possible building blocks of polarization. In this context we need to understand the notion of information in a general sense, wider than, e.g., knowledge or rational belief. Polarization and bipolarization are in fact distinctive features of real-life dynamics, where people form their views (say, decide how to vote or what to buy) by exchanging information with others, or by trusting more or less authoritative channels. Such informational items typically need not to be consistent, nor are acquired via a careful process of individual inquiry and strict rules of belief revision and belief update. Indeed in many such situations individuals deviate from Bayesianism, insofar as they update their beliefs by discarding some available evidence. This happens, e.g., when they display a dogmatic or selective attitude towards the information received.

In the present paper I adopt Argumentation Frameworks [9] as a formalizing tool, which are the most versatile tool to encode the type of informational items we are interested in, as well as the argumentative process of information exchange and update. Indeed, the theoretical tools provided by abstract argumentation will serve to the purpose of

1. Describing both the *total information available* relative to a debated issue and one agent's *partial information.*
2. Provide a measure of the *degree of acceptability* of the debated issue given the available information.
3. Encode the most important policies of *information transmission* between agents and of *information update.*
4. Assess how such policies can impact polarization and bipolarization about the given issue.

Schematically, the argumentative process generating polarization works as in the workflow of Fig. 1. Agent i possesses some information about a given issue a, represented by $I_i(\mathcal{F}, a)$, she transmits some of her information to agent, say, j ($T_{i,j}(\mathcal{F}, a)$), and the latter updates her previous information $I_j(\mathcal{F}, a)$ by combining it with $T_{i,j}(\mathcal{F}, a)$ via some operation $\star$ to be specified.

[2] This explanation assumes that individuals become more convinced of their view when they hear novel and persuasive arguments in favor of their position, and therefore "Group discussion will cause an individual to shift in a given direction to the extent that the discussion exposes that individual to persuasive arguments favoring that direction" [15]. Typically, models inspired by persuasive arguments theory do not assume negative influence of any kind, but presuppose homophily, i.e. stronger interaction with like-minded individuals [23], or biased assimilation of arguments [21].

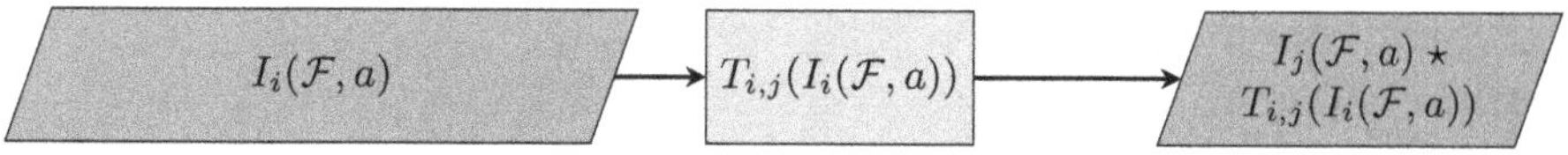

Fig. 1. Schematic flow of information transmission and update between agent i and j

I proceed as follows. Section 2 provides a short introduction to Argumentation Frameworks and shows how to define the scenario of a multi-agent debate. I also show how to apply the acceptability measure defined by [24] to encode the degree of acceptability of a given issue a. Section 3 introduces some relevant policies of information disclosure and update. Section 4 presents two main results that show the impact of these policies on polarization and bipolarization.

2 Argumentation Frameworks and Multi-agent Scenarios

An Argumentation Framework [9], AF for short, consists of a graph where nodes are arguments and a directed edge between a and b is to be read as "argument a attacks argument b". The formal definition is the following

Definition 1 (Pointed Argumentation Framework). *An Argumentation Framework is a 2-ple $\mathcal{F} = (A, R)$ where A is a finite and non-empty set of arguments and $R \subseteq A \times A$. A Pointed Argumentation Framework $\mathcal{F}, a$ consists of an Argumentation Framework $\mathcal{F}$ together with a specified $a \in A$.*

Example 1. Figure 2 provides the graphical representation of a pointed AF, where $A = \{a, b, c, d, e, f\}$, $R = \{(b, a), (c, a), (d, b), (e, c), (f, c)\}$ and the specified argument is a. This will serve us as a running example.

An AF is usually intended to represent a completed debate process. In our specific setting a pointed AF $\mathcal{F}, a$ is meant to encode what is sometimes called a "culturally given pool of arguments" [30] about one issue a, i.e. the full set of arguments and attacks between them that are available to a group of individuals debating over a. *Opinions* held by the participants are represented as sets of arguments they embrace. Conflicts between opinions can be formalised as attacks between sets of arguments. We say that an opinion X attacks an opinion Y if there is an attack $R(x, y) \in X \times Y$. For example, in Fig. 2 it holds that $\{d, e, a\}$ attacks $\{b, c\}$ and viceversa.

One main purpose of argumentation theory is to identify which opinions are intuitively "acceptable". Such opinions are usually called *solutions* (or *extensions*). Typically, one solution should have at least two basic properties, i.e. *conflict-freeness* and *defense* of its own arguments. A set which combines these two properties is said to be *admissible.*

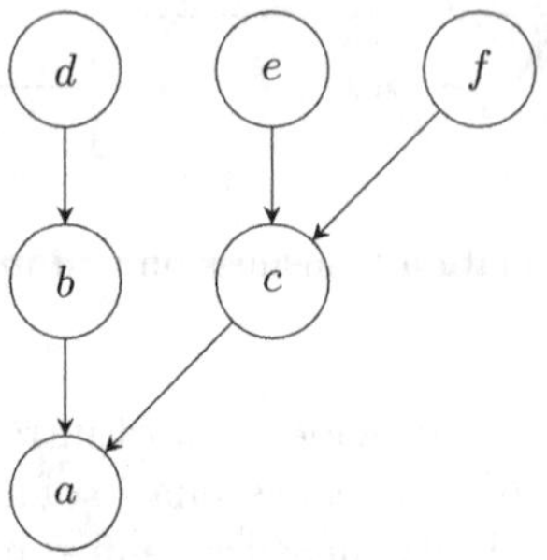

Fig. 2. An example of AF. Labelled nodes represent arguments. Relations of attack between arguments are indicated with an edge.

Definition 2 (Admissibility).

- *A set X is conflict-free if there is no $a, b \in X$ such that $R(a, b)$.*
- *A set X defends an argument a if for all b such that $R(b, a)$ there is a $c \in X$ such that $R(c, b)$.*
- *A set X is admissible iff X is conflict-free and defends all its elements.*

Intuitively, conflict-freeness encodes the internal coherence of an opinion, in the sense that no argument attacks another. The largest conflict-free sets in Fig. 2 are $\{a, d, e, f\}$ and $\{b, c\}$. The second condition (defending all its elements) encodes the fact that for an opinion to be *fully* acceptable it should be able to rebut all its counterarguments.[3] The AF in Fig. 2 has ten admissible sets where the smallest is $\emptyset$ and the largest is $\{a, d, e, f\}$.

The opinions we are interested in for our case are the *pro* and the *contra* opinion about a given issue a. It is straightforward to identify the opinion contra a ($C(a)$) as the set of arguments that attack a, while the opinion pro a ($P(a)$) is the set of arguments that defend a, including a itself. What is left out is the *neutral* opinion $N(a)$, i.e. the set of arguments that neither attack nor defend a. These three opinions are then defined as follows:

Definition 3 (Pro, contra and neutral opinions).

- *$P(a)$ is the set of arguments b such that there is an R-path of even length, including length 0, from b to a.*
- *$C(a)$ is the set of arguments b such that there is an R-path of odd length from b to a.*
- *$N(a)$ is the set of arguments b such that $b \notin P(a)$ and $b \notin C(a)$.*

[3] Admissibility is the basis of most of the solution concepts in the standard Dung's framework such as *preferredness*, *stability* and *groundedness*. For our present purposes we don't need to introduce them.

It is easy to ascertain that, in the pointed AF of Fig. 2, $P(a) = \{a, d, e, f\}$, $C(a) = \{b, c\}$ and $N(a) = \emptyset$. Furthermore, the following holds:

Fact 4. *$P(a) \cap C(a) = \emptyset$ iff both $P(a)$ and $C(a)$ are conflict-free.*

In the following we assume that $P(a)$ and $C(a)$ are conflict-free for all the information frameworks that we consider.

2.1 A Multi-agent Debate

If we regard a specific pointed AF $\mathcal{F}, a$ as the total information available about the issue a, then it is natural to encode the *partial information* available to a participant to the debate as a subgraph of $\mathcal{F}, a$, i.e. a partial representation of the argumentative pool.[4] This captures the fact that an individual may not be aware of some available arguments, or may even not be aware that some argument attacks another.[5] By consequence, the setup of a debate over a can be seen as a multiagent scenario where the information available to each agent i is defined as follows.

Definition 5 (Agent's information). *Given the total information $\mathcal{F}, a = (A, R), a$, the information available to agent i is $I_i(\mathcal{F}, a) = (A_i, R_i), a$ where $A_i \subseteq A$, $a \in A_i$, $R_i \subseteq A_i \times A_i$ and $R_i \subseteq R$.*

Hereafter I denote as $P_i(a)$, $C_i(a)$ and $N_i(a)$ the sets of pro, contra and neutral arguments of agent i about the issue a.

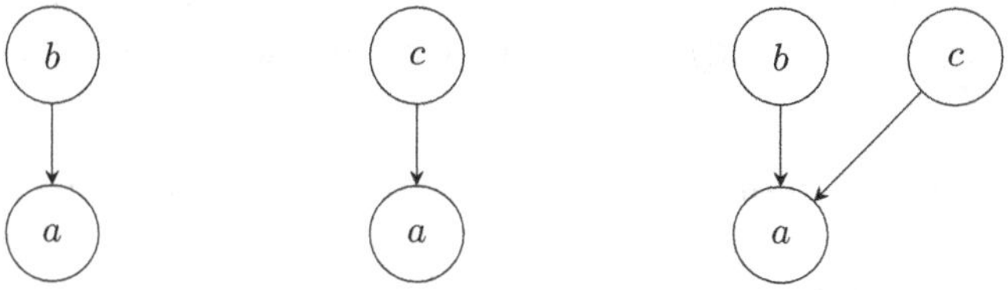

(a) Agent 1 (b) Agent 2 (c) Updated graph

Fig. 3. Information of agents 1 and 2 and their merging

[4] A similar approach is taken by [1,8,25]. There too the information base of an agent is encoded by a subset of a larger *universe* [8] or *universal argumentation framework* [1,25].

[5] Being unaware that b attacks c is the case when one lacks the *warrant* for b to undermine c (see also [29]). To give an example, let c be the argument "Phosphorus is not visible in the sky tonight" and b be the argument "Look, Hesperus is there!". Clearly b constitutes an attack to c only if one is aware that Hesperus and Phosphorus are the same planet.

Example 2. Figure 3(a) (resp. Fig. 3(b)) represents the information $I_1(\mathcal{F}, a)$ available to Agent 1 (resp. $I_2(\mathcal{F}, a)$ available to Agent 2) at the initial state of a debate over a. Both are subgraphs of $\mathcal{F}, a$ in Fig. 2. Here $C_1(a) = \{b\}$ and $C_2(a) = \{c\}$. Therefore both agents have distinct informational items contra a. Intuitively, when they merge such information together, as it happens in discussion, they should both get new information to the effect that $C_1(a) = C_2(a) = \{b, c\}$.

This illustrates polarization on the intuitive level, but does not yet provide a measure of it, which we shall present in the next section.

2.2 Degree of Acceptability

We introduced admissibility as a criterion of full acceptability for an opinion in a debate. However, in many real-life scenarios involving a large number of arguments and rebuttals, full acceptability is a too strict requirement. In most cases the best thing one can do is weighting arguments *pro* and *contra.* This is also what seems to happen in the lab experiments on polarization and risky shift, where people is asked to provide *odds* for a certain decision or opinion [30] and polarization is measured as the shift between the initial and the final odds provided by the participants.

A large literature in abstract argumentation has recently developed to provide measures and weights for arguments and opinions in a debate (see among others [14,20,24]). For our present purposes we shall adopt a measure of acceptability provided by [24], which fulfills a series of useful properties for our case. We first define the set of attacks from set Y to set X in a framework $\mathcal{F}, a$, as $X^{\leftarrow Y}_{\mathcal{F},a} = \{(y, x) \in Y \times X \mid R(y, x)\}$. Then we can define the *degree of acceptability of X w.r.t. Y* as follows:

Definition 6 (degree of acceptability, Matt and Toni 2008).

$$d(X, Y) = \frac{1}{2}(1 + f(|Y^{\leftarrow X}_{\mathcal{F},a}|) - f(|X^{\leftarrow Y}_{\mathcal{F},a}|))$$

where $f : \mathbb{N} \rightarrow \mathbb{N}$ is defined as $f(n) = \frac{n}{n+1}$

We are specially interested in the measures $d(P(a), C(a))$ and $d(C(a), P(a))$. The nice properties of the measure d hang mostly on the fact that f is a monotonic increasing mapping s.t. $f(0) = 0$ and $lim_{n\rightarrow\infty} f(n) = 1$. Such properties are the following:

Fact 7. *The following properties hold for d.*

(a) $0 \leq d(X, Y) \leq 1$
(b) $d(P(a), C(a)) = 1 - d(C(a), P(a))$
(c) $d(P(a), C(a)) < \frac{1}{2}$ *iff* $|P(a)^{\leftarrow C(a)}_{\mathcal{F},a}| > |C(a)^{\leftarrow P(a)}_{\mathcal{F},a}|$
(d) $d(P(a), C(a)) > \frac{1}{2}$ *iff* $|P(a)^{\leftarrow C(a)}_{\mathcal{F},a}| < |C(a)^{\leftarrow P(a)}_{\mathcal{F},a}|$
(e) If $\mathcal{F}, a = (\{a\}, \emptyset), a$ *then* $d(P(a), C(a)) = d(C(a), P(a)) = \frac{1}{2}$

According to property (a) degrees of acceptability are scaled between 0 and 1. Properties (b), (c) and (d) mean that d measures the relative weight of arguments pro and contra a. (e) reflects the fact that a given issue with no arguments pro or contra is to be labeled as undecided.

Example 3. In our argumentative pool of Fig. 2 the degree of acceptability $d(P(a), C(a))$ is $\frac{13}{24}$ and is therefore slightly favorable to the opinion supporting a.

Such measure d can work as the *actual degree of acceptability of* a. In a context where $\mathcal{F}, a$ represents the overall information available and there is no conclusive way of settling the truth, having a measure of this sort is the best we can hope for. In the following we indicate with $d(P_i(a), C_i(a))$ the *degree of acceptability of* a *for agent* i.

Example 4. It is straightforward to ascertain that $d(P_i(a), C_i(a))$ is $\frac{1}{4}$ for both agents in Fig. 3(a) and (b). If we merge the agent's information, as in Fig. 3(c), then $d(P_i(a), C_i(a))$ becomes $\frac{1}{6}$ for each agent, i.e. it lowers. This is particularly interesting when we interpret Fig. 3(c) as the new information available to Agent 1 and 2, as it would result from an information exchange and consequent information update. This is exactly what happens with attitude polarization: both agents radicalize their opinion contra a. Indeed, their degree of acceptability of a ends up being more extreme ($\frac{1}{6}$) than its average before entering the debate ($\frac{1}{4}$).

3 Information Transmission and Information Update

Example 4 shows how polarization may arise in a group when agents share partial information about a given issue. This happened by agents merging their information together as the result of information exchange. However, the mechanisms of information exchange in a debate are way more complex than this. Such mechanisms need to be cut down into their basic components if one wants to capture all the possible ways in which polarization may arise. The basic components are clearly *information transmission* from a sender to a recipient and *information update* by the recipient.

3.1 Information Transmission

Information transmission is the way agents disclose the information they possess. This is encoded as an operation $T^p_{i,j}$ where i is the sender, j the recipient and p is the disclosure policy adopted by the sender. The input of such operation is $I_i(\mathcal{F}, a)$, i.e. the information available to the sender. The disclosure policy p determines the output, which in principle can be any piece of information possessed by i. The latter is then a 2-ple (A', R') with the constraints $A' \subseteq A_i$ and $R' \subseteq R_i$.

There are many possible ways for a sender to transmit information to a recipient. One way is full disclosure o, which consists in one agent disclosing all the available information.

Definition 8 (Open disclosure). *The output of an open disclosure policy for agent i is*

$$T^o_{i,j}(I_i(\mathcal{F}, a)) = A_i, R_i$$

Policy o is typical of a debate where the agents' common interest is to share the best possible information or to settle an issue in the optimal way. But this is not what usually happens in more strategic situations where agents have different goals. People often discloses only the information that is useful for her to win a debate or to fortify her opinion among the audience, e.g. in judicial proceedings or panel discussions. A large set of strategies is available to agents in such contexts. Indeed, most of the situations that can be modeled by our framework are open to *cheap talk* [10], where agents are allowed to lie. In this context, the foundational game-theoretic analysis by [5] shows that, when the interests of the agents are not perfectly aligned, the equilibrium solution requires players not to be fully informative.[6]

Here we define a radical policy s, which consists in delivering only information that speaks in favor of one's opinion.

Definition 9 (Strategic disclosure). *The output of a strategic disclosure policy for agent i is defined by cases as* $T^s_{i,j}(I_i(\mathcal{F}, a)) = A', R'$ *where:*

- *if* $d(P_i(a), C_i(a)) \geq \frac{1}{2}$

$$A' = A_i \setminus \{b \in C_i(a)\}$$
$$R' = R_i \setminus \{(b, c) \in C_i(a) \times P_i(a)\}$$

- *if* $d(P_i(a), C_i(a)) < \frac{1}{2}$

$$A' = A_i \setminus \{b \in P_i(a)\}$$
$$R' = R_i \setminus \{(b, c) \in P_i(a) \times C_i(a)\}$$

3.2 Information Update

When agent j receives new information from i she has to update her informational state in the light of it and of her prior information. This corresponds to an operation $\star^p$ which should output a new information state $I^p_j(\mathcal{F}, a)$ on the basis of her previous information I_j and the information $T_{i,j}$ received by i. Here again, many policies p are available to agents for such an update. For the present purposes I restrict my attention to two of them. The first option is again an open policy $\star^{ou}$, which consists in fully accepting the information received by the sender. This gives rise to the following definition

6 As pointed out by Reviewer 1, modelling information transmission and update in cheap talk situations is a highly interesting venue, which we must leave for future research.

Definition 10 (Open update). *Let $X = (A, R)$, a and $Y = A', R'$ then the output for agent i of $\star^{ou}(X, Y)$ is a pointed AF $(A^{ou}, R^{ou}), a$ where*

$$A^{ou} = A \cup A'$$
$$R^{ou} = (R \cup R') \cap (A^{ou} \times A^{ou})$$

Here the receiver accepts all the new arguments communicated by the sender as well as all the new attacks, provided that both ends are in her updated argument set. In what follows the set of pro (resp. contra and neutral) arguments inherits the apex of the parent argument space, e.g. P^{ou} (resp. C^{ou} and N^{ou}) is the set of the pro (resp. contra and neutral) arguments in A^{ou}.

In most cases, however, information update is more critical than this. Agents often discard evidence that speaks against their prior beliefs, or else devote more scrutiny to it [13]. The latter option is what typically happens when agents try to reduce so-called *cognitive dissonance* [11]. The former is instead a form of what has been called "kripkean dogmatism" [18]. Such update procedures are more articulated. With the next definition we provide an example of one dogmatic policy $\star^d$ of such kind, which is built upon the previously defined $\star^{ou}$.

Definition 11 (Dogmatic update). *Let $X = (A, R), a$, $Y = A', R'$ and let $\star^{ou}(X, Y) = (A^{ou}, R^{ou}), a$ be as in Definition 10. Then the output for agent i of $\star^d(X, Y)$ is a pointed AF (A^d, R^d), a where*

- *If $d(P_i(a), C_i(a)) \geq \frac{1}{2}$*

$$A^d = A^{ou} \setminus (C_i^{ou}(a) \cap (A' \setminus A))$$
$$R^d = R^{ou} \setminus ((C_i^{ou}(a) \times P_i^{ou}(a)) \cap (R' \setminus R))$$

- *If $d(P_i(a), C_i(a)) < \frac{1}{2}$*

$$A^d = A^{ou} \setminus (P_i^{ou}(a) \cap (A' \setminus A))$$
$$R^d = R^{ou} \setminus ((P_i^{ou}(a) \times C_i^{ou}(a)) \cap (R' \setminus R))$$

Under the policy $\star^d$ the agent i updates her framework on the basis of her degree of acceptability of a prior to the exchange of information. If she had a positive degree of accepability about the issue a, she discards all new arguments provided by the sender $(A' \setminus A)$ against a, as well as the new attacks against it $((C_i^{ou}(a) \times P_i^{ou}(a)) \cap (R' \setminus R))$. Otherwise she discards the pro arguments and the new attacks from pro to contra.

4 Results

I show two results holding for a two-agent debate between Agent 1 and 2. Agents follow some combinations of the previously defined policies of information transmission and update. All the following results show what happens after one round of mutual information transmission and update.

Theorem 1. *Let $\mathcal{F}, a = (A, R), a$, $I_1(\mathcal{F}, a) = (A_1, R_1), a$ and $I_2(\mathcal{F}, a) = (A_2, R_2), a$. Let $I_1^{ou}(\mathcal{F}, a) = (A_1^{ou}, R_1^{ou}), a = \star^{ou}(I_1(\mathcal{F}, a), T_{2,1}^{o}(I_2(\mathcal{F}, a)))$ and $I_2^{ou}(\mathcal{F}, a) = (A_2^{ou}, R_2^{ou}), a = \star^{ou}(I_2(\mathcal{F}, a), T_{1,2}^{o}(I_1(\mathcal{F}, a)))$. Suppose further that the distributed information of the agents covers the total information available, i.e. (tot) $A_1 \cup A_2 = A$ and $R_1 \cup R_2 = R$. Then*

(a) $I_1^{ou}(\mathcal{F}, a) = I_2^{ou}(\mathcal{F}, a) = \mathcal{F}, a$
(b) $d(P_1^{ou}(a), C_1^{ou}(a)) = d(P_2^{ou}(a), C_2^{ou}(a)) = d(P(a), C(a))$

Proof. (b) is an immediate consequence of (a). (a) is established as follows. By Definitions 8 and 10 it follows that $A_1^{ou} = A_1 \cup A_2 = A_2^{ou}$ and $R_1^{ou} = R_1 \cup R_2 = R_2^{ou}$. Then, by the condition (*tot*) it follows that $A_1^{ou} = A_2^{ou} = A$ and $R_1^{ou} = R_2^{ou} = R$ and the result is established.

This result shows that by following open policies of information transmission and update all agents can align their opinion to the most reasonable one. However, we must notice that, for this to happen, (*tot*) is a necessary condition. Indeed, Example 4 shows that if such condition fails, the same policies may lead all agents far away from the most reasonable opinion.

Theorem 2. *Let $\mathcal{F}, a = (A, R), a$ with $C(a) \cap P(a) = \emptyset$. Let $I_1(\mathcal{F}, a) = (A_1, R_1), a$. Let $I_1^d(\mathcal{F}, a) = (A_1^d, R_1^d), a = \star^d(I_1(\mathcal{F}, a), T_{2,1}^p(I_2(\mathcal{F}, a)))$ and $T_{2,1}^p(I_2(\mathcal{F}, a)) = A', R'$. Furthermore suppose that $N_1(a) = \emptyset$.*
Then

(a) if $d(P_1(a), C_1(a)) \geq \frac{1}{2}$

$$d(P_1(a), C_1(a)) \leq d(P_1^d(a), C_1^d(a))$$

(b) if $d(P_1(a), C_1(a)) < \frac{1}{2}$

$$d(P_1(a), C_1(a)) \geq d(P_1^d(a), C_1^d(a))$$

Proof. We only prove (a), since (b) follows by the same reasoning. In order to establish (a) it is sufficient to prove that $|C_1(a)^{\leftarrow P_1(a)}| \leq |C_1^d(a)^{\leftarrow P_1^d(a)}|$ and that $|P_1(a)^{\leftarrow C_1(a)}| \geq |P_1^d(a)^{\leftarrow C_1^d(a)}|$ and the result will follow as a consequence of Definition 6. The former is established by ascertaining that $C_1(a)^{\leftarrow P_1(a)} \subseteq C_1^d(a)^{\leftarrow P_1^d(a)}$. The second inequality is established by showing that there are no elements b and c such that $(b, c) \in P_1^d(a)^{\leftarrow C_1^d(a)}$ and $(b, c) \notin P_1(a)^{\leftarrow C_1(a)}$. For reductio we suppose the contrary and then reason by cases.

Case 1. $b \in A_1$, $c \in A_1$ and $(b, c) \in R_1$.
If $c \in P_1(a)$ then $b \in C_1(a)$ and $(b, c) \in P_1(a)^{\leftarrow C_1(a)}$ against the assumption. If $c \in C_1(a)$ then a fortiori $c \in C(a)$. However we also assumed that $c \in P_1^d(a)$ and therefore $c \in P(a)$. This however goes against the assumption that $C(a) \cap P(a) = \emptyset$ and we get a contradiction. The last possibility is $c \in N_1(a)$, but this again is excluded by the assumption that $N_1(a) = \emptyset$. Therefore Case 1 leads to a contradiction.

Case 2. Either $b \notin A_1$ or $c \notin A_1$ or $(b, c) \notin R_1$.
Case 2.1. If $b \notin A_1$ then it must be that $b \in (A' \setminus A_1)$. This excludes the fact that $b \in C_1^d(a)$ since this would imply also that $b \in C_1^{ou}(a)$ and b would therefore have been eliminated by the policy $\star^d$.
Case 2.2 If $c \notin A_1$ then $c \in (A' \setminus A_1)$. This entails that $(b, c) \in (R' \setminus R_1)$. The latter however contradicts the fact that $(b, c) \in P_1^d(a)^{\leftarrow C_1^d(a)}$, since a fortiori $(b, c) \in P_1^{ou}(a)^{\leftarrow C_1^{ou}(a)}$ and therefore (b, c) would have been eliminated by the policy $\star^d$.
Case 2.3 If $(b, c) \notin R_1$ then $(b, c) \in (R' \setminus R_1)$ and a contradiction follows in the same way as in Case 2.2
Therefore Case 2 also leads to a contradiction and the proof is completed.

Under these conditions participants to a debate can only radicalize their prior opinions. This leaves the possibility open for bipolarization to happen. This is what the following example shows.

Example 5. Let the information states of Agent 1 and 2 be as in Fig. 4. Then $d(P_1(a), C_1(a)) = \frac{5}{12}$ and $d(P_2(a), C_2(a)) = \frac{13}{24}$. Here $N_1(a) = N_2(a) = \emptyset$ since both graphs are connected. Suppose that both agents adopt the policies $T_{i,j}^o$ and $\star^d$. Then Agent 1 will update his information state with argument e and attack (e, a). Agent 2 will instead update her state by argument d and the attack (d, c). Therefore $d(P_1^d(a), C_1^d(a)) = \frac{3}{8}$ and $d(P_2^d(a), C_2^d(a)) = \frac{19}{30}$ The group has therefore bipolarized.

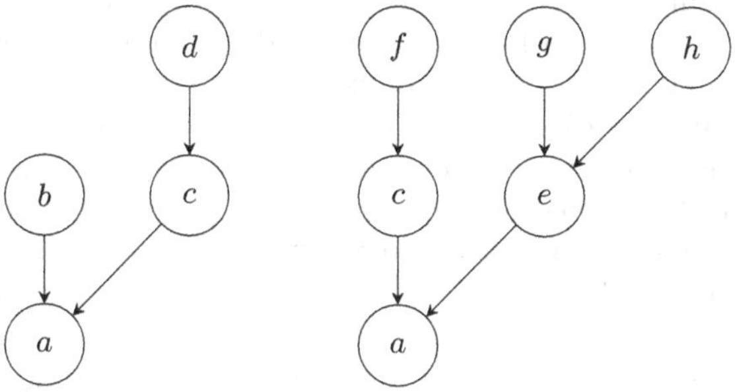

(a) Agent 1 (b) Agent 2

Fig. 4. Bipolarization

Theorem 2 may however not hold when $N_1(a)$ is not empty as the next example shows

Example 6. Let the information states of Agent 1 and 2 be as in Fig. 5. Then $d(P_1(a), C_1(a)) = \frac{1}{4}$. Suppose that Agent 2 adopts the policy $T_{2,1}^o$ and Agent 1 adopts $\star^d$. Then Agent 1 will update his information state with the attack (c, a). As a result $d(P_1^d(a), C_1^d(a)) = \frac{13}{24}$. So Agent 1 shifts her opinion towards the opposite direction.

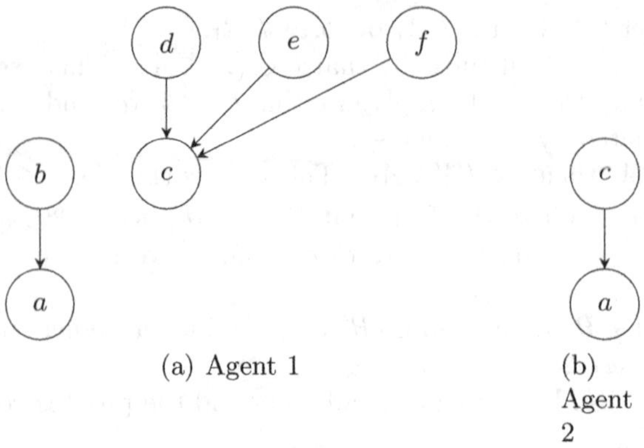

Fig. 5. Opinion shift for Agent 1

This example illustrates how subtly information dynamics can influence opinion change. Indeed, under certain conditions, even providing information contra a may determine a significant positive shift in the degree of acceptability of a. In Example 6 this happens even though the recipient holds a dogmatic attitude in discarding information pro a.

5 Conclusions

One of the main aims of this work is to show how Argumentation Frameworks provide an efficient tool for a formal understanding of polarization and bipolarization dynamics in group discussion. Indeed, they serve to encode both the total information available in a debate about a given issue a, as well as the agents' partial information about it. A large number of policies of information exchange among agents can be defined as specific operations on Argumentation Frameworks. Moreover, the measure provided by [24] provides a way to quantify the degree of acceptability of a, both for the single discussants and on the absolute level. Therefore, it also serves to quantify the distance of the agents' opinion from the actual degree of acceptability of a. More importantly, it quantifies polarization and bipolarization prior and posterior to information exchange.

Theorem 1 of Sect. 4 shows how an open policy of information transmission and information update can help agents to align their views together with the most reasonable opinion about the debated issue. This is often cited as one of the virtues of an open discussion, held by agents without prejudices. Theorem 1 confirms such intuition, but only to a very specific extent. Indeed, the conclusion holds only under the condition (*tot*) that discussants have distributed knowledge of the total information available. Otherwise the possibility is open for polarization to happen and for the agents to be led far away from the most reasonable opinion, as shown in Example 4.

Theorem 2 shows instead that agents who follow a dogmatic policy of information update are bound to radicalize their opinion. This leaves the way open to polarization and bipolarization, as shown in Example 5. Therefore, this result provides an alternative explanation of bipolarization, one which fully lies in the agent's policy of information update and does not recur to standard explanations such as negative influence, as in [16] and [12], or homophily, as in [23]. Nonetheless, in this case too the conclusion holds only under a specific condition. Example 6 illustrates how, when such condition fails, opinion shifts may happen even with a dogmatic policy of information update. This example put the finger on the complexity and the unpredictability of the opinion dynamics generated by information exchange. The study of such dynamics therefore discloses an interesting field of investigation for future research.

As a final consideration, Example 4 illustrates how polarization can happen under conditions of "full rationality", i.e. when agents openly disclose their information and update it on the basis of all the available evidence. On the other hand, Example 5 may seem to suggest that bipolarization requires instead that agents deviate from Bayesian standards, e.g. by being dogmatic and discarding some available evidence. However this conclusion jumps too far. Indeed, Theorem 2 only provides sufficient conditions for (bi)polarization, and not necessary ones. The general question remains open as to whether bipolarization entails such a deviation from Bayesianism. Answering this question goes beyond the scope of this paper, although it deserves the highest priority in this research agenda.[7]

Acknowledgements. This research is financed by the *Riksbankens Jubileumsfond. The Swedish Foundation for Humanities and Social Sciences* (RJ) through the project *Rationality and Group Behavior* (P16-0596:1). Special thanks go to Ilaria Jemos and Roberto Ciuni for their insightful comments and suggestions. I am also indebted to Davide Grossi and Paolo Di Paola for discussion and inspiration.

References

1. Caminada, M., Sakama, C.: On the issue of argumentation and informedness. In: 2nd International Workshop on Argument for Agreement and Assurance (AAA 2015) (2015)
2. Cayrol, C., Lagasquie-Schiex, M.C.: On the acceptability of arguments in bipolar argumentation frameworks. In: Godo, L. (ed.) ECSQARU 2005. LNCS, vol. 3571, pp. 378–389. Springer, Heidelberg (2005). doi:10.1007/11518655_33
3. Cayrol, C., Lagasquie-Schiex, M.C.: Bipolarity in argumentation graphs: Towards a better Understanding. Int. J. Approx. Reason. **54**(7), 876–899 (2013)
4. Coste-Marquis, S., Devred, C., Konieczny, S., Lagasquie-Schiex, M.C., Marquis, P.: On the merging of Dung's argumentation systems. Artif. Intell. **171**, 730–753 (2007)
5. Crawford, V., Sobel, J.: Strategic information transmission. Econom. **50**(6), 1431–1451 (1982)
6. Delobelle, J., Konieczny, S., Vesic, S.: On the aggregation of argumentation frameworks. IJCAI **2015**, 2911–2917 (2015)
7. Delobelle, J., Haret, A., Konieczny, S., Mailly, J., Rossit, J., Woltran, S.: Merging of abstract argumentation frameworks. In: KR 2016, pp. 33–42 (2016)

[7] Thanks to Reviewer 1 for raising this issue.

8. Dupin de Saint-Cyr, F., Bisquert, P., Cayrol, C., Lagasquie-Schiex, M.C.: Argumentation update in YALLA (Yet Another Logic Language for Argumentation). Int. J. Approx. Reason. **75**, 57–92 (2016)
9. Dung, P.M.: On the acceptability of arguments and its fundamental role in nonmonotonic reasoning, logic programming and n-person games. Artif. Intell. **77**(2), 321–357 (1995)
10. Farrell, J.: Cheap talk, coordination, and entry. RAND J. Econ. **18**(1), 34–39 (1987)
11. Festinger, L.: A Theory of Cognitive Dissonance. Stanford University Press, Stanford, CA (1957)
12. Flache, A., Macy, M.W.: Small world and cultural polarization. J. Math. Sociol. **35**, 146–176 (2011)
13. Gilovich, T.: How we know what isnt so. The Free Press, New York (1991)
14. Grossi, D., Modgil, S.: On the graded acceptability of arguments. In: Proceedings of the IJCAI 2017, pp. 868–874 (2015)
15. Isenberg, D.J.: Group polarization: a critical review and a meta-analysis. J. Pers. Soc. Psychol. **50**(6), 1141–1151 (1986)
16. Jager, W., Amblard, F.: Uniformity, bipolarization and pluriformity captured as generic stylized behavior with an agent-based simulation model of attitude change. Comput. Math. Organiz. Theor. **10**, 295–303 (2004)
17. Jern, A., Chang, K.K., Kemp, C.: Belief Polarization is not always irrational. Psychol. Rev. **121**(2), 206–224 (2014)
18. Kelly, T.: Disagreement, dogmatism, and belief polarization. J. Philos. **105**(10), 611–633 (2008)
19. Krizan, Z., Baron, R.S.: Group polarization and choice-dilemmas: How important is self-categorization? Eur. J. Soc. Psychol. **37**, 191–201 (2007)
20. Li, H., Oren, N., Norman, T.J.: Probabilistic argumentation frameworks. In: Modgil, S., Oren, N., Toni, F. (eds.) TAFA 2011. LNCS, vol. 7132, pp. 1–16. Springer, Heidelberg (2012). doi:10.1007/978-3-642-29184-5_1
21. Liu, Q., Zhao, J., Wang, X.: Multi-agent model of group polarisation with biased assimilation of arguments. IET Control Theor. Appl. **9**(3), 485–492 (2014)
22. Lord, C., Ross, L., Lepper, M.: Biased assimilation and attitude polarization: The effects of prior theories on subsequently considered evidence. J. Pers. Soc. Psychol. **37**(11), 2098–2109 (1979)
23. Mäs, M., Flache, A.: Differentiation without distancing. explaining bi-polarization of opinions without negative influence. PLoS ONE **8**(11), e74516 (2013)
24. Matt, P.-A., Toni, F.: A game-theoretic measure of argument strength for abstract argumentation. In: Hölldobler, S., Lutz, C., Wansing, H. (eds.) JELIA 2008. LNCS, vol. 5293, pp. 285–297. Springer, Heidelberg (2008). doi:10.1007/978-3-540-87803-2_24
25. Sakama, C.: Dishonest arguments in debate games. COMMA **2012**(75), 177–184 (2012)
26. Sanders, G.S., Baron, R.S.: Is social comparison irrelevant for producing choice shifts? J. Exper. Soc. Psychol. **13**, 303–314 (1977)
27. Stoner, J.A.: A comparison of individual and group decision involving risk MA thesis, Massachusetts Institute of Technology (1961)
28. Sunstein, C.: Why societies need Dissent. Harvard University Press, Cambridge (2003)
29. Toulmin, S.: The Uses of Argument. Harvard University Press, Cambridge (1958)
30. Vinokur, A., Burnstein, E.: Effects of partially shared persuasive arguments on group-induced shifts. J. Pers. Soc. Psychol. **29**(3), 305–15 (1974)

Doing Without Nature

Frederik Van De Putte[1(✉)], Allard Tamminga[2,3], and Hein Duijf[3]

[1] Ghent University, Blandijnberg 2, 9000 Ghent, Belgium
frederik.vandeputte@ugent.be
[2] University of Groningen, Oude Boteringestraat 52, 9712 GL Groningen, The Netherlands
a.m.tamminga@rug.nl
[3] Utrecht University, Janskerkhof 13, 3512 BL Utrecht, The Netherlands
H.W.A.Duijf@uu.nl

Abstract. We show that every indeterministic n-agent choice model M^i can be transformed into a deterministic n-agent choice model M^d, such that M^i is a bounded morphic image of M^d. This generalizes an earlier result from Van Benthem and Pacuit [16] about finite two-player choice models. It further strengthens the link between STIT logic and game theory, because deterministic choice models correspond in a straightforward way to normal game forms, and choice models are generally used to interpret STIT logic.

1 Introduction

At least since [9], it has become clear that there are strong links between game theory and the model theory of STIT logic. In this paper, we focus on the relation between normal game forms and what we call *choice models*.

A normal game *form* is a strategic game without players' preferences.[1] Choice models will be defined in Sect. 2; they form a specific class of Kripke models for a purely agentive STIT logic, i.e. a logic of (individual and collective) agency that contains no temporal operators. Choice models are e.g. used in [8] to study the complexity of STIT logic for groups. They closely resemble the *choice structures* from [10], the *STIT choice scenarios* from [16], and the *choice Kripke models* from [5].

Normal game forms and choice models are both used to represent the actions of (group) agents, thus providing a basis for the analysis of rational (individual and collective) interaction. The notion of *effectivity for outcomes* is central in both, where an outcome can be thought of as a (non-empty) set of possible worlds. Roughly speaking, an agent is effective for a given outcome in the model if and only if it can ensure that outcome by some action, regardless of what the other agents do.[2]

[1] See e.g. [12] for a solid introduction to the theory of strategic games.
[2] This notion is usually referred to as "α-effectivity" in game theory. We provide a formal definition of it in Sect. 2.

A. Baltag et al. (Eds.): LORI 2017, LNCS 10455, pp. 209–223, 2017.
DOI: 10.1007/978-3-662-55665-8_15

For reasons of space, we cannot provide the full background and history of STIT logic and its relation to game theory in this paper. We refer to [1,9] for general introductions to STIT logic. See [8] for a discussion of the axiomatization and complexity of group STIT interpreted over choice models. [10] appears to be the first paper that explicitly deals with the relation between strategic game forms and STIT. Horty's [9] work in deontic logic, however, was already strongly inspired by the link between game theory and STIT theory. Publications that are directly relevant to this paper are [5,13,14,16].

To explain the aim of this paper, let us first focus on models with only two agents. Like normal game forms, choice models for two agents can be represented using matrices, where the rows represent choices of agent 1 and columns represent choices of agent 2. Figure 1 is a simple example. Each cell[3] in such a matrix represents a combination of actions of each agent – also known as *action profile* – and the corresponding outcome. For instance, if agent 1 chooses row 1 and agent 2 chooses column 2, then there is a unique outcome, viz. c. Note that the action profile (row 1, column 1) allows for two possible worlds, viz. a and b. In this model, agent 1 is i.a. effective for $\{a, b, c\}$ and $\{d, e\}$, whereas agent 2 is effective for $\{a, b, d\}$ and $\{c, e\}$.[4]

a,b	c
d	e

Fig. 1. A choice model for two agents.

Say two models M_1 and M_2 – whether normal game forms or choice models – are *equivalent* if and only if for any given outcome X and every agent j it holds that j is effective for X in M_1 if and only if j is effective for X in M_2. Every normal game form can be translated into an equivalent choice model, where the action profiles in the former correspond to the worlds in the latter. This was first observed by Tamminga in [13, Sect. 3.1]; we will recall the details in Sect. 2. As is shown in [14], the inverse translation works for a specific class of choice models. This class is characterized by the condition known as *determinism*: each choice of the grand coalition, i.e. each action profile, singles out exactly one world. Note that in the above example, this condition is not satisfied: if agent 1 chooses row 1 and agent 2 chooses column 1, then either a or b may result.

A common motivation for determinism is that we can get it "for free" just by moving to a three-agent model, letting "nature" or "the environment" play the role of the third agent (see e.g. [9, p. 91] and [16, p. 300] where this point is made). In other words, nature is an agent that makes its own choices, and in combination

[3] In the two-agent case, cells correspond to the "innermost squares" in the matrix. See Sect. 2 for the general definition of cells in a choice model.

[4] The notion of effectivity is monotonic: whenever an agent is effective for X, it is also effective for every superset of X. In the current example, this means that agent 1 is e.g. also effective for $\{a, b, c, d\}$.

with the choices of the two "real" agents, this determines the outcome. Applying this idea to the example from Fig. 1 yields the model depicted in Fig. 2, where nature gets to choose between the left and right matrix. In this new choice model, agents 1 and 2 are just as effective as they were in the original model given by Fig. 1, but the choices by the group of all agents (including nature) always determine a singleton outcome.

a	c
d	e

b	c
d	e

Fig. 2. A deterministic choice structure for three agents.

Leaving more philosophical issues aside, one may wonder whether this technical trick is really necessary, mathematically speaking. This question will be answered in the present paper: we show that one can indeed do without nature, as long as one does not consider the effectivity of the grand coalition. We prove this by generalizing a proof method from [16], which is discussed in Sect. 3. We generalize this method, first, to an arbitrary finite number of agents and all groups of such agents except the grand coalition (Sect. 4) and second, to infinite models (Sect. 5). We finish with a summary and some questions for future work (Sect. 6).

2 Preliminaries: Group STIT

The notion of effectivity can be made exact and studied formally, using a well-known STIT logic for group agents. In this section, we introduce the formal language of this logic. After that, we give two different semantics for this logic, using choice models and normal game models, and discuss the relation between these types of semantics.

2.1 The Language of Group STIT

Throughout this paper, we assume a fixed, finite set $N = \{1, \ldots, n\} \subset \mathbb{N}$ of agents. We let j range over members of N and we let G range over non-empty subgroups of N. $\mathfrak{P} = \{p_1, p_2, \ldots\}$ is a (countable) set of propositional variables. The formal language $\mathfrak{L}$ is given by the following Backus-Naur form, where p ranges over $\mathfrak{P}$:

$$\varphi ::= p \mid \varphi \wedge \varphi \mid \neg\varphi \mid [G]\varphi \mid \Box\varphi$$

$\mathfrak{L}^{-N}$ denotes the fragment of $\mathfrak{L}$ without the operator $[N]$. Parentheses, brackets, and braces are omitted if the omission does not give rise to ambiguities. The operators $\vee$, $\rightarrow$, $\Diamond$, and $\langle G\rangle$ abbreviate the standard constructions.

A formula $[G]\varphi$ expresses that "the group G sees to it that φ is the case", or alternatively, "given G's choice, φ is necessary".[5] The formula $\Box\varphi$ expresses that "φ is settled true", or equivalently, "whatever the agents choose, φ is the case". In modal logic terminology, $\Box$ corresponds to the universal (or global) modality, since it quantifies over all the worlds in a given model.[6]

2.2 Choice Models

Choice models consist of, on the one hand, a set of possible worlds W, and on the other hand, for every agent $j \in N$, a partition of W that represents the choices of j in the model. The only restriction to these partitions is that they satisfy a specific frame condition, known as *independence of agents*. This condition expresses that no group of agents $G \subseteq N \setminus \{j\}$ can render any of the choices that are available to agent j impossible. In other words, if the agent j has a certain choice, then it can make this choice regardless of what all the other agents do.

To stay in line with standard modal logic terminology, the choices of each agent j will be represented by an equivalence relation $\sim_j$ on W.

Definition 1. *A* choice frame *F is a tuple $\langle W, \langle \sim_j \rangle_{j \in N} \rangle$, where W is a non-empty set (the domain of F), each $\sim_j \subseteq W \times W$ is an equivalence relation, and the* independence of agency *condition obtains:*

(IOA) for all $w_1, \ldots, w_n \in W$, there is a w' such that $w_j \sim_j w'$ for all $j \in N$.

For a given choice frame $F = \langle W, \langle \sim_j \rangle_{j \in N} \rangle$ and non-empty $G \subseteq N$, we define $\sim_G = \bigcap_{j \in G} \sim_j$.[7]

A choice model *M is a triple $\langle W, \langle \sim_j \rangle_{j \in N}, V \rangle$ where $\langle W, \langle \sim_j \rangle_{j \in N} \rangle$ is a choice frame and $V : \mathfrak{P} \to \wp(W)$ is a valuation function.*

We say that M is deterministic *iff $\sim_N = \{(w, w) : w \in W\}$. M is* finite *iff W is finite.*

For every non-empty $G \subseteq N$, the equivalence relation $\sim_G$ in a choice frame induces a partition of W: $Choice_G(M) =_{\mathsf{df}} \{\{w' \mid w' \sim_G w\} \mid w \in W\}$. The members of $Choice_G(M)$ are referred to as the *choices* of the group G in M. Note that $Choice_N(M) = \{X_1 \cap \ldots \cap X_n \mid X_1 \in Choice_1(M), \ldots, X_n \in Choice_n(M)\}$. It is obvious that a choice model M is deterministic iff every member of $Choice_N(M)$ is a singleton. We use the common term *cells* to refer to the members of a $Choice_N(M)$.

---

[5] $[G]$ is also known as the *Chellas STIT*, after the seminal work in the logic of agency by Chellas [4].

[6] Given our semantics, $\Box\varphi$ is definable as $[i][j]\varphi$ for $i \neq j$. We will however treat $\Box$ as primitive for reasons of clarity.

[7] This property may be called the intersection property. In Sect. 6 we briefly mention how it relates to some completeness results.

Definition 2. *Where $M = \langle W, \langle \sim_j \rangle_{j \in N}, V \rangle$ is a choice model and $w \in W$,*

$$\begin{array}{ll} M, w \models p & \textit{iff } w \in V(p) \\ M, w \models \neg\phi & \textit{iff } M, w \not\models \phi \\ M, w \models \phi \wedge \psi & \textit{iff } M, w \models \phi \textit{ and } M, w \models \psi \\ M, w \models \Box\phi & \textit{iff for all } w' \in W \textit{ it holds that } M, w' \models \phi \\ M, w \models [G]\phi & \textit{iff for all } w' \in W \textit{ with } w' \sim_G w \textit{ it holds that } M, w' \models \phi. \end{array}$$

As usual, $\|\varphi\|^M = \{w \in W \mid M, w \models \varphi\}$.

That G is effective for a state of affairs φ in the choice model M can be expressed by means of the formula $\Diamond[G]\varphi$. This formula expresses that, for some world w in the model, $[G]\varphi$ is true. Since $Choice_G(M)$ is a partition of W, this is equivalent to saying that there is a choice $X \in Choice_G(M)$ such that $X \subseteq \|\varphi\|^M$.

2.3 Normal Game Models

In this section, we briefly spell out the semantics for group STIT using normal game forms, following [16]. Subsequently, we discuss the relation between normal game forms and deterministic choice models.

Definition 3. *A* normal game form *for the set of agents $N = \{1, \ldots, n\}$ is a tuple $G = \langle A_i \rangle_{i \in N}$, where each A_i is a non-empty set of actions $a, a', \ldots$ available to agent i. We call $\times_{i \in N} A_i$ the set of* action profiles *of the game, and denote its members by σ, σ', etc. Where $\sigma = \langle a_1, \ldots, a_n \rangle \in \times_{i \in N} A_i$ and $j \in N$, let $\pi^j(\sigma) = a_j$. Where $\emptyset \neq G \subseteq N$, let $\pi^G(\sigma)$ denote G's part in the action profile σ, i.e., $\pi^G(\sigma) = \langle \pi^j(\sigma) \rangle_{j \in G}$.*

A normal game model *is a tuple $S = \langle \langle A_i \rangle_{i \in N}, V \rangle$, where $\langle A_i \rangle_{i \in N}$ is a normal game form and $V : \mathfrak{P} \to \wp(\times_{i \in N} A_i)$ is a valuation function.*

Definition 4. *Where $S = \langle \langle A_i \rangle_{i \in N}, V \rangle$ is a normal game model and $\sigma \in \times_{i \in N} A_i$:*

$$\begin{array}{ll} S, \sigma \models p & \textit{iff } \sigma \in V(p) \\ S, \sigma \models \neg\phi & \textit{iff } S, \sigma \not\models \phi \\ S, \sigma \models \phi \wedge \psi & \textit{iff } S, \sigma \models \phi \textit{ and } S, \sigma \models \psi \\ S, \sigma \models \Box\phi & \textit{iff for all } \sigma' \in \times_{i \in N} A_i, \textit{ it holds that } S, \sigma' \models \phi \\ S, \sigma \models [G]\phi & \textit{iff for all } \sigma' \in W \textit{ with } \pi^G(\sigma) = \pi^G(\sigma') \textit{ it holds that } S, \sigma' \models \phi. \end{array}$$

As was the case for choice models, one can use the language $\mathfrak{L}$ to formalize statements concerning the effectivity of a given (group) agent G for a certain outcome, where outcomes are represented by propositions φ. The formula $\Diamond[G]\varphi$ expresses that there is an action profile σ such that, for all σ' with $\pi^G(\sigma) = \pi^G(\sigma')$, $S, \sigma' \models \varphi$. In other words, the group G has a combined choice $\pi^G(\sigma)$ such that, whatever the other agents do, φ is guaranteed.

2.4 A Correspondence Result

As mentioned in the introduction, there is a well-known correspondence between deterministic choice models on the one hand, and normal game models on the other – see e.g. [5,13,14].[8] To clarify the purpose of our new results, these correspondence results are explicated here.

Definition 5. *Let $S = \langle \langle A_i \rangle_{i \in N}, V \rangle$ be a normal game model. The corresponding choice model $M^S = \langle W, \langle \sim_i \rangle_{i \in N}, V \rangle$ is defined as follows:*

1. $W = \times_{i \in N} A_i$
2. *for all $i \in N$ and $\sigma, \sigma' \in W$, $\sigma \sim_i \sigma'$ iff $\pi^i(\sigma) = \pi^i(\sigma')$.*

Theorem 1. *Let $S = \langle \langle A_i \rangle_{i \in N}, V \rangle$ be a normal game model. Then (a) M^S is a deterministic choice model. Moreover, (b) where $\varphi \in \mathfrak{L}$ and $\sigma \in \times_{i \in N} A_i$: $S, \sigma \models \varphi$ iff $M^S, \sigma \models \varphi$.*

Proof. Suppose the antecedent holds. To obtain (a), note first that each relation $\sim_i$ is an equivalence relation. To see why the condition (IOA) holds for M^S, let $\sigma_1, \ldots, \sigma_n \in \times_{i \in N} A_i$. Let $\sigma = \langle \pi^1(\sigma_1), \ldots, \pi^n(\sigma_n) \rangle$. It can easily be verified that, for all $j \in N$, $\sigma \sim_j \sigma_j$. Finally, to see why M^S is deterministic, note that $\sigma \sim_N \sigma'$ iff for all $j \in N$, $\pi^j(\sigma) = \pi^j(\sigma')$ iff $\sigma = \sigma'$.

The proof of (b) is by a standard induction on the complexity of φ; it suffices to apply the truth conditions from Definitions 2 and 4. □

Definition 6. *Let $M = \langle W, \langle \sim_i \rangle_{i \in N}, V \rangle$ be a deterministic choice model. The corresponding normal game model $S^M = \langle \langle A_i \rangle_{i \in N}, V' \rangle$ is such that the following holds:*

1. *for all $i \in N$, $A_i = Choice_i(M)$*
2. *where $\sigma = \langle X_1, \ldots, X_n \rangle \in \times_{i \in N} A_i$ and $X_1 \cap \ldots \cap X_n = \{w\}$: $\sigma \in V'(p)$ iff $w \in V(p)$.*

Theorem 2. *Let $M = \langle W, \langle \sim_i \rangle_{i \in N}, V \rangle$ be a deterministic choice model and $\varphi \in \mathfrak{L}$. Then (a) $S^M = \langle \langle A_i \rangle_{i \in N}, V' \rangle$ is a normal game model. Moreover, where $\sigma = \langle X_1, \ldots, X_n \rangle \in \times_{i \in N} A_i$, $X_1 \cap \ldots \cap X_n = \{w\}$, and $\varphi \in \mathfrak{L}$: $S^M, \sigma \models \varphi$ iff $M, w \models \varphi$.*

Proof. Suppose the antecedent holds. To obtain (a), it suffices to check that V' is a valuation function. This follows immediately in view of the fact that for every $\sigma \in \times_{i \in N} Choice_i(M)$, there is a world w such that w is the only member of the intersection of all the choices that make up σ. To prove (b), we again apply a standard induction on the complexity of φ, together with the semantic clauses from Definitions 2 and 4. □

Theorems 1 and 2 give us at once:

Corollary 1. *Let $\varphi \in \mathfrak{L}$. Then φ is valid in all deterministic choice models if and only if φ is valid in all normal game models.*

[8] In [13,14], the authors actually establish a correspondence between strategic games and choice models enriched with preference relations $\preceq_i$ for the agents $i \in N$. Ignoring this extra dimension, one obtains exactly the correspondence that we spell out in the present section.

3 Informal Sketch of the Proof

In the remainder we prove that, relative to the fragment $\mathfrak{L}^{-N}$, every (indeterministic) choice model M^i is a bounded morphic image of some deterministic choice model M^d (cf. Theorems 4 and 6 below).[9] In other words, for every world w in M^i there is a world w' in M^d such that, for every formula $\varphi \in \mathfrak{L}^{-N}$, φ is true at w in M^i iff φ is true at w' in M^d. More briefly, M^i and M^d are pointwise equivalent, relative to $\mathcal{L}^{-N}$. In view of the preceding, this result implies that the set of all formulas from $\mathfrak{L}^{-N}$ that are valid in all choice models coincides with the set of all formulas from $\mathfrak{L}^{-N}$ that are valid in all normal game models.[10] This way, we fill an important gap in the comparison of normal game models on the one hand, and choice models (and other traditional semantics of STIT logic) on the other.

Our proof generalizes a construction by Van Benthem and Pacuit [16] that only applies to the case where we have two agents and M^i is finite. The construction by Van Benthem and Pacuit is in turn based on known methods from modal product logics.[11] To guide the reader's intuitions and to explain our own contribution, Van Benthem and Pacuit's construction is explained in the current section.

Consider the model M^i depicted in Fig. 3 – as before, we abstract from the valuation function in our pictures of the models. Note that the cell that contains the highest number of possible worlds is the one containing three worlds: h, i, and j.

a	b,c	d
e,f	g	h,i,j

Fig. 3. An indeterministic choice model.

The proof by Van Benthem and Pacuit basically consists of two steps. The first step is to construct an $m \times m$ matrix $\mathsf{M}(X)$ for every cell X in M^i, where m is the highest number of worlds that occur in one cell of M^i. The points in this matrix are copies of the members of X, and the matrix is constructed in such a way that every $x \in X$ occurs at least once in each row and in each column of $\mathsf{M}(X)$. For instance, the cells in the second row of M^i give us the 3×3 matrices depicted in Fig. 4.

To obtain such $m \times m$ matrices for every cell X in the model, Van Benthem and Pacuit apply a simple arithmetical trick. We give a variant of theirs, that generalizes easily to the case of $n > 2$ agents.

[9] See e.g. [2] for an introduction to the notion of bounded morphisms in modal logic.

[10] For the grand coalition N, determinism obviously makes a difference. That is, within the class of all choice frames, determinism is characterized by the axiom $[N]\varphi \leftrightarrow \varphi$, which is not valid on indeterministic choice frames.

[11] See e.g. [7] for an introduction to modal product logics.

e	f	e
f	e	f
e	f	e

g	g	g
g	g	g
g	g	g

h	i	j
i	j	h
j	h	i

Fig. 4. Some 3×3 matrices.

For every cell X in the model, fix a surjective function $g_m^X : \{1, \ldots, m\} \to X$. Every point in $\mathsf{M}(X)$ is identified by its coordinates $\langle k, l \rangle$, where $k, l \in \{1, \ldots, m\}$. The world $x \in X$ that corresponds to the point $\langle k, l \rangle$ in $\mathsf{M}(X)$ is defined by $f : \{1, \ldots, m\}^2 \to X$ as follows:[12]

$$f(\langle k, l \rangle) = g_m^X(((k + l) \mathsf{mod}\ m) + 1)$$

This way, every row k of $M(X)$ is guaranteed to contain all members of X, and likewise for every column l of $M(X)$.

The second step in the construction from [16] consists in substituting the new, "small" matrices for the cells in the matrix that corresponds to the original model M^i. Applied to the above example, this gives us a single 6×9 matrix (Fig. 5).

a	a	a	b	c	b	d	d	d
a	a	a	c	b	c	d	d	d
a	a	a	b	c	b	d	d	d
e	f	e	g	g	g	h	i	j
f	e	f	g	g	g	i	j	h
e	f	e	g	g	g	j	h	i

Fig. 5. A deterministic choice model.

This new matrix corresponds to a deterministic choice model: rows are again the actions of one agent, columns are the actions of the other agent. The crucial point to note is that, as far as the effectivity of both agents is concerned, M^i and M^d are equivalent, that is, they validate exactly the same formulas in $\mathfrak{L}^{-N}$. Only the effectivity of the grand coalition N is affected.

To see how this new matrix can be accurately defined, we first need to explain how worlds in M^d are defined. For the two-agent case, the worlds in M^d are defined by (a) an index X that refers to a cell in M^i, and (b) the coordinates $k, l \in \{1, \ldots, m\}$ that specify a point in $\mathsf{M}(X)$. Two worlds $\langle X, k, l \rangle$ and $\langle X', k', l' \rangle$ are connected for agent 1, iff (a) the cells X and X' are included in a single choice $Y \in \mathit{Choice}_1(M^i)$, and (b) $k = k'$. Analogously, the new choices of the second agent are defined in terms of $\mathit{Choice}_2(M^i)$ and the indexes l, l' of the new worlds.

[12] Where $i, j \in \mathbb{N}$, $i\ \mathsf{mod}\ j$ is shorthand for "i modulo j", i.e., the remainder after division of i by j.

Even though it does the job, the matrix depicted in Fig. 5 is somewhat large for our purposes: e.g. the third row is superfluous, since it is identical to the first row. Still, the advantage of this construction is that it can easily be generalized to models with n agents. Just as for 2 agents, every cell X in a game for n agents can be replaced with a new, n-dimensional matrix. This is exactly what happens in the proof to which we now turn.

4 Product Construction, Finite Case

In this section and the next one, we prove our main results. In the current section, we will consider the case where M^i is finite and construct an $\mathfrak{L}^{-N}$-equivalent, deterministic *and finite* model M^d from it. In the next section we consider the case where M^i is infinite; using a slightly more complex construction, we construct an $\mathfrak{L}^{-N}$-equivalent deterministic model M^d from M^i. For reasons that will be explained in Sect. 5, the second type of construction will always render an infinite model M^d, even when M^i is finite. Hence, the proof in the present section is not just a special case of the one from the next section.

Recall that we hold the number n of agents fixed in this paper; the construction of M^d will depend in part on this number (see Definition 8). We first lift the equivalence relations $\sim_j^i$ to relations $\approx_j^i$ between the cells in M^i:

Definition 7. *Where $X, Y \in Choice_N(M^i)$: $X \approx_j^i Y$ iff for every $x \in X$ and every $y \in Y$ it holds that $x \sim_j^i y$.*

Each of the following can be easily verified:

Proposition 1. *$X \approx_j^i Y$ iff there are $x \in X$ and $y \in Y$ such that $x \sim_j^i y$.*

Proposition 2. *$\approx_j^i$ is an equivalence relation.*

Definition 8. *Let $M^i = \langle W^i, \langle \sim_j^i \rangle_{j \in N}, V^i \rangle$ be a finite indeterministic choice model. Let m be the number of worlds in the cell in $Choice_N(M^i)$ with the highest cardinality. For every $X \in Choice_N(M^i)$, fix a surjective function $g_m^X : \{1, \ldots, m\} \to X$. Where $X \in Choice_N(M^i)$ and $k_1, \ldots, k_n \in \{1, \ldots, m\}$, let*

$$f(\langle X, k_1, \ldots, k_n \rangle) = g_m^X(((k_1 + \ldots + k_n) \mathsf{mod}\ m) + 1)$$

The model $M^d = \langle W^d, \langle \sim_j^d \rangle_{j \in N}, V^d \rangle$ is defined as follows:

$$\begin{aligned} W^d &= \{\langle X, k_1, \ldots, k_n \rangle : X \in Choice_N(M^i) \text{ and } \{k_1, \ldots, k_n\} \subseteq \{1, \ldots, m\}\} \\ \sim_j^d &= \{(\langle X, k_1, \ldots, k_n \rangle, \langle Y, l_1, \ldots, l_n \rangle) \in (W^d)^2 : X \approx_j^i Y \text{ and } k_j = l_j\} \\ V^d(p) &= \{\langle X, k_1, \ldots, k_n \rangle \in W^d : f(\langle X, k_1, \ldots, k_n \rangle) \in V^i(p)\}. \end{aligned}$$

We first make two basic observations about the set W^d as given by Definition 8. The proofs are safely left to the reader.

Proposition 3. *$f : W^d \to W^i$ is onto.*

If X is a set, let $\mathsf{card}(X)$ denote the number of elements in X.[13]

Proposition 4. $\mathsf{card}(W^d) = m^n \times \mathsf{card}(Choice_N(M^i))$. *Hence, M^d is finite.*

Theorem 3. *M^d is a deterministic choice model.*

Proof. We need to prove a number of things:

1. "$W^d \neq \emptyset$." Immediate, by Proposition 3 and since $W^i \neq \emptyset$.
2. "Every $\sim_j^d$ is an equivalence relation." Immediate in view of Proposition 2 and because of the definition of $\sim_j^d$.
3. "M^d satisfies Independence of Agents." Let $\langle X_j, k_1^j, \ldots, k_n^j \rangle \in W^d$ for all $j \in N$. We have to show that there is a $\langle Y, l_1, \ldots, l_n \rangle \in W^d$ such that for all $j \in N$ it holds that $\langle Y, l_1, \ldots, l_n \rangle \sim_j^d \langle X_j, k_1^j, \ldots, k_n^j \rangle$. First, set $l_j = k_j^j$ for all $j \in N$. Second, fix an arbitrary $x_j \in X_j$ for all $j \in N$. Because of Independence of Agents for M^i, there is a $y \in W^i$ such that $x_j \sim_j^i y$ for all $j \in N$. Let Y be the cell $Y \in Choice_N(M^i)$ that contains y. Then, because of Proposition 1, it must be that $Y \approx_j^i X_j$ for all $j \in N$. By the definition of W^d, it must be that $\langle Y, l_1, \ldots, l_n \rangle \in W^d$. By the definition of $\sim_j^d$, it must be that $\langle Y, l_1, \ldots, l_n \rangle \sim_j^d \langle X_j, k_1^j, \ldots, k_n^j \rangle$ for all $j \in N$. Hence, M^d satisfies Independence of Agents.
4. "$\sim_N^d$ is the identity relation over W^d." Suppose that $\langle X, k_1, \ldots, k_n \rangle \sim_N^d \langle Y, l_1, \ldots, l_n \rangle$. Hence, $\langle X, k_1, \ldots, k_n \rangle \sim_j^d \langle Y, l_1, \ldots, l_n \rangle$ for all $j \in N$. Then $X, Y \in Choice_N(M^i)$ and for all $j \in N$ both $X \approx_j^i Y$ and $k_j = l_j$. Hence it must be that (a) $\langle k_1, \ldots, k_n \rangle = \langle l_1, \ldots, l_n \rangle$. Because $X \approx_j^i Y$ for all $j \in N$, it must be that for all $j \in N$ and for all $x \in X$ and all $y \in Y$ it holds that $x \sim_j^i y$. Hence, for all $x \in X$ and all $y \in Y$ it holds that $x \sim_N^i y$. Because $X, Y \in Choice_N(M^i)$, it must be that (b) $X = Y$. From (a) and (b) we conclude that $\langle X, k_1, \ldots, k_n \rangle = \langle Y, l_1, \ldots, l_n \rangle$.

By (i)-(iv), M^d is a deterministic choice model. □

Theorem 4. *f is a bounded morphism from M^d to M^i in $\mathfrak{L}^{-N}$, i.e.:*

1. *f is onto.*
2. *For all $w, w' \in W^d$ and all non-empty $G \subset N$: if $w \sim_G^d w'$, then $f(w) \sim_G^i f(w')$*
3. *For all $u \in W^d$, all non-empty $G \subset N$, and all $y \in W^i$: if $f(u) \sim_G^i y$, then there is a $v \in W^d$ such that $f(v) = y$ and $u \sim_G^d v$.*
4. *For all $w \in W^d$ and $p \in \mathfrak{P}$: $w \in V^d(p)$ iff $f(w) \in V^i(p)$.*

[13] Note that we apply the card function both to finite and infinite (even uncountable) sets.

Proof. Ad 1. This is Proposition 3.
Ad 2. Let $w = \langle X, k_1, \ldots, k_n\rangle$ and $w' = \langle Y, l_1, \ldots, l_n\rangle$ be arbitrary members of W^d, and suppose that $w \sim^d_G w'$. Let $f(w) = x$ and $f(w') = y$. It follows that $x \in X$, $y \in Y$, and for all $j \in G$, $X \approx^i_j Y$. By Definition 7, for all $j \in G$, $x \sim^i_j y$. Hence, $x \sim^i_G y$.
Ad 3. Suppose the antecedent holds for $u = \langle X, k_1, \ldots, k_n\rangle$. Note that $f(u) \in X$. Let $Y \in Choice_N(M^i)$ be such that $y \in Y$. By Proposition 1 and the supposition, (†) for all $j \in G$, $X \approx^i_j Y$. Fix a $t \in N - G$. For all $j \in N - \{t\}$, let $l_j = k_j$. Let $l_t \in \{1, \ldots, m\}$ be such that $g^X_m(((l_1 + \ldots + l_n) \mathsf{mod}\ m) + 1) = y$. It is a matter of basic arithmetic to check that there is indeed such an l_t. Let now $v = \langle Y, l_1, \ldots, l_n\rangle$. It follows that $f(v) = y$. By (†) and in view of the construction, $u \sim^d_j v$ for all $j \in G$. Hence, $u \sim^d_G v$.
Ad 4. Immediate in view of the definition of V^d. □

It is well-known that, whenever there is a bounded morphism between two models M and M', then these models are pointwise equivalent – see e.g. [2, Proposition 2.14]. Hence, Theorem 4 gives us:

Corollary 2. *For all $\varphi \in \mathfrak{L}^{-N}$ and all $w \in W^d$: $M^d, w \models \varphi$ iff $M^i, f(w) \models \varphi$.*

5 Product Construction, Infinite Case

The proof in Sect. 4 makes essential use of the upper bound m on the cardinality of each $X \in Choice_N(M^i)$. As a result, we can apply well-known arithmetic techniques to construct the n-dimensional matrices that form the core of the construction of M^d. For the infinite case, a slightly different construction is needed.

The idea behind Definition 9 below can be explained as follows. In every world $u \in W^d$, each of the agents gets to choose exactly one world from W^i, and one natural number $k \in \mathbb{N}$. The output for this world, given by f, depends on the one hand on the index X, on the other hand, on which agent $t \in N$ chose the highest number $k_t \in \mathbb{N}$, and the world w_t that this agent t chose.

We first introduce some more notation. Where $\overline{x} = \langle x_1, \ldots, x_k\rangle$ is a k-tuple and $1 \leq j \leq k$, let $\pi^j(\overline{x}) = x_j$. Where $X \subset \mathbb{N}$ is a finite set of natural numbers, let $\mathsf{max}_<(X)$ denote the largest element in X.

Definition 9. *Let $M^i = \langle W^i, \langle \sim^i_j\rangle_{j\in N}, V^i\rangle$ be an infinite indeterministic choice model. For every $X \in Choice_N(M^i)$, fix a surjective function $g^X : W^i \to X$.*[14] *Where $X \in Choice_N(M^i)$, $w_1, \ldots, w_n \in W^i$, and $k_1, \ldots, k_n \in \mathbb{N}$, let*

$$f(\langle X, \langle w_1, k_1\rangle, \ldots, \langle w_n, k_n\rangle\rangle) = g^X(w_l)$$

where $l \in \{1, \ldots, n\}$ is the smallest natural number such that $k_l = \mathsf{max}_<\{k_1, \ldots, k_n\}$. The model $M^d = \langle W^d, \langle \sim^d_j\rangle_{j\in N}, V^d\rangle$ is defined as follows:

[14] Since $X \subseteq W^i$, it can be easily verified that there is at least one such function g^X.

$$W^d = \{\langle X, \langle w_1, k_1\rangle, \ldots, \langle w_n, k_n\rangle\rangle : X \in Choice_N(M^i), w_1, \ldots, w_n \in W^i, \text{ and } k_1, \ldots, k_n \in \mathbb{N}\}$$
$$\sim_j^d = \{(\langle X, \overline{\epsilon}\rangle, \langle Y, \overline{\epsilon'}\rangle) \in (W^d)^2 : X \approx_j^i Y \text{ and } \pi^j(\overline{\epsilon}) = \pi^j(\overline{\epsilon'})\}$$
$$V^d(p) = \{w \in W^d : f(w) \in V^i(p)\}.$$

Note that we use $\overline{\epsilon}, \overline{\epsilon'}, \ldots$ as metavariables for tuples of the form $\langle\langle w_1, k_1\rangle, \ldots, \langle w_n, k_n\rangle\rangle$ that are part of a larger tuple $w \in W^d$.

Theorem 5. *M^d is a deterministic choice model.*

Proof. We need to prove a number of things:

1. "$W^d \neq \emptyset$." Since $W^i \neq \emptyset$, also $Choice_N(M^i) \neq \emptyset$. By Definition 9, $W^d \neq \emptyset$.
2. "Every $\sim_j^d$ is an equivalence relation." Immediate in view of Proposition 2 and by Definition 9.
3. "M^d satisfies Independence of Agents." Consider arbitrary $w_1, \ldots, w_n \in W^d$, where each $w_j = \langle X_j, \langle w_1^j, k_1^j\rangle, \ldots, \langle w_n^j, k_n^j\rangle\rangle$. Fix an arbitrary $x_j \in X_j$ for all $j \in N$. Because of Independence of Agents for M^i, there is a $y \in W^i$ such that $x_j \sim_j^i y$ for all $j \in N$. Let Y be the cell $Y \in Choice_N(M^i)$ that contains y. Then, because of Proposition 1, it must be that (a) $Y \approx_j^i X_j$ for all $j \in N$. Let $w' = \langle Y, \overline{\epsilon}\rangle = \langle Y, \langle w_1^1, k_1^1\rangle, \ldots, \langle w_n^n, k_n^n\rangle\rangle$. By (a) and the definition of $\sim_j^d$, for all $j \in N$, $w_j \sim_j^d w'$.
4. "$\sim_N^d$ is the identity relation over W^d." Suppose that $\langle X, \overline{\epsilon}\rangle \sim_N^d \langle Y, \overline{\epsilon'}\rangle$. Then, for all $j \in N$, $\pi^j(\overline{\epsilon}) = \pi^j(\overline{\epsilon'})$ and hence (a) $\overline{\epsilon} = \overline{\epsilon'}$. Because $X \approx_j^i Y$ for all $j \in N$, it must be that for all $j \in N$ and for all $x \in X$ and all $y \in Y$ it holds that $x \sim_j^i y$. Hence, for all $x \in X$ and all $y \in Y$ it holds that $x \sim_N^i y$. Because $X, Y \in Choice_N(M^i)$, it must be that (b) $X = Y$. From (a) and (b) we conclude that $\langle X, \overline{\epsilon}\rangle = \langle Y, \overline{\epsilon'}\rangle$.

By (i)-(iv), M^d is a deterministic choice model. □

Theorem 6. *f is a bounded morphism from M^d to M^i in $\mathfrak{L}^{-N}$, i.e.:*

1. *f is onto.*
2. *For all $w, w' \in W^d$ and all non-empty $G \subset N$: if $w \sim_G^d w'$, then $f(w) \sim_G^i f(w')$*
3. *Where $u \in W^d$, $G \subset N$, and $f(u) \sim_G^i y$ for a $y \in W^i$: there is a $v \in W^d$ such that $f(v) = y$ and $u \sim_G^d v$.*
4. *For all $w \in W^d$ and $p \in \mathfrak{P}$: $w \in V^d(p)$ iff $f(w) \in V^i(p)$.*

Proof. Ad 1. Let $x \in W^i$ be arbitrary. Let $X \in Choice_N(M^i)$ be such that $x \in X$. Let $y \in W^i$ be such that $g^X(y) = x$. Finally, let $u = \langle X, \langle y, 1\rangle, \ldots, \langle y, 1\rangle\rangle$ be a sequence of length $n + 1$. Note that $u \in W^d$. Moreover, $f(u) = g^X(y) = x$.
Ad 2. Analogous to the proof of Theorem 4.3.
Ad 3. Suppose the antecedent holds for $u = \langle X, \langle w_1, k_1\rangle, \ldots, \langle w_n, k_n\rangle\rangle$. Note that $f(u) \in X$. Let $Y \in Choice_N(M^i)$ be such that $y \in Y$. Note that, by the supposition, (†) for all $j \in G$, $X \approx_j^i Y$. Fix a $t \in N - G$. For all $j \in N - \{t\}$, let $w'_j = w_j$ and $l_j = k_j$. Let l_t be an arbitrary natural number

such that $l_t > l_j$ for all $j \in N - \{t\}$. Fix $w'_t \in Y$ such that $g^Y(w'_t) = y$. Let $v = \langle Y, \langle w'_1, l_1\rangle, \ldots, \langle w'_n, l_n\rangle\rangle$. It follows that $f(v) = g^Y(w'_t) = y$. By (†) and in view of the construction, $u \sim^d_j v$ for all $j \in G$, and hence $u \sim^d_G v$.
Ad 4. Immediate in view of the definition of V^d. □

It is useful, at this point, to check how we can cut down the size of W^d given certain restrictions on M^i. One can e.g. easily observe that, if there is a $Y \in \mathit{Choice}_N(M^i)$ such that, for all $Z \in \mathit{Choice}_N(M^i)$, $\mathsf{card}(Y) \geq \mathsf{card}(Z)$, then we can replace W^i with Y in the definition of g^X and W^d.

A natural follow-up question is: what if W^i is finite? Can we construct one proof that works for both finite and infinite models M^i, and that guarantees that the constructed model M^d is finite whenever M^i is finite? Note that the change that we proposed in the previous paragraph will not do to obtain such a proof. That is, all the natural numbers can still be used for the indices $k_1, \ldots, k_n$, whence W^d is bound to be infinite under the present construction. Moreover, the complication with double indices $\langle k_j, w_j\rangle$ seems necessary in order to ensure that, whatever all the other agents do, any given agent $j \in N$ can still "enforce" every world $x \in X$ for a given cell $X \in \mathit{Choice}_N(M^i)$, by choosing a yet higher index k_j and the world $y \in Y$ with $g^X(y) = x$.

6 Concluding Remarks

In this paper, we have shown that one can retrieve determinism without adding "nature" as an agent. We generalized an earlier result by Van Benthem and Pacuit [16], and showed that every indeterministic n-agent choice model is pointwise equivalent to a deterministic n-agent choice model, as long as we ignore the grand coalition. As a corollary, any (possibly infinite) choice model for n agents can be translated into an $\mathfrak{L}^{-N}$-equivalent normal game form for n agents, where the latter is finite if the former is finite. Our result thus contributes to connecting STIT logic and game theory more generally.

A number of questions should be answered in future work. Let us start with the most technical ones. First, can we rephrase the proof for the infinite case in such a way that it also covers the finite case, ensuring that M^d is finite whenever M^i is? Second, what about STIT logic with infinitely (countably many) agents? Here, the results appear to be mixed. If we only allow for finite groups in the language, we can easily generalize the construction from Sect. 5. However, if we allow for infinite groups $G \subset N$, this construction no longer does the job.

A different issue concerns the axiomatization of the logic we presented. Drawing on earlier results from [8,16], it can be shown that the $\mathfrak{L}^{-N}$-fragment of the logic of deterministic choice models is isomorphic to the modal product logic $\mathbf{S5^n}$. The latter logic is not decidable and cannot be finitely axiomatized for $n > 2$, cf. [8]. A non-standard axiomatization of $\mathbf{S5^n}$ has been presented in [17]. It remains to be seen how this axiomatization can be extended to the full language $\mathfrak{L}$ which includes $[N]$.

There are various ways for retrieving (finite) axiomatizability, decidability and acceptable complexity in the context of group STIT. First, one may restrict the formal language. For instance, it was proven in [11, Sect. 3] that when nesting of STIT operators is not allowed (i) the satisfiability problem becomes decidable in non-deterministic polynomial time (Corollary 1, p. 821), and (ii) the restricted logic becomes finitely axiomatizable (Corollary 2, p. 821).

Second, one may use different models to interpret the STIT language. Most importantly, one may weaken the *intersection property*, which says that $\sim_G = \bigcap_{j\in G} \sim_j$, to the requirement of monotonic effectivity: if $F \subseteq G$, then $\sim_G \subseteq \sim_F$. It has been shown in [3] that complete logics are readily available for these models, typically using Sahlqvist schemes [2].

A third route that was suggested in [16] is to give up the Independence of Agency (IOA) condition. If one does not impose (IOA) on the models, one obtains the non-deterministic counterpart of what Van Benthem and Pacuit call *general game models* [15,16]. Let us call such models *general choice models*. Note that in a general choice model, the choices of one agent may depend on the choices of other agents. The logic of general choice models coincides with the logic of distributed knowledge for arbitrary groups, which is known to be finitely axiomatizable and decidable [6, Chap. 3]. Now, as a matter of fact, our proofs in the current paper do not rely on (IOA), except where we show that the newly constructed model M^d also satisfies (IOA). Hence, our results reduce the problem of axiomatization of the logic of general game models to the axiomatization of the logic of general choice models.

Acknowledgements. Frederik Van De Putte's research for this paper was funded by the Flemish Research Foundation (FWO-Vlaanderen). Allard Tamminga and Hein Duijf gratefully acknowledge financial support from the ERC-2013-CoG project REINS, no. 616512. The research for this paper was facilitated by two research visits of Allard Tamminga to Ghent University that were co-funded by the FWO through the scientific research network for Logical and Methodological Analysis of Scientific Reasoning Processes (LMASRP). We are indebted to Johan van Benthem, Olivier Roy, and Dominik Klein for useful discussions on this paper's topic. We also thank Mathieu Beirlaen and two anonymous referees of LORI for their remarks on previous versions of the paper.

References

1. Belnap, N., Perloff, M., Xu, M., Bartha, P.: Facing the Future: Agents and Choice in Our Indeterminist World. Oxford University Press, Oxford (2001)
2. Blackburn, P., De Rijke, M., Venema, Y.: Modal Logic. Cambridge Tracts in Theoretical Computer Science (2001)
3. Broersen, J., Herzig, A., Troquard, N.: A normal simulation of coalition logic and an epistemic extension. In Samet, D. (ed.) Proceedings of the 11th Conference on Theoretical Aspects of Rationality and Knowledge, pp. 92–101. ACM (2007)
4. Chellas, B.F.: Time and modality in the logic of agency. Stud. Log. **51**(3–4), 485–517 (1992)

5. Ciuni, R., Horty, J.: Stit logics, games, knowledge, and freedom. In: Baltag, A., Smets, S. (eds.) Johan van Benthem on Logic and Information Dynamics. OCL, vol. 5, pp. 631–656. Springer, Cham (2014). doi:10.1007/978-3-319-06025-5_23
6. Fagin, R., Halpern, J.Y., Moses, Y., Vardi, M.Y.: Reasoning About Knowledge. MIT Press, Cambridge (2003)
7. Gabbay, D., Kurucz, A., Wolter, F., Zakharyaschev, M.: Many-dimensional Modal Logics: Theory and Applications. Studies in Logic and the Foundations of Mathematics, vol. 148. North Holland Publishing Company (2003)
8. Herzig, A., Schwarzentruber, F.: Properties of logics of individual and group agency. In: Areces, C., Gobldblatt, R. (eds.) Advances in Modal Logic. College Publications (2008)
9. Horty, J.F.: Agency and Deontic Logic. Oxford University Press, New York (2001)
10. Kooi, B., Tamminga, A.: Moral conflicts between groups of agents. J. Philos. Log. **37**(1), 1–21 (2008)
11. Lorini, E., Schwarzentruber, F.: A logic for reasoning about counterfactual emotions. Artif. Intell. **175**(3), 814–847 (2011)
12. Osborne, M., Rubinstein, A.: A Course in Game Theory, 7th edn. MIT Press, Cambridge (2001)
13. Tamminga, A.: Deontic logic for strategic games. Erkenntnis **78**(1), 183–200 (2013)
14. Turrini, P.: Agreements as norms. In: Ågotnes, T., Broersen, J., Elgesem, D. (eds.) DEON 2012. LNCS, vol. 7393, pp. 31–45. Springer, Heidelberg (2012). doi:10.1007/978-3-642-31570-1_3
15. van Benthem, J.: Logic in Games. MIT Press, Cambridge (2014)
16. Benthem, J., Pacuit, E.: Connecting logics of choice and change. In: Müller, T. (ed.) Nuel Belnap on Indeterminism and Free Action. OCL, vol. 2, pp. 291–314. Springer, Cham (2014). doi:10.1007/978-3-319-01754-9_14
17. Venema, Y.: Rectangular games. J. Symbolic Log. **63**(4), 1549–1564 (1998)

Axiomatizing Epistemic Logic of Friendship via Tree Sequent Calculus

Katsuhiko Sano(✉)

Department of Philosophy, Graduate School of Letters, Hokkaido University, Sapporo, Japan
v-sano@let.hokudai.ac.jp

Abstract. This paper positively solves an open problem if it is possible to provide a Hilbert system to Epistemic Logic of Friendship (EFL) by Seligman, Girard and Liu. To find a Hilbert system, we first introduce a sound, complete and cut-free tree (or nested) sequent calculus for EFL, which is an integrated combination of Seligman's sequent calculus for basic hybrid logic and a tree sequent calculus for modal logic. Then we translate a tree sequent into an ordinary formula to specify a Hilbert system of EFL and finally show that our Hilbert system is sound and complete for an intended two-dimensional semantics.

Keywords: Epistemic logics of friendship · Tree sequent calculus · Hilbert system · Completeness · Cut elimination theorem

1 Introduction

Epistemic Logic of Friendship (**EFL**) is a version of two-dimensional modal logic proposed by [22–24]. Compared to the ordinary epistemic logic [14], one of the key features of their logic is to encode the information of agents into the object language by a technique of hybrid logic [1,3]. Then, a propositional variable p can be read as an indexical proposition such as "I am p" and we may formalize the sentences like "I know that all my friends is p" or "Each of my friends knows that he/she is p". Moreover, the authors of [23,24] provided a dynamic mechanism for capturing public announcements [19], announcements to all the friends, and private announcements [2] and established a relative completeness result (cf. [12,23,24]). This paper focuses on the problem of axiomatizing **EFL** in terms of Hilbert system, i.e., the static part of their framework.

A difficulty of the problem comes from a combination of *modal logic* for agents' knowledge and *hybrid logic* for a friendship relation among agents. If we combine *two hybrid logics* over two-dimensional semantics of [22–24], it is noted that there is an axiomatization of all valid formulas in the semantics by [20, p. 471]. Our approach to tackle the problem is via a sequent calculus, whose idea is originally from Gentzen. In particular, our notion of sequent for **EFL** can be regarded as a combination of a tree or nested sequent [8,15] for modal logic and @-prefixed sequent [7,21] for hybrid logic. One of the merits of our

A. Baltag et al. (Eds.): LORI 2017, LNCS 10455, pp. 224–239, 2017.
DOI: 10.1007/978-3-662-55665-8_16

notion of sequent is that we can translate our sequent into an ordinary formula. This allows us to specify our desired Hilbert system for **EFL**. We note that [9] independently provided a prefixed tableau system for a dynamic extension of **EFL**. There are at least three points we should emphasize on our work. First, our tree sequent system is quite simpler than the tableau system given in [9], i.e., the number of rules of our sequent system is almost half of the number of rules of their system. Second, it is not clear if a prefixed formula in [9] for the tableau calculus can be translated into an ordinary formula. Their result is not concerned with Hilbert system. Third, their syntax contains a special kind of propositional variable (called *feature proposition*) and they include a tableau rule called *propositional cut* to handle such propositions. On the other hand, we can show that our tree sequent calculus enjoys the cut elimination theorem, the most fundamental theorem in proof-theory.

We proceed as follows. Section 2 introduces the syntax and semantics of **EFL**. Section 3 provides a tree sequent calculus for **EFL** and establishes the soundness of the sequent calculus (Theorem 1). Section 4 establishes a completeness result of a cut-free fragment of our sequent calculus (Theorem 2). As a corollary, we also provide a semantic proof of the cut elimination theorem of our sequent calculus (Corollary 1). Section 5 specifies a Hilbert system of **EFL**, and provides a syntactic proof of the equipollence between our proposed Hilbert system and our tree sequent calculus, which implies the soundness and completeness results for our Hilbert system (Corollary 2). Section 6 extends our technical results to cover extensions of **EFL** where a modal operator for states (or a knowledge operator) obeys **S4** or **S5** axioms and a friendship relation satisfies some universal properties (Theorem 5). The result of this section subsumes the logic given in [9], provided we drop the dynamic operator from the syntax of [9]. Section 7 concludes this paper.

2 Syntax and Two-Dimensional Kripke Semantics

Our syntax $\mathcal{L}$ consists of the following vocabulary: a countably infinite set $\mathsf{Prop} = \{p, q, r, \ldots\}$ of propositional variables, a countably infinite set $\mathsf{Nom} = \{n, m, l, \ldots\}$ of agent nominal variables, the Boolean connectives of $\rightarrow$ (the implication) and $\bot$ (the falsum), the satisfaction operators @ and the friendship operator F as well as the modal operator $\Box$. We note that an *agent nominal* $n \in \mathsf{Nom}$ is a syntactic name of an agent or an individual, which amounts to a constant symbol of the first-order logic, while n is read indexically as "I am n." Similarly, we read a propositional variable $p \in \mathsf{Prop}$ also indexically by "I am p," e.g., "I am in danger." The set Form of formulas in $\mathcal{L}$ is defined inductively as follows:

$$\mathsf{Form} \ni \varphi ::= n \,|\, p \,|\, \bot \,|\, \varphi \rightarrow \varphi \,|\, @_n\varphi \,|\, \mathsf{F}\varphi \,|\, \Box\varphi,$$

where $n \in \mathsf{Nom}$ and $p \in \mathsf{Prop}$. Boolean connectives other than $\rightarrow$ or $\bot$ are introduced as ordinary abbreviations. We define the dual of $\Box$ as $\Diamond := \neg\Box\neg$ and the dual of F as $\langle\mathsf{F}\rangle := \neg\mathsf{F}\neg$. Moreover, a formula of the form $@_n\varphi$ is said to be *@-prefixed*. Let us read $\Box$ as "I know that." Here are some examples of how to read formulas:

- $\Box p$, read as “I know that I am p.”
- $@_n\Box p$, read as “n knows that she is p.”
- $\Box @_n p$, read as “I know that agent n is p.”
- $\mathsf{F}p$, read as “all my friends are p.”
- $\mathsf{F}\Box p$, read as “all my friends know that they are p.”
- $\Box\mathsf{F}p$, read as “I know that all my friends are p.”
- $@_n\langle\mathsf{F}\rangle m$, read as “agent m is a friend of agent n.”

We say that a mapping $\sigma : \mathsf{Prop} \cup \mathsf{Nom} \to \mathsf{Form}$ is a *uniform substitution* if σ uniformly substitutes propositional variables by formulas and agent nominals by agent nominals and we use $\varphi\sigma$ to mean the result of applying a uniform substitution σ to φ. In particular, we use $\varphi[n/k]$ to mean the result of substituting each occurrence of agent nominal k in φ uniformly with agent nominal n.

An *model* $\mathfrak{M}$ for our syntax $\mathcal{L}$ is a tuple $(W, A, (R_a)_{a\in A}, (\asymp_w)_{w\in W}, V)$, where W is a non-empty set of possible states, A is a non-empty set of agents, R_a is a binary relation on W ($a \in A$), $\asymp_w$ is a binary relation on A (called a *friendship relation*, $w \in W$), V is a valuation function $\mathsf{Prop} \cup \mathsf{Nom} \to \mathcal{P}(W \times A)$ such that $V(n)$ is a subset of $W \times A$ of the form $W \times \{a\}$, where we denote such unique element a by $\underline{n}$. We do not require any property for R_a and $\asymp_w$ but we will come back to this point in Sect. 6. We say that a tuple $\mathfrak{F} = (W, A, (R_a)_{a\in A}, (\asymp_w)_{w\in W})$ without a valuation is a *frame*.

Let $\mathfrak{M} = (W, A, (R_a)_{a\in A}, (\asymp_w)_{w\in W}, V)$ be a model. Given a pair $(w, a) \in W \times A$ and a formula φ, the satisfaction relation $\mathfrak{M}, (w, a) \models \varphi$ (read “agent a satisfies φ at w in $\mathfrak{M}$ ”) inductively as follows:

$$
\begin{array}{ll}
\mathfrak{M}, (w,a) \models p & \text{iff } (w,a) \in V(p), \\
\mathfrak{M}, (w,a) \models n & \text{iff } \underline{n} = a, \\
\mathfrak{M}, (w,a) \not\models \bot & \\
\mathfrak{M}, (w,a) \models \varphi \to \psi & \text{iff } \mathfrak{M}, (w,a) \models \varphi \text{ implies } \mathfrak{M}, (w,a) \models \psi \\
\mathfrak{M}, (w,a) \models @_n\varphi & \text{iff } \mathfrak{M}, (w,\underline{n}) \models \varphi, \\
\mathfrak{M}, (w,a) \models \mathsf{F}\varphi & \text{iff } (a \asymp_w b \text{ implies } \mathfrak{M}, (w,b) \models \varphi) \text{ for all agents } b \in A, \\
\mathfrak{M}, (w,a) \models \Box\varphi & \text{iff } (wR_a v \text{ implies } \mathfrak{M}, (v,a) \models \varphi) \text{ for all states } v \in W.
\end{array}
$$

Given a class $\mathbb{M}$ of models, we say that a formula φ is *valid* in $\mathbb{M}$ when $\mathfrak{M}, (w, a) \models \varphi$ for all pairs (w, a) in $\mathfrak{M}$ and all models $\mathfrak{M} \in \mathbb{M}$. This paper tackles the question if the set of all valid formulas in the class of all models is axiomatizable.

3 Tree Sequent Calculus of Epistemic Logic of Friendship

A *label* is inductively defined as follows: Any natural number is a label; if α is a label, n is an agent nominal in Nom and i is a natural number, then $\alpha \cdot_n i$ is also a label. When β is $\alpha \cdot_n i$, then we say that β is an *n-child* of α or that α is *an n-parent* of β. A *tree* $\mathcal{T}$ is a set of labels such that the set contains the unique natural number j as the root

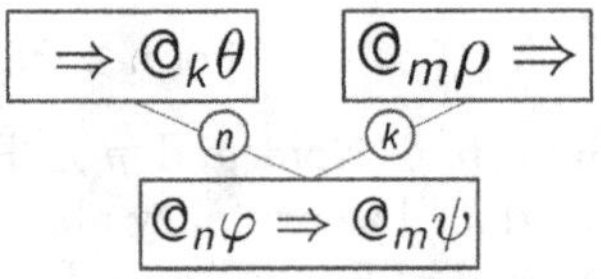

Fig. 1. A tree sequent

label and the set is closed under taking the parent of a label, i.e., $\alpha \cdot_n i \in \mathcal{T}$ implies $\alpha \in \mathcal{T}$ for all labels α, agent nominals n and natural numbers i. For example, all of 0, $0 \cdot_n 1$ and $0 \cdot_k 2$ are labels and they form a finite tree.

Given a label α and an *@-prefixed formula* φ, the expression $\alpha : \varphi$ is said to be a *labelled formula*, where recall that an @-prefixed formula is of the form $@_n\varphi$. A *tree sequent* is an expression of the form

$$\Gamma \overset{\mathcal{T}}{\Rightarrow} \Delta$$

where Γ and Δ are *finite* sets of labelled formulas, $\mathcal{T}$ is a finite tree of labels, and all the labels in Γ and Δ are in $\mathcal{T}$. A tree sequent "$\Gamma \overset{\mathcal{T}}{\Rightarrow} \Delta$" is read as "if we assume all labelled formulas in Γ, then we may conclude some labelled formulas in Δ." A tree sequent $0 : @_n\varphi, 0 \cdot_k 2 : @_m\rho \overset{\mathcal{T}}{\Rightarrow} 0 : @_m\psi, 0 \cdot_n 1 : @_k\theta$ is represented as in Fig. 1, where $\mathcal{T} = \{0, 0 \cdot_n 1, 0 \cdot_k 2\}$. That is, 0, $0 \cdot_n 1$ and $0 \cdot_k 2$ are "addresses" of the root, the left leaf, and the right leaf, respectively.

Table 1. Tree Sequent Calculus **TEFL**

$$(\bot)\ \alpha : @_n\bot, \Gamma \overset{\mathcal{T}}{\Rightarrow} \Delta \qquad (\mathsf{id})\ \alpha : @_n\varphi, \Gamma \overset{\mathcal{T}}{\Rightarrow} \Delta, \alpha : @_n\varphi$$

$$\dfrac{\alpha : @_n m, \alpha : \varphi[n/k], \Gamma \overset{\mathcal{T}}{\Rightarrow} \Delta}{\alpha : @_n m, \alpha : \varphi[m/k], \Gamma \overset{\mathcal{T}}{\Rightarrow} \Delta}\ (\mathsf{rep}_{=1}) \qquad \dfrac{\alpha : @_n m, \alpha : \varphi[m/k], \Gamma \overset{\mathcal{T}}{\Rightarrow} \Delta}{\alpha : @_n m, \alpha : \varphi[n/k], \Gamma \overset{\mathcal{T}}{\Rightarrow} \Delta}\ (\mathsf{rep}_{=2})$$

$$\dfrac{\alpha : @_n n, \Gamma \overset{\mathcal{T}}{\Rightarrow} \Delta}{\Gamma \overset{\mathcal{T}}{\Rightarrow} \Delta}\ (\mathsf{ref}_{=}) \qquad \dfrac{\beta : @_n m, \Gamma \overset{\mathcal{T}}{\Rightarrow} \Delta}{\alpha : @_n m, \Gamma \overset{\mathcal{T}}{\Rightarrow} \Delta}\ (\mathsf{rigid}_{=})$$

$$\dfrac{\alpha : @_n\varphi, \Gamma \overset{\mathcal{T}}{\Rightarrow} \Delta, \alpha : @_n\psi}{\Gamma \overset{\mathcal{T}}{\Rightarrow} \Delta, \alpha : @_n(\varphi \to \psi)}\ ({\to}R) \qquad \dfrac{\Gamma \overset{\mathcal{T}}{\Rightarrow} \Delta, \alpha : @_n\varphi \quad \alpha : @_n\psi, \Gamma \overset{\mathcal{T}}{\Rightarrow} \Delta}{\alpha : @_n(\varphi \to \psi), \Gamma \overset{\mathcal{T}}{\Rightarrow} \Delta}\ ({\to}L)$$

$$\dfrac{\Gamma \overset{\mathcal{T}}{\Rightarrow} \Delta, \alpha : @_m\varphi}{\Gamma \overset{\mathcal{T}}{\Rightarrow} \Delta, \alpha : @_n@_m\varphi}\ (@R) \qquad \dfrac{\alpha : @_m\varphi, \Gamma \overset{\mathcal{T}}{\Rightarrow} \Delta}{\alpha : @_n@_m\varphi, \Gamma \overset{\mathcal{T}}{\Rightarrow} \Delta}\ (@L)$$

$$\dfrac{\alpha : @_n\langle\mathsf{F}\rangle m, \Gamma \overset{\mathcal{T}}{\Rightarrow} \Delta, \alpha : @_m\varphi}{\Gamma \overset{\mathcal{T}}{\Rightarrow} \Delta, \alpha : @_n\mathsf{F}\varphi}\ (\mathsf{F}R)^{*} \qquad \dfrac{\Gamma \overset{\mathcal{T}}{\Rightarrow} \Delta, \alpha : @_n\langle\mathsf{F}\rangle m \quad \alpha : @_m\varphi, \Gamma \overset{\mathcal{T}}{\Rightarrow} \Delta}{\alpha : @_n\mathsf{F}\varphi, \Gamma \overset{\mathcal{T}}{\Rightarrow} \Delta}\ (\mathsf{F}L)$$

$$\dfrac{\Gamma \overset{\mathcal{T}\cup\{\alpha\cdot_n i\}}{\Rightarrow} \Delta, \alpha \cdot_n i : @_n\varphi}{\Gamma \overset{\mathcal{T}}{\Rightarrow} \Delta, \alpha : @_n\Box\varphi}\ (\Box R)^{\dagger} \qquad \dfrac{\beta : @_n\varphi, \Gamma \overset{\mathcal{T}}{\Rightarrow} \Delta}{\alpha : @_n\Box\varphi, \Gamma \overset{\mathcal{T}}{\Rightarrow} \Delta}\ (\Box L)^{\ddagger}$$

$$\dfrac{\Gamma \overset{\mathcal{T}}{\Rightarrow} \Delta}{\Gamma \overset{\mathcal{T}\cup\{\alpha\}}{\Rightarrow} \Delta}\ (w\mathsf{lab})^{\star} \qquad \dfrac{\Gamma \overset{\mathcal{T}}{\Rightarrow} \Delta, \alpha : @_n\varphi \quad \alpha : @_n\varphi, \Pi \overset{\mathcal{T}}{\Rightarrow} \Sigma}{\Gamma, \Pi \overset{\mathcal{T}}{\Rightarrow} \Delta, \Sigma}\ (Cut)$$

$*$: m is a fresh agent nominal in the lower sequent; $\dagger$: $i \in \mathbb{N}$ is fresh in the lower sequent; $\ddagger$: β is an n-child of α; $\star$: $\mathcal{T} \cup \{\alpha\}$ is a tree of labels

Table 1 provides all the initial sequents and all the inference rules of tree sequent calculus **TEFL**, where recall that $\varphi[m/k]$ is the result of substituting each occurrence of agent nominal k in φ with agent nominal m. The system without the cut rule is denoted by TEFL^-. A *derivation* in TEFL (or TEFL^-) is a finite tree generated from initial sequents by inference rules of TEFL (or TEFL^-, respectively). The *height* of a derivation is defined as the maximum length of branches in the derivation from the end (or root) sequent to an initial sequent. A tree sequent $\Gamma \overset{\mathcal{T}}{\Rightarrow} \Delta$ is said to be *provable* in TEFL (or TEFL^-) if there is a derivation in TEFL (or TEFL^-, respectively) such that the root of the tree is $\Gamma \overset{\mathcal{T}}{\Rightarrow} \Delta$.

Let $\mathfrak{M} = (W, A, (R_a)_{a\in A}, (\preceq_w)_{w\in W}, V)$ be a model and $\mathcal{T}$ a tree of labels. A function $f : \mathcal{T} \to W$ is a *$\mathcal{T}$-assignment* in $\mathfrak{M}$ if, whenever β is an n-child of α in $\mathcal{T}$, $f(\alpha)R_{\underline{n}}f(\beta)$ holds. When it is clear from the context, we drop "$\mathcal{T}$-" from "$\mathcal{T}$-assignment". Given any labelled formula $\alpha : @_n\varphi$ with $\alpha \in \mathcal{T}$ and any $\mathcal{T}$-assignment in $\mathfrak{M}$, we define the *satisfaction* for a labelled formula as follows:

$$\mathfrak{M}, f \models \alpha : @_n\varphi \text{ iff } \mathfrak{M}, (f(\alpha), \underline{n}) \models \varphi.$$

where "$\mathfrak{M}, f \models \alpha : @_n\varphi$" is read as "$\alpha : @_n\varphi$ is true at $(\mathfrak{M}, f)$". Given a tree sequent $\Gamma \overset{\mathcal{T}}{\Rightarrow} \Delta$ and a $\mathcal{T}$-assignment in $\mathfrak{M}$, we say that $\Gamma \overset{\mathcal{T}}{\Rightarrow} \Delta$ is true in $(\mathfrak{M}, f)$ (notation: $\mathfrak{M}, f \models \Gamma \overset{\mathcal{T}}{\Rightarrow} \Delta$) if, whenever all labelled formulas of Γ is true in $(\mathfrak{M}, f)$, some labelled formulas of Δ is true in $(\mathfrak{M}, f)$. The following theorem is easy to establish.

Theorem 1 (Soundness of TEFL). *If a tree sequent $\Gamma \overset{\mathcal{T}}{\Rightarrow} \Delta$ is provable in* TEFL *then $\mathfrak{M}, f \models \Gamma \overset{\mathcal{T}}{\Rightarrow} \Delta$ for all models $\mathfrak{M}$ and all assignments f.*

Let us say that an inference rule is *height-preserving admissible* in TEFL^- (or TEFL) if, whenever all uppersequents (premises) of the inference rule is provable by derivations with height no more than n, then the lowersequent (conclusion) of the rule is provable by a derivation whose height is at most n. By induction on height n of a derivation, we can prove the following.

Proposition 1. *The following weakening rules* ($\mathsf{w}R$) *and* ($\mathsf{w}L$) *are height-preserving admissible in* TEFL^- *and* TEFL. *Moreover, the following substitution rule* (sub) *is height-preserving admissible in* TEFL^- *and* TEFL*:*

$$\frac{\Gamma \overset{\mathcal{T}}{\Rightarrow} \Delta}{\Gamma \overset{\mathcal{T}}{\Rightarrow} \Delta, \alpha : @_n\varphi}\ (\mathsf{w}R) \qquad \frac{\Gamma \overset{\mathcal{T}}{\Rightarrow} \Delta}{\alpha : @_n\varphi, \Gamma \overset{\mathcal{T}}{\Rightarrow} \Delta}\ (\mathsf{w}L) \qquad \frac{\Gamma \overset{\mathcal{T}}{\Rightarrow} \Delta}{\Gamma\sigma \overset{\mathcal{T}\sigma}{\Rightarrow} \Delta\sigma}\ (\mathsf{sub}) \quad ,$$

where σ is a uniform substitution, $\mathcal{T}\sigma$ is the resulting tree by substituting agent nominals in $\mathcal{T}$ by σ, $\Theta\sigma := \{\alpha\sigma : \varphi\sigma \in | \alpha : \varphi \in \Theta\}$ and $\alpha\sigma \in \mathcal{T}\sigma$ is the corresponding label to $\alpha \in \mathcal{T}$ by σ.

4 Semantic Completeness of Tree Sequent Calculus of Epistemic Logic of Friendship

In what follows in this section, sets Γ, Δ, etc. of labelled formulas and a tree $\mathcal{T}$ of labels can be possibly (countably) infinite. Following this change, we say that a possibly infinite tree-sequent $\Gamma \overset{\mathcal{T}}{\Rightarrow} \Delta$ is provable in $\mathbf{TEFL}^-$ if there exist finite sets $\Gamma' \subseteq \Gamma$ and $\Delta' \subseteq \Delta$ and finite subtree $\mathcal{T}'$ of $\mathcal{T}$ such that $\Gamma' \overset{\mathcal{T}'}{\Rightarrow} \Delta'$ is provable in $\mathbf{TEFL}^-$.

Definition 1 (Saturated tree sequent). *A possibly infinite tree sequent $\Gamma \overset{\mathcal{T}}{\Rightarrow} \Delta$ is* saturated *if it satisfies the following conditions:*

(rep1) *If $\alpha : @_n m \in \Gamma$ and $\alpha : \varphi[n/k] \in \Gamma$ then $\alpha : \varphi[m/k] \in \Gamma$.*
(rep2) *If $\alpha : @_m n \in \Gamma$ and $\alpha : \varphi[n/k] \in \Gamma$ then $\alpha : \varphi[m/k] \in \Gamma$.*
(ref$_=$) *$\alpha : @_n n \in \Gamma$ for all labels $\alpha \in \mathcal{T}$.*
(rigid$_=$) *If $\alpha : @_n m \in \Gamma$ then $\beta : @_n m \in \Gamma$ for all labels $\beta \in \mathcal{T}$.*
($\to$r) *If $\alpha : @_n(\varphi \to \psi) \in \Delta$ then $\alpha : @_n\varphi \in \Gamma$ and $\alpha : @_n\psi \in \Delta$.*
($\to$l) *If $\alpha : @_n(\varphi \to \psi) \in \Gamma$ then $\alpha : @_n\varphi \in \Delta$ or $\alpha : @_n\psi \in \Gamma$.*
(@r) *If $\alpha : @_n @_m\varphi \in \Delta$ then $\alpha : @_m\varphi \in \Delta$.*
(@l) *If $\alpha : @_n @_m\varphi \in \Gamma$ then $\alpha : @_m\varphi \in \Gamma$.*
(Fr) *If $\alpha : @_n \mathsf{F}\varphi \in \Delta$ then $\alpha : @_n\langle \mathsf{F}\rangle m \in \Gamma$ and $\alpha : @_m\varphi \in \Delta$ for some nominal m.*
(Fl) *If $\alpha : @_n \mathsf{F}\varphi \in \Gamma$ then $\alpha : @_n\langle \mathsf{F}\rangle m \in \Delta$ or $\alpha : @_m\varphi \in \Gamma$ for all nominals m.*
($\Box$r) *If $\alpha : @_n\Box\varphi \in \Delta$ then $\beta : @_n\varphi \in \Delta$ for some n-child β of α.*
($\Box$l) *If $\alpha : @_n\Box\varphi \in \Gamma$ then $\beta : @_n\varphi \in \Gamma$ for all n-children β of α.*

By the standard argument, we can show the following *saturation lemma*.

Lemma 1. *Let $\Gamma \overset{\mathcal{T}}{\Rightarrow} \Delta$ be an unprovable tree sequent in $\mathbf{TEFL}^-$. Then, there exists a saturated (possibly infinite) sequent $\Gamma^+ \overset{\mathcal{T}^+}{\Rightarrow} \Delta^+$ such that it is still unprovable in $\mathbf{TEFL}^-$ and it extends the original tree sequent, i.e., $\Gamma \subseteq \Gamma^+$, $\Delta \subseteq \Delta^+$ and $\mathcal{T} \subseteq \mathcal{T}^+$.*

Lemma 2. *Let $\Gamma \overset{\mathcal{T}}{\Rightarrow} \Delta$ be a saturated and unprovable tree sequent in $\mathbf{TEFL}^-$. Define the derived model $\mathfrak{M} = (\mathcal{T}, A, (R_a)_{a\in A}, (\asymp_\alpha)_{\alpha\in\mathcal{T}}, V)$ from $\Gamma \overset{\mathcal{T}}{\Rightarrow} \Delta$ by:*

- *$A := \{|n| \mid n$ is an agent nominal$\}$, where $|n|$ is an equivalence class of an equivalence relation $\sim$ which is defined as: $n \sim m$* iff *$\alpha : @_n m \in \Gamma$ for some $\alpha \in \mathcal{T}$.*
- *$\alpha R_{|n|}\beta$ iff β is an m-child of α for some $m \in |n|$.*
- *$|n| \asymp_\alpha |m|$ iff $\alpha : @_n\langle \mathsf{F}\rangle m \in \Gamma$.*
- *$(\alpha, |n|) \in V(m)$ iff $\alpha : @_n m \in \Gamma$ ($m \in \mathsf{Nom}$).*
- *$(\alpha, |n|) \in V(p)$ iff $\alpha : @_n p \in \Gamma$ ($p \in \mathsf{Prop}$).*

Then, $\mathfrak{M}$ is a model. Moreover, for every labelled formula $\alpha : @_n\varphi$, we have

(i) *If $\alpha : @_n\varphi \in \Gamma$ then $\mathfrak{M}, (\alpha, |n|) \models \varphi$;*
(ii) *If $\alpha : @_n\varphi \in \Delta$ then $\mathfrak{M}, (\alpha, |n|) \not\models \varphi$.*

Proof. First, let us check that $\mathfrak{M}$ is a model. First of all, note that we can easily verify that $\sim$ is an equivalence relation by the conditions $(\mathbf{ref}_=)$, $(\mathbf{rep}_i)$ and $(\mathbf{rigid}_=)$ of Definition 1. We can also check that if $n \sim m$ then $R_{|n|} = R_{|m|}$ and that if $n \sim n'$ and $m \sim m'$ then $\alpha : @_n\langle \mathsf{F}\rangle m \in \Gamma$ iff $\alpha : @_{n'}\langle \mathsf{F}\rangle m' \in \Gamma$. So both of $R_{|n|}$ and $\asymp_\alpha$ are well-defined. As for the valuation of propositional variables, when $n \sim m$ holds, the equivalence between $\alpha : @_n p \in \Gamma$ and $\alpha : @_m p \in \Gamma$ holds by the saturation conditions $(\mathbf{rep}_1)$ and $(\mathbf{rep}_2)$. For the valuation for agent nominals m, we need to check that $\{(\alpha, |n|) \mid \alpha : @_n m \in \Gamma\}$ is $\mathcal{T} \times \{|m|\}$. But this is clear from the saturation condition $(\mathbf{rigid}_=)$ and the fact that $\sim$ is an equivalence relation.

Now we move to check items (i) and (ii) by induction on φ. We only check the cases where φ is of the form: $\mathsf{F}\varphi$ or $\Box\varphi$, since the other cases are easy to establish by the corresponding saturation conditions of Definition 1.

- Let φ be of the form $\mathsf{F}\varphi$. For (i), assume that $\alpha : @_n\mathsf{F}\varphi \in \Gamma$. We need to show $\mathfrak{M}, (\alpha, |n|) \models \mathsf{F}\varphi$, so let us fix any agent nominal m such that $|n|R_\alpha|m|$. Our goal is to show $\mathfrak{M}, (\alpha, |m|) \models \varphi$. From $|n|R_\alpha|m|$, we get $\alpha : @_n\langle \mathsf{F}\rangle m \in \Gamma$ hence $\alpha : @_n\langle \mathsf{F}\rangle m \notin \Delta$ by the unprovability of $\Gamma \overset{\mathcal{T}}{\Rightarrow} \Delta$. By the condition $(\mathsf{F}\mathrm{l})$, we obtain $\alpha : @_m\varphi \in \Gamma$, which implies our goal by induction hypothesis.
  For (ii), assume that $\alpha : @_n\mathsf{F}\varphi \in \Delta$. By the condition $(\mathsf{R}\mathbf{r})$, $\alpha : @_n\langle \mathsf{F}\rangle m \in \Gamma$ and $\alpha : @_n m \in \Delta$ for some agent nominal m. With the help of induction hypothesis, we have $|n|R_\alpha|m|$ and $\mathfrak{M}, (\alpha, |m|) \not\models \varphi$ for some agent nominal m. Hence $\mathfrak{M}, (\alpha, |n|) \not\models \mathsf{F}\varphi$, as desired.
- Let φ be of the form $\Box\varphi$. To show (i), assume that $\alpha : @_n\Box\varphi \in \Gamma$. We need to show $\mathfrak{M}, (\alpha, |n|) \models \Box\varphi$, so let us fix any label β such that $\alpha R_{|n|}\beta$. Our goal is to show $\mathfrak{M}, (\beta, |n|) \models \varphi$. By $\alpha R_{|n|}\beta$, we can find an agent nominal $m \in |n|$ such that β is an m-child of α. It follows from $m \in |n|$ that $\gamma : @_n m \in \Gamma$ for some label γ. By $\alpha : @_n\Box\varphi \in \Gamma$ and $\gamma : @_n m \in \Gamma$, the saturation condition $(\mathbf{rep}_1)$ implies that $\alpha : @_m\Box\varphi \in \Gamma$. By the saturation condition $(\Box\mathrm{l})$ and the fact that β is an m-child of α, we obtain $\beta : @_m\varphi \in \Gamma$. By induction hypothesis, $\mathfrak{M}, (\beta, |m|) \models \varphi$ hence we obtain our goal by $|m| = |n|$. This finishes to show (i).
  For (ii), assume that $\alpha : @_n\Box\varphi \in \Delta$. By the saturation condition $(\Box\mathbf{r})$, $\beta : @_n\varphi \in \Delta$ for some n-child β of α, i.e., $\alpha R_{|n|}\beta$. By induction hypothesis, $\mathfrak{M}, (\beta, |n|) \not\models \varphi$. So we conclude that $\mathfrak{M}, (\alpha, |n|) \not\models \Box\varphi$. □

Theorem 2 (Completeness of cut-free TEFL$^-$). *If $\mathfrak{M}, f \models \Gamma \overset{\mathcal{T}}{\Rightarrow} \Delta$ for all models $\mathfrak{M}$ and all assignments f, then $\Gamma \overset{\mathcal{T}}{\Rightarrow} \Delta$ is provable in* **TEFL**$^-$.

Proof. Suppose for contradiction that $\Gamma \overset{\mathcal{T}}{\Rightarrow} \Delta$ is unprovable in **TEFL**$^-$. By Lemma 1, we can extend this tree sequent into a saturated (possibly infinite) tree sequent $\Gamma^+ \overset{\mathcal{T}^+}{\Rightarrow} \Delta^+$ which is still unprovable in **TEFL**$^-$. Let $\mathfrak{M}$ be the derived model from $\Gamma^+ \overset{\mathcal{T}^+}{\Rightarrow} \Delta^+$. Let us define $f : \mathcal{T} \to \mathcal{T}$ as the identity mapping. Then it follows from Lemma 2 that $\mathfrak{M}, f \not\models \Gamma \Rightarrow \Delta$, as required. □

By Theorems 1 and 2, the cut elimination theorem of **TEFL** follows.

Corollary 1. *The following are all equivalent:*

1. $\mathfrak{M}, f \models \Gamma \overset{\mathcal{T}}{\Rightarrow} \Delta$ *for all models* $\mathfrak{M}$ *and all assignments* f.
2. $\Gamma \overset{\mathcal{T}}{\Rightarrow} \Delta$ *is provable in* $\mathbf{TEFL}^-$.
3. $\Gamma \overset{\mathcal{T}}{\Rightarrow} \Delta$ *is provable in* **TEFL**.

Therefore, **TEFL** *enjoys the cut-elimination theorem.*

5 Hilbert System of Epistemic Logic of Friendship

This section provides a Hilbert system of the epistemic logic of friendship by "translating" a tree sequent into a formula in $\mathcal{L}$. First of all, let us introduce the notion of *necessity form*, originally proposed in [13] by Goldblatt and used also in [6,11]. Necessity forms are employed to formulate an inference rule of our Hilbert system.

Definition 2 (Necessity form). *Fix an arbitrary symbol # not occurring in the syntax* $\mathcal{L}$. *A* necessity form *is defined inductively as follows:* (i) # *is a necessity form;* (ii) *If* L *is a necessity form and* φ *is a formula, then* $\varphi \to L$ *is also a necessity form;* (iii) *If* L *is a necessity form and* n *is an agent nominal, then* $@_n\Box L$ *is also a necessity form. Given a necessity form* $L(\#)$ *and a formula* φ *of* $\mathcal{L}$, *we use* $L(\varphi)$ *to denote the formula obtained by replacing the unique occurrence of # in* L *by the formula* φ.

When $L(\#)$ is a necessity form of $\psi_0 \to @_n\Box(\psi_1 \to @_m\Box(\psi_2 \to \#))$, then $L(\varphi)$ is $\psi_0 \to @_n\Box(\psi_1 \to @_m\Box(\psi_2 \to \varphi))$. Intuitively, this notion allows us to capture the unique path from a label in a tree of a tree sequent to the root label of the tree.

Table 2 presents our Hilbert system **HEFL**. The underlying idea of the system is: on the top of the propositional part (`Taut` and `MP`), we combine the

Table 2. Hilbert System **HEFL**

(`Taut`) all propositional tautologies	(`MP`) From φ and $\varphi \to \psi$, infer ψ
($\mathtt{K}_\Box$) $\Box(p \to q) \to (\Box p \to \Box q)$	($\mathtt{Nec}_\Box$) From φ, infer $\Box\varphi$
($\mathtt{K}_\mathsf{F}$) $\mathsf{F}(p \to q) \to (\mathsf{F}p \to \mathsf{F}q)$	($\mathtt{Nec}_\mathsf{F}$) From φ, infer $\mathsf{F}\varphi$
($\mathtt{K}_@$) $@_n(p \to q) \to (@_np \to @_nq)$	($\mathtt{Nec}_@$) From φ, infer $@_n\varphi$
(`Ref`) $@_nn$	(`Selfdual`) $\neg @_np \leftrightarrow @_n\neg p$
(`Elim`) $@_np \to (n \to p)$	(`Agree`) $@_n@_mp \to @_mp$
(`Back`) $@_np \to \mathsf{F}@_np$	(`DCom@□`) $@_n\Box @_np \leftrightarrow @_n\Box p$
($\mathtt{Rigid}_=$) $@_nm \to \Box @_nm$	($\mathtt{Rigid}_\neq$) $\neg @_nm \to \Box\neg @_nm$
(`US`) From φ, infer $\varphi\sigma$, where σ is a uniform substitution.	
(`Name`) From $n \to \varphi$, infer φ, where n is fresh in φ.	
(L(`BG`)) From $L(@_n\langle\mathsf{F}\rangle m \to @_m\varphi)$, infer $L(@_n\mathsf{F}\varphi)$, where m is fresh in $L(@_n\mathsf{F}\varphi)$.	

axiomatization of modal logic **K** for the modal operator $\Box$ and the axiomatization of a basic hybrid logic $\mathbf{K}_{\mathcal{H}(@)}$ (see [4,5]) for the modal operator F, with some modification (we need to modify **BG**, the rule of *bounded generalization*, with the help of necessity forms), and then we add three interaction axioms: ($\mathtt{Rigid}_=$), ($\mathtt{Rigid}_{\neq}$), and ($\mathtt{DCom}@\Box$). We note that the axiom ($\mathtt{DCom}@\Box$) is also used for axiomatizing the *dependent product* of two hybrid logics in [20]. Let us define the notion of provability in **HEFL** in as usual. We write $\vdash_{\mathbf{HEFL}} \varphi$ to means that φ is provable in **HEFL**.[1,2]

Proposition 2. *All the following are provable in* **HEFL**.

1. $@_m @_n \varphi \leftrightarrow @_n \varphi$.
2. $n \to (@_n \varphi \leftrightarrow \varphi)$.
3. $@_n m \to (@_n \varphi \leftrightarrow @_m \varphi)$.
4. $@_n m \leftrightarrow @_m n$.
5. $@_n (\varphi \to \psi) \leftrightarrow (@_n \varphi \to @_n \psi)$.
6. $@_n m \to (\varphi[n/k] \leftrightarrow \varphi[m/k])$.

Proof. For the provability of item 1, it suffices to show the right-to-left direction, which is shown by (Agree) and (Selfdual). For the provability of item 2, it suffices to show $n \to (\varphi \to @_n \varphi)$, whose provability is shown by the contraposition of (Elim) and (Selfdual). Then items 3 to 5 are proved similarly as given in [5, p. 293, Lemma 2]. Finally, item 6 is proved by induction on φ. Here we show the case where $\varphi \equiv \Box\psi$ alone, while we note that we need to use item 5 for the case where $\varphi \equiv @_l \psi$. By induction hypothesis, we obtain $\vdash_{\mathbf{HEFL}} @_n m \to (\psi[n/k] \leftrightarrow \psi[m/k])$. By ($\mathtt{K}_\Box$) and ($\mathtt{Nec}_\Box$), we get $\vdash_{\mathbf{HEFL}} \Box @_n m \to (\Box(\psi[n/k]) \leftrightarrow \Box(\psi[m/k]))$. It follows from the axiom ($\mathtt{rigid}_=$) that $\vdash_{\mathbf{HEFL}} @_n m \to ((\Box\psi)[n/k] \leftrightarrow (\Box\psi)[m/k]))$, as desired. $\Box$

The following translation is a key to specify our Hilbert system **HEFL**.

Definition 3 (Formulaic translation). *Given a set Θ of labelled formulas and a label α, we define $\Theta_\alpha := \{\varphi \mid \alpha : \varphi \in \Theta\}$. Let $\Gamma \overset{\mathcal{I}}{\Rightarrow} \Delta$ be a tree sequent. Then the formulaic translation of the sequent at α is defined as:*

$$\left[\!\left[\Gamma \overset{\mathcal{I}}{\Rightarrow} \Delta\right]\!\right]_\alpha := \bigwedge \Gamma_\alpha \to \bigvee \left(\Delta_\alpha, @_{n_1}\Box \left[\!\left[\Gamma \overset{\mathcal{I}}{\Rightarrow} \Delta\right]\!\right]_{\beta_1}, \ldots, @_{n_k}\Box \left[\!\left[\Gamma \overset{\mathcal{I}}{\Rightarrow} \Delta\right]\!\right]_{\beta_k}\right),$$

where β_i is an n_i-child of α, β_is enumerate all children of α, $\bigwedge\emptyset := \top$, and $\bigvee\emptyset := \bot$.

The formulaic translation of a tree sequent of Fig. 1 of Sect. 3 at the root 0 is

$$@_n\varphi \to (@_m\psi \vee @_n\Box(\top \to @_k\theta) \vee @_k\Box(@_m\rho \to \bot)).$$

[1] By (K)-rules and (Nec)-rules for $\Box$, F and $@_n$, the replacement of equivalence holds in **HEFL**.

[2] Given a set $\Gamma \cup \{\varphi\}$ of formulas, we say that φ is *deducible* in **HEFL** from Γ if there exist finite formulas $\psi_1, \ldots, \psi_n \in \Gamma$ such that $(\psi_1 \wedge \ldots \wedge \psi_n) \to \varphi$ is provable in **HEFL**. Then it is easy to see that the deduction theorem holds in **HEFL**.

Theorem 3. *If a tree sequent* $\Gamma \overset{\mathcal{T}}{\Rightarrow} \Delta$ *is provable in* **TEFL** *then the formulaic translation* $[\![\Gamma \overset{\mathcal{T}}{\Rightarrow} \Delta]\!]_i$ *is provable in* **HEFL**, *where a natural number i is the root of* $\mathcal{T}$.

Proof. By induction on height n of a derivation of $\Gamma \overset{\mathcal{T}}{\Rightarrow} \Delta$ in **TEFL**, where i is the root of the tree $\mathcal{T}$. We skip the base case where $n = 0$. Let $n > 0$. It is remarked that, when the sequent is obtained by (rep_l), $(\mathsf{ref}_=)$, $(@L)$, or $(@R)$, respectively, the translation of the sequent at the root is provable by Proposition 2 (6), the axiom (`Ref`), (`Agree`), or Proposition 2 (1), respectively. Here we focus on the cases where $\Gamma \overset{\mathcal{T}}{\Rightarrow} \Delta$ is obtained by $(\Box L)$, $(\mathsf{F}R)$ or $(\mathsf{rigid}_=)$, since these are the cases where we need to be careful and the other cases are easy to establish.

$(\Box L)$ Suppose that $\alpha : @_n\Box\varphi, \Gamma' \overset{\mathcal{T}}{\Rightarrow} \Delta$ is obtained by $(\Box L)$ from $\beta : @_n\varphi, \Gamma' \overset{\mathcal{T}}{\Rightarrow} \Delta$, where $\beta \in \mathcal{T}$ is an n-child of α. By induction hypothesis, we obtain $\vdash_{\mathbf{HEFL}} \left[\!\!\left[\beta : @_n\varphi, \Gamma' \overset{\mathcal{T}}{\Rightarrow} \Delta\right]\!\!\right]_i$. We show that $\vdash_{\mathbf{HEFL}} \left[\!\!\left[\alpha : @_n\Box\varphi, \Gamma' \overset{\mathcal{T}}{\Rightarrow} \Delta\right]\!\!\right]_i$. Let $(\alpha_0, \alpha_1, \ldots, \alpha_l)$ be the unique path from α $(\equiv \alpha_l)$ to the root i $(\equiv \alpha_0)$ of tree $\mathcal{T}$. By induction on $0 \leqslant h \leqslant l$, we show that $\vdash_{\mathbf{HEFL}} \left[\!\!\left[\beta : @_n\varphi, \Gamma' \overset{\mathcal{T}}{\Rightarrow} \Delta\right]\!\!\right]_{\alpha_{l-h}} \rightarrow \left[\!\!\left[\alpha : @_n\Box\varphi, \Gamma' \overset{\mathcal{T}}{\Rightarrow} \Delta\right]\!\!\right]_{\alpha_{l-h}}$.
Let $h = 0$ and so $\alpha_{l-h} = \alpha$. It suffices to show that a formula of the form

$$(\gamma_1 \rightarrow (\delta \vee @_n\Box((\gamma_2 \wedge @_n\varphi) \rightarrow \psi_2)) \rightarrow ((@_n\Box\varphi \wedge \gamma_1) \rightarrow (\delta \vee @_n\Box(\gamma_2 \rightarrow \psi_2))).$$

is provable in **HEFL**. This reduces to the provability of

$$@_n\Box\varphi \wedge @_n\Box((\gamma_2 \wedge @_n\varphi) \rightarrow \psi_2)) \rightarrow @_n\Box(\gamma_2 \rightarrow \psi_2))$$

in **HEFL**. This holds by the axiom (`Dcom□@`) $@_n\Box @_n\varphi \leftrightarrow @_n\Box\varphi$.
Let $h > 0$. But this case is shown with the help of $(\mathsf{Nec}_\Box)$ and $(\mathsf{Nec}_@)$. This completes our induction on h. So we conclude $\vdash_{\mathbf{HEFL}} \left[\!\!\left[\alpha : @_n\Box\varphi, \Gamma' \overset{\mathcal{T}}{\Rightarrow} \Delta\right]\!\!\right]_i$.

$(\mathsf{F}R)$ Suppose that $\Gamma \overset{\mathcal{T}}{\Rightarrow} \Delta', \alpha : @_n\mathsf{F}\varphi$ is obtained by $(\mathsf{F}R)$ from $\alpha : @_n\langle\mathsf{F}\rangle m, \Gamma \overset{\mathcal{T}}{\Rightarrow} \Delta', \alpha : @_m\varphi$ where m is fresh in the conclusion. By induction hypothesis, we have $\vdash_{\mathbf{HEFL}} \left[\!\!\left[\alpha : @_n\langle\mathsf{F}\rangle m, \Gamma \overset{\mathcal{T}}{\Rightarrow} \Delta', \alpha : @_m\varphi\right]\!\!\right]_i$, which is equivalent to $\vdash_{\mathbf{HEFL}} L(@_n\langle\mathsf{F}\rangle m \rightarrow @_m\varphi)$ for some necessitation form L. Fix such necessitation form L. By the inference rule L(`BG`) of **HEFL**, we can obtain $\vdash_{\mathbf{HEFL}} L(@_n\mathsf{F}\varphi)$, which is equivalent to $\vdash_{\mathbf{HEFL}} \left[\!\!\left[\Gamma \overset{\mathcal{T}}{\Rightarrow} \Delta', \alpha : @_n\mathsf{F}\varphi\right]\!\!\right]_i$.

$(\mathsf{rigid}_=)$ Instead of dealing with a general case, we handle a simple example of $\mathcal{T}$ to extract an essence of this case, where we need to use the axioms (`Rigid`$_=$) and (`Rigid`$_{\neq}$). Let $\mathcal{T}$ consists of three labels, i.e., the root i, a

k-child α of i and a k'-child β of i. Let us suppose that $\beta : @_n m, \Gamma' \overset{\mathcal{T}}{\Rightarrow} \Delta$ is obtained by $(\mathsf{rigid}_=)$ from $\alpha : @_n m, \Gamma' \overset{\mathcal{T}}{\Rightarrow} \Delta$. In what follows, for every $\eta \in \mathcal{T}$, let us write $\bigwedge \Gamma'_\eta$ and $\bigvee \Delta_\eta$ by γ_η and δ_η, respectively. Here we note that the following hold:
(α to i) $\vdash_{\mathbf{HEFL}} (@_k\Box((@_n m \wedge \gamma_\alpha) \to \delta_\alpha) \wedge @_n m) \to @_k\Box(\gamma_\alpha \to \delta_\alpha))$.
(i to β) $\vdash_{\mathbf{HEFL}} \neg @_n m \to @_{k'}\Box((@_n m \wedge \gamma_\beta) \to \delta_\beta)$
For (α to i), it suffices to show $\vdash_{\mathbf{HEFL}} @_n m \to @_k\Box @_n m$, which holds by $(\mathtt{Rigid}_=)$, the distribution of @ over the implication and Proposition 2 (1). For (i to β), it suffices to show $\vdash_{\mathbf{HEFL}} \neg @_n m \to @_{k'}\Box\neg @_n m$, which holds by $(\mathtt{Rigid}_{\neq})$, $(\mathtt{Selfdual})$ and Proposition 2 (1). By induction hypothesis, we obtain $\vdash_{\mathbf{HEFL}} \left[\!\left[\alpha : @_n m, \Gamma' \overset{\mathcal{T}}{\Rightarrow} \Delta\right]\!\right]_i$, i.e.,

$$\vdash_{\mathbf{HEFL}} \gamma_i \to (\delta_i \vee @_k\Box((@_n m \wedge \gamma_\alpha) \to \delta_\alpha) \vee @_{k'}\Box(\gamma_\beta \to \delta_\beta)) .$$

It follows from item (α to i) that

$$\vdash_{\mathbf{HEFL}} (@_n m \wedge \gamma_i) \to (\delta_i \vee @_k\Box(\gamma_\alpha \to \delta_\alpha) \vee @_{k'}\Box(\gamma_\beta \to \delta_\beta)) .$$

By this and item (i to β), we can establish:

$$\vdash_{\mathbf{HEFL}} \gamma_i \to (\delta_i \vee @_k\Box(\gamma_\alpha \to \delta_\alpha) \vee @_{k'}\Box((@_n m \wedge \gamma_\beta) \to \delta_\beta)) ,$$

which is equivalent to: $\vdash_{\mathbf{HEFL}} \left[\!\left[\beta : @_n m, \Gamma' \overset{\mathcal{T}}{\Rightarrow} \Delta\right]\!\right]_i$, as desired. □

In what follows in this section, we prove the soundness of **HEFL** for the tree sequent calculus **TEFL** with cut rule. The cut rule is necessary to prove the following.

Lemma 3. *The rules* $(\to R)$, $(\Box R)$, $(@R)$, *and* $(@L)$ *are invertible, i.e., if the lower sequent is provable in* **TEFL** *then the upper sequent is also provable in* **TEFL**.

Proof. We only prove the invertibility of $(\to R)$ and $(\Box R)$. First we deal with $(\to R)$. Suppose that $\Gamma \overset{\mathcal{T}}{\Rightarrow} \Delta, \alpha : @_n(\varphi \to \psi)$ is provable in **TEFL**. This is shown as follows:

$$\frac{\Gamma \overset{\mathcal{T}}{\Rightarrow} \Delta, \alpha : @_n(\varphi \to \psi) \quad \alpha : @_n(\varphi \to \psi), \alpha : @_n\varphi \overset{\mathcal{T}}{\Rightarrow} \alpha : @_n\psi}{\alpha : @_n\varphi, \Gamma \overset{\mathcal{T}}{\Rightarrow} \Delta, \alpha : @_n\psi}\,(Cut) ,$$

where the rightmost tree sequent is provable in **TEFL** by $(\to L)$. Second we move to $(\Box R)$. Suppose that $\Gamma \overset{\mathcal{T}}{\Rightarrow} \Delta, \alpha : @_n\Box\varphi$ is provable in **TEFL**. Then the provability of the upper sequent of $(\Box R)$ is established as follows:

$$\frac{\dfrac{\Gamma \overset{\mathcal{T}}{\Rightarrow} \Delta, \alpha : @_n\Box\varphi}{\Gamma \overset{\mathcal{T}\cup\{\alpha\cdot_n i\}}{\Rightarrow} \Delta, \alpha : @_n\Box\varphi}\,(\mathsf{wlab}) \quad \dfrac{\alpha\cdot_n i : @_n\varphi, \Gamma \overset{\mathcal{T}\cup\{\alpha\cdot_n i\}}{\Rightarrow} \Delta, \alpha\cdot_n i : @_n\varphi}{\alpha : @_n\Box\varphi, \Gamma \overset{\mathcal{T}\cup\{\alpha\cdot_n i\}}{\Rightarrow} \Delta, \alpha\cdot_n i : @_n\varphi}\,(L\Box)}{\Gamma \overset{\mathcal{T}\cup\{\alpha\cdot_n i\}}{\Rightarrow} \Delta, \alpha\cdot_n i : @_n\varphi}\,(Cut) .$$

□

Theorem 4. *If φ is provable in* **HEFL**, *then* $\overset{\mathcal{T}}{\Rightarrow} \alpha : @_n\varphi$ *is provable in* **TEFL** *for all trees* $\mathcal{T}$, $\alpha \in \mathcal{T}$ *and nominals* n *fresh in* φ.

Proof. Suppose that there is a proof $(\varphi_0, \ldots, \varphi_h)$ of φ in **HEFL**. By induction on $0 \leqslant j \leqslant h$, we show that $\overset{\mathcal{T}}{\Rightarrow} \alpha : @_n\varphi_j$ is provable in **TEFL** for all nominals n fresh in φ_j and $\alpha \in \mathcal{T}$. Since the space is limited, we demonstrate some cases. Let us start with (Rigid$_=$), which is shown by the left derivation below. Now we move to (DCom@□). We show the right-to-left direction alone, since the converse direction is shown similarly. Let us see the right derivation below, from which we can obtain the provability of $\overset{\mathcal{T}}{\Rightarrow} \alpha : @_m(@_n\Box @_n p \to @_n\Box p)$ in **TEFL**. Now we deal with some inference rules below.

$$
\begin{array}{cl}
\alpha \cdot_k i : @_n m \overset{\mathcal{T}\cup\{\alpha\cdot_k i\}}{\Rightarrow} \alpha \cdot_k i : @_n m & \\ \hline
\alpha : @_n m \overset{\mathcal{T}\cup\{\alpha\cdot_k i\}}{\Rightarrow} \alpha \cdot_k i : @_n m & (\mathsf{rigid}_=) \\ \hline
\alpha : @_n m \overset{\mathcal{T}\cup\{\alpha\cdot_k i\}}{\Rightarrow} \alpha \cdot_k i : @_k @_n m & (@R) \\ \hline
\alpha : @_n m \overset{\mathcal{T}}{\Rightarrow} \alpha : @_k \Box @_n m & (\Box R) \\ \hline
\alpha : @_k @_n m \overset{\mathcal{T}}{\Rightarrow} \alpha : @_k \Box @_n m & (@L) \\ \hline
\overset{\mathcal{T}}{\Rightarrow} \alpha : @_k(@_n m \to \Box @_n m) & (\to R)
\end{array}
\qquad
\begin{array}{cl}
\alpha \cdot_n i : @_n p \overset{\mathcal{T}\cup\{\alpha\cdot_n i\}}{\Rightarrow} \alpha \cdot_n i : @_n p & \\ \hline
\alpha \cdot_n i : @_n @_n p \overset{\mathcal{T}\cup\{\alpha\cdot_n i\}}{\Rightarrow} \alpha \cdot_n i : @_n p & (@L) \\ \hline
\alpha : @_n \Box @_n p \overset{\mathcal{T}\cup\{\alpha\cdot_n i\}}{\Rightarrow} \alpha : \cdot_n i : @_n p & (\Box L) \\ \hline
\alpha : @_n \Box @_n p \overset{\mathcal{T}}{\Rightarrow} \alpha : @_n \Box p & (\Box R)
\end{array}
$$

(L(BG)) Let $\varphi_j \equiv \Box\psi$ be obtained by (L(BG)). Fix any tree $\mathcal{T}$, $\alpha \in \mathcal{T}$ and fresh nominal k. By induction hypothesis, $\overset{\mathcal{T}}{\Rightarrow} \alpha : @_k L(@_n\langle\mathsf{F}\rangle m \to @_m\varphi)$ is provable in **TEFL**, where m satisfies the freshness condition. By applying Lemma 3 (i.e., the invertibility of the right rules) repeatedly to the consequent of a resulting tree sequent, we obtain the provability of a tree sequent of the form $\Gamma, \beta : @_n\langle\mathsf{F}\rangle m \overset{\mathcal{T}'}{\Rightarrow} \Delta, \beta : @_m\varphi$. Then we apply the right rules in a converse direction of our repeated application of Lemma 3 to conclude that $\overset{\mathcal{T}}{\Rightarrow} \alpha : @_k L(@_n\mathsf{F}\varphi)$ is provable in **TEFL**. To illustrate this argument, let $L \equiv @_n\Box(\psi \to \#)$. By induction hypothesis, $\overset{\mathcal{T}}{\Rightarrow} \alpha : @_k@_n\Box(\psi \to (@_n\langle\mathsf{F}\rangle m \to @_m\varphi))$ is provable in **TEFL**, where m satisfies the freshness condition. By applying Lemma 3 repeatedly, we obtain the provability of $\alpha\cdot_n i : @_n\psi, \alpha\cdot_n i : @_n\langle\mathsf{F}\rangle m \overset{\mathcal{T}\cup\{\alpha\cdot_n i\}}{\Rightarrow} \alpha\cdot_n i : @_m\varphi$ in **TEFL** for some fresh i. Then we proceed as follows:

$$
\begin{array}{cl}
\alpha \cdot_n i : @_n\psi, \alpha \cdot_n i : @_n\langle\mathsf{F}\rangle m \overset{\mathcal{T}\cup\{\alpha\cdot_n i\}}{\Rightarrow} \alpha \cdot_n i : @_m\varphi & \\ \hline
\alpha \cdot_n i : @_n\psi \overset{\mathcal{T}\cup\{\alpha\cdot_n i\}}{\Rightarrow} \alpha \cdot_n i : @_n\mathsf{F}\varphi & (\mathsf{F}R) \\ \hline
\alpha \cdot_n i : @_n\psi \overset{\mathcal{T}\cup\{\alpha\cdot_n i\}}{\Rightarrow} \alpha \cdot_n i : @_n@_n\mathsf{F}\varphi & (@R) \\ \hline
\overset{\mathcal{T}\cup\{\alpha\cdot_n i\}}{\Rightarrow} \alpha \cdot_n i : @_n(\psi \to @_n\mathsf{F}\varphi) & (\to R) \\ \hline
\overset{\mathcal{T}}{\Rightarrow} \alpha : @_n\Box(\psi \to @_n\mathsf{F}\varphi) & (\Box R) \\ \hline
\overset{\mathcal{T}}{\Rightarrow} \alpha : @_k@_n\Box(\psi \to @_n\mathsf{F}\varphi) & (@R)
\end{array}
\quad ,
$$

as required.

(Nec$_\Box$) Let $\varphi_j \equiv \Box\psi$ be obtained by (Nec$_\Box$). Fix any tree $\mathcal{T}$, $\alpha \in \mathcal{T}$ and fresh nominal n. By induction hypothesis, $\overset{\mathcal{T}\cup\{\alpha\cdot_n i\}}{\Rightarrow} \alpha\cdot_n i : @_n\psi$ is provable in **TEFL**, where i is fresh in $\mathcal{T}$. By the rule ($\Box R$) of **TEFL**, the provability of $\overset{\mathcal{T}}{\Rightarrow} \alpha : @_n\Box\psi$ follows, as desired.

(Nec$_\mathsf{F}$) Let $\varphi_j \equiv \mathsf{F}\psi$ be obtained by (Nec$_\mathsf{F}$). Fix any tree $\mathcal{T}$, $\alpha \in \mathcal{T}$ and fresh nominal n. Let m be a fresh nominal in ψ. By induction hypothesis, $\overset{\mathcal{T}}{\Rightarrow} \alpha : @_m\psi$ is provable in **TEFL**. By the admissibility of weakening rule from Proposition 1, we obtain the provability of $\alpha : @_n\langle\mathsf{F}\rangle m \overset{\mathcal{T}}{\Rightarrow} \alpha : @_m\psi$. Since m is fresh in ψ, the rule ($\mathsf{F}R$) enables us to derive the provability of $\overset{\mathcal{T}}{\Rightarrow} \alpha : @_n\mathsf{F}\psi$ in **TEFL**, as desired. □

Corollary 2 (Soudness and Completenss of HEFL). *The following are all equivalent: for every formula φ,*

1. *φ is valid in the class of all models,*[3]
2. *$\overset{\mathcal{T}}{\Rightarrow} \alpha : @_n\varphi$ is provable in* **TEFL**$^-$ *for all $\mathcal{T}$, $\alpha \in \mathcal{T}$ and nominals n fresh in φ,*
3. *$\overset{\mathcal{T}}{\Rightarrow} \alpha : @_n\varphi$ is provable in* **TEFL** *for all $\mathcal{T}$, $\alpha \in \mathcal{T}$ and nominals n fresh in φ,*
4. *φ is provable in* **HEFL**.

Proof. Item 1 is equivalent to the following: $\overset{\mathcal{T}}{\Rightarrow} \alpha : @_n\varphi$ is true for all pairs $(\mathfrak{M}, f)$ of models and assignments, trees $\mathcal{T}$, $\alpha \in \mathcal{T}$ and nominals n fresh in φ. Then the equivalence between items 1, 2 and 3 holds by Corollary 1. The direction from item 4 to item 3 holds by Theorem 4. Finally, the direction from item 3 to item 4 is established as follows. Suppose item 3. Let n be a fresh nominal. By the supposition, $\overset{\{0\}}{\Rightarrow} 0 : @_n\varphi$ is provable in **TEFL**. It follows from Theorem 3 that $\vdash_{\mathbf{HEFL}} [\![\overset{\{0\}}{\Rightarrow} 0 : @_n\varphi]\!]_0$, which implies $\vdash_{\mathbf{HEFL}} @_n\varphi$. By the axiom (Elim), we obtain $\vdash_{\mathbf{HEFL}} n \to \varphi$ hence $\vdash_{\mathbf{HEFL}} \varphi$ by (Name), as required. □

6 Extensions of Epistemic Logic of Friendship

This section outlines how we extend our tree sequent calculus **TEFL** and Hilbert system **HEFL**. In particular, we discuss extensions where $\Box$ follows **S4** or **S5** axioms and/or the friendship relation $\asymp_w$ satisfies some universal properties such as irreflexivity, symmetry, etc. ($w \in W$). We note that [23,24] assume that the friendship relation $\asymp_w$ satisfies irreflexivity and symmetry and that $\Box$ obeys **S5** axioms.

Let us denote a set $\{\Box p \to p, \Box p \to \Box\Box p\}$ by **S4** and a set **S4**$\cup\{p \to \Box\neg\Box\neg p\}$ by **S5**. Let us consider formulas of the form $@_n m$ or $@_n\langle\mathsf{F}\rangle m$, which are denoted by ρ_i, ρ'_i, etc. below. Let us consider a formula φ of the following form:

$$(\rho_1 \wedge \cdots \wedge \rho_h) \to (\rho'_1 \vee \cdots \vee \rho'_l),$$

[3] We do not need to assume that each of our models is *named* in the sense that each agent is named by an agent nominal.

where we note that h and l are possibly zero. We say that a formula of such form is a *regular implication* [17, Sect. 6] (we may even consider a more general class of formulas called *geometric formulas* (cf. [8]), but we restrict our attention to regular implications in this paper for simplicity). The corresponding frame property of a regular implication is obtained by regarding $@_n m$ or $@_n\langle\mathsf{F}\rangle m$ by "$a_n = a_m$" and "$a_n \asymp_w a_m$" and putting the universal quantifiers for all agents and w. For example, irreflexivity and symmetry of $\asymp_w$ are defined by $@_n\langle\mathsf{F}\rangle n \to \bot$ and $@_n\langle\mathsf{F}\rangle m \to @_m\langle\mathsf{F}\rangle n$, respectively. When Λ is one of **S4** and **S5** and Θ is a finite set of regular implications, a Hilbert system $\mathbf{HEFL}(\Lambda \cup \Theta)$ is defined as the axiomatic extension of **HEFL** by new axioms $\Lambda \cup \Theta$.

Now let us move to tree sequent systems. First, we introduce an inference rule for a regular implication. For a regular implication φ displayed above, we can define the corresponding inference rule $(\mathsf{ri}(\varphi))$ for tree sequent calculus as follows (cf. [8,17, Sect. 6]):

$$\frac{\Gamma \overset{\mathcal{T}}{\Rightarrow} \Delta, \alpha : \rho_1 \quad \dots \quad \Gamma \overset{\mathcal{T}}{\Rightarrow} \Delta, \alpha : \rho_h \quad \alpha : \rho'_1, \dots, \alpha : \rho'_l, \Gamma \overset{\mathcal{T}}{\Rightarrow} \Delta}{\Gamma \overset{\mathcal{T}}{\Rightarrow} \Delta}\,(\mathsf{ri}(\varphi))$$

When $\asymp_w$ is irreflexive or symmetric for all $w \in W$, we can obtain the following rule $(\mathsf{irr}_{\asymp})$ or $(\mathsf{sym}_{\asymp})$, respectively:

$$\frac{\Gamma \overset{\mathcal{T}}{\Rightarrow} \Delta, \alpha : @_n\langle\mathsf{F}\rangle n}{\Gamma \overset{\mathcal{T}}{\Rightarrow} \Delta}\,(\mathsf{irr}_{\asymp}) \qquad \frac{\Gamma \overset{\mathcal{T}}{\Rightarrow} \Delta, \alpha : @_n\langle\mathsf{F}\rangle m \quad \alpha : @_m\langle\mathsf{F}\rangle n \Gamma \overset{\mathcal{T}}{\Rightarrow} \Delta}{\Gamma \overset{\mathcal{T}}{\Rightarrow} \Delta}\,(\mathsf{sym}_{\asymp}).$$

Let Λ be one of **S4** and **S5** and Θ be a possibly empty finite set of regular implications. In what follows, we define the tree sequent system $\mathbf{TEFL}(\Lambda;\Theta)$. Recall that the side condition $\ddagger$ of the rule $(\Box L)$ of Table 1. First, depending on the choice of Λ, we change the side condition $\ddagger$ of **TEFL** into the following one:

- $\ddagger_{\mathbf{S4}}$: $\alpha \preceq_n \beta$, where $\preceq_n$ is the reflexive transitive closure of the n-children relation.
- $\ddagger_{\mathbf{S5}}$: $\alpha \sim_n \beta$, where $\sim_n$ is the reflexive, symmetric, transitive closure of the n-children relation.

Second, we extend the resulting system with a set $\{(\mathsf{ri}(\varphi)) \mid \varphi \in \Theta\}$ of inference rules to finish to define the system $\mathbf{TEFL}(\Lambda;\Theta)$. We define $\mathbf{TEFL}(\Lambda;\Theta)^-$ as the system $\mathbf{TEFL}(\Lambda;\Theta)$ without the cut rule.

Given a set Ψ of formulas and a frame $\mathfrak{F} = (W, A, (R_a)_{a\in A}, (\asymp_w)_{w\in W})$ (a model without a valuation), we say that Ψ is *valid* in $\mathfrak{F}$ (notation: $\mathfrak{F} \models \Psi$) if $(\mathfrak{F}, V), (w, a) \models \psi$ for all $\psi \in \Psi$, valuations V and pairs $(w, a) \in W \times A$. We define a class $\mathbb{M}_\Psi$ of models as $\{(\mathfrak{F}, V) \mid \mathfrak{F} \models \Psi\}$. While we omit the detail of the proof, we can obtain the following two theorems by similar arguments to **TEFL** and **HEFL**.

Theorem 5. *Let Λ be one of* **S4** *and* **S5** *and Θ be a possibly empty finite set of regular implications. The following are all equivalent:*

1. $\mathfrak{M}, f \models \Gamma \overset{\mathcal{I}}{\Rightarrow} \Delta$ *for all models* $\mathfrak{M} \in \mathbb{M}_{\Lambda \cup \Theta}$ *and all assignments* f.
2. $\Gamma \overset{\mathcal{I}}{\Rightarrow} \Delta$ *is provable in* $\mathbf{TEFL}(\Lambda; \Theta)^{-}$.
3. $\Gamma \overset{\mathcal{I}}{\Rightarrow} \Delta$ *is provable in* $\mathbf{TEFL}(\Lambda; \Theta)$.

Therefore, $\mathbf{TEFL}(\Lambda; \Theta)$ *enjoys the cut-elimination theorem. Moreover, for every formula* φ, φ *is valid in* $\mathbb{M}_{\Lambda \cup \Theta}$ *iff* φ *is provable in* $\mathbf{HEFL}(\Lambda \cup \Theta)$.

7 Further Directions

This paper positively answered the question if the set of all valid formulas of **EFL** in the class of all models is axiomatizable. We list some directions for further research.

1. Is **HEFL** or **TEFL** decidable?
2. Is it possible to provide a syntactic proof of the cut elimination theorem of **TEFL**?
3. Can we reformulate our sequent calculus into a G3-style calculus, i.e., a contraction-free calculus, all of whose rules are height-preserving invertible?
4. Provide a G3-style labelled sequent calculus for **EFL** based on the idea of doubly labelled formula $(x, y) : \varphi$. This is an extension of G3-style labelled sequent calculus for modal logic in [16,18].
5. Prove the semantic completeness of **HEFL** and its extensions by specifying the notion of *canonical model*.
6. Can we apply our technique of this paper to obtain a Hilbert-system of *Term Modal Logics* which is proposed in [10]?

Acknowledgments. I would like to thank the anonymous reviewers for their careful reading of the manuscript and their many useful comments and suggestions. I presented the contents of this paper first at the 48th MLG meeting at Kaga, Ishikawa, Japan on 6th December 2013 and then at Kanazawa Workshop for Epistemic Logic and its Dynamic Extensions, Kanazawa, Japan on 22nd February 2014. I would like to thank Jeremy Seligman and Fenrong Liu for fruitful discussions of the topic. All errors, however, are mine. The work of the author was partially supported by JSPS KAKENHI Grant-in-Aid for Young Scientists (B) Grant Number 15K21025 and JSPS Core-to-Core Program (A. Advanced Research Networks).

References

1. Areces, C., ten Cate, B.: Hybrid logics. In: Blackburn, P., van Benthem, J., Wolter, F. (eds.) Handbook of Modal Logic, pp. 821–868. Elsevier, Amsterdam (2007)
2. Baltag, A., Moss, L., Solecki, S.: The logic of public announcements, common knowledge and private suspicions. In: Proceedings of TARK, pp. 43–56. Morgan Kaufmann Publishers, Los Altos (1989)
3. Blackburn, P.: Arthur prior and hybrid logic. Synthese **150**(3), 329–372 (2006)
4. Blackburn, P., de Rijke, M., Venema, Y.: Modal Logic. Cambridge Tracts in Theoretical Computer Science. Cambridge University Press, Cambridge (2001)

5. Blackburn, P., ten Cate, B.: Pure extensions, proof rules, and hybrid axiomatics. Studia Logica **84**(3), 277–322 (2006)
6. Blackburn, P., Tzakova, M.: Hybrid completeness. Logic J. IGPL **6**(4), 625–650 (1998)
7. Braüner, T.: Hybrid Logic and Its Proof-Theory. Applied Logic Series, vol. 37. Springer, Dordrecht (2011). doi:10.1007/978-94-007-0002-4
8. Brünnler, K.: Deep sequent systems for modal logic. Arch. Math. Logic **48**, 551–577 (2009)
9. Christoff, Z., Hansen, J.U., Proetti, C.: Reflecting on social influence in networks. J. Logic Lang. Inf. **25**(3), 299–333 (2016)
10. Fitting, M., Thalmann, L., Voronkov, A.: Term-modal logics. Studia Logica **69**(1), 133–169 (2001)
11. Gargov, G., Passy, S., Tinchev, T.: Modal environment for Boolean speculations (preliminary report). In: Skordev, D. (ed.) Mathematical Logic and its Applications. Proceedings of the Summer School and Conference dedicated to the 80th Anniversary of Kurt Gödel, pp. 253–263. Plenum Press, Druzhba (1987)
12. Girard, P., Seligman, J., Liu, F.: General dynamic dynamic logic. In: Ghilardi, S., Bolander, T., Braüner, T., Moss, L.S. (eds.) Advances in Modal Logics, vol. 9, pp. 239–260. College Publications, London (2012)
13. Goldblatt, R.: Axiomatising the Logic of Computer Programming. LNCS, vol. 130. Springer, Heidelberg (1982)
14. Hintikka, J.: Knowledge and Belief: An Introduction to the Logic of the Two Notions. Cornell University Press, Cornell (1962)
15. Kashima, R.: Cut-free sequent calculi for some tense logics. Studia Logica **53**, 119–135 (1994)
16. Negri, S.: Proof analysis in modal logic. J. Philos. Logic **34**, 507–544 (2005)
17. Negri, S., Von Plato, J.: Structural Proof Theory. Cambridge University Press, Cambridge (2001)
18. Negri, S., Von Plato, J.: Proof Analysis. Cambridge University Press, Cambridge (2011)
19. Plaza, J.A.: Logics of public communications. In: Emrich, M.L., Pfeifer, M.S., Hadzikadic, M., Ras, Z.W. (eds.) Proceedings of the 4th International Symposium on Methodologies for Intelligent Systems, pp. 201–216 (1989)
20. Sano, K.: Axiomatizing hybrid products: How can we reason many-dimensionally in hybrid logic? J. Appl. Logic **8**(4), 459–474 (2010)
21. Seligman, J.: Internalization: the case of hybrid logics. J. Logic Comput. **11**(5), 671–689 (2001)
22. Seligman, J., Liu, F., Girard, P.: Logic in the community. In: Banerjee, M., Seth, A. (eds.) ICLA 2011. LNCS, vol. 6521, pp. 178–188. Springer, Heidelberg (2011). doi:10.1007/978-3-642-18026-2_15
23. Seligman, J., Liu, F., Girard, P.: Facebook and the epistemic logic of friendship. In: Proceedings of the 14th Conference on Theoretical Aspects of Rationality and Knowledge (TARK 2013), Chennai, India, pp. 230–238, 7–9 January 2013
24. Seligman, J., Liu, F., Girard, P.: Knowledge, friendship and social announcement. In: van Benthem, J., Liu, F. (eds.) Logic Across the University: Foundations and Applications. Studies in Logic, vol. 47, pp. 445–469. College Publications, London (2013)

The Dynamic Logic of Stating and Asking: A Study of Inquisitive Dynamic Modalities

Ivano Ciardelli(✉)

ILLC, University of Amsterdam, Amsterdam, The Netherlands
i.a.ciardelli@uva.nl

Abstract. Inquisitive dynamic epistemic logic (IDEL) extends public announcement logic incorporating ideas from inquisitive semantics. In IDEL, the standard public announcement action can be extended to a more general *public utterance* action, which may involve a statement or a question. While uttering a statement has the effect of a standard announcement, uttering a question typically leads to new issues being raised. In this paper, we investigate the logic of this general public utterance action. We find striking commonalities, and some differences, with public announcement logic. We show that dynamic modalities admit a set of reduction axioms, which allow us to turn any formula of IDEL into an equivalent formula of static inquisitive epistemic logic. This leads us to establish several complete axiomatizations of IDEL, corresponding to known axiomatizations of public announcement logic.

1 Introduction

Dynamic epistemic logics [2,10] allow us to reason about how an epistemic scenario evolves when certain actions are performed. The simplest kind of action considered in these logics is the *public announcement* of a formula [1,9,11,12]. When φ is publicly announced, all agents learn that φ was true at the time of the announcement, that other agents have also learned this, and so on. The action of announcing a formula φ is associated with a dynamic modality $[\varphi]$, which can be used to relativize a formula to the situation that results from an announcement of φ. The resulting dynamic logic, PAL, admits *reduction axioms*: these are a set of equivalences that allow us to recursively eliminate the dynamic modality, transforming each formula of PAL into a corresponding formula of static epistemic logic. A complete axiomatization of PAL is obtained combining the reduction axioms with a complete axiomatization of epistemic logic.

In recent years, PAL and other dynamic epistemic logics have been fruitfully employed to analyze processes of information exchange [2,10]. However, a typical information exchange process is not a mere sequence of announcements. Rather, it is a process in which certain issues are raised, addressed, and possibly resolved. Typically, issues are raised by asking questions, and resolved by making statements. Starting from this idea, Ciardelli and Roelofsen [8] introduced *inquisitive epistemic logic* (IEL). This framework describes not just the information that agents have, but

A. Baltag et al. (Eds.): LORI 2017, LNCS 10455, pp. 240–255, 2017.
DOI: 10.1007/978-3-662-55665-8_17

also the information that they would like to obtain, i.e., the issues they are interested in. Accordingly, the language of IEL can talk not just about the agent's information, but also about their issues. This is achieved by enriching classical logic with questions (drawing on inquisitive logic [4,5,7]) and allowing modalities to embed both statements and questions. While more expressive than Kripke modalities, the modalities of IEL remain logically very well-behaved. A sound and complete axiomatization of IEL was established in [3].

Ciardelli and Roelofsen [8] also generalized the standard account of public announcements to the inquisitive setting. In the resulting *inquisitive dynamic epistemic logic* (IDEL), agents can not only provide new information by publicly making a statement, but also raise new issues by publicly asking a question. Thus, IDEL provides the tools for a basic modeling of communication as a process in which agents interact by raising and resolving issues. As in PAL, the action of uttering a formula φ is associated with a dynamic modality $[\varphi]$, whose effect is to relativize a formula to the situation that results from the utterance of φ.

In this paper we investigate and axiomatize the resulting dynamic logic. We will show that the key features of PAL are preserved in IDEL, although there are also a few interesting differences. Like in PAL, it is possible to identify a set of logical equivalences by means of which any IDEL-formula can be turned into an equivalent formula of static IEL. In combination with the completeness result for IEL given in [3], this yields a complete axiomatization of IDEL. This provides a proof system that can be used to reason about how a communication scenario evolves when questions are publicly asked, or statements are publicly asserted.

The paper is structured as follows: Sects. 2 and 3 provide an introduction to IDEL. Sections 4 and 5 present novel results on the logic of dynamic modalities, which are used in Sect. 6 to establish a complete axiomatization. Section 7 concludes.

2 Inquisitive Epistemic Logic

In this section, we provide a concise introduction to the framework of inquisitive epistemic logic, following [4]. For discussion and proofs, we refer to [4,8].

The main ingredients for the semantics are the notions of *information states* and *issues*. The former is standard, while the latter stems from work on inquisitive semantics [6,8]. An information state is modeled extensionally by identifying it with the set of worlds compatible with it. Similarly, an issue is modeled extensionally by identifying it with the set of information states where it is *resolved*.

Definition 1 (States and issues). *If $\mathcal{W}$ is a set of possible worlds, then:*

- *an* information state *is a subset $s \subseteq \mathcal{W}$;*
- *an* issue *is a non-empty set I of information states that is* downward closed*: if $s \in I$ and $t \subseteq s$, then $t \in I$. The set of all issues is denoted by $\mathcal{I}$.*

The downward closure condition captures the fact that if an issue I is resolved in s, and if t contains all the information that s contains, then I is resolved in t. The maximal elements in an issue I will be called the *alternatives*. Notice that I can only be truthfully resolved at a world w if some resolving state $s \in I$ contains w, i.e., if $w \in \bigcup I$. We will say that I is an issue *over* the state $\bigcup I$.

Standard epistemic models describe situations which are determined by the truth value of certain primitive facts and by the knowledge of certain agents. Accordingly, a possible world w is fully described by (i) a propositional valuation $V(w)$, specifying which atomic sentences are true at w and (ii) for every agent a, an information state $\sigma_a(w)$, representing the knowledge of a in w. In *inquisitive* epistemic logic, what matters is not only the knowledge that agents have, but also the issues that they entertain. Thus, the description of a possible world w includes, for every agent a, an issue $\Sigma_a(w)$ over $\sigma_a(w)$, called the *inquisitive state* of a in w: s belongs to $\Sigma_a(w)$ iff all the issues entertained by a in w are resolved in s. Intuitively, we think of a as aiming to reach one of the states $s \in \Sigma_a(w)$. Since $\Sigma_a(w)$ is required to be an issue over $\sigma_a(w)$, we have $\sigma_a(w) = \bigcup \Sigma_a(w)$. Hence, the map Σ_a by itself encodes both a's knowledge and a's issues, and we do not need σ_a as a separate component in the model. As in standard epistemic logic, the maps Σ_a may be constrained by specific requirements. Following [8], we build on the strongest version of epistemic logic, which requires factivity and introspection, where the latter now concerns both information and issues.

Definition 2 (Inquisitive epistemic models). *An inquisitive epistemic model for a set $\mathcal{P}$ of atoms and a set $\mathcal{A}$ of agents is a triple $M = \langle \mathcal{W}, \Sigma_{\mathcal{A}}, V \rangle$ where $\mathcal{W}$ is a set (the possible worlds of the model), $V : \mathcal{W} \to \wp(\mathcal{P})$ is a* valuation map, *and $\Sigma_{\mathcal{A}} = \{\Sigma_a \mid a \in \mathcal{A}\}$ is a set of* inquisitive state maps *$\Sigma_a : \mathcal{W} \to \mathcal{I}$, each of which assigns to every world w an issue $\Sigma_a(w)$, in accordance with the following conditions, where $\sigma_a(w) := \bigcup \Sigma_a(w)$ represents the* epistemic state *of a in w.*

- *Factivity: for any $w \in \mathcal{W}$, $w \in \sigma_a(w)$*
- *Introspection: for any $w, v \in \mathcal{W}$, if $v \in \sigma_a(w)$, then $\Sigma_a(v) = \Sigma_a(w)$*

A useful way to draw the state map of an agent is illustrated in the figure below. At a world w, the epistemic state of the agent consists of those worlds included in the same dashed area as w; the solid blocks inside this area are the alternatives for the issue entertained at w—the maximal states in which the issue is resolved.

At w_1 and w_2, the agent's epistemic state is $\{w_1, w_2\}$; the alternatives for the issue entertained are $\{w_1\}$ and $\{w_2\}$, i.e., the agent is interested in whether w_1 or w_2 is actual. At w_3 and w_4, the agent's epistemic state is $\{w_3, w_4\}$, and the unique alternative for the issue entertained is $\{w_3, w_4\}$, i.e., the agent is not interested in acquiring any more specific information. The language of IEL, $\mathcal{L}^{\mathsf{IEL}}$, is given by the following syntax, where $p \in \mathcal{P}$ is an atomic sentence and $a \in \mathcal{A}$ an agent label:[1]

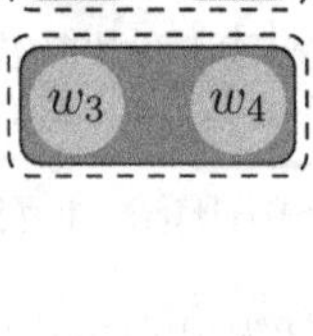

$$\varphi ::= p \mid \bot \mid \varphi \wedge \varphi \mid \varphi \to \varphi \mid \varphi \mathbin{\mathbb{V}} \varphi \mid K_a\varphi \mid E_a\varphi$$

[1] Most previous presentations of IEL ([3,8]) use a *dichotomous* language, in which formulas are divided into two syntactic categories: declaratives and interrogatives. The application of connectives is then subject to syntactic restrictions. Here we follow [4] in using a more general, non-dichotomous language; connectives apply without any restrictions, which leads to a more elegant logic. This difference is not an essential one; the results obtained here can be adapted to the dichotomous setting.

The non-standard items in this language are the connective $\mathbin{\rotatebox[origin=c]{-90}{$\geqslant$}}$, called *inquisitive disjunction* and the modality E_a, which we read as *entertain*. As we shall see, the former allows us to form questions, while the latter allows us to talk about the questions that an agent is interested in. We make use the following abbreviations: $\neg\varphi := \varphi \to \bot$; $\varphi \vee \psi := \neg(\neg\varphi \wedge \neg\psi)$; $\varphi \leftrightarrow \psi := (\varphi \to \psi) \wedge (\psi \to \varphi)$; $?\varphi := \varphi \mathbin{\rotatebox[origin=c]{-90}{$\geqslant$}} \neg\varphi$.

Usually, modal formulas are interpreted in terms of truth-conditions with respect to a possible world. However, the language of IEL comprises not only statements, but also questions, which are not naturally analyzed in terms of truth-conditions. Therefore, following inquisitive semantics [6], the semantics of IEL is given by a relation of *support* between formulas and information states.

Definition 3 (Support). *Let M be a model and s an information state in M.*

1. $M, s \models p \iff p \in V(w)$ *for all worlds* $w \in s$
2. $M, s \models \bot \iff s = \emptyset$
3. $M, s \models \varphi \wedge \psi \iff M, s \models \varphi$ *and* $M, s \models \psi$
4. $M, s \models \varphi \mathbin{\rotatebox[origin=c]{-90}{$\geqslant$}} \psi \iff M, s \models \varphi$ *or* $M, s \models \psi$
5. $M, s \models \varphi \to \psi \iff$ *for every* $t \subseteq s$, $M, t \models \varphi$ *implies* $M, t \models \psi$
6. $M, s \models K_a\varphi \iff$ *for every* $w \in s$, $M, \sigma_a(w) \models \varphi$
7. $M, s \models E_a\varphi \iff$ *for every* $w \in s$ *and every* $t \in \Sigma_a(w)$, $M, t \models \varphi$

The support-set of formula φ in model M is the set $[\varphi]_M := \{s \subseteq \mathcal{W} \mid M, s \models \varphi\}$. A key feature of the support relation is that it is *persistent*: that is, more formulas become supported as information grows: if $M, s \models \varphi$ and $t \subseteq s$, then $M, t \models \varphi$. As a limit case, $\emptyset$ supports all formulas: we refer to $\emptyset$ as the *inconsistent* state.

Although support at a state is the primitive semantic notion in IEL, truth at a world can be defined as support at the corresponding singleton.

Definition 4 (Truth). *We say that φ is* true *at a world w, notation $M, w \models \varphi$, in case $M, \{w\} \models \varphi$. The truth-set of φ in M is the set $|\varphi|_M = \{w \in \mathcal{W} \mid M, w \models \varphi\}$.*

Spelling out Definition 3 for a singleton, one can check that standard formulas receive the usual truth-conditions. The truth-conditions for modal formulas are:

- $M, w \models K_a\varphi \iff M, \sigma_a(w) \models \varphi$
- $M, w \models E_a\varphi \iff$ for every $t \in \Sigma_a(w)$, $M, t \models \varphi$

Notice that the truth-conditions of a modal formula $K_a\varphi$ or $E_a\varphi$ depend crucially on the support conditions of φ, and not just on its truth-conditions.

A formula is said to be *truth-conditional* in case support at an information state s boils down to truth at each world $w \in s$.

Definition 5 (Truth-conditionality). *A formula φ is* truth-conditional *if for all models M and information states s: $M, s \models \varphi \iff M, w \models \varphi$ for all $w \in s$.*

We refer to truth-conditional formulas as *statements* and to non-truth-conditional formulas as *questions* [4]. Intuitively, we read $s \models \varphi$ as "φ is established in s" if φ is a statement, and as "φ is settled in s" if φ is a question.

It is often possible to tell from the form of φ that it is truth-conditional. We define a set $\mathcal{L}_!^{\mathsf{IEL}}$ of *declaratives* as follows, where $\varphi \in \mathcal{L}^{\mathsf{IEL}}$ is any formula:

$$\alpha ::= p \mid \bot \mid K_a\varphi \mid E_a\varphi \mid \alpha \wedge \alpha \mid \varphi \to \alpha$$

In other words, φ is a declarative if the only occurrences of $\mathbin{\rotatebox[origin=c]{180}{$\mathbin{\bigtriangleup}$}}$ in φ, if any, are within the scope of a modality or in a conditional antecedent. Then we have:

Fact 1. *Any* $\alpha \in \mathcal{L}_!^{\mathsf{IEL}}$ *is truth-conditional.*

In particular, $\mathbin{\vee\!\!\!\vee}$−free formulas are truth-conditional; also, modal formulas are always truth-conditional, even when the argument of the modality is a question.

With this background in mind, let us turn to an illustration of the system. First consider a standard propositional formula α. It follows from Fact 1 that the meaning of α is completely determined by its truth-conditions, which are the standard ones. So, the meaning of α is essentially the same as in classical logic.

For an example of a question, consider the formula $?p := p \mathbin{\vee\!\!\!\vee} \neg p$. We have: $s \models ?p \iff s \models p \mathbin{\vee\!\!\!\vee} \neg p \iff s \models p$ or $s \models \neg p \iff s \subseteq |p|_M$ or $s \subseteq |\neg p|_M$. Thus, a state supports $?p$ if it implies either that p is true, or that p is false. Notice that, in any model containing both p-worlds and $\neg p$-worlds, the question $?p$ has two alternatives (two maximal supporting states) as shown in Fig. 1(d).

Now consider the modalities. Since Fact 1 ensures that modal formulas are always truth-conditional, in order to understand the semantics of $K_a\varphi$ and $E_a\varphi$ we need only look at truth-conditions. First, suppose that φ is a statement α. In this case, the two modalities coincide with each other and with the standard box modality of epistemic logic:

$$M, w \models K_a\alpha \iff M, w \models E_a\alpha \iff M, v \models \alpha \text{ for all } v \in \sigma(w)$$

Now consider the case in which φ is a question μ. Recall that, for a question, being supported amounts to being *settled*. Thus, the clauses read as follows:

- $K_a\mu$ is true at w if μ is settled in the epistemic state $\sigma_a(w)$ of agent a at w;
- $E_a\mu$ is true at w if μ is settled in any information state where a's issues at w are settled; that is, if settling the agent's issues implies settling μ.

E.g., $K_a?p$ is true iff the question $?p$ is settled in a's epistemic state—i.e., iff a *knows whether* p. By means of the modalities K_a and E_a, a *wondering* modality is defined: $W_a\varphi := \neg K_a\varphi \wedge E_a\varphi$. The idea is that a wonders about a question μ if her current epistemic state does not settle μ, but she wants to reach a state that does settle μ. As an illustration, consider the model in Fig. 1. At any world, agent a knows whether p ($K_a?p$); agent b doesn't know whether p, but wonders about it ($W_b?p$); agent c doesn't know, and does not wonder ($\neg K_c?p \wedge \neg W_c?p$).

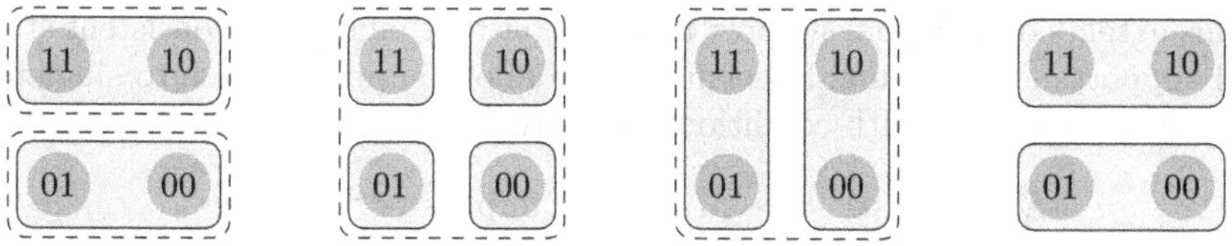

(a) Agent a (b) Agent b (c) Agent c (d) $?p$

Fig. 1. A model for three agents a, b, c, and the alternatives for $?p$.

3 Uttering Statements and Questions

In this section we review how the standard account of public announcements in epistemic logic can be lifted to a general account of *public utterance* in IEL [8].[2] Let us start out by specifying how an model changes as a result of the public utterance of a sentence φ. This works much like in PAL: when φ is uttered, the worlds in which φ is false are dropped from the model, and the state of each agent at a world is restricted by intersecting it with the support set $[\varphi]_M$.

Definition 6. (Updating a model). *The update of* $M = \langle W, \Sigma_{\mathcal{A}}, V\rangle$ *with* $\varphi \in \mathcal{L}^{\mathsf{IDEL}}$ *is the model* $M^\varphi = \langle W^\varphi, \Sigma^\varphi_{\mathcal{A}}, V^\varphi\rangle$*, where* $W^\varphi = W \cap |\varphi|_M$*,* $V^\varphi = V\restriction_{|\varphi|_M}$*, and* $\Sigma^\varphi_a(w) = \Sigma_a(w) \cap [\varphi]_M$*.*

The following fact says that the epistemic state $\sigma^\varphi_a(w) := \bigcup \Sigma^\varphi_a(w)$ of an agent at a world in the updated model is obtained just like in PAL, by restricting the original epistemic state $\sigma_a(w)$ to the set $|\varphi|_M$ of worlds where φ is true.

Fact 2. $\sigma^\varphi_a(w) = \sigma_a(w) \cap |\varphi|_M$

The language $\mathcal{L}^{\mathsf{IDEL}}$ of IDEL extends the static language of IEL by allowing us to conditionalize a formula to the utterance of another formula. Formally, we have:

$$\varphi ::= p \mid \bot \mid \varphi \wedge, \varphi \mid \varphi \to \varphi \mid \varphi \mathbin{\mathbb{V}} \varphi \mid K_a\varphi \mid E_a\varphi \mid [\varphi]\varphi$$

When talking of a formula $[\varphi]\psi$, we will refer to φ as the *label* of the dynamic modality $[\varphi]$, and to ψ as the *argument.* Semantically, assessing a sentence of the form $[\varphi]\psi$ at a model-state pair $\langle M, s\rangle$ requires assessing ψ at the pair $\langle M^\varphi, s \cap |\varphi|_M\rangle$ of the updated model M^φ and the restriction of s to this model. That is, the semantics of IDEL extends Definition 3 with the following clause:

$$M, s \models [\varphi]\psi \overset{def}{\iff} M^\varphi, s \cap |\varphi|_M \models \psi$$

[2] We use the neutral term *utterance* rather than *announcement* (used in [8]) because the latter suggests an informational interpretation. E.g., in IDEL, the utterance of a question such as $?p$ has the effect of raising the issue whether p. This should not be confused with the action of *announcing whether* p, i.e., announcing the true answer to the question $?p$, which is a more standard action of providing information.

For this extended language, support is still persistent, and the inconsistent state $\emptyset$ still supports any formula. Specializing the above support clause to singleton states, we recover the truth-conditions that are familiar from PAL:

$$M, w \models [\varphi]\psi \iff M, w \models \varphi \text{ implies } M^{\varphi}, w \models \psi$$

To familiarize with the effects of the public utterance action, consider first the utterance of a truth-conditional formula α. If I is an issue and s an information state, the *restriction of* I *to* s is the issue $I \upharpoonright s = \{t \cap s \mid t \in I\}$. Then, we have:

Fact 3. *If* $\alpha \in \mathcal{L}^{\mathsf{IDEL}}$ *is truth-conditional, then* $\Sigma_a^{\alpha}(w) = \Sigma_a(w) \upharpoonright_{|\alpha|_M}$

Thus, there is nothing more to the utterance of a statement α than there is in standard PAL: as a consequence of the utterance, worlds where α was false are removed from the model, and the agents' states are restricted accordingly.

Now consider the utterance of a question, say a basic polar question $?p$. Since $?p$ is true at all worlds, no world is removed from the model in the update. Moreover, by Fact 2 we have $\sigma_a^{?p}(w) = \sigma_a(w) \cap |?p|_M = \sigma_a(w)$, i.e., no knowledge is gained from an utterance of $?p$. However, the update changes the inquisitive state of an agent a, from $\Sigma_a(w)$ to $\Sigma_a^{?p}(w) = \Sigma_a(w) \cap [?p]_M = \{s \mid s \in \Sigma_a(w) \text{ and } M, s \models ?p\}$. This means that a's issues after the utterance of $?p$ become more demanding: to settle them, a state must settle a's previous issues, and in addition the question $?p$. In other words, as a result of the utterance, the agent will come to entertain the question $?p$. This illustrates how uttering a question typically results in new issues being raised. The effect of a sequence of utterances on the state of an agent is shown in Fig. 2.

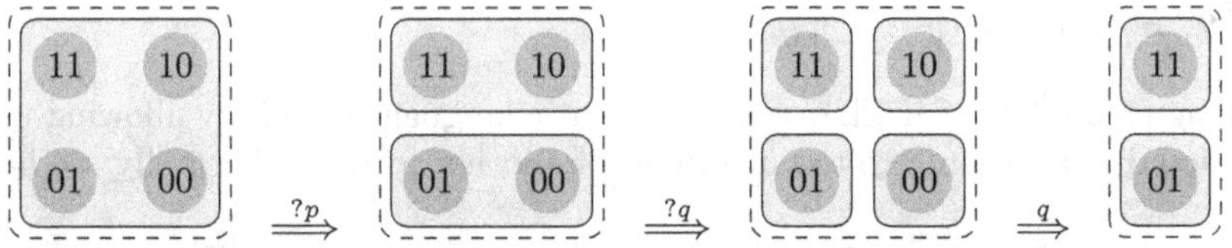

Fig. 2. The effects of a sequence of utterances on the state of an agent.

Let us now look at the significance of a dynamic formula of the form $[\varphi]\psi$. First, consider the special case in which ψ is a statement, i.e., truth-conditional. Then, the following fact ensures that the whole formula $[\varphi]\psi$ is also a statement.

Fact 4. *If* α *is truth-conditional, then so is* $[\varphi]\alpha$ *for any* φ.

Thus, the semantics of $[\varphi]\alpha$ is fully captured by the truth-conditions of $[\varphi]\alpha$. These truth-conditions are standard: $[\varphi]\alpha$ is true if φ is false, or φ is true and α is true in the model resulting from an update with φ. Thus, as in PAL, we can read $[\varphi]\alpha$ as stating that α would be the case after an announcement of φ. In particular, formulas from the language of PAL have the standard truth-conditions.

Now consider the case of a formula $[\varphi]\mu$, where μ is a question. In this case, $[\varphi]\mu$ can be thought of as a "dynamic conditional question", which asks to resolve μ under the assumption not just that φ is true, but that φ were uttered. As an example, consider the formula $[p]?K_a q$.

$$M, s \models [p]?K_a q \iff \begin{array}{l} \text{for all } w \in s \cap |p|_M, \sigma_a(w) \cap |p|_M \subseteq |q|_M \text{ or} \\ \text{for all } w \in s \cap |p|_M, \sigma_a(w) \cap |p|_M \not\subseteq |q|_M \end{array}$$

Thus, $[p]?K_a q$ is settled if it is established either (i) that, if p is true and if a were to learn that p, she would know that q, or (ii) that, if p is true and if a were to learn that p, she would still not know that q. Thus, $[p]?K_a q$ captures the question: "if p were publicly uttered, would a know that q?". And indeed, we have $[p]?K_a q \equiv [p]K_a q \mathbin{\mathbb{V}} [p]\neg K_a q$, which shows that $[p]?K_a q$ can be settled either by establishing that $[p]K_a q$, or by establishing that $[p]\neg K_a q$.

4 Normal Form

In this section, which marks the start of the novel material in the paper, we begin our study of the logic of IDEL by establishing a normal form result which will be very useful in further investigating the logic. The first step is to generalize to IDEL the notion of *declaratives* given above for IEL. The set $\mathcal{L}_!^{\mathsf{IDEL}}$ of declarative formulas of IDEL is defined as follows, where $\varphi \in \mathcal{L}^{\mathsf{IDEL}}$:

$$\alpha ::= p \mid \bot \mid K_a\varphi \mid E_a\varphi \mid [\varphi]\,\alpha \mid \alpha \wedge \alpha \mid \varphi \rightarrow \alpha$$

I.e., α is declarative if all occurrences of $\mathbb{V}$ in α are within (i) the scope of a static modality or (ii) the label of a dynamic modality or (iii) a conditional antecedent. The following can then be shown by means of a straightforward inductive proof.

Proposition 1. *Any declarative formula $\alpha \in \mathcal{L}_!^{\mathsf{IDEL}}$ is truth-conditional.*

Next, we associate with any formula $\varphi \in \mathcal{L}^{\mathsf{IDEL}}$ a set $\mathcal{R}(\varphi)$ of declaratives such that φ is equivalent with the inquisitive disjunction of the $\alpha \in \mathcal{R}(\varphi)$. This shows that any formula of IDEL is equivalent to an inquisitive disjunction of truth-conditional formulas. The inductive proof of the normal form result does not pose particular problems; we omit it here in the interest of space.

Definition 7 (Resolutions).

- $\mathcal{R}(\alpha) = \{\alpha\}$ *if α is of the form $p, \bot, K_a\varphi$ or $E_a\varphi$*
- $\mathcal{R}(\varphi \wedge \psi) = \{\alpha \wedge \beta \mid \alpha \in \mathcal{R}(\varphi)$ *and* $\beta \in \mathcal{R}(\psi)\}$
- $\mathcal{R}(\varphi \mathbin{\mathbb{V}} \psi) = \mathcal{R}(\varphi) \cup \mathcal{R}(\psi)$
- $\mathcal{R}(\varphi \rightarrow \psi) = \{\bigwedge_{\alpha \in \mathcal{R}(\varphi)} \alpha \rightarrow f(\alpha) \mid f : \mathcal{R}(\varphi) \rightarrow \mathcal{R}(\psi)\}$
- $\mathcal{R}([\varphi]\psi) = \{[\varphi]\alpha \mid \alpha \in \mathcal{R}(\psi)\}$

Theorem 1 (Normal form for IDEL).
Let $\varphi \in \mathcal{L}^{\mathsf{IDEL}}$ and let $\mathcal{R}(\varphi) = \{\alpha_1, \ldots, \alpha_n\}$. Then $\varphi \equiv \alpha_1 \mathbin{\mathbb{V}} \ldots \mathbin{\mathbb{V}} \alpha_n$.

5 Reduction

In this section, we will show that the presence of dynamic modalities does not make IDEL more expressive than its static fragment, IEL. As in PAL, any occurrence of a dynamic modality can be paraphrased away, inductively on the structure of the argument. This is easy if the argument is an atom or $\bot$.

Proposition 2. *$[\varphi]p \equiv \varphi \to p$ and $[\varphi]\bot \equiv \neg\varphi$*

Proof. Both $[\varphi]p$ and $\varphi \to p$ are declaratives, and thus truth-conditional by Proposition 1. Thus, to establish the equivalence we just have to show that these formulas have the same truth-conditions. This can be easily checked. The equivalence $[\varphi]\bot \equiv \neg\varphi$ is established by an analogous argument. □

As in PAL, dynamic modalities distribute smoothly over the connectives, which now also include inquisitive disjunction. The distribution over $\wedge$ and $\vee\!\!\!\vee$ is immediate to verify, while the distribution over $\to$ requires a proof.

Proposition 3. $[\varphi](\psi \wedge \chi) \equiv [\varphi]\psi \wedge [\varphi]\chi$

Proposition 4. $[\varphi](\psi \vee\!\!\!\vee \chi) \equiv [\varphi]\psi \vee\!\!\!\vee [\varphi]\chi$

Proposition 5. $[\varphi](\psi \to \chi) \equiv [\varphi]\psi \to [\varphi]\chi$

Proof. We have the following, where the crucial passage from the second to the third line is justified by the set-theoretic fact that the subsets of $s \cap |\varphi|_M$ are all and only the sets of the form $t \cap |\varphi|_M$ for some $t \subseteq s$.

$$\begin{aligned} M, s \models [\varphi](\psi \to \chi) &\iff M^\varphi, s \cap |\varphi|_M \models \psi \to \chi \\ &\iff \forall t \subseteq s \cap |\varphi|_M, \text{ if } M^\varphi, t \models \psi \text{ then } M^\varphi, t \models \chi \\ &\iff \forall t \subseteq s, \text{ if } M^\varphi, t \cap |\varphi|_M \models \psi \text{ then } M^\varphi, t \cap |\varphi|_M \models \chi \\ &\iff \forall t \subseteq s, \text{ if } M, t \models [\varphi]\psi \text{ then } M, t \models [\varphi]\chi \\ &\iff M, s \models [\varphi]\psi \to [\varphi]\chi \end{aligned}$$ □

From Propositions 2 and 5, we also get the standard reduction law for negation.

Corollary 1. $[\varphi]\neg\psi \equiv \varphi \to \neg[\varphi]\psi$

A dynamic modality $[\varphi]$ over a K modality behaves as in PAL: it can be brought within the scope of K, provided that we condition the resulting formula on φ.

Proposition 6. $[\varphi]K_a\psi \equiv \varphi \to K_a[\varphi]\psi$

Proof. Notice that both $\varphi \to K_a[\varphi]\psi$ and $[\varphi]K_a\psi$ are declaratives, and thus truth-conditional. Hence, to establish the equivalence we just have to show that they have identical truth-conditions. Making use of Fact 2, we have:

$$
\begin{aligned}
M, w \models [\varphi]K_a\psi &\iff M, w \models \varphi \text{ implies } M^\varphi, w \models K_a\psi \\
&\iff M, w \models \varphi \text{ implies } M^\varphi, \sigma_a^\varphi(w) \models \psi \\
&\iff M, w \models \varphi \text{ implies } M^\varphi, \sigma_a(w) \cap |\varphi|_M \models \psi \\
&\iff M, w \models \varphi \text{ implies } M, \sigma_a(w) \models [\varphi]\psi \\
&\iff M, w \models \varphi \text{ implies } M, w \models K_a[\varphi]\psi \\
&\iff M, w \models \varphi \to K_a[\varphi]\psi \qquad \square
\end{aligned}
$$

The law that allows us to push a dynamic modality through an E modality is only slightly more complex than the one for K. However, its proof, which relies on the normal form result, is rather difficult, and requires some preliminaries. For this reason, it is given in Appendix Appendix 1..

Proposition 7. $[\varphi]E_a\psi \equiv \varphi \to E_a(\varphi \to [\varphi]\psi)$

One may wonder whether this law could not be simplified. After all, in PAL, it holds generally that $\varphi \to [\varphi]\psi \equiv [\varphi]\psi$, because $[\varphi]\psi$ is always true in a world where φ is false. If this law held in IDEL as well, we could make the reduction law for the E_a modality simpler, and completely analogous to the one for K_a. This is indeed the case if φ is truth-conditional.

Proposition 8. *If* $\varphi \in \mathcal{L}^{IDEL}$ *is truth-conditional, then* $\varphi \to [\varphi]\psi \equiv [\varphi]\psi$

Proof. Suppose φ is truth-conditional, and consider any model M and state s. Using Lemma 3 in Appendix Appendix 1., we have:

$$
\begin{aligned}
M, s \models \varphi \to [\varphi]\psi &\iff M, s \cap |\varphi|_M \models [\varphi]\psi \iff M^\varphi, s \cap |\varphi|_M \cap |\varphi|_M \models \psi \\
&\iff M^\varphi, s \cap |\varphi|_M \models \psi \iff M, s \models [\varphi]\psi \qquad \square
\end{aligned}
$$

However, this does not extend to the case in which φ is a question. E.g., one can check by inspecting the support clauses that $?p \to [?p]?p \equiv \top$, while $[?p]?p \equiv ?p$. Thus, in general $\varphi \to [\varphi]\psi \not\equiv [\varphi]\psi$, and Proposition 7 cannot be simplified.

Equipped with Propositions 2–7, we are now ready to prove that any IDEL-formula is equivalent to some formula in the static language of IEL.

Theorem 2. *For any* $\varphi \in \mathcal{L}^{IDEL}$, $\varphi \equiv \varphi^*$ *for some* $\varphi^* \in \mathcal{L}^{IEL}$.

Proof. By induction on φ. The only non-obvious step is the one for $\varphi = [\psi]\chi$. By i.h., we have $\psi^*, \chi^* \in \mathcal{L}^{IEL}$ with $\psi \equiv \psi^*$, $\chi \equiv \chi^*$. It follows that $\varphi \equiv [\psi^*]\chi^*$. Now we just have to show that $[\psi^*]\chi^* \equiv \varphi^*$ for some $\varphi^* \in \mathcal{L}^{IEL}$. We proceed by induction on χ^*. Since $\chi^* \in \mathcal{L}^{IEL}$, we only have to consider the base cases and the inductive steps for connectives, K_a and E_a. Each of these cases corresponds to one of our equivalences above. As an illustration, we give the step for E.

- Suppose $\chi^* = E_a\xi$. By Proposition 7 we have $\varphi \equiv [\psi^*]\chi^* = [\psi^*]E_a\xi \equiv \psi^* \to E_a(\psi^* \to [\psi^*]\xi)$. Now, ξ is less complex than χ^*, so by induction hypothesis there is a $\xi^* \in \mathcal{L}^{IEL}$ such that $[\psi^*]\xi \equiv \xi^*$. Hence, $\varphi \equiv \psi^* \to E_a(\psi^* \to \xi^*)$. Since both ψ^* and ξ^* are in $\mathcal{L}^{IEL}$, so is the formula $\psi^* \to E_a(\psi^* \to \xi^*)$, which we can thus take to be the desired φ^*. □

6 Axiomatizing IDEL

We can use the reduction of IDEL to IEL to provide a complete axiomatization of IDEL. All we need to do is to enrich a complete system for IEL, like the natural deduction system in [3,4], with inference rules which allow us to perform the reduction. The easiest way to achieve this is to turn the equivalences given by Proposition 2–7 into inference rules, and to equip our system with a rule of replacement of equivalents. These rules are shown in Fig. 3.

!Atom	!⊥	!∧	!→
$\dfrac{[\varphi]p}{\varphi \to p}$	$\dfrac{[\varphi]\bot}{\neg\varphi}$	$\dfrac{[\varphi](\psi \wedge \chi)}{[\varphi]\psi \wedge [\varphi]\chi}$	$\dfrac{[\varphi](\psi \to \chi)}{[\varphi]\psi \to [\varphi]\chi}$
!⩔	**!K**	**!E**	**RE**
$\dfrac{[\varphi](\psi \mathbin{\rotatebox[origin=c]{180}{$\wedge$}\!\!\rotatebox[origin=c]{180}{$\wedge$}} \chi)}{[\varphi]\psi \mathbin{⩔} [\varphi]\chi}$	$\dfrac{[\varphi]K_a\psi}{\varphi \to K_a[\varphi]\psi}$	$\dfrac{[\varphi]E_a\psi}{\varphi \to E_a(\varphi \to [\varphi]\psi)}$	$\dfrac{\varphi \leftrightarrow \psi}{\chi[\varphi/p] \leftrightarrow \chi[\psi/p]}$

Fig. 3. Inference rules for dynamic modalities. All but the last rule are bi-directional.

We will denote derivability in this system by $\vdash_{\mathsf{IDEL}^{\mathsf{RE}}}$, and inter-derivability by $\dashv\vdash_{\mathsf{IDEL}^{\mathsf{RE}}}$. Using the results in the previous section, it is easy to show that this system is sound for IDEL. The next proposition states that this system can prove any formula $\varphi \in \mathcal{L}^{\mathsf{IDEL}}$ to be equivalent to some $\varphi^* \in \mathcal{L}^{\mathsf{IEL}}$. To prove this, we proceed exactly as in the proof of Theorem 2, but with $\equiv$ replaced by $\dashv\vdash_{\mathsf{IDEL}^{\mathsf{RE}}}$.

Proposition 9. *For any $\varphi \in \mathcal{L}^{IDEL}$, $\varphi \dashv\vdash_{IDEL^{RE}} \varphi^*$ for some $\varphi^* \in \mathcal{L}^{IEL}$.*

The completeness of our system follows immediately from this and from the fact that our system includes a complete system for IEL.

Theorem 3. *For any $\Phi \cup \{\psi\} \subseteq \mathcal{L}^{IDEL}$, $\Phi \models \psi \iff \Phi \vdash_{IDEL^{RE}} \psi$.*

The system described in Fig. 3 gives a simple axiomatization for IDEL, similar to the axiomatization of PAL given in [12]. As in the case of PAL, there are some interesting alternatives that we can use instead of replacement of equivalents to ensure that the system can perform the reduction to IEL. One option is to notice that, like in PAL, two dynamic modalities can always be merged into a single, complex one. This gives the following composition law for dynamic modalities.

Proposition 10 (Composition law). $[\varphi][\psi]\chi \equiv [\varphi \wedge [\varphi]\psi]\chi$

To show the validity of the composition law, we start by proving that a sequence of two updates can always be simulated by a single update, just as in PAL.

Lemma 1. *For any $\varphi, \psi \in \mathcal{L}^{\mathsf{IDEL}}$ and for any model M: $(M^\varphi)^\psi = M^{\varphi\wedge[\varphi]\psi}$*

Proof. We start out by showing that $(W^\varphi)^\psi$ and $W^{\varphi\wedge[\varphi]\psi}$ are the same. Notice that, if we take a world $w \in W$ such that $M, w \models \varphi$, then $M, w \models [\varphi]\psi \iff M^\varphi, w \models \psi$. This means that $|\varphi|_M \cap |[\varphi]\psi|_M = |\varphi|_M \cap |\psi|_{M^\varphi}$. So, we have:

$$\begin{aligned} W^{\varphi\wedge[\varphi]\psi} &= W \cap |\varphi \wedge [\varphi]\psi|_M = W \cap |\varphi|_M \cap |[\varphi]\psi|_M \\ &= W \cap |\varphi|_M \cap |\psi|_{M^\varphi} = W^\varphi \cap |\psi|_{M^\varphi} = (W^\varphi)^\psi \end{aligned}$$

This shows that the models $(M^\varphi)^\psi$ and $M^{\varphi\wedge[\varphi]\psi}$ share the same universe of possible worlds. Since the truth-value of a propositional atom at a world is not affected by updates, the two also have the same valuation function. We are left with showing that the two models have the same state map for each agent. Consider an agent a and a world w. Given an info state $s \in [\varphi]_M$, we have $s \subseteq \bigcup[\varphi]_M = |\varphi|_M$, so $s \cap |\varphi|_M = s$. So, for such a state s we have $M, s \models [\varphi]\psi \iff M^\varphi, s \cap |\varphi|_M \models \psi \iff M^\varphi, s \models \psi$. This shows that $[\varphi]_M \cap [[\varphi]\psi]_M = [\varphi]_M \cap [\psi]_{M^\varphi}$. Using this fact, we obtain:

$$\begin{aligned} \Sigma_a^{\varphi\wedge[\varphi]\psi}(w) &= \Sigma_a(w) \cap [\varphi \wedge [\varphi]\psi]_M = \Sigma_a(w) \cap [\varphi]_M \cap [[\varphi]\psi]_M \\ &= \Sigma_a(w) \cap [\varphi]_M \cap [\psi]_{M^\varphi} = \Sigma_a^\varphi(w) \cap [\psi]_{M^\varphi} = (\Sigma_a^\varphi)^\psi(w) \end{aligned}$$ □

Proof of Proposition 10. Using the results in the previous proof, we have:

$$\begin{aligned} M, s \models [\varphi][\psi]\chi &\iff M^\varphi, s \cap |\varphi|_M \models [\psi]\chi \\ &\iff (M^\varphi)^\psi, s \cap |\varphi|_M \cap |\psi|_{M^\varphi} \models \chi \\ &\iff M^{\varphi\wedge[\varphi]\psi}, s \cap |\varphi \wedge [\varphi]\psi|_M \models \chi \\ &\iff M, s \models [\varphi \wedge [\varphi]\psi]\chi \end{aligned}$$ □

We can turn the equivalence $[\varphi][\psi]\chi \equiv [\varphi \wedge [\varphi]\psi]\chi$ into a bidirectional rule, denoted !Comp. Now consider the proof system $\vdash_{\mathsf{IDEL}^{!\mathsf{Comp}}}$ which is like $\vdash_{\mathsf{IDEL}^{\mathsf{RE}}}$, except that the rule RE is substituted by !Comp. Theorem 4 states that this system is sound and complete for IDEL. A proof sketch is given in Appendix Appendix 2..

Theorem 4. *For any $\Phi \cup \{\psi\} \subseteq \mathcal{L}^{\mathsf{IDEL}}$, $\Phi \models \psi \iff \Phi \vdash_{\mathsf{IDEL}^{!\mathsf{Comp}}} \psi$.*

This is analogous to the axiomatization of PAL given in Theorem 7.26 of [10]. Finally, there is yet another alternative that we can use, instead of the rules RE and !Comp. This builds on the next proposition, which ensures that dynamic modalities are monotonic. The straightforward proof is omitted.

Proposition 11 (Monotonicity). *If $\psi \models \chi$, then $[\varphi]\psi \models [\varphi]\chi$.*

We can turn this logical property into an inference rule, that we will call !Mon:

!Mon: given $[\varphi]\psi$ and given a proof of χ from ψ, infer $[\varphi]\chi$.

Let us denote by $\vdash_{\mathsf{IDEL}^{!\mathsf{Mon}}}$ the system which is like $\vdash_{\mathsf{IDEL}^{\mathsf{RE}}}$, except that the rule of replacement of equivalents is substituted by !Mon. The next theorem says that this system, too, is complete for IDEL. A proof sketch is given in Appendix Appendix 2..

Theorem 5. *For any* $\Phi \cup \{\psi\} \subseteq \mathcal{L}^{\mathsf{IDEL}}$, $\Phi \models \psi \iff \Phi \vdash_{\mathsf{IDEL}^{\mathsf{IMon}}} \psi$.

This is analogous to the axiomatization of PAL given by Corollary 12 of [14]. Thus, in this section we have established three different complete axiomatizations for IDEL, each of which is the analogue of one known axiomatization of PAL.

7 Conclusion and Outlook

We have investigated the logic of public utterance in IDEL, and found that it has much in common with standard public announcement logic. Just like in PAL, dynamic modalities can be recursively eliminated, turning each formula into an equivalent static formula. We have exploited this fact to establish three axiomatizations of IDEL, corresponding to existing axiomatizations of PAL. The first one contains, on top of a complete system for inquisitive epistemic logic, the reduction rules and the rule of replacement of equivalents, RE. In the other axiomatizations, RE is not used; instead, we use two other features that IDEL shares with PAL: (i) that a sequence of two dynamic modalities can be reduced to a single dynamic modality and (ii) that dynamic modalities are monotonic.

These results are exciting, as they show that the standard analysis of public announcements, when lifted to an inquisitive semantics framework, can be generalized smoothly to an analysis that deals not only with the effect of making statements, but also with the effect of asking questions. This provides us with a well-behaved logic to reason about how a multi-agent situation evolves not only as a result of new incoming information, but also as a result of new issues being raised. Besides being interesting in its own right, this justifies a more general hope that the wealth of results developed in the field of dynamic epistemic logics can be extended to cover questions (for recent work in this direction, see [13]). The outcome of this would be a more comprehensive analysis of communication as a process in which agents interact by requesting and providing information.

Acknowledgment. Funding from the European Research Council (ERC, grant agreement number 680220) is gratefully acknowledged.

Appendix 1. Proof of the reduction law for *E*

To prove Proposition 7, we first need some lemmata. As a first step, we provide a characterization of the updated inquisitive state $\Sigma_a^{\varphi}(w)$ in terms of resolutions.

Lemma 2. *Let* $\varphi \in \mathcal{L}^{\mathsf{IDEL}}$ *and let* $\mathcal{R}(\varphi) = \{\alpha_1, \ldots, \alpha_n\}$. *Given any* M *and* w*:* $\Sigma_a^{\varphi}(w) = \Sigma_a(w)\!\upharpoonright_{|\alpha_1|_M} \cup \cdots \cup \Sigma_a(w)\!\upharpoonright_{|\alpha_n|_M}$

Proof. Theorem 1 ensures that $\varphi \equiv \alpha_1 \vee\!\!\!\vee \ldots \vee\!\!\!\vee \alpha_n$. Given the support clause for $\vee\!\!\!\vee$, this implies $[\varphi]_M = [\alpha_1]_M \cup \cdots \cup [\alpha_n]_M$. Thus, we have:

$$\begin{aligned}\Sigma_a^{\varphi}(w) &= \Sigma_a(w) \cap [\varphi]_M = \Sigma_a(w) \cap ([\alpha_1]_M \cup \cdots \cup [\alpha_n]_M) \\ &= \bigcup_{1 \le i \le n} (\Sigma_a(w) \cap [\alpha_i]_M) = \Sigma_a^{\alpha_1}(w) \cup \cdots \cup \Sigma_a^{\alpha_n}(w)\end{aligned}$$

Since resolutions are declaratives and thus truth-conditional, we have by Fact 3 that $\Sigma_a^{\alpha_i}(w) = \Sigma_a(w) \upharpoonright_{|\alpha_i|_M}$. Thus, we have $\Sigma_a^{\varphi}(w) = \bigcup_{1 \leq i \leq n} \Sigma_a(w) \upharpoonright_{|\alpha_i|_M}$. □

We will also make use of the next lemma, stating that whenever the antecedent of an implication is a truth-conditional formula α, the clause for implication can be simplified: $\alpha \rightarrow \psi$ is supported at s iff ψ is supported at the state $s \cap |\alpha|_M$.

Lemma 3. *If $\alpha \in \mathcal{L}^{IDEL}$ be truth-conditional. Then for any M, s, and ψ:* $M, s \models \alpha \rightarrow \psi \iff M, s \cap |\alpha|_M \models \psi$

Proof. If α is truth-conditional, then the subsets of s that support α are all and only the subsets of $s \cap |\alpha|_M$. Using this fact, the claim follows straightforwardly by the persistence of support. □

Finally, we will make use of the following equivalence, which can be established simply by spelling out the support conditions for the two formulas.

Lemma 4. *For any $\varphi, \psi, \chi \in \mathcal{L}^{IDEL}$, $(\varphi \vee\!\!\!\vee \psi) \rightarrow \chi \equiv (\varphi \rightarrow \chi) \wedge (\psi \rightarrow \chi)$*

Proof of Proposition 7. Let $\mathcal{R}(\varphi) = \{\alpha_1, \ldots, \alpha_n\}$. First, notice that, by Lemma 2, the information states $s \in \Sigma_a^{\varphi}(w)$ are all and only those of the form $s = t \cap |\alpha_i|_M$ for some $t \in \Sigma_a(w)$ and some $\alpha_i \in \mathcal{R}(\varphi)$. Since $\alpha_i \in \mathcal{R}(\varphi)$, it follows from Theorem 1 that $|\alpha_i|_M \subseteq |\varphi|_M$, whence $t \cap |\alpha_i|_M = t \cap |\alpha_i|_M \cap |\varphi|_M$.

Now suppose that $M, w \models \varphi$, so that w survives in the updated model M^{φ}. Making use of these facts, of Theorem 1, Lemmas 3 and 4, we have:

$$\begin{aligned}
M^{\varphi}, w \models E_a\psi &\iff \forall s \in \Sigma_a^{\varphi}(w), M^{\varphi}, s \models \psi \\
&\iff \forall t \in \Sigma_a(w), \forall \alpha_i \in \mathcal{R}(\varphi), M^{\varphi}, t \cap |\alpha_i|_M \models \psi \\
&\iff \forall t \in \Sigma_a(w), \forall \alpha_i \in \mathcal{R}(\varphi), M^{\varphi}, t \cap |\alpha_i|_M \cap |\varphi|_M \models \psi \\
&\iff \forall t \in \Sigma_a(w), \forall \alpha_i \in \mathcal{R}(\varphi), M, t \cap |\alpha_i|_M \models [\varphi]\psi \\
&\iff \forall t \in \Sigma_a(w), \forall \alpha_i \in \mathcal{R}(\varphi), M, t \models \alpha_i \rightarrow [\varphi]\psi \\
&\iff \forall t \in \Sigma_a(w), M, t \models (\alpha_1 \rightarrow [\varphi]\psi) \wedge \cdots \wedge (\alpha_n \rightarrow [\varphi]\psi) \\
&\iff \forall t \in \Sigma_a(w), M, t \models (\alpha_1 \vee\!\!\!\vee \ldots \vee\!\!\!\vee \alpha_n) \rightarrow [\varphi]\psi \\
&\iff \forall t \in \Sigma_a(w), M, t \models \varphi \rightarrow [\varphi]\psi \\
&\iff M, w \models E_a(\varphi \rightarrow [\varphi]\psi)
\end{aligned}$$

Finally, using this equivalence we get, for any model M and world w:

$$\begin{aligned}
M, w \models [\varphi]E_a\psi &\iff M, w \models \varphi \text{ implies } M^{\varphi}, w \models E_a\psi \\
&\iff M, w \models \varphi \text{ implies } M, w \models E_a(\varphi \rightarrow [\varphi]\psi) \\
&\iff M, w \models \varphi \rightarrow E_a(\varphi \rightarrow [\varphi]\psi)
\end{aligned}$$

We have thus proved that $[\varphi]E_a\psi$ and $\varphi \rightarrow E_a(\varphi \rightarrow [\varphi]\psi)$ have the same truth-conditions. Since both formulas are declaratives, and thus truth-conditional by Proposition 1, this ensures that these formulas are equivalent. □

Appendix 2. Proof of completeness via !Comp and !Mon

Proof of Theorem 4. The proof is analogous to the one in Sect. 7.4 of [10] for PAL. We only provide a proof sketch. We first define a complexity measure as follows:

- $c(p) = c(\bot) = 1$
- $c(\varphi \wedge \psi) = c(\varphi \to \psi) = c(\varphi \vee\!\!\!\vee \psi) = 1 + \max(c(\varphi), c(\psi))$
- $c(K_a\varphi) = c(E_a\varphi) = 1 + c(\varphi)$
- $c([\varphi]\psi) = (4 + c(\varphi)) \cdot c(\psi)$

By recursion on this notion of complexity, we define a map $(\cdot)^* : \mathcal{L}^{\mathsf{IDEL}} \to \mathcal{L}^{\mathsf{IEL}}$:

$$\begin{array}{ll}
p^* = p, \quad \bot^* = \bot & ([\varphi](\psi \wedge \chi))^* = ([\varphi]\psi \wedge [\varphi]\chi)^* \\
(\varphi \wedge \psi)^* = \varphi^* \wedge \psi^* & ([\varphi](\psi \to \chi))^* = ([\varphi]\psi \to [\varphi]\chi)^* \\
(\varphi \to \psi)^* = \varphi^* \to \psi^* & ([\varphi](\psi \vee\!\!\!\vee \chi))^* = ([\varphi]\psi \vee\!\!\!\vee [\varphi]\chi)^* \\
(\varphi \vee\!\!\!\vee \psi)^* = \varphi^* \vee\!\!\!\vee \psi^* & ([\varphi]K_a\psi)^* = (\varphi \to K_a[\varphi]\psi)^* \\
([\varphi]p)^* = (\varphi \to p)^* & ([\varphi]E_a\psi)^* = (\varphi \to E_a(\varphi \to [\varphi]\psi))^* \\
([\varphi]\bot)^* = (\neg\varphi)^* & ([\varphi][\psi]\chi)^* = ([\varphi \wedge [\varphi]\psi]\chi)^*
\end{array}$$

We can then easily prove $\varphi \dashv\vdash_{\mathsf{IDEL}^{!\mathsf{Comp}}} \varphi^*$, using the reduction rules and !Comp. Completeness then follows since $\vdash_{\mathsf{IDEL}^{!\mathsf{Comp}}}$ includes a complete system for IEL. □

Proof of Theorem 5. The proof is similar to the previous one, and to the proof of the analogous result for PAL in [14]. We modify the above definition of $(\cdot)^*$ by setting $([\varphi][\psi]\chi)^* = ([\varphi]([\psi]\chi)^*)^*$. By induction on the complexity of a formula (as defined above), we show that (i) φ^* is well-defined; (ii) if $\varphi \notin \mathcal{L}^{\mathsf{IEL}}$, then $c(\varphi^*) < c(\varphi)$; and (iii) $\varphi \dashv\vdash_{\mathsf{IDEL}^{!\mathsf{Mon}}} \varphi^*$. The only case which is not straightforward is the inductive step for a formula $[\varphi][\psi]\chi$, which I will spell out in detail.

For (i), notice that since $[\psi]\chi$ is less complex than $[\varphi][\psi]\chi$, by induction hypothesis we have that $([\psi]\chi)^*$ is well-defined and less complex than $[\psi]\chi$. It follows that $[\varphi]([\psi]\chi)^*$ is less complex than $[\varphi][\psi]\chi$. So, the induction hypothesis implies that $([\varphi]([\psi]\chi)^*)^*$ is well-defined, i.e., that $([\varphi][\psi]\chi)^*$ is well-defined.

For (ii), as both $[\psi]\chi$ and $[\varphi]([\psi]\chi)^*$ are less complex than $[\varphi][\psi]\chi$, using the induction hypothesis we have $c(([\varphi][\psi]\chi)^*) = c(([\varphi]([\psi]\chi)^*)^*) < c([\varphi]([\psi]\chi)^*) < c([\varphi][\psi]\chi)$. So, $c(([\varphi][\psi]\chi)^*) < c([\varphi][\psi]\chi)$.

For (iii), as $[\psi]\chi$ is less complex than $[\varphi][\psi]\chi$, the induction hypothesis gives $[\psi]\chi \dashv\vdash_{\mathsf{IDEL}^{!\mathsf{Mon}}} ([\psi]\chi)^*$. By two applications of the rule !Mon we get $[\varphi][\psi]\chi \dashv\vdash_{\mathsf{IDEL}^{!\mathsf{Mon}}} [\varphi]([\psi]\chi)^*$. Now, since $[\varphi]([\psi]\chi)^*$ is less complex than $[\varphi][\psi]\chi$, the induction hypothesis applies, and gives $[\varphi]([\psi]\chi)^* \dashv\vdash_{\mathsf{IDEL}^{!\mathsf{Mon}}} ([\varphi]([\psi]\chi)^*)^*$. Putting things together, we have obtained $[\varphi][\psi]\chi \dashv\vdash_{\mathsf{IDEL}^{!\mathsf{Mon}}} ([\varphi]([\psi]\chi)^*)^*$ which is what we need, since by definition $([\varphi]([\psi]\chi)^*)^* = ([\varphi][\psi]\chi)^*$. □

References

1. Baltag, A., Moss, L., Solecki, S.: The logic of public announcements, common knowledge, and private suspicions. In: Proceedings of TARK 7, pp. 43–56. Morgan Kaufmann Publishers (1998)
2. van Benthem, J.: Logical dynamics of information and interaction. Cambridge University Press, Cambridge (2011)
3. Ciardelli, I.: Modalities in the realm of questions: axiomatizing inquisitive epistemic logic. In: Goré, R., Kooi, B., Kurucz, A. (eds.) Advances in Modal Logic, pp. 94–113. College Publications, London (2014)
4. Ciardelli, I.: Questions in logic. PhD thesis, University of Amsterdam (2016)
5. Ciardelli, I., Groenendijk, J., Roelofsen, F.: On the semantics and logic of declaratives and interrogatives. Synthese **192**(6), 1689–1728 (2015)
6. Ciardelli, I., Groenendijk, J., Roelofsen, F.: Inquisitive Semantics. Oxford University Press, Oxford (2017, To appear)
7. Ciardelli, I., Roelofsen, F.: Inquisitive logic. J. Philos. Log. **40**(1), 55–94 (2011)
8. Ciardelli, I., Roelofsen, F.: Inquisitive dynamic epistemic logic. Synthese **192**(6), 1643–1687 (2015)
9. van Ditmarsch, H.: Knowledge Games. PhD thesis, Groningen University (2000)
10. van Ditmarsch, H., van der Hoek, W., Kooi, B.: Dynamic Epistemic Logic. Springer, Netherlands (2007)
11. Gerbrandy, J., Groeneveld, W.: Reasoning about information change. J. log. Lang. Inf. **6**(2), 147–169 (1997)
12. Plaza, J.: Logics of public communications. In: Emrich, M.L., Pfeifer, M.S., Hadzikadic, M., Ras, Z.W. (eds) Proceedings of the Fourth International Symposium on Methodologies for Intelligent Systems, pp. 201–216. Oak Ridge National Laboratory (1989)
13. van Gessel, T.: Action models in inquisitive logic. MSc thesis, University of Amsterdam (2016)
14. Wang, Y., Cao, Q.: On axiomatizations of public announcement logic. Synthese **190**(1), 103–134 (2013)

The Stubborn Non-probabilist—'Negation Incoherence' and a New Way to Block the Dutch Book Argument

Leszek Wroński[1] and Michał Tomasz Godziszewski[2](✉)

[1] Jagiellonian University, Kraków, Poland
leszek.wronski@uj.edu.pl
[2] University of Warsaw, Warsaw, Poland
mtgodziszewski@gmail.com

Abstract. We rigorously specify the class of nonprobabilistic agents which are, we argue, immune to the classical Dutch Book argument. We also discuss the notion of expected value used in the argument as well as sketch future research connecting our results to those concerning incoherence measures.

1 Introduction

Suppose you decide that your first task on a sunny Tuesday morning is to convince your friend who does not subscribe to probabilism (that is, he claims his degrees of belief need not be classical probabilities) of the error of his ways[1]. You decide to try the classical Dutch Book argument first. To your surprise you discover that your friend is not worried about the somewhat pragmatic nature of the argument, allows you to set all the stakes to 1 for convenience, and, while claiming that the set of propositions about which he holds some degree of belief is finite, he is eager to contemplate betting on virtually anything. He also considers a bet to be fair if its expected profit both for the buyer and seller is null, and even accepts the 'package principle', that is, believes a set of bets to be fair if each of the bets in that set is fair. Knowing all that, when telling your friend about how fair betting quotients are connected with the Kolmogorov axioms, and then about the identification of fair betting quotients with degrees of belief, you expect him to be immediately convinced.

To your surprise he shakes his head in opposition, saying 'I agree that fair betting quotients are exactly those which satisfy the axioms of classical probability. Still, even when we set all stakes to 1, I don't believe that these quotients are my degrees of belief.'

'But Alan', you say, 'this is standard. We went through this. We agreed that if your degree of belief in A is $b(A)$, and your degree of belief in $\neg A$ is $b(\neg A)$, then your betting quotient for the bet for A is that particular q for which the expression $b(A) \cdot (1-q) + b(\neg A) \cdot (-q)$, that is, what you expect to be the value

[1] Why Tuesday? See [7].

A. Baltag et al. (Eds.): LORI 2017, LNCS 10455, pp. 256–267, 2017.
DOI: 10.1007/978-3-662-55665-8_18

of the bet, equals 0. And it is a matter of mundane calculation that q is exactly $b(A)$. In general this means that betting quotients are your degrees of belief.'

'Still, look' – your friend responds – 'you're missing one thing. It's just that in my case $b(A) + b(\neg A)$ is in general not equal to 1. My degrees of belief are such that for each proposition A there is a non-zero number r_A for which it holds that $b(A) + b(\neg A) = r_A$; some of those numbers may be equal to 1, but none need be. And so my betting quotient for the bet for a proposition A is in general $^{b(A)}/_{r_A}$. Can you run your argument using such quotients?'

Well, *can you*? It turns out that sometimes you can – but sometimes not. It all depends on the particulars of your friend's belief state. In what follows we will specify the formal details. Notice that the way the story is set up, our friend has granted you the assumptions needed to overcome the well known flaws of the Dutch Book argument (discussed e.g. in [1,17]). Still, it seems that even then he needs not be persuaded by the reasoning. This suggests that we have here a problem for Dutch Book arguments.

The feature of a belief function described above, which we take as suggesting a way in which a nonprobabilist can resist the Dutch Book argument, was first described in print by [8] and called "negation incoherence". In this paper we go further:

- by saying something new regarding why and when a nonprobabilistic agent might not violate the norm of rationality appealed to in Dutch Book arguments;
- more explicitly, by proving a theorem describing the class of those nonprobabilistic agents which are, we believe, immune to Dutch Book arguments,
- and lastly, by discussing some different ways in which an incoherent agent might approach the task of calculating the expected value of some bet.

We take negation incoherence, then, as another reason for which "betting odds and credences come apart" ([2,13]), but which, let us note here, has nothing to do with the issues related to self-location (for the root of a big part of modern literature on that subject see [4]). The key to our idea is that while we see nothing wrong with the classical Dutch Book *theorems*, they concern betting quotients (or odds), while the Dutch Book *argument* tries to establish something about credences. If there are situations in which these two "come apart", that is, should not simply be identified with each other, we should either say that the argument is not applicable (which may be a sensible road to take in the face of self-location problems) or try to first establish a rigorous link between them and then to reevaluate the fate of the Dutch Book argument. The latter is the way we have chosen for this article.[2]

Simply *assuming* that degrees of belief are to be identified with betting quotients would amount to adopting some kind of operational approach to credences, with the details depending on how the understanding of the quotients would be fleshed out. We see little gain from this, aside from a short-lived satisfaction at a spurious connection to empirical matters. We are motivated rather by the spirit

[2] We would like to thank one of the Reviewers for pressing us on this.

of [5]; that is, we try to keep an open mind regarding what degrees of belief *are*, and investigate the relationship between them and betting quotients on a single basic assumption: that whatever they are, they can be expressed by a real number.

2 Details

Notice first that assuming that in general $b(A) + b(\neg A) = 1$ does not amount to assuming the probabilist thesis, that is, the problem is not that of pure *petitio principii.* Still, by doing so we are assuming something with which a nonprobabilist may by no means agree. We just know that by denying it, he has to hold that the additivity axiom or the normalization axiom (stating that the probabilities of tautologies equal 1) is not satisfied by his degrees of belief.

We can arrive at the problem from another direction. The traditional way of looking at the Dutch Book argument for probabilism would have it imply that possessing degrees of belief which violate classical probability axioms is a mark of irrationality. This should be puzzling if we think about the particular form of the 'normalization' axiom used in the classical axiomatization of probability. If we believe tautologies to a degree different from 1, we can apparently be Dutch-booked. Surely there's a mistake here: the choice of the number 1 as the probability of tautologies is purely conventional. The number 2 (say) would do just as well. But if we are careful about setting the betting quotients the way with which our nonprobabilistic friend would agree, then if his degree of belief in countertautologies is 0 and his degree of belief in tautologies is 2, his betting quotient for tautologies is 1, exactly the same as in the classical case.

Let us continue towards the theorem specifying the class of cases in which a nonprobabilist is not Dutch-bookable. In this paper we confine our attention to finite structures.

One of the main points of this paper is to identify/determine the conditions under which it is possible to link betting quotients with credences while arguing for probabilism. We will see that the exact identification of betting quotients and credences is possible when the agent is not "negation-incoherent", and so it should not be surprising that mathematically the betting quotient functions and the credence (degrees of belief) functions are objects of the same type. That is, they are functions from an algebra of events (propositions) defined over a given set (interpreted as a set of possible worlds, sample space or whatnot). The difference between these functions, argumentation-wise, lies in their interpretations and is justified by the way the degrees of belief of a given agent induce betting quotients in betting scenarios via the condition of fairness of bets. All this should be clear by the time the Reader reaches Definition 4 below. Let us start with the basic notions of belief and betting spaces:

Definition 1 (Belief space, betting space). *A belief (betting) space is a tuple* $\langle W, Prop, b\rangle$, ($\langle W, Prop, q\rangle$) *where* W *is a nonempty finite set,* $Prop$ *is a Boolean algebra of subsets of* W *('propositions'), and* b (q) *is a function from* $Prop$ *to* $\mathbb{R}$, *called the* belief function (betting quotient function).

In what follows we always assume that we are given a (finite) set W and a Boolean algebra $Prop \subseteq \mathcal{P}(W)$ of subsets of W ('propositions').

Let us now provide an intuitive description of the concept of betting quotients. We say that a bet on a proposition $A \in Prop$ consists of a stake $s(A)$ and a price $p(A)$ (both real numbers) considered by the agent to be fair for a bet regarding A with that particular stake ('fair' as in 'not favouring either side'), as well as the agent's payoffs: $s(A) - p(A)$ in the case A is true, and $-p(A)$ in the case A is false. Intuitively, the agent's betting quotient for A equals $\frac{p(A)}{s(A)}$, and is simply the price of a bet with the unit stake $s(A) = 1$ considered by the agent as fair. On our account the betting quotient is attached to a proposition, and so it is not the price $p(A)$ that the betting quotient depends on, but rather the other way round: the price that the agent considers fair is determined by her betting quotients and the announced stake. Therefore, it is already here that the Reader might observe that a thing crucial for an accurate interpretation of Dutch Book scenarios is understanding under what conditions the agent considers a given price of a given bet to be fair.

With the interpretation of the betting quotient function in hand, we are in the position to state the formal definition of Dutch Books and recall the Dutch Book Theorems that seem to constitute the main engine of Dutch Book Arguments.

Definition 2 (Dutch Book). *Let W be a non-empty (finite) set and let $\mathcal{F} \subseteq \mathcal{P}(W)$ be a Boolean algebra of its subsets. Let $q : \mathcal{F} \to \mathbb{R}$ be a real-valued function. We say that q is* ***susceptible to a Dutch Book (is Dutch-Bookable; permits a Dutch Book)*** *if there exists a function $s : \mathcal{F} \to \mathbb{R}$ and $\mathcal{F}_0$, a finite non-empty subset of $\mathcal{F}$, such that for any $w \in \bigcup \mathcal{F}$ the following inequality holds:*

$$U(w) = \sum_{E \in \mathcal{F}_0 : w \in E} (1 - q(E))s(E) - \sum_{E \in \mathcal{F}_0 : w \notin E} q(E)s(E) = \sum_{E \in \mathcal{F}_0} (\chi_E(w) - q(E))s(E) < 0,$$

where χ_E is the characteristic function of the set E.

We can also say in such a case that the betting space $\langle W, \mathcal{F}, q \rangle$ permits a Dutch Book, or that it is Dutch-bookable, or that an agent with a betting quotient function q is susceptible to a Dutch Book (is Dutch-bookable).[3]

Definition 3 (Classical probability function (finite)). *A function p from $Prop$ to $\mathbb{R}$ is a* classical probability function *if it satisfies the following three axioms:*

1. *$p(W) = 1$ (the normalization axiom),*
2. *for any A in $Prop$ $p(A) \geq 0$ (the non-negativity axiom),*
3. *for any A and B whose intersection is empty $p(A \cup B) = p(A) + p(B)$ (the additivity axiom).*

Theorem 1 (Dutch Book Theorem - [9,10]). *A betting quotient function is not Dutch-bookable iff it is a classical probability function.*

[3] For a detailed discussion of defining Dutch Books in the more general context of (possibly) nonclassical spaces, including a detailed discussion of the formula for $U(w)$, see [18].

Let us notice that it is crucial to distinguish between:

- the Dutch Book Theorem (DBT), which is an established mathematical result, and
- the Dutch Book Argument (DBA) for probabilism, which only uses the DBT as one of its premises.

Let us consider that direction of the DBA which aims to establish that violating probabilism leads to violating some norm of rationality. The structure of the argument is usually as follows:

1. Assume that a given agent's belief function violates the classical probability axioms.
2. Identify the agent's credences with her betting quotients, that is, the quotients of bets fair according to her[4]—this means that the agent's betting quotient function violates probability axioms.
3. By the Dutch Book Theorem such the agent is guaranteed a sure loss.
4. Ergo, the agent's degrees of belief are irrational.

As the Reader sees in point 2, the argument identifies the degrees of belief with the betting quotients. This might seem close to obvious, as for instance [2] claim[5]:

> "All we need is for there to be a normative link between the belief and the bet. Something like 'Other things being equal (risk-neutral, utility linear with money, ...), an agent who accepts E with 50% certainty is rationally permitted to accept a bet on E that pays twice the stake or better'. This link is broadly accepted, and will be all we need."

What we intend to show in this paper is that the *broad acceptance* reported in the quote above actually deserves serious and careful scrutiny—we hope to demonstrate that it should actually be rejected and although there is a link between the credence and the quotient, it by no means has to be identity in all cases.

The only constraint that we have with respect to the nature of the above-mentioned link is to make sure that the agent expects the value of the bet for A to be 0 (assume all stakes are set to 1; nothing important in the argument depends on that) which is a natural explication of fairness of a given bet: it does not favour any of the sides. Thus, what we need to guarantee, while linking the belief function b to the betting quotient function q, is that for any event A:

$$b(A)(1 - q(A)) - b(\neg A)q(A) = 0.$$

[4] Some variation is possible at this point. Some may prefer to speak instead about bets the agent would accept. This will not be important for the topic of our paper.

[5] Note, though, that the authors talk about and agent being permitted to accept a bet if she does not expect her own loss, and so they do not use the concept of fairness as not favouring either side. This is tangential to our argument.

Notice that as mentioned above, normalization and additivity imply that for any A in *Prop* $b(A) + b(\neg A) = 1$, that is, they imply the assumption we need for the 'classical' connection between degrees of belief and betting quotients, i.e. their identification. In our case we wish to play by our friend's rules, that is, for any A, we want to set the betting quotient for A to $b(A)/(b(A)+b(\neg A))$: this way we will make sure that indeed our friend expects the value of the bet for A to be 0. Therefore, we may then define:

Definition 4 (Induced betting quotient). *A belief space* $\langle W, Prop, b\rangle$ *induces a betting quotient* $q : Prop \to \mathbb{R}$ *if for any* $A \in Prop$*:*

1. $b(A) + b(\neg A) \neq 0$,
2. $q(A) = \frac{b(A)}{b(A)+b(\neg A)}$.

Defined this way, $q(A)$ is the betting quotient which makes a bet for or against A such that an agent with a belief function b expects it to have value 0.[6] It follows that if a belief space induces a betting quotient function, that is, if the first condition of the above definition holds, then that function is unique.

The question which now arises is the following: are there any non-probabilistic epistemic agents (i.e. such that their degrees credences violate the classical probability axioms) that are not susceptible to Dutch Books, i.e. such that their betting quotients (induced by their credences) are not Dutch-Bookable? The answer is given by the following simple theorem:

Theorem 2. *The betting quotient function* q *induced by a belief space* $\langle W, Prop, b\rangle$ *is a classical probability function iff the following conditions hold:*

1. $b(\emptyset) = 0$,
2. *for any* A *in* *Prop* $b(A) \cdot b(\neg A) \geq 0$,
3. *for any* A *and* B *in* *Prop* *with an empty intersection:*

$$\frac{b(A \cup B)}{b(A \cup B) + b(\neg(A \cup B))} = \frac{b(A)}{b(A) + b(\neg A)} + \frac{b(B)}{b(B) + b(\neg B)}.$$

Proof. Let q be the belief quotient function induced by a belief space $\langle W, Prop, b\rangle$.

($\Rightarrow$) Assume q is a classical probability function. By the normalization axiom $q(W) = 1$, and by the additivity axiom $q(\emptyset \cup W) = q(W) = q(W) + q(\emptyset)$, so $q(\emptyset) = 0$. Thus $b(\emptyset) = q(\emptyset) \cdot (b(\emptyset) + b(W)) = 0$.

Let $A \in Prop$. By the definition of the induced betting quotient we have $b(A) \cdot b(\neg A) = [q(A) \cdot (b(A) + b(\neg A))] \cdot [q(\neg A) \cdot (b(A) + b(\neg A))] = q(A) \cdot q(\neg A) \cdot (b(A) + b(\neg A))^2$. As the function q is non-negative and we multiply the square of $(b(A) + b(\neg A))$, the value of the entire expression is ≥ 0.

The last condition holds since it basically says that $q(A) + q(B) = q(A \cup B)$ for disjoint sets in *Prop*, which is guaranteed by the additivity axiom.

($\Leftarrow$) Assume the conditions form the statement of the theorem hold. Trivially, since $b(\emptyset) = 0$ and q is induced by b, it holds that $q(\emptyset) = 0$ and $q(W) = 0 + \frac{b(W)}{b(W)+b(\emptyset)} = 0 + \frac{b(W)}{b(W)+0} = 0 + 1 = 1$, so normalization holds.

[6] For more regarding the notion of expected value see Sect. 4 below.

Let $A \in Prop$. We have $q(A) = \frac{b(A)}{b(A)+b(\neg A)}$, and $b(A)$ and $b(\neg A)$ are of the same sign (or one of them is equal to 0). Thus, both the counter and the denominator of the formula defining $q(A)$ are of the same sign as well (or the counter is equal to 0). Thus, $q(A) \geq 0$.

The additivity of q follows trivially from the formulation of the third condition. □

It is worthwhile to reflect on which steps of the above proof depend on what properties of the classical probability measure on the one hand, and the induced betting quotient on the other. If b satisfies the conditions in the statement of the theorem, then the normalization axiom is implied just by the fact that $b(\emptyset) = 0$. Additivity is immediate in both directions of the reasoning. On the other hand, the same sign of the belief function on complementary events follows from the non-negativity of q, if the latter is the quotient function induced by b. However, the value $b(\emptyset) = 0$ follows from (the conjunction of) normalization and additivity of q. That is, it is not the case that the conditions in the statement of the theorem correspond directly to the respective probability axioms.

3 Discussion

To see an example of a Dutch-bookable, nonprobabilist belief space, consider the space with three atomic propositions depicted in Fig. 1.

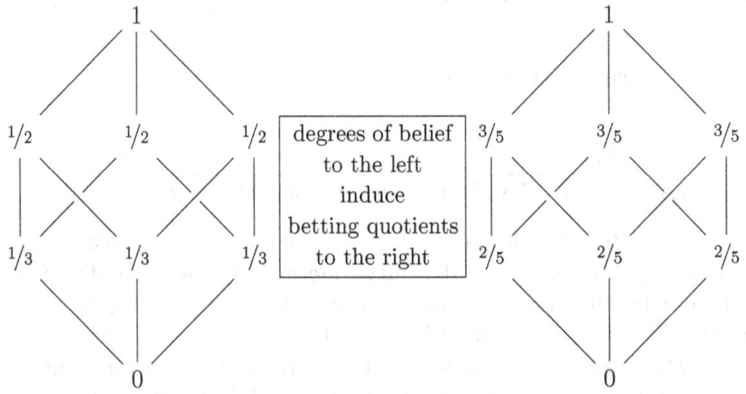

Fig. 1. A nonprobabilist, un-Dutch-bookable belief space and its induced betting quotient function.

That is, consider $\langle W, Prop, b\rangle$ where $W = \{w_1, w_2, w_3\}$, $Prop = \mathcal{P}(W)$, and $b : Prop \to \mathbb{R}$ takes the values $b(\emptyset) = 0$, $b(W) = 1$, $b(\{w_i\}) = \frac{1}{3}$ for each $i \in \{1, 2, 3\}$, and $b(\{w_i, w_j\}) = \frac{1}{2}$ for distinct $i, j \in \{1, 2, 3\}$. Then the induced betting quotient function q is as follows: $q(\emptyset) = 0$, $q(W) = 1$, $q(\{w_i\}) = \frac{2}{5}$ for each $i \in \{1, 2, 3\}$, and $q(\{w_i, w_j\}) = \frac{3}{5}$ for distinct $i, j \in \{1, 2, 3\}$.

(Where a similar illustration appears in the remainder of the paper we shall use the same representational convention, that is, the left algebra represents the function b defined on $\mathcal{P}(\{w_1, w_2, w_3\})$, and the right algebra represents the induced betting quotient function q.)

We can see that the betting quotient function q does not satisfy the classical probability axioms (it is not additive), therefore by Theorem 1 the belief function b is susceptible to a Dutch Book. We can see observe that the belief function b does not satisfy the third condition of Theorem 2, e.g.

$$\frac{b(\{w_1, w_2\})}{b(\{w_1, w_2\}) + b(\{w_3\})} = \frac{3}{5} \neq \frac{4}{5} = \frac{b(\{w_1\})}{b(\{w_1\}) + b(\{w_2, w_3\})} + \frac{b(\{w_2\})}{b(\{w_2\}) + b(\{w_1, w_3\})}.$$

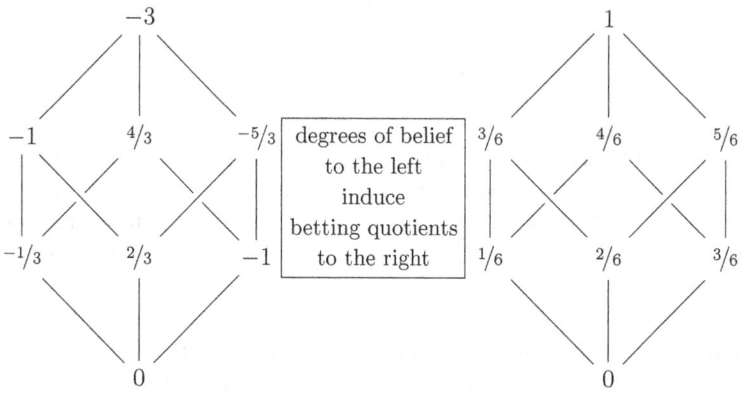

Fig. 2. A nonprobabilist, un-Dutch-bookable belief space with a "wild" belief function.

For a contrasting example, Fig. 2 depicts another nonprobabilistic belief space. As we can see, the induced betting quotient function may be a classical probability function even though the original belief function does not seem to be anything reasonable.

Note that you do need to subscribe to any particular interpretation of belief functions to accept the above argument (for a survey see the already mentioned [5]). The only two things that are needed are that you agree that degrees of belief can be expressed by a real number (so that, for example, you are not a strong operationalist) and agree to our description of the relation between them and the betting quotients.

Notice also that having negative credences—whatever this would mean—does not by itself make you prone to Dutch-Books. You might be exploitable in some ways, and so holding such credences might be irrational. But the main goal of this paper was to distill the essence of the power of the Dutch Book arguments, and from the above Theorem we see that it does not exclude negative credences as irrational.

4 How to Expect Things When You Are Incoherent

We have intentionally used expressions like "the agent expects the bet to have value 0" instead of "according to the agent the expected value of the bet is 0". Consider the following argument (notice that for clarity we have omitted the phrase "according to the agent", but it should be immediate where it is intended to figure):

1. A bet is a random variable;
2. a fair bet is defined as one that has expected value 0;
3. a fair set of bets is defined as one that the expected value of the sum of all the bets in the set is 0;
4. the expected value of a sum of finitely many random variables is equal to the sum of expected values of those random variables;
5. therefore, a finite set of bets all of which are fair is also fair;
6. therefore there are no finite Dutch Books.

Since the conclusion of this argument is false (examples of finite Dutch Books abound), we need to see where it fails. Since a bet outputs a real number (profit) given an element of the sample space (possible world), it can be thought of as a random variable, and so step 1 is true. Steps 2 and 3 are definitions. Step 4 is a basic fact about random variables. Step 5 follows from the previous four, and step 6 is just a reformulation of 5.

So, what is wrong? The culprit is step 2. Yes, it is a definition, but its application to an incoherent agent yields probably unintended results. Compare defining, for an agent with a belief function b, the expected value of a bet for A which pays 1 and costs q as

$$b(A) \cdot (1 - q) + b(\neg A) \cdot (-q) \tag{1}$$

(as is done e.g. by [17]; we have used the same formula in the previous sections) with putting it as

$$\sum_{w \in A} b(\{w\}) \cdot (1 - q) + \sum_{w \notin A} b(\{w\}) \cdot (-q). \tag{2}$$

These two expressions may be different for agents with a nonadditive belief function. (Notice that the former, and not the latter, is, mathematically speaking, the expected value of the bet, considered as a random variable: see e.g. Definition 4.1.1 in [14].) It seems that an incoherent agent may respond to our result in Sect. 2 by saying "that's all very nice, thank you very much for defending me, but really, I expect a bet for A to have value 0 precisedly when (2) is 0, that is, I am using this *atomic* notion of expected value just like it is employed in esteemed publications like [6, p. 615] and [12, Chap. 14]. And so my alleged Dutchbookability is a matter of a different calculation!"

But is that option really available to the agent? It seems to us that the answer is "no". When figuring the relevant betting quotient the agent considers the payoff in the case of winning the bet, the payoff in the case of losing it, and takes into account how probable he or she thinks the two outcomes are, that is, his or hers credences in the proposition (because the bet is won if the proposition is true) and its negation (because it is lost if the proposition is false). The agent's credences in the constituents of the proposition are irrelevant to this; and so, the correct formula to be used is (1). This is of course debatable: but we are willing just to say that at this point the non-probabilist, even though—as pointed out—she could appear to the existing literature, would truly become Stubborn.

4.1 Conclusions

Thus, even if we forget about all the problems of Dutch Book arguments which are usually mentioned in the formal epistemology literature (see e.g. [1,11,17]), it turns out that another one lurks in the basic step of connecting degrees of belief with betting quotients. There is a gap between betting quotients, which theorems in Dutch-Book-inspired formal epistemology are about, and degrees of belief, which those theorems are supposed to be about. This gap prevents the classical Dutch Book argument from being convincing to the target group, that is, nonprobabilists. We propose to bridge that gap using the notion of induced betting quotient; and show that susceptibility to a Dutch Book remains a nontrivial notion: **some nonprobabilists are immune, but others are not.**

5 Relation to Incoherence and Inaccuracy Measures: Some Preliminary Remarks

Since the context of this paper is an argument for probabilism, we have confined our attention to a binary notion: either the agent can be Dutch-booked, and so is irrational, or not, in which case if (s)he indeed is irrational, we need a different argument to show it, since everything seems to be fine about her or his credences at the given moment. This is fine if we are interested in norms of rationality of ideal agents and what exactly it takes to satisfy them. However, if we think of real agents, who—reason would dictate—can only aspire to the probabilistic ideal, or if we would like to compare different violations of probabilism displayed by ideal agents, some graded notion is needed: one which would aim to capture the "distance" between an agent's belief function and some maximally "close" coherent function. (The word "distance" is in quotes since the two-argument functions used in the literature may not be metrics; see e.g. [12].)

One approach would be to use a notion of Dutch Book which would enable us to ask questions regarding "how Dutch-bookable an agent is", for example, intuitively, how much money can be extorted from the agent (assuming some normalization is used). This road is taken by [15]. We have tried to make our paper acceptable to those who think one fault of the classical Dutch Book argument for probabilism from the point view of epistemology was its pragmatic nature; we have thus decided to use a relatively strict notion of a "a bet the agent considers as not favouring any side", and not something similar to "a bet the agent would accept since (s)he considers it to have a nonnegative expected value for her or him", which is the notion Schervish et al. use. We do not know yet how our approach fares if we switch from one notion to the other—this is a task for the future.

The approach by Schervish et al. has been criticised by e.g. [16] on both technical and philosophical grounds, but at least one of their incoherence measures, the "neutral/max" one, stands its ground, and we will consider it in the future. The basic question to be asked is the following. Consider the class N consisting of all nonprobabilist agents which cannot be Dutch-booked (according to a version of our argument which takes into account the class of bets interesting from the point of view of Schervish et al. described above) and the class M consisting of all nonprobabilist agents which can. Are all members of N less incoherent according to the "neutral/max" rule than all members of M?

Another route to consider would be to investigate how members of the classes N and M fare from the standpoint of alethic accuracy (on that notion consult [12]). However, it is not evident what kind of question should be asked in this context. The relationship between graded incoherence and alethic inaccuracy has not been completely worked out and research in that area is ongoing: see e.g. [3]. There seem to be no "simple" theorems to look for in this area; for example, as shown in [3], promoting one virtue does not in general result in promoting the other. Just like in the case of the issues discussed in the previous paragraph, the number of implicit quantifiers involved in researching such issues makes the number of potential formal hypotheses quite high. However, at this moment we are sceptical regarding the outlook of similar endeavours. Consider the nonprobabilist belief space from Fig. 3: on any reasonable inaccuracy measure its "distance" from the closest coherent function will be minimal, and yet it belongs to the class M: it is Dutch-bookable. We will pursue these issues further in [19].

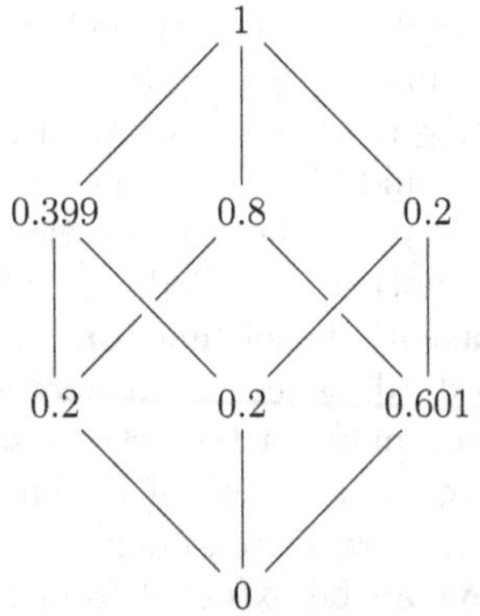

Fig. 3. A belief space which is negation coherent (and so featuring degrees of belief which *are* betting quotients) and Dutch-bookable (since the betting quotients are not additive), but intuitively *very close* to a classical space.

Acknowledgements. Both authors were supported by the Narodowe Centrum Nauki SONATA grant no. UMO-2015/17/D/HS1/01912 'Epistemic inaccuracy and degrees of belief'.

References

1. Bradley, D.J.: A Critical Introduction to Formal Epistemology. Bloomsbury Academic, London (2015)
2. Bradley, D.J., Leitgeb, H.: When betting odds and credences come apart: more worries for Dutch book arguments. Analysis **66**(290), 119–127 (2006)
3. De Bona, G., Staffel, J.: Graded Incoherence for Accuracy-Firsters. Philos. Sci. **84**(April), 1–21 (2017)
4. Elga, A.: Self-locating belief and the Sleeping Beauty problem. Analysis **60**(2), 143–147 (2000)
5. Eriksson, L., Hájek, A.: What are degrees of belief? Studia Logica **86**, 183–213 (2007)
6. Greaves, H., Wallace, D.: Justifying conditionalization: conditionalization maximizes expected epistemic utility. Mind **115**, 607–631 (2006)
7. Hájek, A.: Arguments for-or against-probabilism. Br. J. Philos. Sci. **59**, 793–819 (2008)
8. Hedden, B.: Incoherence without exploitability. Nous **47**(3), 482–495 (2013)
9. Kemeny, J.G.: Fair bets and inductive probabilities. J. Symb. Logic **20**(3), 263–273 (1955)
10. Lehman, R.S.: On confirmation and rational betting. J. Symb. Logic **20**(3), 251–262 (1955)
11. Paris, J.B.: A note on the Dutch Book method. In: Proceedings of the 2nd International Symposium on Imprecise Probabilities and their Applications, Ithaca, New York, pp. 1–16 (2001)
12. Pettigrew, R.: Accuracy and the Laws of Credence. Oxford University Press, Oxford (2016)
13. Rees, O.: Why betting odds and credences come apart. Talk Given at the LOFT 2010 9th Conference on Logic and the Foundations of Game and Decision Theory, University of Toulouse, France, 5–7 July 2010. http://loft2010.csc.liv.ac.uk/papers/20.pdf
14. Schervish, M.J., DeGroot, M.H.: Probability and Statistics. Addison-Wesley (2014)
15. Schervish, M.J., Seidenfeld, T., Kadane, J.B.: Measures of incoherence: how not to gamble if you must. In: Bayesian Statistics 7: Proceedings of the 7th Valencia Conference on Bayesian Statistics, pp. 385–402 (2003)
16. Staffel, J.: Measuring the overall incoherence of credence functions. Synthese **192**(5), 1467–1493 (2015)
17. Vineberg, S.: Dutch book arguments. In: Zalta, E.N. (ed.) The Stanford Encyclopedia of Philosophy. Stanford University, Stanford (2016). Spring 2016 Edition
18. Wroński, L., Godziszewski, M.T.: Dutch Books and nonclassical probability spaces. Eur. J. Philos. Sci. **7**(2), 267–284 (2017)
19. Wroński, L., Godziszewski, M.T.: The Stubborn Non-Probabilist Returns: Incoherence, Non-classical Expectation Values and Dutch Books (in preparation)

Conjunction and Disjunction in Infectious Logics

Hitoshi Omori[1(✉)] and Damian Szmuc[2,3(✉)]

[1] Department of Philosophy, Kyoto University, Kyoto, Japan
hitoshiomori@gmail.com
[2] Department of Philosophy, University of Buenos Aires, Buenos Aires, Argentina
szmucdamian@gmail.com
[3] IIF-SADAF, CONICET, Buenos Aires, Argentina

Abstract. In this paper we discuss the extent to which conjunction and disjunction can be rightfully regarded as such, in the context of infectious logics. Infectious logics are peculiar many-valued logics whose underlying algebra has an absorbing or infectious element, which is assigned to a compound formula whenever it is assigned to one of its components. To discuss these matters, we review the philosophical motivations for infectious logics due to Bochvar, Halldén, Fitting, Ferguson and Beall, noticing that none of them discusses our main question. This is why we finally turn to the analysis of the *truth-conditions* for conjunction and disjunction in infectious logics, employing the framework of plurivalent logics, as discussed by Priest. In doing so, we arrive at the interesting conclusion that —in the context of infectious logics— conjunction *is* conjunction, whereas disjunction *is not* disjunction.

Keywords: Conjunction · Disjunction · Infectious logics · Logics of nonsense · Plurivalent logics

1 Introduction

1.1 Background, Motivation and Aim

The aim of this paper is to discuss the extent to which conjunctions and disjunctions, appearing in the context of what are nowadays called *infectious* logics (cf. [13,22,31]), can be rightfully called *conjunction* and *disjunction*. Infectious logics are, in a nutshell, non-classical many-valued logics that count with a truth-value which is assigned to a compound formula every time it is assigned to at least one of its components. Thus, it is claimed that values behaving in this way exhibit an infectious, contaminating or otherwise absorbing nature.

Salient examples of such logics are the $\{\neg, \wedge, \vee\}$-fragments, also called the "classical" fragments, of Dmitri Bochvar's and Sören Halldén's logic of nonsense, presented in e.g. [5,18] respectively. What differentiates these logics (or, properly speaking, their classical fragments) is that while Bochvar treats the contaminating value as *undesignated*, Halldén (although derivatively, cf. [12, p. 345] and [22]) treats it as *designated*. From this and the absorbing nature of this element

A. Baltag et al. (Eds.): LORI 2017, LNCS 10455, pp. 268–283, 2017.
DOI: 10.1007/978-3-662-55665-8_19

it can be inferred that Bochvar's logic is *paracomplete*, whereas Halldén's logic is *paraconsistent*. In fact, the classical fragment of Bochvar's logic has been also discussed in the relevant literature as Weak Kleene Logic $\mathbf{K_3^w}$, while the classical fragment of Halldén's logic has been also independently discussed in the literature as Paraconsistent Weak Kleene Logic **PWK**. As is easy to notice and as has been already pointed out in many works (e.g. [29]) this logics are such that, respectively, $\vee$-Introduction and $\wedge$-Elimination are invalid in them.

To carry out our present discussion we will scrutinize various motivations given for these infectious logics, in order to determine whether or not the target binary operations are, after all, legitimate disjunctions and conjunctions. To this end, for the case of paracomplete infectious logics, we will consider Bochvar's own nonsense-related account, Melvin Fitting's epistemic semantics [16], Thomas Ferguson's computational interpretation [12,14] and Jc Beall's off-topic reading [2]. Whereas for the case of paraconsistent infectious logics, we will consider Halldén's own nonsense-related account. We will argue that none of these allow to present a cogent reading of disjunction and conjunction, but that an alternative account of the truth and falsity conditions for these connectives, in terms of the discussion of Graham Priest's *plurivalent* semantics carried out in [22,29], indeed does the work.

1.2 Preliminaries

Our language $\mathcal{L}$ consists of a finite set $\{\neg, \wedge, \vee\}$ of propositional connectives and a countable set Prop of propositional parameters. Furthermore, we denote by Form the set of formulas defined as usual in $\mathcal{L}$. We denote formulas of $\mathcal{L}$ by α, β, γ, etc. and sets of formulas of $\mathcal{L}$ by Γ, Δ, Σ, etc.

Definition 1 (Univalent semantics). *A* univalent semantics *for the language $\mathcal{L}$ is a structure $M = \langle \mathcal{V}, \mathcal{D}, \delta \rangle$, where*

- *$\mathcal{V}$ is a non-empty set of truth values,*
- *$\mathcal{D}$ is a non-empty proper subset of $\mathcal{V}$, the designated values,*
- *for every n-ary connective $*$ in the language, $\delta_* : \mathcal{V}^n \to \mathcal{V}$ is the truth function for $*$.*

A univalent interpretation *is a pair $\langle M, \mu \rangle$, where M is such a structure, and μ is an evaluation function from the* Prop *to $\mathcal{V}$. Given an interpretation, μ is extended to a map from all formulas to $\mathcal{V}$ recursively: $\mu(*(\alpha_1, \ldots, \alpha_n)) = \delta_*(\mu(\alpha_1), \ldots, \mu(\alpha_n))$. Finally, $\Sigma \models_u^M \alpha$ iff in every interpretation in which all the formulas of Σ are designated, so is α.*[1]

Note that semantic consequence relations are defined as preservation of designated values, as usual. For an alternative, see Definitions 3 and 4.

[1] We will sometimes omit the subscript u, when contexts disambiguates. Also, we may sometimes make reference of $\models^{\mathbf{L}}$ instead of $\models^{M_{\mathbf{L}}}$.

Definition 2. *The univalent semantics for Weak Kleene and Paraconsistent Weak Kleene for the language $\mathcal{L}$ are the structures $M_{\mathbf{K_3^w}} = \langle \mathcal{V}_{\mathbf{K_3^w}}, \mathcal{D}_{\mathbf{K_3^w}}, \delta_{\mathbf{K_3^w}} \rangle$, and $M_{\mathbf{PWK}} = \langle \mathcal{V}_{\mathbf{PWK}}, \mathcal{D}_{\mathbf{PWK}}, \delta_{\mathbf{PWK}} \rangle$ respectively where*

- $\mathcal{V}_{\mathbf{K_3^w}} = \mathcal{V}_{\mathbf{PWK}} = \{\mathbf{t}, \mathbf{e}, \mathbf{f}\}$,
- $\mathcal{D}_{\mathbf{K_3^w}} = \{\mathbf{t}\}$ *and* $\mathcal{D}_{\mathbf{PWK}} = \{\mathbf{t}, \mathbf{e}\}$,
- $\delta_{\mathbf{K_3^w}} = \delta_{\mathbf{PWK}}$ *and is the set of truth-functions represented by Kleene's 'weak' truth-tables from [19], depicted below.*

	$\delta_\neg$	$\delta_\wedge$	t	e	f	$\delta_\vee$	t	e	f
t	f	t	t	e	f	t	t	e	t
e	e	e	e	e	e	e	e	e	e
f	t	f	f	e	f	f	t	e	f

As is pointed out in [22], $\mathbf{K_3^w}$ can be understood as a logic with gaps endowed with a characterization of logical consequence in terms of truth-preservation, whereas **PWK** can be understood as a logic with gaps endowed with a characterization of logical consequence in terms of non-falsity preservation. The corresponding induced consequence relations being $\models^{\mathbf{K_3^w}}$ and $\models^{\mathbf{PWK}}$.

Nevertheless, it is interesting to notice that these possibilities do not exhaust the way in which we can define logical consequence and, thus, in which we can build logical systems out of the weak truth-tables from Kleene. We define below a q-consequence relation and a p-consequence relation, following the proposals of Grzegorz Malinowksi in [20] and Szymon Frankowski in [17] respectively.

Definition 3 (q-consequence for Kleene's weak truth-tables). $\Sigma \models_q^{\mathbf{WK}} \alpha$ *iff in every interpretation in which all the formulae of Σ are assigned a value in $\{\mathbf{t}, \mathbf{e}\}$, then α is assigned the value* $\mathbf{t}$.

Definition 4 (p-consequence for Kleene's weak truth-tables). $\Sigma \models_p^{\mathbf{WK}} \alpha$ *iff in every interpretation in which all the formulae of Σ are assigned the value* $\mathbf{t}$, *then α is assigned a value in $\{\mathbf{t}, \mathbf{e}\}$.*

2 Infectious Logics: An Overview

As we will briefly see, Weak Kleene and Paraconsistent Weak Kleene are members of a broader family of *infectious logics*. Intuitively, infectious logics are many-valued logics that have an absorbent or infectious truth-value, that is, a truth-value such that it is assigned to a compound formula whenever it is assigned to at least one of its components. More formally:

Definition 5. *A semantics $M = \langle \mathcal{V}, \mathcal{D}, \delta \rangle$ for the language $\mathcal{L}$ is* infectious *iff there is an element $x \in \mathcal{V}$ such that for every n-ary connective $*$ in the language, with an associated truth-function $\delta_* \in \delta$ and for all $v_1, \ldots, v_n \in \mathcal{V}$ it holds that: if $x \in \{v_1, \ldots, v_n\}$, then $\delta_*(v_1, \ldots, v_n) = x$.*

It is easy to see, as has been noticed in e.g. [29], that when the infectious value in question does not belong to the set of designated values, then the logic is paracomplete. By this we mean that there is a valuation such that both A and $\neg A$ are undesignated. Moreover, in these cases, yet another characteristic classical inference is invalid, namely $\vee$-Introduction (sometimes also called 'Addition'), i.e. $\varphi \vDash \varphi \vee \psi$. By this we mean that there is a valuation such that φ is designated, but $\varphi \vee \psi$ is undesignated. This happens, particularly, when ψ receives the infectious undesignated value in question.[2]

Additionally, it is also easy to see, as has been noticed in e.g. [29], that when the infectious value in question belong to the set of designated values, then the logic is paraconsistent. By this we mean that there is a valuation such that both A and $\neg A$ are designated. Moreover, in these cases, yet another characteristic classical inference is invalid, namely $\wedge$-Elimination (sometimes also called 'Simplification'), i.e. $\varphi \wedge \psi \vDash \varphi$. By this we mean that there is a valuation such that $\varphi \wedge \psi$ is designated, but φ is undesignated. This happens, particularly, when ψ receives the infectious designated value in question.[3]

However, these logical behavior could be found to be rather odd and, for this reason, we provide an overview of the philosophical motivations they have received in the literature, in the next subsections. Let us notice that being faithful with the literature will require us reflecting the fact that a considerable amount of motivations have been discussed with regard to the paracomplete case, whereas only a few have been proposed for the paraconsistent case.[4]

2.1 Paracomplete Case

Bochvar's Logic of Nonsense. In the early decades of the last century, paradoxes of set theory devoured the attention of many philosophers and logicians. The first conceptual motivation for an infectious logic relates to these topics. Dmitri Bochvar developed in [5] a three-valued logic to handle the paradoxes of set theory, like Russell's Paradox (cf. [27]).

Bochvar's own take on this issue was that the sentence describing such a paradoxical sets was, properly speaking, *meaningless* or *nonsensical* and —as

[2] Notice that this does *not* suggest that the infectious value does not belong to the set of designated values *if and only if* the logic is paracomplete, for there might well exist paracomplete logics which do not count with an infectious value at all, as in e.g. the well-known Strong Kleene logic $\mathbf{K_3}$ (cf. [19]).

[3] Analogous to the previous footnote, notice that this does *not* suggest that the infectious value does belong to the set of designated values *if and only if* the logic is paraconsistent, for there might well exist paraconsistent logics which do not count with an infectious value at all, as in e.g. the Logic of Paradox due to Priest (cf. [27]).

[4] We should remark that providing a full overview of these motivations will require much more space than we have here. For that reason, we refrained from commenting on some of the motivations for infectious logics, e.g. (the first degree of) Parry systems (cf. [25]) and of Epstein's Dependence and Dual Dependence systems (cf. [11]) discussed in e.g. [12–14,23,24], Deutsch's logic from [9], Daniels' logic from [7], and Priest's logic $\mathbf{FDE}_\varphi$ from [28].

such— it did not deserve to be regarded as either true or false. In more contemporary terms, we would say that Bochvar took such sentences to be truth value *gaps* (cf. [27]).

More importantly, Bochvar was of the idea that sentences or statements containing a meaningless part or subsentence must be, in turn, meaningless themselves. Thus, meaninglessness can be legitimately described as the pathology from which paradoxical sentences suffered, which is indeed itself literally *infectious.* Furthermore, since the meaninglessness of these very sentences is portrayed by Bochvar via the assignment of the corresponding non-classical value we must say that, in terms of Definition 5, the meaningless value is infectious.

These motivations led Bochvar to devise his 'logic of nonsense', which besides the "classical" connectives, has also means to mark nonsensical or meaningless statements. The 'external assertion' operator acts like a characteristic function for true statements, i.e. statements assigned the value and, therefore, not assigned the value false, or the meaningless value. To be precise, then, it is only the $\{\neg, \wedge, \vee\}$-fragment of Bochvar's logic of nonsense that represents an infectious logic (also found in the literature as Weak Kleene Logic $\mathbf{K_3^w}$, on which more below).

Finally, Bochvar took logical consequence as being characterized by *truth-preservation.* That is, necessarily, if the premises are true, the conclusion is true. But, again, if meaningless sentences are neither true nor false, then an inference with true premises but a meaningless conclusion must be *invalid.* This is why, by taking e.g. φ to be true and ψ to be meaningless, $\vee$-Introduction fails.

Fitting's Epistemic Interpretation. With the intention of applying his project (see e.g. [16]) of providing an epistemic interpretation for Kleene logics (cf. [19]) and Belnap-Dunn four-valued logic **FDE**, Melvin Fitting provided in [15] an epistemic interpretation for $\mathbf{K_3^w}$. For this purpose, he recurred to the framework present in e.g. [16], where there is a set of experts expressing their positive and negative opinion on different issues, represented by sentences φ, ψ, etc. In his discussion, and taking φ as an example, Fitting thought of allowing experts to be for (and not against) φ, or both for and against φ, or neither for nor against φ, or against (and not for) φ. These four cases correspond, respectively, (*via* a suitable translation) to the four values $\{\mathbf{t}, \mathbf{b}, \mathbf{n}, \mathbf{f}\}$ of **FDE**.

In this framework, it is possible to ask (when all experts have made their minds about the relevant issues) which is e.g. the set of experts that are in favor of $\varphi \vee \psi$. In this case, we might think of the union of those in favor of φ and those in favor of ψ. However, Fitting notices that in some situations of this sort we might want to *cut-down* the set of experts taken into account to those who have actually expressed an opinion towards both φ and ψ. That is, there can be some situations (Fitting argues) where we may not want to count an expert as being in favor of $\varphi \vee \psi$, if she has no opinion at all with regard to e.g. ψ. In those cases we are interested, in Fitting's terminology, in a 'cut-down' disjunction. (And, by similar remarks, in a 'cut-down' negation and a 'cut-down' conjunction).

The failure of $\vee$-Introduction is, thus, properly understood in epistemic terms by taking disjunction as *cut-down disjunction.* It is not the case that from e.g.

the fact that all experts are in favor of φ it follows that all experts are in favor of $\varphi \vee \psi$, for some experts may have no opinion whatsoever with regard to ψ. If we, additionally, are in a situation where no expert is both in favor and against a certain issue (that is, if no sentence is assigned the truth-value **b**), the logic induced by these cut-down operations is, precisely, $\mathbf{K_3^w}$.

Ferguson's Computational Interpretation. In [12] Thomas Ferguson advances a computational reading of some paracomplete infectious logics by following Belnap's classical remarks about *how a computer should think* (cf. [4]), using **FDE**. Belnap motivates his system by considering a computer retrieving information about certain sentences, where this information can be thought as the truth-value that —the computer is *told*— the given sentences have. Belnap imagines that, with regard to e.g. φ, the computer can be told, i.e. it can retrieve the information that φ is true, or that φ is false, or both, or neither.[5]

Ferguson's 'faulty computer' approach to infectious logics focuses on the idea that a computer may *fail to retrieve the value* of a given sentence. He notices, moreover, that this case must me taken to be essentially different from that where the computer is able to retrieve the value of φ, but it encounters no information regarding its truth or its falsity. The case of a failure retrieving the target value, possibly caused by a memory crash, a physical malfunction, or other problem, is thus different from the case of a successful retrieving attempt, accompanied by the fact that the target value contains no information.

Failures, in Ferguson's approach, must be represented (in an extension of **FDE**) by a fifth value, behaving infectiously and being undesignated. In such a case, $\vee$-Introduction is invalid, for —although a computer might be successful in retrieving the value of φ and, additionally, being told that it is true— it might encounter a critical error or a crash while retrieving the value of ψ and, therefore, an error while retrieving the value of $\varphi \vee \psi$.[6]

Beall's Off-Topic Interpretation. In [2], Jc Beall proposes an alternative interpretation for Weak Kleene Logic, which does not appeal to meaninglessness, as Bochvar's does. Beall focuses on theories formulated in English (or any other natural language). Theories have a distinctive topic, that is, they are *not* about everything, namely, about every concept expressible in English. Intuitively, color theory is about colors, arithmetic is about numbers, and so on and so forth, but color theory is not about numbers and arithmetic is not about colors. To these assumptions Beall adds the intuitive thesis that $\varphi \vee \psi$ is about what φ is about and about what ψ is about, and similarly for $\neg\varphi$ and $\varphi \wedge \psi$.

But this is not all that can be said about theories. Theories, the standard view goes, are sets of sentences closed under logical consequence. But the main question is: *which* logic? Beall remarks that it cannot be a logic that validates

[5] As is noted in [12], while this framework is regarded as a 'single address' approach to Belnap computers, a 'two address' approach can also motivated, with the subtlety that it induces a weaker nine-valued logic.

[6] We shall also mention that in [14] Ferguson discusses another computational interpretation related to McCarthy's logic from [21].

$\vee$-Introduction. For, if that is the case, then any theory that is about what φ is about, will end up being also about what ψ is about, even if ψ is completely off-topic. But this is unintuitive, for then theories will be about every everything, that is, about every concept expressible in English.

To this extent, Beall proposes that the logic under which theories should be closed should be Weak Kleene. By doing this, he also proposes to interpret the infectious value as *off-topic*, thereby understanding validity as *on-topic truth preservation*. It is easy to see how this invalidates $\vee$-Introduction, for even if φ and ψ are both true, φ might well be on-topic while ψ is not. Therefore, $\varphi \vee \psi$ will be true, although off-topic, whence the failure of the corresponding inference.

2.2 Paraconsistent Case

Halldén's Logic of Nonsense. In a similar path than Bochvar, we can find Halldén's own 'logic of nonsense', developed mainly in [18]. Besides the usual set-theoretic paradoxes, Halldén also finds instances of meaningless or nonsensical sentences involved in paradoxes of vagueness (cf. [27]). Again, just like Bochvar, Halldén takes meaningless or nonsensical statements to be neither true nor false, and therefore to be truth-value *gaps*. Halldén also shares the idea that meaninglessness is an infectious feature.

These motivations led him, in turn, to conceive his own 'logic of nonsense', which besides the usual logical connectives $\neg, \wedge, \vee$, has a 'meaningfulness operator' that acts like a characteristic function for meaningful statements, i.e. statements assigned either truth or falsity. To be precise, then, it is only the $\{\neg, \wedge, \vee\}$-fragment of Halldén's logic of nonsense that represents an infectious logic (also found in the literature as **PWK**, on which more below).

Bochvar and Halldén's logics of nonsense have, nevertheless, an important difference. Whereas the first is paracomplete, the latter is paraconsistent. Formally speaking, this means that the truth-value assigned to meaningless sentences is regarded as designated. Thus, there are sentences A (namely, meaningless sentences and sentences containing meaningless statements as subsentences) such that both A and $\neg A$ are designated.

With regard to this, while it is argued in [6] that these renders the meaningless value as being truth-like, a more comprehensive understanding of this feature can be taken into account if we notice that paraconsistency, in Halldén's case, is a derivative phenomenon. This means that Halldén did not think of nonsensical sentences as being both true and false, i.e. as truth-value *gluts* (cf. [27]), much to the contrary he took them to neither true nor false.

The paraconsistent nature of the induced consequence relation is, therefore, better understood if we point out (as [12, p. 345] and [22] do) that Halldén should be regarded as taking validity to be characterized by (forwards) non-falsity preservation, that is, if the premises are non-false, then the conclusion is non-false —instead of the usual (forwards) truth-preservation.[7] In this vein, if

[7] Non-falsity preservation as a motivation for paraconsistency in 'gappy' contexts is discussed in e.g. [1,3].

meaningless sentences are neither true nor false, then an inference with meaningless premises but a false conclusion must be invalid. This is why, by taking e.g. φ to be false and ψ to be meaningless, $\wedge$-Elimination fails.

Dualizing Fitting's Epistemic Interpretation.[8] It is possible to conceive two dualizations of Fitting's epistemic understanding of Weak Kleene logic $\mathbf{K_3^w}$, which will provide an epistemic interpretation of a our target paraconsistent infectious logic, namely, **PWK**. The first one takes the entire framework of Fitting's cut-down operations, but changes the way the consequence relation is defined. Instead of taking validity to be defined by truth-preservation, we change to define it as non-falsity preservation. By this we mean that an inference is valid if and only if the premises are not taken to be false by all the experts, then the conclusion is not taken to be false by all the experts. In such cases, the failure of $\wedge$-Elimination is properly understood in epistemic terms by taking conjunction as cut-down conjunction. Imagine a situation where all experts have no opinion towards φ: in such a situation it is not the case that from e.g. the fact that all experts have no opinion towards $\varphi \wedge \psi$ it follows that all experts have no opinion towards ψ, for all experts may have a negative opinion towards ψ. If we, additionally, are in a situation where no expert is both in favor and against a certain issue (that is, if no sentence is assigned the truth-value **b**), the logic induced by these cut-down operations, taking validity to be defined by non-falsity preservation, is precisely **PWK**.

Alternatively, we could take Fitting's epistemic understanding of **FDE** and build a different interpretation for **PWK**. Again, in this framework, it is possible to ask (when all experts have made their minds about the relevant issues) which is e.g. the set of experts that are in favor of $\varphi \wedge \psi$. In this case, we might think of the intersection of those in favor of φ and those in favor of ψ. However, we might be interested in some situations of this sort we might want to *track-down* those experts who have actually expressed an *inconsistent* opinion towards either φ or ψ. That is, there can be some situations where we may not want to count an expert as having a consistent opinion towards e.g. $\varphi \wedge \psi$, if she has an inconsistent opinion towards e.g. ψ. In those cases we are interested, in analogy with Fitting's terminology, in a 'track-down' disjunction. (And, by similar remarks, in a 'track-down' negation and a 'track-down' conjunction).

Now, imagine a situation where all experts have a negative opinion towards φ and all experts have an inconsistent opinion towards ψ —i.e. they are both for and against ψ. Thus, following the track-down policy we would say that all experts have, therefore, an inconsistent opinion towards $\varphi \wedge \psi$. The failure of $\wedge$-Elimination is, thus, properly understood in epistemic terms by taking conjunction as *track-down conjunction*. In such a situation, it is not the case that from e.g. the fact that all experts are both in favor and against of $\varphi \wedge \psi$ it follows that all experts are in favor of φ, for experts may be both for and against

[8] Unlike the previous interpretations of both the paracomplete and the paraconsistent infectious systems, the following account is our original thought. We would like to thank one of the reviewers for the suggestion to develop further the epistemic readings of infectious logics. For a full technical development of these ideas, see [30].

the conjunction just because they have an inconsistent opinion towards ψ. If we, additionally, are in a situation where no expert is silent regarding all issues (that is, if no sentence is assigned the truth-value **n**), the logic induced by these track-down operations is, precisely, **PWK**.

3 Plurivalent Semantics: Basics

We would like to remark that, even if all of the above formalisms involve infectious connectives that are referred in their respective contexts as *conjunction* and *disjunction*, it is never discussed in these works if the target operations are actual conjunctions and disjunctions, or what makes them be so.

Our discussion in the sections to come is intended to answer this question, by looking at the *truth and falsity conditions* of conjunction and disjunction. We will do this within the framework of plurivalent logics developed by Priest in [29]. Plurivalent logics and their semantics can be thought as an alternative way to look at logical frameworks where instead of a formula's single truth-value coming from an arbitrary set, it is allowed for formulae to have *more than one truth-value*, from a given set. Thus, for example, a setting in which each formulae gets, as a truth-value, a single element of $\{\mathbf{t}, \mathbf{e}, \mathbf{f}\}$, can also be represented in a setting in which every formulae gets, as a truth-value, a subset of $\{\mathbf{t}, \mathbf{f}\}$.

The definitions and results in the first three subsections are all given by Priest in [29] in which proofs are fully spelled out. Moreover, the definitions and results in the last subsection can be found in [22]. Therefore, the results in this section are stated without proofs. Note finally, that our notation as well as the order of the presentation are slightly different from Priest's.[9]

3.1 General Plurivalent Semantics

We begin with the most general case of plurivalent semantics.

Definition 6 (General plurivalent semantics). *Given a univalent interpretation, the corresponding* general plurivalent interpretation *is the same, except that it replaces the evaluation function, μ, with a one-many evaluation relation, $\mathfrak{R}$, between* Prop *and $\mathcal{V}$. Given an interpretation, $\mathfrak{R}$ is extended to a map from* Form *to $\mathcal{V}$ recursively:*

$$*(\alpha_1, \ldots, \alpha_n)\mathfrak{R}\, v \textit{ iff for some } v_1, \ldots, v_n : (\alpha_i \mathfrak{R} v_i \textit{ and } v = \delta_*(v_1, \ldots, v_n)).$$

Finally, $\Sigma \models_g^M \alpha$ iff for all $\mathfrak{R}$, if $\mathfrak{R}$ designates all the formulas of Σ then $\mathfrak{R}$ designates α, where $\mathfrak{R}$ designates α iff $\alpha \mathfrak{R} v$ for some $v \in \mathcal{D}$.

Then, we can again prove a general relation between the univalent semantics and general plurivalent semantics. To this end, we need the following definition.

---

[9] We would like to thank one of the reviewers for the suggestion to restructure the presentation of plurivalent semantics.

Definition 7. *Let $M = \langle \mathcal{V}, \mathcal{D}, \delta \rangle$ be a univalent semantics. Then we can define a univalent semantics $\ddot{M} = \langle \ddot{\mathcal{V}}, \ddot{\mathcal{D}}, \ddot{\delta} \rangle$, where $\ddot{\mathcal{V}} = 2^{\mathcal{V}}$, $\ddot{\mathcal{D}} = \{\ddot{v} \in \ddot{\mathcal{V}} : v \in \ddot{v}$ for some $v \in \mathcal{D}\}$ and*

$$v \in \ddot{\delta}_*(\ddot{v_1}, \ldots, \ddot{v_n}) \textit{ iff for some } v_1, \ldots, v_n : \ (v_i \in \ddot{v}_i \textit{ and } v = \delta_*(v_1, \ldots, v_n)).$$

Proposition 1. *Given any univalent semantics $M = \langle \mathcal{V}, \mathcal{D}, \delta \rangle$, its corresponding general plurivalent semantics can be seen as a univalent semantics $\ddot{M} = \langle \ddot{\mathcal{V}}, \ddot{\mathcal{D}}, \ddot{\delta} \rangle$, i.e. for any $\Sigma \cup \{\alpha\}$: $\Sigma \models_g^M \alpha$ iff $\Sigma \models_u^{\ddot{M}} \alpha$.*

3.2 Positive Plurivalent Semantics

We now turn to the positive plurivalent semantics, which is obtained by adding a constraint to the general plurivalent semantics. Note that the original idea behind the general construction can be found already in [26].

Definition 8 (Positive plurivalent semantics). *Given a univalent interpretation, the corresponding* positive plurivalent interpretation *is the same, except that it replaces the evaluation function, μ, with a one-many evaluation relation, $\mathfrak{R}$, between* Prop *and $\mathcal{V}$ with the following* positivity condition*:*

$$for\ every\ p \in \mathsf{Prop} : p\mathfrak{R}v\ form\ some\ v \in \mathcal{V}.$$

Given an interpretation, $\mathfrak{R}$ is extended to a map from Form *to $\mathcal{V}$ recursively:*

$$*(\alpha_1, \ldots, \alpha_n)\mathfrak{R}v \textit{ iff for some } v_1, \ \ldots, v_n : (\alpha_i \mathfrak{R}\ v_i \textit{ and } v = \delta_*(v_1, \ldots, v_n)).$$

Finally, $\Sigma \models_p^M \alpha$ iff for all $\mathfrak{R}$, if $\mathfrak{R}$ designates all the formulas of Σ then $\mathfrak{R}$ designates α, where $\mathfrak{R}$ designates α iff $\alpha\mathfrak{R}v$ for some $v \in \mathcal{D}$.

Then, we can prove a general relation between the two semantics. To state the result, the following definition will be useful.

Definition 9. *Let $M = \langle \mathcal{V}, \mathcal{D}, \delta \rangle$ be a univalent semantics. Then we can define a univalent semantics $\dot{M} = \langle \dot{\mathcal{V}}, \dot{\mathcal{D}}, \dot{\delta} \rangle$, where $\dot{\mathcal{V}} = 2^{\mathcal{V}} \backslash \emptyset$, $\dot{\mathcal{D}} = \{\dot{v} \in \dot{\mathcal{V}} : v \in \dot{v}$ for some $v \in \mathcal{D}\}$ and*

$$v \in \dot{\delta}_*(\dot{v_1}, \ldots, \dot{v_n}) \textit{ iff for some } v_1, \ldots, v_n : \ (v_i \in \dot{v}_i \textit{ and } v = \delta_*(v_1, \ldots, v_n)).$$

Proposition 2. *Given any univalent semantics $M = \langle \mathcal{V}, \mathcal{D}, \delta \rangle$, its corresponding positive plurivalent semantics can be seen as a univalent semantics $\dot{M} = \langle \dot{\mathcal{V}}, \dot{\mathcal{D}}, \dot{\delta} \rangle$, i.e. for any $\Sigma \cup \{\alpha\}$: $\Sigma \models_p^M \alpha$ iff $\Sigma \models_u^{\dot{M}} \alpha$.*

Remark 1. Let us notice, in passing, that until now our discussion of plurivalent logics has been mainly focused on logical consequence as preservation of "receiving at least one univalently designated value". But that is not the only way, as we can also think of preservation of "not receiving any univalently undesignated value", and more.[10]

[10] As an anonymous reviewer points out, since univalently designated values need not be identified with truth, preserving designated values from premises to conclusion, does *not* collapse with truth-preservation (namely, the preservation of the value **t**).

Once we obtain the plurivalent semantics, we can also characterize the general plurivalent semantic consequence relation in terms of positive plurivalence.

Definition 10. *Let* $M = \langle \mathcal{V}, \mathcal{D}, \delta \rangle$ *be a univalent semantics. Then we can define a univalent semantics* $M^e = \langle \mathcal{V}^e, \mathcal{D}^e, \delta^e \rangle$, *where:* $\mathcal{V}^e = \mathcal{V} \cup \{e\}$, $\mathcal{D}^e = \mathcal{D}$, *and* $\delta^e_*(v^e_1, \ldots, v^e_n) = e$ *iff* $v^e_i = e$ *for some* $v^e_i \in \mathcal{V}^e$. *Otherwise,* $\delta^e_* = \delta_*$.

Proposition 3. *Let* M *be a univalent semantics. Then, for any* $\Sigma \cup \{\alpha\}$: $\Sigma \models^M_g \alpha$ *iff* $\Sigma \models^{M^e}_p \alpha$.

So far we have been looking at the general framework of plurivalent semantics. Here are some examples, obtained by applying plurivalence to the **FDE** family.

Definition 11. *Let* $M = \langle \mathcal{V}, \mathcal{D}, \delta \rangle$ *be a univalent semantics. Then we define a univalent semantics* $M^b = \langle \mathcal{V}^b, \mathcal{D}^b, \delta^b \rangle$, *where:* $\mathcal{V}^b = \mathcal{V} \cup \{b\}$, $\mathcal{D}^b = \mathcal{D} \cup \{b\}$, *and* $\delta^b_* = \delta_*$.

Theorem 1. *Let* M *be a univalent semantics of the* **FDE** *family. Then for any* $\Sigma \cup \{\alpha\}$, *the following hold:* $\Sigma \models^M_p \alpha$ *iff* $\Sigma \models^{M^b}_u \alpha$, *and* $\Sigma \models^M_g \alpha$ *iff* $\Sigma \models^{M^{e,b}}_u \alpha$.

3.3 Yet Another Plurivalent Semantics: Negative Plurivalence

As is well known, Michael Dunn's discovery in [10] offered an intuitive reading of the truth values in the family of **FDE**. Seen in this way, Priest's plurivalent semantics offers yet another way of making sense of truth values in terms of smaller number of truth values. But Priest's construction given in [29] did not give any clue to make sense of Weak Kleene Logic and its paraconsistent variant. It turns out, however, that by considering a rather natural variant of Priest's construction, we obtain an intuitive reading of the truth values for those logics.

Definition 12 (Negative plurivalent semantics). *Given a univalent interpretation, the corresponding* negative plurivalent interpretation *is the same, except that it replaces the evaluation function,* μ, *with a one-many evaluation relation,* $\mathfrak{R}$, *between* Prop *and* $\mathcal{V}$ *with the following* negativity condition*:*

$$for\ every\ p \in \mathsf{Prop}:\ it\ is\ not\ the\ case\ that\ p\mathfrak{R}v\ for\ all\ v \in \mathcal{V}.$$

Given an interpretation, $\mathfrak{R}$ *is extended to a map from* Form *to* $\mathcal{V}$ *recursively:*

$$*(\alpha_1, \ldots, \alpha_n)\mathfrak{R}v\ iff\ for\ some\ v_1, \ldots, v_n : (\alpha_i \mathfrak{R} v_i\ and\ v = \delta_*(v_1, \ldots, v_n)).$$

Finally, we have two definitions of logical consequence in this setting: truth-preservation and non-falsity preservation. In the former case we will say that $\Sigma \models^M_n \alpha$ *iff for all* $\mathfrak{R}$, *if* $\mathfrak{R}$ *designates all the formulas of* Σ *then* $\mathfrak{R}$ *designates* α, *where* $\mathfrak{R}$ *designates* α *iff* $\alpha\mathfrak{R}v$ *for some* $v \in \mathcal{D}$. *For the latter we will say that* $\Sigma \models^M_n \alpha$ *iff for all* $\mathfrak{R}$, *if* $\mathfrak{R}$ *designates all the formulas of* Σ *then* $\mathfrak{R}$ *designates* α, *where* $\mathfrak{R}$ *designates* α *iff it is not the case that* $\alpha\mathfrak{R}v$ *for some* $v \notin \mathcal{D}$.

Then, we can again prove a relation between the univalent semantics and negative plurivalent semantics, but the only case that is allowed for the univalent semantics is the two-valued matrix for classical logic. As [22] shows, negative plurivalence does not define a plurivalent consequence relation if some other matrices from the **FDE** family are taken as the basis. To state the result, the following definition will be useful —alternatively changing the definition of $\dddot{\mathcal{D}}$ to $\dddot{\mathcal{D}} = \{\dddot{v} \in \dddot{\mathcal{V}} : v \notin \ddot{v} \text{ for all } v \notin \mathcal{D}\}$ for the non-falsity preservation case.

Definition 13. *Let $M = \langle \mathcal{V}, \mathcal{D}, \delta \rangle$ be the univalent semantics for classical logic. Then we can define a univalent semantics $\dddot{M} = \langle \dddot{\mathcal{V}}, \dddot{\mathcal{D}}, \dot{\delta} \rangle$, where $\dddot{\mathcal{V}} = 2^{\mathcal{V}} \setminus \mathcal{V}$, $\dddot{\mathcal{D}} = \{\dddot{v} \in \dddot{\mathcal{V}} : v \in \ddot{v} \text{ for some } v \in \mathcal{D}\}$ and*

$$v \in \dddot{\delta}_*(\dddot{v_1}, \ldots, \dddot{v_n}) \textit{ iff for some } v_1, \ldots, v_n : \ (v_i \in \dddot{v_i} \textit{ and } v = \delta_*(v_1, \ldots, v_n)).$$

Proposition 4. *Let M be a univalent semantics for classical logic. Then, for any $\Sigma \cup \{\alpha\}$: $\Sigma \models_u^{\dddot{M}} \alpha$ iff $\Sigma \models^{\mathbf{K_3^w}} \alpha$.*

4 Reflections

4.1 On Infectious Values in Plurivalent Semantics

In the context of both the general and negative plurivalence, following the definitions suggested by Priest, infectious values can *only* be represented by the *empty set*. This is remarkably so, even if we start with univalent semantics other than two-valued classical logic, e.g. the three-valued univalent semantics for $\mathbf{K_3^w}$, or even **FDE**. If we, additionally, think of a truth-value x as being *true* if $\mathbf{t} \in x$, and respectively as being *false* if $\mathbf{f} \in x$, then it is clear that being true or being false implies being non-infectious.[11]

It is for these reasons that if we apply the generalized plurivalence to the e.g. two-valued univalent semantics for classical logic, infectious values cannot be represented with —for instance— the full set $\{\mathbf{t}, \mathbf{f}\}$. We would like to mention, though, that as a remark made by an anonymous reviewer suggests, it will be interesting to discuss definitions of the plurivalent semantics that deviate from Priest's, in order to determine whether or not there is a plurivalent-like setting where, for instance, the full set $\{\mathbf{t}, \mathbf{f}\}$ can represent an infectious value. However, we notice that there is no such thing present in the literature, up to now.

4.2 Addressing the Main Question

In what follows we will present an account of the truth conditions for conjunction and disjunction in the context of plurivalent logics which applies both to general plurivalence and negative plurivalence —something that we take to be an advantage of the present discussion.

[11] Although for an alternative, see [30], where designated infectious values are understood as truth-value *gluts*, i.e. as both-true-and-false.

In the context of infectious logics, interpreted along the lines of Sect. 2, these operations are characterized by the following truth conditions:

$$\varphi \wedge \psi \text{ is true iff } \varphi \text{ and } \psi \text{ are true, and condition } C \text{ applies to } \varphi \text{ and } \psi$$
$$\varphi \vee \psi \text{ is true iff } \varphi \text{ or } \psi \text{ are true, and condition } C \text{ applies to } \varphi \text{ and } \psi$$

where by 'condition C applies to φ and ψ' we mean, respectively, that these sentences are meaningful (in Bochvar's and Halldén's case), that all experts have expressed an opinion toward these sentences (in Fitting's case), that the computer was successful in retrieving the information with regard to these sentences (in Ferguson's case) and that these sentences are on-topic (in Beall's case).

Remark 2. Since truth-values in the plurivalent framework are represented as subsets of some set of univalent truth-values, this implies that the curly brackets act as a meaningful operator (if we take into account Bochvar's and Halldén's interpretation), or as a did-expressed-an-opinion operator (if we take into account Fitting's interpretation), or as a successful-in-retrieving value operator (if we take into account Ferguson's interpretation), or as an on-topic operator (if we take into account Beall's interpretation).

On the more conservative side, the traditional account of conjunction has it that a conjunction is true iff both conjuncts are true, whereas the traditional account of disjunction has it that a disjunction is true iff at least one of both disjunctions is true. We will see, through some technical remarks, that this understanding of conjunction *is* respected in the plurivalent reading of infectious logics, whence we can legitimately say that the operator called 'conjunction' in the context of infectious logic *is conjunction.* However, the standard understanding of disjunction *is not* respected in the plurivalent reading of infectious logics, whence we can legitimately say that the operator called 'disjunction' in the context of infectious logics *is not disjunction.*[12]

By the truth condition for conjunction in the plurivalent semantics, we have:

$$\mathbf{t} \in \ddot{\delta}_{\wedge}(x, y) \text{ iff } \exists x_0, \exists y_0 \in \{\mathbf{t}, \mathbf{f}\} : [x_0 \in x, y_0 \in y \text{ and } \delta_{\wedge}(x_0, y_0) = \mathbf{t}].$$

But the fact that $\delta_{\wedge}(x_0, y_0) = \mathbf{t}$, given the definition of $\delta_{\wedge}$ entails that $x_0 = \mathbf{t}$ and $y_0 = \mathbf{t}$, further implying that none of them is the empty set. Thus we have:

$$\mathbf{t} \in \ddot{\delta}_{\wedge}(x, y) \text{ iff } \mathbf{t} \in x \text{ and } \mathbf{t} \in y.$$

From this we infer that conjunction, in the context of infectious logics represented within the plurivalent semantics, *is conjunction, as traditionally conceived.*

However, we cannot say the same about disjunction, as we now turn to show. By the truth condition for disjunction in the plurivalent semantics, we have:

$$\mathbf{t} \in \ddot{\delta}_{\vee}(x, y) \text{ iff } \exists x_0, \exists y_0 \in \{\mathbf{t}, \mathbf{f}\} : [x_0 \in x, y_0 \in y \text{ and } \delta_{\vee}(x_0, y_0) = \mathbf{t}].$$

[12] Notice that we took the notational liberty of using e.g. $\ddot{\delta}_{\wedge}$ as the paradigmatic case, but nothing really depends on this, and $\dddot{\delta}_{\wedge}$ might be used as well, without any loss.

Nevertheless, the fact that $\delta_\vee(x_0, y_0) = \mathbf{t}$, given the definition of $\delta_\vee$ *does not entail* that both $x_0, y_0 \in \{\mathbf{t}, \mathbf{f}\}$, i.e. it does not imply that none of them is the empty set. From which we can infer that disjunction, in the context of infectious logics represented within the plurivalent semantics, *is not disjunction, as traditionally conceived.*

Let us notice, for some readers might be concerned with the case, that negation (as present in infectious logics represented within plurivalent semantics) is negation as traditionally conceived, that is, it is an operator that flip-flops truth and falsity.[13]

To conclude, we should highlight that the previous remarks about the degree to which the operators called conjunction and disjunction in infectious logics are legitimately called that way did not make any reference to the validity of inference rules or principles where those connectives are features, e.g. of the already mentioned cases of $\wedge$-Elimination and $\vee$-Introduction.

We would like to point out that those are issues that essentially concern the definition of validity. Whether we do that in terms of truth-preservation, or in terms of non-falsity preservation, in terms of q-consequence (cf. [20]) or p-consequence (cf. [17]), following its specific instances defined in Sect. 2, the inferences that are going to be valid or invalid vary, as summarized below.

	$\models^{\mathbf{K_3^w}}$	$\models^{\mathbf{PWK}}$	$\models_q^{\mathbf{WK}}$	$\models_p^{\mathbf{WK}}$
$\psi \models \varphi \vee \neg\varphi$	×	✓	×	✓
$\varphi \wedge \neg\varphi \models \psi$	✓	×	×	✓
$\varphi \models \varphi \vee \psi$	×	✓	×	✓
$\varphi \wedge \psi \models \varphi$	✓	×	×	✓

5 Conclusion

In this paper we discussed the extent to which conjunction and disjunction can be rightfully regarded as logical connectives of those particular sorts, in the context of infectious logics. By turning to the analysis of the *truth-conditions* for these connectives, employing the framework of plurivalent logics, we arrived at the conclusion that —in the context of infectious logics— conjunction *is* conjunction, whereas disjunction *is not* disjunction in the context of infectious logics.

There are a number of directions in which further work related to infectious logics and plurivalent logics can be carried out. Regarding the most prominent and historically salient interpretations of infectious logics, in terms of the 'logics of nonsense' due to Bochvar and Halldén, it will be worth exploring the possibility of defining the semantics of a proper meaningful operator within the plurivalent framework. Both Bochvar and Halldén's logics count with such linguistic devices, although in this work we focused mainly on the $\{\neg, \wedge, \vee\}$-fragment of their systems. We leave these and other discussions for a subsequent paper.

Acknowledgments. We would like to thank the anonymous referees for their helpful (and enthusiastic!) comments that improved our paper. Hitoshi Omori is a Postdoctoral

[13] For further arguments in favor of the traditional account of negation, see [8].

Research Fellow of Japan Society for the Promotion of Science (JSPS). Damian Szmuc is enjoying a PhD fellowship of the National Scientific and Technical Research Council of Argentina (CONICET) and his visits to Kyoto when this collaboration took place were partially supported by JSPS KAKENHI Grant Number JP16H03344.

References

1. Armour-Garb, B., Priest, G.: Analetheism: a pyrrhic victory. Analysis **65**(2), 167–173 (2005)
2. Beall, J.: Off-topic: a new interpretation of weak-kleene logic. Australas. J. Log. **13**(6), 136–142 (2016)
3. Beall, J., Ripley, D.: Analetheism and dialetheism. Analysis **64**(1), 30–35 (2004)
4. Belnap, N.: How a computer should think. In: Ryle, G. (ed.) Contemporary Aspects of Philosophy, pp. 30–55. Oriel Press, Newcastle upon Tyne (1977)
5. Bochvar, D.: On a three-valued calculus and its application in the analysis of the paradoxes of the extended functional calculus. Matamaticheskii Sbornik **4**, 287–308 (1938)
6. Brady, R., Routley, R.: Don't care was made to care. Australas. J. Philos. **51**(3), 211–225 (1973)
7. Daniels, C.: A note on negation. Erkenntnis **32**(3), 423–429 (1990)
8. De, M., Omori, H.: There is more to negation than modality. J. Philos. Log. (2017)
9. Deutsch, H.: Paraconsistent analytic implication. J. Philos. Log. **13**(1), 1–11 (1984)
10. Dunn, M.: Intuitive semantics for first-degree entailments and 'coupled trees'. Philos. Stud. **29**(3), 149–168 (1976)
11. Epstein, R.: The semantic foundations of logic. In: Propositional Logics, 2nd edn., vol. 1. Oxford University Press, New York (1995)
12. Ferguson, T.M.: A computational interpretation of conceptivism. J. Appl. Non Class. Log. **24**(4), 333–367 (2014)
13. Ferguson, T.M.: Logics of nonsense and Parry systems. J. Philos. Log. **44**(1), 65–80 (2015)
14. Ferguson, T.M.: Faulty Belnap computers and subsystems of FDE. J. Log. Comput. **26**(5), 1617–1636 (2016)
15. Fitting, M.: Kleene's three valued logics, their children. Fundamenta Informaticae **20**(1, 2, 3), 113–131 (1994)
16. Fitting, M.: Bilattices are nice things. In: Bolander, T., Hendricks, V., Pedersen, S.A. (eds.) Self-Reference, pp. 53–78. CSLI Publications (2006)
17. Frankowski, S.: Formalization of a plausible inference. Bull. Sect. Log. **33**(1), 41–52 (2004)
18. Halldén, S.: The Logic of Nonsense. Uppsala Universitets Årsskrift (1949)
19. Kleene, S.C.: Introduction to Metamathematics. North-Holland, Amsterdam (1952)
20. Malinowski, G.: Q-consequence operation. Rep. Math. Log. **24**(1), 49–59 (1990)
21. McCarthy, J.: A basis for a mathematical theory of computation. In: Braffort, P., Hirschberg, D. (eds.) Computer Programming and Formal Systems, pp. 33–70. North-Holland Publishing Company, Amsterdam (1963)
22. Omori, H.: Halldén's logic of nonsense and its expansions in view of logics of formal inconsistency. In: Proceedings of DEXA 2016, pp. 129–133. IEEE Computer Society (2016)
23. Paoli, F.: Regressive analytical entailments. Technical report number 33, Konstanzer Berichte zur Logik und Wissenschaftstheorie (1992)

24. Paoli, F.: Tautological entailments and their rivals. In: Béziau, J.Y., Carnielli, W., Gabbay, D. (eds.) Handbook of Paraconsistency, pp. 153–175. College Publications (2007)
25. Parry, W.T.: Ein axiomensystem für eine neue art von implikation (analytische implikation). Ergeb. Eines Math. Kolloqu. **4**, 5–6 (1933)
26. Priest, G.: Hyper-contradictions. Logique et Analyse **27**(107), 237–243 (1984)
27. Priest, G.: An Introduction to Non-classical logic: From If to Is, 2nd edn. Cambridge University Press, Cambridge (2008)
28. Priest, G.: The logic of the catuskoti. Comp. Philos. **1**(2), 24–54 (2010)
29. Priest, G.: Plurivalent logics. Australas. J. Log. **11**(1), 2–13 (2014)
30. Szmuc, D.: An epistemic interpretation of Paraconsistent Weak Kleene. Typescript
31. Szmuc, D.: Defining LFIs and LFUs in extensions of infectious logics. J. Appl. Non Class. Log. **26**(4), 286–314 (2016)

On the Concept of a Notational Variant

Alexander W. Kocurek(✉)

University of California, Berkeley, Berkeley, USA
akocurek@berkeley.edu

Abstract. In the study of modal and nonclassical logics, translations have frequently been employed as a way of measuring the inferential capabilities of a logic. It is sometimes claimed that two logics are "notational variants" if they are translationally equivalent. However, we will show that this cannot be quite right, since first-order logic and propositional logic are translationally equivalent. Others have claimed that for two logics to be notational variants, they must at least be compositionally intertranslatable. The definition of compositionality these accounts use, however, is too strong, as the standard translation from modal logic to first-order logic is not compositional in this sense. In light of this, we will explore a weaker version of this notion that we will call *schematicity* and show that there is no schematic translation either from first-order logic to propositional logic or from intuitionistic logic to classical logic.

Keywords: Translation · Notational variant · Lindenbaum-Tarski algebras · Compositionality · Schematicity

1 Introduction

In the study of modal and nonclassical logics, translations (maps between formulas that faithfully preserve consequence) are frequently employed as a way of measuring the inferential capabilities of a logic. Examples of well-known translations in the literature include:

(a) the double-negation translation of classical logic into intuitionistic logic;
(b) the standard translation of modal logic into first-order logic;
(c) the Gödel translation of intuitionistic logic into classical **S4**.

These translations are often taken to show that the logic being translated can be viewed as a "notational variant" of a fragment of the logic it is translated into. Indeed, a number of authors have conjectured that translational equivalence is a necessary and/or sufficient condition for two logics to be notational variants in the intuitive sense.[1]

Unfortunately, most of these accounts of notational variance are either too weak or too strong. For instance, on any reasonable theory of notational variance,

[1] For claims like this, see [16, p. 67], [5, p. 391], [12, p. 269], [3, p. 108], [13, p. 139], [11, p. 7] and [6, p. 134].

A. Baltag et al. (Eds.): LORI 2017, LNCS 10455, pp. 284–298, 2017.
DOI: 10.1007/978-3-662-55665-8_20

first-order logic and propositional logic are not notational variants. However, we will show in Sect. 3 that first-order logic and propositional logic are translationally equivalent. Thus, any account which says translational equivalence is sufficient for notational variance[2] is too weak.

On the other hand, some have suggested that for two logics to be considered notational variants, they must at least be *compositionally* intertranslatable, in a sense that will be made precise in Sect. 4.[3] Since there is no compositional translation from first-order logic to propositional logic, the former is not a notational variant of the latter in this sense. However, these accounts of notational variance are too strong, since on their definition of compositionality, even the standard translation (in fact, *any* translation) of modal logic into first-order logic is not compositional. Near the end of this paper, a generalization of this notion called *schematicity* that avoids these problems will be proposed, and we will show that there is no schematic translation from first-order logic to propositional logic, or from intuitionistic logic to classical logic.

2 Defining Translations

We start by defining the concept of a logic and a translation in abstract terms.

Definition 1 (Logic). A ***logic*** is a pair $\mathbf{L} = \langle \mathcal{L}, \models \rangle$ where $\mathcal{L}$ is a nonempty class (of ***formulas***) and $\models \subseteq \wp(\mathcal{L}) \times \mathcal{L}$ (the ***consequence relation***) such that:

(i) $\models$ is reflexive, i.e., for all $\phi \in \mathcal{L}$, $\phi \models \phi$
(ii) $\models$ is transitive, i.e., for all $\Gamma, \Delta \subseteq \mathcal{L}$ and all $\phi \in \mathcal{L}$, if $\Gamma \models \phi$ and if $\Delta \models \gamma$ for each $\gamma \in \Gamma$, then $\Delta \models \phi$.

Where $\phi, \psi \in \mathcal{L}$, we will say ϕ and ψ are **L-*equivalent***, written "$\phi \equiv \psi$", if $\phi \models \psi$ and $\psi \models \phi$. We will say that ϕ is **L-*valid***, written "$\models \phi$", if $\varnothing \models \phi$. If $\mathbf{L}$ is a logic, we may write "$\models_{\mathbf{L}}$" and "$\equiv_{\mathbf{L}}$" for the consequence and equivalence relations for $\mathbf{L}$ respectively. We may also write "$\models_i$" instead of "$\models_{\mathbf{L}_i}$", "$\equiv_i$" instead of "$\equiv_{\mathbf{L}_i}$", etc.

This notion of a logic is meant to be fairly general. While it can be generalized even further (allowing for substructural logics, multiple-conclusion logics, etc.), such generalizations will not concern us here. Classical, intuitionistic, modal, and predicate logics can all be viewed as logics in the sense of Definition 1.

Next, we define the concept of a translation.

Definition 2 (Translation). Let $\mathbf{L}_1$ and $\mathbf{L}_2$ be logics. A ***translation*** from $\mathbf{L}_1$ to $\mathbf{L}_2$ is a map $t\colon \mathcal{L}_1 \to \mathcal{L}_2$ such that for all $\Gamma \subseteq \mathcal{L}_1$ and $\phi \in \mathcal{L}_1$, $\Gamma \models_1 \phi$ iff $t[\Gamma] \models_2 t(\phi)$. If t is a translation from $\mathbf{L}_1$ to $\mathbf{L}_2$, we will write "$t\colon \mathbf{L}_1 \rightsquigarrow \mathbf{L}_2$". We will say $\mathbf{L}_1$ is ***translatable*** into $\mathbf{L}_2$, written as "$\mathbf{L}_1 \rightsquigarrow \mathbf{L}_2$", if there is a translation from $\mathbf{L}_1$ to $\mathbf{L}_2$. We will say $\mathbf{L}_1$ and $\mathbf{L}_2$ are ***intertranslatable***, written as "$\mathbf{L}_1 \leftrightsquigarrow \mathbf{L}_2$", if $\mathbf{L}_1 \rightsquigarrow \mathbf{L}_2$ and $\mathbf{L}_2 \rightsquigarrow \mathbf{L}_1$.

[2] E.g., [13, p. 139], [11, p. 7] and [6, p. 134].
[3] E.g., [16, p. 67], [5, p. 391], [12, p. 269] and [3, p. 108].

Example (Double-Negation Translation). Define $\mathtt{At} = \{p_0, p_1, p_2, \dots\}$. Let $\mathcal{L}_{\text{prop}}$ be the set of formulas defined recursively over $\mathtt{At}$ as follows:

$$\phi ::= p \mid \neg\phi \mid (\phi \wedge \phi).$$

Let **CPL** be classical propositional logic over $\mathcal{L}_{\text{prop}}$, and let **IPL** be intuitionistic propositional logic over $\mathcal{L}_{\text{prop}}(\vee, \rightarrow)$, i.e., the result of extending $\mathcal{L}_{\text{prop}}$ with connectives $\vee$ and $\rightarrow$. Define $t(\phi) := \neg\neg\phi$. Then $t\colon \mathbf{CPL} \rightsquigarrow \mathbf{IPL}$.

Example (Standard Translation). Let $\mathtt{Var} = \{x_0, x_1, x_2, \dots\}$ (the set of ***variables***) and for each $n \in \mathbb{N}$, let $\mathtt{Pred}^n = \{P_0^n, P_1^n, P_2^n, \dots\}$ (the set of ***n-place predicates***). Define $\mathcal{L}_{\text{pred}}$ to be the set of formulas defined recursively as follows:

$$\phi ::= P^n(y_1, \dots, y_n) \mid \neg\phi \mid (\phi \wedge \phi) \mid \forall x\, \phi.$$

Let **FOL** be classical first-order logic over $\mathcal{L}_{\text{pred}}$. Define $\mathcal{L}_{\text{prop}}(\Box)$ to be the set of formulas defined recursively over $\mathtt{At}$ as follows:

$$\phi ::= p \mid \neg\phi \mid (\phi \wedge \phi) \mid \Box\phi.$$

Let **K** be the minimal normal modal logic over $\mathcal{L}_{\text{prop}}(\Box)$. Where R is an arbitrarily chosen binary predicate and where $n \in \mathbb{N}$, we define the map ST_n from propositional modal formulas to first-order formulas as follows:

$$\begin{aligned} ST_n(p_i) &= P_i^1(x_n) \\ ST_n(\neg\phi) &= \neg ST_n(\phi) \\ ST_n(\phi \wedge \psi) &= (ST_n(\phi) \wedge ST_n(\psi)) \\ ST_n(\Box\phi) &= \forall x_{n+1}\, (R(x_n, x_{n+1}) \rightarrow ST_{n+1}(\phi)). \end{aligned}$$

Then $ST_n\colon \mathbf{K} \rightsquigarrow \mathbf{FOL}$.

Example (Non-normal Modal Logics). A modal logic over $\mathcal{L}_{\text{prop}}(\Box)$ is said to be ***monotonic*** if it contains all classical tautologies as well as the axiom $\Box(p \wedge q) \rightarrow (\Box p \wedge \Box q)$ and it is closed under uniform substitution, modus ponens, and the rule $\phi \leftrightarrow \psi / \Box\phi \leftrightarrow \Box\psi$. Kracht and Wolter [9, p. 109, Theorem 4.7] showed that the following map is a translation from any monotonic modal logic to a normal bimodal logic (i.e., a modal logic over the language $\mathcal{L}_{\text{prop}}(\Box_1, \Box_2)$, where each $\Box_i$ is a normal modal operator):

$$\begin{aligned} t(p) &= p \\ t(\neg\phi) &= \neg t(\phi) \\ t(\phi \wedge \psi) &= t(\phi) \wedge t(\psi) \\ t(\Box\phi) &= \Diamond_1 \Box_2\, t(\phi). \end{aligned}$$

Thomason [14,15] also shows how to translate any tense logic into a normal (mono)modal logic, though the translation is too complex to state succinctly here.

Logicians have typically taken the existence of such translations to show that the source logic is a mere notational variant of a fragment of the target logic. Gödel [7] (reprinted in [8]) says of the translation from **CPL** to **IPL**:

> If to the primitive notions of Heyting's propositional calculus we let correspond those notions of the classical propositional calculus that are denoted by the same sign and if to absurdity ($\neg$) we let correspond negation ($\sim$), then the intuitionistic propositional calculus H turns out to be a proper subsystem of the ordinary propositional calculus A. With another correlation (translation) of the notions, however, *the classical propositional calculus is*, conversely, *a subsystem of the intuitionistic one* [8, p. 287].

Blackburn et al. [2, p. xi] say of the standard translation from modal logic to first-order logic:

> By adopting the perspective of correspondence theory, modal logic can be regarded as a fragment of first- or second-order classical logic.

Kracht and Wolter [9, p. 100] informally explain the significance of their result that monotonic modal logics are translatable into normal bimodal logics as follows:

> The positive results on simulations [i.e., translations] show that there is no essential difference between the classes of monomodal normal logics, monotonic logics, and polymodal logics.

Finally, Thomason [15, p. 154] summarizes his result that tense logics are translatable into normal (mono)modal logics as follows:

> In general terms, these results would seem to indicate that there is nothing to be gained by considering many modalities rather than just one, except simplicity—anything which can be expressed about the universe in terms of many notions of necessity can be expressed in terms of one, very complex, notion of necessity, by a translation which preserves both the semantic and syntactic consequence relations.

Although we will show that translations between logics are fairly easy to come by, there are non-trivial failures of translatability. For instance, Jeřábek [10, p. 672] showed that there is no translation from **CPL** to the logic of paradox **LP**. As another example, the following is readily verified:

Proposition 3. If $\mathbf{L}_1 \rightsquigarrow \mathbf{L}_2$, then $\mathbf{L}_2$ is compact only if $\mathbf{L}_1$ is.

From this, it follows that second-order logic is not translatable into first-order logic. Moreover, as the next example shows, there are pairs of logics such that neither logic is translatable into the other.[4]

[4] This thereby answers a question posed by Epstein [5, p. 388] in the affirmative. It is also straightforward to generate artificial counterexamples using any two partial orders such that neither is order-embeddable in the other.

Example (Kleene Logic). Let **K3** be the strong Kleene logic over $\mathcal{L}_{\text{prop}}$. Let us write $\mathbf{CPL}^n$ and $\mathbf{K3}^n$ for the logics obtained from **CPL** and **K3** respectively by restricting the set of formulas to those whose atomics are all among $\{p_1, \ldots, p_n\}$. Then neither $\mathbf{CPL}^n$ nor $\mathbf{K3}^n$ is translatable into the other. $\mathbf{CPL}^n$ is not translatable into $\mathbf{K3}^n$ since there are no tautologies in $\mathbf{K3}^n$ (this generalizes to **CPL** and **K3**). And $\mathbf{K3}^n$ is not translatable into $\mathbf{CPL}^n$ since the former has strictly more formulas up to equivalence than the latter (this does not generalize to **CPL** and **K3**; in fact, $\mathbf{K3} \rightsquigarrow \mathbf{CPL}$ by Theorem 14 below).

One might conjecture that two logics are notational variants if they are intertranslatable. However, a number of authors have claimed that intertranslatability is not enough for two logics to be properly called "notational variants". Rather, they must additionally be *translationally equivalent* in the following sense:[5]

Definition 4 (Translational Equivalence). Let $\mathbf{L}_1$ and $\mathbf{L}_2$ be logics. We will say that $\langle t_1, t_2\rangle$ is a ***translation scheme*** between $\mathbf{L}_1$ and $\mathbf{L}_2$ (written as "$t_1, t_2 \colon \mathbf{L}_1 \leftrightsquigarrow \mathbf{L}_2$") if $t_1 \colon \mathbf{L}_1 \rightsquigarrow \mathbf{L}_2$ and $t_2 \colon \mathbf{L}_2 \rightsquigarrow \mathbf{L}_1$ and for all $\phi \in \mathcal{L}_1$ and all $\psi \in \mathcal{L}_2$:

$$t_2(t_1(\phi)) \equiv_1 \phi$$
$$t_1(t_2(\psi)) \equiv_2 \psi.$$

We will say $\mathbf{L}_1$ and $\mathbf{L}_2$ are ***translationally equivalent*** (written "$\mathbf{L}_1 \leftrightsquigarrow \mathbf{L}_2$") if $t_1, t_2 \colon \mathbf{L}_1 \leftrightsquigarrow \mathbf{L}_2$ for some t_1 and t_2.

Translational equivalence is strictly stronger than intertranslatability. In particular, as we will now show, **CPL** and **IPL** are intertranslatable but not translationally equivalent.

Definition 5 (Lindenbaum-Tarski Algebra). Let $\mathbf{L} = \langle \mathcal{L}, \vDash\rangle$ be a logic. The ***Lindenbaum-Tarski algebra*** of **L** is the poset $\mathbb{L}_{\mathbf{L}} = \langle \mathcal{L}/\equiv, \leq\rangle$ where $\mathcal{L}/\equiv$ is the class of $\equiv$-classes on $\mathcal{L}$ and where $[\phi]_{\mathbf{L}}, [\psi]_{\mathbf{L}} \in \mathcal{L}/\equiv$, $[\phi]_{\mathbf{L}} \leq [\psi]_{\mathbf{L}}$ iff $\phi \vDash \psi$ (it is easy to verify this is well-defined since $\vDash$ is transitive).

Proposition 6. $\mathbf{CPL} \leftrightarrows \mathbf{IPL}$ but not $\mathbf{CPL} \leftrightsquigarrow \mathbf{IPL}$.

Proof. We saw above that $\mathbf{CPL} \rightsquigarrow \mathbf{IPL}$ via the double-negation translation. Moreover, by Theorem 14 below, $\mathbf{IPL} \rightsquigarrow \mathbf{CPL}$. Thus, $\mathbf{CPL} \leftrightarrows \mathbf{IPL}$. Suppose $t, s \colon \mathbf{CPL} \leftrightsquigarrow \mathbf{IPL}$. Define $f \colon \mathbb{L}_{\mathbf{CPL}} \to \mathbb{L}_{\mathbf{IPL}}$ and $g \colon \mathbb{L}_{\mathbf{IPL}} \to \mathbb{L}_{\mathbf{CPL}}$ such that $f([\phi]_{\mathbf{CPL}}) = [t(\phi)]_{\mathbf{IPL}}$ and $g([\phi]_{\mathbf{IPL}}) = [s(\phi)]_{\mathbf{CPL}}$ (this is well-defined since translations preserve equivalence). It is easy to check that f and g are order-embeddings such that $f(g([\phi]_{\mathbf{IPL}})) = [\phi]_{\mathbf{IPL}}$ and $g(f([\phi]_{\mathbf{CPL}})) = [\phi]_{\mathbf{CPL}}$. Thus, if $\mathbf{CPL} \leftrightsquigarrow \mathbf{IPL}$, then $\mathbb{L}_{\mathbf{CPL}}$ and $\mathbb{L}_{\mathbf{IPL}}$ would be order-isomorphic, $\lightning$. □

There is an even stronger notion of equivalence between logics, viz., that of isomorphism:

[5] See, e.g., [12, p. 269], [3, p. 108], [13, p. 139] and [6, p. 134].

Definition 7 (Isomorphism). We will say $\mathbf{L}_1$ is ***isomorphic*** to $\mathbf{L}_2$, written as "$\mathbf{L}_1 \cong \mathbf{L}_2$", if there is a bijective $t\colon \mathbf{L}_1 \rightsquigarrow \mathbf{L}_2$.

Observe that if $t\colon \mathbf{L}_1 \rightsquigarrow \mathbf{L}_2$ is bijective, then $t^{-1}\colon \mathbf{L}_2 \rightsquigarrow \mathbf{L}_1$, and therefore $t, t^{-1}\colon \mathbf{L}_1 \leftrightsquigarrow \mathbf{L}_2$. Thus, isomorphism implies translational equivalence. The converse can fail for trivial cardinality reasons. For example, let $\mathbf{CPL}^*$ be the result of adding uncountably many "redundant" unary operators $\bigstar_r$ for each $r \in \mathbb{R}$ such that $\bigstar_r \phi \equiv_{\mathbf{CPL}^*} \phi$. Then $\mathbf{CPL} \leftrightsquigarrow \mathbf{CPL}^*$ but $\mathbf{CPL} \ncong \mathbf{CPL}^*$. Yet intuitively, $\mathbf{CPL}^*$ is a notational variant of $\mathbf{CPL}$. After all, each $\bigstar_r$ is quite straightforwardly *definable* in $\mathbf{CPL}$, and intuitively, adding definable operators to a logic does not yield a new logic. Hence, requiring notational variants to be isomorphic would be unreasonably restrictive. One would prefer a weaker notion of notational variance (such as translational equivalence) on which such artificial cardinality considerations are not deemed essential to a logic.

So suppose we stipulate for a moment that two logics are notational variants just in case they are translationally equivalent. We will now show that $\mathbf{L}_1 \rightsquigarrow \mathbf{L}_2$ just in case $\mathbf{L}_1$ is a notational variant of a fragment of $\mathbf{L}_2$.

Definition 8 (Fragment). Let $\mathbf{L}_1$ and $\mathbf{L}_2$ be logics. We will say $\mathbf{L}_1$ is a ***fragment*** of $\mathbf{L}_2$ (written as $\mathbf{L}_1 \subseteq \mathbf{L}_2$) if (a) $\mathcal{L}_1 \subseteq \mathcal{L}_2$, and (b) for all $\Gamma \subseteq \mathcal{L}_1$ and $\phi \in \mathcal{L}_1$: $\Gamma \vDash_1 \phi$ iff $\Gamma \vDash_2 \phi$.

Proposition 9. Let $\mathbf{L}_1$ and $\mathbf{L}_2$ be logics. Then the following are equivalent:

(a) $\mathbf{L}_1 \rightsquigarrow \mathbf{L}_2$.
(b) There is an $\mathbf{L}_2' \subseteq \mathbf{L}_2$ such that $\mathbf{L}_1 \overset{\leftsquigarrow}{\rightsquigarrow} \mathbf{L}_2'$.
(c) There is an $\mathbf{L}_2' \subseteq \mathbf{L}_2$ such that $\mathbf{L}_1 \leftrightsquigarrow \mathbf{L}_2'$.

Proof. Obviously, (c) implies (b), which implies (a) (since the composition of two translations is also a translation). To show that (a) implies (c), let $t\colon \mathbf{L}_1 \rightsquigarrow \mathbf{L}_2$. Define $\mathbf{L}_{t[1]} = \langle t[\mathcal{L}_1], \vDash_{t[1]} \rangle$ where $t[\Gamma] \vDash_{t[1]} t(\phi)$ iff $t[\Gamma] \vDash_2 t(\phi)$. By definition, $\mathbf{L}_{t[1]} \subseteq \mathbf{L}_2$. Hence, it suffices to show that $\mathbf{L}_1 \leftrightsquigarrow \mathbf{L}_{t[1]}$.

Now, t^{-1} (the inverse of t) may not be a function from $t[\mathcal{L}_1]$ to $\mathcal{L}_1$, since t might not be injective. But since t^{-1} is total on $t[\mathcal{L}_1]$, we can always find a function $t^* \subseteq t^{-1}$ (using the axiom of choice) by selecting a $\psi \in \{\psi' \in \mathcal{L}_1 | t(\psi') = \phi\}$ arbitrarily for each $\phi \in \mathbf{L}_{t[1]}$ and setting $t^*(\phi) = \psi$. Observe that t^* is a right-inverse of t, i.e., for all $\phi \in t[\mathcal{L}_1]$, $t(t^*(\phi)) = \phi$. Using this fact, it is straightforward to verify that $t, t^*\colon \mathbf{L}_1 \leftrightsquigarrow \mathbf{L}_{t[1]}$. □

Hence, if notational variance is translational equivalence, then to show that $\mathbf{L}_1$ is a translatable into $\mathbf{L}_2$ *just is* to show that $\mathbf{L}_1$ is a notational variant of a fragment of $\mathbf{L}_2$.

3 Translating First-Order Logic into Propositional Logic

We will now show that first-order logic is translationally equivalent with propositional logic. In fact, we will show any logic satisfying a few simple properties can be translated into propositional logic.

Definition 10 (Monotonic Logic). We will say a logic $\mathbf{L}$ ***monotonic*** if for all $\Gamma, \Delta \subseteq \mathcal{L}$ such that $\Gamma \subseteq \Delta$ and for all $\phi \in \mathcal{L}$, if $\Gamma \vDash_{\mathbf{L}} \phi$, then $\Delta \vDash_{\mathbf{L}} \phi$.

(Note I am using "monotonic" here in a sense different from the sense of "monotonic" when applied specifically to non-normal modal logics. In what follows, I will only use "monotonic" in the sense of Definition 10.)

Definition 11 (Compact Logic). We will say a logic $\mathbf{L}$ is ***compact*** if for all $\Gamma \subseteq \mathcal{L}$ and $\phi \in \mathcal{L}$, $\Gamma \vDash_{\mathbf{L}} \phi$ only if for some finite $\Gamma_0 \subseteq \Gamma$, $\Gamma_0 \vDash_{\mathbf{L}} \phi$.

The following result is due to Jeřábek [10]:

Theorem 12 (Jeřábek). Let $\mathbf{L}$ be a compact monotonic logic with at most countably many formulas. Then $\mathbf{L} \rightsquigarrow \mathbf{CPL}$.

Jeřábek provides an explicit construction of the translation and shows that the translation is Turing-equivalent to the consequence relation of the source logic. This is quite general, but the details of the proof are quite involved. What is more, the construction is not guaranteed to produce a translation scheme. This raises the question of whether **FOL** and **CPL** are translationally equivalent. We will now show the answer is affirmative. Unlike Jeřábek's constructive proof, our proof will go indirectly via Lindenbaum-Tarski algebras. First, some terminology. A poset $\langle P, \leq \rangle$ is a ***meet-semilattice*** if every finite subset of P has a greatest lower bound.

Definition 13 (Adjunctive Logic). A logic $\mathbf{L}$ is ***adjunctive*** if for any $\Gamma \subseteq \mathcal{L}$, if there is a formula ϕ such that $[\phi]_{\mathbf{L}} = \bigwedge_{\gamma \in \Gamma} [\gamma]_{\mathbf{L}}$, then $\Gamma \vDash_{\mathbf{L}} \phi$. We will often write such a ϕ as "$\bigwedge \Gamma$" given it exists.

Theorem 14. Let $\mathbf{L}_1$ and $\mathbf{L}_2$ be compact monotonic adjunctive logics. Suppose also that $\mathbb{L}_1$ and $\mathbb{L}_2$ are meet-semilattices.

(a) $\mathbf{L}_1 \rightsquigarrow \mathbf{L}_2$ iff there is an order-embedding from $\mathbb{L}_1$ to $\mathbb{L}_2$ that preserves finite meets.
(b) $\mathbf{L}_1 \leftrightsquigarrow \mathbf{L}_2$ iff $\mathbb{L}_1$ is order-isomorphic to $\mathbb{L}_2$.
(c) $\mathbf{L}_1 \cong \mathbf{L}_2$ iff there is an $f\colon \mathbb{L}_1 \cong \mathbb{L}_2$ where $|[\phi]_1| = |f([\phi]_1)|$ for each $\phi \in \mathcal{L}_1$.

Proof. The left-to-right directions are straightforward. For the right-to-left directions:

(a) Let $f\colon \mathcal{L}_1/\equiv_1 \rightarrow \mathcal{L}_2/\equiv_2$ be an order-embedding that preserves finite meets. For each $[\phi]_1 \in \mathcal{L}_1/\equiv_1$, let $f_{[\phi]_1}\colon [\phi]_1 \rightarrow f([\phi]_1)$ be an arbitrary map. Define $t(\phi) = f_{[\phi]_1}(\phi)$. Since $\mathbf{L}_1$ is compact, $\Gamma \vDash_1 \phi$ iff for some finite $\Gamma' \subseteq \Gamma$, $\Gamma' \vDash_1 \phi$. And if Γ' is finite, then $\Gamma' \vDash_1 \phi$ iff $\bigwedge \Gamma' \vDash_1 \phi$ ($\bigwedge \Gamma'$ exists since $\mathbb{L}_1$ is a meet-semilattice). Likewise, $t[\Gamma] \vDash_2 t(\phi)$ iff $t[\Gamma'] \vDash_2 t(\phi)$ for some finite $\Gamma' \subseteq \Gamma$, and $t[\Gamma'] \vDash_2 t(\phi)$ iff $\bigwedge t[\Gamma'] \vDash_2 t(\phi)$. Since f preserves finite meets, $\bigwedge_{\gamma \in \Gamma'} f([\gamma]_1) = f(\bigwedge_{\gamma \in \Gamma'} [\gamma]_1) = f([\bigwedge \Gamma']_1)$. Thus, $\bigwedge t[\Gamma'] \equiv_2 t(\bigwedge \Gamma')$. So to show that t is a translation, it suffices to show that for any $\phi, \psi \in \mathcal{L}_1$, $\phi \vDash_1 \psi$ iff $t(\phi) \vDash_2 t(\psi)$. But $\phi \vDash_1 \psi$ iff $[\phi]_1 \leq_1 [\psi]_1$, iff $f([\phi]_1) \leq_2 f([\psi]_1)$, iff $t(\phi) \vDash_2 t(\psi)$. So $t\colon \mathbf{L}_1 \rightsquigarrow \mathbf{L}_2$.

(b) Let $f\colon \mathcal{L}_1/\equiv_1 \to \mathcal{L}_2/\equiv_2$ be an order-isomorphism. As before, for each $[\phi]_1 \in \mathcal{L}_1/\equiv_1$, let $f_{[\phi]_1}\colon [\phi]_1 \to f([\phi]_1)$ be an arbitrary map. Likewise, for each $[\psi]_2 \in \mathcal{L}_2/\equiv_2$, let $g_{[\psi]_2}\colon [\psi]_2 \to f^{-1}([\psi]_2)$ be arbitrary. Define $t(\phi) = f_{[\phi]_1}(\phi)$ and $s(\psi) = g_{[\psi]_2}(\psi)$. The reasoning above shows that $t\colon \mathbf{L}_1 \rightsquigarrow \mathbf{L}_2$ and $s\colon \mathbf{L}_2 \rightsquigarrow \mathbf{L}_1$. Now, let $\phi \in \mathcal{L}_1$. Then $\phi \equiv_1 s(t(\phi))$ iff $[\phi]_1 = [s(t(\phi))]_1 = f^{-1}([t(\phi)]_2) = f^{-1}(f([\phi]_1)) = [\phi]_1$. So $\phi \equiv_1 s(t(\phi))$ for all $\phi \in \mathcal{L}_1$. Likewise, $\psi \equiv_1 t(s(\psi))$ for all $\psi \in \mathcal{L}_2$. Hence, $t, s\colon \mathbf{L}_1 \leftrightsquigarrow \mathbf{L}_2$.

(c) Under these conditions, we can take each $f_{[\phi]_1}$ to be bijective, making t as a whole bijective. □

Corollary 15. FOL $\cong$ CPL.

Proof. Immediate since $\mathbb{L}_{\mathbf{FOL}}$ and $\mathbb{L}_{\mathbf{CPL}}$ are countable atomless Boolean algebras and any two countable atomless Boolean algebras are isomorphic. □

Note that such an isomorphism is obviously undecidable. One might try to block this result by requiring notational variants to be Turing-equivalent. But this requirement is both too weak and too strong. On the one hand, it is too weak, since monadic first-order logic, which is decidable, would still be deemed to be a notational variant of propositional logic. On the other hand, it is too strong, since it seems plausible that some notational variants of a logic can be more computationally efficient than others. We can illustrate this point with a simple example.

Example. Let $X \subseteq \mathbb{N}$ be a nonrecursive set, and let $\mathcal{L}_{\text{prop}}(\bowtie)$ be the result of adding countably many binary connectives $\bowtie_i$ (where $i \in \mathbb{N}$) to $\mathcal{L}_{\text{prop}}$. We will define the logic $\mathbf{CPL}^{\bowtie}$ semantically. The semantics for atomics and the standard boolean connectives is the same as before. The semantics of $\bowtie_i$ is as follows: if $i \in X$, then $\phi \bowtie_i \psi$ is true on a valuation iff ϕ and ψ are true on that valuation; if $i \notin X$, then $\phi \bowtie_i \psi$ is true on a valuation iff ϕ or ψ is true on that valuation. Finally, $\mathbf{CPL}^{\bowtie} = \langle \mathcal{L}_{\text{prop}}(\bowtie), \vDash_{\bowtie} \rangle$, where $\Gamma \vDash_{\bowtie} \phi$ iff ϕ is true on every valuation on which Γ is true. Intuitively, $\mathbf{CPL}^{\bowtie}$ is a notational variant of **CPL**. After all, each $\bowtie_i$ is definable in terms of connectives in **CPL**: $\mathbf{CPL}(\bowtie)$ is just **CPL** with infinitely many connectives expressing conjunction or disjunction. But $\mathbf{CPL}^{\bowtie}$ is not decidable, since a decision procedure for $\mathbf{CPL}^{\bowtie}$ would generate a decision procedure for X (just check to see if $p \vDash_{\bowtie} p \bowtie_i q$).

Thus, we cannot avoid Theorem 14 by appealing to computability considerations. Something else must explain why **FOL** and **CPL** are not merely notational variants.

Corollary 15 allows us to define a t and s such that $t, s\colon \mathbf{FOL} \leftrightsquigarrow \mathbf{CPL}$ that preserves the boolean connectives exactly:

Proposition 16. There are some $t, s\colon \mathbf{FOL} \leftrightsquigarrow \mathbf{CPL}$ such that $t(\neg\phi) = \neg t(\phi)$ and $t(\phi \wedge \psi) = t(\phi) \wedge t(\psi)$ (and likewise for s).

Proof. Let $i\colon \mathbf{FOL} \rightsquigarrow \mathbf{CPL}$ be bijective. Define t and s as follows:

$$\begin{aligned} t(P^n(y_1,\ldots,y_n)) &= i(P^n(y_1,\ldots,y_n)) & s(p) &= i^{-1}(p) \\ t(\neg\phi) &= \neg t(\phi) & s(\neg\phi) &= \neg s(\phi) \\ t(\phi\wedge\psi) &= t(\phi)\wedge t(\psi) & s(\phi\wedge\psi) &= s(\phi)\wedge s(\psi) \\ t(\forall x\,\phi) &= i(\forall x\, i^{-1}(t(\phi))). \end{aligned}$$

It is straightforward to check by induction that $t(\phi) \equiv_{\mathbf{CPL}} i(\phi)$ for all $\phi \in \mathcal{L}_{\text{pred}}$ and $s(\psi) \equiv_{\mathbf{FOL}} i^{-1}(\psi)$ for all $\psi \in \mathcal{L}_{\text{prop}}$. Hence, for any $\phi \in \mathcal{L}_{\text{pred}}$, $s(t(\phi)) \equiv_{\mathbf{FOL}} i^{-1}(i(\phi)) = \phi$. Likewise, for any $\psi \in \mathcal{L}_{\text{prop}}$, $t(s(\psi)) \equiv_{\mathbf{CPL}} i(i^{-1}(\psi)) = \psi$. So $t, s\colon \mathbf{FOL} \leftrightsquigarrow \mathbf{CPL}$.

It is interesting to note that there is no isomorphism between **FOL** and **CPL** with this property. If there were such an i, then it would have to map both atomic predicate formulas and quantified formulas to atomic propositional formulas (e.g., if $i(F(x)) = \neg\theta$, then $F(x) = i^{-1}(i(F(x))) = i^{-1}(\neg\theta) = \neg i^{-1}(\theta)$, contrary to the fact that $F(x)$ has no negation, $\lightning$). But then $i(\forall x\, F(x))$ and $i(F(x))$ would need to be logically independent atomic formulas, contrary to the fact that $\forall x\, F(x) \vDash_{\mathbf{FOL}} F(x)$, $\lightning$.

4 Compositionality and Schematicity

The notion of a translation as defined in Definition 2 is fairly minimal. In theory, a translation could be quite gerrymandered and complex. In practice, most translations that have been studied are fairly *schematic.* Usually one defines a translation by first defining how to translate the atomic formulas, and then settling how to define the translation of complex formulas in terms of their parts via another schema. And indeed, the translations from **FOL** to **CPL** described in Theorem 14 and Proposition 16 do not have this property. This suggests the thesis that two logics are notational variants just in case they are translationally equivalent via *schematic* translations. In this section, we will explore different ways of fleshing out this idea.

First, to explicate this idea more precisely, we need to build more structure into the definition of a logic. As it stands, a logic is just a class of formulas together with a consequence relation on those formulas. Nothing in Definition 1 demands that the class of formulas a logic is built from must have any underlying compositional structure. Thus, if we want to make use of the notion of schematicity, the definition of a logic must include a specification of its underlying syntactic structure.

Definition 17 (Signature). A ***signature*** is a pair $\Sigma = \langle \mathtt{At}, \mathtt{Op}\rangle$ where $\mathtt{At}$ and $\mathtt{Op}$ are nonempty classes and $\mathtt{Op}$ is a class of pairs $\langle \bigstar, \gamma\rangle$ where $\bigstar$ is a set and γ is an ordinal. The ***Σ-syntax*** is the smallest class $\mathcal{L}_\Sigma$ such that:

(i) for all $\phi \in \mathtt{At}$, $\langle\phi\rangle \in \mathcal{L}_\Sigma$

(ii) for all $\langle\bigstar,\gamma\rangle \in \texttt{Op}$ and all $\rho \in \mathcal{L}_\Sigma^\gamma$ (= the class of γ-sequences of elements of $\mathcal{L}_\Sigma$), $\langle\bigstar,\rho\rangle \in \mathcal{L}_\Sigma$ (we may write "$\bigstar(\rho)$" in place of "$\langle\bigstar,\rho\rangle$").

We call the members of $\mathcal{L}_\Sigma$ the ***Σ-formulas***. A ***Σ-logic*** is a pair $\langle\Sigma,\models\rangle$ where Σ is a signature and $\langle\mathcal{L}_\Sigma,\models\rangle$ is a logic in the sense of Definition 1. A ***translation*** from $\mathbf{L}_1 = \langle\Sigma_1,\models_1\rangle$ to $\mathbf{L}_2 = \langle\Sigma_2,\models_2\rangle$ is just a translation from $\langle\mathcal{L}_{\Sigma_1},\models_1\rangle$ to $\langle\mathcal{L}_{\Sigma_2},\models_2\rangle$.

A number of authors have claimed that for two logics to be notational variants, there need to exist some *compositional* translations between them.[6] To make this precise, we need the following definition:

Definition 18 (Schema)**.** Let $\Sigma = \langle\texttt{At},\texttt{Op}\rangle$ be a signature, and let Π be disjoint from $\mathcal{L}_\Sigma$. A ***Σ-schema*** with parameters in Π is a $\Sigma(\Pi)$-formula where $\Sigma(\Pi) = \langle\texttt{At}\cup\Pi,\texttt{Op}\rangle$. If $\rho \in \mathcal{L}_\Sigma^\gamma$ and if $\Theta(\pi)$ is a Σ-schema where π is a γ-sequence listing the parameters in Θ, we may write "$\Theta(\rho)$" for the Σ-formula obtained by replacing each $\pi(\beta)$ in $\Theta(\pi)$ with $\rho(\beta)$ for $\beta < \gamma$.

Definition 19 (Compositionality)**.** Let $\mathbf{L}_1$ and $\mathbf{L}_2$ be Σ_1- and Σ_2-logics respectively. A translation $t\colon \mathbf{L}_1 \rightsquigarrow \mathbf{L}_2$ is ***compositional*** if for all $\bigstar \in \texttt{Op}_1$, there is an Σ_2-schema $\Theta^\bigstar(\pi)$ such that for all $\rho \in \mathcal{L}_1^\gamma$, $t(\bigstar(\rho)) = \Theta^\bigstar(t\circ\rho)$.

The existence of a translation from one logic to another does not in general imply the existence of a compositional translation from the former to the latter. In particular, there is no compositional translation from **FOL** to **CPL**, nor one from **IPL** to **CPL**.[7]. On the other hand, there is a compositional translation from **CPL** to **FOL** and a compositional translation from **CPL** to **IPL**. Compositional translations can also be used to distinguish **CPL** and most normal modal logics:

Proposition 20. If $\mathbf{L}$ is a normal modal logic and if $t\colon \mathbf{L} \rightsquigarrow \mathbf{CPL}$ is compositional, then $\Box\phi \equiv_{\mathbf{L}} \phi \vee \Box\bot$.

Proof. Suppose $\Theta^\Box(\pi)$ is a $\mathcal{L}_{\mathbf{CPL}}$-schema such that $t(\Box\phi) = \Theta^\Box(t(\phi))$. Observe that $\Theta^\Box(t(\phi)) \equiv_{\mathbf{CPL}} (t(\phi)\wedge\lambda)\vee(\neg t(\phi)\wedge\mu)$, where λ and μ are some $\mathcal{L}_{\mathbf{CPL}}$-formulas. Since $\models_{\mathbf{L}} \Box\top$, we have that $\models_{\mathbf{CPL}} t(\Box\top) \equiv_{\mathbf{CPL}} (t(\top)\wedge\lambda)\vee(\neg t(\top)\wedge\mu) \equiv_{\mathbf{CPL}} \lambda$. Hence, $t(\Box\phi) \equiv_{\mathbf{CPL}} t(\phi)\vee\mu$. Thus, $t(\phi) \models_{\mathbf{CPL}} t(\Box\phi)$, and so $\phi \models_{\mathbf{L}} \Box\phi$, from which it follows that $\Box\phi \equiv_{\mathbf{L}} \phi\vee\Box\bot$. □

Most translations that have been studied in the literature are compositional. So one might suspect we could simply postulate that two logics are notational

[6] [5, p. 391] uses the term "grammatical" instead of "compositional". [9, p. 100], [12, p. 269] and [3, p. 108] build compositionality into the definition of a translation from the start.

[7] This follows from Theorems 23 and 24 below. There are also more direct proofs of these claims. For instance, suppose there were a compositional t: **FOL** $\rightsquigarrow$ **CPL**. Then where Θ is the **CPL**-schema such that $t(\forall x\,\phi) = \Theta(t(\phi))$, we have $t(\forall x\top) = \Theta(t(\top)) \equiv_{\mathbf{CPL}} \Theta(\top)$. Hence, $t(\phi) \models_{\mathbf{CPL}} t(\phi) \leftrightarrow \top \models_{\mathbf{CPL}} \Theta(t(\phi)) \leftrightarrow \Theta(\top) \models_{\mathbf{CPL}} \Theta(t(\phi)) = t(\forall x\,\phi)$ for any $\phi \in \mathcal{L}_{\mathrm{pred}}$. But then $\phi \models_{\mathbf{FOL}} \forall x\,\phi$ for any $\phi \in \mathcal{L}_{\mathrm{pred}}$, $\lightning$.

variants just in case they are compositionally translationally equivalent. But this would be too restrictive. For instance, a number of modal logicians see the van Benthem characterization theorem as showing that modal logic is just (a notational variant of) the bisimulation-invariant fragment of first-order logic via the standard translation.[8] But the standard translation of modal logic into first-order logic is not compositional according to Definition 19.[9] In particular, consider the $\Box$-clause:

$$ST_n(\Box\phi) = \forall x_{n+1}\,(R(x_n, x_{n+1}) \rightarrow ST_{n+1}(\phi)).$$

Since $ST_n(\phi)$ does not occur anywhere as a subformula of $ST_n(\Box\phi)$ (rather, $ST_{n+1}(\phi)$ does), and since compositional translations are required to have the translations of their constituents as subformulas, ST_n is not compositional. In fact, it can be shown that there is no compositional $t\colon \mathbf{K} \rightsquigarrow \mathbf{FOL}$, where $\mathbf{K}$ is the minimal normal modal logic. The following is proved in the appendix:[10]

Theorem 21. Let $\mathbf{L}$ be a normal modal logic. If $t\colon \mathbf{L} \rightsquigarrow \mathbf{FOL}$ is compositional, then $\Box\phi \equiv_{\mathbf{L}} \Box\Box\,\phi$.

Hence, it would be too restrictive to demand that notational variants be compositionally translationally equivalent. Still, arguably there is a sense in which the standard translation is nearly compositional. The problem with the definition of compositionality (Definition 19) is that sometimes a translation can only be defined *simultaneously* with other translations. This is what the standard translation of modal logic into first-order logic illustrates. But intuitively, that should not matter. What is important is not that the translation of a complex formula is strictly a schema of the translation of the parts, but rather that the translation of a complex formula is uniform and fixed solely by its syntactic structure. This motivates a more general notion of compositionality along the following lines:[11]

Definition 22 (Schematicity). Let $\mathbf{L}_1$ and $\mathbf{L}_2$ be Σ_1- and Σ_2-logics respectively, and let T be a class of translations from $\mathbf{L}_1$ to $\mathbf{L}_2$. We will say T is ***compositionally interdependent*** if for each $t \in T$ and for each $\bigstar \in \mathtt{Op}_1$, there is an Σ_2-schema $\Theta^{\bigstar}(\pi)$ with a γ-sequence of distinct parameters π and there is a $\tau \in T^{\gamma}$ such that for all $\rho \in \mathcal{L}_1^{\gamma}$, $t(\bigstar(\rho)) = \Theta^{\bigstar}(\tau \cdot \rho)$, where we define $(\tau \cdot \rho)(\beta) = \tau(\beta)(\rho(\beta))$. We will say a translation is ***schematic*** if it is a member of a compositionally interdependent set.[12]

[8] See, e.g., [1, p. 1] and [2, p. 70].

[9] Mossakowski et al. [11, p. 4] make this observation as well, though they do not offer any alternative notion in its place.

[10] The theorem cannot be extended to all normal modal logics, since there is a compositional translation from **S5** to **FOL** (setting $t(\Box\phi) = \forall x\, t(\phi)$). It is unknown whether the result extends to other logics like **S4** that validate $\Box\phi \leftrightarrow \Box\Box\,\phi$.

[11] The definition is inspired by the definition of "recursive" translations from [6, p. 16], who attributes the definition to Steven Kuhn.

[12] We could also require schematic translations to translate atomic formulas schematically. Such a constraint seems well-motivated, but it was not included in this definition for purposes of generality, as it was not necessary in the results to follow.

If t is compositional, then it is a member of a compositionally interdependent set, but not *vice versa*, as the standard translation from **K** into **FOL** shows. So the fact that no compositional translation from **FOL** to **CPL** exists does not immediately imply that there is no schematic translation from **FOL** to **CPL**. Fortunately, with a little more work, we can achieve this result as well.

Theorem 23. There is no schematic $t\colon \mathbf{FOL} \rightsquigarrow \mathbf{CPL}$.

Proof. Suppose there were such a t. Let $\Theta(\pi)$ be a $\Sigma_{\mathbf{CPL}}$-schema with a single parameter π and let $t'\colon \mathbf{FOL} \rightsquigarrow \mathbf{CPL}$ be such that $t(\exists x\,\phi) = \Theta(t'(\phi))$ (such a schema must exist if there are such schemas for $\forall x$ and $\neg$). Then $\vDash_{\mathbf{CPL}} t(\top) \equiv_{\mathbf{CPL}} t(\exists x\,\top) = \Theta(t'(\top)) \equiv_{\mathbf{CPL}} \Theta(\top)$ (since $t'(\top) \equiv_{\mathbf{CPL}} \top$). Hence, $t'(\phi) \vDash_{\mathbf{CPL}} t'(\phi) \leftrightarrow \top \vDash_{\mathbf{CPL}} \Theta(t'(\phi)) \leftrightarrow \Theta(\top) \vDash_{\mathbf{CPL}} \Theta(t'(\phi)) = t(\exists x\,\phi)$. So $t'(\phi) \vDash_{\mathbf{CPL}} t(\exists x\,\phi)$ for all $\phi \in \mathcal{L}_{\mathbf{FOL}}$.

Now, $\Theta(t'(\phi)), \neg t'(\phi) \vDash_{\mathbf{CPL}} \Theta(\bot)$. Moreover, $\neg t'(\bot) \vDash_{\mathbf{CPL}} t'(\bot) \leftrightarrow \bot \vDash_{\mathbf{CPL}} \Theta(t'(\bot)) \leftrightarrow \Theta(\bot)$. So $t(\exists x\,\phi), \neg t'(\phi), \neg t'(\bot) \vDash_{\mathbf{CPL}} \Theta(t'(\bot))$. But $t'(\bot) \vDash_{\mathbf{CPL}} \Theta(t'(\bot))$, so either way, $t(\exists x\,\phi), \neg t'(\phi) \vDash_{\mathbf{CPL}} \Theta(t'(\bot)) = t(\exists x\,\bot) \equiv_{\mathbf{CPL}} t(\bot)$. Hence, $t(\exists x\,\phi) \vDash_{\mathbf{CPL}} t'(\phi) \vee t(\bot)$. Moreover, the converse holds too, since $t'(\phi) \vDash_{\mathbf{CPL}} t(\exists x\,\phi)$ and $t(\bot) \vDash_{\mathbf{CPL}} t(\exists x\,\phi)$. So $t(\exists x\,\phi) \equiv_{\mathbf{CPL}} t'(\phi) \vee t(\bot)$ for all $\phi \in \mathcal{L}_{\mathbf{FOL}}$. Now, observe that if $s\colon \mathbf{FOL} \rightsquigarrow \mathbf{CPL}$, $s(\phi \wedge \psi) \equiv_{\mathbf{CPL}} s(\phi) \wedge s(\psi)$ for any $\phi, \psi \in \mathcal{L}_{\mathbf{FOL}}$. Thus, we have $t(\exists x\,\phi \wedge \exists x\,\neg\phi) \equiv_{\mathbf{CPL}} t(\exists x\,\phi) \wedge t(\exists x\,\neg\phi) \vDash_{\mathbf{CPL}} (t'(\phi) \wedge t'(\neg\phi)) \vee t(\bot) \equiv_{\mathbf{CPL}} t'(\phi \wedge \neg\phi) \vee t(\bot) \equiv_{\mathbf{CPL}} t'(\bot) \vee t(\bot) \equiv_{\mathbf{CPL}} t(\exists x\,\bot) \equiv_{\mathbf{CPL}} t(\bot)$, $\lightning$. $\square$

We have yet to find a natural example of a pair of logics $\mathbf{L}_1$ and $\mathbf{L}_2$ that are schematically intertranslatable but not schematically translationally equivalent. Given Proposition 6, one might wonder whether **IPL** and **CPL** could witness schematic intertranslatability without schematic translational equivalence. The answer is negative:

Theorem 24. There is no schematic $t\colon \mathbf{IPL} \rightsquigarrow \mathbf{CPL}$.

Proof. Suppose there were a such a t. Let $\Theta(\pi)$ be a $\Sigma_{\mathbf{CPL}}$-schema and let $t'\colon \mathbf{IPL} \rightsquigarrow \mathbf{CPL}$ be such that $t(\neg\phi) = \Theta(t'(\phi))$. Then $\vDash_{\mathbf{CPL}} t(\top) = t(\neg\bot) = \Theta(t'(\bot))$. So $\vDash_{\mathbf{CPL}} \Theta(t'(\bot))$. So $t'(\neg\phi) \vDash_{\mathbf{CPL}} t'(\phi \leftrightarrow \bot) \vDash_{\mathbf{CPL}} t'(\phi) \leftrightarrow t'(\bot) \vDash_{\mathbf{CPL}} \Theta(t'(\phi)) \leftrightarrow \Theta(t'(\bot)) \vDash_{\mathbf{CPL}} \Theta(t'(\phi))$. So $t'(\neg\phi) \vDash_{\mathbf{CPL}} t(\neg\phi)$.

Since $\vDash_{\mathbf{CPL}} t'(\bot) \vee \neg t'(\bot)$ and $\vDash_{\mathbf{CPL}} \Theta(t'(\bot))$, we have that $\vDash_{\mathbf{CPL}} \Theta(\top) \vee \Theta(\bot)$. Now, $t(\bot) \vDash_{\mathbf{CPL}} t(\neg\phi)$; so $\neg t(\neg\phi) \vDash_{\mathbf{CPL}} \neg t(\bot) \equiv_{\mathbf{CPL}} \neg t(\neg\top) = \neg\Theta(t'(\top)) \equiv_{\mathbf{CPL}} \neg\Theta(\top) \vDash_{\mathbf{CPL}} \Theta(\bot) \vDash_{\mathbf{CPL}} \Theta(t'(\neg\phi))$ (since $\neg t(\neg\phi) \vDash_{\mathbf{CPL}} \neg t'(\neg\phi) \vDash_{\mathbf{CPL}} t'(\neg\phi) \leftrightarrow \bot$). Thus, $\neg t(\neg\phi) \vDash_{\mathbf{CPL}} \Theta(t'(\neg\phi)) = t(\neg\neg\phi)$. Hence, $\vDash_{\mathbf{CPL}} t(\neg\phi) \vee \neg t(\neg\phi) \vDash_{\mathbf{CPL}} t(\neg\phi) \vee t(\neg\neg\phi)$. But $t(\phi) \vee t(\psi) \vDash_{\mathbf{CPL}} t(\phi \vee \psi)$. So $\vDash_{\mathbf{CPL}} t(\neg\phi \vee \neg\neg\phi)$, even though $\nvDash_{\mathbf{IPL}} \neg\phi \vee \neg\neg\phi$. $\square$

These results suggest that a more adequate precisification of the concept of notational variance can be stated in terms of schematicity: two logics are notational variants just in case they are schematically translationally equivalent.

This is not to say that schematic translational equivalence is *the* correct precisification of notational variance. Perhaps one will find this particular precisification too restrictive or too general, in which case one might want to explore other notions of notational variance for different purposes. It might turn out that there simply is no unique precisification of this informal concept. Still, schematic translational equivalence at least seems to be an improvement over other notions in the literature in its ability to align more closely with our intuitive judgments.

5 Conclusion

Translations are often employed as a way of determining whether or not one logic is a notational variant of a fragment of another. We saw, however, that most attempts to precisify the concept of a notational variant using translations are either too weak or too strong. If, on the one hand, we stipulate that translational equivalence is sufficient for notational variance, then we will be forced to say that first-order logic and propositional logic are notational variants. If, on the other hand, we require notational variants to be compositionally intertranslatable, then modal logic will not be a notational variant of a fragment of first-order logic. Fortunately, we saw that we could balance between these two proposals by stipulating that two logics are notational variants just in case they are schematically translationally equivalent. Thus, equating notational variance with schematic translational equivalence seems to be a plausible alternative to the previous accounts of notational variance.

A Proof of Theorem 21

Let $\Theta(\xi)$ be a first-order schema such that $t(\Box\phi) = \Theta(t(\phi))$. Without loss of generality, we may assume $\Theta(\xi)$ is in (roughly) prenex normal form, i.e., that:

$$\Theta(t(\phi)) = \mathsf{Q}_1 y_1 \ldots \mathsf{Q}_n y_n \; ((t(\phi) \wedge \lambda) \vee (\neg t(\phi) \wedge \mu))$$

where λ and μ are boolean combinations of atomic **FOL**-formulas and each $\mathsf{Q}_i \in \{\forall, \exists\}$. Observe that:

$$\vDash_{\mathbf{FOL}} t(\Box\top) = \mathsf{Q}_1 y_1 \ldots \mathsf{Q}_n y_n \; ((t(\top) \wedge \lambda) \vee (\neg t(\top) \wedge \mu)) \equiv_{\mathbf{FOL}} \mathsf{Q}_1 y_1 \ldots \mathsf{Q}_n y_n \, \lambda.$$

So $\vDash_{\mathbf{FOL}} \mathsf{Q}_1 y_1 \ldots \mathsf{Q}_n y_n \, \lambda$.

First, we show $\Box\phi \vDash_{\mathbf{L}} \Box\Box\,\phi$. Using the fact that $\vDash_{\mathbf{FOL}} \mathsf{Q}_1 y_1 \ldots \mathsf{Q}_n y_n \, \lambda$:

$$\begin{aligned} t(\Box\phi) &\vDash_{\mathbf{FOL}} t(\Box\phi) \wedge \mathsf{Q}_1 y_1 \ldots \mathsf{Q}_n y_n \, \lambda \\ &\equiv_{\mathbf{FOL}} \mathsf{Q}_1 y_1 \ldots \mathsf{Q}_n y_n \; (t(\Box\phi) \wedge \lambda), \end{aligned}$$

since $y_1, \ldots, y_n$ are already bound in $t(\Box\phi)$. So:

$$\begin{aligned} t(\Box\phi) &\vDash_{\mathbf{FOL}} \mathsf{Q}_1 y_1 \ldots \mathsf{Q}_n y_n \; (t(\Box\phi) \wedge \lambda) \\ &\vDash_{\mathbf{FOL}} \mathsf{Q}_1 y_1 \ldots \mathsf{Q}_n y_n \; ((t(\Box\phi) \wedge \lambda) \vee (\neg t(\Box\phi) \wedge \mu)) \\ &= t(\Box\Box\,\phi). \end{aligned}$$

Hence, $t(\Box\phi) \vDash_{\mathbf{FOL}} t(\Box\Box\,\phi)$, and thus, $\Box\phi \vDash_{\mathbf{L}} \Box\Box\,\phi$.

Next, we show $\Box\Box\,\phi \vDash_{\mathbf{L}} \Box\phi$. Observe that:

$$\Theta(t(\phi)) \equiv_{\mathbf{FOL}} \mathsf{Q}_1 y_1 \ldots \mathsf{Q}_n y_n\, ((t(\phi) \vee \mu) \wedge (\neg t(\phi) \vee \lambda))$$

So:

$$\begin{aligned}
t(\Box\Box\,\phi) &\equiv_{\mathbf{FOL}} \mathsf{Q}_1 y_1 \ldots \mathsf{Q}_n y_n\, ((t(\Box\phi) \vee \mu) \wedge (\neg t(\Box\phi) \vee \lambda))\\
&\vDash_{\mathbf{FOL}} \mathsf{Q}_1 y_1 \ldots \mathsf{Q}_n y_n\, (t(\Box\phi) \vee \mu)\\
&\equiv_{\mathbf{FOL}} t(\Box\phi) \vee \mathsf{Q}_1 y_1 \ldots \mathsf{Q}_n y_n\, \mu\\
&\equiv_{\mathbf{FOL}} t(\Box\phi) \vee (\neg t(\Box\phi) \wedge \mathsf{Q}_1 y_1 \ldots \mathsf{Q}_n y_n\, \mu)\\
&\equiv_{\mathbf{FOL}} (t(\Box\phi) \wedge \mathsf{Q}_1 y_1 \ldots \mathsf{Q}_n y_n\, \lambda) \vee (\neg t(\Box\phi) \wedge \mathsf{Q}_1 y_1 \ldots \mathsf{Q}_n y_n\, \mu)\\
&\equiv_{\mathbf{FOL}} \mathsf{Q}_1 y_1 \ldots \mathsf{Q}_n y_n\, (t(\Box\phi) \wedge \lambda) \vee \mathsf{Q}_1 y_1 \ldots \mathsf{Q}_n y_n\, (\neg t(\Box\phi) \wedge \mu)\\
&\vDash_{\mathbf{FOL}} \mathsf{Q}_1 y_1 \ldots \mathsf{Q}_n y_n\, ((t(\Box\phi) \wedge \lambda) \vee (\neg t(\Box\phi) \wedge \mu))\\
&= t(\Box\Box\,\phi).
\end{aligned}$$

Thus, in particular, $t(\Box\Box\,\phi) \equiv_{\mathbf{FOL}} t(\Box\phi) \vee \mathsf{Q}_1 y_1 \ldots \mathsf{Q}_n y_n\, \mu$. Now, note that $t(\phi \wedge \psi) \equiv_{\mathbf{FOL}} t(\phi) \wedge t(\psi)$. Hence, unpacking $t(\Box(\phi \wedge \psi))$:

$$\begin{aligned}
t(\Box(\phi \wedge \psi)) &\equiv_{\mathbf{FOL}} \mathsf{Q}_1 y_1 \ldots \mathsf{Q}_n y_n\, ((t(\phi \wedge \psi) \wedge \lambda) \vee (\neg t(\phi \wedge \psi) \wedge \mu))\\
&\equiv_{\mathbf{FOL}} \mathsf{Q}_1 y_1 \ldots \mathsf{Q}_n y_n\, ((t(\phi \wedge \psi) \wedge \lambda) \vee (\neg\, (t(\phi) \wedge t(\psi)) \wedge \mu))\\
&\equiv_{\mathbf{FOL}} \mathsf{Q}_1 y_1 \ldots \mathsf{Q}_n y_n\, ((t(\phi \wedge \psi) \wedge \lambda) \vee ((\neg t(\phi) \vee \neg t(\psi)) \wedge \mu))\\
&\equiv_{\mathbf{FOL}} \mathsf{Q}_1 y_1 \ldots \mathsf{Q}_n y_n\, ((t(\phi \wedge \psi) \wedge \lambda) \vee (\neg t(\phi) \wedge \mu) \vee (\neg t(\psi) \wedge \mu))\,.
\end{aligned}$$

Since $\Box(\phi \wedge \psi) \vDash_{\mathbf{L}} \Box\phi$, and since $\mathsf{Q}_1 y_1 \ldots \mathsf{Q}_n y_n\, (\neg t(\psi) \wedge \mu) \vDash_{\mathbf{FOL}} t(\Box(\phi \wedge \psi)$ (given the last equivalence above), that means that $\mathsf{Q}_1 y_1 \ldots \mathsf{Q}_n y_n\, (\neg t(\psi) \wedge \mu) \vDash_{\mathbf{FOL}} t(\Box\phi)$ for any ϕ and ψ. In particular, $\mathsf{Q}_1 y_1 \ldots \mathsf{Q}_n y_n\, (\neg t(\Box\phi) \wedge \mu) \equiv_{\mathbf{FOL}} \neg t(\Box\phi) \wedge \mathsf{Q}_1 y_1 \ldots \mathsf{Q}_n y_n\, \mu \vDash_{\mathbf{FOL}} t(\Box\phi)$. Hence, $\mathsf{Q}_1 y_1 \ldots \mathsf{Q}_n y_n\, \mu \vDash_{\mathbf{FOL}} t(\Box\phi)$. Thus, we have that $t(\Box\Box\,\phi) \equiv_{\mathbf{FOL}} t(\Box\phi)$. □

References

1. Andréka, H., István, N., van Benthem, J.F.A.K.: Modal languages and bounded fragments of predicate logic. J. Philos. Logic **27**(3), 217–274 (1998)
2. Blackburn, P., de Rijke, M., Venema, Y.: Modal Logic. Cambridge University Press, Cambridge (2001)
3. Caleiro, C., Gonçalves, R.: Equipollent logical systems. In: Beziau, J.Y. (ed.) Logica Universalis, pp. 97–109. Springer, Heidelberg (2007). doi:10.1007/978-3-7643-8354-1_6
4. Carnielli, W.A., Coniglio, M.E., D'Ottaviano, I.M.L.: New dimensions on translations between logics. Logica Universalis **3**, 1–18 (2009)
5. Epstein, R.L.: The semantic foundations of logic. In: Epstein, R.L. (ed.) The Semantic Foundations of Logic Volume 1: Propositional Logics. Springer, Dordrecht (1990). doi:10.1007/978-94-009-0525-2_11
6. French, R.: Translational embeddings in modal logic. Ph.D. thesis (2010)

7. Gödel, K.: Zur Intuitionistischen Arithmetik und Zahlentheorie. Ergebnisse eines mathematischen Kolloquiums **4**, 34–38 (1933). Reprinted in Gödel 1986, pp. 286–295
8. Gödel, K.: Collected Works. Oxford University Press, Oxford (1986)
9. Kracht, M., Wolter, F.: Normal monomodal logics can simulate all others. J. Symb. Logic **64**(01), 99–138 (1999)
10. Jeřábek, E.: The ubiquity of conservative translations. Rev. Symb. Logic **5**, 666–678 (2012)
11. Mossakowski, T., Diaconescu, R., Tarlecki, A.: What is a logic translation? Logica Universalis **3**, 95–124 (2009)
12. Pelletier, F.J., Urquhart, A.: Synonymous logics. J. Philos. Logic **32**, 259–285 (2003)
13. Straßburger, L.: What is a logic, and what is a proof? In: Beziau, J.Y. (ed.) Logica Universalis, pp. 135–152. Springer, Basel (2007)
14. Thomason, S.K.: Reduction of tense logic to modal logic. I. J. Symb. Logic **39**(3), 549–551 (1974)
15. Thomason, S.K.: Reduction of tense logic to modal logic II. Theoria **41**(3), 154–169 (1975)
16. Wójcicki, R.: Theory of Logical Calculi: Basic Theory of Consequence Operators. Springer, Dordrecht (1988). doi:10.1007/978-94-015-6942-2

Conditional Doxastic Logic with Oughts and Concurrent Upgrades

Roberto Ciuni(✉)

Department FISPPA, Section of Philosophy, University of Padua, Padua, Italy
roberto.ciuni@unipd.it

Abstract. In this paper, we model the behavior of an epistemic agent that faces a deliberation against a background of oughts, beliefs and information. We do this by introducing a dynamic epistemic logic where ought operators are defined and release of information makes beliefs and oughts co-vary. The static part of the logic extends single-agent Conditional Doxastic Logic by combining dyadic operators for conditional beliefs and oughts that are interpreted over two distinct preorders. The dynamic part of the logic introduces concurrent upgrade operators, which are interpreted on operations that change the two preorders in the same way, thus generating the covariation of beliefs and oughts. The effect of the covariation is that, after receiving new information, the agent will change both her beliefs and her oughts accordingly, and in deliberating, she will pick up the best states among those she takes to be the most plausible.

1 Introduction

Recent works on deontic notions and preferences [8,16,21] have highlighted that deliberation about what ought to be the case or to be done depends on the information available to the agents. However, these works discuss the case where information comes from an infallible source, to the effect that the agent increases her *knowledge*. In real situations, confining ourselves to these cases proves limiting. In particular, we often deliberate on the ground of what we *believe*, simply, and if we come to *change* our beliefs, our deliberation changes accordingly. Take the following example:

Stolen Wallet. Bob has his wallet stolen. He believes that Jones did not steal it, and he knows that Jones ought to be punished if and only if he stole the wallet. Hence, Bob believes Jones ought not to be punished. Ann, whom Bob trusts much, tells Bob that Jones stole his wallet. Bob comes to believe that Jones stole the wallet, and, consequently, he comes to the conclusion that Jones ought to be punished.

The author wishes to thank two anonymous reviewers and Ilaria Canavotto, Davide Grossi, Carlo Proietti for their helpful comments. Research for this paper was carried while the author was a Marie Curie IE Fellow with the WADOXA project at the Institute of Logic, Language and Computation, University of Amsterdam (2015–2016).

A. Baltag et al. (Eds.): LORI 2017, LNCS 10455, pp. 299–313, 2017.
DOI: 10.1007/978-3-662-55665-8_21

In this scenario, the agent assesses what ought to be the case on the ground of a prior set of (conditional) oughts and his beliefs on what is the case. The same mechanism is at stake in any *deliberation scenario.* Also, the example comes with *static* and *dynamic* parts. The static part details what Bob believes and what (he knows) ought to be the case. This includes *conditional oughts*, which provide an instruction on what option turns best in given circumstances. The dynamic part details how received information (from a highly trusted source) *changes* Bob's beliefs on how things are and his deliberation on what ought to be the case. In particular, after Ann's announcement, the beliefs and oughts defined in the static part *co-vary*: Bob comes to assume that what is best must come along with the situation that he now believes—hence, he comes to believe that Jones ought to be punished.

In this paper, we combine *maximality-based semantics* for *beliefs* in the style of [4,7,20] and *oughts* in the style of [11,17,19] in order to model the *statics* of single-agent deliberations scenarios. These semantics interpret conditional belief operators $\mathcal{B}^{\psi}$ on a *plausibility preorder* [4,7,20] and conditional ought operators $\mathcal{O}^{\psi}$ on a *betterness preorder* [11,17,19]. In turn, this combination amounts to extend the setting of (single-agent) *Conditional Doxastic Logic* CDL from [3,4] with ought operators. In order to express the *dynamics* of such scenarios, we extend the framework of DEL (*Dynamic Epistemic Logic*) [2,4,9,10] by defining *concurrent upgrade operations* that change the plausibility and betterness preorder at once. Ideally, these operations capture the effect of information release on the oughts in the system and the beliefs of the agent. The distinctive mark of the new operations is a *Principle of Belief-Ought Covariation* (Sect. 3), which states that a successful information release $\Uparrow \phi$ promotes ϕ-states to best and most plausible states at once.

Deliberation scenarios include some of the most important phenomena of social interaction and individual agency, such as *voting*, *decision-making*, promulgation of a *verdict* in a trial.[1] This calls for the relevance of the logic of plausibility-betterness models (Sect. 2) and the logic of concurrent upgrades (Sect. 3). Indeed, they capture the connection between beliefs, information and oughts, which plays a crucial in the above scenarios. Since the *Stolen Wallet scenario* seems to display the basic features of deliberation scenarios, we use it here as a motivating example, and we test our framework against it (Sect. 5).

One clarification is in order about *deliberation.* This is a complex phenomenon that involves, for instance, cognitive processes, psychological elements, and conditions that trigger a concrete interest in taking side on a given issue. Here, we abstract from these aspects, since we are interested in the role played by (interaction of) beliefs and oughts in deliberation scenarios, and in the impact

[1] The role played by *belief change* in *decision-making* has been investigated by [5] in the context of *epistemic game theory.* Here, we take a more general stance, and we do not aim at modeling game- or decision-theoretical scenarios. Also, the language from [5] does not include ought operators.

of trusted information on the outcome of a deliberation.[2] In this idealized fashion, we say that our agent *deliberates that* ϕ—or *is in a position to deliberate that* ϕ—if the model we use to describe the agent satisfies the set $\{\mathcal{B}^{\top}\psi, \mathcal{O}^{\psi}\phi, \mathcal{O}^{\top}\phi\}$ for some formula ψ. This captures the fact that, in order to deliberate (say) that Jones ought to be punished, the agent needs to believe in the antecedent of a conditional ought that prescribes punishment for Jones. An interesting point is that, while some plausibility-betterness models from Sect. 2 may fail to satisfy set $\{\mathcal{B}^{\top}\psi, \mathcal{O}^{\psi}\phi, \mathcal{O}^{\top}\phi\}$ for some formula ϕ and ψ, a successful *concurrent upgrade* $\Uparrow\psi$ secures that one such set is satisfied: *relevant information from a highly trusted source always puts the agent in a position to deliberate.*

The paper proceeds as follows. Section 2 introduces the logic of plausibility-betterness models and the static language that we need in order to reason about deliberation. Section 3 introduces *concurrent upgrades* and explains their distinctive features and relations with existing upgrades operation from the DEL tradition. Section 4 presents Hilbert-style axiom systems for the two logics, which are easily proved sound and complete on the ground of established results in conditional and dynamic logics. Section 5 applies our framework to the motivating example above. Section 6 briefly discusses a notion of *'ought as' 'norm-abiding'* that must be kept distinct from the notion of *'ought' as 'best'* on which we focus on this paper. Finally, Sect. 7 discusses a possible extension.

2 The Static Logic of Plausibility-Betterness Models

***Syntax*.** Given a non-empty set $\mathcal{P}$ of *atoms*, our language $\mathcal{L}_{\mathcal{P}}$ has the following BNF:

$$\phi ::= p \mid \neg\phi \mid \phi \wedge \psi \mid \phi \vee \psi \mid \phi \rightarrow \psi \mid \top \mid \bot \mid \mathcal{B}^{\psi}\phi \mid \mathcal{O}^{\psi}\phi$$

where $p \in \mathcal{P}$ and $\neg$, $\wedge$, $\vee$, $\rightarrow$, $\top$ and $\bot$ receive their standard interpretation. $\mathcal{B}^{\psi}\phi$ reads 'the agent believes ϕ conditionally on ψ', and $\mathcal{B}^{\psi}$ expresses a belief that is conditional on ψ. *Unconditional beliefs* can be captured by $\mathcal{B}^{\top}$, with $\mathcal{B}^{\top}\phi$ reading 'the agent believes ϕ'. $\mathcal{O}^{\psi}\phi$ reads '(for the agent) it ought to be the case that ϕ conditionally on ψ', and $\mathcal{O}^{\psi}$ expresses an obligation that is conditional on ψ. *Unconditional oughts* can be captured by $\mathcal{O}^{\top}$, with $\mathcal{O}^{\top}\phi$ reading '(for the agent) it ought to be the case that ϕ'. We define the duals of the two operators by $\hat{\mathcal{B}}^{\psi}\phi = \neg\mathcal{B}^{\psi}\neg\phi$ and $\hat{\mathcal{O}}^{\psi}\phi = \neg\mathcal{O}^{\psi}\neg\phi$. We omit reference to $\mathcal{P}$ when possible.

***Semantics*.** We interpret $\mathcal{L}$ on structures combining single-agent *epistemic plausibility models* [4,20] and Hansson-style models [11,19].[3] We use the resulting

[2] Also, we do not presuppose that the deliberating agent has a particular position with respect to the issue in question (for instance, some kind of authority). We attribute deliberation to any agent that can assess what ought to be the case on the ground of believed circumstances.

[3] Similar combinations are defined in [13,14], but the logics defined in those papers differ considerably from the one we are presenting here.

models to define a *maximality-based semantics* for the belief operators—in the style of [4,7,20]—and for the ought operators—in the style of [11,17,19,21].

Definition 1 (Plausibility-betterness models). *A plausibility-betterness model $\mathcal{M}$ is a tuple $(S, R_{\mathcal{B}}, R_{\mathcal{O}}, \mathcal{I})$ where:*

- *S is a nonempty set of* states*;*
- *$R_{\mathcal{B}}$ and $R_{\mathcal{O}}$ are two distinct* preorders *satisfying* connectedness *and* upward well-foundedness.[4]
- *$\mathcal{I} : \mathcal{P} \longmapsto \wp(S)$ is an* interpretation *assigning a truth-set to each atom.*

$R_{\mathcal{B}}(s, s')$ reads 's' is *at least as plausible as* s,' and $R_{\mathcal{O}}(s, s')$ reads 's' is *at least as good as* s.' The interpretation of Boolean constructions is defined recursively in the standard way. A useful notation is this: $[\![\phi]\!]^{\mathcal{M}}$ denotes the truth-set of a formula ϕ in $\mathcal{M}$. Where $\mathcal{R}_{\bullet} \in \{R_{\mathcal{B}}, R_{\mathcal{O}}\}$, we define the set of $\mathcal{R}_{\bullet}$-maximal set of ϕ-states for every formula $\phi \in \mathcal{L}$:

$$\mathsf{max}^{\mathcal{R}_{\bullet}}([\![\phi]\!]^{\mathcal{M}}) = \{s \in [\![\phi]\!]^{\mathcal{M}} \mid \forall s' \in [\![\phi]\!]^{\mathcal{M}} : \mathcal{R}_{\bullet}(s', s)\}$$

More specifically, $\mathsf{max}^{R_{\mathcal{B}}}([\![\phi]\!]^{\mathcal{M}})$ is the set of the *most plausibile* ϕ-states, and $\mathsf{max}^{R_{\mathcal{O}}}([\![\phi]\!]^{\mathcal{M}})$ is the set of the *best* ϕ-states. Upward well-foundedness guarantees that $\mathsf{max}^{\mathcal{R}_{\bullet}}([\![\phi]\!]^{\mathcal{M}})$ is nonempty if $[\![\phi]\!]^{\mathcal{M}}$ is nonempty. Two special cases are $\mathsf{max}^{\mathcal{R}_{\bullet}}([\![\bot]\!]^{\mathcal{M}}) = \emptyset$ and $\mathsf{max}^{\mathcal{R}_{\bullet}}([\![\top]\!]^{\mathcal{M}}) = \{s \in S \mid \forall s' \in S : \mathcal{R}_{\bullet}(s', s)\}$. The latter denotes the most plausible (best) states of the model in question.

Remark 1 (Betterness Preorder). The relation $R_{\mathcal{O}}$ can get different interpretations. One can take it to represent the preferences of the agents, or the ranking of states induced from exogenous sets of norms. Here, we do not need to take side, since the deliberation of an agent can be cast either against the agents' preferences, or against a background of norms dictated by some external authority.

Definition 2 (Satisfaction Relation, static). *Where $\mathcal{M} = (S, R_{\mathcal{B}}, R_{\mathcal{O}}, \mathcal{I})$ is a plausibility-betterness model, the satisfaction relation $\models$ between model and formulas in $\mathcal{L}$ is defined recursively as follows:*[5]

$$\mathcal{M}, s \models p \Leftrightarrow s \in \mathcal{I}(p)$$
$$\mathcal{M}, w \models \mathcal{B}^{\psi}\phi \Leftrightarrow \mathsf{max}^{R_{\mathcal{B}}}([\![\psi]\!]^{\mathcal{M}}) \subseteq [\![\phi]\!]^{\mathcal{M}}$$
$$\mathcal{M}, w \models \mathcal{O}^{\psi}\phi \Leftrightarrow \mathsf{max}^{R_{\mathcal{O}}}([\![\psi]\!]^{\mathcal{M}}) \subseteq [\![\phi]\!]^{\mathcal{M}}$$

---

[4] A preorder $\mathcal{R}$ is *connected* if $\forall s, s' \in S : \mathcal{R}(s, s')$ or $\mathcal{R}(s', s)$. It is *upward well-founded* if, for every $U \subseteq S$, if $U \neq \emptyset$, then $\{s \in U \mid \forall s' \in U : \tilde{\not{\mathcal{R}}}(s, s')\} \neq \emptyset$—the set of the 'most R' among the states in U is nonempty. Here, $\not{\mathcal{R}} = S^2 \setminus \mathcal{R}$, and $\tilde{\mathcal{R}}(s, s')$ is short for '$\mathcal{R}(s, s')$ and $\not{\mathcal{R}}(s', s)$'.

[5] We omit the definition for the Boolean constructions, which is standard.

Satisfaction and validity are defined in the standard way. The agent believes those things that hold in the *most plausible* states available in the system. Similarly, what ought to be the case is what holds in the *best* states in the system. It is clear from the definitions above that beliefs and oughts are *absolute*:[6]

$$\mathcal{M}, s \models \bullet^{\psi}\phi \Leftrightarrow \forall s' \in S : \mathcal{M}, s' \models \bullet^{\psi}\phi$$

where $\bullet^{\psi} \in \{\mathcal{B}^{\psi}, \mathcal{O}^{\psi}\}$. In our semantics, an S5-type *knowledge operator* can be defined as follows:[7]

$$\mathcal{K}\phi = \mathcal{B}^{\neg\phi}\bot$$

This is the standard definition of 'classical' knowledge in conditional doxastic logic [3,4], and it justifies the reading of $\mathcal{K}$ as *unrevisable true belief.* Notice that, in the present setting, $\mathcal{K}$ is a *universal modality* [6], to the effect that knowledge coincides with 'truth in all the possible states': the agent considers possible anything that is not ruled out by the model. *Absoluteness* extends to the knowledge operator. We define the dual of the operator by $\hat{\mathcal{K}}\phi = \neg\mathcal{K}\neg\phi$.

Remark 2 (Conditional Beliefs and Oughts). We read a *conditional belief* as pre-encoding potential belief revision in the sense of [20]: $\mathcal{B}^{\psi}\phi$ means that the agent *would* believe that ϕ was the case, if she came to believe that ψ is the case. We extend this reading to the conditional ought operators, so that $\mathcal{O}^{\psi}\phi$ means that ϕ *would* turn to be the best option if persuasive information that ψ were released. As is well known, conditional beliefs and oughts cannot express the *change* of beliefs or oughts.[8] This can be expressed by the new *upgrade operators* that we introduce in Sect. 3.

This is a plausibility-betterness model—arrows labelled with $R_{\mathcal{B}}$ stay for plausibility relations, black curves for betterness relations, and double arrows are for equally good/plausible states.[9] Reflexive loops are omitted:

The model satisfies $\mathcal{B}^{\top}p$ and $\mathcal{O}^{p}q$, but it does not satisfy $\mathcal{O}^{\top}q$. This makes

$$\mathcal{B}^{\top}\phi \rightarrow (\mathcal{O}^{\phi}\psi \rightarrow \mathcal{O}^{\top}\psi) \qquad (*)$$

invalid: *belief in the antecedent of a conditional ought* does not induce *an unconditional ought.* An equivalent reading is that *the agent is not always in a position to deliberate about a given issue.* Indeed, failure of the above formulas means that some plausibility-betterness model does not satisfy the set $\{\mathcal{B}^{\top}\phi, \mathcal{O}^{\phi}\psi, \mathcal{O}^{\top}\psi\}$ (for some formulas $\phi, \psi \in \mathcal{L}$) and, in our reading, this means that the agent is not deliberating about ψ.[10] Notice that $\mathcal{B}^{\top}\mathcal{O}^{\top}q$ is also not satisfied, due to

[6] That is, their valuations do not vary across states in S of a given model $\mathcal{M}$.

[7] Of course, we could also define $\mathcal{K}\phi$ as $\mathcal{O}^{\neg\phi}\bot$. However, given the conceptual nexus between knowledge and belief, we prefer the definition above.

[8] See for instance [4, pp. 36–37]. [4] discusses the relations between change of beliefs and conditional beliefs, (the crucial point here concerns the so-called *Moore sentences*) but the very same line of reasoning applies to change of oughts and conditional oughts.

[9] Two states s and s' are *equally good* if $R_{\mathcal{O}}(s, s')$ and $R_{\mathcal{O}}(s', s)$. They are *equally plausible* if $R_{\mathcal{B}}(s, s')$ and $R_{\mathcal{B}}(s', s)$.

[10] Unless, of course, there is another formula θ such that $\{\mathcal{B}^{\top}\theta, \mathcal{O}^{\theta}\psi, \mathcal{O}^{\top}\psi\}$.

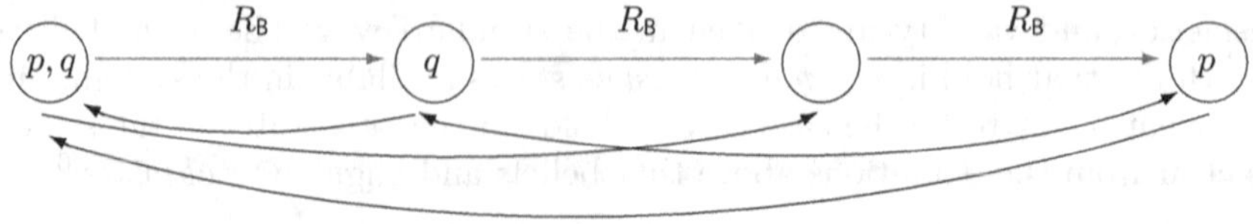

Fig. 1. A plausibility-betterness model.

the absoluteness of $\mathcal{O}^{\psi}$. The following helps understand what kind of agent is described by plausibility-betterness models:

$$\mathcal{O}^{\psi}\phi \rightarrow \mathcal{K}\mathcal{O}^{\psi}\phi \tag{1}$$

$$\mathcal{O}^{\psi}\phi \rightarrow (\mathcal{K}\psi \rightarrow \mathcal{O}^{\top}\phi) \tag{2}$$

that is, *the agent has* perfect knowledge *of the oughts that hold in the model* and *something ought to be the case, if it ought to be the case* conditionally *on a* known *formula.* The latter marks a difference between the way knowledge and belief interact with oughts. Formulas (1) and (2) equally apply to belief operators.

Operators $\mathcal{K}$ and $\mathcal{B}^{\psi}$ obey the standard principles of S5-type knowledge and conditional belief (interpreted on preorders) such as the ones presented in [4, pp. 34–35]. Further principles are presented in the axiomatization from Theorem 1 in Sect. 4.

Remark 3 (Correlation between Beliefs and Oughts). The failure of $\mathcal{B}^{\top}\phi \rightarrow (\mathcal{O}^{\phi}\psi \rightarrow \mathcal{O}^{\top}\psi)$ captures the natural thought that beliefs and oughts are somehow independent from one another. However, a correlation between beliefs and oughts is needed in order to model deontic deliberation. Coming back to the Stolen Wallet scenario, if Bob believes that Jones stole his wallet, and (he knows) it ought to be the case that he is punished in this case, then Bob deliberates that a sanction ought to be enforced. If we cannot draw this conclusion, then we cannot model the connection between what Bob believes and what he deliberates.

In the next section, we introduce dynamic operations that help secure the correlation between beliefs and oughts that is needed in deontic deliberation.

Remark 4 ('Ought' as 'best'). The maximality semantics we are using gives us a semantical reading of 'ought' as 'best'. This reading fits a long-standing intuition that deontic notions involve some kind of 'ideality ordering'—see for instance [11,15,19]. Of course, what is best depends on the circumstances, in a sense: the best scenarios where Jones steals something prescribe an option that is not prescribed by the best scenarios where Jones does not steal anything—Jones ought to be punished conditionally on the former scenario, but not conditionally on the latter scenario.[11] Thus, conditional oughts work as instructions to what

[11] Under this reading, the validity of $\mathcal{O}^{\phi}\phi$ does not cause any concern, even in cases where $\mathcal{O}^{\top}\neg\phi$ is satisfied: the best among the scenarios where Jones steals something are still scenarios where he steals something.

option turns best in given circumstances. In this sense, the oughts we are focusing in this paper are *revisable*, or better *adaptive*: a conditional ought tells us how oughts would be adapted to the circumstances. The dynamics from Sect. 3 will show that information release may induce a change of ought that realizes the potential for adaptation expressed by conditional oughts. Also, there are other (non-adaptive) notions of ought beside the one we are considering. In Sect. 6, we briefly discuss one such notion.

3 The Dynamic Logic of Concurrent Upgrades

Our *dynamic* language $\mathcal{L}_{\Uparrow\!\Uparrow}$ has the following BNF:

$$\phi ::= p \mid \neg\phi \mid \phi \wedge \psi \mid \phi \vee \psi \mid \phi \rightarrow \psi \mid \top \mid \bot \mid \mathcal{B}^{\psi}\phi \mid \mathcal{O}^{\psi}\phi \mid [\Uparrow\!\Uparrow \psi]\phi$$

$[\Uparrow\!\Uparrow \psi]\phi$ reads 'after ψ is announced, ϕ holds true'. The new formulas are interpreted in terms of *model transformers* $\Uparrow\!\Uparrow \psi$ that we call *concurrent upgrades.* These operations are a special kind of *radical upgrade* [4,20] that change the preorders $R_{\mathcal{B}}$ and $R_{\mathcal{O}}$ at once:

Definition 3 (Concurrent Upgrades). *Given a betterness model $\mathcal{M} = (S, R_{\mathcal{B}}, R_{\mathcal{O}}, \mathcal{I})$ and a formula $\psi \in \mathcal{L}$, the upgraded (plausibility-betterness) model $\mathcal{M}^{\Uparrow\!\Uparrow\psi}$ is the tuple $(S, R_{\mathcal{B}}^{\Uparrow\!\Uparrow\psi}, R_{\mathcal{O}}^{\Uparrow\!\Uparrow\psi}, \mathcal{I})$, where:*

$$\begin{aligned} R_{\mathcal{B}}^{\Uparrow\!\Uparrow\psi} &= \{(s,s') \in R_{\mathcal{B}} \mid s,s' \in [\![\psi]\!]^{\mathcal{M}}\} \cup \{(s,s') \in R_{\mathcal{B}} \mid s,s' \notin [\![\psi]\!]^{\mathcal{M}}\} \cup \\ &\quad \cup \{(s,s') \in S \times S \mid s \notin [\![\psi]\!]^{\mathcal{M}} \text{ and } s' \in [\![\psi]\!]^{\mathcal{M}}\} \\ R_{\mathcal{O}}^{\Uparrow\!\Uparrow\psi} &= \{(s,s') \in R_{\mathcal{O}} \mid s,s' \in [\![\psi]\!]^{\mathcal{M}}\} \cup \{(s,s') \in R_{\mathcal{O}} \mid s,s' \notin [\![\psi]\!]^{\mathcal{M}}\} \cup \\ &\quad \cup \{(s,s') \in S \times S \mid s \notin [\![\psi]\!]^{\mathcal{M}} \text{ and } s' \in [\![\psi]\!]^{\mathcal{M}}\} \end{aligned}$$

In words, concurrent upgrades promote ψ-states to both *most plausible* and *best* states, downgrade $\neg\psi$-states accordingly, and leave the orders within the two sets as they are. In particular, *if* $[\![\psi]\!]^{\mathcal{M}} \neq \emptyset$, *then*:

CU1 $\forall s, s' \in S : s \in [\![\neg\psi]\!]^{\mathcal{M}}$ and $s' \in [\![\psi]\!]^{\mathcal{M}} \Rightarrow \tilde{\mathcal{R}}_{\bullet}^{\Uparrow\!\Uparrow\psi}(s,s')$
CU2 $\mathsf{max}^{\mathcal{R}_{\bullet}}([\![\top]\!]^{\mathcal{M}^{\Uparrow\!\Uparrow\psi}}) \subseteq [\![\psi]\!]^{\mathcal{M}}$

where $\tilde{\mathcal{R}}_{\bullet}(s,s')$ is short for $\mathcal{R}_{\bullet}(s,s')$ and $\not{\mathcal{R}}_{\bullet}(s',s)$. As any upgrade, concurrent upgrades do not change facts, thus leaving both S and $\mathcal{I}$ unaltered. It is easy to check that upgraded models are themselves plausibility-betterness models.

Remark 5. Upgrades over *plausibility* preorders are the signature marks of dynamic logics for belief revision [4,20] from the DEL tradition, where they are introduced to capture the effects of *soft* information—that is, information released by a trustworthy yet fallible source. Upgrades over betterness relations are also introduced by [14,22] in order to model the effects of a change of preference. Concurrent upgrades are the first to operate on both preorders at once, to the effect that the two relations change in a uniform way.

Definition 4 (Satisfaction Relation, dynamic). *Where* $\mathcal{M} = (S, R_{\mathcal{B}}, R_{\mathcal{O}}, \mathcal{I})$ *is a plausibility-betterness model, the satisfaction relation* $\models$ *between model and formulas in* $\mathcal{L}$ *extends the satisfaction relation from Definition 2 with:*

$$\mathcal{M}, s \models [\Uparrow \psi]\phi \Leftrightarrow \mathcal{M}^{\Uparrow\psi}, s \models \phi$$

This is the result of $\Uparrow p$ on the model from Fig. 1:[12]

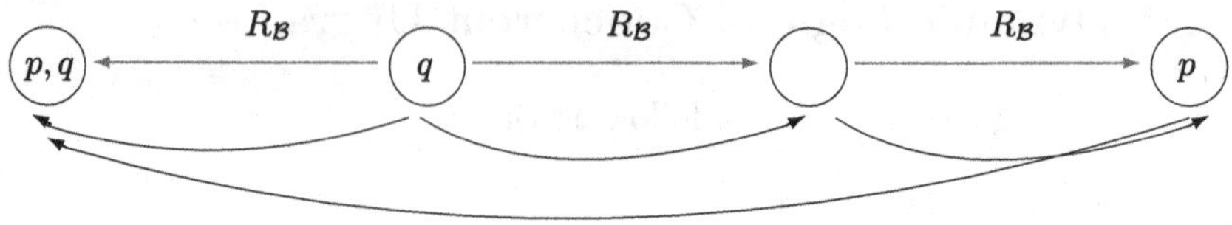

Fig. 2. Upgraded model.

Distinctive features of concurrent upgrades. Along with $\mathcal{B}^{\top}p$ and $\mathcal{O}^{q}p$, the upgraded model satisfies $\mathcal{O}^{\top}p$. Going from concrete to general, the following formula

$$[\Uparrow \phi](\mathcal{B}^{\top}\phi \leftrightarrow \mathcal{O}^{\top}\phi) \tag{3}$$

is valid—it follows from Definition 3. This is a *Principle of Belief-Ought Covariation*, stating that, in presence of new information, the *relevant* oughts and beliefs of the agent *co-vary*. We come back to this in Remark 6 below. By contrast,

$$[\Uparrow \phi](\mathcal{B}^{\phi}\psi \leftrightarrow \mathcal{O}^{\phi}\psi) \tag{*}$$

fails, since concurrent upgrades leave the zones within ϕ- and $\neg\phi$-states as they are in the initial model. Notice that a consequence of formula (3) is the formula

$$[\Uparrow \phi](\mathcal{O}^{\top}\phi \rightarrow \hat{\mathcal{B}}^{\top}\phi) \tag{4}$$

which is a *doxastic* version of the principle of *Ought implies Can*, stating that *after* ϕ *is announced, what ought to be the case is compatible with the agent's belief.* An important principle relating *concurrent upgrades* and *deliberation* is:

$$[\Uparrow \phi](\mathcal{B}^{\top}\phi \rightarrow (\mathcal{O}^{\phi}\psi \rightarrow \mathcal{O}^{\top}\psi)) \tag{5}$$

that follows from formula (3) and the detachment principle $\mathcal{O}^{\top}\phi \rightarrow (\mathcal{O}^{\phi}\psi \rightarrow \mathcal{O}^{\top}\psi)$. *Ideally, an agent deliberates about* ψ *any time she gets relevant information from a highly trusted source.* Thus, our framework does not model *explicit* deliberation processes—that is, the processes that are triggered when the agent asks herself what ought to be the case in the believed circumstances—but rather *implicit* deliberations, which hold as soon as the agent is (from a doxastic point of view) in the position to deliberate about a given issue.

[12] As for the missing arrows, remember that concurrent upgrade $\Uparrow p$ leaves relations within the p- and $\neg p$-zones as they were in Fig. 1.

Strong Beliefs and Oughts. Any formula (atomic, complex, modal, dynamic) can be announced, but our subsequent applications will involve the announcements of *factual statements*—that is, formulas that contain no modal or dynamic operator.[13] As any radical upgrade, concurrent upgrade $\Uparrow\Uparrow \phi$ induces a *strong* belief that ϕ—that is, a belief that is revised only if the agent receives further information that is *inconsistent* with ϕ. Indeed:

Fact 1. *For every $\phi \in \mathcal{L}$, if $[\![\phi]\!]^{\mathcal{M}} \neq \emptyset$, then:*

$$\mathcal{M}^{\Uparrow\Uparrow\phi}, s \models \mathcal{B}^{\psi}\phi \text{ for every } \psi \in \mathcal{L} \text{ such that } [\![\psi]\!]^{\mathcal{M}} \cap [\![\phi]\!]^{\mathcal{M}} \neq \emptyset$$

Indeed, CU1 implies that, for every model $\mathcal{M}$ and $\phi, \psi \in \mathcal{L}$ such that $[\![\psi]\!]^{\mathcal{M}} \cap [\![\phi]\!]^{\mathcal{M}} \neq \emptyset$, if all ϕ-states are more plausible than all $\neg\phi$-states, then all $\psi \wedge \phi$-states are more plausible than all $\psi\wedge\neg\phi$-states for every ψ. Fact 1 equally applies to $\mathcal{O}^{\psi}$, thus pointing at a particular kind of oughts—namely, those that are not dropped until we receive the information that they have been violated.

Remark 6 (Belief-Ought Covariation). The *Principle of Belief-Ought Covariation*—formula (3) above—just tells us that the plausibility and betterness preorders change in the same way. In our intended (DEL-style) interpretation, however, $\Uparrow\Uparrow \phi$ is an information release, and under this interpretation, it is naturally seen as something that operates primarily on the agent's *beliefs*. In line with this, $[\Uparrow\Uparrow \phi]\mathcal{O}^{\top}\psi$ reads 'after the information that ϕ is released, it ought to be the case that ψ'. The natural interpretation is that a formula like $[\Uparrow\Uparrow \phi]\mathcal{O}^{\top}\psi$ captures an *information-induced revision of oughts.* Thus, in Fig. 2 $[\Uparrow\Uparrow p]\mathcal{O}^{\top}q$ tells you *what the agent takes as best after she receives information that* p. Since p is factual and consistent with the initial model (Fig. 1), the formula states *what the agent takes as best after she comes to believe that* p. In general, after a successful announcement that a factual statement ϕ is made, *the agent will revise oughts by picking the best states among those that she takes to be the most plausible.* Thus, change of belief works as a trigger for oughts that were in a sense 'dormant'—that is, oughts that become operative only if the relevant circumstances are believed to be the case.

The principles that we have discussed in the last two sections give us an idea of what kind of agent can be modeled by our framework. In particular, the logic of concurrent upgrades describes an *agent that has* perfect *knowledge of the oughts defined in the system* and is a *cautious* agent—that is an *agent that revises her beliefs and deliberations just after receiving new information from a* highly trusted *source.* This equates with saying that the agent revises oughts in relation to what she comes to (strongly) believe to be the case.

[13] These prove especially interesting. In contrast to announcements of factual formulas, announcements of $\top$,$\bot$, or any (true or false) modal formula in $\mathcal{L}$ cannot change the initial model $\mathcal{M}$. An interesting case holds when *Moore sentences* are (unsuccessfully) announced. We will not face these cases here, and we refer the reader to [9] for them.

4 Axiom System

Axiomatics for the static logic of plausibility-betterness models and the dynamic logic of concurrent upgrades can be easily determined on the ground of established results on static conditional logics and dynamic logics:

Theorem 1. *The static logic of plausibility-betterness models is completely axiomatized by the following Hilbert-style proof-system—where* $\bullet$ *is either* $\mathcal{O}$ *or* $\mathcal{B}$*, and* $\mathcal{K}\phi$ *is short for* $\mathcal{B}^{\neg\phi}\top$*:*

$A1$	*Propositional logic*
$A2$	*S5-axioms for* $\mathcal{K}$
$A3$	$\bullet^{\psi}\phi \rightarrow \mathcal{K} \bullet^{\psi} \phi$
$A4$	$\mathcal{K}\phi \rightarrow \bullet^{\psi}\phi$
$A5$	$\bullet^{\phi}\phi$
$A6$	$\hat{\mathcal{K}}\phi \rightarrow (\bullet^{\psi}\phi \rightarrow \hat{\bullet}^{\psi}\phi)$
$A7$	$\hat{\bullet}^{\psi}\phi \rightarrow (\bullet^{\psi}(\phi \rightarrow \theta) \rightarrow \bullet^{\phi\wedge\psi}\theta)$
$A8$	$\bullet^{\psi}(\phi \rightarrow \theta) \rightarrow (\bullet^{\psi}\phi \rightarrow \bullet^{\psi}\theta)$
$A9$	$\mathcal{K}(\phi \leftrightarrow \psi) \rightarrow (\bullet^{\phi}\theta \leftrightarrow \bullet^{\psi}\theta)$
$A10$	$\bullet^{\phi\wedge\psi}\theta \rightarrow \bullet^{\phi}(\psi \rightarrow \theta)$
$R1$	$\phi, \phi \rightarrow \psi / \psi$
$R2$	$\phi / \mathcal{K}\phi$

The proof is a straightforward adaptation of the completeness proof for system **G** in [17, Theorems 8, 9, 12, 13].

Proposition 1. *The axiom system resulting by replacing A4–A5 in Theorem 1 with:*

$$A^* \quad \mathcal{K}(\psi \rightarrow \phi) \rightarrow \bullet^{\psi}\phi$$

is equivalent with the axiom system consisting of A1–A10 and R1–R2.

Proof. We prove that A* is derivable from A4 and A5 in the axiom system from Theorems 1, and that A4 and A5 are derivable from A* in the alternative axiomatic system.

1. $A^* \vdash A4$. We have $\mathcal{K}\phi \rightarrow \mathcal{K}(\psi \rightarrow \phi)$ from propositional logic, R2, axiom K for $\mathcal{K}$ (A2) and MP. From this and A*, we have $K\phi \rightarrow \bullet^{\psi}\phi$ by transitivity of implication. But this is A4.
2. $A^* \vdash A5$. We have $\mathcal{K}(\phi \rightarrow \phi)$ from propositional logic and R2. From this and $\mathcal{K}(\phi \rightarrow \phi) \rightarrow \bullet^{\phi}\phi$ (instance of A*), we have $\bullet^{\phi}\phi$ by MP.
3. $A4 \wedge A5 \vdash A^*$. We have (1) $\bullet^{\psi}(\psi \rightarrow \phi) \rightarrow (\bullet^{\psi}\psi \rightarrow \bullet^{\psi}\phi)$ (instance of A8). From (1) and propositional logic, we derive (2) $\bullet^{\psi}(\psi \rightarrow \phi) \wedge \bullet^{\psi}\psi \rightarrow \bullet^{\psi}\phi$. From (2), A5 and propositional logic, we derive (3). $\bullet^{\psi}(\psi \rightarrow \phi) \rightarrow \bullet^{\psi}\phi$. From $\mathcal{K}(\psi \rightarrow \phi) \rightarrow \bullet^{\psi}(\psi \rightarrow \phi)$—instance of A4—and (3), we get $\mathcal{K}(\psi \rightarrow \phi) \rightarrow \bullet^{\psi}\phi$ by propositional logic. But this is A*.

Theorem 2. *The dynamic logic of concurrent upgrades is completely axiomatized by a Hilbert-style system including:*

- *axioms A1–A10 and rules R1–R2,*
- *the following reduction axioms:*

$$
\begin{array}{ll}
D1 & [\upuparrows \psi]p \leftrightarrow p \\
D2 & [\upuparrows \psi]\neg\phi \leftrightarrow \neg([\upuparrows \psi]\phi) \\
D3 & [\upuparrows \psi](\phi \wedge \theta) \leftrightarrow [\upuparrows \psi]\phi \wedge [\upuparrows \psi]\theta \\
D4 & [\upuparrows \psi] \bullet^{\theta} \phi \leftrightarrow (\hat{\mathcal{K}}(\psi \wedge [\upuparrows \psi]\theta) \wedge \bullet^{\psi \wedge [\upuparrows \psi]\theta}[\upuparrows \psi]\phi \vee \bullet^{[\upuparrows \psi]\psi}[\upuparrows \psi]\phi \\
D5 & [\upuparrows \psi]\mathcal{K}\phi \leftrightarrow \mathcal{K}[\upuparrows \psi]\phi \\
D6 & [\upuparrows \phi](\mathcal{B}^{\top}\phi \leftrightarrow \mathcal{O}^{\top}\phi)
\end{array}
$$

See [20] for soundness of D1–D5. Soundness of D6 follows from Definition 3. As for completeness of the axiom system in Theorem 2, the proof is similar to the ones in [9]. The reduction axioms D1–D6 are used to inductively simplify any formula until it is reduced to a formula in the static language $\mathcal{L}$.

Notice that, while the logic of plausibility-betterness models is similar to the logic of conditional beliefs on plausibility models with two agents, the latter comes with different information partitions (one per agent), while in our models we have just one information partition, which coincides with the set S of states. Thus, the logic of plausibility-betterness models does not reduce to a special logic of interactive beliefs in the style of [4].

5 Modeling the Stolen Wallet Scenario

Figure 3 displays a model of the static part of the *Stolen Wallet* scenario—that is, the part before Ann talks to Bob. Here, s is 'Jones stole the wallet' and p is 'Jones gets punished'.

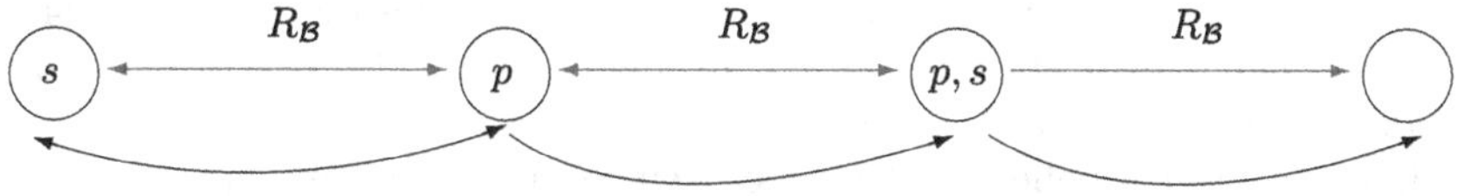

Fig. 3. Stolen Wallet scenario (static).

The model satisfies $\mathcal{O}^{\top}\neg p$: the best states are those where Jones does not get punished, since these are the states where he did not steal the wallet. The model also satisfies $\mathcal{B}^{\top}\neg s$, $\mathcal{K}\mathcal{O}^{\neg s}\neg p$ and $\mathcal{K}\mathcal{O}^{s}p$: the agent knows that Jones ought to be punished only in case he stole the wallet. More in general, Bob has *perfect knowledge* of the oughts defined in the system.

Contrary to the model in Fig. 1, the present model captures a deliberation scenario. Indeed, it also satisfies $\mathcal{B}^{\top}\neg s \rightarrow (\mathcal{O}^{\neg s}\neg p \rightarrow \mathcal{O}^{\top}\neg p)$, which, together with $\mathcal{B}^{\top}\neg s$ and $\mathcal{O}^{\neg s}\neg q$ above, implies that the set $\{\mathcal{B}^{\top}\neg s, \mathcal{O}^{\neg s}\neg p, \mathcal{O}^{\top}\neg p\}$ is satisfied in the model. In our interpretation, this means that Bob is in the position to deliberate that Jones be not punished.

Going to the *dynamic* part of the example, information release by Ann changes Bob's beliefs and deliberation as described by the *Stolen Wallet*—see Introduction. We model Ann conversation with Bob as the announcement $\Uparrow\!\Uparrow s$. This is the effect of the announcement on the beliefs and oughts from Fig. 3:[14]

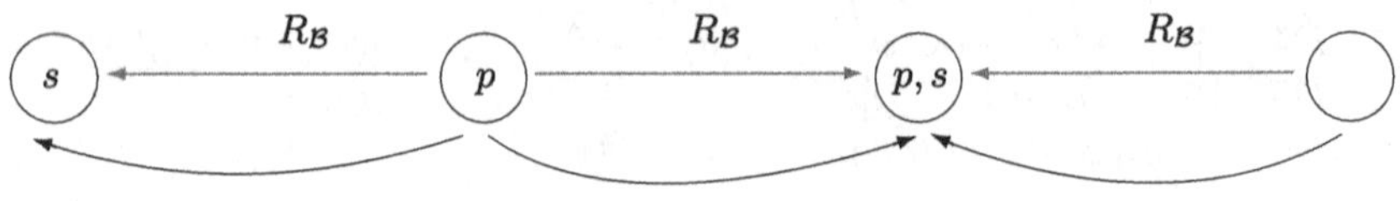

Fig. 4. Stolen Wallet scenario (dynamic).

In particular, the upgraded model satisfies $\mathcal{B}^{\top}s$: Bob has come to believe that Jones stole his wallet. At this point, formula (5) and $\mathcal{O}^{s}p$ suffice for us to realize that $\mathcal{O}^{\top}p$ is also satisfied in the new model: Bob has adjusted the oughts provided by the initial model by taking into account just the best states among the s-states. This also implies that the set $\{\mathcal{B}^{\top}s, \mathcal{O}^{s}p, \mathcal{O}^{\top}p\}$ is now satisfied: in line with his new beliefs, Bob has revised his previous deliberation and has come to deliberate that Jones be punished (Fig. 4).

Notice that, after talking with Ann, Bob has a *strong belief* that Jones stole his wallet. Informally, this means that he revises his beliefs and deliberations just in case the information he receives comes from a highly trusted source: our framework describes Bob as a *cautious* agent. Also, the new models satisfies $\hat{\mathcal{K}}\neg s$—Bob still considers it possible that Jones did not steal the wallet—upgrades do not increase *knowledge*. A natural interpretation of this is that Bob knows that Ann is fallible.

The Stolen Wallet scenario cannot be successfully modeled by traditional upgrade operations. A *radical upgrade* $\Uparrow s$ over the *plausibility relation* [4,20] would just change Bob's beliefs, but not his deliberations. By contrast, a radical upgrade $\Uparrow s$ over the *betterness relation* [14,22] just change the oughts in the model, leaving Bob's beliefs as they are. Neither upgrades can capture a deliberation scenario: the first makes $\{\mathcal{B}^{\top}s, \mathcal{O}^{s}p, \mathcal{O}^{\top}\neg p\}$ satisfied in the new model; the second makes $\{\mathcal{B}^{\top}\neg s, \mathcal{O}^{s}p, \mathcal{O}^{\top}p\}$ satisfied. In our interpretation, this means that Bob is *not* deliberating that p, or that $\neg p$.

6 Different Senses of Oughts

The notion of 'ought' we focus on in this paper is *context-sensitive* and *adaptive*—see Remark 4. A different, non-adaptive sense of 'ought' is at stake when

[14] As for the missing arrows and curves, remember that concurrent upgrade $\Uparrow\!\Uparrow s$ leaves relations within the s- and $\neg s$-zones as they were in Fig. 3. This helps with the missing arrows.

we maintain, for instance, that it *ought* to be the case that Jones does not steal, even in the scenarios where he is indeed stealing something. This example deploys a notion of *'ought' as 'norm-abiding'*, which in turn equates 'what ought to be the case' with 'what follows from the norms'.[15] We maintain that features such as *adaptivity* and *belief-ought covariation* apply to the notion of *'ought' as 'best'*, but should not be applied to the notion of *'ought' as 'norm-abiding'*.

We do not detail a formal semantics for this further notion here, since this would take us too far from our current focus—that is on the *'ought' as 'best'* reading. However, we give a short hint at some possible options.

As for a formal rendering of *'ought' as 'norm-abiding'*, one option is to introduce a propositional constant d in the style of [1] expressing—informally—that all norms are satisfied, and define 'ϕ follows from the norms' ($\Box\phi$) as 'the agent knows that the norms imply ϕ':[16]

$$\Box\phi = \mathcal{K}(d \to \phi)$$

It is easy to see that, if ϕ is a factual formula, then

$$[\Uparrow \phi](\mathcal{B}^{\top}\phi \leftrightarrow \Box\phi) \tag{*}$$

is not valid: *information release does not make 'ought as norm-abiding' co-vary with beliefs*.[17] As for the interaction between the 'adaptive' oughts and the 'non-adaptive' ones, we may want that they coincide *before* any announcement is made and possibly come to diverge only *after* some announcement is made. We can secure the first point by imposing the formula

$$\mathcal{O}^{\top}\phi \leftrightarrow \Box\phi$$

which selects (as 'initial models') only those models $\mathcal{M}$ where $\max^{R_{\mathcal{O}}}(\llbracket\top\rrbracket^{\mathcal{M}}) = \llbracket d\rrbracket^{\mathcal{M}}$. Since the formula

$$[\Uparrow \neg\phi](\mathcal{O}^{\top}\neg\phi \to \Box\neg\phi) \tag{*}$$

is invalid (for ϕ a factual formula), we have that coincidence of the two oughts may be lost after suitable information is announced. We plan to discuss these options in future research, together with other issues concerning the interaction of the two 'oughts' and the formal definition of *'ought' as 'norm-abiding'*.

---

[15] Just to get a concrete feeling of this: the fact (or information) that Jones steals does not make stealing (by Jones) norm-abiding.

[16] This reading relies on the fact that $\mathcal{K}$ is a universal modality in our framework, to the effect that the above definition equates with $\llbracket d\rrbracket^{\mathcal{M}} \subseteq \llbracket\phi\rrbracket^{\mathcal{M}}$ in our framework. The definition extends *perfect knowledge* of the agent to the new deontic component.

[17] Any model $\mathcal{M}$ satisfying $\mathcal{K}(d \to p)$ and $[\Uparrow \neg p]\mathcal{B}^{\top}\neg p$ fails to satisfy $[\Uparrow \neg p](\mathcal{B}^{\top}\neg p \to \Box\neg p)$. Indeed, since d and p contains no modal operators, the interpretation of d and p does not change in the upgraded model, to the effect that $\llbracket d\rrbracket^{\mathcal{M}^{\Uparrow p}} \subseteq \llbracket p\rrbracket^{\mathcal{M}^{\Uparrow p}}$.

7 Multi-agent Extension and Deliberating Societies

In this paper, we have applied our DEL-style framework of concurrent upgrades (Sects. 2 and 3) to a motivating example, which involved a deliberation scenario and information-induced change of deliberation. The framework equally applies to a variety of crucial real-life scenarios such as voting, verdicts in trials, and decision-making. As a consequence, the present framework may have a wide range of applications. Here, we briefly discuss the extension to multi-agent scenarios and its potential for applications.

A multi-agent deliberation framework can be easily obtained along the lines yielding multi-agent plausibility models from single-agents ones in [4]. The resulting framework would include a variety of plausibility relations, each interpreting the conditional beliefs of an agent, and a variety of betterness relations, each interpreting the oughts relative to an agent. *Public announcements* of information from a trusted source would be expressed by concurrent upgrades $\Uparrow\Uparrow \phi$ and would induce a strong belief in ϕ in all agents; at the same time, they would promote ϕ as the best option w.r.t. believed circumstances. Notice that agents would come to different deliberations, depending on their own preference rankings and conditional oughts. By contrast, in a deontically homogeneous community—sharing moral norms, legal system, or social conventions—a public announcement would suffice to create a uniform deliberation. Here, we have at stake a dynamic of *public opinion formation*, but two different dynamics of *deliberation* of a group of agents. A multi-agent version of our framework would provide a straightforward logical tool to model the different dynamics.

An interesting question is whether the dynamics at stake in (deontically) *pluralistic* societies can decrease or block the effect of *information bubbles* [12] on *policy making*. This research perspective would naturally complement formal studies on *pluralistic ignorance* [18] and *cascades* in the *information society* [12], by securing a connection with the crucial issue of policy-making and the efficiency of a deliberating society.

References

1. Anderson, A.R.: The formal analysis of normative concepts. Am. Sociol. Rev. **22**, 9–17 (1967)
2. Baltag, A., Moss, L.S.: Logics for epistemic programs. Synthese **139**, 165–224 (2004)
3. Baltag, A., Smets, S.: Conditional doxastic models: a qualitative approach to dynamic belief revision. Electron. Notes Theor. Comput. Sci. **165**, 5–21 (2006)
4. Baltag, A., Smets, S.: A Qualitative Theory of Dynamic Interactive Belief Revision. Amsterdam University Press, Amsterdam (2008)
5. Baltag, A., Smets, S., Zvesper, J.A.: Keep hoping for rationality: a solution to the backward induction paradox. Synthese **169**(2), 301–333 (2009)
6. Blackburn, P., de Rijke, M., Venema, Y.: Modal Logic. Cambridge University Press, Cambridge (2001)
7. Board, O.: Dynamic interactive epistemology. Games Echonomic Behav. **49**(1), 49–80 (2002)

8. Cariani, F., Kaufmann, M., Kaufmann, S.: Deliberative modality under epistemic uncertainty. Linguist. Philos. **36**, 225–259 (2013)
9. Ditmarsch, H., van der Hoek, W., Kooi, B.: Dynamic Epistemic Logic. Springer, Berlin (2007)
10. Gerbrandy, J., Groeneveld, J.: Reasoning about information change. J. Logic Lang. Inform. **6**(2), 147–169 (1997)
11. Hansson, B.: An analysis of some deontic logics. Noûs **3**, 373–398 (1969)
12. Hendriks, V.F., Hansen, P.G.: Infostorms, 2nd ed. Copernicus/Springer, New York/Berlin (2015)
13. Lang, J., van der Torre, L., Weydert, E.: Hidden uncertainty in the logical representation of desires. In: Proceedings of the XVIII IJCAI, pp. 685–690. Morgan Kaufmann Publisher, Inc., San Francisco (CA) (2003)
14. Liu, F.: Reasoning about Preference Dynamics. Springer, Berlin (2011)
15. Moore, G.E.: Principia Ethica. Cambridge University Press, Cambridge (1903)
16. Pacuit, E., Parikh, R., Cogan, E.: The logic of knowledge based obligations. Synthese **149**, 311–341 (2006)
17. Parent, J.: Maximality vs optimality in dyadic deontic logic. J. Philos. Logic **43**(6), 1101–1128 (2014)
18. Proietti, C., Olsson, E.: A DDL approach to pluralistic ignorance and collective belief. J. Philos. Logic **43**, 499–515 (2014)
19. Spohn, W.: An analysis of Hansson's dyadic deontic logics. J. Philos. Logic **4**, 237–252 (1975)
20. van Benthem, J.: Dynamic logic of belief revision. J. Appl. Non-Classical Logic **17**(2), 129–155 (2007)
21. van Benthem, J., Grossi, D., Liu, F.: Priority structures in deontic logic. Theoria **83**, 116–152 (2014)
22. van Benthem, J., Liu, F.: Dynamic logic of preference upgrade. J. Appl. Non-Classical Logic **17**(2), 157–182 (2007)

On Subtler Belief Revision Policies

Fernando R. Velázquez-Quesada(✉)

Institute for Logic, Language and Computation, Universiteit van Amsterdam, Amsterdam, The Netherlands
F.R.VelazquezQuesada@uva.nl

Abstract. This paper proposes three subtle revision policies that are not propositionally successful (after a single application the agent might not believe the given propositional formula), but nevertheless are not propositionally idempotent (further applications might affect the agent's epistemic state). It also compares them with two well-known revision policies, arguing that the subtle ones might provide a more faithful representation of humans' real-life revision processes.

1 Introduction

Belief revision [1,2] is concerned with belief change fired by incoming information. It emerged as a proper subject in the 80s, when philosophical traditions dealing with both the requirements of rational belief change and the mechanisms by which scientific theories develop [3] converged with computer-science oriented approaches on database updates [4,5] and deontic studies focussed on derogations in legal codes [6]. The seminal [7] is considered the birth of the field; since then, many of its concerns and ideas have been proved relevant not only in philosophy and computer science but also in learning theory [8], among other fields.

Unsurprisingly, there have been many different proposals for representing this process. While one important difference among them has been the representation of beliefs themselves (e.g., syntactically, as either a plain set of formulas or else a deductively closed one [9]; semantically, by means of ordinal functions [10], a system of spheres [11], or a plausibility relation [12]), a more fundamental one has been the actual mechanism though which these beliefs change. Indeed, different revision policies have been proposed over time (see, e.g., [24]).

Despite the number of proposals, most revision policies share one crucial feature: incoming information outweighs current information. Hence, in case of conflict, the new information will prevail. This feature is not accidental, and in fact this *success* requirement is, among the *AGM* postulates [7], the first one: any successful revision act with a consistent formula χ as the new information (a χ-revision) should make χ part of the agent's beliefs.[1] This requirement is reasonable when one considers the situations targeted by the field's origins.

[1] Still, there are proposals rejecting the postulate. In Sect. 3 the reader can find some of them and their relationship with this manuscript's contents.

A. Baltag et al. (Eds.): LORI 2017, LNCS 10455, pp. 314–329, 2017.
DOI: 10.1007/978-3-662-55665-8_22

Indeed, if a new observation contradicts a scientific theory, the theory should be adapted to account for it; if new information is entered into a database, the database should not return old data in future searches; if a law changes, subsequent decisions should be ruled by its most recent version.

However, if one is interested in representing the way actual humans revise their beliefs, giving precedence to the new information might be neither the best nor the most realistic choice. For the first, information simply might not be 100% trustworthy (think of unreliable sources, now infamously called 'fake news'); for the second, we humans tend to be self-righteous [13], assuming most of the time that our beliefs are the correct ones, and thus looking for alternative explanations when something contradicts them.

Is it possible to reconciliate proposed revision policies that give priority to external information with human behaviour which gives priority to what one has? An observation that might prove useful is that revision strategies also tend to be *idempotent*: after a χ-revision, immediate repetitions will not make any difference. But this is not what happens in real life: for human beings, reiterated observations of the same phenomena is likely to have an effect, regardless of whether it contradicts our beliefs or how much we distrust its (possibly different) sources. For human beings, repetition matters.

This work proposes three subtle revision policies that might not lead to a belief after a single revision step, but will lead to a belief after they are repeated enough number of times. In doing so it shows how the effect of a single application of some well-known revision strategies can be reached (and, in some cases, surpassed) by the repetition of the short-term weaker but nevertheless long-term stronger revision policies that will be presented. It starts by providing the tools it will use for depicting beliefs and belief change, then recalling two well-known revision policies (Sect. 2). Then it continues by introducing the subtle revision policies, discussing their short-term and long-term effect, and comparing them with those in the literature (Sect. 3). It ends (Sect. 4) summarising the proposal and discussing further research directions.

2 Preliminaries: Representing Beliefs and Belief Change

In this paper, beliefs are represented by plausibility models [12,14], a qualitative version of the ordinal functions used in [10], and similar to the system of spheres of [11]. In such structures, belief change amounts to plausibility change, for which the layered upgrade of [15] will be used. This section provides the basic definitions the proposal requires. Let $\mathtt{P}$ be a countable set of atomic propositions.

Definition 2.1 (Plausibility model). *A* plausibility model M ($\mathcal{PM}$) *is a tuple* $\langle W, \leq, V \rangle$ *with* $W \neq \varnothing$ *a finite set of worlds,* $\leq \subseteq (W \times W)$ *a total preorder, the agent's* plausibility relation *over* W *(*$u \leq v$ *is read as "*u is at most as plausible as v*") and* $V : \mathtt{P} \to \wp(W)$ *a* valuation function, *indicating the worlds* $V(p) \subseteq W$

in which each $p \in \mathrm{P}$ holds.[2] *Given a plausibility model M, its set of worlds will be also denoted by $\mathcal{D}_M$.*

Intuitively, a $\mathcal{PM}$ represents the order the agent assigns to her epistemic possibilities. Thus, while her knowledge is given by what holds in all possible worlds, her beliefs are given by what holds in the most plausible. In order to provide formal definitions for these notions, the following will be useful.

Definition 2.2. *Let $\leq \subseteq (W \times W)$ be a total preorder over W.*

- $u < v$ ("u is less plausible than v") *iff*$_{def}$ $u \leq v$ *and* $v \not\leq u$.
- $u \sim v$ ("u and v are comparable") *iff*$_{def}$ $u \leq v$ *or* $v \leq u$.
- $u \simeq v$ ("u and v are equally plausible") *iff*$_{def}$ $u \leq v$ *and* $v \leq u$.
- *A* layer *$L \subseteq W$ is a set such that $\leq$ restricted to L is a universal relation on L, but this is not the case for any strict superset of L.*
- *The set of $\leq$-maximum elements in $U \subseteq W$ is given by*

$$\mathrm{Mx}(U) := \{u \in U \mid v \leq u \text{ for all } v \in U\}.$$

A $\mathcal{PM}$ can be understood as one or more layers of equally-plausible worlds, with the layers ordered according to their plausibility.

Definition 2.3 (Induced layers). *A $\mathcal{PM}$ $M = \langle W, \leq, V\rangle$ induces the following sequence of layers:*

$$\mathrm{L}_0(W) := \mathrm{Mx}(W) \qquad\qquad \mathrm{L}_{k+1}(W) := \mathrm{Mx}(W \setminus \bigcup_{\ell=0}^{k} \mathrm{L}_\ell(W))$$

Thus, while $\mathrm{L}_0(W)$ is the set of $\leq$-maximum worlds in W, $\mathrm{L}_1(W)$ is the set of $\leq$-maximum worlds in $W \setminus \mathrm{L}_0(W)$, $\mathrm{L}_2(W)$ is the set of $\leq$-maximum worlds in $W \setminus (\mathrm{L}_0(W) \cup \mathrm{L}_1(W))$ and so on. Note how indeed $\leq$ is the universal relation inside each $\mathrm{L}_k(W)$; moreover, as W is finite, there is $n > 0$ such that $k \geq n$ implies $\mathrm{L}_k(W) = \varnothing$ (denote the smallest such n by nl_M). Finally, the sets $\mathrm{L}_k(W)$ with $k \in \{0, \ldots, \mathrm{nl}_F - 1\}$ form a partition of W.

A formal language. As it will be discussed below, $\mathcal{PM}$s can be described by different formal languages. Here, the following one [15] will be used.

Definition 2.4 (Language $\mathcal{L}$). *Formulas φ, ψ ($\mathcal{L}^f$) and relational expressions π, σ ($\mathcal{L}^r$) of the language $\mathcal{L}$ are given, respectively by*

$$\varphi, \psi ::= \top \mid p \mid \neg\varphi \mid \varphi \vee \psi \mid \langle\pi\rangle\,\varphi \qquad \pi, \sigma ::= \mathbf{1} \mid \leq \mid \geq \mid ?(\varphi, \psi) \mid -\pi \mid \pi \cup \sigma \mid \pi \cap \sigma$$

with $p \in \mathrm{P}$. Propositional constants ($\top, \bot$), other Boolean connectives ($\wedge, \rightarrow, \leftrightarrow$) and dual modal operators $[\pi]$ are defined as usual ($[\pi]\,\varphi := \neg\langle\pi\rangle\,\neg\varphi$ for the latter).

[2] In [14] the plausibility relation is also conversely well-founded, forbidding infinite strictly ascending $\leq$-chains. Here the domain is finite, so this requirement is satisfied automatically.

The set of formulas of $\mathcal{L}$ contains the always true formula ($\top$) and atomic propositions (p), and it is closed under negation ($\neg$), disjunction ($\vee$) and modal operators of the form $\langle \pi \rangle$ with π a relational expression. The set of relational expressions contains a symbol representing the global relation ($\mathbf{1}$), a symbol representing the plausibility relation ($\leq$) and another representing its converse ($\geq$; [16,17]) and an additional construction of the form $?(\varphi, \psi)$ with φ and ψ formulas of the language, and it is closed under Boolean operations over relations (the so called *boolean modal logic*; [18]).

Definition 2.5 (Semantic interpretation). *Let $M = \langle W, \leq, V \rangle$ be a $\mathcal{PM}$. The function $[\![\cdot]\!]^M : \mathcal{L}^f \to \wp(W)$, from formulas in $\mathcal{L}$ to subsets of W, and the function $\langle\!\langle \cdot \rangle\!\rangle^M : \mathcal{L}^r \to \wp(W \times W)$, from relational expressions in $\mathcal{L}$ to binary relations over W, are defined inductively and simultaneously as follows.*

$$\begin{aligned} [\![\top]\!]^M &:= W & [\![\varphi \vee \psi]\!]^M &:= [\![\varphi]\!]^M \cup [\![\psi]\!]^M \\ [\![p]\!]^M &:= V(p) & [\![\langle \pi \rangle \varphi]\!]^M &:= \{w \in W \mid \langle\!\langle \pi \rangle\!\rangle^M_w \cap [\![\varphi]\!]^M \neq \varnothing\} \\ [\![\neg\varphi]\!]^M &:= W \setminus [\![\varphi]\!]^M \end{aligned}$$

with $\langle\!\langle \pi \rangle\!\rangle^M_w$ the set of worlds reachable from w via $\langle\!\langle \pi \rangle\!\rangle^M$,[3] *and*

$$\begin{aligned} \langle\!\langle \mathbf{1} \rangle\!\rangle^M &:= W \times W & \langle\!\langle -\pi \rangle\!\rangle^M &:= (W \times W) \setminus \langle\!\langle \pi \rangle\!\rangle^M \\ \langle\!\langle \leq \rangle\!\rangle^M &:= \leq & \langle\!\langle \pi \cup \sigma \rangle\!\rangle^M &:= \langle\!\langle \pi \rangle\!\rangle^M \cup \langle\!\langle \sigma \rangle\!\rangle^M \\ \langle\!\langle \geq \rangle\!\rangle^M &:= \{(v,u) \mid u \leq v\} & \langle\!\langle \pi \cap \sigma \rangle\!\rangle^M &:= \langle\!\langle \pi \rangle\!\rangle^M \cap \langle\!\langle \sigma \rangle\!\rangle^M \\ \langle\!\langle ?(\varphi,\psi) \rangle\!\rangle^M &:= [\![\varphi]\!]^M \times [\![\psi]\!]^M \end{aligned}$$

As usual, a formula φ is true *at world w in M iff $w \in [\![\varphi]\!]^M$ (in such case, w is called a φ-world). As usual, φ is* valid *($\Vdash \varphi$) iff $[\![\varphi]\!]^M = \mathcal{D}_M$ for every $\mathcal{PM}$ M.*

Relational expressions can be used for defining new modalities:

$$\langle < \rangle \varphi := \langle \leq \cap -\geq \rangle \varphi, \qquad \langle \sim \rangle \varphi := \langle \leq \cup \geq \rangle \varphi, \qquad \langle > \rangle \varphi := \langle \geq \cap -\leq \rangle \varphi.$$

For a sound and complete axiom system characterising formulas of $\mathcal{L}$ valid in $\mathcal{PM}$s, the reader is referred to [15,19].

Knowledge, belief and other epistemic attitudes. As mentioned, while knowledge in $\mathcal{PM}$s corresponds to what holds in all epistemic possibilities, beliefs correspond to what holds in the most plausible ones. But these models can represent further epistemic attitudes. There is the *safe belief* of [20], a belief persistent under revision with any true information, which corresponds to what holds in all worlds $\leq$-reachable from the evaluation point [14]. There is also the *strong belief* of [21,22] (called *robust belief* in [20]), a belief that can only be defeated by evidence (truthful or not) that is known to contradict it, corresponding to those formulas φ for which all φ-worlds are strictly more plausible than all $\neg\varphi$-worlds [14,22]. Another concept that will be useful is that of *very strong belief*, given by those formulas φ for which all φ-worlds appear at the topmost layer. Two further relevant concepts are that of *conditional belief* [12,14,20], what the

[3] Formally, $\langle\!\langle \pi \rangle\!\rangle^M_w := \{u \in W \mid (w,u) \in \langle\!\langle \pi \rangle\!\rangle^M\}$.

agent would have believed it was the case if she would have learnt that some condition was true, and the qualitative *degree of beliefs* [10,11], looking for what holds 'from some level up'. More precisely, at w in M the agent[4]

$$\begin{array}{rll} \textit{knows } \varphi & \text{iff}_{def} & [\![\varphi]\!]^M = W \\ \textit{believes } \varphi & \text{iff}_{def} & \mathrm{L}_0(W) \subseteq [\![\varphi]\!]^M \\ \textit{safely believes } \varphi & \text{iff}_{def} & \langle\!\langle \leq \rangle\!\rangle_w^M \subseteq [\![\varphi]\!]^M \\ \textit{strongly believes } \varphi & \text{iff}_{def} & \text{there is } k < \mathrm{nl}_M \text{ such that } \bigcup_{i=0}^k \mathrm{L}_i(W) = [\![\varphi]\!]^M \end{array}$$

$$\begin{array}{rll} \textit{very strongly believes } \varphi & \text{iff}_{def} & \mathrm{L}_0(W) = [\![\varphi]\!]^M \\ \textit{believes } \varphi \text{ conditionally to } \psi & \text{iff}_{def} & \mathrm{Mx}([\![\psi]\!]^M) \subseteq [\![\varphi]\!]^M \\ \textit{believes } \varphi \text{ with a degree } k & \text{iff}_{def} & \mathrm{Mx}^k(W) \subseteq [\![\varphi]\!]^M \end{array}$$

Note how all these attitudes can be expressed with formulas in $\mathcal{L}$. Indeed, define

$$\begin{array}{rr} \text{knowledge: } K\,\varphi := [\sim]\,\varphi, & \text{strong belief: } St\,\varphi := \langle\leq\rangle\,[\leq]\,\varphi \wedge [\sim](\varphi \rightarrow [\leq]\,\varphi), \\ \text{belief: } B\,\varphi := \langle\leq\rangle\,[\leq]\,\varphi, & \text{very strong belief: } VSt\,\varphi := [\sim](\varphi \leftrightarrow [<]\,\bot), \\ \text{safe belief: } Sf\,\varphi := [\leq]\,\varphi, & \text{conditional belief: } B^{\psi}\,\varphi := \langle\sim\rangle\,\psi \rightarrow \langle\sim\rangle(\psi \wedge [\leq](\psi \rightarrow \varphi)), \\ & \text{degree of belief: } B^k\,\varphi := [\sim]((\lambda_0 \vee \cdots \vee \lambda_k) \rightarrow \varphi), \end{array}$$

with the sequence of formulas λ_k $(k \geq 0)$ for the degree of belief case given by

$$\lambda_0 := [<]\,\bot \text{ and } \lambda_{k+1} := (\neg\lambda_0 \wedge \cdots \wedge \neg\lambda_k) \wedge [<](\lambda_0 \vee \cdots \vee \lambda_k), \text{ so } \mathrm{L}_k(W) = [\![\lambda_k]\!]^M.$$

Then, for every $\mathcal{PM}$ M and world $w \in \mathcal{D}_M$,

$$\begin{array}{rr} w \in [\![K\,\varphi]\!]^M \text{ iff } [\![\varphi]\!]^M = W & w \in [\![St\,\varphi]\!]^M \text{ iff } \exists\, k < \mathrm{nl}_M \text{ s.t. } \bigcup_{i=0}^k \mathrm{L}_i(W) = [\![\varphi]\!]^M \\ w \in [\![B\,\varphi]\!]^M \text{ iff } \mathrm{L}_0(W) \subseteq [\![\varphi]\!]^M & w \in [\![VSt\,\varphi]\!]^M \text{ iff } \mathrm{L}_0(W) = [\![\varphi]\!]^M \\ w \in [\![Sf\,\varphi]\!]^M \text{ iff } \langle\!\langle \leq \rangle\!\rangle_w^M \subseteq [\![\varphi]\!]^M & w \in [\![B^{\psi}\,\varphi]\!]^M \text{ iff } \mathrm{Mx}([\![\psi]\!]^M) \subseteq [\![\varphi]\!]^M \\ & w \in [\![B^k\,\varphi]\!]^M \text{ iff } \mathrm{Mx}^k(W) \subseteq [\![\varphi]\!]^M \end{array}$$

This shows how $\mathcal{L}$ can express several different epistemic attitudes.[5] Moreover, its expressivity is what allows the existence of an axiom system for a modality describing the effects of the *general layered upgrade*, the model operation that will be used here for representing plausibility change.

A plausibility change operation. The 'subtle' policies to be provided in Sect. 3 will be represented with the general layered upgrade of [15, Definition 8]. This operation receives a $\mathcal{PM}$ M and a *layered list* $\mathbb{S}$ over $\mathcal{D}_M$: a finite (possibly empty) list of pairwise disjoint (possible empty) subsets of $\mathcal{D}_M$ together with a default plausibility ordering over $\mathcal{D}_M$. The list's length is denoted by $|\mathbb{S}|$, its kth element is denoted by $\mathbb{S}[k-1]$ (with $0 \leq k < |\mathbb{S}|$, so $\mathbb{S} = \ulcorner \mathbb{S}[0], \cdots, \mathbb{S}[\,|\mathbb{S}|-1\,] \lrcorner$), and $\leq_{\mathbb{S}}^{d}$ is its default plausibility ordering. Intuitively, $\mathbb{S}$ defines layers of elements of $\mathcal{D}_M$ in a new plausibility ordering $\leq_{\mathbb{S}}$, with $\mathbb{S}[0]$ the set of worlds from which the topmost layer will be defined, and $\leq_{\mathbb{S}}^{d}$ used to sort not only each individual set but also those worlds not in $\bigcup_{k=0}^{|\mathbb{S}|-1} \mathbb{S}[k]$.

4 With $\mathrm{Mx}^k(W) := \bigcup_{i=0}^k \mathrm{L}_i(W)$. This is the plausibility-order-is-total version of the standard definition.

5 In particular, it can characterise *all* layers of *any* $\mathcal{PM}$.

As shown in [15], the general layered upgrade can define several well-known order change operations. Still, for this work, a simpler version is enough. Call a layered list *explicit* ($\mathcal{E}$) when its sets define a *quasi-partition* on $\mathcal{D}_W$ (the sets are both pairwise disjoint and collectively exhaustive, but some of them might be empty) and its default ordering is the universal relation ($u \leq^{\mathrm{d}}_{\mathbb{S}} v$ for every $u, v \in \mathcal{D}_W$). Call it *syntactically defined* ($\mathcal{SD}$) when each $\mathbb{S}[k]$ is given by a formula ξ_k that can be evaluated in $\mathcal{PM}$s. Then,

Definition 2.6 (General layered upgrade). *Let M be a $\mathcal{PM}$. If $\mathbb{S}$ is an $\mathcal{E}$-$\mathcal{SD}$ list over $\mathcal{D}_M$ in which each $\mathbb{S}[k]$ is given by a formula ξ_k, then $u \leq_{\mathbb{S}} v$ holds iff either both worlds are in the same set (part 1 below), or else u's set is not above v's (i.e., v's set is strictly above u's; part 2). Formally,*

$$u \leq_{\mathbb{S}(M)} v \quad \mathit{iff}_{def} \quad \underbrace{\bigvee_{k=0}^{|\mathbb{S}|-1} \{u,v\} \subseteq [\![\xi_k]\!]^M}_{1} \vee \underbrace{\bigvee_{k=0}^{|\mathbb{S}|-1} \left(v \in [\![\xi_k]\!]^M \wedge u \notin \bigcup_{l=0}^{k} [\![\xi_l]\!]^M \right)}_{2}$$

It is not difficult to see that $\leq_{\mathbb{S}}$ is indeed a total preorder. Then,

Definition 2.7. *Let $M = \langle W, \leq, V \rangle$ be a $\mathcal{PM}$. If $\mathbb{S}$ is an $\mathcal{E}$-$\mathcal{SD}$ layered list, then the $\mathcal{PM}$ $M_{\mathrm{ly}(\mathbb{S})}$ is defined as $\langle W, \leq_{\mathbb{S}(M)}, V \rangle$. In order to describe the effects of this operation, the language $\mathcal{L}_{\mathrm{ly}}$ extends $\mathcal{L}$ with a modality $\langle \mathrm{ly}(\mathbb{S}) \rangle$ for every $\mathcal{E}$-$\mathcal{SD}$ layered list $\mathbb{S}$. Given a $\mathcal{PM}$ M, define*

$$[\![\langle \mathrm{ly}(\mathbb{S}) \rangle \, \varphi]\!]^M := [\![\varphi]\!]^{M_{\mathrm{ly}(\mathbb{S})}}$$

For a sound and complete axiom system for $\mathcal{L}_{\mathrm{ly}}$ w.r.t. $\mathcal{PM}$s, see [15]. Note that this paper works with $\mathcal{E}$-$\mathcal{SD}$ layered lists because it is both possible and useful. For $\mathcal{E}$, the discussed policies are 'simple', so it is possible (and clearer) to provide explicit definitions for all layers in the resulting plausibility orderings. For $\mathcal{SD}$, $\mathcal{L}$ is expressive enough to characterise syntactically these layers, which yields an (albeit indirect) axiomatisation for the discussed revision policies.

Two well-known revision policies. For a later comparison, here is the general layered upgrade representation of two well-known revision policies. Note, in particular, the discussed crucial properties.

Moderate revision. This policy (see [23] among others) is called *lexicographic* in [24]. A moderate χ-revision makes all χ-worlds more plausible than all $\neg\chi$-worlds, keeping the old ordering within each zone. It can be defined by a general layered upgrade whose ($\mathcal{E}$-$\mathcal{SD}$) layered list uses the following formulas.

Definition 2.8 (Moderate revision). *Let M be a $\mathcal{PM}$ with $n := \mathrm{nl}_M$; let χ be a formula in $\mathcal{L}$. The formulas ξ^{χ}_k for moderate revision, with $0 \leq k \leq (2n-1)$, are defined in the following way (with formulas λ_k as in Page 2).*

$$\xi^{\chi}_0 := \lambda_0 \wedge \chi, \ \ldots \ , \ \xi^{\chi}_{n-1} := \lambda_{n-1} \wedge \chi, \ \xi^{\chi}_n := \lambda_0 \wedge \neg\chi, \ \ldots \ , \ \xi^{\chi}_{2n-1} := \lambda_{n-1} \wedge \neg\chi.$$

With these formulas, a general layered upgrade defines a moderate revision: all former χ-worlds will be above all former $\neg\chi$-worlds, with the relative order within the two zones preserved. Although $[\![\xi_i^\chi]\!]^M = \varnothing$ might hold for some $0 \leq i \leq (2n-1)$, this is not a problem: the sets $[\![\xi_0^\chi]\!]^M, \ldots, [\![\xi_{2n-1}^\chi]\!]^M$ define a quasi-partition of $\mathcal{D}_M$, as required. Here is a modality for this policy.

Definition 2.9. *Let the formulas in* $\{\xi_k^\chi \mid 0 \leq k \leq (2n-1)\}$ *be as in Definition 2.8. Then,*

$$M{\Uparrow_\chi} := M_{\ulcorner \xi_0^\chi, \ldots, \xi_{2n-1}^\chi \lrcorner} \qquad \text{and} \qquad [\![\langle \chi\Uparrow\rangle\, \varphi]\!]^M := [\![\varphi]\!]^{M\Uparrow_\chi}$$

Conservative revision. This revision policy [25] is called *natural* in [26] and *minimal conditional* in [27]. It makes the most plausible χ-worlds the only ones at the top of the ordering, keeping the old order among the rest.

Definition 2.10 (Conservative revision). *Let M be a* $\mathcal{PM}$ *with $n := \mathrm{nl}_M$; let χ be a formula in $\mathcal{L}$. Define $\lambda_\lambda^\chi := \chi \wedge [<]\neg\chi$, characterising the most plausible χ-worlds. Formulas ξ_k^χ for conservative revision, with $0 \leq k \leq n$, are given by*

$$\xi_0^\chi := \lambda_\lambda^\chi, \qquad \xi_1^\chi := \lambda_0 \wedge \neg\lambda_\lambda^\chi, \qquad \ldots, \qquad \xi_n^\chi := \lambda_{n-1} \wedge \neg\lambda_\lambda^\chi.$$

The former most plausible χ-worlds (λ_λ^χ) will move to the top, with the rest keeping their old ordering. Again, the formulas define a quasi-partition of $\mathcal{D}_M$.

Definition 2.11. *Let the formulas in* $\{\xi_k^\chi \mid 0 \leq k \leq n\}$ *be as in Definition 2.10.*

$$M{\uparrow_\chi} := M_{\ulcorner \xi_0^\chi, \ldots, \xi_n^\chi \lrcorner} \qquad \text{and} \qquad [\![\langle \chi\uparrow\rangle\, \varphi]\!]^M := [\![\varphi]\!]^{M\uparrow_\chi}$$

These policies satisfy two important properties. First, they are *propositionally successful*: if the agent considers a propositional χ possible, then after a χ-revision she will believe χ. Second, they are *propositionally idempotent*: after a first revision with a propositional χ, further χ-revisions will not have any effect, not only on the agent's beliefs (the topmost worlds), but also on the agent's plausibility relation.

Proposition 2.1 ((Propositional) success and idempotence). *Let χ be a propositional formula; define $\widehat{K}\,\chi := \neg\, K \neg\chi$. Then*

$$\begin{array}{rll} \textit{Success:} & \Vdash \widehat{K}\,\chi \rightarrow [\chi\Uparrow]\, B\,\chi, & \Vdash \widehat{K}\,\chi \rightarrow [\chi\uparrow]\, B\,\chi. \\ \textit{Idempotence:} & (M{\Uparrow_\chi}){\Uparrow_\chi} = M{\Uparrow_\chi}, & (M{\uparrow_\chi}){\uparrow_\chi} = M{\uparrow_\chi}. \end{array}$$

Proof (Sketch). Success is straightforward. For idempotence, it is enough to show that in both models the ordering is the same. (Since χ is propositional, $[\![\chi]\!]^M = [\![\chi]\!]^{M\Uparrow_\chi} = [\![\chi]\!]^{M\uparrow_\chi}$.) For moderate revision, in $M{\Uparrow_\chi}$ all χ-worlds are already above all $\neg\chi$-worlds; for the conservative, the most plausible χ-worlds in $M{\uparrow_\chi}$ are already the most plausible overall. Thus, repetitions will not have any effect.

These properties might fail when χ involves modalities.[6] Still, as the policies introduced next show, the fact that χ is propositional is not the decisive one.

3 Three Subtle Belief Revision Policies

'All one level up' (ALU). Here is the first subtle revision policy.

Definition 3.1 ('All one level up'). *Let M be a $\mathcal{PM}$ with $n := \mathrm{nl}_M$; let χ be a formula in $\mathcal{L}$. The formulas ξ_k^χ for ALU revision ($0 \leq k \leq n$) are given by*

$$\xi_0^\chi := \lambda_0 \wedge \chi, \qquad \xi_1^\chi := (\lambda_1 \wedge \chi) \vee (\lambda_0 \wedge \neg\chi), \qquad \ldots,$$
$$\ldots, \qquad \xi_{n-1}^\chi := (\lambda_{n-1} \wedge \chi) \vee (\lambda_{n-2} \wedge \neg\chi), \qquad \xi_n^\chi := (\lambda_{n-1} \wedge \neg\chi).$$

From the definition, each $\neg\chi$-world is moved to the layer below. In other words, in the resulting model, all χ-worlds have been pushed 'one level up'. The formulas clearly define a quasi-partition of any model's domain.

Definition 3.2. *Let the formulas in $\{\xi_k^\chi \mid 0 \leq k \leq n\}$ be as in Definition 3.1.*[7]

$$M{\Uparrow}_\chi := M_{\ulcorner \xi_0^\chi, \ldots, \xi_n^\chi \lrcorner} \qquad \text{and} \qquad [\![\langle \chi{\Uparrow} \rangle\, \varphi]\!]^M := [\![\varphi]\!]^{M{\Uparrow}_\chi}$$

Note how ALU does not satisfy the previous two properties.

Fact 3.1. ALU is neither propositionally successful nor propositionally idempotent.

Proof. The following sequence of diagrams (reflexive and transitive arrows omitted) shows (left to right) the effect of a successive application of the ALU policy with p.

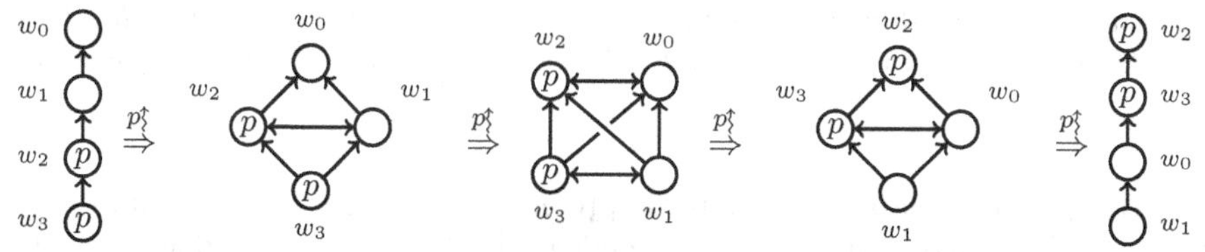

Thus, the ALU policy with a propositional χ might not lead to a belief in χ in one step. Nevertheless, further applications do make a difference.

Proposition 3.1. *For any (mind: finite) $\mathcal{PM}$ M and propositional χ there is $m \geq 0$ s.t.*

---

[6] For success, Moorean phenomena [29] might appear. For idempotence, the $\mathcal{L}$-formula $[>]\bot$ characterises the bottommost elements of the ordering; thus, if the initial ordering is not flat (not all worlds equally plausible), each $[>]\bot$-revision will move to the top different worlds.

[7] Thus, $[\chi{\Uparrow}]\,\varphi := \neg\,\langle \chi{\Uparrow} \rangle\, \neg\varphi$ is such that $\Vdash [\chi{\Uparrow}]\,\varphi \leftrightarrow \langle \chi{\Uparrow} \rangle\, \varphi$.

$$[\![\widehat{K}\,\chi \rightarrow [\chi\Uparrow^{m}]\,B\,\chi]\!]^M = \mathcal{D}_M$$

with $[\chi\Uparrow^{0}]\,\varphi := \varphi$, $[\chi\Uparrow^{1}]\,\varphi := [\chi\Uparrow]\,\varphi$ *and* $[\chi\Uparrow^{k+1}]\,\varphi := [\chi\Uparrow]\,[\chi\Uparrow^{k}]\,\varphi$.

Proof (Sketch). If M has no χ-worlds, any $m \geq 0$ works. Otherwise take $m := \text{lay}^{\chi}_{\curlywedge} + 1$, with $\text{lay}^{\chi}_{\curlywedge}$ the number of the layer in M where the most plausible χ-worlds appear: after $\text{lay}^{\chi}_{\curlywedge}$ applications of $\chi\Uparrow$, these worlds will reach the topmost layer, and after one more, only they will be up there; hence, after $\text{lay}^{\chi}_{\curlywedge} + 1$ steps, the agent will believe χ.[8]

Even though ALU does not have a strong initial effect (it is not propositionally successful), its repetition is not idle (it is not propositionally idempotent), as it eventually leads to a belief in the involved formula. In fact, this policy's long-term effect is stronger than that of the conservative policy, as it becomes idempotent only after all χ-worlds are above all $\neg\chi$-worlds. In other words, its effect become null only when a *strong belief* on χ has been reached.

Proposition 3.2. *Let M be a $\mathcal{PM}$; let χ be propositional. There is $m \geq 0$ s.t.*

$$[\![\widehat{K}\,\chi \rightarrow [\chi\Uparrow^{m}]\,St\,\varphi]\!]^M = \mathcal{D}_M.$$

Proof (Sketch). ALU *has an effect until no χ-world can move up, i.e., until the 'worst' χ-worlds are strictly above the 'best' $\neg\chi$-worlds. Thus, define $m := (\text{lay}^{\chi}_{\curlyvee} - \text{lay}^{\neg\chi}_{\curlywedge}) + 1$, with $\text{lay}^{\chi}_{\curlyvee}$ the number of the layer in M where the least plausible χ-worlds appear. After $\text{lay}^{\chi}_{\curlyvee} - \text{lay}^{\neg\chi}_{\curlywedge}$ applications of $\chi\Uparrow$, the 'worst' χ-worlds and the 'best' $\neg\chi$-worlds will be in the same layer; after one more, a strong belief on χ will be reached.*[9]

'Bottommost one level up' (BLU). The 'all one level up' policy works by pushing all χ-worlds to the upper layer. The 'bottommost one level up' policy below only acts on the least plausible χ-worlds.

Definition 3.3 ('Bottommost one level up'). *Let M be a $\mathcal{PM}$ with $n := \text{nl}_M$; let χ be a formula in $\mathcal{L}$. Define $\lambda^{\chi}_{\curlyvee} := \chi \wedge [>]\neg\chi$, characterising the least plausible χ-worlds. Formulas ξ^{χ}_{k} for BLU revision ($0 \leq k \leq n-1$) are given by*

$$\xi^{\chi}_{i} := (\langle\sim\rangle(\lambda_{i+1} \wedge \lambda^{\chi}_{\curlyvee}) \rightarrow (\lambda_i \vee \lambda^{\chi}_{\curlyvee})) \wedge (\neg\langle\sim\rangle(\lambda_{i+1} \wedge \lambda^{\chi}_{\curlyvee}) \rightarrow (\lambda_i \wedge \neg\lambda^{\chi}_{\curlyvee}))$$

Although more elaborated than the previous case, these formulas fulfil their goal. Indeed, when defining the new model's ith layer (ξ^{χ}_{i}), it is asked whether the next layer in M contains the least plausible χ-worlds (the condition $\langle\sim\rangle(\lambda_{i+1} \wedge \lambda^{\chi}_{\curlyvee})$). If that is the case, this ith layer will contain those worlds in the original ith layer plus the least plausible χ-worlds ($\lambda_i \vee \lambda^{\chi}_{\curlyvee}$); otherwise, it will contain those worlds in the original ith layer that are not the least plausible χ-worlds ($\lambda_i \wedge \neg\lambda^{\chi}_{\curlyvee}$). The formulas certainly define a quasi-partition: the consequent of the implications in each one of them shows how each world satisfies exactly one formula. Then,

---

[8] This relies on the fact that χ is propositional: the set of χ-worlds does not change.

[9] If $\text{lay}^{\chi}_{\curlyvee} - \text{lay}^{\neg\chi}_{\curlywedge} < 0$, then there is already a strong belief on χ at M, and any $m \geq 0$ works.

Definition 3.4. *Let the formulas in* $\{\xi_k^\chi \mid 0 \leq k \leq n-1\}$ *be as in Definition 3.3.*[10] *Then,*

$$M\Uparrow_\chi := M_{\lceil \xi_0^\chi, \ldots, \xi_{n-1}^\chi \rfloor} \qquad \text{and} \qquad [\![\langle \chi\Uparrow \rangle \varphi]\!]^M := [\![\varphi]\!]^{M\Uparrow_\chi}$$

Just as the ALU policy, BLU one does not satisfy the discussed properties.

Fact 3.2. BLU is neither propositionally successful nor propositionally idempotent.

Proof. Here is an example of effect of a successive application of BLU with p.

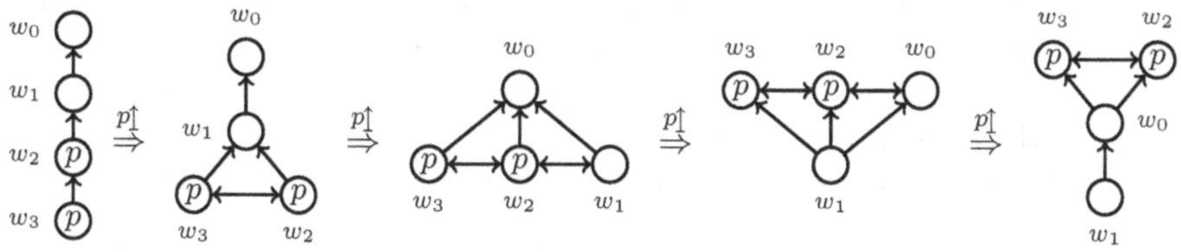

Just as the ALU policy, while a BLU revision with a propositional χ does not need to lead to a belief in χ in one step, further repetitions do make a difference.

Proposition 3.3. *For any* $\mathcal{PM}$ M *and propositional* χ *there is* $m \geq 0$ *s.t.*

$$[\![\widehat{K}\,\chi \rightarrow [\chi\Uparrow^{\,m}]\, B\,\chi]\!]^M = \mathcal{D}_M$$

with $[\chi\Uparrow^{\,0}]\,\varphi := \varphi$, $[\chi\Uparrow^{\,1}]\,\varphi := [\chi\Uparrow]\,\varphi$ *and* $[\chi\Uparrow^{\,k+1}]\,\varphi := [\chi\Uparrow]\,[\chi\Uparrow^{\,k}]\,\varphi$.

Proof (Sketch). Similar to that for the ALU case (Proposition 3.1*), the difference being that it takes* $m := \text{lay}_\curlyvee^\chi + 1$ *BLU steps for the agent to believe* χ*: the most plausible* χ*-worlds do not move up until they become also the least plausible* χ*-ones.*

Once again, even though this revision policy does not have a strong initial impact, its repetition has an effect, as it eventually leads to a belief in the given formula χ. As the proof of the previous proposition shows, this policy is 'slower' (i.e., even subtler) than ALU: it might require more iterations in order to reach the belief point ($\text{lay}_\curlyvee^\chi + 1$ for BLU vs the potentially smaller $\text{lay}_\curlywedge^\chi + 1$ for ALU). Nevertheless, this 'take care of the worst first' approach has its advantages: this policy becomes idle not when the agent has a strong belief on χ but rather when she has a *very strong belief* on it.

Proposition 3.4. *Let* M *be a* $\mathcal{PM}$*; let* χ *be propositional. There is* $m \geq 0$ *s.t.*

$$[\![\widehat{K}\,\chi \rightarrow [\chi\Uparrow^{\,m}]\; VSt\,\chi]\!]^M = \mathcal{D}_M.$$

Proof (Sketch). BLU acts until the 'worst' χ*-worlds cannot be moved up, i.e., until they are at the top. This happens after* $m := \text{lay}_\curlyvee^\chi + 1$ *BLU revision steps.*

[10] Thus, $[\chi\Uparrow]\,\varphi := \neg\langle\chi\Uparrow\rangle\neg\varphi$ is such that $\Vdash [\chi\Uparrow]\,\varphi \leftrightarrow \langle\chi\Uparrow\rangle\,\varphi$.

'Opposite topmost one level down' (OTLD). The 'bottommost one level up' policy shifts the least plausible χ-worlds one level up. The 'opposite topmost one level down' policy is its dual: the most plausible $\neg\chi$-worlds go one level down.

Definition 3.5 ('Opposite topmost one level down'). *Let M be a $\mathcal{PM}$ with $n := \text{nl}_M$; let χ be a formula in $\mathcal{L}$. Define $\lambda_{-1} := \bot$. The formulas ξ_k^χ for OTLD revision ($0 \leq k \leq n-1$) are given by*

$$\xi_i^\chi := (\langle\sim\rangle(\lambda_{i-1} \wedge \lambda_\lambda^{\neg\chi}) \rightarrow (\lambda_i \vee \lambda_\lambda^{\neg\chi})) \wedge (\neg\langle\sim\rangle(\lambda_{i-1} \wedge \lambda_\lambda^{\neg\chi}) \rightarrow (\lambda_i \wedge \neg\lambda_\lambda^{\neg\chi}))$$

When defining the new model's ith layer (ξ_i^χ), it is asked whether the upper layer in M contains the most plausible $\neg\chi$-worlds (the condition $\langle\sim\rangle(\lambda_{i-1}\wedge\lambda_\lambda^{\neg\chi})$). If that is the case, this ith layer will contain those worlds in the original ith layer plus the most plausible $\neg\chi$-worlds ($\lambda_i \vee \lambda_\lambda^{\neg\chi}$); otherwise, it will contain those worlds in the original ith layer that are not the most plausible $\neg\chi$-worlds ($\lambda_i \wedge \neg\lambda_\lambda^{\neg\chi}$).

Definition 3.6. *Let the formulas in $\{\xi_k^\chi \mid 0 \leq k \leq (n-1)\}$ be as in Definition 3.5.*[11] *Then,*

$$M\!\Downarrow_\chi := M_{\lceil \xi_0^\chi,\ldots,\xi_{n-1}^\chi \rfloor} \qquad \text{and} \qquad [\![\langle\chi\Downarrow\rangle\,\varphi]\!]^M := [\![\varphi]\!]^{M\Downarrow_\chi}$$

Again, the OTLD policy is neither (propositionally) successful nor idempotent.

Fact 3.3. OTLD is neither propositionally successful nor propositionally idempotent.

Proof. Here is an example of effect of a successive application of OTLD with p.

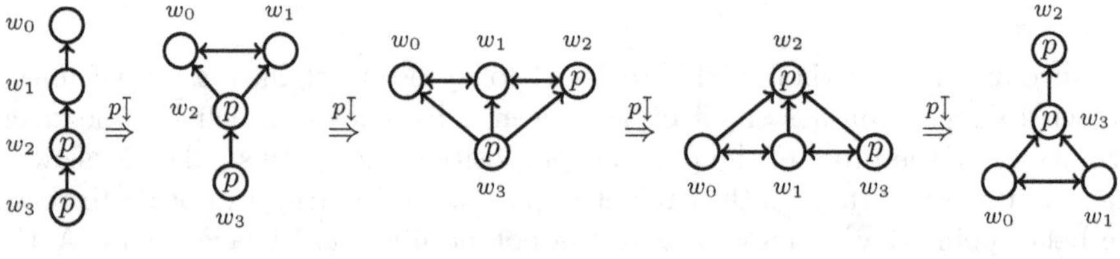

As with the previous cases, OTLD reaches a belief in the involved formula after some repetition.

Proposition 3.5. *For any $\mathcal{PM}$ M and propositional χ there is $m \geq 0$ s.t.*

$$[\![\widehat{K}\,\chi \rightarrow [\chi\Downarrow^m]\,B\,\chi]\!]^M = \mathcal{D}_M$$

with $[\chi\Downarrow^0]\,\varphi := \varphi$, $[\chi\Downarrow^1]\,\varphi := [\chi\Downarrow]\,\varphi$ and $[\chi\Downarrow^{k+1}]\,\varphi := [\chi\Downarrow]\,[\chi\Downarrow^k]\,\varphi$.

[11] Thus, $[\chi\Downarrow]\,\varphi := \neg\langle\chi\Downarrow\rangle\,\neg\varphi$ is such that $\Vdash [\chi\Downarrow]\,\varphi \leftrightarrow \langle\chi\Downarrow\rangle\,\varphi$.

Proof (Sketch). Similar to that for of Proposition 3.3, *here with* $m := (\mathrm{lay}^{\chi}_{\lambda} - \mathrm{lay}^{\neg\chi}_{\lambda}) + 1$: *there will be only* χ*-worlds at the top after the most plausible* $\neg\chi$*-worlds have been pushed below the most plausible* χ-worlds.[12]

This policy requires $(\mathrm{lay}^{\chi}_{\lambda} - \mathrm{lay}^{\neg\chi}_{\lambda}) + 1$ steps to reach a "belief on χ" state, so it is at least as fast as both ALU ($\mathrm{lay}^{\chi}_{\lambda} + 1$) and BLU ($\mathrm{lay}^{\chi}_{\curlyvee} + 1$). In the long term it leads to a strong belief: it keeps changing the plausibility order until all $\neg\chi$-worlds lie at the bottom. Thus, its overall 'belief' effect is stronger than ALU (belief) but weaker than BLU (a very strong belief might not be reached, as the relative position of χ-worlds never changes). Still, by pushing all $\neg\chi$-worlds to the bottom, it reaches what can be called a *very strong disbelief in* $\neg\chi$ (syntactically, $\overline{VSt}\,\chi := [\sim](\chi \leftrightarrow [>]\bot)$).

Proposition 3.6. *Let* M *be a model; let* χ *be propositional. There is* $m \geq 0$ *s.t.*

$$[\![\widehat{K}\,\chi \rightarrow [\chi{↧}^{m}](St\,\chi \wedge \overline{VSt}\,\neg\chi)]\!]^{M} = \mathcal{D}_{M}.$$

Proof (Sketch). OTLD *acts until all* $\neg\chi$*-worlds lie at the bottom, thus yielding a strong belief in* χ *and a very strong disbelief in* $\neg\chi$*. Its works by pushing down the most plausible* $\neg\chi$*-worlds one level at the time, so it reaches its goal after* $m := \mathrm{nl}_M - \mathrm{lay}^{\neg\chi}_{\lambda}$ *steps.*

Comparison. The table below compares the short- and long-term effects of the policies discussed so far (with χ a propositional formula). In fact, informally one can write

$$\text{short-term: } \{{↧},{↥}\} < {⇡} < \uparrow < \Uparrow \qquad \text{and} \qquad \text{long-term: } \uparrow < \{\Uparrow,{⇡}\} < {↧} < {↥},$$

with, in the short-term case, $\{{↧},{↥}\} < {⇡}$ because the latter shifts up *all* χ-worlds. Thus, while in the short-term the subtle policies are weaker, their long-term effect is at least as strong as that of the others.

Policy		Short-term effect on *beliefs*	Long-term effect on *beliefs*
Moderate	$\chi\Uparrow$	Strong belief in χ	Strong belief in χ
Conservative	$\chi\uparrow$	Belief in χ	Belief in χ
ALU	$\chi{⇡}$	—	Strong belief in χ
BLU	$\chi{↥}$	—	Very strong belief in χ
OTLD	$\chi{↧}$	—	Strong belief in χ and very strong disbelief in $\neg\chi$

Other approaches. Some approaches study revision operators that might not satisfy the success postulate. One of the earliest is *non-prioritized* belief revision

[12] If $\mathrm{lay}^{\chi}_{\lambda} - \mathrm{lay}^{\neg\chi}_{\lambda} < 0$, then there is already a belief on χ at M, and any $m \geq 0$ works.

[30], under which new information will be accepted only if it has more 'epistemic value' than the original beliefs it might contradict. This notion of 'epistemic value' is defined sometimes by a set of *core* beliefs (the input will be accepted only if it is consistent with such set: *screened revision* [31]) and sometimes by a set of *credible* formulas (the input will be accepted only if it belongs to such set: *credibility-limited revision* [32]). Neither of the policies proposed here fall into this category. First, the presented setting does not use notions of 'epistemic value' to contrast the input formula with the current beliefs. More importantly, the three policies proposed here always 'accept' the incoming information, albeit with a very high degree of cautiousness (and each one of them in a different way).

A related research line sets the plausibility of the incoming information as that of a second 'reference' formula; thus, the new information might not lie at the top of the ordering after the revision. Approaches following this idea (called *raising* and *lowering* in [33], *revision by comparison* in [34] and *two-dimensional* in [35]) require two formulas as the input for the operation (plus, of course, the representation of the agent's beliefs), and as such differ from the strategy followed by the policies proposed here.

A closer research line is that of *improvement operators* [36]: the incoming χ might not be believed after the improvement, but its plausibility will be increased. A taxonomy of improvement operators is provided in [37], with all of them satisfying **S1** (the ordering within $[\![\chi]\!]^M$ is invariant), and **S2** (so is the ordering within $[\![\neg\chi]\!]^M$). The first subtle policy presented here, ALU, is in fact the unique improvement operator of [36], called *one-improvement* in [37]. However, neither BLU nor OTLD fall in the category: while the first fails to satisfy **S1**, the second fails to satisfy **S2**. Still, lacking such behaviour does not weaken these policies, and in fact it is part of what gives them such a strong long-term effect (Propositions 3.4 and 3.6, respectively).

Other works have proposed policies for small changes in plausibility. As examples, the operation in [38] (called *refinement* in [24]) splits each layer of the original plausibility ordering, placing the χ-worlds above the $\neg\chi$-ones. Refinement is not propositionally successful, but it is propositionally idempotent: once the initial layers have been split, further attempts will not change the ordering. On the other hand, [24] also discusses a *gentle lowering* revision, which pushes the most plausible χ-worlds one level up. Such operation is neither propositionally successful nor propositionally idempotent. Still, its long-term effect is weaker: its iteration will lead to a belief in (a propositional) χ, but neither strong belief nor very strong belief need to be reached.

4 Summary and Future Work

This paper uses $\mathcal{PM}$s for representing beliefs and the general layered upgrade for representing plausibility change. It starts recalling two well-known revision policies, moderate and conservative, both of them being (as many others) propositionally successful (after a χ-revision, the agent will believe χ) and propositionally idempotent (additional χ-revisions are idle). Thus, they have a strong initial effect, but afterwards they do not change the agent's epistemic state.

Then it introduces three 'weaker-yet-stronger' revision policies: 'all one level up', 'bottommost one level up' and 'opposite topmost one level down'. They are weaker in the sense of not being propositionally successful (a single χ-revision might not lead to a belief in χ); they are stronger in the sense of not being propositionally idempotent (further repetitions might have an effect on the agent's epistemic state). In fact, if χ is compatible with the agent's knowledge, ALU eventually leads not only to a belief in χ, but also to a *strong* belief in it. Under the same condition, BLU also leads to an eventual belief in χ. While it might require more iterations to reach such stage, it makes up for its slowness by strengthening the agent's attitude towards χ even more: it eventually leads to a *very strong* belief in it. Finally, OTLD also leads to both a belief and then a strong belief; while it might not reach a very strong belief in χ, it reaches a very strong disbelief in $\neg\chi$. These revision policies are realistic, as indeed there are ideas that are initially refused, only to become eventually accepted after a number of repetitions.[13] Maybe more interesting, such policies could be understood as forms of propositional *inductive* reasoning, as they lead to a given outcome (a belief on χ) not in one single step, but rather after a number of them.

For future work, a first goal is to characterise the presented policies (e.g., in terms of postulates) to provide a more precise comparison.[14] Then one can also explore not only other 'subtle' revision policies, but also 'subtle forms of contraction. Finally, this paper has focussed on the single-agent case, but the technical details will be different in multi-agent scenarios (in particular when considering private revisions).

References

1. Gärdenfors, P., Rott, H.: Belief revision. In: Handbook of Logic in Artificial Intelligence and Logic Programming, vol. 4, pp. 35–132. Oxford University Press, Oxford (1994)
2. Williams, M.A., Rott, H. (eds.): Frontiers in Belief Revision. Kluwer Academic, Dordrecht (2001)
3. Harper, W.L.: Rational conceptual change. In: Proceedings of PSA 1976, vol. 2, pp. 462–494. Philosophy of Science Association (1977)
4. Doyle, J.: A truth maintenance system. Artif. Intell. **12**(3), 231–272 (1979)
5. de Kleer, J.: An assumption-based truth maintenance system. Artif. Intell. **28**(2), 127–162 (1986)
6. Alchourrón, C.E., Makinson, D.: Hierarchies of regulation and their logic. In: Hilpinen, R. (ed.) New Studies in Deontic Logic, pp. 125–148. Reidel, Dordrecht (1981)
7. Alchourrón, C.E., Gärdenfors, P., Makinson, D.: On the logic of theory change: partial meet contraction and revision functions. J. Symbol. Logic **50**(2), 510–530 (1985)
8. Kelly, K.T.: The learning power of belief revision. In Gilboa, I. (ed.) Proceedings of TARK 1998, pp. 111–124. Morgan Kaufmann (1998)

---

[13] Some forms of propaganda are based on this.

[14] Still, the axiom system for the general layered upgrade already provides a syntactic (albeit indirect) characterisation of the presented policies.

9. Rott, H.: Change, Choice and Inference. Oxford Science Publications, Oxford (2001)
10. Spohn, W.: Ordinal conditional functions: a dynamic theory of epistemic states. In: Harper, W.L., Skyrms, B. (eds.) Causation in Decision, Belief Change, and Statistics, pp. 105–134. Kluwer, Dordrecht (1988)
11. Grove, A.: Two modellings for theory change. J. Philos. Logic **17**(2), 157–170 (1988)
12. van Benthem, J.: Dynamic logic for belief revision. J. Appl. Non-Class. Logics **17**(2), 129–155 (2007)
13. Haidt, J.: The Righteous Mind. Vintage (2012)
14. Baltag, A., Smets, S.: A qualitative theory of dynamic interactive belief revision. In: Bonanno, G., van der Hoek, W., Wooldridge, M. (eds.) LOFT7, pp. 13–60. AUP, Amsterdam (2008)
15. Ghosh, S., Velázquez-Quesada, F.R.: A note on reliability-based preference dynamics. In: Hoek, W., Holliday, W.H., Wang, W. (eds.) LORI 2015. LNCS, vol. 9394, pp. 129–142. Springer, Heidelberg (2015). doi:10.1007/978-3-662-48561-3_11
16. Prior, A.N.: Time and Modality. Clarendon Press, Oxford (1957)
17. Burgess, J.P.: Basic tense logic. In: Gabbay, D., Guenthner, F. (eds.) Handbook of Philosophical Logic, vol. II, pp. 89–133. Reidel, Dordrecht (1984)
18. Gargov, G., Passy, S.: A note on boolean modal logic. In: Petkov, P.P. (ed.) Mathematical Logic, pp. 311–321. Plenum Press (1990)
19. Ghosh, S., Velázquez-Quesada, F.R.: Agreeing to agree: reaching unanimity via preference dynamics based on reliable agents. In: Weiss, G., Yolum, P., Bordini, R.H., Elkind, E. (eds.) Proceedings of the AAMAS 2015, pp. 1491–1499. ACM (2015)
20. Stalnaker, R.: Knowledge, belief and counterfactual reasoning in games. Econ. Philos. **12**(2), 133–163 (1996)
21. Battigalli, P., Siniscalchi, M.: Strong belief and forward induction reasoning. J. Econ. Theory **106**(2), 356–391 (2002)
22. Baltag, A., Smets, S.: Group belief dynamics under iterated revision: fixed points and cycles of joint upgrades. In: Heifetz, A. (ed.) TARK, pp. 41–50 (2009)
23. Nayak, A.C., Pagnucco, M., Peppas, P.: Dynamic belief revision operators. Artif. Intell. **146**(2), 193–228 (2003)
24. Rott, H.: Shifting priorities: simple representations for twenty-seven iterated theory change operators. In: Makinson, D., Malinowski, J., Wansing, H. (eds.) Towards Mathematical Philosophy. Trends in Logic, vol. 28, pp. 269–296. Springer, Dordrecht (2009). doi:10.1007/978-1-4020-9084-4_14
25. Rott, H.: Coherence and conservatism in the dynamics of belief II: iterated belief change without dispositional coherence. J. Logic Comput. **13**(1), 111–145 (2003)
26. Boutilier, C.: Revision sequences and nested conditionals. In: Bajcsy, R. (ed.) Proceedings of the 13th IJCAI, pp. 519–525. Morgan Kaufmann (1993)
27. Boutilier, C.: Iterated revision and minimal change of conditional beliefs. J. Philos. Logic **25**(3), 263–305 (1996)
28. Booth, R., Meyer, T.A.: Admissible and restrained revision. J. Artif. Intell. Res. **26**, 127–151 (2006)
29. Holliday, W., Icard, T.: Moorean phenomena in epistemic logic. In: Beklemishev, L., Goranko, V., Shehtman, V. (eds.) AiML, pp. 178–199. College Publications (2010)
30. Hansson, S.: A survey of non-prioritized belief revision. Erkenntnis **50**(2–3), 413–427 (1999)

31. Makinson, D.: Screened revision. Theoria **63**(1–2), 14–23 (1997)
32. Hansson, S., Fermé, E.L., Cantwell, J., Falappa, M.A.: Credibility limited revision. J. Symbol. Logic **66**(4), 1581–1596 (2001)
33. Cantwell, J.: On the logic of small changes in hypertheories. Theoria **63**(1–2), 54–89 (1997)
34. Fermé, E.L., Rott, H.: Revision by comparison. Artif. Intell. **157**(1–2), 5–47 (2004)
35. Rott, H.: Bounded revision: Two-dimensional belief change between conservative and moderate revision. J. Philos. Logic **41**(1), 173–200 (2012)
36. Konieczny, S., Pino Pérez, R.: Improvement operators. In: Brewka, G., Lang, J. (eds.) Proceedings of KR 2008, pp. 177–187. AAAI Press (2008)
37. Konieczny, S., Grespan, M.M., Pino Pérez, R.: Taxonomy of improvement operators and the problem of minimal change. In: Lin, F., Sattler, U., Truszczynski, M. (eds.) Proceedings of KR 2010. AAAI Press (2010)
38. Papini, O.: Iterated revision operations stemming from the history of an agent's observations. In: [2], pp. 279–301

Topo-Logic as a Dynamic-Epistemic Logic

Alexandru Baltag[1], Aybüke Özgün[1,2(✉)], and Ana Lucia Vargas Sandoval[1]

[1] ILLC, University of Amsterdam, Amsterdam, The Netherlands
ozgunaybuke@gmail.com
[2] LORIA-CNRS, Université de Lorraine, Nancy, France

Abstract. We extend the 'topologic' framework [13] with dynamic modalities for 'topological public announcements' in the style of Bjorndahl [5]. We give a complete axiomatization for this "Dynamic Topo-Logic", which is in a sense simpler than the standard axioms of topologic. Our completeness proof is also more direct (making use of a standard canonical model construction). Moreover, we study the relations between this extension and other known logical formalisms, showing in particular that it is co-expressive with the simpler (and older) logic of interior and global modality [1,4,10,14]. This immediately provides an easy decidability proof (both for topologic and for our extension).

1 Introduction

The 'topologic' formalism, introduced by Moss and Parikh [13], and investigated further by Dabrowski et al. [6], Georgatos [8,9] and others, presented a single-agent subset space logic (SSL) for the notions of *knowledge* and *effort* (where "effort" refers to any type of evidence-gathering—via, e.g., measurement, computation, approximation, experiment or announcement—that can lead to an increase in knowledge). They proposed a bimodal language with modalities K and $\Box$, where $K\varphi$ is read as "the agent knows φ (is true)", and the effort modality $\Box\varphi$ says that "φ stays true no matter what further evidence-gathering efforts are made". So $\Box$ captures a notion of *stability under evidence-gathering.* The formulas are interpreted on *subset spaces*, which include the class of topological spaces. In [13] Moss and Parikh gave a sound and complete axiomatization with respect to the class of all subset spaces. The axiomatization for topological spaces has later been studied by Georgatos [8,9], and Dabrowski et al. [6], who independently provided complete axiomatizations and proved decidability. The completeness proofs involve however rather complicated constructions. Moreover, one of the main axioms (the so-called Union Axiom, capturing closure of the topology under binary unions) is extremely complex and looks rather unintuitive.

A different logical formalism with a topological semantics was proposed by Bjorndahl [5], motivated by developments in dynamic epistemic logic. Namely,

A. Özgün—Acknowledges financial support from European Research Council grant EPS 313360.

A. Baltag et al. (Eds.): LORI 2017, LNCS 10455, pp. 330–346, 2017.
DOI: 10.1007/978-3-662-55665-8_23

he proposed a topological semantics (in the style of subset space semantics) for the syntax of Public Announcement Logic (PAL), that assumes as precondition of learning φ the sentence $int(\varphi)$, saying that φ is learnable. Topologically, this corresponds to the interior operator of McKinsey and Tarski [12]. He axiomatized this logic, using natural analogues of the standard reduction axioms of PAL, and showed that this formalism is co-expressive with the simpler (and older) logic of interior $int(\varphi)$ and global modality $K\varphi$ (previously investigated by Bennett, Goranko and Passy, Aiello, and Shehtman [1,4,10,14], extending the work of McKinsey and Tarski [12] on interior semantics).

Another recent development is the work of van Ditmarsch et al. [15], who studied the extension of Bjorndahl's system with a *Topological Arbitrary Public Announcement* modality ('Topo-APAL', for short), that quantifies universally over Bjorndahl-style public announcements (similarly to the way classical APAL modality in [2] quantifies over public announcements). They proved that this is co-expressive with Bjorndahl's logic.

In this paper, we investigate a natural extension of 'topologic', obtained by adding to it Bjorndahl-style dynamic modalities for 'updates' (public announcements). The resulting 'Dynamic Topo-Logic' can be thought of as a logic of evidence-based *knowledge* $K\varphi$, *knowability* $int(\varphi)$, *learning* of new evidence $[\varphi]\psi$ and *stability* $\Box\varphi$ (of some truth φ) under any such evidence-acquisition. We show that this extension is co-expressive with the above-mentioned three logical formalisms: Topo-APAL [15], Bjorndahl's logic of topological public announcements [5], and the logic of interior and global modality [1,4,10,14]. This finally elucidates the relationships between topologic and other modal (and dynamic-epistemic) logics for topology: in particular, topologic is directly interpretable in the simplest logic above (of interior and global modality), which immediately provides a simpler decidability proof (both for topologic and for our extension).

We give a *complete axiomatization* for Dynamic Topo-Logic, which is in a sense simpler than the standard axioms of Topologic: though we have more axioms, each of them is transparent, natural and easily readable, directly reflecting the intuitive meanings of the connectives. Our axiomatization consists of a slightly different version of Bjorndahl's axioms, together with *only two additional proof principles governing the behavior of the topologic "effort" modality* $\Box\varphi$ (which we call "stable truth"): an introduction rule and an elimination axiom. Everything to be said about the effort modality is captured by these two simple principles, which together express the fact that $\Box$ quantifies universally over all updates with any new evidence. In particular, the complicated Union Axiom [8,9] is *not* needed in our system. Our completeness proof is also simpler than the existing completeness proofs for Topologic, making direct use of a *standard canonical topo-model construction.* This shows the advantage of adding dynamic modalities: when considered as a fragment of a dynamic-epistemic logic, Topologic becomes a more transparent and natural formalism, with intuitive axioms and canonical behavior.

In its turn, the effort modality helps to simplify and streamline the axiomatization of Topo-APAL. Indeed, although the two are equivalent, note that the axiomatization of this latter operator in [15] was *essentially infinitary*: it used an inference rule that takes as inputs infinitely many premisses! In contrast, our axiomatization is recursive, being obtained by replacing the infinitary rule with a finitary one, involving the use of "fresh" propositional variables. This rule's soundness is due to the "pure semantical" character of the effort modality $\Box\varphi$ (whose meaning does not depend on the valuation of variables not occurring in φ), in contrast to the "syntactical" character of the APAL modality.[1]

Due to page-limit constraints, this Proceedings version includes only the shortest proofs. The other relevant proofs can be found in the long version of this paper, available online at https://sites.google.com/site/ozgunaybuke/publications.

2 Dynamic Topo-Logic: Syntax, Semantics and Axiomatization

In this section, we present the language of Dynamic Topo-Logic, which is obtained by extending Bjorndahl's logic $\mathbf{L}_{!Kint}$ [5] with the effort modality from Topologic [13]; or equivalently, by extending Topologic with the Tarki-McKinsey interior operator *int* [12] and with Bjorndahl's topological public announcements. As it turns out, interior is in fact *definable* using topological public announcements (since $int(\varphi) = \langle\varphi\rangle\top$). So, although we keep the *int* operator as primitive for technical reasons, in a sense our syntax is essentially given by adding to topologic *only* the dynamic modalities.

Syntax. Given a countable set of propositional variables Prop, the language $\mathcal{L}$ of Dynamic Topo-Logic is defined recursively by the grammar

$$\varphi ::= p \mid \neg\varphi \mid \varphi\wedge\varphi \mid K\varphi \mid int(\varphi) \mid [\varphi]\varphi \mid \Box\varphi,$$

where $p \in$ Prop. The update operator $[\varphi]\psi$ is sometimes denoted by $[!\varphi]\psi$ in Public Announcement Logic literature; we skip the exclamation sign, but we will use the notation [!] for this modality when we do not want to specify the announcement formula φ (so that ! functions as a placeholder for the content of the announcement). We employ the usual abbreviations for propositional connectives $\top, \bot, \vee, \rightarrow, \leftrightarrow$. The dual modalities are defined as $\hat{K}\varphi := \neg K\neg\varphi$, $\Diamond\varphi := \neg\Box\neg\varphi$, $\langle\varphi\rangle\psi := \neg[\varphi]\neg\psi$, and $cl(\varphi) := \neg int\neg\varphi$.

Several fragments of the language $\mathcal{L}$ are of both technical and conceptual interest. For all the fragments studied in this paper, we use a notational convention listing in subscript all the modalities of the corresponding language. For example, $\mathcal{L}_{int}$ denotes the fragment of $\mathcal{L}$ having only the modality *int* (besides propositional connectives); $\mathcal{L}_{Kint}$ has only modalities K and *int* ; $\mathcal{L}_{K\Box}$ has only modalities K and $\Box$, etc.

[1] Indeed, the original paper [2] on "classical" (non-topological) APAL modality contained a similar attempt of converting an infinitary rule into an finitary rule. That was later shown to be flawed: the finitary rule was not sound for the APAL modality (though it is sound for effort)!

Topological Semantics: Intuitions and Motivation. The semantics of this language is over *topological spaces*[2] (X, τ). The points of the space represent "possible worlds" (or *states* of the world). The open sets in the topology are meant to represent *potential evidence*, i.e. facts about the world that are in principle *knowable*, in the sense of being *verifiable*: whenever (in any world in which) they are true, they can be known. In contrast, closed sets represent facts that are in principle *falsifiable* (whenever they are false, their falsity can be known). As it was remarked in [11,16], the closure properties of a topology are intuitively satisfied in this interpretation. First, contradictions ($\emptyset$) and tautologies (X) are in principle verifiable (as well as falsifiable). The conjunction $P \wedge Q$ of two verifiable facts is also verifiable: if $P \wedge Q$ is true, then both P and Q are true, and since both are assumed to be verifiable, they can both be known, and hence $P \wedge Q$ can be known. Finally, if $\{P_i : i \in I\}$ is a (possibly infinite) family of verifiable facts, then their disjunction $\bigvee_{i \in I} P_i$ is verifiable: indeed, if the disjunction is true, then there must exist some $i \in I$ such that P_i is true, and so P_i can be known (since it is verifiable), and as a result the disjunction $\bigvee_{i \in I} P_i$ can also be known (by inference from P_i).

Semantics. A *subset space* is a pair $(X, \mathcal{O})$, where X is a non-empty set and $\mathcal{O} \subseteq \mathcal{P}(X)$ is a non-empty collection of subsets of X. A *subset model* $\mathcal{M} = (X, \mathcal{O}, V)$ is triple where $(X, \mathcal{O})$ is a subset space and $V : \text{Prop} \to \mathcal{P}(X)$ is a valuation function. A *topological model* (or, in short, a *topo-model*) is a subset model $\mathcal{M} = (X, \tau, V)$ where (X, τ) is a topological space. Following the Subset Space Semantics [13], formulas are interpreted on pairs of the form (x, U) where $x \in U \in \mathcal{O}$. Such pairs are called *epistemic scenarios*. The set of all epistemic scenarios of a given topo-model $\mathcal{M}$ is denoted by $ES(\mathcal{M})$. Given a topo-model $\mathcal{M} = (X, \tau, V)$ and an epistemic scenario $(x, U) \in ES(\mathcal{M})$, the semantics for the language $\mathcal{L}$ is given by defining a *satisfaction relation* $(x, U) \models_{\mathcal{M}} \varphi$, as well as a *truth set* (interpretation) $[\![\varphi]\!]^U_{\mathcal{M}} =: \{x \in U \mid (x, U) \models_{\mathcal{M}} \varphi\}$, for all formulas φ. We omit the subscript, writing simply $(x, U) \models \varphi$ and $[\![\varphi]\!]^U$, whenever the model $\mathcal{M}$ is fixed. The definition of satisfaction is by recursion on the complexity of formulas φ:

[2] For a general introduction to topology we refer to [7]. A *topological space* (X, τ) consists of a non-empty set X and a "topology" $\tau \subseteq \mathcal{P}(X)$, i.e. a family of subsets of X (called *open sets*) such that $X, \emptyset \in \tau$, and τ is closed under finite intersections and arbitrary unions. The complements $X \setminus U$ of open sets are called *closed*. The collection τ is called a *topology* on X and elements of τ are called *open sets*. An open set containing $x \in X$ is called an open neighborhood of x. The *interior* $Int(A)$ of a set $A \subseteq X$ is the largest open set contained in A, i.e., $Int(A) = \bigcup\{U \in \tau \mid U \subseteq A\}$, while the closure $cl(A)$ is the smallest closed set containing A. A family $\mathcal{B} \subseteq \tau$ is called a basis for a topological space (X, τ) if every non-empty element of τ can be written as a union of elements of $\mathcal{B}$.

$$\begin{array}{lll}
(x,U) \models p & \text{iff } x \in V(p) & (\text{for } p \in \text{Prop})\\
(x,U) \models \neg\varphi & \text{iff } (x,U) \not\models \varphi & \\
(x,U) \models \varphi \wedge \psi & \text{iff } (x,U) \models \varphi \text{ and } (x,U) \models \psi & \\
(x,U) \models K\varphi & \text{iff } (\forall y \in U)(y,U) \models \varphi & \\
(x,U) \models int(\varphi) & \text{iff } x \in Int([\![\varphi]\!]^U) & \\
(x,U) \models [\varphi]\psi & \text{iff } (x,U) \models int(\varphi) \text{ implies } (x, Int([\![\varphi]\!]^U)) \models \psi & \\
(x,U) \models \Box\varphi & \text{iff } (\forall O \in \tau)\,(x \in O \subseteq U \text{ implies } (x,O) \models \varphi) &
\end{array}$$

We say that a formula φ is *valid in a model* $\mathcal{M}$, and write $\mathcal{M} \models \varphi$, if $(x,U) \models_{\mathcal{M}} \varphi$ for all scenarios $(x,U) \in ES(\mathcal{M})$. We say φ is *valid*, and write $\models \varphi$, if $\mathcal{M} \models \varphi$ for all $\mathcal{M}$.

Intuitive Reading. In an epistemic scenario (x,U), the first component x represents the actual state of the world, while U is the current evidence possessed by the agent; $x \in U$ expresses the *factivity* of evidence. The operator K captures "knowledge" (in the sense of absolute certainty: "infallible knowledge"): in a scenario (x,U), $K\varphi$ holds iff φ is entailed by the agent's evidence U (hence, the above semantic clause). $int(\varphi)$ means that φ is "knowable" in the actual state of the world (though not necessarily knowable in general, in other worlds): there exists some potential evidence (open set containing the actual state) U that entails φ (hence, the actual state is in the interior of φ); the dual closure operator $cl(\varphi)$ means that φ is "unfalsifiable" in the actual state (i.e. it is consistent with all potential evidence at that state). The 'effort' modality $\Box\varphi$ is read as "φ is *stably true*" under evidence-acquisition: i.e. φ is true, and will stay true no matter further evidence is obtained. Finally, we read the dynamic update modalities $[\varphi]\psi$ as "ψ will be true after learning φ". The difference between these Bjorndahl-style updates and the standard update operators (sometimes called public announcements)[3] is that not every truth is automatically "learnable"; so the precondition of updating with φ is the proposition $int(\varphi)$ saying that φ is *knowable* in the actual world (i.e. there exists some true evidence supporting φ).

The above topological semantics for the language $\mathcal{L}$ was in fact previously studied for some of its subfragments. While the semantic clauses for K and $\Box$ were first introduced in [13], the fragment $\mathcal{L}_{!Kint}$ examined in [5]. Bjorndahl provided sound and complete aximatizations for the associated dynamic logic $\mathbf{L}_{!Kint}$ (see Table 1 for the axiomatizations). Moreover, he proved—via a translation using so-called Reduction Axioms (see Table 1)—that the languages $\mathcal{L}_{Kint}$ and $\mathcal{L}_{!Kint}$ are equally expressive under the proposed topological semantics:

Theorem 1 *[5, Proposition 5]. $\mathcal{L}_{Kint}$ and $\mathcal{L}_{!Kint}$ are equally expressive with respect to topo-models.*[4]

---

[3] We prefer to talk about "updates", rather than public announcements, since our setting is single-agent: there is no "publicity" involved. The agent simply learns φ (and implicitly also learns that φ was learnable).

[4] In fact, the modality *int* can be defined in terms of the public announcement modality as $int(\varphi) := \neg[\varphi]\bot$, thus, the language $\mathcal{L}_{!Kint}$ and its fragment $\mathcal{L}_{!K}$ without the modality *int* are also co-expressive.

A Close Relative: Topo-APAL. Yet another relevant variation of $\mathcal{L}$ is the language $\mathcal{L}_{APAL}$ obtained by replacing the effort modality $\Box$ with the so-called *arbitrary announcement modality* $\blacksquare$ that was introduced by Balbiani et al. [2]. Roughly speaking, the arbitrary announcement modality $\blacksquare\varphi$ is read as "φ stays true after any epistemic update". While Balbiani et al. [2] studied this modality on Kripke frames, a topological semantics for $\blacksquare$ was proposed by van Ditmarsch et al. [15]:

$$(x, U) \models \blacksquare\varphi \text{ iff } (\forall\psi \in \mathcal{L}_{!K\,int})((x, U) \models [\psi]\varphi),$$

where [!] is the dynamic operator for topological updates (in the sense above). So $\blacksquare$ quantifies over all Bjorndahl-style topological public announcements of epistemic formulas. In [15], van Ditmarsch et al. proved that:

Theorem 2 *[15, Theorem 19]. $\mathcal{L}_{K\,int}$ and $\mathcal{L}_{APAL}$ are equally expressive expressive with respect to topo-models.*

As stated in the above semantic clause, the arbitrary announcement modality only quantifies over the $\blacksquare$-free formulas, whereas the effort modality quantifies over all open neighborhoods of the actual state x. Intuitively speaking, the effort modality seems stronger than the arbitrary announcement modality. However, quite surprisingly, this is not the case: in this paper, *we will show that the effort modality is in fact semantically equivalent to the arbitrary announcement modality.*

Axiomatizations. Given a formula $\varphi \in \mathcal{L}$, we denote by P_φ the set of all propositional variables occurring in φ (we will use the same notation for the necessity and possibility forms defined below). The axiomatization of our Dynamic Topo-Logic **L** consists of all axioms and rules in Table 1 below.

The intuitive nature of these axioms should be obvious. The first six need no explanation. The Replacement of Equivalents rule (RE) says that *updates are extensional*: learning equivalent sentences gives rise to equivalent updates, while the reduction axiom (R[$\top$]) says that *updating with tautologies is redundant* (nothing changes). The other reduction axioms are the natural analogues of the standard reduction laws in Public Announcement Logic, when one takes into account the fact that the precondition of a Bjorndahl-style update with φ is that φ is (not only true, but also) *knowable*: i.e. $int(\varphi)$. The only essentially new components of our system are the last two: the elimination axiom ([!]$\Box$-elim) and the introduction rule ([!]$\Box$-intro) for the effort modality. Taken together, they say that *θ is a stable truth after learning φ iff θ is true after learning any stronger evidence $\varphi \wedge \rho$.* The left-to-right implication in this statement is directly captured by ([!]$\Box$-elim), while the converse is captured by the rule ([!]$\Box$-intro). The "freshness" of the variable $p \in P$ in this rule ensures that it represents any 'generic' further evidence: this is similar to the introduction rule for the universal quantifier. In essence, the effort axiom and rule express the fact that *the effort modality is in fact a universal quantifier (over potential evidence).*

Simplicity of Our Axioms. One can compare the transparency and simple nature of our axioms with the complexity of the standard axiomatization of

Table 1. The axiomatization **L** of Dynamic Topo-Logic.

	(I) Axioms of system $\mathbf{L}_{Kint}$ for the language $\mathcal{L}_{Kint}$:
(P)	all propositional tautologies and Modus Ponens
($S5_K$)	all $S5$ axioms and rules for knowledge operator K
($S4_{int}$)	all $S4$ axioms and rules for interior operator int
(*K-int*)	*Knowledge implies knowability*: $K\varphi \rightarrow int(\varphi)$
	(II) Additional axioms for $\mathbf{L}_{!Kint}$:
($K_!$)	*Kripke's axiom for* [!]: $[\varphi](\psi \rightarrow \theta) \rightarrow ([\varphi]\psi \rightarrow [\varphi]\theta)$
($Nec_!$)	*Necessitation for* [!]: from $\vdash \theta$, infer $\vdash [\varphi]\theta$.
(RE)	*Replacement of Equivalents*: from $\vdash \varphi \leftrightarrow \psi$, infer $\vdash [\varphi]\theta \leftrightarrow [\psi]\theta$.
	Reduction axioms:
(R[⊤])	$[\top]\varphi \leftrightarrow \varphi$
(R_p)	$[\varphi]p \leftrightarrow (int(\varphi) \rightarrow p)$
($R_\neg$)	$[\varphi]\neg\psi \leftrightarrow (int(\varphi) \rightarrow \neg[\varphi]\psi)$
(R_K)	$[\varphi]K\psi \leftrightarrow (int(\varphi) \rightarrow K[\varphi]\psi)$
($R_{[!]}$)	$[\varphi][\psi]\chi \leftrightarrow [\langle\varphi\rangle\psi]\chi$
	(III) Axioms and rules for effort operator:
([!]□-elim)	$[\varphi]\Box\theta \rightarrow [\varphi \wedge \rho]\theta$ ($\rho \in \mathcal{L}$ arbitrary formula)
([!]□-intro)	From $\vdash \chi \rightarrow [\varphi \wedge p]\theta$, infer $\vdash \chi \rightarrow [\varphi]\Box\theta$ ($p \notin P_\chi \cup P_\theta \cup P_\varphi$ atom).

topologic, containing among others the intricate and opaque Union Axiom, which in our notation reads as:

$$\Diamond\varphi \wedge \hat{K}\Diamond\psi \rightarrow \Diamond(\Diamond\varphi \wedge \hat{K}\Diamond\psi \wedge K\Diamond\hat{K}(\varphi \vee \psi))$$

The fragment $\mathcal{L}_{Kint}$, having only modalities K and int, was first studied in [10] (with a Kripke semantics) and [4] (with the above topological semantics). A complete axiomatization for the topological interpretation was provided by Aiello [1], though a full completeness proof was given later by Shehtman [14]:

Proposition 1 *([14]). The system $\mathbf{L}_{Kint}$, consisting of the axioms and rules in group (I) of Table 1, is sound and complete for the language $\mathcal{L}_{Kint}$.*

The fragment $\mathcal{L}_{!Kint}$, obtained by extending $\mathcal{L}_{Kint}$ with topological update operators ('public announcement'), was introduced and axiomatized by Bjorndahl [5]:

Proposition 2 *([5]). The system $\mathbf{L}_{!Kint}$, consisting of all the axioms and rules in groups (I) and (II) of Table 1, is sound and complete for the language $\mathcal{L}_{!Kint}$.*

Proof. It is easy to see that all the axioms and rules of $\mathbf{L}_{!Kint}$ are sound. A proof of completeness is in [5], for a slightly different, but equivalent, axiomatization. Bjorndahl's system consists of the axioms of $\mathcal{L}_{Kint}$ (i.e. group (I) in our Table), together with our reduction axioms (R_p), ($R_\neg$), (R_K) and ($R_!$), as well as the additional reduction laws, ($R_\wedge$) and (R_{int}) in the Proposition below.[5] Since the latter ones are provable in our system $\mathbf{L}_{!Kint}$, it follows that $\mathbf{L}_{!Kint}$ is complete as well.

Proposition 3. *The first six reduction laws below are provable both in* $\mathbf{L}_{!Kint}$ *and* $\mathbf{L}$ *(for languages* $\mathcal{L}_{!Kint}$ *and* $\mathcal{L}$*, respectively). The seventh schema and the inference rule below can be derived in our full proof system* $\mathbf{L}$*:*

1. ($\langle ! \rangle$) $\langle\varphi\rangle\psi \leftrightarrow (int(\varphi) \wedge [\varphi]\psi)$
2. ($R_\perp$) $[\varphi]\perp \leftrightarrow \neg int(\varphi)$
3. ($R_\wedge$) $[\varphi](\psi \wedge \theta) \leftrightarrow ([\varphi]\psi \wedge [\varphi]\theta)$
4. (R_{int}) $[\varphi]int(\psi) \leftrightarrow (int(\varphi) \rightarrow [\varphi]\psi)$
5. ($R[int]$) $[int(\varphi)]\psi \leftrightarrow [\varphi]\psi$
6. ($R_{[p]}$) $[\varphi][p]\psi \leftrightarrow [\varphi \wedge p]\psi$
7. ($\Box$-elim) $\Box\theta \rightarrow [\rho]\theta$ ($\rho \in \mathcal{L}$ arbitrary formula)
8. ($\Box$-intro) From $\vdash \chi \rightarrow [p]\theta$, infer $\vdash \chi \rightarrow \Box\theta$ ($p \notin P_\chi \cup P_\theta$ atom).

3 Soundness and Expressivity

In this section, we introduce a more general class of models for our language, called *pseudo-models*: these are a special case of the (even more general) Subset Space Semantics introduced in [13]. Pseudo-models include all topo-models, as well as other subset-space models, but they have the nice property that the interior operator *int* can still be interpreted in the standard way. These structures, though interesting enough in themselves, are for us only an auxiliary notion, playing an important technical role in our completeness proof. But for now, we will first prove the soundness of our full system $\mathbf{L}$ from Table 1 with respect to pseudo-models (and thus also over topo-models), and we provide several expressivity results concerning the above defined languages with respect to (both topo-models and) pseudo-models.

Definition 1 (Lattice spaces and Pre-models). *A subset space* $(X, \mathcal{O})$ *is called a* lattice space *if* $\emptyset, X \in \mathcal{O}$*, and* $\mathcal{O}$ *is closed under finite intersections and finite unions. A* pre-model $(X, \mathcal{O}, V)$ *is a triple where* $(X, \mathcal{O})$ *is a lattice space and* $V : \text{Prop} \rightarrow \mathcal{P}(X)$ *is a valuation map.*

Although a lattice space $(X, \mathcal{O})$ is not necessarily a topological space, the family $\mathcal{O}$ constitutes a topological basis over X. Therefore, every pre-model $\mathcal{M} = (X, \mathcal{O}, V)$ has an *associated topo-model* $\mathcal{M} = (X, \tau_\mathcal{O}, V)$, where $\tau_\mathcal{O}$ is the topology generated by $\mathcal{O}$ (i.e., the smallest topology on X such that $\mathcal{O} \subseteq \tau_\mathcal{O}$).

[5] Although Bjorndahl's formulations of ($R_!$) and (R_{int}) are unnecessarily complicated: the first is stated as $[\varphi][\psi]\chi \leftrightarrow [int(\varphi) \wedge [\varphi]int(\psi)]\chi$, while the second as $[\varphi]int(\psi) \leftrightarrow (int(\varphi) \rightarrow int([\varphi]\psi))$. It is easy to see that these are equivalent to our simpler formulations, given the other axioms.

Satisfaction relation in pre-models. Given a pre-model $\mathcal{M} = (X, \mathcal{O}, V)$, the semantics for $\mathcal{L}$ on pre-models can be defined for *all* pairs of the form (x, Y), where $Y \subseteq X$ is an arbitrary subset such that $x \in Y$. It is important to notice that, for a given evaluation pair (x, Y) on pre-models, the set Y is *not necessarily* an element of $\mathcal{O}$. The semantic clauses for the modalities in $\mathcal{L}$ are defined similarly as in Sect. 2, except that $\Box$ quantifies over the elements of $\mathcal{O}$, and *int* is interpreted as the interior operator of the associated topology $\tau_{\mathcal{O}}$. More precisely, given a pre-model $\mathcal{M} = (X, \mathcal{O}, V)$ and (x, Y) with $x \in Y \subseteq X$, we set

$$\begin{array}{lll} (x,Y) \models int(\varphi) & \text{iff} & x \in Int(\llbracket\varphi\rrbracket^Y) \\ (x,Y) \models [\varphi]\psi & \text{iff} & (x,Y) \models int(\varphi) \text{ implies } (x, Int(\llbracket\varphi\rrbracket^Y)) \models \psi \\ (x,Y) \models \Box\varphi & \text{iff} & (\forall O \in \mathcal{O})(x \in O \subseteq Y \text{ implies } (x,O) \models \varphi) \end{array}$$

where *Int* is the interior operator of $\tau_{\mathcal{O}}$.

Validity in pre-models. Although we did not require the neighbourhood Y to be an element of $\mathcal{O}$ in the definition of the satisfaction relation above, the validity on pre-models is defined by *restricting to epistemic scenarios* (x, U) such that $x \in U \in \mathcal{O}$, as in the case for the topo-models. More precisely, we say that a formula φ is *valid in a pre-model* $\mathcal{M}$, and write $\mathcal{M} \models \varphi$, iff $(x, U) \models_{\mathcal{M}} \varphi$ for all *epistemic scenarios* $(x, U) \in ES(\mathcal{M})$. A formula φ is *valid*, denoted by $\models \varphi$, iff $\mathcal{M} \models \varphi$ for all $\mathcal{M}$.

Definition 2 (Pseudo-models for $\mathcal{L}$). *A* pseudo-model $\mathcal{M} = (X, \mathcal{O}, V)$ *is a pre-model such that* $\llbracket int(\varphi)\rrbracket^U \in \mathcal{O}$, *for all* $\varphi \in \mathcal{L}$ *and* $U \in \mathcal{O}$.

It is obvious that the class of pseudo-models includes all topo-models, and that all formulas that are valid on pseudo-models are also valid on topo-models.[6]

Theorem 3. *The system* **L** *is sound with respect to the class of all pseudo-models (and hence also with respect to the class of all topo-models).*

We first prove that $\mathcal{L}_{!K int}$ *and* $\mathcal{L}_{K int}$ *are equally expressive on pseudo-models:*

Lemma 1. $\mathcal{L}_{!K int}$ *and* $\mathcal{L}_{K int}$ *are co-expressive with respect to pseudo-models. In other words, for every formula* $\varphi \in \mathcal{L}_{!K int}$ *there exists a formula* $\psi \in \mathcal{L}_{K int}$ *such that* $\varphi \leftrightarrow \psi$ *is valid in all pseudo-models.*

The proof (in Appendix B.2 of the on-line long version) goes over standard lines, using the reduction laws to push dynamic modalities inside the formula and then eliminate them. The proof uses induction on a non-standard notion of complexity of formulas $<$, given by:

Lemma 2. *There exists a well-founded strict partial order* $<$ *on formulas of* $\mathcal{L}$ *such that*

6 Indeed, this is because the satisfaction relation for epistemic scenarios in any pseudo-model that happens to be a topo-model agrees with the topo-model satisfaction relation.

1. $\varphi \in Sub(\psi)$ *implies* $\varphi < \psi$,
2. $(int(\varphi) \to p) < ([\varphi]p)$,
3. $(int(\varphi) \to \neg[\varphi]\psi) < ([\varphi]\neg\psi)$,
4. $([\varphi]\psi \wedge [\varphi]\chi) < ([\varphi](\psi \wedge \chi))$,
5. $(int(\varphi) \to int([\varphi]\psi)) < ([\varphi]\psi)$,
6. $(int(\varphi) \to K[\varphi]\psi) < ([\varphi]K\psi)$,
7. $[\langle\varphi\rangle\psi]\chi < ([\varphi][\psi]\chi)$.
8. $\varphi \in \mathcal{L}$ *implies* $int(p) < \Box\varphi$,
9. $[p]\varphi < \Box\varphi$,

Next, we prove that $\mathcal{L}$ and $\mathcal{L}_{Kint}$ are equally expressive with respect to the pseudo-models. This result will also be useful in the completeness proof of **L** for topo-models. Toward proving the co-expressivity of $\mathcal{L}$ and $\mathcal{L}_{Kint}$, we follow a similar strategy as in [2,15] and use normal forms in $\mathcal{L}_{Kint}$ as defined in [15, Definition 8].

Definition 3 (Normal form for the language $\mathcal{L}_{Kint}$). *We say a formula $\psi \in \mathcal{L}_{Kint}$ is in* normal form *if it is a disjunction of conjunctions of the form*

$$\delta := \alpha \wedge K\beta \wedge \hat{K}\gamma_1 \wedge \cdots \wedge \hat{K}\gamma_n$$

where $\alpha, \beta, \gamma_i \in \mathcal{L}_{int}$ for all $1 \leq i \leq n$.

Proposition 4. *For any $\varphi, \varphi_i \in \mathcal{L}_{int}$, the following is valid in all pseudo-models:*

$$\Diamond(\varphi \wedge K\varphi_0 \wedge \bigwedge_{1\leq i\leq n} \hat{K}\varphi_i) \leftrightarrow (\varphi \wedge int(\varphi_0) \wedge \bigwedge_{1\leq i\leq n} \hat{K}(int(\varphi_0) \wedge \varphi_i)) \qquad (\mathrm{NF}_n)$$

We now have sufficient machinery to show that $\mathcal{L}$ and $\mathcal{L}_{Kint}$ are equally expressive with respect to pseudo-models.

Theorem 4. *$\mathcal{L}$ and $\mathcal{L}_{Kint}$ are co-expressive with respect to pseudo-models.*

The proof (in Appendix B.4 of the on-line version) uses the above proposition, as well as the co-expressivity of $\mathcal{L}_{!Kint}$ and $\mathcal{L}_{Kint}$, in a similar way to the analogue proofs concerning the arbitrary public announcement logic in [2,15].

Theorem 4 will be used in the completeness proof of **L** for topo-models. Concerning expressivity of $\mathcal{L}$, we also obtain the following result with respect to topo-models.

Theorem 5. *$\mathcal{L}$ and $\mathcal{L}_{Kint}$ are co-expressive with respect to topo-models.*

Theorem 6. *$\mathcal{L}_{K\Box}$ and $\mathcal{L}_{Kint}$ are also co-expressive with respect to topo-models.*

Corollary 1. *$\mathcal{L}$, $\mathcal{L}_{!Kint}$ and $\mathcal{L}_{Kint}$, as well as the language of topologic $\mathcal{L}_{K\Box}$ are all co-expressive with respect to topo-models.*

Proof. The proof follows easily from Theorems 5 and 6, since $\mathcal{L}_{Kint} \subseteq \mathcal{L}_{!Kint} \subseteq \mathcal{L}$ and $\mathcal{L}_{K\Box} \subseteq \mathcal{L}$.

Corollary 2. *Dynamic Topo-Logic $\mathcal{L}$ is decidable and has the Finite Model Property (and thus all its fragments, including in particular topologic, have these properties).*

Proof. This follows from Corollary 1, together with the fact that $\mathcal{L}_{Kint}$ is easily shown to have these properties, by a standard filtration argument. (This is already known, see e.g., [10,14]).

Moreover, not only that $\mathcal{L}_{APAL}$ and $\mathcal{L}$ are co-expressive for topo-models, but also *the effort modality* $\Box$ *and the topological APAL modality* $\blacksquare$ *are in fact equivalent*, in the following sense.

Theorem 7. *Let $t : \mathcal{L}_{APAL} \to \mathcal{L}$ be the map that replaces each instance of $\blacksquare$ with $\Box$. Then for every $\varphi \in \mathcal{L}_{APAL}$, we have that $\varphi \leftrightarrow t(\varphi)$ is valid in all topo-models.*

4 Completeness

In this section we prove the completeness of the proof systems **L** with respect to (both pseudo- and) topo-models. The plan of our proof is as follows. We first prove completeness of **L** with respect to a *canonical pseudo-model*, consisting of *maximally consistent witnessed theories.* Roughly speaking, a theory is witnessed if every $\Diamond\varphi$ (occurring in every "existential context") in the theory is "witnessed" by some atomic formula p, i.e. $\langle p\rangle\varphi$ occurs (in the same existential context) in the theory. Next, we use the co-expressivity of $\mathcal{L}$ and $\mathcal{L}_{Kint}$, as well as the fact that $\mathcal{L}_{Kint}$ cannot distinguish between a pseudo-model and its associated topo-model, to show that **L** is complete with respect to the *canonical topo-model* (associated to the canonical pseudo-model).

The appropriate notion of "existential contexts" is represented by *possibility forms*, in the following sense:

Definition 4 ("Pseudo-modalities": necessity and possibility forms). *For any finite string $s \in (\{\varphi \to \mid \varphi \in \mathcal{L}\} \cup \{K\} \cup \{\psi \mid \psi \in \mathcal{L}\})^* = NF$, we define pseudo-modalities $[s]$ and $\langle s\rangle$, that generalize our dynamic modalities $[\psi]$ and $\langle\psi\rangle$. These pseudo-modalities are functions mapping any formula $\varphi \in \mathcal{L}$ to another formula $[s]\varphi \in \mathcal{L}$ (necessity form), respectively $\langle s\rangle\varphi \in \mathcal{L}$ (possibility form). The definition is by recursion, putting for necessity forms: $[\lambda]\varphi := \varphi$, $[\varphi \to, s]\varphi := \varphi \to [s]\varphi$, $[K, s]\varphi := K[s]\varphi$, $[\psi, s]\varphi := [\psi][s]\varphi$, where λ is the empty string. For possibility forms, we put $\langle s\rangle\varphi := \neg[s]\neg\varphi$.*

Lemma 3. *For every necessity form $s \in NF$, there exist formulas $\theta, \psi \in \mathcal{L}$ such that for all $\varphi \in \mathcal{L}$, we have*

$$\vdash [s]\varphi \text{ iff } \vdash \psi \to [\theta]\varphi.$$

Proof. The proof follows similarly as in [2, Lemma 4.8] and [3, Lemma 2.7].

Lemma 4. *The following rule is admissible in* **L***:*

$$\text{if } \vdash [s][p]\varphi \text{ then } \vdash [s]\Box\varphi, \text{ where } p \notin P_s \cup P_\varphi.$$

Proof. Suppose $\vdash [s][p]\varphi$. Then, by Lemma 3, there exist $\theta, \psi \in \mathcal{L}$ such that $\vdash \psi \to [\theta][p]\varphi$. By the auxiliary reduction law ($R_{[p]}$) in Proposition 3, we get $\vdash \psi \to [\theta \wedge p]\varphi$. By the construction of the formulas ψ and θ, we know that $P_\psi \cup P_\theta \subseteq P_s$, and so $p \notin P_\psi \cup P_\theta \cup P_\varphi$. Therefore, by ([!]$\Box$-intro)), we have $\vdash \psi \to [\theta]\Box\varphi$. Applying again Lemma 3, we obtain $\vdash [s]\Box\varphi$.

Definition 5. *For every countable set of propositional variables P, let $\mathcal{L}^P$ be the language of* **L** *based only on the propositional variables in P. Similarly, let $\mathcal{L}^P_{Kint}$, $\mathcal{L}^P_{!Kint}$ and NF^P denote the corresponding languages restricted by P. A P-*theory *is a* consistent *set of formulas in $\mathcal{L}^P$. Here, "consistent" means consistent with respect to the axiomatization* **L** *formulated for $\mathcal{L}^P$. A* maximal P-theory *is a P-theory Γ that is maximal with respect to $\subsetneq$ among all P-theories; in other words, Γ cannot be extended to another P-theory. A* P-witnessed theory *is a P-theory Γ such that, for every $s \in NF^P$ and $\varphi \in \mathcal{L}^P$, if $\langle s\rangle\Diamond\varphi$ is consistent with Γ then there is $p \in P$ such that $\langle s\rangle\langle p\rangle\varphi$ is consistent with Γ. A* maximal P- witnessed theory *Γ is a P-witnessed theory that is not a proper subset of any P-witnessed theory.*

Lemma 5 (Lindenbaum's Lemma). *Every P-witnessed theory Γ can be extended to a maximal P-witnessed theory T_Γ.*

Lemma 6 (Extension Lemma). *Let P be a set of propositional variables and P' be a countable set of* fresh *propositional variables, i.e., $P \cap P' = \emptyset$. Let $\widetilde{P} = P \cup P'$. Then, every P-theory Γ can be extended to a $\widetilde{P}$-witnessed theory $\widetilde{\Gamma} \supseteq \Gamma$, and hence to a maximal $\widetilde{P}$-witnessed theory $T_\Gamma \supseteq \Gamma$.*

We are now ready to build the canonical pseudo-model. For a fixed countable set P, we define an equivalence relation on maximal P-witnessed theories T and S:

$$T \sim S \text{ iff } (\forall\varphi \in \mathcal{L}^P)(K\varphi \in T \Rightarrow \varphi \in S).$$

Definition 6 (Canonical Pseudo-Model for T_0). *Let T_0 be a maximal P-witnessed theory. The* canonical pseudo-model *for T_0 is a tuple $\mathcal{M}^c = (X^c, \mathcal{O}^c, V^c)$ such that*

- $X^c = \{T \subseteq \mathcal{L}^P \mid T$ *is a maximal P-witnessed theory such that* $T \sim T_0\}$,
- $\mathcal{O}^c = \{\widehat{int(\varphi)} \mid \varphi \in \mathcal{L}^P\}$, *where* $\hat{\theta} = \{T \in X^c \mid \theta \in T\}$ *for any* $\theta \in \mathcal{L}^P$,
- $V^c(p) = \{T \in X^c \mid p \in T\}$.

We let τ^c denote the topology generated by $\mathcal{O}^c$. The associated topo-model $\mathcal{M}^c_\tau = (X^c, \tau^c, V^c)$ is called the canonical topo-model for T_0.

Clearly $\mathcal{M}^c = (X^c, \mathcal{O}^c, V^c)$ is a pre-model, but in fact we prove more, namely:

Lemma 7. *$\mathcal{M}^c = (X^c, \mathcal{O}^c, V^c)$ is a pseudo-model.*

Proof. We need to show that (a) $\mathcal{O}^c$ is closed under finite intersections and finite unions, and (b) for all $\varphi, \alpha \in \mathcal{L}^P$ we have $[\![int(\varphi)]\!]^{\widehat{int(\alpha)}} \in \mathcal{O}^c$. The last item (b) follows from the Truth Lemma (see Appendix C.3 of the on-line version for the proof). We here sketch the proof for the first item: (a.1) closure under finite intersection follows from the normality of *int*, namely from the fact that $\vdash int(\varphi) \wedge int(\psi) \leftrightarrow int(\varphi \wedge \psi)$. (a.2) closure under finite union follows from the fact that $\vdash (int(\varphi) \vee int(\psi)) \leftrightarrow int(int(\varphi) \vee int(\psi))$, and that $int(int(\varphi) \vee int(\psi)) \in \mathcal{L}^P$.

Lemma 8. *Let $T \in X^c$, $\varphi, \alpha \in \mathcal{L}^P$ such that $int(\alpha) \in T$ and $K[\alpha]\varphi \notin T$. Then, there is $S \in X^c$ with $int(\alpha) \in S$ and $[\alpha]\varphi \notin S$.*

Lemma 9 (Truth Lemma). *Let $\mathcal{M}^c = (X^c, \mathcal{O}^c, V^c)$ be the canonical pseudo-model for a maximal P-witnessed theory T_0 and $\varphi \in \mathcal{L}^P$. Then, for all $\alpha \in \mathcal{L}^P$ we have*

$$[\![\varphi]\!]^{\widehat{int(\alpha)}} = \widehat{\langle\alpha\rangle\varphi}.$$

Proof. The proof is by *induction on the well-founded partial order* $<$ on formulas introduced in Lemma 2. We assume the following *Induction Hypothesis* (IH): For $\psi < \varphi$, we have $[\![\psi]\!]^{\widehat{int(\alpha)}} = \widehat{\langle\alpha\rangle\psi}$ for all $\alpha \in \mathcal{L}^P$.

Base case $\varphi = p$:

$$\begin{aligned}
[\![p]\!]^{\widehat{int(\alpha)}} &= \widehat{int(\alpha)} \cap [\![p]\!]^{X^c} && \text{(since } p \text{ is bi-persistent)}\\
&= \widehat{int(\alpha)} \cap V^c(p) && \text{(by the semantics)}\\
&= \widehat{int(\alpha)} \cap \widehat{p} && \text{(by the definition of } V^c)\\
&= \widehat{int(\alpha) \wedge p} && \\
&= \widehat{int(\alpha) \wedge (int(\alpha) \to p)} && \text{(by propositional tautologies)}\\
&= \widehat{int(\alpha) \wedge [\alpha]p} && \text{(by}(R_p))\\
&= \widehat{\langle\alpha\rangle p} && \text{(Proposition 3-}(\langle!\rangle))
\end{aligned}$$

Case $\varphi := \neg\psi$:

$$\begin{aligned}
[\![\neg\psi]\!]^{\widehat{int(\alpha)}} &= \widehat{int(\alpha)} \setminus [\![\psi]\!]^{\widehat{int(\alpha)}} && \text{(by the semantics of } \neg) \\
&= \widehat{int(\alpha)} \setminus \widehat{\langle\alpha\rangle\psi} && \text{(by } \mathbf{IH}) \\
&= \widehat{int(\alpha)} \cap (X^c \setminus \widehat{\langle\alpha\rangle\psi}) \\
&= \widehat{int(\alpha)} \cap \widehat{\neg\langle\alpha\rangle\psi} && \text{(since } X^c \setminus \widehat{\langle\alpha\rangle\psi} = \widehat{\neg\langle\alpha\rangle\psi}) \\
&= \widehat{int(\alpha) \wedge \neg\langle\alpha\rangle\psi} \\
&= \widehat{int(\alpha) \wedge [\alpha]\neg\psi} \\
&= \widehat{\langle\alpha\rangle\neg\psi} && \text{(Proposition 3-}(\langle!\rangle))
\end{aligned}$$

Case $\varphi = \psi \wedge \chi$:

$$\begin{aligned}
[\![\psi \wedge \chi]\!]^{\widehat{int(\alpha)}} &= [\![\psi]\!]^{\widehat{int(\alpha)}} \cap [\![\chi]\!]^{\widehat{int(\alpha)}} && \text{(by the semantics of } \wedge) \\
&= \widehat{\langle\alpha\rangle\psi \wedge \langle\alpha\rangle\chi} && \text{(by } \mathbf{IH}) \\
&= \widehat{\langle\alpha\rangle(\psi \wedge \chi)} && \text{(since } \vdash_{\mathbf{L}} (\langle\alpha\rangle\psi \wedge \langle\alpha\rangle\chi) \leftrightarrow \langle\alpha\rangle(\psi \wedge \chi))
\end{aligned}$$

Case $\varphi = K\psi$:

($\Rightarrow$) Suppose $T \in [\![K\psi]\!]^{\widehat{int(\alpha)}}$. This implies, by the semantic clause of K, that $T \in \widehat{int(\alpha)}$ and $[\![\psi]\!]^{\widehat{int(\alpha)}} = \widehat{int(\alpha)}$. We want to show that $T \in \widehat{\langle\alpha\rangle K\psi}$. By Proposition 3-($\langle!\rangle$) and the reduction axiom (R_K), we obtain $\vdash \langle\alpha\rangle K\psi \leftrightarrow int(\alpha) \wedge K[\alpha]\psi$. We therefore only need to show that $T \in \widehat{int(\alpha)}$ and $T \in \widehat{K[\alpha]\psi}$. We have the former by the assumption. Suppose toward contradiction that $T \notin \widehat{K[\alpha]\psi}$, i.e., $K[\alpha]\psi \notin T$. Then, by Lemma 8, there exists $S \in X^c$ such that $int(\alpha) \in S$ and $[\alpha]\psi \notin S$. Since $\vdash \langle\alpha\rangle\psi \to [\alpha]\psi$, we obtain $\langle\alpha\rangle\psi \notin S$. Therefore, by IH, we have $S \notin [\![\psi]\!]^{\widehat{int(\alpha)}}$. Thus, since $S \in \widehat{int(\alpha)}$, we then conclude $[\![\psi]\!]^{\widehat{int(\alpha)}} \neq \widehat{int(\alpha)}$. By the semantics of K, this means that $[\![K\psi]\!]^{\widehat{int(\alpha)}} = \emptyset$, contradiction our first assumption. Hence, $T \in \widehat{int(\alpha) \wedge K[\alpha]\psi} = \widehat{\langle\alpha\rangle K\psi}$.

($\Leftarrow$) Suppose $T \in \widehat{\langle\alpha\rangle K\psi}$. Then, by the equality $\langle\alpha\rangle K\psi \leftrightarrow int(\alpha) \wedge K[\alpha]\psi$, we have $T \in \widehat{int(\alpha)}$ and $T \in \widehat{K[\alpha]\psi}$. Let $S \in \widehat{int(\alpha)}$. Since $S \sim T$ and $T \in \widehat{K[\alpha]\psi}$, we also have $[\alpha]\psi \in S$. Therefore, by Proposition 3-($\langle!\rangle$), we obtain $\langle\alpha\rangle\psi \in S$. This implies, by IH, that $S \in [\![\psi]\!]^{\widehat{int(\alpha)}}$. Since this holds for all $S \in \widehat{int(\alpha)}$, we have $[\![\psi]\!]^{\widehat{int(\alpha)}} = \widehat{int(\alpha)}$. Hence, $[\![K\psi]\!]^{\widehat{int(\alpha)}} = \widehat{int(\alpha)} \ni T$.

Case $\varphi = int(\psi)$:

($\Rightarrow$) Suppose $T \in [\![int(\psi)]\!]^{\widehat{int(\alpha)}}$. Then, by the semantics of int, there exists $U \in \mathcal{O}^c$ such that $T \in U \subseteq \widehat{int(\alpha)}$ and $U \subseteq [\![\psi]\!]^{\widehat{int(\alpha)}}$ (since $\mathcal{O}^c$ constitutes a basis for τ^c). Then, by IH, we have $U \subseteq \widehat{\langle\alpha\rangle\psi}$. By the construction of $\mathcal{O}^c$, we know that $U = \widehat{int(\gamma)}$ for some $\gamma \in \mathcal{L}^P$. We therefore obtain that $T \in \widehat{int(\gamma)} \subseteq \widehat{\langle\alpha\rangle\psi}$. This means that, for all $S \in \widehat{int(\gamma)}$, we have $S \in \widehat{\langle\alpha\rangle\psi}$. Therefore, $\{\theta \in \mathcal{L}^P \mid K\theta \in T\} \cup \{\neg(int(\gamma) \to \langle\alpha\rangle\psi)\}$ is inconsistent. Then there exists a $\chi \in \{\theta \in \mathcal{L}^P \mid K\theta \in T\}$ such that $\vdash \chi \to (int(\gamma) \to \langle\alpha\rangle\psi)$. Thus, by the normality of K, we have $\vdash K\chi \to K(int(\gamma) \to \langle\alpha\rangle\psi)$. As $K\chi \in T$, we obtain $K(int(\gamma) \to \langle\alpha\rangle\psi) \in T$. Then by axiom ($K$-$int$), we have $int(int(\gamma) \to \langle\alpha\rangle\psi) \in T$. Since int is an $S4$ modality, we have $int(\gamma) \to int(\langle\alpha\rangle\psi) \in T$. Since $T \in \widehat{int(\gamma)}$, we obtain $int(\langle\alpha\rangle\psi) \in T$. Moreover, we have

$$\vdash int(\langle\alpha\rangle\psi) \leftrightarrow int(int(\alpha) \wedge [\alpha]\psi) \qquad \text{(Proposition 3-}(\langle!\rangle))$$
$$\vdash int(int(\alpha) \wedge [\alpha]\psi) \leftrightarrow (int(\alpha) \wedge int([\alpha]\psi))$$
$$\vdash (int(\alpha) \wedge int([\alpha]\psi)) \leftrightarrow (int(\alpha) \wedge (int(\alpha) \to [\alpha]int(\psi))) \qquad \text{(Proposition 3-}(\mathrm{R}_{int}))$$
$$\vdash (int(\alpha) \wedge (int(\alpha) \to [\alpha]int(\psi))) \leftrightarrow (int(\alpha) \wedge [\alpha]int(\psi)))$$
$$\vdash (int(\alpha) \wedge [\alpha]int(\psi))) \leftrightarrow \langle\alpha\rangle int(\psi) \qquad \text{(Proposition 3-}(\langle!\rangle))$$

Therefore, as T is maximal, we obtain $\langle\alpha\rangle int(\psi) \in T$, i.e., $T \in \widehat{\langle\alpha\rangle int(\psi)}$.

($\Leftarrow$) Suppose $T \in \widehat{\langle\alpha\rangle int(\psi)}$. This implies, by the above derivation, that $T \in \widehat{int(\langle\alpha\rangle\psi)}$. By the constraction of $\mathcal{O}^c$, we have $\widehat{int(\langle\alpha\rangle\psi)} \in \mathcal{O}^c$. Moreover, by the T-axiom for int, we have that $\widehat{int(\langle\alpha\rangle\psi)} \subseteq \widehat{\langle\alpha\rangle\psi}$. By IH, we also have that $\widehat{\langle\alpha\rangle\psi} = [\![\psi]\!]^{\widehat{int(\alpha)}}$. Therefore $T \in \widehat{int(\langle\alpha\rangle\psi)} \subseteq \widehat{\langle\alpha\rangle\psi} = [\![\psi]\!]^{\widehat{int(\alpha)}}$, i.e., $T \in [\![int(\psi)]\!]^{\widehat{int(\alpha)}}$.

Case $\varphi = \langle\chi\rangle\psi$:

$$\begin{aligned}
[\![\langle\chi\rangle\psi]\!]^{\widehat{int(\alpha)}} &= \{T \in \widehat{int(\alpha)} \mid (T, Int([\![\chi]\!]^{\widehat{int(\alpha)}})) \models \psi\} \\
&= \{T \in \widehat{int(\alpha)} \mid (T, [\![int(\chi)]\!]^{\widehat{int(\alpha)}}) \models \psi\} && \text{(by the semantics of } int) \\
&= \{T \in \widehat{int(\alpha)} \mid (T, \widehat{\langle\alpha\rangle int(\chi)} \models \psi\} && \text{(by } \mathbf{IH}\text{, since } int(\chi) < \langle\chi\rangle\psi) \\
&= [\![\psi]\!]^{\widehat{\langle\alpha\rangle int(\chi)}} && \text{(since } \widehat{\langle\alpha\rangle int(\chi)} \subseteq \widehat{int(\alpha)}) \\
&= \widehat{\langle(\langle\alpha\rangle int(\chi))\rangle\psi} && \text{(by } \mathbf{IH}\text{, since } \psi < \langle\chi\rangle\psi) \\
&= \widehat{\langle\alpha\rangle\langle\chi\rangle\psi} && (\vdash \langle\alpha\rangle\langle\chi\rangle\psi \leftrightarrow \langle(\langle\alpha\rangle int(\chi))\rangle\psi)
\end{aligned}$$

Case $\varphi = \Box\psi$:

($\Rightarrow$) Suppose $T \in [\![\Box\psi]\!]^{\widehat{int(\alpha)}}$, i.e., $(T, \widehat{int(\alpha)}) \models \Box\psi$. This means that for all $U \in \mathcal{O}$ with $T \in U \subseteq \widehat{int(\alpha)}$, we have $(T, U) \models \psi$. This in particular implies that $(T, \widehat{int(\alpha)}) \models [p]\psi$ for all $p \in P$. To show, let $p \in P$ and suppose

$(T, \widehat{int(\alpha)}) \models int(p)$, i.e., $T \in Int(\llbracket p \rrbracket^{\widehat{int(\alpha)}}) = \llbracket int(p) \rrbracket^{\widehat{int(\alpha)}}$. Since $int(p) < \Box\psi$ (see Lemma 2.8), we know by IH that $\llbracket int(p) \rrbracket^{\widehat{int(\alpha)}} = \widehat{\langle\alpha\rangle int(p)}$. But, as shown in the case for the modality int above, $\vdash \langle\alpha\rangle int(p) \leftrightarrow int(\langle\alpha\rangle p)$, hence, $\llbracket int(p) \rrbracket^{\widehat{int(\alpha)}} = \widehat{int(\langle\alpha\rangle p)}$, thus, $\llbracket int(p) \rrbracket^{\widehat{int(\alpha)}} \in \mathcal{O}^c$. Hence, by the first assumption, we obtain $(T, Int(\llbracket p \rrbracket^{\widehat{int(\alpha)}})) \models \psi$, thus, $(T, \widehat{int(\alpha)}) \models [p]\psi$. Therefore, $T \in \llbracket [p]\psi \rrbracket^{\widehat{int(\alpha)}}$ for all $p \in P$. Then, by IH (since $[p]\psi < \Box\psi$, by Lemma 2.9), we have $\llbracket [p]\psi \rrbracket^{\widehat{int(\alpha)}} = \widehat{\langle\alpha\rangle[p]\psi}$, thus, $\langle\alpha\rangle[p]\psi \in T$. Hence, by Proposition 3-($\langle ! \rangle$), $int(\alpha) \wedge [\alpha][p]\psi \in T$ for all $p \in P$. Since T is P-witnessed and maximal, we then obtain $int(\alpha) \wedge [\alpha]\Box\psi \in T$. Then, by Proposition 3-($\langle ! \rangle$), we conclude $\langle\alpha\rangle\Box\psi \in T$.

($\Leftarrow$) Suppose $T \in \widehat{\langle\alpha\rangle\Box\psi}$. This means (by Proposition 3-($\langle ! \rangle$)) that $T \in \widehat{int(\alpha) \wedge [\alpha]\Box\psi}$, i.e., that $int(\alpha) \in T$ and $[\alpha]\Box\psi \in T$. Then, by axiom ([!]$\Box$-elim), we have that $[\alpha \wedge \chi]\psi \in T$ for all $\chi \in \mathcal{L}^P$. We want to show that $T \in \llbracket \Box\psi \rrbracket^{\widehat{int(\alpha)}}$. Let $U \in \mathcal{O}^c$ such that $T \in U \subseteq \widehat{int(\alpha)}$ and show $T \in \llbracket \psi \rrbracket^U$. By the construction of $\mathcal{O}^c$, we know that $U = \widehat{int(\gamma)}$ for some $\gamma \in \mathcal{L}^P$. We therefore have that $T \in U = \widehat{int(\gamma)} = \widehat{int(\gamma)} \cap \widehat{int(\alpha)} = \widehat{int(\gamma) \wedge int(\alpha)} = \widehat{int(\gamma \wedge \alpha)}$. Hence, $int(\alpha \wedge \gamma) \wedge [\alpha \wedge \gamma]\psi \in T$. Therefore, by Proposition 3-($\langle ! \rangle$) and the fact that T is maximal, we obtain $\langle \alpha \wedge \gamma \rangle\psi \in T$. Thus, by IH (since $\psi < \Box\psi$, by Lemma 2.1), $T \in \llbracket \psi \rrbracket^{\widehat{int(\alpha \wedge \gamma)}}$, i.e., $T \in \llbracket \psi \rrbracket^U$.

Lemma 10. *Let $\mathcal{M} = (X, \mathcal{O}, V)$ be a pseudo-model and $\mathcal{M}_\tau = (X, \tau_\mathcal{O}, V)$ be the associated topo-model. Then, for all $\varphi \in \mathcal{L}_{Kint}$ and $(x, U) \in ES(\mathcal{M})$, we have*

$$(x, U) \models_\mathcal{M} \varphi \text{ iff } (x, U) \models_{\mathcal{M}_\tau} \varphi.$$

Corollary 3. **L** *is complete for canonical pseudo-models and canonical topo-models (and so also complete wrt pseudo-models, as well as wrt topo-models).*

Proof. Let φ be an **L**-consistent formula, i.e., it is a P_φ-theory. Then, by Lemma 6, it can be extended to a maximal Prop-witnessed theory T. Then, by axiom (R[$\top$]), we have $\langle\top\rangle\varphi \in T$, i.e., $T \in \widehat{\langle\top\rangle\varphi}$. Then, by Truth Lemma (Lemma 9), we obtain that $(T, X^c) \models_{\mathcal{M}^c} \varphi$, where $\mathcal{M}^c = (X^c, \mathcal{O}^c, V^c)$ is the canonical pseudo-model for T. This proves the first completeness claim. As for the second: by the co-expressivity of $\mathcal{L}_{Kint}$ and $\mathcal{L}$ on pseudo-models (Theorem 4), there exists a $\psi \in \mathcal{L}_{Kint}$ such that $\varphi \leftrightarrow \psi$ is valid in all pseudo-models. We therefore have $(T, X^c) \models_{\mathcal{M}^c} \psi$. By Lemma 10, we obtain $(T, X^c) \models_{\mathcal{M}_\tau} \psi$ where $\mathcal{M}_\tau$ is the canonical topo-model. Using again the semantic equivalence of φ and ψ (applied to the model $\mathcal{M}_\tau$), we conclude that $(T, X^c) \models_{\mathcal{M}_\tau} \varphi$.

5 Conclusions

This paper throws new light on Topologic and Topo-APAL, elucidating their relations with each other and other modal logics for topology. The addition of dynamic modalities is shown to greatly simplify the axiomatization and completeness proof of Topologic. In on-going work we look at *doxastic* versions of this logic, able to capture learning-theoretic notions; while in future work we plan to investigate *multi-agent* versions.

References

1. Aiello, M.: Theory and practice. Ph.D. thesis, ILLC, Univerisity of Amsterdam (2002)
2. Balbiani, P., Baltag, A., van Ditmarsch, H., Herzig, A., Hoshi, T., Lima, T.D.: 'Knowable' as 'Known after an announcement'. Rew. Symb. Logic **1**, 305–334 (2008)
3. Baltag, A.: To know is to know the value of a variable. In: Proceedings of the 11th Advances in Modal Logic, pp. 135–155 (2016)
4. Bennett, B.: Modal logics for qualitative spatial reasoning. Logic J. IGPL **4**, 23–45 (1996)
5. Bjorndahl, A.: Topological subset space models for public announcements. In: Trends in Logic, Outstanding Contributions: Jaakko Hintikka (2016, to appear)
6. Dabrowski, A., Moss, L.S., Parikh, R.: Topological reasoning and the logic of knowledge. Ann. Pure Appl. Logic **78**, 73–110 (1996)
7. Engelking, R.: General Topology, vol. 6, 2nd edn. Heldermann Verlag, Berlin (1989)
8. Georgatos, K.: Modal logics for topological spaces. Ph.D. thesis, City University of New York (1993)
9. Georgatos, K.: Knowledge theoretic properties of topological spaces. In: Masuch, M., Pólos, L. (eds.) Logic at Work 1992. LNCS, vol. 808, pp. 147–159. Springer, Heidelberg (1994). doi:10.1007/3-540-58095-6_11
10. Goranko, V., Passy, S.: Using the universal modality: gains and questions. J. Log. Comput. **2**, 5–30 (1992)
11. Kelly, K.T.: The Logic of Reliable Inquiry. Oxford University Press, Oxford (1996)
12. McKinsey, J.C.C., Tarski, A.: The algebra of topology. Ann. Math. **2**(45), 141–191 (1944)
13. Moss, L.S., Parikh, R.: Topological reasoning and the logic of knowledge. In: Proceedings of the 4th TARK, pp. 95–105. Morgan Kaufmann (1992)
14. Shehtman, V.B.: "Everywhere" and "Here". J. Appl. Non Class. Logics **9**, 369–379 (1999)
15. Ditmarsch, H., Knight, S., Özgün, A.: Arbitrary announcements on topological subset spaces. In: Bulling, N. (ed.) EUMAS 2014. LNCS (LNAI), vol. 8953, pp. 252–266. Springer, Cham (2015). doi:10.1007/978-3-319-17130-2_17
16. Vickers, S.: Topology via Logic. Cambridge Tracts in Theoretical Computer Science. Cambridge University Press, Cambridge (1989)

Strategic Knowledge of the Past in Quantum Cryptography

Christophe Chareton(✉) and Hans van Ditmarsch

LORIA, CNRS, Université de Lorraine, Nancy, France
{christophe.chareton,hans.van-ditmarsch}@loria.fr

Abstract. We propose an epistemic strategy logic with future and past time operators, called SLKP, for Strategy Logic with Knowledge of the Past. With SLKP we can model mutually observed moves/actions in strategic contexts. In a semantic game, agents may completely or partially observe other agents' moves, their moves may depend on their knowledge of other players' strategies, and their knowledge may depend on the history of their own or other's moves. The logic SLKP also allows us to describe temporal properties involving past, future, and composed tenses such as future perfect or counterfactual assertions. We illustrate SLKP by formalising the quantum cryptography protocol BB84, with the purpose to initiate an integrated epistemic and strategic treatment of agent interactions in quantum systems.

1 Introduction

Strategy Logic (SL) [1,2] provides formal tools to model the ability of agents or coalitions of agents to ensure temporal properties in strategic contexts. In SL, one considers sequences of transitions between possible *states* of a modelled system. In these states agents can concurrently perform *actions*, determining *transitions* to other states.

Strategy logic has powerful modelling possibilities. But more may be needed. In SL, an agent is able to reach a goal if she can perform (play) a conditioned sequence of actions (a strategy) to realise that goal. But how to build this strategy is not addressed. A classical illustration of this restriction is the problem of how to open a strong-box: for any password, any agent is able to compose it. So, the agent has a winning strategy to open the strong-box. But she cannot be called able to open the strong-box.

This problem can be solved by modelling agent's knowledge: in order to open the strong-box, the agent needs to *know* the password. Knowledge is interpreted by identifying, for every agent, an equivalence relation over states. These are the states that the agent cannot distinguish. To realise a goal it is insufficient to have a strategy, the agent must have a *uniform strategy*, indicating the same actions from two states the agent is unable to distinguish. This notion of ability therefore combines knowledge and being able to perform actions. Formal approaches include epistemic multi-agent logics [3,4] and epistemic strategy logics [5,6].

A. Baltag et al. (Eds.): LORI 2017, LNCS 10455, pp. 347–361, 2017.
DOI: 10.1007/978-3-662-55665-8_24

The latter basically enrich the former by use of explicit quantifications over strategies and strategy contexts.

One further issue is how agents observe other agents in the concurrent execution of actions. For example, in a semantic game featuring two agents Alice and Bob, if Alice performs an action, is it (always/sometimes/at some conditions) the case that, after she does, Bob knows this performance has occurred? Public and partial observations are the traditional concern of Dynamic Epistemic Logic (DEL) [7]. In strategy logics, the state of a system is always highly dependent on the strategies played so far. Therefore, SLKP enables to express the knowledge agents may have about the strategies played so far. This affects their knowledge about the current state and about what they can achieve. The logic SLKP enriches epistemic strategy logics of [5,6] with past tense temporal operators. With these operators agents can refer to sequences of past actions, that thus can become conditions for future strategic choices. It can also be seen as an enrichment of DEL with temporal assertions and strategic framework.

We apply our logic SLKP to the modelling of the cryptographic protocol BB84 [8]. This protocol handles information encoded in qubits, held by quantum states of particles. Quantum states cannot be observed without probably being altered. Therefore, eavesdropping may be identified *a posteriori*, by detecting alterations on the sent message. In this paper we use polarised photons as quantum information holders, as in the original description of the protocol [8]. Logical aspects of protocol BB84 have been studied in [9,10]. However, to the best of our knowledge, its multi-agent and epistemic aspects have not yet been investigated. This is what we will do in our contribution.

Modal logical approaches to quantum physics and quantum computing include [11–13]. A recent addition to this corpus is Logic of Quantum Program (LQP) [14,15]. In LQP, actions model the possible or necessary effects of tests for truth of quantum properties. Instead, an action in our framework models quantum measurement by undeterministically bringing one of the mutually incompatible possible results of this measurement. In SLKP, actions are units in a temporal conditional plan: a strategy. Thus we can formalise how an agent reaches a goal after a sequence of actions, depending on other agents' concurrent actions. Quantum epistemic logics [16–18] allow to distinguish between the information contained by a system and the information that an agent may extract from it. Here again, we add a strategic aspect: agents knowledge specifies whether they may acquire further information and reach their goals.

In SLKP, the behaviour of propositional connectives is classical, satisfying distributivity of conjunction over disjunction and the principle of bivalence. In this article, quantum effects are implicitly encoded in an interpretation model that represents the possible transitions in the orthomodular lattice for a Hilbert space. The explicit characterisation of this model class and its axiomatisation is left for future research. This probably also requires an extension of SLKP.

We close the introduction with an overview of our contribution. In Sect. 2 we introduce the protocol BB84. In Sect. 3 we present the syntax and the semantics of SLKP. In Sect. 4 we model the protocol in SLKP and prove the adequation of this modelling. Section 5 contains the conclusions.

2 Protocol BB84

Qubits and measurements. The unit of information in protocol BB84 is the *qubit.* In our description qubits are holded by polarised photons. A qubit is represented by a *projective ray* (or simply a *ray*, that is an element of a vector space quotiented by scalar multiplication) in $\mathbb{R}^2$. The information encoded in a polarised photon is relative to a given *basis* (a pair of orthogonal rays) of $\mathbb{R}^2$, where each element of this basis stands for a value in $\{\mathbf{0}, \mathbf{1}\}$. Here we are using the basis $+ = \{\updownarrow, \leftrightarrow\}$ and $\times = \{\swarrow\!\!\!\nearrow, \nwarrow\!\!\!\searrow\}$. To extract information from a qubit, one can apply a *measurement.* The measurement is, again, relative to a choice of a basis. A measurement is non-deterministic. If the basis chosen for the measurement corresponds to the state of the photon, the result is determined, but if the choice is different the outcome of the measurement may be any state in the qubit's basis that is not orthogonal to the actual state. For example, a $\times$-measurement performed on a photon initially in state $\updownarrow$ of the different basis $+$ may give result $\nwarrow\!\!\!\searrow$ or $\swarrow\!\!\!\nearrow$, whereas the same measurement on a photon in state $\nwarrow\!\!\!\searrow$, of the same basis $\times$ as the measurement, will correctly get $\nwarrow\!\!\!\searrow$. A *wrong* measurement (that is, in a basis different from the one containing the current state of the photon) alters its object. Furthermore, a sequence of two such alterations may result in a bitflip. This enables the detection of eavesdropping in protocol BB84.

Protocol BB84. The protocol enables an agent (Alice in the following) to send an encryption key to another agent (Bob) and detect potential eavesdropping attempt with high probability. It proceeds as follows:

1. Alice chooses a finite sequence of basis, that is an element in $\{\times, +\}^*$. She encodes her message as a series of qubit in this sequence of basis and sends it towards Bob.
2. A potential eavesdropper (Eve), may intercept it. To do so, for each qubit she must choose a measurement basis. If she makes the wrong choice she alters the state of the supporting photon.
3. Bob receives the perhaps altered message. Again, for each qubit he needs to choose a basis. If this qubit has been altered by Eve, then if he chooses the original basis he has a 50% chance to observe the photon in the right basis but with opposite value (as compared to the original encoding).

The communications for the rest of the protocol occur in a public channel:

4. Alice and Bob share the sequence of basis they respectively used. For approximately half the qubits, they used the same basis and the value held by Bob is supposed (if there has not been any eavesdropping attempt) to be conform to Alice's encoding.

In the meanwhile, Eve may also intercept this communication. In this case she also learns which part of the information she got from her measurement was reliable. Then she can read it.

5. Alice and Bob then sacrifice a fraction of the resulting qubits, for which Alice reveals the value she encoded. If they observe, for any sacrificed qubit, that the value measured by Bob differs from Alice's encoding, then someone has altered this qubit by performing a measurement in the wrong basis. In this case they know the transmitted message is not reliable. Otherwise the resulting part of the message constitute their encryption key.

Note that the possibility of a non-detected eavesdropping is not null, depending on the length of the message. A reliable execution of the protocol will use a message sufficiently long so as to keep it below an accepted risk.

We will now introduce SLKP and model the protocol as a run in an interpretation model for SLKP. We will then describe in SLKP the exchanges of informations in the protocol, eavesdropping attempts, and their detection *a posteriori* by Alice and Bob, and verify them in the model.

3 SLKP

First, let us present SLKP syntax. It distinguishes between state and path formulas. In addition to boolean operators, state formulas bring strategic material: for each considered agent a, an existential quantifier $\exists^a x$, over the set of strategies that are available for a and a binder $\downarrow_x$, stating that the (previously quantified and instantiated) strategy x is played in the current semantic game, and an unbinder $\uparrow^x$, by which a strategy is deleted from the current context. This last operator is not used in the present modelling of protocol BB84. Our logic enables strategy refinements for agents: in a given context, an agent may be committed to different strategies at the same time. Then she plays together the actions indicated by these different strategies. In case she cannot (if she is committed to contradictory strategies), the execution stops. For more details about strategy refinement see [19,20].

Path formulas describe the future, with classical LTL operators X (*next*) and U (*until*), or the past, with symmetrical operators P (*previous*) and S (*since*).

Definition 1 (Pseudo-formulas). *Let Ag be a finite set of agents, let At be an enumerable set of propositions, and let X be a set of (strategy) variables. Then the set of* SLKP pseudo-formulas *is defined by the following grammar:*

- *State formulas:* $\psi ::= p \mid \neg\psi \mid \psi \wedge \psi \mid \exists^a x\ \psi \mid \downarrow_x \varphi \mid \uparrow^x \varphi \mid K_a \psi$
- *Path formulas:* $\varphi ::= \zeta \mid \xi$
  - *Future:* $\zeta ::= \psi \mid \neg\zeta \mid \zeta \wedge \zeta \mid \mathsf{X}\zeta \mid \zeta\, \mathsf{U}\, \zeta$
  - *Past:* $\xi ::= \psi \mid \neg\xi \mid \xi \wedge \xi \mid \mathsf{P}\xi \mid \xi\, \mathsf{S}\, \xi$

where $a \in Ag, p \in At$ and $x \in X$.

The universal quantifier $\forall^a x$ and booleans $\vee$ and $\rightarrow$ are introduced in the usual way. The double arrow is written $\Leftrightarrow$ (to distinguish it from state $\leftrightarrow$ for photons). For $\mathsf{O} \in \{\mathsf{X}, \mathsf{S}\}$ and $k \in \mathbb{N}$, we write O^k for a sequence of k successive occurrences of O. As strategy variable *names* are taken into account in the semantics of formulas, some care must be taken when a quantifier is encountered. Thus, well-formed formulas are pseudo-formulas such that every quantifier introduces a fresh strategy variable with regard to the scope in which it appears. Formulas are interpreted, by help of *strategy contexts*, in *Concurrent Epistemic Transition Systems* (CETSs). We first define the latter:

Definition 2 (Concurrent Epistemic Transition Systems). *A CETS is a tuple $\mathcal{G} = \langle St, At, v, Ag, \{\mathcal{R}_a\}_{a \in Ag}, Act, c_0 \rangle$ where:*

- *St is an enumerable non-empty set of* configurations[1].
- *At is a finite non-empty set of* atomic propositions.
- *$v : St \longrightarrow \mathcal{P}(At)$ is a valuation function which maps each configuration c to the set of propositions true at c.*
- *Ag is a finite non-empty set of agents.*
- *For each $a \in Ag$, $\mathcal{R}_a$ is an equivalence accessibility relation. It induces a partition $[St]_a$ of St. For any configuration c, we write $[c]_a$ its equivalence class in $[St]_a$.*
- *For each $a \in Ag, Act_a$ is an enumerable set of actions in $St \times St$. Then, $Act = \bigcup_{a \in Ag} Act_a$. An action $ac_a \in Act_a$ is such that there are C_1 and C_2 in $\mathcal{P}([St]_a)$ such that $dom(ac_a) = \cup C_1$ and $im(ac_a) = \cup C_2$ (where $dom(ac_a)$ and $im(ac_a)$ respectively denote the left and right projection of ac_A).*
- *$c_0 \in St$ is the initial configuration.*

A *strategic context* for a CETS is a pair of an *assignment* for strategy variables and a *commitment* of agents to strategies:

Definition 3 (Strategy, Assignment, Commitment, Context)

- *A* strategy *σ_a for an agent a is a map with domain of definition $dom(\sigma_a) = \bigcup_{ac \in Act_a} dom(ac)$. Given an equivalence class $[s]_a \subseteq dom(\sigma_a)$, it yields an action ac_a for a such that $[s]_a \subseteq dom(ac_a)$.*
- *A* strategy *is the couple $\sigma = \langle \pi_1(\sigma), \pi_2(\sigma) \rangle$ of an agent $\pi_1(\sigma) = a$ and a strategy $\pi_2(\sigma)$ for a.*
- *An* assignment *α is a map which, given a strategy variable x in its domain of definition, yields a strategy $\alpha(x)$.*
- *A* commitment *γ is a set of variables $\gamma \subseteq X$, gathering bindings of strategies to their relative agents.*
- *A* context *κ is a "well-formed" pair $\langle \alpha, \gamma \rangle$ of an assignment α and a commitment γ, that is a pair such that each strategy variable in the commitment is instantiated: $\gamma \subseteq dom(\alpha)$.*

[1] Configurations are commonly referred to as *states*. In this paper we use the word configuration instead, to avoid ambiguities since *state* is also used in its physical sense, to designate the state of a photon in $\{\updownarrow, \leftrightarrow, \searrow, \nearrow\}$.

During the semantic evaluation of a formula, a context $\kappa = \langle \alpha, \gamma \rangle$ must be transformed as we encounter a strategy quantifier $\exists^a x$, a binding $\downarrow_x$ or an unbinding $\uparrow^x$ operator. We write $\alpha[x \mapsto \sigma]$ the assignment of domain $dom(\alpha) \cup \{x\}$ such that $\alpha[x \mapsto \sigma](x) = \sigma$ and for all $y \in dom(\alpha) \backslash \{x\}, \alpha[x \mapsto \sigma](y) = \alpha(y)$. This notation is extended to contexts: $\kappa[x \mapsto \sigma] = \langle \alpha[x \mapsto \sigma], \gamma \rangle$. We also write $\kappa \cup \{x\}$ (respectively $\kappa \setminus \{x\}$) for the context $\langle \alpha, \gamma \cup \{x\} \rangle$ (respectively $\langle \alpha, \gamma \setminus \{x\} \rangle$). Furthermore, if $\gamma = dom(\alpha) = \{x_0, x_1, \ldots, x_i\}$, then we commonly write $\langle \alpha(x_0), \alpha(x_1), \ldots, \alpha(x_i) \rangle$ for the context $\langle \alpha, \gamma \rangle$.

A context κ induces possible *incomes* and *outcomes* from a configuration c. We call *in-outcomes* of κ and c the set of executions that can be if, in c, agents play (and played) according to the strategies stored in κ. To define the in-outcome function, we need to introduce *executions* and the set of possible immediate *successors* of a configuration, given a context.

Definition 4 (Execution, Successor, In-outcome). *Let $\mathcal{G}$ be a CETS and let $\kappa = \langle \alpha, \gamma \rangle$ be a context for $\mathcal{G}$.*

- *An* execution *is a non-empty finite or infinite $]|\lambda|^-, |\lambda|^+[$-indexed sequence of configurations λ_i for $i \in]|\lambda|^-, |\lambda|^+[$. It is such that $|\lambda|^- \in \mathbb{Z}^- \cup \{-\infty\}$ and $|\lambda|^+ \in \mathbb{Z}^+ \cup \{\infty\}$. Given $i \in]|\lambda|^-, |\lambda|^+[$, we write $\lambda_{\leqslant -i}$ for the subsequence of λ ending at index i, and $\lambda_{\geqslant -i}$ for its subsequence starting at index i.*
- *The* successor *function $succ_\kappa : St \longrightarrow \mathcal{P}(St)$ induced by κ characterises, for any configuration c, the set of transitions that are possible if each agent respects the different strategies it is bound to:*

$$succ_\kappa(c) = \{c' \in St \mid there\,is\,x \in \gamma \cap dom(\alpha)\,such\,that$$
$$\alpha(x) = \langle a, \sigma_a \rangle, [c]_a \in dom(\sigma_a)\,and\,(c, c') \in \sigma_a([c]_a)\}$$

- *Let c be a configuration in $\mathcal{G}$. The in-outcomes of κ and c in $\mathcal{G}$ is the set $I/O(\kappa, c)$ of S_λ labelled executions λ in $\mathcal{G}$ such that $\lambda_0 = c$ and, for any $i \in \mathbb{Z}$, iff $\{i, i+1\} \in S_\lambda$, then $\lambda_{i+1} \in succ_\kappa(\lambda_i)$.*

We conclude this section by defining the truth conditions for SLKP formulas:

Definition 5 (Satisfaction). *Let $\mathcal{G}$ be a CEGS, with the notations of Definition 2. Let κ be a context, s be a state in St and λ be an execution. Then:*

- *State formulas*
  - $\mathcal{G}, \kappa, c \models p$ *iff* $p \in v(s)$, *with* $p \in At$
  - $\mathcal{G}, \kappa, c \models \neg\psi$ *iff* $\mathcal{G}, \kappa, c \not\models \psi$
  - $\mathcal{G}, \kappa, c \models \psi_1 \wedge \psi_2$ *iff* $\mathcal{G}, \kappa, c \models \psi_1$ *and* $\mathcal{G}, \kappa, c \models \psi_2$
  - $\mathcal{G}, \kappa, c \models \exists^a x\, \psi$ *iff there is a strategy* σ *for* a *such that* $\mathcal{G}, \kappa[x \mapsto \sigma], c \models \psi$
  - $\mathcal{G}, \kappa, c \models \downarrow_x \varphi$ *iff for any* $\lambda \in I/O(\kappa \cup \{x\}, c), \mathcal{G}, \kappa \cup \{x\}, \lambda \models \varphi$
  - $\mathcal{G}, \kappa, c \models \uparrow^x \varphi$ *iff for any* $\lambda \in I/O(\kappa \setminus \{x\}, c), \mathcal{G}, \kappa \setminus \{x\}, \lambda \models \varphi$
  - $\mathcal{G}, \kappa, c \models K_a\psi$ *iff for any* $c' \in [c]_a, \mathcal{G}, \langle \alpha, \emptyset \rangle, c' \models \psi$

– *Path formulas*
 - $\mathcal{G}, \kappa, \lambda \models \psi$ *iff* $\mathcal{G}, \kappa, \lambda_0 \models \psi$, *if* ψ *is a state formula*
 - $\mathcal{G}, \kappa, \lambda \models \neg\varphi$ *iff* $\mathcal{G}, \kappa, \lambda \not\models \varphi$
 - $\mathcal{G}, \kappa, \lambda \models \varphi_1 \wedge \varphi_2$ *iff* $\mathcal{G}, \kappa, \lambda \models \varphi_1$ *and* $\mathcal{G}, \kappa, \lambda \models \varphi_2$
 - $\mathcal{G}, \kappa, \lambda \models \mathsf{X}\zeta$ *iff* $|\lambda|^+ > 1$ *and* $\mathcal{G}, \kappa, \lambda_{\geqslant 1} \models \zeta$
 - $\mathcal{G}, \kappa, \lambda \models \zeta_1 \mathsf{U} \zeta_2$ *iff there is a number* $i \in \mathbb{N}$ *such that* $|\lambda|^+ > i$, *such that* $\mathcal{G}, \kappa, \lambda_{\geqslant i} \models \zeta_2$ *and such that for any* $0 \leqslant j \leqslant i$, $\mathcal{G}, \kappa, \lambda_{\geqslant j} \models \zeta_1$.
 - $\mathcal{G}, \kappa, \lambda \models \mathsf{P}\xi$ *iff* $|\lambda|^- > 1$ *and* $\mathcal{G}, \kappa, \lambda_{\leqslant -1} \models \xi$
 - $\mathcal{G}, \kappa, \lambda \models \xi_1 \mathsf{S} \xi_2$ *iff there is a number* $i \in \mathbb{N}$ *such that* $|\lambda|^- > i$, *such that* $\mathcal{G}, \kappa, \lambda_{\leqslant -i} \models \xi_2$ *and such that for any* $0 \leqslant j \leqslant i$, $\mathcal{G}, \kappa, \lambda_{\leqslant -j} \models \xi_1$.

Given the empty context $\kappa_\emptyset$ *and a sentence* ψ, *we write* $\mathcal{G}, s \models \psi$ *iff* $\mathcal{G}, \kappa_\emptyset, s \models \psi$, *and* $\mathcal{G} \models \psi$ *iff* $\mathcal{G}, s_0 \models \psi$.

Note that the evaluation of an operator K_a deletes the commitment stored in the current context. If needed, the knowledge an agent has about the context is explicitly stated. For example, let φ be a path formula, then formula $K_a \downarrow_x \varphi$ states that a knows that, if x is played by its relative agent, then φ is ensured: it is true in state c and context κ if and only if, for any $c' \in [c]_a$, for any $\lambda \in \mathsf{I/O}(\langle \kappa \cup \{x\} \rangle, c')$, φ is true in λ.

4 Modelling and Verifying BB84 into SLKP Framework

In Sect. 4.1, we model the BB84 transmission of qubits into CETSs framework. In Sect. 4.2, eavesdropping attempts and detections are characterised by SLKP formulas. They are then verified in Sect. 4.3.

4.1 Modelling the Transmission

Figure 1 depicts a partial execution of protocol BB84. Configurations are given a name c_i. In the logical language we represent a state of a photon by a similarly named proposition describing that the photon is in that state, e.g., $\leftrightarrow$ means "the photon is in state $\leftrightarrow$". The resulting valuation is given in the figure. We also introduce macros for identifying basis and encoding values:

$$\mathbf{0} := \leftrightarrow \vee \swarrow\!\!\!\nearrow \qquad \mathbf{1} := \updownarrow \vee \nwarrow\!\!\!\searrow \qquad \mathbf{C} := \leftrightarrow \vee \updownarrow \qquad \mathbf{D} := \swarrow\!\!\!\nearrow \vee \nwarrow\!\!\!\searrow$$

Transitions are labelled with the actions firing them, each with the mention of its related agent as index. Note that this model happens to be turn-based. This is not a requirement for SLKP but results from modelling choices that were made for sake of simplicity. Except for Bob in the righmost column (configurations whose labels starts with a 3 are called $c_{3\cdot}$-*configurations*), the indistiguishability relations are obtained by the transitive closure of the relation given by dashed (for Eve) and dotted (for Bob) edges. For example, Bob distinguishes between two $c_{3\cdot}$-configurations if and only if the photon is in the same state in them. For sake of readability, we only indicated the case for state $\leftrightarrow$. Since Alice's knowledge is not at stake in our modelling, we did not represent it.

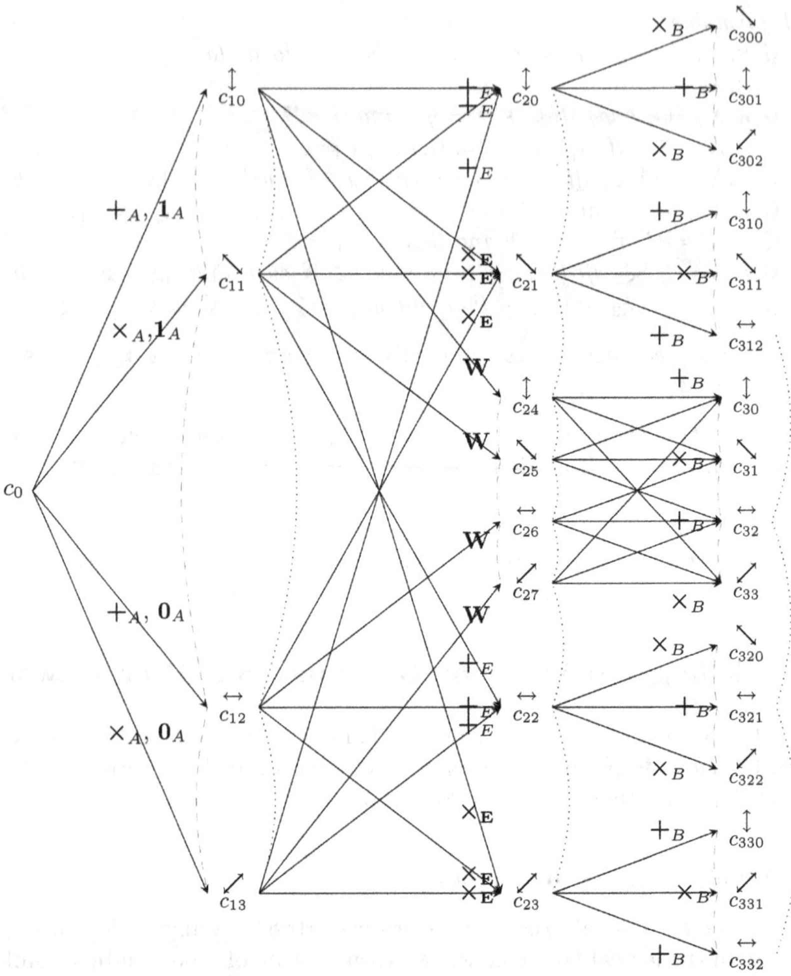

Fig. 1. A CETS for protocol BB84: model $\mathcal{G}_B$

Note also that each agent plays once only in a path leading from c_0 to a c_3.-configuration, and she is never able to distinguish between two configurations where it is her turn. Therefore, for any agent a, a strategy reduces to a single choice of an action ac she plays when it is her turn. We designate it by σ_{ac}. In our modelling, after each sending of a qubit by Alice to Bob, we consider the possibility of a call from Alice to Bob and exchanges of information about this qubit sending. Therefore we only focus on sequences from c_0 to c_3.-configurations. From each c_3.-configuration, the execution goes back to state c_0 to enter a new cycle. For sake of readability of the model, we did not represent the transitions from c_3.-configurations to c_0. Now, let us explain the transitions of the model, referring to the steps of the protocol as described in Sect. 2.

Encoding (step 1). First, Alice plays simultaneously two actions: she chooses an encoding basis (by playing $+_A$ or $\times_A$) and a value to send (action $\mathbf{0}_A$ or $\mathbf{1}_A$). Neither Bob nor Eve is able to distinguish between realisations of any in the four possible resulting states.

Potential eavesdropping (step 2). Then the message comes to Eve' hands.

- She may try to get some information from it, by help of a measurement operation in $\{+_E, \times_{\mathbf{E}}\}$. Doing so she may:
  - choose the wrong measurement basis (the one that Alice did not use for encoding) and alter the photon's state, with two possible results,
  - choose the right basis and observe the photon as it was encoded by Alice.

  In the latter case the execution goes to one of configurations $\{c_i\}_{i\in[20,\ldots,23]}$, where Eve knows the current state of the photon. But she does not know this state was the same in the previous configuration. To know that she lacks the information among which Alice made use of the same basis as her.
- She may also not attempt to read the message. We represent this possibility by a third action *wait* (**W**). In this case she remains unaware of the current state of the photon and the execution goes to a configuration in $\{c_i\}_{i\in[24,\ldots,27]}$.

After Eve's move, Bob does not know either if she made a measurement or played action **W**, nor the current state of the photon.

Reception (step 3). Now, Bob tries to read the message, by performing again a measurement operation randomly chosen in $\{+_B, \times_B\}$. Doing so, he observes the photon in a given state, but he cannot distinguish between configurations resulting from different scenarios ending with the photon in this state.

Communications from Alice to Bob (steps 4 and 5). Suppose, for example, that Alice originally sent a $\updownarrow$ (configuration c_{10}) and Bob gets a $\leftrightarrow$ (configuration c_{312}). Then:

1. the basis used by Alice and Bob are the same.
2. the value obtained by Bob is opposite to the one sent by Alice.

At this stage, Bob does not have these informations. Getting the former at step 4 and the latter at step 5, he will be able to detect an eavesdropping attempt. Note that this identified gap between the state of the photon when sent and its received state could also be due to some noise in the transmission. Then, in practice, the identification of an eavesdropping attempt results from the observation of an unexpected rate of altered photons. We let this consideration for future works and, in the current modelling, we treat any noticed altered transmission of a photon as a potential eavesdropping attempt. Next, we build formulas for characterising eavesdropping attempts, exchanges of informations from Alice to Bob and eavesdropping attempt detections.

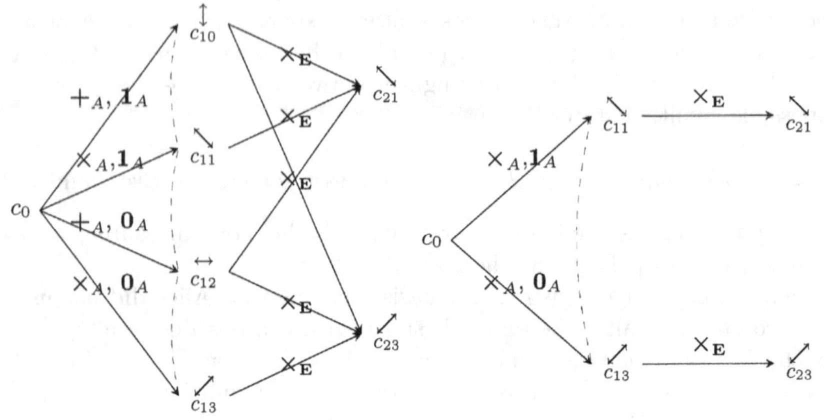

(a) Possible executions.

(b) Executions compatible with the information that Alice played $\times_A$.

Fig. 2. Step 2: Eve plays $\times_{\mathbf{E}}$.

4.2 Formalising Eavesdroppings Attempts and Detection

Eavesdropping attempt. At her turn in the semantic game, Eve does not distinguish between configurations in $\{c_i\}_{i\in[10,\ldots,13]}$. Nevertheless, she knows that by playing $\times_{\mathbf{E}}$, she enforces the ongoing execution to be prefixed by either $c_0{\cdot}c_{10}{\cdot}c_{21}$, $c_0 \cdot c_{10} \cdot c_{23}$, $c_0 \cdot c_{11} \cdot c_{21}$, $c_0 \cdot c_{12} \cdot c_{21}$, $c_0 \cdot c_{12} \cdot c_{23}$ or $c_0 \cdot c_{13} \cdot c_{23}$ (see Fig. 2a). Note that only two of them ($c_0 \cdot c_{11} \cdot {\cdot} c_{21}$ and $c_0 \cdot c_{13} \cdot {\cdot} c_{23}$, represented in Fig. 2b) are compatible with Alice playing $\times_A$. Furthermore, Eve can distinguish between these two since they differ by the ending photon state. Therefore, if Eve plays $\times_{\mathbf{E}}$, then if Alice happens to have played $\times_A$ and in case Eve gets this information, she will be able to infer the state of the photon as encoded by Alice. More generally, an eavesdropping attempt for Eve is a strategy y such that there is a choice of basis x such that if Alice plays x and Eve plays y, then after two steps, these two informations (that Alice plays x and Eve plays y) result in Eve knowing the value originally encoded in the photon. Let us first identify a choice of a basis by Alice, that is a strategy enforcing the photon either to be a ray in basis $+$ or to be a ray in $\times$:

$$\mathsf{B}(x) := \Big(\big(\downarrow_x \mathsf{X}(\mathbf{C})\big) \vee \big(\downarrow_x \mathsf{X}(\mathbf{D})\big)\Big)$$

Now we can define predicate Att, with one free variable. It is true of a strategy y for Eve if and only if y is an eavesdropping attempt:

Definition 6. *Let y stand for a strategy for Eve. Then:*

$$\mathit{Att}(y) := \exists^A x \left[\mathsf{B}(x) \wedge \left(\downarrow_x \downarrow_y \mathsf{X} \bigwedge_{V\in\{\mathbf{0},\mathbf{1}\}} \left(V \rightarrow (\mathsf{X}\, K_E\, (\downarrow_x \downarrow_y (\mathsf{P}\, V)))\right)\right)\right]$$

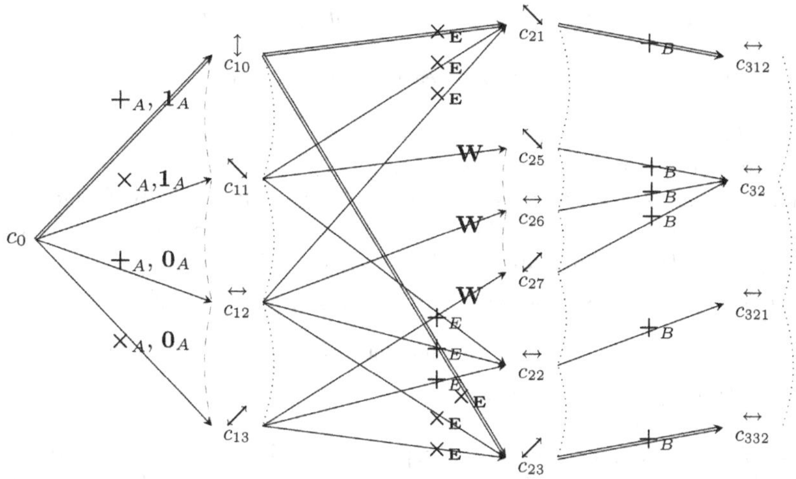

Fig. 3. Step 2: Bob observes $\leftrightarrow$.

Eavesdropping detection. An eavesdropping detection is relative to a pair of strategies (x, x') Alice plays together. Suppose, after step 3 of the protocol, Bob observes the photon in state $\leftrightarrow$. Then he can interpret it as the conclusion of several possible scenarios, consisting in the paths made up of transitions represented in Fig. 3. Now, if Alice has played actions $+_A$ and $\mathbf{0}_A$ in the ongoing semantic game and if, in addition Bob gets this information, then the only scenarios that are still compatible with his knowledge are those represented with double arrows in Fig. 3 ($s_0 \cdot c_{10} \cdot c_{21} \cdot c_{312}$ and $s_0 \cdot c_{10} \cdot c_{23} \cdot c_{332}$). In both Eve has played action $\times_{\mathbf{E}}$: knowing the strategies x and x' played by Alice informed Bob that Eve had played an attempt to eavesdrop the message:

Definition 7. *Let x, x' stand for strategies for Alice. Then:*

$$\mathsf{D}(x, x') := K_B\Big(\forall^E y\ \big(\downarrow_x \downarrow_{x'} \downarrow_y \mathsf{P}^3(\mathsf{Att}(y))\big)\Big)$$

More general, we can formalise that whatever Eve does, if it is an eavedropping attempt, then there are strategies for Alice such that the overall outcome may lead to a detection of this attempt. This is predicate ED:

Definition 8. $\mathsf{ED} := \forall^E y\ \Big(\mathsf{Att}(y) \Leftrightarrow \exists^A x\ \exists^A x'\ \big(\neg \downarrow_x \downarrow_{x'} \downarrow_y \neg(\mathsf{X}^3 \mathsf{D}(x, x'))\big)\Big)$

4.3 Verification

In this section, we verify our modelling for the characterisation of eavesdropping strategies from intruders and their detection. First, we prove that predicate $\mathsf{Att}(y)$ adequately catches eavesdropping attempts:

Proposition 1. *Let y be strategy variable and σ_{ac} be a strategy for Eve. Then:*

$$\mathcal{G}_B, \langle\langle(y \to \sigma_{ac})\rangle, \emptyset\rangle, s_0 \models \mathsf{Att}(y) \; iff \; ac = \times_\mathbf{E} \; or \; ac = +_E$$

Proof. – Case $\times_\mathbf{E}$: from c_0, the outcomes of $\langle \sigma_{\times_A}, \sigma_{\times_\mathbf{E}} \rangle$ all are prefixed by either $c_0 \cdot c_{11} \cdot c_{21}$ or $c_0 \cdot c_{13} \cdot c_{23}$. Furthermore, $[c_{21}]_E = \{c_{21}\}$ and from $\{c_{21}\}$, the single income of $\langle \sigma_{\times_A}, \sigma_{\times_\mathbf{E}} \rangle$ is suffixed by $c_{11} \cdot c_{21}$. Then for any c' in $[c_{21}]_E$, for any $\lambda \in \mathsf{I/O}(\langle \sigma_{\times_A}, \sigma_{\times_\mathbf{E}} \rangle, c')$, $\mathcal{G}_\mathrm{B}, \langle \sigma_{\times_A}, \sigma_{\times_\mathbf{E}} \rangle, \lambda \models \mathsf{P} \searrow$. Similarly for any c' in $[c_{23}]_E$, with state $\nearrow$ instead of $\searrow$. Since $\mathcal{G}_\mathrm{B}, c_{11} \models \searrow$ and $\mathcal{G}_\mathrm{B}, c_{13} \models \nearrow$, for any $\lambda \in \mathsf{I/O}(\langle \sigma_{\times_A}, \sigma_{\times_\mathbf{E}} \rangle, s_0)$ we have that $\mathcal{G}_\mathrm{B}, \langle \sigma_{\times_A}, \sigma_{\times_\mathbf{E}} \rangle, \lambda \models \mathsf{X} \bigwedge_{V \in \{\mathbf{0},\mathbf{1}\}} \Big(V \to (\mathsf{X} K_E (\downarrow_x \downarrow_y (\mathsf{P}\, V))) \Big)$. So $\sigma_{\times_A}$ is an existential witness for the truth of $\mathsf{Att}(y)$ at c_0 with context $\langle\langle(y \to \sigma_{\times_\mathbf{E}})\rangle, \emptyset\rangle$.
- Case $+_E$ is similar.
- Case $\mathbf{W}$: let us write $[\mathbf{W}]_E = \{c_i\}_{i \in [24,\ldots,27]}$. For any $\lambda \in \mathsf{I/O}(\langle \sigma_\mathbf{W} \rangle, s_0)$, $\lambda_2 \in [\mathbf{W}]_E$, so $[\lambda_2]_E = [\mathbf{W}]_E$. Now, check that for any $ac \in \{+_A, \times_A\}$, there are $c, c' \in [\mathbf{W}]_E, \lambda' \in \mathsf{I/O}(\langle \sigma_\mathbf{W}, \sigma_{ac} \rangle, c)$ and $\lambda'' \in \mathsf{I/O}(\langle \sigma_\mathbf{W}, \sigma_{ac} \rangle, c')$ such that $\mathcal{G}_\mathrm{B}, \langle \sigma_\mathbf{W}, \sigma_{ac} \rangle, \lambda' \models \mathsf{P0}$ and $\mathcal{G}_\mathrm{B}, \langle \sigma_\mathbf{W}, \sigma_{ac} \rangle, \lambda'' \models \mathsf{P1}$. So,

$$\mathcal{G}_\mathrm{B}, \langle(y \to \sigma_\mathbf{W})\rangle, \gamma_\emptyset\rangle s_0 \models$$
$$\forall^A x \left[\mathsf{B}(x) \to \left(\downarrow_x \downarrow_y \mathsf{X} \bigvee_{V \in \{\mathbf{0},\mathbf{1}\}} \Big(V \wedge (\neg K_E (\downarrow_x \downarrow_y (\mathsf{P}V))) \Big) \right) \right]$$

Hence $\mathsf{Att}(y)$ is not true at c_0 with context $\langle\langle(y \to \sigma_\mathbf{W})\rangle, \emptyset\rangle$.

Next, we establish that for any strategies x and x' for Alice, $\mathsf{D}(x, x')$ is true only in c_3.-configurations c such that x and x' enforce a state contradicting c.

Proposition 2. *Let σ_B and σ_V be strategies for Alice, and let c be a c_3.-configuration. Then, $\mathcal{G}_B, \langle\langle(x \to \sigma_B), (x' \to \sigma_V)\rangle, \gamma_\emptyset\rangle, c \models \mathsf{D}(x, x')$ iff:*

- *(σ_B, σ_V) is a pair of a choice of a basis and a choice of an encoded value (a strategy in $\{\sigma_{0_A}, \sigma_{1_A}\}$)*
- *and there is $s \in \{\updownarrow, \leftrightarrow, \nearrow, \searrow\}$ such that for all λ in $\mathsf{I/O}(\langle \sigma_B, \sigma_V \rangle, s_0)$, $\mathcal{G}_B, \langle \sigma_B, \sigma_V \rangle, \lambda_1 \models s$ and $\mathcal{G}_B, \lambda_3 \models \overline{s}$, where $\overline{s} \in \{\updownarrow, \leftrightarrow, \nearrow, \searrow\}$ is the state in the same basis as s, with the opposite value.*

Proof (sketch)

- From left to right: if the second item is false then either σ_B does not choose the same basis as c, or σ_V chooses the same value as c. In both cases there are incomes of $\langle \sigma_B, \sigma_V \rangle$ and c where Eve played $\mathbf{W}$. Now, suppose the first item is false. Then either σ_B and σ_V have empty intersection, and $\downarrow_{\sigma_B} \downarrow_{\sigma_V} \downarrow_y \mathsf{P}\varphi$ is false for any configuration c, any strategy y, and any formula φ, or they are equal and, again, there are incomes of $\langle \sigma_B, \sigma_V \rangle$ and c where Eve played $\mathbf{W}$.

- From right to left: Fig. 3 illustrates the case where s is $\updownarrow, \sigma_B = \sigma_{+_A}, \sigma_V = \sigma_{1_A}, \bar{s}$ is $\leftrightarrow$ and, to satisfy the second item in Proposition 2, c must be in $\{s_{312}, s_{332}\}$. Then for any income of $\langle \sigma_B, \sigma_V \rangle$ and $c' \in [c]_B$ (double edges in Fig. 3), Eve played $\times_{\mathbf{E}}$. The other cases are similar.

We can conclude that ED is true at configuration c_0: strategies for Eve that may be detected are the actual eavesdropping attempts.

Proposition 3. $\mathcal{G}_B, s_0 \models \mathit{ED}$

Proof (sketch)

- As established in Proposition 1, possible attempts for Eve are σ_{+_E} and $\sigma_{\times_{\mathbf{E}}}$. Suppose she plays $\sigma_{\times_{\mathbf{E}}}$. Consider path $\lambda = s_0 \cdot c_{10} \cdot c_{21} \cdot c_{312}$. We have that $\lambda \in \mathsf{I/O}(\langle \sigma_{+_A}, \sigma_{1_A}, \sigma_{\times_{\mathbf{E}}} \rangle)$. Since $\mathcal{G}_{\mathrm{B}}, \langle (x \to \sigma_{+_A}), (x' \to \sigma_{1_A}) \rangle, \emptyset \rangle, c_{312} \models \mathsf{D}(x, x')$, $\sigma_{\times_{\mathbf{E}}}$ is detectable by σ_{+_A} and σ_{1_A}. Similarly, σ_{+_E} is detectable by $\sigma_{\times_A}$ and σ_{0_A} (note that σ_{1_A} can be used instead of σ_{0_A}, and reciprocally).
- Now, suppose Eve plays a strategy y such that $\mathsf{Att}(y)$ is not true. Then $y = \sigma_{\mathbf{W}}$ and for all $\lambda \in \mathsf{I/O}(s_0)$, for any $s, s' \in \{\updownarrow, \leftrightarrow, \swarrow\!\!\!\nearrow, \nwarrow\!\!\!\searrow\}$, if $\lambda_1 \models s$ and $\lambda_3 \models s'$, either s and s' share the same encoded value or they belong to different basis. Then for any strategies σ_B and σ_V for Alice, for any $\lambda \in \mathsf{I/O}(\langle \sigma_B, \sigma_V, \sigma_{\mathbf{W}} \rangle, s_0), \mathcal{G}_{\mathrm{B}}, \langle \sigma_B, \sigma_V, \sigma_{\mathbf{W}} \rangle, \lambda \models \neg \mathsf{X}^3 \mathsf{D}(\sigma_B, \sigma_V)$.

5 Conclusion and Future Works

We presented SLKP, an epistemic strategy logic with past time operators. In SLKP we can express the knowledge agents have about the evolution of a system, may it be about the future, about the past or in more structured tense schemes such as future perfect or counterfactual analysis. We illustrated SLKP by modelling the cryptography protocol BB84.

Our work attempts to advance on established research traditions. The logic SLKP is inspired by the strategic reasoning framework [2], by dynamic epistemic logics [7], and by logics for past time as in [21,22]. Featuring strategy refinements, thanks to which an agent may compose her behaviour along different strategies, it can be seen as an epistemic extension of USL [19,20], a non-epistemic extension of SL. In case an agent is committed to contradictory strategies, our semantics adopts the classical interpretation of finite executions [23,24]. The use of epistemic modalities for the analysis of quantum systems has been entertained in many prior publications [16–18,25]. With respect to those our novelty is the usage of a strategic framework and of temporal operators.

We are presently addressing the computational complexity of SLKP. We conjecture that the model-checking problem is PSPACE-complete, and that satisfiability is undecidable, proved by a reduction to the recurrent tiling problem. This line should be continued by search for axiomatized fragments of SLKP.

The expressive power of SLKP should also be further investigated. In particular, since it syntactically enables to discuss the knowledge agents may have about the past, SLKP shines new light on the concept of memory. Although SLKP exclusively uses memoryless strategies, agents are able to determine strategies depending on their knowledge of the past. We also plan to continue the study modelling of quantum systems with SLKP or an adequate extension. This would be, as first steps, by extending the current work on protocol BB84. Such an enrichment would consider Alice sending a whole sequence of photons in a row and would take into account the possible noise in such a transmission. Also, the study could be extended to further quantum cryptography protocols, such as B92 [26] or extensions of BB84 with six states [27]. A further step would be to characterise and axiomatise the class of CETSs that are relevant for the modelling of quantum systems. Then, we would like to model entangled systems and the possible observations and information flows they enable.

Acknowledgement. We thank the reviewers for their helpful comments. We acknowledge financial support from ERC project EPS 313360. Hans van Ditmarsch is also affiliated to IMSc, Chennai, India.

References

1. Chatterjee, K., Henzinger, T.A., Piterman, N.: Strategy logic. Inf. Comp. **208**(6), 677–693 (2010)
2. Mogavero, F., Murano, A., Vardi, M.Y.: Reasoning about strategies. In: Proceedings of FSTTCS, vol. 8, pp. 133–144 (2010)
3. Ågotnes, T., Goranko, V., Jamroga, W., Wooldridge, M.: Knowledge and Ability. Volume Handbook of Epistemic Logic. College Publications (2015)
4. Jamroga, W., Ågotnes, T.: Constructive knowledge: what agents can achieve under imperfect information. J. Appl. Non-Class. Logics **17**, 423–475 (2007)
5. Belardinelli, F.: Reasoning about knowledge and strategies: epistemic strategy logic. In: Proceedings of 2nd International Workshop on Strategic Reasoning, SR, pp. 27–33 (2014)
6. Knight, S., Maubert, B.: Dealing with imperfect information in strategy logic. In: Proceedings of 3rd International Workshop on Strategic Reasoning, SR (2015)
7. Van Ditmarsch, H., van Der Hoek, W., Kooi, B.: Dynamic Epistemic Logic, vol. 337. Springer, Dordrecht (2007)
8. Bennett, C.H., Brassard, G.: Quantum cryptography: public key distribution and coin tossing. In: International Conference on Computers, Systems and Signal Processing, pp. 175–179 (1984)
9. Bergfeld, J.M., Sack, J.: Deriving the correctness of quantum protocols in the probabilistic logic for quantum programs. Soft Comput., 1–21 (2015)
10. Baltazar, P., Chadha, R., Mateus, P.: Quantum computation tree logic, model checking and complete calculus. Int. J. Quantum Inf. **6**, 219–236 (2008)
11. Mittelstaedt, P.: The modal logic of quantum logic. J. Phil. Logic **8**, 479–504 (1979)
12. Dalla Chiara, M.L.: Quantum logic and physical modalities. J. Phil. Logic **6**, 391–404 (1977)
13. Goldblatt, R.I.: Semantic analysis of orthologic. J. Phil. Logic **3**, 19–35 (1974)
14. Baltag, A., Smets, S.: LQP: the dynamic logic of quantum information. Math. Struct. Comput. Sci. **16**(3), 491 (2006)

15. Baltag, A., Smets, S.: The dynamic turn in quantum logic. Synthese **186**, 1–21 (2012)
16. Baltag, A., Smets, S.: Correlated knowledge: an epistemic-logic view on quantum entanglement. Int. J. Theoret. Phys. **49**(12), 3005–3021 (2010)
17. Baltag, A., Smets, S.: Correlated information: a logic for multi-partite quantum systems. Electron. Not. Theoret. Comput. Sci. **270**(2), 3–14 (2011)
18. Beltrametti, E., Chiara, M., Giuntini, R., Sergioli, G.: Quantum teleportation and quantum epistemic semantics. Math. Slovaca **62**(6), 1121–1144 (2012)
19. Chareton, C., Brunel, J., Chemouil, D.: Towards an updatable strategy logic. In: Proceedings of 1st International Workshop on Strategic Reasoning, SR, pp. 91–98 (2013)
20. Chareton, C., Brunel, J., Chemouil, D.: A logic with revocable and refinable strategies. Inf. Comput. **242**, 157–182 (2015)
21. Lichtenstein, O., Pnueli, A., Zuck, L.: The glory of the past. In: Parikh, R. (ed.) Logic of Programs 1985. LNCS, vol. 193, pp. 196–218. Springer, Heidelberg (1985). doi:10.1007/3-540-15648-8_16
22. Vardi, M.Y.: Reasoning about the past with two-way automata. In: Larsen, K.G., Skyum, S., Winskel, G. (eds.) ICALP 1998. LNCS, vol. 1443, pp. 628–641. Springer, Heidelberg (1998). doi:10.1007/BFb0055090
23. De Giacomo, G., Vardi, M.Y.: Linear temporal logic and linear dynamic logic on finite traces. In: Proceedings of 22nd IJCAI, pp. 854–860 (2013)
24. Eisner, C., Fisman, D., Havlicek, J., Lustig, Y., McIsaac, A., Campenhout, D.: Reasoning with temporal logic on truncated paths. In: Hunt, W.A., Somenzi, F. (eds.) CAV 2003. LNCS, vol. 2725, pp. 27–39. Springer, Heidelberg (2003). doi:10.1007/978-3-540-45069-6_3
25. Beltrametti, E., et al.: Epistemic quantum computational structures in a hilbert-space environment. Fundamenta Informaticae **115**, 1–14 (2012)
26. Bennett, C.H.: Quantum cryptography using any two nonorthogonal states. Phys. Rev. Lett. **68**(21), 3121–3124 (1992)
27. Bruß, D.: Optimal eavesdropping in quantum cryptography with six states. Phys. Rev. Lett. **81**(14), 3018–3021 (1998)

Enumerative Induction and Semi-uniform Convergence to the Truth

Hanti Lin(✉)

UC Davis, 1 Shields Ave, Davis, CA 95616, USA
ika@ucdavis.edu

Abstract. I propose a new definition of identification in the limit, also called convergence to the truth, as a new success criterion that is meant to complement, but not replace, the classic definition due to Putnam (1963) and Gold (1967). The new definition is designed to explain how it is possible to have successful learning in a kind of scenario that the classic account ignores—the kind of scenario in which the entire infinite data stream to be presented incrementally to the learner is *not* presupposed to completely determine the correct learning target. For example, suppose that a scientists is interested in whether all ravens are black, and that she will never observe a counterexample in her entire life. This still leaves open whether all ravens (in the universe) are black. From a purely mathematical point of view, the proposed definition of convergence to the truth employs a convergence concept that generalizes net convergence and sits in between pointwise convergence and uniform convergence. Two results are proved to suggest that the proposed definition provides a success criterion that is by no means weak: (i) Between the proposed identification in the limit and the classic one, neither implies the other. (ii) If a learning method identifies the correct target in the limit in the proposed sense, any U-shaped learning involved therein has to be essentially redundant. I conclude that we should have (at least) two success criteria that correspond to two senses of identification in the limit: the classic one and the one proposed here. They are complementary: meeting any one of the two is good; meeting both at the same time, if possible, is even better.

Keywords: Identification in the limit · Convergence to the truth · Enumerative induction · Uniform convergence · Net convergence

1 Introduction

The goal of this paper is to find a new definition of identification in the limit—a new success criterion that is meant to complement, but not replace, the classic definition due to Putnam (1963) and Gold (1967). Let me begin with a motivation. Theoretical computer science has a tradition:

(1) An important part of theoretical computer science is basically the science of *problem solving*; in this science, we define a number of *success criteria* for problem solving.

A. Baltag et al. (Eds.): LORI 2017, LNCS 10455, pp. 362–376, 2017.
DOI: 10.1007/978-3-662-55665-8_25

(2) If a difficult problem cannot be solved by meeting a high success criterion, we try to find out whether it can be solved by meeting a success criterion that is lower—or at least *not as high* if success criteria are partially ordered.
(3) If there is an interesting, tough problem that cannot be solved by meeting any success criterion we have defined, we try to find out whether it is possible to define a new criterion that is low enough to be met for solving the tough problem but, simultaneously, still high enough to deserve to be called a *success* criterion

The kind of situation described by (3) has happened a number of times in the history of theoretical computer science. Classic examples in learning theory include the following: When encountering an interesting problem that cannot be solved by a *decision* procedure, i.e. by an *exact* learning method, Valiant (1984) defined a new success criterion called *probable approximate correctness.* In a similar situation, Putnam (1963) and Gold (1967) defined a new success criterion called *identification in the limit.* I suspect that we are now in a similar situation, too. Let me illustrate with an example.

Consider the following empirical problem. An inquirer is wondering whether all ravens are black. She is to collect more and more ravens, observe their colors, and put forward conjectures after each observation. In standard learning theory, it is typically presupposed that if there are nonblack ravens, then they will be observed by the inquirer sooner or later. With this presupposition, we have **the easy raven problem**, for which there are learning methods that are guaranteed to identify the truth in the limit, meeting the classic success criterion due to Putnam and Gold. But what if that presupposition is not made? Then we have **the hard raven problem**, in which two globally indistinguishable possibilities are both on the table: given that the inquirer only observes ravens that are black in her entire life, it might be that all ravens (in the universe) are black, and it might be that some ravens are not black. The former possibility is a happy possible world; the latter, unfortunate. Nature need not be malicious to make the unfortunate world actual; all it takes is just bad luck, and the bad luck could be generated, for example, by an unknown deterministic process, or by an unknown probabilistic process that involves a sequence of random variables that might or might not be independent and identically distributed. For the hard raven problem, there is no learning that meets the classic success criterion. Then what to do?

I tend to think that the hard raven problem is very interesting. (How interesting is it? More on this below.) If so, by tradition (1)–(3), we should try to look for an alternative to the classic success criterion. Furthermore, the task of *at least trying* to do so is not daunting at all. Indeed, Putnam and Gold only employ the most familiar concept of limit and convergence, namely convergence of a sequence. But in analysis and topology mathematicians have already worked out more elaborated concepts of convergence (and put them to good work):

- convergence of a *sequence* can be generalized to convergence of a so-called *net*;

- convergence of a sequence of functions comes in two flavors: *pointwise* convergence and *uniform* convergence.

The classic success criterion basically requires that, in a sense to be explained in this paper, the learner's convergence to the correct target be "pointwise" and "sequence-like". And, if I am right, it has a very natural variant, which requires that the learner's convergence be "semi-uniform" and "generalized-net-like". Once this idea is developed in rigorous mathematical terms (Definition 6), I will be able to show that this new criterion of identification in the limit is achievable for the hard raven problem (Proposition 2). Two further results are proved to suggest that this new criterion is still high and demanding enough to deserve to be called a success criterion: First, the new criterion is neither stronger nor weaker than the classic criterion (Theorem 1). Second, if the learner can ever solve the problem in the limit in the new sense, then in principle she can do it in pretty much the same way without U-shaped learning (Theorem 2)—namely, without accepting a hypothesis, retracting it, and coming to accept it again. The conclusion to be drawn is that the new success criterion and the classic one are complementary: meeting any one of them is good; meeting both at the same time is even better. In the unfortunate event that only one can be met, let us meet at least one.

The hard raven problem is interesting for a number of reasons. First, it is a paradigmatic case in which enumerative induction is employed. Second, scientists seem to proceed without the presupposition that they are not living in the unfortunate world, or, even if they make that presupposition, they do it only tacitly and probably unconsciously. Whether or not scientists can be said to tacitly make that presupposition, it is interesting to see if they can *do without it.*

It might be worried that the hard raven problem is not really interesting because of the following Kantian objection: *"In order for us to think that the success of science is possible, we have to presuppose that the kind of global underdetermination mentioned above does not hold in the world we actually live in. And, if we presuppose so, we will fall back to the easy raven problem, making the hard problem irrelevant. Granted, this sounds like wishful thinking. But the point is that we have no alternative but to presuppose so—if we really want to think that the success of science is possible."* I used to be sympathetic to this Kantian line of thought until I recognized that it presupposes a lot: as a claim about the possibility of success, it tacitly presupposes that there is no more success criterion than those we have already formulated and put on the table. I propose that we challenge this tacit presupposition, look for a new success criterion, and hold on to the tradition (1)–(3) in theoretical computer science. This paper shows how to do that.

2 Classic Convergence to the Truth

This section presents the Putnam-Gold-style learning model in a very general setting, and defines the hard raven problem. The presentation might appear more complicated than necessary—just look at how long the first definition right below is! But we will gradually see its value: it reveals the moving parts of the classic definition of identification in the limit and, thereby, suggests possible new definitions.

Definition 1 (Problem). A *(learning) problem* is a 5-tuple $\mathcal{P} = (\mathcal{H}, \mathcal{I}, \leq, \mathcal{W}, |\cdot|)$ that satisfies the following conditions (accompanied by interpretations):

- $\mathcal{H}$ is a nonempty set of *hypotheses*, understood as certain theories, languages, concepts, or whatever targets to learn or choose from.
- $\mathcal{I}$ is a nonempty set of *information states*, which are possible inputs into a learning method.
- $\leq$ is a partial order over $\mathcal{I}$. Understand $i \leq i'$ to say that a learner might go from information state i to i' and, when she does, she has more or the same information.
- $\mathcal{W}$ is a nonempty set of possible states of the world, or *possible worlds* for short. $\mathcal{W}$ need not contain all logically possible worlds, but contains exactly those that are logically compatible with the presupposition of the problem—that is, $\mathcal{W}$ represents the content of the presupposition.
- $|\cdot|$ is an *(interpretation) function* defined on $\mathcal{I} \cup \mathcal{H}$, mapping all $i \in \mathcal{I}$ and all $h \in \mathcal{H}$ to subsets of $\mathcal{W}$. $|i|$ is understood to contain exactly the possible worlds in $\mathcal{W}$ at which i is true. Similarly, $|h|$ is understood to contain exactly the possible worlds in $\mathcal{W}$ at which h is true.
- For all $h, h' \in \mathcal{H}$ and $i, i' \in \mathcal{I}$, the following axioms hold:
  - (Nonemptiness) $|h|$ and $|i|$ are nonempty.
  - (Disjointness) $h \neq h'$ implies $|h| \cap |h'| = \varnothing$. (That is, distinct hypotheses are mutually incompatible, competing with each other.)
  - (Monotonicity) $i \leq i'$ implies $|i| \supseteq |i'|$. (That is, there is no loss of information in the state transition from any i to any $i' \geq i$.)
  - (Linearity) Let $\mathcal{I}(w)$ denote $\{i \in \mathcal{I} : w \in |i|\}$, the set of information states that are true at possible world w. Then each partially ordered set $(\mathcal{I}(w), \leq)$ is linear and not longer than the sequence $\omega = \{0, 1, 2, \ldots\}$ of natural numbers. (This linear order is basically the data stream to be presented in possible world w.)

All results in this paper are actually proved in a more general setting, which weakens the axiom of linearity to an axiom called *directedness*.[1] But the above definition is already general enough to capture all Gold-style language learning problems and the easy and hard raven problems.

[1] (Directedness) Each partially ordered set $(\mathcal{I}(w), \leq)$ is directed, namely $i, j \in \mathcal{I}_w$ implies $i, j \leq k$ for some $k \in \mathcal{I}_w$.

Definition 2 (Method). A *(learning) method* for a problem $\mathcal{P} = (\mathcal{H}, \mathcal{I}, \leq, \mathcal{W}, |\cdot|)$ is a function $M : \mathcal{I} \to \mathcal{H} \cup \{?\}$, which is interpreted as follows:[2]

$M(i) = h$ means that M chooses hypothesis h given information/input i;
$M(i) = ?$ means that M suspends judgment given information/input i.

Definition 3 (Classic Identification in the Limit). A method M for a problem $\mathcal{P} = (\mathcal{H}, \mathcal{I}, \leq, \mathcal{W}, |\cdot|)$ is said to *classically converge to the truth*—or *classically identify the correct target in the limit*—just in case:

(1) for each hypothesis $h \in \mathcal{H}$,
(2) for each possible world $w \in |h|$,
(3) there exists an information state $i \in \mathcal{I}(w)$ such that
(4) $M(i') = h$ for all $i' \geq i$ in $\mathcal{I}(w)$.

If a problem admits of such a learning method, it is said to be *classically solvable*—or *learnable*—*in the limit.*[3]

This definition is formulated in a way meant to reveal its moving parts: the order of (1)–(4), and the respective roles played by the five components in a learning problem $(\mathcal{H}, \mathcal{I}, \leq, \mathcal{W}, |\cdot|)$. In Sect. 3, I will exchange the order of (2) and (3) to generate a kind of "semi-uniform" convergence, and then modify here and there to result in the new success criterion promised above.

Imagine a scientist who wonders whether all ravens are black; there are two potential answers: `yes` and `no`. She is going to observe a raven at a time. 0 represents an event of observing a black raven; 1, a nonblack raven. An information state is a binary sequence of finite length. For example, $(0, 0, 1)$ denotes the information state at which the scientist has observed 2 black ravens followed by a nonblack raven. Let $s = (s_1, s_2, \ldots)$ be a binary sequence of infinite length. In a possible world denoted by $(\texttt{yes}, s)$, all ravens are black and the scientist receives data in the order of $s_1, s_2, \ldots$, one at a time—so s only contains 0. Similarly, in a possible world denoted by $(\texttt{no}, s)$, not all ravens are black and the scientist receives data in the order of $s_1, s_2, \ldots$, but this time s can be any infinite binary sequence. For example, let $(0, 0, \ldots, 0, \ldots)$ denote the infinite sequence of 0s. Then $(\texttt{no}, (0, 0, \ldots, 0, \ldots))$ is the unfortunate world in which not all ravens are

[2] For the problems and learning methods defined here to be interesting in computer science, we need to require that the hypotheses in $\mathcal{H}$ and the information states in $\mathcal{I}$ can in principle be encoded by natural numbers. But the results in this paper hold generally whether or not we add this requirement. Furthermore, this definition can be generalized by allowing a learning method to output not just hypotheses in $\mathcal{H}$ but also their Boolean combinations.

[3] Definitions 1–3 are essentially the order-theoretic counterparts of the topologically formulated definitions proposed in Baltag et al. (2015) and in Kelly et al. (2016), provided that we include the generalizations mentioned in preceding footnotes: allowing $(\mathcal{I}(w), \leq)$ to be directed, and allowing a learning method to output Boolean combinations of hypotheses.

black but the scientist will never observe a counterexample in her entire life. Let $\mathcal{W}_{\text{All}}$ denote the set of all possible worlds just defined, or formally:

$$\mathcal{W}_{\text{All}} = \{(\texttt{yes}, (0, 0, \ldots, 0, \ldots))\} \cup \{(\texttt{no}, s) : s \in \{0, 1\}^{\omega}\}.$$

Example 1 (Easy Raven Problem)**.** The easy raven problem $\mathcal{P} = (\mathcal{H}, \mathcal{I}, \leq, \mathcal{W}, |\cdot|)$ is defined by:

- $\mathcal{H} = \{\texttt{yes}, \texttt{no}\}$.
- $\mathcal{I}$ = the set of all binary sequences of finite length.
- $i \leq i'$ iff i is extended by i', for all sequences i and $i' \in \mathcal{I}$.
- $\mathcal{W} = \mathcal{W}_{\text{All}} \smallsetminus \{(\texttt{no}, (0, 0, \ldots, 0, \ldots))\}$, which excludes the unfortunate possible world.
- $|i|$ = the set of possible worlds (h, s) in $\mathcal{W}$ such that s extends i.
  $|\texttt{yes}|$ = the set of possible worlds (h, s) in $\mathcal{W}$ such that $h = \texttt{yes}$.
  $|\texttt{no}|$ = the set of possible worlds (h, s) in $\mathcal{W}$ such that $h = \texttt{no}$.

So this problem presupposes that the scientist does not live in the unfortunate world, and hence that each infinite binary sequence that might be presented to the scientist uniquely determines the true hypothesis.

Example 2 (Hard Raven Problem)**.** The hard raven problem is the same as the easy problem except that:

- $\mathcal{W} = \mathcal{W}_{\text{All}}$, which does contain the unfortunate possible world $(\texttt{no}, (0, 0, \ldots, 0, \ldots))$.

Then we have the following result:

Proposition 1. *The easy raven problem is classically learnable in the limit, but the hard one is not.*

Proof. The claim about the easy raven problem is well known. The non-learnability result of the hard raven problem follows from the fact that any method M is doomed to have this property: $M(i)$ converges to the truth as i travels in the data stream in possible world $(\texttt{yes}, (0, 0, \ldots, 0, \ldots))$ iff it fails to do so in possible world $(\texttt{no}, (0, 0, \ldots, 0, \ldots))$. □

So, given this result and the discussion in the introduction, we should look for a new success criterion that complements the classic one.

3 A New Convergence Concept in Learning Theory

The success criterion to be proposed is based on a new convergence concept. This section will provide the formal definition with a motivating line of thought, which explores possible combinations and/or generalizations of the ideas that lie behind convergence.

The classic definition of identification in the limit can be understood as an implementation of the following informal template:

A. (Pointwise Convergence to the Truth)
No matter which hypothesis $h \in \mathcal{H}$ is true,
no matter which world $w \in |h|$ is actual,
there exists a concept SI of "sufficient information" associated with w such that,
M outputs the truth h as long as the input i is "sufficiently informative" (i.e. in SI) and true at w.

This template gives rise to the classic definition when so-called "concepts of sufficient information associated with w" are defined to be principal upper subsets of $(\mathcal{I}(w), \leq)$. For your reference, a subset S of $(I, \leq)$ is *upper* just in case it has the upward-closure property: for all $i, i' \in I$, if $i \in S$ and $i \leq i' \in I$, then $i' \in S$. A *principal upper* subset of a poset $(I, \leq)$ is a set S that takes this form: $S = \{i' \in I : i \leq i'\}$, for some $i \in I$.

The above template is called "pointwise" because each world—or point—is associated with a concept of sufficient information, i.e. a criterion for convergence. Once this is seen, it is very natural to seek a "more uniform" kind of convergence by exchanging the order of the quantifications over w and over SI. Then we have:

B. (Semi-uniform Convergence to the Truth)
No matter which hypothesis $h \in \mathcal{H}$ is true,
there exists a concept SI of "sufficient information" associated with h such that,
no matter which world $w \in |h|$ is actual,
M outputs the truth h as long as the input i is "sufficiently informative" (i.e. in SI) and true at w.

Note how the existential quantifier is modified: SI is required to be a concept of sufficient information that is associated with, not w, but h. What does that mean? It means that SI has to be a subset of, not $\mathcal{I}(w)$, but $\mathcal{I}(h)$, which is defined as the set of information states that might be true given hypothesis h, namely:

$$\mathcal{I}(h) = \bigcup_{w \in |h|} \mathcal{I}(w).$$

Note that, although $(\mathcal{I}(w), \leq)$ is assumed to be linear, $(\mathcal{I}(h), \leq)$ could be a mere poset. Then the above two informal templates A and B can be made "almost formal" as follows, pending the two blanks to be filled in.

Definition Template 4. The following conditions are two templates for defining *convergence to the truth* or *identification in the limit*:

A. (Pointwise Convergence to the Truth)
For each hypothesis $h \in \mathcal{H}$,
for each possible world $w \in |h|$,
there exists a ______ upper subset SI of linear poset $(\mathcal{I}(w), \leq)$ such that $M(i) = h$ for each information state $i \in SI$ (that is true at w).

B. (Semi-uniform Convergence to the Truth)
For each hypothesis $h \in \mathcal{H}$,
 there exists a ______ upper subset SI of poset $(\mathcal{I}(h), \leq)$ such that,
  (for each possible world $w \in |h|$,)
   $M(i) = h$ for each information state $i \in SI$ (that is true at w).

The parts enclosed in parentheses are redundant, which can be easily proved from the definitions of learning problems and $\mathcal{I}(\cdot)$. But it is important to keep them here—to emphasize the parallel to the informal templates. Note that the term 'upper' has been built into the templates, and rightly so. For if a state i is more informative than some state that has "sufficient information" then i itself should also be taken as a state that has "sufficient information".

It remains to fill in the blank in template B—and this is the crux. In template B, each hypothesis h is required to be associated with a concept SI of sufficient information, which is a subset of $\mathcal{I}(h)$ and serves as a "convergence zone" in which learning methods are required to output the true hypothesis. This suggests that we do *not* want to fill in the blank in template B with 'principal'. Here is why: if a convergence zone in a *partially ordered* set $(\mathcal{I}(w), \leq)$ is allowed to be a principal upper subset, then in effect a convergence zone is allowed to be a *narrow* cone in a *wide* ambient poset $(\mathcal{I}(w), \leq)$. This would make the convergence criterion very weak, too weak to result in what deserves to be called a success criterion. So, for template B, we need to fill in the blank in such a way that requires a convergence zone SI to be reasonably wide.

So I propose that we fill in the blank with 'cofinal', which is defined in order theory as follows:

Definition 5 (Cofinal Subset). A subset S of a poset $(I, \leq)$ is *cofinal* just in case, for each $i \in I$, there exists $i' \in S$ with $i \leq i'$.

A cofinal subset is almost as "wide" as its ambient poset. Furthermore, the concept of cofinal subsets also serves to capture an intuitive idea about sufficient information: *whatever information i we have, in principle it should be always possible for us to obtain more information and (come to) have sufficient information i'*. Now, putting things together, we have:

Definition 6 (Semi-uniform Convergence to the Truth). A method M for a problem $\mathcal{P} = (\mathcal{H}, \mathcal{I}, \leq, \mathcal{W}, |\cdot|)$ is said to *semi-uniformly converge to the truth*—or *semi-uniformly identify the correct target in the limit*—just in case it satisfies the following implementation of template B:

for each hypothesis $h \in \mathcal{H}$,
 there exists a cofinal upper subset SI of $(\mathcal{I}(h), \leq)$ such that,
  (for each possible world $w \in H$,)
   $M(i) = h$ for each information state $i \in SI$ (that is true at w).

If a problem admits of such a learning method, it is called *solvable*—or *learnable—in the limit semi-uniformly.*

This proposal is actually based on a new convergence concept that generalizes standard convergence concepts in analysis and topology. The concept of sequence convergence, which is familiar in analysis, makes use of **principal upper subsets as convergence zones in a linear poset**. Now, by generalizing from 'linear' to 'directed', we obtain net convergence, which is the right convergence concept to employ in general topology. Now, the concept of net convergence uses **principal upper subsets as convergence zones in a directed poset**. That is provably equivalent to using **cofinal upper subsets as convergence zones in a directed poset** (by Proposition 3 in Appendix A). As a further step of generalization, I propose to drop 'directed' and work with any poset. The result is to use **cofinal upper subsets as convergence zones in a poset**, which is exactly the convergence concept that underlies the new, proposed definition of convergence to the truth. See Appendix A for a more rigorous presentation of convergence concepts in analysis and topology.

Now return to learning theory. We have:

Proposition 2 (Hard Raven Problem Solved Semi-uniformly). *The hard raven problem is solvable in the limit semi-uniformly.*

Proof. Consider this learning method: "Say yes when you have not seen any nonblack raven; otherwise say no." This method converges to the truth semi-uniformly for the following two reasons. First, concerning output yes, this method has a "convergence zone" $SI = \mathcal{I}(\texttt{yes})$, which is (trivially) a cofinal upper subset of $(\mathcal{I}(\texttt{yes}), \leq)$. Second, concerning output no, this method has a "convergence zone" $SI' = \{i \in \mathcal{I} : i \text{ extends } 0^n1 \text{ for some } n \in \omega\}$, which is a cofinal upper subset of $(\mathcal{I}(\texttt{no}), \leq)$. This is because $\mathcal{I}(\texttt{no}) = \mathcal{I}$ and any sequence in $\mathcal{I}$ either takes the form of 0^n, which can be extended to a sequence 0^n1 in SI', or takes the form of $0^n1\cdots$, which is already in SI'. □

The learning method employed in the above proof fails to identify the truth in the unfortunate world $(\texttt{no}, (0, 0, \ldots, 0, \ldots))$. Indeed, any method that solves the hard raven problem in the limit semi-uniformly has to fail to do so.[4]

So it might be worried that semi-uniform convergence to the truth is just a way of making the hard raven problem solvable by "ignoring" the unfortunate world, while there are other ways of ignoring worlds. Perhaps the simplest way of ignoring worlds is to adopt the "presupposition" strategy: to presuppose that the inquirer is not living in the unfortunate world and, hence, to change the hard raven problem into the easy raven problem. So, the worry continues, why not just adopt the presupposition strategy?

My reply has two parts, depending on what presupposition really is. First, if presupposition is not something that an inquirer can freely make but is given to the inquirer by the context (such as the context in which the inquirer talks

[4] *Sketch of Proof.* Suppose that M solves the hard raven problem in the limit semi-uniformly. Then, since $\mathcal{I}(\texttt{yes})$ is the set of all finite sequences of 0s linearly ordered by extension, M has to identify the truth yes in the limit at world $(\texttt{yes}, (0, 0, \ldots, 0, \ldots))$. So M has to fail to identify the truth no in the limit at world $(\texttt{no}, (0, 0, \ldots, 0, \ldots))$.

to a certain skeptic), then the presupposition strategy does not make sense. Second, if the inquirer can freely add a presupposition and thereby make the problem easier, then it is not clear why she has to change the problem by making a presupposition. She could, instead, entertain a slightly different but closely related set of competing hypotheses, such as "All ravens observed in my entire life are black" and its negation. In general, a hard problem can be changed into multiple easier ones—but then which one to change into? I propose that, if the inquirer is really interested in addressing the hard raven problem, she should stick to it rather than changing the problem unless there is no achievable success criterion. Is there an achievable success criterion? Yes. Semi-uniform convergence to the truth is achievable for the hard raven problem and, as I shall argue in the next section, it deserves to be taken as a success criterion.

4 Why It's a New Success Criterion

This section provides two reasons for thinking that the new convergence criterion is high enough—demanding enough—to deserve to be called a *success* criterion. Those two reasons are rooted in the two main theorems of this paper.

The first theorem says that, between the two criteria of identification in the limit (the semi-uniform and the classic), neither is strictly more demanding than the other. The following example will be used to prove this claim.

Example 3 (Even-vs-Odd Raven Problem)**.** Now the scientist wonders whether the number of nonblack ravens in the world is even or odd, while presupposing that this number is finite and that data presentations are complete, uniquely determining the truth (as in the easy raven problem). The even-vs-odd raven problem is formally defined as follows:

- $\mathcal{H} = \{\texttt{even}, \texttt{odd}\}$.
- $\mathcal{I} =$ the set of all binary sequences of finite length (as usual).
- $i \leq i'$ iff i is extended by i', for all sequences i and $i' \in \mathcal{I}$ (as usual).
- $\mathcal{W} = \{s \in \{0,1\}^\omega :$ the occurrences of 1 in s are finite in number$\}$
- $|i| = \{s \in \mathcal{W} : s$ extends $i\}$.
  $|\texttt{even}| = \{s \in \mathcal{W} :$ the occurrences of 1 in s are even in number$\}$.
  $|\texttt{odd}| = \{s \in \mathcal{W} :$ the occurrences of 1 in s are odd in number$\}$.

Theorem 1 (Independence Result). *Semi-uniform learnability in the limit and its classical counterpart are independent of each other; that is, neither implies the other. For example, consider the three raven problems defined above: the hard, the easy, and the even-vs-odd. Their respective learnability in the limit is summarized by the following table:*

Raven Problems:	*Hard*	*Easy*	*Even-vs-Odd*
Classically Learnable in the limit?	*No*	***Yes***	***Yes***
Semi-uniformly Learnable in the limit?	***Yes***	***Yes***	*No*

Proof. Thanks to Proposition 2, it suffices to prove the part for the even-vs-odd raven problem. Suppose, for *reductio*, that some learning method M for the even-vs-odd raven problem solves it in the limit semi-uniformly. So there exists a cofinal upper subset SI of $(\mathcal{I}(\texttt{even}), \leq)$ such that $M(i) = \texttt{even}$ for all $i \in SI$. Similarly, there exists a cofinal upper subset SI' of $(\mathcal{I}(\texttt{odd}), \leq)$ such that $M(i) = \texttt{odd}$ for all $i \in SI'$. But here is the key: $\mathcal{I}(\texttt{even}) = \mathcal{I}(\texttt{odd}) = \mathcal{I}$. So, by the order-theoretic result that two cofinal upper subsets SI and SI' of a nonempty poset $(\mathcal{I}, \leq)$ share at least one common element i^*, we have that $M(i^*) = \texttt{even} = \texttt{odd}$, contradiction. The classical learnability of the even-vs-odd problem is witnessed by this learning method: "Say even when you have seen an even number of nonblack ravens; otherwise say odd." □

The second theorem concerns a kind of learning process that cognitive scientists and computational learning theorists call *U-shaped learning* (Carlucci et al. 2005, 2013). U-shaped learning consists of accepting a hypothesis, then retracting it, and then accepting it again:

Definition 7 (U-Shaped Learning). Let M be a learning method for a problem $\mathcal{P} = (\mathcal{H}, \mathcal{I}, \leq, \mathcal{W}, |\cdot|)$. An *instance of U-shaped learning* of M is a sequence (i_1, i_2, i_3) of three information states in $\mathcal{I}$ such that:

1. $i_1 \leq i_2 \leq i_3$,
2. $M(i_1) \neq M(i_2) \neq M(i_3)$,
3. $M(i_1) = M(i_3) \in \mathcal{H}$.

Following Kelly et al. (2016), I submit that such a learning process is epistemically undesirable. So, other things being equal, it had better be avoided or removed whenever possible. Removal of U-shaped learning is defined as follows:

Definition 8 (Removal of U-Shaped Learning). For any learning methods M and M^*, say that M^* can be constructed from M by *removing U-shaped learning* just in case:

1. $M(i) = ?$ implies $M^*(i) = ?$,
2. $M(i) = h$ implies $M^*(i) = h$ or $?$,
3. M^* has no instance of U-shaped learning.

Then we have:

Theorem 2 (Persistence under Removal of U-Shaped Learning). *A method M solves a problem $\mathcal{P}$ in the limit semi-uniformly only if a method M^* can be constructed from M by removing U-shaped learning such that M^* also solves $\mathcal{P}$ in the limit semi-uniformly.*

Proof. Suppose that method M identifies the correct target for problem $\mathcal{P} = (\mathcal{H}, \mathcal{I}, \leq, \mathcal{W}, |\cdot|)$ in the limit semi-uniformly. Then, for each hypothesis $h \in \mathcal{H}$, $(\mathcal{I}(h), \leq)$ has a cofinal upper subset SI_h such that, for all $i \in SI_h$, $M(i) = h$. Construct a learning method M^* from M as follows:

$$M^*(i) = \begin{cases} h & \text{if } i \in SI_h \\ ? & \text{otherwise.} \end{cases}$$

By construction, M^* solves the same problem in the limit semi-uniformly. And, by construction again, it satisfies the first two of the three conditions that jointly define "constructability by removing U-shaped learning". So it suffices to show that M^* satisfies the last of the three conditions, namely that M^* has no instance of U-shaped learning. Suppose, for *reductio*, that M^* has an instance (i_1, i_2, i_3) of U-shaped learning. Then there exists $h^* \in \mathcal{H}$ such that $M^*(i_1) = M^*(i_3) = h^*$. So, by construction, both i_1 and i_3 are in SI_{h^*}. It follows that $i_3 \in \mathcal{I}(h^*)$. So, by the definition of $\mathcal{I}(\cdot)$, there exists $w^* \in \mathcal{W}$ such that $w^* \in |i_3| \cap |h^*|$. Furthermore, since (i_1, i_2, i_3) is an instance of U-shaped learning of M^*, we have $i_1 \leq i_2 \leq i_3$. So, by MONOTONICITY, we have $|i_1| \supseteq |i_2| \supseteq |i_3|$. It follows that $|i_1| \supseteq |i_2| \supseteq |i_3| \ni w^* \in |h^*|$. So $w^* \in |i_2| \cap |h^*|$, and hence $i_2 \in \mathcal{I}(h^*)$. To summarize what we have obtained so far: SI_{h^*} contains i_1 and is an upper subset of $(\mathcal{I}(h^*), \leq)$, and $i_1 \leq i_2 \in \mathcal{I}(h^*)$. It follows from upward closure that SI_{h^*} contains i_2, too. So, by construction, $M^*(i_1) = M^*(i_2) = M^*(i_3) = h^*$. But this contradicts the *reductio* hypothesis that (i_1, i_2, i_3) is an instance of U-shaped learning of M^*. □

So semi-uniform identification in the limit seems to set a standard for problem solving that is by no means low: if a learning method achieves this standard, it has to do so *without* involving U-shaped learning in any essential, unremovable way. We may say that semi-uniform identification in the limit has the property called *persistence under removal of U-shaped learning*. By way of contrast, this property is not shared by classic identification in the limit. Here is a proof sketch: U-shaped learning is provably involved in *all* learning methods that solve the even-vs-odd raven problem in the limit classically. So, when we tackle this problem, removal of U-shaped learning implies no longer having a method that solves it in the limit classically.

Thanks to the above two theorems, semi-uniform identification in the limit is by no means a low criterion. I submit that it is high enough to deserve to be called a success criterion for problem solving. To clarify: I am *not* saying that we should always be satisfied with a solution to a problem that meets only the new success criterion. If a problem can be solved by meeting semi-uniform identification in the limit because it can be solved by meeting a strictly higher criterion—such as the conjunction of classical and semi-uniform identifications in the limit—then we ought to strive for meeting the higher criterion. A success criterion marks an achievement, but does not necessarily mark an achievement that we should be satisfied with. We ought to strive for meeting the highest achievable success criterion.

I also submit that the two kinds of identification in the limit, the classic and the semi-uniform, are *complementary* criteria of success. They set very similar standards of success, but nonetheless each describes a unique way of success—so unique that meeting one does not entail meeting the other. *Meeting any one of these two is good; meeting both at the same time, if possible, is even better.*[5] This is the sense in which these two success criteria are complementary.

That said, it might be possible for their relationship to turn competitive occasionally. Imagine that we have a problem and there are learning methods that solve it in the limit—some do it classically, some do it semi-uniformly, but no single one can do it both classically and semi-uniformly. This raises two questions. *Does there exist such a problem? If yes, which of the two success criteria should be met at the cost of the other?* I do not have an answer at the moment.

Acknowledgements. I thank Kevin Kelly, Konstantin Genin, and two anonymous referees for their helpful comments and suggestions.

A Convergence Generalized

As mentioned in the introduction, the classic definition of identification in the limit only employs the most familiar concept of convergence—pointwise convergence of sequences—while in analysis and topology mathematicians have already worked out more elaborated concepts of convergence. This section provides a quick review of those concepts.

Definition 9 (Sequence Convergence). Let $f : \omega \to Y$ be a sequence of points in a topological space Y. Say that $f(n)$ *converges* to y as n travels upward in $(\omega, \leq)$ just in case:

for any open neighborhood U of y in Y,
there exists $n \in \omega$ such that
$f(m) \in U$ for every $m \geq n$ in ω.

The next step is to go from convergence of a sequence to convergence of a "generalized" sequence, a.k.a. "net". A sequence is defined on a very special set of indices, namely ω, linearly ordered by $\leq$. Let us have a more general set I of indices with a weaker ordering structure:

Definition 10 (Directed Poset). A *poset* $(I, \leq)$ consists of a set I partially ordered by $\leq$. It is *directed* just in case $i, j \in I$ implies $i, j \leq k$ for some $k \in I$.

Definition 11 (Generalized Sequence, or Net). A *generalized sequence* or *net* $f : (I, \leq) \to Y$, is a function from a nonempty directed poset $(I, \leq)$ to a topological space Y.

[5] For example, consider the following problem $\mathcal{P} = (\mathcal{H}, \mathcal{I}, \leq, \mathcal{W}, |\cdot|)$ with $\mathcal{W} = \{w_0, w_1, w_2\}$, $\mathcal{I} = \{0, 1, 2\}$, $0 < 1 < 2$, $|0| = \{w_0, w_1, w_2\}$, $|1| = \{w_1, w_2\}$, $|2| = \{w_2\}$, $\mathcal{H} = \{H_{\text{even}}, H_{\text{odd}}\}$, $|H_{\text{even}}| = \{w_0, w_2\}$, $|H_{\text{odd}}| = \{w_1\}$. Consider this method $M : 0 \mapsto ?, 1 \mapsto H_{\text{odd}}, 2 \mapsto H_{\text{even}}$. This solves the problem in the limit semi-uniformly. But we should not be satisfied with this method, for there is a better one: $M' : 0 \mapsto H_{\text{even}}, 1 \mapsto H_{\text{odd}}, 2 \mapsto H_{\text{even}}$, which solves the problem both semi-uniformly and classically. I thank Konstantin Genin for bringing this example to my attention.

Definition 12 (Net Convergence). Let $f : (I, \leq) \to Y$ be a net. Say that $f(i)$ *converges* to y as i travels in $(I, \leq)$ just in case:

for any open neighborhood U of y in Y,
there exists i in I such that
$f(i') \in U$ for every $i' \geq i$ in I.

Note that the definition of net convergence is formally identical to that of sequence convergence, except that the underlying spaces of indices are generalized. Net convergence has a number of equivalent formulations, some of which are very sophisticated (and stated in terms of, say, filters).[6] Let me introduce the following one:

Proposition 3 (Net Convergence Redefined by "Cofinal Upper"). *Let $f : (I, \leq) \to Y$ be a net. Then net convergence of f can be equivalently redefined in terms of cofinal upper subsets as follows. $f(i)$ converges to y as i travels in $(I, \leq)$ if and only if:*

- *for any open neighborhood U of y,*
  there exists a cofinal upper subset S of $(I, \leq)$ such that
  $f(i) \in U$ for all $i \in S$.

Proof. The ($\Rightarrow$) side follows immediately from the order-theoretic result that every principal upper subset of a nonempty directed poset is cofinal. The ($\Leftarrow$) side follows immediately from the order-theoretic result that every cofinal upper subset of a nonempty directed poset is nonempty (because of cofinality) and thus (by upward closure) includes a principal upper subset. □

With the above formulation of net convergence, we can generalize further by relaxing the restriction to directed posets and allowing any nonempty posets:

Definition 13 (Generalized Convergence and Limit). Let f be a function from a nonempty poset $(I, \leq)$ to an arbitrary topological space Y. Say that $f(i)$ *converges* to y as i travels in $(I, \leq)$ (in the sense of *generalized convergence*) just in case:

for any open neighborhood U of y,
there exists a cofinal upper subset S of $(I, \leq)$ such that
$f(i) \in U$ for all $i \in S$.

If such a y exists uniquely, also say that y is the *generalized limit* of $f(i)$ as i travels in $(I, \leq)$.

The above is the concept of convergence that underlines the definition of semi-uniform identification in the limit: the index i travels in a *partially ordered set* I in general, a convergence zone is required to be cofinal and upper, and the underlying topology of the codomain is discrete.

[6] See Kelley (1991) for a review.

References

Baltag, A., Gierasimczuk, N., Smets, S.: A study in epistemic topology. In: Ramanujam, R. (ed.) Proceedings of the 15th Conference on Theoretical Aspects of Rationality and Knowledge (TARK 2015). ACM 2015 (ILLC Prepublication Series PP-2015-13) (2015)

Carlucci, L., Case, J.: On the necessity of U-shaped learning. Topics Cogn. Sci. **5**(1), 56–88 (2013)

Carlucci, L., Case, J., Jain, S., Stephan, F.: Non u-shaped vacillatory and team learning. In: Jain, S., Simon, H.U., Tomita, E. (eds.) ALT 2005. LNCS, vol. 3734, pp. 241–255. Springer, Heidelberg (2005). doi:10.1007/11564089_20

Gold, E.M.: Language identification in the limit. Inf. Control **10**(5), 447–474 (1967)

Kelley, J.L.: General Topology. Springer, New York (1991)

Kelly, T.K., Genin, K., Lin, H.: Realism, rhetoric, and reliability. Synthese **193**(4), 1191–1223 (2016)

Putnam, H.: Degree of confirmation and inductive logic. In: Schilpp, P.A. (ed.) The Philosophy of Rudolf Carnap. Open Court, La Salle (1963)

Valiant, L.: A theory of the learnable. Commun. ACM **27**(11), 1134–1142 (1984)

How to Make Friends: A Logical Approach to Social Group Creation

Sonja Smets and Fernando R. Velázquez-Quesada(✉)

Institute for Logic, Language and Computation, Universiteit van Amsterdam, Amsterdam, The Netherlands
{S.J.L.Smets,F.R.VelazquezQuesada}@uva.nl

Abstract. This paper studies the logical features of social group creation. We focus on the mechanisms which indicate when agents can form a team based on the correspondence in their set of features (behavior, opinions, etc.). Our basic approach uses a semi-metric on the set of agents, which is used to construct a network topology. Then it is extended with epistemic features to represent the agents' epistemic states, allowing us to explore group-creation alternatives where what matters is not only the agent's differences but also what they know about them. We use tools of dynamic epistemic logic to study the properties of different strategies to network formations.

1 Introduction

It is commonly accepted that our social contacts affect the way we form our opinions about the world. Think, e.g., about socialization (inheriting and disseminating norms, customs, values and ideologies), conformity (changing our attitudes, beliefs and behaviors to match those of others), peer pressure and obedience. These phenomena have been studied not only by empirical sciences (e.g., sociology and social psychology: [1]) but also by theoretical computer science and economy [2]. Within the logic community, epistemic social phenomena have been studied with a diversity of logical tools. Since the birth of dynamic epistemic logic in the late eighties and ninetees, models were first designed to reason about agent's epistemic states in multi-agent environments, and the social dimension has gradually received more attention. As examples we mention the work on communication networks and protocols [3–7], belief change in social networks [8], the analysis of peer pressure [9], the study of informational cascades [10], priority-based peer influence [11], reflective social influence in [12] and the study of diffusion and prediction update in [13]. Still, while the structure of social groups plays an important role in these logical studies, the way the groups are created has till now received much less attention.

This paper focuses on the logical structure behind the creation of social networks. Our basic mechanism for group-creation focusses on agents who become socially connected when the number of features in which they differ is small enough. In line with this idea we propose several group-creation policies, exploring the properties of the resulting networks. In Sect. 2, we introduce a similarity

A. Baltag et al. (Eds.): LORI 2017, LNCS 10455, pp. 377–390, 2017.
DOI: 10.1007/978-3-662-55665-8_26

update operation which generates new reflexive and symmetric social networks. We then discuss alternatives that produce irreflexive and not necessarily symmetric variations. After this, we introduce a version that asks for the agents not only to be 'close enough', but also for the existence of a middleman who can 'connect' them. In Sect. 3, we extend our setting with an epistemic dimension, as in real-life what matters its not only the actual situation, but also what the agents know about it. In this epistemic setting two new operations will be defined, both extending the similarity and middleman similarity operations by asking for the agents to have knowledge of the required condition. In both cases, the *de dicto* and *de re* variations of the epistemic conditions are further explored. The different logical settings in this paper make use of the techniques of *dynamic epistemic logic* (*DEL*; [14–16]) to represent group-creation actions, to define new languages to describe their effects, and to provide sound and complete axiom systems. In Sect. 4 we conclude this paper with a list of topics for future work.

2 Modelling Social Networks

We adopt the basic setting of [13], which is a relational 'Kripke' model in which the domain is interpreted as the set of agents, the accessibility relation represents a social connection from one agent to another, and the atomic valuation describes the features (behavior/opinions) each agent has. Let A denote a finite non-empty set of agents, and P (with $\mathsf{A} \cap \mathsf{P} = \varnothing$) a finite set of features that each agent might or might not have:

Definition 2.1 (Social Network Model). *A* social network model *(SNM) is a tuple $M = \langle \mathsf{A}, S, V \rangle$ where $S \subseteq \mathsf{A} \times \mathsf{A}$ is the* social *relation (Sab indicates that agent a is socially connected to agent b) and $V : \mathsf{A} \to \wp(\mathsf{P})$ is a* feature *function ($p \in V(a)$ indicates that agent a has feature p).*

Relation S is not required to satisfy any specific property (neither irreflexivity nor symmetry). Hence our social relation differs from the friendship relation in e.g. [8,12,13]. Given a social network model, we define a notion of 'distance' between agents based on the number of features in which they differ.

Definition 2.2 (Distance). *Let $M = \langle \mathsf{A}, S, V \rangle$ be a SNM. Define the set of features distinguishing agents $a, b \in \mathsf{A}$ in M as* $\text{msmtch}^M(a,b) := (V(a) \setminus V(b)) \cup (V(b) \setminus V(a))$. *Then, the distance between a and b in M is given by*

$$\text{dist}^M(a,b) := |\text{msmtch}^M(a,b)|$$

Proposition 2.1 *Let $M = \langle \mathsf{A}, S, V \rangle$ be a SNM and take $a, b, c \in \mathsf{A}$. Then,*

- *Non-negativity:* $\text{dist}^M(a,b) \geq 0$.
- *Symmetry:* $\text{dist}^M(a,b) = \text{dist}^M(b,a)$.
- *Reflexivity:* $\text{dist}^M(a,a) = 0$.
- *Subadditivity:* $\text{dist}^M(a,c) \leq \text{dist}^M(a,b) + \text{dist}^M(b,c)$.

Here $\text{dist}^M(a, b)$ is a mathematical distance (satisfying non-negativity, symmetry and reflexivity) and also a semi-metric (a distance satisfying subadditivity).[1] It is not a metric as it does not satisfy *identity of indiscernibles*: $\text{dist}^M(a, b) = 0$ does not imply $a = b$, as two different agents may have exactly the same features.

Static Language $\mathcal{L}$. Following [13], social network models are described by a *propositional* language $\mathcal{L}$ with special atoms describing the agents' features and their social relationship. More precisely, formulas in $\mathcal{L}$ are given by

$$\varphi, \psi ::= p_a \mid \mathrm{S}_{ab} \mid \neg\varphi \mid \varphi \wedge \psi$$

with $p \in \mathsf{P}$ and $a, b \in \mathsf{A}$. We read p_a as *"agent a has feature p"* and S_{ab} as *"agent a is socially connected to b"*. Other Boolean operators ($\vee, \rightarrow, \leftrightarrow, \underline{\vee}$, the latter representing the exclusive disjunction) are defined as usual. Given a SNM $M = \langle \mathsf{A}, S, V \rangle$, the semantic interpretation of $\mathcal{L}$-formulas in M is given by:

$$\begin{array}{llll} M \Vdash p_a & \text{iff}_{\text{def}} \ p \in V(a), & M \Vdash \neg\varphi & \text{iff}_{\text{def}} \ M \not\Vdash \varphi, \\ M \Vdash \mathrm{S}_{ab} & \text{iff}_{\text{def}} \ Sab, & M \Vdash \varphi \wedge \psi & \text{iff}_{\text{def}} \ M \Vdash \varphi \text{ and } M \Vdash \psi. \end{array}$$

A formula $\varphi \in \mathcal{L}$ is valid (notation: $\Vdash \varphi$) when $M \Vdash \varphi$ holds for all models M. Since there are no restrictions on the social relation nor on the feature function, any axiom system for classical propositional logic is fit to characterize syntactically the validities of $\mathcal{L}$ over the class of social network models.

The remainder of this section deals with the creation of new networks by updating the network relation. In contrast to [13], which uses SNMs to study how the fixed network structure leads to changes in features, here we keep the agents' features fixed, focussing instead on the changes in the social structure.

2.1 Similarity Update

There are several ways in which new social relations can be defined. A natural option is to let two agents become friends when they are 'similar' enough. If a *threshold* $\theta \in \mathbb{N}$ is given, with $\theta < |\mathsf{P}|$ (recall: P is finite), then we can define a *similarity update* operation allowing agents to establish connections to others who differ in at most θ features.

Definition 2.3 (Similarity Update). *Let $M = \langle \mathsf{A}, S, V \rangle$ be a SNM; take $\theta \in \mathbb{N}$. The similarity update of M generates a new SNM $M_{\odot_\theta} = \langle \mathsf{A}, S_{\odot_\theta}, V \rangle$ which differs from M only in its social relation, given by*

$$S_{\odot_\theta} := \{(a, b) \in \mathsf{A} \times \mathsf{A} : \ \text{dist}^M(a, b) \le \theta\}$$

Intuitively, each agent defines a circle of ratio θ with herself at the center, and her social contacts will be those agents falling inside it. The social relation of the updated model $M_{\odot_\theta}$ satisfies:

Proposition 2.2. *Let $M = \langle \mathsf{A}, S, V \rangle$ be a SNM, with $M_{\odot_\theta} = \langle \mathsf{A}, S_{\odot_\theta}, V \rangle$ as in Definition 2.3. Then, $S_{\odot_\theta}$ is reflexive and symmetric.*

---

[1] See [17, Chap. 1] for more on mathematical distances.

Proof. Each property follows from its namesake distance property (Proposition 2.1*) and, in the case of reflexivity, from the fact that θ's lower bound is 0.*

One can think of the social network that is generated by a similarity update as representing friends that have mutual access to each others feature-database, allowing every agent to access also her own database. Note that although reflexivity implies that every agent will have at least one friend (i.e., $S_{\odot_\theta}$ is serial), nothing guarantees an agent will have a friend other than herself (as the threshold may be 'too strict').

If a 'friendship' (irreflexive and symmetric) network is required, the update operation has to be adjusted to keep identity pairs out (e.g., $S'_{\odot_\theta}:=\{(a,b) \in \mathsf{A} \times \mathsf{A} : a \neq b \text{ and } \mathrm{dist}^M(a,b) \leq \theta\}$). If, on the other hand, one requires a not necessarily symmetric network of 'informational access' (as in [18]), we can use a *personal threshold* for each agent $a \in \mathsf{A}$ (i.e., a function $\Theta : \mathsf{A} \to \mathbb{N}$). Then each agent can choose 'how different' others may be in order to add them to her social group, and the updated relation is given by $S'_{\odot_\Theta} := \{(a,b) \in \mathsf{A} \times \mathsf{A} : \mathrm{dist}^M(a,b) \leq \Theta(a)\}$. Note how, for example, the distance between a and b, say 2, may be good enough for a to consider b a social contact ($2 \leq \Theta(a)$), but not for b to consider a a social contact ($\Theta(b) < 2$).

We illustrate briefly why other relational properties cannot be guaranteed.

Fact 2.1. The relation $S_{\odot_\theta}$ needs to be neither transitive nor Euclidean.

Proof. Transitivity can fail because, given agents a, b and c, what distinguishes a and b may be *only part of* what distinguishes a and c (i.e., $\mathrm{msmtch}^M(a,b) \subset \mathrm{msmtch}^M(a,c)$). For example, let $\theta = 1$ and consider the updated model below on the right, in which $S_{\odot_\theta}ab$ and $S_{\odot_\theta}bc$, but not $S_{\odot_\theta}ac$.

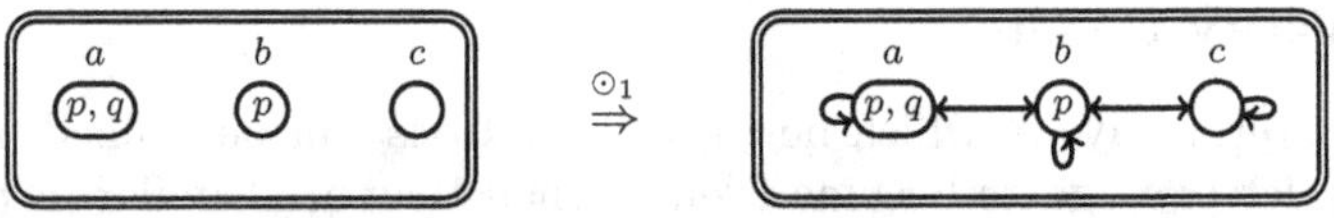

The relation may not be Euclidean because what distinguishes a and b may be *different from* what distinguishes a and c (i.e., $\mathrm{msmtch}^M(a,b) \cap \mathrm{msmtch}^M(a,c) = \varnothing$). For example, take $\theta = 1$: the updated model below on the right is such that $S_{\odot_\theta}ab$ and $S_{\odot_\theta}ac$, but neither $S_{\odot_\theta}bc$ nor $S_{\odot_\theta}cb$.

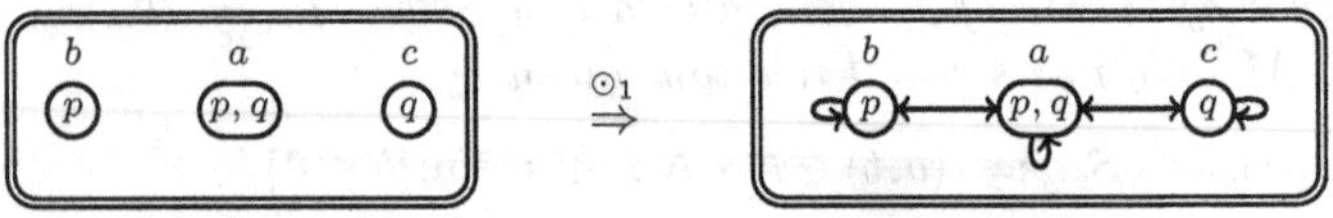

Dynamic Language $\mathcal{L}_{\odot_\theta}$. To express how a social network changes, we use the language $\mathcal{L}_{\odot_\theta}$ which extends the language $\mathcal{L}$ with a 'dynamic' modality $[\odot_\theta]$ to build formulas of the form $[\odot_\theta]\,\varphi$ ("*after a similarity update, φ is the case*"). The semantic interpretation of this modality refers to the similarity-updated model in Definition 2.3 as follows: Let M be a SNM, then

$$M \Vdash [\odot_\theta]\,\varphi \quad \text{iff}_{\text{def}} \quad M_{\odot_\theta} \Vdash \varphi.$$

Different from the well-known case of information updates under public announcements [19,20], no precondition is required for a similarity update of a social network: the operation can take place in any situation. Because of this and the functionality of the model operation, the dual modality $\langle \odot_\theta \rangle \varphi := \neg [\odot_\theta] \neg\varphi$ is such that $\Vdash [\odot_\theta] \varphi \leftrightarrow \langle \odot_\theta \rangle \varphi$.

The following axiom system is build via the *DEL* technique of *recursion axioms.* First, note that, as P is finite, the following $\mathcal{L}$-formula is true in a model M if and only if agents a and b differ in exactly $t \in \mathbb{N}$ features. (The formula states that there is at least one set of features Q, of size t, such that a and b differ in all features in Q and coincide in all features in $\mathsf{P} \setminus \mathsf{Q}$. There can be a most one such set, therefore the formula is true exactly when a and b differ in exactly t features.)

$$\mathrm{Dist}^{t}_{ab} := \bigvee_{\{\mathsf{Q}\subseteq\mathsf{P}:\ |\mathsf{Q}|=t\}} \left(\bigwedge_{p\in\mathsf{Q}} (p_a \veebar p_b) \wedge \bigwedge_{p\in\mathsf{P}\setminus\mathsf{Q}} (p_a \leftrightarrow p_b) \right)$$

The following $\mathcal{L}$-formula is true in M iff a and b differ in at most $\theta \in \mathbb{N}$ features:

$$\mathrm{Dist}^{\leq\theta}_{ab} := \bigvee_{t=0}^{\theta} \mathrm{Dist}^{t}_{ab}$$

Hence the following $\mathcal{L}_{\odot_\theta}$-formula characterizes the social relation in the similarity-updated model: a will consider b as a social contact if and only if, before the operation, a and b differed in at most θ features;

$$\Vdash [\odot_\theta] \mathrm{S}_{ab} \leftrightarrow \mathrm{Dist}^{\leq\theta}_{ab}$$

As only the social relation changes, the reduction axioms and the rules in Table 1 form, together with a propositional system, a sound and strongly complete axiom system characterizing the validities of $\mathcal{L}_{\odot_\theta}$. The here given syntax adapts the work of [13] for threshold-limited influence to the case of similarity update.

Table 1. Axiom system for $\mathcal{L}_{\odot_\theta}$ over social network models ($a, b \in \mathsf{A}$).

$\vdash [\odot_\theta] p_a \leftrightarrow p_a$	From $\vdash \varphi$ infer $\vdash [\odot_\theta] \varphi$
$\vdash [\odot_\theta] \mathrm{S}_{ab} \leftrightarrow \mathrm{Dist}^{\leq\theta}_{ab}$	From $\vdash \psi_1 \leftrightarrow \psi_2$ infer $\vdash \varphi \leftrightarrow \varphi [\psi_2/\psi_1]$ (with $\varphi [\psi_2/\psi_1]$ any formula obtained by replacing one or more occurrences of ψ_1 in φ with ψ_2)
$\vdash [\odot_\theta] \neg\varphi \leftrightarrow \neg [\odot_\theta] \varphi$	
$\vdash [\odot_\theta](\varphi \wedge \psi) \leftrightarrow ([\odot_\theta] \varphi \wedge [\odot_\theta] \psi)$	

If the mentioned 'irreflexive' version of similarity update is used, the axiom characterizing the new social relation should be restricted to cases with $a \neq b$, with a new axiom for the missing case:

$$\vdash [\odot_\theta] \mathrm{S}_{ab} \leftrightarrow \mathrm{Dist}^{\leq\theta}_{ab} \quad \text{for } a \neq b, \qquad\qquad \vdash [\odot_\theta] \mathrm{S}_{aa} \leftrightarrow \bot$$

If the 'personal threshold' option is chosen, then the axiom should state that, after the operation, a includes b as her social contact iff they differ in at most $\Theta(a)$ features:

$$\vdash [\odot_\Theta]\, \mathrm{S}_{ab} \leftrightarrow \mathrm{Dist}_{ab}^{\leq \Theta(a)}$$

2.2 Middleman Similarity Update

The network generated via a similarity update does not depend on the topology of the original network but only on the agent's current 'distance', regardless of whether they were earlier socially connected or not. Yet in most social scenarios, we see that the past network does play a role and that it takes a common acquaintance to introduce new friends to each other who are similar enough.

Definition 2.4. (Middleman Similarity Update). *Let* $M = \langle \mathsf{A}, S, V\rangle$ *be a* SNM*; take* $\theta \in \mathbb{N}$*. The* middleman similarity updated *model* SNM $M_{\widehat{\odot}_\theta} = \langle \mathsf{A}, S_{\widehat{\odot}_\theta}, V\rangle$ *differs from* M *only in its social relation, which is given by*

$$S_{\widehat{\odot}_\theta} := \{(a, b) \in \mathsf{A} \times \mathsf{A} : \mathrm{dist}^M(a, b) \leq \theta \text{ and } \exists\, c \in \mathsf{A} \text{ with } Sac \text{ and } Scb\}$$

In the new network, agent a will include agent b as a social contact if and only if they are similar enough and there is an agent c who belongs to a's social network and who includes b as one of her social contacts. Of course, the social requirements for the middleman c might vary. In some cases, a *symmetric* social relation between him and the two involved agents a and b might be required; while in other cases, an agent who has both a and b in her social network might be enough. Thanks to the formulae describing the social relation S in the syntax, our logical system is capable of dealing with all these cases, and other similar variations. Note also that, in line with our definition, the role of the middleman can be played by agent a or b themselves if they were already friends. Still, requiring a middleman changes the properties of the resulting network:

Fact 2.2. If $M = \langle \mathsf{A}, S, V\rangle$ is a SNM, then $S_{\widehat{\odot}_\theta}$ needs to be neither reflexive nor symmetric. As the diagrams show, a middleman who can establish new relations may not exist (no reflexivity on the left, no symmetry on the right):

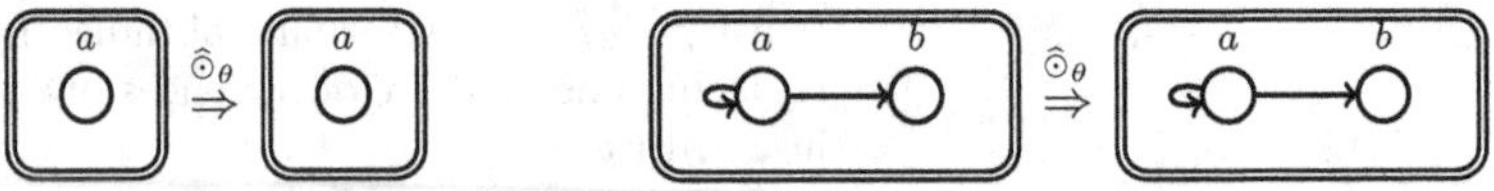

The middleman similarity update is not a monotone operation (i.e., new social links may form and old ones can disappear). To illustrate that the existing social relations are not preserved, take for example a given model with $S = \{(a, b)\}$. The middleman similarity update yields $S_{\widehat{\odot}_\theta} = \{\,\}$ regardless of a and b's similarities: neither Saa nor Sbb holds, hence neither a nor b can now play the role of the middleman. Of course, one can always enforce monotonicity by defining the new social relation in an 'accumulative' way ($S_{\widehat{\odot}_\theta} := S \cup \cdots$), yet this is not always appropriate: in real scenarios, social connections can be created, but unfortunately (and in some cases, fortunately) they can also be dropped. One advantage of not enforcing monotonicity is that it is possible to identify those situations that lead to it in a natural way. In our setting, a *reflexive* S guarantees that old social contacts will be preserved (modulo the agents' distance). In other words, for the agent to preserve her social connections, she should first consider herself 'worthwhile' as a friend.

The middleman similarity update operation *preserves* symmetry and moreover, if the initial M is fully symmetric and the update adds an edge from some a to some b, then it also adds its converse. Of course these preservation properties are no guarantee that $S_{\widehat{\odot}_\theta}$ is symmetric, as even when c plays the role of the middleman 'from left to right', lack of symmetry in the original M may make her unable to play the role 'from right to left'.[2]

Dynamic Language $\mathcal{L}_{\widehat{\odot}_\theta}$. We extend the above language $\mathcal{L}$ with a modality for describing the effect of the middleman similarity update. The resulting language $\mathcal{L}_{\widehat{\odot}_\theta}$ includes a modality $[\widehat{\odot}_\theta]$ for building formulas of the form $[\widehat{\odot}_\theta]\,\varphi$ ("*after a middleman similarity update, φ is the case*"). Its semantics is as follows: let $M = \langle \mathsf{A}, S, V\rangle$ be a SNM, then

$$M \Vdash [\widehat{\odot}_\theta]\,\varphi \quad \text{iff}_{\text{def}} \quad M_{\widehat{\odot}_\theta} \Vdash \varphi.$$

Since both A and P are finite, the axioms and rules in Table 2 (plus a propositional axiom system) characterize the validities of $\mathcal{L}_{\widehat{\odot}_\theta}$ in SNMs. The difference w.r.t. Table 1 is the axiom characterizing the new social relation, asking now for the required middleman. Axiom systems for the variations mentioned before (keeping identity pairs out, personal thresholds) can be obtained as in the 'non-middleman' similarity update case (Page 5).

Table 2. Axiom system for $\mathcal{L}_{\widehat{\odot}_\theta}$ over social network models ($a, b, c \in \mathsf{A}$).

$\vdash [\widehat{\odot}_\theta]\, p_a \leftrightarrow p_a$	$\vdash [\widehat{\odot}_\theta](\varphi \wedge \psi) \leftrightarrow ([\widehat{\odot}_\theta]\,\varphi \wedge [\widehat{\odot}_\theta]\,\psi)$
$\vdash [\widehat{\odot}_\theta]\, \mathrm{S}_{ab} \leftrightarrow \big(\mathrm{Dist}^{\leq\theta}_{ab} \wedge \bigvee_{c \in \mathsf{A}}(\mathrm{S}_{ac} \wedge \mathrm{S}_{cb})\big)$	From $\vdash \varphi$ infer $\vdash [\widehat{\odot}_\theta]\,\varphi$
$\vdash [\widehat{\odot}_\theta]\, \neg\varphi \leftrightarrow \neg\, [\widehat{\odot}_\theta]\,\varphi$	From $\vdash \psi_1 \leftrightarrow \psi_2$ infer $\vdash \varphi \leftrightarrow \varphi\,[\psi_2/\psi_1]$

3 Epistemic Social Networks

The described approach for creating social networks connects agents that are similar enough. However, in real life, two 'identical souls' may never relate to each other, as they may not know about their similarities. Thus, a more realistic representation of social network creation should take into account not only the agents' similarities, but also the knowledge they have about them.

[2] Note that several further constraints can be imposed, for instance one can require that any agent c playing the middleman for a and b should be fully connected to the agents she will 'introduce' (Sac, Sca, Scb, Sbc).

Definition 3.1 (Epistemic Social Network Model). *An* epistemic social network model *(ESNM) is a tuple* $M = \langle W, \mathsf{A}, \sim, S, V\rangle$ *having a set* $W \neq \varnothing$ *of* possible worlds, *a set of agents* A, *an* epistemic *equivalence relation* $\sim: \mathsf{A} \to (W \times W)$ *for each* $a \in \mathsf{A}$, *and at each world the social relation* $S : W \to \wp(\mathsf{A} \times \mathsf{A})$ *and feature function* $V : W \to (\mathsf{A} \to \wp(\mathsf{P}))$.

An ESNM is a standard possible worlds model [21] in which each possible world represents a SNM (Definition 2.1) and the epistemic relation is an equivalence relation. Derived concepts, such as $\text{msmtch}^M(\cdot,\cdot)$ and $\text{dist}^M(\cdot,\cdot)$, can be defined as before *for each possible world* $w \in W$.

Additional constrains can be imposed in the model. For instance, one can ask for the agents to *know themselves* (agent a knows herself at world w if and only if $w \sim_a u$ implies $V_w(a) \subseteq V_u(a)$) or to know who are her contacts (a knows who are her contacts at w if and only if $w \sim_a u$ implies $S_w[a] \subseteq S_u[a]$). For the sake of generality, here no such assumptions will be made.

Epistemic Language $\mathcal{L}^{\mathrm{K}}$. We follow [13] in designing an *epistemic* language $\mathcal{L}^{\mathrm{K}}$ with special atoms to describe the agents' features and social relationship. The formulas φ, ψ of $\mathcal{L}^{\mathrm{K}}$ are given by

$$\varphi, \psi ::= p_a \mid \mathrm{S}_{ab} \mid \neg\varphi \mid \varphi \wedge \psi \mid \mathrm{K}_a\,\varphi$$

with $p \in \mathsf{P}$ and $a, b \in \mathsf{A}$. Formulas of the form $\mathrm{K}_a\,\varphi$ are read as "*agent* a *knows* φ". Given a ESNM model $M = \langle W, \mathsf{A}, \sim, S, V\rangle$, the semantic interpretation of $\mathcal{L}^{\mathrm{K}}$-formulas is standard for Boolean operators and the epistemic modalities, with atoms p_a and S_{ab} interpreted relative to the point of evaluation.

$$(M, w) \Vdash p_a \ \text{iff}_{\text{def}}\ p \in V_w(a), \qquad\qquad (M, w) \Vdash \mathrm{S}_{ab} \ \text{iff}_{\text{def}}\ S_w ab.$$

The definition of (modal) validity ($\Vdash$) is as usual. We adopt here the well-known multi-agent $S5$ axiom system, as no extra restrictions are imposed in the model.

3.1 Knowledge-Based Social Network Creation

Definition 3.2 (Knowledge-based Similarity Update). *Let* $M = \langle W, \mathsf{A}, \sim, S, V\rangle$ *be an ESNM; take* $\theta \in \mathbb{N}$. *The* knowledge-based similarity update *operation generates the ESNM* $M_{\odot_\theta^{\mathrm{K}}} = \langle W, \mathsf{A}, \sim, S_{\odot_\theta^{\mathrm{K}}}, V\rangle$, *differing from* M *only in the social relation at every* $w \in W$, *given by*

$$(S_{\odot_\theta^{\mathrm{K}}})_w := \{(a, b) \in \mathsf{A} \times \mathsf{A} : \forall u \sim_a w\,,\ \text{dist}_u^M(a, b) \leq \theta\}$$

This update operation is based on the operation in Definition 2.3 but asks for an additional epistemic requirement: in world w agent a will add b to her social network if and only if in this world a *knows* that b is similar enough (i.e., b is similar enough in all a's epistemic alternatives from w).

Note how, after this update, the social network at each possible world will be reflexive, even in those cases in which the agent 'does not know herself'.[3]

[3] In any possible world, the distance between any agent and herself is 0.

On the other hand, social relations do not need to be symmetric, as the agents' knowledge might not have such property: at w agent a may know that she and b are similar enough (so $(S_{\odot_\theta^K})_w ab$ will hold), but b may not know this (and thus $(S_{\odot_\theta^K})_w ba$ will fail).

De dicto vs. de re. The knowledge-based similarity update uses a *de dicto* approach: after the operation, a includes b in her network when a knows that b is close enough, even if she does not know exactly which are the features that she shares with b. Indeed, consider the ESNM below on the left:

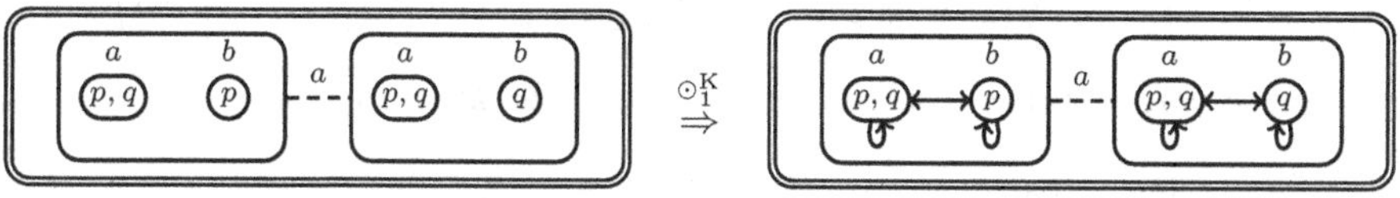

Even though a does not know which features b has, she knows the 'distance' between them is just 1. Thus, after a knowledge-based similarity update with any $\theta \geq 1$, she will add b to her network (the ESNM on the right).

On the other hand, a *de re* approach would ask not for a to know that the number of differences between her and b is 'small enough', but rather for her to point out a 'large enough' set of features on which she and b coincide:

$$(S'_{\odot_\theta^K})_w := \{(a,b) \in \mathsf{A} \times \mathsf{A} : \exists\, \mathsf{Q} \subseteq \mathsf{P} \text{ s.t. } \textit{(i)}\ |\mathsf{Q}| \geq |\mathsf{P}| - \theta \text{ and } \textit{(ii)}\ \forall u \sim_a w\,,\ V_u(a) \cap \mathsf{Q} = V_u(b) \cap \mathsf{Q}\,\}$$

Thus, a will include b in her social network if and only if there is a set of features she knows she and b share, and this set is large enough for their number of differences to be smaller than θ.[4] This variation also highlights an alternative to the basic idea of this proposal: we have related agents when their differences are small enough, but a (perhaps more 'human') alternative is to relate them when their *similarities* are large enough.

The *de dicto* version, asking for the agents to be 'close enough' in all epistemic possibilities, regardless of which one is the set of features in which the agents differ in each one of them, emphasizes that, for the agents, all features are equally important. But one can imagine a more realistic scenario in which certain features are more important than others: take an agent with features $\{p, q, r\}$ choosing an agent with $\{p\}$ over an agent with $\{q, r\}$ because, for her, p is more important than q and r together. Such a setting would require a *de re* approach.

Dynamic Epistemic Language $\mathcal{L}^K_{\odot_\theta^K}$. In order to express the way a knowledge-based similarity update affects a social network, a dynamic modality $[\odot_\theta^K]$ is added to $\mathcal{L}^K$ to yield language $\mathcal{L}^K_{\odot_\theta^K}$. This allows us to express that *"after a knowledge-based similarity update,* φ *is the case"*, $[\odot_\theta^K]\,\varphi$. For its semantic interpretation, let M be an ESNM. Then,

$$(M, w) \Vdash [\odot_\theta^K]\,\varphi \quad \text{iff}_{\text{def}} \quad (M_{\odot_\theta^K}, w) \Vdash \varphi.$$

[4] Both P and θ are commonly known, so a *knows* Q is enough to make her differences with b smaller than θ.

The axiom system is presented in Table 3. Note first the axiom describing the way the agents' knowledge changes: each epistemic modality K_a simply commutes with the dynamic modality, as the epistemic relation is not affected by the update operation. More importantly, the crucial reduction axiom states how, in order for a to 'add' b to her network, it is not enough for a and b to be 'similar enough': a should also know this. The unfolding of $\mathrm{Dist}_{ab}^{\leq\theta}$ in such an axiom makes explicit the *de dicto* approach: a only needs to know that $|\mathrm{msmtch}(a,b)|$ is smaller than θ.

$$\vdash [\odot_\theta^{\mathrm{K}}]\,\mathrm{S}_{ab} \leftrightarrow \mathrm{K}_a \bigvee_{t=0}^{\theta} \bigvee_{\{\mathsf{Q}\subseteq\mathsf{P}:\ |\mathsf{Q}|=t\}} \Big(\bigwedge_{p\in\mathsf{Q}}(p_a \veebar p_b) \wedge \bigwedge_{p\in\mathsf{P}\setminus\mathsf{Q}}(p_a \leftrightarrow p_b)\Big)$$

If the *de re* variation proposed above is chosen, the axiom becomes

$$\vdash [\odot_\theta^{\mathrm{K}}]\,\mathrm{S}_{ab} \leftrightarrow \bigvee_{t=|\mathsf{P}|-\theta}^{|\mathsf{P}|} \bigvee_{\{\mathsf{Q}\subseteq\mathsf{P}:\ |\mathsf{Q}|=t\}} \mathrm{K}_a \bigwedge_{p\in\mathsf{Q}}(p_a \leftrightarrow p_b)$$

Note the differences with the *de dicto* axiom. The fundamental one is the position of the knowledge modality K_a, now under the scope of the disjuctions asking for the existence of the 'large enough' set of features Q ("*there is* Q *of size at least* $|\mathsf{P}|-\theta$ *such that agent* a *knows that ...*"). The other difference is that a does not need to know that Q is exactly what distinguishes her and b; it is enough for her to know that features in Q are common for them. (Note, again how the fact that P and θ are common knowledge implies that a knows Q is large enough.)

Table 3. Axiom system for $\mathcal{L}^{\mathrm{K}}_{\odot_\theta^{\mathrm{K}}}$ over social network models ($a, b \in \mathsf{A}$).

$\vdash [\odot_\theta^{\mathrm{K}}]\,p_a \leftrightarrow p_a$	$\vdash [\odot_\theta^{\mathrm{K}}]\,\mathrm{K}_a\,\varphi \leftrightarrow \mathrm{K}_a\,[\odot_\theta^{\mathrm{K}}]\,\varphi$
$\vdash [\odot_\theta^{\mathrm{K}}]\,\mathrm{S}_{ab} \leftrightarrow \mathrm{K}_a\,\mathrm{Dist}_{ab}^{\leq\theta}$	From $\vdash \varphi$ infer $\vdash [\odot_\theta^{\mathrm{K}}]\,\varphi$
$\vdash [\odot_\theta^{\mathrm{K}}]\,\neg\varphi \leftrightarrow \neg\,[\odot_\theta^{\mathrm{K}}]\,\varphi$	From $\vdash \psi_1 \leftrightarrow \psi_2$ infer $\vdash \varphi \leftrightarrow \varphi\,[\psi_2/\psi_1]$
$\vdash [\odot_\theta^{\mathrm{K}}](\varphi \wedge \psi) \leftrightarrow ([\odot_\theta^{\mathrm{K}}]\,\varphi \wedge [\odot_\theta^{\mathrm{K}}]\,\psi)$	

3.2 Middleman Knowledge-Based Social Network Creation

In the epistemic setting one can ask for a middleman requirement. This again leads to a *de dicto* vs *de re* choice: either a knows there is someone who can link her with the 'similar enough' b (but she might not know who), or else there is someone a knows can link her with b. The definition below follows the *de re* alternative as, intuitively, a should know who is this middleman.

Definition 3.3 (Middleman Knowledge-Based Similarity Update). *Let* $M = \langle W, \mathsf{A}, \sim, S, V\rangle$ *be an ESNM; take* $\theta \in \mathbb{N}$. *The* middleman knowledge-based similarity update *generates ESNM* $M_{\hat{\odot}_\theta^{\mathrm{K}}} = \langle W, \mathsf{A}, \sim, S_{\hat{\odot}_\theta^{\mathrm{K}}}, V\rangle$, *which differs from* M *in its social relation at every* $w \in W$ *as follows:*

$$(S_{\hat{\odot}_\theta^{\mathrm{K}}})_w := \{(a,b) \in \mathsf{A}\times\mathsf{A} :\ \exists\, c \in \mathsf{A} \text{ s.t. } \forall u \sim_a w\,,\ (i)\ \mathrm{dist}_u^M(a,b) \leq \theta \ \text{ and } \ (ii)\ S_u ac \text{ and } S_u cb\,\}$$

Of course, this is not the only alternative in this epistemic-middleman setting. Besides the alternatives to the social requirements for the middleman discussed above, another possibility (suggested by a reviewer) is to shift the epistemic burden to the middleman: a will add agent b to her social network if and only if they are 'close enough' and the middleman knows this (syntactically, $\bigvee_{c \in \mathsf{A}} \mathrm{K}_c(\mathrm{Dist}_{ab}^{\leq \theta})$.

The new relation will be reflexive for an agent a in a world w if the original relation for a was reflexive *in all worlds accessible from* w, but also if somebody else plays the middleman role in all a's epistemic possibilities from w. Additionally, if the original relation is reflexive in all possible worlds, then the operation preserves it (modulo the agents' distance). With respect to symmetry, new relationships may not be 'mutual', as the agents may have 'asymmetric knowledge'.

A weaker *de dicto* requirement on the middleman condition, gives us:

$$(S'_{\widehat{\odot}^{\mathrm{K}}_{\theta}})_w := \{(a,b) \in \mathsf{A} \times \mathsf{A} : \forall\, u \sim_a w\,, \quad \textit{(i)}\ \mathrm{dist}^M_u(a,b) \leq \theta \ \text{ and } \ \textit{(ii)}\ \exists\, c \in \mathsf{A} \text{ s.t. } S_u ac \text{ and } S_u cb\,\}$$

In this variation, it i s enough for a to know there is someone linking her with b, even if she does not know exactly who this middleman is.

A dynamic epistemic language. Our language $\mathcal{L}^{\mathrm{K}}_{\widehat{\odot}^{\mathrm{K}}_{\theta}}$ extends $\mathcal{L}^{\mathrm{K}}$ with a modality $[\widehat{\odot}^{\mathrm{K}}_{\theta}]$ to express formulas of the form $[\widehat{\odot}^{\mathrm{K}}_{\theta}]\,\varphi$ (*"after a middleman knowledge-based similarity update,* φ *is the case"*). For its semantic interpretation, let M be an ESNM, then

$$(M,w) \Vdash [\widehat{\odot}^{\mathrm{K}}_{\theta}]\,\varphi \quad \text{iff}_{\text{def}} \quad (M_{\widehat{\odot}^{\mathrm{K}}_{\theta}}, w) \Vdash \varphi.$$

The axiom characterizing the way the social network changes (see Table 4) reflects both the *de dicto* approach for the knowledge about the distance and the *de re* approach for the knowledge about the middle man. For the alternative *de dicto-de dicto* proposed above (i.e., only knowledge of the existence for both the distance and the middle man), this axiom becomes

$$\vdash [\widehat{\odot}^{\mathrm{K}}_{\theta}]\,\mathrm{S}_{ab} \leftrightarrow \mathrm{K}_a\left(\mathrm{Dist}_{ab}^{\leq \theta} \wedge \bigvee\nolimits_{c \in \mathsf{A}} (\mathrm{S}_{ac} \wedge \mathrm{S}_{cb})\right)$$

Table 4. Axiom system for $\mathcal{L}^{\mathrm{K}}_{\widehat{\odot}^{\mathrm{K}}_{\theta}}$ over social network models ($a, b, c \in \mathsf{A}$).

$\vdash [\widehat{\odot}^{\mathrm{K}}_{\theta}]\,p_a \leftrightarrow p_a$	$\vdash [\widehat{\odot}^{\mathrm{K}}_{\theta}]\,\mathrm{K}_a\,\varphi \leftrightarrow \mathrm{K}_a\,[\widehat{\odot}^{\mathrm{K}}_{\theta}]\,\varphi$
$\vdash [\widehat{\odot}^{\mathrm{K}}_{\theta}]\,\mathrm{S}_{ab} \leftrightarrow \bigvee_{c \in \mathsf{A}} \mathrm{K}_a(\mathrm{Dist}_{ab}^{\leq \theta} \wedge \mathrm{S}_{ac} \wedge \mathrm{S}_{cb})$	From $\vdash \varphi$ infer $\vdash [\widehat{\odot}_{\theta}]\,\varphi$
$\vdash [\widehat{\odot}^{\mathrm{K}}_{\theta}]\,\neg\varphi \leftrightarrow \neg\,[\widehat{\odot}^{\mathrm{K}}_{\theta}]\,\varphi$	From $\vdash \psi_1 \leftrightarrow \psi_2$ infer $\vdash \varphi \leftrightarrow \varphi[\psi_2/\psi_1]$
$\vdash [\widehat{\odot}^{\mathrm{K}}_{\theta}](\varphi \wedge \psi) \leftrightarrow ([\widehat{\odot}^{\mathrm{K}}_{\theta}]\,\varphi \wedge [\widehat{\odot}^{\mathrm{K}}_{\theta}]\,\psi)$	

4 Conclusions

The present proposal explores a threshold approach to social network creation based on the agents' similarities, the key idea being that an agent will add someone to her social network if and only if the distance between them is smaller or equal than the given threshold. In this paper we have studied this idea as well as the *middleman* and *knowledge-based* variations; in each case, the properties of the resulting networks have been explored, and a sound a complete axiom system for the corresponding modality has been presented. The exploration can go deeper: for example, one can look for conditions guaranteeing that the resulting social network will have certain properties (reflexivity, seriality, symmetry, transitivity, Euclideanity). In the middleman case, one can also try to identify those situations in which the update operation will become idempotent, and thus further applications of it will not make a difference.

While this work is an initial exploration of the logical structure behind social group creation, our setting suggests several interesting alternatives. For example, as shown by the *de dicto* understanding of the knowledge requirement about the agent's distances, all features are equally important for all agents: what matters is the number of differences, and not what these differences are. In an alternative setting, we can treat certain features as more important than others, and we can let this 'priority ordering' among features differ from agent to agent. In a similar line, one can imagine situations in which not all features are relevant. Indeed, in [22] the authors use a game theoretic setting to define the agreement and disagreement of agents on a specific feature (or issue), which yields a way for them to update the social relation of agents with respect to one specific feature at a time. A similar idea can be worked out in our setting, which would require that only a subset of all features is relevant for each update operation. Such a structure can be useful when we want agents to control when one of their features becomes visible to other agents. A related alternative is to use an issue-dependent update to define different types of networks or social groups related to agent's different issues; after all, the network of football fans will be rather different from Lady Gaga's fan club.

There are also alternatives to the 'threshold' similarity idea of this proposal. An interesting idea, arising from the cognitive science literature, takes into account the size of the agent's 'social space'. In real life, agents may be willing to keep expanding their social network (even including people who are very different from them) as long as there is still enough space in their social environment. This is famously known as the *Dunbar's number*: a suggested cognitive limit to the number of people with whom one can maintain stable social relationships (see, e.g., [23]). Such a 'group-size' similarity approach produces different social relations, for example the lack of symmetry: in situations in which most agents are 'closer' (i.e., more similar) to agent b than to agent a, while the second may have 'enough space' to include the first in her social group, the first may be 'out of space' before she even considers the second. This group-size approach can be used in combination not only with the presented epistemic models, but also with the above ideas requiring 'some features to be more important than others' or 'not all features to be visible'.

Both the threshold and the mentioned group-size approaches relate agents when they are similar enough, where similarity is based on a distance measure that we have taken to be the standard Hamming distance between two sets of atomic formulae. It would be interesting to compare our obtained results to settings in which other notions of distances between sets are used (such as e.g. the Jaccard distance). An even bigger change is to consider the dual situation in which agents connect when they *complement each other.* In order to deal formally with this *complementary* idea, a more fine-grained setting is needed that takes into account not only the agents' features/behaviors, as in this paper, but also their doxastic state and their preferences (e.g., [11,24]).

In our epistemic setting, an obvious next step is to study also 'knowledge changing operations' within the presented models (e.g., public and private announcements), focussing not only on the changes of the agents' knowledge about each other's features and social connections, but also on the interplay with knowledge-based social network changing operations. There are interesting situations in which agents can learn new facts about each other's features and about each other's social relations from both the knowledge they have about the group-formation rules and the way a network has changed.

Finally, we observe the importance of the interplay between the social network changing operations of this proposal, and the operations that change the features (or behaviour/beliefs) in the proposals mentioned in the introduction. Both ideas deserve to be studied in tandem: the dynamics studied in one can affect the dynamics studied in the other, and our logical setting might be able to capture interesting properties about the interplay of these dynamic mechanisms.[5]

References

1. Crisp, R.J.: Social Psychology: A Very Short Introduction. OUP, Oxford (2015)
2. Easley, D., Kleinberg, J.: Networks, Crowds and Markets: Reasoning about a Highly Connected World. CUP, New York (2010)
3. Wang, Y., Sietsma, F., van Eijck, J.: Logic of information flow on communication channels. In: Grossi, D., Kurzen, L., Velázquez-Quesada, F.R. (eds.) Logic and Interactive RAtionality. Seminar's yearbook 2009. ILLC, pp. 226–245 (2010)
4. van Ditmarsch, H., van Eijck, J., Pardo, P., Ramezanian, R., Schwarzentruber, F.: Epistemic protocols for dynamic gossip. J. Appl. Log. **20**, 1–31 (2017)
5. van Eijck, J., Sietsma, F.: Message-generated kripke semantics. In Sonenberg, L., Stone, P., Tumer, K., Yolum, P. (eds.): AAMAS 2011, IFAAMAS, pp. 1183–1184 (2011)
6. Baltag, A.: Logics for insecure communication. In: van Benthem, J., (ed.) Proceedings of the TARK 2001. Morgan Kaufmann, pp. 111–122 (2001)

[5] In fact, one can see our proposal in this paper as a necessary first step towards that goal, as the formal grounds for both systems need to be settled before looking at their interaction.

7. Apt, K.R., Grossi, D., van der Hoek, W.: Epistemic protocols for distributed gossiping. In: Ramanujam, R. (ed.) Proceedings of the TARK 2015, EPTCS, pp. 51–66 (2015)
8. Liu, F., Seligman, J., Girard, P.: Logical dynamics of belief change in the community. Synthese **191**(11), 2403–2431 (2014)
9. Zhen, L., Seligman, J.: A logical model of the dynamics of peer pressure. Electron. Notes Theoret. Comput. Sci. **278**, 275–288 (2011)
10. Baltag, A., Christoff, Z., Hansen, J.U., Smets, S.: Logical models of informational cascades. In: van Benthem, J., Liu, F. (eds.) Logic Across the University: Foundations and Applications, pp. 405–432. College Publications, London (2013)
11. Ghosh, S., Velázquez-Quesada, F.R.: Agreeing to agree: reaching unanimity via preference dynamics based on reliable agents. In: Weiss, G., Yolum, P., Bordini, R.H., Elkind, E. (eds.) Proceedings of the AAMAS, pp. 1491–1499. ACM (2015)
12. Christoff, Z., Hansen, J.U., Proietti, C.: Reflecting on social influence in networks. J. Logic Lang. Inform. **25**(3–4), 299–333 (2016)
13. Baltag, A., Christoff, Z., Rendsvig, R.K., Smets, S.: Dynamic epistemic logics of diffusion and prediction in social networks (extended abstract). In: Bonanno, G., van der Hoek, W., Perea, A. (eds.) Proceedings of the LOFT 2016 (2016)
14. Baltag, A., Moss, L.S., Solecki, S.: The logic of public announcements, common knowledge, and private suspicions. In: Gilboa, I. (ed.): Proceedings of the TARK-98, pp. 43–56 (1998)
15. van Ditmarsch, H., van der Hoek, W., Kooi, B.: Dynamic Epistemic Logic. Springer, Heidelberg (2008)
16. van Benthem, J.: Logical Dynamics of Information and Interaction. CUP, New York (2011)
17. Deza, M.M., Deza, E.: Encyclopedia of Distances. Springer, Berlin Heidelberg (2009)
18. Carrington, R.: Learning and knowledge in social networks. Master's thesis, ILLC (2013)
19. Plaza, J.A.: Logics of public communications. Synthese **158**(2), 165–179 (2007)
20. Gerbrandy, J., Groeneveld, W.: Reasoning about information change. J. Logic Lang. Inform. **6**(2), 147–169 (1997)
21. Hintikka, J.: Knowledge and Belief. Cornell University Press, Ithaca (1962)
22. Solaki, A., Terzopoulou, Z., Zhao, B.: Logic of closeness revision: challenging relations in social networks. In: Köllner, M., Ziai, R. (eds.) Proceedings ESSLLI 2016 student session, pp. 123–134 (2016)
23. Dunbar, R.I.M.: Neocortex size as a constraint on group size in primates. J. Hum. Evol. **22**(6), 469–493 (1992)
24. Baltag, A., Smets, S.: A qualitative theory of dynamic interactive belief revision. In: Bonanno, G., van der Hoek, W., Wooldridge, M. (eds.): Logic and the Foundations of Game and Decision Theory, pp. 13-60. AUP (2008)

Examining Network Effects in an Argumentative Agent-Based Model of Scientific Inquiry

AnneMarie Borg[1], Daniel Frey[3], Dunja Šešelja[1,2(✉)], and Christian Straßer[1,2(✉)]

[1] Institute for Philosophy II, Ruhr-University Bochum, Bochum, Germany
{dunja.seselja,christian.strasser}@rub.de
[2] Center for Logic and Philosophy of Science, Ghent University, Ghent, Belgium
[3] Faculty of Economics and Social Sciences, Heidelberg University, Heidelberg, Germany

Abstract. In this paper we present an agent-based model (ABM) of scientific inquiry aimed at investigating how different social networks impact the efficiency of scientists in acquiring knowledge. The model is an improved variant of the ABM introduced in [3], which is based on abstract argumentation frameworks. The current model employs a more refined notion of social networks and a more realistic representation of knowledge acquisition than the previous variant. Moreover, it includes two criteria of success: a monist and a pluralist one, reflecting different desiderata of scientific inquiry. Our findings suggest that, given a reasonable ratio between research time and time spent on communication, increasing the degree of connectedness of the social network tends to improve the efficiency of scientists.

1 Introduction

Agent-based models (ABMs) have in recent years been increasingly utilized as a method for tackling socio-epistemic aspects of scientific inquiry [6,18,21,23,24]. The primary value of this approach is that it allows us to tackle questions that are difficult to answer with qualitative methods, such as historical case studies. One such question concerns the impact of different degrees and structures of information flow among scientists on their efficiency in acquiring knowledge. Zollman's pioneering work in this domain [23,24] suggested that a high degree of connectedness of a social network may be epistemically harmful, and that there is a trade-off between the success of agents and the time they need to reach a consensus. Even though it has been shown in [15] that this result, dubbed as "Zollman effect", does not hold for a large portion of the relevant parameter space, structurally different ABMs have come to similar conclusions [9,10].

The research by AnneMarie Borg and Christian Straßer is supported by a Sofja Kovalevskaja award of the Alexander von Humboldt Foundation and by the German Ministry for Education and Research.

A. Baltag et al. (Eds.): LORI 2017, LNCS 10455, pp. 391–406, 2017.
DOI: 10.1007/978-3-662-55665-8_27

The highly idealized nature of these ABMs makes it difficult to assess how relevant their findings are for actual scientific inquiry [15]. On the one hand, idealization and abstraction are necessary for simulations that aim at representing complex real world phenomena [14]. On the other hand, unless we include the most important 'difference making' factors figuring in the target phenomenon, the model might not represent any realistic scenario. Instead, it may represent only a logical possibility, uninformative about the real world.[1] Since due to their high degree of idealization such models operate at a significant representational distance from their intended target, we are faced with the methodological difficulty to judge their adequacy and epistemic merit. What we can do is to compare models with the same intended target that differ in various important respects, such as in the way they structure the phenomenon, in the way some of its elements are represented, in their degree of abstraction and idealization, etc. The possible gain may be that in virtue of this we are able to identify results that are robust under different modeling choices, and, based on that, to identify causes and explanatory relevant properties underlying these commonalities. Similarly, if specific types of models give rise to different outcomes, we may sharpen our understanding of their possible target phenomena and how they apply to them. For example, we may be able to identify distinct sub-classes of phenomena to which the different types of models apply.

The ABM presented in this paper has the same intended target phenomenon as the ones presented in [23,24]. Nevertheless, the representation of social networks, the content of the exchanged information, and the behavior of agents is different, which raises the question whether the "Zollman effect" also occurs in our model. As will be demonstrated, it does not. Given the differences between our models, this result may not be surprising. Nevertheless, it suggests that further investigation is needed to determine to which target phenomena the results of each model apply.

Our model is based on a recently developed argumentative ABM of scientific inquiry (AABMSI) [3]. It represents scientific interaction as argumentative in nature. This means that instead of a simple information flow that lacks critical assessment, in AABMSI agents may refute previously accepted or new information in view of counterarguments. Second, information about a given scientific theory is represented as consisting of a set of arguments (rather than being fully aggregated in a single value, representing one's credence in the given theory [23,24] or the epistemic success of the theory [9,10]). As a result, information sharing is represented as concerning particular parts of the given scientific theory, rather than an aggregated attitude about the whole theory. Third, receiving criticism triggers a search for defense of the given attacked argument. Fourth, the model takes into account that sharing information costs time and hence, that there is a trade-off between time spent on research and time spent on interaction.

[1] Accordingly, we can distinguish between models that provide *how actually* and *how possibly* explanations. While some have suggested that modeling a possibility is epistemically valuable [11], others have argued that if a model merely captures a possibility, it is epistemically and pragmatically idle [19]; instead, the presented possibility has to be understood within a specified context, which makes it relevant for real world phenomena.

In this paper we present an improved and more encompassing variant of AABMSI, aimed at examining the effects of different social networks on the efficiency of scientific inquiry. The model represents a situation in which scientists pursue different scientific theories, with the aim of determining which one is the best, and where they exchange arguments regarding their pursued theories. Compared to the model presented in [3], the current model introduces a number of improvements: (1) it employs the notion of social networks that is typically used in other ABMs of science, such as the complete graph, the wheel and the cycle [9,10,23,24] – this allows for the representation of an increase in information sharing proportional to the population size, which is absent from AABMSI; (2) the heuristic behavior of agents is represented in a more adequate way; (3) an additional criterion of success has been introduced in order to examine the robustness of the results under different standards relevant in scientific practice; (4) the time cost of learning is now proportional to the amount of new information received by the agent; (5) the model has been computationally improved, allowing for statistical analysis of the data and for more reliable results.

Our findings suggest that a higher degree of connectedness leads to a more efficient inquiry, given a reasonable ratio between research time and time spent on communication. While our ABM is still too idealized to draw normative conclusions concerning scientific inquiry that would be useful, for instance, to policy makers,[2] it represents a step further in this direction.

The paper is structured as follows. In Sect. 2 we explicate the main features of our ABM. In Sect. 3 we present the main results of the simulations. In Sect. 4 we compare our model with an argumentation-based ABM introduced in [8], as well as with other ABMs of scientific interaction. We conclude the paper in Sect. 5 by suggesting further enhancements of our model and future research avenues.

2 The Model

Agents in our model represent scientists who inquire into a number of rivaling theories in a given scientific domain. Theories are represented by sets of arguments which scientists can discover and investigate. This is modeled in terms of an argumentative landscape where scientists can investigate arguments by spending time on them, which allows them to discover other arguments or argumentative attacks between arguments of different theories. In certain intervals scientists evaluate theories on the basis of their knowledge about the argumentative landscape and decide which theory to pursue further. Additionally, scientists are situated in communication networks. By communicating with other scientists they enhance their knowledge about the argumentative landscape which may help them to make more informed evaluations.

[2] To this end, one of the tasks for future research is the empirical calibration of the parameters used in the model, as pointed out in [13].

In the remainder of this section we describe the main components of our model in more detail: the underlying argumentative landscape, the behavior of agents and the notion of social networks.[3]

2.1 The Argumentative Landscape

An *abstract argumentation framework*, as introduced by Dung [7], is a directed graph $\mathcal{AF} = \langle \mathcal{A}, \rightsquigarrow \rangle$ where $\mathcal{A}$ is a set of abstract entities called *arguments* and $\rightsquigarrow$ is an *attack relation* between arguments. Where $a, b \in \mathcal{A}$, *a attacks b* if $a \rightsquigarrow b$. A set of arguments $\mathcal{S}$ is *conflict-free* if there are no $a, b \in \mathcal{S}$ such that $a \rightsquigarrow b$.

For our purposes it is useful to add another relation between arguments, the *discovery relation* $\hookrightarrow$. It represents possible paths scientists can take to discover arguments:[4] if $a \hookrightarrow b$, then b can be discovered by the scientists if a has been previously discovered.

A theory is represented by a conflict-free set of arguments connected in a tree-like graph by discovery relations. Argumentative attacks exist only between arguments of different theories. Formally, an *argumentative landscape* is a triple $\mathcal{L} = \langle \mathcal{A}, \rightsquigarrow, \hookrightarrow \rangle$ where the set of arguments $\mathcal{A}$ is partitioned into m theories $\langle \mathcal{A}_1, \ldots, \mathcal{A}_m \rangle$ such that for each $i \in \{1, \ldots, m\}$ the theory $T_i = \langle \mathcal{A}_i, a_i, \hookrightarrow \rangle$ is a tree with root $a_i \in \mathcal{A}_i$ and

$$\rightsquigarrow \subseteq \bigcup_{\substack{1 \leq i,j \leq m \\ i \neq j}} (\mathcal{A}_i \times \mathcal{A}_j) \quad \text{and} \quad \hookrightarrow \subseteq \bigcup_{1 \leq i \leq m} (\mathcal{A}_i \times \mathcal{A}_i).$$

This definition of the attack relation $\rightsquigarrow$ ensures that each theory is conflict-free and that attacks only occur between members of different theories.

At the beginning of the simulation only the roots of each theory (representing a basic underlying hypothesis) are visible to the agents. During the simulation the agents gradually discover arguments and attacks between them (see also Sect. 2.2). Each argument a has a *degree of exploration*: $\mathfrak{expl}(a) \in \{0, \ldots, 6\}$ where 0 means that the argument has not been discovered and 6 means that the argument has been fully researched and it cannot be explored further. By spending time on an argument a, a scientist increases its degree of exploration. The higher the degree of exploration of a, the higher is the likelihood that the connected arguments will become visible. This concerns both arguments connected to a via the discovery relation in the same theory and arguments that attack a or that are being attacked by a. If $\mathfrak{expl}(a) = 0$ [resp. $\mathfrak{expl}(a) = 6$] no [resp. all] relation[s] from a to other arguments have been discovered.

Agents only have subjective knowledge of the landscape, which consists of arguments, their respective degrees of exploration, and attacks. During the simulation agents gain knowledge, on the one hand, by exploring the landscape,

[3] Our ABM is created in NetLogo [22]. The source code is available at: https://github.com/g4v4g4i/ArgABM/tree/LORI-VI-2017.

[4] Other ways of discovering arguments and attacks (via social networks) are discussed below.

and on the other hand, by communicating with other agents. This way two agents may have different knowledge about the degree of exploration of an argument or about discovered relations. See also Sect. 2.3.

2.2 Basic Behavior of Agents

Our model is round based. Each round every agent performs actions that are among the following:

1. the agent investigates the argument a she is currently situated at (i.e. she increases $\mathfrak{expl}(a)$) and while doing so she gradually reveals outgoing discovery relations as well as attacks from and to a;
2. the agent explores her current branch of the theory further by moving along a discovery relation to a neighboring argument (that she can see);
3. the agent leaves her current theory and moves to an argument of a rivaling theory (that she can see).

Every round each agent decides (based on a certain probability) whether to stay on her current argument (option 1) or to move to a new argument in her direct neighborhood (relative to the discovery relation, option 2). If she has reached a leaf of her branch, which is fully explored (i.e., $\mathfrak{expl}(a) = 6$), she backtracks on this branch to find an argument that is not fully explored. In case this fails, she moves to another not fully explored argument in the same theory.

Additionally, every 5 rounds agents consider whether they are still working on the theory they consider the best (with respect to their current subjective knowledge of the landscape). Depending on this decision, they continue with 1 and 2, or move to an alternative theory and start to explore that one (3). Their decision is based on an evaluation of the *degree of defensibility* of a theory.

The degree of defensibility of a theory is the number of defended arguments in this theory, where –informally speaking– an argument a is defended in the theory if it is not attacked or if each attacker b from another theory is itself attacked by some defended argument c in the current theory.

Let us give a more precise formal definition. First, we call a subset of arguments A of a given theory T *admissible* iff for each attacker b of some a in A there is an a' in A that attacks b (we say that a' defends a from the attack by b). Since every theory is conflict-free, it can easily be shown that for each theory T there is a unique maximally admissible subset of T (with respect to set inclusion). An argument a in T is said to be *defended in T* iff it is a member of this maximally admissible subset of T.[5] The *degree of defensibility* of T is equal to the number of defended arguments in T.

[5] Given that theories in our model are conflict-free, our notion of admissibility is actually the same as the one introduced in [7]. In Dung's terminology, our sets of defended arguments correspond to preferred extension (which are exactly the maximally admissible sets), except that we determine these sets relative to given theories.

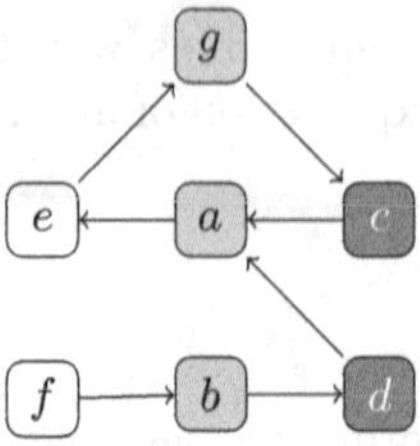

theory	defended	degree of def.
$T_1 = \{e, f\}$ (white)	$\{f\}$	1
$T_2 = \{a, b, g\}$ (gray)	$\{\}$	0
$T_3 = \{c, d\}$ (dark gray)	$\{\}$	0

Fig. 1. Argumentation Framework 1

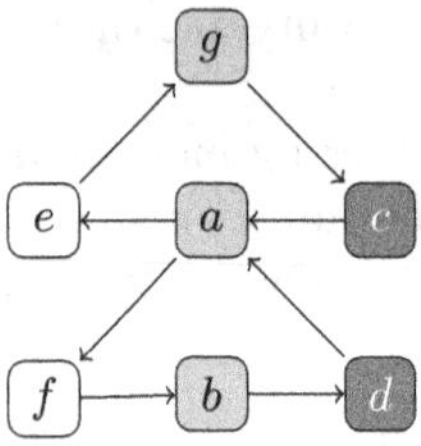

theory	defended	degree of def.
$T_1 = \{e, f\}$ (white)	$\{\}$	0
$T_2 = \{a, b, g\}$ (gray)	$\{a, b, g\}$	3
$T_3 = \{c, d\}$ (dark gray)	$\{\}$	0

Fig. 2. Argumentation Framework 2

Example 1. Figure 1 depicts a situation with three theories as it might occur from the perspective of a given agent: T_1 consisting of arguments e and f (white nodes), T_2 consisting of arguments a, b and g (gray nodes), and T_3 consisting of arguments c and d (dark gray nodes). The arrows represent attacks, we omit discovery relations. We are now interested in the degrees of defensibility our agent would ascribe to the given theories. The table shows which arguments are defended in each theory and their corresponding degree of defensibility. The only defended argument in this situation is f in theory 1. Note for instance that in T_3 the argument d is not defended since no argument *in* T_3 is able to defend it from the attack by b. Although the argument f in T_1 attacks b, it doesn't count as a defender of d for theory T_3 when determining the defended arguments in T_3 since in our account a theory is supposed to defend itself.

Figure 2 depicts the situation after an attack from a to f has been discovered. Consider theory T_2. In this situation a defends b from the attack by f, b defends a from the attack by d, a defends g from the attack by e and g defends a from the attack by c. Hence, all arguments are defended resulting in a degree of defensibility of 3.

A scientist decides to work on another theory T' instead of her current theory T if the degree of defensibility of T' surpasses the one of T by a certain margin. We represent the objectively best theory as a fully defended one.

Moreover, agents are equipped with heuristic abilities. An agent that encounters an attack from an argument b on the argument a, she is currently working on, will try to find a defense for this argument. For this she will consider all the arguments in her subjective knowledge that belong to the same theory as a.

If there is an argument a' among these that can *potentially defend* a from the attack, she will begin to investigate it. That a' potentially defends a from b means that there is an attack from a' to b but this attack has not yet been added to the subjective representation of the landscape of our agent (e.g., since $\mathfrak{expl}(a')$ is too low or since it has not been communicated to her). This means that agents are equipped with 'professional hunches' which help them to tackle problems in their theories.[6]

2.3 Social Networks

Besides discovering the argumentative landscape by exploring it on her own (see Sect. 2.2), an agent can share information about the landscape with other agents.

At the start of a simulation agents are divided into local *collaborative networks*, each consisting of exactly five individuals working on the same theory. During the simulation each agent gathers information (i.e., the degree of exploration of arguments, discovery and attack relations) on her own. Agents of the same collaborative network have the same subjective knowledge of the landscape since whenever an agent learns something new, this is communicated with the other agents in the same collaborative group.

Additionally, the collaborative groups form a *community network*. These have one of the following three structures: a *cycle*, in which each collaborative group is connected to exactly two other groups, a *wheel* which is similar to the cycle, except that a unique group is connected to every other group, and a *complete graph* where each group is connected to all other groups (Fig. 3). Every five rounds, randomly chosen *representative* agents of the collaborative groups communicate along the communication channels of the community network. The different network structures allow us to represent varying degrees of information flow in the scientific community, with the cycle representing the lowest and the complete graph the highest degree of information sharing.[7]

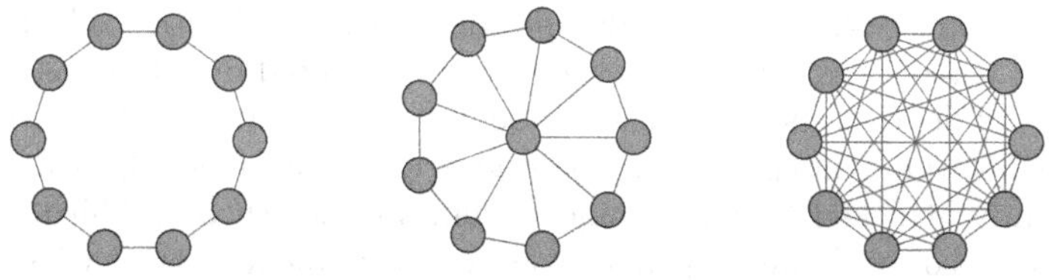

Fig. 3. A cycle, a wheel and a complete graph. Each node is a collaborative group, while the edges represent communication channels

Representative agents do not share their whole knowledge of the landscape with agents from other collaborative networks. Instead, they share the knowledge they have obtained recently by their own exploration of the landscape,

[6] Such hunches are not considered when agents evaluate theories.

[7] In contrast to the current model, in AABMSI [3] network structures are generated probabilistically in specific time intervals.

which consists of the argument they are currently at, its neighboring arguments connected via the discovery relation, their respective degrees of exploration, and attacks to and from their current argument. One way of interpreting this limited knowledge sharing is by considering this to be a situation where the agent writes a paper or gives a presentation on her current research results.

An alternative interpretation of collaborative groups is that they represent one individual working on 5 different arguments (papers, hypotheses, etc.) in a given theory. Whenever this individual communicates with other individuals from the community network it exchanges the information from the neighborhood of one of the 5 arguments she is currently engaged with.

Our model takes into account that information sharing and especially receiving information is time costly: agents who receive information (the representative agents of each collaborative network) are blocked from further exploration for a certain number of rounds. The costs of information sharing are proportional to the amount of new information obtained.[8]

Finally, we distinguish two types of information sharing, characterizing two types of scientists. A *reliable* agent shares all her recent discoveries, whereas a *deceptive* agent withholds the information about attacks on her current theory. This way, agents receiving information from a deceptive agent from theory T might come to a more favorable interpretation of T than if they would have communicated with a reliable agent [4].

3 The Main Findings

We will now specify the parameters used in the simulations and then present our most significant results.

3.1 Parameters Used in Simulations

We have run simulations with a landscape consisting of 3 theories. The landscape is created in three steps: First, each theory is represented by a tree (as explained in Sect. 2.1) of depth 3 such that each node has 4 children (except for the leaves) resulting in 85 arguments per theory. Second, with a chance of 0.3, each argument gets randomly attacked by an argument from another theory. Third, for every argument in theories T_2 and T_3 that attacks an argument in theory T_1 and that is not attacked by an argument from T_1 we add an attack from some random argument in T_1. Thus, it is made sure that T_1 is the objectively best theory and as such is fully defended from all attacks. In this way we wish to represent a scenario in which theories that are rivals to the best one are worse, though not completely problematic (as it would be the case with pseudo-scientific theories).

[8] A representative agent is excluded from research for 1–4 rounds: she always pays the basic cost of information sharing which is 1 round, and in addition, for every 2 fully explored arguments she will pay an additional round. The cost of learning an attack is equivalent to learning one degree of exploration of an argument.

Simulations were run 10.000 times for each of the scenarios with 10, 20, 30, 40, 70 and 100 agents. The scenarios are created by varying:

1. the community network: in the form of a cycle, a wheel and a complete graph;
2. two types of information sharing: reliable and deceptive.[9]

A simulation stops when one of the theories is completely explored. At this point all the agents have one more chance to make their final evaluation and choose their preferred theory. We then evaluate whether the agents have been successful according to the following criteria (where T_1 is the objectively best theory):

1. *monist criterion*, according to which a run is considered successful if, at the end of the run, all agents have converged onto T_1;
2. *pluralist criterion*, according to which a run is considered successful if, at the end of the run, the number of agents working on T_1 is not smaller than the number of agents working on any of the other theories.

The monist criterion is the standard notion of success, often employed in other ABMs of science, e.g. [23,24]. The pluralist criterion, on the other hand, is motivated by the philosophical conception of scientific pluralism, according to which a parallel existence of multiple theories in a given scientific domain is considered epistemically and heuristically beneficial e.g. [5]. This means that the convergence of all scientists onto the objectively best theory isn't a primary epistemic concern for pluralists. Rather, what matters is that the best theory is one of the most actively researched theories.[10]

3.2 Results

In this section we describe the most important findings of the simulations.

[9] Further parameters, with short explanations, are as follows. The *move probability* (set to 0.5) together with the degree of exploration of the argument an agent is situated at, determines the chance that she will move to another argument every 5 rounds (the move incentive is further decreased by $\frac{1}{5}$ for time steps in between). The *visibility probability* (set to 0.5) is the probability with which a new attack is discovered when an agent further explores her argument. The *research speed* (set to 5) determines the number of time steps an agent has to work on an argument a before a reaches its next level of exploration. The *strategy threshold* (set to 0.9) concerns the fact that each theory with a degree of defensibility that is at least 90% of the degree of defensibility of the best theory is considered good enough to be researched by agents. The *jump threshold* (set to 10) concerns the number of evaluations an agent can remain on a theory that is not one of the subjectively best ones.

[10] While our criterion is moderately pluralist, a more radical version would make plurality a necessary condition of success (i.e. populations would be punished for converging on one theory). We leave this consideration for future research.

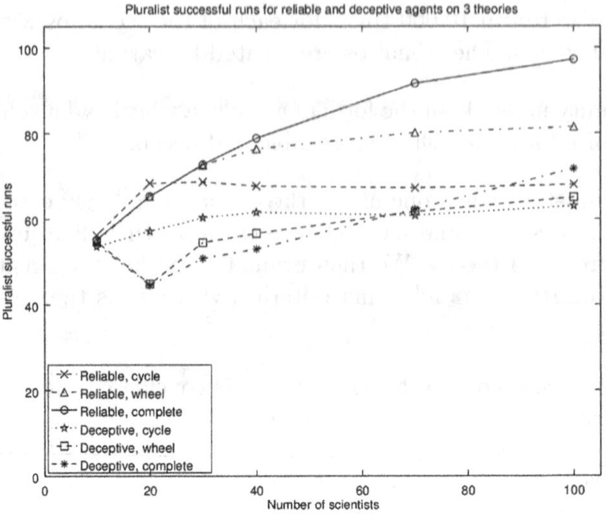

Fig. 4. Pluralist criterion of success, reliable and deceptive agents

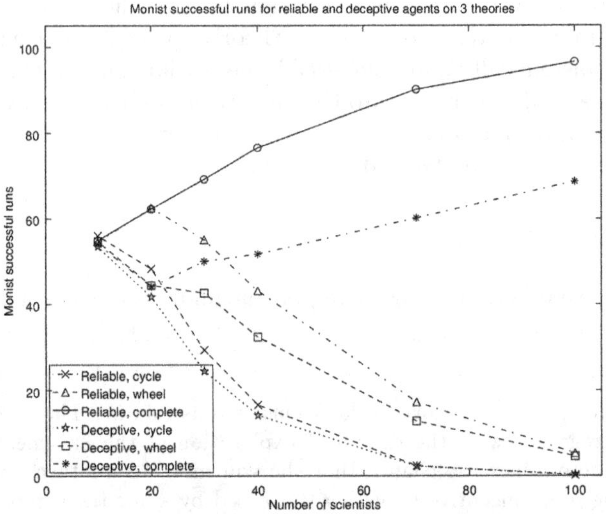

Fig. 5. Monist criterion of success, reliable and deceptive agents

Reliable vs. deceptive agents. With respect to both criteria of success, reliable agents are clearly more successful than the deceptive ones (Figs. 4 and 5), while being only slightly slower (Fig. 6).[11]

[11] The plots concern the landscape consisting of three theories. The results were similar in case of two theories in all the discussed respects, except that the agents were comparatively more efficient.

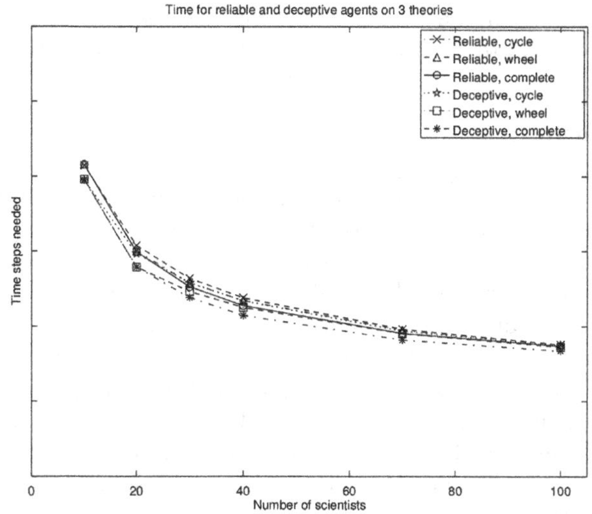

Fig. 6. Time needed

The degree of connectedness. In case of reliable agents, the complete graph tends to outperform the wheel and cycle, with respect to both criteria of success (Figs. 4 and 5), as well as with respect to the speed of exploration (Fig. 6). In other words, a higher degree of connectedness tends to lead to a more efficient inquiry.

In case of deceptive agents, the situation is a bit trickier. On the one hand, higher degrees of connectedness are also beneficial for the success according to the monist criterion (Fig. 5). However, the effect of connectedness is inverse for the pluralist criterion given populations with up to 70 agents (Fig. 4). A possible explanation for this asymmetry between the two success criteria is that deceptive agents in more connected networks share more false positives. As a result, if one theory is explored by a larger number of agents than either of the other theories, for the agents on this theory it will be easier to attract the whole population to it, leading to a fast, possibly wrong, convergence. In contrast, deceptive agents in the less connected networks will spread less information among each other, resulting in fewer cases of wrong convergence. For populations larger than 70 agents these premature convergences of the complete graph are prevented by the fact that each theory has on average enough researchers praising it and pointing out problems with the other theories so that the false positives are debunked as such more often.

Size of the community. Larger populations tend to be more efficient if agents are connected in the complete graph. In contrast, larger populations of agents in less connected networks perform similarly to smaller ones with respect to the pluralist criterion, and drastically worse with respect to the monist criterion.

The latter results from the lack of sufficient information flow, which is why larger populations fail to converge on any (including the best) theory.

4 Discussion

In this section we will first comment on some of our results and then we will compare our model with other ABMs designed to tackle similar questions.

We focus on two particularly interesting results: the impact of deceptive information sharing in contrast to the reliable one, and the impact of the degree of connectedness on reliable agents. Concerning the former, deceptiveness has not been studied in ABMs of science to a large extent.[12] This is, however, an important issue for the efficiency of scientific inquiry. For instance, withholding results that would undermine one's theory, typical for scientific fraud, is an example of deceptive information sharing. Even though deception is clearly problematic from the perspective of ethics of scientific conduct, its effects on the efficiency of the given scientific community aren't immediately clear. For instance, one could assume that presenting one's theory in positive light, in spite of the early problems can attract new researchers and help in developing it further. Our findings suggest that this is in general not the case, i.e. that deception tends to be epistemically harmful. More precisely, assuming that the whole community consists of deceptive scientists, and that scientists prefer theories that have the highest number of defensible results, deception leads to significantly less successful (though slightly faster) inquiry. Whether these results hold also under other assumptions remains to be examined in future research.

Our findings suggest that increasing the degree of connectedness of the communication network tends to be epistemically beneficial. This contrasts with findings obtained in other ABMs [9,10,23,24], according to which agents connected in a cycle perform better than agents connected in a complete graph. In order to see whether and in which sense our results challenge the latter results we first need to highlight some differences between our approach and these ABMs.

First, while in the latter agents are directly connected in the given networks, we employ a more structured approach by distinguishing between collaborative (local research) groups and communal networks between these groups. Note that in the real world it is impossible for each member of a larger scientific community to invest time in communicating with all the other members for the simple fact that communication (such as, e.g. reading papers) costs time, which could otherwise be spent on doing research. Thus, we have started from the assumption that scientists only have limited time for interacting with members of other collaborative groups, while they fully share information within their own collaborative projects. Altogether this means, however, that the highest

[12] One example of an ABM that studies deception in science is [12], which examines the effects of a deceptive agent in a community of epistemically pure agents. The authors show that in general a higher degree of connectivity helps against deceptive information. While our model doesn't examine the case of mixed (reliable and deceiving) agents, our results are, generally speaking, in line with their results.

degree of connectedness examined in our model gives rise to less information flow than in models in which agents are directly linked into a complete graph. Second, the content of the information shared among our agents is more localized and patchy since only representative agents exchange information about local aspects of their theory, namely their current argument and its neighborhood. As a result, agents from different collaborative groups end up having more diverse subjective representations of the landscape than e.g. agents in [23, 24]. Third, our representation of interaction includes a critical component. This is important for a model designed to examine the efficiency of scientific knowledge acquisition due to the fact that criticism has been shown to be truth conducive since it allows for false beliefs to be exposed as such [2]. Finally, similarly to ABMs that employ epistemic landscapes [9, 10, 21], our argumentative landscape allows for the representation of the process of discovery and its timeline. However, unlike the models employing epistemic landscapes, the information that agents encounter via an argumentative landscape is defeasible. This feature allows not only for the representation of critical interaction, but also for specific heuristic behaviors,[13] such as the search for a defense of an argument in case it has been attacked.[14] In this sense, our respective results might concern a different kind of scenario and thus refer to different target phenomena.

Having explicated the differences between our and other ABMs, it is important to notice though that there is no reason to assume that our model introduces more problematic idealizing assumptions than the previous ABMs of scientific interaction when it comes to the representation of a typical scientific inquiry. To the contrary, it includes a number of assumptions directly relevant for its adequate representation. Thus, our findings suggest that the results obtained by others models might not hold for usual cases of scientific inquiry. Instead, they may hold only for some very specific contexts. Which subclass of the phenomenon of scientific inquiry each of these models reliably represents, remains to be tackled in future research.

Finally, let us compare our model with Gabbriellini and Torroni's (G&T) ABM [8]. Their aim is to study polarization effects, e.g., in online debates. Similarly to our approach, their model is based on an abstract argumentation framework. Agents start with an individual partial knowledge of the given framework and enhance their knowledge by means of communication. Since G&T do not model inquiry, their agents cannot discover new parts of the graph by means of 'investigating' arguments. Rather, they exchange information by engaging in

[13] Interestingly, comparing the results of our model that employs the heuristic behavior (HB) and the results produced when HB is removed, shows that HB has hardly any impact on the success of agents, and in some cases it even slightly lowers their success. This seems to suggest that HB, by making agents stay on an undefended argument, waiting to find how to defend it, shields not only the best theory but also the worse ones, leading to an overall less successful inquiry. Examining this issue in more detail remains a task for future research.

[14] Another important difference between our ABM and those in [23, 24] is that the latter examine a fringe case of epistemically similar theories, which makes distinguishing the best one difficult.

a dialogue. This way, agents may learn about new arguments and attacks but also remove attacks. Whether new information is incorporated in the knowledge of an agent depends on the trust relation between the discussants. The beliefs of agents are represented by applying Dung-style admissibility-based semantics to the known part of the argumentation framework of an agent. This is quite different from our model where the underlying graph topology is given by several discovery trees of arguments representing scientific theories and attacks between them. This additional structure of the argumentation graph is essential since we do not model the agents' beliefs in individual arguments but rather evaluative stances of agents that inform their practical decision of which theory to work on. While an admissibility-based semantics would lead to extensions that feature unproblematic sets of arguments from different theories (ones that form conflict-free and fully defensible sets), in our approach agents pick theories to work on. For this, they compare the merits of the given theories, pick the one that is most defended, and employ heuristic behavior to tackle open problems of theories. It will be the topic of future research to include dialogue protocols that are relevant for scientific communication, such as information-seeking, inquiry and deliberation dialogues [20].

5 Conclusion

In this paper we have presented an argumentative ABM aimed at modeling the argumentative nature of scientific inquiry. The model is designed to examine how different kinds of social networks affect the efficiency of scientists in acquiring knowledge. Under the assumption that, in order to conduct their inquiry, scientists only have limited time to spend on communicating with others, our results suggest that a high information flow tends to be epistemically beneficial.

A variety of enhancements can be added to our model in order to make it apt for tackling similar or related questions. First, our current notion of the degree of defensibility represents scientists who prefer theories that exhibit a greater number of defensible results than their rivals. An alternative notion of defensibility would punish theories for having more anomalies (indefensible arguments) than their rivals, thus representing scientists who stick to their theories as long as they are not too anomalous (irrespective of how many positive results they have). Second, adding an explanatory relation and a set of explananda [16] would allow for a more refined representation of the desiderata of scientific theories and evaluative procedures which agents perform when selecting their preferred theory (e.g. in addition to the degree of defensibility, agents can take into account how much their current theory explains, or how well it is supported by evidence). Furthermore, a number of enhancements available from the literature on argumentation frameworks, such as probabilistic semantics [17], values [1], etc. can be introduced in future variants of our ABM. In addition to examining the impact of social networks, the model can be used to examine different heuristic behaviors and evaluations that guide scientific inquiry.

References

1. Bench-Capon, T.J.M.: Value based argumentation frameworks. arXiv:cs.AI/0207059 (2002)
2. Betz, G.: Debate Dynamics: How Controversy Improves Our Beliefs, vol. 357. Springer, Dordrecht (2012)
3. Borg, A., Frey, D., Šešelja, D., Straßer, C.: An Argumentative Agent-Based Model of Scientific Inquiry, pp. 507–510. Springer, Cham (2017). http://dx.doi.org/10.1007/978-3-319-60042-0_56
4. Caminada, M.: Truth, lies and bullshit: distinguishing classes of dishonesty. In: Social Simulation Workshop at the International Joint Conference on Artificial Intelligence (IJCAI) (2009)
5. Chang, H.: Is Water H2O? Evidence, Pluralism and Realism. Springer, Dordrecht (2012)
6. Douven, I.: Simulating peer disagreements. Stud. Hist. Philos. Sci. Part A **41**(2), 148–157 (2010)
7. Dung, P.M.: On the acceptability of arguments and its fundamental role in non-monotonic reasoning, logic programming and n-person games. Artif. Intell. **77**, 321–358 (1995)
8. Gabbriellini, S., Torroni, P.: A new framework for ABMs based on argumentative reasoning. In: Kamiński, B., Koloch, G. (eds.) Advances in Social Simulation. AISC, vol. 229, pp. 25–36. Springer, Heidelberg (2014). doi:10.1007/978-3-642-39829-2_3
9. Grim, P.: Threshold phenomena in epistemic networks. In: AAAI Fall Symposium: Complex Adaptive Systems and the Threshold Effect, pp. 53–60 (2009)
10. Grim, P., Singer, D.J., Fisher, S., Bramson, A., Berger, W.J., Reade, C., Flocken, C., Sales, A.: Scientific networks on data landscapes: question difficulty, epistemic success, and convergence. Episteme **10**(04), 441–464 (2013)
11. Hartmann, S., Reutlinger, A., Hangleiter, D.: Understanding (with) toy models. Br. J. Philos. Sci. (2016)
12. Holman, B., Bruner, J.P.: The problem of intransigently biased agents. Philos. Sci. **82**(5), 956–968 (2015)
13. Martini, C., Pinto, M.F.: Modeling the social organization of science. Eur. J. Philos. Sci. **2**(7), 1–18 (2016)
14. Muldoon, R., Weisberg, M.: Robustness and idealization in models of cognitive labor. Synthese **183**(2), 161–174 (2011)
15. Rosenstock, S., O'Connor, C., Bruner, J.: In epistemic networks, is less really more? Philos. Sci. (2016)
16. Šešelja, D., Straßer, C.: Abstract argumentation and explanation applied to scientific debates. Synthese **190**, 2195–2217 (2013)
17. Thimm, M.: A probabilistic semantics for abstract argumentation. In: ECAI, pp. 750–755 (2012)
18. Thoma, J.: The epistemic division of labor revisited. Philos. Sci. **82**(3), 454–472 (2015)
19. Van Riel, R.: The content of model-based information. Synthese **192**(12), 3839–3858 (2015)
20. Walton, D., Krabbe, E.C.: Commitment in Dialogue: Basic Concepts of Interpersonal Reasoning. SUNY Press, Albany (1995)
21. Weisberg, M., Muldoon, R.: Epistemic landscapes and the division of cognitive labor. Philos. Sci. **76**(2), 225–252 (2009)

22. Wilensky, U.: Netlogo. Center for Connected Learning and Computer Based Modeling. Northwestern University (1999). http://ccl.northwestern.edu/netlogo/
23. Zollman, K.J.S.: The communication structure of epistemic communities. Philos. Sci. **74**(5), 574–587 (2007)
24. Zollman, K.J.S.: The epistemic benefit of transient diversity. Erkenntnis **72**(1), 17–35 (2010)

Substructural Logics for Pooling Information

Vít Punčochář and Igor Sedlár(✉)

Institute of Philosophy, Czech Academy of Sciences, Prague, Czech Republic
{puncochar,sedlar}@flu.cas.cz

Abstract. This paper puts forward a generalization of the account of pooling information – offered by standard epistemic logic – based on intersection of sets of possible worlds. Our account is based on information models for substructural logics and pooling is represented by fusion of information states. This approach yields a representation of pooling related to structured communication within groups of agents. It is shown that the generalized account avoids some problematic features of the intersection-based approach. Our main technical result is a sound and complete axiomatization of a substructural epistemic logic with an operator expressing pooling.

1 Introduction

Alice is visiting her friends, Bob and Cathy. She needs to get to the train station now, and the only option is to take a bus. Alice is not familiar with the bus routes. Bob tells her that it is best to take the bus no. 25, get off at the Main Square and change lines there. However, he does not remember the no. of the connecting line. Cathy does not know this either (she rather bikes), but she takes a look the public transport mobile application and learns that the right bus to take at the Main Square is no. 17.

The information provided by Bob and Cathy needs to be *pooled together* to be helpful for Alice. Similar situations arise on a daily basis. In order to perform even the most rudimentary tasks, agents need to pool information coming from a multitude of sources. While communicating with others, agents pool the received information with the information they already have and possibly send the results further. A good model of pooling is therefore crucial for modelling deliberations, actions and interactions of agents, be they human or artificial.

In epistemic logic the standard representation of information is a set of possible worlds [1]. Information available to an agent (her information state) is modelled as a set of possible worlds "accessible" to the agent [2]. Pooling information states or pieces of information in general is represented by *intersection* of the corresponding sets. This paper puts forward a generalization of the standard

I. Sedlár—This work has been supported by the joint project of the German Science Foundation (DFG) and the Czech Science Foundation (GA ČR) number 16-07954J (*SEGA: From shared evidence to group attitudes*). The authors would like to thank the anonymous reviewers for their comments. The second author's ORCID ID is 0000-0002-1942-7982.

A. Baltag et al. (Eds.): LORI 2017, LNCS 10455, pp. 407–421, 2017.
DOI: 10.1007/978-3-662-55665-8_28

model. Our account, motivated by some unintuitive features of the standard framework, is based on information models for substructural logics.[1] Pooling is modelled by *fusion* of information states, a binary operation that generalizes intersection. This approach yields a more fine-grained representation of pooling; one that allows to model, for example, structured communication within groups of agents ("structured pooling"). Our main technical result is a sound and complete axiomatization of a substructural epistemic logic with an operator expressing outcomes of structured pooling.

Section 2 motivates our approach in more detail. Section 3 introduces the semantics based on information models and Sect. 4 extends the semantics by modalities representing outcomes of structured pooling; our main technical result is established in this section as well. Section 5 shows that the standard intersection-based model is a special case of our framework. Section 6 concludes the paper.

2 Motivation

In general, we may see information states and pieces of information as elements of a partially ordered set, with $x \leq y$ meaning that x supports y. Equivalently, $x \leq y$ means that x "extends" y as $x \leq y$ iff every z supported by y is supported by x. Information pooling can be represented by a binary operation; let us denote as $x \cdot y$ the result of pooling x with y. The representation of information by sets of possible worlds combined with the representation of information pooling in terms of intersection then corresponds to a special case of this framework where the poset at hand is a subset-ordered system of sets closed under intersection.

This special case is intuitively objectionable on several grounds. Firstly, the intersection-based model is *monotonic*; x pooled with y, i.e. $x \cap y$, extends both x and y. This means that support is irrevocable (every piece of information supported by x or y remains supported).

Example 1. As a counterexample to monotonicity, let us consider the following situation. Assume that Ann had believed that her boyfriend David is an honest man (Ann's information state at some point in time, x, supports the information that David is honest, h). Then she dropped that belief as the result of the conversation with Bob who told her that David had an affair with Cathy (the result of pooling x with Bob's information state, y, does not support h; we may assume for the sake of simplicity that it supports $\neg h$).

The standard model would represent the situation incorrectly; Ann would believe that David is honest and that he had an affair with Cathy at the same time.

A related feature of the standard model is that, if x is inconsistent with y (i.e. $x \cap y = \emptyset$), then the result of pooling x with y is the empty set. Hence, pooling

[1] We do not have space to provide an outline of substructural logics. We refer the reader to [7,9,10].

any pair of inconsistent pieces of information gives the same result, and this result supports every piece of information (this feature is known as *explosion*).

Example 2. Let us consider the following counterexample to explosion. Assume that Ann believes that she has free will and that free will is incompatible with physical determinism. Then she talks to Bob who persuades her that the physical world is deterministic. However, as it sometimes happens, she does not abandon the belief that she has free will. This means that the system of her beliefs is inconsistent, but not necessarily that the system supports any information whatsoever. She might hold inconsistent beliefs abut free will without being a right-wing extremist.

A feature of the standard model that comes into play here is that support is closed under classical consequence which validates *ex falso quodlibet.*

Monotonicity and explosion can be avoided by generalizing the standard model so that (i) a pooling operation is used such that $x \cdot y \leq x$ does not hold in general; (ii) mutual inconsistency of x and y is not modelled by $x \cdot y = 0$, where 0 is a trivially inconsistent piece of information; (iii) the support relation between pieces of information is not closed under classical consequence.

In what follows, we provide such a model. Models of this kind are offered, for instance, by various versions of *operational* semantics for substructural logics dating back to [14]. A more complete formulation was provided by Došen [3] and recently by Punčochář [8]. We build on the latter kind of model, extending it with modalities expressing structured communication within groups of agents. To motivate the introduction of such modalities, let us take a look at the connection between the standard model of pooling and communication within groups of agents.

A widespread intuitive interpretation of the standard model of pooling is that $\bigcap_{a \in G} x_a$ is the information state the members of a group G (having information states x_a for $a \in G$) would end up with after *communicating* with each other.[2] From the perspective of this interpretation, the standard framework represents a very special kind of communication within a group of agents, one in which agents pool *all* their information *instantenously* and every piece of information shared by each agent is *equally* considered. This is not how communication within groups usually works. Imagine an office or a research team; members of such groups may exchange partial information sequentially (e.g. Ann talks to Bob and then to Cathy) and some information may not be considered at all (e.g.

[2] For example, "A group has distributed knowledge of a fact φ if the knowledge of φ is distributed among its members, so that by pooling their knowledge together the members of the group can deduce φ, even though it may be the case that no member of the group individually knows φ." [4, p. 3]. This interpretation suffers from well-known problems [5,11,16]; we point out some additional ones. Hence, our paper can be seen as providing an additional argument against considering the standard model to be a good model of communication-related pooling. In the future, we plan to study the "full communication principle" of [5,11] and the dynamic approach of [16] in the context of our framework.

information that contradicts a belief that an agent is not willing to give up). The structure of such *communication scenarios* is often critical when it comes to the outcome of communication.

Before developing this point, we introduce some notation. We may represent the (hypothetical) communication scenario of Ann talking to Bob and then to Cathy (about some issue) by the expression $(a * b) * c$. "Communication within group $G = \{a, b, c\}$" can be seen as being ambiguous between $(a * b) * c$, $(a * c) * b$ etc. Alternatively, the different expressions are related to different ways how agents in G can communicate with each other. If x, y, z are information states of a, b and c, respectively, then the outcome of $(a * b) * c$ should be related to the structured pooling resulting in $(x \cdot y) \cdot z$. When a more specific formulation is preferred, we may say that $(x \cdot y) \cdot z$ represents a's information state after the scenario $(a * b) * c$ has been realized.

It turns out that, given the link between communication scenarios and pooling, some algebraic properties of intersection are problematic. Take commutativity and associativity, for example.[3]

Example 3. Assume that Ann has not yet formed an opinion about a new colleague, Bob. She has the tendency to accept the first strong opinion she hears from others. Cathy likes Bob very much but David does not like him at all. When it comes to her eventual opinion about Bob, it is obviously important to whom she talks first.

In general, assume that a's information state is partial with respect to p (it supports neither p nor $\neg p$), b's state supports p and c's state supports $\neg p$. Assume that when communicating with other agents, a accepts only information that is consistent with her state. Now if a communicates with b and then with c – that is scenario of the type $(a * b) * c$ – then her resulting state supports p; if a communicates with c and then with b – that is scenario of the type $(a * c) * b$ – her state supports $\neg p$.

Example 4. Assume that Cathy has read recently in a newspaper that Ms. X, Bob's favourite politician, obtained some money from Mr. Y, a man involved in organized crime. Consider two communication scenarios about the credibility of Ms. X. In the first scenario, which is of the type $a * (b * c)$, Cathy first talks to Bob and Bob is subsequently discussing the same issue with Ann. Since Bob trusts Ms. X, he does not believe the information conveyed by Cathy about the problematic money from Mr. Y, and he does not pass this information to Ann. In the second scenario, which is of the type $(a * b) * c$, Ann is discussing Ms. X's credibility with Bob first and subsequently with Cathy. Unlike in the first scenario, she ends up believing that Ms. X obtained some money from Mr. Y, which she learned from Cathy.

These examples show that scenarios $(a * b) * c$ and $(a * c) * b$, and $(a * b) * c$ and $a * (b * c)$, respectively, might actually lead to different outcomes. In addition, some communication scenarios may be more effective or leading to more desirable

[3] For associativity, see also [13].

results than others. It makes sense, therefore, to extend the formal language at hand with modalities indexed by communication scenarios; e.q. $\Box_{(a*b)*c}\alpha$ meaning that, after $(a * b) * c$ is realized, a's information state supports α. This language can then be used to formalize reasoning of agents about communication scenarios. Such reasoning is, of course, a vital part of reasoning about agent interactions.

Example 5. Going back to Example 1, we may assume that if David's information state supports $\Box_a h$ and $\Box_{a*b}\neg h$, then he would try to prevent $a * b$ from realizing.

As another example, consider the situation where a team leader's information state supports both $\Box_{(a*b)*c}\alpha$ and $\Box_{a*c}\alpha$, where α is necessary for a to perform some task. It is then reasonable for the team leader to suggest $a * c$ and not $(a * b) * c$ as the former requires less resources (team members, time) than the former to reach the same goal (a' having information α).

Our framework, introduced in the next two sections, combines a generalization of the intersection-based model of pooling, based on operational substructural semantics, with a modal language allowing to express reasoning about hypothetical communication scenarios.

3 Information Models

In this section, we reconstruct the semantic framework for substructural logics introduced in [8] and summarize some of the results needed in the next section. For the sake of brevity, the results are presented without proofs; the interested reader is referred to [8].

Let us fix a set of atomic formulas At. The variables $p, q, \ldots$ range over elements of At. An *information model* is a structure of the following type:

$$\mathcal{M} = \langle S, +, \cdot, 0, 1, C, V\rangle.$$

S is an arbitrary nonempty set, informally construed as a set of information states[4]; $+$ and $\cdot$ are binary operations on S (addition and fusion of states); 0 and 1 are two distinguished elements of S (the trivially inconsistent state and the logical state); and V is a valuation, that is a function assigning to every atomic formula a subset of S. The following conditions are assumed:

1. $\langle S, +, 0\rangle$ is a join-semilattice with the least element 0, i.e. $+$ is idempotent, commutative and associative, and $x + 0 = x$ for every $x \in S$. The semilattice determines an ordering of S: $a \leq b$ iff $a + b = b$.
2. The operation $\cdot$ is distributive in both directions over $+$, i.e. $x \cdot (y + z) = (x \cdot y) + (x \cdot z)$ and $(y + z) \cdot x = (y \cdot x) + (z \cdot x)$
3. $1 \cdot x = x$ and $0 \cdot x = 0$

[4] By "information states" we mean bodies of information that might be "available to" agents, but we do not assume that every information state is an information state of an agent.

4. C satisfies the following conditions: (a) there is no x such that $0Cx$, (b) if xCy, then yCx, (c) $(x+y)Cz$ iff xCz or yCz
5. V assigns to every atomic formula an ideal in $\mathcal{M}$, that is a subset $I \subseteq S$ satisfying: (a) $0 \in I$, (b) $x + y \in I$ iff $x \in I$ and $y \in I$

Information models derive from Došen's grupoid models for substructural logics [3]. We extend Došen's models with 0, allowing us to have a simpler semantic clause for disjunction. Moreover, these structures are enriched with the compatibility relation C that allows us to introduce a paraconsistent negation avoiding the principle of explosion (*ex falso quotlibet*).[5]

Informally, information states $x \in S$ represent *bodies of information* that can be said to *support* specific pieces of information. For example, the beliefs of an agent or the evidence produced during a criminal trial can be seen as bodies of information supporting information that is not explicitly part of the respective body. The state 1 represents the "logical" state supporting all the logically valid formulas and 0 represents the trivially inconsistent state supporting every formula. The relation C represents compatibility between information states. Informally, xCy means that y does not support any information that contradicts the information supported by x; for more details, see [6]. The operation $+$ yields the *common content* of the states x, y. The state $x + y$ supports any piece of information supported by both x and y. The operation $+$ will correspond to intersection of the sets of supported formulas (see the construction of canonical models in this section). Dually, in the specific models in which the states x and y are represented as sets of possible worlds, $+$ corresponds to union (see Sect. 5). The operation $\cdot$ yields a *fusion* $x \cdot y$ of information states x, y. Importantly, a fusion of two information states may involve far more (or less) than the intersection of sets of possible worlds. None of the following are assumed:

- $x \cdot y \leq x$ (monotonicity)
- if not xCy, then $x \cdot y = 0$ (explosion)
- $(x \cdot y) \cdot z = (x \cdot z) \cdot y$ (commutativity)
- $x \cdot (y \cdot z) = (x \cdot y) \cdot z$ (associativity)

Formulas of the language L are defined as follows:

$$\alpha ::= p \mid \bot \mid t \mid \neg\alpha \mid \alpha \rightarrow \alpha \mid \alpha \wedge \alpha \mid \alpha \otimes \alpha \mid \alpha \vee \alpha.$$

With respect to a given information model $\mathcal{M}$ a relation of support $\Vdash_{\mathcal{M}}$ between information states from S and L-formulas is defined recursively by the following clauses (we drop the subscript):

- $x \Vdash p$ iff $p \in V(x)$.
- $x \Vdash \bot$ iff $x = 0$.
- $x \Vdash t$ iff $x \leq 1$.
- $x \Vdash \neg\alpha$ iff for any y, if yCx then $y \nVdash \alpha$.
- $x \Vdash \alpha \rightarrow \beta$ iff for any y, if $y \Vdash \alpha$, then $x \cdot y \Vdash \beta$.
- $x \Vdash \alpha \wedge \beta$ iff $x \Vdash \alpha$ and $y \Vdash \beta$.

[5] An alternative extension of Došen's semantics is due to Wansing [15] who adds to Došen's models a constructive negation based on positive and negative valuation.

- $x \Vdash \alpha \otimes \beta$ iff there are y, z such that $y \Vdash \alpha$, $z \Vdash \beta$, and $x \leq y \cdot z$.
- $x \Vdash \alpha \vee \beta$ iff there are y, z such that $y \Vdash \alpha$, $z \Vdash \beta$, and $x \leq y + z$.

If $x \Vdash \alpha$, we say that x supports α. The proposition $||\alpha||_{\mathcal{M}}$ expressed by α in $\mathcal{M}$ is the set of states of $\mathcal{M}$ that support α.

Theorem 1. *For any information model $\mathcal{M}$ and any L-formula α, $||\alpha||_{\mathcal{M}}$ is an ideal in $\mathcal{M}$.*

Accordingly, $y \leq x$ only if every α supported by x is supported by y ("information state y *extends* x"). We say that an L-formula α is valid in an information model $\mathcal{M}$ if the logical state 1 supports α in $\mathcal{M}$. An L-formula is valid in a class of information models if it is valid in every model in the class. Let α be an L-formula and Δ a nonempty set of L-formulas. We say that α is semantically FL-valid ($\vDash_{FL} \alpha$) if α is valid in every information model. α is a semantic FL-consequence of Δ ($\Delta \vDash_{FL} \alpha$) if for any state x of any information model, if x supports every formula from Δ, then x supports α.

Lemma 1. *An implication $\alpha \to \beta$ is valid in $\mathcal{M}$ iff, for all $x \in S$, $x \Vdash \alpha$ only if $x \Vdash \beta$.*

The logic of all information models is a non-distributive, non-associative, and non-commutative version of Full Lambek calculus with a paraconsistent negation. The logic can be axiomatized by a Hilbert-style axiomatic system (that we call FL) containing the following axiom schemata and inference rules:

*A*1	$\alpha \to \alpha$
*A*2	$\bot \to \alpha$
*A*3	$(\alpha \wedge \beta) \to \alpha$
*A*4	$(\alpha \wedge \beta) \to \beta$
*A*5	$\alpha \to (\alpha \vee \beta)$
*A*6	$\beta \to (\alpha \vee \beta)$
*A*7	$(\alpha \otimes (\beta \vee \gamma)) \to ((\alpha \otimes \beta) \vee (\alpha \otimes \gamma))$

*R*1	$\alpha, \alpha \to \beta / \beta$
*R*2	$\alpha \to \beta / (\beta \to \gamma) \to (\alpha \to \gamma)$
*R*3	$\gamma \to \alpha, \gamma \to \beta / \gamma \to (\alpha \wedge \beta)$
*R*4	$\alpha \to \gamma, \beta \to \gamma / (\alpha \vee \beta) \to \gamma$
*R*5	$\alpha \to \neg\beta / \beta \to \neg\alpha$
*R*6	$\alpha \to \beta / (\gamma \otimes \alpha) \to (\gamma \otimes \beta)$
*R*7	$\alpha \to (\beta \to \gamma) / (\alpha \otimes \beta) \to \gamma$
*R*8	$(\alpha \otimes \beta) \to \gamma / \alpha \to (\beta \to \gamma)$
*R*9	$t \to \alpha / \alpha$
*R*10	$\alpha / t \to \alpha$

Lemma 2. *Every axiom of FL is semantically FL-valid and all the rules preserve semantic FL-validity in all information models.*

A proof in the system FL is defined in the standard way as a finite sequence of L-formulas such that every formula in the sequence is either an instance of an axiom schema, or a formula that is derived by applying an inference rule to formulas that occur earlier in the sequence. We say that α is FL-provable ($\vdash_{FL} \alpha$), if there is a proof $\beta_1, \ldots, \beta_n$ such that $\alpha = \beta_n$. The expression $\alpha_1, \ldots, \alpha_n \vdash_{FL} \beta$ is an abbreviation for $\vdash_{FL} (\alpha_1 \wedge \ldots \wedge \alpha_n) \rightarrow \beta$, and if Δ is a set of L-formulas, $\Delta \vdash_{FL} \beta$ means that there are $\alpha_1, \ldots, \alpha_n \in \Delta$ such that $\alpha_1, \ldots, \alpha_n \vdash_{FL} \beta$.

Definition 1. *A set of L-formulas λ is a logic over FL iff (a) λ contains all the axioms of FL, (b) λ is closed under the rules of FL, and (c) λ is closed under uniform substitutions of L-formulas.*

For any logic over FL we construct a canonical model. For a given logic λ the canonical model of λ is constructed out of λ-theories.

Definition 2. *Let λ be a logic over FL. A nonempty set of L-formulas Δ is an λ-theory if it satisfies the following two conditions:*

(a) if $\alpha \in \Delta$ and $\beta \in \Delta$, then $\alpha \wedge \beta \in \Delta$,
(b) if $\alpha \in \Delta$ and $\alpha \rightarrow \beta \in \lambda$, then $\beta \in \Delta$.

Definition 3. *Let λ be a logic over FL. The canonical model of λ is the structure $\mathcal{M}^\lambda = \langle S^\lambda, +^\lambda, \cdot^\lambda, 0^\lambda, 1^\lambda, C^\lambda, V^\lambda \rangle$, where*

- *S^λ is the set of all λ-theories,*
- *$\Gamma +^\lambda \Delta = \Gamma \cap \Delta$,*
- *$\Gamma \cdot^\lambda \Delta = \{\alpha;$ for some $\gamma \in \Gamma$ and $\delta \in \Delta, (\gamma \otimes \delta) \rightarrow \alpha \in \lambda\}$,*
- *0^λ is the set of all L-formulas,*
- *$1^\lambda = \lambda$,*
- *$\Gamma C^\lambda \Delta$ iff for all α, if $\neg\alpha \in \Gamma$, then $\alpha \notin \Delta$,*
- *$\Gamma \in V^\lambda(p)$ iff $p \in \Gamma$.*

Theorem 2. *$\mathcal{M}^\lambda$ is an information model.*

Theorem 3. *For any L-formula α and λ-theory Γ, the following holds:*
$\Gamma \Vdash \alpha$ in $\mathcal{M}^\lambda$ iff $\alpha \in \Gamma$.

Assume that λ is given by an axiomatic system that is sound with respect to a class of information models that contains the canonical model of λ. The following direct corollary of Theorem 3 guarantees that the system must be also complete with respect to the class.

Corollary 1. *$\alpha \in \lambda$ iff α is valid in $\mathcal{M}^\lambda$.*

In particular, using Lemma 2 and Corollary 1 we obtain completenes of FL.

Corollary 2. *$\Delta \vDash_{FL} \beta$ iff $\Delta \vdash_{FL} \beta$.*

Since the construction leads to strong completeness of FL, we obtain compactness immediately.

Corollary 3. *If $\Delta \vDash_{FL} \beta$, then there is a finite $\Gamma \subseteq \Delta$ such that $\Gamma \vDash_{FL} \beta$.*

4 Communication Models

This section extends information models by a representation of information about the results of communications scenarios.

Let us fix a set of expressions Ag, representing a set of agents. We will define inductively a set of expressions $CS(Ag)$ (structured communication scenarios over Ag), or just CS, in the following way: 1. every $a \in Ag$ is in CS. 2. if G and H are in CS, then the expression $(G * H)$ is also in CS. 3. Nothing else is in CS. A communication scenario can be viewed as a binary tree whose leaves represent agents. The *principal agent of* $G \in CS$ is the agent denoted by the leftmost occurrence of an agent variable in G. For example, the principal agent of both $a * (b * c)$ and $(a * b) * c$ is a.

Variables $G, H, \ldots$ range over elements of CS. A *communication model* is any tuple

$$\mathcal{M} = \langle S, +, \cdot, 0, 1, C, \{f_G\}_{G \in CS}, V \rangle,$$

where $\mathcal{M} = \langle S, +, \cdot, 0, 1, C, V \rangle$ is an information model and $\{f_G\}_{G \in CS}$ is a collection of unary functions on S satisfying for every $G, H \in CS$:

$$\begin{array}{ll} (f0) & f_G(0) = 0 \\ (f+) & f_G(x + y) = f_G(x) + f_G(y) \\ (f\cdot) & f_{G*H}(x) \leq f_G(x) \cdot f_H(x) \end{array}$$

Informally, $f_a(x)$ is the information state of agent a, *according to* the body of information x; see [12]. $f_{a*b}(x)$ is the information state of a, according to x, after the communication scenario $a * b$ is realized.[6] In general, $f_G(x)$ is the information state of the principal agent of scenario G after G has been carried out, according to x. The result of pooling the information of the principal agent of G after realizing G with the information state of the principal agent of H after realizing H is represented by f_{G*H}. We call this the information state of the scenario $G * H$.

Our three "frame conditions" represent the following informal assumptions about communication scenarios. First, every G has an inconsistent information state according to 0, $(f0)$. This is straightforward as 0 supports every piece of information, i.e. it supports every piece of information about every G. Second, the information state of G according to the common content of x and y is the common content of $f_G(x)$ and $f_G(y)$, $(f+)$. This is a consequence of the interpretation of $x + y$ as the intersection of the information provided by x and y. Third, the information state of $G * H$ according to x extends the fusion of $f_G(x)$ and $f_H(x)$, $(f\cdot)$. This represents the fact that structured pooling is based on fusion of information states.

---

[6] The body of information x consists of information on a number of topics, including agents a and b, and how they react to receiving specific information in communication. $f_{a*b}(x)$ is the information constituting a's information state after receiving information from b, according to what x says about a and b.

Note, however, that on the level of information states we do not define pooling *as* fusion, that is we do not require

$$(f{=})\quad f_{G*H}(x) = f_G(x) \cdot f_H(x)$$

The reason will become clear in the next section where we explain how our framework generalizes the standard intersection-based model of pooling (i.e. epistemic logic with distributed knowledge). A spoiler: in the particular cases where our models correspond to standard models, the information states are sets of possible worlds, $*$ is union and $\cdot$ is intersection. In these cases $(f{=})$ naturally fails; it is only the case that $f_{G \cup H}(x) \subseteq f_G(x) \cap f_H(x)$.

The language $L_\Box$ is obtained by adding to L a modality $\Box_G$ for every communication scenario G:

$$\alpha ::= p \mid \bot \mid t \mid \neg\alpha \mid \alpha \to \alpha \mid \alpha \wedge \alpha \mid \alpha \otimes \alpha \mid \alpha \vee \alpha \mid \Box_G\alpha$$

The semantic clauses for the language $L_\Box$ extend the semantic clauses for L with the following clause for the group modalities:

$x \Vdash \Box_G\alpha$ iff $f_G(x) \Vdash \alpha$.

The following result extends Theorem 1. The result shows that complex formulas and atomic formulas express propositions of the same kind. This will guarantee that the logic of all communication models is closed under uniform substitution.

Theorem 4. *For any communication model $\mathcal{M}$ and any $L_\Box$-formula α, $||\alpha||_\mathcal{M}$ is an ideal in $\mathcal{M}$.*

Proof. Let $\mathcal{M}$ be a communication model and α an $L_\Box$-formula. We have to show that $||\alpha||_\mathcal{M}$ is an ideal in $\mathcal{M}$. This can be proved by induction. We will show the inductive step for $\Box_G$. The inductive assumption is that for a given $L_\Box$-formula β, $||\beta||_\mathcal{M}$ is an ideal in $\mathcal{M}$. We will show that $||\Box_G\beta||_\mathcal{M}$ is also an ideal. First, since $0 \Vdash \beta$ and $f_G(0) = 0$, we have $f_G(0) \Vdash \beta$, i.e. $0 \Vdash \Box_G\beta$. Second, $x+y \Vdash \Box_G\beta$ iff $f_G(x+y) \Vdash \beta$ iff $f_G(x)+f_G(y) \Vdash \beta$ iff $f_G(x) \Vdash \beta$ and $f_G(y) \Vdash \beta$ iff $x \Vdash \Box_G\beta$ and $y \Vdash \Box_G\beta$.

Let α be an $L_\Box$-formula and Δ a nonempty set of $L_\Box$-formulas. We say that α is semantically *PFL*-valid ($\vDash_{PFL} \alpha$) if α is valid in every communication model; α is a semantic *PFL*-consequence of Δ ($\Delta \vDash_{PFL} \alpha$) if for any state x of any communication model, x supports every formula from Δ only if x supports α.

The axiomatic system *PFL* is given by axioms and rules of *FL* plus the axiom *A*8, and the rules *R*11 and *R*12.

*A*8	$(\Box_G\alpha \wedge \Box_G\beta) \to \Box_G(\alpha \wedge \beta)$
*R*11	$\alpha \to \beta / \Box_G\alpha \to \Box_G\beta$
*R*12	$(\alpha \otimes \beta) \to \gamma / (\Box_G\alpha \wedge \Box_H\beta) \to \Box_{G*H}\gamma$.

The following claim extends Lemma 2.

Lemma 3. *Every axiom of PFL is semantically PFL-valid and all the rules preserve semantic PFL-validity in all communication models.*

Proof. (a) Let x be an arbitrary state of a communication model such that $x \Vdash \Box_G\alpha \wedge \Box_G\beta$. Then $f_G(x) \Vdash \alpha$ and $f_G(x) \Vdash \beta$, i.e. $f_G(x) \Vdash \alpha \wedge \beta$. So $x \Vdash \Box_G(\alpha \wedge \beta)$. It follows that $1 \Vdash (\Box_G\alpha \wedge \Box_G\beta) \to \Box_G(\alpha \wedge \beta)$, so we have proved that $A8$ is semantically PFL-valid.

(b) Assume that $1 \Vdash \alpha \to \beta$ in an arbitrary communication model. Let x be a state of that model such that $x \Vdash \Box_G\alpha$. Then $f_G(x) \Vdash \alpha$ and it follows from our assumption that $f_G(x) \Vdash \beta$. So $x \Vdash \Box_G\beta$. It follows that $1 \Vdash \Box_G\alpha \to \Box_G\beta$, so we have proved that $R11$ preserves semantic PFL-validity.

(c) Assume that in a communication model $\mathcal{M}$, $1 \Vdash (\alpha \otimes \beta) \to \gamma$. We will prove that $1 \Vdash (\Box_G\alpha \wedge \Box_H\beta) \to \Box_{G*H}\gamma$. Assume $x \Vdash \Box_G\alpha \wedge \Box_H\beta$. Then $f_G(x) \Vdash \alpha$ and $f_H(x) \Vdash \beta$. It follows that $f_G(x) \cdot f_H(x) \Vdash \alpha \otimes \beta$, and so $f_G(x) \cdot f_H(x) \Vdash \gamma$. Since $f_{G*H}(x) \leq f_G(x) \cdot f_H(x)$, it holds $f_{G*H}(x) \Vdash \gamma$, due to Theorem 4. As a consequence, $x \Vdash \Box_{G*H}\gamma$.

Definition 4. *A set of $L_\Box$-formulas λ is called a logic over PFL if the following three conditions are satisfied: (a) λ contains all the axioms of PFL, (b) λ is closed under the rules of PFL, (c) λ is closed under uniform substitutions of $L_\Box$-formulas.*

Theories related to logics over PFL are defined in the same way as theories related to logics over FL (see Definition 2) with the difference that if λ is a logic over PFL then λ-theories are sets of $L_\Box$-formulas. The construction of the canonical model for a given logic over PFL extends the construction from the previous section.

Definition 5. *Let λ be a logic over PFL. The canonical model of λ is the structure $\mathcal{M}^\lambda = \langle S^\lambda, +^\lambda, \cdot^\lambda, 0^\lambda, 1^\lambda, C^\lambda, \{f_G^\lambda\}_{G \in CS}, V^\lambda\rangle$, where $S^\lambda, +^\lambda, \cdot^\lambda, 0^\lambda, 1^\lambda, C^\lambda$ and V^λ are defined as in Definition 3 and for any $G \in CS$ and any λ-theory Γ we define:*

$$f_G^\lambda(\Gamma) = \{\alpha \in L_\Box; \Box_G\alpha \in \Gamma\}.$$

Let us fix a logic λ over PFL. We will write just S, $+$, $\cdot$, 0, 1, C, f_G, V instead of S^λ, $+^\lambda$, $\cdot^\lambda$, 0^λ, 1^λ, C^λ, f_G^λ, V^λ.

Lemma 4. *If Γ is a λ-theory, then $f_G(\Gamma)$ is also a λ-theory.*

Proof. (a) Assume that $\alpha \in f_G(\Gamma)$ and $\beta \in f_G(\Gamma)$, i.e. $\Box_G(\alpha) \in \Gamma$ and $\Box_G(\beta) \in \Gamma$. Since Γ is a λ-theory, $\Box_G\alpha \wedge \Box_G\beta \in \Gamma$. Since λ contains all the axioms of PFL, $\Box_G(\alpha \wedge \beta) \in \Gamma$, due to $A8$. It follows that $\alpha \wedge \beta \in f_G(\Gamma)$.

(b) Assume that $\alpha \in f_G(\Gamma)$ and $\alpha \to \beta \in \lambda$. Then $\Box_G\alpha \in \Gamma$ and since λ is closed under the rules of PFL, $\Box_G\alpha \to \Box_G\beta \in \lambda$, due to $R11$. It follows that $\Box_G\beta \in \Gamma$. So, $\beta \in f_G(\Gamma)$.

Lemma 5. *For any λ-theories Γ, Δ the following conditions are satisfied:*

(a) $f_G(0) = 0$,
(b) $f_G(\Delta + \Gamma) = f_G(\Delta) + f_G(\Gamma)$,
(c) $f_{G*H}(\Delta) \leq f_G(\Delta) \cdot f_H(\Delta)$.

Proof. The case (a) is immediate. (b) We are proving $f_G(\Delta \cap \Gamma) = f_G(\Delta) \cap f_G(\Gamma)$. It holds that $\alpha \in f_G(\Delta \cap \Gamma)$ iff $\Box_G \alpha \in \Delta \cap \Gamma$ iff $\Box_G \alpha \in \Delta$ and $\Box_G \alpha \in \Gamma$ iff $\alpha \in f_G(\Delta)$ and $\alpha \in f_G(\Gamma)$ iff $\alpha \in f_G(\Delta) \cap f_G(\Gamma)$.

(c) We are proving $f_G(\Delta) \cdot f_H(\Delta) \subseteq f_{G*H}(\Delta)$. Assume $\alpha \in f_G(\Delta) \cdot f_H(\Delta)$. That means that there are $\beta \in f_G(\Delta)$ and $\gamma \in f_H(\Delta)$ such that $(\beta \otimes \gamma) \to \alpha \in \lambda$. The rule $R12$ gives us $(\Box_G \beta \wedge \Box_H \gamma) \to \Box_{G*H} \alpha \in \lambda$. Moreover, $\Box_G \beta \in \Delta$ and $\Box_H \gamma \in \Delta$, so $\Box_G \beta \wedge \Box_H \gamma \in \Delta$. It follows that $\Box_{G*H} \alpha \in \Delta$, and, consequently, $\alpha \in f_{G*H}(\Delta)$.

Lemmas 4 and 5 lead to the following strengthening of Theorem 2.

Theorem 5. *$\mathcal{M}^\lambda$ is a communication model.*

Theorem 5 allows us to express the following "truth-lemma" as a meningful statement. In addition, we will show that the statement is true.

Theorem 6. *For any $L_\Box$-formula α and any λ-theory Γ:*

$\Gamma \Vdash \alpha$ *in* $\mathcal{M}^\lambda$ *iff* $\alpha \in \Gamma$.

Proof. We can proceed by induction on the complexity of α. The base of the induction and the inductive steps for $\neg$, $\to$, $\wedge$, $\otimes$, and $\vee$ are the same as in the proof of Theorem 3. Let us consider the case of $\Box_G$. The induction hypothesis is that the claim holds for an $L_\Box$-formula β. To see that then the claim holds also for $\Box_G \beta$, we can observe that the following equivalences hold: $\Gamma \Vdash \Box_G \beta$ iff $f_G(\Gamma) \Vdash \beta$ iff $\beta \in f_G(\Gamma)$ iff $\Box_G \beta \in \Gamma$.

Corollary 4. *$\alpha \in \lambda$ iff α is valid in $\mathcal{M}^\lambda$.*

Corollary 5. *$\Delta \vDash_{PFL} \beta$ iff $\Delta \vdash_{PFL} \beta$.*

Corollary 6. *If $\Delta \vDash_{PFL} \beta$, then there is a finite $\Gamma \subseteq \Delta$ such that $\Gamma \vDash_{PFL} \beta$.*

5 The Standard Framework as a Special Case

We show in this section that every model of standard epistemic logic with distributed knowledge (i.e. every standard intersection-based model) corresponds to a particular communication model. The modality $\Box_G$ will boil down to standard distributed knowledge in these specific cases. The language L_D is a basic language of classical propositional logic enriched with a modality of distributed knowledge for every set of agents $A \subseteq Ag$:

$\alpha ::= p \mid \neg\alpha \mid \alpha \wedge \alpha \mid D_A \alpha.$

A *standard model* is a tuple $\mathfrak{M} = \langle W, \{R_a\}_{a \in Ag}, V \rangle$, where W is a non-empty set (of possible worlds), $R_a : W \to \mathcal{P}(W)$, for every $a \in Ag$, and $V : At \to \mathcal{P}(W)$. Moreover, for every set of agents $A \subseteq Ag$, we define a function $R_A : W \to \mathcal{P}(W)$ in the following way:

$R_A(w) = \bigcap_{a \in A} R_a(w).$

In a given standard model $\langle W, R, V \rangle$, a relation of truth between worlds and L_D-formulas is defined in the following way:

- $w \models p$ iff $w \in V(p)$,
- $w \models \neg\alpha$ iff $w \not\models \alpha$,
- $w \models \alpha \wedge \beta$ iff $w \models \alpha$ and $w \models \beta$,
- $w \models D_A\alpha$ iff for every $v \in R_A(w)$, $v \models \alpha$.

An L_D-formula is valid in a standard model iff it is true in every world of that model. We can assign to every communication scenario G a set of agents $s(G)$ by the following recursive equations:

$$s(a) = \{a\}, \text{ for every } a \in Ag, \text{ and } s(G * H) = s(G) \cup s(H).$$

So, $s(G)$ is the set of agents occurring in G. Now we will construct for any given standard model $\mathfrak{M} = \langle W, \{R_a\}_{a \in Ag}, V \rangle$ a communication model $\mathfrak{M}^i = \langle S, +, \cdot, 0, 1, C, \{f_G\}_{G \in CS}, V^\dagger \rangle$ in the following way:

- $S = \mathcal{P}(W)$,
- $x + y = x \cup y$ and $x \cdot y = x \cap y$,
- $0 = \emptyset$ and $1 = W$,
- xCy iff $x \cap y \neq \emptyset$,
- $f_G(x) = \bigcup_{w \in x} R_{s(G)}(w)$,
- $x \in V^\dagger(p)$ iff $x \subseteq V(p)$.

Lemma 6. *For all $\mathfrak{M}$, $\mathfrak{M}^i$ is a communication model.*

Proof. We will verify only that the three conditions for the group functions are satisfied. In $\mathfrak{M}^i$ these conditions boil down to the following claims:

- $\bigcup_{w \in \emptyset} R_A(w) = \emptyset$,
- $\bigcup_{w \in x \cup y} R_A(w) = (\bigcup_{w \in x} R_A(w)) \cup (\bigcup_{w \in y} R_A(w))$,
- $\bigcup_{w \in x} R_{A \cup B}(w) \subseteq (\bigcup_{w \in x} R_A(w)) \cap (\bigcup_{w \in x} R_B(w))$.

The first two claims are obvious. We will prove the third one. Assume that $v \in \bigcup_{w \in x} R_{A \cup B}(w)$. Then there is $w \in x$ such that $v \in R_{A \cup B}(w)$, i.e. for all $a \in A \cup B$, $v \in R_a(w)$. It follows that there is $w \in x$ such that for all $a \in A$, $v \in R_a(w)$, and there is $w \in x$ such that for all $a \in B$, $v \in R_a(w)$. In other words, there is $w \in x$ such that $v \in R_A(w)$, and there is $w \in x$ such that $v \in R_B(w)$, i.e. $v \in (\bigcup_{w \in x} R_A(w)) \cap (\bigcup_{w \in x} R_B(w))$.

Now it can be explained why we did not require $f_{G*H}(x) = f_G(x) \cdot f_H(x)$. This equation does not hold even in the most simple models generated by standard epistemic models. In particular, it does not generally hold that

$$(\bigcup_{w \in x} R_A(w)) \cap (\bigcup_{w \in x} R_B(w)) \subseteq \bigcup_{w \in x} R_{A \cup B}(w).$$

For the sake of simplicity, assume that $A = \{a\}$, $B = \{b\}$, and $x = \{w_1, w_2\}$. Consider for example the situation described by this table:

	R_a	R_b	$R_{\{a,b\}}$
w_1	$\{w_1, v\}$	$\{w_1\}$	$\{w_1\}$
w_2	$\{w_2\}$	$\{w_2, v\}$	$\{w_2\}$

In this situation, $v \in (\bigcup_{w \in x} R_A(w)) \cap (\bigcup_{w \in x} R_B(w))$ but $v \notin \bigcup_{w \in x} R_{A \cup B}(w)$. (Nevertheless, $(f\!=)$ holds in $\mathfrak{M}^i$ for singleton states x.) Dually speaking, suppose that v is the only world in which p is false. Then, the state $f_{A \cup B}(x) = \{w_1, w_2\}$ supports the information that p but the state $f_A(x) \cap f_B(x) = \{v, w_1, w_2\}$ does not support p.

As the last step, let us introduce a recursive translation tr of $L_\Box$ into L_D. For every atomic formula p, we define $tr(p) = p$. Moreover, $tr(\bot) = q \wedge \neg q$ and $tr(t) = \neg(q \wedge \neg q)$, for a selected atomic formula q. The translation operates on complex formulas according to these equations:

$$tr(\neg\alpha) = \neg tr(\alpha) \qquad tr(\Box_G \alpha) = D_{s(G)} tr(\alpha)$$
$$tr(\alpha \wedge \beta) = tr(\alpha) \wedge tr(\beta) \qquad tr(\alpha \otimes \beta) = tr(\alpha) \wedge tr(\beta)$$
$$tr(\alpha \to \beta) = \neg(tr(\alpha) \wedge \neg tr(\beta)) \qquad tr(\alpha \vee \beta) = \neg(\neg tr(\alpha) \wedge \neg tr(\beta))$$

Theorem 7. *For every $\mathfrak{M}$, every set x of its worlds, and every $L_\Box$-formula α, the following holds:*

$$x \Vdash \alpha \text{ in } \mathfrak{M}^i \text{ iff for all } w \in x,\ w \vDash tr(\alpha) \text{ in } \mathfrak{M}.$$

Proof. Induction on the complexity of α. We will show just the case of $\Box_G$. As the induction hypothesis we assume that our claim holds for an $L_\Box$-formula β. The following equivalences show that then it must hold also for $\Box_G\beta$.

$$\begin{aligned} x \Vdash \Box_G\beta &\text{ iff } f_G(x) \Vdash \beta \\ &\text{ iff } \textstyle\bigcup_{w \in x} R_{s(G)}(w) \Vdash \beta \\ &\text{ iff for all } v \in \textstyle\bigcup_{w \in x} R_{s(G)}(w),\ v \vDash tr(\beta) \\ &\text{ iff for all } w \in x \text{ and for all } v \in R_{s(G)}(w),\ v \vDash tr(\beta) \\ &\text{ iff for all } w \in x,\ w \vDash D_{s(G)} tr(\beta) \\ &\text{ iff for all } w \in x,\ w \vDash tr(\Box_G\beta). \end{aligned}$$

Corollary 7. *For every $L_\Box$-formula α, $tr(\alpha)$ is valid in $\mathfrak{M}$ iff α is valid in $\mathfrak{M}^i$.*

6 Conclusion

In this paper, we have formulated a generalization of the standard semantics for distributed knowledge. The standard modality of distributed knowledge/belief that is indexed by sets of agents has been generalized to an epistemic modality which is relative to structured communication scenarios. Our general framework allows to add this modality to a large class of non-classical logics extending a weak, non-associative, non-distributive and non-commutative Full Lambek Calculus with a paraconsistent negation.

References

1. van Benthem, J., Martinez, M.: The stories of logic and information. In: van Benthem, J., Adriaans, P. (eds.) Philosophy of Information, chap. 7, pp. 217–280. Elsevier (2008)
2. van Ditmarsch, H., Halpern, J.Y., van der Hoek, W., Kooi, B.: An introduction to logics of knowledge and belief. In: van Ditmarsch, H., Halpern, J.Y., van der Hoek, W., Kooi, B. (eds.) Handbook of Epistemic Logic, chap. 1, pp. 1–51. College Publications (2015)
3. Došen, K.: Sequent-systems and grupoid models II. Stud. Logica **48**, 41–65 (1989)
4. Fagin, R., Halpern, J.Y., Moses, Y., Vardi, M.Y.: Reasoning About Knowledge. MIT Press, Cambridge (1995)
5. van der Hoek, W., van Linder, B., Meyer, J.-J.: Group knowledge is not always distributed (neither is it always implicit). Math. Soc. Sci. **38**, 215–240 (1999)
6. Mares, E.: Relevant Logic: A Philosophical Interpretation. Cambridge University Press, Cambridge (2004)
7. Paoli, F.: Substructural Logics: A Primer. Springer, Dordrecht (2002)
8. Punčochář, V.: Substructural inquisitive logics. Under review (201X)
9. Read, S.: Relevant Logic: A Philosophical Examination of Inference. Basil Blackwell, Oxford (1988)
10. Restall, G.: An Introduction to Substructural Logics. Routledge, London (2000)
11. Roelofsen, F.: Distributed knowledge. J. Appl. Non-Class. Logics **16**, 255–273 (2007)
12. Sedlár, I.: Substructural epistemic logics. J. Appl. Non-Class. Logics **25**, 256–285 (2016)
13. Sequoiah-Grayson, S.: Epistemic closure and commutative, nonassociative residuated structures. Synthese **190**, 113–128 (2013)
14. Urquhart, A.: Semantics for relevant logics. J. Symbol. Logic **37**, 159–169 (1972)
15. Wansing, H.: An informational interpretation of substructural propositional logics. J. Logic Lang. Inform. **2**, 285–308 (1993)
16. Ågotnes, T., Wáng, Y.N.: Resolving distributed knowledge. In: Proceedings of TARK 2015, pp. 31–50 (2015)

Logical Argumentation Principles, Sequents, and Nondeterministic Matrices

Esther Anna Corsi and Christian G. Fermüller(✉)

Institut für Computersprachen, Vienna University of Technology,
Favoritenstr. 9-11, 1040 Vienna, Austria
{esther,chrisf}@logic.at

Abstract. The concept of "argumentative consequence" is introduced, involving only the attack relations in Dung-style abstract argumentation frames. Collections of attack principles of different strength, referring to the logical structure of claims of arguments, lead to new characterizations of classical and nonclassical consequence relations. In this manner systematic relations between structural constraints on abstract argumentation frames, sequent rules, and nondeterministic matrix semantics for corresponding calculi emerge.

Keywords: Argumentation · Sequent calculus · Nondeterministic matrices

1 Introduction

In a seminal paper Dung [7] demonstrated that various concepts of nonmonotonic reasoning, logic programming, and game theory can be modeled profitably using so-called abstract argumentation frameworks. The latter are just directed graphs, where the vertices are identified with arguments and the edges represent an attack relation between arguments. A lot of research is devoted to different 'semantics' of abstract argumentation and to corresponding reasoning mechanisms as documented, e.g., in [5,15]. While the technical progress as well as application oriented developments in this area are impressive, certain foundational issues, in particular regarding logical principles guiding the search for consequence relations that lurk behind frameworks of (explicit and implicit) arguments, remain less well explored.

We are interested in "logics of argumentation". Of course, quite different concepts can be associated with this term. Here, we refer to consequence relations extracted solely from the structure of attack relations, thus providing an alternative semantics of formal logic that does not involve Tarski's classic concept of truth in a model, but rather takes (potential) counter-arguments as central. We call a proposition G an *argumentative consequence* of propositions $F_1, \ldots, F_n$ if every argument that attacks G also attacks at least one of the premises F_i (see Sect. 4). Admittedly, the identification of arguments with single propositions, that is at play here, calls for further explanation. We take up this issue

A. Baltag et al. (Eds.): LORI 2017, LNCS 10455, pp. 422–437, 2017.
DOI: 10.1007/978-3-662-55665-8_29

in Sect. 2, but right away declare that by a *semi-abstract argumentation frame* we mean a directed graph, where each vertex carries a propositional formula that can be understood as the *claim* of an argument and where the edges represent *attacks* between corresponding arguments. In other words, a semi-abstract argumentation frame is just like an ordinary abstract argumentation frame in the sense of Dung [7], where additionally each vertex is labeled by a formula representing its claim.

The endeavor to extract a consequence relation from given semi-abstract argumentation frames gets off the ground by realizing that the logical structure of (claims of) arguments poses constraints on the structure of attack relations. For example, it seems natural to stipulate that every argument that attacks an argument with claim F also attacks (at least implicitly) all arguments that feature the stronger claim $F \wedge G$. Similarly, if an argument X attacks a disjunctive claim $F \vee G$ then it seems reasonable to assume that X also attacks any (bolder) argument with either claim F or claim G. Note that the mentioned "attack principles" are intuitively justified, independently of whether we interpret claims classically, intuitionistically or according to an even weaker logic. We discuss such principles in Sect. 3 and investigate in the rest of the paper the prospects of extracting a corresponding logic of argumentation. In particular, we are interested in determining a collection of attack principles that yields classical logic as the corresponding argumentative consequence relation. To this aim we employ Gentzen's classical sequent calculus **LK** [11] and relate its inference rules to argumentation in a systematic manner in Sect. 5. Perhaps not surprisingly, it turns out that some of the principles, that have to be imposed on the attack relation in order to induce classical consequence, are intuitively very demanding and in fact unjustified in a specific sense, that will be made precise in Sect. 6. This triggers the question whether a collection of attack principles, that are justifiable intuitively as well as formally, gives raise to some interesting nonclassical logic. A simple positive answer arises in Sects. 7 and 8 as a straightforward application of the theory of canonical sequent calculi [3,4]: disregarding problematic attack principles corresponds to a *fragment* of the classical sequent calculus in which some of the logical rules are missing. The resulting logic can be characterized by nondeterminstic matrices. In Sect. 9 we will hint at related literature, before concluding with some questions for future research in Sect. 10.

2 Semi-abstract Argumentation

As already indicated, we look at arguments as logical propositions. Although one can find several examples in the literature, where arguments indeed consist in single (possibly logically complex) propositions (see, e.g., many articles in [15]), it is more common to distinguish explicitly between the *support* and the *claim* of an argument. Whereas the latter usually is indeed a single sentence, the support may consist of several statements, rules, or even of small theories, including default rules (see, e.g., [5]). Here we focus on *semi-abstract argumentation*, which identifies claims of arguments with (interpreted) logical formulas, but ignores supports.[1] More formally, let PV be an infinite set of propositional variables and define the set of propositional formulas $\mathcal{PL}$ over PV by

$$\mathcal{F} ::= \mathcal{F} \vee \mathcal{F} \mid \mathcal{F} \wedge \mathcal{F} \mid \mathcal{F} \supset \mathcal{F} \mid \neg \mathcal{F} \mid \mathcal{PV}$$

where $\mathcal{F}$ and $\mathcal{PV}$ are used as meta-variables for formulas and propositional variables, respectively.

Definition 1. *A* semi-abstract argumentation frame (SAF) *is a directed graph* $(\mathcal{A}, R_{\rightarrow})$*, where each vertex* $a \in \mathcal{A}$ *is labeled by a formula of* $\mathcal{PL}$*, representing the* claim *of argument* a*, and the edges* ($R_{\rightarrow}$) *represent the* attack relation *between arguments.*

Note that an SAF is like an ordinary abstract argumentation frame as introduced by Dung [7], except for attaching a formula (its claim) to each argument. We say that F *attacks* G and write $F \longrightarrow G$ if there is an edge from an argument labeled by F to one labeled by G.[2] But the reader should be aware of the fact that, in general, $F \longrightarrow G$ stands for "an argument with claim G is attacked by some argument with claim F". We abbreviate "not $F \longrightarrow G$" by $F \not\longrightarrow G$.

Example 1. Consider the following statements:

- "The overall prosperity in society increases." (P)
- "Warlike conflicts about energy resources arise." (W)
- "The level of CO_2 emissions is getting dangerously high." (C)
- "Awareness about the need of environmental protection increases." (E)

Consider an argumentation frame containing arguments, where the claims consist in some of these statements or in some simple logical compounds thereof. Using the indicated abbreviations and identifying vertices with theirs labels, a concrete corresponding SAF $S_E = (\mathcal{A}, R_{\rightarrow})$ is given by $\mathcal{A} = \{P, E, W, P \supset C, E \vee C, P \wedge C\}$ and $R_{\rightarrow} = \{E \longrightarrow P \supset C,\ W \longrightarrow E \vee C,\ W \longrightarrow P \wedge C,\ E \vee C \longrightarrow P\}$.

3 Logical Attack Principles

It seems reasonable to expect that an argument that attacks G also attacks claims that logically entail G. In our notation, this amounts to the following *general attack principle*:

(A) If $F \longrightarrow G$ and $G' \models G$ then $F \longrightarrow G'$.

[1] Of course, the idea to label arguments in an argumentation frame with formulas is not new (see, e.g., [6,12,13]). We use the labels to highlight the logical form of concrete claims, while abstracting away from their particular support or any specific form of attack.

[2] The same formula may occur as claim of different arguments. Thus we (implictly) refer to *occurrences* of formulas, rather than to formulas themselves when talking about attacks in a given SAF.

Applied naively, principle **(A)** is problematic for at least two reasons. (1) We have not specified which logic the consequence relation $\models$ refers to. Classical logic may be a canonical choice, but we should not dismiss weaker logics, that are potentially more adequate in the context of defeasible reasoning, too quickly. (2) Even for classical propositional logic deciding logical consequence is computationally intractable, in general. Arguably, a realistic model of argumentation might insist on constraining **(A)** to arguments G that *immediately* follow from G' in some appropriate sense. This motivates our focus on principles that follow already from *simple and transparent instances* of **(A)**:

(A.$\wedge$) If $F \longrightarrow A$ or $F \longrightarrow B$ then $F \longrightarrow A \wedge B$.
(A.$\vee$) If $F \longrightarrow A \vee B$ then $F \longrightarrow A$ and $F \longrightarrow B$.
(A.$\supset$) If $F \longrightarrow A \supset B$ then $F \longrightarrow B$.

These specific instances of **(A)** involve only very basic consequence claims, that are already valid in *minimal logic* [16], i.e. in the positive fragment of intuitionistic logic (and in fact in even weaker logics).

Proposition 1. *If "$\models$" refers to minimal logic, then the general attack principle* **(A)** *entails the specific attack principles* **(A.$\wedge$)**, **(A.$\vee$)**, **(A.$\supset$)**.

Proof. Immediate from the fact that $A \wedge B \models A$, $A \wedge B \models B$, $A \models A \vee B$, $B \models A \vee B$, $B \models A \supset B$ in minimal logic. □

Principle **(A.$\supset$)** might be considered intuitively less obvious than **(A.$\wedge$)** and **(A.$\vee$)**. Indeed, the above justification of **(A.$\supset$)** depends on the fact that $B \models A \supset B$ according to minimal logic and thus involves a logical principle that may be disputed, e.g., from the point of view of "relevant entailment" (see [8]). Therefore we prefer to replace **(A.$\supset$)** by the following attack principle:

(B.$\supset$) If $F \longrightarrow B$ and $F \not\longrightarrow A$ then $F \longrightarrow A \supset B$.

As we will see in Sects. 5 and 6, **(B.$\supset$)** relates to a basic inference principle about implicative premises in logical consequence claims, that holds in a very wide range of logics.

We have not yet specified any principle involving negation. Negation is often defined by $\neg F =_{df} F \supset \bot$, where $\bot$ is an atomic formula that signifies an elementary contradiction. But minimal logic treats $\bot$ just like an arbitrary propositional variable and thus does not give rise to specific attack principle for negation. However the following principle seems intuitively plausible: If an argument attacks (an argument with claim) A then it does not simultaneously also attack the negation of A. In symbols: dummy

(B.$\neg$) If $F \longrightarrow A$ then $F \not\longrightarrow \neg A$.

We will show in Sect. 8 that the (weak) attack principles mentioned so far give raise to a logic that arises from dropping some logical rules from Gentzen's classical sequent calculus **LK**. But we are also interested in the question, which (stronger) attack principles have to be imposed on semi-abstract argumentation frames in order to recover ordinary classical logic. To this aim we introduce the following additional attack principles that are inverse to **(A.**$\wedge$**)**, **(A.**$\vee$**)**, **(B.**$\supset$**)**, and **(B.**$\neg$**)**, respectively.

(C.$\wedge$**)** If $F \longrightarrow A \wedge B$ then $F \longrightarrow A$ or $F \longrightarrow B$.
(C.$\vee$**)** If $F \longrightarrow A$ and $F \longrightarrow B$ then $F \longrightarrow A \vee B$.
(C.$\supset$**)** If $F \longrightarrow A \supset B$ then $F \longrightarrow B$ and $F \not\longrightarrow A$.
(C.$\neg$**)** If $F \not\longrightarrow A$ then $F \longrightarrow \neg A$.

Conditions like **(A.**$\wedge$**)** seem to entail that the corresponding SAFs are infinite. However, we may *relativize* the attack principles to sets of formulas Γ (usually the set of claims of arguments of some finite argumentation frame). E.g.,

(A.$\wedge$**)** For every $A, B, F \in \Gamma$: $F \longrightarrow A$ or $F \longrightarrow B$ implies $F \longrightarrow A \wedge B$, if $A \wedge B \in \Gamma$.

In the following we will tacitly assume that attack principles are relativized to some (finite) set of formulas that will always be clear from the context.

4 Argumentative Consequence

Viewing an argument attacking a certain claim F as a kind of counter-model to F suggests the following definition of consequence, that, in contrast to usual definitions of logical consequence, neither refers to truth values nor to interpretations in the usual (Tarskian) sense.

Definition 2. *F is an* argumentative consequence *of (the claims of) arguments $A_1, \ldots, A_n$ with respect to an SAF S ($A_1, \ldots, A_n \models^S_{arg} F$) if all arguments in S that attack F also attack A_i for some $i \in \{1, \ldots, n\}$.*[3] *For a set of SAFs $\mathcal{S}$ $A_1, \ldots, A_n \models^{\mathcal{S}}_{arg} F$ if $A_1, \ldots, A_n \models^S_{arg} F$ for all $S \in \mathcal{S}$.*

To render this notion of consequence plausible from a logical perspective, the underlying SAFs should be rich enough to contain (potential) arguments that feature also the subformulas of occurring formulas as claims. Moreover, we want these SAFs to satisfy at least some of our logical attack principles.

Definition 3. *An SAF S is* logically closed *with respect to the set of formulas Γ if all subformulas of formulas in Γ occur as claims of some argument in S.*

Let $\mathcal{AP}$ be a set of attack principles, then F is an argumentative $\mathcal{AP}$*-consequence of $A_1, \ldots, A_n$ ($A_1, \ldots, A_n \models^{\mathcal{AP}}_{arg} F$) if $A_1, \ldots, A_n \models^S_{arg} F$ for every SAF S that is logically closed with respect to $\{A_1, \ldots, A_n, F\}$ and moreover satisfies all (appropriately relativized) attack principles in $\mathcal{AP}$.*

[3] Note that, if we identify arguments with counter-models and if S contains all relevant counter-models, then argumentative consequence coincides with ordinary logical consequence: every counter-model of the conclusion must invalidate some premise.

We do *not* suggest that argumentation frames should always be logically closed. We rather view logical closure as an operation that augments a given SAF by "potential claims of arguments", which are *implicit* according to logical attack principles. Argumentative consequence thus refers to a "logical completion" of interpreted argumentation frames, rather than directly to arbitrarily given collections of arguments.

Example 2. Continuing Example 1 of Sect. 2, we observe the SAF S_E is almost, but not yet fully, logically closed with respect to the statements that appear as claims of arguments: we just have to add one more vertex with label C (for "Awareness about the need of environmental protection increases") to obtain logical closure.

More interestingly, we may check which additional (implicit) attacks are induced by which of our attack principles: Since there is an argument with claim $E \vee C$ that attacks an argument with claim P, principle **(A.$\wedge$)** stipulates that there is also an attack from $E \vee C$ to an argument with the the stronger claim $P \wedge C$. In other words **(A.$\wedge$)** induces the addition of the edge $E \vee C \longrightarrow P \wedge C$. Note that this corresponds to the plausible assumption that an argument that attacks the claim that "the overall prosperity in society increases" also attacks the statement "The overall prosperity in society increases and (moreover) the level of CO_2 emissions is getting dangerously high." Similarly the principle **(A.$\vee$)** stipulates that an argument with claim C (added for logical closure, as explained above) should be attacked by an argument with claim W, since such an argument already attacks the weaker claim $E \vee C$. The addition of $W \longrightarrow C$, likewise induces the additional attack edge $W \longrightarrow E$. Moreover, since we have $E \longrightarrow P \supset C$, but $E \not\longrightarrow P$, the principle **(B.$\supset$)** induces the addition of $E \longrightarrow C$ to $R_{\rightarrow}$. The stronger principle **(C.$\supset$)** would call for $E \longrightarrow C$ even without $E \not\longrightarrow P$. The strong conjunction principle **(C.$\wedge$)** demands that either $W \longrightarrow P$ or $W \longrightarrow C$. Both seem reasonable with respect to the intended interpretation of S_E. However, $W \longrightarrow C$ is already present anyway if **(A.$\vee$)** is imposed, as explained above. Likewise the strong disjunction principle **(C.$\vee$)** is already satisfied.

In the next section we will investigate which collection $\mathcal{AP}$ of attack principles allows one to recover classical logical consequence as argumentative $\mathcal{AP}$-consequence. Gentzen's classical sequent calculus **LK** turns out to be a perfect tool for this task. Hence, we generalize the consequence relation to (disjunctive) sets of premises, as usual in proof theory. Moreover, we adopt the convention to identify finite lists and sets of formulas, and write, e.g., Γ, F for $\Gamma \cup \{F\}$.

Definition 4. *Let Γ and Δ be finite sets of formulas and let S be an SAF. Δ is an* argumentative consequence *of Γ with respect to an SAF S ($\Gamma \models^S_{arg} \Delta$) if all arguments in S that attack every $F \in \Delta$ attack at least some $G \in \Gamma$.*

The generalization to sets of SAFs and sets of attack principles is just as indicated above.

With respect to an SAF $S = (\mathcal{A}, R_{\rightarrow})$ we define:

- $\mathrm{atts}_S(F) =_{df} \{A \mid A \longrightarrow F, A \in \mathcal{A}\}$,

- $\overline{\text{atts}}_S(\Gamma) =_{df} \bigcup_{F \in \Gamma} \text{atts}_S(F)$,
- $\underline{\text{atts}}_S(\Gamma) =_{df} \bigcap_{F \in \Gamma} \text{atts}_S(F)$.

The following simple facts will be useful below:

(a) $\Gamma \models^S_{\text{arg}} \Delta$ iff $\underline{\text{atts}}_S(\Delta) \subseteq \overline{\text{atts}}_S(\Gamma)$.
(b) $\overline{\text{atts}}_S(\Gamma, F) = \overline{\text{atts}}_S(\Gamma) \cup \text{atts}_S(F)$.
(c) $\underline{\text{atts}}_S(\Gamma, F) = \underline{\text{atts}}_S(\Gamma) \cap \text{atts}_S(F)$.

We will drop the index S if no ambiguity arises.

5 Relating Sequent Rules and Attack Principles

In our version of Gentzen's classical sequent calculus **LK** [11], sequents are pairs of sets of formulas, written as $\Gamma \vdash \Delta$. Initial sequents (*axioms*) are of the form $A, \Gamma \vdash \Delta, A$. The logical rules (for introducing logical connectives) are as follows:

$$\frac{A, \Gamma \vdash \Delta}{\Gamma \vdash \Delta, \neg A}\ (\neg, r) \qquad \frac{\Gamma \vdash \Delta, A}{\neg A, \Gamma \vdash \Delta}\ (\neg, l)$$

$$\frac{\Gamma \vdash \Delta, A \qquad \Gamma \vdash \Delta, B}{\Gamma \vdash \Delta, A \wedge B}\ (\wedge, r) \qquad \frac{A, B, \Gamma \vdash \Delta}{A \wedge B, \Gamma \vdash \Delta}\ (\wedge, l)$$

$$\frac{\Gamma \vdash \Delta, A, B}{\Gamma \vdash \Delta, A \vee B}\ (\vee, r) \qquad \frac{A, \Gamma \vdash \Delta \qquad B, \Gamma \vdash \Delta}{A \vee B, \Gamma \vdash \Delta}\ (\vee, l)$$

$$\frac{A, \Gamma \vdash \Delta, B}{\Gamma \vdash \Delta, A \supset B}\ (\supset, r) \qquad \frac{\Gamma \vdash \Delta, A \qquad B, \Gamma \vdash \Delta}{A \supset B, \Gamma \vdash \Delta}\ (\supset, l)$$

Note that we do not need to use structural rules: weakening is redundant because of the more general form of axioms compared to Gentzen's $A \vdash A$; contraction is eliminated because we treat sequents as pairs of *sets* of formulas. Moreover the calculus is cut-free complete with respect to the classical consequence relation $\models_{cl}$ (generalized to disjunctions of premises, as usual). We rely on the following well known facts (see, e.g., [16]).

Proposition 2. *$\Gamma \vdash \Delta$ is derivable in* **LK** *iff $\Gamma \models_{cl} \Delta$.*

Proposition 3 (e.g., [16], Proposition 3.5.4). *The rules of* **LK** *are invertible; i.e., if a sequent $\Gamma \vdash \Delta$ is derivable and $\Gamma \vdash \Delta$ is an instance of a lower sequence of an* **LK**-*rule then the corresponding instance(s) of the upper sequent(s) is (are) derivable, too.*

Let CAP consist of the attack principles **(A.∧)**, **(A.∨)**, **(B.⊃)**, **(B.¬)**, **(C.∧)**, **(C.∨)**, **(C.⊃)**, and **(C.¬)**. We first show that **LK** is sound with respect to argumentative CAP-consequence.

Theorem 1. *If $\Gamma \vdash \Delta$ is derivable in* **LK** *then* $\Gamma \models^{\mathsf{CAP}}_{arg} \Delta$.

Proof. Clearly, $A, \Gamma \models^{\mathsf{CAP}}_{\mathrm{arg}} \Delta, A$. It remains to check that the inference rules of **LK** preserve CAP-consequence. We only present two cases.

$(\neg, r)$ We have to show that $A, \Gamma \models^{\mathsf{CAP}}_{\mathrm{arg}} \Delta$ implies $\Gamma \models^{\mathsf{CAP}}_{\mathrm{arg}} \Delta, \neg A$. The premise states that $\overline{\mathrm{atts}}_S(\Gamma, A) \supseteq \underline{\mathrm{atts}}_S(\Delta)$ for every SAF S that is logically closed with respect to $\Gamma \cup \Delta \cup \{A\}$ and satisfies the CAP-principles. The conclusion states that $\overline{\mathrm{atts}}_{S'}(\Gamma) \supseteq \underline{\mathrm{atts}}_{S'}(\Delta, \neg A)$, where S' now ranges over the CAP-complying SAFs that are closed with respect to $\Gamma \cup \Delta \cup \{\neg A\}$. Since every SAF S' of the second kind reduces to one of the first kind (without $\neg A$) we may argue over any such SAF and drop the reference. We obtain:

$$
\begin{aligned}
& \overline{\mathrm{atts}}(\Gamma, A) \supseteq \underline{\mathrm{atts}}(\Delta) \\
\Leftrightarrow\ & \overline{\mathrm{atts}}(\Gamma) \cup \mathrm{atts}(A) \supseteq \underline{\mathrm{atts}}(\Delta) \\
\Rightarrow\ & (\overline{\mathrm{atts}}(\Gamma) \cup \mathrm{atts}(A)) \cap \mathrm{atts}(\neg A) \supseteq \underline{\mathrm{atts}}(\Delta) \cap \mathrm{atts}(\neg A) \\
\Leftrightarrow\ & (\overline{\mathrm{atts}}(\Gamma) \cap \mathrm{atts}(\neg A)) \cup (\mathrm{atts}(A) \cap \mathrm{atts}(\neg A)) \supseteq \underline{\mathrm{atts}}(\Delta) \cap \mathrm{atts}(\neg A) \\
\Leftrightarrow\ & \overline{\mathrm{atts}}(\Gamma) \cap \mathrm{atts}(\neg A) \supseteq \underline{\mathrm{atts}}(\Delta) \cap \mathrm{atts}(\neg A) \quad [\text{using } (\mathbf{B}.\neg)] \\
\Rightarrow\ & \overline{\mathrm{atts}}(\Gamma) \supseteq \underline{\mathrm{atts}}(\Delta) \cap \mathrm{atts}(\neg A) \\
\Leftrightarrow\ & \overline{\mathrm{atts}}(\Gamma) \supseteq \underline{\mathrm{atts}}(\Delta, \neg A).
\end{aligned}
$$

Crucially, $(\mathbf{B}.\neg)$ amounts to $\mathrm{atts}(A) \cap \mathrm{atts}(\neg A) = \emptyset$.

$(\supset, l)$ We show that, if $\Gamma \models^{\mathsf{CAP}}_{\mathrm{arg}} \Delta, A$ and $B, \Gamma \models^{\mathsf{CAP}}_{\mathrm{arg}} \Delta$, then $A \supset B, \Gamma \models^{\mathsf{CAP}}_{\mathrm{arg}} \Delta$. Like above, the premises amount to $\overline{\mathrm{atts}}(\Gamma) \supseteq \underline{\mathrm{atts}}(\Delta, A)$ and $\overline{\mathrm{atts}}(B, \Gamma) \supseteq \underline{\mathrm{atts}}(\Delta)$, respectively; whereas the conclusion is $\overline{\mathrm{atts}}(A \supset B, \Gamma) \supseteq \underline{\mathrm{atts}}(\Delta)$. We use $(\cdot)^c$ to denote the complement with respect to the set of arguments in question.

$$
\begin{aligned}
& \underline{\mathrm{atts}}(\Delta) \cap \mathrm{atts}(A) \subseteq \overline{\mathrm{atts}}(\Gamma) \text{ and } \underline{\mathrm{atts}}(\Delta) \subseteq \overline{\mathrm{atts}}(\Gamma) \cup \mathrm{atts}(B) \\
\Leftrightarrow\ & \underline{\mathrm{atts}}(\Delta) \cap \mathrm{atts}(A) \cap (\overline{\mathrm{atts}}(\Gamma))^c = \emptyset \text{ and } \underline{\mathrm{atts}}(\Delta) \cap (\overline{\mathrm{atts}}(\Gamma) \cup \mathrm{atts}(B))^c = \emptyset \\
\Leftrightarrow\ & \underline{\mathrm{atts}}(\Delta) \cap (\overline{\mathrm{atts}}(\Gamma))^c \cap \mathrm{atts}(A) = \emptyset \text{ and } \underline{\mathrm{atts}}(\Delta) \cap (\overline{\mathrm{atts}}(\Gamma))^c \cap (\mathrm{atts}(B))^c = \emptyset \\
\Leftrightarrow\ & \underline{\mathrm{atts}}(\Delta) \cap (\overline{\mathrm{atts}}(\Gamma))^c \cap (\mathrm{atts}(A) \cup (\mathrm{atts}(B))^c) = \emptyset \\
\Leftrightarrow\ & \underline{\mathrm{atts}}(\Delta) \cap (\overline{\mathrm{atts}}(\Gamma))^c \cap ((\mathrm{atts}(A))^c \cap \mathrm{atts}(B))^c = \emptyset \\
\Leftrightarrow\ & \underline{\mathrm{atts}}(\Delta) \cap (\overline{\mathrm{atts}}(\Gamma) \cup (\mathrm{atts}(A))^c \cap \mathrm{atts}(B))^c = \emptyset \\
\Leftrightarrow\ & \underline{\mathrm{atts}}(\Delta) \subseteq \overline{\mathrm{atts}}(\Gamma) \cup (\mathrm{atts}(B) \setminus \mathrm{atts}(A)) \quad [\text{using } (\mathbf{B}.\supset)] \\
\Rightarrow\ & \underline{\mathrm{atts}}(\Delta) \subseteq \overline{\mathrm{atts}}(\Gamma) \cup \mathrm{atts}(A \supset B)
\end{aligned}
$$

Note that $(\mathbf{B}.\supset)$ amounts to $\mathrm{atts}(B) \setminus \mathrm{atts}(A) \subseteq \mathrm{atts}(A \supset B)$.

The other cases are analogous. For later reference, we list which of the attack principles are used for which further rules: $(\mathbf{C}.\neg)$ for $(\neg, l)$, $(\mathbf{C}.\wedge)$ for $(\wedge, r)$, $(\mathbf{A}.\wedge)$ for $(\wedge, l)$, $(\mathbf{A}.\vee)$ for $(\vee, r)$, $(\mathbf{C}.\vee)$ for $(\vee, l)$, and $(\mathbf{C}.\supset)$ for $(\supset, r)$. □

To show the completeness of **LK** with respect to argumentative CAP-consequence, we rely on the invertibility of the logical rules (Proposition 3).

Theorem 2. *If* $\Gamma \models^{\mathsf{CAP}}_{arg} \Delta$ *then* $\Gamma \vdash \Delta$ *is derivable in* **LK**.

Proof. We have to check the inverse directions of the implications in the proof of Theorem 1. Again, we just present two cases, since the others are similar.

($\neg, r$) We show that $A, \Gamma \not\models^{\mathsf{CAP}}_{\mathrm{arg}} \Delta$ implies $\Gamma \not\models^{\mathsf{CAP}}_{\mathrm{arg}} \Delta, \neg A$. To this aim assume that $\overline{\mathrm{atts}}_S(A, \Gamma) \not\supseteq \underline{\mathrm{atts}}_S(\Delta)$ for some CAP-complying SAF $S = (\mathcal{A}^S, R^S_{\rightharpoonup})$ that is logically closed with respect to $\Gamma \cup \Delta \cup \{A\}$. In other words there is an $F \in \underline{\mathrm{atts}}_S(\Delta)$, such that $F \notin \overline{\mathrm{atts}}_S(A, \Gamma)$. The latter implies $F \not\longrightarrow A$ in S. Now let S' be an CAP-complying SAF that is logically closed with respect to $\Gamma \cup \Delta \cup \{\neg A\}$, where the attack relation restricted to those (claims of) arguments that already occur in S coincides with $R^S_{\rightharpoonup}$. Since $F \not\longrightarrow A$ also in S', we obtain $F \longrightarrow \neg A$ from (**C.**$\neg$). Since $F \in \underline{\mathrm{atts}}_{S'}(\Delta)$ we conclude that $F \in \underline{\mathrm{atts}}_{S'}(\Delta, \neg A)$. On the other hand, $F \notin \overline{\mathrm{atts}}_{S'}(\Gamma)$, since otherwise we already had $F \in \overline{\mathrm{atts}}_S(A, \Gamma)$. Thus we have shown that $\overline{\mathrm{atts}}_{S'}(\Gamma) \not\supseteq \underline{\mathrm{atts}}_{S'}(\Delta, \neg A)$, which entails $\Gamma \not\models^{\mathsf{CAP}}_{\mathrm{arg}} \Delta, \neg A$.

($\supset, l$) We again proceed indirectly and show that (1) $\Gamma \not\models^{\mathsf{CAP}}_{\mathrm{arg}} \Delta, A$ implies $A \supset B, \Gamma \not\models^{\mathsf{CAP}}_{\mathrm{arg}} \Delta$, and (2) $B, \Gamma \not\models^{\mathsf{CAP}}_{\mathrm{arg}} \Delta$ implies $A \supset B, \Gamma \not\models^{\mathsf{CAP}}_{\mathrm{arg}} \Delta$. For (1) assume that $\overline{\mathrm{atts}}_S(\Gamma) \not\supseteq \underline{\mathrm{atts}}_S(\Delta, A)$ for some CAP-complying SAF $S = (\mathcal{A}^S, R^S_{\rightharpoonup})$ that is logically closed with respect to $\Gamma \cup \Delta \cup \{A\}$. Thus there is an $F \in \underline{\mathrm{atts}}_S(\Delta, A)$, such that $F \notin \overline{\mathrm{atts}}_S(\Gamma)$. In particular $F \in \underline{\mathrm{atts}}_S(\Delta)$. Let S' be an CAP-complying SAF that is logically closed with respect to $\Gamma \cup \Delta \cup \{A \supset B\}$, where the attack relation restricted to those (claims of) arguments that already occur in S coincides with $R^S_{\rightharpoonup}$. Then (**C.**$\supset$) implies $F \not\longrightarrow A \supset B$ in S'. Therefore $\overline{\mathrm{atts}}_{S'}(A \supset B, \Gamma) \not\supseteq \underline{\mathrm{atts}}_{S'}(\Delta)$, which in turn entails $A \supset B, \Gamma \not\models^{\mathsf{CAP}}_{\mathrm{arg}} \Delta$.

For (2) assume that $\overline{\mathrm{atts}}_S(\Gamma, B) \not\supseteq \underline{\mathrm{atts}}_S(\Delta)$ for some CAP-complying SAF $S = (\mathcal{A}^S, R^S_{\rightharpoonup})$ that is logically closed with respect to $\Gamma \cup \Delta \cup \{B\}$. Thus there is an $F \in \underline{\mathrm{atts}}_S(\Delta)$, such that $F \notin \overline{\mathrm{atts}}_S(\Gamma, B)$. Let S' be an SAF like in case (1). Then $F \not\longrightarrow B$ also in S'. Therefore (**C.**$\supset$) implies $F \not\longrightarrow A \supset B$ in S', which entails $A \supset B, \Gamma \not\models^{\mathsf{CAP}}_{\mathrm{arg}} \Delta$, like in case (1). □

6 Sorting Out Attack Principles

We have seen that the collection CAP of attack principles leads to a characterization of classical logical consequence, that replaces model theoretic (Tarskian) semantics by a reference to specific structural properties of logically closed semi-abstract argumentation frames. Remember that classical logic is the top element in the lattice of all possible logics over the language $\mathcal{PL}$. (As usual, we identify a logic with a set of formulas that is closed under substitution and modus ponens, here.) Therefore it is hardly surprising that some of the principles in CAP might be considered too demanding to be adequate for models of logical argumentation. Consider for example (**C.**$\neg$): it says that any argument that does not attack a claim A can be understood as an argument that attacks $\neg A$. In other words every argument has to attack either A or $\neg A$. This is hardly plausible and should be contrasted with the inverse principle (**B.**$\neg$), which just stipulates that no argument attacks A and $\neg A$, simultaneously. We have justified (**A.**$\vee$) and (**A.**$\wedge$) in Sect. 3 as immediate instances of the general principle that, if an argument attacks a claim A, then it (implicitly) also attacks claims from which A logically follows. But the plausibility of the inverse principles (**C.**$\vee$) and (**C.**$\wedge$) remains in question. Intuitively, it seems justifiable to stipulate that an argument that attacks both, A and B, attacks also $A \vee B$ (**C.**$\vee$). However the requirement that any argument attacking a conjunction must attack also at least one of the conjuncts intuitively seems too strong: think of the instance $A \wedge \neg A$, against which an agent presumably may have a reasonable (general) argument, without knowing an argument that attacks either A or $\neg A$.

Rather than to simply appeal to pre-theoretic intuitions, as just outlined, we want to present a simple formal interpretation of attacks involving logically compound claims, that supports some, but not all of the attack principles in CAP. To this aim we employ standard modal logic and refer to *Kripke interpretations* $\langle W, R, V\rangle$, where W is a non-empty set of *states*, $R \subseteq W \times W$ the *accessibility relation*, and V a *valuation* $V : W \times PV \longrightarrow \{\mathbf{t}, \mathbf{f}\}$ that assigns a truth value to each propositional variable in each state. The language $\mathcal{PL}$ is enriched by a modal operator $\Box$ and its dual $\Diamond = \neg\Box\neg$. The valuation V is extended from propositional variables to classical formulas (elements of $\mathcal{PL}$) as usual. For $\Box$ we have

$$V(w, \Box F) = \mathbf{t} \text{ iff } \forall v\text{: } wRv \text{ implies } V(v, F) = \mathbf{t}.$$

A Kripke interpretation $\langle W, R, V\rangle$ is a *model* of formula F if $V(w, F) = \mathbf{t}$ for all $w \in W$. F is a $\mathcal{K}$-consequence of $G_1, \ldots, G_n$ with respect to a class $\mathcal{K}$ of Kripke interpretations if every model of $G_1, \ldots, G_n$ is also a model of F.

We view the states W as possible states of affairs and interpret wRv as "v is a possible alternative from the viewpoint of w". An attack is considered to involve the claims of the involved arguments in two ways: (1) the claim of the attacking argument is asserted to hold in all alternatives; (2) the negation of the claim of the attacked argument is asserted to hold there. Accordingly, we define:

Definition 5. *For all $F, G \in \mathcal{PL}$:*

- $\iota(F \longrightarrow G) =_{df} \Box(F \wedge \neg G)$;
- $\iota(F \not\longrightarrow G) =_{df} \Diamond(\neg F \vee G)$ *(or, equivalently, $\neg\Box(F \wedge \neg G)$).*

Recall that attack principles are implications between (disjunctions or conjunctions) of assertions of the form $F \longrightarrow G$ or $F \not\longrightarrow G$. We call an attack principle $\mathcal{K}$*-justified* if the implication translates into a valid $\mathcal{K}$-consequence claim via ι.

We have not yet imposed any restriction on Kripke interpretations. The intended interpretation of the accessibility relation R as "possible alternative" might suggest that R is an equivalence relation, or at least reflective. However we will only impose the weaker condition of seriality. Let $\mathcal{D}$ be the corresponding class of Kripke interpretations, where for every $w \in W$ there is a $v \in W$ such that wRv.

Theorem 3. *The attack principles* **(A.∧)**, **(A.∨)**, **(C.∨)**, **(C.⊃)**, *and* **(B.¬)** *are all $\mathcal{D}$-justified.*

Proof. To show that **(A.∧)** is $\mathcal{D}$-justified we have to check that $\Box(F \wedge \neg A) \models_{\mathcal{D}} \Box(F \wedge \neg(A \wedge B))$ as well as $\Box(F \wedge \neg B) \models_{\mathcal{D}} \Box(F \wedge \neg(A \wedge B))$, which is obvious. The cases for **(A.∨)**, **(C.∨)** and **(C.⊃)** are similar.

(B.¬) is $\mathcal{D}$-justified because $\Box(F \wedge \neg A) \models_{\mathcal{D}} \Diamond(\neg F \vee \neg A)$, where the seriality of $\mathcal{D}$-models is used. □

Theorem 4. *The attack principles* **(C.∧)**, **(C.¬)**, *and* **(B.⊃)** *are not $\mathcal{D}$-justified.*

Proof. It is straightforward to find counter-models for the following consequence claims:

- $\Box(F \wedge \neg(A \wedge B)) \models_{\mathcal{D}} \Box(F \wedge \neg A)$,
- $\Box(F \wedge \neg(A \wedge B)) \models_{\mathcal{D}} \Box(F \wedge \neg B)$,
- $\Diamond(\neg F \vee \neg A) \models_{\mathcal{D}} \Box(F \wedge \neg A)$,
- $\Diamond(\neg F \vee A),\ \Box(F \wedge \neg B) \models_{\mathcal{D}} \Box(F \wedge (A \wedge \neg B))$. □

To sum up, we have seen that our (admittedly rather unsophisticated and coarse) modal interpretation of the attack relation supports a formal justification of the collection of attack principles MAP ={ **(A.∧)**, **(A.∨)**, **(C.∨)**, **(C.⊃)**, **(B.¬)**}. Moreover, the modal interpretation allows us to reject the attack principles **(C.∧)** and **(C.¬)**, thus suggesting that a corresponding "logic of argumentation" should be weaker than classical logic.

Of course, the interpretation of the attack relation using modalities is not unique. We briefly discuss three alternatives.

Definition 6. *For all $F, G \in \mathcal{PL}$:*

- $\iota_1(FG) =_{df} \Diamond(F \wedge \neg G)$;	- $\iota_1(F \not\longrightarrow G) =_{df} \Box(\neg F \vee G)$.
- $\iota_2(FG) =_{df} \Diamond(\neg F \vee \neg G)$;	- $\iota_2(F \not\longrightarrow G) =_{df} \Box(F \wedge G)$.
- $\iota_3(FG) =_{df} \Box(\neg F \vee \neg G)$;	- $\iota_3(F \not\longrightarrow G) =_{df} \Diamond(F \wedge G)$.

The interpretation ι_1 is similar to ι, but less demanding because (1) the claim of the attacking argument is asserted to hold in just one of the alternatives and (2) the negation of the claim of the attacked argument is asserted to hold only there. Interpretation ι_2 suggests that $F \longrightarrow G$ means that there is at least one possible state in which $F \supset \neg G$ (equivalently: $\neg F \vee \neg G$) holds; whereas according to ι_3 all possible states must be of this form. These alternative interpretations of the attack relation (abstracted to the claims of the argument) are arguably more problematic than the interpretation ι suggested in Definition 5. This is also witnessed by the attack principles that are justified or rejected by the respective interpretations: The set of attack principles justified by ι_1 is MAP_1 ={**(A.∧)**, **(A.∨)**, **(B.⊃)**, **(C.∧)**}; ι_2 justifies MAP_2 ={**(A.∧)**, **(A.∨)**, **(B.⊃)**, **(C.∧)**, **(C.¬)**}; whereas ι_3 is extremely demanding and rejects all of our attack principles except **(A.∧)** and **(A.∨)**.

7 A 'Logic of Argumentation'

We have seen that the attack principles in MAP are more plausible than the collection CAP that induces classical logical consequence. Thus the question arises, whether argumentative consequence relative to MAP can be characterized in a similar manner. We provide a positive by showing that $\models^{\mathsf{MAP}}_{\mathrm{arg}}$ matches the sequent calculus **LM**, that arises from dropping the rules $(\neg, l)$, $(\wedge, r)$, and $(\supset, l)$ from **LK**.

Theorem 5. *$\Gamma \vdash \Delta$ is derivable in* **LM** *iff $\Gamma \models^{\mathsf{MAP}}_{arg} \Delta$.*

Proof. For the left-to-right direction (soundness of **LM** with respect to $\models^{\mathsf{MAP}}_{\mathrm{arg}}$) it suffices to observe that only attack principles in MAP are used to show the soundness of those **LM**-rules in the proof of Theorem 1.

For the other direction (completeness of **LM** with respect to $\models^{\mathsf{MAP}}_{\mathrm{arg}}$) we can no longer rely on the invertibility of rules, as we did for **LK**. We rather show that for every sequent $\Gamma \vdash \Delta$ that is not derivable in **LM** there is an MAP *counter model*; i.e., an SAF S that is logically closed with respect to $\Gamma \cup$ and satisfies all principles in MAP, such that $\Gamma \not\models^{S}_{\mathrm{arg}} \Delta$.

A sequent $\Gamma_0 \vdash \Delta_0$ is called **LM**-*irreducible* if it is neither an axiom nor an instance of a lower sequent of an **LM**-rule. Note that every sequent that is not derivable in **LM**, results from applying **LM**-rules backwards until one hits an **LM**-irreducible sequent $\Gamma_0 \vdash \Delta_0$. It follows from the soundness of **LM** with respect to $\models^{\mathsf{MAP}}_{\mathrm{arg}}$ that $\Gamma \vdash \Delta$ has an MAP-counter-model if $\Gamma_0 \vdash \Delta_0$ has one, too. It thus remains to show that every **LM**-irreducible sequent has an MAP-counter-model.

First observe that every **LM**-irreducible sequent $\Gamma_0 \vdash \Delta_0$ is of the following form:

- $\Gamma_0 \cap \Delta_0 = \emptyset$;
- every $F \in \Gamma_0$ is atomic, a negation or an implication;
- every $F \in \Delta_0$ is atomic or a conjunction.

We construct an SAF $S = (\mathcal{A}, R_{\rightarrow})$ where the set of claims of arguments in $\mathcal{A}$ consists of all subformulas occurring in $\Gamma_0 \cup \Delta_0$. Additionally there is special argument in $\mathcal{A}$ with a claim x that is a new atomic formula (i.e., x does not occur as a subformula in $\Gamma_0 \cup \Delta_0$). The attack relation $R_{\rightarrow}$ is obtained by setting $x \rightarrow F$ for all $F \in \Delta_0$. Moreover, because of **(C.**$\vee$**)**, we have to add an attack from x to a disjunction $G \vee H$, if x attacks G as well as H (and if $G \vee H$ occurs as a subformula of some formula in $\Gamma_0 \cup \Delta_0$). It is easy to check that S satisfies all attack principles in MAP and that $\Gamma_0 \not\models^{S}_{\mathrm{arg}} \Delta_0$. (Remember in particular, that no disjunctions can occur in an irreducible sequent.) I.e., S is an MAP counter-model of $\Gamma_0 \vdash \Delta_0$. □

8 Nondeterministic Matrices

We have identified a 'logic of argumentation' with a consequence relation arising from certain plausible principles about logically closed collections of (claims of) arguments and managed to characterize this logic in terms of a variant of Gentzen's classical sequent calculus **LK**, where some of the logical rules have been discarded. We finally ask whether our logic appears already in a different context, pointing to a different type of semantics. A positive answer is provided by the theory of *canonical signed calculi* and *nondeterministic matrix semantics* (see [3,4]).

Definition 7. *A* classical Nmatrix $\mathcal{N}$ *consists in a function* $\tilde{\neg} : \{\mathbf{t}, \mathbf{f}\} \to 2^{\{\mathbf{t},\mathbf{f}\}} \backslash \emptyset$ *and a function* $\tilde{\circ} : \{\mathbf{t}, \mathbf{f}\}^2 \to 2^{\{\mathbf{t},\mathbf{f}\}} \backslash \emptyset$ *for each* $\circ \in \{\wedge, \vee, \supset\}$.

A corresponding dynamic valuation *is a function* $\tilde{v}_{\mathcal{N}} : \mathcal{PL} \to \{\mathbf{t}, \mathbf{f}\}$ *such that* $\tilde{v}_{\mathcal{N}}(\neg A) \in \tilde{\neg}(\tilde{v}_{\mathcal{N}}(A))$ *and* $\tilde{v}_{\mathcal{N}}(A \circ B) \in \tilde{\circ}(\tilde{v}_{\mathcal{N}}(A), \tilde{v}_{\mathcal{N}}(B))$ *for* $\circ \in \{\wedge, \vee, \supset\}$. $\tilde{v}_{\mathcal{N}}$ *is a* model *of* A *if* $\tilde{v}_{\mathcal{N}}(A) = \mathbf{t}$; *it is a model of* Γ *if it is a model of every* $A \in \Gamma$.

Δ *is a* dynamical consequence *of* Γ *with respect to* $\mathcal{N}$, *written* $\Gamma \models^{\mathcal{N}}_{dyn} \Delta$ *if every model of* Γ *is a model of some* $A \in \Delta$.

Consider the following classical Nmatrix $\mathcal{M}_{\mathsf{MAP}}$:

	$\tilde{\wedge}$	$\tilde{\vee}$	$\tilde{\supset}$
t t	{t, f}	{t}	{t}
t f	{f}	{t}	{t, f}
f t	{f}	{t}	{t}
f f	{f}	{f}	{t}

	$\tilde{\neg}$
t	{t, f}
f	{t}

The following is just an instance of Theorem 62 of [4].

Corollary 1. $\Gamma \vdash \Delta$ *is derivable in* **LM** *iff* $\Gamma \models^{\mathcal{M}_{\mathsf{MAP}}}_{dyn} \Delta$.

Combing this observation with Theorem 5, we have thus connected argumentative consequence with respect to the attack principles in MAP with logical consequence defined with respect to a specific nondeterministic valuation. (This type of analysis can straightforwardly be extended to, e.g., the collection MFAP of principles, Sect. 6).

9 Remarks on Related Work

To our best knowledge, the idea to saturate argumentation frames with respect to principles that make logically implicit arguments explicit, engendering a notion of logical consequence that is based on (attacking) arguments in place of classical (counter-)models, is new. However there are a number of interesting approaches that also aim at connecting Dung-style argumentation with logical inference. We briefly review some of this work.

Arieli and Straßer [2] provide an analysis of logical argumentation that is based on the identification of sequents with logical arguments. They analyze and characterize various forms of attack in a corresponding proof theoretic framework. Note however, that our approach is more general in the sense that we do not restrict attention to (purely) *logical* arguments, i.e., to arguments were the claim logically follows from its support. Moreover we deliberately abstract away from the internal structure of arguments and focus on the logical forms of their claims in order to obtain an alternative consequence relation not between arguments, but, as usual, between propositions.

Grossi [13,14] has shown that one can compactly characterize a wide range of Dung-style semantics by viewing argumentation frames as Kripke frames and introducing a modal operator that refers to the inverse of the attack relation. Also Caminada and Gabbay [6] exploit connections between modal logic and argumentation. This might be somewhat reminiscent of our interpretation of the attack relation in semi-abstract argumentation frames in Sect. 6. However, rather that identifying accessibility with (inverted) attack or arguments with states, we use the notions of standard normal modal logics to obtain a simple model of argumentative attack that enables one to distinguish plausible and implausible logical attack principles.

Yet further interesting foundational approaches to what can be viewed as 'logic of argumentation', albeit in a very different sense than the one outlined in this paper can be found, e.g., in [1,9,10].

10 Conclusion

We have specified some simple principles about attacking logically compound claims. The collection CAP of such principles leads to a characterization of classical logical consequence that only involves argumentation frames. But only a proper subset MAP $\subset$ CAP is justified with respect to a simple and coarse, but still useful modal model of attack. We characterized the corresponding 'argumentative' consequence relation $\models^{\mathsf{MAP}}_{\mathrm{arg}}$ by a variant of Gentzen's sequent calculus, where some of the logical rules are missing. This, in turn, triggers an alternative characterization in terms of nondeterministic valuations. We emphasize that do not claim technical sophistication or originality; to the contrary, we consider the technical simplicity of our approach a virtue, rather than an obstacle, in view of potential applications.

The interplay between logical principles about argumentation, on the one hand, and inference principles as studied in proof theory, on the other hand, certainly deserves further studies. Among many possible directions for further research, we single out the following questions:

- Can one extend the approach to cover also bi-polar argumentation frameworks, where in addition to attacks also a support relation between arguments and corresponding logical argument principles are considered?
- Is it possible to define a modal interpretation of support that justifies the corresponding support principles of MAP?
- How to generalize from semi-abstract frameworks to fully interpreted argumentation models, where also the support part of arguments is instantiated?
- Is it possible to relate the nondeterministic matrix semantics of Sect. 8 to logical argument principles in a more direct manner?

- Do more subtle modal interpretations of argument attack (or support) than the one of Sect. 6 lead to interesting alternative 'logics of argumentation'?
- Is there a relation between relevance logic and logics of argumentation?
- Can these theoretical insights into ('semi-abstract') logical argumentation be usefully employed to shed light on more practical (e.g., computational) problems in formal argumentation?

We think that the answer to all these question is positive. In particular, we already have promising results regarding bipolar argumentation frameworks that we plan to include in an extended version of this paper.

Acknowledgments. We like to thank Stefan Woltran and the referees for interesting suggestions regarding related work.

References

1. Amgoud, L., Besnard, P., Hunter, A.: Foundations for a logic of arguments. In: Logical Reasoning and Computation: Essays dedicated to Luis Fariñas del Cerro, pp. 95–108 (2016)
2. Arieli, O., Straßer, C.: Sequent-based logical argumentation. Argument Comput. **6**(1), 73–99 (2015)
3. Avron, A., Lev, I.: Non-deterministic multiple-valued structures. J. Logic Comput. **15**(3), 241–261 (2005)
4. Avron, A., Zamansky, A.: Non-deterministic semantics for logical systems. In: Gabbay, D., Guenthner, F. (eds.) Handbook of Philosophical Logic, vol. 16, pp. 227–304. Springer, Dordrecht (2011). doi:10.1007/978-94-007-0479-4_4
5. Besnard, P., Hunter, A.: Elements of Argumentation. MIT Press, Cambridge (2008)
6. Caminada, M.W.A., Gabbay, D.M.: A logical account of formal argumentation. Stud. Logica **93**(2), 109–145 (2009)
7. Dung, P.M.: On the acceptability of arguments and its fundamental role in non-monotonic reasoning, logic programming and n-person games. Artif. Intell. **77**(2), 321–357 (1995)
8. Dunn, J.M., Restall, G.: Relevance logic. In: Gabbay, D.M., Guenthner, F. (eds.) Handbook of Philosophical Logic, vol. 6, pp. 1–128. Springer, Dordrecht (2002). doi:10.1007/978-94-017-0460-1_1
9. Dyrkolbotn, S.: On a formal connection between truth, argumentation and belief. In: Colinet, M., Katrenko, S., Rendsvig, R.K. (eds.) ESSLLI Student Sessions 2013. LNCS, vol. 8607, pp. 69–90. Springer, Heidelberg (2014). doi:10.1007/978-3-662-44116-9_6
10. Gabbay, D.M.: Dungs argumentation is essentially equivalent to classical propositional logic with the peirce-quine dagger. Logica Universalis **5**(2), 255–318 (2011)
11. Gentzen, G.: Untersuchungen über das Logische Schließen I & II. Mathematische Zeitschrift **39**(1), 176–210, 405–431 (1935)
12. Gorogiannis, N., Hunter, A.: Instantiating abstract argumentation with classical logic arguments: Postulates and properties. Artif. Intell. **175**(9–10), 1479–1497 (2011)
13. Grossi, D.: Argumentation in the view of modal logic. In: McBurney, P., Rahwan, I., Parsons, S. (eds.) ArgMAS 2010. LNCS (LNAI), vol. 6614, pp. 190–208. Springer, Heidelberg (2011). doi:10.1007/978-3-642-21940-5_12

14. Grossi, D.: On the logic of argumentation theory. In: van der Hoek, W., Kaminka, G., Lesperance, Y., Luck, M., Sandip, S. (eds.) Proceedings of the 9th International Conference on Autonomous Agents and Multiagent Systems, vol. 1, pp. 409–416. International Foundation for Autonomous Agents and Multiagent Systems (2010)
15. Rahwan, I., Simari, G.R.: Argumentation in Artificial Intelligence. Springer, New York (2009)
16. Troelstra, A.S., Schwichtenberg, H.: Basic Proof Theory, 2nd edn. Cambridge University Press, Cambridge (2000)

Non-triviality Done Proof-Theoretically

Rohan French[1] and Shawn Standefer[2](✉)

[1] Monash Unversity, Melbourne, Australia
[2] University of Melbourne, Melbourne, Australia
sstandefer@unimelb.edu.au

1 Introduction

It is well known that naive theories of truth based on the three-valued schemes $K3$ and LP are non-trivial. This is shown by the fixed-point model construction of Kripke (1975). Kremer (1988) presents sequent systems for some fixed-point theories of truth, proves a completeness result, and provides an inferentialist interpretation of these systems. Kremer's model constructions show that the systems are non-trivial. Yet, there has been little work done to obtain a proof-theoretic explanation for why these systems are non-trivial, whereas a similar *classical* system is trivial.

Our goal is to gain some insight into *why* systems like $K3$ and LP are, when endowed with a truth predicate and enough syntactic machinery to get us into trouble, non-trivial by examining how to prove this in a purely proof-theoretic manner—attending to sequent calculi formulations of the two systems. For the sake of simplicity we focus on $K3$ for the bulk of the paper; the considerations for LP are largely dual. We begin with the basic sequent system and some results and problems in Sect. 2. These problems motivate a different sequent formulation in Sect. 3, which we show to be non-trivial. We then close in Sect. 4 with directions for future work.

2 Opening Moves

The kind of the sequent calculus we have in mind is found in Fig. 1. Throughout, $\pm A$ stands for A or $\neg A$, depending on whether the sign is $+$ or $-$. We note that to obtain a sequent calculus for LP, one would replace the axiom [$K3$] with [LP].

$$\frac{}{\Gamma \succ \Delta, p, \neg p}\ [LP]$$

We will focus on the quantifier-free fragment that includes no variables and no function symbols. There are two reasons. First, the addition of quantifiers changes little with respect to the question with which we are concerned, namely why are certain naive truth theories non-trivial. There is little interaction between the quantifiers and the syntactic theory that we will adopt, namely quotation names. Quantifiers will be more involved with a more complex syntactic theory, such as classical Peano arithmetic, and further issues will be raised

A. Baltag et al. (Eds.): LORI 2017, LNCS 10455, pp. 438–450, 2017.
DOI: 10.1007/978-3-662-55665-8_30

$$\frac{}{\Gamma, \pm p \succ \pm p, \Delta}\,[Id] \qquad \frac{}{\Gamma, p, \neg p \succ \Delta}\,[K3]$$

$$\frac{\Gamma \succ \Delta, A \quad \Gamma, A \succ \Delta}{\Gamma \succ \Delta}\,[Cut]$$

$$\frac{\Gamma, A \succ \Delta}{\Gamma, \neg\neg A \succ \Delta}\,[DNL] \qquad \frac{\Gamma \succ A, \Delta}{\Gamma \succ \neg\neg A, \Delta}\,[DNR]$$

$$\frac{\Gamma A, B \succ \Delta}{\Gamma, A \wedge B \succ \Delta}\,[\wedge L] \qquad \frac{\Gamma \succ A, \Delta \quad \Gamma \succ B, \Delta}{\Gamma \succ A \wedge B, \Delta}\,[\wedge R]$$

$$\frac{\Gamma, \neg A \succ \Delta \quad \Gamma, \neg B \succ \Delta}{\Gamma, \neg(A \wedge B) \succ \Delta}\,[\neg\wedge L] \qquad \frac{\Gamma \succ \neg A, \neg B, \Delta}{\Gamma \succ \neg(A \wedge B), \Delta}\,[\neg\wedge R]$$

$$\frac{\Gamma, A \succ \Delta}{\Gamma, T\langle A\rangle \succ \Delta}\,[TL] \qquad \frac{\Gamma \succ \Delta, A}{\Gamma \succ \Delta, T\langle A\rangle}\,[TR]$$

$$\frac{\Gamma, \neg A \succ \Delta}{\Gamma, \neg T\langle A\rangle \succ \Delta}\,[\neg TL] \qquad \frac{\Gamma \succ \Delta, \neg A}{\Gamma \succ \Delta, \neg T\langle A\rangle}\,[\neg TR]$$

Fig. 1. A Sequent Calculus for $K3$

there, such as ω-consistency. The second reason is that quantifiers bring some additional complexity that we think detracts from the overall aim of this project, getting a proof-theoretic grip on non-triviality. We expect that the arguments that we develop can be adapted to include quantifiers with standard rules.

Proposition 1. *Suppose that the sequent* $\succ\Delta$ *is provable in the above sequent calculus. Then there are some formulas* Δ'*, where each formula* $A \in \Delta'$ *is a subformula or a negated subformula of a formula in* Δ *such that* $\succ\Delta'$*.*

Proof. By induction on the construction of derivations, noting that (i) each of the right-rules have the feature in question, and (ii) that there are no rules which move formulas from the antecedant to the succedent of sequents. □

For the moment, we will say that a system is *trivial* if the sequent $\emptyset \succ \emptyset$ is derivable. Since the systems under consideration will have admissible weakening rules, all sequents will be derivable in a trivial proof system. A system is *non-trivial* just in case it is not trivial.

Theorem 2. *The above system is non-trivial.*

Proof. Suppose that we can derive $\emptyset \succ \emptyset$. By inspection of the above rules it is easy to see that it must be that the final rule in such a derivation was $[Cut]$, which means that we have, for some A, (i) $A\succ$, and (ii) $\succ A$. By the above Proposition and (ii) it follows that we have some instance of one of our initial sequents of the form

$$\overline{\succ A_1, \ldots, A_n}$$

for some formulas $A_1, \ldots, A_n$. This is impossible, though, so there is no such derivation. □

The basic idea of the consistency proof is simple. In order for $\emptyset \succ \emptyset$ to be derivable, we need the sequents $\emptyset \succ A$ and $A \succ \emptyset$ to be derivable. The sequents are unbalanced, in the sense that they each have an empty side. Either the antecedents or the succedents have been cleared out. If one can show that one cannot derive sequents that are unbalanced on each side, then non-triviality follows. A similar argument can be run for LP, changing Δ and Δ' to be in antecedent, rather than succedent position in Proposition 1 and adjusting the argument in the proof of Theorem 2 accordingly.

The system, as it stands, does not incorporate any syntactic theory. As a result, equivalences, such as a liar Ta being equivalent to $\neg Ta$, or even the weaker co-entailments, are not derivable. For the syntactic theory, we will use quotation names.[1] We start with a classical base language L and extend it to a language, L^+ with truth and quotation names, $\langle A \rangle$, for each sentence A. The names and formulas of L^+ are defined by simultaneous induction, so that $T\langle Fb \rangle$ is a formula, as is $T\langle T\langle Gc \rangle \rangle$. The use of quotation names does not force there to be, for example, liar sentences in the language, although we will assume there are such. To use the syntactic theory, we require that there are new axioms or rules in the language for quotation names and identity. The rules we would like to add are in Fig. 2.[2] These rules, however, will encounter serious difficulties that will lead us to proceed in a different direction.

$$\frac{\langle A \rangle \neq \langle B \rangle, \Gamma \succ \Delta}{\Gamma \succ \Delta}\,[QNL] \qquad \frac{\Gamma \succ \Delta, \langle A \rangle = \langle B \rangle}{\Gamma \succ \Delta}\,[QNR]$$

In [QNR] and [QNL], $A \neq B$.

$$\frac{\Gamma, s = t, A(t), A(s) \succ \Delta}{\Gamma, s = t, A(t) \succ \Delta}\,[=L1] \qquad \frac{\Gamma, s = t \succ \Delta, A(t), A(s)}{\Gamma, s = t, \succ \Delta, A(t)}\,[=L2]$$

$$\frac{\Gamma, A(t), A(s) \succ \Delta, s \neq t}{\Gamma, A(t) \succ \Delta, s \neq t}\,[\neq R1] \qquad \frac{\Gamma \succ \Delta, A(t), A(s), s \neq t}{\Gamma, \succ \Delta, A(t), s \neq t}\,[\neq R2]$$

$$\frac{a = a, \Gamma \succ \Delta}{\Gamma \succ \Delta}\,[=Ref] \qquad \frac{\Gamma \succ \Delta, a \neq a}{\Gamma \succ \Delta}\,[\neq Ref]$$

Fig. 2. Identity and quotation name rules for $K3$

[1] See Gupta (1982), Kremer (1988), or Ripley (2012).

[2] These two rules are based on those from Kremer (1988) and from Negri and von Plato (2001).

Identity is not treated partially, as truth is. The classical treatment of identity extends to identities between terms not in the base language. More generally, we also want to consider languages that have predicates in addition to the truth predicate and identity. These base language expressions, including identity, are, in the $K3T$ theory, treated classically. To accommodate them in the sequent calculus, we can add the following axioms, where p is a T-free atom.[3]

$$\frac{}{\Gamma \succ p, \neg p, \Delta}\ [Cl]$$

An alternative, which we will not adopt, is to use the following rule.

$$\frac{\Gamma, p \succ \Delta \quad \Gamma, \neg p \succ \Delta}{\Gamma \succ \Delta}\ [Cl]$$

Let us call the system with the axiom $[Cl]$ and the rules above $K3TL^{=}$, where L is the classical base language. The addition of $[Cl]$ adds another way to obtain derivable sequents of the form $\succ A_1, \ldots, A_n$. This, along with the identity rules that delete formulas raise problems for extending the balance argument in Theorem 2 to work for the extended system.

With identity in the system, the definition of triviality has to be modified, since it may be the case, for example, that while $\emptyset \succ \emptyset$ isn't derivable, $a = \langle \neg Ta \rangle \succ \emptyset$ is.[4] A more general definition of triviality is needed. A system is *trivial*, in the extended sense, if $\Gamma_0 \succ \Delta_0$ is derivable, where Γ_0 is a multiset of equalities, Δ_0 is a multiset of inequalities, and at least one contains a formula with a quotation name on one side and a non-quotation name on the other. Apart from the new definition of triviality, the addition of the identity rules requires modifying the non-triviality proof, since sequents of the form $\succ A_1, \ldots, A_n$ are now derivable. Apart from the preceding issue, the balance argument does not seem to guarantee that the system isn't trivial in the extended sense. While it guarantees that both sides do not end up empty, it does not appear to guarantee non-triviality in the extended sense.

We would like to prove the system $K3TL^{=}$ non-trivial, but the proof-theoretic methods for doing so run into difficulties. We have not seen these pointed out before, so we will briefly indicate some of the hurdles. One way to prove the non-triviality of $K3TL^{=}$ would be to give a cut elimination argument. For a sequent calculus with the structural rules of contraction and weakening absorbed, a common route to eliminating cuts proceeds via a few lemmas, including the *inversion lemma* showing that all the rules are invertible, which says, roughly, that if a derivable sequent could be the conclusion of a rule, then the premiss of that rule is also derivable.[5] In particular, it would require that if $T\langle A \rangle, \Gamma \succ \Delta$ is derivable in n steps, then $A, \Gamma \succ \Delta$ is also derivable in n steps. Our diagnosis of the problem arises is the following: atoms using the truth predicate can occur as the conclusions of rules and as atoms in axioms. Inversion requires

[3] The appropriate $[Cl]$ axiom for LP would be $\Gamma, p, \neg p \succ \Delta$.

[4] This problem was pointed out by Kremer (1988).

[5] Negri and von Plato (2001, 32).

that for arbitrary A, $\Gamma, A \succ T\langle A\rangle, \Delta$ be derivable in 1 step, since $\Gamma, T\langle A\rangle \succ T\langle A\rangle, \Delta$ is. But, that is not generally the case.

One might try to force the inversion lemma, by adding as an axiom $\Gamma, A \succ T\langle A\rangle, \Delta$, but this, in turn, would require that an arbitrary A could be inverted. For example, if A is $B \vee C$, then $\Gamma, B \succ T\langle A\rangle, \Delta$ and $\Gamma, C \succ T\langle A\rangle, \Delta$ would have to be axioms. It appears, then, that obtaining the inversion lemma for these rules would require the addition of a rule with no premises that allows one to infer any derivable sequent. While that would make derivations shorter, it would not be insightful, even if the rest of the argument for the cut elimination theorem worked.

There is one additional hurdle for giving a direct cut elimination argument that we will highlight, as it is important for the approach adopted in the next section. In this style of sequent system, there is usually only an identity substitution rule on the left, in our case the $[= L1]$ and $[\neq R1]$ rules. Substitution on the right is achieved indirectly, starting with the desired term in a formula on the right, e.g. Fb and then proceeding to replace it on the left by means of the rule to obtain, e.g. $a = b, Fa \succ Fb$. This appears, however, to be inadequate in the case of truth as the truth rules can introduce distinguished terms, namely quotation names, in either the antecedent or succedent. None of the other rules introduce terms that are new to the proof. The use of identity substitution rules on the right, as well as the left, creates the same sort of trouble for the elimination proof that truth case did above. There appears, then, not to be any way to obtain $a = \langle Fb\rangle, Fb \succ Ta$ without using cut, as the truth rule would yield $T\langle Fb\rangle$ in the succedent, requiring the use of a term substitution. We will, then, move to a different setting for proving non-triviality.

3 Sequent Systems with Annotations

We will use an alternative sequent system to deal with some of the indicated issues related to truth and quotation names. Specifically, we will use a modified form of $K3TL$, without identity rules and without identity in the object language. Suppose one is given a language L^+. Let a *syntax set* $\mathscr{E}$ for L^+ be a set of identities, each of which is of one of the following three forms: $\langle A\rangle = \langle B\rangle$, $a = \langle B\rangle$, or $b = c$, where a, b, c are names that are not quotation names.[6] An *identity set* $\mathscr{E}$ is a syntax set obeying the following closure conditions:

1. for all sentences A, $\langle A\rangle = \langle A\rangle$ is in $\mathscr{E}$,
2. if $s = t$ is in $\mathscr{E}$, then $t = s$ is in $\mathscr{E}$, and
3. if $s = t$ and $t = u$ are in $\mathscr{E}$, then $s = u$ is in $\mathscr{E}$.

Finally, an *annotation set* $\mathscr{E}$ is an identity set not containing $\langle A\rangle = \langle B\rangle$, where A and B are distinct formulas. Given an annotation set $\mathscr{E}$, say that two terms, s, t, appearing in identities in $\mathscr{E}$ are equivalent in $\mathscr{E}$ just in case $s = t$ is in $\mathscr{E}$. We will consider the proof systems $K3TL\mathscr{E}$, for each $\mathscr{E}$. So, a given proof will have a particular annotation set $\mathscr{E}$ that affects its rules.

[6] Note that we are assuming there are no function symbols in the language.

Taking the interpretations of the identities in an annotation set $\mathscr{E}$ to be fixed to a standard interpretation, then, $\mathscr{E}$ contains the syntactic information of L^+. A particular ground model for L^+ may interpret the non-quotation names, and names outside of $\mathscr{E}$ differently, as well as the predicates, but that is fine, since we are only interested in the syntactic theory, as captured by $\mathscr{E}$. From an inferentialist point of view, the use of the annotation sets presents no philosophical problems.

Rather than use multisets, the systems $K3TL\mathscr{E}$ will use sequences on either side of the turnstile. The sequences Γ, Δ are permitted to be empty. We also add a permutation rule for both sides, although we will generally suppress it in what follows. The purpose of the switch from multisets to sequences is to facilitate the definition of a trace and an ancestor, which are used for the proof of the elimination theorem.

The next definition is used to integrate $\mathscr{E}$ into the proof system. Say that two formulas A and B are equivalent in $\mathscr{E}$ just in case there are sequences of terms $c_1, \ldots, c_n$, and $d_1, \ldots, d_n$, not occurring in quotation names in A, such that B can be obtained from A by replacing one or more occurrences of c_i in A with d_i, where for each i, c_i and d_i are equivalent in $\mathscr{E}$.

The axioms, $[Id]$ and $[K3]$, are generalized to include the following instances, where $\pm p$ is of the form $\pm Tb$. In $[K3]$, the antecedent formulas may be in any order.

$$\frac{}{\Gamma, \pm Tb, \Sigma \succ \Theta, \pm Tc, \Delta}\,[Id] \qquad \frac{}{\Gamma, Tb, \neg Tc, \Sigma \succ \Delta}\,[K3]$$

In these axioms, b and c must be equivalent in $\mathscr{E}$. The axiom form of $[Cl]$ does not need to be changed.

The truth rules are similarly modified.

$$\frac{\Gamma, A, \Sigma \succ \Delta}{\Gamma, Tb, \Sigma \succ \Delta}\,[TL] \qquad \frac{\Gamma \succ \Delta, A, \Sigma}{\Gamma \succ \Delta, Tb, \Sigma}\,[TR]$$

$$\frac{\Gamma, \neg A, \Sigma \succ \Delta}{\Gamma, \neg Tb, \Sigma \succ \Delta}\,[\neg TL] \qquad \frac{\Gamma \succ \Delta, \neg A, \Sigma}{\Gamma \succ \Delta, \neg Tb, \Sigma}\,[\neg TR]$$

In these rules, b and $\langle A \rangle$ must be equivalent in $\mathscr{E}$.

We add the following rules to $K3TL\mathscr{E}$.

$$\frac{\Gamma, A, A, \Sigma \succ \Delta}{\Gamma, A, \Sigma \succ \Delta}\,[WL] \qquad \frac{\Gamma \succ \Delta, A, A, \Sigma}{\Gamma \succ \Delta, A, \Sigma}\,[WR]$$

$$\frac{\Gamma, A, B, \Sigma \succ \Delta}{\Gamma, B, A, \Sigma \succ \Delta}\,[CL] \qquad \frac{\Gamma \succ \Delta, A, B, \Sigma}{\Gamma \succ \Delta, B, A, \Sigma}\,[CR]$$

Finally, $K3TL\mathscr{E}$ does not take the rule $[Cut]$ as primitive, although, as we will show, this does not affect what sequents are provable.

An upshot of internalizing the syntactic theory, in the manner that we have done, is that it permits us to return to the simple definition of triviality, namely the derivability of $\emptyset \succ \emptyset$. The reason that we had to move to a more complicated definition of triviality in Sect. 2 was that we wanted to permit the use of syntactic resources in the derivation of triviality, as the syntactic resources, in a sense, come for free. The use of any syntactic resources, however, would preclude the derivation of $\emptyset \succ \emptyset$. Since we no longer record appeal to syntactic resources with identities and negations of identities in sequents, the complications are no longer needed.

We will state two lemmas concerning equivalence in $\mathscr{E}$, to be used later.

Lemma 3. *If A and A' are equivalent in $\mathscr{E}$ and B and B' are equivalent in $\mathscr{E}$, then the following are equivalent in $\mathscr{E}$.*

- $\neg A$ *and* $\neg A'$
- $A \wedge B$ *and* $A' \wedge B'$
- $\neg(A \wedge B)$ *and* $\neg(A' \wedge B')$

Proof. This is proved by induction on the complexity of A and B. □

It is not the case that if A and B are equivalent in $\mathscr{E}$ then $T\langle A\rangle$ and $T\langle B\rangle$ will be. This is because A and B may be distinct sentences, in which case, one will not have $\langle A\rangle = \langle B\rangle$ in $\mathscr{E}$. This is, however, as it should be. We can say something about relations between formulas equivalent in $\mathscr{E}$.

Lemma 4. *Suppose A and A' are equivalent in $\mathscr{E}$. Then, if $\Gamma \succ \Delta, A, \Sigma$ is derivable, then $\Gamma \succ \Delta, A', \Sigma$ is derivable, and if $\Gamma, A, \Sigma \succ \Delta$ is derivable, then $\Gamma, A', \Sigma \succ \Delta$ is derivable. Furthermore, if the original sequent was derivable in n steps, then the new sequent is derivable in at most n steps.*

Proof. The proof is by induction on the construction of the proof. If A is principle in an axiom, then the result of replacing A with A' in the axiom will still be an axiom, and similarly if A is parametric.

The structural rules are taken care of by the induction hypothesis. We will present [*WL*], [*WR*] being similar. Suppose $\Gamma, A, A, \Sigma \succ \Delta$ is derivable. By the inductive hypothesis, then $\Gamma, A', A', \Sigma \succ \Delta$ is derivable. By [*WL*], $\Gamma, A', \Sigma \succ \Delta$ is derivable. The connective rules are immediate from the induction hypothesis.

Let us look at the truth rules. If A is principle in one of the truth rules, then it is of the form $\pm Tb$. Since A' is equivalent in $\mathscr{E}$, then A' is of the form $\pm Tc$ and c and b are equivalent in $\mathscr{E}$. It follows that the sequent replacing A with A' is also a conclusion of a truth rule. □

The contraction rule we use does not permit contraction across formulas equivalent in $\mathscr{E}$. This is, however, shown to be admissible by the previous proof.

Corollary 1. *Fix $\mathscr{E}$ and let A and A' be equivalent in $\mathscr{E}$. If $\Gamma, A, A', \Sigma \succ \Delta$ is derivable, then so is $\Gamma, A, \Sigma \succ \Delta$. If $\Gamma \succ \Delta, A, A', \Sigma$ is derivable, then so is $\Gamma \succ \Delta, A, \Sigma$.*

The form the cut rule that we will show is admissible is the following.

$$\frac{\Gamma \succ \Delta, M, \Xi \quad \Phi, M', \Sigma \succ \Theta}{\Phi, \Gamma, \Sigma \succ \Delta, \Theta, \Xi}\ [Cut]$$

In this rule, M and M' must be equivalent in $\mathscr{E}$. In light of Lemma 3, we could require that M is identical to M', but we will not do so here.

We need to define the notions of being parametric in a rule and parametric ancestor. We use the definitions of Bimbó (2015, 34–35) modified in the obvious way for our rules, which for reasons of space we will leave slightly informal here. The non-displayed formulas in the axioms are parametric in the axioms. In the connective and structural rules, the non-displayed formulas are parametric. A formula occurrence A in the premiss of a rule is a *parametric ancestor* of a formula occurrence B in the conclusion of that rule iff they are related by the transitive closure of the following relation: Both are occurrences of the same formula and either (i) they are both parametric in the rule and occur in the same position, (ii) they are displayed in $[CL]$ or $[CR]$ and not in the same position, or (iii) they are displayed in $[WL]$ or $[WR]$. Note that in some rules, such as $[\neg \wedge L]$ and $[WR]$, formulas in the conclusion can have more than one parametric ancestor in the premises, e.g. each formula in Γ in $[\neg \wedge L]$ and A in $[WR]$. The *contraction count* of a formula occurrence is the number of applications of $[WL]$ or $[WR]$ in which one of its parametric ancestors is displayed. The contraction count of an application of cut is the sum of the contraction counts of the two occurrences of the cut formula.

Define the *trace tree* of an occurrence of a formula as follows. If A is parametric in an inference, then A's trace is extended with a branch containing the corresponding occurrence of A in the premises. If A is principle in a $\neg$ rule, so is of the form $\neg\neg B$, then its trace tree is extended with a branch containing B. If A is principle in a $\wedge$ rule, then it is of the form $B \wedge C$ and its trace tree is extended with a branch containing B and one containing C. If A is principle in a $\neg\wedge$ rule, then it is of the form $\neg(B \wedge C)$ and its trace tree is extended with a branch containing $\neg B$ and one containing $\neg C$. If A is principle in a truth rule, then it is of the form Tb, and its trace is the displayed B in the premiss. The negated truth rules are similar. If A is principle in a contraction rule, then its trace tree is extended with a branch containing one occurrence of A and one containing the other.

Define the *trace weight* $t(A)$ of an occurrence of a formula A as the number of truth rules featuring nodes of A's trace tree as principle in their conclusions. Define the *grade* $g(A)$ of A as the number of logical connectives appearing in A outside the scope of quotation names. Define the *complexity* of a cut as being $\omega \cdot (t(M) + t(M')) + g(M)$, where M and M' are, respectively, the occurrences of M and M' displayed in the left and right premises of [Cut]. The *rank* of the cut is defined as the sum of the *left rank*, which is the number of steps in which a parametric ancestor of M occurs in the succedent of the left premiss, plus the *right rank*, which is the number of steps in which a parametric ancestor of M' occurs in the antecedent of the right premiss. We will follow Bimbó's triple induction proof technique, modified in the indicated ways to account for the differing rules.[7]

Proposition 5. *Let* $\mathscr{E}$ *be an annotation set. If* $\Gamma \succ \Delta, M, \Xi$ *and* $\Phi, M', \Sigma \succ \Theta$ *are derivable, where* M, M' *are equivalent in* $\mathscr{E}$*, then* $\Phi, \Gamma, \Sigma \succ \Delta, \Theta, \Xi$ *is derivable without cut.*

Proof. It is sufficient to show that uppermost cuts can be eliminated from derivations. The proof proceeds by triple induction on the cut complexity, rank, and contraction count. The left rank is lowered, and then the right rank is lowered, then the complexity is lowered, lowering the contraction count as needed. As usual, we can break the cases into groups, depending on the rank of the left cut premiss and the rank of the right cut premiss. We will present a selection of the cases, generally presenting instances where the cuts are simpler than the general case due to the ordering of formulas.

We will start with the cases in which both cut premises come via axioms.

Case: Both premises are from $[Id]$. This splits into subcases, depending on whether either cut formula is parametric in the axiom. In the case in which both cut formulas are principle, we may have one premiss as $\Gamma, Ta \succ Tb, \Delta$ and the other as $\Sigma, Tc \succ Td, \Theta$. Since we know that the following pairs are equivalent in $\mathscr{E}$, $\langle a, b\rangle$, $\langle c, d\rangle$, and $\langle b, c\rangle$, it follows that a and d are equivalent in $\mathscr{E}$. This means that the sequent $\Gamma, \Sigma, Ta \succ Td, \Delta, \Theta$ is an axiom.

The case in which one premiss comes from $[K3]$ and one from $[Cl]$ is straightforward. Similarly, the case in which one comes from $[Id]$ and one from either $[K3]$ or $[Cl]$ is straightforward.

The permutation cases are taken care of by the induction hypothesis on rank.

Case: The left premiss comes via $[WR]$. This breaks into subcases depending on the complexity of the cut formula. Here we will assume that it is greater than 1. The proof then looks like the following.

$$\dfrac{\dfrac{\Gamma \succ \Delta, M, M'}{\Gamma \succ \Delta, M} \qquad M'', \Sigma \succ \Theta}{\Gamma, \Sigma \succ \Delta, \Theta}$$

Since M and M' are equivalent in $\mathscr{E}$, as are M and M'', it follows that M' and M'' are. We can then permute the cut upwards as follows.

$$\dfrac{\dfrac{\Gamma \succ \Delta, M, M' \quad M'', \Sigma \succ \Theta}{\Gamma, \Sigma \succ \Delta, \Theta, M} \qquad M'', \Sigma \succ \Theta}{\Gamma, \Sigma, \Sigma \succ \Delta, \Theta, \Theta}$$

The new cuts can be eliminated by the induction hypothesis on the contraction count. The desired endsequent can then be obtained by repeated use of contraction and permutation rules.

The other cases involving $[WL]$ and $[WR]$ are similar.

[7] See Bimbó (2015, 36-51) for details.

The cases in which one or both cut formulas are parametric in their respective inferences are handled by the usual induction hypothesis on rank.

We will do a few cases in which both cut formulas are principle in their inferences.

Case: $[\wedge]$. Both cut formulas are principle, so the proof looks like the following.

$$\dfrac{\dfrac{\Gamma \succ \Delta, A \quad \Gamma \succ \Delta, B}{\Gamma \succ \Delta, A \wedge B} \quad \dfrac{A', B', \Sigma \succ \Theta}{A' \wedge B', \Sigma \succ \Theta}}{\Gamma, \Sigma \succ \Delta, \Theta}$$

In this proof, A and A', as well as B and B', are, respectively, equivalent in $\mathscr{E}$.

This is transformed into the following.

$$\dfrac{\Gamma \succ \Delta, B \quad \dfrac{\Gamma \succ \Delta, A \quad A', B', \Sigma \succ \Theta}{B', \Gamma, \Sigma \succ \Delta, \Theta}}{\Gamma, \Gamma, \Sigma \succ \Delta, \Delta, \Theta}$$

The new cuts can be eliminated using the induction hypothesis on complexity. The desired endsequent can then be obtained by repeated use of contraction and permutation rules.

The $[\neg\wedge]$ and double negation cases are similar.

Case: $[T]$. The proof ends with the following.

$$\dfrac{\dfrac{\Gamma \succ \Delta, A}{\Gamma \succ \Delta, Tb} \quad \dfrac{A, \Sigma \succ \Theta}{Tc, \Sigma \succ \Theta}}{\Gamma, \Sigma \succ \Delta, \Theta}$$

In this proof, b and c are equivalent in $\mathscr{E}$.

This can be transformed into the following.

$$\dfrac{\Gamma \succ \Delta, A \quad A, \Sigma \succ \Theta}{\Gamma, \Sigma \succ \Delta, \Theta}$$

The new cut can be eliminated by appeal to the induction hypothesis on complexity, since the trace weight of the cut formula has been reduced.

Case: $[\neg T]$. The proof ends with the following.

$$\dfrac{\dfrac{\Gamma \succ \Delta, \neg A}{\Gamma \succ \Delta, \neg Tb} \quad \dfrac{\neg A, \Sigma \succ \Theta}{\neg Tc, \Sigma \succ \Theta}}{\Gamma, \Sigma \succ \Delta, \Theta}$$

In this proof, b and c are equivalent in $\mathscr{E}$.

This can be transformed into the following.

$$\dfrac{\Gamma \succ \Delta, \neg A \quad \neg A, \Sigma \succ \Theta}{\Gamma, \Sigma \succ \Delta, \Theta}$$

The new cut can be eliminated by appeal to the induction hypothesis on complexity, since the trace weight of the cut formula has been reduced.

Finally, we observe that in no case did the trace weight of a cut formula increase from the original cut to the new cuts in the transformed proof. □

$$\frac{\Gamma, \pm Ts \succ \Delta}{\Gamma, t = s, \pm Tt \succ \Delta}\,[=L1] \qquad \frac{\Gamma \succ \Delta, \pm Ts}{\Gamma, t = s, \succ \Delta, \pm Tt}\,[=L2]$$

$$\frac{\Gamma, \pm Ts \succ \Delta}{\Gamma, s = t, \pm Tt \succ \Delta}\,[=L3] \qquad \frac{\Gamma \succ \Delta, \pm Ts}{\Gamma, s = t, \succ \Delta, \pm Tt}\,[=L4]$$

$$\frac{\Gamma, \pm Ts \succ \Delta}{\Gamma, \pm Tt \succ \Delta, t \neq s}\,[\neq R1] \qquad \frac{\Gamma \succ \Delta, \pm Ts}{\Gamma \succ \Delta, \pm Tt, t \neq s}\,[\neq R2]$$

$$\frac{\Gamma, \pm Ts \succ \Delta}{\Gamma, \pm Tt \succ \Delta, s \neq t}\,[\neq R3] \qquad \frac{\Gamma \succ \Delta, \pm Ts}{\Gamma \succ \Delta, \pm Tt, s \neq t}\,[\neq R4]$$

$$\frac{\Gamma, A \succ \Delta}{\Gamma, T\langle A\rangle \succ \Delta}\,[TL] \qquad \frac{\Gamma \succ \Delta, A}{\Gamma \succ \Delta, T\langle A\rangle}\,[TR]$$

$$\frac{\Gamma, \neg A \succ \Delta}{\Gamma, \neg T\langle A\rangle \succ \Delta}\,[\neg TL] \qquad \frac{\Gamma \succ \Delta, \neg A}{\Gamma \succ \Delta, \neg T\langle A\rangle}\,[\neg TR]$$

Fig. 3. Identity and truth rules in KR

Since this proof system is somewhat non-standard, we will demonstrate its adequacy by showing that it is equivalent to a fragment of the sequent system for Strong Kleene truth from Kremer (1988). Rather than give Kremer's rules, we will use the rules from $K3TL^{=}$, with restrictions that we will indicate. These rules are admissible in Kremer's system. The fragment in which we will be interested here is the quantifier-free fragment without identity axioms and whose identity rules are restricted to operate only on literals using the truth predicate (Fig. 3). Further, the proofs are required to be *syntax consistent*, in the sense that for a given derivation, the equalities on the left and negations of equalities on the right of the end sequent do not imply, using classical equational logic, $\langle A\rangle = \langle B\rangle$, for any distinct formulas A and B.[8] Call this fragment KR. For a given $\mathscr{E}$ and an instance of a truth rule whose displayed premiss formula is $\pm A$ and whose conclusion is $\pm Tb$, say that a set of identities Ξ *underwrites* the application of the rule just in case Ξ contains the identity $\langle A\rangle = b$. Say that a set of identities underwrites a truth axiom of one of the followings forms,

- $\Gamma, \pm Tb, \Sigma \succ \Delta, \pm Tc, \Theta$, or
- $\Gamma, Tb, \neg Tc, \Sigma \succ \Delta$,

just in case Ξ contains $b = c$. A set of identities Ξ underwrites a proof just in case Ξ underwrites each truth rule and truth axiom in the proof.

The axioms in KR can be used for arbitrary formulas, rather than being restricted to atoms. This is not a problem, since $K3TL\mathcal{E}$ allows one to prove that the axioms hold for arbitrary formulas.

[8] Syntax consistency says, roughly, that the set of equalities on the left and negations of equalities on the right can be extended to an annotation set.

Lemma 6. *The axioms [Id] and [K3] are derivable with an arbitrary formula A replacing p. The axioms [Cl] are derivable with an arbitrary T-free formula B replacing p.*

The proof is by induction on the construction of A, or B. The proof is routine, so we omit it.

With that lemma in hand, we can now state the equivalence between the systems.

Theorem 7. *1. Let Π be a $K3TL\mathcal{E}$ derivation of $\Gamma \succ \Delta$. Then there is a derivation of $\Gamma, \Xi \succ \Delta$ in KR, where Ξ is a set of identities that underwrites each truth rule and truth axiom used in Π.*

2. Let Π be a proof of $\Gamma, \Sigma \succ \Delta, \Theta$ in KR, where Σ is a set of identities introduced via identity rules and Θ is a set of negated identities introduced via identity rules. Then $\Gamma \succ \Theta$ is derivable in $K3TL\mathcal{E}$, provided $\mathcal{E}$ contains $\Sigma \cup \Theta^*$, where $\Theta^* = \{s = t : s \neq t \in \Theta\}$.

For reasons of space, we will sketch the proof. For 1, one translates a proof Π of $\Gamma \succ \Delta$ in $K3TL\mathcal{E}$ into a proof Π' in KR. We briefly describe some of the cases. Whenever a truth rule is used in Π, a truth rule is used in Π', followed by an appropriate identity rule if the principal formula in the rule in Π was of the form $\pm Ta$ rather than $\pm T\langle A\rangle$. These additional identities make up Ξ.

For 2, one translates a proof Π' of $\Gamma, \Sigma \succ \Delta, \Theta$ into a proof Π of $\Gamma \succ \Delta$ in $K3TL\mathcal{E}$. The important cases for this proof are the identity cases. These are taken care of by global transformations on trace trees in Π.

We will note that the theorem, and proof, carry over to LP, using Kremer's sequent calculus for LP.

The restriction in KR to use identity rules only on literals using the truth predicate is to facilitate the proof of Theorem 7, and it is not a great restriction. Nothing additional is provable if the identity rules are allowed to substitute into truth atoms in complex formulas.

Proposition 8. *Let A be a formula with at least one occurrence of Ta and let A' be A with one or more occurrences of Ta replaced by Tb. The rules*

$$\frac{\Gamma, A \succ \Delta}{a = b, \Gamma, A' \succ \Delta}\,[=L] \qquad \frac{\Gamma, A \succ \Delta}{b = a, \Gamma, A' \succ \Delta}\,[=L]$$

$$\frac{\Gamma, A \succ \Delta}{\Gamma, A' \succ \Delta, a \neq b}\,[\neq R] \qquad \frac{\Gamma, A \succ \Delta}{\Gamma, A' \succ \Delta, b \neq a}\,[\neq R]$$

are admissible in KR.

The proof is by induction on the construction of A. It is straightforward, so we omit it. We will turn to the conclusion.

4 Conclusion

We began with the aim of providing some proof-theoretic explanation of why the fixed-point theories of truth based on $K3$ and LP are non-trivial. Focusing on $K3$, we proved that a very basic system, one with no additional syntactic theory, is non-trivial by a balancing argument. The argument extends to LP. Enriching this system with a modest syntactic theory and identity, leads to difficulties showing that cut is eliminable. Indeed, the enrichment requires a more complicated definition of triviality. We proposed non-standard systems that internalize the syntactic theory in the annotation sets and their effect on the truth rules. A cut elimination argument can be carried out for those systems, showing that the systems are non-trivial in the original sense that $\emptyset \succ \emptyset$ is not derivable. Finally, we showed that the systems are intertranslatable with fragments of Kremer's system.

There is still work to be done. It would be good to obtain a completeness result of some kind for the systems with annotation sets. We also hope to generalize the initial balancing argument to work more broadly. But, we have made some progress towards our initial goal of getting a proof-theoretic justification for the non-triviality of naive truth in $K3$ and LP.

Acknowledgments. We would like to thank the members of audiences at the Otago Logic Seminar and the Australasian Association for Logic Conference 2016 for feedback on this material. Shawn Standefer's research was supported by the Australian Research Council, Discovery Grant DP150103801.

References

Bimbó, K.: Proof Theory: Sequent Calculi and Related Formalisms. CRC Press, Boca Raton (2015)

Gupta, A.: Truth and paradox. J. Philos. Logic **11**(1), 1–60 (1982)

Kremer, M.: Kripke and the logic of truth. J. Philos. Logic **17**, 225–278 (1988)

Kripke, S.: Outline of a theory of truth. J. Philos. **72**, 690–716 (1975)

Negri, S., von Plato, J.: Structural Proof Theory. Cambridge University Press, Cambridge (2001)

Ripley, D.: Conservatively extending classical logic with transparent truth. Rev. Symb. Logic **5**(2), 354–378 (2012)

Sette's Logics, Revisited

Hitoshi Omori(✉)

Department of Philosophy, Kyoto University, Kyoto, Japan
hitoshiomori@gmail.com

Abstract. One of the simple approaches to paraconsistent logic is in terms of three-valued logics. Assuming the standard behavior with respect to the "classical" values, there are only two possibilities for paraconsistent negation, namely the negation of the Logic of Paradox and the negation of Sette's logic $\mathbf{P}^1$. From a philosophical perspective, the paraconsistent negation of $\mathbf{P}^1$ is less discussed due to the lack of an intuitive reading of the third value. Based on these, the aim of this paper is to fill in the gap by presenting a semantics for $\mathbf{P}^1$ à la Jaśkowski which sheds some light on the intuitive understanding of Sette's logic. A variant of $\mathbf{P}^1$ known as $\mathbf{I}^1$ will be also discussed.

1 Introduction

1.1 Background and Aim

Paraconsistent logics are characterized by the failure of *ex contradictione quodlibet* (ECQ hereafter). Since the modern birth of paraconsistency, infinitely many systems of paraconsistent logic have been devised and studied through various approaches based on various motivations.

One of the simple approaches to paraconsistent logic is in terms of three-valued logics. Assuming the standard behavior with respect to the "classical" values, namely two truth values except the third or intermediate truth value, there are only two possibilities for paraconsistent negation, namely the negation of the Logic of Paradox (cf. [20], **LP** hereafter) and the negation of $\mathbf{P}^1$. $\mathbf{P}^1$ is one of the oldest and famous systems of paraconsistent logic devised by Antonio Sette in [22]. The logic has been studied in depth by e.g. [15,21]. However, these examinations are all guided by more mathematical/technical interest, and its *philosophical* interest has been rather small. This is not a surprise since the three-valued semantics for $\mathbf{P}^1$ seems extremely artificial. This is in sharp contrast with **LP**. Indeed, due to the two-valued *relational* semantics devised by Michael Dunn in [9] for logics in the **FDE** family, to which **LP** belongs, we now have a very clear understanding that the truth and falsity conditions for the logical connectives in **FDE** family are identical with those for classical logic. The crucial difference lies in the relation between the truth and falsity. And thus we can discuss the philosophical issues in view of the relation between the truth and falsity.

For the case with $\mathbf{P}^1$, however, alternative semantics are not completely non-existent. There are two previous attempts to the best of author's knowledge.

A. Baltag et al. (Eds.): LORI 2017, LNCS 10455, pp. 451–465, 2017.
DOI: 10.1007/978-3-662-55665-8_31

First, there is a Kripke semantics for the system, developed in [1], but unfortunately, the semantics does not give us too much philosophical insight. Second, there is a semantics devised by Walter Carnielli and Mamede Lima-Marques in [5]. However, their semantics also leaves some room for improvement, especially its semantics not being compositional, and its philosophical implications have been rather limited.

The main motivation behind the paper is to offer a two-valued semantics for $\mathbf{P}^1$ that would shed some new light on the understanding of the third value in its semantics as well as negation of $\mathbf{P}^1$, very much in the same spirit of Dunn's discovery. And, the aim of this paper is to present a new semantics for $\mathbf{P}^1$ along the line of Stanisław Jaśkowski's discussive logics. I will also deal with the system $\mathbf{I}^1$ which is a dual of $\mathbf{P}^1$, introduced in [23].

1.2 Preliminaries

Our language $\mathcal{L}$ throughout this paper, unless otherwise stated, consists of a set $\{\sim, \rightarrow\}$ of propositional connectives and a countable set Prop of propositional variables. We denote by Form the set of formulas defined as usual in $\mathcal{L}$. Moreover, we denote a formula by A, B, C, etc. and a set of formulas by Γ, Δ, Σ, etc.

2 The Paraconsistent Case

I first review the basic results well-known for the system $\mathbf{P}^1$. I then briefly introduce discussive logic of Jaśkowski, and this will be followed by a presentation of the new semantics which is equivalent to the original three-valued semantics.

2.1 Basics

First, we begin with the three-valued semantics.

Definition 1. *A* three-valued $\mathbf{P}^1$-valuation *is a homomorphism from* Form *to* $\{\mathbf{1}, \mathbf{i}, \mathbf{0}\}$*, induced by the following truth tables:*

A	$\sim A$
1	**0**
i	**1**
0	**1**

$A \rightarrow B$	**1**	**i**	**0**
1	**1**	**1**	**0**
i	**1**	**1**	**0**
0	**1**	**1**	**1**

Note that the designated values are **1** *and* **i***, and we define the semantic consequence relation* $\models_3$ *as usual in terms of the preservation of designated values.*

Remark 1. Conjunction and disjunction may be defined by $\sim(A \rightarrow \neg B)$ and $\neg A \rightarrow B$ respectively where $\neg A =_{\text{def.}} \sim(\sim A \rightarrow A)$, and have the following truth tables.[1]

[1] These definitions can be found in [16].

$A \wedge B$	1	i	0
1	1	1	0
i	1	1	0
0	0	0	0

$A \vee B$	1	i	0
1	1	1	1
i	1	1	1
0	1	1	0

However, for the sake of simplicity, I will not consider these connectives in the rest of this paper.

We now turn to the proof theory in terms of a Hilbert-style calculus.

Definition 2. *The system* $\mathbf{P}^1$ *consists of the following axiom schemata and a rule of inference.*[2]

(Ax1) $A \to (B \to A)$

(Ax2) $(A \to (B \to C)) \to ((A \to B) \to (A \to C))$

(Ax3) $({\sim}A \to {\sim}B) \to (({\sim}A \to {\sim}{\sim}B) \to A)$

(Ax4) $(A \to B) \to {\sim}{\sim}(A \to B)$

(MP) $$\frac{A \quad A \to B}{B}$$

Finally, we write $\Gamma \vdash_{\mathbf{P}^1} A$ *iff there is a sequence of formulas* $B_1, \ldots, B_n, A$ $(n \geq 0)$*, called* a derivation, *such that every formula in the sequence either (i) belongs to* Γ*; (ii) is an axiom of* $\mathbf{P}^1$*; (iii) is obtained by (MP) from formulas preceding it in the sequence.*

Then, Sette established two results in [22]. First one is the following soundness and completeness result.

Theorem 1 (Sette). *For all* $\Gamma \cup \{A\} \subseteq \mathsf{Form}$, $\Gamma \models_3 A$ *iff* $\Gamma \vdash_{\mathbf{P}^1} A$.

In order to state the second result, it is convenient to introduce the following definition.

Definition 3. *Let* $\mathbf{L}_1$ *and* $\mathbf{L}_2$ *be logics taken as sets of formulas closed under an appropriate relation of deducibility. Then,* $\mathbf{L}_1$ *is said to be* maximal relative to $\mathbf{L}_2$ *if the following holds:*

- *The languages of* $\mathbf{L}_1$ *and* $\mathbf{L}_2$ *are the same;*
- $\mathbf{L}_1 \subseteq \mathbf{L}_2$*;*
- $\mathbf{L}_1 \cup \{G\} = \mathbf{L}_2$ *for any theorem* G *of* $\mathbf{L}_2$ *which is not a theorem of* $\mathbf{L}_1$.

With this definition in mind, the second result may be stated as follows.

Theorem 2 (Sette). $\mathbf{P}^1$ *is maximal relative to classical logic.*

Remark 2. Note that in view of a general result established by Arnon Avron, Ofer Arieli and Anna Zamansky, it follows that $\models_3$ is maximally paraconsistent in the strong sense by [2, Corollary 3.6].

[2] The original axiomatization had one more axiom ${\sim}(A \to {\sim}{\sim}A) \to A$ which was later proved to be redundant. See, for example, [16].

Remark 3. Here are a few more facts that are well-known in relation to $\mathbf{P}^1$.

- Even though $A \to (\sim A \to B)$ is not provable, $\sim A \to (\sim\sim A \to B)$ *is* provable. Other systems with the same feature include the system **Z** of Jean-Yves Béziau (cf. [4]), $\mathbf{TCC}_\omega$ of Gordienko (cf. [10]) and **PCL1** of Toshiharu Waragai and Tomoki Shidori (cf. [24,25]).
- $\mathbf{P}^1$ is obtained by adding the following axioms to the so-called $\mathbf{C}_n$-systems devised by Newton da Costa (cf. [8]): $(\sim A)^{(n)}$, $(A \wedge B)^{(n)}$, $(A \vee B)^{(n)}$ and $(A \to B)^{(n)}$. In other words, not only negation but every complex formulas behave as in classical logic in $\mathbf{P}^1$.

2.2 Discussive Logics: An Interlude

Our new semantics for Sette's $\mathbf{P}^1$ is a variant of the semantics for discussive logics, originally developed by Jaśkowski in [12,13] (note that [12] is a new translation of [11]). In this subsection, we quickly review the semantics for discussive logics.

Definition 4. *The discussive language $\mathcal{L}_d$ consists of a set $\{\sim, \vee, \wedge_d, \to_d\}$ of propositional connectives and a countable set* Prop *of propositional variables. We denote by* Form_d *the set of formulas defined as usual in $\mathcal{L}_d$. Moreover, we denote a formula by A, B, C, etc. and a set of formulas by Γ, Δ, Σ, etc.*

Then, Jaśkowski's definition of discussive validities relies on the modal logic **S5** as follows.

Definition 5. *Let $\mathcal{L}_d$ and $\mathcal{L}_M$ be the discussive language and the language for* **S5** *respectively. Then, the translation τ from $\mathcal{L}_d$ to $\mathcal{L}_M$ is defined as:*

- $\tau(p) = p$ *for* $p \in \mathsf{Prop}$,
- $\tau(\sim A) = \neg\tau(A)$,
- $\tau(A \vee B) = \tau(A) \vee \tau(B)$,
- $\tau(A \wedge_d B) = \tau(A) \wedge \Diamond\tau(B)$,
- $\tau(A \to_d B) = \Diamond\tau(A) \supset \tau(B)$.

Then, a discussive formula A is $\mathbf{D}_2$-valid iff $\vdash_{\mathbf{S5}} \Diamond\tau(A)$.

Remark 4. Note that there are several translations discussed in the literature of discussive logics. The difference among translations lies in how the discussive conjunction is translated. For the details, see [19].

Remark 5. One may wonder what happens if we replace **S5** by other modal logics. In fact, this question has been one of the main topics for the Polish school of paraconsistency. For example, [17] explores the weakest modal logic that defines $\mathbf{D}_2$, and [18] explores the weakest normal and regular modal logics that defines $\mathbf{D}_2$.

We may also state the semantics of discussive logic *without* the help of translation, as suggested by Janucz Ciuciura in [7].

Definition 6. *A discussive model is a pair* $\langle W, v\rangle$ *where* W *is a non-empty set and* $v : W \times \mathsf{Prop} \longrightarrow \{0,1\}$*, an assignment of truth values to state-variable pairs. Valuations* v *are then extended to interpretations* I *to state-formula pairs by the following conditions.*

- $I(w,p) = v(w,p)$,
- $I(w,{\sim}A) = 1$ *iff* $I(w,A) \neq 1$,
- $I(w, A{\wedge_d}B) = 1$ *iff* $I(w,A) = 1$ *and for some* $x \in W$: $1 \in I(x,B)$,
- $I(w, A{\vee}B) = 1$ *iff* $I(w,A) = 1$ *or* $I(w,B) = 1$,
- $I(w, A{\to_d}B) = 1$ *iff for all* $x \in W$: $I(x,A) \neq 1$ *or* $I(w,B) = 1$.

Then, $\models_j A$ *iff for every discussive model* $\langle W, v\rangle$, $I(w,A) = 1$ *for some* $w \in W$.

Remark 6. Note the unusual definition of the valid formulas which is not a mistake but reflects the "diamond effect" of the original definition of $\mathbf{D}_2$-validities. Note also that elements of W may be regarded as discussants in a discussion.

2.3 Discussive Semantics

We now present a new semantics. Since it may be seen as a variation of the semantics for discussive logic, we refer to the semantics as discussive semantics.

Definition 7 ($\mathbf{P}^1$-discussive-model). $\mathbf{P}^1$*-discussive-model is a pair* $\langle W, v\rangle$*, where* W *is a nonempty set and* $v : W \times \mathsf{Prop} \longrightarrow \{0,1\}$*, an assignment of truth values to state-variable pairs. Valuations* v *are then extended to interpretations* I *to state-formula pairs by the following conditions.*

- $I(w,p) = v(w,p)$,
- $I(w,{\sim}A) = 1$ *iff for some* $x \in W : I(x,A) = 0$,
- $I(w, A{\to}B) = 1$ *iff for all* $x{\in}W : I(x,A) = 0$ *or for some* $y{\in}W : I(y,B) = 1$.

Then, $\models_d A$ *iff for every* $\mathbf{P}^1$*-discussive-model* $\langle W, v\rangle$, $I(w,A) = 1$ *for some* $w \in W$.

Remark 7. Here are a few remarks. First, note the "diamond feature" of the definition of the validity. Again, this is not a mistake but reflects the idea of Jaśkowski. Second, note that the above truth condition for negation is also considered by Ciuciura in [6] in terms of a variant of $\mathbf{D}_2$ based on a comment (as a translator) by Jerzy Perzanowski (cf. [13, p. 59]).[3] Finally, note that the truth condition for the conditional is even more "modalised" than that for the conditional in $\mathbf{D}_2$. More specifically, in order to evaluate the conditional to be true at a state w, it suffices for the consequent of the conditional to be true at *some* state, not necessarily at the state w.

Now, by considering a special case of the above semantics in which there are only two states (or discussants), we obtain the following four-valued semantics.

[3] Note however that the main result of Ciuciura contains a mistake, as pointed out in [19]. Therefore, the variant of $\mathbf{D}_2$ remains to be explored further.

Definition 8. *A four-valued* $\mathbf{P}^1$*-valuation is a homomorphism from* Form *to* $\{\mathbf{1}, \mathbf{i}, \mathbf{j}, \mathbf{0}\}$, *induced by the following matrices:*

A	$\sim A$
$\mathbf{1}$	$\mathbf{0}$
$\mathbf{i}$	$\mathbf{1}$
$\mathbf{j}$	$\mathbf{1}$
$\mathbf{0}$	$\mathbf{1}$

$A \to B$	$\mathbf{1}$	$\mathbf{i}$	$\mathbf{j}$	$\mathbf{0}$
$\mathbf{1}$	$\mathbf{1}$	$\mathbf{1}$	$\mathbf{1}$	$\mathbf{0}$
$\mathbf{i}$	$\mathbf{1}$	$\mathbf{1}$	$\mathbf{1}$	$\mathbf{0}$
$\mathbf{j}$	$\mathbf{1}$	$\mathbf{1}$	$\mathbf{1}$	$\mathbf{0}$
$\mathbf{0}$	$\mathbf{1}$	$\mathbf{1}$	$\mathbf{1}$	$\mathbf{1}$

Note that the designated values are $\mathbf{1}$, $\mathbf{i}$ *and* $\mathbf{j}$, *and we define the semantic consequence relation* $\models_4$ *as usual in terms of the preservation of designated values.*

Remark 8. In this four-valued semantics, the intermediate values are representing the two possibilities depending on which of the two participants are saying false. We can in fact "merge" these two possibilities, and let the third value stand for the case in which the two participants disagree. As a result, we obtain the three-valued semantics presented in Definition 1. The precise relation between the two semantics will be established in Remark 9.

Since $\models_4$ may be seen as a special case of $\models_d$, we obtain the following lemma.

Lemma 1. *For all* $A \in$ Form, *if* $\models_d A$ *then* $\models_4 A$.

2.4 The Main Result

We now turn to the main result that the three-valued semantics and discussive semantics are equivalent. To this end, we need the following key lemma.

Lemma 2. *If* $\Gamma \models_4 A$ *then* $\Gamma \models_3 A$.

Proof. We prove the contrapositive. Assume $\Gamma \not\models_3 A$. Then there is a three-valued-$\mathbf{P}^1$-valuation I_0 such that $I_0(\Gamma) \in \mathcal{D}$ and $I_0(A) \notin \mathcal{D}$. Then let I_1 be a four-valued-$\mathbf{P}^1$-valuation such that $I_1(p) = I_0(p)$. Then, we have that $I_1(A) = \mathbf{1}$ iff $I_0(A) = \mathbf{1}$ and $I_1(A) = \mathbf{0}$ iff $I_0(A) = \mathbf{0}$. This can be proved by a simple induction on the complexity of A as follows:

- The base case when A is a propositional variable, it is obvious by definition.
- For induction step, we consider the following two cases.
  - If A is of the form $\sim B$, then by IH, we have that
    - * $I_1(B) = \mathbf{1}$ iff $I_0(B) = \mathbf{1}$ and
    - * $I_1(B) = \mathbf{0}$ iff $I_0(B) = \mathbf{0}$.

    Then, by the truth-table, we obtain that

$$\begin{aligned} I_1(\sim B) = \mathbf{0} &\text{ iff } I_1(B) = \mathbf{1} && \text{By the truth table}\\ &\text{ iff } I_0(B) = \mathbf{1} && \text{IH}\\ &\text{ iff } I_0(\sim B) = \mathbf{0} && \text{By the truth table}\end{aligned}$$

$$\begin{aligned} I_1(\sim B) = \mathbf{1} &\text{ iff } I_1(B) \neq \mathbf{1} && \text{By the truth table}\\ &\text{ iff } I_0(B) \neq \mathbf{1} && \text{IH}\\ &\text{ iff } I_0(\sim B) = \mathbf{1} && \text{By the truth table}\end{aligned}$$

- If A is of the form $B \to C$, then by IH, we have that
  * $I_1(B) = \mathbf{1}$ iff $I_0(B) = \mathbf{1}$, $I_1(B) = \mathbf{0}$ iff $I_0(B) = \mathbf{0}$, and
  * $I_1(C) = \mathbf{1}$ iff $I_0(C) = \mathbf{1}$, $I_1(C) = \mathbf{0}$ iff $I_0(C) = \mathbf{0}$.

  Then, by the truth-table, we obtain that

$$\begin{aligned} I_1(B \to C) = \mathbf{0} &\text{ iff } I_1(B) \neq \mathbf{0} \text{ and } I_1(C) = \mathbf{0} && \text{By the truth table} \\ &\text{ iff } I_0(B) \neq \mathbf{0} \text{ and } I_0(C) = \mathbf{0} && \text{IH} \\ &\text{ iff } I_0(B \to C) = \mathbf{0} && \text{By the truth table} \end{aligned}$$

$$\begin{aligned} I_1(B \to C) = \mathbf{1} &\text{ iff } I_1(B) = \mathbf{0} \text{ or } I_1(C) \neq \mathbf{0} && \text{By the truth table} \\ &\text{ iff } I_0(B) = \mathbf{0} \text{ and } I_0(C) \neq \mathbf{0} && \text{IH} \\ &\text{ iff } I_0(B \to C) = \mathbf{1} && \text{By the truth table} \end{aligned}$$

  This completes the proof.

Once this is established, it is easy to see that the desired result holds since $I_1(A) = \mathbf{0}$ iff $I_0(A) = \mathbf{0}$ implies that $I_1(A) \notin \mathcal{D}$ iff $I_0(A) \notin \mathcal{D}$. □

We are now ready to prove the main result.

Theorem 3 (Main Theorem). *For all* $A \in \mathsf{Form}$, $\models_3 A$ *iff* $\models_d A$.

Proof. For the left-to-right direction, if $\models_3 A$ then $\vdash_{\mathbf{P^1}} A$ by Theorem 1. And one may check the soundness with respect to the Kripke semantics, i.e. that if $\vdash_{\mathbf{P^1}} A$ then $\models_d A$. The details are left as an exercise. For the other direction, if $\models_d A$ then $\models_4 A$ by Lemma 1. Thus, together with Lemma 2, we obtain the desired result. □

Corollary 1. *For all* $A \in \mathsf{Form}$, $\vdash_{\mathbf{P^1}} A$ *iff* $\models_d A$.

Proof. By Theorems 1 and 3. □

Remark 9. Note that by Lemma 2, we obtain that $\models_4 A$ implies $\models_3 A$. Moreover, in view of Theorem 3, we obtain that $\models_3 A$ implies $\models_d A$ and since $\models_d A$ implies $\models_4 A$ by Lemma 1, we obtain $\models_3 A$ implies $\models_4 A$. Therefore we reach that $\models_4 A$ iff $\models_3 A$ for all $A \in \mathsf{Form}$.

3 The Paracomplete Case

We now turn to the dual case. Namely, we deal with the system $\mathbf{I^1}$ developed by Sette and Carnielli in [23]. The section is structured as in the previous section.

3.1 Basics

Again, we begin with the three-valued semantics.

Definition 9. *A three-valued* $\mathbf{I}^1$-valuation *is a homomorphism from* Form *to* $\{\mathbf{1}, \mathbf{i}, \mathbf{0}\}$, *induced by the following matrices:*

A	$\sim A$
1	**0**
i	**0**
0	**1**

$A \to B$	**1**	**i**	**0**
1	**1**	**0**	**0**
i	**1**	**1**	**1**
0	**1**	**1**	**1**

Note that the designated value is **1**, *and we define the semantic consequence relation* $\models_3^i$ *as usual in terms of the preservation of the designated value.*

We now turn to the proof theory, again in terms of a Hilbert-style calculus.

Definition 10. *The system* $\mathbf{I}^1$ *consists of the following axiom schemata and a rule of inference.*

(Ax1) $A \to (B \to A)$

(Ax2) $(A \to (B \to C)) \to ((A \to B) \to (A \to C))$

(Ax3) $(\sim\sim A \to \sim B) \to ((\sim\sim A \to B) \to \sim A)$

(Ax4) $\sim\sim(A \to B) \to (A \to B)$

(MP) $\dfrac{A \quad A \to B}{B}$

Finally, we define $\Gamma \vdash_{\mathbf{I}^1} A$ *iff there is a sequence of formulas* $B_1, \ldots, B_n, A$ $(n \geq 0)$, *called* a derivation, *such that every formula in the sequence either (i) belongs to* Γ; *(ii) is an axiom of* $\mathbf{I}^1$; *(iii) is obtained by (MP) from formulas preceding it in the sequence.*

Then, the following results, together with some further results, were established by Sette and Carnielli in [23].

Theorem 4 (Sette & Carnielli). *For all* $\Gamma \cup \{A\} \subseteq$ Form, $\Gamma \models_3^i A$ *iff* $\Gamma \vdash_{\mathbf{I}^1} A$.

Theorem 5 (Sette & Carnielli). $\mathbf{I}^1$ *is maximal relative to classical logic.*

3.2 Kripke Semantics

We now present a new semantics for $\mathbf{I}^1$ à la Kripke.

Definition 11 ($\mathbf{I}^1$-Kripke-model). $\mathbf{I}^1$-*Kripke-model is a pair* $\langle W, v \rangle$, *where* W *is a nonempty set and* $v : W \times \mathsf{Prop} \longrightarrow \{0, 1\}$, *an assignment of truth values to state-variable pairs. Valuations* v *are then extended to interpretations* I *to state-formula pairs by the following conditions.*

- $I(w, p) = v(w, p)$,
- $I(w, \sim A) = 1$ *iff for all* $x \in W : I(x, A) = 0$,
- $I(w, A \to B) = 1$ *iff for some* $x \in W : I(x, A) = 0$ *or for all* $y \in W : I(y, B) = 1$.

Finally, $\models_k A$ *iff for every* $\mathbf{I}^1$-*Kripke-model* $\langle W, v \rangle$, $I(w, A) = 1$ *for all* $w \in W$.

Remark 10. Unlike the discussive semantics for $\mathbf{P}^1$, the semantic consequence relation requires to check *all* worlds/states to make the concerned sentence valid. Moreover, compared to the Kripke semantics for intuitionistic logic, the negation has the same truth condition of the form "necessarily not" (of course the frame condition is different, so not completely the same), but this is not the case for the conditional.

Now, by considering a special case of the Kripke semantics in which there are only two participants, we obtain the following four-valued semantics.

Definition 12. *A* four-valued $\mathbf{I}^1$-valuation *is a homomorphism from* Form *to* $\{\mathbf{1}, \mathbf{i}, \mathbf{j}, \mathbf{0}\}$, *induced by the following matrices:*

A	$\sim A$
1	0
i	0
j	0
0	1

$A \to B$	1	i	j	0
1	1	0	0	0
i	1	1	1	1
j	1	1	1	1
0	1	1	1	1

Note that the designated value is $\mathbf{1}$, *and we define the semantic consequence relation* $\models_4^i$ *as usual in terms of the preservation of the designated value.*

Remark 11. Note first the difference in the set of designated values between $\models_4$ and $\models_4^i$ are due to the different definitions of validities in $\models_d$ and $\models_k$ respectively. Intuitively speaking, being more strict in the modal semantics corresponds to less designated values in the many-valued semantics. Note too that as with the case for $\mathbf{P}^1$, we obtain the three-valued semantics by "merging" the intermediate values in the above four-valued semantics.

Since $\models_4^i$ may be seen as a special case of $\models_k$, we obtain the following lemma.

Lemma 3. *For all* $A \in$ Form, *if* $\models_k A$ *then* $\models_4^i A$.

3.3 The Main Result

Now we turn to the main observation. Once again, we need the following lemma.

Lemma 4. *If* $\Gamma \models_4^i A$ *then* $\Gamma \models_3^i A$.

Proof. We prove the contrapositive. Assume $\Gamma \not\models_3^i A$. Then there is a three-valued-$\mathbf{I}^1$-valuation I_0 such that $I_0(\Gamma) \in \mathcal{D}$ and $I_0(A) \notin \mathcal{D}$. Then let I_1 be a four-valued-$\mathbf{I}^1$-valuation such that $I_1(p) = I_0(p)$. Then, we have that $I_1(A) = \mathbf{1}$ iff $I_0(A) = \mathbf{1}$ and $I_1(A) = \mathbf{0}$ iff $I_0(A) = \mathbf{0}$. This can be proved by a simple induction on the complexity of A. Once this is established it is easy to see that the desired result holds since $I_1(A) = \mathbf{1}$ iff $I_0(A) = \mathbf{1}$ implies that $I_1(A) \neq \mathbf{1}$ iff $I_0(A) \neq \mathbf{1}$, namely $I_1(A) \notin \mathcal{D}$ iff $I_0(A) \notin \mathcal{D}$. □

We are now ready to prove the main result.

Theorem 6 (Main Theorem). *For all* $A \in \mathsf{Form}$, $\models^i_3 A$ *iff* $\models_k A$.

Proof. For the left-to-right direction, if $\models^i_3 A$ then $\vdash_{\mathbf{I}^1} A$ by Theorem 1. And one may check the soundness with respect to the Kripke semantics, i.e. that if $\vdash_{\mathbf{I}^1} A$ then $\models_k A$. The details are left as an exercise. For the other direction, if $\models_k A$ then we obtain $\models^i_4 A$ by Lemma 3. Thus, together with Lemma 4, we obtain the desired result. □

Corollary 2. *For all* $A \in \mathsf{Form}$, $\vdash_{\mathbf{I}^1} A$ *iff* $\models_k A$.

Proof. By Theorems 1 and 6. □

4 Reflections

In this section, we first observe some differences from the two alternative semantics for $\mathbf{P}^1$ offered in the literature. These will be followed by a brief discussion in relation to the results established by Barteld Kooi and Allard Tamminga. Finally, we sketch some of the philosophical implications of the discussive semantics presented in this paper.

4.1 A Comparison to the Semantics by Araujo, Alves, and Guerzoni

The semantics developed by Araujo, Alves, and Guerzoni in [1] takes the form of the more familiar Kripke semantics, as follows.

Definition 13 (Araujo & Alves & Guerzoni). *An* AAG-$\mathbf{P}^1$-model *is a triple* $\langle W, R, v\rangle$ *where* W *is a non-empty set,* R *is a binary reflexive relation on* W *and* $v : W \times \mathsf{Prop} \longrightarrow \{0, 1\}$, *an assignment of truth values to state-variable pairs. Valuations* v *are then extended to interpretations* I *to state-formula pairs by the following conditions:*

- $I(w, p) = v(w, p)$,
- $I(w, {\sim}p) = 1$ *iff for some* $w' \in W$: *(*wRw' *and* $I(w', A) \neq 1$*) if* $p \in \mathsf{Prop}$,
- $I(w, {\sim}A) = 1$ *iff* $I(w, A) \neq 1$ *if* $A \notin \mathsf{Prop}$,
- $I(w, A{\to}B) = 1$ *iff* $I(w, A) \neq 1$ *or* $I(w, B) = 1$.

Based on this model, the validity is defined in the usual manner.

Definition 14. *For all* $A \in \mathsf{Form}$, $\models_{\mathbf{AAG}} A$ *iff for every AAG-*$\mathbf{P}^1$*-model* $\langle W, R, I\rangle$, $I(w, A) = 1$ *for all* $w \in W$.

Then, Araujo, Alves, and Guerzoni established the following result (cf. [1, Proposition 4.8]).

Theorem 7 (Araujo & Alves & Guerzoni). *For all* $A \in \mathsf{Form}$, $\vdash_{\mathbf{P}^1} A$ *iff* $\models_{\mathbf{AAG}} A$.

Remark 12. Compared to the $\mathbf{P}^1$-discussive-model, we may observe the following differences:

- the binary relation R on W is taken to be reflexive. This is a reasonable option since the validity of the law of the excluded middle corresponds to the reflexivity of R when negation is seen as negative modality of the form "possibly not".
- the truth condition for negation is divided into two cases depending on the negated formula being atomic or not. This is also a reasonable option since after some calculations, one realize that the paracosistent behavior is observed only with respect to the atomic level.
- the truth condition for conditional is stated point-wise, exactly like in modal logics based on classical logic.

The main problem with this semantics, however, is that we do *not* have a uniform understanding of negation since the truth condition for negation depends on the negated formula. This is not the case in the discussive semantics presented in this paper in which we have a clear understanding of paraconsistent negation as a negative modality.

4.2 A Comparison to the Semantics by Carnielli and Lima-Marques

We now turn to compare the new semantics with the so-called Society Semantics devised by Carnielli and Lima-Marques in [5].

Definition 15. *A* society *S is a denumerable set of agents $S = \{Ag_1, Ag_2, \dots, \}$ where an* agent *Ag_i is a pair $\langle C_i, L_i \rangle$ formed by a collection C_i of propositional variables in a formal language and an underlying logic L_i.*

Remark 13. For our purposes, we assume that the underlying logic is classical logic for all agents.

Definition 16. *An agent Ag* accepts *a formula A iff all classical valuations which satisfy the variables of Ag also satisfy A. Moreover, a society is* open *if it accepts a formula in case any of its agent do. We denote open societies by S^+.*

Definition 17. *Let S^+ be an open society. Then the satisfiability relation $\Vdash$ between S^+ and an arbitrary formula A is inductively defined as follows.*

- *$S^+ \Vdash p$ iff for some agent $Ag \in S^+$: $p \in Ag$,*
- *$S^+ \Vdash {\sim}p$ iff for some agent $Ag \in S^+$: $p \notin Ag$,*
- *$S^+ \Vdash {\sim}A$ iff $S^+ \nVdash A$ for $A \notin \mathsf{Prop}$,*
- *$S^+ \Vdash A \to B$ iff $S^+ \nVdash A$ or $S^+ \Vdash B$.*

Definition 18. *A formula A is* satisfiable *in an open society iff for some open society S^+, $S^+ \Vdash A$. Moreover, A is an* open-tautology *(notation: $\models_o A$) iff A is satisfiable in every open society.*

Based on these definitions, Carnielli and Lima-Marques established the following result.

Theorem 8 (Carnielli & Lima-Marques). *For all* $A \in \mathsf{Form}$, $\vdash_{\mathbf{P}^1} A$ *iff* $\models_o A$.

Remark 14. Note that Carnielli and Lima-Marques introduces a variant of the Kripke semantics of Araujo, Alves, and Guerzoni based on their semantics. Note also that Carnielli and Lima-Marques also introduced *close society* to characterize the semantic consequence relation of $\mathbf{I}^1$. For the details, see [5].

Remark 15. Compared to the $\mathbf{P}^1$-discussive-model, we may observe the following differences. First, as Carnielli and Lima-Marques admits, the truth condition for negation is not compositional. This is also the feature of the Kripke semantics of Araujo, Alves, and Guerzoni which seems to be not so plausible from a philosophical perspective. However, this problem is solved in our discussive semantics. Second, the truth condition for the conditional is exactly as in classical modal logic which was also the case with the Kripke semantics of Araujo, Alves, and Guerzoni. However, by having a variant of discussive conditional, we reach a more uniform perspective on Sette's logics, and this seems to be quite surprising as well as interesting.

In sum, compared to the previous two attempts in offering alternative semantics for $\mathbf{P}^1$, the notable feature of the discussive semantics presented in this paper is the finding of the truth condition for the conditional. This turns out to be the key to afford a compositional semantics along the path that Jaśkowski paved already a while ago.

4.3 Yet Another Comparison

When it comes to the relation between three-valued logics and modal logics, there is an extremely interesting and general result established by Barteld Kooi and Allard Tamminga in [14].[4] In brief, Kooi and Tamminga observe that every truth-functional three-valued propositional logics can be conservatively translated into the modal logic **S5**. Since $\mathbf{P}^1$ is also a three-valued logic, it follows from the general result of Kooi and Tamminga that there is a close connection between $\mathbf{P}^1$ and **S5**, and so the observation of the paper may seem to be not so surprising.[5]

The general translation manual offered by Kooi and Tamminga is quite complex. Indeed, as they point out, the length of the translations produced by their manual is exponential. This gives rise to a question if there is a shorter translation into **S5** for some three-valued logics. In order to show that the answer is affirmative, Kooi and Tamminga present translations for **LP** and the Strong Kleene logic whose translations are quite distinct from the one in the manual.

Now, back to the worry of one of the reviewers. Is the close connection between $\mathbf{P}^1$ and **S5** surprising? To my surprise, it turns out that the application of the translation manual for $\mathbf{P}^1$ and $\mathbf{I}^1$ will give rise to the discussive

[4] I would like to thank one of the reviewers for directing my attention to the result by Kooi and Tamminga reported in [14].

[5] This was a worry of one of the reviewers.

and Kripke semantics, presented in this paper, respectively.[6] Thus, one of the reviewer's insight is justified.

Still, some of the results in this paper are *not* immediate corollaries of the general result of Kooi and Tamminga. For example, the four-valued semantics which bridges the three-valued and discussive semantics is not an immediate corollary. Moreover, we may raise the following question: what is the relation between the translation manual of Kooi and Tamminga and discussive logics in general? For example, is it possible to characterize three-valued logics that have discussive semantics? I will leave these questions for another occasion.

4.4 A Brief Philosophical Sketch

Finally, I wish to briefly discuss some philosophical implications of the discussive semantics.

One of the obvious problems of the three-valued semantics of $\mathbf{P}^1$ is that it is not clear at all how we can make sense of the third truth value. Another problem, related to the first problem is that it is hard to grasp the intuitive meaning of negation and the conditional. Negation is particularly problematic since it plays an important role in defining paraconsistent logics.

In face of these problems, the discussive semantics does offer one of the many ways to make sense of the third truth value and understand the unary connective intended to be negation. Indeed, for the truth values, we obtain an intuitive reading of the truth values. More specifically, the values may be read discussively within the context of a discussion by two person: **1** stands for evaluated as true by both discussants, **0** for evaluated as false by both discussants, and **i** for evaluated as true by one of the discussants and false by another.[7] Moreover, we may understand the paraconsistent negation of $\mathbf{P}^1$ as a negative modality "possibly not". In this way, we may assess $\mathbf{P}^1$ in view of motivations along discussive logic, although this was obviously not the intended reading of Sette.

Of course, discussive semantics is not the only semantics that makes sense of the three-valued semantics. For example, we may consider the following semantics which may be seen as a variant of Dunn semantics.

Definition 19. *A* $\mathbf{P}^1$-interpretation *is a left-total relation, r, between propositional parameters and the values 1 and 0. More precisely, $r \subseteq \mathsf{Prop} \times \{1, 0\}$ such that $pR1$ or $pR0$ for all $p \in \mathsf{Prop}$. Given an interpretation, r, this is extended to a left-total relation between all formulas and truth values by the following clauses:*

$$\begin{array}{llll} {\sim}Ar1 & \textit{iff } Ar0, & {\sim}Ar0 & \textit{iff not } Ar1, \\ A{\to}Br1 & \textit{iff not } Ar1 \textit{ or } Br1, & A{\to}Br0 & \textit{iff } Ar1 \textit{ and not } Br1. \end{array}$$

Based on this, A is a $\mathbf{P}^1$-relational semantic consequence of Γ ($\Gamma \models_r A$) iff for every $\mathbf{P}^1$-interpretation r, if $Br1$ for all $B \in \Gamma$ then $Ar1$.

Then, as expected, we may observe the following fact.

[6] The details are left for interested readers.

[7] Recall here that we may obtain the four-valued semantics in which the disagreement case will be split into two cases depending on which of the two discussants is seeing the proposition as true.

Proposition 1. *For all* $\Gamma \cup \{A\} \subseteq \mathsf{Form}$, $\Gamma \models_r A$ *iff* $\Gamma \models_3 A$.

Now, the above semantics may suggest that $\mathbf{P}^1$ may fit the dialetheic approach to paraconsistency as well. Indeed, the above semantics relies only on two truth values, truth and falsity, and compatible with the view that some propositions are both truth and false. Note, of course, that the above falsity conditions for negation and the conditional are quite distinctive, and how dielatheists should set up the falsity condition is quite independent of the non-exclusiveness of truth and falsity. This seems to be an interesting implication for dialetheists.

In the end, there are at least four semantics for the proof system of $\mathbf{P}^1$: three-valued, four-valued, discussive and Dunn semantics. This is somewhat similar to the case with **FDE** in which there are at least three kinds of semantics four-valued semantics, Dunn semantics, and Routleys' star semantics. Further details towards more satisfactory assessment of $\mathbf{P}^1$, especially contrasting some differences between $\models_d$ and $\models_r$, will be a topic for further discussion.

5 Concluding Remarks

In this paper, I offered a new discussive semantics for Sette's paraconsistent logic $\mathbf{P}^1$ and a new Kripke semantics for Sette and Carnielli's paracomplete logic $\mathbf{I}^1$. I hope this will be of help to deepen our philosophical understanding or assess the philosophical importance of both systems.

For future directions, there are some technical questions that might be of interest on its own. For example, one may ask the same question explored in the context discussive logic to see if we can replace **S5**-like set up by more weaker set up. Another question that is more challenging is to find an alternative semantics for C-systems of da Costa. Thanks to the extensive work by Avron and his colleagues, we now have non-deterministic semantics for C-systems. However, when it comes to Kripke semantics, it is still unclear if we can devise one beside the "limit" system $\mathbf{C}_\omega$ which was explored by Matthias Baaz in [3]. Based on our tour in this paper, we should perhaps keep some distance from Kripke and get closer to Jaśkowski. I do not have any idea more than this, unfortunately, and I must leave the problem for interested readers.

Acknowledgments. Hitoshi Omori is a Postdoctoral Research Fellow of Japan Society for the Promotion of Science (JSPS). I would like to thank the anonymous referees for their kind and helpful comments that improved the paper substantially. I would also like to thank audiences at *Prague Seminar on Paraconsistent Logic* for useful comments and discussions.

References

1. de Araujo, A.L., Alves, E.H., Guerzoni, J.A.D.: Some relations between modal and paraconsistent logic. J. Non-Class. Logic **4**(2), 33–44 (1987)
2. Arieli, O., Avron, A., Zamansky, A.: Maximal and premaximal paraconsistency in the framework of three-valued semantics. Stud. Logica. **97**, 31–60 (2011)

3. Baaz, M.: Kripke-Type Semantics for Da Costa's Paraconsistent Logic C_ω. Notre Dame J. Formal Logic **27**(4), 523–527 (1986)
4. Béziau, J.-Y.: The paraconsistent Logic Z-A possible solution to Jaśkowski's problem. Logic Logical Philos. **15**, 99–111 (2006)
5. Carnielli, W.A., Lima-Marques, M.: Society semantics and multiple-valued logics. In: Contemporary Mathematics, vol. 235, pp. 33–52. American Mathematical Society (1999)
6. Ciuciura, J.: A quasi-discursive system ND_2^+. Notre Dame J. Formal Logic **47**, 371–384 (2006)
7. Ciuciura, J.: Frontiers of the discursive logic. Bull. Sect. Logic **37**(2), 81–92 (2008)
8. da Costa, N.C.A.: On the theory of inconsistent formal systems. Notre Dame J. Formal Logic **15**, 497–510 (1974)
9. Dunn, M.: Intuitive semantics for first-degree entailments and 'coupled trees'. Philos. Stud. **29**(3), 149–168 (1976)
10. Gordienko, A.B.: A paraconsistent extension of Sylvan's logic. Algebra Logic **46**(5), 289–296 (2007)
11. Jaśkowski, S.: Propositional calculus for contradictory deductive systems. Stud. Logica **24**, 143–157 (1969)
12. Jaśkowski, S.: A propositional calculus for inconsistent deductive systems. Logic Logical Philos. **7**, 35–56 (1999)
13. Jaśkowski, S.: On the discussive conjunction in the propositional calculus for inconsistent deductive systems. Logic Logical Philos. **7**, 57–59 (1999)
14. Kooi, B., Tamminga, A.: Three-valued logics in modal logic. Stud. Logica **101**(5), 1061–1072 (2013)
15. Lewin, R.A., Mikenberg, I.F., Schwarze, M.G.: Algebraization of paraconsistent logic P^1. J. Non-Class. Logic **7**(1/2), 145–154 (1990)
16. Marcos, J.: On a problem of da costa. In: Essays on the Foundations of Mathematics and Logic, pp. 53–69. Polimetrica (2005)
17. Nasieniewski, M., Pietruszczak, A.: On the weakest modal logics defining Jaśkowski's logic D_2 and the D_2-consequence. Bull. Sect. Logic **41**(3–4), 215–232 (2012)
18. Nasieniewski, M., Pietruszczak, A.: On modal logics defining Jaśkowski's D2-consequence. In: Tanaka, K., Berto, F., Mares, E., Paoli, F. (eds.) Paraconsistency: Logic and Applications. Logic, Epistemology, and the Unity of Science, vol. 26, pp. 141–161. Springer, Dordrecht (2013). doi:10.1007/978-94-007-4438-7_9
19. Omori, H., Alama, J.: Axiomatizing Jaśkowski's Discussive Logic D_2 (under review)
20. Priest, G.: The logic of paradox. J. Philos. Logic **8**(1), 219–241 (1979)
21. Pynko, A.P.: Algebraic study of Sette's maximal paraconsistent logic. Stud. Logica **54**(1), 89–128 (1995)
22. Sette, A.: On the propositional calculus P^1. Math. Japonicae **16**, 173–180 (1973)
23. Sette, A., Carnielli, W.: Maximal weakly-intuitionistic logics. Stud. Logica **55**, 181–203 (1995)
24. Waragai, T., Omori, H.: Some new results on PCL1 and its related systems. Logic Logical Philos. **19**(1–2), 129–158 (2010)
25. Waragai, T., Shidori, T.: A system of paraconsistent logic that has the notion of "behaving classically" in terms of the law of double negation and its relation to S5. In: Béziau, J.Y., Carnielli, W.A., Gabbay, D. (eds.) Handbook of Paraconsistency, pp. 177–187. College Publications (2007)

Multi-agent Belief Revision Using Multisets

Konstantinos Georgatos(✉)

Department of Mathematics and Computer Science, John Jay College,
City University of New York, 524 West 59th Street, New York, NY 10019, USA
kgeorgatos@jjay.cuny.edu

Abstract. Revising a belief set K with a proposition a results in a theory that entails a. We consider the case of a multiset of beliefs, representing the beliefs of multiple agents, and define its revision with a multiset of desired beliefs the group of agents should have. We give graph theoretic semantics to this revision operation and we postulate two classes of distance-based revision operators. Further, we show that this multiset revision operation can express the merging of the beliefs of multiple agents.

Keywords: Belief revision · Reasoning with multisets · Graph-based reasoning · Geodesic reasoning · Reasoning with similarity · Distance-based reasoning · Belief merging

1 Introduction

When there is agent interaction, we need to keep track of the beliefs of several agents simultaneously. The belief of a single agent may result in an action that modifies the belief of other agents. Such an action can be simply a disclosure of an agent belief to another agent. In addition, we may need to keep track of the beliefs that agents have of each other. For example, when an agent makes an announcement to a group of other agents, then the belief of the announcer becomes common knowledge among the receiving group of agents. It is imperative, therefore, that the logical system we use to describe the interaction of agents is able to express the resulting epistemic scenarios. Traditionally, the language employed to describe the interaction of multiple agents makes use of an indexed modality K_i to describe the knowledge of an agent i (see [4]); that is, the index varies over the agents, and we can express, for example, that agent i knows that agent j knows a with the formula $\mathsf{K}_i\mathsf{K}_j a$. This representation, however, is not always appropriate. Consider for example an announcement from an agent that is only partially successful, e.g. in the case of a faulty channel. We do not know which agents have received it and, therefore, it is not clear how to represent this uncertain epistemic state using indices. Similarly, if the group of agents is anonymous, e.g. in the case of anonymous polling, indexical modalities is not an appropriate representation.

Support for this project was provided by a PSC-CUNY Award, jointly funded by The Professional Staff Congress and The City University of New York.

A. Baltag et al. (Eds.): LORI 2017, LNCS 10455, pp. 466–479, 2017.
DOI: 10.1007/978-3-662-55665-8_32

This paper builds on the observation that multiple agent modeling may benefit by the use of multisets. In the case of multiple agents, we can use a multiset A of propositions to represent the beliefs of each individual. As two or more agents may have the same belief, a belief can appear more than once in the epistemic state and therefore the epistemic state denoted by A should express this multiplicity. For example, there are three reviewers of an article. Two of them believe that the paper should be accepted, denoted by a, and one believes that the paper should be rejected, denoted by r. Their group belief corresponds to the multiset $\{a, a, r\}$. Notice that a multiset representation is not sensitive to order, so beliefs in a multiset cannot be indexed. Therefore, this representation is appropriate when the set of multiple agents is anonymous.

The representation of belief states solely with multisets is limited, too. To see that, suppose there are three reviewers of an article as above. The paper is accepted so *at least* two of them believe it should be accepted. There is no multiset of propositions that can represent the majority. We need an additional construct such as the following

$$(a, a, r) \vee (a, a, a)$$

Now, consider the revision $A * b$ of a multiset A representing the beliefs of a group of agents with a formula b. It is natural to assume that b is a single proposition, as it could represent the common hypothesis that the agents need to assume or even the common constraint that the agents need to agree with. This view of revision is limited. Revision is not always completely successful. For example, suppose five agents revise their beliefs with b with 60% sucess. In this case it will be more appropriate to revise with the multiset

$$(b, b, b, \top, \top)$$

instead of a single formula b. Therefore, multisets of formulas should combine to express group epistemic change.

Finally, consider four agents that meet to agree on either a or $\neg a$. How would we represent their agreement? For example

$$\{a \vee \neg a, a \vee \neg a, a \vee \neg a, a \vee \neg a\}$$

is not appropriate as it is equivalent to a multiset of tautologies. The following expression seems more appropriate

$$\{a, a, a, a\} \vee \{\neg a, \neg a, \neg a, \neg a\}.$$

which is not expressible using a single multiset of formulas.

To summarize:

- single formulas cannot convey the quantitative information (such as majority or probability judgments) inherent in a multiset
- single multisets cannot be used to reason with cases, or represent conditional or disjunctive information.

In this paper, we define a framework that employs a more general language that allows boolean combinations of multisets. In other words, a belief change operation, say revision, $A * B$, will combine A and B which are boolean combinations of multisets of formulas and are interpreted over multisets of worlds but they are not necessarily *single* multisets of formulas. We will use a single language for A, B and $A * B$ and therefore several results from the traditional propositional belief change theory apply at both the syntactic and semantic level. Our approach is very similar to the jump from real numbers to vectors of real numbers of a fixed dimension.

In the next section, we will define the extended language. In Sect. 3, we will introduce and characterize geodesic belief revision in the extended language and in Sect. 4 we will show how some belief merging formalism can be expressed using the new framework. In the last section, we conclude.

2 Language and Semantics of Multisets

We will use a propositional language $\mathcal{L}$ with a finite set of atomic propositions closed under the usual classical connectives $\neg, \wedge$ and $\vee$ and with a finite set of atomic propositions. An interpretation w is a function from atomic propositions to $\{T, F\}$. An interpretation extends to a map from $\mathcal{L}$ to $\{T, F\}$ and will be called a *model* of ϕ if it maps ϕ to T. We write W for the set of all models. If A is a set of formulas then we write $v(A)$ to denote the set of all models of A.

In order to express theories based on multisets, we will consider a new language $\mathcal{L}^{\text{multi}}$ closed under the boolean connectives, but whose atomic formulas are multisets of formulas in $\mathcal{L}$ and formulas I^n_ϕ, for each $\phi \in \mathcal{L}$ and $n \in \{1, 2, \ldots\}$, that are not expressible using multisets of formulas in $\mathcal{L}^n$, called *diagonal* formulas. In the following, we will draw attention to fragments $\mathcal{L}^n$, for a given positive integer n, of $\mathcal{L}^{\text{multi}}$, that are formed like $\mathcal{L}^{\text{multi}}$ but the atomic formulas are diagonals I^n_ϕ and multisets of size n. We have

$$\mathcal{L}^{\text{multi}} = \bigcup^{n} \mathcal{L}^n.$$

We will use curly brackets { and } to write multisets. For example

$$\{a, b, b\} \vee \{c, d, f\} \vee I^3_a$$

belongs to $\mathcal{L}^3$, for $a, b, c \in \mathcal{L}$, and

$$\{a, b, b\} \vee \{c, d, f, d\} \vee I^5_a$$

does not belong to $\mathcal{L}^3$ but belongs to $\mathcal{L}^{\text{multi}}$.

To interpret $\mathcal{L}^{\text{multi}}$ we use finite multisets from W, the set of binary valuations that interpret $\mathcal{L}$. Let M be the set of finite multisets from W, i.e.,

$$M = \{\{w_1, \ldots, w_n\} : w_i \in W, n > 1\}.$$

We will write M^n for the subset of M that contains all multisets of cardinality n for some positive integer n. A *state* is a subset of M. A *product state* is a state of the form

$$\{A_1, A_2, \ldots, A_n\} = \{\{w_1, w_2, \ldots, w_n\} : w_i \in A_i, \text{ for } i = 1, \ldots, n\},$$

where $A_i \subseteq W$ and $n > 1$. A product state is obviously a state but a state is not necessarily a product state. For this reason, multisets of propositions correspond to product states but general formulas in $\mathcal{L}^{\text{multi}}$ are interpreted as states.

The interpretation of $\mathcal{L}^{\text{multi}}$ on M is straightforward. Suppose that $v(\phi)$ returns all valuations that make $\phi \in \mathcal{L}$ true. This map extends to atomic formulas with

$$v(\{a_1, a_2, \ldots, a_n\}) = \{v(a_1), v(a_2), \ldots, v(a_n)\}$$

and

$$v(I_a^n) = \{\{w, w, ..., w\} \in M^n : w \in v(a)\},$$

and to the whole of $\mathcal{L}^{\text{multi}}$ using intersection, union and complement. We will use $a, b, c, \ldots$ for propositions in $\mathcal{L}^{\text{multi}}$ and we will use $A, B, C, \ldots$ for theories.

Consequence and consistency in $\mathcal{L}^{\text{multi}}$ is defined semantically using the above interpretation. We have that a implies b iff $v(a) \subseteq v(b)$, and a and b are consistent iff $v(a) \cap v(b) \neq \varnothing$, where $a, b \in \mathcal{L}^{\text{multi}}$.

In particular, for multisets of propositions we have

$$\{a_1, a_2, \ldots, a_n\} \text{ implies } \{b_1, b_2, \ldots, b_n\}$$

if a_i implies $b_{\pi(i)}$ for all i, where $1 \leqslant i \leqslant n$, and some π permutation on n. The opposite direction does not hold. That is, we may have

$$\{a_1, a_2, \ldots, a_n\} \text{ implies } \{b_1, b_2, \ldots, b_n\}$$

but there is no permutation π such that a_i implies $b_{\pi(i)}$ for all i. As a counterexample, we have

$$\{a \wedge b, a \vee b\} \text{ implies } \{a, b\},$$

where $v(a) = \{w_1, w_2\}$ and $v(b) = \{w_1, w_3\}$.

It is known (see [2]) that

$$\{a_1, a_2, \ldots, a_n\} \text{ is equivalent to } \{b_1, b_2, \ldots, b_n\},$$

if and only if, a_i is equivalent to $b_{\pi(i)}$, for all i, where $1 \leqslant i \leqslant n$, and some π permutation on n. Similarly, it is easy to see that $\{a_1, a_2, \ldots, a_n\}$ and $\{b_1, b_2, \ldots, b_n\}$ are consistent iff a_i and $b_{\pi(i)}$ are consistent, for all i, where $1 \leqslant i \leqslant n$, and some π permutation on n.

3 Geodesic Revision

We will study geodesic revision operators on multisets of fixed cardinality n, where $n > 2$. Considering a fixed cardinality n will allow us to work on metric spaces whose metric is generated by a geodesic metric on worlds.

The use of geodesic metric rests on a novel view of similarity as a derived concept. Traditionally, similarity has been conceived as a primitive concept usually represented by distance; that is, the following identification is made:

$$\text{similarity} = \text{distance}$$

Our idea [5] is that similarity does not have to be primitive but it can be generated by a relation of indistinguishability. In particular, we do not need quantitative data, such a measurements, to generate a distance function. This idea can be summarized by the following maxim: two objects are similar when there is a context within which they are indistinguishable. Further, similarity can be *measured* with degrees of indistinguishability.

For example, although two similar houses might appear different in various details when we stand in front of them, they will appear identical if we observe them from an appropriate distance x. Thus, indistinguishability at distance x implies similarity. The smaller the distance x, the more similar the objects are.

A simple representation of indistinguishability by a reflexive symmetric non-transitive relation goes back to [15]. Such relations have been studied together with a set under various names such as tolerance spaces [18], proximity spaces [1], and others, but the best way to describe a set of worlds with an indistinguishability relation is simply a graph. Similarity now will be the distance map defined on the graph defined by the shortest path. Given a relation R the distance from y to x is the least number of times we need to apply R in order to reach y from x. Traditionally, this kind of relation has been called *geodesic*. We have

$$\text{similarity} = \text{geodesic distance (of a graph)}$$

Note that similarity for us is a distance rather than a relation taking values to the interval $[0, 1]$ as in the representation of similarity in fuzzy reasoning that usually assumes transitivity. Using graphs with their geodesic metric generalizes several popular formalisms such as threshold and integer metrics as well as hamming distance (see [7]).

Geodesic semantics have been successfully developed for a variety of belief change operators such as revision, update, conditionalization, and contraction [6–9] and this paper is an effort towards extending geodesic semantics to multisets.

Definition 1. *Let W be a set and $R \subseteq W \times W$ a relation on W. Then (W, R) is called a* (connected) tolerance space *when R is reflexive, symmetric, and (W is) connected, i.e., for all $x, y \in W$ there is a non negative integer n such that $xR^n y$.*

In the above definition, we assume $R^0 = \mathrm{id}_W$, $R^n = R^{n-1} \circ R$ for $n > 0$.

Given a tolerance space (W, R) we can define a metric called *geodesic* with a map d from $W \times W$ to Z^+ (the set of non-negative integers) where

$$d(x, y) = \min\{n \mid xR^n y\}.$$

Note that a geodesic metric is not any integer metric. The values of the geodesic metric are determined by adjacency. This property can be described with: for all $x, y \in W$ such that $d(x, y) = n$ with $1 < n < \infty$ there is $z \in W$ with $z \neq x, y$ such that $d(x, y) = d(x, z) + d(z, y)$. In particular we can choose z so that $d(x, z) = 1$. Note here that a geodesic metric is a topological metric, that is, it satisfies identity, symmetry and triangle inequality.

The geodesic distance extends to distance between non-empty subsets with

$$d(A, B) = \min\{d(x, y) \mid x \in A, y \in B\}. \tag{1}$$

We shall also write $d(x, A)$ for $d(\{x\}, A)$. Similarly for $d(A, x)$. We will write A^c for the complement of A and A^n for the set $\{x \in W : d(A, x) \leqslant n\}$ (where $n = 0, 1, \ldots$).

The distance d between models lifts to a distance between subsets of models using (1). We can now define a revision operator in terms of distance. Let

$$A * B = \begin{cases} \{y \in B : d(A, y) = d(v(A), v(B)\} \text{ if } A, B \neq \varnothing \\ B \qquad\qquad\qquad\qquad\qquad\qquad\quad \text{otherwise} \end{cases} \tag{2}$$

or, equivalently,

$$A * B = \begin{cases} A^{d(A,B)} \cap B \text{ if } A, B \neq \varnothing \\ B \qquad\qquad\quad \text{otherwise.} \end{cases}$$

The operator $*$ is a revision operator because it is defined through distance minimization as in [7,13]. We illustrate the above definition with the following

Example 1. In Fig. 1 and the rest of the figures graph edges represent the (irreflexive part of the) indistinguishability relation R. Let $A = \{a\}$, $B = \{b\}$, and $C = \{c_1, c_2\}$. Then $A * (B * C) = \{c_1\} \neq \{c_1, c_2\} = (A * B) * C$ (this also shows that revision does not satisfy association).

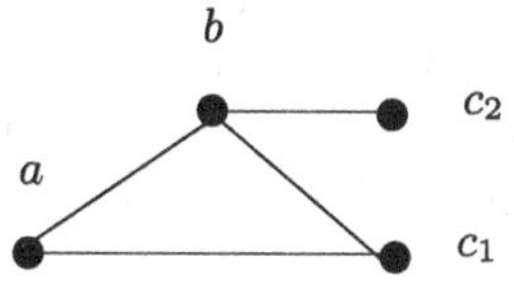

Fig. 1. Geodesic revision example

Table 1. Geodesic revision rules

1. $\phi * \psi \vdash \psi$
2. If ψ is consistent, then $\phi * \psi$ is consistent
3. If ϕ is inconsistent, then $\phi * \psi \leftrightarrow \psi$
4. If $\phi \wedge \psi$ is consistent, then $\phi * \psi \leftrightarrow \phi \wedge \psi$
5. If $\psi_1 \leftrightarrow \psi_2$ and $\phi_1 \leftrightarrow \phi_2$ then $\phi_1 * \psi_1 \leftrightarrow \phi_2 * \psi_2$
6. If $\psi \vdash \neg\phi$ then $\phi * \psi \leftrightarrow (\phi * \neg\phi) * \psi$
7. If $\psi \vdash \neg\phi$ then $\phi * \psi \leftrightarrow (\neg\phi * \phi) * \psi$
8. If $\phi * \psi \leftrightarrow \chi * \psi$ then $\phi * \psi \leftrightarrow (\phi \vee \chi) * \psi$
9. If $\psi \vdash \neg\phi$ then $\phi * \psi \vdash \neg\psi * \psi$
10. If $\phi \vdash \neg\psi$ then $\phi * \neg\phi \vdash \neg\psi$ iff $\psi * \neg\psi \vdash \neg\phi$

Call a revision operator *geodesic* if it satisfies the properties of Table 1. (For an explanation of of the postulates appearing in Table 1 we refer the reader to [7].)

The following has been proved in [7].

Proposition 1. *If $*$ is an operator that satisfies (2), then $*$ is a geodesic revision operator. Conversely, given a geodesic revision operator $*$, then there exists a binary relation R such that (W, R) is a tolerance space, where W is the set of models, and $*$ satisfies (2).*

The above framework applies to multisets as well. Let (W, R) be a tolerance space. Then, let Let M^n be the set of multisets from W with a fixed length n, i.e.,

$$M^n = \{\{w_1, \ldots, w_n\} : w_i \in W\}.$$

The relation of indistinguishability R can be extended to a relation R' of indistinguishability between multisets of worlds in two ways as follows

1. We have $(x_1, \ldots, x_n)R'(y_1, \ldots, y_n)$ if there exists a permutation π and $j \in \{1, \ldots, n\}$ such that $x_j R y_{\pi(j)}$ and $x_i = y_{\pi(i)}$ for $i = 1, \ldots, n$ and $i \neq j$. Now that we have an indistinguishability relation among multisets we can define a tolerance space with a geodesic metric. The geodesic metric is defined as the shortest path (of R') between the two multisets but one may show that it can be reduced to the geodesic metric d of the original space W:
$$d_{\Sigma}(\{w_1, w_2, \ldots, w_n\}, \{w'_1, w'_2 \ldots, w'_n\}) = \min_{\pi}\{\Sigma_i d(w_i, w'_{\pi(i)})\},$$
where π varies over the set of permutations of $\{1, 2, \ldots, n\}$.
2. We have $(x_1, \ldots, x_n)R'(y_1, \ldots, y_n)$ if there exists a permutation π such that $x_i R y_{\pi(i)}$ for all $i = 1, \ldots, n$. As in (1) above, we can define a tolerance space with a geodesic metric using the shortest path (of R') between the two multisets and again one may show that it can be reduced to the geodesic metric d of the original space W:
$$d_m((x_1, \ldots, x_n), (y_1, \ldots, y_n)) = \min_{\pi}\{\max_{i}\{d(x_i, y_{\pi(i)})\}\}.$$

In both cases, the geodesic metric lifts to a metric between subsets of multisets and therefore a revision operator can be defined the same way as in (2). We will denote the revision operators based on d_{Σ} and d_m with $*_{\Sigma}$ and $*_m$, respectively. Both operators $*_{\Sigma}$ and $*_m$ are defined using Eq. (2) so by Proposition 1 both operators are geodesic and, therefore, satisfy the postulates of Table 1 (where formulas in the table belong to $\mathcal{L}^{\text{multi}}$). To illustrate the process and difference of those two operators consider the following example.

Example 2. Consider the model of Fig. 2 consisting of five worlds and four propositions a, b, c, and d where $v(a) = \{w_1\}$, $v(b) = \{w_5\}$, $v(c) = \{w_2\}$, and $v(d) = \{w_4, w_5\}$.

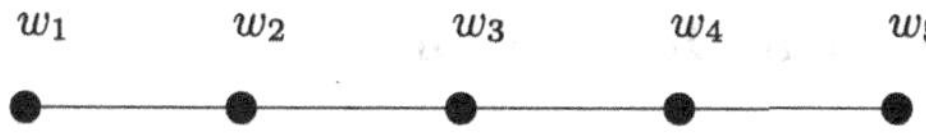

Fig. 2. Multiset revision

Now consider the revision of the multiset $\{a, a, b\}$ with the multiset $\{c, c, d\}$. We have that

$$v(\{a, a, b\}) = \{\{w_1, w_1, w_5\}\}$$

and

$$v(\{c, c, d\}) = \{\{w_2, w_2, w_4\}, \{w_2, w_2, w_5\}\}.$$

We have that

$$v(\{a, a, b\} *_\Sigma \{c, c, d\}) = \{c, c, b\}$$

because the multiset $\{w_2, w_2, w_5\}$ is closer to $\{w_1, w_1, w_5\}$ than $\{w_2, w_2, w_4\}$ since $d_\Sigma(\{w_1, w_1, w_5\}, \{w_2, w_2, w_5\}) = 2$ and $d_\Sigma(\{w_1, w_1, w_5\}, \{w_2, w_2, w_4\}) = 3$.

In contrast, we have that

$$v(\{a, a, b\} *_m \{c, c, d\}) = \{c, c, d\}$$

because the multisets $\{w_2, w_2, w_4\}$ and $\{w_2, w_2, w_5\}$ are equidistant to $\{w_1, w_1, w_5\}$ since $d_m(\{w_1, w_1, w_5\}, \{w_2, w_2, w_5\}) = d_m(\{w_1, w_1, w_5\}, \{w_2, w_2, w_5\}) = 1$.

Our aim is to give a logical characterization of the operators $*_\Sigma$ and $*_m$. To this end:

Definition 2. *Let $*$ and $*'$ be binary operators on $\mathcal{L}$ and $\mathcal{L}^n$, respectively, then*

1. *$*'$ will be called locally generated (by $*$) if*

$$v(\{a_1, \ldots, a_n\} *' \neg\{a_1, \ldots, a_n\}) = v(\bigvee_{j=1}^{n} \{a_1, \ldots, a_j * \neg a_j, \ldots, a_n\})$$

2. *$*'$ will be called globally generated (by $*$) if*

$$v(\{a_1, \ldots, a_n\} *' \neg\{a_1, \ldots, a_n\}) = v(\{a_1 * \neg a_1, \ldots, a_n * \neg a_n\}).$$

The above properties although restricted to multisets of $\mathcal{L}$ are enough to characterize the operators as the following shows.

Theorem 1. *Let $*$ and $*'$ be binary operators on $\mathcal{L}$ and $\mathcal{L}^n$, respectively, then*

1. *$*'$ is coordinate generated by $*$, if and only if, $*' = *_\Sigma$, and*
2. *$*$ is globally generated by $*$, if and only if, $*' = *_m$,*

where $*_\Sigma$ and $*_m$ are induced by the geodesic metric d corresponding to $*$.

The above shows that globally generated revision operators are associated with the standard distance of multisets of fixed cardinality called metric distance of roots (akin to Frechet distance–see [3]). The coordinate generated revision operators are associated with a minimization of the so-called Manhattan or ℓ^1 distance over all possible permutations. If the above metrics are restricted to the distance between a multiset and a single element then one obtains the Σ and Max operators of Sect. 4 in [12].

4 Comparison with Other Work

In this section, we will express a few merging operators studied in the literature using a multiset revision operator.

We start with the traditional notion of merging [14,16], that was perceived as a commutative version of revision. When we perform this sort of binary merging, we choose the models of the belief sets to be merged that are the most similar. We will confine ourselves to the geodesic framework; therefore we will pick the closest models with respect to the geodesic distance.

Example 3. We illustrate the process with the following example (edges represent the reflexive symmetric tolerance relation).

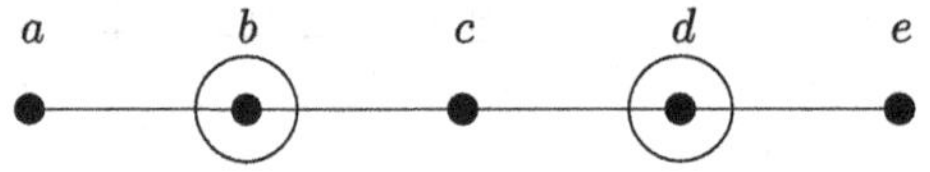

Fig. 3. Non-commutative revision

In Fig. 3, let $A = \{a, b\}$ and $B = \{d, e\}$. Then the merging, denoted by $A \otimes B$, of A with B equals the subset $\{b, d\}$ containing the elements of A and B whose distance is the least among the elements of the two sets: the distance of b from d is 2 while the distance of a from d and e from d is 3. This form of merging corresponds to arbitration of [14], and is a special case of the distance-based merging operator of [17].

This merging operator is defined on subsets of the tolerance space as follows:

$$A \otimes B = \begin{cases} \{x \in A, y \in B : d(x, y) = d(A, B)\} & \text{if } A, B \neq \varnothing \\ A \cup B & \text{otherwise.} \end{cases}$$

We can show that

Theorem 2. *Assume $v(\phi), v(\psi) \neq \varnothing$. Then, if $\otimes$ is a geodesic merging operator on $\mathcal{L}$, then there exists a coordinate generated revision operator on $\mathcal{L}^2$ such that*

$$v(\phi \otimes \psi) = \{w : \{w, w\} \in v(\{\phi, \psi\} * I_{\phi \vee \psi})\},$$

The merging operator defined has an important property, namely, it implies disjunction:

$$\phi \otimes \psi \vdash \phi \vee \psi.$$

There are cases, however, where this is not possible or not desirable. Next, we will express two forms of binary merging (introduced in [10]), that do not imply disjunction.

Example 4. Suppose that we count the pennies saved in a jar. An initial count finds 112 pennies. A second count finds 114 pennies. It seems plausible that the merge of these two counts is the set $\{112, 113, 114\}$ as one or both counts could have been wrong. Using the propositions of Example 3, we would like that the extension of $A \otimes_{\mathsf{c}} B$ is the set $\{b, c, d\}$ (see Fig. 4). This form of merging is called *convex.*

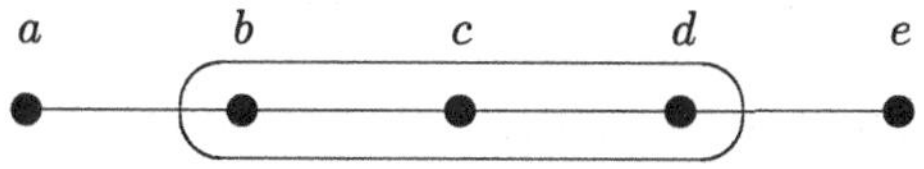

Fig. 4. Convex merging

Example 5. Suppose now that we need to classify submitted articles into three groups: *accept*, *reject* and *borderline*. For a given paper, we receive reviews from two referees. One thinks it belongs to the accept group and the other to the reject group. It seems to me that the merge of those two opinions is borderline. Obviously this notion of merging seems more appropriate when beliefs have different sources such as the case of voting. Using the propositions of Example 3, we would like that $A \odot B$ is modeled by the set $\{b\}$ (see Fig. 5). This form of merging is called *barycentric.*

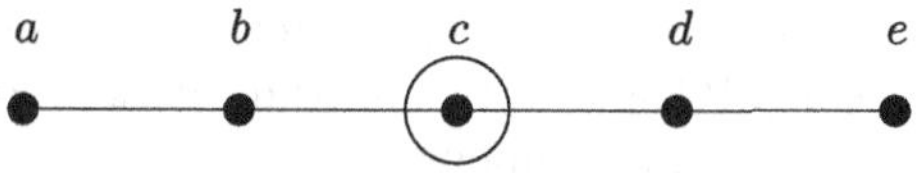

Fig. 5. Barycentric merging

The above merging operators are defined on subsets with

$$A \otimes_{\mathsf{c}} B = \begin{cases} \{x : d(A, x) + d(B, x) = d(A, B)\} & \text{if } A, B \neq \varnothing \\ A \cup B & \text{otherwise.} \end{cases}$$

and

$$A \odot B = \begin{cases} \{x \in \text{mid}(A, B) : d(A, x) + d(B, x) = d(A, B)\} & \text{if } A, B \neq \varnothing \\ A \cup B & \text{otherwise.} \end{cases}$$

where

$$\text{mid}(A, B) = A^k \cap B^k, \quad k = \min\{l : A^l \cap B^l \neq \varnothing\}.$$

Theorem 3. *Assume $v(\phi), v(\psi) \neq \varnothing$. Then,*

1. *if $\otimes_{\mathsf{c}}$ is a convex merging operator on $\mathcal{L}$, then there exists a coordinate generated revision operator on $\mathcal{L}^2$ such that*

$$v(\phi \otimes_{\mathsf{c}} \psi) = \{w : \{w, w\} \in v(\{\phi, \psi\} * I_{\top})\}, \text{ and}$$

2. *if $\odot$ is barycentric merging operator on $\mathcal{L}$, then there exists a pair generated revision operator on $\mathcal{L}^2$ such that*

$$v(\phi \odot \psi) = \{w : \{w, w\} \in v(\{\phi, \psi\} * I_{\top})\}.$$

Finally, we show how multiset revision can be used to express belief aggregation of multiple agents using the Integrity Constraints (IC) merging operator. An IC merging operator is a map from a multiset E of bases called profile and a formula to a base. A base is a finite set of propositional formulas. As IC merging operators allow bases in the multisets, we will assume that profiles, as well as the output of an IC merging operator, are made out of single propositional formulas. Further, we will also assume that profiles are restricted to a fixed size group of agents. Therefore, such profiles cannot be combined into a single one, as the group of agents represented by the resulting profile will double in size. In other words, we will assume that an IC merging operator Δ is a map from $\mathcal{L}^n \times \mathcal{L} \to \mathcal{L}$. Now, we have that

$$v(\Delta_\mu(E)) = \{w : \{w, w, \ldots, w\} \in v(E * I^n_\mu)\},$$

where $*$ can be any multiset merging operator introduced earlier.

In order to simulate fully IC operators, our framework needs to define merging revision over multisets of arbitrary dimensions, that is on $\mathcal{L}^{\text{multi}}$ rather than $\mathcal{L}^n$. We think that such an extension will also express more sophisticated merging operators that deal with possibly inconsistent bases such as the DA^2 operators (see [11]). Nevertheless, multiset revision shows that merging can be conceived through a *unified* process of minimization, at least within geodesic metric spaces. In other words, we only need to logically describe a single minimization operation and the various belief change operators arise from this minimization operation by varying the underlying language.

5 Further Work

The assumption in Sect. 3 that multisets have a fixed size n is critical—not only because this assumption underlies the technical development of the results of Sect. 3, but also because it corresponds to a very specific epistemic scenario; namely, the case where the population is fixed and we receive information from all individuals in the population, albeit anonymously. However, this is not always the case: we may have access to only proper subgroups of all agents. For example consider two belief multisets taken from two different college classes. Those multisets not only may differ in size, but they may not be combined into a single multiset as there might be students who belong to both classes.

The above framework can be *made* to accommodate multisets of varying size. In order to retain the validity of the results of Sect. 3, we will reinterpret an extended language of multisets of formulas with varying sizes on a tolerance space of multisets of valuations of a large enough *fixed* size. First, we will allow multisets of formulas of arbitrary size bounded by a fixed number N that represents the population size. That is, our language is

$$\mathcal{L}^{\downarrow N} = \bigcup_{n=1}^{N} \mathcal{L}^n.$$

We will alter the semantics of $\mathcal{L}^{\downarrow N}$ by interpreting them on the set M^N of multisets of binary valuations of fixed size N (see Sect. 2) as follows

$$\begin{aligned} &v(\{a_1, a_2, \ldots, a_m\}) \\ &\quad = \{\{w_1, w_2, \ldots, w_N\} : w_i \in M \text{ for } 1 \leqslant i \leqslant N, w_i \in v(a_i) \text{ for } 1 \leqslant j \leqslant m\} \end{aligned} \quad (3)$$

where $1 \leqslant m \leqslant N$, so $\{a_1, a_2, \ldots, a_m\} \in \mathcal{L}^{\downarrow N}$ and $v(\{a_1, a_2, \ldots, a_m\}) \subseteq M^N$. In effect, we interpret multisets of size less than N with multisets of size N using the following equivalence:

$$\{a_1, a_2, \ldots, a_m\} \leftrightarrow \{a_1, a_2, \ldots, a_m, \top, \ldots, \top\},$$

where $1 \leqslant m \leqslant N$, $\{a_1, a_2, \ldots, a_m\} \in \mathcal{L}^m$ and $\{a_1, a_2, \ldots, a_m, \top, \ldots, \top\} \in \mathcal{L}^N$. Interpreting $\mathcal{L}^{\downarrow N}$ on M^N allow us to replace $\mathcal{L}^n$ with $\mathcal{L}^{\downarrow N}$ in both Definition 2 and Theorem 1.

Although the above extension allows us to express incomplete information from subgroups of agents, by way of smaller size multisets, it does not fully address the complex issues arising from the employment of varying size multisets. One such issue is the way we combine multisets. Combining multisets of beliefs can be accomplished through a sophisticated merging mechanism (such as the ones discussed in Sect. 4), or can be as simple as concatenation. For example, if we combine two multisets that may include agents that are common in both, a *with-replacement* choice of sorts, conjunction of the multisets seems adequate:

$$v(\{a_1, a_2, \ldots, a_k\} \wedge \{b_1, b_2, \ldots, b_l\}) = v(\{a_1, a_2, \ldots, a_k\}) \cap v(\{b_1, b_2, \ldots, b_l\}).$$

This intersection is well defined as both multisets are interpreted over the same space. If, instead, we combine two multisets that represent disjoint groups of agents (*without-replacement* choice) then concatenation $\circ$ is more appropriate:

$$v(\{a_1, a_2, \ldots, a_k\} \circ \{b_1, b_2, \ldots, b_l\}) = v(\{a_1, a_2, \ldots, a_k, b_1, b_2, \ldots, b_m\},$$

where $k + l \leqslant N$. A comprehensive approach to multisets should include logical operators that correspond to all natural ways we can combine them.

Finally, the tolerance space itself may need to be adjusted according to the way multisets arise. An interesting reading of a multiset of beliefs arises when the set of agents, whose beliefs are represented in the multiset, is randomly selected. In this scenario, projecting the multiset to a population multiset via Eq. (3) is not appropriate. As an example, suppose that political leanings of agents can be conservative (c), liberal (l) and mixed (m) and those are mutually exclusive. Also, suppose that the basic tolerance relation among the resulting binary valuations is $\{c\}R\{m\}$ and $\{m\}R\{l\}$ but not $\{c\}R\{l\}$. For simplicity, assume that the population is of size 6 and we observe the multiset $\{c, c, c\}$ of a random sample of size 3. According to Eq. (3), the multiset $\{c, c, c\}$ is equivalent to $\{c, c, c, \top, \top, \top\}$. Therefore,

$$d(\{c, c, c\}, \{c, c, c, l, l, l\}) = d(\{c, c, c\}, \{c, c, c, c, c, c\}) = 0.$$

However, observing $\{c, c, c\}$ through a random sample indicates a slight preference for a population that is all conservative, and the tolerance space of multisets should reflect this preference. Therefore, our framework is not developed enough to account for all epistemic scenarios that may arise through the use of multisets, but it is useful as it help us express those scenarios naturally.

6 Conclusion

We have defined and axiomatized two revision operators over multisets of propositional formulas, representing the belief state of a group of multiple agents. We have shown that these revision operators can express a variety of merging operations, thereby reducing the process of merging to revision. We think that these operators, including their extensions to richer language and variants to diverse metric spaces, merit further study. The belief change of multiple agents is more complicated and includes changes that do not necessarily arise in the single agent case, but this framework shows that many central tools for the single agent case, such as minimization, carry over to the multiple agent case.

References

1. Bell, J.L.: A new approach to quantum logic. Br. J. Philos. Sci. **37**, 83–99 (1986)
2. Blizard, W.D., et al.: Multiset theory. Notre Dame J. Formal Logic **30**(1), 36–66 (1988)
3. Deza, M.M., Deza, E.: Encyclopedia of Distances, pp. 1–583. Springer, Heidelberg (2009). doi:10.1007/978-3-642-00234-2
4. Fagin, R., Halpern, J.Y., Moses, Y., Vardi, M.Y.: Reasoning About Knowledge. MIT Press, Cambridge (1995)
5. Georgatos, K.: On indistinguishability and prototypes. Logic J. IGPL **11**(5), 531–545 (2003)
6. Georgatos, K.: Belief update using graphs. In: Proceedings of the Twenty-First International Florida Artificial Intelligence Research Society Conference, pp. 649–654. AAAI Press (2008)

7. Georgatos, K.: Geodesic revision. J. Logic Comput. **19**(3), 447–459 (2009)
8. Georgatos, K.: Conditioning by minimizing accessibility. In: Bonanno, G., Löwe, B., Hoek, W. (eds.) LOFT 2008. LNCS (LNAI), vol. 6006, pp. 20–33. Springer, Heidelberg (2010). doi:10.1007/978-3-642-15164-4_2
9. Georgatos, K.: Iterated contraction based on indistinguishability. In: Artemov, S., Nerode, A. (eds.) LFCS 2013. LNCS, vol. 7734, pp. 194–205. Springer, Heidelberg (2013). doi:10.1007/978-3-642-35722-0_14
10. Georgatos, K.: Graph-based belief merging. In: Hoek, W., Holliday, W.H., Wang, W. (eds.) LORI 2015. LNCS, vol. 9394, pp. 102–115. Springer, Heidelberg (2015). doi:10.1007/978-3-662-48561-3_9
11. Konieczny, S., Lang, J., Marquis, P.: DA2 merging operators. Artif. Intell. **157**(12), 49–79 (2004)
12. Konieczny, S., Pérez, R.P.: Merging with integrity constraints. In: Hunter, A., Parsons, S. (eds.) ECSQARU 1999. LNCS (LNAI), vol. 1638, pp. 233–244. Springer, Heidelberg (1999). doi:10.1007/3-540-48747-6_22
13. Lehmann, D.J., Magidor, M., Schlechta, K.: Distance semantics for belief revision. J. Symb. Log. **66**(1), 295–317 (2001)
14. Liberatore, P., Schaerf, M.: Arbitration (or how to merge knowledge bases). IEEE Trans. Knowl. Data Eng. **10**(1), 76–90 (1998)
15. Poincaré, H.: La Valeur de la Science. Flammarion, Paris (1905)
16. Revesz, P.Z.: On the semantics of theory change: arbitration between old and new information. In: Proceedings of the Twelfth ACM SIGACT-SIGMOD-SIGART Symposium on Principles of Databases, pp. 71–82 (1993)
17. Schlechta, K.: Non-prioritized belief revision based on distances between models. Theoria **63**(1–2), 34–53 (1997)
18. Zeeman, E.C.: The topology of the brain and visual perception. In: Fort, M.K. (ed.) The Topology of 3-Manifolds, pp. 240–256. Prentice Hall, Englewood Cliffs (1962)

Boosting Distance-Based Revision Using SAT Encodings

Sébastien Konieczny, Jean-Marie Lagniez, and Pierre Marquis(✉)

CRIL, Université d'Artois & CNRS, Lens, France
{konieczny,lagniez,marquis}@cril.fr

Abstract. Belief revision has been studied for more than 30 years, and the theoretical properties of the belief revision operators are now well-known. Contrastingly, there are almost no practical applications of these operators. One of the reasons is the computational complexity of the corresponding inference problem, which is typically NP-hard and coNP-hard. Especially, existing implementations of belief revision operators are capable to solve toy instances, but are still unable to cope with real-size problem instances. However, the improvements achieved by SAT solvers for the past few years have been very impressive and they allow to tackle the solving of instances of inference problems located beyond NP. In this paper we describe and evaluate SAT encodings for a large family of distance-based belief revision operators. The results obtained pave the way for the practical use of belief revision operators in large-scale applications.

1 Introduction

Propositional belief revision has received much attention for the past thirty years [1,9], and the theoretical properties of belief revision operators are nowadays well-known. Contrastingly, far less studies have focused so far on the *computational aspects* of propositional belief revision. An explanation of this is that the inference problem for belief revision operators (i.e., the problem of deciding whether $\varphi \circ \mu \models \alpha$ holds, given three formulae φ, μ and α) is typically intractable. Indeed, the complexity of this problem has been identified for many operators, and it is typically both NP-hard and coNP-hard [14] and lies at the first or even at the second level of the polynomial hierarchy [8,14]. Existing implementations [7,18] are able to handle very small instances, but are far from being able to deal with real-size belief revision instances.

Interestingly, the improvements achieved by SAT solvers for the past few years have been huge. A current research direction is to leverage them to address the solving of instances of inference problems located beyond NP. Following this research line, we describe and evaluate SAT encodings for a large family of distance-based belief revision operators. For such operators, the models of the revised base are the models of the new piece of information μ which are at a minimal distance of the belief base φ. Among them, Dalal revision operator,

A. Baltag et al. (Eds.): LORI 2017, LNCS 10455, pp. 480–496, 2017.
DOI: 10.1007/978-3-662-55665-8_33

based on the Hamming distance between propositional worlds, is probably the best known [3].

In this paper, we define, give SAT encodings, and do experiments on the family of topic-decomposable distance-based revision operators. Topic-decomposable distances are complex distances, obtained by aggregating simpler distances defined on topics, which are (possibly non-disjoint) subsets of variables. The family includes as specific cases standard distances considered in belief revision, especially the Hamming distance and the drastic distance.

We present and evaluate SAT encoding schemes $E_{\circ_d}(\varphi, \mu)$ for such operators $\circ_d$. The encodings are CNF formulae which are query-equivalent to the revised bases $\varphi \circ_d \mu$ corresponding to the distance-based revision operator $\circ_d$ under consideration. Roughly, the idea underlying the encoding schemes is to make independent the languages used for the belief base φ and for the new piece of information μ, then to define defaults aiming to reconcile these languages. These defaults are used to compute the minimal distance *min* between the belief base and the new information (using a weighted partial MAXSAT solver). A constraint ensuring that the distance between any model of μ and φ is equal to *min* is finally added. The resulting encoding can thus be viewed as a query-equivalent compilation of the revised base. Indeed, in order to determine whether $\varphi \circ_d \mu \models \alpha$ holds, it is enough to check whether $E_{\circ_d}(\varphi, \mu) \models \alpha$ holds, which can be solved by checking the (un)satisfiability of $E_{\circ_d}(\varphi, \mu) \wedge \neg\alpha$. Empirically, our approach is efficient enough to compute encodings for belief revision instances based on thousands of variables.

The contributions of this work are:

- the definition of the family of topic-decomposable distance-based revision operators,
- the proposal of SAT encoding schemes for several topic-decomposable distance-based revision operators,
- the description of a set of benchmarks for distance-based belief revision,
- and the experimental evaluation of our encodings on these benchmarks.

2 Some Background on Belief Revision

Let $\mathcal{L}_\mathcal{P}$ be a propositional language built up from a finite set of propositional variables $\mathcal{P}$ and the usual connectives. $\bot$ (resp. $\top$) is the Boolean constant always false (resp. true). An interpretation (or world) is a mapping from $\mathcal{P}$ to $\{0, 1\}$, denoted by a bit vector whenever a strict total order on $\mathcal{P}$ is specified. The set of all interpretations is denoted $\mathcal{W}$. An interpretation ω is a model of a propositional formula $\alpha \in \mathcal{L}_\mathcal{P}$ if and only if it makes it true in the usual truth functional way. $Mod(\alpha)$ denotes the set of models of φ, i.e., $Mod(\alpha) = \{\omega \in \mathcal{W} \mid \omega \models \alpha\}$. $\models$ denotes logical entailment and $\equiv$ logical equivalence, i.e., $\alpha \models \beta$ iff $Mod(\alpha) \subseteq Mod(\beta)$ and $\alpha \equiv \beta$ iff $Mod(\alpha) = Mod(\beta)$. $Var(\alpha)$ denotes the set of variables occurring in α.

Let X be any subset of $\mathcal{P}$, the X-projection of an interpretation ω on X, noted $\omega^{\downarrow X}$, is the restriction of ω on the variables in X. For instance, with

$\mathcal{P} = \{a, b, c, d, e, f\}$ (ordered in this way), if $X = \{a, b, c\}$, and $\omega = 101001$, then $\omega^{\downarrow X} = 101$.

A *belief base* is a propositional formulae (or equivalently a finite set of propositional formulae interpreted conjunctively) φ, that represents the current beliefs of an agent.

A *belief revision scheme* $\circ$ is a mapping from $\mathcal{L}_\mathcal{P} \times \mathcal{L}_\mathcal{P}$ to $\mathcal{L}_\mathcal{P}$, associating with a belief base φ and a formula (a new piece of information) μ a belief base $\varphi \circ \mu$ called the revised base. Belief revision operators are the belief revision schemes satisfying the following postulates:

Definition 1 ([9]). *A belief revision scheme $\circ$ is a belief revision operator satisfying the following postulates. For every formula $\mu, \mu_1, \mu_2, \varphi, \varphi_1, \varphi_2$:*

(R1) $\varphi \circ \mu \models \mu$
(R2) *If $\varphi \wedge \mu$ is consistent, then $\varphi \circ \mu \equiv \varphi \wedge \mu$*
(R3) *If μ is consistent, then $\varphi \circ \mu$ is consistent*
(R4) *If $\varphi_1 \equiv \varphi_2$ and $\mu_1 \equiv \mu_2$, then $\varphi_1 \circ \mu_1 \equiv \varphi_2 \circ \mu_2$*
(R5) $(\varphi \circ \mu_1) \wedge \mu_2 \models \varphi \circ (\mu_1 \wedge \mu_2)$
(R6) *If $(\varphi \circ \mu_1) \wedge \mu_2$ is consistent, then $\varphi \circ (\mu_1 \wedge \mu_2) \models (\varphi \circ \mu_1) \wedge \mu_2$*

Belief revision operators can be characterized in terms of total preorders over interpretations. Indeed, each belief revision operator corresponds to a faithful assignment [9]:

Definition 2 (faithful assignment). *A* faithful assignment *is a mapping which associates with every base φ a preorder $\leq_\varphi$ over interpretations such that for every base φ, φ_1, φ_2, it satisfies the following conditions:*

(1) *If $\omega \models \varphi$ and $\omega' \models \varphi$, then $\omega \simeq_\varphi \omega'$*
(2) *If $\omega \models \varphi$ and $\omega' \not\models \varphi$, then $\omega <_\varphi \omega'$*
(3) *If $\varphi_1 \equiv \varphi_2$, then $\leq_{\varphi_1} = \leq_{\varphi_2}$*

where $<_\varphi$ is the strict part of $\leq_\varphi$ and $\simeq_\varphi$ is the indifference relation induced by $\leq_\varphi$.

Theorem 1 ([9]). *A belief revision scheme $\circ$ is a* belief revision operator *if and only if there exists a faithful assignment associating every base φ with a total preorder $\leq_\varphi$ over $\mathcal{W}$ such that for every formula μ, $Mod(\varphi \circ \mu) = \min(Mod(\mu), \leq_\varphi)$.*

3 Topic-Decomposable Distance-Based Revision Operators

Among revision operators are *distance-based* operators, which select as result the models of μ that are the closest ones to φ:

Definition 3 (pseudo-distance, distance). *Let X be a subset of $\mathcal{P}$. A* pseudo-distance *d between X-interpretations is a mapping $d : \mathcal{W}_X \times \mathcal{W}_X \rightarrow \mathbb{N}$ such that for any X-interpretations ω_1 and ω_2:*

- $d(\omega_1, \omega_2) = 0$ *if and only if* $\omega_1 = \omega_2$
- $d(\omega_1, \omega_2) = d(\omega_2, \omega_1)$

d is a distance *when it satisfies in addition the triangular inequality, i.e., for any interpretations ω_1, ω_2, and ω_3:*

- $d(\omega_1, \omega_3) \leq d(\omega_1, \omega_2) + d(\omega_2, \omega_3)$

Usual distances are the drastic distance ($d_D(\omega, \omega') = 0$ if $\omega = \omega'$ and 1 otherwise), which corresponds to the infinity-norm distance, also known as Chebyshev distance, and the Hamming distance ($d_H(\omega, \omega') = n$ if ω and ω' differ on n variables), which corresponds to the 1-norm distance, also referred to as the Manhattan distance. One can also consider weighted versions of these distances, where each propositional variable x is associated with a (non-null) weight $\rho(x)$, and then the weighted Hamming distance is given by $d_{H\rho}(\omega, \omega') = \sum_{\{x | \omega(x) \neq \omega'(x)\}} \rho(x)$. Similarly, a weighted drastic distance is defined as $d_{D\rho}(\omega, \omega') = \max_{\{x | \omega(x) \neq \omega'(x)\}} \rho(x)$.

Sometimes one can identify different topics, on which formulae and interpretations can be evaluated. Some of these topics can be more important than others, so having conflicts on some topics can be more problematic than on some others. See [11] for a criticism of (simple) Hamming distance, and a justification of the use of weights or topics.

Let f be an aggregation function, i.e., a mapping associating an integer $i = f(\boldsymbol{v}_n)$ with any finite vector $\boldsymbol{v}_n = (i_1, \ldots, i_n)$ of integers. Let us recall the definition of topic-decomposable distance from [11]:

Definition 4 (topic-decomposable distance). *Let $\mathcal{T} = \{T_1, \ldots, T_m\}$ be a collection of non-empty subsets of $\mathcal{P}$ (topics) such that $\bigcup_{i=1}^{m} T_i = \mathcal{P}$. A pseudo-distance d between interpretations is* $\mathcal{T}$-decomposable *if and only if there exist m pseudo-distances $d_1, \ldots, d_m$ and an aggregation function f such that each d_i ($i \in \{1, \ldots, m\}$) is between T_i-interpretations, and for all ω, $\omega' \in \mathcal{W}$:*

$$d(\omega, \omega') = f(d_1(\omega^{\downarrow T_1}, \omega'^{\downarrow T_1}), \ldots, d_m(\omega^{\downarrow T_m}, \omega'^{\downarrow T_m})).$$

Note that distinct topics from a topic decomposition $\mathcal{T}$ of X may share some variables of X.

In [11] Lafage and Lang do not specify the properties they expect for the aggregation function. In this work we require the following:

Definition 5 (aggregation function). *An* aggregation function *f is a mapping associating an integer $i = f(\boldsymbol{v}_n)$ with any finite vector $\boldsymbol{v}_n = (i_1, \ldots, i_n)$ of integers. It is assumed that whatever the integer n, $f(\boldsymbol{v}_n) = 0$ if and only if $\boldsymbol{v}_n = \boldsymbol{0}_n$ where $\boldsymbol{0}_n$ is the vector of size n containing only null coordinates. f should also be non-decreasing in each argument. We finally assume that if $\boldsymbol{v}_m$ ($m \geq n$) is any vector containing the same coordinates as $\boldsymbol{v}_n$ but completed with $m - n$ zeroes, then $f(\boldsymbol{v}_m) = f(\boldsymbol{v}_n)$.*

Note that standard aggregation functions, as Σ (sum), max, Leximax or Leximin satisfy these requirements.

In order to define $\mathcal{T}$-decomposable distances from their components, we will take advantage of the following structure:

Definition 6 (decomposition distance). *Let $\delta = \{\mathcal{T}, \mathcal{D}, f\}$, with $\mathcal{T} = \{T_1, \ldots, T_m\}$ be a collection of non-empty subsets of $\mathcal{P}$ (topics) such that $\bigcup_{i=1}^{m} T_i = \mathcal{P}$, $\mathcal{D} = \{d_1, \ldots, d_m\}$ be a collection of pseudo-distances such that each d_i ($i \in \{1, \ldots, m\}$) is between T_i-interpretations, and f be an aggregation function. We call such a δ a* composition frame, *and d^δ the* decomposition distance *induced by δ (or simply the δ-decomposable distance).*

Let us now introduce topic-decomposable distance-based revision operators:

Definition 7 (topic-decomposable distance-based revision operator). *Let δ be a composition frame. A* topic-decomposable distance-based revision operator *$\circ_{\mathrm{d}}^{\delta}$ is defined as $Mod(\varphi \circ_{\mathrm{d}}^{\delta} \mu) = \min(Mod(\mu), \leq_{\varphi}^{\delta})$, where*

- $\omega \leq_{\varphi}^{\delta} \omega'$ *iff* $d^\delta(\omega, \varphi) \leq d^\delta(\omega', \varphi)$
- $d^\delta(\omega, \varphi) = \min_{\omega' \models \varphi} d^\delta(\omega, \omega')$
- d^δ *is the δ-decomposable distance*

We can easily show that:

Proposition 1. *Any topic-decomposable distance-based revision operator $\circ_{\mathrm{d}}^{\delta}$ is a belief revision operator.*

It is easy to check that the drastic distance d_D and the Hamming distance d_H (and, similarly, their weighted counterparts d_{D_ρ} and d_{H_ρ}) are (somewhat trivial) topic-decomposable pseudo-distances. Indeed, each of them is the decomposition distance induced by the composition frame $\{\mathcal{T} = \{\{\mathcal{P}\}\}, \mathcal{D}, f\}$ where $\mathcal{D}$ is the singleton consisting of the distance itself, and f is the identity function.

Several new, yet interesting belief revision operators can be defined as members of this family. For instance, a revision operator that first looks at Hamming distance between interpretations (like Dalal revision $\circ_{d_H}$), but in case of equality, focuses on some specific variables. The corresponding topic-decomposable distance can be built up using Σ as the aggregation function on a first topic equal to $\mathcal{P}$, and then on other topics containing the variables of interest. Formally:

Definition 8 (DVI revision operators). *Let $Y = \langle x_1, \ldots, x_k \rangle$ be a vector of propositional variables of $\mathcal{P}$. The Dalal revision with Variables of Interests operator $\circ_{\mathrm{d}}^{\delta_{DVI(Y)}}$ is the topic-decomposable distance-based revision operator defined by the decomposition frame $\delta_{DVI(Y)} = \{\mathcal{T}, \mathcal{D}, \Sigma^{DVI(Y)}\}$ such that:*

- $\mathcal{T} = \{\mathcal{P}, \{x_1\}, \ldots, \{x_k\}\}$
- $\mathcal{D} = \{d_H, d_D, \ \ldots, \ d_D\}$
- $\Sigma^{DVI(Y)}(i_0, i_1, \ldots, i_k) = 2^{k+1}.i_0 + \Sigma_{j=1}^{k} 2^j.i_j$

Here is an illustrative example.

Example 1. Suppose that $\mathcal{P} = \{x_1, x_2, x_3\}$ and $Y = \langle x_1, x_2, x_3 \rangle$. Let $\varphi \equiv (x_1 \leftrightarrow x_2) \wedge (x_2 \leftrightarrow \neg x_3)$ and let $\mu = (x_1 \wedge x_3) \vee (x_2 \wedge x_3)$. Assuming $x_1 < x_2 < x_3$, we have $Mod(\varphi) = \{001, 110\}$ and $Mod(\mu) = \{011, 101, 111\}$. Every model of μ is at Hamming distance 1 from φ. Accordingly, $\varphi \circ_{d_H} \mu$ is equivalent to μ. Contrastingly, the distance of 011 (resp. 101, 111) to φ for the Dalal revision with Variables of Interests operator defined above is 20 (resp. 18, 24). Thus, we have that $\varphi \circ_d^{\delta_{DVI(Y)}} \mu$ is equivalent to $x_1 \wedge \neg x_2 \wedge x_3$.

Proposition 2. *For any set Y, for any φ, μ, $\varphi \circ_d^{\delta_{DVI(Y)}} \mu \models \varphi \circ_{d_H} \mu$.*

Many other refinements of (and variations around) Dalal revision operator can be figured out from topic-decomposable distances.

4 SAT Encodings

We now describe SAT encodings for the topic-decomposable distance-based revision operators based on the Hamming distance or the drastic distance on each topic, and on the aggregation functions $w\Sigma, w\mathsf{Leximax}, w\mathsf{Leximin}$, which are weighted versions of the standard aggregation function Σ, Leximax, Leximin, where w is a weight function on topics (it associates an integer $w(T_i)$ with each topic), and a topic T_i of weight $w(T_i)$ is duplicated $w(T_i)$ times before the aggregation.

Our SAT encodings for topic-decomposable distance-based belief revision mainly use the same techniques as those considered in our previous work [10] on belief merging.

Given a topic-decomposable pseudo-distance d, a belief base φ represented as a CNF formula, a change formula μ represented as a CNF formula, we are going to show that our encoding scheme E generates a CNF formula $E_{\circ_d}(\varphi, \mu)$ of size polynomial in $|\varphi| + |\mu|$ which is query-equivalent to $\varphi \circ_d \mu$. Let us first make precise what query-equivalent means.

Definition 9 (query-equivalence).

- *A propositional formula α is said to be* query-equivalent *to a propositional formula β whenever α has the same logical consequences as β over $Var(\beta)$, i.e., for ever formula γ over $Var(\beta)$, we have $\alpha \models \gamma$ if and only if $\beta \models \gamma$.*
- *A mapping τ associating a* CNF *formula α with a given propositional formula β is* query-equivalence preserving *if and only if α is query-equivalent to β.*

In our approach, both φ and μ are supposed to be CNF formulae. This is not a limitation of the framework, since any formula can be transformed in linear time into a query-equivalent CNF formula using Tseitin or Plaisted/Greenbaum translation functions [15,17]. Indeed, Tseitin and Plaisted/Greenbaum translation functions τ_T and τ_{PG} (respectively) [15,17] are query-equivalence preserving mappings from propositional circuits to CNF, and they can be computed in linear time in the size of the input β.

Example 2. As a matter of example, let us consider again φ represented by $(x_1 \leftrightarrow x_2) \wedge (x_2 \leftrightarrow \neg x_3)$ and $\mu = (x_1 \wedge x_3) \vee (x_2 \wedge x_3)$. Using Tseitin translation function, we get $\tau_T(\varphi) = a_0 \wedge (\neg a_0 \vee a_1) \wedge (\neg a_0 \vee a_2) \wedge (a_0 \vee \neg a_1 \vee \neg a_2) \wedge (\neg a_1 \vee \neg x_1 \vee x_2) \wedge (\neg a_1 \vee x_1 \vee \neg x_2) \wedge (a_1 \vee x_1 \vee x_2) \wedge (a_1 \vee \neg x_1 \vee \neg x_2) \wedge (\neg a_2 \vee \neg x_2 \vee \neg x_3) \wedge (\neg a_2 \vee x_2 \vee x_3) \wedge (a_2 \vee x_2 \vee \neg x_3) \wedge (a_2 \vee \neg x_2 \vee x_3)$.

The auxiliary, fresh variables a_0, a_1, a_2 correspond respectively to φ, and its two main subformulae $x_1 \leftrightarrow x_2$ and $x_2 \leftrightarrow \neg x_3$. The unit clause x_0 expresses that β holds, the next three clauses that it is equivalent to $a_1 \wedge a_2$, the next three clauses that a_1 is equivalent to $x_1 \leftrightarrow x_2$, and finally the last three clauses that a_2 is equivalent to $x_2 \leftrightarrow \neg x_3$.

Similarly, we get $\tau_T(\mu) = b_0 \wedge (\neg b_0 \vee b_1 \vee b_2) \wedge (b_0 \vee \neg b_1) \wedge (b_0 \vee \neg b_2) \wedge (\neg b_1 \vee x_1) \wedge (\neg b_1 \vee x_3) \wedge (b_1 \vee \neg x_1 \vee \neg x_3) \wedge (\neg b_2 \vee x_2) \wedge (\neg b_2 \vee x_3) \wedge (b_2 \vee \neg x_2 \vee \neg x_3)$. This time, the auxiliary variables which are introduced are b_0, b_1, b_2.

Plaisted/Greenbaum translation function is a bit lighter (it leads to less clauses). Here, $\tau_{PG}(\varphi) = a_0 \wedge (\neg a_0 \vee a_1) \wedge (\neg a_0 \vee a_2) \wedge (\neg a_1 \vee \neg x_1 \vee x_2) \wedge (\neg a_1 \vee x_1 \vee \neg x_2) \wedge (\neg a_2 \vee \neg x_2 \vee \neg x_3) \wedge (\neg a_2 \vee x_2 \vee x_3)$. $\tau_{PG}(\mu) = b_0 \wedge (\neg b_0 \vee b_1 \vee b_2) \wedge (\neg b_1 \vee x_1) \wedge (\neg b_1 \vee x_3) \wedge (\neg b_2 \vee x_2) \wedge (\neg b_2 \vee x_3)$.

We can observe that each of $\tau_T(\varphi)$ and $\tau_{PG}(\varphi)$ is query-equivalent to φ. And similarly for $\tau_T(\mu)$ and $\tau_{PG}(\mu)$ w.r.t. μ Especially, the clause $\neg x_1 \vee \neg x_3$ is a logical consequence of φ, so it is also a logical consequence of $\tau_T(\varphi)$ and of $\tau_{PG}(\varphi)$.

Whatever the used translation function τ, let us denote by $A(\tau(\beta)) = Var(\tau(\beta)) \setminus Var(\beta)$ the set of auxiliary variables introduced in $\tau(\beta)$. In the case when φ and/or μ are not given as CNF formula(e), one can always take advantage of $\tau = \tau_T$ and/or $\tau = \tau_{PG}$ to turn them into query-equivalent formulae. The point is that this translation is safe as to the solving of the (inference problem associated to) revision. To be more precise:

Proposition 3. *Let $X = Var(\varphi) \cup Var(\mu)$ and let d_X be a topic-decomposable distance over $\mathcal{W}_X$ induced by a topic decomposition $\mathcal{T}_X = \{T_1, \ldots, T_m\}$ of X, an aggregation function f, and m pseudo-distances $d_1, \ldots, d_m$ where each d_i ($i \in \{1, \ldots, m\}$) is between T_i-interpretations. Let $Y = X \cup A(\tau(\varphi)) \cup A(\tau(\mu))$. Then, provided that $A(\tau(\varphi)) \cap A(\tau(\mu)) = \emptyset$ (which is harmless, since the names given to the auxiliary variables do not matter), one can associate with d_X a topic-decomposable pseudo-distance d_Y over $\mathcal{W}_Y$ induced by a topic decomposition $\mathcal{T}_Y = \{T_1, \ldots, T_m, T_{m+1}\}$ of Y, the aggregation function f, and the $m+1$ pseudo-distances $d_1, \ldots, d_m, d_{m+1}$, with $T_{m+1} = A(\tau(\varphi)) \cup A(\tau(\mu))$ and d_{m+1} any pseudo-distance between T_{m+1}-interpretations. By construction, d_Y is such that $\tau(\varphi) \circ_{d_Y} \tau(\mu)$ is query-equivalent to $\varphi \circ_{d_X} \mu$.*

Let us now explain how SAT encoding schemes can be exploited to compute polynomial-size encodings, given by CNF formulae which are query-equivalent to the revised bases $\varphi \circ_d \mu$ for the topic-decomposable distance-based revision operators.

Formally, the objective is to associate with each φ and μ a CNF propositional formula noted $E_{\circ_d}(\varphi, \mu)$ which is query-equivalent to $\varphi \circ_d \mu$; thus, $E_{\circ_d}(\varphi, \mu)$

must have the same logical consequences φ as those of $\varphi \circ_d \mu$, provided that the queries φ are built up from the variables occurring in φ or μ. Furthermore, one expects the size of the encoding $E_{\circ_d}(\varphi, \mu)$ to be polynomial in the size of φ plus the size of μ.

Such encodings $E_{\circ_d}(\varphi, \mu)$ are computed via a two-step compilation process:

(1) using a solver for weighted partial MAXSAT, one first computes the value *min*, which is the distance of μ to φ, i.e., the minimal value of $\{d(\omega, \varphi) \mid \omega \models \mu\}$,
(2) once *min* has been computed, one generates the encoding $E_{\circ_d}(\varphi, \mu)$ which states (among other things) that the distance of μ to φ must be equal to *min*.

The generated encoding $E_{\circ_d}(\varphi, \mu)$ is a CNF formula, enabling to take advantage of the power of SAT solvers for solving the inference problem when the queries φ are also given as CNF formulae.

From now on, we suppose that $Var(\varphi) \cup Var(\mu) = \{x_1, \ldots, x_n\}$. All the encodings $E_{\circ_d}(\varphi, \mu)$ described in the following share a common part $C(\varphi, \mu)$ given by

$$\mu \wedge \varphi' \wedge \bigwedge_{j=1}^{n} (d_j \vee \neg x_j \vee x'_j) \wedge (d_j \vee x_j \vee \neg x'_j).$$

φ' is a clone of φ obtained by renaming in it every occurrence of a variable x_j by an occurrence of the fresh variable x'_j. Such a renaming of the bases enables it to freeze any conflict which would exist in the conjunction of μ and φ. This is reminiscent to the consistency-based approach to belief merging reported in [6]. The last conjunct of $C(\varphi, \mu)$ is a constraint based on *discrepancy variables* d_j, such that d_j must be set to true whenever it is not possible to assume that $x_j \leftrightarrow x'_j$ holds without violating $C(\varphi, \mu)$.

Distances. Taking into account the distance d under consideration (d_D or d_H) requires to add a further constraint of the form $\bigwedge_{i=1}^{m} D^i$ to $C(\varphi, \mu)$ where m is the number of topics of $\mathcal{T}$. For each topic $T_i \in \mathcal{T}$, D^i is a CNF formula over the variables $d_1, \ldots, d_n$ plus a number of additional fresh variables. Some (actually r_i) of them give the binary representation $b^i_{r_i}, \ldots, b^i_1$ of $max_{x_j \in T_i} d_j$ (resp. $\Sigma_{x_j \in T_i} d_j$) when the drastic distance (resp. the Hamming distance) is considered (see [10] for more details). For each model ω of $C(\varphi, \mu) \wedge \bigwedge_{i=1}^{m} D^i$, the bit vector obtained by projecting ω over those $\Sigma_{i=1}^{m} r_i$ additional variables is the binary representation of the distance of the projection of ω over the variables of μ with the projection of ω over the variables of φ'.

Aggregators. The objective is now to find *min*, the minimal distance of μ to φ. Let us first focus on the easiest case $f = \Sigma$. In this case, the value we look for is the minimal value *min* which can be taken by $\Sigma_{i=1}^{m} w_{T_i} \times (\Sigma_{j=1}^{r_i} 2^{j-1} \times b^i_j)$. Since this objective function is linear, in order to compute *min*, we take advantage of a weighted partial MAXSAT solver. Once this is done, to get $E_{\circ_d}(\varphi, \mu)$, it is enough to conjoin with $C(\varphi, \mu) \wedge \bigwedge_{i=1}^{m} D^i$ a CNF formula query-equivalent to the constraint $\Sigma_{i=1}^{m} k_i \times (\Sigma_{j=1}^{r_i} 2^{j-1} \times b_j^i) = min$. A polynomial-size CNF formula query-equivalent to this last constraint can be computed using a weighted parallel binary counter [16].

Let us now consider the harder cases f = Leximax and f = Leximin. Let r be $max_{i=1}^{m} r_i$. First of all, for aligning the binary representations $b_{r_i}^i, \ldots, b_1^i$ over r bits when i varies from 1 to m, we introduce some fresh variables assigned to false. Then we generate an additional CNF constraint $P(\varphi)$ which requires the introduction of m^2 additional variables $p_{i,j}$. This constraint is used to "sort" the bit representations associated with the topics (i.e., to associate with each j a position i) depending on the respective values of their bit vectors $b_r^j \ldots b_1^j$. As in [10], $P(\varphi)$ requires $(5 \times r + 2) \times m^3$ clauses: $2 \times m^3$ clauses are used to express the fact that each j is associated with a unique i (a pigeonhole instance) and $5 \times r \times m^3$ clauses are used to ensure (thanks to a standard comparator) that for every $j, k \in \{1, \ldots, m\}$, $i \in \{1, \ldots, m\text{-}1\}$, if $p_{i,j}$ and $p_{i+1,k}$ are set to true, then $b_r^j \ldots b_1^j$ is greater than or equal to (resp. lower than or equal to) $b_r^k \ldots b_1^k$ when f = Leximax (resp. f = Leximin). Thus, the only j such that $p_{1,j}$ is true is such that the value of $b_r^j \ldots b_1^j$ is maximal (resp. minimal) when f = Leximax (resp. f = Leximin), and so on.

The following step aims at taking account for the weights w_{T_i} ($i \in \{1, \ldots, m\}$). We determine the positions of the binary representations associated with the topics for which the corresponding bit vectors take the same values (they are necessarily pairwise adjacent because of the constraint $P(\varphi)$). To do so, we add a further CNF constraint $A(\varphi)$ requiring the introduction of m fresh variables e^i, so that e^1 is set to true and for every $i \in \{1, \ldots, m-1\}$, e^i is set to true precisely when the binary representations corresponding to the topics associated with positions i and $i-1$ correspond to different bit vectors. $A(\varphi)$ requires $(r+1) \times m^3$ additional clauses.

The next step is to add a constraint $K(\varphi)$ which is used to make the sums of the weights w_{T_i} which are associated with equal bit vectors (indeed, unlike for the case $f = \Sigma$, multiplying by w_{T_i} the value of the corresponding bit vector is not convenient when a lexicographic comparison is to be achieved). Let $s = \lceil log_2(\Sigma_{i=1}^{m} w_{T_i}) \rceil$. Constraint $K(\varphi)$ requires the introduction of $m \times s$ fresh variables, i.e., m bit vectors $t_s^i \ldots t_1^i$, and it ensures that for each $i \in \{1, \ldots, m\}$, $t_s^i \ldots t_1^i$ is the binary representation of w_{T_i} when e^i is true, and $t_s^i \ldots t_1^i$ is the binary representation of the sum of the value of $t_s^{i-1} \ldots t_1^{i-1}$ with w_{T_i} when e^i is false. $K(\varphi)$ is based on a half-adder and requires $6 \times m \times s$ clauses. Then one needs to add a further constraint $O(\varphi)$ which is used to "sort" the bit vectors $b_r^i, \ldots, b_1^i$ for $i \in \{1, \ldots, m\}$. This constraint requires the introduction of $m \times r$ fresh variables, i.e., m bit vectors $o_r^i \ldots o_1^i$. It ensures that for every $i, j \in \{1, \ldots, m\}$, if $p_{i,j}$ is set to true, then $b_r^j \ldots b_1^j$ is equal to $o_r^i \ldots o_1^i$. This constraint requires $2 \times r \times m^2$ additional clauses.

Now, in order to compute min (which can be viewed here as a sorted list of ordered pairs of integers, where the second element of each pair is the number of repetitions of the first element that must be considered), one needs first to compute a model which minimizes the value v_o^1 of $o_r^1 \ldots o_1^1$, and then minimizes (resp. maximizes) the value v_t^1 of $t_s^1 \dots t_1^1$ when $f = \mathsf{Leximax}$ (resp. $f = \mathsf{Leximin}$). We achieve the two optimization processes in one step, using a weighted partial MAXSAT solver on the instance given by the hard constraint $E_{\circ_d}(\varphi, \mu) = C(\varphi, \mu) \wedge \bigwedge_{i=1}^{m} D^i \wedge P(\varphi) \wedge A(\varphi) \wedge K(\varphi) \wedge O(\varphi)$ and the objective function $2^s \times \Sigma_{i=1}^{r} 2^{i-1} \times o_i^1 + \Sigma_{i=1}^{s} 2^{i-1} \times t_i^1$ (resp. $2^s \times \Sigma_{i=1}^{r} 2^{i-1} \times o_i^1 + \Sigma_{i=1}^{s} 2^{i-1} \times \neg t_i^1$) when $f = \mathsf{Leximax}$ (resp. $f = \mathsf{Leximin}$).

Once an optimal solution is found, we add to the hard constraint $s + r \times v_t^1$ unit clauses in order to set the variables $t_s^1, \dots, t_1^1$, as well as the variables $o_r^j, \dots, o_1^j$ ($j \in \{1, \dots, v_t^1\}$) to the truth values they have in this solution. We iterate this process by considering then the second greatest (resp. least) value of the bit vectors $o_r^{v_t^1+1}, \dots, o_1^{v_t^1+1}$ for $i \in \{1, \dots, m\}$, and so on. The number of iterations is upper bounded by m. The computation of *min* is achieved when all the iterations have been done. Then $E_{\circ_d}(\varphi, \mu)$ is equal to $C(\varphi, \mu) \wedge \bigwedge_{i=1}^{m} D^i \wedge P(\varphi) \wedge A(\varphi) \wedge K(\varphi) \wedge O(\varphi)$ conjoined with all the unit clauses which have been generated during the optimization step.

By construction of the encodings, all the belief revision operators under consideration are *query-compactable* [2]:

Proposition 4. *For each topic-decomposable distance d induced by $f \in \{w\Sigma$, $w\mathsf{Leximax}, w\mathsf{Leximin}\}$ and local distances which are Hamming or drastic ones, the size of $E_{\circ_d}(\varphi, \mu)$ is polynomial in the size of φ plus the size of μ and $E_{\circ_d}(\varphi, \mu)$ is query-equivalent to $\varphi \circ_d \mu$.*

A direct consequence of the previous proposition is that the inference problems for the topic-decomposable distance-based belief revision operators under consideration can be reduced to the classical entailment problem by taking advantage of our encoding schemes. Since the size of $E_{\circ_d}(\varphi, \mu)$ is in every case polynomial in the size of φ plus the size of μ, we get that the corresponding inference problems (when queries φ are unrestricted propositional formulae) are compilable to coNP, and are among the hardest ones (see [12] for more details on the compilability classes):

Corollary 1. *For each topic-decomposable distance d induced by $f \in \{w\Sigma$, $w\mathsf{Leximax}, w\mathsf{Leximin}\}$ and local distances which are Hamming or drastic ones, the inference problem for $\circ_d$ is $\mathsf{compcoNP}$-complete.*

Accordingly, our results extend some compilability results known for Dalal revision operator [12]. From the practical side, the computational effort required to generate $E_{\circ_d}(\varphi, \mu)$ is spent only once (during the compilation phase), independently of the number of queries. Since the complexity of the inference problem falls to coNP once the preprocessing has been done, this effort can be easily balanced by considering sufficiently many queries.

5 Empirical Evaluation

Benchmarks. The non-availability of belief revision benchmarks corresponding to an actual application was a difficulty we had to face. To deal with it, we started with 295 unsatisfiable CNF instances used as benchmarks for the MUS competition in 2011 (http://www.cril.fr/SAT11/). We filtered from those benchmarks 220 CNF instances, precisely the ones which can be solved in less than 300 s by the weighted partial MAXSAT solver MaxHS [4,5] (the objective was to remove the most difficult MAXSAT instances). The number of variables of the selected instances varies from 26 to 4426259, with an average of 83240 variables. The number of clauses varies from 70 to 15983633 with an average of 279887 clauses.

From each such CNF formula Σ, we selected at random (following a uniform distribution and using a generate-and-test approach) a satisfiable subset φ_Σ of clauses containing 80% of the number of clauses of Σ. For generating μ_Σ we followed a similar generation methodology, but limited the number of selected clauses to (approximately) 5%, 15%, 35%, or 50% of the number of clauses of Σ. Those 4 thresholds are intended to capture different revision scenarios, from a "light" revision where the revision formula μ_Σ consists of only a few clauses to a more "severe" revision situation, where μ_Σ is quite huge. The generation process ensures that μ_Σ is a satisfiable CNF formula and that $\varphi_\Sigma \wedge \mu_\Sigma$ is unsatisfiable. Indeed, one wants to avoid trivial cases of belief revision, i.e., the ones when $\varphi_\Sigma \wedge \mu_\Sigma$ is satisfiable (in this case, **(R2)** requires the revised base to be equivalent to $\varphi_\Sigma \wedge \mu_\Sigma$). This explains why the retained thresholds are only approximate ones (sometimes additional clauses must be added to μ_Σ for guaranteeing the unsatisfiability of $\varphi_\Sigma \wedge \mu_\Sigma$). Following this approach, we derived $220 \times 4 = 880$ belief revision instances $(\varphi_\Sigma, \mu_\Sigma)$.

As to the topics, we considered sets $\mathcal{T}$ consisting of 1, 2, 5, 10, 15 and 20 elements. The case when only one topic is considered amounts to "standard" distance-based revision. Each topic T_i of $\mathcal{T}$ is obtained by selecting at random (following a uniform distribution) 30% of the variables of $Var(\varphi_\Sigma) \cup Var(\mu_\Sigma)$. When necessary, an additional topic is added to $\mathcal{T}$ for ensuring that $\bigcup_{T_i \in \mathcal{T}} T_i = Var(\varphi_\Sigma) \cup Var(\mu_\Sigma)$. Each topic T_i is associated with a weight $w(T_i)$ between 1 and 10 and chosen at random. $w(T_i)$ represents the significance of T. From the aggregation point of view, when T_i ($i \in \{1, \ldots, n\}$) has a weight $w(T_i)$, in the computation of the distance between two worlds ω and ω', the argument $d_i(\omega^{\downarrow T_i}, \omega'^{\downarrow T_i})$ is repeated $w(T_i)$ times. Clearly enough, this is the same as multiplying $d_i(\omega^{\downarrow T_i}, \omega'^{\downarrow T_i})$ by $w(T_i)$ when the global aggregation function f is $w\Sigma$, but it leads to distinct distances in general when f is $w\mathsf{Leximin}$ or $w\mathsf{Leximax}$. Considering 6 possible sizes for $\mathcal{T}$ led to $880 \times 6 = 5280$ topic-decomposable instances $(\mathcal{T}, \varphi_\Sigma, \mu_\Sigma)$. The instances, their generator (and the whole set of empirical results) are available at http://www.cril.fr/KC/br2cnf.html.

Setting. For each of the 5280 topic-decomposable belief revision instances, we have considered 2 candidate distances for each local distance d_i: the Hamming distance and the drastic distance. Finally, as to the global aggregation function f needed to define the topic-decomposable distance d inducing the belief revision operator under consideration, we have considered 3 functions: $w\Sigma$, $w\mathsf{Leximin}$, and $w\mathsf{Leximax}$. This finally led to $5280 \times 2 \times 3 = 31680$ topic-decomposable distance-based belief revision instances $(\mathcal{T}, \varphi_\Sigma, \mu_\Sigma, \circ_d)$.

For each instance, we took advantage of the SAT encoding schemes $E_{\circ_d}(\varphi_\Sigma, \mu_\Sigma)$ as reported in Sect. 4 to generate a query-equivalent CNF formula. Our experiments have been conducted on Intel Xeon E5-2643 (3.30 GHz) processors with 32 GiB RAM on Linux CentOS. We allocated 900 s CPU time and 8 GiB of memory per instance.

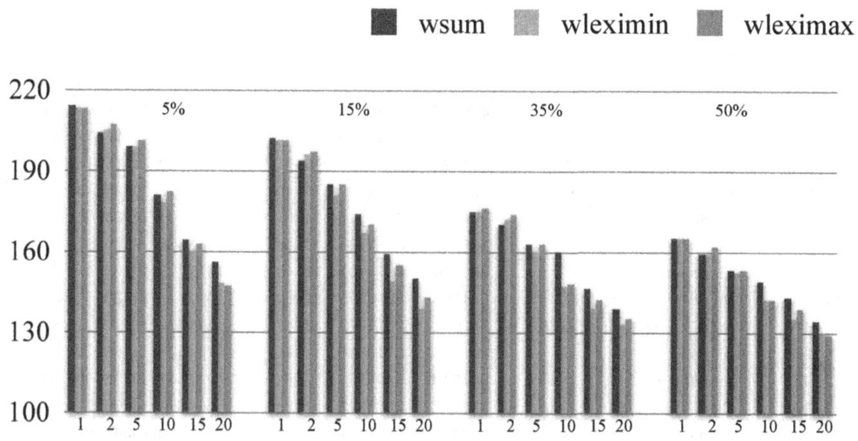

Fig. 1. Number of solved instances for different sizes of μ and different numbers of topics. The distance used is the Hamming one d_H.

Empirical results. Let us first focus on the drastic distance d_D which turned out to be the easiest case, computationally speaking. Given the computational resources allocated, we have been able to generate the encodings for all the 31680 instances but 336 (288 of them coming from the same 6 CNF instances). This represents (approx.) 99% of the topic-decomposable distance-based belief revision instances we have considered. For the instances for which the generation was feasible, the average generation time was 22.43 s, the worst case was 785 s. As to the number of variables (resp. clauses), the worst case was equal to (approx.) 3.6 million (resp. 14 million).

Let us then focus on the Hamming distance d_H. In Fig. 1 are indicated the numbers of solved instances (out of the 220) for different sizes of μ and numbers of topics when d_H is considered. One can easily see in this figure that both parameters have an impact on the difficulty of generating the encoding.

We now give more detailed results for the case $f = w\mathsf{Leximin}$ and $d = d_H$, which proved to be the most difficult scenario. In Table 1, for each size of μ (i.e., $5\%, 15\%, 35\%, 50\%$) and each number of topics (i.e., $1, 2, 5, 10, 15, 20$), we report the number of solved instances (out of 220) within the time and memory bounds,

Table 1. Results for $f = w\mathsf{Leximin}$ and $d = d_H$.

%mu	#T	#solved	avg. time	max. time	min. time	avg. #var	max. #var	min. #var	avg. #cl	max. #cl	min. #cl
5	1	213	38.4808	899.46	0	386026	5907706	226	778634	11811657	437
5	2	205	45.7588	756.69	0	566257	5084802	391	1236960	11097772	984
5	5	199	82.2348	887.36	0	879919	8737708	628	2019340	19966128	2203
5	10	178	122.551	859.89	0	1024460	9783278	1349	2414570	22654048	7040
5	15	160	121.378	676.36	0.01	800255	8947349	2062	1911580	21226404	13322
5	20	148	172.762	892.44	0.14	726893	7835375	3047	1769200	18978889	25198
15	1	201	54.7176	876.86	0	289224	2743707	226	588028	5255155	438
15	2	196	76.058	875.09	0	444427	4947647	391	980266	10731927	985
15	5	181	86.3864	856.39	0	583383	8737708	628	1342630	20065116	2204
15	10	167	132.848	834.85	0	688920	8380732	1349	1632140	20248744	7041
15	15	149	129.767	893.95	0.01	534537	8947349	2062	1285300	21281274	13323
15	20	139	196.685	858.48	0.18	527622	9125590	3047	1292340	21866643	25199
35	1	175	61.0705	803.32	0	126799	1689229	226	267289	3449743	452
35	2	172	80.4879	829.02	0	245727	4947647	391	551284	10929903	999
35	5	160	90.1234	896.61	0	198657	3817739	628	465638	8860787	2218
35	10	147	111.083	872.56	0	280196	6865407	1349	671725	16267341	7055
35	15	139	124.084	605.08	0	284763	5856095	2062	697729	14120161	13337
35	20	133	189.07	883.93	0.18	329617	7316079	3047	820222	17616391	25213
50	1	165	57.246	837.68	0	70207.7	716066	226	156578	1465372	463
50	2	159	51.0458	628.34	0	94154.6	1101822	391	217551	2396056	1010
50	5	152	79.0841	872.84	0	132565	2000729	628	314010	4609591	2229
50	10	142	101.54	897.53	0	194128	6865407	1349	471072	16348489	7066
50	15	135	136.915	877.36	0.01	199578	783143	2062	495474	1909354	13348
50	20	130	191.321	825.83	0.18	302605	7316079	3047	757717	17655136	2522

and the average *avg* and the maximum and the minimum of the values of the following measurements: the compilation time (in seconds) needed to compute the encoding (*time*), the number *#var* of variables in the encoding, and the number *#cl* of clauses in it.

From these experiments, one can make the following observations. First, one can note that the size of the formula μ has an impact on the difficulty (for instance, for a unique topic, 213 instances have been solved for a size of μ of 5%, and "only" 165 instances for a size of μ of 50%). But the greatest source of difficulty seems to be the number of topics (213 instances solved for one topic vs. 148 instances for 20 topics for a size of μ of 5%, and from 165 to 130 instances solved for a size of μ of 50%).

Table 2 (resp. Table 3) reports the same kind of measurements for $f = w\mathsf{Leximax}$ (resp. $f = w\Sigma$). Similar observations as the ones made for $f = w\mathsf{Leximin}$ about the impact of the size of μ and the number of topics can also be done for those two aggregation functions. Unsurprisingly, looking at the number of instances "solved", the "hardness" of the instances obtained for $f = w\mathsf{Leximax}$ appears as similar to the ones of the corresponding instances for $f = w\mathsf{Leximin}$.

Table 2. Results for $f = w\mathsf{Leximax}$ and $d = d_H$.

%mu	#T	#solved	avg. time	max. time	min. time	avg. #var	max. #var	min. #var	avg. #cl	max. #cl	min. #cl
5	1	213	39.931	868.64	0	386026	5907706	226	778634	11811657	437
5	2	207	39.5016	691.14	0	569724	5084802	391	1245750	11097772	984
5	5	201	74.6523	760.76	0	885310	8737708	628	2031650	19966128	2203
5	10	182	111.554	804.17	0	1026790	9783278	1349	2419830	22654050	7042
5	15	163	123.695	824.42	0.01	827095	8947349	2062	1976280	21226417	13322
5	20	147	177.478	876.36	0.08	708300	9125590	3047	1722970	21833514	25209
15	1	201	53.4791	762.6	0	289224	2743707	226	588028	5255155	438
15	2	197	54.9609	796.77	0	452246	4947647	391	995154	10731927	985
15	5	185	80.3196	839.36	0	658773	8737708	628	1517230	20065116	2204
15	10	170	126.352	845.67	0	784672	8751179	1349	1854490	20848972	7043
15	15	155	142.257	816.41	0.01	762194	8947349	2062	1825040	21281287	13323
15	20	143	193.719	893.73	0.08	555254	7373477	3047	1358460	17699333	25210
35	1	176	65.2532	899.82	0	137931	2086001	226	290681	4384428	452
35	2	174	79.3148	870.4	0	249685	4947647	391	560567	10929903	999
35	5	163	86.8453	858.17	0	279900	6083327	628	653163	14037642	2218
35	10	148	106.351	867.95	0	328283	6865407	1349	786258	16267343	7057
35	15	142	133.89	890.64	0	375744	8221280	2062	913011	19275067	13337
35	20	135	230.019	894.36	0.1	373476	7373477	3047	926447	17781160	25224
50	1	165	57.7481	848.88	0	70207.7	716066	226	156578	1465372	463
50	2	162	61.9183	859.48	0	103994	1101822	391	240676	2396056	1010
50	5	153	74.1976	891.83	0	151718	2745305	628	359986	6433006	2229
50	10	142	92.8003	892.33	0	171436	3639701	1349	416643	8615495	7068
50	15	139	146.263	770.45	0	343132	8380242	2062	837112	19729492	13348
50	20	129	221.471	703.56	0.11	324223	7316079	3047	809519	17655148	25235

Furthermore, the instances obtained for $f = w\Sigma$ appears as slightly easier than the ones of the corresponding instances for $f = w\mathsf{Leximax}$ (especially when the size of μ and the number of topics are high).

In Tables 1, 2, and 3, the case when $\#T = 1$ corresponds precisely to Dalal revision. We can observe on Table 3 that for a small size of μ (5%) most instances have been solved (214 out of 220), with a reasonable average time of 43 s. The average number of variables in the instances is 83240, and the average number of clauses is 279887. This shows that undoubtedly Dalal revision can be computed efficiently for large-size instances thanks to the encoding we point out.

These results should be contrasted with previous implementations of belief revision operators, for which solving instances of such a size was clearly out of reach. Note that those implementations do not correspond to distance-based operators: [18] encodes revision operators based on transmutation, [7] encodes revision operators based on language reconciliation, and [13] encodes partial-meet and kernel contraction. But noticeably in each of these three cases, no empirical evaluation was reported, or the instances under consideration were limited to be built up from a few dozens of variables.

Table 3. Results for $f = w\Sigma$ and $d = d_H$.

%mu	#T	#solved	avg. time	max. time	min. time	avg. #var	max. #var	min. #var	avg. #cl	max. #cl	min. #cl
5	1	214	43.5343	866.03	0	388336	5907706	226	784187	11811657	437
5	2	204	48.5623	514.68	0	555211	5086581	487	1213040	11101977	1052
5	5	199	94.8606	892.67	0	883152	8741391	713	2026030	19973570	1576
5	10	181	99.4129	760.91	0	1029270	9789091	1422	2420080	22660787	3268
5	15	164	97.8304	729.02	0	824714	8955449	2223	1955560	21228391	5176
5	20	156	117.773	825.72	0	748603	9135718	3118	1791345	21824355	7289
15	1	202	57.6379	870.47	0	293132	2743707	226	596101	5255155	438
15	2	194	70.8754	897.13	0	438054	4949485	487	967042	10736273	1053
15	5	185	91.2105	747.01	0	682350	8741391	713	1572110	20072558	1577
15	10	174	111.604	791.15	0	877804	9377697	1422	2069170	22358992	3269
15	15	159	113.972	864.67	0	793179	8955449	2223	1884540	21283261	5177
15	20	150	128.309	882.78	0	580689	7845670	3118	1392370	18998942	7290
35	1	175	60.3497	790.45	0	126799	1689229	226	267289	3449743	452
35	2	170	73.1915	813.47	0	238452	4949485	487	535214	10934249	1067
35	5	163	98.3538	834.7	0	326413	6086818	713	759278	14044635	1591
35	10	160	120.138	896.45	0	589290	7158010	1422	1396320	16958646	3283
35	15	146	108.166	844.91	0	393917	8387940	2223	940866	19662489	5191
35	20	139	133.365	886.58	0	348501	7049833	3118	838498	17026971	7304
50	1	165	59.0681	886.05	0	70207.7	716066	226	156578	1465372	463
50	2	159	63.7416	681.17	0	94895.7	1103118	487	219139	2399027	1078
50	5	153	84.6987	854.33	0	198170	3888204	713	466058	9157465	1602
50	10	149	94.5133	848.7	0	348109	6871077	1422	831604	16355040	3294
50	15	143	121.752	859.69	0	346522	8387940	2223	830872	19730597	5202
50	20	134	145.103	838.1	0	318841	7049833	3118	769852	17067544	731

6 Conclusion

We have introduced a general family of revision operators, based on topic-decomposable distances. This family captures well-known distance-based operators, but contains as well new interesting variations of previous operators. We have presented SAT encoding schemes for operators of this family. Based on them, one can compute polynomial-size encodings which are query-equivalent to the corresponding revised bases. This shows that the inference problem for belief revision for those operators is compilable to coNP.

We have evaluated our encoding schemes on non-trivial instances; leveraging the power of SAT solvers, we have shown that the resulting encodings can be computed within reasonable time and space limits, for instances based on thousands of variables which are out of reach of previous implementations.

We would like to insist on the fact that these instances have been defined from benchmarks from the 2011 MUS competition, which are non-trivial formulae. Being able to compute the result of the revision process for most of them shows that our approach can be used for real, large-scale applications where belief revision is required. Dalal revision being a specific operator of our family (among the easiest ones), this paper is the first one (as far as we know) presenting a convincing implementation of Dalal revision for practical applications.

By showing how SAT solvers can be exploited for solving revision problems located higher than coNP, this work also contributes to the recent Beyond NP initiative (beyondnp.org). As a perspective for further research, other distances and other aggregation functions will be targeted.

References

1. Alchourrón, C.E., Gärdenfors, P., Makinson, D.: On the logic of theory change: partial meet contraction and revision functions. J. Symb. Logic **50**(2), 510–530 (1985)
2. Cadoli, M., Donini, F., Liberatore, P., Schaerf, M.: The size of a revised knowledge base. Artif. Intell. **115**(1), 25–64 (1999)
3. Dalal, M.: Investigations into a theory of knowledge base revision. In: Proceedings of the Seventh AAAI Conference on Artificial Intelligence (AAAI 1988), pp. 475–479 (1988)
4. Davies, J., Bacchus, F.: Exploiting the power of MIP solvers in MAXSAT. In: Järvisalo, M., Van Gelder, A. (eds.) SAT 2013. LNCS, vol. 7962, pp. 166–181. Springer, Heidelberg (2013). doi:10.1007/978-3-642-39071-5_13
5. Davies, J., Bacchus, F.: Postponing optimization to speed up MAXSAT solving. In: Schulte, C. (ed.) CP 2013. LNCS, vol. 8124, pp. 247–262. Springer, Heidelberg (2013). doi:10.1007/978-3-642-40627-0_21
6. Delgrande, J.P., Schaub, T.: A consistency-based framework for merging knowledge bases. J. Appl. Logic **5**(3), 459–477 (2007)
7. Delgrande, J.P., Liu, D.H., Schaub, T., Thiele, S.: COBA 2.0: a consistency-based belief change system. In: Mellouli, K. (ed.) ECSQARU 2007. LNCS (LNAI), vol. 4724, pp. 78–90. Springer, Heidelberg (2007). doi:10.1007/978-3-540-75256-1_10
8. Eiter, T., Gottlob, G.: On the complexity of propositional knowledge base revision, updates, and counterfactuals. Artif. Intell. **57**(2–3), 227–270 (1992)
9. Katsuno, H., Mendelzon, A.O.: Propositional knowledge base revision and minimal change. Artif. Intell. **52**(3), 263–294 (1991)
10. Konieczny, S., Lagniez, J.M., Marquis, P.: SAT encodings for distance-based belief merging operators. In: Proceedings of the Thirty-First AAAI Conference on Artificial Intelligence (AAAI 2017), pp. 1163–1169 (2017)
11. Lafage, C., Lang, J.: Propositional distances and preference representation. In: Benferhat, S., Besnard, P. (eds.) ECSQARU 2001. LNCS (LNAI), vol. 2143, pp. 48–59. Springer, Heidelberg (2001). doi:10.1007/3-540-44652-4_6
12. Liberatore, P.: Compilation of intractable problems and its application to artificial intelligence. Ph.D. thesis, Università di Roma "La Sapienza" (1998)
13. Lundberg, R.U., Ribeiro, M.M., Wassermann, R.: A Framework for empirical evaluation of belief change operators. In: Barros, L.N., Finger, M., Pozo, A.T., Gimenénez-Lugo, G.A., Castilho, M. (eds.) SBIA 2012. LNCS, pp. 12–21. Springer, Heidelberg (2012). doi:10.1007/978-3-642-34459-6_2
14. Nebel, B.: How hard is it to revise a belief base? In: Dubois, D., Prade, H. (eds.) Belief Change. Handbook of Defeasible Reasoning and Uncertainty Management Systems, vol. 3, pp. 77–145. Kluwer Academic, Netherlands (1998)
15. Plaisted, D.A., Greenbaum, S.: A structure-preserving clause form translation. J. Symb. Comput. **2**(3), 293–304 (1986)
16. Sinz, C.: Towards an optimal CNF encoding of Boolean cardinality constraints. Technical report, Symbolic Computation Group, University of Tübingen (2005)

17. Tseitin, G.: Structures in constructive mathematics and mathematical logic. In: On the Complexity of Derivation in Propositional Calculus, pp. 115–125. Steklov Mathematical Institute (1968)
18. Williams, M.A., Sims, A.: Saten: an object-oriented web-based revision and extraction engine. CoRR cs.AI/0003059 (2000)

Counterfactuals in Nelson Logic

Andreas Kapsner[1(✉)] and Hitoshi Omori[2]

[1] Department of Philosophy, LMU Munich, Munich, Germany
Andreas.Kapsner@lrz.uni-muenchen.de
[2] Department of Philosophy, Kyoto University, Kyoto, Japan

Abstract. We motivate and develop an extension of Nelson's constructive logic **N3** that adds a counterfactual conditional to the existing setup. After developing the semantics, we will outline how our account will be able to give a nice analysis of natural language counterfactuals. In particular, the account does justice to the intuitions and arguments that have lead Alan Hájek to claim that most conditionals are false, but assertable, without actually forcing us to endorse that rather uncomfortable claim.

1 Introduction

1.1 Aim

In this paper, we will motivate and develop an extension of Nelson's constructive logic **N3** that adds a subjunctive or counterfactual conditional to the existing setup. We will mostly be concerned with somewhat technical questions about the formal semantics, but we will outline how our account will be able to give a nice analysis of natural language counterfactuals. In particular, we will be able to do justice to the intuitions and arguments that have lead Alan Hájek to claim that most conditionals are false, but assertable (cf. [3]) without having to actually endorse that rather strange claim. This will, due to restrictions of space, remain a very rough outline of a theory of natural language conditionals. A more full-blooded theory will appear as a companion paper that focuses more on the linguistic evidence than the technical ideas, problems and solutions that are at the center of this paper.

1.2 Preliminaries

Throughout this paper, our languages $\mathcal{L}_{\sqsupset}$, $\mathcal{L}_{\rightarrow}$ and $\mathcal{L}_{\sqsupset,\rightarrow}$ consist of finite sets $\{-, \wedge, \vee, \sqsupset\}$, $\{-, \wedge, \vee, \rightarrow\}$ and $\{-, \wedge, \vee, \sqsupset, \rightarrow\}$ of propositional connectives, respectively, and a countable set Prop of propositional variables which we denote by p, q, etc. Furthermore, we denote by $\mathsf{Form}_{\sqsupset}$, $\mathsf{Form}_{\rightarrow}$ and $\mathsf{Form}_{\sqsupset,\rightarrow}$ the set of formulas defined in the languages $\mathcal{L}_{\sqsupset}$, $\mathcal{L}_{\rightarrow}$ and $\mathcal{L}_{\sqsupset,\rightarrow}$ respectively as follows.

$$A ::= p \mid -A \mid (A \wedge B) \mid (A \vee B) \mid (A \sqsupset B),$$
$$A ::= p \mid -A \mid (A \wedge B) \mid (A \vee B) \mid (A \rightarrow B),$$
$$A ::= p \mid -A \mid (A \wedge B) \mid (A \vee B) \mid (A \sqsupset B) \mid (A \rightarrow B).$$

We denote a formula of the language by A, B, C, etc. and a set of formulas of the language by Γ, Δ, Σ, etc.

A. Baltag et al. (Eds.): LORI 2017, LNCS 10455, pp. 497–511, 2017.
DOI: 10.1007/978-3-662-55665-8_34

1.3 Revisiting Nelson Logics

Our starting point are the so-called Nelson logics, named after David Nelson. There are two main variants of these logics called **N3** and **N4**. **N3** allows for gaps in the semantics, and **N4** features both gaps and gluts.

Definition 1. *A* **N3***-model for* $\mathcal{L}_{\rightarrow}$ *is a structure* $\langle W, \leq, \Vdash \rangle$, *W being a non-empty set of partially ordered* ($\leq$) *worlds and* $\Vdash : W \times \mathsf{Prop} \longrightarrow \{\emptyset, \{0\}, \{1\}\}$ *is an assignment of truth values to state-variable pairs with the following condition:*

- *for all p and all worlds w and w′, if* $w \leq w'$ *and* $w \Vdash_1 p$, *then* $w' \Vdash_1 p$, *and*
- *for all p and all worlds w and w′, if* $w \leq w'$ *and* $w \Vdash_0 p$, *then* $w' \Vdash_0 p$.

Valuations $\Vdash$ *are then extended by the following conditions:*

$$
\begin{array}{ll}
w \Vdash_1 -A & \textit{iff } w \Vdash_0 A \\
w \Vdash_0 -A & \textit{iff } w \Vdash_1 A \\
w \Vdash_1 A \wedge B & \textit{iff } w \Vdash_1 A \textit{ and } w \Vdash_1 B \\
w \Vdash_0 A \wedge B & \textit{iff } w \Vdash_0 A \textit{ or } w \Vdash_0 B \\
w \Vdash_1 A \vee B & \textit{iff } w \Vdash_1 A \textit{ or } w \Vdash_1 B \\
w \Vdash_0 A \vee B & \textit{iff } w \Vdash_0 A \textit{ and } w \Vdash_0 B \\
w \Vdash_1 A \rightarrow B & \textit{iff for all } x \geq w,\ x \nVdash_1 A \textit{ or } x \Vdash_1 B \\
w \Vdash_0 A \rightarrow B & \textit{iff } w \Vdash_1 A \textit{ and } w \Vdash_0 B
\end{array}
$$

Remark 1. Here are some intuitive explanations of **N3**-models. Worlds are intuitively to be understood as stages of investigation, and the accessibility relation marks that one stage is an epistemically possible development from one stage to another. We give both of the values 1 and 0 a substantive reading: 1 stands for verifiable, 0 for falsifiable. This is in contrast to the semantics of intuitionistic logic, in which value 1 marks the constructive notion of being provable and the other the mere absence of that notion.

Moreover, for **N3** we allow $\Vdash$ to be a partial function, so that statements might not receive either value at a given world. This reflects the fact that at a stage of investigation, a statement might be neither verifiable nor falsifiable. Note that $w \Vdash_0 p$ is not equivalent to $w \nVdash_1 p$ any more, and that the same of course goes for $w \Vdash_1 p$ and $w \nVdash_0 p$.

Note also that hereditary constraints for both 1 and 0 reflect that we assume that verifications and falsifications are conclusive.

Finally, the verification and falsification conditions are guided by the BHK-style clauses. The philosophical plausibility of these clauses is discussed at length in [5], where it is argued that these clauses give a much more faithful formal representation of the constructive thoughts in semantics that Michael Dummett has campaigned for throughout his career (even though Dummett himself was championing intuitionistic logic).[1] In this connection, of special interest is the negation, which combines a strong claim to being constructive with a more natural behavior than intuitionistic logic shows. All double negation laws, for example, are valid in **N3**; on the other hand, the Law of Excluded Middle fails.

[1] For more technical discussions related to Nelson's logics, see [4,7].

Remark 2. The logic **N4** gives even more options: It allows $\Vdash$ to assign 1, 0, neither or both values to a statement at a world. That is, we are not dealing with a valuation *function* any more, but with a valuation *relation*.[2] This is the only difference between the two logics, and everything else that follows in this section applies to both of them.

The most significant difference between **N3** and **N4** is that the Law of Explosion, $(A \wedge -A) \models B$, holds in the former, but not in the latter. In light of the intended interpretation, the choice between the two comes down to this question: Do we want to consider it possible that the very same proposition is verified and falsified at the same time. The intuitive answer, and the answer that both Dummett and [5] give, is "no". Therefore, we consider **N3** to be our preferred logic, even though there might be some reason to move to **N4** once we add counterfactual conditionals, as we will see below.

The second item of interest in the stock of propositional connectives, after negation, is the conditional. It combines, in its positive clause, the intuitionistic account with the classical account in its negative clause. The intuitionistic idea is that $A \rightarrow B$ is verified at a world iff in every world that presents a conceivable extension of our knowledge at the original world, either A is not (yet) verified, or B is verified. The classical idea is that $A \rightarrow B$ is falsified in just one kind of situation: One where A is verified and B falsified. Again, [5] presents an extended discussion that concludes that these clauses are both in considerable harmony with what Dummett had to say about conditionals, as well as a good model of our actual intuitions about verifiability and falsifiability conditions for conditionals.

However, the account can only claim to be one of conditionals in the indicative mood, such as "If it rains tomorrow, I will stay at home." Counterfactual or subjunctive conditionals, such as "If kangaroos had no tails, they would topple over." will clearly need to be analyzed differently. It is the aim of this paper to show how that can be done.

2 Expanding N3 with a Counterfactual Conditional

In our search for a suitable account for semantic clauses that we might add to Nelson logic to obtain a counterfactual conditional, we turn to one of the most influential accounts available, namely David Lewis's. In a nutshell, he considers counterfactuals to be about possible worlds that bear a particular relation to the actual world: They are worlds in which the antecedent of the counterfactual conditional is true and which, moreover, are maximally similar to the actual world. The question whether the conditional is true then comes down to the question whether the consequent is true in all of these possible worlds. Talk of similarity is notoriously slippery here, even though Lewis has a point when he argues that our intuitions about conditionals tend to be slippery to just the same degree (see [6, p. 91]).

[2] In other words, it is like the Michael Dunn's semantics (cf. [1]) for the so-called Belnap-Dunn logic or **FDE**.

We first review some of the basics of conditional logics, by following the presentation given by Graham Priest in [12].[3]

Definition 2. *A Chellas-model for $\mathcal{L}_{\sqsupset}$ is a structure $\langle W, \{R_A : A \in \mathsf{Form}\}, \Vdash\rangle$, where W is a nonempty set, $\{R_A : A \in \mathsf{Form}\}$ is a collection of binary relations on W, R_A, one for every formula, A, and $\Vdash : W \times \mathsf{Prop} \longrightarrow \{0,1\}$ is an assignment of truth values to state-variable pairs.*[4] *Valuations $\Vdash$ are then extended by the following conditions:*

$$\begin{array}{ll} (-) \quad w \Vdash_1 -A & \textit{iff } w \nVdash_1 A \,; \\ (\vee)\ w \Vdash_1 A \vee B & \textit{iff } w \Vdash_1 A \textit{ or } w \Vdash_1 B \,; \\ (\wedge)\ w \Vdash_1 A \wedge B & \textit{iff } w \Vdash_1 A \textit{ and } w \Vdash_1 B \,; \\ (\sqsupset)\ w \Vdash_1 A \sqsupset B & \textit{iff for all } y \in W \textit{ such that } wR_Ay; y \Vdash_1 B. \end{array}$$

Furthermore, $\models_l A$ iff for every Chellas-model $\langle W, \{R_A : A \in \mathsf{Form}\}, \Vdash\rangle$, $w \Vdash_1 A$ for all $w \in W$.

Remark 3. The above system is like the modal logic **K** in the sense that it has no constraints on the binary relations R_A ($A \in \mathsf{Form}$). And as in standard modal logic, we may consider some additional conditions on R_A. Let us first fix additional notations:

- $f_A(w) := \{x \in W : wR_Ax\}$
- $[A] := \{x \in W : w \Vdash_1 A\}$

Following Priest in [12], we may consider the following six conditions.

1. $f_A(w) \subseteq [A]$
2. If $w \in [A]$ then $w \in f_A(w)$
3. If $[A] \neq \emptyset$ then $f_A(w) \neq \emptyset$
4. If $f_A(w) \subseteq [B]$ and $f_B(w) \subseteq [A]$ then $f_A(w) = f_B(w)$
5. If $f_A(w) \cap [B] \neq \emptyset$ then $f_{A\wedge B}(w) \subseteq f_A(w)$
6. If $w \in [A]$ and $w' \in f_A(w)$ then $w = w'$.[5]

Now, the adoption and implementation of these ideas into the Nelson framework presents us with some fascinating questions and puzzles. First and foremost, we must decide what the verification and falsification clauses of the counterfactual conditional will look like, a highly non-trivial task, as we will see. Then, we need to consider in which ways the two accessibility relations that will show

[3] Note, however, that for the notation we follow Krister Segerberg's notation in [13]. More specifically, we use $\sqsupset$ and $>$ for would- and might-conditionals respectively. This has an intuitive appeal since these symbols are half-box and diamond which reflect the truth conditions for those conditionals.

[4] Note that we have $w \Vdash_0 A$ iff $w \nVdash_1 A$ here since we are reviewing the conditional logic based on classical logic.

[5] See [6, p.15] for the original discussion of these requirements and their motivation. If, instead of the last requirement, we add
- If $x \in f_A(w)$ and $y \in f_A(w)$ then $x = y$,

we get the system preferred by Robert Stalnaker.

up in our models, R_A and $\leq$, should interact. Lastly, we may wonder what our account let's us say about so-called "might conditionals", which are normally taken to be the duals of the "would conditionals". As we will see, upholding this duality might not be the most natural option to take in our setting.

Let us start, then, by considering the ways in which we might want to say that a counterfactual can be verified, and what we should ask for it to be falsified. Here is what we take to be the most natural first response to this question: We leave the positive clause as we found it in Lewis. That is, we say that a counterfactual is verified iff the consequent is verified at all the maximally close worlds in which the antecedent is verified.

We will take this as our starting point in judging when a counterfactual is verified. There is a small but important difference in our setup, in that we will have the same kinds of worlds in our similarity relation as in our model generally. That is, we are not, as Lewis is, dealing with complete worlds. Our valuation function stays partial. Otherwise, however, we will take the clause on board unchanged, at least to characterize our positive notion. We will say that a counterfactual is verified in the same circumstances (modulo the incomplete worlds) in which Lewis would call it true. But unlike him, we will not call it falsified (or false) in all other cases. The Nelson set-up allows us to think of how to falsify a counterfactual independently. As we will see, that opens up a wide space of counterfactuals that are neither verifiable nor falsifiable (and, as it will further unfold, these are really the most interesting and useful counterfactuals in normal conversations).

So, what would a falsified counterfactual look like? First, are there such things at all? We believe so. How about "If Kennedy had not been shot, Obama would have been sworn into office as president of the United States three decades earlier"? That, we hope you agree, would not have happened. Anyone who made such a claim would not be seen to have made a correct assertion, unless it was a very unusual context in which he uttered it (or, maybe, unless he succeeded, in the course of the conversation, in making the context so unusual that it would have been appropriate to say what he did).

Taking this as an intuitively clear case of what we are out to model, we propose the following in analogy to the Lewisian positive clause:

> *A counterfactual is falsified iff in all the closest, most similar worlds in which the antecedent is verified, the consequent is falsified.*

We will have occasion to slightly tweak this clause for technical reasons, but we start here as the most intuitive first attempt, at least according to our intuitions. Here is the intuition spelled out: In every world that is otherwise as similar as possible as ours, but in which Kennedy was not shot, it would be possible to verify that Obama was not elected into office three decades earlier. For starters, in those worlds, without further prompts from the context that time travel or some such device is under consideration, Obama would simply have been too young to run for president. That is actually enough to note, even though there are other reasons why we should regard this counterfactual as falsified.

There are other possibilities we might consider. As there now is a space between falsifiable and unverifiable statements, we could think of utilizing the latter in our clause, i.e.:

A counterfactual is falsified iff in all the closest, most similar worlds in which the antecedent is verified, the consequent is unverified.

We think this less plausible, but leave more detailed discussion of this and other alternatives for another time.[6] What we will try in this paper is to find a formal account that captures the above idea of falsifiability for counterfactuals as fully as possible. If we hook up all the preceding ideas, formally we will get the following:

Definition 3 (Lewis-Nelson model). *A Lewis-Nelson-model for $\mathcal{L}_{\sqsupset,\rightarrow}$ is a structure $\langle W, \leq, \{R_A : A \in \mathsf{Form}\}, \Vdash\rangle$, where W is a non-empty set (of states); $\leq$ is a partial order on W; $\{R_A : A \in \mathsf{Form}\}$ is a collection of binary relations on W, R_A, one for every formula, A with the above constraints; and $\Vdash: W \times \mathsf{Prop} \longrightarrow \{\emptyset, \{0\}, \{1\}\}$ is an assignment of truth values to state-variable pairs with the condition that $w_1 \Vdash_i p$ and $w_1 \leq w_2$ only if $w_2 \Vdash_i p$ for all $p \in \mathsf{Prop}$, all $w_1, w_2 \in W$ and $i \in \{0, 1\}$. Valuations $\Vdash$ are then extended by the following conditions in addition to those in* **N3***-model:*

- $w \Vdash_1 A \sqsupset B$ *iff for all* $y \in W$ *such that* $wR_Ay, y \Vdash_1 B$,
- $w \Vdash_0 A \sqsupset B$ *iff for all* $y \in W$ *such that* $wR_Ay, y \Vdash_0 B$.

Finally, the semantic consequence is now defined as follows: $\Sigma \models_{ln} A$ iff for all Lewis-Nelson-models $\langle W, \leq, \{R_A : A \in \mathsf{Form}\}, \Vdash\rangle$, and for all $w \in W$: $w \Vdash_1 A$ if $w \Vdash_1 B$ for all $B \in \Sigma$.

Remark 4. This way of fixing the clauses for the counterfactual brings a certain limitation in expressivity: We have not given a semantics for indicatives embedded into counterfactuals, i.e., statements like $A \sqsupset (B \rightarrow C)$. The reason for that is that we are only relating single worlds in R_A. To get an account of statements like the one just mentioned, we should rather relate worlds to pointed Nelson models. The modification is straightforward, but it makes the following exposition rather messy, and, from a natural language point of view, the loss seems not too great. "If A had been the case, then, if B, then C" seems something one would utter in the rarest of circumstances. We have, at any rate, bigger problems than such contrived statements to worry about, as we will see presently.

3 Bad News: Triviality

While the clauses above look, we believe, quite compelling at first blush, we immediately run into dire trouble with them.

[6] One reason why this alternative might seem tempting is that it leaves our original condition free to serve as the falsification clause for an added might-conditional. Might-conditionals, however, are yet another topic we don't have the space to cover in this piece.

3.1 The Collapse

Theorem 1. *$\models_{ln}$ is inconsistent.*

Proof. Just note that both $\models_{ln} (A \wedge -A) \sqsupset A$ and $\models_{ln} -((A \wedge -A) \sqsupset A)$ holds in the concerned semantics. Indeed, we have that $wR_{A\wedge -A}x$ does not hold for all $x \in W$ in view of the constraint 1, and so by *ex contradictione quodlibet*, we obtain the desired result. □

Remark 5. This is a similar observation to Wansing's observation in [17] for a connexive variant of **N3** being inconsistent.

Now, in order to overcome the problem of inconsistency, there are at least two avenues to pursue. We will outline those avenues in the following two subsections.

3.2 Route 1: From N3 to N4

The first option is to step back from **N3** to **N4**, a contradiction-friendly (i.e., paraconsistent) subsystem of **N3** mentioned at the beginning of our piece.[7] We refer to the new model with four-valued valuation instead of three-valued valuation as *Lewis-N4 model.* Then, from a technical viewpoint, we still obtain the following result.

Proposition 1. *Both $(A \wedge -A) \sqsupset A$ and $-((A \wedge -A) \sqsupset A)$ are valid.*

Proof. For the first formula, we use the fact that $A \wedge -A$ is satisfiable in **N4**, and that we have the constraint 5. For the second formula, we also use the falsity condition for $\sqsupset$. □

Of course, there is a worry of collapsing again. Thankfully, this is not the case due to the following result.

Proposition 2. *$(A \wedge -A) \sqsupset B$ is invalid, and thus it is not trivial.*

Proof. Let us consider a one-world Lewis-N4 model with wR_Ax identified with $w = x$ & $x \Vdash_1 A$. We can then see that the six conditions are satisfied as follows.

- For the first condition, it is obvious that if wR_Ax then $x \Vdash_1 A$.
- For the second condition, it is also obvious that if $w \Vdash_1 A$ then $w = w$ & $w \Vdash_1 A$.
- For the third condition, if $x_0 \Vdash_1 A$ for some $x_0 \in W$, then since W is one-element world, we obtain that $w = x_0$ and thus we obtain that wR_Ax_0.
- For the fourth condition, if wR_Ax implies $x \Vdash_1 B$, then by the definition of R_A, we obtain that wR_Ax implies wR_Bx. Similarly, if wR_Bx implies $x \Vdash_1 A$, then we obtain that wR_Bx implies wR_Ax, as desired.
- For the fifth condition, it is immediate that $wR_{A\wedge B}x$ implies wR_Bx in view of the definition of R_A and the truth condition for conjunction.
- For the final condition, it is obvious by the definition of R_A.

[7] This also works for the case with Wansing's connexive logic **C**, a variant of **N4**.

Now, let A and B be such that $w \Vdash_1 A$, $w \Vdash_0 A$ and $w \nVdash_1 B$. Then it follows that $w \nVdash_1 (A \wedge -A) \sqsupset B$, as desired. □

Remark 6. Note that one-world model is the "classical" model in which the indicative conditional will be not constructive, but classical. This might be of interest for those who are more sympathetic with classical logic rather than intuitionistic logic.

Therefore, from a purely technical perspective, a shift from **N3** to **N4** works well, in that it avoids the triviality result.[8] However, it is not so straightforward from a more philosophical perspective. In particular, the understanding of the verification and falsification becomes less intuitive once we allow these kinds of overlaps between them (see [5, pp. 150–153] for discussion). And this motivates us to consider the second avenue to which we turn in the next subsection.

3.3 Route 2: Different Verification and Falsification Conditions

As we have observed, the move to the **N4**-based logic works well technically, but philosophically it is a doubtful one. So, we here seek for options to keep **N3**.

A quick reflection on our proof of the inconsistency reveals that we are relying on the fact that none of the worlds are $(A \wedge -A)$-world. In order to fix this feature, one simple and natural option is to require that there is at least one A-world in the truth and falsity conditions which is maximally similar to the present world and in which the antecedent holds. This is an idea that can already be found in Lewis's original discussion ([6, p. 25]); of course in his case he only considered adding the requirement to the truth condition of the counterfactual.[9] He was brought to ponder this alternative because he was wondering how natural it is to consider a counterfactual with impossible antecedent to be true. In the end, he opted for the earlier, simpler variant, but he was far from displaying strong feelings on the matter. He thought that our intuitions about such impossible counterfactuals were malleable and not indicative of much of substance, so he more or less left the readers free to choose their favorite option. We cannot afford an equally leisurely stance; for us, the consistency of the system is at stake, so we should better explore the alternative clauses to make sure they give us a well-behaved system. To be specific, we want to replace the truth and falsity conditions in the Chellas-Nelson model as follows:

- $w \Vdash_1 A \sqsupset B$ iff $\begin{cases} \text{for some } x \in W, wR_A x \text{ and} \\ \text{for all } x \in W \text{ such that } wR_A x, x \Vdash_1 B, \end{cases}$
- $w \Vdash_0 A \sqsupset B$ iff $\begin{cases} \text{for some } x \in W, wR_A x \text{ and} \\ \text{for all } x \in W \text{ such that } wR_A x, x \Vdash_0 B. \end{cases}$

[8] We emphasize again that $(A \wedge -A) \rightarrow B$ is invalid in **N4**.

[9] Basically the same idea is applied by Priest in [11] in which he discusses the cancellation account of negation.

Once this adjustment is made, the same proof for the inconsistency will not work any more.[10] In particular, we obtain both $\not\models_{ln} (A \wedge -A) \sqsupset A$ and $\not\models_{ln} -((A \wedge -A) \sqsupset A)$.

Proposition 3. *$(A \wedge -A) \sqsupset A$ is not valid, and thus the system is non-trivial.*

Proof. Let us again consider a one-world Lewis-Nelson model with wR_Ax identified with $w = x$ & $x \Vdash_1 A$. We can then see that the six conditions are satisfied as in the proof of Proposition 2. Then the desired result follows since we have $w \nVdash_1 A \wedge -A$ for all A. The same model shows that $-((A \wedge -A) \sqsupset A)$ is not valid. □

Remark 7. Once fears of triviality are out of the way, it deserves to be highlighted that we obtain the following equivalence: $-(A \sqsupset B) \leftrightarrow (A \sqsupset -B)$. In view of this equivalence, we obtain the following characteristic thesis of connexive logics[11] with respect to $\sqsupset$:

- Boethius's thesis: $(A \sqsupset B) \rightarrow -(A \sqsupset -B)$

Note that there are some connections between conditional logics and connexive logics discussed since [10], and more recently in [15]. The interesting feature in the present context is that some rather natural considerations led us to connexivity, which is a highly nonclassical property. In particular, it led us down an intensional route to connexivity, one of the approaches suggested by Wansing in [17].[12]

These reflections will have to suffice, at least in this piece, to give an idea of the wealth of interesting questions and connections we tapped into. There are, on closer inspection, many more combinations that might be plausible alternatives to our choice that would be worth exploring. And crucially, we have yet to give our account of might conditionals, conditionals of the form "If A had been the case, B might have been the case". This important part of our story will have to be told another time when more space is available. Instead, we close our technical discussion with some reflections on the two accessibility relations and then turn at the end of the paper to the question of how well our account can actually make sense of assertions of counterfactuals.

[10] One of the reasons why Lewis was not too concerned about the "right" choice between the two clauses for the truth of conditionals is this: He realized that, in the classical case he was considering, he could define either one of the two conditionals in terms of the other ([6, p. 26]). Here is how to define the old condition in terms of the new one: $A \sqsupset_{old} B =_{def} (A \sqsupset_{new} A) \supset (A \sqsupset_{new} B)$. Now, if this was possible in our setting as well, of course, then we would be faced with disaster again. Luckily, the equivalence does not hold in our system if we replace the material conditional with the constructive Nelson conditional. The same is true of $-(A \sqsupset_{new} A) \vee (A \sqsupset_{new} B)$.

[11] For more on connexive logics in general, see [19].

[12] For another interesting case for connexive logics through a natural consideration, see [18]. Note also that Wansing's approach to connexive logics can be applied to other systems than Nelson logics. For some examples, see [8,9].

4 Issues of Accessibility

As we have two accessibility relations in our model, a natural question to ask is how they interact. As it stands, any assumptions we make here will only affect formulas in which both indicative and subjunctive conditionals occur embedded into each other, which is something rarely observed in natural language. Nonetheless, it would be nice to have our models the most plausible interpretation that we can find.

The first thing to notice here is that, unless we impose rather *ad hoc* restrictions, we will lose the hereditary property of the Nelson model, at least for statements with counterfactual conditionals in them.[13] This should not be seen as cause for alarm. For starters, this is known to happen in other cases of modal extensions of Nelson logic.[14] But more importantly, on reflection it really strikes us as a desirable property. Think of the so-called Sobel sequences that are of central interest in the discussions of counterfactuals. These are sequences of counterfactuals with more and more elaborate antecedents that go back and forth from being true to being false. An example that is often discussed is:

> If the U.S. threw all its nuclear weapons into the sea, there would be war; but if all nations with nuclear weapons threw them into the sea, there would be peace.

Now, let's assume that these are true counterfactuals. Consider what change the single counterfactual

> If the U.S. threw all its nuclear weapons into the sea, there would be war.

will go through if we utter it before and after we learn that all nations except the U.S. have indeed thrown their nuclear weapons into the sea. Before we learned that fact, it would seem that the counterfactual should be counted as true, while afterwards, it should be seen as false. In other words, thinking in terms of verifications and falsifications in Nelson logics, the counterfactual should be verifiable at x, but falsifiable (or not verifiable) at y, where $x < y$. That is, the heredity constraint must be watered down if we want to accommodate these analogues to Sobel sequences; all is well with our model in that regard.

A different question here is which combinations of worlds that stand in the Nelson relation, $x < y$, such that y is a true expansion of x, should we possibly want to find in the relation R_A to our world. That is, should it be possible that (1) neither wR_Ax nor wR_Ay, (2) wR_Ax but not wR_Ay, (3) wR_Ay but not wR_Ax or that (4) both wR_Ax and wR_Ay? (1) is obviously a combination that we need. But to get an account that is of any interest, we will in addition need to countenance at least one of the other possibilities. Which one(s)?

---

[13] We would like to thank Massimiliano Carrara for directing our attention to this issue.

[14] See, for example, [16].

(2) looks like a possibility that we would like to keep. Is a world in which it is neither verifiable nor falsifiable that Bigfoot exists closer to our world than one in which it is verifiable that he exists? To us, it seems so.

It also seems that we will want that possibility (3), mostly in cases in which y exceeds x in information that we in the actual world have available also. If x is a world in which it is uncertain whether the moon is made of green cheese and y is one in which it is verifiably not, then y is closer to ours. (A little care is needed here, because factual similarity is actually not necessarily at the top of the list for judging the counterfactual similarity of worlds, see the discussion about "If Nixon had pushed the button..." in the literature about conditional logics. Nonetheless, there are clearly cases in which factual similarity will make the difference, and that is enough to answer our question here).

Finally, (4) seems to be the most disposable option, but we think it is plausible to keep it, as well. If the things that make the difference between x and y are completely unrelated to the content of the counterfactual, it seems unnecessary to ban either one from the circle of closest worlds if the other one is in it.

These are tentative conclusions that we are happy to reconsider. At this point, however, we will not impose any restrictions and allow all of the combinations (1) to (4) in our models.[15]

5 Unfalsifiability as an Assertability Condition

What we have achieved so far is an account of the verification and falsification conditions of counterfactual conditionals in a **N3** setting, and we showed that the resulting system is non-trivial. In this sense, we have offered a constructive alternative to the conditional logics of a more classical provenance. However, the account allows us to do more, namely to give a new and, we believe, very attractive analysis of what it is to correctly assert a counterfactual. The key move is to question the natural correlation between verifiability and assertability of counterfactuals.

The idea behind this account is one developed for other kinds of statements in [5] (drawing on and greatly expanding corresponding ideas in Dummett's own writing). Namely, that for a certain number of assertions, the question whether they have been made correctly does not come down to whether they are verifiable, but rather whether they are *unfalsifiable.*

[15] A more intricate condition on the two relation was proposed by a reviewer, and we thank her or him for the inspiration:

If $w \leq x$ and xR_Ax' then there is a $w' \in W$ such that wR_Aw' and $w' \leq x'$.

This condition is analogous to what in many-dimensional modal logics is called left-commutativity (see [2, p. 221]). Once parsed, this condition indeed seems eminently plausible. With the vocabulary we have introduced so far, however, it seems that no difference to the consequence relation is made by imposing the condition. This will change when, in later work, we will introduce a suitable might-conditional, a topic we have to leave out for reasons of space. When we will address this, we will be sure to come back to the reviewer's condition.

Take, for example, future tense statements. If I had to be able to supply a verification of every statement in the future tense I correctly assert, I would not be able to say much of interest. The better way to deal with this is to say that such statements are correct as long as they are unfalsifiable. The day might come when the audience finds out that what the statement had claimed did not come to pass. In that case, the speaker will have to take back his assertion. But until that day, what he has said will stand.

In the present context, we would like to suggest that many counterfactuals are of the same kind. In many cases, we seem to be able to assert them without it being clear at all that they are verifiable (or true, in the classical case). For reasons of space, this is not the place in which we will convince the reader fully of that claim. However, we can give some gestures in the right direction that, we believe, shows this to be a path worth investigating.

Let us start by giving an example that illustrates what we have in mind (and one which many readers might be familiar with):[16]

> Years ago, a few of us were at a restaurant in NY–Red Smith, Frank Graham, Allie Reynolds, Yogi [Berra] and me. At about 11.30 p.m., Ted [Williams] walked in helped by a cane. Graham asked us what we thought Ted would hit if he were playing today. Allie said, "due to the better equipment probably about .350." Red Smith said. "About .385." I said, "due to the lack of really great pitching about .390." Yogi said, ".220." We all jumped up and I said, "You're nuts, Yogi! Ted's lifetime average is .344." "Yeah," said Yogi, "but he is 74 years old."
>
> Buzzie Bavasi, baseball executive

This story nicely illustrates several points. First, the speakers in it are disagreeing, and we would presumably have some trouble convincing ourselves that one of them is really, clearly, objectively at fault. This is very typical for statements made under the unfalsifiability norm: They easily accommodate cases of faultless disagreement, a topic which over the last years has attracted a lot of attention from philosophers of language (see [5]). However, the example seems, under closer inspection, also to suggest an alternative resolution: Maybe it would be best to say that the original conditional was ambiguous, and there was no fact of the matter whether what it expressed concerned possible worlds in which Ted Williams was playing as a young man today and worlds in which he was playing as an old one. This is, we believe, very likely to be correct, and our account is not meant to supplant an unusual story for a much more recognizable one.[17] On that account, many counterfactuals are ambiguous, and without disambiguation will leave space for apparent disagreement that will dissolve once it is agreed upon which sense of the counterfactual is talked about. Lewis's classic example "If kangaroos had no tails, they would topple over" is a case in point. Yes, they would, if the existing animals were all of a sudden stripped of their appendage.

[16] A quote from [14, p. 202].

[17] In this sense, the example serves to show that we are not overly and unnecessarily ambitious here.

No, they wouldn't, for evolution would not have allowed a species that falls over all the time, so they would have adopted a different posture if they had come up without tails. This is no deep disagreement, for we can spell out which counterfactual we are really interested in discussing, and once we do it, it seems clear which answer is the right one.

But note something about the baseball story: There is more disagreement in the story than just the one between Yogi Berra and the others (a disagreement that, if the speakers were feeling serious about it in the first place, could be easily resolved by disambiguation in just the way the kangaroo case can be dealt with). The *others* were disagreeing with each other, as well. And here, it is clear that they had the same counterfactual in mind.

Now, what do we want to say about that (very real, even if good-humored) disagreement? That there is a precise average that would be the right answer to the question, and thus, that at most one of them could have been right to say what he said (and much more likely, none of them)? And that, if someone had been there to point that fact out, they all would have had to retract their statements?

Or would we rather say that they, somehow, all were making correct assertions, at least correct in the sense that none of them could have rightly been required by the others to retract his statement?

We think that the latter is much more appealing, and it is just what the unfalsifiability account allows us to say. Among the closest possible worlds in which Ted Williams, as a young man, would play today, there is a world in which he hits .350, one where he hits .385, and one in which he manages .390. There are many worlds in between, but there is no closest world in which he hits .220, at least not as a young man. That is, there are indeed things in this context that are falsifiable in our sense. It's just the things the three first speakers were saying were not among them: They all said something unfalsifiable, and thus they all made a correct assertion, their disagreement notwithstanding.

They all make correct assertions, at least ones that could under normal circumstances be uttered seriously and without reprimand. In [5], we argue at length for this weak understanding of correctness. Smith, Graham, Reynolds and Bavasi were in *faultless disagreement*, a rather typical dialectical pattern also discussed in [5].

6 Updating Hájek: Most Counterfactuals Are Unverifiable

Our analysis points to a possible solution to a potentially uncomfortable theoretical situation that Hájek has diagnosed. He has (in unpublished but much read work) argued at length that while many counterfactuals are assertable, most of them are false. His arguments come in many different guises, but they all have a similar basic pattern. In each case in which

(1) If A had been the case, not-B might have been the case?
is true, and there are many of these cases,

(2) If A had been the case, B would have been the case?
cannot be true.

It is simply in their semantical nature that the truth of the two statements are incompatible. Then he goes through many examples in which we would like to say that a would-counterfactual of form (2) is true and points out that the corresponding might-conditional of the form (1) is true. Even if that might-conditional speaks of the remotest possibility ("If I had jumped just now, I might not have come down, because I might have quantum-tunnelled to China"), it rules out the truth of the sane-sounding would-conditional ("If I had jumped just now, I would have come down"). To insist on the truth of the second is, so Hájek, tantamount to insisting on the falsity of the first, which in turn is to deny the lessons of quantum physics. So, Hájek argues, if you would like to hold on to quantum physics, you will have to consider "If I had jumped just now, I would have come down" to be false, but of course it seems like something you should surely be allowed to say in normal conversation. Hence, even though most counterfactuals are false, many of them are assertable. Hájek himself acknowledges that this is a rather strange conclusion to draw, but he sees no way of avoiding it.

Here is how our story gives a twist to the situation that makes it, in our view, much more palatable. First of all, we will speak of verifiability instead of truth, but that is just for consistency with the general story we have presented so far. Hájek might reject all that and keep speaking of realistic truth and still be enamored by our account; he only would have to accept realistic truth value gaps, at least for counterfactuals.

So, where Hájek says that most counterfactuals are not true, we agree to the corresponding claim: Most counterfactuals are not verifiable. But where he goes on to conclude that most counterfactuals are false, we resist the corresponding conclusion: It is not the case that most counterfactuals are falsifiable. For that would mean, on our account, that in all the nearest A-worlds, B is actually falsified. What we say is that most counterfactuals are neither verifiable nor falsifiable, but that many of them are assertable nonetheless. Again, that is just what we are saying about future tense statements, and it seems to us a conclusion that is much easier to find our peace with than Hájek's conclusion.

To fully convince you (and us) of the viability of our alternative, we will have to give the examples Hájek gives much more detailed attention, and, even more importantly, we will have to submit something we have not yet given you: Our explanation of might-conditionals in our framework. The simple solution of classicists to assume might- and would-conditionals to be duals does not service us well, and our alternative takes more space to motivate than we have available here.

7 Conclusion

The last section gives a hint of the direction in which our further research is going to move. In this paper, we have laid the groundwork in showing how a constructive counterfactual conditional can be added to Nelson logic. This is of interest to the growing community of researchers working on Nelson logic, whether or not they see the same potential we see in the idea that assertions of counterfactuals are often to be judged by the standard of unfalsifiability.

Acknowledgments. Hitoshi Omori is a Postdoctoral Research Fellow of Japan Society for the Promotion of Science (JSPS). We would like to thank the anonymous referees for their helpful comments that improved our paper. We would also like to thank Massimiliano Carrara, Roberto Ciuni, and the participants of *Kyoto Philosophical Logic Workshop I* for useful comments and discussions.

References

1. Dunn, M.: Intuitive semantics for first-degree entailments and 'coupled trees'. Philos. Stud. **29**(3), 149–168 (1976)
2. Gabbay, D., Kurucz, A., Wolter, F., Zakharyaschev, M.: Many-Dimensional Modal Logics: Theory and Applications (2003)
3. Hájek, A.: Most counterfactuals are false. Unpublished draft http://philrsss.anu.edu.au/people-defaults/alanh/papers/MCF.pdf
4. Kamide, N., Wansing, H.: Proof Theory of N4-Related Paraconsistent Logics. Studies in Logic, vol. 54. College Publications, London (2015)
5. Kapsner, A.: Logics and Falsifications. Trends in Logic, vol. 40. Springer, Heidelberg (2014)
6. Lewis, D.K.: Counterfactuals. Blackwell, Boston (1973)
7. Odintsov, S.P.: Constructive Negations and Paraconsistency. Trends in Logic, vol. 26. Springer, Heidelberg (2008)
8. Omori, H.: A simple connexive extension of the basic relevant logic BD. IfCoLog J. Logics Appl. **3**(3), 467–478 (2016)
9. Omori, H.: From paraconsistent logic to dialetheic logic. In: Andreas, H., Verdée, P. (eds.) Logical Studies of Paraconsistent Reasoning in Science and Mathematics. TL, vol. 45, pp. 111–134. Springer, Cham (2016). doi:10.1007/978-3-319-40220-8_8
10. Pizzi, C.: Boethius' thesis and conditional logic. J. Philos. Logic **6**, 283–302 (1977)
11. Priest, G.: Negation as cancellation and connexive logic. Topoi **18**(2), 141–148 (1999)
12. Priest, G.: An Introduction to Non-Classical Logic: From If to Is, 2nd edn. Cambridge University Press, Cambridge (2008)
13. Segerberg, K.: Notes on conditional logic. Stud. Logica **48**(2), 157–168 (1989)
14. Sider, T.: Logic for Philosophy. Oxford University Press, Oxford (2010)
15. Unterhuber, M.: Beyond system P - Hilbert-style convergence results for conditional logics with a connexive twist. IfCoLog J. Logics Appl. **3**(3), 377–412 (2016)
16. Wansing, H.: Semantics-based nonmonotonic inference. Notre Dame J. Formal Logic **36**(1), 44–54 (1995)
17. Wansing, H.: Connexive modal logic. In: Schmidt, R., Pratt-Hartmann, I., Reynolds, M., Wansing, H. (eds.) Advances in Modal Logic, vol. 5, pp. 367–383. King's College Publications, London (2005)
18. Wansing, H.: A note on negation in categorial grammar. Logic J. IGPL **15**, 271–286 (2007)
19. Wansing, H.: Connexive logic. In: Zalta, E.N. (ed.) The Stanford Encyclopedia of Philosophy (2014). http://plato.stanford.edu/archives/fall2014/entries/logic-connexive/. Fall 2014 edition

A Dynamic Approach to Temporal Normative Logic

Fengkui Ju[1(✉)] and Gianluca Grilletti[2]

[1] School of Philosophy, Beijing Normal University, Beijing, China
fengkui.ju@bnu.edu.cn
[2] Institute for Logic, Language and Computation, University of Amsterdam, Amsterdam, Netherlands
grilletti.gianluca@gmail.com

Abstract. State commands refer to states, not actions. They have a temporal dimension explicitly or implicitly. They indirectly change what we are permitted, forbidden or obligated to do. This paper presents DTNL, a deontic logic meant to handle state commands based on the branching-time temporal logic PCTL^*. The models of DTNL are trees with *bad states*, which are identified by a propositional constant $\mathfrak{b}$ introduced in the language. To model state commands, a dynamic operator that adds states to the extension of $\mathfrak{b}$ is introduced.

Keywords: Deontic logic · Temporal logic · Commands · Deadlines

1 Background

There are two types of commands that refer to *actions* and *states* respectively. The former can be called action commands and the latter state ones. For example *never touch the button* is an action command since it imposes a restriction on which actions can be legally performed, while *ensure that the table is clean before the meeting*; *everything be in order until I get back* [12]; *nobody sit in the first row* [12] are state commands as they impose conditions on the future state of affairs.

Commands of both types can change what the agent is *permitted*, *forbidden* or *obligated* to do. After the action command *never touch the button*, the agent is not allowed to touch the button any more. After the state command *ensure that the table is clean*, the agent has the obligation to make it true that the table is clean.

Unlike action commands, state commands change what the agent can do indirectly. The state command *ensure that the table is clean* is not ordering the agent to clean the table himself. All it requires is that the agent makes it the case that the table is clean. The agent can execute the command by letting someone else clean the table.

A. Baltag et al. (Eds.): LORI 2017, LNCS 10455, pp. 512–525, 2017.
DOI: 10.1007/978-3-662-55665-8_35

Commands have a temporal dimension. *Ensure that the table is clean before the meeting* imposes a deadline conditional on future events: the command is fulfilled if the table is clean just before the meeting; *everything be in order until I get back* imposes a certain obligation *until* some other condition is fulfilled; *nobody sit in the first row* imposes a permanent condition: the agent should prevent *from now on* that someone sits in the first row.

There are two perspectives concerning how the agent should behave after a command: the perspective of the agent himself and the external perspective. They make a difference for the commands without a *deadline*. We take *ensure that the table is clean some day* as an example. From the external perspective, after the command, the agent should make it true that the table is clean, although he can do it in whichever day he wants. But from the perspective of the agent, the command creates no obligation, as it will not be violated at any point in time. This difference was already pointed out by [6] in the context of *deadlines of norms.*

Computation Tree Logic (CTL), proposed in [5], is a branching-time temporal logic broadly used in computer science. Examples of properties expressible by CTL are *it is sure that* ϕ *will happen in the next moment*; *it is sure that* ϕ *will eventually happen*; *it is possible that* ϕ *will hold until* ψ. By generalizing CTL, [1] presents a deontic logic, called Normative Temporal Logic (NTL). The models of this logic are transition systems plus illegal transitions. An example of properties expressible in NTL is *the agent is allowed to act to make* ϕ *true in the next moment.* Compared with conventional deontic logics, NTL has two interesting features. Firstly, it makes the idea explicit that the agent acts to make things true. Secondly, normative notions expressible by it have a temporal dimension.

NTL is conceptually suitable to handle state commands, but technically has two problems. Firstly, NTL, as CTL, has a syntactic restriction on applications of temporal operators: they have to be immediately preceded by path quantifiers. This implies that the temporal formulas such as ϕ *will eventually happen* are not well-formed formulas and iterations of temporal operators are not allowed. So it is hard to express and interpret state commands within NTL. Secondly, NTL uses transition systems with illegal transitions to handle normative notions, but these models are *historyless* in the sense that whether a transition is illegal or not has nothing to do with what the agent has done in the past. However, normative notions with a temporal dimension essentially involve past actions. For example, assume a scenario in which a child has to collect 100 coins in a piggy bank, and only then retrieve them by crashing the container. As the difference made by putting a coin into the bank can not be seen, the child has to rely on his memory of the past actions to know when to break the bank.

Full Computation Tree Logic (CTL*), introduced in [7], is an extension of CTL that does not have the syntactic restriction mentioned previously. PCTL* is a further extension of CTL* with two past operators whose completeness is shown in [9]. In what follows we present DTNL ("Dynamic Temporal Normative Logic"), a deontic logic based on PCTL*. This logic takes *trees* with *bad states* as its models, instead of general transition systems with *bad transitions.* A special propositional constant $\mathfrak{b}$ is introduced to indicate bad states. Using this constant, the normative notions of permission, prohibition and obligation can be defined as in [3,8]. A dynamic operator representing state commands is also introduced. Its function is to update the model, adding those states that violate the command to the extension of $\mathfrak{b}$. The logic follows the agent's perspective, and commands without a deadline might not create the corresponding obligations. Our way of viewing commands follows [2,10] which think that the meaning of commands lies in how they change the agent's internal state.

2 Language

Let Φ_0 be a countable set of atomic propositions and p range over Φ_0. Define the language Φ_{DTNL} as the following:

$$\phi ::= p \mid \top \mid \mathfrak{b} \mid \neg\phi \mid (\phi \wedge \phi) \mid \mathrm{Y}\phi \mid (\phi \mathrm{S} \phi) \mid \mathrm{X}\phi \mid (\phi \mathrm{U} \phi) \mid \mathbf{A}\phi \mid [!\phi]\phi$$

The featured formulas of this language are read as follows:

1. $\mathfrak{b}$: this is a *bad* state.
2. $\mathrm{Y}\phi$: ϕ was the case in the *last* moment.
3. $\phi\mathrm{S}\psi$: ϕ has been the case *since* ψ.
4. $\mathrm{X}\phi$: ϕ will be the case in the *next* moment.
5. $\phi\mathrm{U}\psi$: ϕ will be the case *until* ψ.
6. $\mathbf{A}\phi$: no matter how the agent will act in the future, ϕ is the case *now.*
7. $[!\phi]\psi$: ψ is the case after the command *make* ϕ *true* is given.

Note that the language Φ_{DTNL} is the language of PCTL* plus the propositional constant $\mathfrak{b}$ and the dynamic operator $[!\phi]$.

It seems strange to say that no matter how the agent will act in the future, ϕ is the case now, but in fact this is fine. Whether a sentence that involves time relations is true or not now might be dependent on how the agent will act in the future. For example, whether a student will pass an exam is dependent on how he will study. In order to make a sentence true now, the agent has to act in a certain way in the future.

The other usual propositional connectives and the falsum $\bot$ are defined in the usual way:

1. $\mathfrak{f} := \neg\mathfrak{b}$: this is a fine state.
2. $\mathrm{P}\phi := (\top\mathrm{S}\phi)$: ϕ was the case.
3. $\mathrm{H}\phi := \neg\mathrm{P}\neg\phi$: ϕ has been the case.

4. $\mathrm{F}\phi := (\top \mathrm{U}\phi)$: ϕ will be the case.
5. $\mathrm{G}\phi := \neg\mathrm{F}\neg\phi$: ϕ will always be the case.
6. $\mathbf{E}\phi := \neg\mathbf{A}\neg\phi$: the agent has a way to act in the future s.t. ϕ is the case now; $\mathbf{E}\phi$ intuitively means that making ϕ true is *achievable.*

Let $\mathrm{X}^n\phi$ denote $\mathrm{X}\ldots\mathrm{X}\phi$ where n is the number of occurrences of X. The state commands mentioned in the beginning can be expressed as follows:

1. *Ensure that the table is clean before the meeting*: $!\mathrm{X}^{k-1}c$
2. *Everything be in order until I get back*: $!(o\mathrm{U}b)$
3. *Nobody sit in the first row*: $!\mathrm{G}\neg s$

For the first example, we suppose that the starting time of the meeting is already fixed and there are k units of time from now to it.

3 Models

Let W be a nonempty set of states and R a binary relation on it. A sequence $w_0 \ldots w_n$ of states (possibly of length one) is called an *R-sequence* if $w_0 R \ldots R w_n$. (W, R) is a *tree* if there is a $r \in W$, called the *root of the tree* s.t. for any w, there is a unique R-sequence from r to w. Immediate consequences of the definition are that the root is unique and R is irreflexive. R is *serial* if for any w, there is a u s.t. Rwu (there are no end points).

A serial tree (W, R) is understood as a time structure encoding an agent's actions (the transitions) and states in time (the nodes). At any state w, the *history* of the agent up to that point is represented by the path connecting the root to w (the actions performed). The seriality condition corresponds to the fact that the agent can always perform an action at any given time, while a branching in the tree is interpreted as a situation in which the agent *can choose between different possible actions.*

Fix a serial tree (W, R). Here are some auxiliary notations. An R-sequence $w_0 \ldots w_n$ starting from the root is an *history* of w_n. For any w and u, u is a *historical* state of w if there is an R-sequence $u_0 \ldots u_n$ s.t. $0 < n$, $u_0 = u$ and $u_n = w$. w is a *future* state of u if u is a historical state of w. Note that a state can not be a historical or future state of itself.

An infinite R-sequence is a *path.* A path starting at the root is a *timeline.* A path $w_0 \ldots$ *passes by* a state x if $x = w_i$ for some $i > 0$. Let π be a path. We use $\pi(i)$ to denote the $i+1$-th element of π, ${}^i\pi$ the prefix of π to the $i+1$-th element and π^i the suffix of π from the $i+1$-th element. For example, if $\pi = w_0 \ldots$, then $\pi(2) = w_2$, ${}^2\pi = w_0 w_1 w_2$ and $\pi^2 = w_2 \ldots$. For any history $w_0 \ldots w_n$ and path $u_0 \ldots$ s.t. $w_n = u_0$, let $w_0 \ldots w_n \otimes u_0 \ldots$ denote the timeline $w_0 \ldots w_n u_1 \ldots$.

A tuple $\mathfrak{M} = (W, R, r, B, V)$ is a *model* if

1. (W, R) is a serial tree with r as the root
2. B is a subset of W meeting the following conditions:
   (a) if $w \in B$, then $u \in B$ for any u s.t. Rwu
   (b) if $w \notin B$, then there is a u s.t. Rwu and $u \notin B$
3. V is a function from Φ_0 to 2^W

B is called the set of *bad* states and $W - B$ the set of *fine* ones. Intuitively, a transition (w, u) is *illegal* if u is a bad state. The first constraint on B is called *persistency of liability*; it indicates that if we reached a state in which we failed to fulfill a command, then this holds for all its successors. This constraint implies that if a state is bad, then all of its future states are bad, and if a state is fine, then all of its historical states are fine. The second constraint on B is called *seriality of legality*; it means that if a state is fine, then at least a successor of it is fine. The conjunction of the two constraints is called *normative coherence.* Figure 1 illustrates a model.

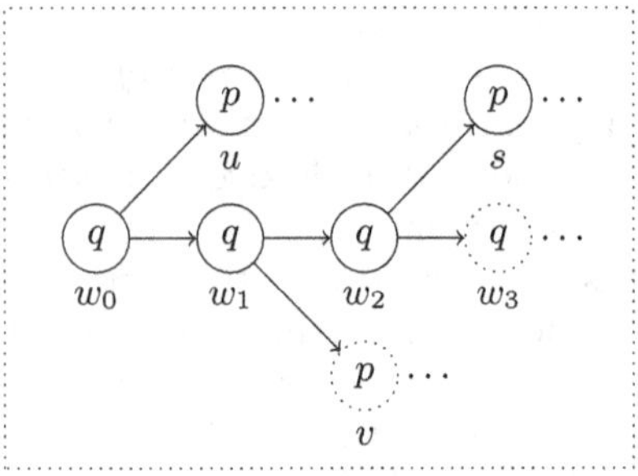

Fig. 1. This figure indicates a model. w_0 is the root. Dotted circles denote bad states and solid circles fine states. Arrows are transitions.

The following is an intuitive interpretation of the models: an agent's possible actions are encoded by a serial tree (as mentioned above). At any moment, the agent has a set of rules he should respect, and these rules are encoded by the bad states: the agent is allowed to travel to the fine states but not to the bad ones. If the agent has not done anything illegal, there is always something legal for him to do, and if he has done something illegal, there is nothing legal for him to do. We will get back to the last point again in Sect. 6.

4 Semantics

Following the dynamic approach, we define by mutual recursion the truth of formulas with respect to a model and an update operation to interpret the dynamic operator $!\phi$.

$\mathfrak{M}, \pi, i \Vdash \phi$, the formula ϕ being true at the state $\pi(i)$ relative to the timeline π in the model $\mathfrak{M}$, is defined as follows:

$$
\begin{array}{lll}
\mathfrak{M}, \pi, i \Vdash p & \Leftrightarrow & \pi(i) \in V(p) \\
\mathfrak{M}, \pi, i \Vdash \top & & \\
\mathfrak{M}, \pi, i \Vdash \mathfrak{b} & \Leftrightarrow & \pi(i) \in B \\
\mathfrak{M}, \pi, i \Vdash \neg\phi & \Leftrightarrow & \text{not } \mathfrak{M}, \pi, i \Vdash \phi \\
\mathfrak{M}, \pi, i \Vdash \phi \wedge \psi & \Leftrightarrow & \mathfrak{M}, \pi, i \Vdash \phi \text{ and } \mathfrak{M}, \pi, i \Vdash \psi \\
\mathfrak{M}, \pi, i \Vdash \mathrm{Y}\phi & \Leftrightarrow & i > 0 \text{ and } \mathfrak{M}, \pi, i-1 \Vdash \phi \\
\mathfrak{M}, \pi, i \Vdash \phi \mathrm{S} \psi & \Leftrightarrow & \text{there is a } j \leq i \text{ s.t. } \mathfrak{M}, \pi, i-j \Vdash \psi \\
& & \text{and } \mathfrak{M}, \pi, i-k \Vdash \phi \text{ for any } k < j \\
\mathfrak{M}, \pi, i \Vdash \mathrm{X}\phi & \Leftrightarrow & \mathfrak{M}, \pi, i+1 \Vdash \phi \\
\mathfrak{M}, \pi, i \Vdash \phi \mathrm{U} \psi & \Leftrightarrow & \text{there is a } j \text{ s.t. } \mathfrak{M}, \pi, i+j \Vdash \psi \\
& & \text{and } \mathfrak{M}, \pi, i+k \Vdash \phi \text{ for any } k < j \\
\mathfrak{M}, \pi, i \Vdash \mathbf{A}\phi & \Leftrightarrow & \text{for any path } \rho \text{ starting at } \pi(i), \mathfrak{M}, {}^{i}\pi \otimes \rho, i \Vdash \phi \\
\mathfrak{M}, \pi, i \Vdash [!\phi]\psi & \Leftrightarrow & \mathfrak{M}^{!\phi}_{\pi(i)}, \pi, i \Vdash \psi
\end{array}
$$

$\mathbf{A}$ is a universal quantifier over possible timelines. The evaluation of some formulas does not depend on the whole path but only on the point of the path with the selected index (e.g., p, $\top$, $\mathfrak{b}$ and $\mathbf{A}\phi$). We will call a formula for which this property holds in every model a *state formula*, while a formula whose semantical interpretation depends on other points of the path (e.g., $\mathrm{Y}\phi$, $\phi\mathrm{S}\psi$, $\mathrm{X}\phi$ and $\phi\mathrm{U}\psi$) will be called a *temporal formula*. In particular, if ψ is a state formula then $[!\phi]\psi$ is a state formula too.

A path $w_0 \ldots$ is *legal* if w_i is a fine state for every $i > 0$. The update model $\mathfrak{M}^{!\phi}_{\pi(i)}$ is defined as follows.

Definition 1 (Update with commands). *Let $\mathfrak{M} = (W, R, r, B, V)$ be a model, ϕ a formula and w a state. Let $w_0 \ldots w_i$ be the history of w. Define a set $X^{!\phi}_w$ of states as follows: for any $x \in W$, $x \in X^{!\phi}_w \Leftrightarrow$ (i) x is a future state of w, (ii) x is a fine state, and (iii) there is no legal path ρ starting at w and passing by x s.t. $\mathfrak{M}, w_0 \ldots w_i \otimes \rho, i \Vdash \phi$. Let $\mathfrak{M}^{!\phi}_w = (W, R, r, B \cup X^{!\phi}_w, V)$ if $B \cup X^{!\phi}_w$ is normatively coherent, or else $\mathfrak{M}^{!\phi}_w = \mathfrak{M}$. $\mathfrak{M}^{!\phi}_w$ is called the result of updating $\mathfrak{M}$ at w with the command $!\phi$.*

Proposition 1. *Fix $\mathfrak{M} = (W, R, r, B, V)$, π, i and ϕ. $B \cup X^{!\phi}_{\pi(i)}$ is not normatively coherent $\Leftrightarrow$ $\pi(i)$ is a fine state in $\mathfrak{M}$ and there is no legal path ρ starting at $\pi(i)$ s.t. $\mathfrak{M}, {}^{i}\pi \otimes \rho, i \Vdash \phi$.*

Proof. ($\Leftarrow$) Assume that $\pi(i)$ is as in the hypothesis. Since $\pi(i)$ is not reachable from $\pi(i)$, by definition of $X^{!\phi}_{\pi(i)}$ it follows that $\pi(i) \notin B \cup X^{!\phi}_{\pi(i)}$.

Consider now a successor u of $\pi(i)$. If $u \notin B$, by hypothesis we know that there is no legal path ρ starting at $\pi(i)$ s.t. $\mathfrak{M}, {}^{i}\pi \otimes \rho, i \Vdash \phi$ and passing by u. By definition of $X^{!\phi}_{\pi(i)}$ this implies $u \in X^{!\phi}_{\pi(i)}$, and because u was arbitrary, it follows that all the successors of $\pi(i)$ lie in $B \cup X^{!\phi}_{\pi(i)}$, and so the set is not normatively coherent.

($\Rightarrow$) The result is proved by contraposition. We want to show that $B \cup X^{!\phi}_{\pi(i)}$ is normatively coherent assuming that $\pi(i)$ is a bad state or that there is a legal path ρ starting at $\pi(i)$ s.t. $\mathfrak{M}, {}^{i}\pi \otimes \rho, i \Vdash \phi$.

The former case: assume that $\pi(i)$ is a bad state in $\mathfrak{M}$. Then $X^{!\phi}_{\pi(i)} = \emptyset$ and $B \cup X^{!\phi}_{\pi(i)} = B$. Then $B \cup X^{!\phi}_{w}$ is trivially normatively coherent.

The latter case: assume that there is a legal path ρ as described above. Spelling out the definition of normative coherence, we have to show that the successors of a state in $B \cup X^{!\phi}_{\pi(i)}$ are again in $B \cup X^{!\phi}_{\pi(i)}$ (persistency of liability) and that given a state $w \notin B \cup X^{!\phi}_{\pi(i)}$ then it has a successor u s.t. $u \notin B \cup X^{!\phi}_{\pi(i)}$ (seriality of legality).

The first condition it's easily checked. If $w \in B$ then all its successors are in B because we assumed that $\mathcal{M}$ is normative coherent. Otherwise if $w \in X^{!\phi}_{\pi(i)}$, consider u s.t. Rwu; clearly we have

$$\begin{aligned} w \in X^{!\phi}_{\pi(i)} &\Rightarrow R\pi(i)w \text{ and for all } \rho \text{ starting at } \pi(i) \\ &\quad\; \text{and passing by } w\text{: } \mathcal{M}, {}^{i}\pi \otimes \rho, i \nVdash \phi \\ &\Rightarrow R\pi(i)u \text{ and for all } \rho \text{ starting at } \pi(i) \\ &\quad\; \text{and passing by } u\text{: } \mathcal{M}, {}^{i}\pi \otimes \rho, i \nVdash \phi \\ &\Rightarrow u \in B \text{ or } u \in X^{!\phi}_{\pi(i)} \\ &\Rightarrow u \in B \cup X^{!\phi}_{\pi(i)} \end{aligned}$$

For the second condition, consider $w \notin B \cup X^{!\phi}_{\pi(i)}$ and suppose that $R\pi(i)w$ (otherwise the result trivially follows). We have, by definition of $X^{!\phi}_{\pi(i)}$, that there exists a *legal* path ρ starting at $\pi(i)$ passing by w such that $\mathcal{M}, {}^{i}\pi \otimes \rho, i \Vdash \phi$. And now it easily follows that for any successor u of w in the path ρ we have $u \notin B$ (as ρ is legal) and $u \notin X^{!\phi}_{\pi(i)}$ (as the path $\pi(i) \otimes \rho$ witness this). □

The proof of this proposition is omitted due to limit of space. That there is no legal path ρ starting at $\pi(i)$ s.t. $\mathfrak{M}, \pi \otimes \rho, i \Vdash \phi$ means that making ϕ true at $\pi(i)$ is forbidden. $\mathfrak{M}^{!\phi}_{w}$ is understood as follows. Assume that the agent is at the state w and the command *make ϕ true* is given to him. If w is a fine state but it is not allowed to make ϕ true, then the agent considers the command strange and ignores it. Assume otherwise. Then the agent scans the fine states that he can reach from w one by one. He marks a state bad if he finds this: *if he travels to it, there would be no legal way to make ϕ true at w,* no matter where he goes afterwards. $X^{!\phi}_{w}$ is the collection of the states that he marks bad. After marking, the agent behaves by taking the new bad states into consideration. Figure 2 illustrates how a command updates a model.

Note that the set $X^{!\phi}_{w}$ is defined w.r.t. $\mathfrak{M}$. Updating $\mathfrak{M}$ at w with $!\phi$ only changes the future states of w.

A formula ϕ is *valid* if for any $\mathfrak{M}$, π and i, $\mathfrak{M}, \pi, i \Vdash \phi$. Let Γ be a set of formulas and ϕ a formula. $\Gamma \models \phi$, Γ *entails* ϕ, if for any $\mathfrak{M}$, π and i, if $\mathfrak{M}, \pi, i \Vdash \Gamma$, then $\mathfrak{M}, \pi, i \Vdash \phi$. We in the sequel use DTNL to denote the set of valid formulas.

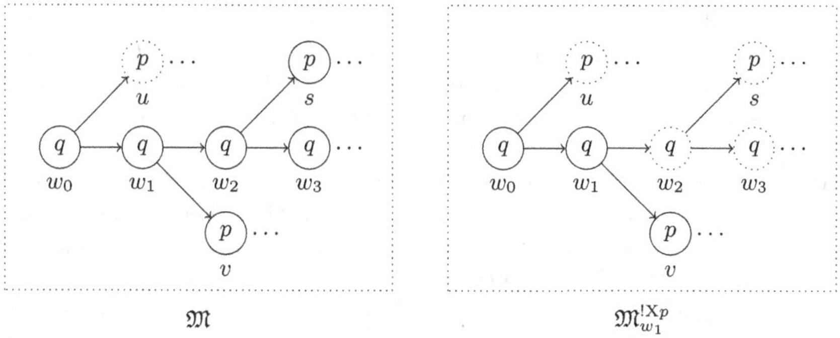

Fig. 2. This figure illustrates how a model is updated by a command. The valuation of two propositions p and q is depicted. $\mathfrak{M}_{w_1}^{!Xp}$ is the result of updating $\mathfrak{M}$ at w_1 with the command *make p true in the next moment* (!Xp).

5 A Static Deontic Logic by Reduction

Without the dynamic operator, the static part of DTNL is just PCTL* plus the propositional constant $\mathfrak{b}$. As mentioned in the introduction, there is already a complete axiomatization of PCTL* in the literature. We can get a complete axiomatization for the static part of DTNL by adding the axioms $\mathfrak{b} \rightarrow \mathbf{AX}\mathfrak{b}$ and $\mathfrak{f} \rightarrow \mathbf{EX}\mathfrak{f}$ to PCTL*. The two formulas respectively express the two constraints on the models of DTNL: persistency of liability and seriality of legality.

The formula XG$\mathfrak{f}$ indicates that from the next moment on, the state will always be fine. It expresses legal paths in the following sense: for any model $\mathfrak{M}$ and history $w_0 \dots w_i$, a path ρ is legal iff $\mathfrak{M}, w_0 \dots w_i \otimes \rho, i \Vdash \mathrm{XG}\mathfrak{f}$. Using this formula and the path quantifiers $\mathbf{A}$ and $\mathbf{E}$, we can define some deontic notions.

1. $\mathcal{P}\phi := \mathbf{E}(\mathrm{XG}\mathfrak{f} \wedge \phi)$: the agent has a legal way to act in the future s.t. ϕ is the case now. That is, the agent is *permitted* to make ϕ true now.
2. $\mathcal{F}\phi := \mathbf{A}(\mathrm{XG}\mathfrak{f} \rightarrow \neg\phi)$: no matter how the agent will legally act in the future, ϕ is not the case now. That is, the agent is *forbidden* to make ϕ true now.
3. $\mathcal{O}\phi := \mathbf{A}(\mathrm{XG}\mathfrak{f} \rightarrow \phi)$: no matter how the agent will legally act in the future, ϕ is the case now. That is, the agent is *obligated* to make ϕ true now.

The truth conditions of the normative formulas can be specified by use of legal paths.

$$\mathfrak{M}, \pi, i \Vdash \mathcal{P}\phi \Leftrightarrow \text{there is a legal path } \rho \text{ from } \pi(i) \text{ s.t. } \mathfrak{M}, {}^i\pi \otimes \rho, i \Vdash \phi$$
$$\mathfrak{M}, \pi, i \Vdash \mathcal{F}\phi \Leftrightarrow \text{there is no legal path } \rho \text{ from } \pi(i) \text{ s.t. } \mathfrak{M}, {}^i\pi \otimes \rho, i \Vdash \phi$$
$$\mathfrak{M}, \pi, i \Vdash \mathcal{O}\phi \Leftrightarrow \text{for any legal path } \rho\text{from } \pi(i), \mathfrak{M}, {}^i\pi \otimes \rho, i \Vdash \phi$$

The deontic operators can be treated as quantifiers over legal paths.

We obtained a deontic logic, namely the static part of DTNL, for which normative formulas have a temporal dimension. For example, $\mathcal{O}\mathrm{F}p$ says that the agent ought to make p true at some point in the future.

A quite different case is $\mathcal{O}p$. This formula does not mean that the agent ought to make p true, but that the agent ought to *act in the future to make p true at the present moment.* As the condition is independent from the future events, this should count as a trivial obligation, and in fact it can be verified that $\mathfrak{f} \rightarrow (p \leftrightarrow \mathcal{O}p)$ is a valid formula.

As defined above, a path is legal if it consists only of legal transitions, but there are other properties of paths expressible in DTNL that are interesting in normative contexts. One of them is *containing finitely many illegal transitions*, which is expressed by $\mathrm{FG}\mathfrak{f}$. This property can be intuitively understood as *mainly* consisting of legal transitions and is *slightly* weaker than legal paths. Using this property, we can define different normative notions in a similar fashion as the one presented above. This issue deserves a closer look.

6 Explanations

The models of DTNL are trees with bad states. Note that in trees, defining bad states or bad transitions results in equivalent semantics and there is no real difference between the two approaches. We mentioned in Sect. 1 that bad transitions in general transition systems are *historyless* and not suitable to handle normative notions with a temporal dimension. A concrete example is given in Fig. 3.

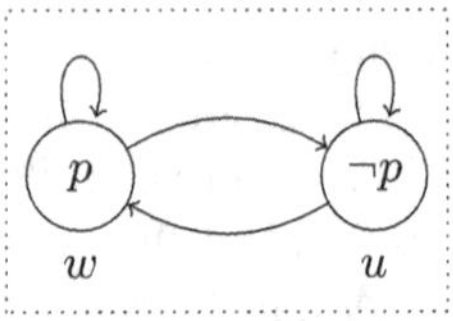

Fig. 3. The structure in this figure represents a conceivable scenario. Let $\phi = \mathcal{O}(\mathrm{X}\neg p \wedge \mathrm{XXG}p)$. This formula says what follows is obligatory: firstly make p false; then make p true; then keep p true forever. In fact, it can be checked that there is no way to arrange illegal transitions in this structure s.t. ϕ is true at w. This shows the importance of keeping track of the past actions.

Note that technically, trees with bad states are equivalent to general transition systems with *history-dependent* bad transitions. By history-dependent, we mean whether a transition is bad or not is dependent on specific histories, that is, finite transition sequences. Roughly, the arguments for the equivalence can go as follows: *point generated submodels* preserve truth; pointed transition systems with history-dependent bad transitions can be safely *unwound* to trees with bad states. We refer to [4] for the details of generated submodels and unwinding. However, if we work with transition systems with history-dependent bad transitions, the definitions, especially the definition for the update with commands, would be very complicated. Working with trees plus bad states makes things much easier.

The special constraint persistency of liability on the models of DTNL denotes that if a state is bad, then all of its successors are bad as well. We use this constraint for two reasons. One reason concerns offering state permissions such as *you may let the prisoners go today or tomorrow.* Offering permissions tends to make bad states fine. This paper focuses on giving commands and does not deal with offering permissions, but once we want to handle it in this framework, it becomes clear why the constraint is needed. Another reason is conceptual. If a command has been violated, then this fact will remain true also in the future. The constraint *persistency of liability* is coherent with this.

7 Commands Without a Deadline

The formula $[!\phi]\mathcal{O}\phi$ is not generally valid. For example, $[!\mathrm{F}p]\mathcal{O}\mathrm{F}p$ is not valid, as shown in Fig. 4. Therefore, commands do not always cause the expected effects, sometimes they *fail.* Note that the command $!\mathrm{F}p$ does not have a *deadline* and so it can not be checked if it has been violated after a finite amount of time. Actually, this failure only happens for this kind of commands.

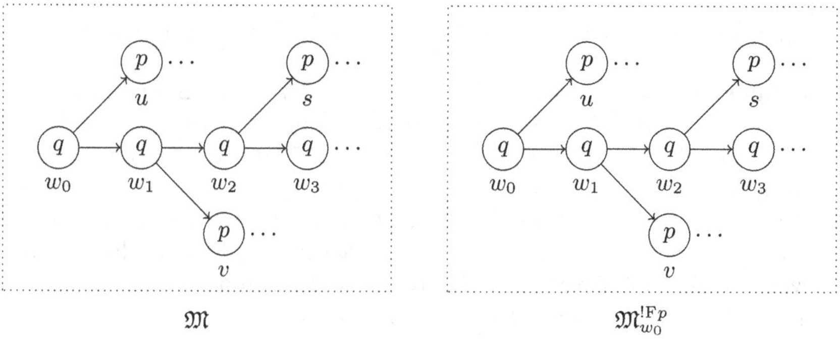

Fig. 4. This figure explains why $[!\mathrm{F}p]\mathcal{O}\mathrm{F}p$ is not valid. The model $\mathfrak{M}$ does not contain any bad state and all the paths in it are legal. The command $!\mathrm{F}p$ at w_0 in $\mathfrak{M}$ does not change anything, as no matter where the agent goes from w_0, there is a way to make $\mathrm{F}p$ true at w_0. Therefore, $w_0w_1\ldots$ is a legal path of $\mathfrak{M}^{!\mathrm{F}p}_{w_0}$. However, it can be seen that not $\mathfrak{M}^{!\mathrm{F}p}_{w_0}, w_0w_1\ldots, 0 \Vdash \mathrm{F}p$. Then not $\mathfrak{M}^{!\mathrm{F}p}_{w_0}, w_0w_1\ldots, 0 \Vdash \mathcal{O}\mathrm{F}p$, that is, not $\mathfrak{M}, w_0w_1\ldots, 0 \Vdash [!\mathrm{F}p]\mathcal{O}\mathrm{F}p$.

We say that a formula ϕ is *colorless* if it contains no occurrence of $\mathfrak{b}$. Colorless formulas are not sensitive to badness of states.

Proposition 2. *Let* ϕ *be a colorless formula. Then* $\mathfrak{M}, \pi, i \Vdash [!\phi]\mathcal{O}\phi \Leftrightarrow$ *for any path* ρ *starting at* $\pi(i)$*, if* $\mathfrak{M}, {}^{i}\pi \otimes \rho, i \Vdash \{\mathrm{X}^n\mathbf{E}\mathrm{Y}^n(\mathrm{XG}\mathfrak{f} \wedge \phi) \mid n \in \mathbb{N}\}$*, then* $\mathfrak{M}, {}^{i}\pi \otimes \rho, i \Vdash \phi$.

Proof. ($\Leftarrow$) By contraposition, assume $\mathfrak{M}, \pi, i \nVdash [!\phi]\mathcal{O}\phi$. By spelling out the semantic clauses, this means that there exists a legal path ρ of $\mathfrak{M}^{!\phi}_{\pi(i)}$ starting at $\pi(i)$ s.t. $\mathfrak{M}^{!\phi}_{\pi(i)}, {}^{i}\pi \otimes \rho, i \nVdash \phi$. It is straightforward that ρ is also a legal path of $\mathfrak{M}$ as well. Moreover, as ϕ is colorless, $\mathfrak{M}, {}^{i}\pi \otimes \rho, i \nVdash \phi$.

It remains to show that $\mathfrak{M}, {}^{i}\pi \otimes \rho, i \Vdash \mathrm{X}^n\mathbf{E}\mathrm{Y}^n(\mathrm{XG}\mathfrak{f} \wedge \phi)$ for every value of $n \in \mathbb{N}$. Fix the notation $\rho = u_0u_1 \ldots$ and $j \geq 0$. As u_j is a fine state in $\mathfrak{M}^{!\phi}_{\pi(i)}$, there exists a legal path τ in $\mathfrak{M}$ starting at $\pi(i)$ and passing by u_j s.t. $\mathfrak{M}, {}^{i}\pi \otimes \tau, i \Vdash \phi$ (notice that τ and ρ coincide up to the n-th state because our structure is a *tree*). Then clearly $\mathfrak{M}, {}^{i}\pi \otimes \tau, i \Vdash \mathrm{XG}\mathfrak{f} \wedge \phi$ and consequently $\mathfrak{M}, {}^{i}\pi \otimes \rho, i \Vdash \mathrm{X}^j\mathbf{E}\mathrm{Y}^j(\mathrm{XG}\mathfrak{f} \wedge \phi)$. Since j was arbitrary, it follows $\mathfrak{M}, {}^{i}\pi \otimes \rho, i \Vdash \{\mathrm{X}^n\mathbf{E}\mathrm{Y}^n(\mathrm{XG}\mathfrak{f} \wedge \phi) \,|\, n > 0\}$, as wanted.

($\Rightarrow$) By contraposition, assume there is a path ρ starting at $\pi(i)$ s.t. $\mathfrak{M}, {}^{i}\pi \otimes \rho, i \Vdash \{\mathrm{X}^n\mathbf{E}\mathrm{Y}^n(\mathrm{XG}\mathfrak{f} \wedge \phi) \,|\, n \in \mathbb{N}\}$ but $\mathfrak{M}, {}^{i}\pi \otimes \rho, i \nVdash \phi$. Note that, as ϕ is colorless, this is equivalent to $\mathfrak{M}^{!\phi}_{\pi(i)}, {}^{i}\pi \otimes \rho, i \nVdash \phi$. So in order to show $\mathfrak{M}, \pi, i \nVdash [!\phi]\mathcal{O}\phi$, it suffices to prove that ρ is a fine path of $\mathfrak{M}^{!\phi}_{\pi(i)}$.

Let $\rho = u_0u_1 \ldots$ and $j \geq 0$. It is easy to show, using again that ϕ is colorless, that the hypothesis $\mathfrak{M}, {}^{i}\pi \otimes \rho, i \Vdash \mathrm{X}^j\mathbf{E}\mathrm{Y}^j(\mathrm{XG}\mathfrak{f} \wedge \phi)$ implies that every u_j is a fine in $\mathfrak{M}$ and is not an element of $X^{!\phi}_{\pi(i)}$, and this means exactly that u_j is a fine state of $\mathfrak{M}^{!\phi}_{\pi(i)}$. Since j was arbitrary, it follows that ρ is a fine path, as wanted. □

A formula ϕ is *co-compact* if $\{\mathrm{X}^n\mathbf{E}\mathrm{Y}^n\phi \,|\, n \in \mathbb{N}\} \models \phi$. An example of such a formula is Gp, while Fp is not. Co-compact formulas are exactly the formulas that can be *falsified in finite steps*: if the formula is not true at a path, this is attested by an initial segment. For example, if Gp is false at π, then in order to know this, we just have to scan π up to a state in which p is false. Note that if Gp is true, we might need to scan the whole path: it is possible that the agent will never execute $!\phi$ but we are unable to attest this. In other words, if ϕ is not co-compact, then $!\phi$ *is a command without a deadline.*

Proposition 3. *Let ϕ be a colorless formula. Then $[!\phi]\mathcal{O}\phi$ is valid $\Leftrightarrow$ ϕ is co-compact.*

Proof. By Proposition 2, it suffices to show that $\{\mathrm{X}^n\mathbf{E}\mathrm{Y}^n(\mathrm{XG}\mathfrak{f} \wedge \phi) \,|\, n \in \mathbb{N}\} \models \phi$ $\Leftrightarrow \{\mathrm{X}^n\mathbf{E}\mathrm{Y}^n\phi \,|\, n \in \mathbb{N}\} \models \phi$. As $\mathrm{X}^n\mathbf{E}\mathrm{Y}^n(\mathrm{XG}\mathfrak{f} \wedge \phi)$ implies $\mathrm{X}^n\mathbf{E}\mathrm{Y}^n\phi$ for any n, the direction from right to left is trivial.

For the other direction, reasoning by contraposition, assume $\{\mathrm{X}^n\mathbf{E}\mathrm{Y}^n\phi \,|\, n \in \mathbb{N}\} \not\models \phi$. Then there is $\mathfrak{M}, \pi$ and i s.t. $\mathfrak{M}, \pi, i \Vdash \{\mathrm{X}^n\mathbf{E}\mathrm{Y}^n\phi \,|\, n \in \mathbb{N}\}$ but $\mathfrak{M}, \pi, i \nVdash \phi$. Note that, as ϕ is colorless, we can assume that *all the states in $\mathfrak{M}$ are fine*, and so it follows trivially $\mathfrak{M}, \pi, i \Vdash \{\mathrm{X}^n\mathbf{E}\mathrm{Y}^n(\mathrm{XG}\mathfrak{f} \wedge \phi) \,|\, n \in \mathbb{N}\}$, thus showing that $\{\mathrm{X}^n\mathbf{E}\mathrm{Y}^n(\mathrm{XG}\mathfrak{f} \wedge \phi) \,|\, n \in \mathbb{N}\} \not\models \phi$ as wanted. □

Something worth mentioning is that a command containing a co-compact formula might have *different deadlines in different situations* and so a bound to the length of the initial segment falsyfing the formula can not be given. An example is !Gp. These commands just have implicit deadlines. We say that a formula ϕ is *explicitly co-compact* if there is a n s.t. $X^n\mathbf{E}Y^n\phi \rightarrow \phi$ is valid. The commands containing this sort of formulas have explicit deadlines. For example, $!(Xp \vee XXp)$ has an explicit deadline, that is, two steps; if p is not true in two steps, then the command is violated.

It can be seen that past operators let us easily state the condition of $[!\phi]\mathcal{O}\phi$ being true and the definition of co-compact formulas. This is the reason that we introduce them.

8 Connections with Other Work

We mentioned in Sect. 1 the deontic logic NTL which is based on CTL and presented in [1]. The language of NTL, Φ_{NTL}, is defined as follows[1]:

$$\phi ::= p \mid \top \mid \neg\phi \mid (\phi \wedge \phi) \mid \mathcal{P}X\phi \mid \mathcal{P}(\phi U\phi) \mid \mathcal{O}X\phi \mid \mathcal{O}(\phi U\phi)$$

Note that Φ_{NTL} is a fragment of Φ_{DTNL} that contains only state formulas.

A model of NTL is a structure (W, R, η, V) where W and V are as usual, R is a serial relation on W, and η is a subset of R whose complement is serial, called the *set of illegal transitions*[2]. A path $w_0w_1\dots$ is called *legal* if $(w_i, w_{i+1}) \notin \eta$ for any i. The semantics of Φ_{NTL} is a special case of the semantics of Φ_{DTNL}, but now $\mathcal{P}$ and $\mathcal{O}$ can be regarded as an existential and universal quantifier over legal paths respectively.

Say that a formula ϕ in Φ_{DTNL} is *f-valid* if for any $\mathfrak{M}$, π and i, if $\pi(i)$ is a fine state, then $\mathfrak{M}, \pi, i \Vdash \phi$. NTL can be embedded into DTNL under the notion of f-validity.

Let $\mathfrak{M} = (W, R, \eta, V)$ be a NTL model s.t. (W, R) is a tree. Let r be the root of (W, R). Let $B = \{y \in W \mid (x, y) \in \eta \textit{ for some } x\}$ and B' be the smallest set containing B and closed under R. It can be verified that for any w, if $w \notin B'$, then there is a u s.t. Rwu and $u \notin B'$. Define $\mathfrak{M}^{\times}$ as the structure (W, R, r, B', V), which is a DTNL model.

Lemma 1. *Let $\mathfrak{M} = (W, R, \eta, V)$ be a* NTL *model s.t. (W, R) is a tree. Let w be a fine state of $\mathfrak{M}^{\times}$. Then for any $\phi \in \Phi_{\mathsf{NTL}}$, $\mathfrak{M}, w \Vdash \phi \Leftrightarrow \mathfrak{M}^{\times}, w \Vdash \phi$.*

The lemma follows immediately by noticing that for any fine state u of $\mathfrak{M}^{\times}$, a path π starting at u is legal in $\mathfrak{M}$ iff it is legal in $\mathfrak{M}^{\times}$.

Proposition 4. *For any ϕ of Φ_{NTL}, ϕ is valid in* NTL *$\Leftrightarrow$ ϕ is f-valid in* DTNL.

Proof. Assume that ϕ is not valid in NTL. Then there is an NTL model $\mathfrak{M} = (W, R, \eta, V)$ and a state w s.t. $\mathfrak{M}, w \nVdash \phi$. Let $\mathfrak{M}' = (W', R', \eta', V')$ be the *unwinding* of $\mathfrak{M}$ from w. It follows immediately that $\mathfrak{M}'$ is also an NTL model,

---

[1] In [1], every deontic operator has a parameter referring to a specific set of illegal transitions. Here we ignore the parameter, but this is not crucial for the comparison between NTL and DTNL.

[2] The models of NTL in [1] have initial states that are omitted here.

that $\mathfrak{M}', w \not\Vdash \phi$ and that w is a fine state of $(\mathfrak{M}')^\times$. By Lemma 1, $(\mathfrak{M}')^\times, w \not\Vdash \phi$ and so ϕ is not f-valid in DTNL.

Assume that ϕ is not f-valid in DTNL. Then there is a DTNL model $\mathfrak{M} = (W, r, R, B, V)$ and a fine state w s.t. $\mathfrak{M}, w \not\Vdash \phi$. Let $\mathfrak{M}' = (W, R, \eta, V)$ where $\eta = \{(x, y) \mid y \in B\}$. Then $\mathfrak{M}'$ is an NTL model and it can be seen that $(\mathfrak{M}')^\times = \mathfrak{M}$. By Lemma 1, $\mathfrak{M}', w \not\Vdash \phi$ and so ϕ is not valid in NTL. □

Standard Deontic Logic (SDL), proposed in [11], is one of the most well-known deontic logics. SDL is a typical normal modal logic, where $\Box\phi$ is interpreted as *it ought to be that* ϕ, and $\Diamond\phi$ as *it may be that* ϕ. The frames of SDL are *serial* relational structures.

As mentioned in [1], SDL is a sublogic of NTL under the translation σ defined as follows:

$$\begin{aligned} p^\sigma &= p \\ (\neg\phi)^\sigma &= \neg\phi^\sigma \\ (\phi \wedge \psi)^\sigma &= \phi^\sigma \wedge \psi^\sigma \\ (\Box\phi)^\sigma &= \mathcal{O}\mathrm{X}\phi^\sigma \\ (\Diamond\phi)^\sigma &= \mathcal{P}\mathrm{X}\phi^\sigma \quad \text{(derived)} \end{aligned}$$

Under this translation, the formulas of Φ_{SDL} have temporal reading: $\Box\phi$ means that *it ought to be that* ϕ *is true in the next moment.* Note that for any $\phi \in \Phi_{\mathsf{SDL}}$, ϕ^σ is a state formula.

Actually, this previous fact can be stronger: SDL can be embedded to NTL. As NTL can be embedded to DTNL, SDL can be embedded to DTNL too.

9 Future Work

The expressive power of DTNL needs further study. It is still not known whether the dynamic operator can be reduced: if this is the case, then the completeness of DTNL follows by the completeness of PCTL*. Another point worth mentioning is that in defining the commands with a deadline, we use the inference $\{\mathrm{X}^n\mathbf{E}\mathrm{Y}^n\phi \mid n \in \mathbb{N}\} \models \phi$. It's still not known if this entailment can be expressed by a formula of DTNL.

Offering state permissions is another important way to change the normative state of an agent, but it is harder to capture as it raises some interesting issues such as *free-choice* permission. Formalizing giving permissions is another future work.

The constraint *persistency of liability* on the models of DTNL says that if the agent has done something illegal, then there will be nothing legal for him to do. This implies that DTNL only works for ideal agents who always comply, so this logic is far from being realistic. One way to solve this issue is to introduce more shades, instead of a simple distinction between bad and fine states.

Acknowledgment. We would like to thank the audience of the Workshop of Language and Logic in China 2016, the Tsinghua-Bayreuth Logic Workshop, the LIRa seminar at University of Amsterdam, the seminar at Osnabrück University and the 3rd PIOTR project meeting. Thanks also go to Maria Aloni, Johan van Benthem, Ilaria Canavotto,

Jan van Eijck, Davide Grossi, Frank Veltman and Yanjing Wang. Fengkui Ju was supported by the National Social Science Foundation of China (No. 12CZX053), the Fundamental Research Funds for the Central Universities (No. SKZZY201304) and China Scholarship Council. Gianluca Grilletti was supported by the European Research Council (ERC) under the European Union's Horizon 2020 research and innovation programme (No. 680220).

References

1. Ågotnes, T., van der Hoek, W., Rodríguez-Aguilar, J., Sierra, C., Wooldridge, M.: A temporal logic of normative systems. In: Makinson, D., Malinowski, J., Wansing, H. (eds.) Towards Mathematical Philosophy, pp. 69–106. Springer, Netherlands (2009)
2. Aloni, M.: Free choice, modals, and imperatives. Nat. Lang. Seman. **15**(1), 65–94 (2007)
3. Anderson, A.: Some nasty problems in the formal logic of ethics. Noûs **1**(4), 345–360 (1967)
4. Blackburn, P., de Rijke, M., Venema, Y.: Modal Logic. Cambridge University Press, Cambridge (2002)
5. Clarke, E.M., Emerson, E.A.: Design and synthesis of synchronization skeletons using branching time temporal logic. In: Kozen, D. (ed.) Logic of Programs 1981. LNCS, vol. 131, pp. 52–71. Springer, Heidelberg (1982). doi:10.1007/BFb0025774
6. Dignum, F., Broersen, J., Dignum, V., Meyer, J.-J.: Meeting the deadline: why, when and how. In: Hinchey, M.G., Rash, J.L., Truszkowski, W.F., Rouff, C.A. (eds.) FAABS 2004. LNCS, vol. 3228, pp. 30–40. Springer, Heidelberg (2004). doi:10.1007/978-3-540-30960-4_3
7. Emerson, E., Halpern, J.: "Sometimes" and "not never" revisited: on branching versus linear time temporal logic. J. ACM **33**(1), 151–178 (1986)
8. Kanger, S.: New foundations for ethical theory. In: Hilpinen, R. (ed.) Deontic Logic: Introductory and Systematic Readings, pp. 36–58. Springer, Netherlands (1971)
9. Reynolds, M.: An axiomatization of PCTL*. Inf. Comput. **201**(1), 72–119 (2005)
10. Veltman, F.: Imperatives at the borderline of semantics and pragmatics. Manuscript (2010)
11. von Wright, G.: Deontic logic. Mind **60**(237), 1–15 (1951)
12. Zanuttini, R., Pak, M., Portner, P.: A syntactic analysis of interpretive restrictions on imperative, promissive, and exhortative subjects. Nat. Lang. Linguist. Theory **30**(4), 1231–1274 (2012)

Labelled Sequent Calculus for Inquisitive Logic

Jinsheng Chen(✉) and Minghui Ma

Institute of Logic and Cognition, Sun Yat-Sen University, Guangzhou, China
jinsheng.chen@foxmail.com, mamh6@mail.sysu.edu.cn

Abstract. A contraction-free and cut-free labelled sequent calculus GInqL for inquisitive logic is established. Labels are defined by a set-theoretic syntax. The completeness of GInqL is shown by the equivalence between the Hilbert-style axiomatic system and sequent system.

Keywords: Inquisitive logic · Labelled sequent calculus · Cut-elimination

1 Introduction

The natural language meaning processing has been more conversational and dynamic since Stalnaker [9] proposed to treat the meaning of a sentence to be its potential to update the common ground (shared information) of conversational participants. Stalnaker's approach is dynamic in the sense that a sentence provides information to eliminate possibilities from the common ground of participants. This kind of pragmatic theory is generalized to inquisitive semantics in Groenendijk [3] and Mascarenhas [4], which characterizes the assertions and questions in conversations by informative and inquisitive semantic terms. The associated logic was axiomatized by Mascarenhas [4], and a sound and complete tree sequent calculus was independently established by Sano [8].

Ciardelli and Roelofsen [1] propose a sound and complete Hilbert-style axiomatic system InqL for the generalized inquisitive semantics. It is the extension of the intermediate logic KP with the axiom $\neg\neg p \to p$ where only propositional variables are allowed to substitute for the variable p in the axiom. The set of all theorems of the logic InqL is hence not closed under uniform substitution. The semantics for InqL is not based on the Kripke semantics for the intermediate logic KP but given in a fixed set of states. The basic semantic notion is the support relation between a state and a formula.

The labelled sequent calculi for normal modal logics are systematically established by Negri [5] as an uniform approach to the proof analysis of modal logics. The basic idea is that the Kripke semantics of a modal operator is transformed into sequent rules. The meaning of a modal operator is written into the right introduction rule of that operator, and the inversion principle provides the left

M. Ma—The work is supported by Chinese national foundation of social sciences (grant no. 16CZX049).

A. Baltag et al. (Eds.): LORI 2017, LNCS 10455, pp. 526–540, 2017.
DOI: 10.1007/978-3-662-55665-8_36

introduction rule. This is a very uniform approach to the proof theory of normal modal logics. Moreover, Dyckhoff and Negri [2] extend the labelled approach to intermediate logics, where the intermediate logic KP was not investigated.

Given the inquisitive semantics of propositional logic, the present paper shall extend the labelled approach to the inquisitive logic Inq. A contraction-free and cut-free labelled sequent calculus GInqL is established. Compared with labelled sequent calculi for normal modal logics and intermediate logics, there is a kind of complexity about labels used in the labelled sequent calculus GInqL. This is partially due to the validity of the special axiom $\neg\neg p \to \neg p$, and partially due to the semantics of implication. The present paper starts from an introduction to inquisitive logic in Sect. 2. The labelled sequent calculus GInqL is established in Sect. 3. Section 4 proves the admissibility of structural rules, and finally the soundness and completeness of GInqL is shown in Sect. 5.

2 Inquisitive Logic

Let $\mathcal{P}$ be a denumerable set of propositional variables. The set of all formulas $\mathcal{L}$ is defined inductively by

$$\mathcal{L} \ni \alpha ::= p \mid \bot \mid (\alpha \wedge \beta) \mid (\alpha \vee \beta) \mid (\alpha \to \beta), \text{ where } p \in \mathcal{P}.$$

Define $\top := \bot \to \bot$ and $\neg\alpha := \alpha \to \bot$. The *length* of a formula is the number of connectives occuring in it.

An *index*, denoted by w, u, v etc., is a subset of $\mathcal{P}$. A *state*, denoted by s, t etc., is a set of indices. The empty set $\emptyset$ is called the *inconsistent state*. The set of all states is denoted by $S_{\mathcal{P}}$.

Definition 1. *The* support relation *between a state s and a formula α, notation $s \models \alpha$, is defined inductively as follows:*

(1) *$s \models p$ iff $p \in X$ for all $X \in s$.*
(2) *$s \models \bot$ iff $s = \emptyset$.*
(3) *$s \models \alpha \wedge \beta$ iff $s \models \alpha$ and $s \models \beta$.*
(4) *$s \models \alpha \vee \beta$ iff $s \models \alpha$ or $s \models \beta$.*
(5) *$s \models \alpha \to \beta$ iff for every state $t \subseteq s$, if $t \models \alpha$, then $t \models \beta$.*

A formula α is valid, *notation $\models \alpha$, if $s \models \alpha$ for every state $s \in S_{\mathcal{P}}$.*

Lemma 1. *For every formula α and states s, t, the following hold:*

(1) *$\emptyset \models \alpha$.*
(2) *if $s \models \alpha$, then $t \models \alpha$ for every $t \subseteq s$.*
(3) *if $s \models \neg\alpha$ and $t \models \neg\alpha$, then $s \cup t \models \neg\alpha$.*

Proof. The condition (1) is obvious by definition. The persistency condition (2) is shown in [1]. For (3), assume $s \models \neg\alpha$ and $t \models \neg\alpha$. Let $t' \subseteq s \cup t$ and $t' \models \alpha$. Then $t' = t' \cap (s \cup t) = (t' \cap s) \cup (t' \cap t)$. By (2), $t' \cap s \models \alpha$ and $t' \cap t \models \alpha$. By assumption, $t' \cap s = \emptyset = t' \cap t$. Hence $t' = \emptyset$. □

Definition 2. *The Hilbert-style axiomatic system* HInqL *for inquisitive logic consists of the following axiom schemata and inference rules:*

- *Axiom schemata:*
  (A1) $\alpha \to (\beta \to \alpha)$
  (A2) $(\alpha \to (\beta \to \gamma)) \to ((\alpha \to \beta) \to (\alpha \to \gamma))$
  (A3) $\alpha \wedge \beta \to \alpha$
  (A4) $\alpha \wedge \beta \to \beta$
  (A5) $\alpha \to \alpha \vee \beta$
  (A6) $\beta \to \alpha \vee \beta$
  (A7) $(\alpha \to \gamma) \to ((\beta \to \gamma) \to (\alpha \vee \beta \to \gamma))$
  (A8) $\alpha \to (\beta \to \alpha \wedge \beta)$
  (A9) $\bot \to \alpha$
  (KP) $(\neg\alpha \to \beta \vee \gamma) \to (\neg\alpha \to \beta) \vee (\neg\alpha \to \gamma)$
  (N) $\neg\neg p \to p$, *for every propositional variable* $p \in \mathcal{P}$.
- *Inference rule (MP): from* α *and* $\alpha \to \beta$ *infer* β.

By $\vdash_{\mathsf{HInqL}} \alpha$ *we mean that* α *is provable in the system* HInqL.

Theorem 1 ([1]). *For any formula* α, $\vdash_{\mathsf{HInqL}} \alpha$ *if and only if* $\models \alpha$.

3 A Labelled Sequent Calculus

Let $\mathcal{A} = \{a_i \mid i \in \mathbb{N}\}$ be a set of variables ranging over the set of all singleton states. Let $\mathcal{V} = \{x_i \mid i \in \mathbb{N}\}$ be a set of variables ranging over the set of all states $S_{\mathcal{P}}$. We use a, b, c etc. as metavariables for variables in $\mathcal{A}$, and x, y, z etc. as metavariables for $\mathcal{V}$. In particular, let ϵ be a constant symbol standing for the state $\emptyset$. Moreover, let $\cup$ and $\cap$ be set-theoretic symbols of union and intersection.

Definition 3. *The set of* labels *is defined inductively by the following rule:*

$$L \ni s ::= \epsilon \mid a \mid x \mid (s \cup s) \mid (s \cap s),$$

where $a \in \mathcal{A}$ *and* $x \in \mathcal{V}$. *Let* R *be a binary relational symbol. A* relational atom *is an expression of the form* sRt *where* $s, t \in L$.

A *labelled formula* is of the form $s : \alpha$ where α is a formula and s is a label. A *term* is a labelled formula or a relational atom. We use A, B, C etc. to denote terms. A *sequent* is an expression of the form $\Gamma \Rightarrow \Delta$ where Γ and Δ are finite (possibly empty) multisets of terms.

Definition 4. *The labelled sequent calculus* GInqL *consists of the following axioms and rules:*

(1) *Axioms:*

$$(Id_{\mathcal{P}})\ a : p, \Gamma \Rightarrow \Delta, a : p \quad (Id_R)\ sRt, \Gamma \Rightarrow \Delta, sRt \quad (Id_\bot)\ s : \bot, \Gamma \Rightarrow \Delta, s : \bot$$

(2) *Rules:*

$$(L\wedge)\ \frac{s:\alpha, s:\beta, \Gamma \Rightarrow \Delta}{s:\alpha\wedge\beta, \Gamma \Rightarrow \Delta} \qquad (R\wedge)\ \frac{\Gamma \Rightarrow \Delta, s:\alpha \quad \Gamma \Rightarrow \Delta, s:\beta}{\Gamma \Rightarrow \Delta, s:\alpha\wedge\beta}$$

$$(L\vee)\ \frac{s:\alpha, \Gamma \Rightarrow \Delta \quad s:\beta, \Gamma \Rightarrow \Delta}{s:\alpha\vee\beta, \Gamma \Rightarrow \Delta} \qquad (R\vee)\ \frac{\Gamma \Rightarrow \Delta, s:\alpha, s:\beta}{\Gamma \Rightarrow \Delta, s:\alpha\vee\beta}$$

$$(L_p)\ \frac{a:p, s:p, sRa, \Gamma \Rightarrow \Delta}{\Gamma, s:p, sRa \Rightarrow \Delta} \qquad (R_p)\ \frac{sRa, \Gamma \Rightarrow \Delta, a:p}{\Gamma \Rightarrow \Delta, s:p}$$

$$(L{\rightarrow})\ \frac{sRt, s:\alpha\rightarrow\beta, \Gamma \Rightarrow \Delta, t:\alpha \quad t:\beta, sRt, s:\alpha\rightarrow\beta, \Gamma \Rightarrow \Delta}{sRt, s:\alpha\rightarrow\beta, \Gamma \Rightarrow \Delta}$$

$$(R{\rightarrow})\frac{sRx, x:\alpha, \Gamma \Rightarrow \Delta, x:\beta}{\Gamma \Rightarrow \Delta, s:\alpha\rightarrow\beta} \qquad (tran)\ \frac{sRt, sRu, uRt, \Gamma \Rightarrow \Delta}{sRu, uRt, \Gamma \Rightarrow \Delta}$$

$$({}^{\cup}R)\ \frac{(s\cup t)Rs, \Gamma \Rightarrow \Delta}{\Gamma \Rightarrow \Delta} \qquad (R^{\cup})\ \frac{sR(u\cup v), sRu, sRv, \Gamma \Rightarrow \Delta}{sRu, sRv, \Gamma \Rightarrow \Delta}$$

$$(\bot_1)\frac{}{a:\bot, \Gamma \Rightarrow \Delta} \qquad (\bot_2)\frac{\epsilon Rt, \Gamma \Rightarrow \Delta \quad \Gamma \Rightarrow \Delta, t:\bot}{\Gamma \Rightarrow \Delta} \qquad (\bot_3)\frac{\epsilon Rt, t:\alpha, \Gamma \Rightarrow \Delta}{\epsilon Rt, \Gamma \Rightarrow \Delta}$$

$$(\bot_4)\frac{\Gamma \Rightarrow \Delta, \epsilon Rt, t:\bot}{\Gamma \Rightarrow \Delta, t:\bot} \qquad (\bot_5)\frac{\Gamma \Rightarrow \Delta, \epsilon Rt, t:\bot}{\Gamma \Rightarrow \Delta, \epsilon Rt}$$

$$(S)\frac{aRt, t:\bot, \Gamma \Rightarrow \Delta \quad aRt, tRa, \Gamma \Rightarrow \Delta}{aRt, \Gamma \Rightarrow \Delta}$$

$$(\cap_1)\frac{sR(s\cap t), \Gamma \Rightarrow \Delta}{\Gamma \Rightarrow \Delta} \qquad (LE)\frac{sRs, \Gamma \Rightarrow \Delta}{\Gamma \Rightarrow \Delta} \qquad (RE)\frac{\Gamma \Rightarrow \Delta, \epsilon Ra}{\Gamma \Rightarrow \Delta}$$

$$(R_1)\frac{\Gamma \Rightarrow \Delta, \epsilon Rs, \epsilon R(s\cup t) \quad \Gamma \Rightarrow \Delta, \epsilon Rt, \epsilon R(s\cup t)}{\Gamma \Rightarrow \Delta, \epsilon R(s\cup t)}$$

$$(R_2)\frac{(t_1\cup t_2)Rt, \Gamma \Rightarrow \epsilon R((t_1\cap t)\cup(t_2\cap t)), \epsilon Rt, \Delta}{(t_1\cup t_2)Rt, \Gamma \Rightarrow \epsilon Rt, \Delta}$$

In the rule ($\mathrm{R_p}$)*, $a \in A$ does not occur in the conclusion. In the rule* ($R \rightarrow$)*, $x \in V$ does not occur in the conclusion.*

A labelled formula $s : \alpha$ in the lower sequent of a rule in GInqL is called *principal.* The *height* of a derivation in GInqL is the greatest number of successive applications of rules in it, where an axiom has height 0. By GInqL $\vdash_k \Gamma \Rightarrow \Delta$ we mean that $\Gamma \Rightarrow \Delta$ is derivable with a height of derivation at most k. We say a sequent rule

$$\frac{\Gamma_1 \Rightarrow \Delta_1 \quad \ldots \quad \Gamma_n \Rightarrow \Delta_n}{\Gamma_0 \Rightarrow \Delta_0}\,(\mathrm{R})$$

is *height-preserving admissible* in a sequent calculus if the conclusion $\Gamma_0 \Rightarrow \Delta_0$ has a derivation with the height not larger than the maximal height of derivations of premisses. We say that (R) is *height-preserving invertible* if GInqL $\vdash_k \Gamma_0 \Rightarrow \Delta_0$ implies GInqL $\vdash_k \Gamma_i \Rightarrow \Delta_i$ for all $1 \leq i \leq n$.

Lemma 2. *The following rules are admissible in* GInqL*:*

(1) *The rule* $(\cup\bot)$*: if* $\vdash_{\mathsf{GInqL}} \Gamma \Rightarrow \Delta, \epsilon Rs, \epsilon R(s \cup t), s : \bot, (s \cup t) : \bot, (s \cup t) : \bot$ *and* $\vdash_{\mathsf{GInqL}} \Gamma \Rightarrow \Delta, \epsilon Rt, \epsilon R(s \cup t), t : \bot, (s \cup t) : \bot, (s \cup t) : \bot$*, then* $\vdash_{\mathsf{GInqL}} \Gamma \Rightarrow \Delta, (s \cup t) : \bot$.
(2) *The rule* $(\cup\cap)$*: if* $\vdash_{\mathsf{GInqL}} (t_1 \cup t_2)Rt, \Gamma \Rightarrow \epsilon R((t_1 \cap t) \cup (t_2 \cap t)), \epsilon Rt, ((t_1 \cap t) \cup (t_2 \cap t)) : \bot, t : \bot, t : \bot, \Delta$*, then* $\vdash_{\mathsf{GInqL}} (t_1 \cup t_2)Rt, \Gamma \Rightarrow t : \bot, \Delta$.

Proof. For $(\cup\bot)$, assume $\vdash_{\mathsf{GInqL}} \Gamma \Rightarrow \Delta, \epsilon Rs, \epsilon R(s \cup t), s : \bot, (s \cup t) : \bot, (s \cup t) : \bot$. By two applications of $(\bot_5)$, we have $\Gamma \Rightarrow \Delta, \epsilon Rs, \epsilon R(s \cup t), (s \cup t) : \bot$. Similarly, we get $\Gamma \Rightarrow \Delta, \epsilon Rt, \epsilon R(s \cup t), (s \cup t) : \bot$. By (R_1), we have $\Gamma \Rightarrow \Delta, \epsilon(s \cup t), (s \cup t) : \bot$. By $(\bot_4)$, we have $\Gamma \Rightarrow \Delta, (s \cup t) : \bot$.

For $(\cup\cap)$, assume $\vdash_{\mathsf{GInqL}} (t_1 \cup t_2)Rt, \Gamma \Rightarrow \epsilon R((t_1 \cap t) \cup (t_2 \cap t)), \epsilon Rt, ((t_1 \cap t) \cup (t_2 \cap t)) : \bot, t : \bot, t : \bot, \Delta$. By two applications of $(\bot_5)$, we have $(t_1 \cup t_2)Rt, \Gamma \Rightarrow \epsilon R((t_1 \cap t) \cup (t_2 \cap t)), \epsilon Rt, t : \bot, \Delta$. By (R_2), we have $(t_1 \cup t_2)Rt, \Gamma \Rightarrow \epsilon Rt, t : \bot, \Delta$. By $(\bot_4)$, we have $(t_1 \cup t_2)Rt, \Gamma \Rightarrow t : \bot, \Delta$. □

Proposition 1. *The following sequents are derivable in* GInqL*:*

(1) $s : \alpha, \Gamma \Rightarrow \Delta, s : \alpha$.
(2) $sRt, s : \alpha, \Gamma \Rightarrow \Delta, t : \alpha$.
(3) $sRt, s : \alpha \to \beta, t : \alpha, \Gamma \Rightarrow \Delta, t : \beta$.
(4) $s : \bot, \Gamma \Rightarrow \Delta, s : \alpha$.
(5) $t_1 : \neg\alpha, t_2 : \neg\alpha, \Gamma \Rightarrow \Delta, t_1 \cup t_2 : \neg\alpha$.

Proof. We sketch only the proofs of (1), (2) and (5), and the proofs of (3) and (4) are similar. For (1), we show it by induction on the length of α. Consider the case $\alpha = p \in \mathcal{P}$. By (Idp), we have $a : p, sRa, s : p, \Gamma \Rightarrow \Delta, a : p$. By $(\mathrm{L_p})$, we have $sRa, s : p, \Gamma \Rightarrow \Delta, a : p$. By (R_p), we have $s : p, \Gamma \Rightarrow \Delta, s : p$.

The case $\alpha = \bot$ follows from $(Id_{\bot_1})$.

The cases for $\vee$ and $\wedge$ are easily done. Suppose $\alpha = \beta \to \gamma$. By inductive hypothesis, we have $x : \beta, sRx, s : \beta \to \gamma, \Gamma \Rightarrow \Delta, x : \beta, x : \gamma$ and $x : \gamma, x : \beta, sRx, s : \beta \to \gamma, \Gamma \Rightarrow \Delta, x : \gamma$. Then by $(L \to)$, we have $x : \beta, sRx, s : \beta \to \gamma, \Gamma \Rightarrow \Delta, x : \gamma$. By $(R \to)$, we have $s : \beta \to \gamma, \Gamma \Rightarrow \Delta, s : \beta \to \gamma$.

(2) By induction on the length of α. Consider the case $\alpha = p \in \mathcal{P}$. From (1), we have $a : p, sRt, tRa, sRa, s : p, \Gamma \Rightarrow \Delta, a : p$. By $(\mathrm{L_P})$, we have $sRt, tRa, sRa, s : p, \Gamma \Rightarrow \Delta, a : p$. By (tran), we have $sRt, tRa, s : p, \Gamma \Rightarrow \Delta, a : p$. By $(\mathrm{R_P})$, we have $sRt, s : p, \Gamma \Rightarrow \Delta, t : p$.

Consider the case $\alpha = \bot$. By (Id_R), we have $\epsilon Rt, \Gamma, \epsilon Rs, sRt, s : \bot \Rightarrow \Delta, \epsilon Rt, t : \bot$. By $(tran)$, we have $\Gamma, \epsilon Rs, sRt, s : \bot \Rightarrow \Delta, \epsilon Rt, t : \bot$. By (1), we have $\Gamma, \epsilon Rs, sRt, s : \bot \Rightarrow \Delta, \epsilon Rt, t : \bot, s : \bot$. Applying $(\bot_2)$ to the last two sequents, we have $\Gamma, sRt, s : \bot \Rightarrow \Delta, \epsilon Rt, t : \bot$. By $(\bot_4)$, we have $\Gamma, sRt, s : \bot \Rightarrow \Delta, t : \bot$.

The cases for $\vee$ and $\wedge$ are easily done. Suppose $\alpha = \beta \to \gamma$. From (1), we have $\Gamma, sRt, tRx, sRx, s : \beta \to \gamma, x : \beta \Rightarrow x : \gamma, \Delta, x : \beta$ and $x : \gamma, \Gamma, sRt, tRx, sRx, s : \beta \to \gamma, x : \beta \Rightarrow x : \gamma, \Delta$. By $(L \to)$, we have $\Gamma, sRt, tRx, sRx, s : \beta \to \gamma, x : \beta \Rightarrow$

$x : \gamma, \Delta$. By $(tran)$, we have $\Gamma, sRt, tRx, s : \beta \to \gamma, x : \beta \Rightarrow x : \gamma, \Delta$. By $(R \to)$, we have $\Gamma, sRt, s : \beta \to \gamma \Rightarrow t : \beta \to \gamma, \Delta$.

(5) From (2), we have $t_1 : \neg\alpha, t_2 : \neg\alpha, t_1R(t_1 \cap x), xR(x \cap t_1), x : \alpha, \Gamma' \Rightarrow \Delta', (t_1 \cap x) : \alpha, (t_1 \cap x) : \bot$. Then by $(\cap_1)$, we have $t_1 : \neg\alpha, t_2 : \neg\alpha, t_1R(t_1 \cap x), x : \alpha, \Gamma' \Rightarrow \Delta', (t_1 \cap x) : \alpha, (t_1 \cap x) : \bot$. From (4), we have $t_1 : \neg\alpha, t_2 : \neg\alpha, t_1R(t_1 \cap x), (t_1 \cap x) : \bot, x : \alpha, \Gamma' \Rightarrow \Delta', (t_1 \cap x) : \bot$. Applying $(L \to)$ to the last two sequents for $t_1 : \neg\alpha$ and $t_1R(t_1 \cap x)$, we have $t_1 : \neg\alpha, t_2 : \neg\alpha, t_1R(t_1 \cap x), x : \alpha, \Gamma' \Rightarrow \Delta', (t_1 \cap x) : \bot$. Then by $(\cap_1)$, we have $t_1 : \neg\alpha, t_2 : \neg\alpha, x : \alpha, \Gamma' \Rightarrow (t_1 \cap x) : \bot, \Delta'$.

Let $\Delta' = \{\epsilon R(t_1 \cap x), \epsilon R((t_1 \cap x) \cup (t_2 \cap x)), ((t_1 \cap x) \cup (t_2 \cap x)) : \bot, ((t_1 \cap x) \cup (t_2 \cap x)) : \bot, \Delta''\}$, we have $t_1 : \neg\alpha, t_2 : \neg\alpha, x : \alpha, \Gamma' \Rightarrow (t_1 \cap x) : \bot, \epsilon R(t_1 \cap x), \epsilon R((t_1 \cap x) \cup (t_2 \cap x)), ((t_1 \cap x) \cup (t_2 \cap x)) : \bot, ((t_1 \cap x) \cup (t_2 \cap x)) : \bot, \Delta''$. Similarly, starting from $t_1 : \neg\alpha, t_2 : \neg\alpha, t_2R(t_2 \cap x), xR(x \cap t_2), x : \alpha, \Gamma' \Rightarrow \Delta', (t_2 \cap x) : \alpha, (t_2 \cap x) : \bot$, we have $t_1 : \neg\alpha, t_2 : \neg\alpha, x : \alpha, \Gamma' \Rightarrow (t_2 \cap x) : \bot, \epsilon R(t_2 \cap x), \epsilon R((t_1 \cap x) \cup (t_2 \cap x)), ((t_1 \cap x) \cup (t_2 \cap x)) : \bot, ((t_1 \cap x) \cup (t_2 \cap x)) : \bot, \Delta''$. Applying Lemma 2 $(\cup\bot)$, we have $t_1 : \neg\alpha, t_2 : \neg\alpha, x : \alpha, \Gamma' \Rightarrow ((t_1 \cap x) \cup (t_2 \cap x)) : \bot, \Delta''$. Let $\Gamma' = \{(t_1 \cup t_2)Rx, \Gamma\}$ and $\Delta'' = \{\epsilon R((t_1 \cap x) \cup (t_2 \cap x)), \epsilon Rx, x : \bot, x : \bot, \Delta\}$ and apply Lemma 2$(\cup\cap)$, we have $(t_1 \cup t_2)Rx, t_1 : \neg\alpha, t_2 : \neg\alpha, \Gamma, x : \alpha \Rightarrow \Delta, x : \bot$. Finally by $(R{\to})$, we have $t_1 : \neg\alpha, t_2 : \neg\alpha, \Gamma \Rightarrow \Delta, t_1 \cup t_2 : \neg\alpha$. □

Proposition 2. *For any axiom α in* HInqL, *$\Rightarrow s : \alpha$ is derivable in* GInqL.

Proof. We sketch only the proof of (KP) and (N), and the remaining axioms are shown similarly. For (KP), by Proposition 1(2) for $(x_2 \cup x_3)Rx_3$, we have $sRx_1, x_1Rx_2, x_1Rx_3, x_1R(x_2 \cup x_3), (x_2 \cup x_3)Rx_3, x_1 : \neg\alpha \to \beta \vee \gamma, x_2 : \neg\alpha, x_3 : \neg\alpha, x_2 \cup x_3 : \gamma \Rightarrow x_2 : \beta, x_3 : \gamma$. By $(^{\cup}R)$, we have $sRx_1, x_1Rx_2, x_1Rx_3, x_1R(x_2 \cup x_3), x_1 : \neg\alpha \to \beta \vee \gamma, x_2 : \neg\alpha, x_3 : \neg\alpha, x_2 \cup x_3 : \gamma \Rightarrow x_2 : \beta, x_3 : \gamma$. Similarly, we have $sRx_1, x_1Rx_2, x_1Rx_3, x_1R(x_2 \cup x_3), x_1 : \neg\alpha \to \beta \vee \gamma, x_2 : \neg\alpha, x_3 : \neg\alpha, x_2 \cup x_3 : \beta \Rightarrow x_2 : \beta, x_3 : \gamma$. Applying $(L\vee)$ to above two sequents, we have $sRx_1, x_1Rx_2, x_1Rx_3, x_1R(x_2 \cup x_3), x_1 : \neg\alpha \to \beta \vee \gamma, x_2 : \neg\alpha, x_3 : \neg\alpha, x_2 \cup x_3 : \beta \vee \gamma \Rightarrow x_2 : \beta, x_3 : \gamma$. By Proposition 1(5), we have $sRx_1, x_1Rx_2, x_1Rx_3, x_1R(x_2 \cup x_3), x_1 : \neg\alpha \to \beta \vee \gamma, x_2 : \neg\alpha, x_3 : \neg\alpha \Rightarrow x_2 : \beta, x_3 : \gamma, x_2 \cup x_3 : \neg\alpha$. Taking this to be the left premiss and the above one as the right premiss, by $(L \to)$ for $x_1R(x_2 \cup x_3)$ and $x_1 : \neg\alpha \to \beta \vee \gamma$, we have: $sRx_1, x_1Rx_2, x_1Rx_3, x_1R(x_2 \cup x_3), x_1 : \neg\alpha \to \beta \vee \gamma, x_2 : \neg\alpha, x_3 : \neg\alpha \Rightarrow x_2 : \beta, x_3 : \gamma$. Applying $(R^{\cup})$ and $(R{\to})$ two times, and then $(R\vee)$ and $(R{\to})$, we have the desired result.

(N) By Proposition 1(1), we have $sRx, xRa, aRy, x : \neg\neg p, y : p, y : \bot \Rightarrow a : p, y : \bot$. By (Id_R), we have $a : p, yRa, sRx, xRa, aRy, x : \neg\neg p, y : p \Rightarrow a : p, y : \bot$. Then by (L_p), we have $yRa, sRx, xRa, aRy, x : \neg\neg p, y : p \Rightarrow a : p, y : \bot$. Applying (S) to these two sequents, we have $sRx, xRa, aRy, x : \neg\neg p, y : p \Rightarrow a : p, y : \bot$. Then by $(R{\to})$, we have $sRx, xRa, x : \neg\neg p \Rightarrow a : p, a : p \to \bot$. By $(\bot_1)$, we have $a : \bot, sRx, xRa, x : \neg\neg p \Rightarrow a : p$. By $(L \to)$, we have $xRa, sRx, x : \neg\neg p \Rightarrow a : p$. By (R_p), we have $sRx, x : \neg\neg p \Rightarrow x : p$. By $(R \to)$, we have $\Rightarrow s : \neg\neg p \to p$. □

4 Admissibility of Structural Rules

In this section, we shall show the height-preserving admissibility of substitution, weakening and contraction rules, and the admissibility of the cut rule.

Definition 5 (Label substitution). *Suppose labels $s, t \in \mathcal{A}$ or [$s, t \neq \epsilon$ and $s, t \notin \mathcal{A}$]. For every term A, define $A[s/t]$ inductively as follows:*

$$
\begin{aligned}
sRt[s'/t'] &= sRt \text{ if } t' \notin \{s,t\}; & sRt[s'/s] &= s'Rt \text{ if } s \neq t;\\
sRt[s'/t] &= sRs' \text{ if } s \neq t; & sRs[s'/s] &= s'Rs';\\
(s:\alpha)[s'/t] &= s:\alpha \text{ if } s \neq t; & (s:\alpha)[s'/s] &= s':\alpha.
\end{aligned}
$$

For any finite multiset of terms Γ, let $\Gamma[s/t] = \{A[s/t] \mid A \in \Gamma\}$.

Lemma 3. *The following rule of substitution*

$$\frac{\Gamma \Rightarrow \Delta}{\Gamma[t/s] \Rightarrow \Delta[t/s]}\,(sub)$$

is height-preserving admissible in GInqL.

Proof. By induction on the height of the derivation of $\Gamma \Rightarrow \Delta$ in GInqL. Suppose that $n = 0$ and $[t/s]$ is not a vacuous substitution. Then $\Gamma \Rightarrow \Delta$ is an axiom or the conclusion of $(\bot_1)$. Obviously, $\Gamma[t/s] \Rightarrow \Delta[t/s]$ is also an axiom or the conclusion of $(\bot_1)$. Suppose $n > 0$ and the substitution is not vacuous. We have the following cases:

Case 1. The last rule is $(L\wedge)$, $(R\wedge)$, $(L\vee)$, $(R\vee)$, $(L\!\to)$, (L_p), $({}^{\cup}R)$, $(R^{\cup})$, (tran), $(\bot_2)$, $(\bot_3)$, $(\bot_4)$,$(\bot_5)$ (S), $(\cap_1)$, (R_1), (R_2), (LE) or (RE). These cases are similar. First apply the (sub) to the premisses and then apply the rule. We specify only the case for $(L\wedge)$ where the derivation ends with

$$\frac{\vdash_{n-1} s:\alpha, s:\beta, \Gamma' \Rightarrow \Delta}{\vdash_n s:\alpha\wedge\beta, \Gamma' \Rightarrow \Delta}\,(L\wedge)$$

By the induction hypothesis, we have $\vdash_{n-1} t:\alpha, t:\beta, \Gamma'[t/s] \Rightarrow \Delta[t/s]$. By $(L\wedge)$, one gets $\vdash_n t:\alpha\wedge\beta, \Gamma'[t/s] \Rightarrow \Delta[t/s]$.

Case 2. The last rule is $(R\!\to)$ or (R_p). These two cases are similar and we specify only $(R\!\to)$. Let the derivation be

$$\frac{\vdash_{n-1} sRt', t':\alpha, \Gamma \Rightarrow \Delta', t':\beta}{\vdash_n \Gamma \Rightarrow \Delta', s:\alpha\to\beta}\,(R\!\to)$$

where $t' \neq s$ and t' is a state variable not in the conclusion. By the induction hypothesis, we obtain the following derivation:

$$\frac{\vdash_{n-1} tRt', t':\alpha, \Gamma[t/s] \Rightarrow \Delta'[t/s], t':\beta}{\vdash_n \Gamma[t/s] \Rightarrow \Delta'[t/s], t:\alpha\to\beta}\,(R\!\to)$$

If t is the eigenvariable, the derivation ends with

$$\frac{\vdash_{n-1} sRt, t : \alpha, \Gamma \Rightarrow \Delta', t : \beta}{\vdash_n \Gamma \Rightarrow \Delta', s : \alpha \to \beta} (R\to)$$

By the induction hypothesis, we have $\vdash_{n-1} sRu, u : \alpha, \Gamma \Rightarrow \Delta', u : \beta$ where u is a fresh state variable. Again by the induction hypothesis, one gets $tRu, u : \alpha, \Gamma[t/s] \Rightarrow \Delta'[t/s], u : \beta$. By $(R\to)$, one gets $\vdash_n \Gamma[t/s] \Rightarrow \Delta'[t/s], t : \alpha \to \beta$. □

Proposition 3. *The following rules of weakening*

$$\frac{\Gamma \Rightarrow \Delta}{A, \Gamma \Rightarrow \Delta}(wl) \qquad \frac{\Gamma \Rightarrow \Delta}{\Gamma \Rightarrow \Delta, A}(wr)$$

are height-preserving admissible in GInqL.

Proof. By induction on the height of the derivation of the premiss. If $\Gamma \Rightarrow \Delta$ is an axiom or the conclusion of $(\bot_1)$, so are $A, \Gamma \Rightarrow \Delta$ and $\Gamma \Rightarrow \Delta, A$. If the last rule (L_p), $(L\wedge)$, $(R\wedge)$, $(L\vee)$, $(R\vee)$,$(L\to)$, $({}^{\cup}R)$, $(R^{\cup})$, (tran), $(\bot_2)$, $(\bot_3)$, $(\bot_4)$,$(\bot_5)$, (S), $(\cap_1)$, (R_1), (R_2), (LE) or (RE), it is done by straightforward induction. If the last rule is $(R\to)$ or (R_p), by applying (sub) to the premiss, one gets a sequent in which the eigenvariable does not occur in A or Γ, Δ. The conclusion is obtained by induction hypothesis and the rule $(R\to)$ or (R_p). □

Proposition 4. *All rules in* GInqL *are height-preserving invertible.*

Proof. The height-preserving invertibility of rules for $\wedge$ and $\vee$ is done as for classical propositional logic in [7]. The invertibility of (L_p), $(L\to)$, $({}^{\cup}R)$, $(R^{\cup})$, (tran),$(\bot_2)$, $(\bot_3)$,$(\bot_4)$,$(\bot_5)$,(S), $(\cap_1)$, (R_1), (R_2), (LE) and (RE) is obtained by height-preserving weakening. For $(R\to)$, we do it by induction on the height n of the derivation of the premiss $\Gamma \Rightarrow \Delta, s : \alpha \to \beta$. When $n = 0$, it is an axiom or the conclusion of $(\bot_1)$. But then $sRt, t : \alpha, \Gamma \Rightarrow \Delta, t : \beta$ is also an axiom or the conclusion of $(\bot_1)$. If $\Gamma \Rightarrow \Delta, s : \alpha \to \beta$ is obtained by a rule (R) other than $(R\to)$ and (R_P), then we apply the induction hypothesis to the premiss(es) of (R) and then apply (R). Now consider the case that the last rule is (R_p). Let the derivation be:

$$\frac{\vdash_{n-1} sRa, \Gamma \Rightarrow \Delta, a : p, s : \alpha \to \beta}{\vdash_n \Gamma \Rightarrow \Delta, s : p, s : \alpha \to \beta} (R_P)$$

The sequent we have to derive is $sRx, sRa, \Gamma, x : \alpha \Rightarrow \Delta, a : p, x : \beta$. Since x cannot be equal to a, the process goes as other rules. If $\alpha \to \beta$ is principal, then the premiss of $(R\to)$ gives the desired sequent. The proof of invertibility of (R_p) is similar to that of $(R\to)$. □

Theorem 2. *The rules of contraction*

$$(cl)\ \frac{A, A, \Gamma \Rightarrow \Delta}{A, \Gamma \Rightarrow \Delta} \qquad (cr)\ \frac{\Gamma \Rightarrow \Delta, A, A}{\Gamma \Rightarrow \Delta, A}$$

are height-preserving admissible in GInqL.

Proof. By induction on the height n of the derivation of $A, A, \Gamma \Rightarrow \Delta$ and $\Gamma \Rightarrow \Delta, A, A$. If $n = 0$, then $A, A, \Gamma \Rightarrow \Delta$ and $\Gamma \Rightarrow \Delta, A, A$ are axioms or the conclusions of $(\bot_1)$, and so are $A, \Gamma \Rightarrow \Delta$ and $\Gamma \Rightarrow \Delta, A$. Assume $n > 0$. Consider the last rule (R). If the contraction term A is not principle, then by the induction hypothesis and (R) we get $\vdash_n A, \Gamma \Rightarrow \Delta$ and $\vdash_n \Gamma \Rightarrow \Delta, A$. Suppose that the contraction term A is principal. The proof is quite similar to the admissibility of contraction in [2]. For instance, the derivation ends with

$$\frac{\vdash_{n-1} sRx_1, x_1 : \alpha, \Gamma \Rightarrow \Delta, x_1 : \beta, s : \alpha \to \beta}{\vdash_n \Gamma \Rightarrow \Delta, s : \alpha \to \beta, s : \alpha \to \beta}\ (R\to)$$

By the invertibility of $(R\to)$, one gets $\vdash_{n-1} sRx_1, sRx_2, x_1 : \alpha, x_2 : \alpha, \Gamma \Rightarrow \Delta, x_1 : \beta, x_2 : \beta$. By (sub), one gets $\vdash_{n-1} sRx_1, sRx_1, x_1 : \alpha, x_1 : \alpha, \Gamma \Rightarrow \Delta, x_1 : \beta, x_1 : \beta$. By the induction hypothesis, one gets $\vdash_{n-1} sRx_1, x_1 : \alpha, \Gamma \Rightarrow \Delta, x_1 : \beta$. By $(R\to)$, one gets $\vdash_{n-1} \Gamma \Rightarrow \Delta, s : \alpha \to \beta$. □

Theorem 3. *The cut rule*

$$(cut)\ \frac{\Gamma \Rightarrow \Delta, s : \alpha \quad s : \alpha, \Gamma' \Rightarrow \Delta'}{\Gamma, \Gamma' \Rightarrow \Delta', \Delta}$$

is admissible in GInqL.

Proof. The proof is organized as follows. We consider first the case that at least one premiss in a cut is an axiom or conclusion of $(\bot_1)$ and show how cut is eliminated. For the rest there are three cases: (1) The cut term is not principal in either premiss of cut. (2) The cut term is principal in just one premiss of cut. (3) The cut term is principal in both premisses of cut.

(1) Cut with an axiom or conclusion of $(\bot_1)$ as premiss. If at least one of the premisses of cut is an axiom or conclusion of $(\bot_1)$, we distinguish two cases:

(1.1) The left premiss $\Gamma \Rightarrow \Delta, s : \alpha$ of cut is an axiom or conclusion of $(\bot_1)$. If the cut term $s : \alpha$ is in Γ, we derive $\Gamma, \Gamma' \Rightarrow \Delta', \Delta$ from $s : \alpha, \Gamma' \Rightarrow \Delta'$ by weakening. If $a : \bot$ is a term in Γ, then $\Gamma, \Gamma' \Rightarrow \Delta', \Delta$ is a conclusion of $(\bot_1)$.

(1.2) The right premiss $s : \alpha, \Gamma' \Rightarrow \Delta'$ is an axiom or conclusion of $(\bot_1)$. The conclusion is an axiom except the case that the cut term is $a : \bot$. In this case, either the first premiss $\Gamma \Rightarrow \Delta, a : \bot$ is an axiom and $\Gamma, \Gamma' \Rightarrow \Delta', \Delta$ follows as in case 1, or $\Gamma \Rightarrow \Delta, a : \bot$ has been derived. There are many cases according to the rule used. Theses are transformed into derivations with cuts of lesser cut-height. The transformations are special cases of the transformations below.

(2) Cut with neither premiss an axiom. We have three cases.

(2.1) Cut term $s : \alpha$ is not principal in the left premiss. We have subcases according to the rule used to derive the left premiss. For $(L\vee)$, $(R\vee)$, $(L\wedge)$ and $(R\wedge)$, the transformations are easily done. Now we consider other cases.

(2.1.1) (L_p) with $\Gamma = a : p, s : p, sRa, \Gamma''$. The derivation

$$\frac{\dfrac{a : p, s : p, sRa, \Gamma'' \Rightarrow \Delta, s : \alpha}{\Gamma'', s : p, sRa \Rightarrow \Delta, s : \alpha}\ (L_p) \qquad s : \alpha, \Gamma' \Rightarrow \Delta'}{\Gamma', \Gamma'', s : p, sRa \Rightarrow \Delta, \Delta'}\ (Cut)$$

is transformed into the derivation with a cut of lower height:

$$\cfrac{\cfrac{a:p, s:p, sRa, \Gamma'' \Rightarrow \Delta, s:\alpha \qquad s:\alpha, \Gamma' \Rightarrow \Delta'}{a:p, s:p, sRa, \Gamma', \Gamma'' \Rightarrow \Delta, \Delta'}\,(Cut)}{\Gamma', \Gamma'', s:p, sRa \Rightarrow \Delta, \Delta'}\,(L_p)$$

(2.1.2) (R_p) with $\Gamma = \Gamma'', sRa$ and $\Delta = \Delta'', a:p$. The derivation

$$\cfrac{\cfrac{sRa, \Gamma'' \Rightarrow \Delta'', a:p, s:\alpha}{\Gamma'' \Rightarrow \Delta'', s:p, s:\alpha}\,(R_p) \qquad s:\alpha, \Gamma' \Rightarrow \Delta'}{\Gamma', \Gamma'' \Rightarrow \Delta', \Delta'', s:p}\,(Cut)$$

is transformed in to the derivation with a cut of lower height:

$$\cfrac{\cfrac{sRa, \Gamma'' \Rightarrow \Delta'', a:p, s:\alpha \qquad s:\alpha, \Gamma' \Rightarrow \Delta'}{sRa, \Gamma', \Gamma'' \Rightarrow \Delta', \Delta'', a:p}\,(Cut)}{\Gamma', \Gamma'' \Rightarrow \Delta', \Delta'', s:p}\,(R_p)$$

(2.1.3) $(L\rightarrow)$ with $\Gamma = sRt, s:\alpha \rightarrow \beta, \Gamma''$. The left premiss $sRt, s:\alpha \rightarrow \beta, \Gamma'' \Rightarrow \Delta, s:\alpha$ is obtained by $(L\rightarrow)$ from $sRt, s:\alpha \rightarrow \beta, \Gamma'' \Rightarrow \Delta, s:\alpha, t:\alpha$ and $t:\beta, sRt, s:\alpha \rightarrow \beta, \Gamma'' \Rightarrow \Delta, s:\alpha$. The cut is

$$\cfrac{sRt, s:\alpha \rightarrow \beta, \Gamma'' \Rightarrow \Delta, s:\alpha \qquad s:\alpha, \Gamma' \Rightarrow \Delta'}{sRt, s:\alpha \rightarrow \beta, \Gamma', \Gamma'' \Rightarrow \Delta, \Delta'}\,(Cut)$$

We have the following two cuts of lower height:

$$\cfrac{sRt, s:\alpha \rightarrow \beta, \Gamma'' \Rightarrow \Delta, s:\alpha, t:\alpha \qquad s:\alpha, \Gamma' \Rightarrow \Delta'}{sRt, s:\alpha \rightarrow \beta, \Gamma', \Gamma'' \Rightarrow \Delta, \Delta', t:\alpha}\,(Cut)$$

and

$$\cfrac{t:\beta, sRt, s:\alpha \rightarrow \beta, \Gamma'' \Rightarrow \Delta, s:\alpha \qquad s:\alpha, \Gamma' \Rightarrow \Delta'}{t:\beta, sRt, s:\alpha \rightarrow \beta, \Gamma', \Gamma'' \Rightarrow \Delta, \Delta'}\,(Cut)$$

By $(L\rightarrow)$, we have $sRt, s:\alpha \rightarrow \beta, \Gamma', \Gamma'' \Rightarrow \Delta, \Delta'$.

(2.1.4) $(R\rightarrow)$ with $\Gamma = sRx, x:\alpha, \Gamma''$ and $\Delta = s:\alpha \rightarrow \beta, \Delta''$. The derivation

$$\cfrac{\cfrac{sRx, x:\alpha, \Gamma'' \Rightarrow \Delta'', x:\beta, s:\alpha}{\Gamma'' \Rightarrow \Delta'', s:\alpha \rightarrow \beta, s:\alpha}\,(R\rightarrow) \qquad s:\alpha, \Gamma' \Rightarrow \Delta'}{\Gamma', \Gamma'' \Rightarrow \Delta', \Delta'', s:\alpha \rightarrow \beta}\,(Cut)$$

is transformed into the derivation with a cut of lower height

$$\cfrac{\cfrac{sRx, x:\alpha, \Gamma'' \Rightarrow \Delta'', x:\beta, s:\alpha \qquad s:\alpha, \Gamma' \Rightarrow \Delta'}{sRx, x:\alpha, \Gamma', \Gamma'' \Rightarrow \Delta', \Delta'', x:\beta}\,(Cut)}{\Gamma', \Gamma'' \Rightarrow \Delta', \Delta'', s:\alpha \rightarrow \beta}\,(R\rightarrow)$$

(2.1.5) Transformations of other rules are similar. We apply the cuts of lower height then apply the rule.

(2.2) Cut term $s:\alpha$ is principal in the left premiss only. The derivation is transformed in one with a cut of lower cut-height according to derivation of the right premiss. We have a few subcases according to the rule used. For $L\vee$, $R\vee$, $L\wedge$ and $R\wedge$, the transformations are easily done. Now we consider other cases:

(2.2.1) (L_p) with $\Gamma' = a:p, s:p, sRa, \Gamma''$. The derivation

$$\dfrac{\Gamma \Rightarrow \Delta, s:\alpha \qquad \dfrac{s:\alpha, a:p, s:p, sRa, \Gamma'' \Rightarrow \Delta'}{s:\alpha, \Gamma'', s:p, sRa \Rightarrow \Delta'}\,(L_p)}{\Gamma, \Gamma'', s:p, sRa \Rightarrow \Delta, \Delta'}\,(Cut)$$

is transformed into the derivation with a cut of lower cut-height

$$\dfrac{\dfrac{\Gamma \Rightarrow \Delta, s:\alpha \qquad s:\alpha, a:p, s:p, sRa, \Gamma'' \Rightarrow \Delta'}{a:p, s:p, sRa, \Gamma, \Gamma'' \Rightarrow \Delta, \Delta'}\,(Cut)}{\Gamma, \Gamma'', s:p, sRa \Rightarrow \Delta, \Delta'}\,(L_p)$$

(2.2.2) (R_p) with $\Gamma' = \Gamma'', sRa$ and $\Delta' = \Delta'', a:p$. The derivation

$$\dfrac{\Gamma \Rightarrow \Delta, s:\alpha \qquad \dfrac{s:\alpha, sRa, \Gamma'' \Rightarrow \Delta'', a:p}{s:\alpha, \Gamma'' \Rightarrow \Delta'', s:p}\,(R_p)}{\Gamma, \Gamma'' \Rightarrow \Delta, \Delta'', s:p}\,(Cut)$$

is transformed in to the derivation with a cut of lower height

$$\dfrac{\dfrac{\Gamma \Rightarrow \Delta, s:\alpha \qquad s:\alpha, sRa, \Gamma'' \Rightarrow \Delta'', a:p}{sRa, \Gamma, \Gamma'' \Rightarrow \Delta, \Delta'', a:p}\,(Cut)}{\Gamma, \Gamma'' \Rightarrow \Delta, \Delta'', s:p}\,(R_p)$$

(2.2.3) $(L\to)$ with $\Gamma' = sRt, s:\alpha \to \beta, \Gamma''$. Let the left premiss of cut be $\Gamma \Rightarrow \Delta, s:\alpha$, and let the right premiss end with

$$\dfrac{s:\alpha, sRt, s:\alpha \to \beta, \Gamma'' \Rightarrow \Delta, t:\alpha \quad s:\alpha, t:\beta, sRt, s:\alpha \to \beta, \Gamma'' \Rightarrow \Delta'}{s:\alpha, sRt, s:\alpha \to \beta, \Gamma'' \Rightarrow \Delta'}\,(L\to)$$

and let the conclusion of cut be $sRt, s:\alpha \to \beta, \Gamma, \Gamma'' \Rightarrow \Delta, \Delta'$. Then we have the following cuts

$$\dfrac{\Gamma \Rightarrow \Delta, s:\alpha \quad s:\alpha, sRt, s:\alpha \to \beta, \Gamma'' \Rightarrow \Delta', t:\alpha}{sRt, s:\alpha \to \beta, \Gamma, \Gamma'' \Rightarrow \Delta, \Delta', t:\alpha}\,(Cut)$$

and

$$\dfrac{\Gamma \Rightarrow \Delta, s:\alpha \quad s:\alpha, t:\beta, sRt, s:\alpha \to \beta, \Gamma'' \Rightarrow \Delta'}{t:\beta, sRt, s:\alpha \to \beta, \Gamma, \Gamma'' \Rightarrow \Delta, \Delta'}\,(Cut)$$

Then by $(L\to)$, we have $sRt, s:\alpha \to \beta, \Gamma, \Gamma'' \Rightarrow \Delta, \Delta'$.

(2.2.4) $(R\rightarrow)$, with $\Gamma' = sRx, x : \alpha, \Gamma''$ and $\Delta' = s : \alpha \rightarrow \beta, \Delta''$. The derivation

$$\dfrac{\Gamma \Rightarrow \Delta, s:\alpha \qquad \dfrac{s:\alpha, sRx, x:\alpha, \Gamma'' \Rightarrow \Delta'', x:\beta}{s:\alpha, \Gamma'' \Rightarrow \Delta'', s:\alpha\rightarrow\beta}(R\rightarrow)}{\Gamma, \Gamma'' \Rightarrow \Delta, \Delta'', s:\alpha\rightarrow\beta}(Cut)$$

is transformed into the derivation with a cut of lower cut-height

$$\dfrac{\dfrac{\Gamma \Rightarrow \Delta, s:\alpha \qquad s:\alpha, sRx, x:\alpha, \Gamma'' \Rightarrow \Delta'', x:\beta}{sRx, x:\alpha, \Gamma, \Gamma'' \Rightarrow \Delta, \Delta'', x:\beta}(Cut)}{\Gamma, \Gamma'' \Rightarrow \Delta, \Delta'', s:\alpha\rightarrow\beta}(R\rightarrow)$$

(2.2.5) Transformations of other rules are similar: apply the cut of lower cut-height then apply the rule.

(2.3) Cut formula $s : \alpha$ is principal in both premisses, and we have a few subcases. We prove by considering different forms of $s : \alpha$. Conjunction and disjunction are easily done.

(2.3.1) α is $\bot$. Then the left premiss of (Cut) is derived by $(\bot_4)$ and the right premiss is derived by $(\bot_1)$. We transform the derivation

$$\dfrac{\dfrac{\Gamma \Rightarrow \Delta, a:\bot, \epsilon Ra}{\Gamma \Rightarrow \Delta, a:\bot}(\bot_4) \qquad \dfrac{}{a:\bot, \Gamma' \Rightarrow \Delta'}(\bot_1)}{\Gamma, \Gamma' \Rightarrow \Delta, \Delta'}(Cut)$$

into

$$\dfrac{\dfrac{\Gamma \Rightarrow \Delta, a:\bot, \epsilon Ra \qquad a:\bot, \Gamma' \Rightarrow \Delta'}{\Gamma, \Gamma' \Rightarrow \Delta, \Delta', \epsilon Ra}(Cut)}{\Gamma, \Gamma' \Rightarrow \Delta, \Delta'}(RE)$$

(2.3.2) $s : \alpha$ is $s : p$. The derivation is

$$\dfrac{\dfrac{sRa, \Gamma \Rightarrow \Delta, a:p}{\Gamma \Rightarrow \Delta, s:p}(R_p) \qquad \dfrac{b:p, s:p, sRb, \Gamma'' \Rightarrow \Delta'}{\Gamma'', s:p, sRb \Rightarrow \Delta'}(L_p)}{sRb, \Gamma, \Gamma'' \Rightarrow \Delta, \Delta'}(Cut)$$

If $s \notin \mathcal{A}$ and $s \neq \epsilon$, the derivation can be transformed into

$$\dfrac{\dfrac{sRb, \Gamma \Rightarrow \Delta, b:p \qquad \dfrac{\Gamma \Rightarrow \Delta, s:p \qquad b:p, s:p, sRb, \Gamma'' \Rightarrow \Delta'}{b:p, sRb, \Gamma, \Gamma'' \Rightarrow \Delta, \Delta'}(Cut)}{sRb, sRb, \Gamma, \Gamma, \Gamma'' \Rightarrow \Delta, \Delta, \Delta'}(Cut)}{sRb, \Gamma, \Gamma'' \Rightarrow \Delta, \Delta'}(Ctr^*)$$

where the upper cut is of smaller derivation height and the lower on a smaller cut formula (the reason why $a : p$ is shorter than $s : p$ is that we can define the *weight* of a labelled formula such that the weight of $a : p$ is smaller that the weight of $s : p$ where $\epsilon \neq s \notin \mathcal{A}$), Ctr^* denotes repreated applications of contraction rules, and the leftmost premiss is obtained by the substitution (b/a) from $sRa, \Gamma \Rightarrow \Delta, a : p$.

If $s \in \mathcal{A}$, applying substitution to the upper sequent of left premiss, we have $sRs, \Gamma \Rightarrow \Delta, s : p$. Cutting with $\Gamma'', s : p, sRb \Rightarrow \Delta'$, we have $sRs, sRb, \Gamma, \Gamma'' \Rightarrow \Delta, \Delta'$. By (LE), we have $sRb, \Gamma, \Gamma'' \Rightarrow \Delta, \Delta'$.

If $s = \epsilon$, the derivation is transformed into

$$\dfrac{\dfrac{\Gamma \Rightarrow \Delta, \epsilon : p \qquad b : p, \epsilon : p, \epsilon Rb, \Gamma'' \Rightarrow \Delta'}{b : p, \epsilon Rb, \Gamma, \Gamma'' \Rightarrow \Delta, \Delta'}\,(Cut)}{\epsilon Rb, \Gamma, \Gamma'' \Rightarrow \Delta, \Delta'}\,(\bot_3)$$

where the cut is of smaller derivation height.

(2.3.3) $s : \alpha$ is $s : \alpha \to \beta$. Let the left premiss of cut end with

$$\dfrac{sRx_1, x_1 : \alpha, \Gamma \Rightarrow \Delta, x_1 : \beta}{\Gamma \Rightarrow \Delta, s : \alpha \to \beta}\,(R{\to})$$

and let the right premiss of cut end with

$$\dfrac{sRx_2, s : \alpha \to \beta, \Gamma'' \Rightarrow \Delta', x_2 : \alpha \quad x_2 : \beta, sRx_2, s : \alpha \to \beta, \Gamma'' \Rightarrow \Delta'}{sRx_2, s : \alpha \to \beta, \Gamma'' \Rightarrow \Delta'}\,(L{\to})$$

and let the conclusion of cut be $sRx_2, \Gamma, \Gamma'' \Rightarrow \Delta, \Delta'$. Suppose the premisses from left to right are of height m, n, k, respectively. The (Cut) above is of cut-height $m + max(n + k) + 2$. The transformed derivation contains four cuts

$$\dfrac{\Gamma \Rightarrow \Delta, s : \alpha \to \beta \qquad sRx_2, s : \alpha \to \beta, \Gamma'' \Rightarrow \Delta', x_2 : \alpha}{sRx_2, \Gamma, \Gamma'' \Rightarrow \Delta, \Delta', x_2 : \alpha}\,(Cut)$$

$$\dfrac{\Gamma \Rightarrow \Delta, s : \alpha \to \beta \qquad x_2 : \beta, sRx_2, s : \alpha \to \beta, \Gamma'' \Rightarrow \Delta'}{sRx_2, x_2 : \beta, \Gamma, \Gamma'' \Rightarrow \Delta, \Delta'}\,(Cut)$$

$$\dfrac{sRx_2, \Gamma, \Gamma'' \Rightarrow \Delta, \Delta', x_2 : \alpha \qquad sRx_2, x_2 : \alpha, \Gamma \Rightarrow \Delta, x_2 : \beta}{sRx_2, sRx_2, \Gamma, \Gamma, \Gamma'' \Rightarrow \Delta, \Delta, \Delta', x_2 : \beta}\,(Cut)$$

$$\dfrac{sRx_2, sRx_2, \Gamma, \Gamma, \Gamma'' \Rightarrow \Delta, \Delta, \Delta', x_2 : \beta \quad sRx_2, x_2 : \beta, \Gamma, \Gamma'' \Rightarrow \Delta, \Delta'}{sRx_2, sRx_2, sRx_2, \Gamma, \Gamma, \Gamma, \Gamma'', \Gamma'' \Rightarrow \Delta, \Delta, \Delta, \Delta', \Delta'}\,(Cut)$$

where $sRx_2, x_2 : \alpha, \Gamma \Rightarrow \Delta, x_2 : \beta$ is obtained by the substitution (x_2/x_1) from $sRx_1, x_1 : \alpha, \Gamma \Rightarrow \Delta, x_1 : \beta$. The first two cuts are of less cut height. Others are on shorter formulas. Then we apply (Ctr) to obtain $sRx_2, \Gamma, \Gamma'' \Rightarrow \Delta, \Delta'$. □

Corollary 1. *The following rule is admissible in* GInqL*:*

$$\dfrac{\Rightarrow s : \alpha \qquad \Rightarrow s : \alpha \to \beta}{\Rightarrow s : \beta}$$

Proof. Assume $\Rightarrow s : \alpha$ and $\Rightarrow s : \alpha \to \beta$. By (sub), $\Rightarrow t : \alpha$. By the invertibility of $(R{\to})$, from $\Rightarrow s : \alpha \to \beta$, we have $sRt, t : \alpha \Rightarrow t : \beta$. By (cut), $sRt \Rightarrow t : \beta$. By (sub), $sRs \Rightarrow s : \beta$. By (LE), one gets $\Rightarrow s : \beta$. □

5 Soundness and Completeness

In this section, we shall prove the completeness of GInqL with respect to the Hilbert-style axiomatic system HInqL. First of all, we need to interpret labelled formulae and sequents in the state space $\wp\wp(\mathcal{P})$.

Definition 6. *A interpretation I is a surjective function from the set of labels L to $\wp\wp(\mathcal{P})$ satisfying the following conditions:*

(1) $I(\epsilon) = \emptyset$.
(2) *For any variable $x \in V$, $I(x) \in \wp\wp(\mathcal{P})$.*
(3) *For any variable $a \in A$, $I(a) = \{w\}$ for some $w \in \wp(\mathcal{P})$.*
(4) $I((s \cup t)) = I(s) \cup I(t)$.
(5) $I((s \cap t)) = I(s) \cap I(t)$.

Given an interpretation I, we say that a labelled formula $s : \alpha$ is *true* under I, notation $I \models s : \alpha$, if $I(s) \models \alpha$. A relational atom sRt is *true* under I, notation $I \models sRt$, if $I(s) \supseteq I(t)$. A sequent $\Gamma \Rightarrow \Delta$ is *true* under I, notation $I \models \Gamma \Rightarrow \Delta$, if the following condition holds: if $I \models A$ for all A in Γ, then there is at least one term B in Δ such that $I \models B$. We say that a sequent $\Gamma \Rightarrow \Delta$ is *valid*, notation $\models \Gamma \Rightarrow \Delta$ if $I \models \Gamma \Rightarrow \Delta$ for all interpretations I.

Theorem 4. *If a sequent $\Gamma \Rightarrow \Delta$ is derivable in* GInqL*, then $\Gamma \Rightarrow \Delta$ is valid.*

Proof. By induction on the height of derivation of $\Gamma \Rightarrow \Delta$ in GInqL. Obviously all axioms are valid. It suffices to show that every rule preserves validity. Here we sketch the proof of some cases and the remaining cases are easily shown.

(R_p) Suppose $I \models \Gamma$. Suppose that no term in Δ is true under I. To show that $I(s) \models p$, we have to show that for every $w \in I(s)$, $p \in w$. For a specific w, since I is surjective, we can pick a singleton variable a which does not occur in the conclusion such that $I(a) = \{w\}$. Then $I(a) \models p$. Hence $I(s) \models p$.

($L\!\rightarrow$) Suppose $I \models sRt, s : \alpha \rightarrow \beta, \Gamma$. Then $I(s) \supseteq I(t)$ and $I(s) \models \alpha \rightarrow \beta$. By the validity of the first premiss, at least one term in Δ is satifiable or I satisfies $t : \alpha$. If it is the former case, this complete the proof. If it is the latter case, since $I(s) \supseteq I(t)$ and $I(s) \models \alpha \rightarrow \beta$, we have $I(t) \models \beta$. Then we can conclude that at least one term in Δ is true.

($R\!\rightarrow$) Suppose that $I \models \Gamma$ and no term in Δ is true under I. To show that $I(s) \models \alpha \rightarrow \beta$, we have to show that for any $t' \subseteq I(s)$ and $t' \models \alpha$, $t' \models \beta$. For a specific t', since I is surjective, we can pick a label x which does not occur in Γ, Δ and $s : \alpha \rightarrow \beta$ such that $I(x) = t'$. Then $I(x) \models \beta$, *i.e.*, $t' \models \beta$. Since t' is arbitrary, $I(s) \models \alpha \rightarrow \beta$.

($R^{\cup}$) Suppose $I \models sRu, sRv, \Gamma$. Then we have $I(s) \supseteq I(u)$ and $I(s) \supseteq I(v)$. It follows that $I(s) \supseteq (I(u) \cup I(v))$, *i.e.*, $I(s) \supseteq (I(u \cup v))$. By the validity of the premiss, we have the desired result.

(R_1) Suppose that $I \models \Gamma$ and no term in Δ is true under I. Then $I \models \epsilon R(s \cup t)$. Suppose not, by the validity of left premiss, I satisfies ϵRs. By the validity of right premiss, $I \models \epsilon Rt$. Then $\emptyset \supseteq I(s)$ and $\emptyset \supseteq I(t)$. It follows that

$\emptyset \supseteq (I(s) \cup I(t))$. Then $\emptyset \supseteq I(s \cup t)$. Hence, I satisfies $\epsilon R(s \cup t)$. Contradiction. Hence, I satisfies $\epsilon R(s \cup t)$.

(R_2) Suppose that $I \models (t_1 \cup t_2)Rt, \Gamma$ and no term in Δ is true under I. By the validity of the premiss, I satisfies $\epsilon R((t \cap t_1) \cup (t \cap t_2))$ or ϵRt. Since $I(t_1 \cup t_2) \supseteq I(t)$, $I((t \cap t_1) \cup (t \cap t_2)) = I(t)$. Hence $I \models \epsilon Rt$. □

Lemma 4. *Suppose Γ is the set of labelled formulas with the same label x and $\Gamma' = \{\beta \mid x : \beta \in \Gamma\}$. If $\Gamma \Rightarrow x : \alpha$ is valid, then $\Gamma' \models \alpha$.*

Proof. Assume that $\Gamma \Rightarrow x : \alpha$ is valid. Then for any interpretation I, if $I(x) \models \Gamma$, then $I(x) \models \alpha$. If follows that x can range over $\wp\wp(\mathcal{P})$. Hence $\Gamma' \models \alpha$. □

Theorem 5. *Suppose Γ is the set of labelled formulas with the same label x and $\Gamma' = \{\beta \mid x : \beta \in \Gamma\}$. For every formula α, $\Gamma' \vdash_{\mathsf{HInqL}} \alpha$ iff $\Gamma \Rightarrow x : \alpha$ is derivable in* GInqL.

Proof. Assume that $\Gamma \Rightarrow x : \alpha$ is derivable in GInqL. By the soundness of GInqL, $\Gamma \Rightarrow x : \alpha$ is valid. By Lemma 4, $\Gamma' \models \alpha$. By the completeness of HInqL, $\Gamma' \vdash_{\mathsf{HInqL}} \alpha$. Conversely, assume $\Gamma' \vdash_{\mathsf{HInqL}} \alpha$. By Proposition 2 and Corollary 1, $\Gamma \Rightarrow x : \alpha$ is derivable in GInqL. □

6 Concluding Remarks

The labelled sequent calculus GInqL for inquisitive logic seems not to be pure in the sense that the semantics of logical connectives is integrated into the calculus. The surfacial difference with non-labelled sequent calucli for normal modal logics is not essential because it is a real proof system consisting of right introduction rules and left introduction rules for all logical connectives. The meaning of a logical connective is indicated by its right introduction rule. The calculus GInqL is a proof system for the complete inquisitive logic InqL, and it is designed according to the inquisitive semantics.

References

1. Ciardelli, I., Roelofsen, F.: Inquisitive logic. J. Philosoph. Logic **40**, 55–94 (2011)
2. Dyckhoff, R., Negri, S.: Proof analysis in intermediate logics. Arch. Math. Logic **51**, 71–92 (2012)
3. Groenendijk, J.: Inquisitive semantics: two possibilities for disjunction. In: Bosch, P., Gabelaia, D., Lang, J. (eds.) TbiLLC 2007. LNCS, vol. 5422, pp. 80–94. Springer, Heidelberg (2009). doi:10.1007/978-3-642-00665-4_8
4. Mascarenhas, S.: Inquisitive Semantics and Logic. Master Thesis, University of Amsterdam (2009)
5. Negri, S.: Proof analysis in modal logic. J. Philosoph. Logic **34**, 507–544 (2005)
6. Negri, S.: Contraction-free sequent calculi for geometric theories with an application to Barr's theorem. Arch. Math. Logic **42**, 389–401 (2003)
7. Negri, S., von Plato, J.: Structural Proof Theory. Cambridge University Press, Cambridge (2001)
8. Sano, K.: Sound and complete tree-sequent calculus for inquisitive logic. In: Ono, H., Kanazawa, M., Queiroz, R. (eds.) WoLLIC 2009. LNCS, vol. 5514, pp. 365–378. Springer, Heidelberg (2009). doi:10.1007/978-3-642-02261-6_29
9. Stalnaker, R.: Assertion. Syntax Semant. **9**, 315–332 (1978)

Testing Minimax for Rational Ignorant Agents

Marc-Kevin Daoust(✉) and David Montminy

University of Montreal, Montreal, Canada
{marc-kevin.daoust,david.montminy}@umontreal.ca

Abstract. Richard Pettigrew [13,14] defends the following theses: (1) epistemic disutility can be measured with strictly proper scoring rules (like the Brier score) and (2) at the beginning of their credal lives, rational agents ought to minimize their worst-case epistemic disutility (Minimax). This leads to a Principle of Indifference for ignorant agents. However, Pettigrew offers no argument in favour of Minimax, suggesting that the epistemic conservatism underlying it is a "normative bedrock." Is there a way to test Minimax? In this paper, we argue that, since Pettigrew's Minimax is *impermissive*, an argument against credence permissiveness constitutes an argument in favour of Minimax, and that arguments for credence permissiveness are arguments against Minimax.

Keywords: Rationality · Minimax · Uniqueness · Permissiveness · Scoring rule · Objective bayesianism

1 Introduction

Meet Pria, a perfectly rational ignorant agent.[1] Pria wants some coins to gamble with her friends. She visits four coin factories, where she is told the following:

1. *Fair Coin Factory* produces unbiased coins. The objective probability that their coins will land heads is 0.5.
2. *Mystery Coin Factory* is short on specifics. The objective probability that their coins will land heads is kept secret.

Now, consider the proposition "when tossed, the coin will land on heads" (H). For each of the coins produced in these factories, what epistemically rational credence in H can Pria have? Clearly, if she got a coin from *Fair Coin Factory*, Pria is rationally required to assign a credence of 0.5 in H. However, if she got a coin from *Mystery Coin Factory*, things get complicated.

Richard Pettigrew [13, Chap. 12–13]; [14] defends the claim that, at the beginning of their credal lives, rational agents ought to minimize their worst-case disutility (as measured by strictly proper scoring rules). Call this the Minimax rule. If Pettigrew is right, it leads to a Principle of Indifference for prior credence assignments. Such a principle would state that, if a rational ignorant agent entertains only P and $\sim P$, then his or her credence in P should be equal

[1] We borrowed this example from Schoenfield [18, p. 640].

A. Baltag et al. (Eds.): LORI 2017, LNCS 10455, pp. 541–553, 2017.
DOI: 10.1007/978-3-662-55665-8_37

to his or her credence in $\sim P$. Suppose that an agent has no evidence concerning H's objective probability. In such a case, an agent minimizes his or her worst-case disutility by assigning a credence of 0.5 in H and a credence of 0.5 in $\sim H$. Therefore, in a case like *Mystery Coin Factory*, Pria's credence in H should be equal to his or her credence in $\sim H$, and this is precisely what the Principle of Indifference would recommend.[2]

Is Minimax a genuine rationality requirement? Could Pria follow a different decision rule and yet be rational? Pettigrew has little to say in favour of the Minimax rule. According to him, accepting or rejecting Minimax is a matter of "normative bedrock," as he indicates in the following:

> Minimax makes all but the most risk-averse behaviour irrational; Maximax makes all but the most risk-seeking behaviour irrational ... Why favour Minimax over Maximax? The answer, I think, lies in cognitive conservatism ... At this point, it seems to me, we have reached normative bedrock: one cannot argue for cognitive conservatism from more basic principles. [14, p. 45–46]

We think that Pettigrew is partly right: a proof of the Minimax rule cannot be offered. However, such a definitive answer is unsatisfactory to us. Moreover, there is an indirect way to defend Minimax, namely by rejecting competing decision rules. In this paper, we argue that Minimax is the only impermissive decision rule for ignorant agents. Conversely, if credence permissiveness is true, then Pettigrew's account of epistemic norms is compromised, since Minimax is the only impermissive decision rule for ignorant agents. This means that there is a test for Minimax: in defending credence uniqueness, one argues for the rejection of all decision rules for ignorant agents *except Minimax.*

In addition to providing an evaluation of the Minimax rule as a rationality requirement, our paper aims at showing that maximal risk-aversion under ignorance and the denial of permissiveness go hand in hand. Moreover, our results could be relevant in other contexts. For example, the argument of this paper could provide grounds for constraints on priors probabilities assigned to hypotheses in science, machine learning or rational reasoning in economics. After all, whether we are concerned with doxastic states, scientific hypotheses or members of a partition in a database, the problem is essentially the same: we want to determine the rational constraints on the initial probabilities assigned to propositions.

In Sect. 2, we introduce strictly proper scoring rules, the Minimax rule and credence permissiveness. In Sect. 3, we argue that Minimax is the only impermissive decision rule for ignorant agents. In Sect. 4, we reply to various objections.

[2] In such a case Pria is facing a situation akin to the one an agent faces in the famous Ellsberg paradox. In our framework, however, there is no need to make a distinction between risk and uncertainty, since in all cases, Pria is maximally ignorant. Indeed, since she has no prior information, she is always dealing with uncertainty. Minimax does not in fact solve the Ellsberg paradox: it simply avoids it when it is used by a maximally ignorant agent.

2 Scoring Rules, Minimax and Credence Permissiveness

In this section, we introduce Pettigrew's account using strictly proper scoring rules, the Minimax rule and credence permissiveness.[3] A scoring rule is a function measuring the inaccuracy of a prediction. The more accurate an agent's prediction is, the better his or her score is. Some scoring rules have particular properties. For example, strictly proper scoring rules are uniquely optimized when agents report objective (or true) probabilities. Specifically, only one credence assignment optimizes the expected score of a strictly proper scoring rule. A well-known strictly proper scoring rule is the Brier score, a quadratic scoring rule.[4] When an agent assigns a credence X in P, the Brier score is measured by the following:

1. If P is true, then the agent's score is $(1 - X)^2$.
2. If P is false, then the agent's score is $(0 - X)^2$.

The lower an agent's total score is, the more accurate his or her credence assignments are.

Here is an example of how the Brier score works. Suppose that Pria, our perfectly rational ignorant agent, wonders what her credence in P should be. She has evidence that P's objective probability is 0.6. She assigns a credence X in P. If P happens to be true, then her Brier score will be $(1 - X)^2$. If P happens to be false, then her Brier score will be $(0 - X)^2$. What is her expected Brier score?[5] Since Pria has evidence that P's objective probability is 0.6, her expected score is $(0.6 \cdot (1 - X)^2) + (0.4 \cdot X^2)$. So if she assigned a credence in P of 0.8, her expected score would be $(0.6 \cdot 0.2^2) + (0.4 \cdot 0.8^2) = 0.28$.

Since the Brier score is a strictly proper scoring rule, Pria will optimize her expected score by satisfying the following: X must be equal to 0.6. In other words, Pria will optimize her expected score by assigning a credence in P that is equal to P's objective probability (relative to her evidence).

Pettigrew [13, Chap. 12–13], [13] also defends the claim that, at the beginning of their credal lives, rational agents ought to minimize their worst-case disutility. This is the Minimax rule. Pettigrew argues that, since epistemic disutility is measured by strictly proper scoring rules, these requirements lead to a Principle of Indifference for ignorant agents. According to such a principle, if a rational ignorant agent entertains only P and $\sim P$, then his or her credence in P should be equal to his or her credence in $\sim P$.

---

[3] Readers interested in the formal aspects of strictly proper scoring rules and the Principle of Indifference may consult [13–15], and also [19] for a discussion on the relation between the Principle of Indifference and inductive inference.

[4] See [3].

[5] The expected result is sometimes called the weighted mean result. For example, suppose that, in a fair lottery, 5 participants each have 1 chance in 5 to win a single prize of \$50. In that lottery, 4 participants won't win a prize, and 1 participant will win \$50. Since $(40 + 150)/5 = 10$, the weighted mean value of this lottery is \$10. This means that \$10 is the expected prize to each participant.

In the next sections, we will argue that Minimax is the only impermissive decision rule for ignorant agents. In other words, we will soon argue that there is a strong connection between Minimax and credence uniqueness.[6] Here is how White defines Uniqueness:

1. *Epistemic uniqueness*: "If an agent whose total evidence is E is fully rational in taking doxastic attitude D to P, then necessarily, any subject with total evidence E who takes a different attitude to P is less than fully rational" [21, p. 312].

Since, in this paper, we are concerned with rational credences, we will only address the following types of Uniqueness and permissiveness:

1. *Credence uniqueness*: relative to an agent's total evidence E, there is a unique rational credence assignment function towards any proposition P;
2. *Permission to have incompatible epistemic standards*: relative to a body of evidence, it is possible for a rational agent to choose between incompatible rational standards governing credence assignments, such as prior functions;
3. *Permission within some epistemic standards*: relative to a body of evidence and some rational standards, it is possible for a rational agent to assign incompatible credences towards a proposition P.

Uniqueness is controversial. Kelly in [8, pp. 175–76]; [9, pp. 120–22]; [10] thinks that Uniqueness is implausible since it does not admit of any exceptions. Following [1], he acknowledges that two rational agents with common prior probabilities who update their beliefs by conditionalization may never come to hold incompatible credences. What Kelly rejects, however, is that there is a unique set of rational prior probabilities relative to a body of evidence. Another argument against credence uniqueness comes from [17], who thinks that distinct rational agents are entitled to hold incompatible epistemic standards, such as distinct priors or conditional probability functions. According to her, " two people with the same evidence reasonably have different opinions about whether P, it is because these people have each adopted a different set of reasonable epistemic standards" [17]. If she is right, the Minimax rule is merely permitted, not required.[7]

Even if there are intuitive reasons to defend permissiveness, there are some serious objections against such a view. A plausible argument against credence permissiveness comes from White [20]. We normally think of rational agents as holding prior updating functions and responding to new evidence in accordance with these functions. Responding to evidence is a one-way process: once a rational agent gathers new evidence, that rational agent updates his or her credences accordingly. However, according to White, credence permissiveness implies that,

[6] For various recent arguments against credence uniqueness, one may consult [2,9,10, 12,15,16]. For various recent arguments in favour of credence uniqueness, one may consult [5,6,20,21].

[7] Specifically, while every agent could be bound to a specific impermissive scoring rule and a specific impermissive decision rule, different agents could be bound to distinct incompatible scoring rules or distinct incompatible decision rules.

relative to a body of evidence, incompatible prior updating functions are rationally permitted. Therefore, it would mean that, even when gathering new evidence in favour of P, a rational agent might not have to change his or her credence in P. Instead, that agent could simply change his or her prior updating function. That is incompatible with the claim that rational agents are responsive to new evidence (by updating their credences). White explains his argument in the following:

> Suppose that you and I share our total evidence E. My subjective probability for P is x, and yours is lower at y. We each now obtain additional evidence E', which supports P. My confidence in P rises to x' and yours to y', which happens to be equal to the x that I held prior to obtaining E'. We have each updated our convictions appropriately in response to the new evidence. But now let's suppose that we were each fully rational in holding our different degrees of belief x and y given just evidence E ... Why then shouldn't I just keep my confidence in P at x, if it suits me? [20, pp. 454–55]

3 A Test for Minimax

3.1 Minimax as the only Impermissive Decision Rule for Ignorant Agents

We will now argue that Minimax is the only risk-based impermissive decision rule for ignorant agents. By risk-based decision rules, we simply mean decision rules that are primarily based on risk considerations. Before offering a formal proof of that connection, we will offer an informal explanation of why this is the case.

First, it is important to mention that we are interested in decidable and risk-based decision rules. We are interested in decision rules for which there exists a known effective method based on risk factors for determining which decisions are warranted. For example, suppose that Pria is in front of two glasses. One of them contains petrol, and one of them contains gin. She is ignorant of which glass contains gin. Suppose that Pria is subject to the following rule: she must choose the glass containing gin. Surely, that rule would be impermissive, since there is only one glass containing gin. But since Pria is ignorant, she cannot follow such a rule. From her perspective, it is equally risky to drink from either glass. This means that there is no effective risk-based method available to her for determining which glass to choose.

Now, suppose that there is an impermissive and decidable decision rule A for ignorant agents such that A is not Minimax. To make things simpler, we will focus on the folllowing carving of two mutually exclusive and exhaustive possibilities:

1. *Two-propositions carving*: if the set of propositions contains only "the coin will land on heads" (H) and "the coin will land on tails" (T), then Pria should assign a credence of X and $(1 - X)$ in each proposition.[8]

Minimax would recommend assigning a credence of 0.5 in each of the propositions. Since A is different from Minimax, X must be different from 0.5. Now, suppose that A is the rule assigning a credence of 0.6 in one proposition and 0.4 in the other. Surely, A is impermissive, since there is a unique credence assignment towards each proposition. Unfortunately, Pria is ignorant, which makes it difficult for her to apply the rule correctly. Pria could assign a credence of 0.6 in H, but she could also assign a credence of 0.6 in T. Surely, she knows that, if she assigns a credence of 0.6 in H, she has to assign a credence of 0.4 in T, since H and T form the mutually exclusive and exhaustive carving of logical possibilities. But how is she supposed to decide between assigning a credence of 0.6 in H or assigning a credence of 0.6 in T? This situation is similar in fashion to the situation where Pria ought to choose between the gin and the petrol glasses. As with the glasses, there is no way to decide how the rule should be applied.

Besides, if she is permitted to assign a credence of 0.6 in one of these propositions while having no information at all concerning H and T, there is no difference for her *in terms of risk* between assigning a higher credence in H and assigning a higher credence in T. In such a context, a risk-based decision rule should recommend assigning a credence of 0.6 in *either* proposition, since no risk-based distinction can be made between H and T.

3.2 A Formal Proof in Favour of that Connection

We will now argue that, necessarily, if A is an impermissive, risk-based, decidable decision rule for an ignorant agent entertaining a carving of mutually exclusive and exhaustive logical possibilities $(L_1, L_2, ..., L_n)$ and a set of credences $(1/m_1, 1/m_2, ..., 1/m_n)$, then A is Minimax.

Definitions and Assumptions

(i) Decidability: a situation is decidable when, relative to a relation between a set of inputs and a set of outputs, there exists a known effective method for determining which output is associated with an input.[9]

(ii) Ignorance*: an agent is ignorant* when, except for carvings of logical possibilities and rational risk factors, he or she has no background knowledge about the world. His or her body of evidence is maximally uninformative.

[8] What we here call a "carving" is sometimes called a partition elsewhere. The reason for this terminological choice is that, in the literature from which our paper draws, "carving" is a much more popular term.

[9] Decidability should be understood in its most basic sense: given an input (evidential carving), there is *always* an effective method for returning an output (credence assignment).

(iii) Risk Factor: Suppose that $(Y_1, Y_2, ..., Y_n)$ are some possible truth values of propositions that are part of the carving of mutually exclusive and exhaustive logical possibilities $(L_1, L_2, ..., L_n)$. A risk factor is an interval $[B_n, W_n]$ such that, relative to possibilities $(L_1, L_2, ..., L_n)$ and credence assignments $(1/m_1, 1/m_2, ..., 1/m_n)$, B_n is the best-case sum of $(Y_1 - (1/m_1))^2 + (Y_2 - (1/m_2))^2, ..., +(Y_n - (1/m_n))^2$ and W_n is the worst-case sum of $(Y_1 - (1/m_1))^2 + (Y_2 - (1/m_2))^2, ..., +(Y_n - (1/m_n))^2$.

(iv) Minimax: Minimax is a decidable and impermissive decision rule such that, when entertaining the carving of mutually exclusive and exhaustive logical possibilities $(L_1, L_2, ..., L_n)$, agents assign credences $(1/m_1, 1/m_2, ..., 1/m_n)$ such that the worst-case sum of $(Y_1 - (1/m_1))^2 + (Y_2 - (1/m_2))^2, ..., +(Y_n - (1/m_n))^2$ is minimized. This also means that Minimax has a risk factor of $[W_n, W_n]$.

(v) Principle of Indifference: a decision rule such that, when entertaining a carving of mutually exclusive and exhaustive logical possibilities $(L_1, L_2, ..., L_n)$, an ignorant* agent assigns the credences $(1/n, 1/n, ..., 1/n)$.

(vi) Minimax implies the Principle of Indifference: as established in Pettigrew [13,14], Minimax leads to the Principle of Indifference. So when ignorant* agents assign credences $(1/m_1, 1/m_2, ..., 1/m_n)$, Minimax is satisfied if and only if $1/m_1 = 1/n, 1/m_2 = 1/n, ..., 1/m_n = 1/n$.

(vii) Risk-based decision rule: a risk-based decision rule relies on a risk factor. For example, Minimax is based on a risk factor of $[W_n, W_n]$.

(viii) Permissive decision rule: a decision rule is permissive if and only if distinct incompatible credence assignments satisfy the rule. For example, if A permits (L_1, L_2) to be in relation with credence assignments $(1/m_1, 1/m_2)$ and $(1/m_2, 1/m_1)$ while $1/m_2 \neq 1/m_1$, then A is permissive.

Theorem

Necessarily, if A *is an impermissive, risk-based, decidable decision rule for an ignorant* agent entertaining a carving of mutually exclusive and exhaustive logical possibilities* $(L_1, L_2, ..., L_n)$ *and a set of credences* $(1/m_1, 1/m_2, ..., 1/m_n)$, *then* A *is Minimax.*

Proof

(1) Assume for *reductio*: A is not Minimax and A is a risk-based, decidable and impermissive decision rule for an ignorant* agent entertaining a carving of mutually exclusive and exhaustive logical possibilities $(L_1, L_2, ..., L_n)$ and a unique set of credences $(1/m_1, 1/m_2, ..., 1/m_n)$.

(2) Given (vi), (vii) and (1), A is not Minimax, which means that, relative to the carving $(L_1, L_2, ..., L_n)$ in relation with the credences $(1/m_1, 1/m_2, ..., 1/m_n)$, there is the smallest credence assignment $1/m_i$ and the greatest credence assignment $1/m_j$ such that $1/m_i \neq 1/m_j$.

(3) Given (iii) and (2), since $1/m_i \neq 1/m_j$, A is based on a risk factor $[B_n, W_n]$ where $B_n \neq W_n$.

(4) Relative to a carving of two possibilities, the carving (L_1, L_2) in relation with either $(1/m_i, 1/m_j)$ or $(1/m_j, 1/m_i)$, where $1/m_i \neq 1/m_j$.

(5) First possibility: $(L_1,\ L_2)$ is in relation with $(1/m_i, 1/m_j)$. Since $1/m_i$ is the smallest credence assignment and $1/m_j$ the greatest credence assignment, this means that $B_2 = (1-(1/m_j))^2 + (0-(1/m_i))^2$ and $W_2 = (1-(1/m_i))^2 + (0-(1/m_j))^2$.

(6) Second possibility: (L_1, L_2) is in relation with $(1/m_j, 1/m_i)$. Since $1/m_i$ is the smallest credence assignment and $1/m_j$ the greatest credence assignment, this means that $B_2 = (1-(1/m_j))^2 + (0-(1/m_i))^2$ and $W_2 = (1-(1/m_i))^2 + (0-(1/m_j))^2$.

(7) So given (4), (5) and (6), $B_2 = (1-(1/m_j))^2 + (0-(1/m_i))^2$ and $W_2 = (1-(1/m_i))^2 + (0-(1/m_j))^2$, regardless of whether (L_1, L_2) is in relation with $(1/m_j, 1/m_i)$ or in relation with $(1/m_i, 1/m_j)$.

(8) Assume that, for a carving containing $n-1$ elements, there are at least two credence assignments satisfying a risk factor $[B_{n-1},\ W_{n-1}]$ where $B_{n-1} \neq W_{n-1}$.

(9) Relative to a carving of n possibilities, the carving $(L_1, L_2, ..., L_c,$ $L_d, ..., L_n)$ can be in relation with either $(1/m_1, 1/m_2, ..., 1/m_i, 1/m_j, ..., 1/m_n)$ or $(1/m_1,\ 1/m_2, ..., 1/m_j, 1/m_i, ..., 1/m_n)$, where $1/m_i \neq 1/m_j$.

(10) We can rearrange the credence assignments $(1/m_1, 1/m_2..., 1/m_n)$ from the smallest to the greatest credence assignment. This new set can be denoted by $(R_1, R_2, R_3, ..., R_n)$. Naturally, if $1/m_i$ is the smallest credence assignment and $1/m_j$ the greatest credence assignment, this means that $(R_1, R_2, R_3, ..., R_n)$ amounts to $(1/m_i, R_2, R_3, ..., 1/m_j)$.

(11) Given (9) and (10), $B_n = (0-(1/m_i))^2 + (0-R_2)^2 + (0-R_3)^2 ... + (1-(1/m_j))^2$ and $W_n = (1-(1/m_j))^2 + (0-R_2)^2 + (0-R_3)^2 ... + (0-(1/m_i))^2$, regardless of whether (L_c, L_d) is in relation with $(1/m_j, 1/m_i)$ or in relation with $(1/m_i, 1/m_j)$.

(12) By induction from (7), (8) and (11), for any carving containing $n \geq 2$ elements, there are at least two credence assignments satisfying a risk factor $[B_n, W_n]$ where $B_n \neq W_n$.

(13) Given (3) and (12), for A and any carving of n possibilities, an agent cannot determine if (L_c, L_d) is in relation with $(1/m_j, 1/m_i)$ or in relation with $(1/m_i, 1/m_j)$ by making a risk-based distinction.

(14) Following (viii), if A permits (L_c, L_d) to be in relation with $(1/m_i, 1/m_j)$ **and** $(1/m_j, 1/m_i)$, then A is permissive.

(15) However, given (1), A is impermissive.

(16) So (L_c, L_d) is not in relation with $(1/m_i, 1/m_j)$ **and** $(1/m_j, 1/m_i)$.

(17) Given (13), we cannot explain why (L_c, L_d) is not in relation with $(1/m_i, 1/m_j)$ **and** $(1/m_j, 1/m_i)$ by making a risk-based distinction.

(18) Given (1) and (15), (L_c, L_d) is in relation with $(1/m_i, 1/m_j)$ if and only if it is false that (L_c, L_d) is in relation with $(1/m_j, 1/m_i)$.

(19) Following (i), if A is decidable, the agent has an effective method for determining if (L_c, L_d) is in relation with $(1/m_i, 1/m_j)$.

(20) Given (ii), if the agent is ignorant*, he or she has no knowledge of an effective method for determining if (L_c, L_d) is in relation with $(1/m_i, 1/m_j)$.

(21) Following (1), the agent is ignorant*.

(22) So given (20) and (21), the agent has no effective method for determining if (L_c, L_d) is in relation with $(1/m_i, 1/m_j)$.

(23) If A is decidable, the agent has an effective method for determining if (L_c, L_d) is in relation with $(1/m_j, 1/m_i)$.

(24) But, once again, given (1), the agent is ignorant*.

(25) So following (23) and (24), the agent also has no effective method for determining if (L_c, L_d) is in relation with $(1/m_j, 1/m_i)$.

(26) Given (i), (17), (22) and (25), A is undecidable.

(27) By *reductio* from (1), (2) and (26), necessarily, if A is not Minimax and A is a risk-based decidable and impermissive relation between an ignorant* agent's carving $(L_1, L_2, ..., L_n)$ and a set of credences $(1/m_1, 1/m_2, ..., 1/m_n)$, then there is a counterexample to A.

(28) Given (iv) and (vii), Minimax is a risk-based, decidable and impermissive decision rule.

(29) Given (27) and (28), necessarily, if A is an impermissive, risk-based and decidable decision rule between an ignorant* agent's carving $(L_1, L_2, ..., L_n)$ and a set of credences $(1/m_1, 1/m_2, ..., 1/m_n)$, then A is Minimax.

□

4 Objections and Replies

In this section, we answer some objections one could raise against our test.

4.1 Endorsing Versus Conforming to Minimax

Here is a first objection against our model. Minimax is a rule governing an agent's attitudes towards epistemic risk at the beginning of his or her credal life. Pettigrew, for example, argues that "Minimax makes all but the most risk-averse behaviour irrational; Maximax makes all but the most risk-seeking behaviour irrational" [14, p. 45]. Even if Uniqueness is true, that does not provide a reason to think that agents ought to be risk-averse at the beginning of their credal lives. This would mean only that, regardless of how we interpret an ignorant agent's attitude towards epistemic risk, his or her credences will be compatible with Minimax. Interestingly, some defenders of Uniqueness reject Pettigrew's defense of Minimax. For example, Sophie Horowitz thinks that Pettigrew's defense of Minimax is puzzling, as she indicates in the following:

> Maximin yields such unreasonable recommendations under normal circumstances. A maximally risk-averse believer would never form any opinions unless she could be sure she was right; a maximally risk-averse actor would never leave the house. So can we give any further argument for why superbabies should follow Maximin? ... Why should superbabies use a special decision rule? Without a significant backstory, it seems arbitrary to treat an agent's very first decision differently from others [7, pp. 5–6].[10]

This is a perfectly good objection. Agents could hold credences in conformity with Minimax without having a genuine aversion towards epistemic risk. We have not offered reasons to think that, if Uniqueness is true, agents ought to be risk-averse at the beginning of their credal lives. There could be another explanation of why agents ought to have credences in accordance with Minimax, and this is a limit to our testing procedure. However, Minimax could also be a sound consequence of credence uniqueness. So while such an objection is relevant to shed light on what we proved exactly, it does not affect our conclusion that, if credence permissiveness is false, the only decision rule left for rational ignorant agents is Minimax.

4.2 Minimax and Strictly Proper Scoring Rules

Here is another objection. We claimed that the Minimax rule is impermissive. However, one reason why Minimax is impermissive is that an agent's epistemic disutility is measured with strictly proper scoring rules. If epistemic disutility could be measured differently, Minimax could be permissive. At least, since Pettigrew's Principle of Indifference would not necessarily hold, satisfying the Minimax rule would not necessarily lead to impermissive credence assignments. In such a context, we need a reason to think that epistemic disutility ought to be measured with strictly proper scoring rules.

According to Pettigrew in [13], five principles offer grounds in favour of strictly proper scoring rules. *Alethic Vindication* states that the omniscient credence function is the ideal function; *Perfectionism* states that a function's inaccuracy is its divergence from the ideal one; *Divergence Additivity* sums up local functions; and *Divergence Continuity* makes them continuous; finally, *Decomposition*, which is the *ceteris paribus* version of *Calibration*, states that a legitimate inaccuracy measure is determined both by the divergence of a function c from its

[10] Horowitz refers to Pettigrew's book *Accuracy and the Laws of Credence* [13], where he uses the expression "Maximin". We refer to his paper "Accuracy, Risk and the Principle of Indifference" [14] where he uses the expression "Minimax". Both terms refer to the same decision rule.

well-calibrated counterpart c' and by the divergence of c' from the ideal credence function.[11]

We can easily see why a rational agent's epistemic disutility ought to be measured with strictly proper scoring rules. Improper scoring rules can be optimized through inconsistent combinations of credences. Surely, consistency is a minimal rationality requirement.[12] Therefore, there is an independent justification in favour of strictly proper scoring rules.

To begin with, consider the following expected scoring rule:

(1) Variable function: $Z \cdot ((1 - X)^{(n+1)}) + (1 - Z) \cdot ((0 - X)^{(n+1)})$ for $n \geq 0$

Z is P's objective probability. X is an agent's credence in P. This function is a polynomial generalization of the quadratic scoring rule. Indeed, when n equals 1, this expected scoring rule is the expected Brier score, a strictly proper scoring rule. However, when n is not equal to 1, this expected scoring rule is improper and recommends either "liberal" or "conservative" credences. If it is possible to measure an agent's epistemic disutility through distinct incompatible scoring rules, some distinct values of n could be rationally permitted, depending on the epistemic objectives of agents.

Here are two examples of distinct scoring rules. Suppose that n equals 0.5. We get the following expected scoring rule function:

(2) Liberal scoring rule: $Z \cdot ((1 - X)^{3/2}) + (1 - Z) \cdot ((0 - X)^{3/2})$

This scoring rule rewards "liberal" credences. If an agent has evidence that P's objective probability is 0.8, his or her expected score would be optimized

---

[11] See [13], Sect. 4.3] for formal definitions and the importance of calibration. *Decomposition* is crucial to Pettigrew's account of accuracy: it allows him to exclude uncalibrated divergence measures. Roughly, an agent has a calibrated credence X in P if, relative to a body of evidence, X is the proportion of all propositions of a certain type (P) that appear to be true. This is the distinctive feature of strictly proper scoring rules: an agent's expected score is maximized by assigning calibrated credences in P. For example, the following improper scoring rule respects *Alethic vindication, Perfectionism, Divergence Additivity* and *Divergence Continuity* but violates *Decomposition*:

1. Relative to an agent's credence X in P, if P is true, then the agent's score is $(1 - X)$;
2. Relative to an agent's credence X in P, if P is false, then the agent's score is $|0 - X|$.

See also [13, pp. 37–40, 48].

[12] Some authors have suggested that there are no distinct consistency requirements of rationality. Specifically, it is possible that process requirements of rationality, which govern how rational agents form and revise beliefs, secure consistency [11]. What matters in this paper is that inconsistent agents violated at least one rationality requirement. See also [4, sect. 9.2] on consistency requirements.

by assigning a credence of ≈ 0.941 in P. In general, this scoring rule rewards credences that are closer to 0 or 1.[13]

Now, by way of contrast, suppose that n equals 2. We get the following expected scoring rule function:

(3) Conservative scoring rule: $Z \cdot ((1 - X)^3) + (1 - Z) \cdot ((0 - X)^3)$

This scoring rule rewards "conservative" credences. If an agent has evidence that P's objective probability is 0.8, his or her expected score would be optimized by assigning a credence of ≈ 0.666 in P. In general, this scoring rule is optimized by credences that are closer to 0.5.[14]

What's wrong with these scoring rules (either liberal or conservative)? The problem is that they are sometimes optimized by irrational credence combinations. Indeed, in many situations, these scoring rules would recommend holding inconsistent combinations of credences. Here is an example of such inconsistencies. Suppose that Pria is about to roll a fair six-sided die. She believes that the objective probability of rolling any of the 6 numbers is 1. She also believes that the objective probability of rolling one of the 6 numbers is ≈ 0.17. First, consider the liberal scoring rule ($n = 0.5$). If the objective probability of rolling a one is ≈ 0.17, her expected score would be optimized by assigning a credence of ≈ 0.04 in that proposition. The same goes for rolling a two, a three, and so forth. Since all 6 numbers form the collectively exclusive and exhaustive possibilities, Pria should believe that the sum of her credences in each of the propositions equals 1. However, she assigned a credence of ≈ 0.04 in each outcome. Since $\approx 0.04 \cdot 6$ does not equal 1, that scoring rule would lead to inconsistent credence assignments. The same goes for the conservative scoring rule ($n = 2$). If the objective probability of rolling a one is ≈ 0.17, her expected score would be optimized by assigning a credence of ≈ 0.31 in that proposition. Since all 6 numbers are the collectively exclusive and exhaustive possibilities, Pria should believe that the sum of her credences in each of the propositions equals 1. But again, since $\approx 0.31 \cdot 6$ does not equal 1, that scoring rule leads to inconsistent credence assignments.[15] Therefore, measuring an agent's epistemic disutility with strictly proper scoring rules appears uncontroversial.

[13] We simply calculated the derivative of this function to find its optimum. The derivative of the function $f(X) = 0.8 \cdot ((1 - X)^{3/2}) + 0.2 \cdot ((0 - X)^{3/2})$ is equal to 0 when $X \approx 0.941$.

[14] The derivative of the function $f(X) = 0.8 \cdot ((1 - X)^3) + 0.2 \cdot ((0 - X)^3)$ is equal to 0 when $X \approx 0.666$.

[15] While we focused on two instances of improper scoring rules, the same problem would arise with any improper scoring rule. The fact that a scoring rule is not uniquely optimized when $Z = X$ explains why an agent could end up with an inconsistent combination of attitudes. Only improper scoring rules are not uniquely optimized when $Z = X$.

5 Conclusion

This paper started with the following problem: how can we offer an argument in favour of Pettigrew's Minimax? We suggested that, since Minimax is the only impermissive decision rule for rational ignorant agents, analyzing the plausibility of credence uniqueness is an indirect way to test Minimax. Conversely, if credence permissiveness is true, then there are rational alternatives to Minimax.

References

1. Aumann, R.J.: Agreeing to disagree. Annals Stat. **4**, 1236–1239 (1976)
2. Ballantyne, N., Coffman, E.J.: Uniqueness, evidence, and rationality. Philos. Impr. **11**(18), 1–13 (2011)
3. Brier, G.W.: Verification of forecasts expressed in terms of probability. Mon. Weather Rev. **78**(1), 1–3 (1950)
4. Broome, J.: Rationality Through Reasoning. Wiley, Oxford (2013)
5. Christensen, D.: Conciliation, Uniqueness and Rational Toxicity. Noûs. doi:10.1111/nous.12077 (2014)
6. Horowitz, S.: Immoderately Rational. Philos. Stud. **167**(1), 41–56 (2014)
7. Horowitz, S.: Laws of credence and laws of choice. Episteme **14**(1), 31–37 (2017). doi:10.1017/epi.2016.49
8. Kelly, T.: The Epistemic Significance of Disagreement. In: Gendler, T.S., Hawthorne, J. (eds.) Oxford Studies in Epistemology, pp. 167–96. Clarendon Press, Oxford (2005)
9. Kelly, T.: Peer disagreement and higher-order evidence. In: Warfield, T., Feldman, R. (eds.) Disagreement, pp. 111–72. Oxford University Press, Oxford (2010)
10. Kelly, T.: Evidence can be permissive. In: Steup, M., Turri, J., Sosa, E. (eds.) Contemporary Debates in Epistemology, pp. 298–312. Wiley, Boston (2014)
11. Kolodny, N.: State or process requirements? Mind **116**(462), 372–385 (2007)
12. Meacham, C.G.: Impermissive Bayesianism. Erkenntnis **79**(6), 1185–1217 (2014)
13. Pettigrew, R.: Accuracy and the Laws of Credence. Oxford University Press, Oxford (2016a)
14. Pettigrew, R.: Accuracy, risk, and the principle of indifference. Philos. Phenomenol. Res. **92**(1), 35–59 (2016b). doi:10.1111/phpr.12097
15. Predd, J.B., Seiringer, R., Lieb, E.H., Osherson, D.N., Vincent Poor, H., Kulkarni, S.R.: Probabilistic coherence and proper scoring rules. IEEE **55**(10), 4786–4792 (2009)
16. Sharadin, N.: A paritial defense of permissivism. Ratio **30**, 57–71 (2015). http://onlinelibrary.wiley.com/doi/10.1111/rati.12115/full
17. Schoenfield, M.: Permission to believe: why permissivism is true and what it tells us about irrelevant influences on belief. Noûs. **48**(2), 193–218 (2014)
18. Schoenfield, M.: Bridging rationality and accuracy. J. Philos. **112**(12), 633–57 (2015). doi:10.5840/jphil20151121242
19. Smithson, R.: The principle of indifference and inductive scepticism. Brit. J. Phil. Sci. **68**, 253–272 (2017)
20. White, R.: Epistemic Permissiveness. Philos. Perspect. **19**(1), 445–459 (2005)
21. White, R.: Evidence cannot be permissive. In: Steup, M., Turri, J., Sosa, E. (eds.) Contemporary Debates in Epistemology, pp. 312–23. Wiley, Boston (2014)

A Reconstruction of Ex Falso Quodlibet via Quasi-Multiple-Conclusion Natural Deduction

Yosuke Fukuda[1(✉)] and Ryosuke Igarashi[2]

[1] Graduate School of Informatics, Kyoto University, Kyoto, Japan
yfukuda@fos.kuis.kyoto-u.ac.jp
[2] Graduate School of Letters, Kyoto University, Kyoto, Japan
igarashi.r0922@gmail.com

Abstract. This paper is intended to offer a philosophical analysis of the propositional intuitionistic logic formulated as *NJ*. This system has been connected to Prawitz and Dummett's proof-theoretic semantics and its computational counterpart. The problem is, however, there has been no successful justification of *ex falso quodlibet* (EFQ): "From the absurdity '$\bot$', an arbitrary formula follows." To justify this rule, we propose a novel intuitionistic natural deduction with what we call quasi-multiple conclusion. In our framework, EFQ is no longer an inference deriving everything from '$\bot$', but rather represents a "jump" inference from the absurdity to the other possibility.

Keywords: Ex Falso Quodlibet · Intuitionistic logic · Proof-theoretic semantics · Curry–Howard correspondence · Catch/throw mechanism

1 Introduction

The aim of this paper is to provide a better understanding of the logical system of the propositional fragment of intuitionistic logic formulated as *NJ*. As the logic of constructivism, *NJ* has been regarded as being best explained by Prawitz and Dummett's proof-theoretic semantics and its computational counterpart through the Curry–Howard correspondence. We agree with this understanding as a basic direction. At the same time, however, it should also be pointed out that such understanding has certain inadequacies. The subject matter of the problem is about *ex falso quodlibet* (EFQ), which is usually formalized as '$\bot$'-elimination:

$$\frac{\bot}{A}\ \mathrm{EFQ}$$

As various previous studies pointed out, it cannot be said that there is a good justification for EFQ to dispel the counter-intuitive look of EFQ. A point of skepticism regarding EFQ is, of course, from its inference: *"From the absurdity (i.e., falsum or contradiction) '$\bot$', an arbitary formula A follows."* The problem is, there seems to be no relevance between '$\bot$' and arbitary A. As a matter of fact, people who are considered the founders of intuitionism have adopted such

A. Baltag et al. (Eds.): LORI 2017, LNCS 10455, pp. 554–569, 2017.
DOI: 10.1007/978-3-662-55665-8_38

a position of skepticism, as van Dalen's article points out [19]. Thus, we justify EFQ in such a way as to avoid as much as possible the skepticism that has been presented so far, by using what we call *quasi-multiple-conclusion natural deduction.*

As a thread of guidance in such an attempt, we will incorporate two ideas that come from logical and computational points of view. One is Tennant's consideration [17] on logical consequence that '$\bot$' represents *deadend* in reasoning. The second is an intuition, which is probably shared by some computer scientists, that EFQ expresses some sort of *exception handling* in computation. According to these views, we abandon the policy of justifying EFQ, appealing to the semantical value of '$\bot$'. Rather, we characterize EFQ as a structural rule that expresses *jump inference*, which is particularly characteristic of the computational interpretation of classical logic, e.g., as Griffin and Parigot studied [4, 11]. In our view, EFQ is not an inference deriving everything from '$\bot$', but represents a "jump" from *deadend* to *another possibility.* To formalize this, we think a system with a catch/throw mechanism is most appropriate. Regarding such a system, one formulated by Nakano [8] already exists. However, because the system of Nakano is minimal logic and does not contain an inference on '$\bot$', we extend it to intuitionistic. Moreover, the logic obtained in this manner takes a form that naturally expands the constructive semantics originally associated with *NJ*, such as proof-theoretic semantics of Prawitz and Dummett, and the computational explanation, in a way that avoids the difficulties already pointed out.

2 Backgrounds

2.1 Theory of Meaning and the Computational Interpretation

According to the use theory of meaning, proof-theoretic semantics claims that the introduction rule in natural deduction, as itself, explains the meaning of a connective. This is because the introduction rule defines the usage of the connective, i.e., under what condition we can assert it. For example, the introduction rule of conjunction stipulates the usage that from proofs of propositions A and B, one can conclude $A \wedge B$. From a computational point of view, it means from constructions a of type A and b of type B, it is possible to construct a pair $\langle a, b \rangle$ of type $A \wedge B$ in simply-typed λ-calculus. This correspondence between natural deduction and typed λ-calculus relies on a general framework, the Curry–Howard correspondence [15], which has substantially been studied by various computer scientists. On the other hand, the elimination rule is understood just as the "consequence" of the introduction rule. To stipulate appropriate condition for this "consequence" relation between introduction and elimination, Dummett [2] introduced the notion of "harmony," that is, the *reducibility* of proof detours. This condition, as a global property of the whole system, corresponds to the *normalizability* of deductions, which is that of terms in typed λ-calculus.

Furthermore, the correspondence has now been studied for what was considered to be a particular concept of computation so far. The catch/throw mechanism is one of them, and it is used in some practical programming languages, such as *Common Lisp* [16], for *exception handling*. For instance, given a list of natural numbers $\langle n_1, n_2, \cdots, n_m \rangle$, let us consider calculating their product $\prod_{i=1}^{m} n_i$. Without the mechanism, we have to calculate "all" of the products one by one from n_1, even if there is 0 in the list. However, by using the catch/throw, we can immediately "throw" 0 as the result at the time that 0 appears in the list; namely, the mechanism is used for an evacuation from redundant computations.

2.2 Previous Validations for EFQ and Their Inadequacies

There are some validations for EFQ by Dummett [2,3] and Prawitz [13]. However, it has been pointed out that both of the works are inadequate (cf., Tennant [17], Hand [5], Onishi [10], Tranchini [18], and Cook and Cogburn [1]). First, Prawitz seems to think that because there is no proof detour of '⊥' due to the lack of the introduction rule, '⊥'-elimination is *vacuously* in "harmony." The problem here is, as Tennant [17] pointed out, that this justification apparently appeals to the meta-level '⊥'-elimination or something equivalent, and is not so good of an explanation as to avoid the skepticism of EFQ. Second, Dummett proposed two kinds of justification: One identifies '⊥' with "0 = 1", and the other considers an implicit introduction rule of '⊥' that has all atomic formulae as assumptions:

$$\frac{0 = 1}{A}\ \bot E \qquad\qquad \frac{p_0 \qquad p_1 \qquad p_2 \qquad \cdots}{\bot}\ \bot I$$

where p_i is an atomic formula for all natural numbers i.

In the former, although every proposition indeed can be derived from "0 = 1" in some mathematical theories, this solution works only when the theory is of arithmetic; in the latter, the introduction and the elimination is harmonious, but the meaning of '⊥' relies on a particular language (cf., Onishi [10])[1]. In a nutshell, depending on particular theories, both lack a kind of generality that is expected for logical constants.

3 A Quasi-Multiple-Conclusion Natural Deduction: *NQ*

The first step toward giving an intuitive meaning to EFQ is to define our natural deduction *NQ*, whose main feature is to handle EFQ as a kind of structural inference rule via what we call *quasi-multiple conclusion*. Its main part is based on Nakano's constructive minimal logic [8,9], but because his setting is for minimal logic, we extend it with '⊥' to deal with EFQ.

[1] There is another formalization by using second-order propositional logic, but it seems that the formalization has the same problem.

3.1 Syntax and Semantics

The syntax and semantics of *NQ* are defined in what follows.

Definition 1 (Formulae). *The* formulae *are inductively defined as follows:*

$$A, B ::= p \mid \bot \mid A \vee B \mid A \wedge B \mid A \supset B$$

p denotes a propositional variable. '⊥' is the absurdity, which represents a logical deadend. *The connectives of '∨', '∧', and '⊃' are for disjunction, conjunction, and implication respectively. We define the* negation *of A, written* $\neg A$*, as* $A \supset \bot$.

The formulae of *NQ* are exactly the same as the propositional fragment of intuitionistic natural deduction *NJ*.

Definition 2 (Judgments). *A* judgment *is a triple, written* $\Gamma \vdash A; \Delta$*, where A is a formula; each of* Γ *and* Δ *is a (possibly empty) set of formulae.*

A judgment $\Gamma \vdash A; \Delta$ means that: there exists at least one *construction* (or *witness*) of the formulae of $\{A\} \cup \Delta$ under the constructions of Γ. It also represents that A is the *current formula* of concern within deductions, and Δ is a *container* holding the other formulae. Intuitively speaking, in $\Gamma \vdash A; \Delta$, propositions in Δ can be understood as *other possibilities* that are temporary "put aside," and 'A' is a formula that is *provisionally* asserted. From this point of view, single-conclusion natural deduction can be regarded as a special case where there is only one possibility.

For example, let us consider a case where one is looking for one's wallet. One knows that the wallet is in the desk or the chest or possibly the fridge. In such a situation, one may guess that the wallet is in the desk and confirm this. In this seeking process, "in the desk" corresponds to 'A' and the others to 'Δ'. Note that although according to Theorem 1 below, quasi-multiple-conclusion "$A; \Delta$" is logically equivalent to a disjunction of all formulae in $\{A\} \cup \Delta$, it cannot be identified with an assertion of the disjunction when they are viewed as speech acts. This is because in the former case, only A is in the scope of the assertion, whereas in the latter, the assertion concerns with all disjuncts[2].

Definition 3 (Inference rules). *The* inference rules *of NQ are inductively defined as follows (Note that here, we use '∅' to denote the empty set):*

Structural rules

$$\frac{}{A \vdash A; \varnothing}\ Ax \qquad \frac{\Gamma \vdash A; \Delta}{\Gamma, B \vdash A; \Delta}\ LW \qquad \frac{\Gamma, B, B \vdash A; \Delta}{\Gamma, B \vdash A; \Delta}\ LC$$

$$\frac{\Gamma \vdash \bot; A, \Delta}{\Gamma \vdash A; \Delta}\ EFQ \qquad \frac{\Gamma \vdash A; A, \Delta}{\Gamma \vdash A; \Delta}\ Catch \qquad \frac{\Gamma \vdash A; \Delta}{\Gamma \vdash B; A, \Delta}\ Throw$$

[2] At this point, our system avoids the problem of multiple conclusion pointed out by Dummett [2]. However, we will not go further into this debate.

Logical rules

$$\frac{\Gamma \vdash A; \Delta \qquad \Gamma' \vdash B; \Delta'}{\Gamma, \Gamma' \vdash A \wedge B; \Delta, \Delta'} \wedge I \qquad \frac{\Gamma \vdash A \wedge B; \Delta}{\Gamma \vdash A; \Delta} \wedge E_0 \qquad \frac{\Gamma \vdash A \wedge B; \Delta}{\Gamma \vdash B; \Delta} \wedge E_1$$

$$\frac{\Gamma \vdash A \vee B; \Delta \qquad \Gamma', A \vdash C; \Delta' \qquad \Gamma'', B \vdash C; \Delta''}{\Gamma, \Gamma', \Gamma'' \vdash C; \Delta, \Delta', \Delta''} \vee E \qquad \frac{\Gamma \vdash A; \Delta}{\Gamma \vdash A \vee B; \Delta} \vee I_0$$

$$\frac{\Gamma \vdash B; \Delta}{\Gamma \vdash A \vee B; \Delta} \vee I_1 \quad \frac{\Gamma, A \vdash B; \Delta}{\Gamma \vdash A \supset B; \Delta} (*), \supset I \qquad \frac{\Gamma \vdash A \supset B; \Delta \qquad \Gamma' \vdash A; \Delta'}{\Gamma, \Gamma' \vdash B; \Delta, \Delta'} \supset E$$

The structural rules are described as follows. Ax states that there exists a trivial witness of A. The EFQ's premise expresses that the current formula '$\bot$' is *deadend*[3], so one considers the inference with A as current by taking it from the container. In this setting, EFQ is no longer an inference that derives everything from absurdity[4]. Hence, it avoids the skepticism mentioned earlier.

The Throw and Catch are rules that correspond to the catch/throw mechanism of a typed λ-calculus explained later, but logically, they are structural rules for quasi-multiple conclusion. It may seem that this solution is just passing the problem of EFQ to Throw, as there appears to be an arbitary formula in the conclusion. This suspicion is, however, pointless. This is because Throw is a kind of weakening and intuitively corresponds to a process that, "putting aside" the proposition currently considered, starts to think about a new possibility. In this sense, Throw is not an inference that derives an arbitary conclusion from the premise, which is considered problematic. On the other hand, Catch is a kind of contraction and corresponds to a process "going back" to the possibility considered before.

The logical rules are defined in the same manner of Nakano's formalization [8] because NQ is just an extension of Nakano's logic with '$\bot$', and the difference does not affect the logical rules. Note that there is a side condition for $\supset I$, marked $(*)$, to keep NQ constructive. We omit the precise definition of the condition because it is the same as Nakano's and is a little difficult to explain due to limitations of space. The condition is also needed for the compositionality of derivations explained in Sect. 4.1.

According to the framework of proof-theoretic semantics, each I-rule successfully explains the meaning of the connective. It is also obvious that all structural rules are valid in terms of the reading of judgments explained above. With regard to EFQ, there are no constructions of '$\bot$', because it represents a logical deadend. Thus, from the assumption of EFQ, it follows that at least one formula in $\{A\} \cup \Delta$ has a construction. We can also provide a rigorous definition of *proof-theoretic validity* [12] for NQ instead of the naive notion of "construction" by expanding the original definition. However, we will not go into detail about this

[3] An intuition behind here is Tennant's remark [17]: "'$\bot$' represents a *logical deadend*, and thus there are no further inferences after it appears within deductions."

[4] In this sense, this rule is not well expressed its name, i.e., "from contradiction, anything" (ex falso quodlibet). In the following, however, we will continue to use this name for the sake of simplicity.

because it requires further technical and philosophical consideration, and the space is limited.

3.2 Examples and Properties of *NQ*

In this section, we are going to see characteristic examples of *NQ* and discuss some desired properties of *NQ*.

First, whereas there are no structural rules as primitive for the succedent of judgments, all of the common structural rules follow from the Throw and Catch rules.

Lemma 1 (Structural rules for the succedent). *The following* exchange, weakening, *and* contraction *rules are derivable:*

$$\dfrac{\Gamma \vdash A; B, \Delta}{\Gamma \vdash B; A, \Delta}\,Ex \qquad \dfrac{\Gamma \vdash A; \Delta}{\Gamma \vdash A; B, \Delta}\,W \qquad \dfrac{\Gamma \vdash A; B, B, \Delta}{\Gamma \vdash A; B, \Delta}\,C$$

Proof. By the following derivations, respectively:

$$\dfrac{\dfrac{\Gamma \vdash A; B, \Delta}{\Gamma \vdash B; A, B, \Delta}\,\text{Throw}}{\Gamma \vdash B; A, \Delta}\,\text{Catch} \qquad \dfrac{\dfrac{\Gamma \vdash A; \Delta}{\Gamma \vdash B; A, \Delta}\,\text{Throw}}{\Gamma \vdash A; B, \Delta}\,\text{Ex} \qquad \dfrac{\dfrac{\dfrac{\Gamma \vdash A; B, B, \Delta}{\Gamma \vdash B; A, B, \Delta}\,\text{Ex}}{\Gamma \vdash B; A, \Delta}\,\text{Catch}}{\Gamma \vdash A; B, \Delta}\,\text{Ex}$$

□

Second, in our framework, we can use another form of disjunction rules, which are defined as direct inference rules by using quasi-multiple-conclusion.

Theorem 1 (Another formalization of disjunction). *The following I-rule and E-rule for disjunction are derivable from the original rules.*

$$\dfrac{\Gamma \vdash A; B, \Delta}{\Gamma \vdash A \vee B; \Delta}\,\vee I' \qquad \dfrac{\Gamma \vdash A \vee B; \Delta}{\Gamma \vdash A; B, \Delta}\,\vee E'$$

Proof. By the following derivations respectively:

$$\dfrac{\dfrac{\dfrac{\dfrac{\Gamma \vdash A; B, \Delta}{\Gamma \vdash A \vee B; B, \Delta}\,\vee I_0}{\Gamma \vdash B; A \vee B, \Delta}\,\text{Ex}}{\Gamma \vdash A \vee B; A \vee B, \Delta}\,\vee I_1}{\Gamma \vdash A \vee B; \Delta}\,\text{Catch} \qquad \dfrac{\Gamma \vdash A \vee B; \Delta \quad \dfrac{}{A \vdash A; \varnothing}\,\text{Ax} \quad \dfrac{\dfrac{}{B \vdash B; \varnothing}\,\text{Ax}}{B \vdash A; B}\,\text{Throw}}{\Gamma \vdash A; B, \Delta}\,\vee E$$

□

This formalization is intuitively described as follows: for a disjunctive formula that expresses plural possibilities, the elimination rule enables us to focus on one possibility (i.e., one disjunct) within reasoning, and the introduction is the reverse inference rule to combine such possibilities.

Finally, the following is an instructive example that shows us how EFQ works in inferences.

Theorem 2 (Disjunctive Syllogism). *$\neg A, A \vee B \vdash B; \varnothing$ is derivable.*

Proof. By the following derivation:

$$\dfrac{\dfrac{\dfrac{}{\neg A \vdash \neg A; \varnothing}\,\text{Ax} \qquad \dfrac{\dfrac{}{A \vee B \vdash A \vee B; \varnothing}\,\text{Ax}}{A \vee B \vdash A; B}\,\vee E'}{\neg A, A \vee B \vdash \bot; B}\,\supset E}{\neg A, A \vee B \vdash B; \varnothing}\,\text{EFQ}$$

□

Interestingly, the proof reflects an intuition of disjunctive syllogism such that: *"Since we have $\neg A$ and $A \vee B$, there is no possibility that A holds. Hence B holds."* Namely, we first focus on A as the current concerned formula from $A \vee B$ by $\vee E'$. Noticing that there will be a deadend '$\bot$' from A and $\neg A$, we abort proving '$\bot$' but assert B by EFQ. This inference demonstrates EFQ's behavior in deductions as "jump" from the deadend to another possibility.

In the above, we defined *NQ* as an appropriate system for justifying intuitionistic logic in terms of proof-theoretic semantics. To accomplish this purpose, however, there are two things left that we have to show:

- Harmony, or normalizability of deductions
- Soundness and completeness with regard to NJ

To show this fact, we will construct a corresponding λ-calculus in the next section.

4 The Corresponding Typed λ-Calculus: λ_{NQ}

In this section, we define λ_{NQ} as a tool for investigating the properties of *NQ*.

Definition 4 (Syntax). *A countably infinite set of* type variable, *written V_{ty}, is assumed to be given. Similarly, V_{ind} and V_{tag} are for individual variables and tag variables. We also assume that all of the V_{ty}, V_{ind}, and V_{tag} are disjoint. Then,* types *and* terms *are defined by the following grammar:*

$$\begin{array}{ll} \textit{Types} & A, B ::= p \mid \bot \mid A \vee B \mid A \wedge B \mid A \supset B \\ \textit{Terms} & M, N, L ::= x \mid \lambda x.M \mid MN \mid \langle M, N \rangle \mid \pi_0 M \mid \pi_1 M \mid \iota_0 M \mid \iota_1 M \\ & \qquad \mid \textbf{\textit{case}}\ M\ \textbf{\textit{of}}\ [x]N\ \textbf{\textit{or}}\ [y]L \mid u \mid \textbf{\textit{throw}}\ u\ M \mid \textbf{\textit{catch}}\ u\ M \end{array}$$

where p, x, and u range over V_{ty}, V_{ind}, and V_{tag} respectively.

The type constructors of '$\vee$', '$\wedge$', and '$\supset$' are for the cartesian product, disjoint sum, and function space, respectively. Under the Curry–Howard isomorphism, we may identify them with connectives: disjunction, conjunction, and implication.

The term constructors are all standard as defined in the literature[5], except for the terms of u, **throw**, and **catch**. Namely, (1) x is a variable; (2) $(\lambda x.M)$ and (MN) are for function abstraction and application respectively; (3) $\langle M, N\rangle$ is for pairing and $\pi_0 M$ (resp., $\pi_1 M$) is the left (resp., right) projection; (4) $\iota_0 M$ (resp., $\iota_1 M$) is the left (resp., right) injection to make a disjoint sum, and $(\mathbf{case}\ M\ \mathbf{of}\ [x]N\ \mathbf{or}\ [y]L)$ is a pattern maching of the disjoint sum. The rest of terms, are in the same way as Nakano's $L_{c/t}$, used to deal with the catch/throw mechanism. Namely, (5) u is a tag variable for labeling a pair of catch and throw constructs; (6) $(\mathbf{throw}\ u\ M)$ is an exception operation that throws M as a result to the corresponding "catch" labeled by the same u; and (7) $(\mathbf{catch}\ u\ M)$ catches an exception raised by some "throw," labeled by u within M.

Definition 5 (Substitution). *The* substitution *of a term N for a variable x in a term M, written $M[x := N]$, is defined as usual, that is, it is the term obtained by replacing all of the free occurrences of x in M with N. The* simultaneous substitution, *written as $M[x_0 := N_0, \cdots, x_{n-1} := N_{n-1}]$, is also defined similarly.*

Definition 6 (Reduction). *An* evaluation context *C is, used for "global reduction," defined to be a* pseudo-term *M, which has a hole $[]$ within the term, which is no different from the usual term except it must have exactly one hole so as to be substituted with some term. For a context C and a term M, $C[M]$ is defined to be the term obtained by replacing the $[]$ in C with M.*

Then, the one-step reduction relation for redex, *written $\mapsto$, is defined to be the least relation closed under the following rules:*

$$\begin{aligned}
(\lambda x.M)N &\mapsto M[x := N] && \\
\pi_0(\langle M, N\rangle) &\mapsto M && \\
\pi_1(\langle M, N\rangle) &\mapsto N && \\
\textit{\textbf{case}}\ (\iota_0 M)\ \textit{\textbf{of}}\ [x]N\ \textit{\textbf{or}}\ [y]L &\mapsto N[x := M] && \\
\textit{\textbf{case}}\ (\iota_1 M)\ \textit{\textbf{of}}\ [x]N\ \textit{\textbf{or}}\ [y]L &\mapsto L[y := M] && \\
C[\textit{\textbf{throw}}\ u\ M] &\mapsto \textit{\textbf{throw}}\ u\ M && \textit{if } (\dagger) \textit{ holds} \\
\textit{\textbf{catch}}\ u\ M &\mapsto M && \textit{if } u \notin FV(M) \\
\textit{\textbf{catch}}\ u\ (\textit{\textbf{throw}}\ u\ M) &\mapsto M && \textit{if } u \notin FV(M)
\end{aligned}$$

where $(\dagger)$ is defined as:

$$C \not\equiv []\ \textit{and}\ C\ \textit{do not capture any free variables occurring in}\ (\textit{\textbf{throw}}\ u\ M).$$

Then, the one-step reduction *relation, written $\rightarrow_\beta$, is defined to be the least relation closed under the following rule:*

$$C[M] \rightarrow_\beta C[N] \qquad \textit{if } M \mapsto N$$

[5] We recommend some text (e.g., Sørensen and Urzyczyn's [15]) for people who are not familiar with these definitions, although the precise one is determined by Definition 6.

The multi-step reduction *relation, written* $\rightarrow^{+}_{\beta}$, *is defined to be the transitive closure of* $\rightarrow_{\beta}$. *Namely, if* $M_1 \rightarrow_{\beta} \cdots \rightarrow_{\beta} M_n$ *holds, then so does* $M_1 \rightarrow^{+}_{\beta} M_n$.

Example 1. $C_1 = (M\,[])$ and $C_2 = \langle [], L\rangle$ are both evaluation contexts. Then, $C_1[N] = (MN)$ and $C_2[N] = \langle N, L\rangle$ are terms.

Example 2. The following are reduction examples. Note that the last two show that the throw operation cannot throw a variable to the outside of its bound scope.

1. $\mathbf{catch}\ u\ (\pi_0\langle x, \mathbf{throw}\ u\ y\rangle) \rightarrow_{\beta} \mathbf{catch}\ u\ x \rightarrow_{\beta} x$
2. $\mathbf{catch}\ u\ (\pi_0\langle x, \mathbf{throw}\ u\ y\rangle) \rightarrow_{\beta} \mathbf{catch}\ u\ (\mathbf{throw}\ u\ y) \rightarrow_{\beta} y$
3. $(\lambda x.(\mathbf{throw}\ u\ x)) \not\rightarrow_{\beta} \mathbf{throw}\ u\ x$
4. $\mathbf{catch}\ u\ (\mathbf{case}\ M\ \mathbf{of}\ [x]N\ \mathbf{or}\ [y](\mathbf{throw}\ u\ y)) \not\rightarrow_{\beta} \mathbf{catch}\ u\ (\mathbf{throw}\ u\ y)$

Then, the type judgment and typing rules are defined in what follows.

Definition 7 (Type judgment). *An* expression *is a pair of a term M and a type A, written as* $M : A$. *We say an expression is* individual (resp. tag) *if its term is a individual (resp. tag) variable. Then, a* type judgment $\Gamma \vdash M : A; \Delta$ *is defined to be a triple that consists of a set of individual expressions* Γ, *an expression* $M : A$, *and a set of tag expressions* Δ.

Definition 8 (Typing rules). *The* typing rules *are defined as follows:*

Structural typing rules

$$\frac{}{x : A \vdash x : A; \varnothing}\ Ax \qquad \frac{\Gamma \vdash M : A; \Delta}{\Gamma, x : B \vdash M : A; \Delta}\ LW \qquad \frac{\Gamma \vdash M : A; \Delta}{\Gamma \vdash M : A; u : B, \Delta}\ RW$$

$$\frac{\Gamma, x : B, y : B \vdash M : A; \Delta}{\Gamma, z : B \vdash M[x := z, y := z] : A; \Delta}\ LC \qquad \frac{\Gamma \vdash M : \bot; u : A, \Delta}{\Gamma \vdash \textit{\textbf{catch}}\ u\ M : A; \Delta}\ EFQ$$

$$\frac{\Gamma \vdash M : A; u : A, \Delta}{\Gamma \vdash \textit{\textbf{catch}}\ u\ M : A; \Delta}\ Catch \qquad \frac{\Gamma \vdash M : A; \Delta}{\Gamma \vdash \textit{\textbf{throw}}\ u\ M : B; u : A, \Delta}\ Throw$$

Logical typing rules

$$\frac{\Gamma \vdash M : A; \Delta \quad \Gamma' \vdash N : B; \Delta'}{\Gamma, \Gamma' \vdash \langle M, N\rangle : A \wedge B; \Delta, \Delta'}\ \wedge I \qquad \frac{\Gamma \vdash M : A \wedge B; \Delta}{\Gamma \vdash \pi_0 M : A; \Delta}\ \wedge E_0$$

$$\frac{\Gamma \vdash M : A \wedge B; \Delta}{\Gamma \vdash \pi_1 M : B; \Delta}\ \wedge E_1 \qquad \frac{\Gamma \vdash M : A; \Delta}{\Gamma \vdash \iota_0 M : A \vee B; \Delta}\ \vee I_0 \qquad \frac{\Gamma \vdash M : B; \Delta}{\Gamma \vdash \iota_1 M : A \vee B; \Delta}\ \vee I_1$$

$$\frac{\Gamma \vdash M : A \vee B; \Delta \quad \Gamma', x : A \vdash N : C; \Delta' \quad \Gamma'', y : B \vdash L : C; \Delta''}{\Gamma, \Gamma', \Gamma'' \vdash \textit{\textbf{case}}\ M\ \textit{\textbf{of}}\ [x]N\ \textit{\textbf{or}}[y]L : C; \Delta, \Delta', \Delta''}\ \vee E$$

$$\frac{\Gamma, x : A \vdash M : B; \Delta}{\Gamma \vdash (\lambda x.M) : (A \supset B); \Delta}\ (*), \supset I \qquad \frac{\Gamma \vdash M : A \supset B; \Delta \quad \Gamma' \vdash N : A; \Delta'}{\Gamma, \Gamma' \vdash (MN) : B; \Delta, \Delta'}\ \supset E$$

Among the structural typing rules, Ax and LW are standard, and LC is a natural rule for formalizing the notion of contraction. The Catch and Throw rules, in the same manner as Nakano's $L_{c/t}$, reflect their mechanism. The term ($\mathbf{throw}\ u\ M$) in the conclusion of Throw is intended to "jump" to the corresponding "catch" (of type B), and thus, the Catch rule requires that M and u have the same type, so as to catch an exception raised by the throw operation. The rule EFQ also constructs a "catch" term because the term M will get stuck in the computation, as '$\bot$' represents the type of "deadend," but it may also "jump" as an evacuation from the deadend, and hence EFQ wraps the term M with catch. We will discuss this rule further in a later section.

Note that there is, compared with those of *NQ*, one additional rule RW. It is logically equivalent to *NQ* even if we remove RW, but we adopt it to capture the fine-grained computational structures of λ_{NQ}. In other words, as a technical reason, we need this rule to prove the subject reduction theorem explained later.

All of the logical typing rules are standard w.r.t terms, namely, they are the same as simply-typed λ-calculus, and are defined in the same manner as Nakano [8]. We also require the same side condition of $\supset I$ mentioned in Definition 3.

4.1 Properties of λ_{NQ}

In this section, we prove and discuss some properties of λ_{NQ}. The main results of λ_{NQ} are proofs of the so-called subject reduction theorem and strong normalization theorem, which are famous criteria for well-definedness of calculi. Through these theorems and the Curry–Howard correspondence, we will give a computational meaning to the reasoning of *NQ* w.r.t. '$\bot$' and EFQ.

First, as we mentioned already, there exists a correspondence between logical system *NQ* and typed λ-calculus λ_{NQ}. Namely, the provability of *NQ* and the typability of λ_{NQ} coincide as the following theorem states.

Theorem 3 (Curry–Howard isomorphism). *$\Gamma \vdash A; \Delta$ is derivable in NQ iff $\Gamma \vdash M : A; \Delta$ is derivable in λ_{NQ} for some M.*

Proof. Straightforward. □

To solve the normalizability, we show the following lemma and theorem.

Lemma 2 (Substitution). *If $\Gamma, x_0 : A, \ldots, x_{n-1} : A \vdash M : B; \Delta$ and $\Gamma' \vdash N : A; \Delta'$ are derivable, then so is $\Gamma, \Gamma' \vdash M[x_0 := N, \cdots, x_{n-1} := N] : B; \Delta, \Delta'$.*

Proof. By induction on $\Gamma, x_0 : A, \ldots, x_{n-1} : A \vdash M : B; \Delta$. □

Theorem 4 (Subject reduction). *If $\Gamma \vdash M : A; \Delta$ is derivable and $M \to_\beta M'$, then $\Gamma \vdash M' : A; \Delta$ is also derivable.*

Proof. Because we formalized the typing rules of λ_{NQ} as multiplicative, the proof is a little bit different from one used in Nakano's additive system, but together with Lemma 2, it readily holds in almost the same manner presented in [9]. □

From a logical viewpoint, both the lemma and the theorem state that the compositionality of derivations and proof normalization are well-defined in *NQ* respectively through the Curry–Howard correspondence.

Now, the strong normalization theorem is proved by embedding from λ_{NQ} into Kameyama and Sato's $L'_{c/t}$ [6][6], which is known to be a strong normalizing calculus and is an extension of Nakano's $L_{c/t}$ with '$\bot$' and the following rule:

$$\frac{\Gamma \vdash M : \bot; \Delta}{\Gamma \vdash \textbf{abort}\ M : A; \Delta}\ \text{Abort}$$

At a glance, this Abort rule is a kind of EFQ rule, and, in fact, their $L'_{c/t}$ logically corresponds to intuitionistic logic. However, this Abort, a kind of EFQ rule, causes the same problems as mentioned in Sect. 2.2. Moreover, they did not give any intuitive meanings of EFQ because it is not in their scope. We use $L'_{c/t}$ just as a technical tool for proving properties of λ_{NQ}.

Then, we define a translation map from λ_{NQ} into $L'_{c/t}$, that preserves their typabilities, and prove a lemma to show that the map also preserves reductions.

Definition 9 (Translation map). *The* translation map $[\![-]\!]$ *is a function which maps* λ_{NQ}*'s derivations* $\Gamma \vdash M : A; \Delta$ *to* $L'_{c/t}$*'s derivations* $[\![\Gamma \vdash M : A; \Delta]\!]$. *It is defined to be a one-to-one mapping by induction on* $\Gamma \vdash M : A; \Delta$.

We show only the translation of EFQ, and the other translations are defined so as to be one-to-one mapping. Note that we use $[\![M]\!]$ *to denote the resulting term of* $[\![\Gamma \vdash M : A; \Delta]\!]$ *for some presupposed derivation* $\Gamma \vdash M : A; \Delta$.

$$\left[\!\!\left[\frac{\Gamma \vdash M : \bot; u : A, \Delta}{\Gamma \vdash \textbf{catch}\ u\ M : A; \Delta}\ \mathit{EFQ}\right]\!\!\right] \overset{def}{=} \cfrac{\cfrac{\Gamma \vdash [\![M]\!] : \bot; u : A, \Delta}{\Gamma \vdash \textbf{abort}\ [\![M]\!] : A; u : A, \Delta}\ \mathit{Abort}}{\Gamma \vdash \textbf{catch}\ u\ (\textbf{abort}\ [\![M]\!]) : A; \Delta}\ \mathit{Catch}$$

Then, the well-definedness of $[\![-]\!]$*:*

"If $\Gamma \vdash M : A; \Delta$ *is derivable in* λ_{NQ}*, then so is* $[\![\Gamma \vdash M : A; \Delta]\!]$ *in* $L'_{c/t}$*."*

can be proved by straightforward induction.

Lemma 3. *If* $\Gamma \vdash M : A; \Delta$ *is derivable and* $M \to_\beta M'$*, then* $[\![M]\!] \to^+_\beta [\![M']\!]$.

Proof. $M \to_\beta M'$ is derived from $N \mapsto N'$ for some terms N, N' s.t. $M \equiv C[N]$ and $M' \equiv C[N']$ for some context C. Then, it is enough to show that $[\![N]\!] \mapsto^+ [\![N']\!]$ by cases on the redex N.

Case 1: $N \equiv \textbf{catch}\ u\ (\textbf{throw}\ u\ L)$ for some L. If N is derived by EFQ, $[\![N]\!] \equiv \textbf{catch}\ u\ (\textbf{abort}\ (\textbf{throw}\ u\ [\![L]\!])) \mapsto \textbf{catch}\ u\ (\textbf{throw}\ u\ [\![L]\!]) \mapsto [\![L]\!] \equiv [\![N']\!]$. Otherwise, N from Catch, and $[\![N]\!] \equiv \textbf{catch}\ u\ (\textbf{throw}\ u\ [\![L]\!]) \mapsto [\![L]\!] \equiv [\![N']\!]$.

Case 2: N is one of the rest. In this case, the redex $[\![N]\!]$ one-to-one corresponds to the original N, namely, $[\![N]\!] \mapsto [\![N']\!]$ holds because $N \mapsto N$ holds. □

Finally, we can now obtain the following two main results for *NQ*.

[6] $L'_{c/t}$ was named as $L_{c/t}$ in their paper, which is the same name as Nakano's calculus. In this paper, we use $L'_{c/t}$ to denote Kameyama and Sato's calculus.

Theorem 5 (Soundness and completeness w.r.t. *NJ*). *$\Gamma \vdash_{NQ} A; \Delta$ if and only if $\Gamma \vdash_{NJ} \bigvee(\{A\} \cup \Delta)$, where $\bigvee \Sigma$ means the disjunction of all the formulae in Σ. In particular, $\Gamma \vdash_{NQ} A; \varnothing$ if and only if $\Gamma \vdash_{NJ} A$.*

Proof. For the only-if part, suppose that $\Gamma \vdash_{NQ} A; \Delta$. Then, $\Gamma \vdash_{NQ} \bigvee(\{A\} \cup \Delta); \varnothing$ follows from $\vee E'$ and hence $[\![\Gamma \vdash \bigvee(\{A\} \cup \Delta); \varnothing]\!]$ holds in $L_{c/t}$. Therefore, $\Gamma \vdash_{NJ} \bigvee(\{A\} \cup \Delta)$ follows from a fact that $L_{c/t}$ logically corresponds to *NJ* [6].

For the if part, we first show that $\Gamma \vdash_{NJ} \bigvee(\{A\} \cup \Delta)$ implies $\Gamma \vdash_{NQ} \bigvee(\{A\} \cup \Delta); \varnothing$ by induction on the derivation. The only non-trivial case is EFQ, but it can be dealt with the following translation from *NJ* derivations into *NQ* derivations:

$$\dfrac{\Gamma \vdash \bot}{\Gamma \vdash A}\ \text{EFQ} \qquad \Longrightarrow \qquad \dfrac{\dfrac{\Gamma \vdash \bot; \varnothing}{\Gamma \vdash \bot; A}\ \text{W}}{\Gamma \vdash A; \varnothing}\ \text{EFQ}$$

Then, $\Gamma \vdash_{NQ} A; \Delta$ follows from $\Gamma \vdash_{NQ} \bigvee(\{A\} \cup \Delta); \varnothing$ by using $\vee E'$ and Ex. □

Theorem 6 (Strong normalization). *For every typable term M in λ_{NQ}, there is no infinite reduction sequence starting from M.*

Proof. Suppose that there exists an infinite reduction sequence starting from M. Considering the subject reduction property of λ_{NQ} and Lemma 3, there will be an infinite reduction sequence starting from $[\![M]\!]$, in $L'_{c/t}$, but this contradicts the strong normalization property of $L'_{c/t}$. □

Thanks to these theorems, we can conclude that *NQ* is logically equivalent to *NJ* and successfully represents intuitionistic logic, and *NQ* satisfies the normalizability of deduction, which is a criterion for "harmony" as mentioned in Sect. 2.

4.2 '⊥' as a Type of Computational Deadend

As a consequence of the strong normalizability of λ_{NQ}, we show that '⊥' actually represents the type of "computational deadend," that is, every term of type '⊥' will "get stuck in computation" eventually. Here, we will explain what is intended by the word "computation" before explaining the intuitive meaning of '⊥'.

Firstly, let us imagine an extension of λ_{NQ} with natural numbers, the type **Nat** for the numbers, and addition operator in order to compute mathematical expressions. Then, for instance, an expression of type **Nat** is computed as follows:

$$(\lambda x.(x+4))(1+2) \to_\beta (\lambda x.(x+4))\,3 \to_\beta (3+4) \to_\beta 7$$

In this example, we eventually get 7 as the final result of the program. The important point here is that: when we want to do "computation" of a program, we intend to obtain some "concrete" result, such as the above 7 in the final analysis, whereas there may be "auxiliary" computations, such as value propagation through variables, as $(\lambda x.(x+4))$ did in the above example.

For the above view of computation, we say that a program will get stuck in computation when the program does not produce any kind of concrete results, and we can see that programs of type '$\bot$' are indeed in computational deadend, as we discuss it hereafter.

For the sake of simplicity, we consider the judgments of form "$\Gamma \vdash M : \bot; \varnothing$" in what follows, but the same story holds in general, i.e., for judgments of form "$\Gamma \vdash M : \bot; \Delta$". Then, it is easy to see, through the Curry–Howard correspondence, that there is no derivable judgment of form "$\varnothing \vdash M : \bot; \varnothing$" because of the consistency of *NJ* and Theorem 5. It also means that if a judgment "$\Gamma \vdash M : \bot; \varnothing$" is derivable, the term M must be constructed from some variable in Γ, whose type has '$\bot$' as a subformula[7]. Therefore, considering the strong normalization theorem, M will eventually get stuck in some normal form[8] because no further reductions occur in such variables. In particular, unlike in the case of the variables of type **Nat**, there is no "concrete" term instantiation to the variables of type '$\bot$'. This is why we say that every program of type '$\bot$' will get stuck and hence is in computational deadend.

To see the above intuition more precisely, let us consider the following:

Example 3. A typing of a kind of disjunctive syllogism is derived as follows. Note that, here we use M^A to express $M : A$ for space convenience.

$$
\dfrac{\dfrac{\dfrac{\vdots}{\Gamma \vdash M^{\neg A}; \varnothing} \qquad \dfrac{\dfrac{\vdots}{\Gamma' \vdash N^{A \vee B}; \varnothing} \quad \dfrac{}{x^A \vdash x^A; \varnothing}\,\text{Ax} \quad \dfrac{\dfrac{}{y^B \vdash y^B; \varnothing}\,\text{Ax}}{y^B \vdash (\mathbf{throw}\ u\ y)^A; u^B}\,\text{Throw}}{\Gamma' \vdash (\mathbf{case}\ N\ \mathbf{of}\ [x]x\ \mathbf{or}\ [y](\mathbf{throw}\ u\ y))^A; u^B}\,\vee E}{\Gamma, \Gamma' \vdash (M(\mathbf{case}\ N\ \mathbf{of}\ [x]x\ \mathbf{or}\ [y](\mathbf{throw}\ u\ y)))^{\bot}; u^B}\,\supset E}{\Gamma, \Gamma' \vdash (\mathbf{catch}\ u\ (M(\mathbf{case}\ N\ \mathbf{of}\ [x]x\ \mathbf{or}\ [y](\mathbf{throw}\ u\ y))))^B; \varnothing}\,\text{EFQ}
$$

The last term derived by EFQ is a construction that represents the proof of the disjunctive syllogism. The computation (i.e., proof normalization) of the term captures how the actual reasoning of the proposition works, and its computation proceeds depending on the form of N as follows: if N was constructed from $\vee I_1$, i.e., $N \equiv \iota_1 L$ for some L^B, then the last term reduces to L via the throw operation; however, if N is constructed from $\vee I_0$, i.e., $N \equiv \iota_0 L$ for some L^A, then the last term will reduce to ($\mathbf{catch}\ u\ (ML)$), and because the type of (ML) is '$\bot$', the whole computation will gets stuck.

In the sequel, if a term of type '$\bot$' appears within a deduction, then the reasoning process of the derivation will be described depending on the following two cases: (i) if the construction actually needs the term of '$\bot$', then it will get stuck; (ii) otherwise, that is, if the reasoning does not need such terms, then a concrete construction will appear eventually.

---

[7] It is equal to say, also through the Curry–Howard correspondence, that: "if $\Gamma \vdash \bot$ holds in *NJ*, then the proof must depend on some assumption consisted of '$\bot$'."

[8] A term M is said to be in *normal form* if there is no further reduction from M.

5 Related Work

Sørensen and Urzyczyn proposed a *dialogue semantics* (i.e., *game semantics*) to model the implicational fragment of intuitionistic logic, in their textbook [15]. The semantics deals with a game with two players, prover and skeptic, to prove propositions by the players' dialogue. In their setting, when a player is supposed to prove '$\bot$', the player aborts the dialogue's process but asserts another formula with knowledge gained during the dialogue. Thus, their formalization seems to rely on the same intuition as ours regarding '$\bot$' and EFQ, whereas their model is not a natural deduction and is not proper based on our motivation.

It is worth noting that there is a justification of classical logic by using multiple-conclusion natural deduction [14]. It is also well known that Parigot's $\lambda\mu$-calculus [11] for classical logic has multiple-conclusion judgments, and $\lambda\mu$ had already involved essentially the same intuition that EFQ as a structural rule. Because intuitionistic logic is a subsystem of classical logic, we can also appeal to these results to justify EFQ. These are, however, too strong because our subject is an analysis of intuitionistic logic. Although we adopted Nakano's $L_{c/t}$ as a basis because our aim is to justify intuitionistic logic, it is another interesting direction to give a novel justification of classical logic with $\lambda\mu$.

6 Conclusion

We have proposed *NQ* to justify EFQ by using quasi-multiple-conclusion natural deduction and "jump inference." In a nutshell, the problem of the justifications of EFQ appealing to the meaning of '$\bot$' is that there is no semantic connection between '$\bot$' and arbitrary propositions. In *NQ*, on the other hand, EFQ is considered a structural rule that expresses a "jump" inference from the deadend to another possibility. In terms of this reinterpretation, EFQ is no longer a counter-intuitive rule that derives every proposition from the meaning of '$\bot$', but rather one that retrieves another possibility on the grounds that the curernt reasoning reaches the deadend.

The above result also indicates that it is indeed minimal logic that is most fitting for the standard proof-theoretic semantics via single-conclusion natural deduction and simply-typed λ-calculus. Together with the result in Parigot [11], we can now obtain a rough yet instructive sketch of the correspondence among the logics, natural deduction systems, and calculi in Table 1. Although the correspondence between logics and systems is well known in the literature [7,15], our study supports this view from the philosophical viewpoint of proof-theoretic semantics.

In spite of the substantial change of the underlying systems, intuitionistic logic represented as *NQ* is still constructive in the sense that the disjunctive property still holds. On the other hand, it can also be said that intuitionic logic, truly viewed, takes a step toward *unconstructivity* in that it essentially contains the function of "jump inference," which has been considered a characteristic of classical logic. This fact suggests that its status of the logic of "constructivism" should be reconsidered.

Table 1. Logics, systems, and calculi

Logic	System	Calculus
Classical	Multiple	$\lambda\mu$
Intuitionistic	Quasi-multiple	λ_{NQ}
Minimal	Single	$\lambda_{\rightarrow}$

Acknowledgment. We appreciate the helpful comments of Atsushi Igarashi and Takuro Onishi that improved the presentation of this paper. We are also grateful to the anonymous referees for their helpful comments.

References

1. Cook, R.T., Cogburn, J.: What negation is not: intuitionism and '0=1'. Analysis **60**(265), 5–12 (2000)
2. Dummett, M.: The Logical Basis of Metaphysics. Harvard University Press, Cambridge (1991)
3. Dummett, M.: Elements of Intuitionism. Oxford Logic Guides. Clarendon Press, Oxford (2000)
4. Griffin, T.G.: A formulae-as-type notion of control. In: Proceedings of the 17th ACM SIGPLAN-SIGACT Symposium on Principles of Programming Languages, pp. 47–58 (1990)
5. Hand, M.: Antirealism and falsity. In: Gabbay, D.M., Wansing, H. (eds.) What is Negation?, pp. 185–198. Springer, Dordrecht (1999). doi:10.1007/978-94-015-9309-0_9
6. Kameyama, Y., Sato, M.: Strong normalizability of the non-deterministic catch/throw calculi. Theor. Comput. Sci. **272**(1–2), 223–245 (2002)
7. Maehara, S.: Eine darstellung der intuitionistischen logik in der klassischen. Nagoya Math. J. **7**, 45–64 (1954)
8. Nakano, H.: A constructive logic behind the catch and throw mechanism. Ann. Pure Appl. Logic **69**(2), 269–301 (1994)
9. Nakano, H.: Logical structures of the catch and throw mechanism. Ph.D. thesis, The University of Tokyo (1995)
10. Onishi, T.: Proof-theoretic semantics and bilateralism. Ph.D. thesis, Kyoto University (2012). (Written in Japanese)
11. Parigot, M.: $\lambda\mu$-Calculus: an algorithmic interpretation of classical natural deduction. In: Voronkov, A. (ed.) LPAR 1992. LNCS, vol. 624, pp. 190–201. Springer, Heidelberg (1992). doi:10.1007/BFb0013061
12. Prawitz, D.: Meaning approached via proofs. Synthese **148**(3), 507–524 (2006)
13. Prawitz, D.: Pragmatist and verificationist theories of meaning. In: Auxier, R.E., Hahn, L.E. (eds.) The Philosophy of Michael Dummett. Open Court Publishing Company (2007)
14. Read, S.: Harmony and autonomy in classical logic. J. Philos. Logic **29**(2), 123–154 (2000)
15. Sørensen, H., Urzyczyn, P.: Lectures on the Curry-Howard Isomorphism. Elsevier, Amsterdam (2006)
16. Steele, G.L.: Common LISP: The Language, 2nd edn. Digital Press, Newton (1990)

17. Tennant, N.: Negation, Absurdity and Contrariety. In: Gabbay, D.M., Wansing, H. (eds.) What is Negation?, pp. 199–222. Springer, Dordrecht (1999)
18. Tranchini, L.: The role of negation in proof-theoretic semantics: a proposal. Fuzzy Logics Interpret. Logics Resour. **9**, 273–287 (2008)
19. van Dalen, D.: Kolmogorov and Brouwer on constructive implication and the Ex Falso rule. Russ. Math. Surv. **59**(2), 247–257 (2004)

A Nonmonotonic Modal Relevant Sequent Calculus

Shuhei Shimamura(✉)

Nihon University, Funabashi, Japan
shimamura.shuuhei@nihon-u.ac.jp

Abstract. Motivated by semantic inferentialism and logical expressivism proposed by Robert Brandom, in this paper, I submit a nonmonotonic modal relevant sequent calculus equipped with special operators, $\Box$ and **R**. The base level of this calculus consists of two different types of atomic axioms: material and relevant. The material base contains, along with all the flat atomic sequents (e.g., Γ_0, p $|\!\sim_0$ p), some non-flat, defeasible atomic sequents (e.g., Γ_0, p $|\!\sim_0$ q); whereas the relevant base consists of the local region of such a material base that is sensitive to relevance. The rules of the calculus uniquely and conservatively extend these two types of nonmonotonic bases into logically complex material/relevant consequence relations and incoherence properties, while preserving *Containment* in the material base and *Reflexivity* in the relevant base. The material extension is supra-intuitionistic, whereas the relevant extension is stronger than a logic slightly weaker than R. The relevant extension also avoids the fallacies of relevance. Although the extended material consequence relation is defeasible and insensitive to relevance, it has local regions of indefeasibility and relevance (the latter of which is marked by the relevant extension). The newly introduced operators, $\Box$ and **R**, codify these local regions within the same extended material consequence relation.

Keywords: Nonmonotonicity · Defeasibility · Relevance · Semantic inferentialism · Logical expressivism

1 Introduction

1.1 Background

There are two features of inference that may motivate us to reject monotonicity: relevance and defeasibility. On one hand, we usually demand, in everyday inferential practice, that the premise and the consequence be relevant. For example, it sounds inappropriate to say, "The grass is green; *therefore*, the snow is white." Relevance logicians take this feature of our folk concepts of inference seriously and bring it into the context of formal inference by imposing a certain constraint on the validity of their conditionals. Attention to defeasibility, on the other hand, is raised on rather different grounds. While formal inferences, such as *Modus Ponens*, are safely assumed to be indefeasible, once we go beyond that well-domesticated realm of inferences, indefeasibility no longer appears to be the norm. For instance, an everyday inference—such as, "The road is wet; therefore, it rained"—is easily defeated by a new piece of

A. Baltag et al. (Eds.): LORI 2017, LNCS 10455 pp. 570–584, 2017.
DOI: 10.1007/978-3-662-55665-8_39

information, such as, "A street cleaner has just passed by." Such fragility is broadly observed even in inferences in more sophisticated areas, such as law and sciences (except, perhaps, for microscopic physics). People interested in reasoning in these areas have sought for a system to deal with defeasible inferences.

These two nonmonotonic features of inference, relevance and defeasibility, have mainly been investigated separately in the contexts of formal and non-formal inferences. However, there seems to be no reason this must be so. After all, sensitivity to relevance also appears to be an important feature of *non-formal* inferences. For example, it would be strange to say, "The road is wet; and *the snow is white*; *therefore*, it rained." What I offer in this paper is a system capable of dealing with inferences that are both defeasible *and* relevant. Such a system is not only rarely investigated, but also philosophically motivated. In particular, my system is motivated by two distinctive philosophical ideas.[1]

1.2 Philosophical Motivations

Semantic Inferentialism. One such philosophical motivation is semantic inferentialism advocated by Robert Brandom [3, 4]. Semantic inferentialism is a radical generalization of inferentialist semantics in logics. According to the latter, more modest view, the meaning of a logical connective is explained, roughly, by the rules governing its use in formal inferences. The former view radicalizes this idea by claiming that even the meaning of a *non-logical* expression can be explained by the rules governing its inferential use. If such inferential use is restricted to the use in formal inferences, however, this view is hopeless. After all, the validity of a formal inference tells us nothing about the meanings of its non-logical components, as the inference is always valid, by its formal nature, no matter what those components are. From a wider perspective, however, there appear to be those inferences in which non-logical expressions are involved *materially*—in that the correctness of those inferences depends on the involvement of particular non-logical expressions. Remember the inference from "The road is wet" to "It rained." Here, "The road" is materially involved, since this inference turns incorrect if "The road" is replaced by, say, "The water." According to semantic inferentialism, the meaning of "The road" is explained by the rule governing its use in this or many other inferences of the same sort. Those inferences are called *material inferences*. As illustrated by the above example, material inferences are typically defeasible. Thus, semantic inferentialists who care about relevance need a system that can deal with *defeasible*, relevant inferences.

Logical Expressivism. The other philosophical background motivating my system is logical expressivism, proposed, again, by Brandom [5]. For semantic inferentialists, meanings of both logical and non-logical expressions are explained by their inferential rules. Then, what distinguishes the logical vocabulary from the non-logical vocabulary? Logical expressivists answer this question by offering, roughly, the following two

---

[1] This is a product of our collaborative work with Robert Brandom's research group. The technical results reported here are mine.

requirements for logicality: (i) the usage of logically complex expressions must be *elaborated from* (or determined by) the usage of non-logical expressions in the underlying material inferential practice; (ii) the inferential role of a piece of logical vocabulary must be to *express* (or codify at the level of the object language) various features of the underlying material inferential practice.

These two requirements of logical expressivism put several constraints on my logical system. On one hand, (i) demands that an expressivist logical system must provide a set of rules that uniquely extend the underlying material (therefore, defeasible) inferential relation over non-logical (i.e., atomic) sentences into the ones over logically complex sentences. On the other hand, (ii) imposes two more constraints on such a system. First, if adding a bit of logical vocabulary changed a feature of the underlying material inferential practice that it is supposed to express, it would not count as expressing that feature. Therefore, the logical extension required by (i) must be done conservatively. Second, (ii) also requires that for a piece of logical vocabulary to play an expressive role, it must codify some distinctive feature of the underlying material inferential relation. A typical instance of such expressive vocabulary is the conditional. As long as the deduction theorem holds (i.e., $\Gamma \mid\sim A \rightarrow B$ *iff* $\Gamma, A \mid\sim B$, where $\mid\sim$ stands for the material implication), the conditional codifies a piece of information on which sentence follows from which in a given context.

Now, what other features of the underlying material inferential relation are worth expressing? Indefeasibility and relevance should naturally be included. Although the material inferential relation is defeasible in general, there is a local region in which inferences hold indefeasibly. *Modus Ponens* (i.e., $A, A \rightarrow B \mid\sim B$), for instance, typically belong to that region. In my system, such local indefeasibility is expressed by a box operator such that $\Gamma \mid\sim \Box A$ *iff* $\forall\Delta$ $(\Delta, \Gamma \mid\sim A)$.[2] Similarly, although semantic inferentialists do not explicitly require that the underlying material inferential relation must be relevant, it would be helpful to delineate a local region in which inferences are sensitive to relevance. After all, that "The road is wet" is relevant to "It rained," while "The snow is white" is not also seems to be an important aspect of the usage of these non-logical expressions involved, and therefore, according to semantic inferentialism, to their meanings. (It is no wonder that instances of material inferences that semantic inferentialists like to cite are almost always relevant ones.) In my system, such a local region of relevant inferences is expressed by a new operator, **R**, such that $\Gamma \mid\sim \mathbf{R}A$ *iff* $\Gamma \mid\sim_r A$, where $\mid\sim_r$ stands for the implication sensitive to relevance.

1.3 Prospects

Here is what I offer in the rest of this paper. As semantic inferentialists, we begin with the underlying material consequence relation and incoherence property over atomic language, $\mid\sim_0$, which I call the *material base*. Although the material base is defeasible (and therefore nonmonotonic), we stipulate that it satisfies *Containment* (i.e., $\forall\Gamma_0\forall p$ $(\Gamma_0, p \mid\sim_0 p)$) and therefore has a local region of indefeasibility. We also stipulate that although the entire base is not relevant, it has a local region of relevance, $\mid\sim_{r0} \subseteq \mid\sim_0$,

[2] This indefeasibility box and the mechanism that makes it work are proposed by Hlobil [6].

which I call *the relevant base*, and this region satisfies *Reflexivity* (i.e., $\forall p\ (p \mid\!\sim_{r0} p)$). Now, our sequent calculus uniquely extends such a material/relevant base into the logically complex material/relevant consequence and incoherence, $\mid\!\sim / \mid\!\sim_r$. Such logical extensions are conservative and preserve *Containment* in $\mid\!\sim$ and *Reflexivity* in $\mid\!\sim_r$. Moreover, $\mid\!\sim$ is supra-intuitionistic, whereas $\mid\!\sim_r$ is stronger than a logic slightly weaker than logic R. $\mid\!\sim_r$ also avoids the fallacies of relevance. Finally, our logical connectives include, among others, $\rightarrow$, $\Box$, and **R**. These connectives jointly allow us to codify, within the object language (more specifically in $\mid\!\sim$), which sentence is indefeasibly and/or relevantly implied by which in a given context.

2 The System

2.1 The Material Base

Our atomic language (L_0) consists of a set of atomic sentences (L_0^-) and $\bot$. The underlying material consequence relation and incoherence over L_0, or the material base, is defined as $\mid\!\sim_0 \subseteq P(L_0^-) \times L_0$. Since we treat premises as a set of formulas, *Permutation* and *Contraction* are stipulated. We also stipulate that the material base obeys (at least) the following two principles:

Containment: $\forall \Gamma_0 \subseteq L_0^- \forall p \in L_0^- (\Gamma_0,\ p \mid\!\sim_0 p)$.
Ex Falso Fixo Quodlibet (*ExFF*): $\forall \Gamma_0 \subseteq L_0^- \forall p \in L_0^- (\forall \Delta_0 \subseteq L_0^- (\Delta_0,\ \Gamma_0 \mid\!\sim_0 \bot \Rightarrow \Gamma_0 \mid\!\sim_0 p))$.

ExFF is our restricted version of explosion principle, according to which explosion occurs only if the premise-set is *indefeasibly* incoherent (i.e., remains incoherent under the arbitrary addition of extra premises). Since we consider the material base as defeasible, however, *Weakening* is not a part of our stipulations.

2.2 The Relevant Base

There is no guarantee that the material base is sensitive to relevance. That is, when $\Gamma_0 \mid\!\sim_0 p$, Γ_0 may contain a premise that is irrelevant to the implication of p. To make the material base sensitive to relevance, we must rule out such an irrelevant implication from a redundant premise-set. This means that the underlying consequence and incoherence that are sensitive to relevance, or the relevant base ($\mid\!\sim_{r0}$), must be a subset of the material base.

Subset: $\Gamma_0 \subseteq L_0^- \forall p \in L_0 (\Gamma_0 \mid\!\sim_{r0} p \Rightarrow \Gamma_0 \mid\!\sim_0 p)$.

In the relevant base, any versions of the explosion principle should not be stipulated, as they, by their nature, ignore the relevance of implications. *Containment* should also be rejected on the same grounds. Instead of *Containment*, however, the following weaker stipulation should be acceptable even in the relevant base:

$$\textit{Reflexivity}: \forall p \in L_0^-(p \mid\sim_{r0} p).$$

Now, it is an open question whether we should make any other stipulations in the relevant base. The answer to this question depends on how we understand the concept of relevance. In the context of the formal inference, for example, the concept of relevance is understood to involve the variable sharing principle.[3] According to this principle, A → B is a theorem of (i.e., formally valid in) a given relevance logic only if A and B share at least one propositional variable. Is there any similar principle that captures an aspect of relevance in the context of the non-formal, material inference? Arguably, the following principle might be a candidate:

$$\textit{No Redundancy}: \forall \Gamma_0 \subseteq L_0^- \forall p \in L_0(\Gamma_0,\ p \mid\sim_{r0} p \Rightarrow \Gamma_0 = \varnothing).$$

In any case, it is a new and important task to analyze the concept of relevance in the material inference. To make room for different conceptions of such material relevance, in this paper, I do not impose any constraints, except for *Subset* and *Reflexivity*.

2.3 Logical Extension

Next, let us extend these material/relevant bases over atomic language ($\mid\sim_0 / \mid\sim_{r0}$) into logically complex material/relevant consequence relations and incoherence properties ($\mid\sim / \mid\sim_r \subseteq P(L^-) \times L$). Our logically complex language (L) is defined as $\{\bot\} \cup L^-$, where L^- has the following syntax:

$$\textbf{Syntax of } \mathbf{L^-}: \ \varphi ::= p \mid \varphi \rightarrow \varphi \mid \neg\, \varphi \mid \varphi \,\&\, \varphi \mid \varphi \vee \varphi \mid \Box\, \varphi \mid R\, \varphi.$$

Now, we are going to take in the material/relevant bases as axioms of our logical system and extend them via our logical rules. Let us call this system *nonmonotonic modal relevant sequent calculus* (NMMR for short). NMMR must deal with quite a number of different snake turnstiles (to be exact, the number is four times the cardinality of the power set of the atomic language). To avoid confusion, let me provide a quick overview of the intended readings and mutual relationship of these turnstiles. First, NMMR distinguishes four types of consequence relations and incoherence properties: That is, (i) the extended material consequence and incoherence ($\mid\sim$), (ii) the *flat* region of (i) ($\mid\sim_f$); (iii) the *relevant* region of (i) ($\mid\sim_r$); and the *flat* region of (iii) ($\mid\sim_{rf}$). The *flat* snake turnstile ($\mid\sim_f$) is supposed to mark, within the material consequence and incoherence ($\mid\sim$), the logical ramifications from the tautological implications of *Containment*; while the *relevant flat* snake turnstile ($\mid\sim_{rf}$) is supposed to mark, within the relevant material consequence and incoherence ($\mid\sim_r$), the logical ramifications from the tautological implications of *Reflexivity*.

Furthermore, in NMMR, each of these four snake turnstiles is superscripted by an indexed upward arrow ($^{\uparrow X}$), where X is a set of sets of atomic formulas. The intended reading of, say, "$\Gamma \mid\sim^{\uparrow X} A$" is that for any atomic $\Delta_0 \in X$, $\Delta_0 \cup \Gamma$ implies A. In other

[3] See, for instance, Anderson & Belnap [1]. Also Mares [7].

words, X is supposed to keep track of sets of atomic sentences that do *not* jointly defeat the implication of A from Γ. If X = {Ø}, "$\Gamma \mid\sim^{\uparrow X} A$" is abbreviated as "$\Gamma \mid\sim A$"; while if X = P(L$_0$), "$\Gamma \mid\sim^{\uparrow X} A$" is abbreviated as "$\Gamma \mid\sim^{\uparrow} A$". The same reading and abbreviations apply to all the other types of snake turnstiles with "$^{\uparrow X}$". The unindexed upward arrow ($^{\uparrow}$) plays a pivotal role in introducing the nonmonotonicity box (□) into our system.

It is important to keep in mind, however, that the principal consequence relation and incoherence property in NMMR is the one represented by the plain snake turnstile without the upward arrow (i.e., $\mid\sim$). After all, all the other turnstiles are simply means for marking those properties of the principal turnstile in which we are interested, such as relevance ($\mid\sim_r$), logical derivability ($\mid\sim_f$ and $\mid\sim_{rf}$), and robustness under the augmentation of the premise-set ($\mid\sim^{\uparrow X}$). In the end, some pieces of such information are encoded in the principal turnstile by means of our special operators, such as **R** and □.

Now, let us turn to the logical extension. First, there are four types of axioms, which correspond to the four types of consequence and incoherence, as mentioned above.

Axioms of NMMR

Ax1: If $\Gamma \mid\sim_0 p$, then $\Gamma \mid\sim p$ is an axiom.
Ax2: $\Gamma, p \mid\sim_f p$ is an axiom.
Ax3: If $\Gamma \mid\sim_{r0} p$, then $\Gamma \mid\sim_r p$ is an axiom.
Ax4: $p \mid\sim_{rf} p$ is an axiom.

Recall that *Containment* is stipulated in $\mid\sim_0$, and *Reflexivity* and *Subset* are stipulated in $\mid\sim_{r0}$. Thus, it is obvious that at this stage of the construction, $\mid\sim\ \supseteq\ \mid\sim_f$, $\mid\sim_r\ \supseteq\ \mid\sim_{rf}$, and $\mid\sim\ \supseteq\ \mid\sim_r$.

Next, let us set down the connective rules that extend these four consequence relations and incoherence properties over the atomic language into the ones over the logically complex language as defined above. (Note that since we treat the premise(s) as a set, *Permutation* and *Contraction* are automatically built into our system.)

Connective rules of NMMR

$$\frac{\Gamma \mid\sim_{[r][f]}{}^{\uparrow X} B \qquad \Delta,\ C \mid\sim_{[r]f}{}^{\uparrow Y} A}{\Gamma, \Delta, B \rightarrow C \mid\sim_{[r][f]}{}^{\uparrow X[/Y]} A}\text{LLC1} \qquad \frac{\Gamma,\ A \mid\sim_{[r][f]}{}^{\uparrow X} B}{\Gamma \mid\sim_{[r][f]}{}^{\uparrow X} A \rightarrow B}\text{RC}$$

(where **R** does not appear in Δ, C, or A)

$$\frac{\Gamma \mid\sim_{r[f]}{}^{\uparrow X} B \qquad \Delta,\ C \mid\sim_{[r]f}{}^{\uparrow Y} A}{\Gamma,\ \Delta,\ B \rightarrow C \mid\sim_{[r][f]}{}^{\uparrow X[/Y]} A}\text{LLC2}$$

(where **R** appears in Δ, C, or A)

$$\frac{\Gamma \mid\sim_{[r][f]}{}^{\uparrow X} A}{\Gamma, \neg A \mid\sim_{[r][f]}{}^{\uparrow X} \bot}\text{LN} \qquad \frac{\Gamma,\ A \mid\sim_{[r][f]}{}^{\uparrow X} \bot}{\Gamma \mid\sim_{[r][f]}{}^{\uparrow X} \neg A}\text{RN}$$

$$\frac{\Gamma, \mathrm{B} \mid\sim_{[r]f}{}^{\uparrow X} \mathrm{A}}{\Gamma, \mathrm{C} \,\&\, \mathrm{B}[/\mathrm{B} \,\&\, \mathrm{C}] \mid\sim_{[r][f]}{}^{\uparrow X} \mathrm{A}}\mathrm{LL\&} \qquad \frac{\Gamma \mid\sim_{[r][f]}{}^{\uparrow X} \mathrm{A} \qquad \Gamma \mid\sim_{[r][f]}{}^{\uparrow X} \mathrm{B}}{\Gamma \mid\sim_{[r][f]}{}^{\uparrow X} \mathrm{A} \,\&\, \mathrm{B}}\mathrm{R\&}$$

$$\frac{\Gamma, \mathrm{A} \mid\sim_{[r][f]}{}^{\uparrow X} \mathrm{C} \qquad \Gamma, \mathrm{B} \mid\sim_{[r][f]}{}^{\uparrow X} \mathrm{C}}{\Gamma, \mathrm{A} \vee \mathrm{B} \mid\sim_{[r][f]}{}^{\uparrow X} \mathrm{C}}\mathrm{LV} \qquad \frac{\Gamma \mid\sim_{[r][f]}{}^{\uparrow X} \mathrm{A}[/\mathrm{B}]}{\Gamma \mid\sim_{[r][f]}{}^{\uparrow X} \mathrm{A} \vee \mathrm{B}}\mathrm{RV}$$

$$\frac{\Gamma, \mathrm{A} \mid\sim_{[r][f]}{}^{\uparrow X} \mathrm{B}}{\Gamma, \Box\mathrm{A} \mid\sim_{[r][f]}{}^{\uparrow X} \mathrm{B}}\mathrm{LB} \qquad \frac{\Gamma \mid\sim_{[r][f]}{}^{\uparrow} \mathrm{A}}{\Gamma \mid\sim_{[r][f]}{}^{[\uparrow]} \Box\mathrm{A}}\mathrm{RB}$$

$$\frac{\Gamma, \mathrm{A} \mid\sim_{[r][f]}{}^{\uparrow X} \mathrm{B}}{\Gamma, \mathbf{R}\mathrm{A} \mid\sim_{[r][f]}{}^{\uparrow X} \mathrm{B}}\mathrm{LR} \qquad \frac{\Gamma \mid\sim_{r[f]}{}^{\uparrow X} \mathrm{A}}{\Gamma \mid\sim_{[r][f]}{}^{\uparrow X} \mathbf{R}\mathrm{A}}\mathrm{RR}$$

$$\frac{\Gamma \mid\sim_{[f]}{}^{\uparrow} \mathrm{A}}{\Gamma, [\mathrm{B}] \mid\sim_{[f]}{}^{[\uparrow]} \mathrm{A}}\mathrm{pW}$$

(where **R** does not appear in A)

$$\frac{\Gamma, \mathrm{p}_1, \ldots, \mathrm{p}_n \mid\sim_{[f]}{}^{\uparrow X} \mathrm{A} \qquad \Gamma \mid\sim_{[f]}{}^{\uparrow Y} \mathrm{A}}{\Gamma \mid\sim_{[f]}{}^{\uparrow \{\{p1, \ldots, pn\}\} \cup X \cup Y} \mathrm{A}}\mathrm{PushUpUN}$$

(where **R** does not appear in A, and $\mathrm{p}_1, \ldots, \mathrm{p}_n \in \mathrm{L}_0^-$)

$$\frac{\Gamma \mid\sim_{[f]}{}^{\uparrow} \bot}{\Gamma \mid\sim_{[f]}{}^{[\uparrow]} \mathrm{A}}\mathrm{ExFF}$$

(where **R** does not appear in A)

These rules are systematically ambiguous with respect to the square-bracketed elements. First, if the same element is square-bracketed at *both* top *and* bottom sequents, it must be shared by both sequents. Take RC as an example. RC says that for any of the four types of consequence and incoherence, if Γ together with A imply B, then Γ implies A → B in *that same* consequence and incoherence. The same reading also applies to all the other rules. Second, if an element is square-bracketed *only* at the *bottom* sequent, it is optional at the bottom. For instance, take RR. Since there is no square-bracket around "r" at the top sequent of RR, the top sequent must be relevant. However, since "r" at the bottom sequent is square-bracketed, the bottom sequent can be either relevant or non-relevant. The similar reading goes for LLC1, LLC2, LL&, RB, pW, and ExFF. Finally, if an element is square-bracketed *and* prefixed with a backslash, it can replace the adjoining element. For example, LL& says that from $\Gamma, \mathrm{B} \mid\sim_{[r]f}{}^{\uparrow X} \mathrm{A}$, both $\Gamma, \mathrm{C} \,\&\, \mathrm{B} \mid\sim_{[r][f]}{}^{\uparrow X} \mathrm{A}$ and $\Gamma, \mathrm{B} \,\&\, \mathrm{C} \mid\sim_{[r][f]}{}^{\uparrow X} \mathrm{A}$ are derivable. LLC1, LLC2, and RV should be read in the similar manner.

Setting these complications aside, most of the rules are, more or less, familiar from Gentzen's LJ. However, there are several rules that may need further explanations.

First, pW and PushUpUN are supposed to guarantee that $|\sim^{\uparrow}$ marks what it is supposed to mark, namely, the indefeasibility of implications (for technical details, see the proof of Lemma 4 below). Then, RB allows us to encode, within the object language, such indefeasibility of implications by means of $\Box$. Note that since "$^{\uparrow}$" of the bottom sequent of RB is optional, this encoding can also be done in $|\sim$, our principal turnstile.

Second, RR does a similar job with respect to the relevance of implications, which is supposed to be marked by $|\sim_{r}$. RR allows us to encode this by means of our newly introduced operator, **R**. Again, since "$_{r}$" at the bottom sequent of RR is optional, such encoding can also be done in $|\sim$.

Third, it may also be noted that the left rules for the conditional and the conjunction (i.e., LLC1, LLC2, and LL&) are remarkably weaker than the other left rules, as they demand that (one of) the top sequent(s) must be flat (i.e., $|\sim_{f}$ or $|\sim_{rf}$). This restriction is forced by one of our motivating ideas for the entire project, that we are dealing with consequence relations and incoherence properties that are defeasible. Consider LL&, for instance. Without the restriction at issue, LL& would allow us to derive p&q $|\sim$ r from p $|\sim$ r. However, q is arbitrary and may be a defeater of the implication of r from p (e.g., consider the case where p: the match is struck, q: the match is wet, and r: the match lights). Thus, assuming that the conjunction means what it is supposed to mean on the left-hand side (i.e., p *and* q), LL& should not be applied across the board. Then, under what condition is LL& applied safely? Our answer is that it is when LL& is applied in logical derivations—namely those derivations that are traced back to the tautological implications of *Containment* or *Reflexivity*, as we are happy to acknowledge that tautological implications and their logical ramifications are indefeasible (although implications are defeasible across the board). Thus, the restriction above follows. The same justification, mutatis mutandis, can be applied to LLC1 and LLC2.[4]

Finally, remember that at the atomic level, it is stipulated that $|\sim \supseteq |\sim_{f}$, $|\sim_{r} \supseteq |\sim_{rf}$, and $|\sim \supseteq |\sim_{r}$. These inclusion relations between different types of snake turnstiles are preserved by our logical rules. In other words, $|\sim_{f}$, $|\sim_{r}$, and $|\sim_{rf}$ can be understood as marking a special region of our main turnstile, $|\sim$. Given our rules, this intuitively makes sense. After all, most of our rules (i.e., RC, LN, RN, R&, LV, RV, LB, LR) uniformly extend all the four types of turnstiles. For the rules that require the top sequent(s) of having a special property (i.e., LLC1, LLC2, LL&, RB, RR, pW, and

[4] Two more comments are in order for LLC1 and LLC2. First, it may be unnoticeable how LLC2 differs from LLC1. The only formal difference is that the left top sequent of LLC2 must be relevant, while the corresponding sequent of LLC1 need not. This requirement on LLC2 is crucial for **R** to codify what it is supposed to codify (see Proposition 12 below). Second, given that the indexed upward arrow is supposed to mark sets of non-defeaters of a given implication, one may find the upward arrow of the bottom sequent of LLC1 and LLC2 (i.e., $^{\uparrow X[/Y]}$) counterintuitive. Upon closer look, however, there is no substantial harm here. After all, if the bottom sequent is non-relevant, then it is supposed to hold indefeasibly (remember that the right top sequent must be flat, and therefore is supposed to hold indefeasibly). If the bottom sequent is relevant, on the other hand, both right and left top sequents must also be relevant. Then, both indices of those top sequents (i.e., $^{\uparrow X}$ and $^{\uparrow Y}$) must be the empty set, since PushUpUN has no application in relevant sequents. The technical advantage of the current formulations of LLC1 and LLC2 are substantial. They enable us to prove the admissibility of restricted versions of Cut in NMMR (see Proposition 7 below).

ExFF), it is always optional whether to keep that special property at the bottom sequent. (For a rigorous proof of this point, see Proposition 6 below.)

3 Properties of the System

In this section, we show that NMMR has several desirable properties, as stated in Sect. 1.3.

3.1 Conservativeness

First, NMMR does not add any atomic relevant consequence or incoherence that does not hold at the relevant base. That is, NMMR conservatively extends $|\sim_{r0}$.

Proposition 1. $\forall \Gamma_0 \subseteq L_0^- \forall p \in L_0(\Gamma_0 |\sim_r p \Leftrightarrow \Gamma_0 |\sim_{r0} p)$.

Proof. The right-to-left direction. Straightforward from Ax3. *The left-to-right direction.* Straightforward from the fact that all the connective rules concerning $|\sim_r$ are complicating. ■

For the Conservativeness proof for the material base, we first need to prove the following lemma.[5]

Lemma 1. $\forall \Gamma_0 \subseteq L_0^- \forall p \in L_0(\Gamma_0 |\sim^{\uparrow X} p \Rightarrow \forall \Delta_0 \in X\ (\Gamma_0, \Delta_0 |\sim_0 p))$.

Proof. By induction on the proof height of the antecedent sequent. The base case is trivial. In induction step, the antecedent sequent can only come by PushUpUN, pW, or ExFF. In any of these cases, given the Induction Hypothesis, the consequent sequent follows. ■

Now, we are in a position to prove that NMMR also conservatively extends $|\sim_0$.

Proposition 2. $\forall \Gamma_0 \subseteq L_0^- \forall p \in L_0(\Gamma_0 |\sim p \Leftrightarrow \Gamma_0 |\sim_0 p)$.

Proof. The right-to-left direction. Straightforward from Ax1. *The left-to-right direction.* Straightforward from Lemma 1. ■

3.2 Preservation of Reflexivity and Containment

On the one hand, NMMR preserves *Reflexivity* in the relevant base with respect to the ϒ-free fragment of the extended relevant (flat) consequence and incoherence.

Proposition 3. $\forall A \in L^- \left(A |\sim_{r[f]} A\right)$, where no □ appears in A.

Proof. By induction on the complexity of the formula on the right-hand side. ■

On the other hand, NMMR also preserves *Containment* in the material base with respect to the **R**-free fragment of the extended material (or flat) consequence and incoherence. We first need to prove the following lemma.

[5] Recall that it is syntactically stipulated that $X \subseteq P(L_0)$.

Lemma 2. $\forall\Gamma \subseteq L^{-}\forall A \in L^{-}(\Gamma, A \mid\sim_{[f]}{}^{[\uparrow]}A)$, where no **R** appears in Γ or A.

Proof. By induction on the complexity of the formula on the right-hand side. ■

Now, we are in a position to prove our target proposition.

Proposition 4. $\forall\Gamma \subseteq L^{-}\forall A \in L^{-}(\Gamma, A \mid\sim_{[f]}{}^{[\uparrow]}A)$, where no **R** appears in A.

Proof. Take arbitrary Γ, and let Γ' be the largest **R**-free subset of Γ. By Lemma 2, for any **R**-free A, $\Gamma', A \mid\sim_{[f]}{}^{\uparrow} A$. By (repeated) applications of pW, $\Gamma, A \mid\sim_{[f]}{}^{[\uparrow]} A$ ■

3.3 Avoiding Fallacies of Relevance

One may well wonder in what sense the extended "relevant" consequence and incoherence ($\mid\sim_r$) of NMMR are relevant. Here is one answer: In $\mid\sim_r$, the fallacies of relevance[6] are not derivable as a theorem.

Proposition 5. None of the following formulas are a theorem in $\mid\sim_r$ of NMMR: (1) $(A \,\&\, \neg A) \to B$; (2) $A \to (B \to A)$; (3) $A \to (B \to B)$; (4) $\neg A \to (A \to B)$; (5) $(A \to B) \vee (B \to C)$; (6) $A \to (B \vee \neg B)$.

Proof. By counterexamples. Suppose that A = p, B = q, and C = r. Pick up a relevant base in which (i) *No Redundancy* holds, (ii) neither $p \mid\sim_{r0} q$ nor $q \mid\sim_{r0} r$, and (iii) neither $p \mid\sim_{r0} q$ nor $p, q \mid\sim_{r0} \bot$. (1): $\mid\sim_r (p \,\&\, \neg p) \to q$ can only come (by RC) from $p \,\&\, \neg p \mid\sim_r q$. But the latter sequent is not derivable by any of our rules. (2): $\mid\sim_r p \to (q \to p)$ can only come (by RC) from $p, q \mid\sim_r p$. But the latter sequent does not hold because of (i). (3): $\mid\sim_r p \to (q \to q)$ can only come (by RC) from $p, q \mid\sim_r q$. But the latter sequent does not hold because of (i). (4): $\mid\sim_r \neg p \to (p \to q)$ can only come (by RC) from $p, \neg p \mid\sim_r q$. But given that $q \neq \bot$, the latter sequent is not derivable from any of our rules. (5): $\mid\sim_r (p \to q) \vee (q \to r)$ can come from either $p \mid\sim_r q$ or $q \mid\sim_r r$. Neither of them holds by (ii). (6): $\mid\sim_r p \to (q \vee \neg q)$ can come from either $p \mid\sim_r q$ or $p, q \mid\sim_r \bot$. Neither of them holds by (iii). ■

3.4 Strength of the System

Inclusion Relations Between Different Snake Turnstiles. It is supposed to be the case that subscripted snake turnstiles (i.e., $\mid\sim_f$, $\mid\sim_r$, and $\mid\sim_{rf}$) mark a special region of corresponding non-subscripted snake turnstiles (i.e., $\mid\sim$, $\mid\sim$, and $\mid\sim_r$, respectively). This is shown to be the case by the following proposition.

Proposition 6. The following inclusion relations hold between $\mid\sim^{\uparrow X}$, $\mid\sim_f{}^{\uparrow X}$, $\mid\sim_r{}^{\uparrow X}$, and $\mid\sim_{rf}{}^{\uparrow X}$:

(1) If $\Gamma \mid\sim_{rf}{}^{\uparrow X} A$, then $\Gamma \mid\sim_r{}^{\uparrow X} A$;
(2) If $\Gamma \mid\sim_{r[f]}{}^{\uparrow X} A$, then $\Gamma \mid\sim_{[f]}{}^{\uparrow X} A$;
(3) If $\Gamma \mid\sim_f{}^{\uparrow X} A$, then $\Gamma \mid\sim^{\uparrow X} A$.

[6] See, for instance, Anderson & Belnap [1] and Mares [7].

Proof. Induction on the proof height of the antecedent sequents. (Note that the inductive proof of (3) relies on (1) for some special cases (i.e., the cases for LLC2 and RR) in its induction step.) ■

These inclusion relations (especially, (1) and (3)) enable us to estimate the strength of $|\sim$ and $|\sim_r$ on the basis of the strength of $|\sim_f$ and $|\sim_{rf}$, which we are going to estimate below in this section.

Admissibility of (Restricted Versions of) Cut. Remember that NMMR does not have any version of Cut as its rule. However, it is provable that the following restricted versions of Cut are admissible in NMMR:

(1) $$\frac{\Gamma |\sim_f{}^{\uparrow X} M \qquad \Delta, M |\sim_f{}^{\uparrow Y} A}{\Gamma, \Delta |\sim_f{}^{[\uparrow]} A}\,\text{fCut}$$

(where **R** does not appear in Γ, Δ, M, or A)

(2) $$\frac{\Gamma |\sim_{rf}{}^{\uparrow X} M \qquad \Delta, M |\sim_{rf}{}^{\uparrow Y} A}{\Gamma, \Delta |\sim_{rf} A}\,\text{rfCut}$$

(where $\Box$ does not appear in Γ, Δ, M, or A)

Proposition 7. Both fCut and rfCut are admissible in NMMR.

Proof. The proposition is equivalent to the conjunction of the following two claims: (1) Every proof tree of NMMR + fCut such that fCut is applied only at the last step can be transformed into a proof tree that shares the same top and bottom sequents but has no fCut application in it; (2) every proof tree of NMMR+rfCut such that rfCut is applied only at the last step can be transformed into a proof tree that shares the same top and bottom sequents but has no rfCut application in it. Both (1) and (2) can be shown by double induction on the rank of the proof tree at issue and the depth of M as usually defined. ■

Before turning to the estimating job, we need to show one more lemma.

Lemma 3. $\Gamma |\sim_{[r][f]}{}^{\uparrow X} A \to B \Rightarrow \Gamma, A |\sim_{[r][f]}{}^{\uparrow X} B$

Proof. By induction on the proof height of the left-hand side sequent. ■

The Extended Material Consequence and Incoherence. Now, let us estimate how strong the extended material consequence and incoherence of NMMR are. The following proposition answers this question.

Proposition 8. $|\sim_f$ is as strong as the intuitionistic logic.

Proof. By Proposition 7 and Lemma 3, in the **R**-free fragment of $|\sim_f$, *Modus Ponens* is admissible (i.e., ($|\sim_f A$ and $|\sim_f A \to B$) $\Rightarrow$ $|\sim_f B$). For any **R**-free formula, the following eight axioms of Hilbert-style axiomatization of the intuitionistic logic are all derivable as theorems with respect to the same fragment of $|\sim_f$: (1) $(A \to B) \to (A \to (B \to C))$ $(A \to C))$; (2) $A \to (B \to A)$; (3) $A \to (A \vee B)$ and $B \to (A \vee B)$; (4) $(A \to C)$ $((B \to C) \to ((A \vee B) \to C)))$; (5) $B \to (A \to (A \vee B))$; (6) $(A \& B) \to A$ and $(A \& B) \to B$; (7) $(A \to B) \to ((A \to \neg B) \to \neg A)$; (8) $\neg A \to (A \to B)$. ■

Since $|\!\sim\ \supseteq\ |\!\sim_f$ (see (3) of Proposition 6), we conclude that the extended material consequence and incoherence are supra-intuitionistic.

The Extended Relevant Material Consequence and Incoherence. Similarly, one may wonder how strong the logic of the extended relevant consequence and incoherence of NMMR is. The following proposition provides us with an answer.

Proposition 9. In the □-free fragment of $|\!\sim_{rf}$, nine of eleven axioms of a Hilbert-style axiomatization of logic R[7] are derivable as theorems, and the two inference rules of that axiomatization are admissible.

Proof. Adjunction is equivalent to R&. By Proposition 7 and Lemma 3, in the □-free fragment of $|\!\sim_{rf}$, *Modus Ponens* is admissible (i.e., ($|\!\sim_{rf}$ A and $|\!\sim_{rf}$ A → B) ⇒ $|\!\sim_{rf}$ B). Among the following eleven axioms of Hilbert-style axiomatization of logic R, all but (9) and (11) are derivable as theorems with respect to the same fragment of $|\!\sim_{rf}$: (1) A → A; (2) (A → B) → ((B → C) → (A → C)); (3) A → ((A → B) → B); (4) (A → (A → B)) → (A → B); (5) (A & B) → A and (A & B) → B; (6) A → (A ∨ B) and B → (A ∨ B); (7) ((A → B) & (A → C)) → (A → (B & C)); (8) (((A → B) → C) → ((A → C) & (B → C))) & (((A → C) & (B → C)) → ((A → B) → C)); (9) (A & (B ∨ C)) → ((A & B) ∨ (A & C)); (10) (A → ¬ B) → (B → ¬ A); (11) ¬¬ A → A. ■

Since $|\!\sim_r\ \supseteq\ |\!\sim_{rf}$ (see (1) of Proposition 6), we conclude that the extended relevant consequence and incoherence are stronger than a logic that is slightly weaker than logic R.

3.5 The Expressive Power of Logical Connectives

As logical expressivists, we want our logical connectives to express (i.e., codify within the object language) various features of the material/relevant consequence and incoherence. Let us take the conditional as an example. A conditional implied by a given premise-set is supposed to express that the antecedent implies the consequent in the context of that premise-set. Similarly, a negation implied by a given premise-set is supposed to express that the negated formula is incompatible with that premise-set. Such expressive readings of our connectives are supported by the following proposition:

Proposition 10

(1) $\Gamma\,|\!\sim_{[r][f]}{}^{\uparrow X} A \rightarrow B \Leftrightarrow \Gamma,\ A\,|\!\sim_{[r][f]}{}^{\uparrow X} B$;

(2) $\Gamma\,|\!\sim_{[r][f]}{}^{\uparrow X} \neg A \Leftrightarrow \Gamma,\ A\,|\!\sim_{[r][f]}{}^{\uparrow X} \bot$;

(3) $\Gamma\,|\!\sim_{[r][f]}{}^{\uparrow X} A \& B \Leftrightarrow (\Gamma\,|\!\sim_{[r][f]}{}^{\uparrow X} A$ and $\Gamma\,|\!\sim_{[r][f]}{}^{\uparrow X} B)$.

7 See Anderson & Belnap [1] and Anderson, Belnap, & Dunn [2].

Proof. The right-to-left direction: Straightforward from RC, R&, and RN, respectively; *The left-to-right direction*: (1) is already shown as Lemma 3. (2) and (3) are similarly shown by induction on the proof height of the left-hand side sequents. (Note that the inductive proof of (3) relies on (2) of Proposition 6 for the case for LLC2 in its induction step). ■

Our box operator (i.e., □) is introduced to express that a boxed formula is indefeasibly implied by a given premise-set. This is shown to be the case in two steps. First, the following lemma guarantees that the unindexed upward arrow (i.e., "↑") indicates that an implication is indefeasible:

Lemma 4. $\forall \Delta(\Gamma, \Delta | \sim A) \Leftrightarrow \Gamma | \sim^{\uparrow} A$.

Proof. The left-to-right direction. Suppose $\Gamma, \Delta | \sim A$, for any Δ. Thus, $\Gamma, \Delta_0 | \sim A$, for any atomic Δ_0. By repeated applications of PushUpUN, $\Gamma | \sim^{\uparrow} A$. *The right-to-left direction.* Suppose $\Gamma | \sim^{\uparrow} A$. Take an arbitrary Δ. Enumerate Δ as $\{B_1, \ldots, B_n\}$. By repeated applications of pW, $\Gamma, B_1, \ldots, B_{n-1} | \sim^{\uparrow} A$. Then, by the last application of pW, $\Gamma, B_1, \ldots, B_n | \sim A$. Since Δ is arbitrary, $\forall \Delta (\Gamma, \Delta | \sim A)$. ■

Next, via the following sublemma and lemma, the following proposition guarantees that such indefeasibility of an implication is expressed by our modality box.

Sublemma 1. $\Gamma | \sim_{[f]}{}^{\uparrow X} \Box A \Rightarrow \Gamma | \sim_{[f]}{}^{\uparrow} A$.

Proof. By induction on the proof height of the left-hand side sequent. ■

Lemma 5. $\Gamma | \sim^{\uparrow} A \Leftrightarrow \Gamma | \sim \Box A$.

Proof. The left-to-right direction. Straightforward from RB. *The right-to-left direction.* Straightforward from Sublemma 1. ■

Proposition 11. $\Gamma | \sim \Box A \Leftrightarrow \forall \Delta(\Gamma, \Delta | \sim A)$.

Proof. Straightforward from Lemmas 4 and 5. ■

Finally, how do we codify, within the object language, the relevance of implications? Here, our new operator, **R**, plays a crucial role. Remember that the material consequence and incoherence that the snake turnstile subscripted with "r" represents are sensitive to *relevance* in that the fallacies of relevance do not hold there (see Proposition 5). To show that such relevant consequences are expressed by the new operator, we first need to prove the following slightly stronger lemma.

Lemma 6. $\Gamma | \sim_{[r][f]}{}^{\uparrow X} \mathbf{R}A \Leftrightarrow \Gamma | \sim_{r[f]}{}^{\uparrow X} A$.[8]

Proof. The right-to-left direction. Straightforward from RR. *The left-to-right direction.* By induction on the proof height of the left-hand side sequent. ■

Now, our target proposition immediately follows from this lemma.

Proposition 12. $\Gamma | \sim \mathbf{R}A \Leftrightarrow \Gamma | \sim_r A$.

Proof. Straightforward from Lemma 6. ■

[8] A note on the intended reading of this biconditional: It is optional whether the left-hand sequent is relevant, while the right-hand sequent must be relevant.

Given the expressive power of the logical vocabulary at hand, we are now in a position to express, within the object language, various interesting features of the material consequence relation and incoherence property. For instance, that A *indefeasibly and relevantly implies* B in the context of Γ is expressed by " □ (A → B) & **R** (A → B)" implied by Γ.

4 Conclusion

Motivated by semantic inferentialism and logical expressivism, in this paper, I propose a nonmonotonic sequent calculus equipped with special logical operators, □ and **R**. The base level of this calculus includes two different types of atomic axioms: material and relevant. The material base comprises defeasible atomic sequents, as well as indefeasible (including formally valid) ones, whereas the relevant base consists of the local region of such a material base that is sensitive to relevance. The rules of the calculus uniquely and conservatively extend these two sorts of nonmonotonic bases into logically complex consequence relations and incoherence properties, while preserving *Containment* in the material base and *Reflexivity* in the relevant base. The material extension is supra-intuitionistic, whereas the relevant extension is stronger than a logic slightly weaker than R. The relevant extension also avoids the fallacies of relevance. Although the extended material consequence relation, as a whole, is defeasible and insensitive to relevance, it has local regions of indefeasibility and relevance (the latter of which is marked by the relevant extension). The newly introduced operators, □ and **R**, codify these local regions within the same extended material consequence relation.

Given these properties, it would be fair to conclude that this system can be a common platform for people interested in different sorts of inferences, such as defeasible/indefeasible (including formal) inferences and relevant/non-relevant inferences. The system can not only deal with these (combinatorially four) different types of inferences at the same time, but can also allow us to "talk about", within its object language and the same extended material consequence relation, those different inferences.

References

1. Anderson, A.R., Belnap, N.D.: Entailment: The Logic of Relevance and Necessity, vol. I. Princeton University Press, Princeton (1975)
2. Anderson, A.R., Belnap, N.D., Dunn, J.M.: Entailment: The Logic of Relevance and Necessity, vol. II. Princeton University Press, Princeton (1992)
3. Brandom, R.: Making It Explicit. Harvard University Press, Cambridge (1994)
4. Brandom, R.: Articulating Reasons. Harvard University Press, Cambridge (2001)
5. Brandom, R.: Between Saying and Doing: Towards an Analytic Pragmatism. Oxford University Press, Oxford (2008)

6. Hlobil, U.: A nonmonotonic sequent calculus for inferentialist expressivists. In: Arazim, P., Dančák, M. (eds.) The Logica Yearbook 2015, pp. 87–105. College Publications, London (2016)
7. Mares, E.: Relevance logic. In: Zalta, E.N., Nodelman, U., Allen, C. (eds.) Stanford Encyclopedia of Philosophy (2012). http://plato.stanford.edu/archives/spr2014/entries/logic-relevance/. Last accessed 27 Feb 2016

A Formalization of the Greater Fools Theory with Dynamic Epistemic Logic

Hanna S. van Lee(✉)

Center for Information and Bubble Studies, University of Copenhagen, Copenhagen, Denmark
hannavanlee@hum.ku.dk

Abstract. The greater fools explanation of financial bubbles says that traders are willing to pay more for an asset than they deem it worth, because they anticipate they might be able to sell it to someone else for an even higher price. As agents' beliefs about other agents' beliefs are at the heart of the greater fools theory, this paper comes to formal terms with the theory by translating the phenomenon into the language and models of dynamic epistemic logic. By presenting a formalization of greater fools reasoning, structural insights are obtained pertaining to the structure of its higher-order content and the role of common knowledge.

1 Introduction

A financial bubble describes specific scenarios in which asset prices rise way beyond fundamental value and in which eventually the market crashes. Such scenarios are unwelcome because they make the market unpredictable and uncontrollable, and because they create inequality at the risk of ruining individuals, firms and even nations. To understand the circumstances under which prices may deviate from their fundamental value to subsequently prevent the occurence of bubbles – or at least limit their consequences – financial bubbles have been extensively studied both theoretically and empirically.[1] Nevertheless, there are contrasting explanations for the occurrence of financial bubbles. Some explain the sudden rise of prices by irrational or noise traders [10]; others suggest that traders' herding behavior lead the price astray from the asset's fundamental value [16]; yet others point to traders who are rationally willing to pay more for an asset than they deem it worth, because they anticipate they might be able to sell it to someone else for an even higher price, known as the *greater fools* explanation for bubbles [1,8,14]. The greater fools theory sounds like a crisp and clear explanation for a mismatch between price and value, but the variety of results on the theory shows that it is a surprisingly difficult theory to model and analyze [5]. This paper aims for a structural epistemic understanding of the

My research is financed by the Carlsberg Foundation. I would like to thank Thomas Bolander, Vincent F. Hendricks, Rasmus K. Rensdvig and two anonymous referees for their valuable comments on this paper.

[1] See [7] for an extensive overview of studies of financial bubbles.

A. Baltag et al. (Eds.): LORI 2017, LNCS 10455, pp. 585–597, 2017.
DOI: 10.1007/978-3-662-55665-8_40

greater fools theory by focussing on micro-economic features as investor information and behavior rather than on the macro-economic perspective of regulation and market conditions.

The following describes a greater fools scenario that will be repeatedly referred to in this paper. Imagine a market with one orange tree and a boy called Arthur interested in buying the tree. Based on certain predictions of the tree's harvest, Arthur believes ownership of the orange tree is worth 2 euro. Would it be rational for him to pay 3 euro for the orange tree? Not if he is buying the orange tree exclusively for the tree's harvest. However, imagine that instead of owning the orange tree, Arthur is interested in reselling the orange tree for 4 euro to his friend Barbara, making a profit of 1 euro. The story continues with Barbara, who agrees that the orange tree is worth 2 euro but is willing to pay 4 euro because she expects to sell the orange tree for 5 euro to her friend Chris. When the orange tree is traded further based on similar reasoning, the price may rise far beyond it's assumed value of 2 euro. In this story, Arthur seems to act like a fool by paying more for the tree than he deems it worth, but his behavior is justified by his belief in a greater fool named Barbara, who believes in the even greater fool Chris, etc.

Although agents' beliefs about other agents' beliefs are at the heart of the greater fools theory, existing studies typically do not formally include higher-order reasoning. As an epistemic approach is currently lacking from the literature, epistemic (including doxastic) logic seems like a suitable candidate framework for obtaining a novel understanding of the theory. On top of the framework's language and model representation appropriate for studying higher-order epistemic structures, *dynamic* epistemic logics introduce action models to describe changes due to interaction between agents. This paper comes to formal terms with the greater fools theory by translating the phenomenon into the language and models of public announcement logic, thereby unfolding the higher-order content of the theory. By presenting an epistemic formalization of a greater fools bubble, structural insights are obtained pertaining to the pattern of it's higher-order beliefs and the role of common knowledge in the burst.

A proper understanding of investors' reasoning about the market and about each other potentially plays a key role in preventing future crises. The ambition to apply dynamic epistemic logics to strategic reasoning in finance is still quite young. In 2012 [9] and 2014 [11], authors use probabilistic dynamic epistemic logics to model Aumann's agreement theorem. The current paper adds to the application of dynamic epistemic logics to social interaction and rationality in general and in specific to studies of the bubble-fueling herding behavior, such as models of informational cascades [3,17].[2]

[2] The greater fools explanation of a bubble must not be confused with herding phenomena as the two are fundamentally different: where in a herding bubble investors act the same because of an incentive to follow the crowd, investors in a greater fools bubble simply act the same as a result of similar reasoning, as will be elaborately discussed in this paper.

The next section discusses essential concepts of the greater fools theory, while at the same time motivating the abstractions made in the models in Sect. 3, where a semantic formalization of greater fools bubbles is introduced by means of a public announcement logic, illustrated by three cases of greater fools reasoning. In Sect. 4 some results of the formalization are observed and proven, and the final section concludes and mentions directions for further work.

2 Central Concepts

Market Place. Theoretical models of finance offer a great variety of descriptions of the highly complex market places. As models of dynamic epistemic logic focus on interacting individuals, rather than on the crowd as a whole, a market in this paper is a place where individuals sequentially trade one asset at a time with each other.

Traders in a market may have different motivations for buying and selling. For instance, agents may choose to trade to spread risks, to stimulate liquidity in the market, to mislead others, to profit by speculation, or to enjoy dividend pay out. As only the last two motivations are relevant to greater fools reasoning, all other possible motivations are ignored in the forthcoming analyses.

Furthermore, in this paper the price is a variable on which agents' attitudes are defined, as will be shown in Sect. 3. When a seller and a buyer agree on the price, the trade may take place, thereby letting the price be determined by supply and demand only on an individual level.

Fundamental Value. A bubble is typically defined by referring to the deviation of the price from the so-called *fundamental value*, also *instrinsic* or *natural value*. As the concept of fundamental value often gives rise to many questions - e.g. when confused for the *market* value and in discussions related to the efficient market hypothesis - a few notes on it's nature must be given. The fundamental value is derived from expected (discounted) cash flows that the asset pays out, such as it's future dividend.[3] This value is objective in the sense that *given the same information* about the asset, all agents would agree on the fundamental value. To ensure the existence of an objective fundamental value in the most simple way, it is in the forthcoming assumed that resources are efficiently allocated, such that traders expect to profit exclusively at the expense of others.[4] As a consequence, the fundamental value is never based on personal preferences or needs.

Reasoning under Uncertainty. In practise, though, agents typically do not agree on the fundamental value, because they have access to different information

[3] See [19] for a more eleborate definition of the fundamentals that influence the fundamental value.

[4] This eliminates the situation where Arthur, who owns an apple tree, and Barbara, who owns an orange tree, trade apples for oranges and mutually benefit from the trade.

about an asset's fundamentals. As will be shown in Sect. 4, the possibility of a disagreement about the asset's value is essential to the rise of greater fools bubbles: as soon as all uncertainties about other traders' beliefs are eliminated, the price will fall back to the fundamental value.

A distinction must be made here between uncertainty about the fundamental value of the asset on one hand and uncertainty about the (future) beliefs and behavior of other traders on the other hand. Having no crucial role in the reasoning driving a greater fools bubble, the models in the forthcoming formalization do not include informative updates on the asset's value itself: that is, agents are not directly informed about the asset's fundamentals, nor is information about the asset somehow indirectly revealed through prices or actions that traders take.

Rational Bubbles. As said before, there is no consensus about the circumstances under which a bubble may occur. There is one topic in particular that nourishes much debate: the relation between rational traders and the possibility of a bubble, e.g. [6]. Some claim that under specific assumptions, rational bubbles will not form: as traders are immediately aware that they are being exploited, they will refuse to buy an overpriced asset, even under presence of asymmetric information [18]. Others suggest that enough rational traders will guarantee that any potential mispricing induced by behavioral traders (or noise) will be corrected [12]. Contrarily to those two theories, others argue that rational traders will not necessarily prevent a greater fools bubble from occurring, because rational traders prefer to ride the bubble rather than attack it [2,8]. That is, because they can profit from less informed traders only if they exit the market just prior to the crash, which seems to capture aspects of what often happens during real episodes of bubbles [1].

The next sections try to answer the question under which epistemic circumstances greater fools reasoning may lead to overpricing of the asset and what may cause the bubble to crash.[5]

3 A Formalization of the Greater Fools Theory

Even though the intuition behind the greater fools theory is simple, coming to formal terms with the idea it represents reveals some obscurities. To focus on the epistemic structures of the greater fools theory, only the necessary ingredients are used in the forthcoming qualitative analysis: a finite set of traders,[6] an asset (an orange tree), traders' attitudes towards trading the asset, and communication that reveals traders' first- and higher-order beliefs about the value of the asset. For simplicity, all agents are assumed to have unlimited financial resources. In addition, when agents are indifferent with respect to trading or not, agents refrain from trading.

[5] A rational model of greater fools bubble should however not rule out the important role played by irrational behavior and mass psychology.

[6] Although the models are a representation of the epistemic states of a few individuals, these individuals can be interpreted to represent a group of homogeneous traders.

3.1 Language and Plausibility Models

The semantic formalization of the greater fools theory comes down to unfolding the meaning of being willing to buy an asset for a price that is higher than the assumed value of the asset. To construct a translation of this meaning in epistemic logic, the following language is used:

Definition 1 *(Language)*

$$\varphi ::= v{=}n \mid \neg\varphi \mid \varphi \vee \varphi \mid K_i\varphi \mid B_i\varphi \mid CK\varphi \mid \mathsf{sell}_i(p) \mid \mathsf{buy}_i(p) \mid [!\varphi]\varphi$$

$$\textit{for } 0 \leq n, p \leq \mathbf{L} \in \mathbb{N} \textit{ and } i \in \mathcal{A}$$

Here, $v{=}n$ says "the value of the orange tree is n", $\neg$ and $\vee$ translate to "not" and "or", $K_i\varphi$ expresses that agent i knows φ while $B_i\varphi$ expresses that agent i believes φ. Furthermore, $CK\varphi$ says "it is common knowledge among all agents that φ". The expression $\mathsf{sell}_i(p)$ says: "if agent i owns the asset, then she wants to sell it for price p" and similarly, $\mathsf{buy}_i(p)$: "if agent i does not own the asset, then she wants to buy it for p". The dynamic sentence $[!\varphi]\varphi$ means "after a public announcement of φ, φ is true". The set $\mathcal{A}$ is the set of agents, currently defined as $\{a, b, c, f\}$ existing of Arthur, Barbara, Chris and Farmer Flora. Due to constructions later, n and p are limited to an unspecified large finite natural number $\mathbf{L}$. Furthermore, let $v > n$ ("v is strictly higher than n") be short for $\bigvee_{n<n'\leq \mathbf{L}} v{=}n'$ and $v < n$ ("v is strictly lower than n") short for $\bigvee_{n'<n} v{=}n'$. The language is intepreted on standard plausibility models, cf. [4].

Definition 2 *(Plausibility model). Given a set of agents $\mathcal{A}$ and a limit $\mathbf{L} \in \mathbb{N}$, a plausibility model is a tuple $\mathcal{M} = \langle \mathcal{W}, \mathcal{V}, \preceq \rangle$ with $\mathcal{W}$ a set of worlds, $\mathcal{V} : \mathcal{W} \to \{v{=}n\}_{0\leq n\leq \mathbf{L}}$ a valuation map assigning a value of the asset to each world, and $\preceq : \mathcal{A} \to \mathscr{P}(\mathcal{W} \times \mathcal{W})$ a plausibility relation such that for all $i \in \mathcal{A}$, the relation $\preceq(i)$ is reflexive, transitive, conversely well-founded and locally connected.*[7] *A pointed plausibility model $(\mathcal{M}, w)$ designates one real world $w \in \mathcal{W}$.*

Note that by definition of the valuation map, at every world $w \in \mathcal{W}$ the asset is assigned exactly one value n (while there may be values that are assigned to none or more worlds). The expression $(w, u) \in \preceq(i)$ is usually written $w \preceq_i u$ and says "u is at least as plausible as w". Because $\preceq_i$ is conversely well-founded there exists necessarily a set of worlds that are considered most plausible. Let the indistinguishability relation $\sim : \mathcal{A} \to \mathscr{P}(\mathcal{W} \times \mathcal{W})$ be defined by $w \sim_i u$ only if $w \preceq_i u$ or $u \preceq_i w$. This implies that $\sim_i$ is an equivalence relation. Finally, let $\sim^\circ := \bigcup_{i\in\mathcal{A}} \sim_i$. At the end of the next section, Fig. 1 shows two examples of a plausibility model and satisfaction in a model (given in Definition 4).

---

[7] The relation $\preceq_i$ being locally connected means that for all $w, u \in \mathcal{W}$ whenever they are related by the symmetric closure of $\preceq_i$, then $w \preceq_i u$ or $u \preceq_i w$.

3.2 A Meaningful Translation

In order to characterize the reasoning used in a greater fools episode, the meaning of $\mathsf{sell}_i(p)$ and $\mathsf{buy}_i(p)$ need introduction. The most generalized justification for selling an asset for a certain price is to believe that the asset is worth less than that price. For a price equal or lower than the value of the asset, the agent would rather keep the asset and enjoy the dividend cash flow that corresponds to the value. In addition, an agent wants to sell the asset for at least the highest price she believes she can get for it,[8] as characterized below:

$$\mathsf{sell}_i(p) \leftrightarrow \left(B_i(v < p) \wedge B_i \neg \left(\bigvee_{j \neq i} \bigvee_{p' > p} \mathsf{buy}_j(p') \right) \right)$$

A trader may have two different kinds of motivation for buying the orange tree for price p: either the agent believes the tree is worth more than the price, or the agent believes she can sell the asset to another trader $j \in \mathcal{A}$ for a higher price $p' > p$, as characterized below:

$$\mathsf{buy}_i(p) \leftrightarrow \left(B_i\,(v > p) \vee B_i \left(\bigvee_{j \neq i} \bigvee_{p' > p} \mathsf{buy}_j(p') \right) \right)$$

The meaning of $\mathsf{buy}_j(p')$ is that agent j either believes $v > p'$, or that she can sell the asset for $p'' > p'$ to another agent j', etc. Hence $\mathsf{buy}_i(p)$ gives rise to an up to k-th order belief about the value of the asset. The k-th order belief refers to the asset being traded k times for a rising price at each next trade $p < p_2 < ... < p_k$. Note that this is a static encoding of being willing to buy the asset. In the next section it is argued by means of an example that one might intuitively prefer a dynamic encoding of buying (see Definition 5). For now, let the static interpretations of selling and buying be as in Definition 4.

Definition 3 *(Information update). Given a plausibility model $\mathcal{M}$ and a public announcement φ, the updated model $\mathcal{M}^\varphi$ is the restriction of $\mathcal{M}$ to all worlds where φ is true, i.e. $\langle \mathcal{W}^\varphi, \mathcal{V}^\varphi, \preceq^\varphi \rangle$ such that $\mathcal{W}^\varphi := \{w \in \mathcal{W} | (\mathcal{M}, w) \models \varphi\}$ and $\mathcal{V}^\varphi$ and $\preceq^\varphi$ are defined as $\mathcal{V}$ and $\preceq$ restricted to $w \in \mathcal{W}^\varphi$. When $\mathcal{W}^\varphi$ is empty, $\mathcal{M}^\varphi$ is undefined.*

Through the dynamic operator $[!\varphi]$ agents as well as third person modellers may reason about what will happen after an announcement of φ, cf. [13]. The models to follow will specifically reason about announcements of an agent's (un)willingness to buy or sell.

---

[8] The second condition has the realistic consequence that when for example Arthur believes he can sell the asset for 10, after which he learns no other agent in fact wants to buy the asset for 10, he will lower the price for which he offers to sell. The price will continue to drop until he either has an agreement to trade, or Arthur realizes he cannot sell the asset for more than the value he deems it worth.

Definition 4 *(Truth). Given a plausibility model $\mathcal{M}$ and a world $w \in \mathcal{W}$, truth is defined:*

$$
\begin{array}{lll}
(\mathcal{M}, w) \models \top & & \textit{always} \\
(\mathcal{M}, w) \models v{=}n & \textit{iff} & v{=}n \in \mathcal{V}(w) \\
(\mathcal{M}, w) \models \neg\varphi & \textit{iff} & (\mathcal{M}, w) \not\models \varphi \\
(\mathcal{M}, w) \models \varphi \vee \psi & \textit{iff} & (\mathcal{M}, w) \models \varphi \textit{ or } (\mathcal{M}, w) \models \psi \\
(\mathcal{M}, w) \models K_i\varphi & \textit{iff} & (\mathcal{M}, w') \models \varphi \textit{ for all } w' \sim_i w \\
(\mathcal{M}, w) \models B_i\varphi & \textit{iff} & (\mathcal{M}, w') \models \varphi \textit{ for all } w' \in \mathsf{max}_{\preceq_i}\{u \in \mathcal{W} | u \sim_i w\} \\
(\mathcal{M}, w) \models CK\varphi & \textit{iff} & (\mathcal{M}, w') \models \varphi \textit{ for all } w' \sim^\circ w \\
(\mathcal{M}, w) \models \mathsf{sell}_i(p) & \textit{iff} & (\mathcal{M}, w) \models B_i(v < p) \textit{ and } (\mathcal{M}, w) \models B_i \neg \bigvee_{j \neq i} \bigvee_{p' > p} \mathsf{buy}_j(p') \\
(\mathcal{M}, w) \models \mathsf{buy}_i(p_1) & \textit{iff} & (\mathcal{M}, w) \models B_i(v > p_1) \textit{ or there is a } k \geq 1 \\
 & & \textit{such that } p_n < p_{n+1} \textit{ for all } 1 \leq n < k \textit{and} \\
 & & (\mathcal{M}, w) \models B_i B_{j_1} ... B_{j_k}(v > p_k) \\
(\mathcal{M}, w) \models [!\varphi]\psi & \textit{iff} & (\mathcal{M}^\varphi, w) \textit{ is defined implies } \mathcal{M}^\varphi, w \models \psi
\end{array}
$$

Here, $\mathsf{max}_{\preceq_i}\mathcal{U} := \{u \in \mathcal{U} | w \preceq_i u \text{ for all } w \in \mathcal{U}\}$. The meaning of belief thus refers to the most plausible states that are consistent with the agent's knowledge at state w. This implies that belief (and also $\mathsf{sell}_i(p)$ and $\mathsf{buy}_i(p)$) is universally true within an agent's knowledge set: if $(\mathcal{M}, w) \models B_i\varphi$ then also $(\mathcal{M}, u) \models B_i\varphi$ for all u such that $u \sim_i w$.

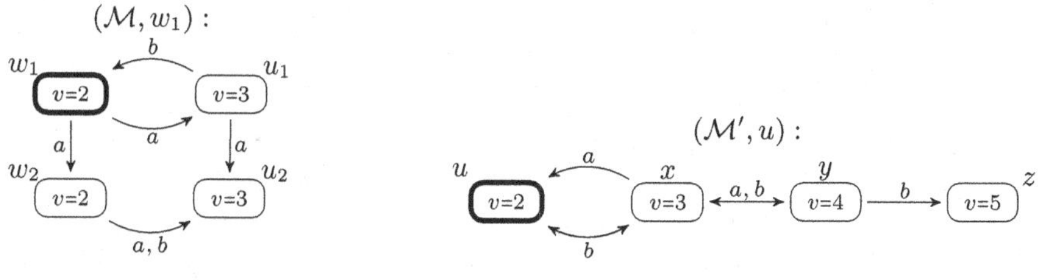

Fig. 1. Two examples of plausibility models

Figure 1 presents two different pointed plausibility models where the true world is signed bold and $w \rightarrow_i u$ represents that $w \preceq_i u$. Reflexive, transitive and locally connected arrows are omitted from all figures. In the left model, both Arthur and Barbara know that the orange tree is worth either 2 or 3 euro. Barbara believes the asset is worth 2 euro while Arthur believes it is worth 3 euro, as $(\mathcal{M}, w) \models v{=}2$ for all $w \in \mathsf{max}_{\preceq_b}\{u \in \mathcal{W} | u \sim_b w_1\}$ and similarly for Arthur. Furthermore, note that Barbara knows Arthur's beliefs, while Arthur incorrectly believes that $B_b(v{=}3)$. In the right model, Arthur is indifferent whether $v{=}3$ or $v{=}4$ while he does believe that $v < 3$. As Arthur however believes that that $\mathsf{buy}_b(4)$, it is false that $\mathsf{sell}_a(3)$, while $\mathsf{sell}_a(4)$ and $\mathsf{sell}_a(10)$ are true.

3.3 Three Cases of Greater Fools Reasoning

Inspired by reality and theoretical models of trade, the following protocol for trading is chosen: first the agent holding the orange tree publicly announces that she wants to sell it for price p (given that she wants to). Then, if possible, another agent publicly replies that he is willing to buy the tree for that price and only if two agents publicly agree on a price, a trade will take place. If this is not possible, all agents will announce that they are not willing to buy the tree for that price. In that case the selling agent lowers the price of her offer, unless the price is equal to her believed fundamental value. To establish this,[9] a trade between i and j for price p is formally defined as a public announcement:

$$!\mathsf{trade}_{i,j}(p) :=!CK\left(\mathsf{sell}_i(p) \wedge \mathsf{buy}_j(p)\right)$$

To get a better intuition about the formal interpretation of buying, and to conclude that greater fools bubbles can exist (see Proposition 2), three different scenarios will now be modelled, each with a different justification for Arthur buying the orange tree for 2 euro. To focus on the mismatch between traders' beliefs about value of the asset and the price it is traded for, all agents in the examples correctly agree that the value is 2. This is generalised in Sect. 4.

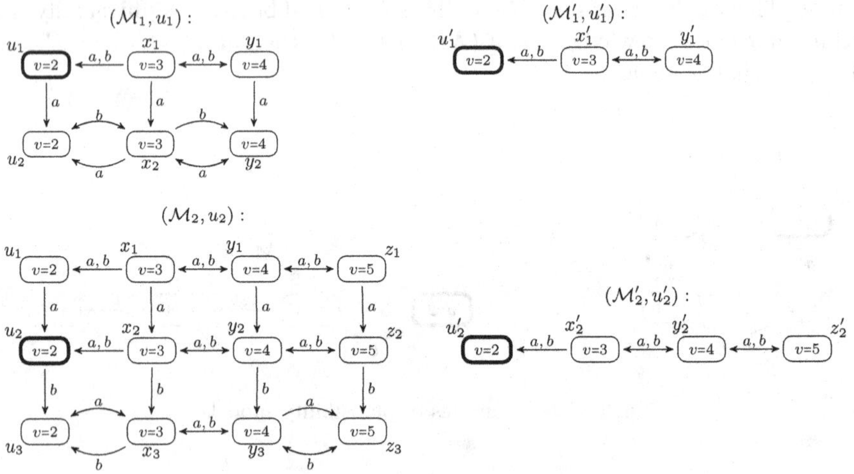

Fig. 2. Models of Scenario 1 and 2

Scenario 1: $B_a B_b(v > 3)$ Suppose Farmer Flora (whose epistemic and doxastic relations are not represented in the models for sake of simplicity) has offered to sell the orange tree for 2 euro. Consider pointed plausibilty model $(\mathcal{M}_1, u_1)$

[9] Following the protocol, communication should precede the trade such that the trade does not envoke any epistemic change. Moreover, the framework does not formally keep track of who owns the tree, thus a trade should not envoke any atomic change. The chosen encoding of trading fulfills both desiderata.

in Fig. 2, representing Arthur's and Barbara's epistemic and doxastic attitudes. Note that $B_aB_b(v > 3)$ indeed holds. By the truth definition of $\mathsf{buy}_i(p)$, this means that Arthur wants to buy the tree for 2 euro. After Arthur's announcement $!\mathsf{buy}_a(2)$ the trade takes place: $!\mathsf{trade}_{f,a}(2)$. Unfortunately for Arthur, his belief about Barbara is wrong: $(\mathcal{M}_1, u_1) \not\models B_b(v > 3)$ because $(\mathcal{M}_1, u_1) \models B_b(v{=}2)$. Thus, after Arthur announces he wants to sell the tree for 3, Barbara rejects his offer by announcing $\neg\mathsf{buy}_b(3)$, resulting in model $(\mathcal{M}_1', u_1')$.

This scenario demonstrates a miniature bubble scenario where Arthur bought the orange tree for a price higher than he believed it was worth because he expected to sell the tree to Barbara for even more. As his expectation turned out to be false, he could not sell the tree and the bubble crashed softly in the sense of leaving Arthur with the unsellable asset.

Scenario 2: $B_aB_bB_a(v > 4)$ Again, suppose Farmer Flora offers to sell the orange tree for 2 euro. Now, Arthur knows that $B_b(v{=}2)$ but he still wants to buy the tree for 2 euro because now $B_aB_bB_a(v > 4)$. This can be seen in $(\mathcal{M}_2, u_2)$ in Fig. 2 where $(\mathcal{M}_2, u_3) \models B_a(v > 4)$, and therefore $(\mathcal{M}_2, u_2) \models B_bB_a(v > 4)$, and therefore $(\mathcal{M}_2, u_2) \models B_aB_bB_a(v > 4)$. Like in Scenario 1, announcements of $!\mathsf{buy}_a(2)$ and $!\mathsf{trade}_{f,a}(2)$ will be executed. Note that this time, Arthur is right about Barbara: $B_bB_a(v > 4)$ is true at u_2. However, when Arthur announces to sell for 3 euro, Barbara learns that Arthur believes the value is lower than 3 such that no longer $B_bB_a(v > 4)$ as can be seen in the resulting model $(\mathcal{M}_2', u_2')$.[10] As a result, Barbara does not want to buy the orange tree for 3 euro, because she knows she will not be able to sell it back to Arthur for 4 euro. As in the previous scenario, Arthur is left with the unsellable asset.

This raises the question whether Arthur could foresee this series of events. Under the current definition of $\mathsf{buy}_i(p)$, agents do not look forward in time. However, Arthur can reason about what would happen after announcing $\mathsf{sell}_a(3)$ as $(\mathcal{M}_2, u_2) \models K_a[!\mathsf{sell}_a(3)]K_b\neg B_a(v > 4)$.

In general, a trader (e.g. a) that is willing to buy an asset for price p because he believes in the existence of a greater fool who believes in an even greater fool, etc., must not be able to predict that the trader he anticipates to resell the asset to (e.g. b) will refrain from buying at some point in the expected future. Under this dynamic interpretation of rationality, the meaning of buy has to be altered as following:

Definition 5 *(Truth predictive rationality)*

$$(\mathcal{M}, w) \models \mathsf{buy}_i(p_1) \quad \textit{iff} \quad (\mathcal{M}, w) \models B_i(v > p_1) \textit{ or, there is a } k \geq 1 \textit{ such that } p_n < p_{n+1} \textit{ for all } 1 \leq n < k \textit{ and } (\mathcal{M}, w) \models B_i[!\mathsf{sell}_i(p_2)]...B_{j_{k-1}}[!\mathsf{sell}_{j_{k-1}}(p_{k-1})]B_{j_k}(v > p_k)$$

10 After the update, worlds u_1', x_1', y_1' and z_1' become redundant by their bisimulation to worlds u_1, x_1, y_1 and z_1 respectively. Such bisimulation contraction is harmless in epistemic logic.

This definition is essentially a model restriction revealing agents' predictive abilities. In Scenario 2, this means that Arthur initially does not want to buy the orange tree for 2 from Farmer Flora, because he can predict that Barbara will learn that she cannot resell the tree to Arthur for 4 euro.

Notice the difference between Arthur's predictive abilities in Scenario 1 and Scenario 2. In the former, Arthur's justification for buying was Barbara's belief pertaining to the value – which he can only learn *after* he bought the asset and announces to Barbara that he wants to sell it. In the latter, his justification for buying was Barbara's belief about his belief – which he can predict to change before buying the asset.

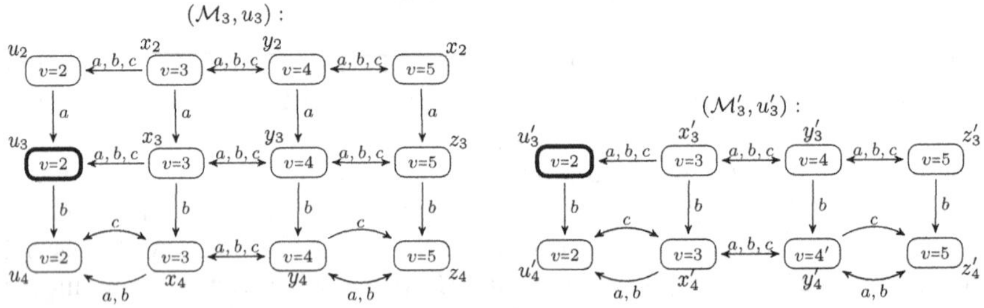

Fig. 3. The first two models of Scenario 3

Scenario 3: $B_aB_bB_c(v > 4)$ Consider plausibility model $(\mathcal{M}_3, u_3)$ in Fig. 3, where Farmer Flora offered to sell the orange tree for 2 euro and Arthur is willing to buy the tree because $B_aB_bB_c(v > 4)$. As Barbara indeed believes that $B_c(v > 4)$, Barbara is willing to buy the tree for 3 euro. The announcements $!\mathsf{buy}_b(3)$ and $!\mathsf{trade}_{a,b}(3)$ result in $(\mathcal{M}'_3, u'_3)$ in Fig. 3. Notice that this model resembles $(\mathcal{M}_1, u_1)$ from Scenario 1, where in this case Barbara offers to sell the orange tree for 4 euro and Chris will reply that he does not want to buy, resulting in a model similar to $(\mathcal{M}'_1, u'_1)$ from Fig. 2. This time, Barbara is left with a tree she can only sell for 1 euro. Note that by same reasoning as Arthur's in Scenario 1, Barbara could not foresee the bubble to crash in her hands. Given her beliefs about Chris, Barbara was rational to buy the tree from Arthur for 3 euro.

4 Results

The last scenario may be extended with more agents following the same or similar patterns. However, under predictive rationality (Definition 5), a justification like $B_aB_bB_a(v > 4)$ becomes invalid for $\mathsf{buy}_a(2)$. The consequence of predictive rationality for the structure of higher-order beliefs can be generalized as follows:

Proposition 1. *Given* $(\mathcal{M}, w)$, $\langle i, j_1, ..., j_k\rangle$ *and* $p_n < p_{n+1}$ *for all* $1 \leq n < k$, *it holds that:* $(\mathcal{M}, w) \models B_i B_{j_1} ... B_{j_k}(v > p_k)$ *and all agents in* $\langle i, j_1, ..., j_k\rangle$ *are pairwise distinct, if and only if,*
$(\mathcal{M}, w) \models B_i[!\mathsf{sell}_i(p_2)]...B_{j_{k-1}}[!\mathsf{sell}_{j_{k-1}}(p_{k-1})]B_{j_k}(v > p_k)$.

Proof. By definition of $\mathsf{sell}_i(p)$, an announcement of selling can only reveal information about the beliefs of the announcing agent. Hence whenever all traders in the sequence are *fresh*, nothing can be derived from the announcements of selling about agents who are expected to trade in the future. However if there are some agents $j_n = j_{n'}$ for $n \neq n'$, the announcement $!\mathsf{sell}_{j_n}(p_n)$ after j_n's first trade will inform all other traders that she will not want to buy the asset for $p_{n'} > p_n$ further down the line. In particular, $(\mathcal{M}, w) \models K_i[!\mathsf{sell}_{j_n}(p_n)]\neg\mathsf{buy}_{j_n}(p_{n'})$. By definition of $\mathsf{buy}_i(p)$, this implies $(\mathcal{M}, w) \not\models B_i[!\mathsf{sell}_i(p_2)]...B_{j_{k-1}}[!\mathsf{sell}_{j_{k-1}}(p_{k-1})]B_{j_k}(v > p_k)$.

Next, the examples from the previous section are used to prove that greater fools bubbles may occur[11] and the observations are generalized into a conclusion regarding the role of common knowledge in the burst. But first, some concepts need clarification.

Definition 6 (Highest Fundamental Value). *Given a* $(\mathcal{M}, w)$, *let the unique highest fundamental value* v_{max} *be such that there exists an agent* $i \in \mathcal{A}$ *and a world* $u \in \mathsf{max}_{\preceq_i}$ *such that* $(\mathcal{M}, u) \models v{=}v_{\mathsf{max}}$ *and there is no agent* $i \in \mathcal{A}$ *and a world* $u \in \mathsf{max}_{\preceq_i}$ *where* $(\mathcal{M}, u) \models v > v_{\mathsf{max}}$.

Definition 7 (Overpricing). *An asset is overpriced in* $(\mathcal{M}, w)$ *when* $(\mathcal{M}, w) \models \bigvee_{i,j\in\mathcal{A}} \bigvee_{p>v_{\mathsf{max}}} [!\mathit{trade}_{i,j}(p)]\top$.

Definition 8 (Bubble). *A sequence of events* $\langle\alpha_1, ..., \alpha_n\rangle$ *of type* $!\mathsf{sell}_i(p)$, $!\mathsf{buy}_i(p)$, $!\neg\mathsf{buy}_i(p)$ *or* $!\mathsf{trade}_{i,j}(p)$ *is called a bubble if for some* $(\mathcal{M}, w)$ *and* $l < n$ *it holds that* $(\mathcal{M}, w) \models [\alpha_1]...[\alpha_l] \bigvee_{i,j\in\mathcal{A}} \bigvee_{p>v_{\mathsf{max}}} [\mathit{trade}_{i,j}(p)]\top$ *while* $(\mathcal{M}, w) \models [\alpha_1]...[\alpha_l]...[\alpha_n]\neg \bigvee_{i\in\mathcal{A}} \bigvee_{p>v_{\mathsf{max}}} \mathsf{buy}_i(p)$.

That is, a bubble describes a sequence of events in which the asset can be traded for a price that is mutually believed to be too high and in the end the price falls back to the fundamental value: the crash.[12]

Proposition 2. *A bubble may form under the current greater fools encoding of buying and selling, even when all agents agree about the value of the asset.*

---

[11] This is not evident, as e.g. [1] merely shows – by simulating a growing bubble with a price that runs up automatically in every period – that *if* a bubble exists, people are willing to ride the bubble.

[12] Note that while a bubble is typically described by referring to an *extreme* overpricing, this definition allows for a minimal overpricing of $v_{\mathsf{max}} + 1$. It is here chosen to refrain from an ad hoc specification of "*extreme*", but if preferred one can easily adjust the definition and examples accordingly.

Proof. The scenarios in Sect. 3.3 constitute proof by example.

Firstly, note that Proposition 2 is proven for both the static and dynamic interpretation of buying. The difference between the two is that under the dynamic interpretation an asset can only be traded as most as many times as there are agents in the model. Secondly, note that greater fools bubbles are thus robust to symmetric atomic information. Instead, it is the higher-order uncertainty that sustains the asset to be traded for prices exceeding the fundamental value. It follows that the price drops to v_{max} when it is common knowledge that nobody believes the asset is worth more than v_{max}:

Proposition 3. *Given* $(\mathcal{M}, w)$, $(\mathcal{M}, w) \models CK\neg \bigvee_{i\in\mathcal{A}} B_i(v > v_{\mathsf{max}})$ *implies* $(\mathcal{M}, w) \models \neg \bigvee_{i\in\mathcal{A}} \bigvee_{p>v_{\mathsf{max}}} \mathsf{buy}_i(p)$.

Proof. Suppose $(\mathcal{M}, w) \models CK\neg \bigvee_{i\in\mathcal{A}} B_i(v > p)$. This means, there is no sequence of agents $\langle j_1, ..., j_k\rangle$ such that $B_{j_1}...B_{j_k}(v > v_{\mathsf{max}})$. By definition of $\mathsf{buy}_i(p)$ this implies that nobody wants to buy the asset for $p > v_{\mathsf{max}}$.

Proposition 3 demonstrates the informational transparancy of common knowledge that leads to the burst of a bubble. As illustrated in the three scenarios, the beliefs of all traders are sequentially revealed by all traders offering to sell the asset for a gradually rising price and finally all traders reject to buy. This conclusion supports Conlon [8] who presents a finite horizon "nth order" rational asset price bubble, where greater fools bubbles can only grow when there is a higher order possibility of a trader who does not believe the asset is overpriced.

5 Conclusions and Further Work

The greater fools explanation of an asset bubble is a theory that explains why an asset may be traded at prices far exceeding the fundamental value of an asset, even when all traders are rational. This paper provided a formal epistemic interpretation of being willing to sell or buy during greater fools episodes, which shows that epistemic logic is a natural setting in which to formalize the greater fools theory. It has been shown that under two different interpretations of rationality a greater fools bubble may arise even though all agents agree about the value. The literature is not clear about whether or not predictive rationality is assumed, even though it appears that it has consequences for the structure of the beliefs that form greater fools bubbles. It has finally been shown that common knowledge, which is achieved through communication, is an informational bubble buster.

This first formalization creates venues for future research, continuing the persuit of a deep understanding and a detailed description of the financial market. Firstly, it should be noted that the models rely on the assumption that all trade is publicly executed and all traders know each others' identity. Opening the model to anonymous traders and belief upgrades about the fundamental value – thereby shifting focus slighty away from purely greater fools reasoning – may change the conclusions. Furthermore, the used framework of standard epistemic logic can be enriched with for example probabilities (cf. [15]) to specify the risk of actions or the deeper justifications of agents' beliefs. The scenarios can be further formalized by translating the natural language protocol to transition rules cf. [17]. Furthermore, it would be interesting to study the interplay of different types of agents (based on e.g. strategy, level of rationality, mutual trust, expertise) in a greater fools bubble.

References

1. Abreu, D., Brunnermeier, M.K.: Bubbles and crashes. Econometrica **71**(1), 173–204 (2003)
2. Allen, F., Morris, S., Postlewaite, A.: Finite bubbles with short sale constraints and asymmetric information. J. Econ. Theor. **61**, 206–229 (1993)
3. Baltag, A., Christoff, Z., Hansen, J.U., Smets, S.: Logical models of informational cascades. In: van Benthem, J., Liu, F. (eds.) Logic Across the University: Foundations and Applications - Proceedings of the Tsinghua Logic Conference, vol. 47, pp. 405–432. College Publications, London (2013)
4. Baltag, A., Smets, S.: Dynamic belief revision over multi-agent plausibility models. In: Bonanno, G., Wooldridge, M. (eds.) Proceedings of the 7th Conference on Logic and the Foundations of Game and Decision (LOFT 2006), pp. 11–24. University of Liverpool (2006)
5. Barlevy, G.: Bubbles and fools. Econ. Perspect. **39**(2), 54–76 (2015)
6. Brunnermeier, M.K.: Asset Pricing Under Asymmetric Information: Bubbles, Crashes, Technical Analysis and Herding. Oxford University Press, New York (2001)
7. Brunnermeier, M.K.: Bubbles, 2nd edn. In New Palgrave Dictionary of Economics. Palgrave Macmillan, London (2008)
8. Conlon, J.R.: Simple finite horizon bubbles robust to higher order knowledge. Econometrica **72**(3), 927–936 (2004)
9. Dégremont, C., Roy, O.: Agreement theorems in dynamic-epistemic logic. J. Philos. Logic **41**, 735–764 (2012)
10. De Long, J.B., Shleifer, A., Summers, L.H., Waldmann, R.J.: Noise trader risk in financial markets. J. Polit. Econ. **98**, 703–738 (1990)
11. Demey, L.: Agreeing to disagree in probabilistic dynamic epistemic logic. Synthese **191**, 409–438 (2014)
12. Fama, E.F.: The behavior of stock-market prices. J. Bus. **38**, 34–105 (1965)
13. Gerbrandy, J., Groeneveld, W.: Reasoning about information change. J. Logic Lang. Inform. **6**, 147–169 (1997)
14. Kindleberger, C.P., Aliber, R.Z.: Manias, Panics, and Crashes: a history of financial crises, 5th edn. John Wiley and Sons Inc., USA (2005)
15. Kooi, B.: Probabilistic dynamic epistemic logic. J. Logic Lang. Inform. **12**, 381–408 (2003)
16. Levine, S.S., Zajac, E.J.: The institutional nature of price bubbles. SSRN Electroni. J. **11**(1), 109–126 (2007)
17. Rendsvig, R.K.: Aggregated beliefs and informational cascades. In: Grossi, D., Roy, O., Huang, H. (eds.) LORI 2013. LNCS, vol. 8196, pp. 337–341. Springer, Heidelberg (2013). doi:10.1007/978-3-642-40948-6_29
18. Tirole, J.: On the possibility of speculation under rational expectations. Econometrica **50**(5), 1163–1181 (1982)
19. Vogel, H.L.: Financial Market: Bubbles and Crashes. Cambridge University Press, New York (2010)

On Axiomatization of Epistemic GDL

Guifei Jiang[1(✉)], Laurent Perrussel[2], and Dongmo Zhang[3]

[1] Software College, Nankai University, Tianjin, China
g.jiang@nankai.edu.cn
[2] IRIT, University of Toulouse, Toulouse, France
[3] SCEM, Western Sydney University, Penrith, Australia

Abstract. The Game Description Language (GDL) has been introduced as an official language for specifying games in the AAAI General Game Playing Competition since 2005. It was originally designed as a declarative language for representing rules of arbitrary games with perfect information. More recently, an epistemic extension of GDL, called EGDL, has been proposed for representing and reasoning about imperfect information games. In this paper, we develop an axiomatic system for a variant of EGDL and prove its soundness and completeness with respect to the semantics based on the epistemic state transition model. With a combination of action symbols, temporal modalities and epistemic operators, the completeness proof requires novel combinations of techniques used for completeness of propositional dynamic logic and epistemic temporal logic. We demonstrate how to use the proof theory for inferring game properties from game rules.

1 Introduction

General Game Playing (GGP) is concerned with creating intelligent agents that can play previously unknown games by just being given their rules [6]. To specify a game played by autonomous agents, a formal game description language, called GDL, has been introduced as an official language for GGP since 2005. GDL is defined as a high-level, machine-processable language for representing the rules of arbitrary games with perfect information [16]. Originally designed as a logic programming language, GDL has been recently adapted as a logical language for game specification and strategic reasoning [25]. Based on this, the epistemic extension of GDL, called EGDL, has been developed for representing and reasoning about imperfect information games [14].

Syntactically, EGDL extends GDL with the standard epistemic operators to specify the rules of imperfect information games and capture the epistemic status of agents. For example, an EGDL-formula $\mathsf{K}_r(does(a) \wedge \bigcirc wins(r)) \rightarrow does(a)$ specifies that if an agent knows that taking an action leads to win at the next state, then she takes that action at the current state. Semantically, EGDL is interpreted over epistemic state transition models which are used to

Most of the work was done while the first author was a postdoc at IRIT, University of Toulouse.

A. Baltag et al. (Eds.): LORI 2017, LNCS 10455, pp. 598–613, 2017.
DOI: 10.1007/978-3-662-55665-8_41

represent synchronous and deterministic games with imperfect information. The expressive power and computational efficiency of EGDL have been investigated in [14]. In this paper, we address the fundamental logical question of a complete axiomatization for EGDL.

The axiomatic system for EGDL that we present is composed of axiom schemes and inference rules that capture the logical properties of the semantical models. With action, temporal and epistemic operators, the completeness proof of EGDL, however, is non-trivial. It requires novel combinations and extensions of techniques from both propositional dynamic logic (PDL) [15] and epistemic temporal logics (ETLs) [7].

To achieve the completeness of EGDL, we first construct a pre-model for a consistent EGDL-formula φ out of maximal consistent subsets of a finite set of formulas, called the *closure* of φ. Similar to [7], we define a number of different distinct levels of closure so as to deal with the epistemic operators. The techniques to construct the pre-model has been strongly influenced by two sources which gave us valuable insights: [15] providing an elementary proof of the completeness of PDL, and [7] presenting a general framework for completeness proofs of ETLs. Unfortunately, the pre-model is non-deterministic and thus is not an epistemic state transition model. To fill this gap, we then transform the pre-model into an epistemic state transition model with an equivalent satisfiability of EGDL-formulas. Such transformation is inspired by the method used in [18] to transform a non-deterministic automata into a deterministic one. From the completeness proof, we derive the *finite model property* for EGDL: every EGDL-formula that is satisfiable in some epistemic state transition model is satisfiable in a finite epistemic state transition model. We also demonstrate how to use the proof theory for inferring game properties from game rules.

The rest of this paper is structured as follows: Sect. 2 establishes the syntax and the semantics of EGDL. Section 3 provides a sound and complete axiomatic system for EGDL and demonstrates how to use the proof theory for reasoning about game rules. Section 4 discusses the related work. Finally we conclude with future work.

2 The Framework

All games are assumed to be played in multi-agent environments. Each game is associated with a game signature. A *game signature* $\mathcal{S}$ is a triple $(N, \mathcal{A}, \Phi)$, where $N = \{1, 2, \cdots, m\}$ is a non-empty finite set of agents; $\mathcal{A} = \bigcup_{r \in N} A^r$, where A^r consists of a non-empty finite set of *actions* for agent r such that different agents have different actions, i.e., $A^{r_1} \cap A^{r_2} = \emptyset$ if $r_1 \neq r_2$ and each agent has an action without effects, i.e., $noop^r \in A^r$, and $\Phi = \{p, q, \cdots\}$ is a finite set of propositional atoms for specifying individual features of a game state.

Through the rest of the paper, we will fix a game signature $\mathcal{S}$ and all concepts will be based on this game signature unless otherwise specified.

2.1 Epistemic State Transition Models

We consider synchronous imperfect information games where all players move simultaneously and may have partial information of the game states. The structures of these games may be specified by epistemic state transition models defined as follows:

Definition 1. *An* epistemic state transition (EST) model *M is a tuple $(W, I, T, \{R_r\}_{r\in N}, \{L_r\}_{r\in N}, U, g, \pi)$, where*

- *W is a non-empty set of* possible states.
- *$I \subseteq W$, representing a set of* initial *states.*
- *$T \subseteq W \backslash I$, representing a set of* terminal *states.*
- *$R_r \subseteq W \times W$ is an equivalence relation for agent r, indicating the states that are indistinguishable for r.*
- *$L_r \subseteq W \times A^r$ is a* legality *relation for agent r, describing the legal actions of agent r at each state. Let $L_r(w) = \{a \in A^r : (w, a) \in L_r\}$ be the set of all legal actions of agent r at state w. To make a game playable, we assume that (i) each agent has at least one available action at each state: $L_r(w) \neq \emptyset$ for all $r \in N$ and $w \in W$, and (ii) at all terminal states each agent can only take action noop: $L_r(w) = \{noop^r\}$ for any $r \in N$ and $w \in T$.*
- *$U : W \times \prod_{r\in N} A^r \hookrightarrow W \backslash I$ is a partial* update *function, specifying the state transformations, such that $U(w, \langle noop^r\rangle_{r\in N}) = w$ for any $w \in T$.*
- *$g : N \to 2^W$ is a* goal *function, specifying the winning states for each agent.*
- *$\pi : W \to 2^\Phi$ is a standard valuation function.*

Note that different from [14], (i) we consider a general case without assuming a unique initial state; (ii) the update function is partial, as not all joint actions are possible in all states due to the legality relation. In particular, there is no semantical condition to guarantee that all joint legal actions lead to valid next states. Such a condition is not easy to provide, giving the legal conditions are defined for individual agents. Besides this, we do not require each agent knows her own legal actions, since in GGP it may occur that an agent fails to figure out her legal actions given the limited time. In that case, the game master assigns a random legal action for her. For convenience, let D denote the set of all joint actions $\prod_{r\in N} A^r$. For $d \in D$, let $d(r)$ denote agent r's action in the joint action d. We write $R_r(w)$ for the set of all states that agent r cannot distinguish from w, i.e., $R_r(w) = \{u \in W \ : \ wR_r u\}$. We now define the notion of *a path* to specify the set of all possible ways in which a game can develop.

Definition 2. *Given an EST-model $M = (W, I, T, \{R_r\}_{r\in N}, \{L_r\}_{r\in N}, U, g, \pi)$, a path δ is an infinite sequence of states and actions $w_0 \xrightarrow{d_1} w_1 \xrightarrow{d_2} w_2 \cdots \xrightarrow{d_j} \cdots$ such that for all $j \geq 1$ and for any $r \in N$,*

1. *$w_j = U(w_{j-1}, d_j)$ (state update);*
2. *$(w_{j-1}, d_j(r)) \in L_r$ (that is, any action that is taken must be legal);*
3. *if $w_{j-1} \in T$, then $w_{j-1} = w_j$ (that is, a loop after reaching a terminal state).*

It follows that only the first state may be initial, i.e., $w_j \notin I$. Let $\mathcal{P}(M)$ denote the set of all paths in M. When M is fixed, we simply write $\mathcal{P}$. For a path $\delta \in \mathcal{P}$ and a position $j \geq 0$, we use $\delta[j]$, $\delta[0,j]$ and $\delta[j,\infty]$ to denote the j-th state of δ, the finite prefix $w_0 \xrightarrow{d_1} w_1 \xrightarrow{d_2} \cdots \xrightarrow{d_j} w_j$ of δ and the infinite suffix path $w_j \xrightarrow{d_{j+1}} w_{j+1} \xrightarrow{d_{j+2}} \cdots$ of δ, respectively. Finally, we write $\theta_r(\delta, j)$ for the action of agent r taken at stage j of δ.

The following definition, by extending equivalence relations over states to paths, characterizes precisely what an agent with *imperfect recall and perfect reasoning* can in principle know during a game.

Definition 3. *Two paths δ, $\delta' \in \mathcal{P}$ are imperfect recall (also called memoryless) equivalent for agent r, written $\delta \approx_r \delta'$, iff $\delta[0] R_r \delta'[0]$.*

That is, imperfect recall requires an agent to be only aware of the present state but forget everything that happened. This is similar to the notion of imperfect recall in ATL [20].

2.2 The Syntax

Let us now introduce an epistemic extension of the game description language GDL [25] to represent games with imperfect information. We further provide a semantics for the language based on the epistemic state transition model. In the following, we call this resulting framework EGDL for short.

Definition 4. *The language $\mathcal{L}$ of EGDL is generated by the following BNF:*

$$\varphi ::= p \mid initial \mid terminal \mid legal(a^r) \mid wins(r) \mid does(a^r) \mid$$

$$\neg\varphi \mid \varphi \wedge \psi \mid \bigcirc\varphi \mid \mathsf{K}_r\varphi \mid \mathsf{C}\varphi$$

where $p \in \Phi$, $r \in N$ and $a^r \in A^r$.

Other connectives $\vee$, $\rightarrow$, $\leftrightarrow$, $\top$, $\bot$ are defined by $\neg$ and $\wedge$ in the standard way. Intuitively, *initial* and *terminal* specify the initial state and the terminal states of a game, respectively; $does(a^r)$ asserts that agent r takes action a at the current state; $legal(a^r)$ asserts that action a is available to agent r at the current state; and $wins(r)$ asserts that agent r wins at the current state. The formula $\bigcirc\varphi$ means "φ holds in the next state". All these components are inherited from GDL. The epistemic operators K and C are taken from the Modal Epistemic Logic [5,9]. The formula $\mathsf{K}_r\varphi$ is read as "agent r knows φ", and $\mathsf{C}\varphi$ as "φ is common knowledge among all the agents in N". As usual, we write $\widehat{\mathsf{K}}_r$ for the dual of K_r and $\mathsf{E}\varphi$ for $\bigwedge_{r \in N} \mathsf{K}_r\varphi$, saying that every agent in N knows φ.

To illustrate the intuition of the language, let us consider a variant of the Tic-Tac-Toe, called Krieg-Tictactoe in [19].

Example 1. Krieg-Tictactoe is played by two players, cross x and naught o, who take turns marking cells in a 3×3 board. Different from standard Tic-Tac-Toe,

1. $initial \leftrightarrow turn(\mathsf{x}) \wedge \neg turn(\mathsf{o}) \wedge \bigwedge_{i,j=1}^{3} \neg(p^{\mathsf{x}}_{i,j} \vee p^{\mathsf{o}}_{i,j})$
2. $tried(a^r_{i,j}) \rightarrow p^{-r}_{i,j}$
3. $wins(r) \leftrightarrow (\bigvee_{i=1}^{3}\bigwedge_{l=0}^{2} p^r_{i,1+l}) \vee (\bigvee_{j=1}^{3}\bigwedge_{l=0}^{2} p^r_{1+l,j}) \vee (\bigwedge_{l=0}^{2} p^r_{1+l,1+l}) \vee (\bigwedge_{l=0}^{2} p^r_{1+l,3-l})$
4. $teminal \leftrightarrow wins(\mathsf{x}) \vee wins(\mathsf{o}) \vee \bigwedge_{i,j=1}^{3} (p^{\mathsf{x}}_{i,j} \vee p^{\mathsf{o}}_{i,j})$
5. $turn(r) \wedge \neg terminal \rightarrow \bigcirc\neg turn(r) \wedge \bigcirc turn(-r)$
6. $legal(noop^r) \leftrightarrow turn(-r) \vee terminal$
7. $legal(a^r_{i,j}) \leftrightarrow turn(r) \wedge \neg p^r_{i,j} \wedge \neg tried(a^r_{i,j}) \wedge \neg terminal$
8. $\bigcirc p^r_{i,j} \leftrightarrow p^r_{i,j} \vee (does(a^r_{i,j}) \wedge \neg(p^{\mathsf{x}}_{i,j} \vee p^{\mathsf{o}}_{i,j}))$
9. $\bigcirc tried(a^r_{i,j}) \leftrightarrow tried(a^r_{i,j}) \vee (does(a^r_{i,j}) \wedge p^{-r}_{i,j})$
10. $does(a^r_{i,j}) \rightarrow \mathsf{K}_r(does(a^r_{i,j}))$
11. $initial \rightarrow \mathsf{E}initial$
12. $(turn(r) \rightarrow \mathsf{E}turn(r)) \wedge (\neg turn(r) \rightarrow \mathsf{E}\neg turn(r))$
13. $(p^r_{i,j} \rightarrow \mathsf{K}_r p^r_{i,j}) \wedge (\neg p^r_{i,j} \rightarrow \mathsf{K}_r \neg p^r_{i,j})$
14. $(tried(a^r_{i,j}) \rightarrow \mathsf{K}_r tried(a^r_{i,j})) \wedge (\neg tried(a^r_{i,j}) \rightarrow \mathsf{K}_r \neg tried(a^r_{i,j}))$

Fig. 1. An EGDL description of Krieg-Tictactoe.

each player can see her own marks, but not those of her opponent, just like the chess variant *Kriegspiel* [17].

To represent the Krieg-Tictactoe, we first describe its game signature, written $\mathcal{S}_{KT}$, as follows: $N_{KT} = \{\mathsf{x}, \mathsf{o}\}$ where x denotes the player who marks the symbol cross and o denotes the player who marks the symbol naught; $A^r_{KT} = \{a^r_{i,j} : 1 \leq i,j \leq 3\} \cup \{noop^r\}$, where $a^r_{i,j}$ denotes the action that player r marks cell (i,j) with her symbol; $\Phi_{KT} = \{p^r_{i,j}, tried(a^r_{i,j}), turn(r) : r \in \{\mathsf{x}, \mathsf{o}\}$ and $1 \leq i,j \leq 3\}$, where $p^r_{i,j}$ represents the fact that cell (i,j) is marked with player i's symbol, $tried(a^r_{i,j})$ represents the fact that player r has tried to mark cell (i,j) but failed before, and $turn(r)$ says that it is player r's turn now. The rules of Krieg-Tictactoe are specified by EGDL in Fig. 1 (where $1 \leq i,j \leq 3$, $r \in \{\mathsf{x}, \mathsf{o}\}$ and $-r$ represents r's opponent).

Rules 1–5 specify the initial state, each player's winning states, the terminal states and the turn-taking. In particular, Rule 2 specifies that if a player has tried to mark a cell, then the corresponding cell is marked by the opponent.

The preconditions of each action (legality) are specified by 6 and 7. *The player who has the turn can mark any non-terminal cell such that (i) it is not marked by herself, and (ii) she has never tried to mark it before. A player can only do action noop at the terminal states or the states where it is not her turn.*

Rules 8 and 9 are the combination of the frame axioms and the effect axioms. Rule 8 states that *a cell is marked with a player's symbol in the next state if the player takes the corresponding action at the current state or the cell has been marked by her symbol before.* Similarly, Rule 9 says that *an action is tried by a player in the next state if the action is ineffective while still taken by the player at the current state, or this action has been tried before.*

The rest of the rules specify the epistemic status of the game. Rule 10 states each player knows which action she is taking. Rule 11 and Rule 12 say both players know the initial state and their turns, respectively. Rule 13 says that *each player knows which cell is marked or not with her symbol.* Similarly, Rule 14 states that *each player knows which cell is tried or not by herself.*

Note that rules 12–14 together specify the epistemic relations for each player: *two states are indistinguishable for a player if their configurations are the same from her point of view.* Finally, let Σ_{KT} be the set of rules 1–14.

2.3 The Semantics

The semantics of EGDL-formulas is based on the epistemic state transition models.

Definition 5. *Let M be an EST-model. Given a path δ in M and a formula $\varphi \in \mathcal{L}$, we say φ is true at δ under M, denoted by $M, \delta \models \varphi$, according to the following definition:*

$M, \delta \models p$	*iff*	$p \in \pi(\delta[0])$
$M, \delta \models \neg\varphi$	*iff*	$M, \delta \not\models \varphi$
$M, \delta \models \varphi_1 \wedge \varphi_2$	*iff*	$M, \delta \models \varphi_1$ *and* $M, \delta \models \varphi_2$
$M, \delta \models initial$	*iff*	$\delta[0] \in I$
$M, \delta \models terminal$	*iff*	$\delta[0] \in T$
$M, \delta \models wins(r)$	*iff*	$\delta[0] \in g(r)$
$M, \delta \models legal(a^r)$	*iff*	$(\delta[0], a^r) \in L_r$
$M, \delta \models does(a^r)$	*iff*	$\theta_r(\delta, 0) = a^r$
$M, \delta \models \bigcirc\varphi$	*iff*	$M, \delta[1, \infty] \models \varphi$
$M, \delta \models \mathsf{K}_r\varphi$	*iff*	*for any* $\delta' \in \mathcal{P}$, *if* $\delta \approx_r \delta'$, *then* $M, \delta' \models \varphi$
$M, \delta \models \mathsf{C}\varphi$	*iff*	*for any* $\delta' \in \mathcal{P}$, *if* $\delta \approx_N \delta'$, *then* $M, \delta' \models \varphi$

where $\approx_N$ is the transitive closure of $\bigcup_{r \in N} \approx_r$.

A formula φ is *globally true or valid* in an EST-model M, written $M \models \varphi$, if $M, \delta \models \varphi$ for any $\delta \in \mathcal{P}$. A formula φ is *valid*, written $\models \varphi$, if $M \models \varphi$ for any EST-model M. Let Σ be a set of formulas in $\mathcal{L}$, then M is a *model* of Σ if $M \models \varphi$ for all $\varphi \in \Sigma$.

The following result specifies some generic game properties.

Proposition 1. *For any $r \in N$, $\varphi \in \mathcal{L}$ and $a^r, b^r \in A^r$,*

1. $\models \neg \bigcirc initial$
2. $\models terminal \to \bigwedge_{a^r \in A^r \setminus \{noop^r\}} \neg legal(a^r) \wedge legal(noop^r)$
3. $\models \bigvee_{a^r \in A^r} does(a^r)$
4. $\models \neg(does(a^r) \wedge does(b^r))$ *for* $a^r \neq b^r$
5. $\models does(a^r) \to legal(a^r)$
6. $\models \bigvee_{a^r \in A^r} legal(a^r)$
7. $\models terminal \wedge \varphi \to \bigcirc\varphi$

The first formula says that a game would never go back to its initial state once it starts. The second formula specifies that all players can only take action "noop" at the terminal states. The third and forth formulas prescribe that there is a unique action for each player at all game states. The fifth formula asserts that any action that is taken should be legal. The sixth formula specifies that each player has at least one legal action at each state. The last formula requires that a terminal state leads to a self-loop.

Besides those generic game properties, EGDL is also able to specify epistemic properties of a game. For instance, whether each player always knows her own legal actions in the course of the game. This property as well as some other well-known properties have been discussed in [14]. It is worth of mentioning that although these properties are expressible in EGDL, different from the generic game properties, they are not valid for any game model.

3 Axiomatization

In this section, we develop an axiomatic system for the logic EGDL, and provide its soundness and completeness with respect to the epistemic state transition models.

3.1 The Axiomatic System

EGDL consists of the following axiom schemas and inference rules: For any a^r, $b^r \in A^r$, $r \in N$ and $\varphi, \psi \in \mathcal{L}$,

- Axiom Schemas:
  1. All tautologies of classical propositional logic.
  2. $\neg \bigcirc\, initial$
  3. $terminal \rightarrow \bigwedge_{a^r \in A^r \setminus \{noop^r\}} \neg legal(a^r) \wedge legal(noop^r)$
  4. $\bigvee_{a^r \in A^r} does(a^r)$
  5. $\neg(does(a^r) \wedge does(b^r))$ for $a^r \neq b^r$.
  6. $\bigcirc(\varphi \rightarrow \psi) \rightarrow (\bigcirc\varphi \rightarrow \bigcirc\psi)$
  7. $\neg \bigcirc \varphi \leftrightarrow \bigcirc\neg\varphi$
  8. $does(a^r) \rightarrow legal(a^r)$
  9. $\varphi \wedge terminal \rightarrow \bigcirc\varphi$
  10. $\mathsf{K}_r(\varphi \rightarrow \psi) \rightarrow (\mathsf{K}_r\varphi \rightarrow \mathsf{K}_r\psi)$
  11. $\mathsf{K}_r\varphi \rightarrow \varphi$
  12. $\mathsf{K}_r\varphi \rightarrow \mathsf{K}_r\mathsf{K}_r\varphi$
  13. $\neg\mathsf{K}_r\varphi \rightarrow \mathsf{K}_r\neg\mathsf{K}_r\varphi$
  14. $\mathsf{E}\varphi \leftrightarrow \bigwedge_{r=1}^{m} \mathsf{K}_r\varphi$
  15. $\mathsf{C}\varphi \rightarrow \mathsf{E}(\varphi \wedge \mathsf{C}\varphi)$
- Inference Rules:

  (R1) From φ, $\varphi \rightarrow \psi$ infer ψ.

  (R2) From φ infer $\bigcirc\varphi$.

  (R3) From φ infer $\mathsf{K}_r\varphi$.

  (R4) From $\varphi \rightarrow \mathsf{E}(\varphi \wedge \psi)$ infer $\varphi \rightarrow \mathsf{C}\psi$.

Besides the axioms mentioned in Proposition 1, the axioms for temporal and epistemic operators are well-known. Note that since we focus on games with imperfect recall, thus there is no interaction properties between epistemic and temporal operators. Let $\vdash$ denote the provability in EGDL. The notion of *the syntactic consequence (derivation)* is defined in the standard way.

With the proof theory, we are now able to derive the following formulas from the rules of Krieg-Tictactoe specified in Fig. 1.

Proposition 2. *For any* $r \in N_{KT}$ *and* $a^r_{i,j} \in A^r_{KT}$ *,*

1. $\vdash_{\Sigma_{KT}} initial \rightarrow \mathsf{C} initial$
2. $\vdash_{\Sigma_{KT}} legal(a^r_{i,j}) \rightarrow \mathsf{K}_r(legal(a^r_{i,j}))$
3. $\vdash_{\Sigma_{KT}} does(a^r_{i,j}) \rightarrow \bigcirc \mathsf{K}_r(p^r_{i,j} \vee tried(a^r_{i,j}))$
4. $\vdash_{\Sigma_{KT}} \mathsf{K}_r tried(a^r_{i,j}) \rightarrow \mathsf{K}_r p^{-r}_{i,j}$

That is, in Krieg-Tictactoe, the turn-taking is common knowledge (Clause 1). Each player knows her own available actions (Clause 2). If an agent marks a cell at the current state, then she will knows either this cell has been marked or been tried by herself at the next state (Clause 3). Moreover, if a player knows that she has tried to mark a cell, then she knows the corresponding cell has been marked by the opponent (Clause 4).

3.2 Completeness Proofs

The completeness result is achieved in two step. First, we construct a pre-model for a consistent formula φ out of consistent subsets of a finite set of formulas, called the *closure of* φ. The construction resembles those previously used for completeness of propositional dynamic logic [15] and epistemic temporal logics [7]. Next we transform the pre-model into an epistemic state transition model and show that the satisfiability of EGDL-formulas is invariant under such transformation. This idea is captured in Figure 2 (where $r \in N$ and $k \in \mathbb{N}$).

Let us now fix a formula $\varphi \in \mathcal{L}$, which is consistent in EGDL, i.e., not $\vdash \neg\varphi$. We define $ad(\varphi)$ to be the greatest number of alternations of distinct K_r's along any branch in φ's parse tree. If φ involves the common knowledge operator C, let $ad(\varphi) = 0$. For instance, $ad(\mathsf{K}_{r_1}\mathsf{K}_{r_2}\mathsf{K}_{r_1}p) = 3$; $ad(\mathsf{K}_{r_1}\mathsf{K}_{r_1}\mathsf{K}_{r_2}p) = 2$; $ad(\mathsf{C}\mathsf{K}_{r_1}\mathsf{K}_{r_2}p) = 0$; temporal operators are not considered, so that $ad(\mathsf{K}_{r_1}\mathsf{K}_{r_2}\bigcirc \mathsf{K}_{r_1}p) = 3$.

A finite sequence $\sigma = r_1 r_2 \cdots r_k$ of agents, possibly equal to the null sequence ϵ, is called an *index* if $r_i \neq r_{i+1}$ for all $i < k$. We write $|\sigma|$ for the length k of such a sequence. In particular, $|\epsilon| = 0$.

Let N^* be the set of all finite sequences over N, we define the absorptive concatenation function $\#$ from $N^* \times N$ to N^* as follows: Given a sequence $\sigma \in N^*$ and an agent $r \in N$,

$$\sigma\#r = \begin{cases} \sigma & \text{if the final element of} \, \sigma \, \text{is} \, r; \\ \sigma r & \text{otherwise.} \end{cases}$$

Given $\varphi \in \mathcal{L}$, for each $k \geq 0$, we define the k-closure $cl_k(\varphi)$, and for each agent $r \in N$, we define the k, r-closure $cl_{k,r}(\varphi)$. The definitions of these sets proceeds by mutual recursion:

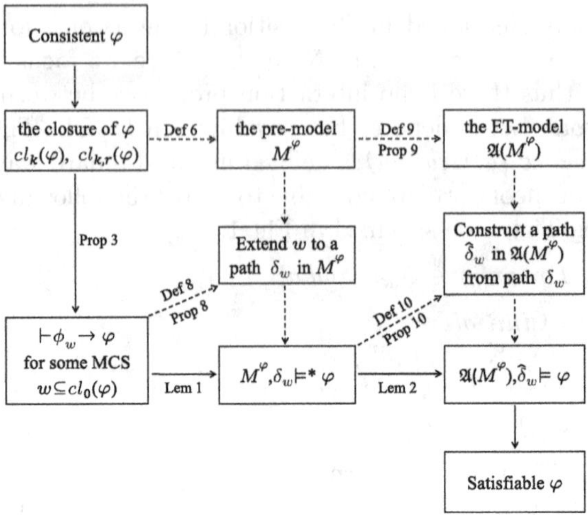

Fig. 2. The roadmap of the completeness proof for EGDL. Note that solid arrows denote the process to achieve the completeness, and dashed arrows denote the notions and their properties to obtain the intermediate results. The abbreviation "MCS" denotes the maximal consistent set.

1. *The basic closure* $cl_0(\varphi)$ is the smallest set containing φ such that
   (a) it is closed under subformulas.
   (b) if $\mathsf{E}\psi \in cl_0(\varphi)$, then $\mathsf{K}_{r_1}\psi, \cdots, \mathsf{K}_{r_m}\psi \in cl_0(\varphi)$.
   (c) if $\mathsf{C}\psi \in cl_0(\varphi)$, then $\mathsf{EC}\psi \in cl_0(\varphi)$.
   (d) if $\psi \in cl_0(\varphi)$ and ψ is not of the form $\neg\psi'$, then $\neg\psi \in cl_0(\varphi)$.
2. Let $cl_{k,r}(\varphi)$ be the union of $cl_k(\varphi)$ with the set of formulas of the form $\mathsf{K}_r(\psi_1 \vee \cdots \psi_n)$ or $\neg\mathsf{K}_r(\psi_1 \vee \cdots \psi_n)$, where the ψ_i are distinct formulas in $cl_k(\varphi)$.
3. $cl_{k+1}(\varphi) = \bigcup_{r=1}^{m} cl_{k,r}(\varphi)$.

If X is a finite set of formulas, we write ϕ_X for the conjunction of all the formulas in X. A finite set X of formulas is said to be consistent if ϕ_X is consistent. A finite set Cl of formulas is said to be *negation-closed* if, for all $\psi \in Cl$, either $\neg\psi \in Cl$ or ψ is of the form $\neg\psi'$ and $\psi' \in Cl$. Note that the sets $cl_k(\varphi)$ and $cl_{k,r}(\varphi)$ are negation-closed. We define an *atom* of Cl to be a maximal consistent subset of Cl. The set of all atoms of Cl is denoted as $\mathcal{A}_{Cl}$. We have the following properties.

Proposition 3. *Suppose that X is a finite set of formulas and Cl is a negation-closed set of formulas. For any $\varphi_1, \varphi_2 \in \mathcal{L}$,*

1. *if $\vdash \phi_X \to \varphi_1$ and $\vdash \varphi_1 \to \varphi_2$, then $\vdash \phi_X \to \varphi_2$.*
2. *if X is an atom of Cl and $\psi \in Cl$, then either $\vdash \phi_X \to \psi$ or $\vdash \phi_X \to \neg\psi$.*
3. *$\vdash \bigvee_{X \in \mathcal{A}_{Cl}} \phi_X$.*

The construction of the pre-model of φ is based on the atoms of the closures of φ. Let $d = ad(\varphi)$.

Definition 6. *The pre-model of φ, denoted by $M^\varphi = (W^\varphi, I^\varphi, T^\varphi, \{R_r^\varphi\}_{r\in N}, \{L_r^\varphi\}_{r\in N}, U^\varphi, g^\varphi, \pi^\varphi)$, is constructed as follows:*

1. *W^φ consists of all the pairs (σ, X) such that σ is an index, $|\sigma| \leq d$, and*
   (a) if $\sigma = \epsilon$ then X is an atom of $cl_d(\varphi)$, and
   (b) if $\sigma = \tau r$ then X is an atom of $cl_{k,r}(\varphi)$, where $k = d - |\sigma|$.
2. *$I^\varphi = \{(\sigma, X) \in W^\varphi \ : \ \vdash \phi_X \rightarrow initial\}$.*
3. *$T^\varphi = \{(\sigma, X) \in W^\varphi \ : \ \vdash \phi_X \rightarrow terminal\}$.*
4. *$(\sigma, X)R_r^\varphi(\tau, Y)$ iff $\sigma\#r = \tau\#r$ and $\{\psi : \mathsf{K}_r\psi \in X\} = \{\chi : \mathsf{K}_r\chi \in Y\}$.*
5. *$((\sigma, X), a^r) \in L_r^\varphi$ iff $\vdash \phi_X \rightarrow legal(a^r)$.*
6. *$U^\varphi((\sigma, X), d) = (\tau, Y)$ iff $\vdash \phi_X \rightarrow \bigwedge_{r\in N} legal(d(r))$, $\sigma = \tau$ and the formula $\phi_X \wedge \bigcirc\phi_Y$ is consistent.*
7. *$g^\varphi(r) = \{(\sigma, X) \in W^\varphi \ : \ \vdash \phi_X \rightarrow wins(r)\}$.*
8. *$\pi^\varphi((\sigma, X)) = \{p \in \Phi \ : \ \vdash \phi_X \rightarrow p\}$.*

It is easy to see that the update function in M^φ is non-deterministic, and thus M^φ is not an EST-model. We redefine the notion of a path as follows:

Definition 7. *Given the pre-model $M^\varphi = (W^\varphi, I^\varphi, T^\varphi, \{R_r^\varphi\}_{r\in N}, \{L_r^\varphi\}_{r\in N}, U^\varphi, g^\varphi, \pi^\varphi)$ of φ, a path δ of M^φ is an infinite sequence of states and actions $w_0 \xrightarrow{d_1} w_1 \xrightarrow{d_2} \cdots$ such that the conditions are the same as those in Definition 2 except changing Condition 1 to $w_j \in U^\varphi(w_{j-1}, d_j)$ due to nondeterminacy.*

Similarly, we generalize the indistinguishable relation to the paths. We say that two paths δ, δ' of M^φ are indistinguishable for agent r, denoted by $\delta \approx_r^\varphi \delta'$, iff $\delta[0]R_r^\varphi\delta'[0]$. The truth conditions for all EGDL-formulas under the pre-model are exactly the same as those in Definition 5. In particular, we use $M^\varphi, \delta \models^* \varphi$ to denote that φ is true at path δ under M^φ.

If $w = (\sigma, X)$ is a state, we define ϕ_w to be the formula ϕ_X. Following [7], we say that the state w *directly decides* a formula ψ if either $\psi \in w$, $\neg\psi \in w$, or $\psi = \neg\psi'$ and $\psi' \in w$. We say that w *decides* ψ if either $\vdash \phi_w \rightarrow \psi$, or $\vdash \phi_w \rightarrow \neg\psi$. Clearly, if w directly decides ψ, then w decides ψ. Note that if $\sigma = \tau r$, then each σ-state directly decides every formula in $cl_{k,r}(\varphi)$. Also, every ϵ-state directly decides every formula in $cl_d(\varphi)$. In particular, we have the following results about formulas with K-operators.

Proposition 4. *Given two states $w = (\sigma, X)$ and $u = (\tau, Y)$, if $\sigma\#r = \tau\#r$, then the same formulas of the form $\mathsf{K}_r\psi$ are directly decided by w and u.*

Given a σ-state w, we use $\Phi_{w,r}$ for the disjunction of all the formulas ϕ_u, where u is a σ-state satisfying $wR_r^\varphi u$, and we use $\Phi_{w,r}^+$ for the disjunction of all the formulas ϕ_u, where u is a $\sigma\#r$-state satisfying $wR_r^\varphi u$.

Proposition 5. *1. If w is a σ-state and u is a σ-state or $\sigma\#r$-state such that not $wR_r^\varphi u$, then $\vdash \phi_w \rightarrow \mathsf{K}_r\neg\phi_u$.*
2. For all σ-states w, $\vdash \phi_w \rightarrow \mathsf{K}_r\Phi_{w,r}$.
3. For all σ-states w, if $|\sigma\#r| \leq d$, then $\vdash \phi_w \rightarrow \mathsf{K}_r\Phi_{w,r}^+$.

The following two propositions show that the pre-model has properties resembling those for the truth conditions for formulas in the basic closure.

Proposition 6. *For all σ-states w and $\mathsf{K}_r\psi \in cl_0(\varphi)$, the following are equivalent.*

1. $\vdash \phi_w \rightarrow \neg \mathsf{K}_r\psi$.
2. *There is some σ-state u such that $wR_r^\varphi u$ and $\vdash \phi_u \rightarrow \neg\psi$.*

Please recall that when the formula φ contains the common knowledge operator, we take $d = 0$, so that all states are ϵ-states.

Proposition 7. *Given $\mathsf{C}\psi \in cl_0(\varphi)$, the following are equivalent.*

1. $\vdash \phi_w \rightarrow \neg \mathsf{C}\psi$
2. *there is a state u reachable from w through the relation R_r^φ such that $\vdash \phi_u \rightarrow \neg\psi$.*

The next definition specifies how to extend a σ-state for φ to a path in the pre-model.

Definition 8. *Given an arbitrary σ-state w, we define a sequence δ_w of states and actions $w_0 \xrightarrow{d_1} w_1 \xrightarrow{d_2} \cdots$ as follows: for any $r \in N$ and $j \geq 1$,*

1. *w_j is a σ-state in W^φ, and $d_j \in D$.*
2. *$w_0 = w$.*
3. *$\phi_{w_{j-1}} \wedge \bigcirc \phi_{w_j}$ is consistent.*
4. *$d_j(r) = a_j^r$ iff $\vdash \phi_{w_{j-1}} \rightarrow does(a_j^r)$.*

The following result shows such generated sequence is indeed a path of M^φ.

Proposition 8. *Given an arbitrary σ-state w, the sequence δ_w is a path of M^φ.*

Proof. We first show that δ_w is infinite. Suppose not that there is some state w_l with no successor. By Axiom 4 and Axiom 5, such an action a_j^r for each agent always exists. Then it is only the case that $\vdash \phi_{w_l} \rightarrow \neg \bigcirc \phi_s$ for all atoms s of $cl_{k,r}(\varphi)$ where $k = d - |\sigma|$ if $\sigma = \tau r$, or $cl_d(\varphi)$ if $\sigma = \epsilon$. But by Proposition 3.3 and (R2) we have $\vdash \bigcirc \bigvee_{X \in \mathcal{A}_{cl_{k,r}(\varphi)}} \phi_X$ where $k = d - |\sigma|$ if $\sigma = \tau r$; $\vdash \bigcirc \bigvee_{X \in \mathcal{A}_{cl_d(\varphi)}} \phi_X$ if $\sigma = \epsilon$, which contradicts that w_l is consistent. With this, it remains to show that δ_w satisfies the conditions of a path in Definition 7.

Condition 1 holds directly by the definition of U^φ and Axiom 8. Regarding Condition 2, by Clause 3 and Axiom 8, we have $\vdash \phi_{w_{j-1}} \rightarrow legal(d_j(r))$ for any $r \in N$ and $j \geq 1$, so $d_j(r) \in L_r^\varphi(w_{j-1})$. Regarding Condition 3, assume $w_{l-1} \in T^\varphi$ and $w_{l-1} \neq w_l$ for some $l \geq 1$. Without loss of generalization, say $w_{l-1} = (\sigma, X_{l-1})$ and $w_l = (\sigma, X_l)$. Then $\vdash \phi_{X_{l-1}} \rightarrow terminal$, and there is some $\alpha \in cl_{k,r}(\varphi)$ where $k = d - |\sigma|$ if $\sigma = \tau r$, or $\alpha \in cl_d(\varphi)$ if $\sigma = \epsilon$, such that either ($\alpha \in X_{l-1}$ and $\alpha \notin X_l$) or ($\alpha \notin X_{l-1}$ and $\alpha \in X_l$). By symmetry, it suffices to show the case $\alpha \notin X_{l-1}$ and $\alpha \in X_l$. Then $\vdash \phi_{X_{l-1}} \rightarrow \neg\alpha$ and $\vdash \phi_{X_l} \rightarrow \alpha$. From the former and by Axiom 9, we get $\vdash \phi_{X_{l-1}} \rightarrow \bigcirc\neg\alpha$, so $\vdash \phi_{X_{l-1}} \rightarrow \neg \bigcirc \alpha$ by Axiom 7. While from $\vdash \phi_{X_l} \rightarrow \alpha$ and by (R2), $\vdash \bigcirc\phi_{X_l} \rightarrow \bigcirc\alpha$, which contradicts that $\phi_{X_{l-1}} \wedge \bigcirc\phi_{X_l}$ is consistent. Thus, for all $j \geq 1$, if $w_{j-1} \in T^\varphi$, then $w_{j-1} = w_j$. This completes the proof. (□)

We now come to one of the main intermediate results.

Lemma 1. *For every $\alpha \in cl_0(\varphi)$ and every ϵ-state w,*

$$M^{\varphi}, \delta_w \models^* \alpha \text{ iff } \vdash \phi_w \to \alpha.$$

It is routine to prove this by induction on the complexity of α. As we noted before, the pre-model M^{φ} of φ is not an EST-model. To achieve the completeness result of EGDL, it suffices to transform the pre-model of ϕ into a deterministic model with an equivalent satisfiability. Inspired by [18], we redefine states as a subset of atoms and treat all the successors as a single state in the new model. The transformation is given as follows:

Definition 9. *Let $M^{\varphi} = (W^{\varphi}, I^{\varphi}, T^{\varphi}, \{R_r^{\varphi}\}_{r\in N}, \{L_r^{\varphi}\}_{r\in N}, U^{\varphi}, g^{\varphi}, \pi^{\varphi})$ be the pre-model of φ. Then $\mathfrak{A}(M^{\varphi})$ is a model $(S, I, T, \{R_r\}_{r\in N}, \{L_r\}_{r\in N}, U, g, \pi)$ based on M^{φ} such that*

1. *S consists of all the pairs (σ, Γ) such that σ is an index, $|\sigma| \leq d$, and*
   (a) if $\sigma = \epsilon$ then Γ is a non-empty subset of $\mathcal{A}_{cl_d(\varphi)}$, and
   (b) if $\sigma = \tau r$ then Γ is a non-empty subset of $\mathcal{A}_{cl_{k,r}(\varphi)}$, where $k = d - |\sigma|$.
2. *$I = \{(\sigma, \Gamma) \in S \ : \ \Gamma \subseteq \{X \ : \ (\sigma, X) \in I^{\varphi}\}\}$.*
3. *$T = \{(\sigma, \Gamma) \in S \ : \ \Gamma \subseteq \{X \ : \ (\sigma, X) \in T^{\varphi}\}\}$.*
4. *$(\sigma, \Gamma) R_r (\tau, \Delta)$ iff $\sigma\#r = \tau\#r$ and $\{\psi : \mathsf{K}_r\psi \in \bigcup \Gamma\} = \{\chi : \mathsf{K}_r\chi \in \bigcup \Delta\}$.*
5. *$L_r((\sigma, \Gamma)) = \bigcup_{X\in\Gamma} L_r^{\varphi}((\sigma, X))$.*
6. *$U((\sigma, \Gamma), d) = (\sigma, \Delta)$ where $\Delta = \{Y \ : \ (\sigma, Y) \in \bigcup_{X\in\Gamma} U^{\varphi}((\sigma, X), d)\}$.*
7. *$g(r) = \{(\sigma, \Gamma) \in S \ : \ \Gamma \subseteq \{X \ : \ (\sigma, X) \in g^{\varphi}(r)\}\}$.*
8. *$\pi((\sigma, \Gamma)) = \bigcup_{X\in\Gamma} \pi^{\varphi}((\sigma, X))$.*

The following result shows the associated model $\mathfrak{A}(M^{\varphi})$ is just what we want.

Proposition 9. *Given a pre-model M^{φ} of φ, the model $\mathfrak{A}(M^{\varphi})$ is an EST-model.*

Proof. Clearly, $S \neq \emptyset$ and $I \cap T = \emptyset$ follows from $\vdash \neg(initial \wedge terminal)$. It is straightforward that the epistemic relation R_r is equivalent. Regarding L_r, for any $(\sigma, \Gamma) \in S$, since $\Gamma \neq \emptyset$ and $L_r^{\varphi}((\sigma, X)) \neq \emptyset$ for any $(\sigma, X) \in W^{\varphi}$, so by definition $L_r((\sigma, \Gamma)) \neq \emptyset$ (Condition (i)). Assume $(\sigma, \Gamma) \in T$, then by the definition we have $(\sigma, X) \in T^{\varphi}$ for any $X \in \Gamma$, so $\vdash \phi_X \to terminal$. And by Axiom 3, we have $\vdash \phi_X \to \bigwedge_{a^r \in A^r \setminus \{noop^r\}} \neg legal(a^r) \wedge legal(noop^r)$, so $L_r^{\varphi}((\sigma, X)) = \{noop^r\}$ for any $X \in \Gamma$. Thus, $L_r((\sigma, \Gamma)) = \{noop^r\}$ (Condition (ii)). It remains to show that the update function U satisfies the assumption.

We first show that for any state $(\sigma, \Gamma) \in S$ and joint action $d \in D$, $U((\sigma, \Gamma), d)$ is non-initial. This follows from the fact that for any $d \in D$ and $(\sigma, X) \in W^{\varphi}$, $U^{\varphi}((\sigma, X), d) \cap I^{\varphi} = \emptyset$ (by Axiom 2).

We next show that $U((\sigma, \Gamma), d)$ is unique if exists. Suppose not, then there are (σ, Δ) and (σ, Δ') such that $U((\sigma, \Gamma), d) = (\sigma, \Delta)$, $U((\sigma, \Gamma), d) = (\sigma, \Delta')$ and $\Delta \neq \Delta'$. But by the definition we have $\Delta = \Delta' = \{Y \ : \ (\sigma, Y) \in \bigcup_{X\in\Gamma} U^{\varphi}((\sigma, X), d)\}$: a contradiction. Thus, $U(s, d)$ is unique.

The last assumption follows from the fact that for any $(\sigma, X) \in T^\varphi$, $U^\varphi((\sigma, X), \langle noop^r\rangle_{r\in N}) = (\sigma, X)$. This is proved by a similar method of Proposition 8. (□)

To complete the transformation, we next show how to generate a path in $\mathfrak{A}(M^\varphi)$ from a given path in the pre-model M^φ of φ.

Definition 10. *Let M^φ be the pre-model of φ. For any path $\delta := (\sigma, X_0) \xrightarrow{d_1} (\sigma, X_1) \xrightarrow{d_2} \cdots$ of M^φ, we define a sequence of states and joint actions $\widehat{\delta} := (\tau, \Gamma_0) \xrightarrow{d'_1} (\tau, \Gamma_1) \xrightarrow{d'_2} \cdots$ with respect to δ as follows: for any $j \geq 1$,*

1. *$\sigma = \tau$ and $d_j = d'_j$,*
2. *$\Gamma_0 = \{X_0\}$, and*
3. *$\Gamma_j = \{X_j : (\sigma, X_j) \in \bigcup_{X_{j-1}\in\Gamma_{j-1}} U^\varphi((\sigma, X_{j-1}), d_j)\}$.*

Proposition 10. *For any path δ of M^φ, the sequence $\widehat{\delta}$ is a path of $\mathfrak{A}(M^\varphi)$.*

Proof. Let $\delta := (\sigma, X_0) \xrightarrow{d_1} (\sigma, X_1) \xrightarrow{d_2} \cdots$ and $\widehat{\delta} := (\tau, \Gamma_0) \xrightarrow{d_1} (\tau, \Gamma_1) \xrightarrow{d_2} \cdots$. Clearly, $\widehat{\delta}$ is infinite as δ is infinite. It suffices to show that $\widehat{\delta}$ satisfies all the conditions of a path in Definition 2. Let us first consider Condition 1. Suppose not that there is some $k \geq 1$ such that $\widehat{\delta}[k] \in I$, then by the definition of $\mathfrak{A}(M^\varphi)$, we have for all $(\sigma, X) \in \widehat{\delta}[k]$, $(\sigma, X) \in I^\varphi$. In particular, $(\sigma, X_k) \in I^\varphi$, so $\vdash X_k \rightarrow initial$. Then by (R2) we have $\vdash \bigcirc X_k \rightarrow \bigcirc initial$. But by Axiom 2 we have $\vdash X_{k-1} \rightarrow \neg\bigcirc initial$, contradicting that the formula $X_{k-1} \wedge \bigcirc X_k$ is consistent. Thus, $\widehat{\delta}[j] \notin I$ for all $j \geq 1$. Condition 2 holds directly by the last two clauses of Definition 10. Regarding Condition 3, for any $r \in N$, we have $d_j(r) \in L^\varphi_r((\sigma, X_{j-1}))$ by Definition 7. Since $X_{j-1} \in \Gamma_{j-1}$, so we have $L^\varphi_r((\sigma, X_{j-1})) \subseteq L_r((\sigma, \Gamma_{j-1}))$ by Definition 9. Thus, $d_j(r) \in L_r((\sigma, \Gamma_{j-1}))$. Regarding Condition 4, it suffices to show the following fact that for any $(\sigma, X) \in T^\varphi$ and $d \in D$,

$$U^\varphi((\sigma, X), d) = \begin{cases} \{(\sigma, X)\} & \text{if } d = \langle noop^r\rangle_{r\in N}; \\ \emptyset & \text{otherwise.} \end{cases}$$

By Axiom 3, $\vdash \phi_X \rightarrow \bigwedge_{r\in N}(\bigwedge_{a^r\in A^r\setminus\{noop^r\}} \neg legal(a^r) \wedge legal(noop^r))$. And by Axiom 8, we have $\vdash \phi_X \rightarrow \bigwedge_{r\in N}(\bigwedge_{a^r\in A^r\setminus\{noop^r\}} \neg does(a^r))$. Thus, $U^\varphi((\sigma, X), d) = \emptyset$ for any $d \neq \langle noop^r\rangle_{r\in N}$. Then by Axiom 4, we have $\vdash \phi_X \rightarrow \bigwedge_{r\in N} does(noop^r)$. Since $\phi_X \wedge \bigcirc\phi_X$ is consistent, so $(\sigma, X) \in U^\varphi((\sigma, X), \langle noop^r\rangle_{r\in N})$. Suppose there is another state $(\tau, Y) \in W^\varphi$ such that $(\tau, Y) \in U^\varphi((\sigma, X), \langle noop^r\rangle_{r\in N})$ and $(\sigma, X) \neq (\tau, Y)$. Then there is some $\alpha \in cl_{k,r}(\varphi)$ where $k = d - |\sigma|$ if $\sigma = \tau r$, or $\alpha \in cl_d(\varphi)$ if $\sigma = \epsilon$, such that either $(\alpha \in X$ and $\alpha \notin Y)$ or $(\alpha \notin X$ and $\alpha \in Y)$. By symmetry, it suffices to show the case $\alpha \in X$ and $\alpha \notin Y$. Then $\vdash \phi_X \rightarrow terminal \wedge \alpha$. And by Axiom 9, we get $\vdash \phi_X \rightarrow \bigcirc\alpha$. By Proposition 3.2, we have either $\vdash \phi_Y \rightarrow \alpha$ or $\vdash \phi_Y \rightarrow \neg\alpha$. The former contradicts with the assumption $\alpha \notin Y$. Thus, $\vdash \phi_Y \rightarrow \neg\alpha$. Then by (R2), we have $\vdash \bigcirc(\phi_Y \rightarrow \neg\alpha)$, so $\vdash \bigcirc\phi_Y \rightarrow \neg\bigcirc\alpha$, contradicting that $\phi_X \wedge \bigcirc\phi_Y$ is consistent. Thus, $U^\varphi((\sigma, X), \langle noop^r\rangle_{r\in N}) = \{(\sigma, X)\}$. For any

$(\sigma, \Gamma_{j-1}) \in T$, by the definition we have $(\sigma, X) \in T^{\varphi}$ for any $X \in \Gamma_{j-1}$, then by the fact for any $X \in \Gamma_{j-1}$, $U^{\varphi}((\sigma, X), d_j) = \{(\sigma, X)\}$ and $d_j = \langle noop^r \rangle_{r \in N}$, so $\{X_j : (\sigma, X_j) \in \bigcup_{X \in \Gamma_{j-1}} U^{\varphi}((\sigma, X), \langle noop^r \rangle_{r \in N})\} = \Gamma_{j-1}$. Thus, $\Gamma_j = \Gamma_{j-1}$. This completes the proof of the proposition. (□)

Then we have the following equivalent result in terms of the transformations.

Lemma 2. *Let M^{φ} be the pre-model of φ and δ be a path of M^{φ}. Then for any $\alpha \in \mathcal{L}$,*

$$M^{\varphi}, \delta \models {}^{*}\alpha \quad \textit{iff} \quad \mathfrak{A}(M^{\varphi}), \widehat{\delta} \models \alpha.$$

We are now in the position to prove the soundness and completeness results of EGDL with respect to the epistemic state transition models.

Theorem 1. *The logic EGDL is sound and complete with respect to the class of epistemic state transition models, i.e., for every $\varphi \in \mathcal{L}$, $\models \varphi$ iff $\vdash \varphi$.*

Proof. We only show the completeness. Assume $\nvdash \varphi$, then $\neg\varphi$ is consistent, so there is an ϵ-state w such that $\vdash \phi_w \to \neg\varphi$. By Lemma 1, $M^{\neg\varphi}, \delta_w \models {}^{*}\neg\varphi$. And by Lemma 2, we have $\mathfrak{A}(M^{\neg\varphi}), \widehat{\delta_w} \models \neg\varphi$. Thus, $\not\models \varphi$. This completes the proof. (□)

Then we have the following result saying that EGDL has the *finite model property*.

Theorem 2. *Let φ be a formula in EGDL. If φ is satisfiable, then it is satisfiable in a finite epistemic state transition model.*

4 Related Work

To deal with imperfect information games, many logics, mostly epistemic extensions of Alternating-time Temporal Logic, Strategy Logic and PDL, have been developed [1,2,10–13]. Different from them, as shown in [14], EGDL uses a bottom-up approach in order to create a balance between expressive power and computational efficiency. It is a conservative extension of a simple and practical logical language GDL. Besides the literature discussed in Introduction, the following is also worth mentioning.

Zhang and Thielscher propose a dynamic extension of GDL for reasoning about game strategies, and develop a sound and complete axiomatic system for this logic [24]. With different languages and semantics, their axiomatization and techniques to prove the completeness are different from ours. In particular, they make use of forgetting techniques while we combine techniques used for completeness of PDL and ETLs.

As a logic programming language, GDL has recently been extended to GDL-II and GDL-III so as to incorporate imperfect information games [22,23]. They are different from EGDL in two aspects: (i) GDL-II and GDL-III are purely logic programming languages and do not provide a reasoning facility to reason about epistemic game rules. While as a logic EGDL is able to represent and reason about rules of imperfect information. Moreover, we have developed an axiomatic system for EGDL. (ii) GDL-II and GDL-III considers games with perfect recall players and randomness, such as dice rolling and card shuffling. While EGDL focuses on imperfect recall games without randomness. Yet EGDL is flexible enough to specify perfect recall as well as the state-based memory and the action-based memory [4].

Finally, it is worth mentioning that EGDL has similarities with ETLs such as CKL_m [7], but they are significantly different in the following ways: (i) With *does*(.) operator, EGDL can express actions and their effects, thus it can be used for reasoning about actions, while epistemic temporal logics are not. Moreover, with action operator, the completeness proof of EGDL is different from those of epistemic temporal logics; (ii) EGDL contains a single temporal operator ("next"), and can only represent finite steps of time. (iii) Model checking for EGDL is in Δ_2^p, while, for epistemic temporal logics, it is at least PSPACE-hard [21].

5 Conclusion

We have developed a sound and complete axiomatic system for a variant of EGDL. From the completeness proof, we have derived the finite model property of this logic. We have also demonstrated how to use the proof theory for inferring game properties from game rules.

Directions of future research are manifold. We intend to investigate the satisfiability problem of EGDL. The hardness of the satisfiability problem for EGDL follows from the fact that EGDL is a conservative extension of $S5_n^C$, and the satisfiability problem for $S5_n^C$ is EXPTIME-complete [8]. We also want to study the definability problem of EGDL [3]: which properties of games are definable by means of EGDL-formulas? For instance, this paper shows that EGDL is able to provide a description for Krieg-Tictactoe. It would be interesting to consider the other direction: whether Krieg-Tictactoe is completely or even uniquely specified by such a description.

Acknowledgments. We are grateful to Prof. Thomas Ågotnes and A/Prof. Yi Wang for their valuable suggestions, and special thanks are due to two anonymous referees for their insightful comments.

References

1. van Benthem, J.: Games in dynamic-epistemic logic. Bull. Econ. Res. **53**(4), 219–248 (2001)
2. van Benthem, J.: Logic in Games. MIT Press, Cambridge (2014)
3. Blackburn, P., De Rijke, M., Venema, Y.: Modal Logic. Cambridge University Press, Cambridge (2002)

4. van Ditmarsch, H., Knight, S.: Partial information and uniform strategies. In: Bulling, N., Torre, L., Villata, S., Jamroga, W., Vasconcelos, W. (eds.) CLIMA 2014. LNCS (LNAI), vol. 8624, pp. 183–198. Springer, Cham (2014). doi:10.1007/978-3-319-09764-0_12
5. Fagin, R., Moses, Y., Halpern, J.Y., Vardi, M.Y.: Reasoning About Knowledge. MIT press, Cambridge (2003)
6. Genesereth, M., Love, N., Pell, B.: General game playing: overview of the AAAI competition. AI Mag. **26**(2), 62–72 (2005)
7. Halpern, J.Y., van Der Meyden, R., Vardi, M.Y.: Complete axiomatizations for reasoning about knowledge and time. SIAM J. Comput. **33**(3), 674–703 (2004)
8. Halpern, J.Y., Moses, Y.: A guide to completeness and complexity for modal logics of knowledge and belief. Artif. Intell. **54**(3), 319–379 (1992)
9. Hintikka, J.: Knowledge and Belief: An Introduction to the Logic of the Two Notions. Cornell University Press, Ithaca (1962)
10. van der Hoek, W., Pauly, M.: Modal logic for games and information. In: Blackburn, P., van Benthem, J., Wolter, F. (eds.) Handbook of Modal Logic, vol. 3, pp. 1077–1148. Elsevier, Amsterdam (2006)
11. van der Hoek, W., Wooldridge, M.: Cooperation, knowledge, and time: alternating-time temporal epistemic logic and its applications. Stud. Log. **75**(1), 125–157 (2003)
12. Huang, X., van der Meyden, R.: An epistemic strategy logic. In: SR 2014, pp. 35–41 (2014)
13. Jamroga, W., van der Hoek, W.: Agents that know how to play. Fundam. Inform. **63**(2), 185–219 (2004)
14. Jiang, G., Zhang, D., Perrussel, L., Zhang, H.: Epistemic GDL: a logic for representing and reasoning about imperfect information games. In: IJCAI 2016, pp. 1138–1144 (2016)
15. Kozen, D., Parikh, R.: An elementary proof of the completeness of PDL. Theor. Comput. Sci. **14**(1), 113–118 (1981)
16. Love, N., Hinrichs, T., Haley, D., Schkufza, E., Genesereth, M.: General game playing: Game description language specification. Stanford Logic Group (2006). http://logic.stanford.edu/reports/LG-2006-01.pdf
17. Pritchard, D.B.: The Encyclopedia of Chess Variants. Games & Puzzles, UK (1994)
18. Rabin, M.O., Scott, D.: Finite automata and their decision problems. IBM J. Res. Dev. **3**(2), 114–125 (1959)
19. Schiffel, S., Thielscher, M.: Reasoning about general games described in GDL-II. In: AAAI 2011, pp. 846–851 (2011)
20. Schobbens, P.Y.: Alternating-time logic with imperfect recall. Electron. Notes Theor. Comput. Sci. **85**(2), 82–93 (2004)
21. Sistla, A.P., Clarke, E.M.: The complexity of propositional linear temporal logics. J. ACM **32**(3), 733–749 (1985)
22. Thielscher, M.: A general game description language for incomplete information games. In: AAAI 2010, pp. 994–999 (2010)
23. Thielscher, M.: GDL-III: a proposal to extend the game description language to general epistemic games. In: ECAI 2016, pp. 1630–1631 (2016)
24. Zhang, D., Thielscher, M.: A logic for reasoning about game strategies. In: AAAI 2015, pp. 1671–1677 (2015)
25. Zhang, D., Thielscher, M.: Representing and reasoning about game strategies. J. Philos. Log. **44**(2), 203–236 (2015)

Putting More Dynamics in Revision with Memory

Sébastien Konieczny[1(✉)] and Ramón Pino Pérez[2]

[1] CRIL - CNRS & Université d'Artois, Lens, France
konieczny@cril.fr
[2] Facultad de Ciencias, Universidad de los Andes, Mérida, Venezuela
pino@ula.ve

Abstract. We have proposed in previous works [14,15] a construction that allows to define operators for iterated revision from classical AGM revision operators. We called these operators *revision operators with memory* and show that the operators obtained have nice logical properties. But these operators can be considered as too conservative, since the revision policy of the agent, encoded as a faithful assignment, does not change during her life. In this paper we propose an extension of these operators, that aims to add more dynamics in the revision process.

1 Introduction

The predominant approaches for modelling belief change was proposed by Alchourrón, Gärdenfors and Makinson and is known as the AGM framework [1,10]. The core of this framework is a set of logical properties that a revision operator has to satisfy to guarantee a nice behaviour. A drawback of AGM definition of revision is that it is a static one, which means that, with this definition of revision operators, one can have a rational one step revision but the conditions for the iteration of the process are very weak. The problem is that AGM postulates state conditions only between the initial knowledge base, the new evidence and the resulting knowledge base. But the way to perform further revisions on the new knowledge base does not depend on the way the old knowledge base was revised.

Numerous proposals have tried to state a logical characterization that adequately models iterated belief change behaviour [6,8,9,14,17,19,20][1]. The core work on iterated revision is the proposal of Darwiche and Pearl [9] and its developments [4,11,13,16]. The main idea that is common to all of these works is that the belief base framework is not sufficient to encompass iterated revision, since one needs some additional information for coding the revision policy of the agent. So the need of *epistemic states* to encode the agent's "state of mind" is widely accepted. An epistemic state allows to code the agent's beliefs but also to code her relative confidence in alternative possible states of the world. Epistemic

[1] See also [22–24]. We do not adress this kind of operators in this paper since they require an additional numerical information with the new evidence.

A. Baltag et al. (Eds.): LORI 2017, LNCS 10455, pp. 614–627, 2017.
DOI: 10.1007/978-3-662-55665-8_42

states can be represented by several means: pre-orders on interpretations [9,17], conditionals [6,9], epistemic entrenchments [19,23], prioritized belief bases [2,3], etc. In this work we will focus on the representation of epistemic states in terms of pre-orders on interpretations.

In [14,15], we define a family of revision operators that we have called *revision operators with memory*. These operators can be defined from any classical AGM revision operator [1,12] and they have good properties for iterated revision.

In fact revision operators with memory use the faithful assignment provided by the classical AGM revision operator as an *a priori* information. This *a priori* information is attached to the new evidence, and the completed information obtained is then incorporated to the old epistemic state with the usual *primacy of update* requirement. The ontology for this pre-processing step, associating an additional information to the incoming new evidence is the following. Suppose that the agent has no information (no belief) about the world and learns a (first) new evidence. Then, this new evidence alone can provide more change in the agent's mind than just the addition of a belief.

As an example, suppose that the agent learns $\varphi = a \wedge b \wedge c \wedge d$, where a, b, c, d are atomic formulae. Then her preferred worlds (the ones she finds the more plausible) will be the ones where the four atomic formulae are true. But it can be sensible for her to find the worlds where three of the atomic formulae are true more plausible than the ones where only two are, etc.

So the new evidence does not simply imply a partition between the believed worlds and the unbelieved ones, but defines several stratas, depending of the plausibility of each world, given the new evidence. We call this property *strong primacy of update*. This induced preferential information was given here by a "Dalal distance" policy[2], but more complex or realistic policies can be also used depending on the particular context.

The point in the definition of revision operators with memory, is that this *a priori* information, carried by a new evidence depends only on the new evidence by itself, and does not depend on the current agent's beliefs. Going back to the previous example, the fact that the worlds where three of the four atomic formulae are true are preferred to the ones where only two are, does not depend on any other information than the new evidence itself. So this *a priori* information has to be added to the new evidence before incorporating it in the agent's epistemic state.

More precisely, the revision policy of revision operators with memory is the following: the revision of the current epistemic state Φ – represented by a pre-order over possible worlds – by a new piece of information α – a formula – is the epistemic state (pre-order) obtained after the following two steps:

- First, take the pre-order $\leq_\alpha$ associated to α by the AGM revision operator (faithful assignment [12]) given at the beginning of the process.
- Second, take the lexicographical pre-order associated to $\leq_\alpha$ and Φ. The pre-order obtained in this way is the new epistemic state.

Note that there is a very static feature in this process: the way in which we associate a pre-order to the new piece of information is always the same; it is

[2] The Dalal distance [7] is a Hamming distance between interpretations.

given by the *fixed* AGM operator from which we start all the process. In some sense this is contrary to the principle of *priority of the new information*.

In this work we solve this problem. In order to do that we take the revision policy as an epistemic state and naturally this revision policy will change progressively with the successive revisions. The new process can be described in the following manner: first of all, an epistemic state is composed by a faithful assignment, say f, and a distinguished formula ϕ. When α, the new evidence, arrives, we revise as follows:

(i) the new distinguished formula ϕ' will be a formula having as models the minimal models of α with respect to the $f(\phi)$ pre-order.
(ii) The new assignment f' will coincide with f on the formulas not equivalent to ϕ'. On formulas equivalent to ϕ' it will be the lexicographical pre-order associated to $f(\alpha)$ and $f(\phi)$.

Thus, this method allows to incorporate the changes step by step in a very natural way. This process agrees with the postulate of primacy of the new information. Unlike our original revision operators with memory that mix the new piece of information with the oldest information (which is static), our present operators mix the new piece of information with the current epistemic state.

The rest of the paper is organized as follows: in Sect. 2, we recall the logical characterization of iterated revision operators of Darwiche and Pearl. In Sect. 3, we recall the definition of revision operators with memory and state the general logical results. Then, in Sect. 4, we show how to add more dynamics to revision operators with memory. We conclude in Sect. 5 with some general remarks.

2 Iterated Revision Postulates

We give here a formulation of AGM postulates for belief revision *à la* Katsuno and Mendelzon [12]. More exactly, we give a formulation of these postulates in terms of epistemic states [9]. The epistemic states framework is an extension of the belief bases one. Intuitively an epistemic state can be seen as a composed information: the beliefs of the agent, plus all the information that the agent needs about how to perform revision (preference ordering, conditionals, etc.). Then we give the additional iteration postulates proposed by Darwiche and Pearl [9].

2.1 Formal Preliminaries

We will work in the finite propositional case. A belief base φ is a finite set of formulae, which can be considered as the formula that is the conjunction of its formulae. The set of all interpretations is denoted $\mathcal{W}$. Let φ be a formula, $Mod(\varphi)$ denotes the set of models of φ, *i.e.* $Mod(\varphi) = \{I \in \mathcal{W} :\ I \models \varphi\}$.

A pre-order $\leq$ is a reflexive and transitive relation, and $<$ is its strict counterpart, i.e. $I < J$ if and only if $I \leq J$ and $J \not\leq I$. As usual, $\simeq$ is defined by $I \simeq J$ iff $I \leq J$ and $J \leq I$. A pre-order is total if and only if $\forall I, J, I \leq J$ or $J \leq I$.

To each epistemic state Ψ is associated a belief base $Bel(\Psi)$ which is a propositional formula representing the objective (logical) part of Ψ. The models of Ψ are the models of its associated belief base, thus $Mod(\Psi) = Mod(Bel(\Psi))$. Let Ψ be an epistemic state and μ be a sentence denoting the new information. $\Psi \circ \mu$ denotes the epistemic state resulting of the revision of Ψ by μ. For reading convenience we will write respectively $\Psi \vdash \mu$, $\Psi \wedge \mu$ and $I \models \Psi$ instead of $Bel(\Psi) \vdash \mu$, $Bel(\Psi) \wedge \mu$ and $I \models Bel(\Psi)$.

Two epistemic states are equivalent, noted $\Psi \equiv \Psi'$, if and only if their objective parts are equivalent formulae, *i.e.* $Bel(\Psi) \leftrightarrow Bel(\Psi')$. Two epistemic states are equal, noted $\Psi = \Psi'$, if and only if they are identical. Thus equality is stronger than equivalence.

2.2 AGM Postulates for Epistemic States

Let Ψ be an epistemic state and μ and φ be formulae. An operator $\circ$ that maps an epistemic state Ψ and a formula μ to an epistemic state $\Psi \circ \mu$ is said to be a revision operator on epistemic states if it satisfies the following postulates [9]:

(R*1) $\Psi \circ \mu \vdash \mu$
(R*2) If $\Psi \wedge \mu \nvdash \bot$, then $\Psi \circ \mu \leftrightarrow \Psi \wedge \mu$
(R*3) If $\mu \nvdash \bot$, then $\Psi \circ \mu \nvdash \bot$
(R*4) If $\Psi_1 = \Psi_2$ and $\mu_1 \leftrightarrow \mu_2$, then $\Psi_1 \circ \mu_1 \equiv \Psi_2 \circ \mu_2$
(R*5) $(\Psi \circ \mu) \wedge \varphi \vdash \Psi \circ (\mu \wedge \varphi)$
(R*6) If $(\Psi \circ \mu) \wedge \varphi \nvdash \bot$, then $\Psi \circ (\mu \wedge \varphi) \vdash (\Psi \circ \mu) \wedge \varphi$

This is nearly the Katsuno and Mendelzon formulation of AGM postulates [12]; the only differences are that we work with epistemic states instead of belief bases and that postulate (R*4) is weaker than its AGM counterpart. See [9] for a full motivation of this definition.

A representation theorem states how revisions can be characterized in terms of pre-orders on interpretations. In order to give such a semantical representation, the concept of faithful assignment on epistemic states is defined.

Definition 1. *A function that maps each* epistemic state Ψ *to a pre-order* $\leq_\Psi$ *on interpretations is called a* faithful assignment over epistemic states *if and only if: 1. If* $I \models \Psi$ *and* $J \models \Psi$*, then* $I \simeq_\Psi J$
2. If $I \models \Psi$ *and* $J \not\models \Psi$*, then* $I <_\Psi J$
3. If $\Psi_1 = \Psi_2$*, then* $\leq_{\Psi_1} = \leq_{\Psi_2}$

Now the reformulation of the Katsuno and Mendelzon [12] representation theorem in terms of epistemic states is:

Proposition 1 ([9]). *A revision operator* $\circ$ *satisfies postulates (R*1-R*6) if and only if there exists a faithful assignment (over epistemic states) that maps each epistemic state* Ψ *to a total pre-order* $\leq_\Psi$ *such that*

$$Mod(\Psi \circ \mu) = \min(Mod(\mu), \leq_\Psi)$$

Notice that this theorem gives information only on the objective part of the resulting epistemic state, but does not allow to know what is the pre-order associated with $\Psi \circ \mu$, *i.e.* we can not identify the new epistemic state, but only its associated belief base $Mod(\Psi \circ \mu)$. Making the parallel with the classical Katsuno and Mendelzon representation theorem (*cf* Definition 2 and [12]), that allows to define exactly what is the belief base $Mod(\Psi \circ \mu)$[3], the last theorem is only a weak representation theorem.

2.3 Darwiche and Pearl Postulates

A strong limitation of AGM revision postulates is that they impose very weak constraints on the iteration of the revision process. Darwiche and Pearl [8,9] proposed postulates for iterated revision. The aim of these postulates is to keep as much as possible of conditional beliefs (a conditional belief can be expressed as "if μ would be the case, then φ must be true") of the old belief base. These conditional beliefs are encoded in the total pre-orders on interpretations. So, besides postulates (R*1-R*6), a revision operator has to satisfy:

(C1) If $\varphi \vdash \mu$, then $(\Psi \circ \mu) \circ \varphi \equiv \Psi \circ \varphi$
(C2) If $\varphi \vdash \neg\mu$, then $(\Psi \circ \mu) \circ \varphi \equiv \Psi \circ \varphi$
(C3) If $\Psi \circ \varphi \vdash \mu$, then $(\Psi \circ \mu) \circ \varphi \vdash \mu$
(C4) If $\Psi \circ \varphi \nvdash \neg\mu$, then $(\Psi \circ \mu) \circ \varphi \nvdash \neg\mu$

These postulates can be explained as follows: (C1) states that if two pieces of information arrive and if the second implies the first, the second alone would give the same belief base. (C2) says that when two contradictory pieces of information arrive, the second alone would give the same belief base. (C3) states that an information should be retained after revising by a second information such that, when revising the current belief base by it, the first one holds. (C4) says that no piece of information can contribute to its own denial.

3 Revision Operators with Memory

A "classical" AGM revision operator is equivalent to a faithful assignment over belief bases as stated in the following theorem [12].

Definition 2. *A function that maps each* belief base φ *to a pre-order* $\leq_\varphi$ *on interpretations is called a* faithful assignment over belief bases *if and only if:*

1. *If* $I \models \varphi$ *and* $J \models \varphi$*, then* $I \simeq_\varphi J$
2. *If* $I \models \varphi$ *and* $J \not\models \varphi$*, then* $I <_\varphi J$
3. *If* $\varphi_1 \leftrightarrow \varphi_2$*, then* $\leq_{\varphi_1} = \leq_{\varphi_2}$

[3] Recall that classical AGM operators are functions that map a belief base and a formula to a belief base, which is (completely) defined by the theorem, whereas Proposition 1 concerns operators that are functions which map an epistemic state and a formula to an epistemic state, that is not completely defined by the theorem.

It is important to note that, in what follows, we have two distinct kinds of faithful assignments: one over belief bases and one over epistemic states.

Proposition 2 ([12]). *A revision operator $\circ$ satisfies classical AGM postulates (R1-R6)*[4] *if and only if there exists a faithful assignment (over belief bases) that maps each belief base φ to a total pre-order $\leq_\varphi$ such that: $Mod(\varphi \circ \mu) = \min(Mod(\mu), \leq_\varphi)$.*

So one can define a revision operator directly by defining the corresponding faithful assignment over belief bases. It is the case for most distance-based revision operators such as Dalal operator [7,12].

More precisely we say that a revision operator $\circ$ is defined from a distance d iff the following conditions hold:

- d is a (pseudo-)distance, that is d is a function $d : \mathcal{W} \times \mathcal{W} \mapsto \mathbb{N}$ which satisfies: $d(I, J) = d(J, I)$ and $d(I, J) = 0$ iff $I = J$.
- The distance between an interpretation I and a belief base φ is defined as:

$$d(I, \varphi) = \min \{d(I, J) : J \models \varphi\}$$

- This distance induces a faithful assignment: $I \leq_\varphi J$ iff $d(I, \varphi) \leq d(J, \varphi)$
- And the revision operator is defined by $Mod(\varphi \circ \mu) = \min(Mod(\mu), \leq_\varphi)$

One can check that the assignment obtained like this is a faithful assignment and thus that all operators defined in this way satisfy AGM postulates. It can also be easily checked that operators defined in this way do not satisfy many of the iterated revision postulates.

Now we will give a construction that allows, from a given faithful assignment (*i.e.* from a given classical AGM revision operator), to define another revision operator that satisfies AGM postulates but also most of the iterated revision postulates.

First, let us notice that an epistemic state can be represented by a total pre-order on interpretations as suggested by Proposition 1 and by several related works (*cf e.g.* [3,9]). So, with this particular representation (identifying the epistemic state Ψ with a pre-order $\leq_\Psi$), the belief base $Bel(\Psi)$ is simply the formula whose models are minimal for the pre-order, that is $Bel(\Psi) = \min(\mathcal{W}, \leq_\Psi)$. And the other interpretations are ordered according to their relative plausibility for the agent. For example, $I \leq_\Psi J$ means that the agent that is in the epistemic state Ψ considers I as at least as plausible as J. It is this preferential information that can be used to encompass the iterated revision behaviour, by considering revision operators as functions that map a pre-order (epistemic state) and a formula (new information) into a new pre-order (epistemic state). This idea is the mainstay in most of iterated revision works [4,9,11,13,16,19,23].

So, using this representation by means of pre-orders on interpretations and Proposition 1 we will define a familly of revision operators as follows:

[4] It is the same set of postulates than (R*1-R*6) but expressed for belief bases instead of epistemic states (*cf* [12]).

Definition 3. *Suppose that we have a function that maps each belief base φ to a pre-order $\leq_\varphi$. Then, we define the epistemic state (the pre-order) $\Psi \circ \varphi$ resulting of the revision of Ψ by the new information φ as:*

$$I \leq_{\Psi \circ \varphi} J \text{ iff } I <_\varphi J \text{ or } (I \simeq_\varphi J \text{ and } I \leq_\Psi J)$$

Then one can check that:

Proposition 3 ([15]). *If the function that maps each belief base φ to a total pre-order $\leq_\varphi$ is a faithful assignment over belief bases, then the revision operator on epistemic states defined in Definition 3 satisfies postulates (R*1-R*6). We will call such operators* revision operators with memory.

So, with Definition 3, one can start from any epistemic state (total pre-order over interpretations) and carry on iterated revisions. A particular epistemic state we can mention is the "empty" epistemic state, where the agent has no belief and no preferential information, that is, such that $\forall I, J \in \mathcal{W}\ I \simeq J$. We will denote by Ξ this epistemic state. So, the objective part of this epistemic state is $Bel(\Xi) = \top$. It can be considered as the epistemic state generalisation of $\top$ for the belief base framework, since they are both neutral elements for the corresponding operators: $\Xi \circ \varphi \equiv \varphi$ (as $\top \circ \varphi \equiv \varphi$ in the belief base framework). One can consider that all agents start with this epistemic state (we will consider this in the examples). Concerning iteration postulates:

Proposition 4 ([15]). *Revision operators with memory satisfy postulates (C1), (C3) and (C4).*

It can be also easily checked that (C2) is satisfied by a unique revision operator with memory, since it demands (in the presence of the other revision postulates), that the pre-order associated to a belief base by the faithful assignment on belief base used in Definition 3 is a two-level pre-order with the models of the belief base at the lowest level and the counter-models at the higher one. This operator will be presented in the next section.

So most of our revision operators with memory do not satisfy (C2). But we do not consider this as a drawback. We rather think that it is (C2) that is not fully satisfactory. In fact C2 demands that a piece of information that is accepted but later contradicted is completely discarded. One could argue for a more subtle behavior where only the contradicted part (so not all the formula) is discarded. See [15,17] for more explanations on this point. For instance suppose that you learn a big conjunction $a \wedge b \wedge \ldots \wedge z$ and that later you learn $\neg a$. Couldn't be natural to try to keep $b \wedge \ldots \wedge z$? Or should we discard it completely as required by (C2) ? According to us (C2) should not be regarded as a first class requirement (conversely to other postulates), but as an optional property that makes a distinction between two kinds of revision operators: the ones that consider that contradicting a piece of information amounts to discrediting its source, and then to discard it completely, and the ones that have a more subtle behavior and that only remove the contradicting parts of the pieces of information.

For a more complete logical characterization of this family of operators see [14].

3.1 Basic Memory Operator

Let us define the assignment that maps each belief base to a pre-order in the following way:

Definition 4. *Let φ be a belief base, the basic pre-order $\leq_\varphi^b$ associated to φ is defined as: $I \leq_\varphi^b J$ if and only if $I \models \varphi$ or ($I \not\models \varphi$ and $J \not\models \varphi$)*

So we have what we call a basic order, which is a two-level order (at most), with the models of φ at the lowest level and the other worlds at the highest level.

Definition 5. *The* basic memory operator *is the memory operator obtained from this assignment (i.e. the operator obtained by Definitions 4 and 3).*

It is worthy to note that if one uses this faithful assigment (Definition 4) to define a classical AGM operator (Proposition 2), one obtains the *full meet revision operator* which is not a good operator. But, even with this basic order on belief bases in the revision with memory framework, one can build very complex epistemic states. This is due to revision memory. The assignment of Definition 4 is a faithful assignment on belief bases; with Propositions 3 and 4, it is easy to show that:

Proposition 5 ([15]). *The only revision operator with memory that satisfies (R*1-R*6) and (C1–C4) is the basic memory revision operator.*

This operator has been already studied in the literature under different particular representations: in [19] with epistemic entrenchments, in [2,21] with polynomials and syntactic belief bases. Finally, we can note that Liberatore has shown [18] that several problems are computationally simpler for the basic memory operator than for the other iterated belief revision proposals (including Boutilier's natural revision [5], Lehmann's ranking revision [17] and Williams' transmutations [23]).

3.2 Dalal Memory Operator

We use in this section the Hamming distance d_H between interpretations[5]. The Dalal distance between an interpretation I and a belief base φ is defined as $d_D(I, \varphi) = \min_{J \models \varphi}(d_H(I, J))$.

Let's define the assignment that maps each belief base to a pre-order in the following way:

Definition 6. *Let φ be a belief base, the pre-order $\leq_\varphi^d$ associated to φ is defined as: $I \leq_\varphi^d J$ if and only if $d_D(I, \varphi) \leq d_D(J, \varphi)$*

So we have a pre-order with the models of φ at the lowest level and the other worlds in the higher levels, according to their Dalal distance.

[5] The Hamming distance between two interpretations is the number of propositional letters on which the two interpretations differ.

Definition 7. *The* Dalal memory operator *is the memory operator obtained from this assignment (i.e. the operator obtained by Definitions 6 and 3).*

We can show through a simple example that this operator differs from the classical Dalal revision operator [7,12]. Let a and b be two propositional letters and consider for example the sequence $\Psi = \Xi \circ a \circ b \circ \neg(a \wedge b)$. The classical Dalal operator gives $Bel(\Psi) = (a \wedge \neg b) \vee (\neg a \wedge b)$, whereas Dalal memory operator gives $Bel(\Psi) = (\neg a \wedge b)$. This behaviour seems more natural since at the next to last step we learned that b was true, and it is normal to keep some credit for this evidence in the following step. It is in this way, that our operators use revision with "memory".

4 Dynamical Revision Operators with Memory

For revision operators with memory, the revision policy is fixed once the operator is chosen. For example for the Dalal memory operator, the way to associate a pre-order to each new evidence is completely determined at the beginning of the process by the Dalal distance.

So, whereas the aim of revision operators with memory is to give a strong preference to the new evidence, one can object that the faithful assignment used to associate a pre-order to the new evidence does not change and so, that an old information is used in each revision step.

The solution to cope with this objection is to find a way to change the faithful assignment during the course of revisions. Such a solution will be given in this section. So, first, let's sum up the way revision operators with memory work:

- The definition of a particular operator lies in the chosen faithful assignment over belief bases. Let's call f such an assignment. So, for each formula φ, f associates a total pre-order $f(\varphi)$ (also noted $\leq_{f(\varphi)}$) satisfying the conditions of Definition 2.
- Each time a new evidence φ comes, the operator associates to it its corresponding pre-order $f(\varphi)$.
- The new epistemic state is the result of incorporating the pre-order in the old epistemic state, giving preference to the new evidence (*i.e.* to the pre-order) by using a lexicographical order: $\leq_{\Phi \circ \varphi} = \leq_{lex(f(\varphi), \Phi)}$ [6].

So, what we want now is to be able to change f during the agent's life. That is, to dynamically change the revision policy of the agent, so that when a new evidence comes, it is not always associated to the same pre-order.

The idea is to start from an *a priori* faithful assignment over belief bases such as for revision by memory operators, but then to modify it at each revision step. To be able to do that, we have to use a more general definition of epistemic states. The (representation of) epistemic states we use for revision operators with memory are pre-orders on interpretations $\leq_\Phi$, from which we can extract the corresponding belief base $Bel(\Phi) = min(\mathcal{W}, \leq_\Phi)$.

[6] Where $I \leq_{lex(\leq_1, \leq_2)} J$ means $I <_1 J$ or ($I \simeq_1 J$ and $I \leq_2 J$).

For dynamical revision operators with memory, the representation of an epistemic state we use is a couple $\Phi = (\varphi, f)$, where φ is the current belief base and f is the current faithful assignment (So, with this representation, we can extract the pre-order corresponding to the belief base: $f(\varphi)$, and straightforwardly $Bel(\Phi) = \varphi$).

As for classical revision operators with memory, to define a particular dynamical revision with memory operator, one needs an initial, a *priori* faithful assignment over belief bases (*i.e.* a classical AGM revision operator), that will encode the initial revision policy of the agent.

So let's define dynamical revision operators with memory:

Definition 8. *Let $\Phi = (\varphi, f)$ be an epistemic state and let μ be a formula denoting a new evidence. We define the new epistemic state $\Phi \circ \mu$, resulting of the* dynamical revision with memory *of Φ by μ, as $\Phi \circ \mu = (\varphi', f')$, where φ' is a formula whose models are $\min(Mod(\mu), f(\varphi))$, and f' is a function (faithful assignment over belief bases) that is identical to f for each belief base ψ except when $\psi \leftrightarrow \varphi'$. In this case $f'(\psi)$ is defined as:*

$$I \leq_{f'(\psi)} J \textit{ iff } I <_{f(\mu)} J \textit{ or } (I \simeq_{f(\mu)} J \textit{ and } I \leq_{f(\varphi)} J)$$

So, pointwise, the dynamical operators work exactly the same way as memory operators. The difference is that they also change the given faithful assignment over belief bases at each step. One could believe that the difference between the two families of operators is not huge, since the corresponding pre-orders (faithful assignment) used change only for one value at each step. But as we will see next the dynamical revision operators with memory satisfy the following postulate that the revision operators with memory do not (always) satisfy (*cf* Example 1):

(C5) If $Bel(\Phi) \leftrightarrow \mu$ then $\Phi \circ \mu = \Phi$.

This axiom says that the current epistemic state does not change in all cases where the new piece of information coincides with the observable part of this epistemic state. Note that this axiom is almost trivial in the classical AGM framework[7]. But in the framework of complex epistemic states it is not the case. In fact, as we already mentioned, the revision operators with memory do not (always) satisfy (C5) as can be seen in the following example.

Example 1. *We are reasoning about an electronic circuit with two components, the left one and the right one. The propositional variable l means that the left component is working, and r encodes the fact the right component is working. Suppose we start from the Dalal classical AGM revision operator. Let Φ be the epistemic state with observable part being the following formula:* "only one of the two components is working" *($Bel(\Phi) = (l \wedge \neg r) \vee (\neg l \wedge r)$). Let μ be the formula expressing that* "the component on the left is not working" *($\mu = \neg l$). The beliefs of the epistemic state after the revision $\Phi \circ \mu$ using Dalal memory operator is* "only the component on the right is working". *The other (conditional)*

[7] In that framework, it is a consequence of the other axioms.

information of this epistemic state can be described by the conditionals "if the component on the right is not working then the two components are bad" *and* "if the components on the left is working then only the component on the left is working". *Now if we revise this current epistemic state by the fact that* "only the component on the right is working" *($\varphi = \neg l \wedge r$), which is indeed the beliefs of the current epistemic state, we obtain a different epistemic state in which for instance we have the conditional* "if the component on the left is working then the two components are working". *On the contrary, the current epistemic state does not change after revision by this new information when the operator is the dynamical Dalal revision with memory operator.*

We illustrate this example below. In order to do that consider a language $\mathcal{L}$ with only two propositional letters l and r. We will denote interpretations simply by the truth assignment, i.e. 10 *denotes the interpretation mapping l to true and r to false. Two interpretations are equivalent, with respect to the pre-order, if they appear at the same level. An interpretation I is better than another interpretation J ($I < J$) if it appears at a lower level. $\circ_{MD}$ denotes the Dalal revision with memory operator and $\circ_{DMD}$ the dynamical Dalal with memory operator.*

Let's see the pre-order associated to some belief bases by the faithful assignment over belief bases given by the Dalal distance:

$$\leq^D_\Phi = \begin{array}{c} 11\ 00 \\ 01\ 10 \end{array} \qquad \leq^D_\mu = \begin{array}{c} 10\ 11 \\ 00\ 01 \end{array} \qquad \leq^D_\varphi = \begin{array}{c} 10 \\ 11\ 00 \\ 01 \end{array}$$

And the epistemic states reached by the operators are:

$$\leq_\Phi = \begin{array}{c} 11\ 00 \\ 01\ 10 \end{array} \qquad \leq_{\Phi \circ_{MD} \mu} = \begin{array}{c} 11 \\ 10 \\ 00 \\ 01 \end{array} \qquad \leq_{\Phi \circ_{MD} \mu \circ_{MD} \varphi} = \begin{array}{c} 10 \\ 11 \\ 00 \\ 01 \end{array}$$

$$\leq_\Phi = \begin{array}{c} 11\ 00 \\ 01\ 10 \end{array} \qquad \leq_{\Phi \circ_{DMD} \mu} = \begin{array}{c} 11 \\ 10 \\ 00 \\ 01 \end{array} \qquad \leq_{\Phi \circ_{DMD} \mu \circ_{DMD} \varphi} = \begin{array}{c} 11 \\ 10 \\ 00 \\ 01 \end{array}$$

Note that the idea here is that, when the agent receives a new evidence that she has met before, the repetition of this evidence suggests that the old beliefs of the agent were correct, and so she holds on to the last pre-order that corresponds to this evidence.

In fact, if one considers the definition of iterated revision operators according to Darwiche and Pearl (cf Sect. 2.2), it amounts to say that we change the revision operators at each step, since the corresponding faithful assignment over epistemic states changes at each step. So, in a sense, dynamical revision operators with memory are definable by a family of revision operators with memory (that corresponds to the set of all faithful assignments over belief bases reached by the course of revisions).

Concerning the logical properties of this family of operators, it is easy to check the following:

Theorem 6. *A dynamical revision operator with memory satisfies (R*1)-(R*6). It satisfies (C1), (C3), (C4) and (C5) but it never satisfies (C2).*

Finally, as another example, let's see the behaviour of the full meet revision operator $\circ_B$, the basic memory operator $\circ_{MB}$ and the dynamical basic memory operator $\circ_{DMB}$ (they are all built from the same faithful assignment over belief bases) on the same situations.

Example 2. *Consider a language $\mathcal{L}$ with only two propositional letters a and b (considered in that order for the valuations). Let's see the pre-order associated to some belief bases by the faithful assignment over belief bases given by the Basic distance:*

$$\leq^B_a = \begin{matrix} 00\ 01 \\ 10\ 11 \end{matrix} \qquad \leq^B_b = \begin{matrix} 00\ 10 \\ 01\ 11 \end{matrix}$$

$$\leq^B_{a \wedge b} = \begin{matrix} 00\ 01\ 10 \\ 11 \end{matrix} \qquad \leq^B_{\neg a} = \begin{matrix} 10\ 11 \\ 00\ 01 \end{matrix}$$

And the epistemic states reached by the operators are:

$$\leq_{a \circ_{MB} b} = \leq_{a \circ_{DMB} b} = \begin{matrix} 00 \\ 10 \\ 01 \\ 11 \end{matrix} \qquad \leq_{a \circ_{MB} b \circ_{MB} \neg a} = \leq_{a \circ_{DMB} b \circ_{DMB} \neg a} = \begin{matrix} 10 \\ 11 \\ 00 \\ 01 \end{matrix}$$

$$\leq_{a \circ_{MB} b \circ_{MB} \neg a \circ_{MB} a \wedge b} = \begin{matrix} 10 \\ 00 \\ 01 \\ 11 \end{matrix} \qquad \leq_{a \circ_{DMB} b \circ_{DMB} \neg a \circ_{DMB} a \wedge b} = \begin{matrix} 00 \\ 10 \\ 01 \\ 11 \end{matrix}$$

So we have that:

$$a \circ_B b \circ_B \neg a \circ_B a \wedge b \circ_B \neg b \equiv \neg b$$
$$a \circ_{MB} b \circ_{MB} \neg a \circ_{MB} a \wedge b \circ_{MB} \neg b \equiv \neg a \wedge \neg b$$
$$a \circ_{DMB} b \circ_{DMB} \neg a \circ_{DMB} a \wedge b \circ_{DMB} \neg b \equiv a \wedge \neg b$$

As noted previously, the full meet revision operator $\circ_B$ does not have a very good behaviour: each time the new evidence contradicts the current beliefs, the new beliefs are only the logical consequences of the new evidence. So, it absolutely does not consider the previous revisions. With the revision operator with memory $\circ_{MB}$, the agent is able to build complex epistemic states (pre-orders), that lead to a satisfactory behaviour for iterated revision. With this operator, the two evidences $\neg a$ and $\neg b$ recently learned lead to this belief base. With the dynamical revision operator with memory $\circ_{DMB}$, the evidence learned at the next to last step ($a \wedge b$) recalls the agent the last time she had this belief (after $a \circ_{DMB} b$), and this modifies her epistemic state.

5 Conclusion

It is worthy to note that the two families of operators defined, revision with memory and dynamical revision with memory, are revision operators in the sense of Darwiche and Pearl, that is, they map an epistemic state and a formula (new evidence) to an epistemic state. We have shown that one can use any standard AGM revision operator and turns it to a DP iterated revision operator using revision with memory and dynamical revision with memory (C2 is not satisfied, but this postulate is criticizable).

Note that [3] considers revision of epistemic states by epistemic states. In this work, even if at the end of the process we work with two pre-orders, the second one is obtained from the input, that is a single formula (as usual in AGM/DP framework), by a pre-processing step.

It is interesting also to note that our definition of epistemic states for dynamical revision with memory is more complicated than usual DP ones: so this work illustrates that one can encode sublter behaviours with more complicated epistemic states. Studying this kind of generalized epistemic states and its application seems to be an interesting research issue.

References

1. Alchourrón, C.E., Gärdenfors, P., Makinson, D.: On the logic of theory change: partial meet contraction and revision functions. J. Symbolic Logic **50**, 510–530 (1985)
2. Benferhat, S., Dubois, D., Papini, O.: A sequential reversible belief revision method based on polynomials. In: Proceedings of the Sixteenth National Conference on Artificial Intelligence (AAAI 1999) 1999
3. Benferhat, S., Konieczny, S., Papini, O., Pino Pérez, R.: Iterated revision by epistemic states: axioms, semantics and syntax. In: Proceedings of the Fourteenth European Conference on Artificial Intelligence (ECAI 2000), pp. 13–17 (2000)
4. Booth, R., Meyer, T.: Admissible and restrained revision. J. Artif. Intell. Res. **26**, 127–151 (2006)
5. Boutilier, C.: Revision sequences and nested conditionals. In: Proceedings of the Thirteenth International Joint Conference on Artificial Intelligence (IJCAI 1993) 1993
6. Boutilier, C.: Iterated revision and minimal change of conditional beliefs. J. Philos. Logic **25**(3), 262–305 (1996)
7. Dalal, M.: Investigations into a theory of knowledge base revision: preliminary report. In: Proceedings of the National Conference on Artificial Intelligence (AAAI 1988), pp. 475–479 (1988)
8. Darwiche, A., Pearl, J.: On the logic of iterated belief revision. In: Theoretical Aspects of Reasoning about Knowledge: Proceedings of the 1994 Conference (TARK 1994), pp. 5–23. Morgan Kaufmann (1994)
9. Darwiche, A., Pearl, J.: On the logic of iterated belief revision. Artif. Intell. **89**, 1–29 (1997)
10. Gärdenfors, P.: Knowledge in Flux. MIT Press, Cambridge (1988)
11. Jin, Y., Thielscher, M.: Iterated belief revision, revised. Artif. Intell. **171**, 1–18 (2007)

12. Katsuno, H., Mendelzon, A.O.: Propositional knowledge base revision and minimal change. Artif. Intell. **52**, 263–294 (1991)
13. Konieczny, S., Grespan, M. M., Pino Pérez, R.: Taxonomy of improvement operators and the problem of minimal change. In: Proceedings of the 12th International Conference on Principles of Knowledge Representation and Reasoning (KR 2010), pp. 161–170 (2010)
14. Konieczny, S., Pino Pérez, R.: A framework for iterated revision. J. Appl. Non-Classical Logics **10**(3–4), 339–367 (2000)
15. Konieczny, S., Pérez, R.P.: Some operators for iterated revision. In: Benferhat, S., Besnard, P. (eds.) ECSQARU 2001. LNCS (LNAI), vol. 2143, pp. 498–509. Springer, Heidelberg (2001). doi:10.1007/3-540-44652-4_44
16. Konieczny, S., Pino Pérez, R.: Improvement operators. In: Principles of the 11th International Conference on Principles of Knowledge Representation and Reasoning (KR 2008), pp. 177–186 (2008)
17. Lehmann, D.: Belief revision, revised. In: Proceedings of the Fourteenth International Joint Conference on Artificial Intelligence (IJCAI 1995), pp. 1534–1540 (1995)
18. Liberatore, P.: The complexity of iterated belief revision. In: Afrati, F., Kolaitis, P. (eds.) ICDT 1997. LNCS, vol. 1186, pp. 276–290. Springer, Heidelberg (1997). doi:10.1007/3-540-62222-5_51
19. Nayak, A.: Iterated belief change based on epistemic entrenchment. Erkenntnis **41**, 353–390 (1994)
20. Nayak, A. C., Foo, N. Y., Pagnucco, M., Sattar, A.: Changing conditional beliefs unconditionally. In: Proceedings of the Sixth Conference of Theoretical Aspects of Rationality and Knowledge (TARK 1996), pp. 119–135. Morgan Kaufmann, Netherlands (1996)
21. Papini, O.: Iterated revision operators stemming from the history of an agent's observations, Frontiers of Belief Revision, pp. 281–303. Kluwer (2001).
22. Spohn, W.: Ordinal conditional functions: a dynamic theory of epistemic states. In: Harper, W.L., Skyrms, B. (eds.) Causation in Decision, Belief Change, and Statistics, vol 2, pp. 105–134 (1987)
23. Williams, M. A.: Transmutations of knowledge systems. In: Proceedings of the Fourth International Conference on the Principles of Knowledge Representation and Reasoning (KR 1994), pp. 619–629 (1994)
24. Williams, M. A.: Iterated theory base change: a computational model. In: Proceedings of the Fourteenth International Joint Conference on Artificial Intelligence (IJCAI 1995), pp. 1541–1550 (1995)

Short Papers

An Empirical Route to Logical 'Conventionalism'

Eugene Chua[(✉)]

Wolfson College, University of Cambridge, Cambridge CB3 9BB, UK
eugene.chua@cantab.net

Abstract. The laws of classical logic are taken to be logical truths, which in turn are taken to hold objectively. However, we might question our faith in these truths: why are they true? One general approach, proposed by Putnam [8] and more recently Dickson [3] or Maddy [5], is to adopt empiricism about logic. On this view, logical truths are true because they are true of the world alone – this gives logical truths an air of objectivity. Putnam and Dickson both take logical truths to be true in virtue of the world's structure, given by our best empirical theory, quantum mechanics. This assumes a determinate logical structure of the world given by quantum mechanics. Here, I argue that this assumption is false, and that the world's logical structure, and hence the related 'true' logic, are underdetermined. This leads to what I call empirical conventionalism.

Keywords: Philosophy of logic · Philosophy of physics · Conventionalism

1 Empiricism

Consider the classical distributive law over conjunctions for all sentences *p*, *q*, and *r*:

$$(\textbf{DIST}): p \text{ and } (q \text{ or } r) \leftrightarrow (p \text{ and } q) \text{ or } (p \text{ and } r)$$

As with other 'laws' of classical logic (**CL**), **DIST** is often taken as a *logical truth* – regardless of *p*, *q* or *r*, it *objectively* holds. Braving heresy, one might ask: *why* are they true? Logical conventionalism, a once-popular approach, takes logical truths to follow from the meanings of subsentential operators. In short, logical truths, e.g. **DIST**, are true 'in virtue of meaning' or 'true by convention'.[1] However, such conventionalism appears intuitively unsatisfactory for explaining **DIST**'s objective truth given its dependence on (at best) intersubjective conventions.[2]

One attractive alternative is *empiricism*: facts about the world *alone* determine the logical truths, independent of conventions. This *prima facie* avoids the aforementioned problem, since a logic is true in virtue of facts about a mind-independent world. How is a logic validated by the world? Maddy [5] proposes[3] that 'logic is true of the world because of its underlying structural features'. The difficulty, then, is determining the

[1] Warren [12], p. 120.
[2] Quine [9] remains the starting point against *explicit* conventionalism. However, see [12] who argues for *implicit* conventionalism.
[3] Maddy [5], p. 226.

A. Baltag et al. (Eds.): LORI 2017, LNCS 10455, pp. 631–636, 2017.
DOI: 10.1007/978-3-662-55665-8_43

world's actual logical structure. Here, the empiricist strategy is to 'read off' logic from our best fundamental sciences,[4] often taken to be quantum mechanics (**QM**). Notably, Putnam [8] and Dickson [3] employ this strategy to argue that **DIST** *is* false, in light of the logic of quantum mechanics, quantum logic (**QL**). Instead, **QL** is the 'true' logic.[5] The objectivity of empiricism thus comes at a cost: **DIST** was, after all, a logical law whose objective truth we hoped to establish. The empiricist bites the bullet here: forsaking **DIST** is a worthwhile price for reclaiming objectivity.

Here, I offer a critique of this strategy, specifically its presupposition of a determinate world-structure prescribed by **QM**, by showing that the world-structure in **QM** is *empirically conventional*: nothing within **QM**'s formalism, from which all empirical results are derived, can determine the choice of world-structure. Since logic is tied to world-structure, the world alone fails to decide our logic. Putnam's challenge to the conventionalist fails: the relevant empirical facts determining 'true' logic are still conventional, leaving us with another form of conventionalism.

Going forward, I assume a basic understanding of **QL** and **QM**. In Sect. 2, I introduce the disagreement between **QL** and **CL** regarding **DIST**. In Sect. 3, I present two well-known interpretations of **QM** with different prescriptions of world-structure. In Sect. 4, I argue that the choice between these prescription is empirically conventional. *A fortiori*, so is the 'true' logic. This leads to a form of conventionalism even *within* empiricism.

2 Distribution

The distinguishing feature of **QL** is the sort of propositions it governs, viz. *experimental propositions* about quantum systems, e.g. 'the system passes a test for some possible state $\boldsymbol{P}$ with probability 1', with a bijection to subspaces in Hilbert space. While disjunction in **CL** is equivalent to classical union $\cup$, it is well-known that the disjunction in **QL**, $\mathbf{A} \vee_{QL} \mathbf{B}$, is *not* $\mathbf{A} \cup \mathbf{B}$. Rather, it is **A** and **B**'s *span*.[6] Indeed, two subspaces' union is generally not itself a subspace in **QL**. Recall the empiricist claim: **QL**, based off the structure of **QM**, shows that **DIST** *is* false. Consider **QL**'s equivalent of **DIST**:

$$(\mathbf{DIST}^{\dagger}):\ \boldsymbol{R} \wedge \left(\boldsymbol{P} \vee_{QL} \boldsymbol{Q}\right) = (\boldsymbol{R} \wedge \boldsymbol{P}) \vee_{QL} (\boldsymbol{R} \wedge \boldsymbol{Q})$$

Here, $\boldsymbol{P}$, $\boldsymbol{Q}$ and $\boldsymbol{R}$ are arbitrary distinct subspaces, $\wedge$ is interpreted as intersection, and $\vee_{QL}$ is interpreted as span. By considering **P**, **Q** and **R** as any three distinct coplanar subspaces we can trivially demonstrate that $\mathbf{DIST}^{\dagger}$ is *not* a logical truth in **QL**.

Objection: we have not yet shown **DIST**'s failure, using classical 'and' and 'or'. I merely demonstrated $\mathbf{DIST}^{\dagger}$'s falsehood using '$\wedge$' and '$\vee_{QL}$', on a restricted class of experimental propositions. Hence Maudlin's complaint: "quantum 'logic' isn't logic,

[4] Putnam [8], p. 179.
[5] Dickson [3], p. 2.
[6] The span of two subspaces is the plane containing them and their superpositions.

i.e. isn't an account of conjunction and disjunction".[7] In response, proponents of **QL** must show that, in some sense, the world's apparent **CL**-structure, in virtue of which we adopted **CL**, is 'really' a **QL**-structure: we were simply *mistaken* about the world's logical structure underlying our use of 'or'.

One strategy comes to mind. First, in the context of experimental propositions, show that there is **(a)** no meaningfully definable classical 'or', and **(b)** the best replacement for 'or' is '$\vee_{QL}$'. Next, **(c)** show that the experimental propositions of QL exhaust all propositions about the world. Without **(c)**, a possibility that the world's logical structure is really classical remains: the non-classical nature of **QL** only arises in specific contexts, e.g. measurements. Given **(a)** – **(c)**, the empiricist can assert that there is *no other way* to 'read off' an appropriate notion of disjunction from the world's structure without '$\vee_{QL}$'. The empiricist takes this to mean that what we meant by 'or' was *really* '$\vee_{QL}$', and so what we thought was **DIST** was *really* **DIST**†. Since **DIST**† is false, so is **DIST**.

In the context of experimental propositions, there is justification for both **(a)** and **(b)**. Regarding **(a)**, there is no clear way to introduce classical 'or' into the logic of experimental propositions, **QL**, since there is, in general, no experimental proposition or subspace in Hilbert space corresponding to the classical '***P*** or ***Q***'.[8] This gives us reason to claim that we cannot even speak of the classical 'or' meaningfully in terms of experimental propositions. Dickson [3] argues further for **(b)**: plausibly, '$\vee_{QL}$' is the only other candidate[9] in this framework for replacing the classical 'or', in the sense that '$\vee_{QL}$' share many similarities in terms of logical behavior with classical disjunction, apart from distribution.[10]

3 Interpretation

It seems clear that **DIST**† cannot be 'read off' the structure of experimental propositions in **QM**. However, do experimental propositions exhaust all propositions about the world? There are at least two interpretations of **QM** – Bohmian mechanics (**BM**) and Everettian mechanics (**EM**) – which are well-known for capturing the same empirical results as the standard **QM** algorithm. They are *empirically equivalent*, despite prescribing radically different world-structures and thereby answers to **(c)**.

The relevant difference here turns on the status of superpositions in the world, and hence '$\vee_{QL}$': under **BM**, the wave-function of a superposed system is a physical pilot wave 'piloting' particles with *classically determinate* positions and trajectories to *one of the superposed states* with a stochastic distribution constrained by the Born rule.[11] Importantly, the particles themselves are *either* in 'support' of one state *or* another in the *sense of* ***CL***: "the [position/trajectory] configuration of the system is located only in

[7] Maudlin [6], p. 479.
[8] Bacciagaluppi [1], p. 19.
[9] Dickson [3], p. 4.
[10] For an exposition of the logical behavior of '$\bigvee$**QL**', see Humberstone [4], pp. 913-917.
[11] See Bohm [2].

one of these different components, and this is already a matter of classical logic".[12] On this view, **(c)** fails to obtain since the world is *fundamentally* classical: we are not justified to replace *all* propositions about the world with experimental propositions, and the more fundamental propositions obey **CL**. The use of **QL** reflects not the world-structure but *our epistemic inability* to access the hidden position variables.

Contra **BM**, **EM** claims that the experimental propositions of **QL**, *does* completely describe the world (as a sum of many dynamically-decohered 'worlds'[13]). Regarding quantum disjunction: **EM** takes superposed states as descriptions of *single* systems. On **EM**, systems which are in superposed states *stay so* after measurement: each component state of superpositions *actually* obtains (in dynamically isolated 'worlds'). Hence, on this view, all propositions about the world *really* do behave like experimental propositions of **QL**: we use quantum disjunction not as a result of some epistemic limitations *a la* **BM** but as a true description of world-structure. Thus **(c)** obtains under **EM**.

4 Empirical Conventionalism

I began by asking why logical truths are true. Hoping to avoid the conventionalist path, I turned to empiricism. Yet, as demonstrated with **BM** and **EM**, there is no determinate answer to one key part of the empiricist strategy, viz. **(c)**. This situation leads us back to a form of conventionalism distinct from traditional logical conventionalism, *empirical conventionalism*, best described by Sklar [10]: "insofar as the two theories have the same predictive content with regard to the directly observable facts, they ought to be viewed as merely conventional alternatives to one another and not as genuinely alternative theories about the nature of the world."[14] In other words, **QM** does not enforce a single interpretation. *A fortiori*, **QM** does not enforce a single 'true' logic.

To break the purported symmetry between interpretations of **QM**, we might invoke some 'further fact' to decide on one interpretation over another. Putnam [7] famously distinguished two types of facts constraining total science:[15] *internal coherence constraints* (**ICC**) demand that science must cohere with simplicity, agree with intuition, and so on, while *external coherence constraints* (**ECC**) demand that science must agree with experimental checks, i.e. empirical facts. Hence, an interpretation is chosen not only because it coheres with all possible empirical facts, viz. **ECC**, but also because of simplicity, intuitiveness, etc., viz. **ICC**. Putnam suggests that **ICC** provides a further fact which decides between empirically conventional choices; in considering **ICC** one may 'refute' conventionalism. However, I see two problems with this strategy.

Firstly, **ICC** simply makes the *un*objective elements involved in interpretational choice more obvious. While something *can* be a determinate fact of the matter *given* **ICC**, these considerations of simplicity, intuitiveness, etc., are *exactly* what appears to

[12] Bacciagaluppi [1], p. 31.
[13] For more on decoherence or the status of 'worlds' in **EM**, see Wallace [11].
[14] Sklar [10], p. 958.
[15] Putnam [7], p. 33.

be less-than-objective. Even if there *could* be a decisive fact of the matter given **ICC**, I am not sure there are objective grounds for the **ICCs** themselves.

Secondly, it is unclear whether there even *is* a fact of the matter under **ICC** as to whether **BM** or **EM** is better. It is simply *not apparent* which is simpler and so on. Given the complicated nature of **QM**, and the technical and conceptual apparatus required for both **BM** (hidden level of phenomena, physical 'pilot-wave', non-locality) and **EM** (a world of infinitely many decohered 'worlds'), neither **BM** nor **EM** obviously satisfies **ICC** better than the other. One is left to their metaphysical predilections.

In any case, the empiricist has already lost much in adopting **ICC**. Empiricism aimed to place logical truths on objective grounds, *contra* logical conventionalism, by appealing to empirical facts about a mind-independent world. The situation I have briefly sketched, however, is one where there is *no determinate interpretation* for **QM**. **ECC** alone does not determine the true world-structure and 'true' logic; we must appeal to considerations about **ICC**. The resulting decision seems to turn on something about *us*, as rational beings, as scientists, and so on. Empiricism thus fails to obtain objectivity for logic, leading instead to empirical conventionalism in the context of **QM**.

5 Conclusion

Empiricists hoping to recover the 'true' logic by appealing to the world alone should see that the world-structure in our best theory, **QM**, is underdetermined: a choice between them, and hence a choice of 'true' logic, is *empirically conventional* – we have no empirical reason to think that **DIST** (and **CL**) is true of the world or otherwise. Relying on **ICC** to resolve this only ameliorates the situation by relying on less-than-objective facts. Thus, empiricism fares no better than logical conventionalism in accounting for the objectivity of logical truths: something broadly conventional lurks.

References

1. Bacciagaluppi, G.: Is Logic Empirical? (2009). http://philsci-archive.pitt.edu/3380/. Accessed 13 June 2017
2. Bohm, D.: A suggested interpretation of the quantum theory in terms of "hidden" variables I. Phys. Rev. **85**(2), 166–179 (1952)
3. Dickson, M.: Quantum Logic Is Alive ∧ (It Is True ∨ It Is False) (2001). http://mdickson.net/pubs/Quantum_Logic.pdf. Accessed 13 June 2017
4. Humberstone, L.: The Connectives. MIT Press, Cambridge (2011)
5. Maddy, P.: Second Philosophy: A Naturalistic Method. Oxford University Press, Oxford (2007)
6. Maudlin, T.: Distilling metaphysics from quantum physics. In: Loux, M., Zimmerman, D. (eds.) The Oxford Handbook of Metaphysics, pp. 461–487. Oxford University Press, Oxford (2003)
7. Putnam, H.: The refutation of conventionalism. Noûs **8**(1), 25–40 (1974)
8. Putnam, H.: "The Logic of Quantum Mechanics", in Mathematics, Matter and Method, vol. 1, pp. 174–197. Cambridge University Press, Cambridge (1975)

9. Quine, W.V.O.: Truth by convention. In: Benacerraf, P., Putnam, H. (eds.) Philosophy of Mathematics: Selected Readings, 2nd edn., pp. 329–354. Cambridge University Press, Cambridge (1984)
10. Sklar, L.: Spacetime and conventionalism. Philos. Sci. **71**(5), 950–959 (2004)
11. Wallace, D.: Decoherence and Ontology (or: How I learned to stop worrying and love FAPP). In: Saunders, S., Barrett, J., Kent, A., Wallace, D. (eds.) Many Worlds?: Everett, Quantum Theory & Reality, pp. 53–72. Oxford University Press, Oxford (2010)
12. Warren, J.: Revisiting quine on truth by convention. J. Philos. Log. **46**(2), 119–139 (2017)

Beating the Gatecrasher Paradox with Judiciary Narratives

Rafal Urbaniak[1,2(✉)]

[1] Centre for Logic and Philosophy of Science, Ghent University, Ghent, Belgium
rfl.urbaniak@gmail.com
[2] Institute of Philosophy, Sociology and Journalism, University of Gdansk, Gdańsk, Poland
https://ugent.academia.edu/RafalUrbaniak/Papers

Abstract. A probabilistic model for the narrative approach to reasoning in legal fact-finding is developed and applied to the gatecrasher paradox.

According to *Classical Legal Probabilism* CLP, the fact-finders' degrees of beliefs are to be modeled by standard probability distributions, and criminal standard of proof beyond reasonable doubt should be equated with a certain high threshold probability of guilt [3]. The *paradox of the gatecrasher* [4] was formulated as a criticism of CLP.[1]

> Suppose our guilt threshold is high, say at 0.99. Consider the situation in which 1000 fans enter a football stadium, and 999 of them avoid paying for their tickets. A random spectator is tried for not paying. The probability that the spectator under trial did not pay exceeds 0.99. Yet, intuitively, a spectator cannot be considered guilty on the sole basis of the number of people who did and did not pay.

CLP has also been criticized for not being a sensible or useful model of judiciary fact-finding reasoning from various angles (see for example [2,4–6,9,10,13–18]). The critics of CLP argue that the view is blind to various phenomena that an adequate philosophical account of legal fact-finding should explain. Some of them pertain to procedural issues [14]: proceedings are back-and-forth between opposing parties, cross-examination is crucial, and yet CLP seems to take no notice of this dynamics. Some have to do with reasoning methods which are not only evidence-to-hypothesis, but also hypotheses-to-evidence [2,18] and involve inference to the best explanation [6].

The main competing approach—the no plausible alternative story (NPAS) theory [1]—is one in which the fact-finding process is seen as an interplay of evidence and various explanations (often called *narratives*) presented by opposing

The research has been supported by Research Foundation Flanders.
The research was funded by National Centre for Science grant number 2016/22/E/HS1/00304.

[1] The paradox is mathematically the same as the *prisoners in a yard scenario* [13], where a group of prisoners commits a group killing, and it's impossible to identify the single innocent prisoner.

A. Baltag et al. (Eds.): LORI 2017, LNCS 10455, pp. 637–642, 2017.
DOI: 10.1007/978-3-662-55665-8_44

parties [10]. The most plausible one with no plausible alternative is to be the decisive one. From this perspective, intuitively, the gatecrasher is a no-started, for it provides no real narration of what happened.

Alas, the notion of plausibility as used in NPAS [1] is a primitive notion with no explication. If formal, especially probabilistic, methods are of any value, the view is a step back in this respect, for it "solves" a selection of problems with the probabilistic tools by giving up on them.

New Legal Probabilism (NLP) [7] improves on NPAS by developing an informal and real-case-based analysis of the conditions on conviction beyond reasonable doubt. The key conditions are:

(Evidential support)	The defendant's guilt probability on the evidence should be sufficiently supported by the evidence, and a successful accusing narration should explain the relevant evidence
(Evidential completeness)	The evidence available at trial should be complete as far as a reasonable fact-finders' expectations are concerned
(Resiliency)	The prosecutor's narrative, based on the available evidence, should not be susceptible to revision given reasonably possible future arguments and evidence
(Narrativity)	The narrative offered by the prosecutor should answer all the natural or reasonable questions one may have about what happened, given the content of the prosecutor's narration and the available evidence.

The analysis is informal, but it's specific enough to be a part of departure for a formal explication. The goal of this paper is to use probabilistic tools to provide a formal theory of narratives as used in judiciary fact-finding, inspired by NLP.

The *object language* is a standard propositional language $\mathcal{L}$ extended to a language $\mathcal{L}^+$ with primitive unary operators $E, N_1^A, \ldots, N_k^A, N_1^D, \ldots, N_k^D$, and the guilt statement constant G.[2] The intended interpretation of Ep is p *is part of evidence*, and the operator is needed because in judiciary fact-finding the information what is and what isn't evidence plays an important role (the same point will hold for narratives).

$N_i^A p$ means p *is part of an accusing narration* $\mathtt{N}_i^A$ and $N_i^D p$ means p *is part of a defending narration* $\mathtt{N}_i^D$. Each narration $\mathtt{N}_i$ is taken to be a finite set of sentences $n_{i1}, n_{i2}, \ldots, n_{ik_i}$ of $\mathcal{L}^+$.[3]

---

[2] The content of the guilt statement is G which has the form of $G \equiv g_1 \wedge \cdots \wedge g_l$ for appropriate $g_1, \ldots, g_l \in \mathcal{L}$.

[3] In contexts in which it is irrelevant whether a narration is an accusing one or not, I will suppress the superscripts.

E stands ambiguously for the set of all sentences constituting evidence and for the conjunction thereof. Which reading is meant will always be clear from the context (this convention applies to all finite sets of sentences considered in this paper). $\mathtt{E}^d$ stands for $\mathtt{E}^d = \{E\varphi | \varphi \in \mathtt{E}\}$ and $\mathtt{E}^-$ is $\{\neg E\varphi | \varphi \notin \mathtt{E}\}$. For any narration $\mathtt{N}_i$, symbols $\mathtt{N}_i$, $\mathtt{N}_i^d$, and $\mathtt{N}_i^-$ are to be understood analogously to E, $\mathtt{E}^d$ and $\mathtt{E}^-$. $\mathtt{N}^d$ is the (positive) *description* of all the narrations, $\bigcup_i \mathtt{N}_i^d$, and $\mathtt{N}^- = \bigcup_i \mathtt{N}_i^-$ adds that this description is complete.

Priors are modeled as a partial probabilistic distribution (to be able to model ignorance and the suspension of illegitimate bias) P, which (partially) maps $\mathcal{L}^+ \times \mathcal{P}(\mathcal{L}^+)$ to $[0,1]$ [11,12]. P has to have an extension to a total conditional probability distribution over $\mathcal{L}^+$ satisfying the standard axioms of conditional probability, and it has to satisfy the following:

$$\mathtt{P}(\top|\Gamma) = 1 \qquad \mathtt{P}(\bot|\Gamma) = 0 \tag{1}$$

$$\varphi \in \Gamma \Rightarrow \mathtt{P}(\varphi|\Gamma)\downarrow \tag{2}$$

$$\mathtt{P}(\varphi|\Gamma)\downarrow \Leftrightarrow \mathtt{P}(\neg\varphi|\Gamma)\downarrow \qquad \mathtt{P}(\varphi\wedge\psi|\Gamma)\downarrow \Leftrightarrow \mathtt{P}(\psi\wedge\varphi|\Gamma)\downarrow \tag{3}$$

$$\mathtt{P}(\varphi\wedge\psi|\Gamma) > 0 \Rightarrow \mathtt{P}(\varphi|\Gamma)\downarrow, \mathtt{P}(\psi,|\Gamma)\downarrow \qquad \mathtt{P}(\varphi|\Gamma) = 0 \Rightarrow \mathtt{P}(\varphi\wedge\psi|\Gamma)\downarrow \tag{4}$$

$$\mathtt{P}(\varphi|\Gamma)\uparrow \Rightarrow \mathtt{P}(\varphi\wedge\psi|\Gamma)\uparrow \qquad \text{unless } \mathtt{P}(\psi|\Gamma) = 0 \tag{5}$$

$$\text{If } \mathtt{P}(\varphi|\Gamma) > 0, \mathtt{P}(\varphi\wedge\psi|\Gamma) = 0, \qquad \text{then } \mathtt{P}(\psi|\Gamma) = 0 \tag{6}$$

Four types of stances that a fact-finder might take towards a claim will be considered: *uncontroversial acceptability threshold*, a, *negligibility threshold*, $\mathtt{n} =_{df} 1 - \mathtt{a}$, *strong plausibility* s and *rejectability*, $\mathtt{r} =_{df} 1 - \mathtt{s}$. We require $\mathtt{a} > \mathtt{s} > \mathtt{r} > \mathtt{n}$. Updates used in what follows are:

name	notation	meaning
full	$\mathtt{P}^f(\varphi\|\Gamma)$	$\mathtt{P}(\varphi\|\mathtt{E}, \mathtt{E}^d, \mathtt{E}^-, \mathtt{N}^d, \mathtt{N}^-, \mathtt{G}, \Gamma)$
n-full	$\mathtt{P}^{nf}(\varphi\|\Gamma)$	$\mathtt{P}(\varphi\|\mathtt{E}, \mathtt{E}^d, \mathtt{N}^d, \mathtt{N}^-, \mathtt{G}, \Gamma)$
informed	$\mathtt{P}^i(\varphi\|\Gamma)$	$\mathtt{P}(\varphi\|\mathtt{E}, \mathtt{E}^d, \mathtt{N}^d, \mathtt{G}, \Gamma)$
evidential	$\mathtt{P}^e(\varphi\|\Gamma)$	$\mathtt{P}(\varphi\|\mathtt{E}, \mathtt{E}^d, \mathtt{E}^-, \mathtt{G}, \Gamma)$
argued	$\mathtt{P}^a(\varphi\|\Gamma)$	$\mathtt{P}(\varphi\|\mathtt{N}^d, \mathtt{G}, \Gamma)$
play-along	$\mathtt{P}^{N_j}(\varphi\|\Gamma)$	$\mathtt{P}(\varphi\|\mathtt{N}_j, \mathtt{N}^d, \mathtt{N}^-, \mathtt{G}, \Gamma)$
n-extended play-along	$\mathtt{P}^{nN_j}(\varphi\|\Gamma)$	$\mathtt{P}(\varphi\|\mathtt{N}_j, \mathtt{E}, \mathtt{E}^d, \mathtt{N}^d, \mathtt{N}^-, \mathtt{G}, \Gamma)$
e-extended play-along	$\mathtt{P}^{eN_j}(\varphi\|\Gamma)$	$\mathtt{P}(\varphi\|\mathtt{N}_j, \mathtt{E}, \mathtt{E}^d, \mathtt{E}^-, \mathtt{N}^d, \mathtt{G}, \Gamma)$
f-extended play-along	$\mathtt{P}^{fN_j}(\varphi\|\Gamma)$	$\mathtt{P}(\varphi\|\mathtt{N}_j, \mathtt{E}, \mathtt{E}^d, \mathtt{E}^-, \mathtt{N}^d, \mathtt{N}^-, \mathtt{G}, \Gamma)$

The conditions on narratives are:

$$\mathtt{P}^f(\neg(\mathtt{N}_i \wedge \mathtt{N}_j)) \geq \mathtt{a}, \text{ for } i \neq j \qquad \text{(Exclusion)}$$

$$\mathtt{P}^{f\mathtt{N}_i^A}(G) \geq \mathtt{a} \wedge \mathtt{P}^{f\mathtt{N}_k^D}(\neg G) \geq \mathtt{a} \qquad \text{(Decision)}$$

$$\mathtt{P}^e(\mathtt{N}_k) \geq \mathtt{n} \qquad \text{(Initial plausibility)}$$

$$\mathtt{P}^f(\mathtt{N}_1 \vee \cdots \vee \mathtt{N}_k) \geq s \qquad \text{(Exhaustion)}$$

In general, we'd normally expect the persecution to accept any claim relevant to the case that, given the evidence and the persecution's narration, is at least as likely as the guilt statement that they're putting forward. This motivates the following requirement:[4]

(Commitment) For any φ relevant to the case, if $\mathtt{P}^{f\mathtt{N}_i^A}(\varphi) \geq \mathtt{P}^{f\mathtt{N}_i^A}(\mathtt{G})$, then $\mathtt{P}^{eN_i}(N_i(\varphi)) \geq \mathtt{s}$.

Now we turn to assessment criteria. An accusing narration has to satisfy (Explaining Evidence A) and a defending narration has to satisfy (Explaining Evidence D). A narration shouldn't *miss any evidence*: there shouldn't be any piece of evidence that is expected, given the narration and the background knowledge, but isn't. We say that $\mathtt{N}_i$ is *gappy* ($\mathbf{G}(\mathtt{N}_i)$) just in case there are claims that the narration should choose from and yet it doesn't. An accusing narration $\mathtt{N}_i^A$ *dominates* the set of all accusing narrations $\mathbb{N}^A$ just in case it doesn't miss any evidence, it doesn't contain any gap, in light of all available information and evidence it is at least as likely any other accusing narration, and it is strongly plausible, given all available information. A dominating narration $\mathtt{N}_i^A$ is *resilient* ($\mathbf{R}(\mathtt{N}_i^A)$) just in case there is no non-negligible potential evidence that might undermine it, at least in light of all we know (minus the negative description of the evidence, to avoid triviality)—that is, no φ with $\mathtt{P}^{nf}(E\varphi) \geq \mathtt{n}$ – such that if $\mathtt{E}$ was modified to $\mathtt{E} \cup \{\varphi\}$, $\mathtt{N}_i^A$ would no longer dominate. A defense narration $\mathtt{N}_k^D$ *raises reasonable doubt* ($\mathbf{RD}(\mathtt{N}_k^D)$) if it has no gaps, and hasn't been rejected given all that we know.

[4] A set of sentences is relevant for the case if it is consistent with the background knowledge and there is a narration such that its posterior probability given all background knowledge together with that set is different from its posterior probability given all background knowledge only. A set of sentences is a minimal relevant set if no proper subset thereof is a relevant set. A sentence is relevant if it or its negation is a member of a minimal relevant subset.

$$\text{For any } e \in \mathtt{E}, [\neg \mathtt{P}^{\mathtt{N}_i^A}(\neg Ee) \geq \mathtt{s} \Rightarrow \mathtt{P}^{\mathtt{N}_i^A}(e) \geq \mathtt{s}] \qquad \text{(Explaining evidence A)}$$

$$\text{For any } e \in \mathtt{E}, \text{ if there is } N_i^A \text{ such that } \mathtt{P}(\mathtt{N}_i^A|e) > \mathtt{P}(\mathtt{N}_i^A), \text{ then } \mathtt{P}^{\mathtt{N}_k^D}(e) \geq \mathtt{r}. \qquad \text{(Explaining evidence D)}$$

$$\mathbf{ME}(\mathtt{N}_i) \Leftrightarrow \text{ for some } \varphi_1, \ldots, \varphi_u \notin \mathtt{E}: [\mathtt{P}^{nN_i}(E(\varphi_1) \vee \cdots \vee E(\varphi_u)) \geq \mathtt{s}].. \qquad \text{(Missing evidence)}$$

$$\mathbf{G}(\mathtt{N}_i) \Leftrightarrow \text{ for some } \varphi_1, \ldots, \varphi_u \notin \mathtt{N}_i \; \mathtt{P}^{f\mathtt{N}_i}(\varphi_1 \vee \cdots \vee \varphi_u) \geq \mathtt{s} \wedge \mathtt{P}^{eN_i}(N_i(\varphi_1) \vee \cdots \vee N_i(\varphi_u)) \geq \mathtt{s}. \qquad \text{(Gap)}$$

$$\mathbf{D}(\mathtt{N}_i^A) \Leftrightarrow \neg\mathbf{ME}(\mathtt{N}_i^A) \wedge \neg\mathbf{G}(\mathtt{N}_i^A) \wedge \mathtt{P}^f(\mathtt{N}_i^A) \geq \mathtt{P}^f(\mathtt{N}_j^A) \text{ for all } j \neq i \wedge \mathtt{P}^f(\mathtt{N}_i^A) \geq \mathtt{s}. \qquad \text{(Domination)}$$

$$\mathbf{RD}(\mathtt{N}_k^D) \Leftrightarrow \neg\mathbf{G}(\mathtt{N}_k^D) \wedge \mathtt{P}^f(\mathtt{N}_k^D) \geq \mathtt{r} \qquad \text{(Reasonable doubt)}$$

We say that a conviction is *beyond reasonable doubt* if it is justified by a resilient dominating narration and no defense narration raises reasonable doubt.

Let's now look back at the gatecrasher. Name the suspects $1, 2, \ldots, 1000$. $\mathtt{g_i}$ means "i is guilty of gatecrashing" and $\mathtt{e}$ is the evidence available in the gatecrasher. Take the accusing narration "1, together with 998 other people crashed the gates" to be $\mathtt{N}_1 = \{\mathtt{g}_1, \mathtt{e}\}$. Given the relevance of all g_i $(i \neq 1)$, (Commitment) entails that for any $i \neq 1$ we have $\mathtt{P}^{e\mathtt{N}_1}(\mathtt{N}_1(\mathtt{g}_i)) \geq \mathtt{s}$. By (Decision) $\mathtt{P}^{f\mathtt{N}_1}(\mathtt{g}_i) \geq \mathtt{a} > \mathtt{s}$. It's also obvious that $\mathtt{g}_i \notin \mathtt{N}_1$. These taken together, however, mean that $\mathtt{N}_1$ fails to satisfy (Gap) and so it fails to justify a conviction beyond reasonable doubt. If $\mathtt{N}_1$ is replaced with $\mathtt{N}_1^+ = \{\mathtt{g}_1, \mathtt{e}\} \cup \{\mathtt{g}_k | k \neq 1\}$, so that (Gap) is satisfied, the resulting narration simply becomes highly implausible, because it contradicts evidence.

References

1. Allen, R.J.: No plausible alternative to a plausible story of guilt as the rule of decision in criminal cases. In: Cruz, L.L.J. (ed.) Proof and Standards of Proof in the Law, Northwestern University School of Law, pp. 10–27 (2010)
2. Allen, R.J., Pardo, M.S.: The problematic value of mathematical models of evidence. J. Legal Stud. **36**(1), 107–140 (2007). doi:10.1086/508269
3. Bernoulli, J.: Ars Conjectandi (1713)
4. Jonathan Cohen, L.: The Probable and the Provable. Oxford University Press, Oxford (1977). doi:10.2307/2219193
5. Jonathan Cohen, L.: Subjective probability and the paradox of the gatecrasher. Ariz. State Law J. 627–634 (1981)
6. Dant, M.: Gambling on the truth: the use of purely statistical evidence as a basis for civil liability. Columbia J. Law Soc. Probl. **22**, 31–70 (1988)
7. Di Bello, M.: Statistics and probability in criminal trials. Ph.D. thesis, University of Stanford (2013)

8. Haack, S.: Evidence Matters: Science, Proof, and Truth in the Law. Cambridge University Press, Cambridge (2014)
9. Haack, S.: Legal probabilism: an epistemological dissent. In: [8], pp. 47–77. Cambridge University Press, Cambridge (2014)
10. Ho, H.L.: A Philosophy of Evidence Law: Justice in the Search for Truth. Oxford University Press, Oxford (2008)
11. Lepage, F.: Partial probability functions and intuitionistic logic. Bull. Sect. Logic **41**(3/4), 173–184 (2012)
12. Lepage, F., Morgan, C.: Probabilistic canonical models for partial logics. Notre Dame J. Formal Logic **44**(3), 125–138 (2003)
13. Nesson, C.R.: Reasonable doubt and permissive inferences: the value of complexity. Harv. Law Rev. **92**(6), 1187–1225 (1979). doi:10.2307/1340444
14. Stein, A.: Foundations of Evidence Law. Oxford University Press, Oxford (2005)
15. Tribe, L.H.: A further critique of mathematical proof. Harv. Law Rev. **84**, 1810–1820 (1971)
16. Tribe, L.H.: Trial by mathematics: precision and ritual in the legal process. Harv. Law Rev. **84**(6), 1329–1393 (1971)
17. Underwood, B.D.: The thumb on the scale of justice: burdens of persuasion in criminal cases. Yale Law J. **86**(7), 1299–1348 (1977). doi:10.2307/795788
18. Wells, G.L.: Naked statistical evidence of liability: Is subjective probability enough? J. Pers. Soc. Psychol. **62**(5), 739–752 (1992). doi:10.1037/0022-3514.62.5.739

Distributed Knowledge Whether
(Extended Abstract)

Jie Fan(✉)

School of Philosophy, Beijing Normal University, Beijing, China
fanjie@bnu.edu.cn

1 Introduction

As is known, by putting their knowledge together, agents can obtain distributed knowledge. However, by pooling their non-ignorance, agents can only obtain *distributed knowledge as to whether* something holds, rather than distributed knowledge (of something). Take a simple example. Suppose that Ann knows whether φ and Bob knows whether ψ. Then by 'sharing' their non-ignorance, what Ann and Bob obtain is not distributed knowledge of $\varphi \wedge \psi$, but distributed knowledge as to whether $\varphi \wedge \psi$.

Despite so many results about the notion of distributed knowledge in the literature (see e.g. [2,6,8,12,15]), the notion of 'distributed knowledge whether' has not yet received its deserved attention. A natural question arises: what does it mean when one says "a group **G** *distributedly knows whether* φ"? To answer this question, let us first look at the following statement:

> $\cdots$ *Although Alice and Bob do not have distributed knowledge as to whether p is true or false before any messages are sent, if p is true then they can obtain distributed knowledge that p is true by communication, and if p is false, then they can obtain distributed knowledge that p is false by communication.* [2, p. 331]

When omitted the modifier 'can', the above sentence contains an idea: "A group has distributed knowledge (as to) whether p (is true or false)" is equivalent to "if p is true then the group has distributed knowledge that p (is true), and if p is false, then the group has distributed knowledge that p is false". Since distributed knowledge is always true (cf. e.g. [1, p. 34]), this is equivalent to "the group has distributed knowledge that p, or the group has distributed knowledge that $\neg p$".

Inspired by the above-mentioned intuitive meaning of 'distributed knowledge whether', we will understand this notion in the following way:

$$\text{distributedly knows whether } \varphi \iff \text{distributedly knows } \varphi \text{ or distributedly knows } \neg\varphi.$$

Due to the intuition of distributed knowledge (see e.g. [2]), the intuition of 'distributed knowledge whether' can be understood as follows: a group *distributedly*

This research is funded by China Postdoctoral Science Foundation (2017T100050). The author thanks two anonymous referees for their insightful comments. Xingchi Su [13] also proposed the 'distributed knowledge whether' operator independently.

A. Baltag et al. (Eds.): LORI 2017, LNCS 10455, pp. 643–647, 2017.
DOI: 10.1007/978-3-662-55665-8_45

knows whether φ, if by putting their non-ignorance together, the members of the group could know whether φ.

Section 2 introduces the language **PLKwDw** of 'distributed knowledge whether' logic and its extension with public announcements, and their semantics, both of which are fragments of a much larger logic. After proposing a suitable bisimulation notion for **PLKwDw** in Sect. 3, we compare the expressive hierarchy for this logic and some variations in Sect. 4, on both $\mathcal{K}$ and $\mathcal{S}5$. Section 5 presents proof systems for **PLKwDw** and its public announcements extension, and shows the soundness and the completeness. Finally we conclude in Sect. 6.

2 Preliminaries

In this paper, we will assume **P** and **I**, respectively, to be a fixed nonempty set of propositional variables and a finite nonempty group of agents. Let $\mathbf{Gr} = \mathcal{P}(\mathbf{I})\backslash\{\emptyset\}$, i.e., the set of all nonempty subgroup of **I**. We here list the logical languages involved in this paper, with the left being their respective notation and the right their respective original sources, where $p \in \mathbf{P}$, $i \in \mathbf{I}$ and $\mathbf{G} \in \mathbf{Gr}$.

Definition 1 (Languages).

EL	$\varphi ::= p \mid \neg\varphi \mid \varphi \wedge \varphi \mid K_i\varphi$	*[9]*
PLKw	$\varphi ::= p \mid \neg\varphi \mid \varphi \wedge \varphi \mid Kw_i\varphi$	*[4,5]*
PAL	$\varphi ::= p \mid \neg\varphi \mid \varphi \wedge \varphi \mid K_i\varphi \mid [\varphi]\varphi$	*[11]*
PLKwA	$\varphi ::= p \mid \neg\varphi \mid \varphi \wedge \varphi \mid Kw_i\varphi \mid [\varphi]\varphi$	*[5]*
ELD	$\varphi ::= p \mid \neg\varphi \mid \varphi \wedge \varphi \mid K_i\varphi \mid D_{\mathbf{G}}\varphi$	*[7,8]*
PLKwDw	$\varphi ::= p \mid \neg\varphi \mid \varphi \wedge \varphi \mid Kw_i\varphi \mid Dw_{\mathbf{G}}\varphi$	*this paper*
PALD	$\varphi ::= p \mid \neg\varphi \mid \varphi \wedge \varphi \mid K_i\varphi \mid [\varphi]\varphi \mid D_{\mathbf{G}}\varphi$	*[14]*
PLKwADw	$\varphi ::= p \mid \neg\varphi \mid \varphi \wedge \varphi \mid Kw_i\varphi \mid [\varphi]\varphi \mid Dw_{\mathbf{G}}\varphi$	*this paper*

Intuitively, $K_i\varphi$ means "agent i knows that φ", $Kw_i\varphi$ means "agent i knows whether φ" (or equivalently, "i is not ignorant about φ" [10]), $[\psi]\varphi$ means "after every announcement of ψ, φ holds", $D_{\mathbf{G}}\varphi$ means "group **G** has distributed knowledge of φ", and $Dw_{\mathbf{G}}\varphi$ means "group **G** distributedly knows whether φ".

Although the knowledge operator K has mostly been interpreted on epistemic models, it can also be investigated in the absence of any condition. In the sequel, we mainly focus on models without any constraints.

Definition 2 (Models and pseudo models). *A* model *is a triple* $\mathcal{M} = \langle S, \{\to_i \mid i \in \mathbf{I}\}, V\rangle$, *where* S *is a nonempty set of states, for every* $i \in \mathbf{I}$, $\to_i \subseteq S \times S$ *is an accessibility relation over* S, *and* $V : \mathbf{P} \to \mathcal{P}(S)$ *is a valuation. In models, for each* $\mathbf{G} \in \mathbf{Gr}$, *we define* $\to_{D_G}$ *as* $\bigcap_{i \in G} \to_i$. *We use* $\mathcal{K}$ *and* $\mathcal{S}5$, *respectively, to mean the class of all models and the class of reflexive, transitive and symmetric models.*

$\mathcal{M} = \langle S, \{\to_i \mid i \in \mathbf{I}\}, \{\to_{D_G} \mid \mathbf{G} \in \mathbf{Gr}\}, V\rangle$ *is said to be a* pseudo model, *if* $\langle S, \{\to_i \mid i \in \mathbf{I}\}, V\rangle$ *is a model, and* $\to_{D_G}$ *is a binary relation on* S *such that for any* $\mathbf{G} \in \mathbf{Gr}$, *any* $i \in \mathbf{I}$, $\to_{D_i} = \to_i$ *and, for any* $\mathbf{G}, \mathbf{G}' \in \mathbf{Gr}$, $\mathbf{G} \subseteq \mathbf{G}'$ *implies* $\to_{D_{G'}} \subseteq \to_{D_G}$.

Let $\mathcal{M} = \langle S, \{\to_i \mid i \in \mathbf{I}\}, \{\to_{D_\mathbf{G}} \mid \mathbf{G} \in \mathbf{Gr}\}, V\rangle$ be a pseudo model and $s \in S$. The semantics for Boolean formulas are as usual, and

$\mathcal{M}, s \vDash K_i\varphi$	$\Longleftrightarrow$	for all $t \in S$ such that $s \to_i t : \mathcal{M}, t \vDash \varphi$
$\mathcal{M}, s \vDash D_\mathbf{G}\varphi$	$\Longleftrightarrow$	for all $t \in S$ such that $s \to_{D_\mathbf{G}} t : \mathcal{M}, t \vDash \varphi$
$\mathcal{M}, s \vDash Kw_i\varphi$	$\Longleftrightarrow$	for all $t_1, t_2 \in S$ such that $s \to_i t_1, s \to_i t_2$: $\mathcal{M}, t_1 \vDash \varphi \Longleftrightarrow \mathcal{M}, t_2 \vDash \varphi$
$\mathcal{M}, s \vDash Dw_\mathbf{G}\varphi$	$\Longleftrightarrow$	for all $t_1, t_2 \in S$ such that $s \to_{D_\mathbf{G}} t_1, s \to_{D_\mathbf{G}} t_2$: $\mathcal{M}, t_1 \vDash \varphi \Longleftrightarrow \mathcal{M}, t_2 \vDash \varphi$
$\mathcal{M}, s \vDash [\psi]\varphi$	$\Longleftrightarrow$	$\mathcal{M}, s \vDash \psi$ implies $\mathcal{M}\|\psi, s \vDash \varphi$

Where $\mathcal{M}|\psi$ is the model of $\mathcal{M}$ restricted to ψ-states.

Satisfaction in a pseudo model is defined as usual. The semantics on a *model* is similar, except that in that case, $\to_{D_\mathbf{G}}$ is a defined notation rather than a primitive one. We say that φ is *valid (with respect to models)*, written $\vDash \varphi$, if $\mathcal{M}, s \vDash \varphi$ for any pointed model $(\mathcal{M}, s)$. Given any language $\mathcal{L}$, we use $(\mathcal{M}, s) \equiv_\mathcal{L} (\mathcal{M}', s')$ to denote that $(\mathcal{M}, s)$ and $(\mathcal{M}', s')$ satisfy the same $\mathcal{L}$-formulas.

3 Bisimulation

This section proposes a suitable bisimulation notion for **PLKwDw**. The 'distributed knowledge whether' operators are not invariant under *Kw*-bisimulation (called 'Δ-bisimulation' in [4, Definition 3.3]), as illustrated by two models $(\mathcal{M}, s)$ and $(\mathcal{M}', s')$, where p-state s ij-accesses to a p-state and a $\neg p$-state, and p-state s' i-access to a p-state and a $\neg p$-state, but j-access to another p-state and another $\neg p$-state.

The following notion of bisimulation is inspired by the similarity between 'distributed knowledge whether' and 'knowledge whether', and *Kw*-bisimulation.

Definition 3 (*DKw*-**bisimulation**). *Let $\mathcal{M} = \langle S, \{\to_i \mid i \in \boldsymbol{I}\}, \{\to_{D_G} \mid \boldsymbol{G} \in \mathbf{Gr}\}, V\rangle$ be a pseudo model. We say that a nonempty binary relation $Z \subseteq S \times S$ is a DKw*-bisimulation *on $\mathcal{M}$, if for all $i \in \boldsymbol{I}$, all $\boldsymbol{G} \in \mathbf{Gr}$, $(s, t) \in Z$ implies*

(INV) $s \in V(p)$ iff $t \in V(p)$ for all $p \in \boldsymbol{P}$,

(Kw-ZIG) for all s', if $s \to_i s'$ and $s \to_i s_1$ and $s \to_i s_2$ for some s_1, s_2 such that $(s_1, s_2) \notin Z$, then there is a t' such that $t \to_i t'$ and $(s', t') \in Z$,

(Kw-ZAG) for all t', if $t \to_i t'$ and $t \to_i t_1$ and $t \to_i t_2$ for some t_1, t_2 such that $(t_1, t_2) \notin Z$, then there is a s' such that $s \to_i s'$ and $(s', t') \in Z$,

(Dw-ZIG) for all s', if $s \to_{D_G} s'$ and $s \to_{D_G} s_1$ and $s \to_{D_G} s_2$ for some s_1, s_2 such that $(s_1, s_2) \notin Z$, then there is a t' such that $t \to_{D_G} t'$ and $(s', t') \in Z$,

(Dw-ZAG) for all t', if $t \to_{D_G} t'$ and $t \to_{D_G} t_1$ and $t \to_{D_G} t_2$ for some t_1, t_2 such that $(t_1, t_2) \notin Z$, then there is a s' such that $s \to_{D_G} s'$ and $(s', t') \in Z$.

Models $(\mathcal{M}, s)$ and $(\mathcal{M}', s')$ are DKw-bisimilar, *notation: $(\mathcal{M}, s) \leftrightarrows_{DKw} (\mathcal{M}', s')$, if there exists a DKw-bisimulation Z on the disjoint union of $\mathcal{M}$ and $\mathcal{M}'$ such that $(s, s') \in Z$.*

Proposition 1. *Given pseudo models $(\mathcal{M}, s)$ and $(\mathcal{M}', s')$, if $(\mathcal{M}, s) \leftrightarrows_{DKw} (\mathcal{M}', s')$, then $(\mathcal{M}, s) \equiv_{\boldsymbol{PLKwDw}} (\mathcal{M}', s')$, and the converse holds when both $\mathcal{M}$ and $\mathcal{M}'$ are image-finite pseudo models.*

4 Expressivity

This section compares the relative expressivity of the languages introduced in Definition 1, on both $\mathcal{K}$ and $\mathcal{S}5$. We adopt the definition of expressivity in [3].

The expressivity hierarchy on $\mathcal{K}$ is as follows, where the arrow from a language A to another B denotes $B > A$, and arrows are transitive, whereas two languages connected with *neither* direct *nor* indirect arrows are incomparable.

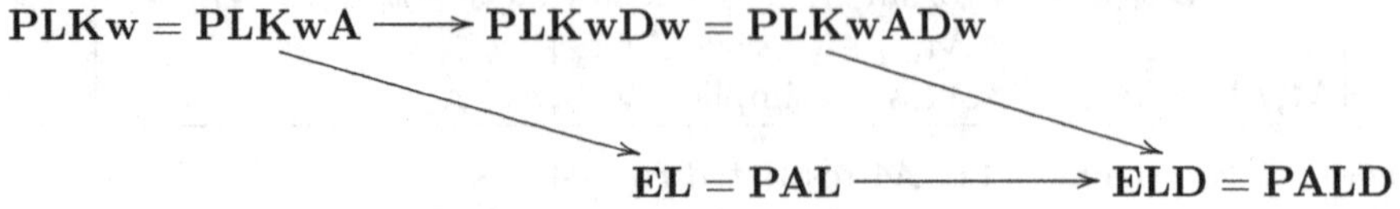

PLKwDw $>$ **PLKw** follows from the fact that Dw operators are not invariant under Kw-bisimulation; **PLKwDw** $=$ **PLKwADw** is immediate by the validity $\vDash [\psi]Dw_{\mathbf{G}}\varphi \leftrightarrow (\psi \to Dw_{\mathbf{G}}[\psi]\varphi \vee Dw_{\mathbf{G}}[\psi]\neg\varphi)$; **PLKwDw** $\asymp$ **EL** is because Dw operators are not invariant under K-bisimulation and K operators are not invariant under DKw-bisimulation. **ELD** $>$ **PLKwDw** follows from the fact that Kw and Dw operators are definable with K and D operators and that **PLKwDw** $\not\succeq$ **EL**; and other expressivity results can be found in [5,11,12,14].

However, on $\mathcal{S}5$, since operators K and D are definable with operators Kw and Dw, respectively, we thus get the following expressive hierarchy:

$$\mathbf{PLKw} = \mathbf{PLKwA} = \mathbf{EL} = \mathbf{PAL} \longrightarrow \mathbf{PLKwDw} = \mathbf{PLKwADw} = \mathbf{ELD} = \mathbf{PALD}.$$

5 Axiomatizations

The proof system $\mathbb{PLKWADW}$ is the extension of $\mathbb{CLA}$ in [5] with the following axioms and rules. By deleting all reduction axioms, we obtain $\mathbb{PLKWDW}$.

A-Dw	$[\psi]Dw_{\mathbf{G}}\varphi \leftrightarrow (\psi \to Dw_{\mathbf{G}}[\psi]\varphi \vee Dw_{\mathbf{G}}[\psi]\neg\varphi)$
Dw$\wedge$	$Dw_{\mathbf{G}}(\chi \to \varphi) \wedge Dw_{\mathbf{G}}(\neg\chi \to \varphi) \to Dw_{\mathbf{G}}\varphi$
Dw $\leftrightarrow$	$Dw_{\mathbf{G}}\varphi \leftrightarrow Dw_{\mathbf{G}}\neg\varphi$
Dw1	$Kw_i\varphi \leftrightarrow Dw_i\varphi$
Dw2	$Dw_{\mathbf{G}}\varphi \to Dw_{\mathbf{G}'}\varphi$, where $\mathbf{G} \subseteq \mathbf{G}'$
REDw	from $\varphi \leftrightarrow \psi$ infer $Dw_{\mathbf{G}}\varphi \leftrightarrow Dw_{\mathbf{G}}\psi$

Theorem 1. *$\mathbb{PLKWDW}$ and $\mathbb{PLKWADW}$ are both sound and strongly complete with respect to both the class of all pseudo models and the class of all models.*

The completeness of $\mathbb{PLKWADW}$ reduces to that of $\mathbb{PLKW}$, which in turn consists of two steps. First, every consistent set of formulas is satisfiable in a pseudo model. For this, we define a canonical pseudo model for the system, where the domain and valuation are as usual, the canonical relation $\to_i^c$ is defined such that $s \to_i^c t$ iff there exists χ such that for all φ, if $Kw_i(\chi \to \varphi) \in s$, then $\varphi \in t$, and $\to_{D_{\mathbf{G}}}^c$ is such that $s \to_{D_{\mathbf{G}}}^c t$ iff there exists χ such that for all φ, if $Dw_{\mathbf{G}}(\chi \to \varphi) \in s$, then $\varphi \in t$. Second, if a set of formulas is satisfiable in a pseudo model, then it is also satisfiable in a model. The proof strategy is to think of each $D_{\mathbf{G}_j}$ as an agent and add it to $\mathbf{G}_j$, which extends an idea in [2, pp. 357–358]. We omit the proof details due to space limitations.

6 Conclusion

In this paper, we defined an intuitive formal semantics for the notion of 'distributed knowledge whether', proposed a suitable bisimulation for the knowing whether logic **PLKwDw** with this notion. We compared the relative expressivity of **PLKwDw** and some variations, on both $\mathcal{K}$ and $\mathcal{S}5$. We presented complete axiomatizations for **PLKwDw** and its extension with public announcements.

For future work, we can study other group notions of 'knowledge whether', for example, common knowledge whether.

References

1. van Ditmarsch, H., van der Hoek, W., Kooi, B.: Dynamic Epistemic Logic: Synthese Library, vol. 337. Springer, Heidelberg (2007)
2. Fagin, R., Halpern, J.Y., Vardi, M.Y.: What can machines know? On the properties of knowledge in distributed systems. J. ACM **39**(2), 328–376 (1992)
3. Fan, J.: Removing your ignorance by announcing group ignorance: A group announcement logic for ignorance. Stud. Log. **9**(4), 4–33 (2016)
4. Fan, J., Wang, Y., van Ditmarsch, H.: Almost necessary. Adv. Modal Log. **10**, 178–196 (2014)
5. Fan, J., Wang, Y., van Ditmarsch, H.: Contingency and knowing whether. Rev. Symbolic Log. **8**(1), 75–107 (2015)
6. Hakli, R., Negri, S.: Proof Theory for Distributed Knowledge. In: Sadri, F., Satoh, K. (eds.) CLIMA 2007. LNCS, vol. 5056, pp. 100–116. Springer, Heidelberg (2008). doi:10.1007/978-3-540-88833-8_6
7. Halpern, J.Y., Moses, Y.: Knowledge and common knowledge in a distributed environment. In: Proceedings of the 3rd ACM Conference on Distributed Computing. pp. 50–61 (1984), a revised and expanded version appears as: IBM Research Report RJ 4421. IBM, Yorktown Heights, N.Y. (1988)
8. Halpern, J.Y., Moses, Y.: Knowledge and common knowledge in a distributed environment. J. ACM **37**(3), 549–587 (1990)
9. Hintikka, J.: Knowledge and Belief. Cornell University Press, Ithaca (1962)
10. van der Hoek, W., Lomuscio, A.: A logic for ignorance. Electron. Notes Theoret. Comput. Sci. **85**(2), 117–133 (2004)
11. Plaza, J.: Logics of public communications. In: Proceedings of the 4th ISMIS. pp. 201–216. Oak Ridge National Laboratory (1989)
12. Roelofsen, F.: Distributed knowledge. J. Appl. Non Class. Log. **17**(2), 255–273 (2007)
13. Su, X.: Distributed knowing whether. In: Proceedings of Sixth International Conference on Logic, Rationality and Interaction (LORI) (2017, to appear)
14. Wáng, Y.N., Ågotnes, T.: Public Announcement Logic with Distributed Knowledge. In: Ditmarsch, H., Lang, J., Ju, S. (eds.) LORI 2011. LNCS, vol. 6953, pp. 328–341. Springer, Heidelberg (2011). doi:10.1007/978-3-642-24130-7_24
15. Wáng, Y.N., Ågotnes, T.: Public announcement logic with distributed knowledge: expressivity, completeness and complexity. Synthese **190**, 135–162 (2013)

A Note on Belief, Question Embedding and Neg-Raising

Michael Cohen(✉)

Department of Philosophy, Stanford University, Stanford 94305, USA
micohen@stanford.edu

Abstract. The epistemic verb to believe does not embed polar questions, unlike the verb to know. After reviewing this phenomenon, I propose an explanation which connects the neg-raising behavior of belief with its embedding patterns (following [14]). I use dynamic epistemic logic to model the presuppositions and the effects associated with belief assertions.

1 Introduction

This note is concerned with the failure of the English verb to believe to embed polar questions, as opposed to to know, a phenomenon exemplified in the following pairs of sentences (infelicitous sentences are marked with *):

(1) (a) * Michael believes whether it is raining in Munich.
(b) Michael knows whether it is raining in Munich.
(2) (a) * Michael does not believe whether it is raining in Munich.
(b) Michael does not know whether it is raining in Munich.

(1) (a) exemplifies the failure of embedding whether complements in to believe; (2) (a) exemplifies the failure of embedding whether complements under negation in to believe. The behavior of to believe differs from predicates such as to be certain and to be sure, which allow for whether complements under negation (but not otherwise), and *responsive* verbs like to know and to see, which embed all interrogatives (wh-complements). While a full theory should be able to explain the embedding patterns of a wide class of predicates, this short note will only be concerned with the behavior of to believe.[1]

The focus on the behavior of to believe and to know has an epistemological motivation. The above phenomenon points to a certain tension between two tendencies within epistemology: one tendency is to tightly connect the notions of knowledge and belief by some logical relations; a different tendency, albeit less common, connects knowledge with questions.[2] Explaining the above phenomenon can resolve the apparent tension between these two tendencies.

[1] For an overview of the problem of the embedding patterns of questions, see e.g. [5].
[2] For a few examples of this tendency, within both epistemology and epistemic logic, see e.g. [3,11].

A. Baltag et al. (Eds.): LORI 2017, LNCS 10455, pp. 648–652, 2017.
DOI: 10.1007/978-3-662-55665-8_46

In [9], Hintikka argues that the difference between verbs like to know and verbs like to believe lies in *veridicality*: the former verb is veridical, the latter is not.[3] However, this hypothesis fails to explain why to be certain and to believe behave differently, given that they are both non-veridical.[4]

A different approach, going back to Zuber [14], connects the embedding behavior of to believe with its neg-raising (NR) behavior. Neg-raising predicates are predicates in which the wide scope of the negation tends to be interpreted with a narrow scope. For an example with to believe consider (3):

(3) Michael does not believe that it is raining $\rightsquigarrow$ Michael believes that it is not raining.

Here, I follow Zuber's approach: I will try to account for the infelicity of 'believe whether' sentences using patterns such as the one in (3). NR can be formally stated as the material implication $\neg Vp \rightarrow V\neg p$ (where V is some epistemic verb). Call the latter formula **Strong NR**. **Strong NR** is materially equivalent to the disjunction $Vp \vee V\neg p$. If **Strong NR** is always true for the verb to believe in natural language, the disjunction is always true for belief. But the disjunction $Vp \vee V\neg p$ should be equivalent to the construction V *whether* p. Thus, if **Strong NR** is valid for some verb V, then V *whether* p is valid. It follows that assertions of the form V *whether* p are necessarily uninformative, and therefore defective.[5] I will expand and refine this basic proposal below.

2 The Present Proposal

One major weakness of the basic proposal above is that the principle **Strong NR** for belief seems to be too strong. **Strong NR** predicts that sentences that express agnosticism using the verb to believe are contradictory, which clearly is not the case. Consider (4):

(4) Noor does not believe that God exists, but she does not believe that God does not exist either.

The surface form of (4) is $\neg B_n p \wedge \neg B_n \neg p$, which is inconsistent with **Strong NR** for belief, stated as $\neg B_n p \rightarrow B_n \neg p$.[6] My account below aims to avoid this problem.

In what follows, I start by presenting the logical framework, then explain the way I model presuppositions and assertions in the logic, and finally explain the linguistic phenomena.

[3] A verb is veridical if it entails the truth of its complement.

[4] See [5] for a recent overview and a development of Hintikka's proposal.

[5] Similar arguments have been proposed. Zuber's [14] original argument uses strong assumptions about factivity and concludes that the construction '* believing whether' is *contradictory*. Egré ([5], footnote 3) acknowledges the possibility of this proposal, but does not develop it. [12] endorses a similar kind of argument and develops it using inquisitive semantics (on grounds independent from mine). [10] proposes a similar solution but develops it differently.

[6] The account in [10] is open to this kind of criticism.

I set my account in dynamic doxastic logic, a version of dynamic epistemic logic (DEL) based on plausibility models (see [2] for the appropriate formal definitions). I present a model in which: (i) we have a neg-raising behavior for beliefs, and (ii) the construction 'believe whether' is in some way defective, by virtue of (i).

The doxastic part of the logic allows me to represent agents' beliefs, taken as listeners of an assertion. The dynamic part allows me to represent the effect of assertions on the information states of the agents. The language $\mathcal{L}$ is given by

$$\varphi := p \mid \varphi \vee \psi \mid \neg\varphi \mid B_a\varphi \mid \triangle_a\varphi \mid C_G\varphi \mid [\varphi!]\psi$$

$B_a\varphi$ reads 'agent a believes that φ'; C_G reads 'φ is a common belief among group G'; $[\varphi!]\psi$ reads 'after the update with φ, ψ is the case'. I assume a reductive approach to interrogative complements (see [13] for a defense), which predicts that the meaning of V *whether* p is equivalent to V *that* p *or* V *that not* p. Thus, in the context of doxastic logic, the non-contingency operator $\triangle_a\varphi$ is taken to be equivalent to $B_a\varphi \vee B_a\neg\varphi$ and represents the ungrammatical construction 'a believes whether φ'.

The language is interpreted over epistemic plausibility models (see [2]). An epistemic plausibility model is a Kripke model $M = (W, \geq_a, \geq_b, ...V)$, where W is a non-empty finite set of possible worlds, V is a valuation function, and $\geq_a$ is the *epistemic plausibility* order of agent a, a reflexive transitive relation on W. As usual, a believes φ in w iff the set of epistemically *best worlds* (worlds ranked highest in the ordering) is a subset of φ. The C_G operator represents the transitive closure of the derived belief relations of a group of agents.

The semantics of the $[\varphi!]\psi$ operator is understood in terms of elimination of possible worlds (i.e. as a *hard* update). We follow the standard treatment in DEL and assume that updates are truthful. This means that for a model and a world (M, w), updating with φ results in (M_φ, w), a model in which all the $\neg\varphi$ worlds are eliminated, *if* φ is true in (M, w). If φ is false in (M, w), we say that the update *aborts*, in which case it does not change the model (see [6] for the latter terminology).

With this logical framework, we can model the presuppositions of assertions. I take NR to be an effect on the doxastic states of the listeners of an assertion. This can be stated informally as:

(5) In standard contexts, after the assertion that $\neg B_c p$, the result is $B_a B_c \neg p$.

This formulation of NR is much weaker than **Strong NR**: it takes NR to be an effect of an assertion, and not a semantic validity. Similarly to [8], I take NR predicates to have an excluded middle presupposition. In the case of belief, this amounts to an *opinionatedness presupposition* $B_a\varphi \vee B_a\neg\varphi$. However, in line with (5), I treat this presupposition as a *soft presupposition*: a presupposition that arises in default contexts but is easily cancelable.[7]

To do so, let us first define a ***pre*** function on the logical language that takes a formula and returns its presupposition, s.t. $\textit{pre}(\neg\varphi) = \textit{pre}(\varphi)$ and $\textit{pre}(B_a\varphi) =$

[7] See [1] for more about the notion of soft presuppositions.

$pre(\triangle_a\varphi) = B_a\varphi \vee B_a\neg\varphi$. I require that $\mathtt{pre}(\varphi) = \mathtt{pre}(\neg\varphi)$ to capture the idea that presuppositions survive under negation. Moreover, I specify that $pre(\triangle\varphi) = pre(B_a\varphi) = B_a\varphi \vee B_a\neg\varphi$ to capture the idea that NR verbs (like to believe) have an opinionatedness presupposition. Combining the two restrictions, we get that $pre(\neg B_a\varphi) = B_a\varphi \vee B_a\neg\varphi$.

A crucial point of my proposal pertains to the way I represent assertions and their presuppositions in DEL. Here, I follow the ideas in [6] with some crucial modifications. The idea is to represent the assertion of φ in the logical language in the following way:

(6) Asserting $\varphi \approx [C_G(pre(\varphi))!][\varphi!]$

In other words, asserting φ formally amounts to a sequence of two updates: the first is with the information that the presupposition of φ is common belief within the group, the second is with φ itself. This has the following (desirable) effect: if the presupposition of φ is not accepted as common belief before the assertion, then the first update aborts (as it is false) and does not change the model. This models presupposition failure.

We are now in a position to answer our main question. I start by explaining NR in the current system. In line with (5), we formalize NR as:

(7) $[C_G(pre(\neg B_c\varphi))!][\neg B_c\varphi!]B_a(B_c\neg\varphi)$

(7) represents the assertion that $\neg B_c\varphi$, divided into an update with the presupposition and an update with the proposition asserted, and the effect that it has: $B_a(B_c\neg\varphi)$. Now let us suppose that the assertion $\neg B_c\varphi$ is accepted in a context, i.e. the two updates do not abort. Then agent a's conclusion about agent c's beliefs is a result of a simple disjunctive syllogism performed by a: the move from $B_a(B_c\varphi \vee B_c\neg\varphi)$ (which is the case since the first update did not abort) and $B_a\neg B_c\varphi$ (which holds due to the second update) to $B_aB_c\neg\varphi$. We thus have a dynamic representation of the NR behavior of belief.

What is the effect of asserting $\triangle_a\varphi$? Given (6), asserting $\triangle_a\varphi$ will amount to the following sequence of updates: $[C_G(\triangle_a\varphi)!][\triangle_a\varphi!]$. It is easy to see that such a sequence of updates cannot be informative: if the first update does not abort then everybody believes that $\triangle_a\varphi$. In that case, the second update with $\triangle_a\varphi$ makes no difference on the agents' beliefs. The framework therefore predicts that assertions of $\triangle_a\varphi$ are necessarily uninformative. This gives us an answer to our main question: the construction 'believe whether' is ungrammatical since it is necessarily uninformative.[8] A similar explanation predicts that an assertion of $\neg\triangle_a\varphi$ will always contradict background context.

The present proposal also accounts for the felicity of sentences like (4), which express agnosticism. Unlike the original proposal, the assertion of (4) does not result in contradiction, rather in *belief revision.*

---

[8] The exact connection between uninfomativness and ungrammaticality should be spelled out. [7] offers a possible approach.

I model (4) as a sequence of two assertions, $\neg B_n p$ and then $\neg B_n \neg p$. This gives a us a sequence of four updates:

(8) $[C_G(B_n p \vee B_n \neg p)!][\neg B_n p!][C_G(B_n p \vee B_n \neg p)!][\neg B_n \neg p!]B_a(\neg B_n p \wedge \neg B_n \neg p)$

Assuming for simplicity that agent a is the only hearer of (4), consider an initial context in which a believes that n is opinionated: $B_a(B_n p \vee B_n \neg p)$. Thus, the first update of (8) does not abort. The second update eliminates worlds in which $B_n p$ is the case. After that, $B_a(B_n \neg p)$ is the case. Hence, the third update does not abort (a still believes n is opinionated). The fourth update eliminates all worlds in which $B_n \neg p$ is the case. This forces a to revise her old belief that $B_n \neg p$, resulting in a model in which $B_a(\neg B_n p \wedge \neg B_n \neg p)$ holds. This shows that sentences like (4) can be informative under the current proposal.[9]

References

1. Abusch, D.: Presupposition triggering from alternatives. J. Semant. **27**(1), 37–80 (2009)
2. Baltag, A., Renne, B.: Dynamic epistemic logic. In: Zalta, E.N. (ed.) The Stanford Encyclopedia of Philosophy. Springer, Netherlands (2016)
3. van Benthem, J., Minică, Ş.: Toward a dynamic logic of questions. J. Philos. Log. **41**(4), 633–669 (2012)
4. Cohen, M.: Belief: neg-raising and question embedding. Manuscript (2017)
5. Egré, P.: Question-embedding and factivity. Grazer Philosophische Stud. **77**, 85–125 (2008)
6. van-Eijck, J., Unger, C.: The epistemics of presupposition projection. In: Aloni, M., Dekker, P., Roelofsen, F. (eds.) Proceedings of the 16th Amsterdam Colloquium, pp. 235–240, Palteam ILLC. University of Amsterdam (2007)
7. Gajewski, J.: L-analyticity and natural language. Manuscript (2002)
8. Gajewski, J.: Neg-raising and polarity. Linguist. Philos. **30**(3), 289–328 (2007)
9. Hintikka, J.: Different Constructions in terms of the basic epistemological verbs. In: The Intentions of Intensionality, pp. 1–25. Kluwer, Dordrecht (1975)
10. Mayr, C.: Predicting Polar question embedding. In: Truswell, R. (ed.) Proceedings of Sinn und Bedeutung, vol. 21, pp. 1–18. University of Edinburgh, Edinburgh (2017)
11. Schaffer, J.: Knowing the Answer. Philos. Phenomenol. Res. **75**(2), 383–403 (2007)
12. Theiler, N., Roelofsen, F., Aloni, M.: A uniform semantics for declarative and interrogative complements. Manuscript, 29 November, 2016 version (2016)
13. Spector, B., Egré, P.: A uniform semantics for embedded interrogatives: An answer, not necessarily the answer. Synthese **192**(6), 1729–1784 (2015)
14. Zuber, R.: Semantic restrictions on certain complementizers. In: Proceedings of the 12th International Congress of Linguists, Tokyo, pp. 434–436 (1982)

[9] A longer version of this submission, [4], develops the above proposal more formally and considers some special cases in which to believe embeds wh-questions.

Distributed Knowing Whether
(Extended Abstract)

Xingchi Su(✉)

Department of Philosophy, Peking University, Beijing, China
1501211414@pku.edu.cn

Abstract. Standard epistemic logic studies reasoning patterns about 'knowing that', where interesting group notions of 'knowing that' arise naturally, such as distributed knowledge and common knowledge. In recent research, other notions of knowledge are also studied, such as 'knowing whether', 'knowing how', and so on. It is natural to ask what are the group notions of these non-standard knowledge expressions. This paper makes an initial attempt in this line, by looking at the notion corresponding to distributed knowledge in the setting of 'knowing whether'. We introduce the distributed know-whether operator, and give complete axiomatizations of the resulting logics over arbitrary or $\mathcal{S}5$ frames, based on the corresponding axiomatizations of 'knowing whether'.

Keywords: Epistemic logic · Distributed knowledge · Knowing whether · Completeness

1 Introduction

The concept of *distributed knowledge* was proposed in [1] by computer scientists to ascribe knowledge to machines in distributed systems. It is one of the most important notions of group knowledge discussed in [2] besides *general knowledge* (everyone knows) and *common knowledge* (everyone knows that everyone knows that everyone knows...). Intuitively, a group has the distributed knowledge of ϕ means that ϕ can be known to an 'wise man' if he has all the knowledge of the group members. Compared to other kinds of group knowledge, the distributed knowledge of a group may not be known by any individual in the group.

Since the work of [1,2], distributed knowledge has attracted a lot of attention in epistemic logic and computer science. A complete axiom system of epistemic logic with distributed knowledge is given in [3]. Adding distributed knowledge to epistemic language also brings new technical questions, e.g., the language is no longer invariant under bisimulation as discussed in [4], which also leads to some alternative semantics for distributed knowledge operator [5].

Besides group knowledge, some new notions of knowledge are also drawing attention in recent years (cf. [6] for a survey). For example, 'agent i *knows whether* ϕ' [7], 'agent i *knows how* to achieve ϕ' [8], and 'agent i *knows why* ϕ is true' [9]. Clearly, 'knowing whether' can be expressed by 'knowing that', but

A. Baltag et al. (Eds.): LORI 2017, LNCS 10455, pp. 653–657, 2017.
DOI: 10.1007/978-3-662-55665-8_47

the logic with 'knowing whether' operator only is less expressive than standard epistemic logic [10]. A natural question is to ask what are the intuitive group knowledge notions of such non-standard notions of knowledge. In this line of work, we make the first attempt by considering the counterpart of distributed knowledge in the setting of 'knowing whether'.[1]

Informally, a group G *distributedly knows whether* ϕ if we know whether ϕ given all information from all members of G. For example, if A knows whether p and B knows whether q, and A and B share their information with C, then C should know whether $p \vee q$ is true. In this case we can say group $\{A, B\}$ distributedly knows whether $p \vee q$.

In the rest of this introduction we review the standard distributed knowledge logic and the logic of 'knowing whether'.

Given a set of agents A, and a set of propositional variable P, the language ELD of epistemic logic with distributed knowledge is given by:

$$\phi := p \mid \phi \wedge \phi \mid \neg\phi \mid K_i\phi \mid D\phi$$

where $p \in P$, $i \in A$. The model for ELD is the classical multi-agent Kripke model: $M =< W, \{ \rightarrow_i \mid i \in A\}, V >$.

The semantics of ELD is given as usual in modal logic with $K_i\phi$ and $D\phi$ defined as follows where the interaction relation $\bigcap_{i\in A} \rightarrow_i$ captures the intuition that the agents put their information together:
$M, w \models K_i\phi \iff M, v \models \phi$ for all v such that $w \rightarrow_i v$.
$M, w \models D\phi \iff M, v \models \phi$ for all v such that $(w, v) \in \bigcap_{i\in A} \rightarrow_i$.

On the other hand, knowing whether ϕ ($K_i^w\phi$) means knowing that ϕ is true or knowing that ϕ is false. Formally:
$M, s \models K_i^w\phi \iff$ for all t_1, t_2 with $s \rightarrow_i t_1, s \rightarrow_i t_2 : (M, t_1 \models \phi \Leftrightarrow M, t_2 \models \phi)$.

A complete axiom system called NCL^2 for the modal language containing K_i^w as the only primitive modality is given in [10]:

TAUT and all instances of tautologies	$K_i^w\phi \leftrightarrow K_i^w\neg\phi$
$(K^wCon) : K_i^w(\chi \rightarrow \phi) \wedge K_i^w(\neg\chi \rightarrow \phi) \rightarrow K_i^w\phi$	$(MP) : \phi, \phi \rightarrow \psi\ /\psi$
$(K^wDis) : K_i^w\phi \rightarrow K_i^w(\phi \rightarrow \psi) \vee K_i^w(\neg\phi \rightarrow \chi)$	$(NEC) :\vdash \phi \Rightarrow\vdash K_i^w\phi$
$(REK^w) :\vdash \phi \leftrightarrow \psi \Rightarrow\vdash K_i^w\phi \leftrightarrow K_i^w\psi$	

Note that NCL is not a *normal* modal logic, where the K axiom for K_i^w is not admissible.

2 Syntax and Semantics of *DKW*

In this section, we give the language and semantics of distributed knowing whether logic DKW with both knowing whether operator and distributed knowing whether operator.

[1] Jie Fan also proposed the *distributed-knowing-whether* operator independently, with a citation to [11] in the same proceedings.

[2] The name NCL comes from non-contingency logic, cf. [10].

Definition 1. *Let P be a set of propositional variables and A be a set of agents. The language L^{D^w} for DKW is defined as:*

$$\phi := \top \mid p \mid \neg\phi \mid \phi \wedge \phi \mid K_i^w \phi \mid D^w \phi$$

where $p \in P$ and $i \in A$. We use the operator D^w to refer to the distributed knowing whether.

Definition 2. *Given a model $M = \langle S, \{\rightarrow_i \mid i \in A\}, V\rangle$, the semantics of DKW is defined as follows (semantics of K_i^w is defined as before): $M, s \models D^w\phi \iff M, t \models \phi$ for all $(s,t) \in \bigcap_{i\in A} \rightarrow_i$ or $M, t \models \neg\phi$ for all $(s,t) \in \bigcap_{i \in A} \rightarrow_i$.*

Note that the semantics of D_i^w is in the line of the semantics of the standard distributed knowledge operator D_i.

We propose the axiom system of DKW based on all of axioms and rules in NCL and the following extra axioms:

$$(D^wCon): D^w(\chi \rightarrow \phi) \wedge D^w(\neg\chi \rightarrow \phi) \rightarrow D^w\phi \qquad (K^wD)\ K_i^w\phi \rightarrow D^w\phi$$
$$(D^wDis): D^w\phi \rightarrow D^w(\phi \rightarrow \psi) \vee D^w(\neg\phi \rightarrow \chi) \qquad D^w\phi \leftrightarrow D^w\neg\phi$$
$$RED^w: \vdash \phi \leftrightarrow \psi \Rightarrow \vdash D^w\phi \leftrightarrow D^w\psi$$

Similar to the axiomatization of epistemic logic with distributed knowledge [3], the distributed knowing whether operator has the axioms for the knowing whether operator in NCL, and (K^wD) captures the interaction of K_i^w and D^w. Intuitively we can think D^w captures the knowledge-whether of the wise man who have all the information of the group members.

3 Completeness to the Class of $\mathcal{K}$-frames

In order to prove the completeness of DKW, we take a method proposed by Fagin, Halpern and Vardi in [3]. The basic proof idea consists of 3 steps: 1. we show that DKW is *pseudo-satisfied* in *pseudo-model*; 2. we unravel the pseudo-model to make it tree-like and prove that all formulas are invariant on the corresponding nodes of two models; 3. finally we transform the pseudo-model into a real model by replacing every relation $\rightarrow_D$ with $\rightarrow_i$ for each $i \in A$ and also prove that all the formulas are invariant on every node.

The main idea of the pseudo-model is just regarding the operator D^w as a knowing-whether operator. And the construction of the pseudo-model M^* is same as the construction of the canonical model given in [10].

Lemma 1. *Truth Lemma*: $M^*, s \models^* \phi$ iff $\phi \in S$ for all $\phi \in L^{D^w}$ where $\models^*$ represents pseudo-satisfaction.*

The approach to unravel M^* has been given in [3]. The tree-like model is $M_T^* = \langle T, \{\rightarrow_i^T \mid i \in A\}, \rightarrow_D^T, V^T\rangle$. We can show:

Proposition 1. *$M_T^*, s \models^* \psi$ iff $M^*, g(s) \models^* \psi$ for all $\psi \in L^{D^w}$.*

However, M_T^* is still a pseudo-model and the relations $\rightarrow_D$ actually should not exist in a real model. The key process to construct the real model is changing the relations in the tree-like model M_T^*. The state space and the valuation inherited from M_T^* directly.

Definition 3. *Let* $M = \langle T, \{\rightarrow_i \mid i \in A\}, V^T \rangle$. *We set* $w \rightarrow_i^+ v$ *if* $w \rightarrow_D v$ *in* M_T^* *for each* $i \in A$. *Let* $\rightarrow_i = \rightarrow_i^T \cup \rightarrow_i^+$ *for any* $i \in A$.

Because M and M_T^* have the same state space T, we should show that the set of formulas Φ satisfied in M is exactly the set of formulas pseudo-satisfied in the corresponding node in M_T^*.

Proposition 2. *For any* $\psi \in L^{D^w}$, *there is* $M, s \models \psi$ *iff* $M_T^*, s \models^* \psi$.

Theorem 1. *The logic DKW is strongly complete with respect to the class* $\mathcal{K}$ *of all frames.*

4 Axiomatization for DKWS5 and Completeness

The axiom system of DKWS5 is formed by the system of DKW and the following axioms:

$$(K^wT)\ K_i^w\phi \wedge K_i^w(\phi \rightarrow \psi) \wedge \phi \rightarrow K_i^w\psi \qquad (wK^w5)\ \neg K_i^w\psi \rightarrow K_i^w\neg K_i^w\phi$$
$$(D^wT)\ D^w\phi \wedge D^w(\phi \rightarrow \psi) \wedge \phi \rightarrow D^w\psi \qquad (wD^w5)\ \neg D^w\psi \rightarrow D^w\neg D^w\phi$$

To prove the completeness to S5-frames, the basic idea is similar. Construct the new S5-pseudo-model M° by reflexive closure of $\rightarrow_i^*$ in M^*. We can show that the $\{\rightarrow_i^\circ \mid i \in A\}$ is Euclidean. The truth lemma can be proved in the same way. Then, the way to change M° into a tree-like model M_T° is almost same to that mentioned above. But after unraveling, we have to do reflexive, symmetric and transitive closure step by step. We can prove:

$M_T^\circ, s \models^* \psi$ iff $M^\circ, g(s) \models^* \psi$ for all $\psi \in L^{D^w}$

Finally, construct the real model as following definition:

Definition 4. *Let* $M_E = \langle T, \{\rightarrow_i^E \mid i \in A\}, V^T \rangle$. *We set* $w \rightarrow_i^{E+} v$ *if* $w \rightarrow_D^{\circ T} v$ *in* M_T° *for each* $i \in A$. *Let* $\rightarrow_i^E$ *be the transitive closure of* $\rightarrow_i^{\circ T} \cup \rightarrow_i^{E+}$ *for every* $i \in A$.

It is trivial that $\rightarrow_i^E$ is reflexive. Then we can prove that it preserved symmetry after doing transitive closure. Finally, we can show:

For any $\psi \in L^{D^w}$, there is $M^E, s \models \psi$ iff $M_T^\circ, s \models^* \psi$.

Theorem 2. *The logic DKWS5 is strongly complete with respect to the class of all S5 frames.*

5 Conclusion

Reconstructing the distributed knowledge of a group with the non-classical operator *knowing whether* is the main contribution of the paper. In DKW, we give the axioms (D^wCon) and (D^wDis) which are similar to (K^wCon) and (K^wDis) in form. Recalling the original idea of distributed knowledge, we can regard it as the knowledge owned by an outsider who has exactly all the information from every members in the group. In this sense, the outsider has the same ability to infer as any other agent.

When proving the completeness, we take the method of *pseudo-model*. Regard operator D^w as another K_i^w to construct a pseudo-model. Then transform it into a tree-like model with better properties. Following that, construct the real model based on the tree-like model by replacing these $\rightarrow_D$ with $\rightarrow_i$ for each $i \in A$.

As mentioned above, this research focus on one kind of group knowledge (distributed knowledge) and one type of knowledge expression (knowing whether). Further questions for consideration include:

- Some other kinds of group knowledge can also be combined with 'knowing whether'. For example, 'commonly knowing whether'.
- Apart from 'knowing whether', there are also other expressions of knowledge including 'knowing what', 'knowing why', 'knowing how', etc. Combining these expressions of knowledge with group knowledge may be of continued interest.

References

1. Halpern, J.Y.: Using reasoning about knowledge to analyze distributed systems. Annu. Rev. Comput. Sci. **2**, 37–68 (1987)
2. Halpern, J.Y., Moses, Y.: Knowledge and common knowledge in a distributed environment. Annu. Rev. Comput. Sci. **37**(3), 549–587 (1990)
3. Fagin, R., Halpern, J.Y., Vardi, M.Y.: What can machines know?: on the properties of knowledge in distributed systems. J. Assoc. Comput. Mach. **39**(2), 328–376 (1992)
4. Roelofsen, F.: Distributed knowledge. J. Appl. Non Class. Logics **17**(2), 255–273 (2007)
5. Gerbrandy, J.D.: Bisimulations on planet kripke (2006)
6. Wang, Y.: Beyond knowing that: a new generation of epistemic logics. In: Jaakko Hintikka on knowledge and game theoretical semantics (2016). arXiv.org/abs/1605.01995. (forthcoming)
7. Fan, J., Wang, Y., van Ditmarsch, H.: Almost neccessary. Proceedings of AiML **10**, 178–196 (2014)
8. Wang, Y.: A logic of goal-directed knowing how, Synthese (2017). forthcoming
9. Xu, C., Wang, Y., Studer, T.: A logic of knowing why (2016). arXiv:1609.06405
10. Fan, J., Wang, Y., Ditmarsch, H.V.: Contingency and knowing whether. Rev. Symbol. Logic **8**(1), 1–33 (2015)
11. Fan, J.: Distributed knowledge whether to appear. In: proceedings of Sixth International Conference on Logics, Rationality and Interaction (LORI) (2017)

A Causal Theory of Speech Acts

Chiaki Sakama(✉)

Department of Computer and Communication Sciences,
Wakayama University, 930 Sakaedani, Wakayama 640-8510, Japan
sakama@sys.wakayama-u.ac.jp

Abstract. In speech acts, a speaker utters sentences that might affect the belief state of a hearer. To formulate causal effects in assertive speech acts, we introduce a logical theory that encodes causal relations between speech acts, belief states of agents, and truth values of sentences. We distinguish trustful and untrustful speech acts depending on the truth value of an utterance, and distinguish truthful and untruthful speech acts depending on the belief state of a speaker. Different types of speech acts cause different effects on the belief state of a hearer, which are represented by the set of models of a causal theory. Causal theories of speech acts are also translated into logic programs, which enables one to represent and reason about speech acts in answer set programming.

1 Introduction

An *assertive speech act* commits a speaker to the truth of the expressed proposition [5]. Although assertive sentences are either true or false, a speaker generally does not have complete knowledge of the world. It may happen that a speaker utters a sentence that is believed to be true by herself while it is actually false. In this case, a speaker acts truthfully but a hearer would consider the speaker untrustful. On the other hand, there is a case that a speaker utters a sentence disbelieved by himself while the statement happens to be true. In this case, a speaker acts untruthfully but a hearer would consider the speaker trustful. Whether a speech act is truthful or not depends on the belief state of a speaker, while whether a speech act is trustful or not is judged by the truth of information conveyed by the utterance. At this point, four different combinations of speech acts are considered—(i) both truthful and trustful, (ii) truthful but untrustful, (iii) untruthful but trustful, and (iv) both untruthful and untrustful.

In this paper, we distinguish truthful/untruthful and trustful/untrustful speech acts and consider their performative effects on hearers. Different from previous studies based on modal logic [2,3] or dynamic epistemic logic [6], we formulate assertive speech acts using a *causal logic* introduced in [4]. The logic can simply encode causal relations in speech acts and a causal theory is implemented by logic programming.

2 Causal Theory

We first review a causal logic of [4]. Let $\mathcal{L}$ be a language of propositional logic and $\mathcal{A}$ a finite set of *atoms* in the language. *Formulas* (or *sentences*) in $\mathcal{L}$ are

A. Baltag et al. (Eds.): LORI 2017, LNCS 10455, pp. 658–663, 2017.
DOI: 10.1007/978-3-662-55665-8_48

defined as follows: (i) If $p \in \mathcal{A}$, then p is a formula. (ii) If φ and ψ are formulas, then $\neg\varphi, \varphi \wedge \psi, \varphi \vee \psi, \varphi \supset \psi$, and $\varphi \equiv \psi$ are all formulas. In particular, $\top$ and $\bot$ represent valid and contradictory formulas, respectively. We often use parentheses "()" in a formula as usual. A finite set T of formulas is identified with the conjunction of all formulas in T. The set of all formulas in $\mathcal{L}$ is represented by $\mathcal{F}$. A *literal* is an atom A or its negation $\neg A$. An *interpretation* I is a complete and consistent (finite) set of literals, i.e., $L \in I$ iff $\neg L \notin I$ for any literal L appearing in a theory. A literal L is *true* in an interpretation I iff $L \in I$. The truth value of a formula φ in I is defined based on the usual truth tables of propositional connectives. An interpretation I *satisfies* a formula φ (written $I \models \varphi$) iff φ is true in I. Given formulas φ and ψ, $\varphi \Rightarrow \psi$ is called a *causal rule* meaning that "ψ is caused if φ is true." In particular, the rule $(\top \Rightarrow \psi)$ is a *fact* representing that ψ is true, and $(\varphi \Rightarrow \bot)$ is a *constraint* representing that φ cannot be true.

A *causal theory* is a finite set of causal rules. Given a causal theory T and an interpretation I, define $T^I = \{\psi \mid (\varphi \Rightarrow \psi) \in T \text{ for some } \varphi \text{ and } I \models \varphi\}$. Then I is a *model* of T if and only if $I = \{L \mid T^I \models L\}$ where $T^I \models L$ means that T^I entails L in classical logic. We say that I *satisfies* every rule in T if I is a model of T. By definition, I is a model of T iff I is the unique model of T^I. A causal theory T is *consistent* if it has a model; otherwise, T is *inconsistent*. Actions and their effects are represented by a causal theory as follows. First, atoms of the language are expressions of the forms: a_t and f_t where a, f and t are action, fluent, and time names, respectively. a_t means that an action a occurs at time t, and f_t means that a fluent f holds at time t. In this paper we consider actions as assertive speech acts by an agent. An utterance of a sentence σ by an agent x at time t is represented by the atom $Utter_t(x, \sigma)$.[1] The truth value of a sentence is represented by a fluent. When a sentence σ is true (resp. false) at time t, we write $Hold_t(\sigma)$ (resp. $\neg Hold_t(\sigma)$). A belief state of an agent is also represented by a fluent. When an agent x believes (resp. disbelieves) a sentence σ at time t, it is represented by the literal $Bel_t(x, \sigma)$ (resp. $\neg Bel_t(x, \sigma)$). We define $Hold_t(\top) \equiv Bel_t(x, \top) \equiv \top$, $Hold_t(\bot) \equiv Bel_t(x, \bot) \equiv \bot$ and $Hold_t(\neg\sigma) \equiv \neg Hold_t(\sigma)$ for any x, σ and t.

A causal theory must specify conditions that are sufficient for every fact to be caused. To this end, a causal theory of action contains *action rules*: $(a_t \Rightarrow a_t)$ and $(\neg a_t \Rightarrow \neg a_t)$ which represent that the occurrence (resp. non-occurrence) of an action a at a time t is caused whenever a occurs (resp. does not occur) at t. Likewise, to explain facts at the moment t when a fluent f comes into existence, a causal theory of action contains *fluent rules*: $(f_t \Rightarrow f_t)$ and $(\neg f_t \Rightarrow \neg f_t)$ which represent that the state of a fluent f at a time t is determined outside the theory. Finally, fluents that do not change by an action are represented by *inertia rules*: $(f_t \wedge f_{t+1} \Rightarrow f_{t+1})$ and $(\neg f_t \wedge \neg f_{t+1} \Rightarrow \neg f_{t+1})$ which represent that if the truth value of f at time t is identical with the value at time $t+1$ then the truth value at time $t+1$ is caused by virtue of its persistence. Note that $\Rightarrow$ is not identical to material implication in classical logic. In fact, the above

[1] $Utter_t(x, \sigma)$ is represented by the proposition $utter\_x\_\sigma_t$ where $utter\_x\_\sigma$ is an action name. We write $utter\_x\_\sigma_t$ by $Utter_t(x, \sigma)$ for notational convenience.

rules become tautologies if $\Rightarrow$ is replaced by $\supset$. Those rules are *not* tautologies in causal logic. For notational convenience, we use $L^{\pm}$ meaning L or $\neg L$. For instance, $(a_t^{\pm} \Rightarrow a_t^{\pm})$ means $(a_t \Rightarrow a_t)$ and $(\neg a_t \Rightarrow \neg a_t)$, and $(f_t^{\pm} \wedge f_{t+1}^{\pm} \Rightarrow f_{t+1}^{\pm})$ means $(f_t \wedge f_{t+1} \Rightarrow f_{t+1})$ and $(\neg f_t \wedge \neg f_{t+1} \Rightarrow \neg f_{t+1})$.

3 Causal Theories of Speech Acts

Definition 3.1 (causal theory of speech acts). Let x be an agent, $\sigma \in \mathcal{F}$, and t a parameter representing time. A *causal theory of speech acts* $\mathcal{CT}^t_{x\sigma}$ consists of rules:

$$Utter_t^{\pm}(x,\sigma) \Rightarrow Utter_t^{\pm}(x,\sigma), \tag{1}$$
$$Hold_t^{\pm}(\sigma) \Rightarrow Hold_t^{\pm}(\sigma), \tag{2}$$
$$Hold_t^{\pm}(\sigma) \wedge Hold_{t+1}^{\pm}(\sigma) \Rightarrow Hold_{t+1}^{\pm}(\sigma), \tag{3}$$
$$Bel_t^{\pm}(x,\sigma) \Rightarrow Bel_t^{\pm}(x,\sigma), \tag{4}$$
$$Bel_t^{\pm}(x,\sigma) \wedge Bel_{t+1}^{\pm}(x,\sigma) \Rightarrow Bel_{t+1}^{\pm}(x,\sigma). \tag{5}$$

The rules (1) are action rules, the rules (2) and (4) are fluent rules, and the rules (3) and (5) are inertia rules. The rules (1) represent that an agent x utters (or does not utter) a sentence σ at time t. (2) represent that a sentence σ is true (or false) at time t. (4) represent that an agent x believes (or disbelieves) a sentence σ at time t.

A speech act by an agent is *trustful* (resp. *untrustful*) if the agent utters a true (resp. *false*) sentence. They are represented by causal theories as follows.

Definition 3.2 (trustful/untrustful speech acts). A *trustful* (or *untrustful*) *speech act* of a sentence σ by an agent a at time t is defined as follows.

$$\mathbf{Trustful}(a,\sigma,t) := \mathcal{CT}^t_{a\sigma} \cup \{Utter_t(a,\sigma) \wedge \neg Hold_t(\sigma) \Rightarrow \bot\}.$$
$$\mathbf{Untrustful}(a,\sigma,t) := \mathcal{CT}^t_{a\sigma} \cup \{Utter_t(a,\sigma) \wedge Hold_t(\sigma) \Rightarrow \bot\}.$$

By contrast, a speech act by an agent is *truthful* (resp. *untruthful*) if the agent utters a sentence believed (resp. *disbelieved*) to be true.

Definition 3.3 (truthful/untruthful speech acts). A *truthful* (or *untruthful*) *speech act* of a sentence σ by an agent a at time t is defined as follows.

$$\mathbf{Truthful}(a,\sigma,t) := \mathcal{CT}^t_{a\sigma} \cup \{Utter_t(a,\sigma) \wedge \neg Bel_t(a,\sigma) \Rightarrow \bot\}.$$
$$\mathbf{Untruthful}(a,\sigma,t) := \mathcal{CT}^t_{a\sigma} \cup \{Utter_t(a,\sigma) \wedge Bel_t(a,\sigma) \Rightarrow \bot\}.$$

Whether a speech act is trustful or not is determined by the truth of an utterance, while whether a speech act is truthful or not is determined by the belief state of a speaker. Any logical combination of trustfulness and truthfulness is consistent, for instance, $\mathbf{Trustful}(a,\sigma,t) \wedge \mathbf{Untruthful}(a,\sigma,t)$

has the model:[2] $\{U_t(a,\sigma),\ \neg B_t(a,\sigma), \neg B_{t+1}(a,\sigma),\ H_t(\sigma),\ H_{t+1}(\sigma)\}$ which represents that a speaker a utters a disbelieved sentence σ that happens to be true. **Untrustful**$(a,\sigma,t) \wedge$ **Truthful**(a,σ,t) has the model $\{U_t(a,\sigma),\ B_t(a,\sigma),\ B_{t+1}(a,\sigma),\ \neg H_t(\sigma),\ \neg H_{t+1}(\sigma)\}$ which represents that a speaker a utters a believed-true sentence σ that is in fact false.

Next we consider the effect of a speech act on a hearer. Suppose that a speaker a utters a sentence σ at time t, which brings about a hearer b's believing σ at time $t+1$. It is represented by the causal rule:

$$Utter_t(a,\sigma) \Rightarrow Bel_{t+1}(b,\sigma). \tag{6}$$

On the hearer's side, she would believe an utterance only when it is consistent with her own belief. The situation is represented by the constraint:

$$Bel_t(b,\neg\sigma) \wedge Bel_t(b,\sigma) \Rightarrow \bot. \tag{7}$$

Prepare rules (4) and (5) for b and $\neg\sigma$, and put them together with (6) and (7). Let $\Delta^t_{ab\sigma} = \{Bel^{\pm}_t(b,\sigma) \Rightarrow Bel^{\pm}_t(b,\sigma),\ \ Bel^{\pm}_t(b,\sigma) \wedge Bel^{\pm}_{t+1}(b,\sigma) \Rightarrow Bel^{\pm}_{t+1}(b,\sigma),\ \ Bel^{\pm}_t(b,\neg\sigma) \Rightarrow Bel^{\pm}_t(b,\neg\sigma),\ \ Bel^{\pm}_t(b,\neg\sigma) \wedge Bel^{\pm}_{t+1}(b,\neg\sigma) \Rightarrow Bel^{\pm}_{t+1}(b,\neg\sigma),\ \ Utter_t(a,\sigma) \Rightarrow Bel_{t+1}(b,\sigma),\ \ Bel_\tau(b,\neg\sigma) \wedge Bel_\tau(b,\sigma) \Rightarrow \bot \text{ (for } \tau = t,\, t+1)\}$.

Definition 3.4 ((mis)inform/(in)sincere). Let a and b be two agents, $\sigma \in \mathcal{F}$, and t a parameter representing time. Then define

$$\textbf{Inform}(a,b,\sigma,t) := \textbf{Trustful}(a,\sigma,t) \cup \Delta^t_{ab\sigma}.$$
$$\textbf{Misinform}(a,b,\sigma,t) := \textbf{Untrustful}(a,\sigma,t) \cup \Delta^t_{ab\sigma}.$$
$$\textbf{Sincere}(a,b,\sigma,t) := \textbf{Truthful}(a,\sigma,t) \cup \Delta^t_{ab\sigma}.$$
$$\textbf{Insincere}(a,b,\sigma,t) := \textbf{Untruthful}(a,\sigma,t) \cup \Delta^t_{ab\sigma}.$$

If the speech act is trustful (resp. untrustful), the speaker brings true (resp. false) information to the hearer. In this case, we say that *a informs* (resp. *misinforms*) *b* of σ. On the other hand, if the speech act is truthful (resp. untruthful) a speaker a communicates a believed-true (resp. disbelieved) sentence σ to a hearer b. In this case, we say that *a sincerely* (resp. *insincerely*) *communicates* σ to b. The effect of an utterance is observed, for instance, by comparing two models: $M_1 = \{U_t(a,\sigma),\ \neg B_t(b,\sigma),\ B_{t+1}(b,\sigma)\} \cup N$ and $M_2 = \{\neg U_t(a,\sigma),\ \neg B_t(b,\sigma),\ \neg B_{t+1}(b,\sigma)\} \cup N$ of **Inform**(a,b,σ,t) where $N = \{B_t(a,\sigma),\ B_{t+1}(a,\sigma),\ \neg B_t(b,\neg\sigma),\ \neg B_{t+1}(b,\neg\sigma),\ H_t(\sigma),\ H_{t+1}(\sigma)\}$. When a hearer b disbelieves the sentence σ at time t, an utterance of σ changes the belief state of the hearer at time $t+1$ as far as b disbelieves $\neg\sigma$. Such a belief change does not happen if b believes $\neg\sigma$ at t. When a speaker a utters a believed-true sentence σ that is actually false, it would *mislead* a hearer b's acquiring the false belief σ. The situation is represented by the model $\{U_t(a,\sigma), B_t(a,\sigma),\ \neg B_t(b,\sigma),\ \neg B_t(b,\neg\sigma),\ \neg H_t(\sigma), B_{t+1}(a,\sigma), B_{t+1}(b,\sigma), \neg B_{t+1}$

[2] U means *Utter*, B means *Bel*, and H means *Hold*.

$(b, \neg\sigma), \neg H_{t+1}(\sigma)\,\}$ of **Sincere**(a, b, σ, t). On the other hand, if a speaker a utters a disbelieved sentence σ that is actually false and it causes a hearer b's acquiring the false belief σ, the speaker *deceives* the hearer. The situation is represented by the model $\{\, U_t(a, \sigma),\ \neg B_t(a, \sigma),\ \neg B_t(b, \sigma), \neg B_t(b, \neg\sigma), \neg H_t(\sigma), \neg B_{t+1}(a, \sigma), B_{t+1}(b, \sigma),\ \neg B_{t+1}(b, \neg\sigma),\ \neg H_{t+1}(\sigma)\,\}$ of **Insincere** (a, b, σ, t).

Some formal properties are addressed as follows.

Proposition 3.1. *Let a and b be two agents, $\sigma \in \mathcal{F}$, and t a parameter representing time. Also let* **Comm** *be either* **Inform**, **Misinform**, **Sincere**, *or* **Insincere**. *It holds that*

(i) **Trustful**$(a, \sigma, t) \wedge$ **Untrustful**$(a, \sigma, t) \supset \neg\, Utter_t(a, \sigma)$.
(ii) **Truthful**$(a, \sigma, t) \wedge$ **Untruthful**$(a, \sigma, t) \supset \neg\, Utter_t(a, \sigma)$.
(iii) **Trustful**$(a, \sigma, t) \wedge$ **Truthful**$(a, \sigma, t) \supset (Utter_t(a, \sigma) \supset Hold_t(a, \sigma) \wedge Bel_t(a, \sigma))$.
(iv) **Trustful**$(a, \sigma, t) \wedge$ **Untruthful**(a, σ, t) $\supset (Utter_t(a, \sigma) \supset Hold_t(a, \sigma) \wedge \neg Bel_t(a, \sigma))$.
(v) **Untrustful**$(a, \sigma, t) \wedge$ **Truthful**(a, σ, t) $\supset (Utter_t(a, \sigma) \supset \neg Hold_t(a, \sigma) \wedge Bel_t(a, \sigma))$.
(vi) **Untrustful**$(a, \sigma, t) \wedge$ **Untruthful**(a, σ, t) $\supset (Utter_t(a, \sigma) \supset \neg Hold_t(a, \sigma) \wedge \neg Bel_t(a, \sigma))$.
(vii) **Comm**$(a, b, \sigma, t) \supset (Utter_t(a, \sigma) \wedge Bel_{t+1}(b, \sigma) \supset \neg Bel_t(b, \neg\sigma))$.
(viii) **Comm**$(a, b, \sigma, t) \supset (Utter_t(a, \sigma) \wedge \neg Bel_{t+1}(b, \sigma) \supset Bel_t(b, \neg\sigma))$.

4 Encoding in Logic Programs

A causal rule $L_1 \wedge \cdots \wedge L_n \Rightarrow L_0$ where L_i $(0 \leq i \leq n)$ is a literal, is translated into the logic programming rule: $L_0 \leftarrow not\,\neg L_1, \ldots, not\,\neg L_n$ where *not* represents *negation as failure*. Let Π_T be the logic program that is obtained from a causal theory T by translating each causal rule in T into the logic programming rule in Π_T. Then an interpretation I is a model of T iff I is a consistent and complete *answer set* of Π_T [4]. By this fact, a causal theory of speech acts is represented by a logic program as follows.

Definition 4.1 (logic program of speech acts). Let $CT^t_{x\sigma}$ be a causal theory of speech acts of Definition 3.1. Then a *logic program* $\Pi^t_{x\sigma}$ *associated with* $CT^t_{x\sigma}$ consists of rules:

$$Utter_t(x, \sigma) \leftarrow not\,\neg\, Utter_t(x, \sigma), \qquad \neg\, Utter_t(x, \sigma) \leftarrow not\ Utter_t(x, \sigma),$$
$$Hold_t(\sigma) \leftarrow not\,\neg Hold_t(\sigma), \qquad \neg Hold_t(\sigma) \leftarrow not\ Hold_t(\sigma),$$
$$Hold_{t+1}(\sigma) \leftarrow not\,\neg Hold_t(\sigma),\ not\,\neg Hold_{t+1}(\sigma),$$
$$\neg Hold_{t+1}(\sigma) \leftarrow not\ Hold_t(\sigma),\ not\ Hold_{t+1}(\sigma),$$
$$Bel_t(x, \sigma) \leftarrow not\,\neg Bel_t(x, \sigma), \qquad \neg Bel_t(x, \sigma) \leftarrow not\ Bel_t(x, \sigma),$$
$$Bel_{t+1}(x, \sigma) \leftarrow not\,\neg Bel_t(x, \sigma),\ not\,\neg Bel_{t+1}(x, \sigma),$$
$$\neg Bel_{t+1}(x, \sigma) \leftarrow not\ Bel_t(x, \sigma),\ not\ Bel_{t+1}(x, \sigma).$$

By the correspondence between a causal theory T and its logic programming translation Π_T, we have the next result.

Proposition 4.1. *Let $\Pi^t_{x\sigma}$ be a logic program associated with a causal theory $\mathcal{C}T^t_{x\sigma}$. Then, I is a model of $\mathcal{C}T^t_{x\sigma}$ iff I is an answer set of $\Pi^t_{x\sigma}$.*

(Un)trustful or (un)truthful speech acts, and (mis)inform or (in)sincere communication are also represented by logic programs. In this way, we can compute the effect of assertive speech acts in *answer set programming* [1].

References

1. Brewka, G., Eiter, T., Truszczyński, M.: Answer set programming at a glance. CACM **54**, 92–103 (2011)
2. Cohen, P.R., Levesque, H.: Speech acts and rationality. In: Proceedings of the 23rd Annual Meeting of ACL, pp. 49–59 (1985)
3. Demolombe, R.: Reasoning about trust: a formal logical framework. In: Jensen, C., Poslad, S., Dimitrakos, T. (eds.) iTrust 2004. LNCS, vol. 2995, pp. 291–303. Springer, Heidelberg (2004). doi:10.1007/978-3-540-24747-0_22
4. Giunchiglia, E., Lee, J., Lifschitz, V., McCain, N., Turner, H.: Nonmonotonic causal theories. Artif. Intell. **153**, 49–104 (2004)
5. Searle, J.R.: Expression and Meaning. Cambridge University Press, Cambridge (1979)
6. Yamada, T.: Logical dynamics of some speech acts that affect obligations and preferences. Synthese **165**, 295–315 (2008)

An Axiomatisation for Minimal Social Epistemic Logic

Liang Zhen(✉)

Department of Philosophy, University of Auckland, Auckland, New Zealand
soul_lz@hotmail.com

Abstract. A two-dimensional modal logic, intended for applications in social epistemic logic, with one dimension for agents and the other for epistemic states is given. The language has hybrid logic devices for agents, as proposed in earlier papers by Seligman, Liu and Girard. We give an axiomatisation and a proof of its completeness.

We start with a minimal language, with no restrictions on the epistemic and social relations. Even the denotation of names is not assumed to be the same in every state. Like standard hybrid logic, our language has a mixture of devices from modal propositional logic (modalities) and predicate logic (names and predication). But unlike hybrid logic the names do not refer to points of evaluation (worlds, in the case of hybrid logic) but to agents. This imbalance upsets the canonical model method of proving completeness in a way that is hard to restore using the technique of "witnesses" familiar from predicate logic.

1 Language, Semantics and Axiomatisation

Definition 1. The language of social epistemic logic consists of a set Prop of propositional variables p, a set Nom of agent names n and the set of formulas φ constructed from them as follows:

$$\varphi \quad ::= \quad p \mid n \mid \neg\varphi \mid (\varphi \wedge \varphi) \mid \langle K\rangle\varphi \mid \langle S\rangle\varphi \mid @_n\varphi$$

Standard abbreviations for $\vee$, $\rightarrow$ and $\leftrightarrow$ and duals for $[K] = \neg\langle K\rangle\neg$ and $[S] = \neg\langle S\rangle\neg$.

Formulas express agent-indexical propositions: they must be evaluated from the perspective of an agent to receive a truth value. From my perspective, $\langle S\rangle\varphi$ means that I stand in some relationship to someone else, from whose perspective φ is true. $\langle K\rangle\varphi$ means that it is possible for me that φ is true. Shifts in perspective are achieved with the @ operator: $@_n\varphi$ means that φ is true from n's perspective.

Definition 2. A social epistemic model M is a tuple $\langle W, A, k, s, g, V\rangle$ where W is a set (of *states*), A is a non-empty set (of *agents*), k_a is a binary relation on

A. Baltag et al. (Eds.): LORI 2017, LNCS 10455, pp. 664–669, 2017.
DOI: 10.1007/978-3-662-55665-8_49

W for each agent a, s_w is a binary relation on A for each state w, $g_w\colon \mathsf{Nom} \to A$ assigns an agent $g_w(n)$ to each name n in each state w, and $V\colon \mathsf{Prop} \to \mathrm{pow}(W \times A)$ assigns a subset of $W \times A$ (i.e., an agent-indexical proposition) to each propositional variable.

Definition 3. Given a model $M = \langle W, A, k, s, g, V\rangle$, the relation $\models$ (*satisfies*) is defined recursively as follows:

$$
\begin{array}{lll}
M, w, a \models p & \text{iff} & \langle w, a\rangle \in V(p) \\
M, w, a \models n & \text{iff} & g_w(n) = a \\
M, w, a \models \neg\varphi & \text{iff} & M, w, a \not\models \varphi \\
M, w, a \models (\varphi \wedge \psi) & \text{iff} & M, w, a \models \varphi \text{ and } M, w, a \models \psi \\
M, w, a \models \langle K\rangle\varphi & \text{iff} & M, v, a \models \varphi \text{ for some } v \text{ such that } k_a(w, v) \\
M, w, a \models \langle S\rangle\varphi & \text{iff} & M, w, b \models \varphi \text{ for some } b \text{ such that } s_w(a, b) \\
M, w, a \models @_n\varphi & \text{iff} & M, w, g_w(n) \models \varphi
\end{array}
$$

As usual, φ is *valid* in M if $M, w, a \models \varphi$ for all w and a; it is *valid* if it is valid in every social epistemic model.

$\vdash \varphi$ if φ is a tautology	Tautology
$\vdash [S](\varphi \to \psi) \to ([S]\varphi \to [S]\psi)$	K_S
$\vdash [K](\varphi \to \psi) \to ([K]\varphi \to [K]\psi)$	K_K
$\vdash @_n(\varphi \to \psi) \to (@_n\varphi \to @_n\psi)$	$K_@$
$\vdash @_n\varphi \leftrightarrow \neg @_n \neg\varphi$	Selfdual$_@$
$\vdash @_n n$	Ref$_@$
$\vdash @_n @_m \varphi \leftrightarrow @_m\varphi$	Agree
$\vdash @_m n \to @_n m$	Symmetric
$\vdash n \to (\varphi \leftrightarrow @_n\varphi)$	Introduction
$\vdash @_n\varphi \to [S]@_n\varphi$	Back
if $\vdash \varphi$ and $\vdash \varphi \to \psi$, then $\vdash \psi$	Modus ponens
if $\vdash \varphi$ then $\vdash [S]\varphi$	Necessitation of S
if $\vdash \varphi$ then $\vdash @_n\varphi$	Necessitation of @
if $\vdash \varphi$ then $\vdash [K]\varphi$	Necessitation of K
if $\vdash @_n\varphi \to (n_1 \vee n_2 \vee \ldots \vee n_k)$ then $\vdash \neg @_n\varphi$	Nominal Rule

Fig. 1. The system $\mathsf{H}_{\mathsf{SEL}}$ of minimal social epistemic logic.

Although $\mathsf{H_{SEL}}$ is closely based on the usual axiomatisation for hybrid logic, the Back Axiom is restricted to the social modality and we have an additional rule: the Nominal Rule. This is sound in hybrid logic and so must be admissible, but it is not needed to proof the completeness of hybrid logic. We need to know that is it sound:

Proposition 1. *The Nominal Rule is sound.*

Theorem 1 (Soundness $\mathsf{H_{SEL}}$). *If* $\mathsf{H_{SEL}} \vdash \varphi$ *then* φ *is a* SEL*-validity.*

2 Completeness

We assume standard concepts, such as consistency, maximal consistency, logical closure, etc. First, fix an enumeration e for the set of formulas. By a standard Linenbaum argument we can prove:

Proposition 2. *There is a function* $\Gamma \mapsto \Gamma^e$ *over the sets of formulas such that if* Γ *is consistent,* Γ^e *is maximal consistent and* $\Gamma \subseteq \Gamma^e$.

Definition 4. A *network* $\mathcal{N} = \langle W, A, \mathsf{k}, \mathsf{s}, \delta \rangle$ consists of non-empty sets $W, A \subseteq \mathbb{N}$, a relation $\mathsf{k}_a \subseteq W \times W$ for each $a \in A$, a relation $\mathsf{s}_w \subseteq A \times A$ for each $w \in W$, and for each *node* $\langle w, a \rangle$, maximal consistent set $\delta(w, a)$.

Definition 5. A network $\mathcal{N}$ is *coherent* iff for every $w, v \in W, a, b \in A$

(k) if $\mathsf{k}_a(w, v)$ then if $[K]\varphi \in \Sigma$, then $\varphi \in \Gamma$, for any formula φ
(n) $\delta(w, a) \cap \delta(w, b) \cap \mathsf{Nom} = \emptyset$ if $a \neq b$
(s) if $\mathsf{s}_w(a, b)$ then if $[S]\varphi \in \Sigma$, then $\varphi \in \Gamma$, for any formula φ
(@) $@\delta(w, a)^1 = @\delta(w, b)$

$\mathcal{N}$ is *saturated* iff

(S) if $\langle S\rangle\varphi \in \delta(w, a)$ then there is a $b \in A$ such that $\mathsf{s}_w(a, b)$ and $\varphi \in \delta(w, b)$
($@_n$) if $@_n\varphi \in \delta(w, a)$ then there is a $b \in A$ such that $n \in \delta(w, b)$
(K) if $\langle K\rangle\varphi \in \delta(w, a)$ then there is a $v \in W$ such that $\mathsf{k}_a(w, v)$ and $\varphi \in \delta(v, a)$
$\mathcal{N}$ is *perfect* iff it is both coherent and saturated.

The saturation conditions all impose existential demands on a network.

Proposition 3. *For any perfect network* $\mathcal{N} = \langle W, A, \mathsf{k}, \mathsf{s}, \delta \rangle$ *the structure* $M_{\mathcal{N}} = \langle W, A, \mathsf{k}, \mathsf{s}, g, V \rangle$ *where* $g_w(n) = a$ *iff* $n \in \delta(w, a)$
$V(p) = \{\langle w, a \rangle | p \in \delta(w, a), p \in \mathit{Prop}\}$
is a social network model.

Lemma 1 (Truth). *Given any perfect network* $\mathcal{N}$*, for any* $w \in W, a \in A$ *and any formula* φ*,* $M_{\mathcal{N}}, w, a \models \varphi$ *iff* $\varphi \in \delta(w, a)$.

[1] In this paper, $@X = \{@_n\varphi | @_n\varphi \in X\}$.

We will show at the end of paper that every consistent set Γ has a perfect network $\mathcal{N}$ with $\Gamma \subseteq \delta(0,0)$. The Truth Lemma then implies that $M_{\mathcal{N}}, 0, 0 \models \varphi$ for every $\varphi \in \Gamma$. As it conventionally, this, together with soundness (Theorem 1) is enough to prove:

Theorem 2 (Completeness). *φ is derivable from formulas in Γ in H_{SEL} iff φ is a consequence of Γ, i.e. φ is satisfied by every M, w, a that satisfies Γ.*

3 A Perfect Network

The goal of this section is to build a perfect network by removing defect, one-by-one.

Definition 6. Given network $\mathcal{N}$ with node $\langle w, a\rangle$ and a formula φ,

$[w, a, \langle S\rangle\varphi]$ is an *S defect* iff $\langle S\rangle\varphi \in \delta(w, a)$ but no b has $\mathsf{s}_w(a, b)$ and $\varphi \in \delta(w, b)$
$[w, a, @_n\varphi]$ is an $@_n$ *defect* iff $@_n\varphi \in \delta(w, a)$ but no b has $n \in \delta(w, b)$
$[w, a, \langle K\rangle\varphi]$ is a *K defect* iff $\langle K\rangle\varphi \in \delta(w, a)$ but no v has $\mathsf{k}_a(w, v)$ and $\varphi \in \delta(v, a)$

We say that $\mathcal{N}' = \langle W', A', \mathsf{k}_{'}, \mathsf{s}_{'}, \delta'\rangle$ is an *extension* of network $\mathcal{N} = \langle W, A, \mathsf{k}, \mathsf{s}, \delta\rangle$ and write $\mathcal{N} \trianglelefteq \mathcal{N}'$ if all the obvious inclusions hold: $W \subseteq W'$, $A \subseteq A'$, $\mathsf{k}_a \subseteq \mathsf{k}_{a'}$, $\mathsf{s}_w \subseteq \mathsf{s}_{w'}$, and $\delta(w, a) \subseteq \delta'(w, a)$ for each $w \in W$ and $a \in A$. In the final case, we actually have an equality, $\delta(w, a) = \delta'(w, a)$, since they are both maximal.

Definition 7. If D is a defect of network $\mathcal{N}$ then network $\mathcal{N}'$ is a *repair* of D in $\mathcal{N}$ iff

$\mathcal{N}'$ is a coherent extension of $\mathcal{N}$ and neither have D nor new defects at nodes of $\mathcal{N}$.

Due to space constraint, we just give the constructions for Lemmas 2, 3 and 4.

Lemma 2. *For any coherent network $\mathcal{N}$ and $w \in W$, $a \in A$, $[w, a, \langle S\rangle\varphi]$ has a repair.*

Proof. **Case One**: $\langle S\rangle(m \wedge \varphi) \in \delta(w, a)$ and $m \in \delta(w, b)$ for some $m \in \mathsf{Nom}$ and $b \in A$. Here is a formal definition of $\mathcal{N}' = \langle W, A, \mathsf{k}, \mathsf{s}', \delta\rangle$ where

$$\mathsf{s}'_v = \begin{cases} \mathsf{s}_v & \text{if } v \neq w \\ \mathsf{s}_w \cup \{\langle a, c\rangle\} & \text{Otherwise} \end{cases}$$

where $c \in A$ is the least b satisfying the condition of Case One.

Case Two: Otherwise.
$\mathcal{N}' = \langle W, A', \mathsf{k}, \mathsf{s}', \delta' \rangle$ such that
$A' = A \cup \{|A|\}$[2]

$$\mathsf{s}'_v = \begin{cases} \mathsf{s}_v & \text{if } v \neq w \\ \mathsf{s}_w \cup \{\langle a, |A| \rangle\} & \text{Otherwise} \end{cases}$$

$$\delta'(v,b) = \begin{cases} [\{\varphi\} \cup \{\psi | [S]\psi \in \delta(w,a)\}]^e & \text{if } v = w, b = |A| \\ [@\delta(v,a) \cup \{\neg n | n \in \mathsf{Nom}\}]^e & \text{if } v \neq w, b = |A| \\ \delta(v,b) & \text{Otherwise} \end{cases}$$

Lemma 3. *For any coherent network $\mathcal{N}$ and $w \in W$, $a \in A$, $[w, a, @_n\varphi]$ has a repair.*

Proof. Define $\mathcal{N}' = \langle W, A', \mathsf{k}, \mathsf{s}, \delta' \rangle$ as follows:

$A' = A \cup \{|A|\}$

$$\delta'(v,b) = \begin{cases} [\{n\} \cup @\delta(w,a)]^e & \text{if } v = w, b = |A| \\ [@\delta(v,a) \cup \{\neg n | n \in \mathsf{Nom}\}]^e & \text{if } v \neq w, b = |A| \\ \delta(v,b) & \text{Otherwise} \end{cases}$$

Lemma 4. *For any coherent network $\mathcal{N}$ and $w \in W$, $a \in A$, $[w, a, \langle K \rangle \varphi]$ has a repair.*

Proof. Define $\mathcal{N}' = \langle W', A, \mathsf{k}', \mathsf{s}, \delta' \rangle$ as follows:
$W' = W \cup \{|W|\}$

$$\mathsf{k}'_b = \begin{cases} \mathsf{k}_b & \text{if } b \neq a \\ \mathsf{k}_b \cup \{\langle w, |W| \rangle\} & \text{Otherwise} \end{cases}$$

$\delta'(|W|, a) = [\{\varphi\} \cup \{\psi | [K]\psi \in \delta(w,a)\}]^e$

For all $b \neq a$, $\delta'(v,b) = \begin{cases} [@\delta'(|W|,a) \cup \{\neg n | n \in \mathsf{Nom}\}]^e & \text{if } v = |W| \\ \delta(v,b) & \text{Otherwise} \end{cases}$

Let d be an enumeration of all potential defects.

Definition 8. Given a consistent set Γ, define a sequence of networks $\mathcal{N}_0 \trianglelefteq \mathcal{N}_1 \trianglelefteq \dots$ as follows: $\mathcal{N}_0 = \langle W_0, A_0, \mathsf{k}_0, \mathsf{s}_0, \delta_0 \rangle$ is defined by $W_0 = \{0\}$, $A_0 = \{0\}$, $\mathsf{k}_0 = \emptyset$, $\mathsf{s}_0 = \emptyset$ and $\delta_0(0,0) = \Gamma^e$. Then $\mathcal{N}_{i+1}$ is the repair of $d(j)$ in $\mathcal{N}_i$, where j is the least number such that $d(j)$ is a defect of $\mathcal{N}_i$.

Then define the *network of Γ* to be $\mathcal{N}_\Gamma = \langle \bigcup_i W_i, \bigcup_i A_i, \bigcup_i \mathsf{k}_i, \bigcup_i \mathsf{s}_i, \bigcup_i \delta_i \rangle$.

Lemma 5. *$\mathcal{N}_\Gamma$ is perfect.*

Lemma 6. *For any consistent set Γ, there is a perfect network $\mathcal{N}$ for which $\Gamma \subseteq \delta(0,0)$*

---

[2] $|W|$ and $|A|$ are the successors of the largest number in W and A respectively since $W, A \subseteq \mathbb{N}$.

4 Conclusion and Further Work

We have provided a complete axiomatisation for the minimal social epistemic logic. Several obvious extensions present themselves:

1. Rigid social epistemic models
2. Strengthening the epistemic logic to S4, S5, KD45
3. Interactions between K and S
4. Downarrow. The binder $\downarrow_n \varphi$ from hybrid logic.

References

1. Blackburn, P., De Rijke, M., Venema, Y.: Modal Logic. Cambridge tracts in theoretical computer science, vol. 53. Cambridge University of Press, Chennai (2001)
2. Areces, C., ten Cate, B.: 14 Hybrid logics. Stud. Log. Pract. Reasoning **3**, 821–868 (2007)
3. Seligman, J., Liu, F., Girard, P.: Logic in the Community. In: Banerjee, M., Seth, A. (eds.) ICLA 2011. LNCS, vol. 6521, pp. 178–188. Springer, Heidelberg (2011). doi:10.1007/978-3-642-18026-2_15
4. Seligman, J., Liu, F., Girard, P.: Facebook and the epistemic logic of friendship. arXiv:1310.6440 (2013)

Relief Maximization and Rationality

Paolo Galeazzi[1(✉)] and Zoi Terzopoulou[2]

[1] CIBS, Copenhagen University, Copenhagen, Denmark
pagale87@gmail.com
[2] ILLC, University of Amsterdam, Amsterdam, The Netherlands
zoiterzopoulou@yahoo.com

Abstract. This paper introduces the concept of relief maximization in decisions and games and shows how it can explain experimental behavior, such as asymmetric dominance and decoy effects. Next, two possible evolutionary explanations for the survival of relief-based behavior are sketched.

Keywords: Relief maximization · Rationality · Choice principles

1 Introduction: Asymmetric Dominance

A popular book by Dan Ariely [Ari08], entitled *Predictably Irrational*, begins with the case of an advertisement for a yearly subscription to the *Economist*. The customer was offered the following three alternatives: (1) online subscription, \$59; (2) print subscription, \$125; (3) print-and-online subscription, \$125. The second option is obviously dominated by the third. When Ariely wanted to test the effect of including the dominated alternative (called the "decoy option"), he asked one hundred MIT students to choose one of the options in the menu and found that the students decided as follows: 16 students chose (1), 0 students chose (2), 84 students chose (3). Then Ariely removed the decoy option from the menu and asked the students again. In this case 68 students opted for (1), while only 32 chose (3). The mere existence of a dominated option, that nobody opted for, caused a substantial difference in the choice between the other two alternatives. The first chapter of his book is full of similar examples, where the decoy effect is evidently playing a considerable role in the individuals' decisions.

The subscription case presented by Ariely is an example of decision under certainty. Decoy effects, also called "asymmetric dominance" effects, have been recently investigated in the context of decisions under uncertainty too, and in game theory in particular. In [CPB07], the authors conducted an experimental analysis on a set of two-player games with three possible actions, one of which (the decoy action E) was dominated by just one of the remaining two (action C), and with no other dominance relation between other actions. Their experiments were performed both on symmetric and asymmetric games, with similar results.

A. Baltag et al. (Eds.): LORI 2017, LNCS 10455, pp. 670–675, 2017.
DOI: 10.1007/978-3-662-55665-8_50

For the purposes of this work, however, it will be enough to present the symmetric case. The symmetric games used in [CPB07] are the following.[1]

I	C	D	E
C	40	20	60
D	60	40	20
E	20	0	40

II	C	D	E
C	60	20	40
D	0	80	0
E	20	0	20

III	C	D	E
C	80	20	40
D	40	80	0
E	60	0	20

IV	C	D	E
C	60	20	40
D	0	80	0
E	40	0	20

V	C	D	E
C	80	20	20
D	40	80	0
E	60	0	0

Players first had to play the 3×3 games, and later, as control condition, they had to play the 2×2 games without the dominated action E. In short, the conclusions were that asymmetric dominance effects were significantly exhibited.

2 Relief Maximization

We can formulate a general choice principle, that we call *relief maximization*, giving rise and possible explanation to asymmetric dominace effects. Formally, a game is a tuple $G = \langle I, (A_i, u_i)_{i \in I} \rangle$, with I the set of players, and action set A_i and utility function $u_i : \times_{i \in I} A_i \to \mathbb{R}$ for each $i \in I$. Let us denote $A_{-i} := \times_{j \neq i} A_j$, and define the following quantity as the *relief* of action $a_i \in A_i$:

$$\min_{a_{-i} \in A_{-i}} \{u_i(a_i, a_{-i}) - \min_{a'_i \in A_i} u_i(a'_i, a_{-i})\}. \tag{1}$$

Intuitively, the relief of action a_i is the minimal gain that a_i can secure with respect to the worst choices in all action profiles $a_{-i} \in A_{-i}$. Relief maximization is the principle that "cares" about this quantity and wants to maximize it:

$$\arg\max_{a_i \in A_i} \min_{a_{-i} \in A_{-i}} \{u_i(a_i, a_{-i}) - \min_{a'_i \in A_i} u_i(a'_i, a_{-i})\}.$$

To the best of our knowledge, the concept of relief appeared in decision theory only as an exercise in Problem 2.5.7 in [Jef90], and has gone unused since then.[2] However, the link between relief maximization and decoy effects is apparent. Relief maximization would capture asymmetric dominance effects in all games presented before: since only actions that are never worst responses have nonzero relief, an agent who maximizes relief will always choose the dominant action, in line with the behavior largely observed in the experiments.

[1] The symmetric games presented there are actually six, but one of them is irrelevant for our study. As usual, since games are symmetric, it suffices to specify row player's payoffs.

[2] For a (compact convex) set of probabilities $\Gamma \subseteq \Delta(A_{-i})$, it is also straightforward to generalize the definition to the "multiple-prior version" of relief maximization: $\arg\max_{a_i} \min_{P \in \Gamma} \{\mathbb{E}_P [u_i(a_i, a_{-i})] - \min_{a'_i} \mathbb{E}_P [u_i(a'_i, a_{-i})]\}$.

3 On the Rationality of Maximizing Relief

Mostly due to the axiomatization given in [Sav54], the notion of rationality is usually associated with the maximization of expected utility: an agent is rational if and only if she maximizes her expected utility. In this perspective, the agent is required to form beliefs in terms of a single probability distribution $P \in \Delta(A_{-i})$, and to choose an action that maximizes the expected utility $\mathbb{E}_P[u_i(a_i, a_{-i})]$. Are then asymmetric dominance effects irrational? First of all, it is easy to check that relief maximization violates the independence of irrelevant alternatives and Savage's axiom P1 (see [CPB07, Sav54]). Relief maximization and asymmetric dominance effects are hence at odds with expected utility maximization, and they would be irrational according to standard rational choice theory.

According to a more ecological perspective instead (e.g., [Gig08]), the qualities and the rationality of a choice cannot be evaluated independently of the environment in which it takes place. To study the evolutionary performance of relief maximization we can use a *multi-game* model where different principles compete with each other in an environment that consists of a variety of decision problems (see [GF17]). Consider for example a population living in an environment composed of the five games from Sect. 1. Since in all those games the Nash equilibria are in the main diagonal, they incentivize coordination between the players. To have a more diverse environment, let us also consider their "anti-coordination" versions, where rows C and D are swapped for each other in all five payoff matrices. Notice that E is still dominated by C, and no other dominance relations exist in the games. So C is still the relief maximizing action in all ten games. For the sake of example suppose also that each game has the same probability of occurring, and that the only two principles in the population are relief maximization *Rel* and maxmin Mm, that is defined as follows:

$$\arg\max_{a_i} \min_{a_{-i}} u_i(a_i, a_{-i}).$$

When agents from this population are repeatedly matched to play the games in the environment, each agent chooses actions according to his or her own choice principle in all games he or she is involved. The mutual choices of the agents, together with the occurrence probability of each game, define an expected utility for any of the choice principles in the multi-game, as specified by the next table. Each number denotes the expected utility of the principle listed in the row when the co-player is choosing according to the principle in the column.

	Mm	*Rel*
Mm	44	44
Rel	42	42

Relief maximization (and, therefore, asymmetric dominance effects) turns out to be strictly dominated by maxmin. In general, relief maximization has no bite unless there is an action that is a never-worst response and seldom prescribes a unique choice consequently. Because of this weakness, relief maximizers would often lose the evolutionary competition in many environments that are not explicitly tailored to relief maximization.

4 Combining Principles

One might then wonder why asymmetric dominance effects are still consistently observed in actual behavior, given that agents who fall prey to decoys should go extinct in most reasonable circumstances. We sketch two possible explanations here. The first is centered on the notion of *context dependency*. A choice principle is context-dependent if the evaluation of each action is not independent of the menu of actions available to the decision maker (DM henceforth). Think of maxmin: each action is evaluated by looking at the minimal utility possibly achievable, which does not vary across different menus and is thus context-independent. Instead, the relief of an action a_i, as defined in (1), essentially depends on the other actions a_i' in the menu of possible alternatives. It is then possible that relief-based reasoning survived as a side effect of the evolutionary success of other context-dependent principles, such as regret minimization (see [GF17]). Humans might have simply retained a general propensity for choosing in a context-dependent manner (by comparing options with each other), that can be evolutionarily beneficial when implemented in terms of regret, but much less beneficial when implemented in terms of relief. This direction would be in line with the observation in [Ari08] that "humans rarely choose things in absolute terms. [...] Rather, we focus on the relative advantage of one thing over another, and estimate value accordingly" (p. 2). The human attitude of judging things relative to the context might then be much more rational than it is normally believed in the literature. Not all context-dependent principles are equally good, though.

A second explanation is based on the possibility that different choice principles could also be mixed towards a final decision. Indeed, different principles may be triggered by different (features of) choice situations, as it is the case for relief maximization in the presence of an asymmetrically dominated action, but agents might also combine, in a lexicographic or in a weighted way, different principles in the same decision problem. In some decision situations in fact two or more actions may be equally optimal according to a given principle. How will DM choose between them? An option is to assume that DM will simply pick one at random, but an alternative would be to reconsider the equally optimal actions in the light of a different principle, i.e., to use a second principle to break the tie. In the following we investigate the possibility of using relief maximization as tie-breaker.[3]

[3] Note that in general a tie-breaker is needed for a principle to always output a single action.

Let us introduce two more sophisticated variants of maximinimizers Mm and regret minimizers Reg, which is defined as follows:

$$\arg\max_{a_i}\min_{a_{-i}}\{u_i(a_i, a_{-i}) - \max_{a_i'} u_i(a_i', a_{-i})\}.$$

So, type Mm^+ first discards all actions that do not maximize the minimum, and then, among the surviving actions, discards all those that do not maximize relief; and mutatis mutandis for Reg^+. If we consider the multi-game environment consisting of the ten games introduced before, the resulting expected utilities are in the next table. It turns out that reasoning in terms of relief, in combination with other principles, is not necessarily detrimental to the player's utility.

	Mm	Mm^+	Reg	Reg^+
Mm	44	44	44	44
Mm^+	42	42	42	42
Reg	46	46	46	46
Reg^+	46	46	46	46

When combined with maxmin, relief maximization performs rather poorly, but when combined with the other context-dependent principle presented here, namely regret minimization, then relief-based reasoning can survive the evolutionary competition. In the example considered, principles Mm and Mm^+ are dominated by both Reg and Reg^+, and the evolutionarily stable states (see [Wei95]) are all population states with the two principles Reg and Reg^+ only. If we allow the agents to evaluate options according to multiple choice principles, that is, to tackle the decision problem from many different perspectives and combine them into a final decision, then it is evolutionarily less implausible to observe the survival of relief-maximizing behaviors, such as asymmetric dominance effects.

5 Conclusion

This paper presented a choice principle, relief maximization, that is able to capture asymmetric dominance and decoy effects. We next sketched two possible explanations for the survival of relief-based behaviors and considered a more ecological and multi-principled view on rational choice, where DM combines different principles to evaluate choices, and different combinations are evolutionarily selected over others depending on the decision problems in the environment.

References

[Ari08] Ariely, D.: Predictably Irrational. HarperCollins, New York (2008)
[CPB07] Colman, A.M., Pulford, B.D., Bolger, F.: Asymmetric dominance and phantom decoy effects in games. Organ. Behav. Hum. Decis. Process. **104**(2), 193–206 (2007)

[GF17] Galeazzi, P., Franke, M., Representations, S.: Rationality and evolution in a richer environment. Philos. Sci. (2017, forthcoming)
[Gig08] Gigerenzer, G.: Why heuristics work. Perspect. Psychol. Sci. **3**(1), 20–29 (2008)
[Jef90] Jeffrey, R.C.: The Logic of Decision. University of Chicago Press, Chicago (1990)
[Sav54] Savage, L.J.: The Foundations of Statistics. Dover, New York (1954)
[Wei95] Weibull, J.W.: Evolutionary Game Theory. MIT Press, Cambridge (1995)

Reason to Believe

Chenwei Shi[1(✉)] and Olivier Roy[2]

[1] ILLC, University of Amsterdam, Amsterdam, the Netherlands
shichenwei88@gmail.com
[2] Universität Bayreuth, Bayreuth, Germany
olivier.roy@bayreuth.de

Abstract. In this paper we study the relation between nonmonotonic reasoning and belief revision. Our main conceptual contribution is to suggest that nonmonotonic reasoning guides but does not determine an agent's belief revision. To be adopted as beliefs, defeasible conclusions should remain stable in the face of certain bodies of information. This proposal is formalized in what we call a two-tier semantics for nonmonotonic reasoning and belief revision. The main technical result is a sound and complete axiomatization for this semantic.

1 Introduction

Nonmonotonic reasoning and belief revision, although structurally very similar [12], are conceptually distinct. Paraphrasing Gärdenfors [6], nonmonotonic reasoning consists in drawing defeasible conclusions from what you believe. In this paper we view these as *permissible* conclusions. You may, but do not have to draw them. Belief revision, on the other hand, is a theory of how you should change your mind in the face of new information. It describes how this "change in view" [8] *ought* to take place. Belief revision is thus more normatively demanding than nonmonotonic reasoning. The first consists in keeping track of what "normally" or "commonsensically" follows from your beliefs. The second requires you to come to new beliefs and retract old ones.

How should nonmonotonic reasoning and belief revision be related? How, in other words, should permissible but defeasible conclusions relate to the way you should update your belief upon learning? In this paper we study an answer to these questions in terms of cautiousness. As belief revision is more normatively demanding than nonmonotonic reasoning, agents may use caution in forming new beliefs in the face of incoming information, refraining from drawing all defeasible conclusions from their original beliefs.

To be more specific, the agent circumscribes the range of what she may come to believe upon learning new information by nonmonotonic reasoning. Among the defeasible conclusions derived from her nonmonotonic reasoning, the agent filters out the ones which are not entrenched enough. Conclusion A is more entrenched than conclusion B if, when more information is taken into account in the nonmonotonic reasoning, conclusion A still follows while conclusion B does not.

The paper presents a formal framework in which we characterize the relation between nonmonotonic reasoning and belief revision as described above.

A. Baltag et al. (Eds.): LORI 2017, LNCS 10455, pp. 676–680, 2017.
DOI: 10.1007/978-3-662-55665-8_51

The framework we use i interesting in itself in the sense of integrating two heterogeneous parts, one is qualitative and the other is quantitative. The interaction between them leads to the formation and revision of the agent's beliefs.

Furthermore, we provide a sound and complete logic for reasoning about the key notions involved in our framework. However, due to the page limit, we decide not to include the details. The interested readers are referred to the long version of this paper.

2 Two-Tier Mechanism for Belief Revision

In this section, we aim to characterize the notion of belief revision based on the agent's nonmonotonic reasoning and cautiousness. To achieve this, we introduce a frame which includes a set of possible worlds W. All subsets of W are taken as propositions. The usual power set algebra on W gives us the boolean connectives: $P \wedge Q := P \cap Q$, $P \vee Q := P \cup Q$, $\neg P := W - P$, $P \to Q := \neg P \vee Q$.

Moreover, we impose a transitive and reflexive order $\succeq \subseteq W \times W$ and an equivalence relation $R \subseteq W \times W$ on W. $\succeq$ is interpreted as a normality order [3,9] and $u \succeq v$ means u is at least as normal as v to the agent. For simplicity, we will assume that $\succ := \succeq \setminus \approx$ is well-founded, where $\approx := \succeq \cap \preceq$. Then $\mathsf{Max}_{\succeq}(X) = \{w \in X \mid \neg\exists v \in X : v \succ w\}$ denotes the most normal worlds the agent would take given the proposition X. As in the literature on epistemic logic [5], R is taken as the epistemic accessibility relation and Ruv means that the agent cannot epistemically distinguish v from u. If w is the actual world where the agent is located, then $R(w) = \{v \in W \mid Rwv\}$ denotes the set of possible worlds the agent takes epistemically possible.

The agent's nonmonotonic reasoning is expressed by an operator $D^P Q$ which says that the agent would expect Q to be the case given the information that P is the case. In the frame, given $\succeq$ and R, the operator $D^P Q$ is defined as follows.

$$D^P Q \text{ is the case in the world } w \text{ iff } \mathsf{Max}_{\succeq}(P \cap R(w)) \subseteq Q$$

where $P, Q \subseteq W$. This definition says that the agent would take those propositions which hold on all the most normal worlds in $P \cap R(w)$ as permissible conclusions. Note that in the definition, $R(w)$ represents what the agent knows. The agent's nonmonotonic reasoning is made by applying those default rules (represented by the normality order in the frame) to the given information P and what she has already known.

The way we characterize the agent's nonmonotonic reasoning follows the line of [3,9]. The difference is that we model the agent's knowledge base and how it is involved in the agent's nonmonotonic reasoning explicitly in the frame.

The same characterization can also be applied to the agent's belief revision as did in [2,4] by interpreting the normality order as a plausibility order and the operator $D^P Q$ as the agent's conditional belief. This way of modeling belief revision implies that belief revision and nonmonotonic reasoning are governed by the same logic.

Our approach to belief revision is different from the model using plausibility order. As described in the Introduction, we think that the agent's belief revision is guided but not determined by her nonmonotonic reasoning. So the remaining part of this section will show how the agent's nonmonotonic reasoning guides her belief revision and what extra process contributes to the agent's belief revision. Moreover, we will investigate whether belief revision and nonmonotonic reasoning in our setting follow the same logic.

As we have said in the Introduction, by nonmonotonic reasoning, the agent collects some defeasible conclusions which are not yet qualified as her belief. These conclusions need to pass a further test called "stability test": when taking more information into account, whether the agent would still expect those conclusions to hold. Next we will address the question of what information should be taken into consideration in the stability test to evaluate the conclusions. The idea is that the agent would consider those propositions credible enough. To model the credibility of each proposition, we use a probability function.

Definition 1 ***(Two-tier frame).*** *A two-tier frame* $\mathcal{F} = (W, R, \succeq, \mu)$ *is a structure where* $W, R, \succeq$ *are a set of possible worlds, an equivalence relation on* W *and a transitive and reflexive order on* W *as explained above. And* $\mu\colon W \to \mathbf{P}$ *is a function assigning to each possible world* $w \in W$ *a probability function* $\mu_w\colon 2^{R(w)} \to [0,1]$ *satisfying Kolmogorov's Axioms such that*

- $\mu_w(w) > 0$;
- $\mu_v = \mu_w$ *for any* $v \in R(w)$.

Given $w \in W$ *and* $X \subseteq W$, *we define the probability function conditional on* X *with* $X \cap R(w) \neq \emptyset$ *as*

$$P_w(Y|X) := \frac{\mu_w(Y \cap X \cap R(w))}{\mu_w(X \cap R(w))}$$

where $Y \subseteq W$.

Then how is the agent's probability function applied in the stability test? It is used to measure the entrenchment of the conclusions derived in the non-monotonic reasoning. The details are spelled out in the following definition of the operator $\rightsquigarrow^r$.

Definition 2. *For any* $r \in [0,1]$, $\mathcal{F}, w \models X \rightsquigarrow^r Y$[1] *if and only if for any* $Q \subseteq W$, *if* $P_w(Q \mid X) \geq r$, *then* $\mathsf{Max}_{\succeq}(Q \cap X \cap R(w)) \subseteq Y$.

To get a feeling for this definition, take X in $X \rightsquigarrow^r Y$ as W. The formula $W \rightsquigarrow^r Y$ says that conditional on any proposition $P \subseteq W$ whose probability is not less than r, the agent would draw the conclusion Y from W. To put it another way, taking into account each item of information which is sufficiently probable, i.e. not lower than r, the agent would still expect the conclusion Y.

Therefore, the smaller is the number r, the more entrenched is the conclusion Y. Note that

$$\mathcal{F}, w \models W \rightsquigarrow^0 Y \textbf{ iff } R(w) \subseteq Y\ ,$$

which can be interpreted as the knowledge operator (the agent knows Y) and

$$\mathcal{F}, w \models X \rightsquigarrow^1 Y \textbf{ iff } \mathsf{Max}_{\succeq}(X \cap R(w)) \subseteq Y\ ,$$

which is the same as the operator $D^X Y$ for nonmonotonic reasoning.

[1] We write $w \in P \subseteq W$ as $\mathcal{F}, w \models P$.

With this in hand, we can state our definition of belief revision based on cautious nonmonotonic inferences.

Definition 3. *Given a two-tier frame* $\mathcal{F}$ *and* $w \in W$,

$$\mathcal{F}, w \models B^X_{0.5}Y \textit{ if and only if } \mathcal{F}, w \models X \leadsto^{0.5} Y \ .$$

This definition definitely ought to remind some readers of the notion of belief which only requires a sufficiently high probability (beyond 0.5 for example). Indeed, our notion of belief implies belief with high probability. If $\mathcal{F}, w \models B^X_{0.5}Y$, then $P_w(Y|X) > 0.5$. But it is more than high probability alone. It relies even more heavily on the agent's nonmonotonic reasoning. This ensures that under $\models$, $B^X_{0.5}Y$ is deductively closed: for any two-tier frame $\mathcal{F}$ and any world w in it, $\mathcal{F}, w \models B^X_{0.5}(P \wedge Q) \leftrightarrow B^X_{0.5}P \wedge B^Y_{0.5}Q$.[2]

We next show that our notion of belief revision $B^X_{0.5}Y$ is not governed by the same logic of nonmonotonic reasoning D^PQ. Several results about valid and invalid formulas are summarized in the following table, including those in the class of two-tier frames where $\succeq$ is connected.

In any $\mathcal{F}$ where $\succeq$ is not necessarily connected		
	$O = D$	$O = B_{0.5}$
CM:$O^PQ \wedge O^PY \rightarrow O^{P\wedge Q}Y$	✓	✓
Cut: $O^PQ \wedge O^{P\wedge Q}Y \rightarrow O^PY$	✓	×
Or:$O^PY \wedge O^QY \rightarrow O^{P\vee Q}Y$	✓	×
And: $O^XP \wedge O^XQ \rightarrow O^X(P \wedge Q)$	✓	✓
In any $\mathcal{F}$ where $\succeq$ is connected		
	$O = D$	$O = B_{0.5}$
NR: $\neg O^{P\wedge Q}Y \wedge \neg O^{P\wedge\neg Q}Y \rightarrow \neg O^PY$	✓	✓
RM: $(O^PY \wedge \neg O^{P\wedge Q}Y) \rightarrow O^P\neg Q$	✓	×
CV: $\neg O^PQ \rightarrow (O^PY \rightarrow O^{P\wedge\neg Q}Y)$	✓	×
DR:$\neg O^PY \wedge \neg O^QY \rightarrow \neg O^{P\vee Q}Y$	✓	×

3 Future Work

The mechanism presented in Sect. 2 can also be seen as a proposal for how categorical, i.e. normality- or plausibility-based reasoning should relate to quantitative, graded reasoning. We did not, however, impose any restrictions on the relationship between the agent's normality order and credence function. In recent years there has been a renewal of interest in such bridges between categorical and graded reasoning [10, 11]. One obvious next step would be to compare the present proposal with these, especially in view of the dynamic interaction between the normality order and the probability function.

Our notion of belief revision as cautious nonmonotonic reasoning uses the idea that the agent's defeasible conclusions should remain stable in the face of

[2] The validity of a formula in this framework is defined in the usual way.

certain bodies of information. This of course raises the question of the relation between our proposal and the stability theory of knowledge [13] and of belief [10].

The former requires that the agent's belief remains stable in the face of any true information. The latter requires that for the agent the probability of a certain proposition remains beyond a particular threshold in face of any information consistent with the proposition. What is then the relation between our stability theory of belief and the other two stability theories? It is worth mentioning that the stability theory of belief [10] aims to achieve a notion of belief which can imply a sufficiently high probability while still validating the closure under conjunction. This aim is also achieved by our notion of belief.

Our definition of the operator $\rightsquigarrow^r$ provides a way of measuring the entrenchment of the agent's defeasible conclusions. How does this measurement of entrenchment relate to the discussion of entrenchment in the literature of belief revision [7]? To be more concrete, how is our method of measurement different from and connected to Sphon's "degree of belief" [14] which is also studied in [1,15]?

References

1. Aucher, G.: A combined system for update logic and belief revision. Master's thesis, ILLC, University of Amsterdam (2003)
2. Baltag, A., Smets, S.: A qualitative theory of dynamic interactive belief revision. Texts Log. Games **3**, 9–58 (2008)
3. Boutilier, C.: Conditional logics of normality: a modal approach. Artif. Intell. **68**, 87–154 (1994)
4. Boutilier, C.: Unifying default reasoning and belief revision in a modal framework. Artif. Intell. **68**, 33–85 (1994)
5. Fagin, R., Halpern, J.Y., Moses, Y., Vardi, M.: Reasoning About Knowledge. MIT press, Cambridge (2004)
6. Gärdenfors, P.: Belief revision and nonmonotonic logic: two sides of the same coin? In: Logics in AI, pp. 52–54 (1991)
7. Gärdenfors, P. (ed.): Belief Revision, vol. 29. Cambridge University Press, Cambridge (2003)
8. Gilbert, H.: Change in View. MIT Press, Cambridge (1986)
9. Kraus, S., Lehmann, D., Magidor, M.: Nonmonotonic reasoning, preferential models and cumulative logics. Artif. Intell. **44**(1), 167–207 (1990)
10. Leitgeb, H.: The stability theory of belief. Philos. Rev. **123**(2), 131–171 (2014)
11. Lin, H., Kelly, K.T.: Propositional reasoning that tracks probabilistic reasoning. J. Philos. Log. **41**(6), 957–981 (2012)
12. Makinson, D.: Five faces of minimality. Stud. Log. **52**(3), 339–379 (1993)
13. Rott, H.: Stability, strength and sensitivity: converting belief into knowledge. Erkenntnis **61**(2–3), 469–493 (2004)
14. Spohn, W.: Ordinal conditional functions: a dynamic theory of epistemic states. In: Harper, W.L., Skyrms, B. (eds.) Causation in Decision, Belief Change, and Statistics, pp. 105–134. Springer, Netherlands (1988)
15. Van Ditmarsch, H.P.: Prolegomena to dynamic logic for belief revision. In: van der Hoek, W. (ed.) Uncertainty, Rationality, and Agency, pp. 175–221. Springer, Netherlands (2005)

Justification Logic with Approximate Conditional Probabilities

Zoran Ognjanović[1], Nenad Savić[2], and Thomas Studer[2](✉)

[1] Mathematical Institute SANU, Belgrade, Serbia
zorano@mi.sanu.ac.rs
[2] Institute of Computer Science, University of Bern, Bern, Switzerland
{savic,tstuder}@inf.unibe.ch

Abstract. The importance of logics with approximate conditional probabilities is reflected by the fact that they can model non-monotonic reasoning. We introduce a new logic of this kind, CPJ, which extends justification logic and supports non-monotonic reasoning with and about evidences.

1 Introduction

Justification logic [1] is a variant of modal logic that 'unfolds' the $\Box$-modality into justification terms, i.e., justification logics replace modal formulas $\Box\alpha$ with formulas of the form $t{:}\alpha$ that mean *t is a justification for the agent's belief (or knowledge) in* α. This interpretation of justification logic has many applications and has been successfully employed to analyze many different epistemic situations including certain forms of defeasible knowledge [2–5,12].

In a general setting, justifications need not to be certain. Milnikel [14] was the first to approach this problem with his logic of uncertain justifications. Kokkinis et al. [8–10] study probabilistic justification logic, which provides a very general framework for uncertain reasoning with justifications that subsumes Milnikel's system.

In the present paper we extend probabilistic justification logic with operators for approximate conditional probabilities. Formally, we introduce formulas $\mathsf{CP}_{\approx r}(\alpha, \beta)$ meaning *the probability of* α *under the condition* β *is approximately* r. This makes it possible to express defeasible inferences for justification logic. For instance, we can express

if x justifies that Tweety is a bird, then *usually* $t(x)$ justifies that Tweety flies

as $\mathsf{CP}_{\approx 1}(t(x){:}\mathsf{flies}, x{:}\mathsf{bird})$.

Z. Ognjanović—Supported by the Ministry of education, science and technological development grants 174026 and III44006.
N. Savić—Supported by the SNSF project 200021_165549 *Justifications and non-classical reasoning*.

A. Baltag et al. (Eds.): LORI 2017, LNCS 10455, pp. 681–686, 2017.
DOI: 10.1007/978-3-662-55665-8_52

Our paper builds on previous work on probabilistic logics and non-monotonic reasoning. Logics with probability operators are important in artificial intelligence and computer science in general [6,15]. They are interpreted over Kripke-style models with probability measures over possible worlds. Ognjanović and Rašković [16] develop probability logics with infinitary rules to obtain strong completeness results. The recent [17] provides an overview over the topic of probability logics.

Kraus et al. [11] propose a hierarchy of non-monotonic reasoning systems. In particular, they introduce a core system P for default reasoning and establish that P is sound and complete with respect to preferential models. Lehmann and Magidor [13] propose a family of non-standard ($^*\mathbb{R}$) probabilistic models. A default $\alpha \rightarrowtail \beta$ holds in a model of this kind if either the probability of α is 0 or the conditional probability of β given α is infinitesimally close to 1. Using this interpretation, they show that system P is also sound and complete with respect to $^*\mathbb{R}$-probabilistic models. Rašković et al. [18] present a logic with approximate conditional probabilities, $\mathsf{LPP}^{\mathsf{S}}$, whose models are a subclass of non-standard $^*\mathbb{R}$-probabilistic models. They prove the following: for any finite default base Δ and for any default $\alpha \rightarrowtail \beta$

$$\Delta \vdash_{\mathsf{P}} \alpha \rightarrowtail \beta \quad \text{iff} \quad \Delta \vdash_{\mathsf{LPP^S}} \alpha \rightarrowtail \beta.$$

We will introduce operators for approximate conditional probabilities to justification logic. This makes it possible to formalize non-monotonic reasoning with and about evidences.

2 Basic Justification Logic J

Let C be a countable set of constants, V a countable set of variables, and Prop a countable set of atomic propositions. Justification terms and formulas are given as follows:

$$t ::= c \mid x \mid (t \cdot t) \mid (t + t) \mid !t \qquad \text{and} \qquad \alpha ::= p \mid \neg\alpha \mid \alpha \wedge \alpha \mid t : \alpha$$

where $c \in C$, $x \in V$, and $p \in \mathsf{Prop}$. We denote the set of all justification formulas by $\mathsf{Fml_J}$. Other classical Boolean connectives, $\vee$, $\rightarrow$, $\leftrightarrow$, as well as $\bot$ and $\top$, are defined as usual.

The axioms of the logic J are following:

all propositional tautologies	$u : (\alpha \rightarrow \beta) \rightarrow (v : \alpha \rightarrow u \cdot v : \beta)$
$u : \alpha \rightarrow u + v : \alpha$	$v : \alpha \rightarrow u + v : \alpha$

A set $\mathsf{CS} \subseteq \{(c, \alpha) \mid c \in C, \alpha \text{ is an instance of any axiom of } \mathsf{J}\}$ is called *constant specification.* For a given constant specification CS, we define the Hilbert-style *deductive system* $\mathsf{J_{CS}}$ by adding the following two rules to the axioms of J:

1. For $(c, \alpha) \in \mathsf{CS}$, $n \in \mathbb{N}$, infer $!^n c :!^{n-1} c : \cdots :!c : c : \alpha$
2. From α and $\alpha \rightarrow \beta$ infer β

A *basic evaluation* for $\mathsf{J_{CS}}$, where CS is any constant specification, is a function $*$ such that $* : \mathsf{Prop} \to \{\text{true}, \text{false}\}$ and $* : \mathsf{Term} \to \mathcal{P}(\mathsf{Fml_J})$, and for $u, v \in \mathsf{Term}$, any constant c and $\alpha \in \mathsf{Fml_J}$ we have:

1. if there is $\beta \in v^*$ with $\beta \to \alpha \in u^*$, then $\alpha \in (u \cdot v)^*$
2. $u^* \cup v^* \subseteq (u + v)^*$
3. if $(c, \alpha) \in \mathsf{CS}$, then $\alpha \in c^*$ and for each $n \in \mathbb{N}$, $!^n c :!^{n-1} c : \cdots :!c : c : \alpha \in (!^{n+1} c)^*$.

Instead of writing $*(t)$ and $*(p)$, we write t^* and p^* respectively. Now, we are ready to define the notion of truth under a basic evaluation. The binary relation $\Vdash$ is defined by:

$$* \Vdash p \;\text{ iff }\; p^* = \text{true} \qquad\qquad * \Vdash \neg\alpha \;\text{ iff }\; * \nVdash \alpha$$
$$* \Vdash \alpha \wedge \beta \;\text{ iff }\; * \Vdash \alpha \text{ and } * \Vdash \beta \qquad\qquad * \Vdash t : \alpha \;\text{ iff }\; \alpha \in t^*$$

3 The Logic CPJ

Consider a non-standard elementary extension $^*\mathbb{R}$ of the real numbers. An element ϵ of $^*\mathbb{R}$ is called an infinitesimal iff $|\epsilon| < \frac{1}{n}$ for every $n \in \mathbb{N}$. Let S be the unit interval of the Hardy field $\mathbb{Q}[\epsilon]$, which contains all rational functions of a fixed positive infinitesimal ϵ of $^*\mathbb{R}$, for details see, e.g., [7].

The set of probabilistic formulas, denoted by $\mathsf{Fml_P}$, is the smallest set that contains all the formulas of the form

$$\mathsf{CP}_{\geq s}(\alpha, \beta) \qquad \mathsf{CP}_{\leq s}(\alpha, \beta) \qquad \mathsf{CP}_{\approx r}(\alpha, \beta)$$

for $\alpha, \beta \in \mathsf{Fml_J}$, $s \in S$, and $r \in \mathbb{Q} \cap [0, 1]$ and that is closed under negation and conjunction. We use $\varphi, \psi, \ldots$ to denote $\mathsf{Fml_P}$-formulas. The set of all formulas, Fml, of the logic CPJ is defined by $\mathsf{Fml} = \mathsf{Fml_J} \cup \mathsf{Fml_P}$. Elements of Fml will be denoted by $\theta, \theta_1, \theta_2, \ldots$. We use the following standard abbreviations, see [18]:

$$\mathsf{CP}_{<s}(\alpha, \beta) \quad \mathsf{CP}_{>s}(\alpha, \beta) \quad \mathsf{CP}_{=s}(\alpha, \beta) \quad \mathsf{P}_{\rho s}\alpha \text{ with } \rho \in \{\geq, \leq, >, <, =, \approx\}.$$

The semantics for the logic CPJ is based on possible worlds models. Let CS be the constant specification. A $\mathsf{CPJ_{CS}}$-*model* (or just model) is a tuple $M = \langle W, H, \mu, * \rangle$ where:

- W is a non-empty set of objects called worlds
- H is an algebra of subsets of W
- μ is a finitely additive probability measure on H
- $*$ is a function from W to all basic J_{CS}-evaluations. We write $*_w$ for $*(w)$.

Let $M = \langle W, H, \mu, * \rangle$. We put $[\alpha]_M := \{w \in W \mid *_w \Vdash \alpha\}$. Whenever M is clear from the context, we will write $[\alpha]$ instead of $[\alpha]_M$.

A $\mathsf{CPJ_{CS}}$-model M is *measurable* if and only if $[\alpha]_M \in H$, for every $\alpha \in \mathsf{Fml_J}$. A $\mathsf{CPJ_{CS}}$-model M is *neat* if and only if the empty set has the zero probability and no other set has. The class of all measurable and neat $\mathsf{CPJ_{CS}}$ models is denoted by $\mathsf{CPJ_{CS,Meas,Neat}}$.

Let CS be any constant specification. The *satisfaction relation* $\models \subseteq \mathsf{CPJ}_{\mathsf{CS,Meas,Neat}} \times \mathsf{Fml}$ is defined, for any $M \in \mathsf{CPJ}_{\mathsf{CS,Meas,Neat}}$, as follows:

1. $M \models \alpha$ if for every $w \in W$, $*_w \Vdash \alpha$,
2. $M \models \mathsf{CP}_{\leq s}(\alpha, \beta)$ if either $\mu([\beta]) = 0$ and $s = 1$, or $\mu([\beta]) > 0$ and $\frac{\mu([\alpha\wedge\beta])}{\mu([\beta])} \leq s$,
3. $M \models \mathsf{CP}_{\geq s}(\alpha, \beta)$ if either $\mu([\beta]) = 0$, or $\mu([\beta]) > 0$ and $\frac{\mu([\alpha\wedge\beta])}{\mu([\beta])} \geq s$,
4. $M \models \mathsf{CP}_{\approx r}(\alpha, \beta)$ if either $\mu([\beta]) = 0$ and $r = 1$, or $\mu([\beta]) > 0$ and for each $n \in \mathbb{N}$, $\frac{\mu([\alpha\wedge\beta])}{\mu([\beta])} \in [\max\{0, r - \frac{1}{n}\}, min\{1, r + \frac{1}{n}\}]$,
5. $M \models \neg\varphi$ iff it is not the case that $M \models \varphi$,
6. $M \models \varphi \wedge \psi$ iff $M \models \varphi$ and $M \models \psi$.

We assume that the conditional probability is by default 1, whenever the condition has the probability 0, which explains the formulation of case 3 in the above definition.

We introduce the following axiom system for $\mathsf{CPJ}_{\mathsf{CS}}$, where we set $f(s,t) := \min\{1, s+t\}$, $r^- := \mathbb{Q} \cap [0, r)$, and $r^+ := \mathbb{Q} \cap (r, 1]$:

Axiom schemes

1) all J_{CS}-provable formulas
2) all $\mathsf{Fml}_{\mathsf{P}}$-instances of classical tautologies
3) $\mathsf{CP}_{\geq 0}(\alpha, \beta)$
4) $\mathsf{CP}_{\leq s}(\alpha, \beta) \rightarrow \mathsf{CP}_{<t}(\alpha, \beta)$, $t > s$
5) $\mathsf{CP}_{<s}(\alpha, \beta) \rightarrow \mathsf{CP}_{\leq s}(\alpha, \beta)$
6) $\mathsf{P}_{\geq 1}(\alpha \leftrightarrow \beta) \rightarrow (\mathsf{P}_{=s}\alpha \rightarrow \mathsf{P}_{=s}\beta)$
7) $\mathsf{P}_{\leq s}\alpha \leftrightarrow \mathsf{P}_{\geq 1-s}\neg\alpha$
8) $(\mathsf{P}_{=s}\alpha \wedge \mathsf{P}_{=t}\beta \wedge \mathsf{P}_{\geq 1}\neg(\alpha \wedge \beta)) \rightarrow \mathsf{P}_{=f(s,t)}(\alpha \vee \beta)$
9) $\mathsf{P}_{=0}\beta \rightarrow \mathsf{CP}_{=1}(\alpha, \beta)$
10) $(\mathsf{P}_{=t}\beta \wedge \mathsf{P}_{=s}(\alpha \wedge \beta)) \rightarrow \mathsf{CP}_{=\frac{s}{t}}(\alpha, \beta)$, $t \neq 0$
11) $\mathsf{CP}_{\approx r}(\alpha, \beta) \rightarrow \mathsf{CP}_{\geq r_1}(\alpha, \beta)$, $r_1 \in r^-$
12) $\mathsf{CP}_{\approx r}(\alpha, \beta) \rightarrow \mathsf{CP}_{\leq r_1}(\alpha, \beta)$, $r_1 \in r^+$.

Inference Rules

1. From θ_1 and $\theta_1 \rightarrow \theta_2$ infer θ_2.
2. From α infer $\mathsf{P}_{\geq 1}\alpha$.
3. From the set of premises $\{\varphi \rightarrow \mathsf{P}_{\neq s}\alpha \mid s \in S\}$ infer $\varphi \rightarrow \bot$.
4. Let $r \in \mathbb{Q} \cap [0,1]$. From the two sets of premises $\{\varphi \rightarrow \mathsf{CP}_{\geq r - \frac{1}{n}}(\alpha, \beta) \mid n \geq \frac{1}{r}, n \in \mathbb{N}\}$ and $\{\varphi \rightarrow \mathsf{CP}_{\leq r + \frac{1}{n}}(\alpha, \beta) \mid n \geq \frac{1}{1-r}, n \in \mathbb{N}\}$ infer $\varphi \rightarrow \mathsf{CP}_{\approx r}(\alpha, \beta)$.

Axiom 3, putting $\top$ instead of β, says that the probability of each formula being satisfied in some set of worlds is at least 0, and we can easily infer (using $\neg\alpha$ instead of α) that the upper bound is 1, i.e. $\mathsf{P}_{\leq 1}\alpha$. Axioms 4 and 5 say that we can weaken the degree of confidence of truth, while Axiom 6 says that equivalent formulas have the same probability. Axiom 8 corresponds to finite addivity of a measure. Axiom 9 ensures that the conditional probability is equal to 1 whenever the condition has probability 0. Axiom 10 is the formula that states the standard definition of the conditional probability. Finally, the Axioms 11 and 12 (together with Inference Rule 4) give us the relationship between the conditional probability infinitesimally close to the some rational number $r \in [0,1]$ and the standard conditional probability.

Note that there are two bottom elements in Fml, namely $\bot_{\mathsf{J}} \in \mathsf{Fml}_{\mathsf{J}}$ and $\bot_{\mathsf{P}} \in \mathsf{Fml}_{\mathsf{P}}$. Accordingly we say that a set T of Fml-formulas is CS-*consistent* if $T \nvdash_{\mathsf{CS}} \bot_{\mathsf{J}}$ and $T \nvdash_{\mathsf{CS}} \bot_{\mathsf{P}}$.

Similar to [18], we can establish an extended completeness result.

Theorem 1. *Let* CS *be any constant specification. A set* T *of* Fml*-formulas is* CS*-consistent if and only if* T *has a* $\mathsf{CPJ}_{\mathsf{CS,Meas,Neat}}$*-model, i.e., there exists a* $\mathsf{CPJ}_{\mathsf{CS,Meas,Neat}}$*-model* M *with* $M \models \theta$ *for each* $\theta \in T$.

4 Conclusion

We extended probabilistic justification logic with operators for approximate conditional probabilities, which makes it possible to express defaults in justification logic. In particular:

$$\mathsf{CP}_{\approx 1}(t(x){:}\mathsf{flies}, x{:}\mathsf{bird}) \tag{1}$$

means *if* x *justifies that Tweety is a bird, then usually* $t(x)$ *justifies that Tweety flies*;

$$\mathsf{CP}_{\approx 1}(\neg t(x){:}\mathsf{flies}, x{:}\mathsf{penguin}) \tag{2}$$

means *if* x *justifies that Tweety is a penguin, then usually it is not the case that* $t(x)$ *justifies that Tweety flies*;

$$\mathsf{CP}_{\approx 1}(x{:}\mathsf{bird}, x{:}\mathsf{penguin}) \tag{3}$$

means *if* x *justifies that Tweety is a penguin, then usually* x *also justifies that Tweety is a bird.*

Similar to [13,18], it is possible to show that (the corresponding translations) of the axioms and rules of system P are sound with respect to CPJ. In particular we can apply the rule of cautious monotonicity to (2) and (3) in order to infer

$$\mathsf{CP}_{\approx 1}(\neg t(x){:}\mathsf{flies}, x{:}\mathsf{penguin} \wedge x{:}\mathsf{bird}),$$

which is consistent with (1).

Besides the possibility of expressing defaults, CPJ also features non-monotonic versions of classical operations on justifications. Let us consider the sum operator with its defining axiom

$$u : \alpha \vee v : \alpha \rightarrow u + v : \alpha. \tag{4}$$

This axiom states that justifications are monotone: if u justifies α, then the combination of u with v still justifies α. Often the sum operation is motivated as follows. Think of u and v as two volumes of book collection and $u + v$ as the set of those two volumes. Imagine that volume u contains a justification for a proposition α, i.e., $u : \alpha$ is the case. Then the larger set $u + v$ also contains a justification for α, i.e., $u + v : \alpha$. This idea is reflected in the provability semantics of justification logic where the sum operation is interpreted as proof concatenation, which, of course, is monotone.

This motivational example can also be read in another way. It is possible that the second volume v contains a retraction of α, i.e., it withdraws the justification given for α in volume u. To model situations of this kind, one could introduce a non-monotonic sum operation, $\dot{+}$, with

$$\mathsf{CP}_{\approx 1}(u \dot{+} v{:}\alpha, u{:}\alpha) \quad \text{and} \quad \mathsf{CP}_{\approx 1}(u \dot{+} v{:}\alpha, v{:}\alpha).$$

Using the (Or) rule of system P we get $\mathsf{CP}_{\approx 1}(u \dot{+} v{:}\alpha, u{:}\alpha \vee v{:}\alpha)$, which is a non-monotonic version of (4).

References

1. Artemov, S.N.: Explicit provability and constructive semantics. BSL **7**(1), 1–36 (2001)
2. Artemov, S.N.: The logic of justification. RSL **1**(4), 477–513 (2008)
3. Baltag, A., Renne, B., Smets, S.: The logic of justified belief, explicit knowledge, and conclusive evidence. APAL **165**(1), 49–81 (2014)
4. Bucheli, S., Kuznets, R., Studer, T.: Justifications for common knowledge. Appl. Non Class. Log. **21**(1), 35–60 (2011)
5. Bucheli, S., Kuznets, R., Studer, T.: Realizing public announcements by justifications. J. Comput. Syst. Sci. **80**(6), 1046–1066 (2014)
6. Fagin, R., Halpern, J.: Reasoning about knowledge and probability. J. ACM **41**(2), 340–367 (1994)
7. Ikodinović, N., Rašković, M., Marković, Z., Ognjanović, Z.: A first-order probabilistic logic with approximate conditional probabilities. Log. J. IGPL **22**(4), 539–564 (2014)
8. Kokkinis, I.: The Complexity of non-iterated probabilistic justification logic. In: Gyssens, M., Simari, G. (eds.) FoIKS 2016. LNCS, vol. 9616, pp. 292–310. Springer, Cham (2016). doi:10.1007/978-3-319-30024-5_16
9. Kokkinis, I., Maksimović, P., Ognjanović, Z., Studer, T.: First steps towards probabilistic justification logic. Log. J. IGPL **23**(4), 662–687 (2015)
10. Kokkinis, I., Ognjanović, Z., Studer, T.: Probabilistic justification logic. In: Artemov, S., Nerode, A. (eds.) LFCS 2016. LNCS, vol. 9537, pp. 174–186. Springer, Cham (2016). doi:10.1007/978-3-319-27683-0_13
11. Kraus, S., Lehmann, D.J., Magidor, M.: Nonmonotonic reasoning, preferential models and cumulative logics. Artif. Intell. **44**(1–2), 167–207 (1990)
12. Kuznets, R., Studer, T.: Update as evidence: belief expansion. In: Artemov, S., Nerode, A. (eds.) LFCS 2013. LNCS, vol. 7734, pp. 266–279. Springer, Heidelberg (2013). doi:10.1007/978-3-642-35722-0_19
13. Lehmann, D.J., Magidor, M.: What does a conditional knowledge base entail? Artif. Intell. **55**(1), 1–60 (1992)
14. Milnikel, R.S.: The logic of uncertain justifications. APAL **165**(1), 305–315 (2014)
15. Nilsson, N.: Probabilistic logic. Artif. Intell. **28**, 7187 (1986)
16. Ognjanović, Z., Rašković, M.: Some probability logics with new types of probability operators. J. Log. Comput. **9**(2), 181–195 (1999)
17. Ognjanović, Z., Rašković, M., Marković, Z.: Probability Logics. Springer, Heidelberg (2016)
18. Rašković, M., Marković, Z., Ognjanović, Z.: A logic with approximate conditional probabilities that can model default reasoning. Int. J. Approx. Reason. **49**(1), 52–66 (2008)

From Concepts to Predicates Within Constructivist Epistemology

Farshad Badie(✉)

Center for Linguistics, Aalborg University,
Rendsburggade 14, 9000 Aalborg, Denmark
badie@id.aau.dk

Abstract. In this research constructivist epistemology provides a ground for conceptual analysis of concept construction, conception production, and concept learning processes. Relying on a constructivist model of knowing, this research will make an epistemological and logical linkage between concepts and predicates.

1 Introduction

Constructivist epistemology, as a style of thinking about knowledge, focuses on the question of whether, and under which conditions, human beings may construct their own knowledge structures and produce their understanding of the world. It holds that humans can know [about] their personal built up constructions. Actually constructivist epistemology focuses on HowNess of meaning construction with regard to humans' own experiences of the world. It assumes that humans reflect their constructed knowledge on their experiences and pre-conceptions of the world. Regarding [8, 9], I assume that knowledge is not a representation of objective entities (e.g., objective facts, objective procedures), but a compendium (and construction) of concepts, conceptual relationships, and rules that have proven to be useful in expressing humans' experiential world. According to [5, 6], Bruner believed that knowing is how human beings get beyond the information given. According to Bruner, knowing the world is not just perceiving something, but it's constructing it. I shall therefore conclude that the active process of knowledge construction includes information selection, information transformation, decision making, hypothesis generation, and meaning construction from information and experiences. My central focus is on the assumption that we can, reasonably and logically, employ a constructivist model of knowing in order to describe knowledge construction over concepts as a kind of conditional reasoning. This research will focus on conceptual analysis of concepts, concept constructions, conceptions, and concept learning in order to make a logical junction between conceptions and predications. Relying on a constructivist model of knowing, this research will make an epistemological and logical linkage between concepts and predicates.

A. Baltag et al. (Eds.): LORI 2017, LNCS 10455, pp. 687–692, 2017.
DOI: 10.1007/978-3-662-55665-8_53

2 Concepts in the Framework of Constructivism

Over the years, the concept of '*concept*' has been quite imprecise and concepts have not been used consistently, see [11, 13, 15]. More specifically, it is not always transparent if what is meant by the expression 'concept' is some mental representation of phenomena in the world, for example as mental pictures of 'red dog', or whether a concept always has to be bound up with some linguistic expression, e.g., the words 'dog' and 'red' in the concept 'red dog', see [10]. In my opinion, it seems to be plausible that concepts could be the primary units of knowledge—the basic materials, it is often said—out of which humans' thoughts are built and developed. In order to express my conception of 'concepts', I need to focus on the concept of 'meaning'. In the framework of constructivism, a meaning might be interpreted as a 'conceptual structure' and, as such, meanings, to a large extent, influence any individual human being's constructions and developments of her/his individual experiential reality. Therefore, meanings could be interpreted to be constructed over conceptual entities. Thus, any conceptual entity can be interpreted as a building block of a conceptual structure. Note that we can have different perceptions of conceptual entities and, of course, there is no absolute schema for conceptual entities. In my opinion, conceptual entities might be labelled 'concepts' and, subsequently, a conceptual entity, as a representation of a piece of reality, or even fiction, in an individual's mind, can be interpreted to be a basic material of [to-be-constructed] meanings. This could be in line with Bartlet's idea that concepts might be realised as representations of [pieces of] reality in the minds of humans, see [3].

A *concept construction* process is structured over the union of three sub-processes consisting of (i) *concept formation*, (ii) *concept trans-formation*, and (iii) *concept re-formation*, see [1, 2]. Concept reformation happens either after transformation, or at the more specific levels of conceptualisations (of the formed concepts) and as the outcome of conceptual changes (see [12]). For example, the concept 'red dog' could be constructed by Bob based on his intra-psychological and inter-psychological processes. More specifically, (a) some processes, like, thinking about red dogs and studying the topic 'red dogs' are intra-psychological. Accordingly, the concept 'red dog' could be reflected and epitomised in Bob's mind in order to be developed; (b) according to inter-psychological processes, the concept 'red dog' can become transformed from Bob into other agents, and vice-versa. For example, his interactions with other humans, animals, and dogs are inter-psychological. Bob may have conversational exchanges with other humans about their conceptions of red dogs. He can watch and touch red dogs. Bob—by moving from (a) to (b) and vice-versa and, by modifying his conception of 'red dog' over time—becomes concerned with the development of his mental construction of the concept 'red dog'. He can, at anytime, make inferences based on his most modified conception of 'red dogs'. Note that his most modified conception(s) is/are produced based on his most specified constructed 'red dog'. This conclusion is in line with Vygotsky's theory of social constructivism. Vygotsky in [16] stated that "Every function in the child's cultural development appears twice: first, on the social level, and later, on the individual level; first, between people (inter-psychological) and then, inside the child (intra-psychological)".

3 Constructivist Concept Learning

Bruner believed that "to perceive is to categorize, to conceptualize is to categorize, to learn is to form categories, to make decisions is to categorize". In the framework of constructivism, a *categorisation* (or *classification*) can be interpreted as a process of 'constructing', which is dependent on 'representations'. In addition, a classification corresponds with an 'assignment'. It is worth mentioning that the term 'construction' can also express the creation of an abstract entity, see [7, 14]. Actually any individual human being is capable of classifying a phenomenon as belonging to one or to multiple classes with her/his determined and specified labels. I shall claim that humans form and produce multiple classes of different phenomena in order to construct knowledge over those phenomena. It seems to be undeniable that classifications are strongly dependent on mental representations of any individual's personal to-be-constructed constructions based on her/his own conceptions (as the products of her/his constructed concepts). Taking the concepts of 'classification' and 'construction' into account, we could realise that there is a strong logical bi-conditional relationship between the categories (i) classification, regarding hierarchical viewpoints, and (ii) mental construction, as well as mental representation of a phenomenon.

Concept learning can, by seeing concepts as classes, be regarded as (a) the developmental process of concept construction and, accordingly, as (b) the specification of the conceptualisation of the constructed concepts. Concept learning is activable with regard to humans' specification of the conceptualisations of the characteristics and properties of concepts and through experiencing various collections of examples of those concepts. Through concept learning, the phenomena of 'classification', 'representation', 'construction', and 'abstraction' are linked together. In addition, I want to stress that the conceptual interrelationships between 'concepts' (as basic materials of meanings) and 'meanings' (as conceptual structures) establish a semantics based upon humans' concept constructions within their constructivist concept learning process.

4 From Conceptions to Predicates

How could we establish a logical junction between humans' constructed concepts and their expressed conceptions? By taking the phenomenon of 'classification' into account, we can, logically, consider any conception as a class (and, in fact, as a mathematical set). Subsequently, by representing that set in the form of a predicate, that conception can be expressed and stated. More specifically, a unary set and its superset make a *class inclusion* relationship in the form of SubClass $\subseteq$ SuperClass. For example, `Dog` $\subseteq$ `Animal` and `Red` $\subseteq$ `Colour`. Accordingly, a class inclusion relationship can be expressed in the form of a [*unary*] *predicate inclusion*. For example, `Dog` $\subseteq$ `Animal` and `Red` $\subseteq$ `Colour` are representable in the form of `Dog` $\sqsubseteq$ `Animal` and `Red` $\sqsubseteq$ `Colour`, respectively. Subsequently, a unary predicate inclusion can be expressed in the form of a *predicate assertion*. For example, `Dog` $\sqsubseteq$ `Animal` and `Red` $\sqsubseteq$ `Colour` are expressible by `Animal(Dog)` and `Colour(Red)`, respectively. The logical expression '`Animal(Dog)`' expresses that the class `Animal` covers the class `Dog` and, in fact, all `Dog`s are interpreted and understood as `Animal`s.

Therefore, the class Dog is subsumed under the class Animal. Furthermore, relying on inductive rules, the class inclusions Dog $\sqsubseteq$ Animal and Red $\sqsubseteq$ Colour can be merged in order to produce RedDog $\sqsubseteq$ ColouredAnimal and, subsequently, to produce ColouredAnimal(RedDog). I shall conclude that representing the logical description 'Dog $\sqsubseteq$ Animal' is strongly dependent on the following processes in one's mind:

(I) Constructing Dog and Animal and, respectively, producing personal conceptions of them. (II) Scheming some individual dogs and animals (as the instances of Dog and Animal, respectively) and, accordingly, producing personal conceptions of them. (III) Constructing logical relationships between Dog and Animal and, subsequently, producing personal conceptions of them. It shall be emphasised that any semantic interpretation is given sense over the collection of (I), (II), and (III). Logically, the triple (I, II, III) is equivalent to: (classes, inclusions and memberships in classes, the interrelationships between the members of various classes).

Assessed by formal logic, a predicate is an expression that makes an 'assignment'. Human beings, by their semantic interpretations, assign their own conceptions and the interrelationships between those conceptions into logical values. A predicate is capable of connecting with one or more singular terms to make a proposition. Predicates express the conditions that the entities referred to may satisfy, see [4]. More specifically, predicates, by employing semantic interpretations (i.e., generated interpretation functions from words and symbols into truth values), transmit the characteristics as well as the properties of conceptions into either statements or values. Consequently, unary predicates could stand in the place of conceptions and n-ary predicates (for $n \geq 2$) could stand in the place of conceptions' interrelationships. It can, therefore, be concluded that 'a predicate is an assignment function from characteristics and properties of a conception (and in fact, of a constructed concept) into subjects'. A subject is something which is—in a situation/setting—the conceptual entity (i.e., a configuration) of the act of linguistic communication (i.e., transfer of information) or cognition (i.e., transformation of information) of the interlocutor uttering the statement. Therefore, subjection is an assertive predication, see [10]. Figure 1 represents the analysed conceptual relationship between concepts and predicates.

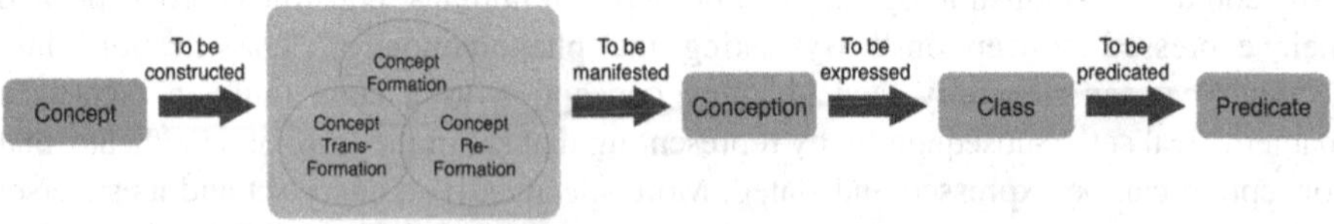

Fig. 1. From concepts to predicates in the framework of constructivism

5 Concluding Remarks

Predication of a conception (or, equivalently: a to-be-created class) is concerned with the question 'what is it to state something about that conception?' and, thus, a predication tackles to find an answer for describing and expressing the question 'what is there for a produced conception?'. Heuristically, the latter question focuses on the existence of a constructed concept. This question is concerned with ontological descriptions of a constructed concept, while the first question is concerned with a description of that concept. I shall acknowledge that there is a strong correlation between predication of a conception and that concept's ontology. Relying on constructivist ontology, any individual human being has her/his personal constructed concepts, and hence, it is reasonable to assume that the constructed concepts are valid. Let me conclude that a predication is, indirectly, concerned with a kind of ontological underpinning of a conception and, in fact, of a constructed concept. In addition, it must be assumed that the background knowledge has strong interrelatedness with humans' ontological conceptions that are generated with regard to the structures of the pieces of reality in their minds. So, there are strong correlations between 'pre-concept descriptions and pre-conceptions' and 'ontological conceptions'. Accordingly, the pre-concept descriptions and pre-conceptions could be realised as the outcomes of ontological conceptions. I shall therefore conclude that there is a triangle covering (a) ontological realisation of a concept, (b) concept construction, and (c) predication of a conception. In my opinion, the realisation of characteristics and properties of concepts, tackle to deal with their ontologies (note that the predications are generated in order to express the properties of concepts and in order to reflect those concepts in multiple subjects). Therefore, the predication functions are from those characteristics and properties into subjects. I shall also conclude that the philosophy of constructivism is a kind of comprehensive and explanatory ontology of human beings, and the constructivist epistemology provides a model of knowing over this ontology. The central focus of constructivist epistemology is the origin of an individual's constructed knowledge.

References

1. Badie, F.: A semantic basis for meaning construction in constructivist interactions. In: Proceedings of the International Conference On Cognition And Exploratory Learning In Digital Age, IADIS, pp. 369–373 (2015)
2. Badie, F.: Towards a semantics based framework for meaning construction in constructivist interactions. In: Proceedings of ICERI15 in Research Innovation International Conference of Education, IATED, pp. 7995–8002 (2015)
3. Bartlett, F.C.: A Study in Experimental and Social Psychology. Cambridge University Press, Cambridge (1932)
4. Blackburn, S.: The Oxford Dictionary of Philosophy (2016)
5. Bruner, J.: Going Beyond the Information Given. Norton, New York (1974)
6. Bruner, J.: Acts of Meaning. Harvard University Press, Cambridge (1990)
7. Cambridge Online Dictionary: Cambridge University Press (2017) http://dictionary.cambridge.org/dictionary/english
8. von Foerster, H.: Observing Systems. Intersystems Publications, Seaside (1981)
9. von Glasersfeld, E.: An introduction to radical constructivism. In: Watzlawick, P. (ed.) The Invented Reality. Norton, New York (1984)
10. Götzsche, H.: Deviational Syntactic Structures. New York, Bloomsbury Academic (2013)

11. Hampton James, A., Moss, E.H.: Concepts and Meaning: Introduction to the Special Issue on Conceptual Representation, Language and Cognitive Processes, pp. 505–512, Taylor & Francis, London (2003)
12. Limon, M.: Conceptual change in history. In: Limon, M., Mason, L. (eds.) Reconsidering Conceptual Change: Issues in Theory and Practice, pp. 259–289. Kluwer, Dordrecht (2002)
13. Margolis, E., Laurence, S.: The Conceptual Mind: New Directions in the Study of Concepts. MIT Press, Cambridge (2015)
14. Oxford Online Dictionaries: Oxford University Press (2017) https://en.oxforddictionaries.com
15. Peacocke, C.: A Study of Concepts. MIT Press, Cambridge (1992)
16. Vygotsky, L.: Interaction between learning and development. In: Gauvain., Cole. (eds.) Readings on the Development of Children, pp. 34–40. Scientific American Books, New York (1978)

Author Index

Printed in the United States
By Bookmasters